International Association for Hydraulic Research
International Union of Theoretical and Applied Mechanics

Practical Experiences with Flow-Induced Vibrations

Symposium Karlsruhe/Germany
September 3-6, 1979
University of Karlsruhe

Editors
E. Naudascher and **D. Rockwell**

With 429 Figures and 217 Charts

Springer-Verlag
Berlin Heidelberg New York 1980

Prof. Dr. EDUARD NAUDASCHER

Institut für Hydromechanik, Universität Karlsruhe
Kaiserstraße 12, 7500 Karlsruhe

Prof. Dr. DONALD ROCKWELL

Department of Mechanical Engineering, Lehigh University
Bethlehem, Pa. 18015, USA

ISBN-13: 978-3-642-81530-0 e-ISBN-13: 978-3-642-81528-7
DOI: 10.1007/ 978-3-642-81528-7

Library of Congress Cataloging in Publication Data
Main entry under title:
Practical experiences with flow-induced vibrations.
(IUTAM-Symposien)
At head of title: International Association for Hydraulic Research, International Union of Theoretical and Applied Mechanics.
Bibliography: p.
Includes index. 1. Vibration-Congresses. 2. Fluid dynamics-Congresses. I. Naudascher, Eduard. II. Rockwell, Donald. III. International Association for Hydraulic Research. IV International Union of Theoretical and Applied Mechanics. V. Series: International Union of Theoretical and Applied Mechanics. IUTAM-Symposien.
TA355.P68 620.3 80-20244

Softcover reprint of the hardcover 1st edition 1980

2061/3020/543210

SCIENTIFIC COMMITTEE

R.E.D.Bishop, UK
M. Hug, France
S.J. Crandall, USA
G.V. Parkinson, Canada
L.I. Sedov, USSR

E. Naudascher and D.O. Rockwell, Chairmen

LOCAL ORGANIZING COMMITTEE

R. Böser, D. Danzeisen, V. Gottwald
B. Herrling, R. Kleine, M. Leschziner, A. Richter
P.Schaupp, N.Thang, B. Treiber, R. Zschernitz

R. Ermshaus, Chairman

SYMPOSIUM SPONSORS

Deutsche Forschungsgemeinschaft
Land Baden Württemberg
Department of the US Navy, ONR
Department of the US Army, HQ Europe

Badenwerk Aktiengesellschaft, Karlsruhe
Bethlehem Steel Corporation, Bethlehem, PA, USA
Hochschulvereinigung, Universität Karlsruhe
Klein, Schanzlin & Becker-Stiftung, Stuttgart
Lufthansa Airlines, Frankfurt
Sulzer Aktiengesellschaft, Winterthur, Switzerland
Schluchseewerk Aktiengesellschaft, Freiburg
J.M. Voith GmbH, Heidenheim

Participants of the IAHR/IUTAM Symposium 1979 in front of the Karlsruhe Palace

PREFACE

Symposia in the areas of fluid mechanics and hydraulics are, in many cases, strongly oriented toward the researcher, with only peripheral participation of the practicing engineer. The possible danger therein is an increased emphasis on those problems that are theoretically or experimentally attractive, rather than those that would most effectively lead to improved design criteria. To bring the practical aspects of flow-induced vibrations to the forefront, and to enhance interchange of information on shortcomings of design criteria, it was decided to extend invitations to field engineers, designers and researchers. The leading theme of the Symposium would be experiences from practice, encompassing information on failures, fatigue, and wear; and wherever possible, the associated cures would be addressed as well. Moreover, to promote interaction between disciplines, a variety of specialities, ranging from tall structures to heat exchanger and reactor components, was included; but to keep the undertaking within a reasonable scope, topics dealing with slender body aerodynamics, and high-speed internal and external flow, were not incorporated. During this planning phase, many of our colleagues encouraged us to follow through with the spirit of the Symposium, and provided a number of helpful suggestions. It was gratifying to receive the endorsement of the International Association for Hydraulic Research and the International Union for Theoretical and Applied Mechanics, as the variety of interests represented by these organizations enhanced the theme of the Symposium. Particularly satisfying was the balance achieved in the number of partipants from practice (118) and research organizations (123). This spectrum of representatives from 23 countries provided an international perspective that furtherenriched cross-communication.

The collection of contributions included herin consists of a range of papers in each of the basic disciplines, selected discussions, as well as Invited Papers and General Lectures. These latter two types of contributions provided focal points for each Session, as well as overviews of areas of practical importance, thereby establishing a framework for each Session. To emphasize the practical features of all contrubutions, authors were asked to summarize each case history using a chart format, in which the essential aspects of a given case are depicted, thereby highlighting symptoms and possible cures. These Charts, designed to be self-explanatory and included in these Proceedings, were enlarged and displayed in the Symposium hall, providing participants with a stimulus for information exchange before and between Sessions.

To bring together the various desciplines represented at the Symposium, and provide a common ground for discussion, there was a Workshop on Prediction and Modelling, a Film Session (on observed vibrations), and a Panel Discussion on Critical Unresolved Problems. The Workshop addressed capabilities and limitations of nonlinear oscillator modelling, discrete vortex methods, and simulation of incident urbulence; corresponding overview papers, representing the state of the art, are included in these Proceedings. During the Film Session, a number of unsteady fluid-dynamic and fluid-elastic phenomena were screened, setting the stage for subsequent discussion. A listing of these films, available from the authors, is given herein. The Panel Discussion, held at the end of the Symposium, involved assessment of unsolved aspects of several practical classes of problems, with substantial input from consulting and industrial engineers from the Panel, and participants as well. Six position papers on these unresolved problems set forth shortcomings of design criteria and suggest lines of future research.

As Chairman of the Scientific Committee, we should like to express our appreciation to all Those who helped make this undertaking a success. In the initial stages, members of the Scientific Committee played a key role in the paper selection process, which involved careful review of a large number of manuscripts. Presidents of the sponsoring organizations, Dr. M. Hug (IAHR) and Dr. F. Niordson (IUTAM) supported our efforts in the planning phase, and Prof. H. Schlichting, Prof. L.I. Sedov, as well as Dr. Hug, enriched our opening and closing festivities with their addresses.

During the three years of preparation for the Symposium, a number of our colleagues at the Institute of Hydromechanics, University of Karlsruhe, contributed in various ways. Here we mention a few of them, who represent the spirit and efforts of many. As Chairman of the Local Organizing Committee, Dr. Rolf Ermshaus served invaluably not only in this capacity, but also in several other essential roles before, during, and after the Symposium; preparation of the final form of the Proceedings involved an immense effort on his part. Although it is impossible to note here the key contributions of all of the Local Organizing Committee, Dr. B.Herrling energetically coordinated many of the local organizing tasks, Dr. A.Richter effectively guided financial aspects, and Mrs. R.Boeser capably assumed many of the administrative responsibilities. Of course, an undertaking such as this would not be possible without generous sponsorship of the organizations listed above; their support is gratefully acknowledged.

Finally, we should like to note that the success of each of the Technical Sessions is a credit not only to the Chairman and Co-Chairman, who effectively presided over the presentations and discussion, but also to the participants who, through discussions, provided suggestions and ideas, many of which have been incorporated in the Proceedings. A number of our colleagues painstakingly produced Invited Papers and General Lectures that set the tone for the Symposium, and will be of importance for years to come. To mention just three that characterize many, the works of Dr. P.Kolkman, Prof. M.Paidoussis and Mr. R. Wardlaw provide a fascinating panorama of case studies. These contributions, together with the others, will hopefully give the reader an overview of capabilities and shortcomings of present design criteria, possible cures, and meaningful paths of research.

Karlsruhe
February 1980

E. Naudascher and D. Rockwell

PARTICIPANTS

Argentina: M.S. Barchilon, A.A. Urbano

Australia: W.H. Melbourne

Austria: K. Giersig

Belgium: C. Arnold, A. D'Haenens, J. de Keuster, G. v.Langenhove

Brazil: E.V. Blotta, F.L.S. Futuro, E.Koelle, B.Petry, H.U. Schmider

Bulgaria: V.A. Dzhupanov, W. Gheroff

Canada: A.G. Davenport, R.T. Hartlen, J.H. Lever, M.P. Païdoussis, G.V. Parkinson, M.J. Pettigrew, B.J. Vickery, R.L. Wardlaw, D.S. Weaver

Czechoslovakia: M. Pichal, M. Pirner, L. Půlpitel, E. Šeda, A. Skalicky

Ecuador: D.M. Castro

France: D. Abrassart, F. Axisa, D. Bai, B. Barthelet, F. Boulot, A. Daubert, L. Fabre, J.C. Guilloud, M. Hug, J.C. Jouannet, D. Milan, I.J.P. Mollet, F.N. Remy, M. Sagner, B. Villard, M. Watremetz

Germany D.R.: K. Hennig

Germany F.R.: H. Annemüller, H. Bardowicks, V. Bauernfeind. E.Berger, R. Bouecke, E. Brandt, R. Buchwald, F. Büsching, A. Dinkelacker, F. Durst, O. Eichler, R. Ermshaus, L. Fink, W.J. Gerasch, U.Goss, K. Haffke, W. Hammer, B. Herrling, P. Holz, D. Holzdeppe, G. Jodelsberger, H. Kirchner, R. Kleine, J. Knaul, M. Leschziner, C. Liess, O. Mahrenholtz, T. Merkle, C. Miller, L. Möcks, E. Mosonyi, C. Nath, H.G. Natke, E. Naudascher, J. Osterwalder, K. Petrikat, E. Plate, A. Richter, G. Rouvê, H. Ruscheweyh, U. Simon, H. Sompek, B. Schäfer, N. Schauki, H. Schlichting, G. Schmausser, U. Schumann, M. Topal-Gökceli, B. Treiber, P. Ulith, G. Valtinat, U. Vogel, H.J. Wehling, G. Wonik

India: P.C. Saxena

Italy: R. Galetto, F. de Leva, G. Rossi

Japan: H. Aoki, T. Fujikawa, N. Hagiwara, F. Hara, T. Hiramatsu, N. Ishii, H. Ishizaki, T. Kobayashi, T. Kawase, T. Kubota, M. Matsumoto, T. Miyata, K. Ogihara, E. Outa, N. Suzuki, Y. Ueda

Kuwait: M.A. Hassouneh

Mexico: G.A. Echavez

Netherlands: C.H. v.Deelen, G.J.M. Hertogh, A.B.M. Hoff, R.J. de Jong, P.A. Kolkman, H.v.Koten, H.K. Kwee, J.W.G.v.Nunen, A. Vrijer, H. v.d.Weijde

Norway: E. Hjorth-Hansen, K.O. Holden, K.A. Nyhus, L. Vinnogg

Portugal: C.A.M. Ramos

Sweden: G.B. Benkö, B. Forsblom, J.E.V. Nilsson, G. Svensson, K.E. Widell

Switzerland: U. Bolleter, G.M. Capitaine, B. Chaix, Y.N. Chen, P. Doerfler, M.P. Escudier, M. Gerodetti, A.H. Glattfelder, H. Grein, C. Gremli, H. Hinz, H.K. Höller, C.S. Jacobsen, N. Jaussi, R. Jenny, A. Kirschner, R. Miksicek, H.E. Minor, E. Muller, M. Staehle, T. Staubli, A. Temler, D. Vischer, G. Vuilloud

United Kingdom: R. Ali, J. Armitt, J. Bokota, W.B. Bryce, A.E. Collinson, R.N. Cox, J.C. Duthie, J.A. Fitzpatrick, M.E. Ginniff, J.D. Hardwick, P. Holmes, C.H. Jones, M.J. Kenn, R. King, M.J. Lambert, B.E. Lee, A. Marshall, D.J. Maull, B.C. Murray, M. Owen, M.W. Parkin, J.N. Pinder, T.L. Shaw, B.M.H. Soper, G.D. Thomas, D.E.J. Walshe, M.C. Welbourne, P.N. Whitton, H.Y. Wong, L.R. Wootton, T.A. Wyatt, G.A.I. Young, M. Zdravkovich

United States: W.T. Ashurst, R.D. Blevins, J.P. Breslin, J.H.Douma, E.H. Dowell, H.T. Falvey, F.J. Fischer, H.M. Fitzpatrick, G.E. Hecker, R.E. Henderson, K. Karamcheti, B.T. Lubin, W.J. McCroskey, F. Neilson, F.E. Reed, D.O. Rockwell, D.W.Sallet, S.D. Savkar, R.H. Scanlan, D.A. Skinner, G.C.C. Smith, D.T. Tsahalis, I.L. Vêr, S. Vigander, M.W. Wambsganss J.R. Weske, S.R. Ziada

U.S.S.R.: G. Fomine, V. Grin, A.G. Kostjuk, K.S. Kolesnikov, G.K. Mikhailov, L.E. Sedov

Venezuela: K. Zagustin

CONTRIBUTORS

Aoki, H.	JAP	Fuji Electric Co. Ltd., Kawasaki Factory, Tanabe-Shinden 1-1 Kawasaki - 210
Ashurst, W.T.	USA	Sandia Laboratories, Livermore, California 94450
Bai, D.	F	E.D.F. - D.E.R., 6, quai Watier, 78400 Chatou
Bardowicks, H.	D	Inst. für Mechanik, Universität Hannover, Appelstr. 11, 3000 Hannover
Baylac, G.	F	Electricité de France, Laboratoire National d'Hydraulique, 6, quai Watier, 78400 Chatou
Berger, E.	D	Inst. für Fluiddynamik, TU, Straße des 17.Juni 135, 1000 Berlin 12
Bertholon, G.	F	NEYRPIC, Rue Général Mangin, 38041 Grenoble
Blevins, R.D.	USA	General Atomic Co., P.O.Box 81608, San Diego, California 92138
Bolleter, U.	CH	Sulzer Brothers Ltd., 8401 Winterthur
Boulot, F.	F	Electricité de France, Laboratoire National d'Hydraulique, 6, quai Watier, 78400 Chatou
Breslin, J.P.	USA	Davidson Lab., Stevens Inst. of Technology, 711 Hudson St. Hoboken, NJ 07030
Büsching, F.	D	Lehrstuhl für Hydromechanik, TU Braunschweig, Beethovenstr. 51a, 3300 Braunschweig
Campagna, A.O.	CDN	Chalk River Nuclear Laboratories, Chalk River, Ontario K0J 1J0
Cassell, A.C.	GB	Civil Engineering Dept., Imperial College, London SW7 2BU
Chaix, B.	CH	Inst. für Flüssigkeitstechnik, ETH, Sonneggstr. 3, 8092 Zürich
Chen, Y.N.	CH	Sulzer Brothers Ltd., 8401 Winterthur
Chow, W.Y.	USA	Hydro Research Science, 3334 Victor Court, Santa Clara, CA 95050
Cox, R.N.	GB	Dept. of Aeronautics and Fluid Mechanics, University of Glasgow, Glasgow G12 8QQ, Scotland
Dalgliesh, W.A.	CDN	Div. of Building Research, National Res. Council of Canada, Ottawa, Ontario K1A 0R6
Davenport, A.G.	CDN	The University of Western Ontario, London, Ontario N6A 5B9
Douma, J.H.	USA	1001 Manning Street, Great Falls, Virginia 22066
Dowell, E.H.	USA	Mechanical and Aerospace Engineering, Princeton University, Princeton, New Jersey 08544
Drake, M.E.	GB	C.E.G.B., Berkeley Nuclear Laboratories, Berkeley, Glos.
Durst, F.	D	Inst. für Hydromechanik, Universität Karlsruhe, Kaiserstrasse 12, 7500 Karlsruhe 1
Eichler, O.	D	J.M. Voith GmbH, Postfach 1940, 7920 Heidenheim
Elder, R.A.	USA	Bechtel Inc., P.O.Box 3965, San Francisco, CA 94119
Ellis, B.R.	GB	Building Research Establishment, Sheffield
Escudier, M.P.	CH	Brown Boveri & Cie.AG, BBC Forschungszentrum, 5404 Baden
Evans, R.A.	GB	Dept. of Building Science, Sheffield University, Sheffield S10 2TN
Falvey, H.T.	USA	US Bureau of Reclamation, P.O.Box 25007, Denver, CO 80225
Fêtizon, F.	F	CETIM, 52, avenue Félix-Louat, 60300 Senlis Cedex
Fischer, F.J.	USA	Shell Development Company, P.O.Box 1380, Houston, TX 77001
Florjancic, D.	CH	Sulzer Brothers Ltd., 8401 Winterthur
Gerodetti, M.	CH	Versuchsanstalt für Wasserbau, ETH, 8092 Zürich
Glattfelder, A.H.	CH	Escher Wyss AG, Postfach, 8023 Zürich
Gorelov, D.	SU	Inst. of Hydrodynamics, Siberian Branch of the USSR Academy of Sciences, Novosibirsk - 630090
Gregoire, J.	F	Electricité de France, Laboratoire National d'Hydraulique, 6 quai Watier, 78400 Chatou
Grein, H.	CH	Escher Wyss, 8023 Zürich
Grunwald, G.	DDR	Zentralinstitut für Kernforschung, Akademie der Wissenschaften der DDR, Berlin
Grootenhuis, P.	GB	Civil Engineering Dept., Imperial College, London, SW7 2BU
Guerrero, H.N.	USA	C-E Power Systems Eng., Inc., 100 Prospect Hill Rd.,Windsor,CT 06095
Halle, H.	USA	Argonne National Laboratory, Argonne, IL 60439
Hara, F.	JAP	Tokyo University of Science, 1-3 Kagurazaka, Shinjuku-ku, Tokyo
Hardwick, J.D.	GB	Civil Engineering Dept., Imperial College, London SW7
Hart, E.D.	USA	Waterways Experiment Stn., Corps of Engineers, P.O.Box 631, Vicksburg, MS 39180

Hartlen, R.T.	CDN	Ontario Hydro Research Div., 800 Kipling Ave.,Toronto,Ontario M8Z 5S4
Hassouneh, M.A.		Kuwait National Petroleum Co., P.O.Box 19499, Kheitan, Kuwait
Hay, J.S.	GB	Transport and Road Research Lab., Crowthorne, Berkshire
Hennig, K.	DDR	Zentralinstitut für Mathematik und Mechanik, Mohrenstraße 39 108 Berlin
Hirose, A.	JAP	Osaka University, Toyonaka, Osaka-560
Hite, J.E.	USA	Waterways Experiment Stn.,Corps of Engineers,P.O.Box 631, Vicksburg, MS 39180
Huntley, I.D.	GB	Dept. of Mech. Eng.,Paisley College of Technology, High Street, Paisley, Renfrewshire PA1 2BE Scotland
Huser, L.	CH	Escher Wyss AG, Postfach, 8023 Zürich
Imaichi, K.	JAP	Osaka University, Toyonaka, Osaka-560
Ishii, N.	JAP	Osaka Electro-Communication University, Neyagawa, Osaka-572
Ishizaki, H.	JAP	Disaster Prevention Res. Inst., Kyoto University, Uji,Kyoto-611
Jaster, W.	CDN	Ontario Hydro Res. Div., 800 Kipling Ave., Toronto, Ont., M8Z 5S4
Jeary, A.P.	GB	Building Res. Establishment, Sheffield
Jenny, R.J.	CH	Sulzer Brothers Ltd., Postfach, 8023 Zürich
Johns, D.J.	GB	Loughborough University, Loughborough, Leics., LE11 3TU
Jong, de, R.J.	NL	Delft Hydraulics Laboratory, P.O.Box 177, 2600 MH Delft
Jones, R.	GB	B.H.R.A. Fluid Engineering, Cranfield, Bedford
Jones, W.T.	USA	Shell Development Co., P.O.Box 1380, Houston, TX 77001
Jouannet, J.C.	F	CETIM, 52, avenue Fêlix-Louat, 60300 Senlis Cedex
Katinas, V.	SU	Lietuvos TSR Mokslu Akademija, Linino prosp. Nr. 3, 232600 Vilnius, MTP-1, Litauische SSR
Kantayya, R.C.	USA	C-E Power Systems Eng., Inc., 100 Prospect Hill Rd.,Windsor,CT 06095
Kawase, T.	JAP	Waseda University, 3-4-1 Okubo, Shinjuku, Tokyo 160
Kenn, M.J.	GB	Civil Eng. Dept., Imperial College, London SW7 2BU
King, R.	GB	B.H.R.A. Fluid Engineering, Cranfield, Bedford
Kirchner, H.	D	J.M. Voith GmbH, Postfach 1940, 7920 Heidenheim
Kolkman, P.A.	NL	Delft Hydraulics Lab., P.O.Box 177, 2600 MH Delft
Konishi, J.	JAP	Dept. of Civil Eng., Kyoto University, Yoshida Hommachi, Sakyo-ku, Kyoto
Koten van, H.	NL	TNO, P.O. Box 49, 2600 AA Delft
Kovalenko, V.A.	SU	Turbine Bldg. Association, Leningradskij Metallitsheskij Zavod
Kubota, T.	JAP	Fuji Electric Co.Ltd., Kawasaki Factory, Tanabe-Shinden 1-1, Kawasaki-210
Kumai, T.	JAP	Mitsui Shipbuilding Co.Ltd., Tsukiji 5-6-4, Tokyo
Lee, B.E.	GB	Dept. of Building Science, Sheffield University, Sheffield S10 2TN
Ljatkher, V.M.	SU	D 362, P.O. Box 393, Hydroproject, Moscow
Lubin, B.T.	USA	C-E Power Systems Eng., Inc., 100 Prospect Hill Rd.,Windsor,CT 06095
Machiyama, T.	JAP	Waseda University, 3-4-1 Okubo, Shinjuku, Tokyo 160
Mahrenholtz, O.	D	Inst. for Mechanics, University of Hannover, Appelstrasse 11, 3000 Hannover
Makihata, T.	JAP	Disaster Prevention Res. Inst., Kyoto University, Uji Kyoto-611
Malyshev, V.M.	SU	Turbine Bldg. Association, Leningradskij Metallitsheskij Zavod
March, P.A.	USA	TVA Engineering Laboratory, Norris, TN 37828
Matsumoto, M.	JAP	Dept. of Civil Eng., Kyoto University, Yoshida Hommachi,Sakyo-ku, Kyoto 606
Maull, D.J.	GB	Dept. of Engineering, Cambridge University, Trumpington St., Cambridge CB2 1PZ
McCroskey, W.J.	USA	U.S.Army Aero-Mechanics Lab., Ames Res.Center, Moffett Field, CA 94035
Mee, W.T.	GB	Sir Alexander Gibb & Partners, Early House, Reading
Melbourne, W.H.	AUS	Dept. of Mech.Engineering, Monash University, Wellington Rd., Clayton - 3168
Merkle, T.	D	J.M. Voith GmbH, Postfach 1940, 7920 Heidenheim
Milan, D.	F	NEYRPIC, Rue Gênêral Mangin, 38041 Grenoble
Mitchell, W.R.	AUS	Hydro-Electric Commission of Tasmania, Elizabeth Street, Hobart-7000
Miyata, T.	JAP	Dept. of Civil Eng.,University of Tokyo, Bunkyo-ku, Tokyo 113
Mulcahy, T.M.	USA	Argonne National Laboratory, Argonne, IL 60439
Nath, C.	D	Germanischer Lloyd, P.O.Box 111606, 2000 Hamburg 11

Naudascher, E.	D	Institut of Hydromechanics, University of Karlsruhe, 7500 Karlsruhe
Neilson, F.M.	USA	USAE Waterways Experiment Stn.,P.O.Box 631, Vicksburg, MS 39180
Nunen van, J.W.G.	NL	(DNW) German-Dutch-Wind Tunnel, Voorsterweg 31, 8316 PR Marknesse
Ogihara, K.	JAP	Toyo University, Civil Eng. Dept., 2100 Kujirai,Kawagoe,Saitama 350
Okauchi, I.	JAP	Dept. of Civil Eng., Chuo University, Tokyo 112
Païdoussis, M.P.	CDN	Dept. of Mech. Eng., McGill University, 817 Sherbrooke St. W., Montreal, Quebec H3A 2K6
Parkin, M.W.	GB	UKAEA, Windscale Works, Seascale, Cumbria CA20 1PF
Parkinson, G.V.	CDN	Dept. of Mech. Eng.,University of British Columbia, 2075 Wesbrook Place, Vancouver, B.C. V6T 1W5
Pasche, E.	D	Inst. für Wasserbau, TH Aachen, Templergraben 55 5100 Aachen
Payer, H.G.	D	Germanischer Lloyd, P.O.Box 111606, 2000 Hamburg 11
Petrie, A.M.	GB	Dept. of Mech. Eng., Paisley College, High St., Paisley, Renfrewshire, PA1 2BE, Scotland
Petrikat, K.	D	Autenbrunnstrasse 6, 7022 Leinfelden-Echterdingen
Pettigrew, M.J.	CDN	Atomic Energy of Canada, Chalk River Nuclear Lab.,Chalk River, Ontario K0J 1J0
Pirner, M.	CS	TZUS, Praha 3, Lucemburska 29
Planchard, J.	F	Elêctricitê de France, Laboratoire National d'Hydraulique, 6, quai Watier, 78400 Chatou
Platen, E.	DDR	Zentralinstitut für Mathematik und Mechanik, Mohrenstr.39, 108 Berlin
Pless, E.A.	D	Germanischer Lloyd, P.O.Box 111606, 2000 Hamburg 11
Pugnet, L.	F	Elêctricitê de France, Laboratoire National d'Hydraulique, 6, quai Watier, 78400 Chatou
Půlpitel, L.	CS	ČKD Blansko, 678 18 Blansko
Ramanathan, V.	IND	P.O. Khadakwasla, Research Station, Poona-411 024
Reed, R.E.	USA	Littleton Res. and Eng. Corp.,95 Russel St., Littleton,MA 01460
Rockwell, D.O.	USA	Dept. of Mech.Eng., Lehigh University, Bethlehem, PA 18015
Rohde, F.G.	D	TH Aachen, Institut für Wasserbau, Templergraben 55, 5100 Aachen
Rouvê, G.	D	TH Aachen, Institut für Wasserbau, Templergraben 55, 5100 Aachen
Rudavsky, A.B.	USA	Hydro Res. Science, 3334 Victor Court, Santa Clara, CA 95050
Ruscheweyh, H.	D	Inst. für Leichtbau, RWTH Aachen, Wüllnerstr. 7, 5100 Aachen
Sallet, D.W.	USA	University of Maryland, Dept. of Mech.Eng.,College Park, Md 20742
Saxena, P.C.	IND	P.O. Khadakwasla, Reseach Stn., Poona-411 024
Scanlan, R.H.	USA	Princeton University, Princeton, NJ 08544
Schachenmann, A.	CH	Sulzer Brothers Ltd., 8401 Winterthur
Schumann, U.	D	Kernforschungszentrum Karlsruhe, Postfach 3640, 7500 Karlsruhe
Shaw, T.L.	GB	Rock Cottage, Blagdon, Bristol BS18 6TW
Sheinin, I.S.	SU	B.E.Wedenejew-WNIIG, 21 Gshatskaja-Str., 195220 Leningrad
Shiraishi, N.	JAP	Dept. of Civil Eng., Kyoto University, Yoshida Hommachi, Sakyo-ku,Kyoto
Sinnarkar, R.M.	IND	P.O. Khadakwasla, Research Station, Poona-411 024
Skladnev, M.F.	SU	B.E.Wedenejew-WNIIG, 21 Gshatskaja-Str., 195220 Leningrad
Solana, P.	YV	Instituto de Mecanica de Fluidos, Universidad Central, Caracas
Stürchler, R.	CH	Sulzer Brothers Ltd., 8401 Winterthur
Tangemann, H.	D	Inst. for Mechanics, University of Hannover, Appelstr. 11,3000 Hannover
Toueille, R.	F	S.M.F. Company, Tarbes
Traut, F.-J.	D	TH Aachen, Inst. für Wasserbau, Templergraben 55, 5100 Aachen
Tsui, Y.T.	CDN	IREQ, Institut de recherche de l'Hydro-Quêbec,P.O.Box 1000, Varennes, Quêbec J0L 2P0
Ueda, S.	JAP	Ishikawajima Harima Heavy Industry Co.Ltd., 2018-24, Gumisawa-cho, Totsuka-ku, Yokohama
Ulith, P.	D	J.M. Voith GmbH, 7920 Heidenheim
Venkataraman, C.P.	IND	P.O. Khadakwasla, Research Station, Poona-411 024
Vickery, B.J.	CDN	University of Western Ontario, London, Ontario
Vigander, S.	USA	TVA Engineering Laboratory, Norris, TN 37828
Vrijer, A.	NL	Delft Hydraulics Laboratory, P.O.Box 177, 2600 Delft
Wambsganss, M.W.	USA	Argonne Nat. Laboratory, 9700 South Cass Ave., Argonne IL 60439

Wang, J.C.	USA	Hydro Res. Science, 3334 Victor Court, Santa Clara,CA 95050
Wardlaw, R.L.	CDN	National Res. Council of Canada, Ottowa, Ontario KIA OR6
Watremetz, M.	F	CETIM, 52, avenue Fèlix-Louat, 60300 Senlis Cedex
Watson, A.G.	GB	C.E.G.B., Berkeley Nuclear Lab., Berkeley, Glos.
Weaver, D.S.	CDN	McMaster University, Dept. of Mech. Eng.,Hamilton, Ontario L8S 4L7
Whitton, P.N.	GB	C.E.G.B., Berkeley Nuclear Lab., Berkeley, Glos.
Widell, K.-E.	S	Stal-Laval-Turbin AB, 61220 Finspong
Wong, H.Y.	GB	University of Glasgow, Dept. of Aeronautics and Fluid Mech., Glasgow G12 8QQ, Scotland
Wootton, L.R.	GB	Atkins Research & Development, Woodcote Grove, Ashley Road, Epsom, Surrey
Wyatt, T.A.	GB	Dept. of Civil Eng., Imperial College, London SW7 2BU
Yavits, S.N.	SU	Turbine Bldg. Association, Leningradskij Metallitsheskij Zavod
Zagustin, K.	YV	Instituto de Mecanica de Fluidos, Universidad Central, Caracas
Zdravkovich, M.	GB	Dept. Aeromech. Eng., University of Salford, Salford M5 4WT
Žukauskas, A.	SU	Lietuvos TSR Mokslu Akademija, Linino prosp. Nr. 3, 232600 Vilnius, MTP-1, Litauische SSR

CONTENTS

Technical Session A : REACTOR AND HEAT-EXCHANGER COMPONENTS

Technical Session D : SHIP STRUCTURES

Technical Session E : OCEAN STRUCTURES

Technical Session F : BRIDGE DECKS, BEAMS, AND CABLES

Technical Session G : HIGH-RISE BUILDINGS AND STRUCTURES

Workshop : PREDICTION AND MODELLING OF FLOW-INDUCED VIBRATIONS

Panel Discussion : CRITICAL UNRESOLVED PROBLEMS

A1 FLOW-INDUCED VIBRATIONS IN NUCLEAR REACTORS AND HEAT EXCHANGERS
Practical experiences and state of knowledge

Michael P. Païdoussis
Professor and Chairman
Department of Mechanical Engineering
McGill University
Montreal, Québec, Canada

SUMMARY

This paper presents 50-odd practical cases of flow-induced vibration problems in heat exchangers and nuclear reactors, sufficient to make it clear that such problems are of great technical and financial importance and of safety concern. Broadly grouped according to flow configuration, the vibration problems are distinguished as having been induced by (a) cross-flow, (b) internal axial (pipe) flow, (c) external axial flow, and (d) annular and leakage flow. Those in (a), the most numerous, are further classified according to the most likely underlying mechanism: (i) vortex shedding, with or without acoustic resonance, (ii) fluid-elastic instability, and (iii) buffeting and impingement by high velocity fluid streams.

Then follows a brief review of the state of the art in each of these categories. It is shown that the mechanism of vibrations induced by axial flow, either internal or external, is well understood, but means of prediction of the excitation field are not yet adequate. Similarly, the fundamental mechanism of annular-flow-induced vibrations is fairly well understood, but complexities of geometry and flow field make vibration prediction sometimes difficult. In contrast, the mechanism of cross-flow-induced vibration of cylinder arrays, as opposed to single cylinders, by vortex shedding is not sufficiently clear, and its distinction from buffeting is fuzzy; moreover, prediction methods are inadequate. Also, the underlying mechanism of fluidelastic instability is not fully understood even though methods of prediction are available. Finally, although some of the building blocks for comprehensive design guides do not exist, a great deal of research, testing and refining remains to be done.

1. INTRODUCTION

The purpose of this paper is twofold: to present an assortment of reported cases of flow-induced vibration problems in heat exchangers and nuclear reactors, and to discuss the state of the art of theoretical and practical knowledge presently available, and variously used, to prevent such occurrences.

The list of practical experiences presented here is by no means exhaustive, nor is the information relating to each necessarily complete. Manufacturers, utilities and other users, and state boards are not always forthcoming in disclosing problems in full blown detail, even when they have been able to successfully solve them. Several factors come into play, amongst which are maintenance of commercial advantage, preservation of technical secrets, corporate image and national pride, and even fear of litigation. In the nuclear industry there is the added need for circumspection to guard against the emotional distortions of fact, of which the antinuclear zealots have shown themselves capable.

In this light, two points should be made at the outset. Firstly, the apparently large number of flow-induced vibration problems presented here which are associated with nuclear power plants, *vis-à-vis* conventional ones, does not imply that the former are more prone to this kind of failure, but merely reflects the availability of this information, as nuclear regulatory boards demand that even the smallest of operating problems be reported. Secondly, the degree of accessibility of this information varies from country to country and, hence, if incidence of failure in some reactor systems, or in some countries, would appear to be higher than in others on the basis of the cases discussed here, this should not be construed to mean that the latter are necessarily free of such difficulties - perhaps, quite the contrary. In this connection, the list of cases to be presented also reflects, to some extent, the author's abilities as a detective, no less than the extent of his contacts in the industries involved, which cannot be claimed to be uniformly spread throughout the world. The author is most grateful to the individuals and industries who responded to his appeal for help in assembling the cases of flow-induced vibration to be discussed here; one purpose of the foregoing is to ensure that this compilation of flow-induced problems is not made the object of facile, and hence distorted, conclusions based merely on appearances.

Some statistics will put the problem of flow-induced vibration in heat exchangers and nuclear reactors in perspective. Although overall statistics are not available, some information relating to nuclear plants does exist. Thus, in all plants considered by the U.S. Nuclear Regulatory Commission, there were 261 shutdowns and power reductions between May 1974 and December 1975 for other than refueling and maintenance purposes, of which 23 were due to flow-induced vibration problems [1]. More particularly, considering reported tube failures in steam generators of water-cooled nuclear power plants around the world, of the 19 in-service failures reported to the end of 1971, two were due to flow-induced vibration (fretting) [2]; in 1973, of the 13 reported cases, again two were related to flow-induced vibration [3]; in 1974 most failures were caused by corrosion [4]; and in 1976 of the 3,763 steam generator tubes plugged, 2.6% failed due to fretting and 0.1% due to fatigue [5]. More particularly still, of the 62 reported PWR† steam generator problems [6], five were caused by flow-induced vibration and ten by the allied mechanism of water hammer. Thus, it is quite clear that flow-induced vibration is a real, though not overwhelming problem in nuclear plant. This is also the case with conventional plants, in boilers and heat exchangers alike. Concerning the latter, although it is generally accepted that "the vast majority of shell-and-tube heat exchangers apparently do not have flow-induced tube vibration problems" [7], there is no shortage of reportable cases; some spectacular failures come to mind, such a heat exchanger bundle in a petroleum refinery being completely destroyed within 24 hours of start-up.

The financial implications of power plants being put out of service for repairs are quite impressive, and in the case of nuclear plants quite formidable. In the U.S.A. replacement power costs alone, for a 740 MWe plant, were \$1.7 million for 17 days in 1973 [8]. Estimates of steam generator replacement, including replacement power costs during outage, range from \$100 million to \$300 million per plant [6]. For one particular reactor being out of service in U.K., replacement generator costs were £100,000 per day in 1977, and if the boiler involved had to be replaced, the plant would be out of service for five years, at an estimated total cost of £200 million [9]. In the nuclear industry flow-

---

† Pressurized Water Reactor

induced vibration problems can also have a health dimension. Even minor repairs involve handling of at least mildly radioactive materials. Major repairs to reactor internals sometimes necessitate the use of frogmen in the still radioactive pressure vessel [10]. The duration of one such vibration-induced shutdown, involving divers, was nearly three years.

Why do flow-induced vibration failures continue to occur? The following are some of the reasons: (i) flow-induced vibration is usually a secondary design parameter, as compared to heat transfer or, if applicable, reactor physics imperatives; (ii) the state of the art is insufficiently advanced to permit fully successful design in all cases; (iii) under the pressure of the enormous outage costs, interim "solutions", based on incomplete understanding of the mechanism involved, themselves fail in turn; (iv) feedback from the field is sometimes very sketchy, and not widely accessible; (v) there is insufficient lateral communication with other engineering fields where flow-induced vibration problems are encountered. Some of these points will be expanded further, below, and will be amply illustrated in Section 2.

In connection with (ii) above, it has oft been said that one of the problems is that researchers communicate their findings to designers neither in sufficiently understandable form, nor sufficiently quickly, if at all. Perhaps in an effort to correct this, researchers are now apt to propose design guidelines and computerized models, based nevertheless on insufficiently tested theory. In 1955, in a paper dealing with flow about bluff bodies, Roshko remarked [11] that not enough attention had been paid to the accumulated experimental information, which is so indispensable for the realistic formulation of theoretical models. Alas, nowadays too much theorizing is made based on too little experimental information; some researchers charge ahead to propose all-encompassing theories without cautioning the unwary reader of their tentativeness. Compounding the problem is the fact that, unlike unpretentious old fashioned empirical design guides, modern theories come with sophisticated façades and computerized formats, possessing an aura of authority not sufficiently frequently questioned by designers.

Feedback from the field, referred to in (iv) above, can be frustrating to researcher and designer alike, mainly because operators' first priority is to get things working again rather than find out why they did not work in the first place. The following report from a field engineer illustrates the point [12]: "There is a severe vibration on this unit. The forced draft duct, gas duct, and superheater-economizer sections all vibrate. The frequency I would guess to be 60-100 cps. It feels about like one of these 'ease tired feet' vibration machines. Vibration starts at approximately 700,000 lb of steam per hr and continues up to and above full load on gas". Even this information, however, is usually never published; for, once the problem is solved, often heuristically, the file is closed and the case remains known to but a few.

The above discussion makes it abundantly clear that there is a need for a Symposium such as this, where field experiences from many domains will be brought to the attention of designers and researchers working in different fields. It will become clear that the easy problems have already been solved. Alas, the days are long past when waving a three-pronged fork in the air and noting the difference in tone between row-array and column-array waving was sufficiently new to merit publication in the Proceedings of the Royal Society (London). The quality of the quantitative information demanded nowadays, no less than the complexity of the phenomena involved, requires consistent effort and commitment by the researcher, if enlightenment is to be achieved. Therein lies the challenge.

2. A COMPENDIUM OF PRACTICAL EXPERIENCES

The practical experiences of flow-induced vibration to be related here have been assembled after a two-year quest which, at the time of writing, has resulted in a 50 cm thick pile of documentation - some of which, to be frank, required a great deal of ferreting out.

The Cases, mostly field experiences, are presented in the appended 42 Charts. Charts I-XXIII contain the Cases related to cross-flow-induced vibrations; Charts XXIV-XXVIII and XXIX-XXXII, those related to internal and to external axial flow, respectively; Charts XXXIII-XLII, those related to annular and leakage flow. As some Charts contain more than one Case, the Cases have also been numbered, sequentially, for easy reference; thus, Chart I contains Case 1, Chart II contains Cases 2 and 3, *et seq*. In each case, a brief description of the flow-induced vibration is given, together with a likely cause of the occurrence and the solution implemented. The description given is sometimes a highly compacted statement of a great deal of information, and sometimes just about all that is available.

Several comments should be made.

(i) Manufacturers' and proprietors' names have been suppressed in the Charts, since they need not concern us here.

(ii) A great deal of information, essential for the presentation of fully documented cases, is proprietary and inaccessible to the author. Consequently, the interpretations and explanations given here will likely differ from those to be eventually put forth by those directly involved. It is hoped that their inclusion here will serve as a catalyst to publication.

(iii) The likely cause of the flow-induced vibration problem given in the primary sources referred to is sometimes no longer valid, in view of present knowledge, and differs from that put forth by the author. In this connection, the dates of the references involved are most important.†

(iv) Some of the cases presented have been taken directly from papers with similar aims to this one; notably those compiled by Hartlen [13], Nelms and Segaser [14], and Pettigrew [15]. Other rich sources of cases of flow-induced vibration in heat exchangers, namely the data banks of the Tubular Exchanger Manufacturers Association (TEMA) and Heat Transfer Research Inc. (HTRI) of the U.S.A., were unfortunately not accessible.

(v) Cases relating to pumps, valves and the like have not been included, as they are in the realm of Session C of this Symposium.

The following abbreviations and symbols are used in the Tables:

AGR	= advanced gas-cooled reactor	BWR	= boiling water reactor
GCR	= gas cooled reactor	LMFBR	= liquid metal fast breeder reactor

---

† It is recalled that until the late 1960's it was not widely known that the Strouhal number for tube arrays in cross-flow differed from the classical S=0.2 (for subcritical Re), and that the existence of fluid-elastic (or 'whirling') instability in such systems was not recognized till the end of 1970. This period, before the pioneering papers of Y.N. Chen [16] and H.J. Connors [17], may be referred to as B.C., without implying any metaphysical qualities to the authors. Other benchmark contributions, such as Owen's [18] and Savkar and So's [19], similarly influenced contemporary thinking, each in its own way.

LWCHWR = light-water cooled, heavy-water-moderated reactor

PWR = pressurized water reactor

D = diameter

f_n = natural frequency

K = constant in Connors' fluid-elastic expression

p = pitch, or intertube distance, in cylindrical arrays

S = Strouhal number

S_C = Y.N. Chen Strouhal number

δ = logarithmic decrement

PHWR = pressurized heavy-water-moderated and -cooled reactor

f_a = acoustic frequency

f_{vs} = vortex-shedding frequency

m = mass of the component per unit length

s_t, s_ℓ = distance between columns and rows, respectively, divided by D

S_{FH} = Fitz-Hugh Strouhal number

U = flow velocity

ρ = fluid density

The classification of the Cases into four classes, according to whether the vibration is induced by cross, axial internal, axial external, or annular flow is in many ways arbitrary, as in some situations the flow often has both axial and cross components and is probably unsteady too; also, since this paper deals exclusively with contained flows, the distinction between external axial and annular flows is artificial, depending on the annular gap. Nevertheless, this classification serves two important purposes. Firstly, it provides a framework for the discussion of the state of knowledge, the great majority of fundamental work being compartmentalized according to whether it deals with cross or axial external flow, or with internal flow. Secondly, it serves one, after even casual perusal of the Charts, to arrive at the immediate and important conclusion that most serious problems, so far, have been caused by annular and cross-flow.

Concerning the Cases of cross-flow-induced vibration presented in Charts I-XXIII, it is noted that those involving solitary cylinders in cross-flow are rather rare, indicating that design information is perhaps adequate in this area. This is not the case with groups of cylinders, where problems abound, notably cases of fluid-elastic instability.

Mechanism	Certain or very likely	Possible
Vortex shedding	4	6
Vortex shedding and acoustic resonance	1	1
Fluid-elastic instability	10	6
Buffeting and impingement	7	4
Unknown	2	-

Table 1: Classification of the 33 Cases of cross-flow-induced vibration

Table 1 categorizes the 33 Cases presented; where there is more than one plausible mechanism, the Case appears as "possible" under two categories. It is noted that the many cases of fluid-elastic instability reflects the dates of design. Before 1970 there was a great deal of Procrusteanism in trying to explain everything in terms of vortex shedding (e.g. Cases 16-20, 22; Charts XI-XIV). Even after this instability was put on the map by Connors, however, design criteria remained too liberal for quite a time. Another major source of cross-flow-induced vibration is impingement of high velocity streams, e.g., on tubing close to the inlet nozzle of heat exchangers (*vide* Cases 9,11,12; Charts VII, IX).

The Cases associated with internal and external axial flow (Charts XXIV-XXXII) are quite varied. The mechanism is sometimes associated with excessive turbulence, noise or acoustic coupling, swirl, or more exotic excitations. Finally, the Cases of annular and leakage flow (Charts XXXIII-XLII) are quite impressive by their destructiveness; it would appear that, in contrast to the previous classes of flows, the possibility of flow-induced vibrations here was often entirely overlooked.

Further discussion of the Cases will be presented in conjunction with the state of the art reviews, which follow. One last general observation, however, is that the cures implemented are mostly of the "run of the mill" variety, as opposed to the ingenious type; e.g., stiffening tubing by reducing spans or by lacing, introducing deresonating baffles, providing added or stiffer supports, reducing upstream noise, diminishing leakage flow, reducing local flow velocities, etc. Moreover, in the great majority of Cases, the cures did not result in reduced efficiency of operation; in other words, they could have been made part of the original design, had problems been anticipated - although this was sometimes impossible because of the poor state of knowledge at the time.

3. VIBRATION INDUCED BY CROSS FLOW: STATE OF THE ART

3.1 Solitary cylinders

The forces exerted on a solitary bluff body, and especially on a cylinder, and the vibration induced thereby have been studied extensively for a long time. Several reviews exist to which the reader is referred [20-25], as well as the proceedings of the previous IUTAM/IAHR Symposium at Karlsruhe [26].

In the context of heat-exchanger and nuclear-reactor components, two mechanisms of excitation of solitary cylinders in cross-flow are of interest: vortex shedding and buffeting.

The mechanism of vortex-excited vibration is well known. In the subcritical and transcritical flow régimes characterized by $Re \lesssim 2x10^5$ and $Re \gtrsim 3.5 x10^6$, respectively, periodic shedding of vortices may excite large amplitude oscillation, if system damping is small, and especially if coincidence between vortex-shedding frequency, f_{vs}, and natural frequency, f_n, should occur. Typical Strouhal numbers for the two régimes are S=0.2 and S=0.3, respectively. In the intermediate, critical and supercritical Re régime, S can be as large as 0.45, but vortex-shedding periodicity is not sharply defined. All the foregoing are quantitatively highly dependent on such parameters as confinement (blockage), incident free-stream turbulence, aspect ratio and two-dimensionality of flow. Lift and drag coefficients, required for computation of the excitation forces are available [27].

An important phenomenon associated with vortex-shedding excitation is lock-in [28] (more appropriately conveyed by the French *accrochage*). Suppose that resonance ($f_{vs}=f_n$) in a system is associated with a flow velocity U_0; through vibration of the cylinder at f_n, lock-in occurs at $U_1<U_0$, and vortex shedding is forced to also occur at f_n; this continues to $U_2>U_0$, before vortex shedding snaps back to the value prescribed by U_2S/D. The phenomenon occurs provided that the parameter $m\delta/\rho D^2$ is not overly large [25]. In addition to normal resonance, harmonic and sub-harmonic resonances have also been observed, e.g. [29]; moreover, oscillation predominantly in the streamwise direction at $2f_{vs}$ has also been observed, in contrast to the usual case when it is mainly normal to the flow.

Finally, it was recently reported [30] that lock-in of vortex shedding to a transverse acoustical frequency, f_a, of the flow channel is possible; also, vortex shedding may be induced to occur at twice the normal frequency, to equal f_a.

A great deal of work is currently under way to fill the gaps in our knowledge, e.g. in the critical and supercritical ranges of Re. Mathematical modelling of the dynamical behaviour of the cylinder (*vide* [25]) is progressing.

The second mechanism of interest is turbulent buffeting by the flow. It is operative at all flow velocities and involves a much wider frequency spectrum and smaller amplitudes, in comparison to resonant vortex excitation; the response to buffeting alone is proportional roughly to U^2. Recent work by Savkar & So [19] renders the distinction between vortex-shedding and buffeting excitation rather delicate. For a fixed cylinder they found the fluctuating lift coefficient is diminished by almost an order of magnitude when the free-stream turbulence level† is increased from 1% to >2-4%; the force spectrum on the cylinder is then fairly broad-band with a maximum at a frequency somewhat higher than f_{vs}. It may be expected, nevertheless, that if the cylinder were free to vibrate at a U such that $f_{vs} \simeq f_n$, the resonance effect described earlier would still occur even at large levels of turbulence (albeit with reduced energy input) - for motion correlates vortex shedding spatially and reinforces periodicity.

In general, knowledge in this area for design purposes is adequate and improving. However, it has been traditional to consider fluctuating forces as not posing any serious problems, provided that resonance is avoided, since their magnitude is considerably less than the steady drag force. This may be sound provided that f_{vs} and f_n are truly far apart and buffeting generates very small amplitudes; otherwise fatigue may occur even in the absence of resonance (Case 33).

The state of the art for cross-flow-induced vibration of solitary cylinders is quite advanced in comparison to that for cylinder arrays (e.g. tube banks or bundles of cylinders) in cross-flow. In this case, in addition to buffeting and vortex excitation, jet switching [31] and fluid-elastic instability [17], and acoustic coupling also, are possible. These phenomena will be discussed next.

3.2 Vortex shedding and buffeting in tube banks

Traditionally, vortex shedding in tube banks is presumed to be similar to that for a solitary cylinder, vortices being shed alternately from each tube, with some kind of symmetric or antisymmetric synchronization for tubes in the same row and in adjacent columns. This is rather difficult to visualize in closely spaced tube arrays, where there is obviously insufficient room for the shear layers to roll up in the classical manner. To overcome this difficulty, several alternative mechanisms differing from classical vortex shedding, yet possessing periodicity, have been proposed, notably Chen's "jet switch", "wake swing" and "jet instability" models [32,33] — but remain unsubstantiated by measurement.

The Strouhal numbers associated with flow through tube banks have been compiled for different layout angles by Fitz-Hugh [34,25] and by Chen [35]. The two sets of S, to be referred to as S_{FH} and S_C, are compared in Fig.1. The

† In these experiments, turbulence intensity was increased to as high as 9%, with an integral scale $\ell/D \simeq 1.3$.

spacing parameters s_t, s_ℓ, and for equilateral triangular arrays, p/D are defined in Fig.2, as is the flow velocity $U=U_\infty[s_t/(s_t-1)]$, where U_∞ is the free stream velocity upstream of the tube bank. It is noted that although both Chen and Fitz-Hugh draw their information from mostly the same sources, large discrepancies exist (Fig.1); the main differences are associated with the very high values of S_C as $s_t \to 1$ [35], values as high as 4.2. Moreover, it is not made clear if in the collected data there was acoustic reinforcement or not.

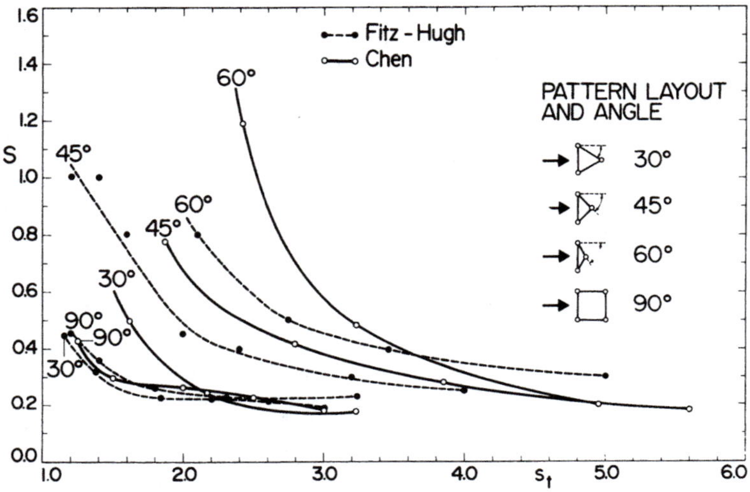

FIG. 1. COMPARISON OF STROUHAL NUMBERS GIVEN BY FITZ-HUGH AND REVISED CHEN PLOTS FOR TUBE BANKS IN CROSS FLOW

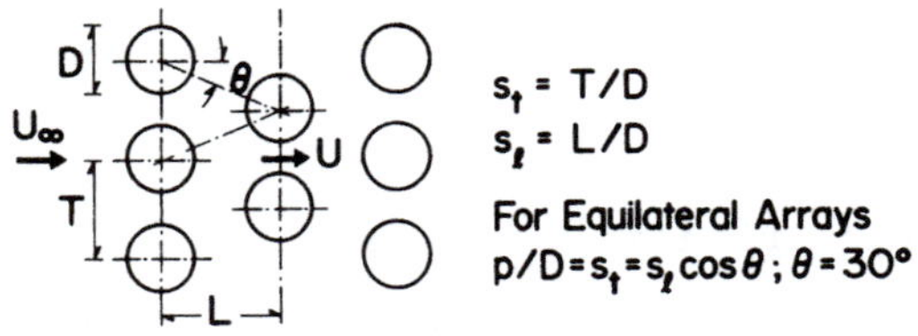

FIG. 2. DEFINITION DIAGRAM FOR CROSS FLOW IN TUBE ARRAYS

There is considerable skepticism about the existence of identifiable vortex shedding and discrete periodicity deep in a tube bank [18], although accepted as present in at least the first few rows [36]; Grover and Weaver [37] report discrete vortex shedding in the first 15 rows of a tube array and find that flow-field periodicity is greatly enhanced by the correlating influence of tube motions. For liquids, according to [36], vortex-shedding excitation may occur if f_n coincides, or is close to, f_{vs} up to $(U/f_nD)^{-1}=S<1$, and, according to another study [38], provided that the damping parameter $m\delta^{\frac{1}{2}}/\rho D^2<30$. Vortex-shedding excitation in two-phase flows appears to be absent, whether by increased fluid damping, decreased speed of sound, or another mechanism, is not known.

For computing the vortex-shedding response in tube banks, data on lift coefficients, C_L, are rather scarce (e.g.[36]), especially in view of the large map of s_t and s_ℓ that needs be filled. Some of the C_L's quoted are very small, ~0.05, and this fact is utilized by Chen [33] to separate regions of likely and unlikely vortex-excited vibration.

Buffeting in tube banks is essentially of the same nature as for a solitary cylinder, except that upstream cylinders in this case add to the general turbulent excitation of those downstream. Deep enough in a tube bank, one may presume that a steady state is reached, i.e., a stationary spectrum of turbulent pressure fluctuations is achieved. In a remarkable study [18], Owen goes beyond this point and proposes a model in which buffeting is responsible for the observed periodicity, casting doubt on the very existence of vortex shedding as such, or other regular wake patterns. He states: "Deep enough within a bank of tubes, the cumulative growth of random irregularities in the labyrinthine-like, high-Reynolds-number flow must lead to a state of almost complete incoherence on which it is difficult to imagine any superposed regular pattern....to be discernible." By physical reasoning he then goes on to predict that in a given tube bank there is a predominant frequency, obtainable from

$$(f_bD/U)s_\ell s_t = 3.05(1-1/s_t)^2 + 0.28, \qquad (1)$$

where f_b is the peak of the turbulent energy spectrum.

The differences between vortex-shedding advocates and Owen disciples, although fundamental, may not be as irreconcilable as might appear. They reflect the same lack of clarity in the distinction between classical vortex shedding and buffeting brought to the fore by Savkar and So, in their work on solitary cylinders discussed earlier. From the *practical* viewpoint the two precepts lead to essentially the same Strouhal numbers, as may be seen in Fig.3, where $S=f_bD/U$ from equation (1) is compared with S_{F-H} and S_C — except that Owen's data turn downwards for sufficiently small s_t*. In this light, the observed "vortex-shedding peaks" in diagrams of tube vibration amplitude *versus* U [36,37] may equally well be referred to as "buffeting resonance peaks"; then

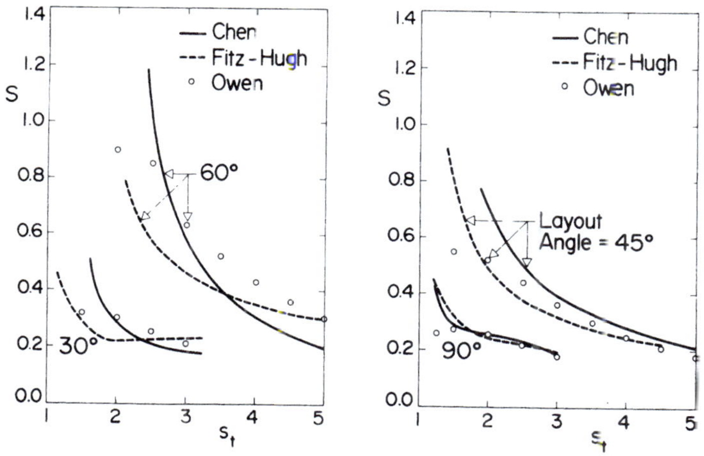

FIG. 3. VORTEX-SHEDDING STROUHAL NUMBERS ACCORDING TO CHEN AND FITZ-HUGH COMPARED TO OWEN'S DOMINANT BUFFETING FREQUENCY STROUHAL NUMBER

* In this connection, Owen himself suggested that it would be prudent to place reservations on (1) for $s_t \to 1$.

the peak is not due to the coincidence of f_{vs} with f_n (f_{vs} being meaningless in this theory) but to coincidence of f_n with f_b. *In practice*, since $f_b \simeq f_{vs}$, it makes no difference. Since Owen's work is based on sound physical reasoning, one is inclined to lean towards that model of excitation, rather than the "vortex shedding" models. Finally, one might be tempted to explain the absence of "vortex shedding" in two-phase flow in terms of the much broader, wide band energy spectrum of such flows.

In the discussion of the Cases, as well as in most of the following discussion, the conventional "vortex shedding" nomenclature has been retained. Clearly, in view of the above, one could read "buffeting resonance", where vortex shedding resonance is mentioned. In what follows, where f_{vs} is mentioned, one should read "f_{vs} or f_b, depending on one's point of view".

In the foregoing, no acoustical coupling was assumed to exist; its presence may radically alter the picture. Let f_a be the natural frequency of a standing wave in the conduit containing the tube array, perpendicular both to flow and to the tube axes. Then, if f_a coincides with f_b given by (1), or f_{vs} from Fig.1, this provides a mechanism of energy transfer from the mean flow to acoustical wave motion; moreover, acoustical vibration has a correlating effect on vorticity, enhancing the energy content at f_{vs}. Indeed, acoustical coupling then becomes the dominant mechanism, and the previous reservations about how deep in a tube bank may vortex shedding occur vanish. The resultant sound amplification may lead to deafening levels (e.g. Cases 21,23). If a triple coincidence of f_a, f_{vs}, and f_n of any component in a heat exchanger should occur, the heat exchanger may be expected to be destroyed within hours (as mentioned in Case 23). A good analytical foundation is given by Baylac & Grégoire [39]. This phenomenon is associated mainly with gas flows. In water, the speed of sound is higher and equipment dimensions generally smaller, implying high f_a; at the same time flow velocities are smaller, and f_{vs} tend to be lower; hence, resonances are more difficult to achieve. For steam raising units, Chen & Young [40] propose the empirical guide that in-line arrays will not develop serious acoustical resonances if $(Re/S)(1-1/s_\ell)^2(1/s_t) \leqslant 2{,}000$, i.e. if viscous dissipation in the fluid is relatively high.

3.3 Fluidelastic instability of tube banks

At sufficiently high reduced flow velocities, U/f_nD, tube banks are subject to a so-called "fluidelastic" or "whirling" instability, resulting in large amplitude limit-cycle motion, with frequency f_n. Although elliptic paths of individual cylinders and organized motions of groups of cylinders are observable from time to time, the overall pattern appears to be incoherent, perhaps due to nonlinear effects of amplitude. The mathematical models in existence, the original one by Connors [17], as well as Blevins' more refined one [41,42], are attractively simple. They propose a quasi-static, position-dependent mechanism associated with energy input/output arguments; if the tubes execute small motions with prescribed relationships to those of adjacent tubes, then energy may be extracted from the fluid flow over a cycle of motion; if this is larger than the energy lost by dissipation, then the motion is amplified, i.e. the system is unstable. This mechanism is related to that proposed earlier by Roberts [31], where energy is fed into the vibrating cylinders through the motion-induced switching of paired "jets" in the interstitial passages, the switching implying changes in the wake size and, hence, in the drag forces; because of the finite time required for the jet switch to be effected, this mechanism is expected to be operative only for $U/f_nD>75$.

The underlying mechanism of fluidelastic instability cannot be said to be well and truly understood. A number of key questions remain unanswered. One is why the cylinders oscillate at f_n [24,43], when there is no *a priori* reason for this; theory simply takes this as given. Theory also proposes a specific whirling pattern, involving at least three cylinders; yet Weaver & Grover have shown [44] that even if all cylinders but one in a matrix are immobile, then that one becomes fluidelastically unstable at the prescribed flow; this is supported by Gorman's work with dissimilar tubes [45]. Also, the monotonic effect predicted by theory in which detuning the tubes stabilizes the system [41], is similarly not borne out by experiments [47], disappearing if detuning is larger than ~9%.

The original Connors expression for the threshold of instability of a single row of tubes

$$(U/f_nD) = K(m\delta/\rho D^2)^{\frac{1}{2}} \tag{2}$$

with K=9.9, has unfortunately been misused, for tube banks. However, modified with lower K's - e.g. K=3.3, the rock-bottom safe value proposed by Pettigrew [36,46] - it has served well as a design tool. The question remains why.

The reason, and it is surprising no one has stumbled on this before, is that it is similar to "well-founded" empirical expressions rather than because it is based necessarily on sound physical principles. Thus, Coit, Peake & Lohmeier's [48] empirical "severity number", extant since 1966, may be re-written in "Connor's form" as $(U/f_nD) = 15.4\ (m\delta/\rho D^3)^{\frac{1}{2}}$ for condensers, where D is in inches - the expression not being dimensionless; since typically, D≃0.75-1.25 in. for condensers, one obtains

$$(U/f_nD) = 15.4(m\delta/\rho D^2)^{\frac{1}{2}}\ , \tag{3}$$

remarkably similar to (2). It is nonconservative, by the way; hence, the condenser failures reported in the Charts (e.g. Case 25). Significantly, Peake, Gestenkorn & Arnold [49] in 1975 recommended reduction of the severity factor by 1/1.7, which leads to K=9.06!

Also, Thorngren's "collision damage number" [50] may be re-written in the form

$$(U/f_nD) = 2.76(m/\rho D^2)^{\frac{1}{2}}[(p/D)-1]^{\frac{1}{2}}\ ; \tag{4}$$

here, taking for instance p/D=1.5 in (4) and the conservative values δ=0.1, K=3.3 in the Pettigrew safe version of (2), one obtains

$$(U/f_nD)_{THOR} \simeq 1.95(m/\rho D^2)^{\frac{1}{2}} \quad \text{vs} \quad (U/f_nD)_{CON/PET} \simeq 1.04(m/\rho D^2)^{\frac{1}{2}}\ . \tag{5}$$

Finally, it should also be noted that it is possible to obtain (2) by very simple means, not taking recourse to the geometric and position-dependent arguments of the theory, as demonstrated by Heinecke [51].

The single-row expression originally proposed by Connors has been adhered to tenaciously, in the face of contradictory evidence, much as Aristotelian physics was in the Middle Ages - in both cases, and with no such wish by the authors, retarding progress. Thus, the evidence of [52] suggests that the critical U/f_nD is proportional to $\delta^{0.1}$ to $\delta^{0.3}$, rather than $\delta^{0.5}$; also Weaver's work [44,53] suggests $\delta^{0.21}$. Moreover, experiments by Nicolet *et al.* in which ρ was varied [52] suggest $U/f_nD \propto (m/\rho D^2)^{0.3}$, rather than $\propto (m/\rho D^2)^{0.5}$. In Figs. 4(a) - (c) are plotted the results from several sources taking $(U/f_nD) \propto \delta^0$, $\delta^{0.25}$ and $\delta^{0.5}$. It is noted that Fig.4(b) gives perhaps the narrowest band; moreover, $\delta^{0.25}$ gives good agreement with Weaver's data. These figures show how much of an open question this matter is. It is strongly suggested that consideration be given to adopting a relationship such as

$$(U/f_nD) = K[(m/\rho D^2)^{\frac{1}{2}}\delta^{\frac{1}{4}}], \qquad \text{or even } (U/f_nD) = K[(m/\rho D^2)^{\alpha}\delta^{\beta}]; \tag{6}$$

the first, i.e. that of Fig.4(b), with some safety margin on K, could be a start.

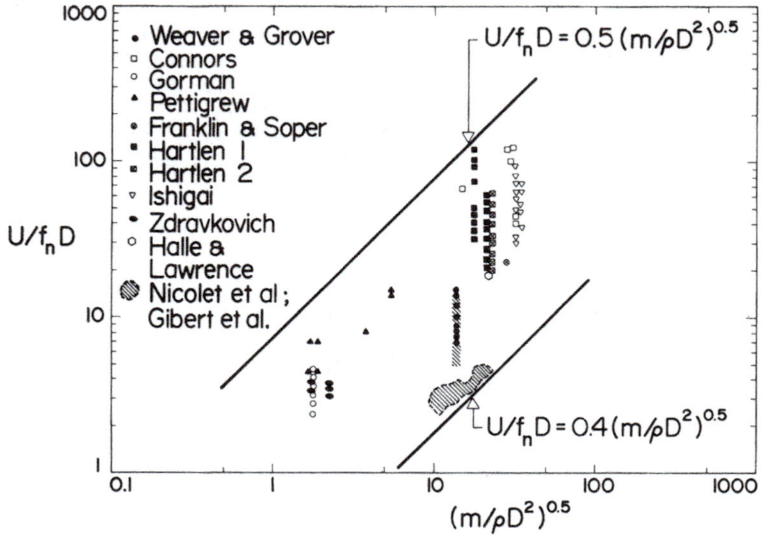

FIG. 4(a). VARIATION OF MASS PARAMETER WITH REDUCED VELOCITY FOR FLUIDELASTIC INSTABILITY (δ ABSENT FROM CORRELATION)

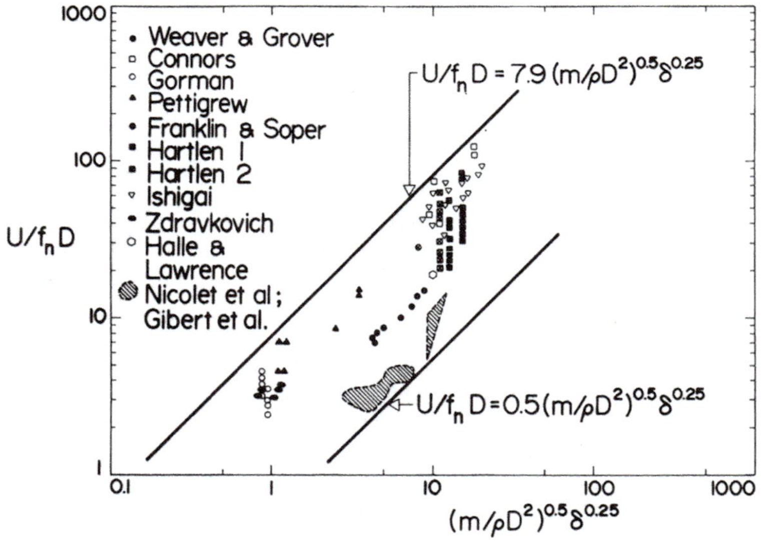

FIG. 4(b). VARIATION OF MASS PARAMETER AND DAMPING WITH REDUCED VELOCITY FOR FLUIDELASTIC INSTABILITY (WITH $\delta^{0.25}$ IN THE CORRELATION)

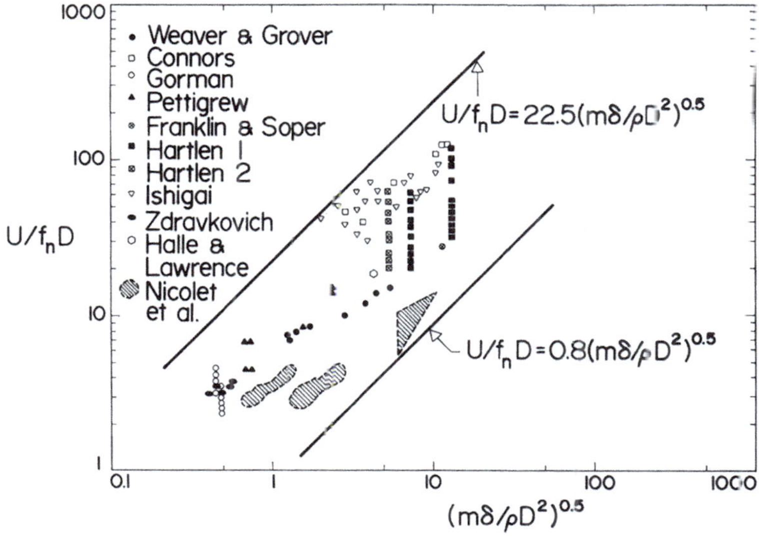

FIG 4(c). VARIATION OF MASS PARAMETER AND DAMPING WITH REDUCED VELOCITY FOR FLUIDELASTIC INSTABILITY (WITH $\delta^{0.5}$ IN THE CORRELATION)

Finally, it would instinctively seem right that an expression such as (6) should depend explicitly on s_t and s_ℓ, much as Thorngren's did (cf. [35]). An attempt to decide on this is made in Fig.5 only for data involving a single row. If the original form (involving $\delta^{\frac{1}{2}}$) is retained, then there seems to be a reduction of U_{crit} with s_t; however, in the expression proposed here, there is no clear dependence on s_t. For tube banks the situation is even less clear, since most experimental points are for tube arrays such that $1.2 \leq p/D \leq 1.5$, hardly a wide enough range to permit empirical correlation.

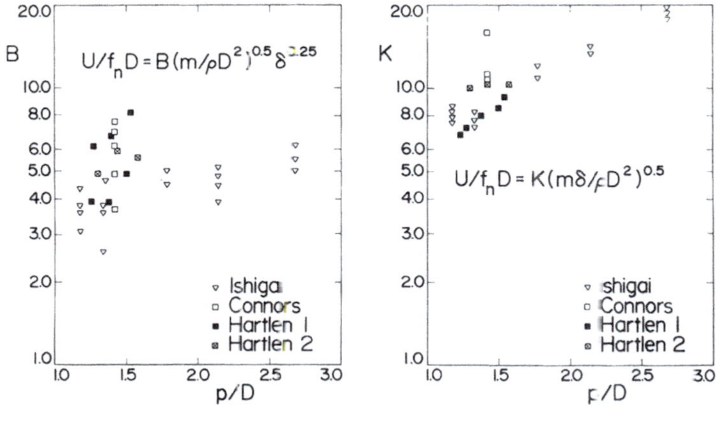

FIG. 5. VARIATION OF STABILITY COEFFICIENTS WITH PITCH RATIO FOR FLUIDELASTIC INSTABILITY OF SINGLE ROWS OF TUBES

In the foregoing, the U utilized is as shown in Fig.2. Weaver, on the other hand, proposes that another flow velocity be used corresponding to the minimum flow area, which is different than U for small s_ℓ. Clearly, unless more is understood on the mechanism of fluidelastic instability, one cannot resolve which U is the "right" one.

3.4 Tools for the designer for tube bank vibration

Several attempts have been made to provide a unified approach for dealing with at least some of the mechanisms discussed above, e.g. by Nicolet *et al.*[52], Gibert *et al.*[38], Pettigrew *et al.*[36,46], Y.N. Chen [35,33], Zdravkovich *et al.*[54,55]. Also, several review papers have appeared, e.g. [56-59,24] and Blevins' book [25]. Finally, a number of more practical papers, perhaps of more direct interest to the designer, have appeared from time to time, e.g. [7,14,48-50,60-63].

It may be considered that for steam generators, but also more widely if acoustical coupling is absent, Pettigrew's *et al.* approach is a sound and well-tested conservative design guide. For that reason it has been adopted for analysing the Cases of the Charts with respect to fluidelastic instability. This is shown in Fig.6, where all the cases pronounced as fluidelastically unstable are seen to lie in the unstable region if K=3.3 is adopted - even for the unrealistically high value of δ=0.1 assumed for some of the Cases.†

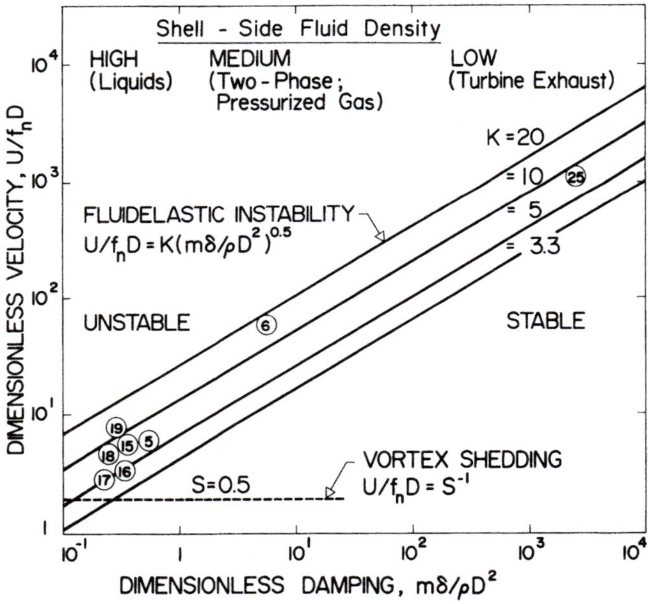

FIG. 6. PRACTICAL CASES OF FLUIDELASTIC INSTABILITY, FROM THE CHARTS OF THIS PAPER (CIRCLES) COMPARED TO PREDICTED INSTABILITY THRESHOLDS WITH VARIOUS VALUES OF K. NUMBERS IN CIRCLES ARE THE CASE NUMBERS

†It is noted that the value of f_n used to evaluate U/f_nD in some of the Cases was the tube single-span frequency, in the absence of detailed information. Strictly, the lowest f_n should have been utilized, treating the whole tube as a multiply supported beam (cf. Case 1).

Hence, the work of [36] may be used as a safe design guide for fluidelastic instability - and for vortex-shedding/buffeting excitation as well. However, in terms of fluidelastic instability, Pettigrew's expression may be *overly* conservative, because of the use of $U/f_nD \propto \delta^{0.5}$. A further manifestation of this is that a number of stably operating heat exchangers lie in the unstable régime in Pettigrew's stability chart [36], although the converse has never been found to occur.

Finally, it should be stressed once more that, although interim design guides are a necessity, there is a wide gap between the state of knowledge and the state of the art. Definitive design guides cannot be constructed before a great deal of further fundamental work is done.

4. VIBRATION INDUCED BY INTERNAL AXIAL FLOW: STATE OF THE ART

Internal flow influences the dynamics of piping (a) by means of the gross momentum transport associated with the mean flow, (b) through substantial perturbations of the mean velocity or mass flux, and (c) by the effect of near-field and far-field noise transmitted through the mean flow.

4.1 Mean flow effects

The dynamical effect of uniform steady flow on straight piping supported at both ends is to reduce its natural frequencies, f_n. This effect, which increases with flow velocity U, is associated with the centrifugal force exerted by the flow on the curved portion of a momentarily and arbitrarily deformed pipe - acting in a similar way to a compressive load. For sufficiently high U, this force overcomes the flexural restoring force, and the pipe buckles in its first mode (*vide*, e.g. [64]). In cantilevered pipes the mean flow also generates a damping effect, which for sufficiently high U changes into negative damping, causing the pipe to lose stability by flutter [65]. The mechanism here is more complex and is associated with a balance of work done, per cycle, by the centrifugal and Coriolis forces. Fig.7(a) shows the nondimensional critical flow velocity for the onset of buckling of pipes simply supported at both ends, $u_{cb}=\pi$, and the corresponding velocity for flutter of cantilevered pipes, u_{cf}. Here $u=(M/EI)^{\frac{1}{2}}UL$, where M is the mass per unit length of the conveyed fluid, EI the pipe flexural rigidity and L its length, or length between supports; m is the mass of the empty pipe, per unit length; u_{cb} and u_{cf} are the critical values of u, as defined above. As shown in Fig.7(a), cantilevered pipes are surprisingly more stable than those simply supported at both ends.

Other interesting features of the system are that (i) with flow there are no classical normal modes, as all eigenmodes involve travelling wave components [65] - this also implies peculiar wave transmission characteristics [66]; (ii) internal pressurization has a destabilizing effect on the system and may itself become the primary cause of buckling of pipes supported at both ends [67].

If the pipe is very thin, so that it may be considered to be a cylindrical shell, then in addition to the beam-type instabilities discussed above, shell-mode instabilities may also occur [68,69]; indeed for short shells, it is the latter which are the critical modes. The essential character of the dynamics of the system is nevertheless the same as for thicker pipes [70].

The behaviour of curved pipes is essentially the same, but with two main differences: (i) the dynamics of in-plane and out-of-plane motions are distinctly different (loss of stability being associated with the latter), and

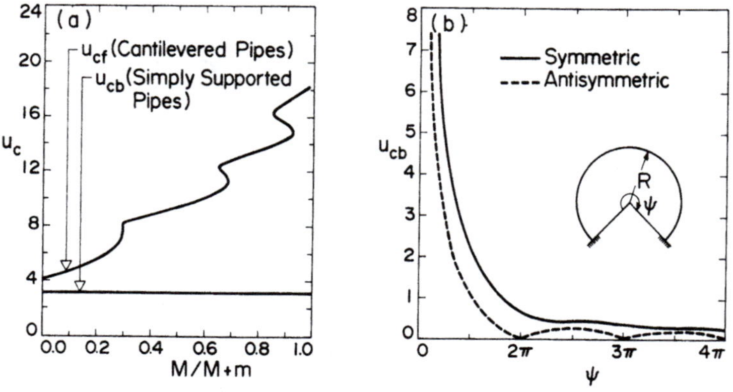

FIG. 7. (a) THE DIMENSIONLESS CRITICAL FLOW VELOCITIES FOR STRAIGHT PIPES CONVEYING FLUID [65]
(b) THE CORRESPONDING FLOW VELOCITIES FOR CURVED SIMPLY-SUPPORTED PIPES [72]

(ii) the in-plane shape of the pipe may be continuously altered with increasing flow as a result of the large centrifugal forces present [71-73]. The dimensionless critical flow velocity for out-of-plane buckling of simply supported pipes, u_{cb}, is shown in Fig.7(b) as a function of the subtended angle ψ.

Finally, if the flow has an appreciable swirl component, it may cause the piping to respond as a whirling shaft in its beam modes, or cause more complex shell-mode instabilities (Case 39).

The state of knowledge on the effect of the mean flow on piping is excellent and the literature very rich. The interested reader is referred to the pertinent review papers [74-77]. However, although fundamentally important, the fluidelastic instabilities referred to above rarely materialize in practice, mainly because in most common engineering systems the rigidity of the piping is too large as compared to the centrifugal forces attainable†; i.e., for most engineering systems $u<1$.

The following may be proposed as very conservative design guides, after a ~100% safety margin has been imposed to account for the effect of imperfections on u_{cb}. For straight piping, fluidelastic instabilities will not occur if $u<\frac{1}{2}\pi$. For shells with thickness-to-radius ratio, $h/a<10^{-2}$, $u' = U/\{E/\rho_s(1-\nu^2)\}^{\frac{1}{2}}<0.2$, as well as $u<\frac{1}{2}\pi$, may be proposed; however, in this case the results depend sensitively on h/a. For curved pipes subtending an angle of 180^0 without intermediate support, a rough design guide is that $u<\frac{1}{2}$ [72]. The foregoing apply provided that $\bar{p}AL^2/EI$ and $\bar{P}L^2/EI$ are appreciably <1, $\bar{p}$ and $\bar{P}$ being the mean pressure and the compressive force, respectively, and A being the flow cross-sectional area. These criteria have been adequately tested for straight piping [65,68,79-81]; however, the theoretical findings for curved pipes remain untested.

Finally, if a pipe should fracture in a way that the fluid escapes asymmetrically to the pipe axis, the lateral reaction force will cause the pipe to deform or move and perhaps hit and damage other components. This is the "pipe whip" problem, which has received considerable attention recently, e.g. [25,82,83].

---

† This is not always true; some interesting cases of such instabilities, partly in combination with pressurization effects, are reported by Den Hartog [78].

4.2 Gross velocity and mass-flux perturbations

It was shown that if the flow velocity in a pipe is disturbed harmonically with a circular frequency Ω, so that $U=U_0(1+\mu\cos\Omega t)$, μ being a small parameter, the piping may develop parametric oscillations (instabilities) in the neighbourhood of $\Omega/\Omega_n=2/k$, k=1, 2, 3 .... [84,76,85]. $\Omega_n=2\pi f_n$ are the circular eigenfrequencies of the pipe at a given mean flow velocity, U_0. The most important oscillations occur for k=1, i.e. where $\Omega=2\Omega_n$. Fig.8 shows some typical results for a simply supported and a cantilevered pipe. For the former, it is seen that there are unstable regions around $\Omega/\Omega_1=2$ and 1; similarly for the cantilevered pipe, about $\Omega/\Omega_2=2$ and 1. It is noted that Ω_2 and Ω_1 are the frequencies at the given u_0; these are lower than Ω_{02} and Ω_{01}, the values associated with $u_0=0$, as noted in §4.1 and seen on the scales of the right-hand ordinates of the figures. It is noted in Fig.8(a) that dissipation reduces the instability zones considerably. Theory was confirmed by experiments [84].

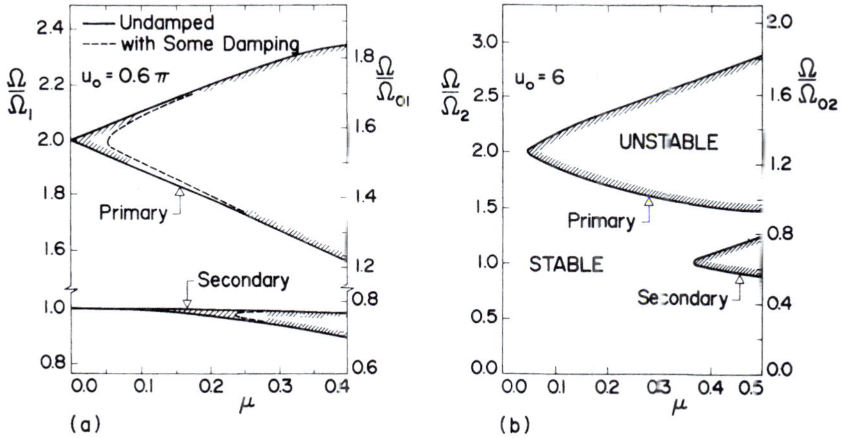

FIG 8 TYPICAL PARAMETRIC INSTABILITY REGIONS FOR (a) A SIMPLY SUPPORTED PIPE, AND (b) A CANTILEVERED PIPE, CONVEYING PULSATILE FLOW [76]

A similar approach was taken by Hara [86] in a theoretical-experimental study of vibrations induced by two-phase flow (in the slug- and plug-flow régimes) in horizontal piping. A number of dominant frequencies were found to exist in the otherwise random "liquid slug arrival" spectrum. Significantly, strong vibrations could be excited when the ratio of one of these frequencies, $\Omega_{s\ell}$, to the pipe first-mode frequency, Ω_1, was in the neighbourhood of $\Omega_{s\ell}/\Omega_1=2$, 1 and 2/3 - i.e., the first three classical parametric instability ratios.

4.3 Vibration induced by near- and far-field noise

Vibration of piping may be induced by relatively small pressure fluctuations generated in the turbulent boundary layer and by singularities in the flow conduit. The local pressure fluctuations associated with the boundary layer and the non-propagating part of disturbances associated with singularities are usually referred to as the near field; propagating disturbances in the form of acoustic waves constitute the far field. The current trend is to analyse the two together by (i) computing the total stochastic excitation field, usually in the form of a power spectral density and appropriate axial and circumferential correlation functions, usually presuming stationarity and ergodicity, and (ii) calculating the response of the piping to this excitation by means of appropriate acceptance, transfer matrix or finite element techniques. (In all this the effect of the mean flow is usually either neglected altogether or is presumed to have been approximately considered, e.g. in lowering the piping f_n; this is not too unreasonable for $u<0.25$, although the effect of neglecting the non-existence of classical modes has never been tested.) Clearly, knowledge of the following three is essential: (i) source signatures; (ii) propagation characteristics, and (iii) structural characteristics.

Considering the sources of excitation, it is noted that boundary-layer pressure fluctuations are of small scale and relatively low intensity in straight piping. The characteristics of these fluctuations are fairly well known, e.g. [87-90]. However, singularities such as pipe expansions and contractions, bends and tee-junctions, protuberances, pumps and blowers, generate disturbances of larger correlation lengths and higher intensities; moreover, they act as acoustical sources in the far field. The signatures of these components are less well documented and predictable - *vide* [91,92] - and a great deal of work remains to be done here.

The propagation characteristics of acoustical disturbances are quite complex. Some key features will be mentioned. Wave propagation is generally frequency-dispersive; i.e., the speed of sound, c, depends on the wavelength. Moreover, not all frequencies can propagate, and there are "cut-off" frequencies for some of the acoustical modes [92]. Acoustical waves are not always attenuated through dissipation and radiation; depending on the form of the conduit, local acoustical resonances may develop, amplifying the sound, e.g. [94,95] and Case 35. Further, although often only plane waves are considered, higher-order acoustic modes may be important also [96-98], especially through modal shape "coincidence" with shell modes of the pipe. Finally, it is not always possible to consider the propagation of acoustical disturbances independently of the excitation of the pipe, i.e. as if the pipe were inert; this is so when there is insufficient mismatch between the specific acoustic impedances† of the fluid and pipe materials [99],e.g. in liquid-filled metal piping. In such cases the pipe response modifies the sound field and *vice versa* - a simple manifestation being the well known dependence of c in a pipe on the flexibility of the pipe wall. The result is that acoustical-structural interactions become strongly coupled, and the acoustic field may no longer be viewed as simply providing a forcing function; wave propagation characteristics in the piping itself come into play, and the problem becomes related in some ways to that of waterhammer - which will not be discussed here (*vide* [100] for a review).

The field is too complex, and really deserves a special review paper of its own, especially the subject of far-field excitation. The interested reader is referred to Refs.[90,92,93,101-104], where further references may be found on the fundamentals of the problem. Finally, reference will be made to some recent work of direct interest to heat-exchanger and nuclear-reactor piping. Typical studies considering mainly near-field excitation are those of [105-107]. The characteristics of singularities, their contribution to near- and far-field noise, propagation characteristics in piping components, and the response of piping to such excitations, have been studied extensively in France [108-110,91, 94]. Analytical methods and computer codes have been developed for the design of piping of gas-cooled reactors (cf. Case 35); the analytical methods have been tested by experiment, and agreement between the two is remarkably good, as may be seen for a typical case in Fig.9. Similar studies have been undertaken in the U.K. and U.S.A. in conjunction with the GCR and LMFBR programmes [111-118], dealing both with gas and liquid-filled piping.

---

† Defined as equal to ρc, ρ being the density and c the speed of sound.

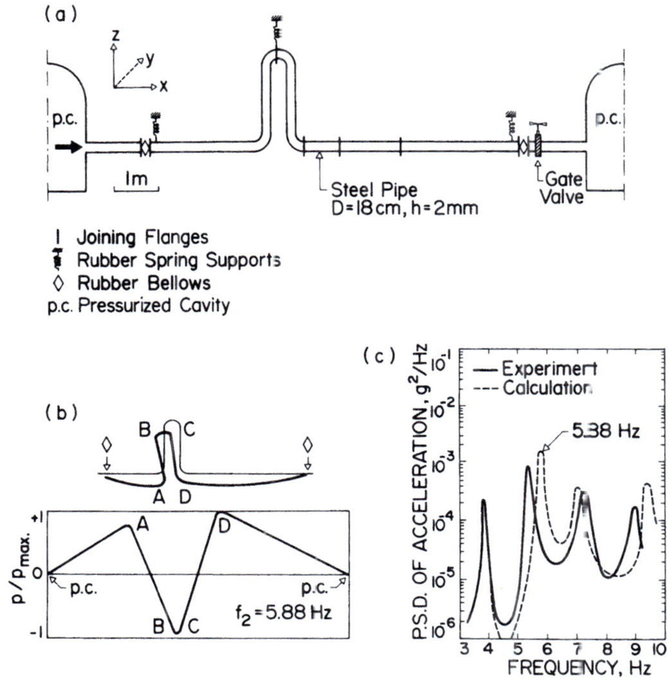

FIG. 9. (a) SKETCH OF TEST APPARATUS: PIPING CONVEYING WATER
(b) MODAL SHAPE OF 2nd MODE, AND CORRESPONDING FLUCTUATING PRESSURE DISTRIBUTION
(c) TYPICAL PIPE ACCELERATION POWER SPECTRAL DENSITY (PARTIALLY OPEN VALVE) [91]

4.4 Tools for the designer for flow-induced vibrations of piping

It may be concluded that the state of the art in this field is quite advanced, yet not always adequately good for achieving trouble-free design of piping. The phenomena involved are mostly well understood, and techniques for calculating piping response due to near- and far-field noise are rapidly improving [91,94,105-118]. Unfortunately, some of these techniques and the computer programs based thereupon are not available in full detail in the open literature.

5. VIBRATION INDUCED BY EXTERNAL AXIAL FLOW: STATE OF THE ART

The vibration characteristics of cylindrical structures in axial flow, e.g. heat exchanger tubing or nuclear reactor fuel bundles, are in many ways similar to those of pipes with internal flow. This permits a rather compact review of this topic, and justifies *a posteriori* the fairly long discussion in §4.

5.1 General dynamical behaviour

Considering mean flow effects, for sufficiently high u_e† fluidelastic instabilities are possible [119,120] - generally both buckling and flutter -

† See footnote on next page

the critical flow velocities being $u_{ec}>\pi$ for a simply supported cylinder in unconfined fluid, and $u_{ec}>2\pi$ for one with clamped ends. Recently, the effect of both internal and external axial flow acting simultaneously has been studied [121], and a typical combined stability map is shown in Fig.10(a). Here u_i and u_e are the dimensionless internal and external flow velocities, respectively; the discrepancy with the experimental points obviates the need for a safety factor, as noted previously in §4.1. If there is an appreciable pulsating component in the flow, then parametric instabilities may also occur [122].

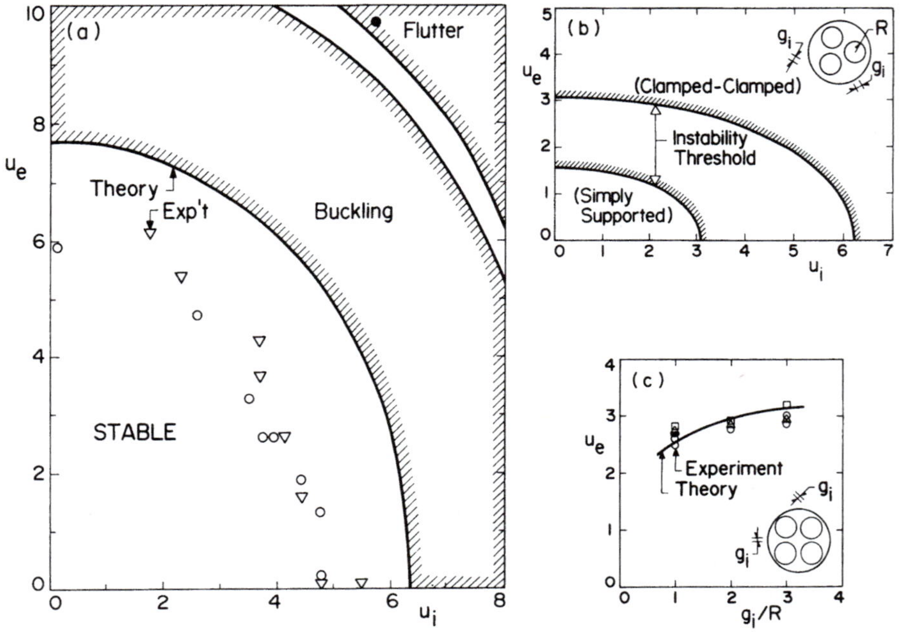

FIG.10. (a) STABILITY MAP FOR SOLITARY CYLINDER WITH CLAMPED ENDS, WITH BOTH INTERNAL AND EXTERNAL AXIAL FLOW [121]
(b) SAME FOR A THREE-CYLINDER CLUSTER [127]
(c) EFFECT OF CONFINEMENT ON STABILITY WITH EXTERNAL FLOW ONLY [126]

For multi-cylinder arrays, especially tightly spaced ones, hydrodynamic coupling becomes important, and motions of one cylinder influence those of all other proximate ones, an effect first examined by S.-S. Chen [123]. The effect on stability of confinement by the flow channel and of hydrodynamic coupling to adjacent cylinders has been studied theoretically and experimentally [120,124-127], both effects lowering the stability threshold, besides generating complex coupled motion effects [123,126]; a stability map for buckling of clusters of tubes with both internal and external flow is shown in Fig.10(b), for the geometry shown, where all interstitial gaps, g_i, are ¼ of a cylinder radius. This figure may be used as a design guide in conjunction with a safety factor to account for imperfections. Finally, Fig.10(c) shows the effect (for external flow only) on stability of proximity between cylinders and of cylinders to the channel wall, and the degree of agreement between theory and experiment.

† $u=(\rho_e A_e/EI)^{1/2}U_e L$, where ρ_e and U_e are, respectively, the density and flow velocity of the external flow, and A_e the external cross-sectional area of the tube; EI and L are as in §4.1. Similarly, $u_i=(M_i/EI)^{1/2}U_i L$ is identical to the u defined in §4.1.

The hydrodynamic coupling effect mentioned above plays an important rôle at all flow velocities, i.e., it is not only important for stability considerations; it has been studied intensively for cylinders in still fluid, both theoretically and experimentally [123,127-129]. This effect is directly related to the existence of an added mass matrix, each element of which plays a rôle similar to that of a structural influence coefficient; i.e., it defines the force imparted on any given cylinder in a given direction, when another undergoes a unit acceleration. This coupling effect is shown clearly in Fig.11, where when one cylinder is given an oscillatory excitation, all others are seen to respond. The closely spaced multiple resonance peaks in Fig.11(a) are a manifestation of the eight (2x4) closely spaced first-mode natural frequencies clustered around where f_1 would have been if the cylinders were sufficiently far apart, cf. Fig.11(b), i.e. if the coupling were very weak; in this latter case there are actually still eight first-mode eigenfrequencies, but they are essentially coincident and $\simeq f_1$ [123,130]. Both classical [123,130-132] and finite-element techniques [130,133] are available for determining the added mass matrix of complex cylinder arrays, as well as for taking viscous coupling into account [125,133]. Recently, structural coupling was also shown to be important [134].

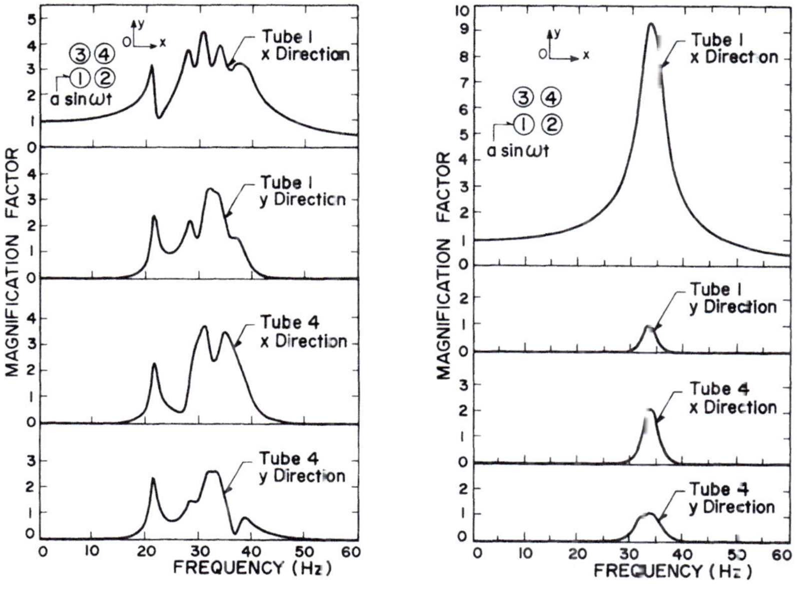

FIG. 11. RESONANCE AND COUPLING EFFECTS IN STATIONARY LIQUID (a) FOR A FOUR-CYLINDER TIGHTLY SPACED CLUSTER, AND (b) FOR A MORE WIDELY SPACED CLUSTER, WHEN CYLINDER 1 IS EXCITED SINUSOIDALLY [123]

Some work has also been done in connection with gross motions of a cluster (bundle) of fuel-element rods, as distinct from vibrations of the individual fuel elements, e.g. [135, 136] - *vide* Case 42.

5.2 Sub-critical vibration

Apart from their inherent technical importance, the studies referred to above [119-136] serve to lay the foundation for understanding and predicting the small-amplitude, "sub-critical" vibrations which are the paramount concern of the designer, insofar as axial-flow-induced vibration is concerned. "Sub-critical" vibrations, so called because they lie below the threshold of instabilities, are random vibrations predominantly in the first axial mode of the cylinder, $f_n=f_1$, and of very small amplitude - typically, $10^{-3}<y/D<10^{-2}$, where y is the r.m.s. amplitude and D the cylinder diameter. The vibration amplitude increases roughly as U^2. Such small amplitude vibration would ordinarily be of little concern, were it not for the close spacing of adjacent cylinders or other structures, so that even small amplitude vibration may cause fretting wear or damage (cf. Cases 40-42).

Several mechanisms of sub-critical vibration have been proposed; they have been classified and discussed in detail by the author [137,138]. Thus, sub-critical vibration was supposed to be a self-excited phenomenon [139], a forced vibration excited mainly by near-field noise [140-142], or a parametrically excited vibration [143]. It should be noted here that all these studies predate the important findings that motions of cylinders in a cluster are hydrodynamically strongly coupled, and that one deals in such situations with a clustered group of first-mode frequencies, rather than a single f_1. Accordingly, in most of the above studies [139-143] it is tacitly assumed that f_1 is *the* first-mode frequency of the cylinder, as if it were a solitary cylinder in unconfined flow; moreover, in all cases, it is supposed that each cylinder vibrates essentially independently of all others; both assumptions are wrong, as is now known. As a result, the methods of predicting vibration amplitudes, to be described next, no doubt suffer.

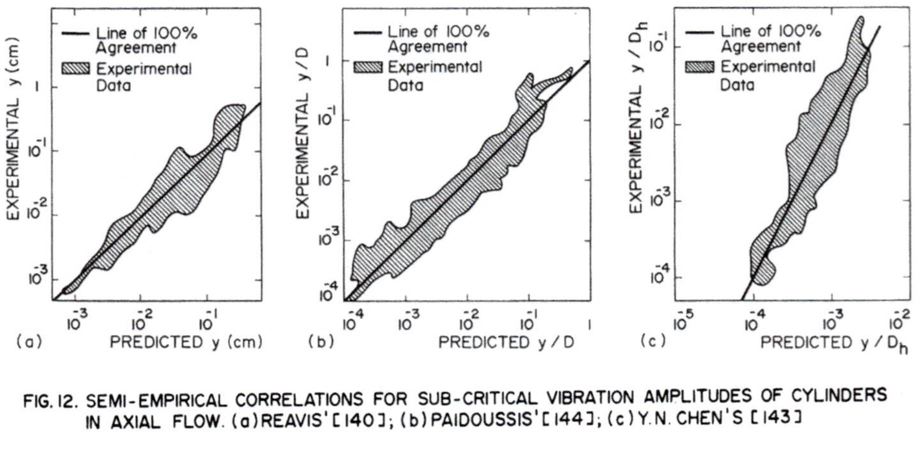

FIG. 12. SEMI-EMPIRICAL CORRELATIONS FOR SUB-CRITICAL VIBRATION AMPLITUDES OF CYLINDERS IN AXIAL FLOW. (a) REAVIS' [140]; (b) PAIDOUSSIS' [144]; (c) Y.N. CHEN'S [143]

Methods for predicting sub-critical vibration amplitudes are either semi-empirical [144,140,143] or analytical [141,142]. The degree of success achieved by the semi-empirical methods is shown in Fig.12, where it is seen that prediction within one order of magnitude is possible; (D_h in Fig.12(c) is the hydraulic diameter). Of the mechanisms proposed, that of forced vibration seems the most likely to be correct [137]. Both Reavis' semi-empirical correlation of Fig.12(a) and the analytical methods are based thereon; the difference lies in that the former resorts to the use of an arbitrary multiplicative factor (5<C<200 roughly) for the amplitude so as to obtain a reasonable correlation, while the latter presume to predict vibration amplitude completely analytically.

The analytical methods [141,142] in general achieve no better results than Reavis did, *before* applying his empirical factor. Nevertheless, under one of the following special provisions, they can achieve remarkable success: (i) the excitation spectrum and other statistical correlations, used in the analysis, are measured in the same apparatus where vibration amplitudes are measured [141], or (ii) one is dealing with very "quiet" laboratory circulation systems or with two-phase flows [141,142]. This is illustrated in Fig.13. It is noted that in the analytical models no account is taken of the effect of singularities discussed in §4.3 on the excitation field; in effect, in the foregoing only boundary-layer pressure fluctuations were normally taken into account, since they were the only ones one could predict. However, if the *in situ* excitation field is used, as suggested in (i), then the near-field and far-field effects of singularities are taken into account, and accuracy of prediction improves, which partly explains the degree of agreement achieved in Fig.13; that other

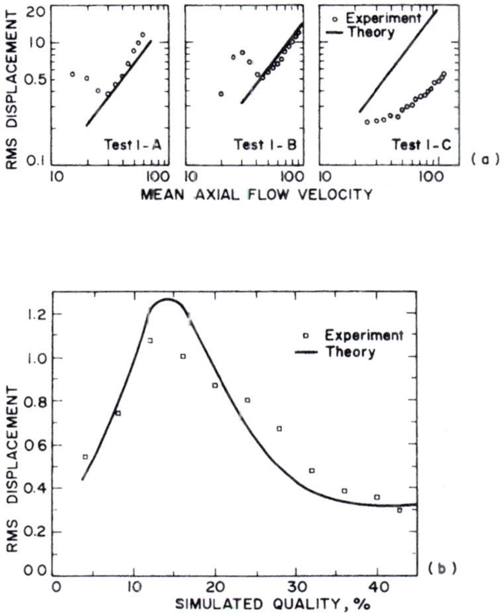

FIG 13 SUB-CRITICAL VIBRATION AMPLITUDE OF CYLINDERS IN AXIAL FLOW BY MEANS OF THE ANALYTICAL METHODS, WHEN USING "MEASURED" EXCITATION FIELD IN THEORY (a) FOR A SINGLE PHASE FLOW [141] (b) FOR A SIMULATED TWO-PHASE FLOW [142]

than boundary-layer pressure fluctuations are important was convincingly demonstrated by Kadlec & Ohlmer [145], by conducting the same experiment (with the same test cylinders) in different circulation systems and obtaining large differences in vibration amplitudes. In addition to this, the experiments of Fig. 13(a) were conducted in a well designed, quiet hydraulic loop with a minimum of loop disturbances; similarly, the experiments of Fig.13(b) were conducted with air-water mixtures, simulating two-phase flows, which display a remarkable lack of memory; i.e., in both cases, local perturbations make up a relatively more important part of the overall excitation field than is normally the case. Thus, (i) and (ii) together explain the unusually good results shown in Fig.13.

Clearly, the current status is, in general, unsatisfactory. The semi-empirical methods, on the one hand, predict vibration amplitude only within one order of magnitude (cf. also [146]). The analytical methods, on the other hand, although capable of much better accuracy, for industrial systems require measurment of the excitation field *in situ*; one might as well, and more easily, measure the response *in situ*, by-passing the need of analysis altogether. As noted in a previous review "It is obvious that what is missing is the characterization of the pressure field in realistic flow conditions. Such information as power-spectral densities of pressure fluctuations induced by valves, bends, etc., and their decay characteristics with distance .... all this would be useful to know and it is one possible direction that future research should take" [137]. Well, with the recently developed data banks for characterization of singularities in piping (§4.3), some of this "missing" information is now at hand, readily offering means for improvement of prediction of vibration amplitude in any given system. Moreover, some studies on far-field excitation have now been made [147,148], and on the near-field effect of singularities and convected turbulence [149]. Also, a set of newer reviews and overviews have appeared [74,150,151].

5.3 Concluding remarks

All the foregoing dealt mostly with cylindrical structures in axial flow. There is, of course, a very rich literature on vibration and flutter of panels, which was developed in response to aeronautical problems: the now classical field of aeroelasticity. The interested reader is referred to [152-154] for very useful information on plate and shell problems. Additonally, a considerable amount of work has been done on the dynamics and on stability of parallel multi-plate assemblies, which are the flat-plate equivalents of clusters of fuel rods [155-158]; fuel assemblies of this type "collapse", i.e. buckle, if the dimensionless flow velocity u', defined in §4.1, becomes high enough, the critical u' depending on plate thickness and width, and on separation between plates.

It may be concluded that although the state of the art for axial-flow-induced vibration is generally not quite satisfactory, the potential for improvement is great, indeed greater than for cross-flow-induced vibration. The reasons for this are that (i) the phenomena involved are fairly well understood, and (ii) in the absence of wakes and other aero- or hydrodynamically messy features, the problem lends itself to mathematical modelling much more readily. This is reflected by the appearance of a tentative design guide for axial-flow-induced vibration as early as 1971 [159]. The relative rarity of serious problems, however, as compared to cross-flow-induced vibration has consistently resulted in a lower priority being placed on research in this area.

6. VIBRATION INDUCED BY ANNULAR AND LEAKAGE FLOWS

As mentioned earlier, the distinction between axial contained flow and annular flow is rather artificial: it could all be considered as annular; especially some of the work described in §5 relating to narrow annuli, e.g. [124,133,149], could easily have been included here. The designation "annular-flow-induced vibration" has nevertheless been retained in categorizing the Cases for (i) problems in which shell-like cylindrical structures are immersed in annular flow (e.g., Cases 46,49,50, involving the thermal shield of PWR's); (ii) problems where the flow characteristics depend inherently on the annular geometry (e.g. Case 52); (iii) problems involving constricted passages, associated with leakage flows (e.g. Case 48). Moreover, "annular" is used here in its hydraulically generic sense, to include not only cylindrical geometries, but also flat plates in narrow rectangular conduits, as in Cases 43 and 44.

6.1 Cylindrical structures subjected to annular flow

A considerable amount of work has been done on the dynamics of cylinders containing, or being immersed in, annular flow. The impetus for most of these studies derives from attempts to model the dynamics of two major components of some types of nuclear reactors, namely the thermal shield and the core barrel, which lie concentrically within the pressure vessel, separated from one another by a liquid annulus.

Most studies consider only the free vibration characteristics, modelling the system in different ways, e.g. as a cylindrical beam within a rigid cylindrical container [160-162], as a cylindrical beam within a cylindrical shell [163] and *vice versa* [164], or as two coaxial shells [106,165-170]. Moreover, in most of these studies, the fluid in the annulus is taken not to be flowing. Hence, the main distinctive feature of the dynamics of these systems is associated with the large added mass effect, which results from the large accelerations suffered by the fluid in the narrow annular passages when the structure vibrates. As a result, the *in vacuo* structural frequencies may be diminished by as much as a factor of 2 to 5; moreover, strong hydrodynamic coupling in the structural motions develops, leading to distinct sets of coupled symmetric and antisymmetric modes. Motions of one cylindrical component may generate enormous unsteady pressures and forces on the annulus walls (*vide* Case 46). In these studies fluid motions are normally described by means of ideal flow theory, although some studies take viscous effects into account [168-169]. Agreement between measured and predicted free vibration characteristics is generally good [106,165,167,170].

Mean flow effects on the dynamics of such systems have not been studied as extensively, perhaps because the hydrodynamics of flow in real-life annular conduits can be very complex, as demonstrated for example by studies of multi-inlet flow into the annular region [171,172] and flow in an annular diffuser [173]. Hence, in studying *flow-induced* vibrations, the device mentioned in §4 and §5 is used here also: mean flow effects are neglected, and the only manifestations of flow taken into account are the flow-induced near- and far-field excitations [149,174-177], or pressure changes associated with "blowdown" situations, e.g. [187-189]. The methods used are similar to those of §4.3. The state of the art in this area is quite advanced; there exist computer codes for predicting the excitation field and the response of all major components of reactor internals, e.g. [176]; moreover, the degree of agreement achieved between theoretical and experimental vibration-amplitude p.s.d.'s can be spectacularly good. Once again, however, the specialized computer codes involved are not available in full in the open literature.

6.2 Annular- and leakage-flow-induced instabilities

The essentials of the mechanism of leakage-flow-induced vibration may be obtained from Cases 48 and 43. Briefly, whenever a structure is so shaped as to severely restrict the annular flow around it, and especially if the constriction is at its upstream end, then the system is potentially unstable; both divergence and flutter instabilities of the structure may be induced [178-180]. A downstream constriction, on the other hand, tends to stabilize the system. A simplified account of how these effects may arise is offered by Miller [181]. This mechanism is familiar in other fields, giving rise, for example, to vibrations of pistons [182], seals [183] and valves at small openings [184-185].

There is a superficial similarity between these instabilities and the hydroelastic instabilities of cylinders in narrow conduits described in §5.1 that may be worth clarifying. The essence of the difference in mechanisms lies in that in the latter case, cf.[120], small motions of the cylindrical structure do not induce gross changes in the axial flow field, although they do perturb the flow laterally, in accordance with slender-body theory. The leakage flow mechanism, on the other hand, depends functionally on gross mass-flux disturbances associated with partial asymmetric closure of the annular gap at the constriction and on attendant mass-influx asymmetry in the annular flow beyond. In the absence of a constriction the leakage mechanism effectively becomes inoperative. Furthermore, the hydroelastic instabilities of §5.1 require that the body be flexible, although rigid body instabilities are also possible but are then governed by the leading- and trailing-edge geometry [186].

6.3 Some concluding comments

From the above discussion it is evident that (i) the dynamical effects associated with axial flow are greatly magnified in annular geometries by the effect of severe fluid confinement; (ii) flow-induced motions in narrow annuli involving constrictions are prone to leakage-flow-induced instabilities, quite apart from possible hydroelastic instabilities.

7. CONCLUSION

This paper offers a *tour d'horizon* of the many-faceted problems of flow-induced vibration in heat exchangers and nuclear reactors. Some original and hopefully useful comments have been made with regard to the mechanisms of excitation of vibration induced by cross flow in §3.2 and 3.3. The practical Cases described and the state-of-the-art reviews make it abundantly clear that a great deal of progress has been made in the past decade; yet, physical understanding remains inadequate in a number of key areas, especially in the case of cross flow. The fluid mechanics of the fluid-structure interaction present the greatest difficulties, reflected by the broad simplifications sometimes introduced in the analysis, which contrast to the refinement of the structural formulation; e.g. sometimes the fluid is assumed to be stationary when it is not; worse still, some analyses treat what might be termed *stationary* fluid flows, and some even *stationery* flows; the latter shall remain anonymous.

In some cases, although the computerized models available have demonstrated remarkable predictive capabilities, the essential mechanisms involved remain physically obscure. It is still to be hoped that once the underlying mechanism becomes understood then one may use simplified analytical models which will provide the required information more meaningfully and cheaply - much as the complexities of the epicyclical theories of Ptolemy vanished with Copernicus' revival of Aristarchus' heliocentric principle for planetary motion.

In writing this paper, it became painfully obvious how general is the lack of familiarity by most workers in this area with similar work in other fields; perhaps, this Symposium will succeed in breaking down some of this artificial compartmentalization. More surprising was the evident lack of familiarity, in some cases, of parallel work in the *same* field, despite the great advances in computerized bibliographical retrieval; whether this signifies that the individual researcher has been overwhelmed by the shear volume of technical paper production, or whether it reflects the maxim that "the function of a scientist is to generate papers, not to read them", one can only guess. Of course, the state-of-the-art reviews cannot hope to rectify this situation, as it was found impossible to refer to all pertinent studies, in the interests of brevity; thus, for the subject of §4.1 alone, one could easily have cited 75 papers. Accordingly, some significant contributions have gone unmentioned, and the author apologizes to those colleagues whose work was not covered adequately, or not at all.

It is hoped that this paper will serve the twin goal of inspiring researchers to contribute in areas of great practical need, and stimulating designers and operators to avail themselves of the fundamental and practical knowledge accumulated to date through research.

8. ACKNOWLEDGEMENTS

The financial support of the Natural Sciences and Engineering Research Council of Canada is hereby gratefully acknowledged, as is the assistance of many colleagues all over the world - too numerous to mention them all, but especially O.S. Tatone, S.J. Price and M.J. Pettigrew.

9. REFERENCES

1. Deddens, J.C. & Clark, R.H., "Increasing reliability and availability of B&W nuclear steam systems", Paper PGT76-25, American Power Conference, Chicago, Apr 1976.

2. Stevens-Guille, P.D., "Steam Generator Tube Failures: A World Survey of Water-Cooled Nuclear Power Reactors to the End of 1971", Atomic Energy of Canada Ltd., Report AECL-4449, Apr 1973.

3. Stevens-Guille, P.D. & Hare, M.G., "Steam Generator Tube Failures: World Experience in Water-Cooled Nuclear Power Reactors in 1973", AECL-5013, Jan 1975.

4. Hare, M.G., "Steam Generator Tube Failures: World Experience in Water-Cooled Nuclear Power Reactors in 1974", AECL-5242, Apr 1976.

5. Tatone, O.S., & Pathania, R.S., "Steam Generator Tube Failures: Experience in Water-Cooled Nuclear Reactors during 1976", AECL-6095, Feb 1978.

6. Levenson, M., "R&D status report. Nuclear Power Division", EPRI Journal, Vol.3, 1978, pp.51-56.

7. Chenoweth, J.M., & Kistler, R.S., "Tube Vibrations in Shell-and-Tube Heat Exchangers", Heat Transfer Research Inc., Tech. Report, Sep 1976.

8. "Nuclear Power Experience", Section PWR-2, II, Item 3; Verna, B. (ed.), Encino, California.

9. Crimmin, W.R.C., Gray, I., Richards, D.J.W., & Dell, J.D., "Wylfa Nuclear Boiler Repair", Central Electricity Generating Board (U.K.) Report G823, Sep 1977.

10. "Nuclear Power Experience", Section PWR-2, II, Items 2 and 21.

11. Roshko, A., "On the wake and drag of bluff bodies", J.Aeronautical Sciences Vol.22, 1955, pp.124-132.

12. Cohan, L.J., & Deane, W.J., "Elimination of destructive, self-excited vibrations in large gas and oil-fired utility units", Trans ASME, J.Engng for Power, Vol.87, 1965, pp.223-228.

13. Hartlen, R.T., "Recent field experience with flow-induced vibration of heat exchanger tubes", Paper 611, Int'l Symposium on Vibration Problems in Industry, Keswick, U.K., Apr 1973.

14. Nelms, H.A., & Segaser, C.L., "Survey of Nuclear Reactor System Primary Circuit Heat Exchangers", Oak Ridge National Laboratory Report ORNL-4399, Apr 1969.

15. Pettigrew, M.J., "Flow-Induced Vibration of Nuclear Power Station Components", AECL-5852, Sep 1977.

16. Chen, Y.N., "Flow-induced vibration and noise in tube bank heat exchangers due to von Kármán streets", Trans ASME, J.Engng for Industry, Vol.89, 1968, pp. 134-146.

17. Connors, H.J., Jr., "Fluidelastic vibration of tube arrays excited by cross-flow", in "Flow-Induced Vibration in Heat Exchangers", ASME, New York, 1970.

18. Owen, P.R., "Buffeting excitation of boiler tube vibration", J.Mechanical Engineering Science, Vol.7, 1965, pp.431-439.

19. Savkar, S.D., & So, R.M.C., "On the Buffeting Response of a Cylinder in a Turbulent Cross Flow", General Electric Co. Report GEAP-24149, Sep 1978.

20. Marris, A.W., "A review on vortex streets, periodic wakes and induced vibration phenomena", Trans ASME, J.Basic Engng, Vol.86, 1964, pp.185-196.

21. Morkovin, M.V., "Flow around a circular cylinder - a kaleidoscope of challenging fluid phenomena", ASME Symposium on Fully Separated Flows, 1964, pp.102-118.

22. Berger, E., & Wille, R., "Periodic flow phenomena" in Annual Review of Fluid Mechanics, Vol.4, 1972, pp.313-340.

23. Mulcahy, T.M., & Chen, S.S., "Annotated Bibliography on Flow Induced Vibration", Argonne National Laboratory, Tech.Mem.ANL-CT-74-05, Jan 1974.

24. Savkar, S.D., "A survey of flow-induced vibrations of cylindrical arrays in cross-flow", ASME Paper 76-WA/FE-21, 1976.

25. Blevins, R.D., "Flow-Induced Vibration"; Van Nostrand Rheinhold Co., New York, 1977.

26. "Flow-Induced Structural Vibrations", ed. E. Naudascher; Springer-Verlag, Berlin, 1974.

27. Richter, A., & Naudascher, E., "Fluctuating forces on a rigid circular cylinder in a confined flow", J.Fluid Mechanics, Vol.78, 1976, pp.561-576.

28. Bishop, R.E.D., & Hassan, A.Y., "The lift and drag forces on a circular cylinder in a flowing-field", Proc. Roy. Soc. (London), A, Vol.277, 1964, pp.51-75.

29. King, R., Prosser, M.J., & Johns, D.J., "On vortex excitation of model piles in water", J.Sound & Vibration, Vol.29, 1972, pp.169-188.

30. Welsh, M.C., & Gibson, D.C., "Acoustic resonances generated by flow past a flat plate in a duct", Paper presented in the 1979 Joint ASME/CSME Applied Mechanics-Fluids Engineering-Biomechanics Conference, Niagara Falls, Jun 1979.

31. Roberts, B.W., "Low Frequency, Aeroelastic Vibrations in a Cascade of Circular Cylinders", Mechanical Engineering Science Monograph No.4, Sep 1966.

32. Chen, Y.N., "Excitation sources of the flow-induced vibrations and noise in tube bank heat exchangers", in 'Noise and Fluid Engineering" (ed.R.Hickling), pp.239-246; ASME, New York, 1977.

33. Chen, Y.N., "Criteria for the cross-flow-induced tube vibrations in tube bank heat exchangers", Paper 2.6, Int'l Conf. on Vibration in Nuclear Plant (ViNP), Keswick, 1978.

34. Fitz-Hugh, J.S., "Flow-induced vibration in heat exchangers", Paper 427, Int'l Symp. Vibration Problems in Industry, Keswick, 1973.

35. Chen, Y.N., "The sensitive tube spacing region of tube bank heat exchangers for fluid-elastic coupling in cross-flow", in "Fluid-Structure Interaction Phenomena in Pressure Vessel and Piping Systems" (ed. M.K. Au-Yang & S.J. Brown); ASME, 1977.

36. Pettigrew, M.J., Sylvestre, Y., Campagna, A.O., "Vibration analysis of heat exchanger and steam generator designs", Nuclear Engineering & Design, Vol. 48, 1978, pp.97-115.

37. Grover, L.K., & Weaver, D.S., "Cross-flow induced vibrations in a tube bank-Vortex shedding", J.Sound & Vibration, Vol.59, 1978, pp.263-276.

38. Gibert, R.J., Chabrerie, J. & Sagner, M., "Vibrations of tube arrays in transversal flow", Paper F4/h, 4th Int'l Conf. on Structural Mechanics in Reactor Technology (SMiRT), San Francisco, 1977.

39. Baylac, G., & Grégoire, J.P., "Etude des écoulements et des phénomènes acoustiques dans un faisceau de tubes", in "Aéro-hydro-élasticité", pp.279-379; Eyrolles, Paris, 1973.

40. Chen, Y.N., & Young, W.C., "The orbital movement and the damping of the fluidelastic vibration of tube banks due to vortex formation. Part 3", Trans.ASME, J.Engineering for Industry, Vol.96, 1974, pp.1072-1075.

41. Blevins, R.D., "Fluidelastic whirling of a tube row", Trans ASME, J.Pressure Vessel Technology, Vol.96, 1974, pp.263-267.

42. Blevins, R.D., "Fluidelastic whirling of tube rows and tube arrays", ASME Paper 76-WA/FE-26, Dec 1976.

43. Savkar, S.D., "A brief review of flow-induced vibrations of tube arrays in crossflow", Trans ASME, J. Fluids Engng, Vol.99, 1977, pp.517-519.

44. Weaver, D.S., & Grover, L.K., "Cross flow induced vibrations in a tube bank", ASME Paper 78-PVP-25, Jun 1978.

45. Gorman, D.J., "Experimental study of peripheral problems related to liquid flow induced vibration in heat exchangers and steam generators", Paper F4/g, SMiRT, San Francisco, 1977.

46. Pettigrew, M.J., & Gorman, D.J., "Vibration of heat exchanger components in liquid and two-phase cross flow", Paper 2.3, ViNP, Keswick, 1978.

47. Weaver, D.S., & Lever, J., "Tube frequency effects on cross flow induced vibrations in tube arrays", Paper IV.4, 5th Biennial Symposium on Turbulence, Dept Chem. Engng, University of Missouri-Rolla, Oct 1977.

48. Coit, R.L., Peake, C.C., & Lohmeier, A., "Design and manufacture of large surface condensers - problems and solutions", Proc. American Power Conference, Vol.28, 1966, pp.469-483.

49. Peake, C.C., Gestenkorn, G.F., & Arnold, T.R., "Some reliability considerations for large surface condensers", Proc. American Power Conference, Vol. 37, 1975, pp.562-574.

50. Thorgren, J.T., "Predict exchanger tube damage", Hydrocarbon Processing, Vol.49, 1970, pp.129-131.

51. Heinecke, E.P., "Fluidelastic vibrations in heat exchangers with tubes in cross-flow", Paper 2.4, ViNP, Keswick, 1978.

52. Nicolet, J.B., Sagner, M., & Régis, G., "Vibrations de faisceaux de tubes sous excitation aérodynamique", Revue Générale de Thermique, No.128, 1976, pp.869-891.

53. Weaver, D.S., & Grover, L.K., "Cross-flow induced vibrations in a tube bank-turbulent buffeting and fluid elastic instability", J.Sound & Vibration, Vol.59, 1978, pp.277-294.

54. Southworth, P.J., & Zdravkovich, M.M., "Cross-flow-induced vibrations of finite tube banks in in-line arrangements", J.Mechanical Engineering Science, Vol.17, 1975, pp.190-198.

55. Zdravkovich, M.M., & Namork, J.E., "Structure of interstitial flow between closely spaced tubes in staggered array", in "Flow-Induced Vibrations" (ed. S.-S. Chen & M.D. Bernstein), pp.41-46; ASME, New York, 1979.

56. Chen, S.S., "Flow-induced vibration of circular cylindrical structures. Part II: cross-flow considerations", The Shock & Vibration Digest, Vol.9, 1977, pp.21-27.

57. Moretti, P.M., "Critical review of the literature and research in flow-induced vibrations in heat exchangers", AIChE Symposium on Flow Induced Vibrations in Heat Exchangers, Vol.70, 1973, pp.185-189.

58. "Flow-Induced Vibration in Heat Exchangers", ASME, New York, 1970.

59. Blevins, R.D., "Flow-Induced vibration in nuclear reactors: a review", to appear in Progress in Nuclear Energy.

60. Erskine, J.B., & Waddington, W., "A review of some tube vibration failures in shell and tube heat exchangers and failure prediction methods", Paper 428, Int'l Symp. on Vibration Problems in Industry, Keswick, 1973.

61. Schwartz, Jr., G.W., "Preventing vibration in shell-and-tube heat exchangers", Chemical Engineering, Vol.83, 1976, pp.134-140.

62. Ananthanarayanan, K.S., "Preliminary Design Handbook for Flow-Induced Vibration of Light Water Reactors", General Electric Report GEAP-24158, Nov 1978.

63. Eisinger, F.L., "Prevention and cure of flow-induced vibration problems in tubular heat exchangers", in "Flow-Induced Vibrations" (ed. S.-S.Chen & M.D. Bernstein), pp.47-55; ASME, New York, 1979.

64. Housner, G.W., "Bending vibrations of a pipe when liquid flows throught it", J.Applied Mechanics, Vol.19, 1952, pp.205-208.

65. Gregory, R.W., & Paidoussis, M.P., "Unstable oscillation of tubular cantilevers conveying fluid. I. Theory; II. Experiments", Proc.Royal Society (London), 293, A, 1966, pp.512-527 and 528-542.

66. Chen, S.-S., "Vibrations of continuous pipes conveying fluid", in "Flow-Induced Structural Vibrations" (ed. E.Naudascher), pp.663-675; Springer-Verlag, Berlin, 1974.

67. Naguleswaran, S., & Williams, C.J.H., "Lateral vibration of pipe conveying fluid", J.Mechanical Engineering Science, Vol.10, 1968, pp.228-238.

68. Paidoussis, M.P., & Denise, J.-P., "Flutter of thin cylindrical shells conveying fluid", J.Sound & Vibration, Vol.20, 1972, pp.9-26.

69. Weaver, D.S., & Unny, T.E., "Dynamic stability of fluid-conveying pipes", J.Applied Mechanics, Vol.40, 1973, pp.48-52.

70. Paidoussis, M.P., "Flutter of conservative systems of pipes conveying incompressible fluid", J.Mechanical Engineering Science, Vol.17, 1975, pp.19-25.

71. Chen, S.-S., "Flow-induced in-plane instabilities of curved pipes", Nuclear Engineering & Design, Vol.23, 1972, pp.29-38.

72. Chen, S.-S., "Out-of-plane vibration and stability of curved tubes conveying fluid", J.Applied Mechanics, Vol.40, 1973, pp.362-368.

73. Doll, R.W., & Mote, C.D., Jr., "On the dynamic analysis of curved and twisted cylinders transporting fluids", Trans ASME, J.Pressure Vessel Technology, Vol.98, 1976, pp.143-150.

74. Chen, S.-S., "Parallel flow-induced vibrations and instabilities of cylindrical structures", Shock & Vibration Digest, Vol.6, 1974, pp.1-11.

75. Paidoussis, M.P., "Vibrations of tubes containing flowing fluid", in "Aéro-hydro-élasticité", pp.441-500; Eyrolles, Paris, 1973.

76. Paidoussis, M.P., & Issid, N.T., "Dynamic stability of pipes conveying fluid", J.Sound & Vibration, Vol.33, 1974, pp.267-294.

77. Chen, S.-S., "Flow-induced vibrations of circular cylindrical structures. Part I: Stationary fluids and parallel flow", Shock & Vibration Digest, Vol.9, 1977, pp.25-38.

78. Den Hartog, J.P., "John Orr Memorial Lecture: Recent cases of mechanical vibration", The South African Mechanical Engineer, Vol. 19, 1969, pp.53-68.

79. Paidoussis, M.P., "Dynamics of tubular cantilevers conveying fluid", J. Mechanical Engineering Science, Vol.12, 1970, pp.85-103.

80. Dodds, H.L., & Runyan, H.L., "Effect of high velocity fluid flow on the bending vibrations and static divergence of a simply supported pipe", NASA Technical Note D-2870, 1965.

81. Liu, H.-S., & Mote, C.D.,Jr., "Dynamic response of pipes transporting fluids", Trans ASME, J.Engineering for Industry, Vol.96, 1974, pp.591-596.

82. Esswein, G., *et al.*,"Pipe whip dynamics", in "Dynamic Analysis of Pressure Vessel & Piping Components" (ed. C. Sundararajan), ASME, New York, 1977.

83. Haktamir, S.A., & Tolle, G.C., "An approximate method for the determination of the response frequency of pipe whip", ASME Paper 79-PVP-123, 1979.

84. Chen, S.-S., "Dynamic stability of a tube conveying fluid", Proc. ASCE, J.of the Engineering Mechanics Division, Vol.97, 1971, pp.1469-1485.

85. Paidoussis, M.P., & Issid, N.T., "Experiments on parametric resonance of pipes containing pulsatile flow", J.Applied Mechanics, Vol.43, 1976, pp. 198-202.

86. Hara, F., "Two-phase-flow-induced vibrations in a horizontal piping system", Bulletin of the JSME, Vol.20, 1977, pp.419-427.

87. Willmarth, W.W., "Wall pressure fluctuations in a turbulent boundary layer", J. Acoustical Society of America, Vol.28, 1956, pp.1048-1053.

88. Bakewell, H.P.,Jr., "Narrow-band investigations of the longitudinal space-time correlation function in turbulent airflow", J.Acoust.Society of America, Vol.36, 1964, pp.146-148.

89. Clinch, J.M., "Measurements of the wall pressure field at the surface of a smooth-walled pipe containing turbulent waterflow", J.Sound & Vibration, Vol.9, 1969, pp.398-420.

90. Willmarth, W.W., "Pressure fluctuations beneath turbulent boundary layers", Annual Review of Fluid Mechanics, Vol.7, 1975, pp.13-38.

91. Gibert, R.J., Axisa, F., Villard, B., "Flow induced vibrations of piping system (Vibration sources - Mechanical response of the pipes)", Paper 6.2, ViNP, Keswick, 1978.

92. Reethof, G., "Turbulence-generated noise in pipe flow", Annual Reviews of Fluid Mechanics, Vol.10, 1978, pp.333-367.

93. Lin, T.C., & Morgan, G.W., "Wave propagation through fluid contained in a cylindrical, elastic shell", J.Acoustical Society of America, Vol.28, 1956, pp.1165-1176.

94. Baylac, G.,*et al.*, "Calculation of acoustical resonances in irregular cavities with application to noise-induced stress in expansion joints", Trans. ASME, J.Engineering for Industry, Vol.98, 1976, pp.101-108.

95. Rockwell, D., & Naudascher, E., "Self-sustained oscillations of impinging free shear layers", Annual Reviews of Fluid Mechanics, Vol.11, 1979, pp. 67-94.

96. Rattaya, J.V., & Junger, M.C., "Flow excitation of cylindrical shells and associated coincidence effects", J.Acoustical Society of America, Vol.36, 1964, pp.878-884.

97. Bull, M.K., & Norton, M.P., "Higher-order acoustic modes due to internal flow disturbances in relation to acoustic radiation from pipes", Proc.9th Int'l Congress on Acoustics, Vol.II, p.726; Madrid, 1977.

98. Walter, J.L., McDaniel, O.H., Reethof, G., "Coincidence of higher-order acoustic modes in pipes with pipe vibrational modes", J.Acoustical Society of America, Vol.62, 1977, pp.S84.

99. Firth, D., "Acoustic vibration of structures in liquids", Shock & Vibration Digest, Vol.9, 1977, pp.3-7.

100.Streeter, V.L., & Wylie, E.B., 'Waterhammer and surge control", Annual Reviews of Fluid Mechanics, Vol.6, 1974, pp.57-73.

101.Rubinow, S.I., & Keller, J.B., "Wave propagation in a fluid-filled tube", J.Acoustical Society of America, Vol.50, 1971, pp.198-223.

102.Khosrovani, H., Cohen, R., Charaud, R.C., "Response of thin-walled cylindrical duct to internal airflow", J.Acoustical Society of America, Vol.50, 1971, pp.241-247.

103.Harel, P., & Perulli, M., "Influence of a stationary uniform axial flow on propagation of acoustic modes of vibration in cylindrical duct", J.Sound & Vibration, Vol.15, 1971, pp.455-474.

104. Skalak, R., "Wave propagation in blood flow", in "Biomechanics" (ed.Y.C. Fung), pp.20-46; ASME, New York, 1966.

105. Clinch, J.M., "Prediction and measurement of the vibrations induced in thin-walled pipes by the passage of internal turbulent waterflow", J.Sound & Vibration, Vol.12, 1970, pp.429-451.

106. Levin,L., Milan, D., Payan, G., "Recherches sur la réponse de coques cylindriques à des excitations hydrodynamiques", La Houille Blanche,Vol.26,1971, pp.505-512.

107. Lakis, A.A., Paidoussis, M.P., "Prediction of the response of a cylindrical shell to arbitrary or boundary-layer-induced random pressure fields", J. Sound & Vibration, Vol.25, 1972, pp.1-27.

108. Baylac, G., Gibert, R.J., Livolant, M., "Calcul de l'amplification acoustique dans les circuits complexes: application aux vibrations des échangeurs du reacteur CHINON 3", La Houille Blanche, Vol.26, 1971, pp.407-428.

109. Baylac, G., & Copin, A., "Vibration studies of the new primary circuit of the CHINON 3 reactor", Paper F4/3, 3rd SMiRT, London, 1975.

110. Gibert, R.J., "Etude des fluctuations de pression dans les circuits parcourus par des fluides", Commissariat à l'Energie Atomique (France), Note CEA No.1925, 1976.

111. Fahy, F.J., & Firth, D., "Acoustic excitation of flexural modes in a pipe which incorporates a 90 degree radiussed bend", Paper 6.1, ViNF, Keswick, 1978.

112. Spanton, J.H., & Ashton, M.W., "Measurements of fluctuating forces generated by internal flow in pipe line components", Paper 6.3, ViNP, Keswick, 1978.

113. Matthews, L.J., "The prediction of sound pressure levels in the primary coolant circuit of Dungeness 'B' AGR", Paper 8.2, ViNP, Keswick, 1978.

114. Bentley, P.G., & Rowley, R., "Acoustic vibration measurements in PFR", Paper 8.3, ViNP, Keswick, 1978.

115. Firth, D., *et al.*, "An acoustic vibration model of a fast reactor", Paper 8.4, ViNP, Keswick, 1978.

116. Wilkinson, D.H., "Acoustic and mechanical vibrations of liquid-filled pipework systems", Paper 8.5, ViNP, Keswick, 1978.

117. A-Moneim, M.T., "Numerical fluid-structure interaction analysis of piping systems", in "Fluid Structure Interaction Phenomena in Pressure Vessel and Piping Systems" (ed. M.K. Au-Yang & S.J. Brown), pp.107-121, ASME, New York, 1977.

118. A-Moneim, M.T., Chang, Y.W., Fistedis, S.H., "Computer modelling of piping components for transient hydrodynamic-structural response", Nuclear Engineering & Design, Vol.49, 1978, pp.175-186.

119. Paidoussis, M.P., "Dynamics of flexible slender cylinders in axial flow. Part 1: Theory; Part 2: Experiments", J.Fluid Mechanics, Vol.26, 1966, pp.717-736 and 737-751.

120. Paidoussis, M.P., "Dynamics of cylindrical structures subjected to axial flow", J.Sound & Vibration, Vol.29, 1973, pp.365-385.

121. Hannoyer, M.J., & Paidoussis, M.P., "Instabilities of tubular beams simultaneously subjected to internal and external axial flows", Trans. ASME, J.Mechanical Design, Vol.100, 1978, pp.328-336.

122. Paidoussis, M.P., "Stability of flexible slender cylinders in pulsatile axial flow", J.Sound & Vibration, Vol.42, 1975, pp.1-11.

123. Chen, S.-S., "Vibration of nuclear fuel bundles", Nuclear Engineering & Design, Vol.35, 1975, pp.399-422.

124. Paidoussis, M.P., & Pettigrew, M.J., "Dynamics of flexible cylinders in axisymmetrically confined flow", J.Applied Mechanics, Vol.46, 1978, pp.37-44.

125. Paidoussis, M.P., & Suss, S., "Stability of a cluster of flexible cylinders in bounded axial flow", J.Applied Mechanics, Vol.44, 1977, pp.401-408.

126. Paidoussis, M.P., "The dynamics of clusters of flexible cylinders in axial flow: theory and experiments", J.Sound & Vibration, Vol.65, 1979, pp.391-417.

127. Besançon, P., "Dynamics of a Cluster of Pipes Conveying Fluid in Bounded Axial Flow", M.Eng. Thesis, McGill University, Jul 1978.

128. Chen, S.-S., "Vibrations of a row of circular cylinders in a liquid", Trans.ASME, J.Engineering for Industry, Vol.97, 1975, pp.1212-1217.

129. Chen, S.-S., & Jendrzejczyk, J.A., "Experiments on fluidelastic vibration of cantilevered tube bundles", Trans.ASME, J.Mechanical Design, Vol.100, 1978, pp.540-548.

130. Paidoussis, M.P., Suss, S., Pustejovsky, M., "Free vibration of cylinders in liquid-filled channels", J.Sound & Vibration, Vol.55, 1977, pp.443-459.

131. Chung, H., & Chen, S.-S., "Vibration of a group of circular cylinders in confined fluid", J.Applied Mechanics, Vol.44, 1977, pp.213-217.

132. Yamamoto, T., "Hydrodynamic forces on multiple circular cylinders", Proc. ASCE, J. of the Hydraulic Division, Vol.102, 1976, pp.1193-1210.

133. Yang, C.I., & Moran, T.J., "Calculations of added mass and damping coefficients for hexagonal cylinders in confined viscous fluid", in 'Flow Induced Vibrations" (ed. S.-S. Chen & M.D. Bernstein), pp.97-103; ASME, New York, 1979.

134. Jagannath, D.V., & Paidoussis, M.P., "Solid-fluid interactions in the vibration of nuclear fuel bundles", Paper 1.2, ViNP, Keswick, 1978.

135. Paidoussis, M.P., "Dynamics of fuel strings in axial flow", Annals of Nuclear Energy, Vol.3, 1976, pp.19-30.

136. Mason, V.A., *et al.*,"Dynamics of Nuclear Fuel Assemblies in Vertical Flow Channels: Computer Modelling and Associated Studies", Atomic Energy of Canada Ltd., Report AECL-5976, Oct 1978.

137. Paidoussis, M.P., "Vibration of cylindrical structures induced by axial flow", Trans. ASME, J.Engineering for Industry, Vol.96, 1974, pp.547-552.

138. Paidoussis, M.P., "The dynamical behaviour of cylindrical structures in axial flow", Annals of Nuclear Energy, Vol.1, 1974, pp.33-106.

139. Quinn,E.P., "Vibration of fuel rods in parallel flow", General Electric Report GEAP-4059, 1962.

140. Reavis, J.R., "Vibration correlation for maximum fuel-element displacement in parallel turbulent flow", Nuclear Science & Engineering, Vol.38, 1969, pp.63-69.

141. Chen, S.-S., & Wambsganss, M.W., Jr., "Parallel-flow-induced vibration of fuel rods", Nuclear Engineering & Design, Vol.18, 1972, pp.253-278.

142. Gorman, D.J., "Analytical and experimental investigation of the vibration of cylindrical reactor fuel elements in two-phase parallel flow", Nuclear Science & Engineering, Vol.44, 1971, pp.277-290.

143. Chen, Y.N., "Turbulence-induced vibration of tube bundle heat exchangers with cross and parallel flow. Part 1: Parallel flow", in Ref.[58],pp.57-66.

144. Paidoussis, M.P., "An experimental study of vibration of flexible cylinders induced by nominally axial flow", Nuclear Science & Engineering, Vol.35, 1969, pp.127-138.

145. Kadlec, J., & Ohlmer, E., "On the reproducibility of the parallel-flow induced vibration of fuel pins", Nuclear Engineering & Design, Vol.17, 1971, pp.355-360.

146. Henry, C.D., Lee, G.E., Wambsganss, M.W., "Flow-induced vibrations in GCFR core components", in "Flow Induced Vibrations" (ed. S.-S. Chen & M.D. Bernstein), pp.83-90; ASME, New York, 1979.

147. Hine, M.J., "Acoustically induced vibrations of slender rods in a cylindrical duct", J.Applied Mechanics, Vol.40, 1973, pp.459-463.

148. Lin, H.-C., & Chen, S.-S., "Acoustically induced vibration of circular cylindrical rods", J.Sound & Vibration, Vol.51, 1977, pp.89-96.

149. Mulcahy, T.M., Yeh, T.T., & Miskevics, A.J., "Turbulence and Rod Vibrations in an Annular Region with Upstream Disturbances", General Electric Report GEAP-24173, Jan 1979.

150. Lane, A.D., *et al.*,"The measurement and prediction of vibration in CANDU-PHW fuel and channel assemblies", Paper 1.1, ViNP, Keswick, 1978.

151. Shin, Y.W., "Two-phase Flow-Induced Vibrations of Rods in Parallel Flow: A State-of-the-Art Review", General Electric Report GEAP-24148, Oct 1978.

152. Bisplinghoff, R.L., Ashley, H., & Halfman, R.L., "Aeroelasticity", Addison-Wesley Publ.Co., Reading (Mass.), 1955.

153. Dowell, E.H., "Panel flutter: A review of the aeroelastic stability of plates and shells", AIAA Journal, Vol.8, 1970, pp.385-399.

154. Dowell, E.H., "Aeroelasticity of Plates and Shells", Noordhoff Int'l Publ.; Leyden, 1975.

155. Miller, D.R., "Critical flow velocities for collapse of reactor parallel-plate assemblies", Trans. ASME, J.Engineering for Power, Vol.82, 1960 pp.83-95.

156. Rosenberg, G.S., & Youngdahl, C.K., "A simplified dynamic model for the vibration frequencies and coolant flow velocities for reactor parallel plate fuel assemblies", Nuclear Science & Engineering, Vol.13, 1962, pp. 91-102.

157. Scavuzzo, R.J., "Hydraulic instability of flat parallel plate assemblies", Nuclear Science & Engineering, Vol.21, 1965, pp.463-472.

158. Wambsganss, M.W., "Second-order effects as related to critical coolant flow velocities and reactor parallel plate fuel assemblies", Nuclear Engineering & Design, Vol.5, 1967, pp.268-276.

159. Wambsganss, M.W., & Chen, S.-S., "Tentative Design Guide for Calculating the Vibration Response of Flexible Cylindrical Elements in Axial Flow", Argonne National Laboratory Report ANL-ETD-71-07, 1971.

160. Lazzeri, L., & Dini, D., "Vibrations of a pendular structure of circular cross section inside a concentric pipe", Ingegneria Nucleare, Vol.10, 1969, pp.48-57.

161. Bowers, G., & Horvay, G., "Beam modes of vibration of a thin cylindrical shell flexibly supported and immersed in water inside of a coaxial cylindrical container of slightly larger radius", Nuclear Engineering & Design, Vol.26, 1974, pp.291-298.

162. Hennig, K., & Grunwald, G., "Treatment of flow-induced pendulum oscillations", Paper 4.9, ViNP, Keswick, 1978.

163. Chen, S.-S., "Free vibration of a coupled fluid/structural system", J. Sound & Vibration, Vol.21, 1972, pp.387-398.

164. Turula, P., & Mulcahy, T.M., "Computer modelling of flow-induced in-reactor vibrations", Proc. ASCE, J. of the Power Division, Vol.103, 1977, pp.37-50.

165. Levin, L., & Milan, D., "Coupled breathing vibrations of two thin cylindrical coaxial shells in a fluid", Paper 616, Int'l Symposium Vibration Problems in Industry, Keswick, 1973.

166. Krajcinovic, D., "Vibrations of two coaxial cylindrical shells containing fluid", Nuclear Engineering & Design, Vol.30, 1974, pp.242-248.

167. Au-Yang, M.K., "Free vibration of fluid-coupled coaxial cylindrical shells of different lengths", J.Applied Mechanics, Vol.43, 1976, pp.480-484.

168. Chen, S.-S., Wambsganss, M.W., & Jendrzejczyk, J.A., "Added mass and damping of a vibrating rod in confined viscous fluids", J.Applied Mechanics, Vol.43, 1976, pp.325-329.

169. Yeh, T.T., & Chen, S.-S., "Dynamics of a cylindrical shell system coupled by viscous fluid", J.Acoustical Society of America, Vol.62, 1977, pp.262-270.

170. Chu, M., *et al.*,"An experimental study of gap and thickness influence on the vibration of flexible fluid coupled cylinders", in "Dynamics of Fluid-Structure Systems in the Energy Industry" (ed. M.K. Au-Yang, S.J. Brown & J.R. McDermott), pp.1-19; ASME, New York, 1979.

171. Daubert, A., Pugnet, L., Boulot, F., "On a global hydraulic instability in the vessel of a pressurized water reactor", in Ref.[26], pp.75-80.

172. Chevalier, J., "Some hydrodynamical problems encountered in nuclear power plants", IAHR J. of Hydraulic Research, Vol.15, 1977, pp.125-149.

173. Fricke, F.R., Davies, P.O.A.L., Parkin, M.W., "Periodicity in annular diffuser flow", in Ref.[26], pp.68-74.

174. Au-Yang, M.K., "Pump-induced acoustic pressure distribution in an annular cavity bounded by rigid walls", ASME Paper 78-PVP-69, Jun 1978.

175. Au-Yang, M.K., & Connelly, W.H., "A computerized method for flow-induced random vibration analysis of nuclear reactor internals", ASME Paper 76-WA/PVP-12, Dec 1976.

176. Epstein, A., *et al.*, "Hydroelastic model of PWR reactor internals Safran 1 - validation of a vibration calculation method", Paper 4.3, ViNP, Keswick, 1978.

177. Jaudet, A., & Sol, J.C., "Study of the vibrational behaviour of the vessel and internals of the Fessenheim 1 PWR - Application to vibrational monitoring", Paper 7.3, ViNP, Keswick, 1978.

178. Miller, D.R., & Kennison, R.G., "Theoretical analysis of flow-induced vibration of a blade suspended in a flow channel", ASME Paper 66-WA/NE-1, Nov 1966.

179. Zanoni, E.A., & Lowe, P.A., "Stability analysis of a rigid blade in a parallel fluid flow", in "Proc. Conference on Flow-Induced Vibrations in Reactor System Components", pp.321-334; Argonne National Lab. Report ANL-7685, May 1970.

180. Corr, J.E., "Big Rock Point Vibration Analysis", pp.272-289; Argonne National Lab. Report ANL-7685, May 1970.

181. Miller, D.R., "Generation of positive and negative damping with a flow restrictor in axial flow", pp.304-307, Argonne National Lab. Report ANL-7685, May 1970.

182. Boyd, J., "The influence of fluid forces on the sticking and lateral vibration of pistons", J.Applied Mechanics, Vol.31, 1964, pp.397-401.

183. Petrikat, K., "Seal Vibration" Paper in Session F of this Conference, pp. 117-140 of Preprints III.

184. Kolkman, P.A., "Development of vibration-free gate design: learning from experience and theory", General Lecture in Session F of this Conference, pp.80-116 of Preprints III.

185. Weaver, D.S., "Flow-induced vibrations in valves operating at small openings", Paper in Session C of this Conference, pp.11-25 of Preprints II.

186. Paidoussis, M.P., "Dynamics of submerged towed cylinders", Proc. 8th Symposium on Naval Hydrodynamics", Pasadena 1970, pp.981-1016; U.S. Office of Naval Research, ARC-179.

187. Belytschko, T., "Methods and programs for analysis of fluid-structure systems", Nuclear Engineering & Design, Vol.42, 1977, pp.41-52.

188. Schumann, U., "Dynamics of a nuclear-reactor shell structure in an incompressible fluid", in "Recent Developments in Theoretical and Experimental Fluid Mechanics" (ed. U.Müller *et al.*), pp.577-586; Springer-Verlag, Berlin, 1979.

189. Papers B6/1-B6/7 of "Session B6: PWR Problems" of 5th Int'l Conf. on Structural Mechanics in Reactor Technology, Berlin, 1979.

CHART I: TUBING & ANTIVIBRATION BARS IN PWR STEAM GENERATORS

Case 1 Class: cross-flow-induced vibration Date: 1970-1977

(1)

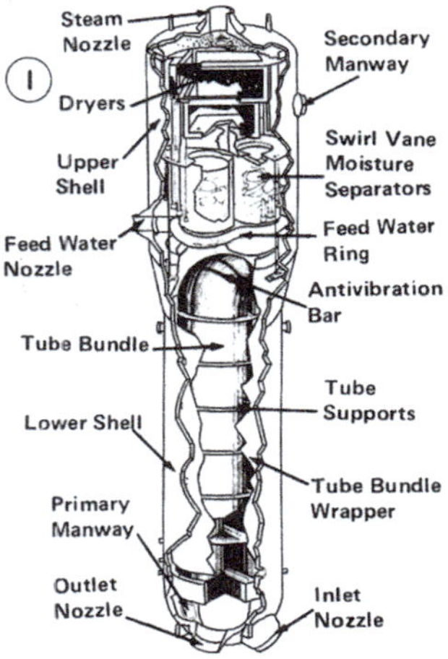

Problem: Several occurrences of tubes leaking in steam generators, from 1970 on, were localized by inspection in the U-bend region (Sch.1). Several tubes had to be plugged on many occasions: e.g. 17 in Jul 72, 57 in Oct 76-Feb 77. The leaking was attributed to thinning of the tube wall by vibration-induced wear and impact with anti-vibration bars (Sch.2); indentation and wear of both tubes and bars was noted. Extensive eddy-current testing revealed many tubes had been affected (of 140 tubes tested in Sep/Oct 77, 25 had >50% wall thinning).

Mechanism: Analysis showed that the mechanism involved was fluid-elastic instability of the inadequately supported tubes in cross-flow in the U-bend region (support case 2, Sch.3). Mode shape for 48th row shows large amplitude at the U bend (Sch.5).

Cure: Installation of additional anti-vibration bars between the others to provide further support points (Sch.4). The new bars were of chrome-plated Inconel 600 to resist wear, and of rectangular cross-section, while the old ones were round carbon steel. The new design was analysed and found to be below threshold for fluid-elastic instability (support case 1, Sch.3; Sch.6).

(2)

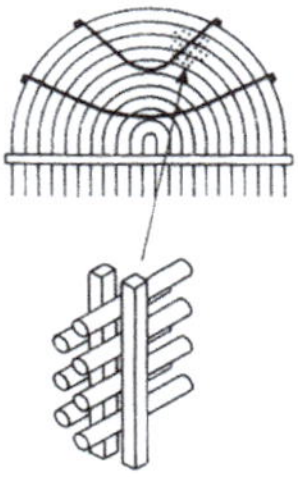

References: 1. Hare, M.G., *Nuclear Safety*, Vol.18, 1977, pp.355-364.

2. Stevens-Guille, P.D., *Atomic Energy of Canada AECL-4753*, Mar 1974.

3. Verna, B.(ed.), *Nuclear Power Experience*, Section PWR-2, V.D., Items 128,155,162.

4. Ottoson, H.L. & Gould, W.R., Proc. *IAEA Symposium on Experience from Operating and Fuelling Nuclear Power Plants;* Vienna 1974, pp.151-166.

5. Anonymous, *Steam Generator Inspections, San Onofre Nuclear Generating Station, Unit 1*, Ch.4; S.Calif. Edison Co., Jan 1977.

6. Anonymous, *Power Reactor Docket Information*, Docket 50266-788, Oct 1977.

(3)

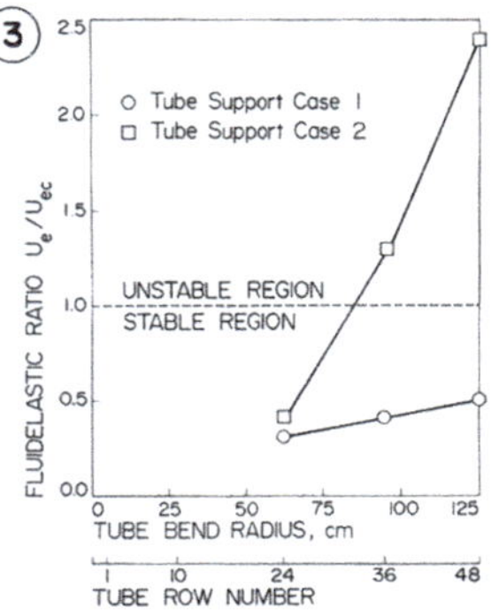

(4)

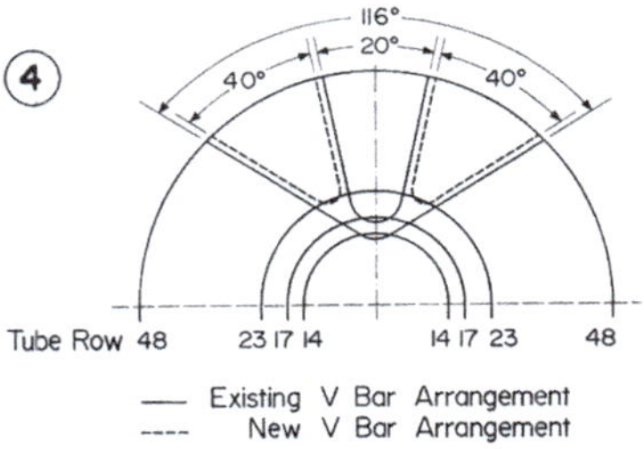

(5)

(6)

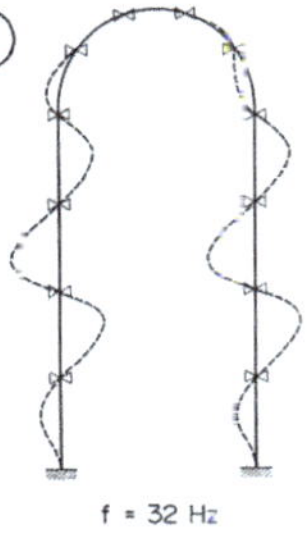

CHART II: TUBING IN PWR STEAM GENERATORS

Case 2 Class: cross-flow-induced vibration Date: 1976

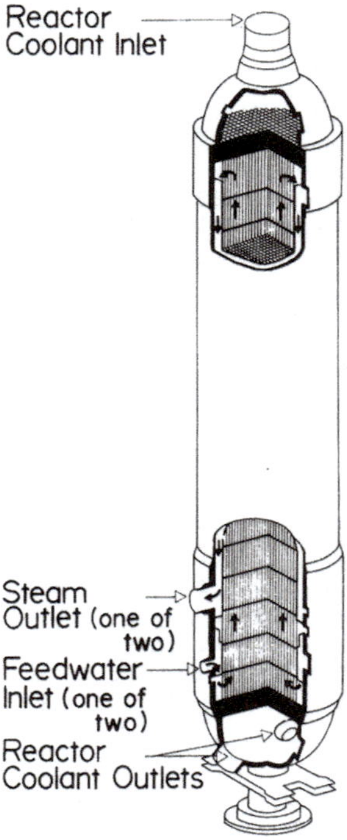

Problem: In all, six tubes developed leaks, in three shell-and-tube "once-through" steam generators (see schematics), each of which has an "open lane" of tubes; this lane is one tube row wide and extends from the periphery of the tube bundle to the centerline. Steam velocities in this lane are higher. Most tube leaks were in the open lane and were due to high cycle fatigue.

Mechanism: Vibration measurements during start-up and in laboratory tests indicated amplitudes insufficiently high to cause failure by flow-induced vibration *alone*. Thermal stresses through use of cold auxiliary feed water may have contributed to failures.

Cure: Minimizing use of auxiliary feed-water system and stiffening the tubes in the open lane.

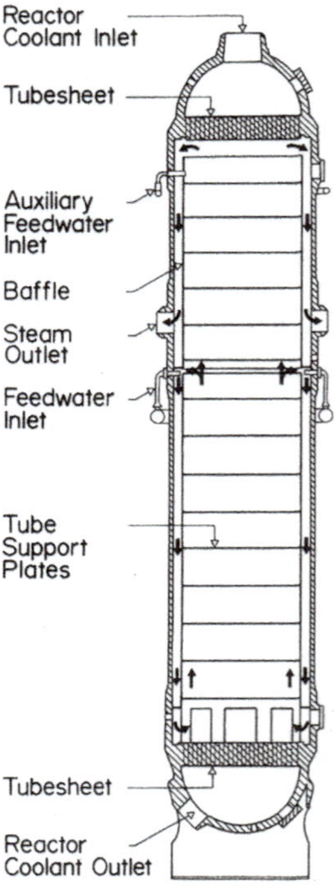

References: 1. Tatone, O.S., & Pathania, R.S., *Atomic Energy of Canada AECL-6095*, Feb 1978.

2. Deddens, J.C., & Fehland, F.R., Paper PGTP 77-38, *American Power Conference;* Chicago, Apr 1977.

Case 3 Class: cross-flow-induced vibration Date: 1976

Problem: Six tubes had to be plugged in these once-through steam generators because of fretting wear at the top support plate.

Mechanism: Most probably cross-flow-induced vibration; details and precise mechanism involved unknown.

Cure: Plugging operation.

Reference: Tatone, O.S. & Pathania, R.S., *op. cit.*

CHART III: (a) TUBING IN PHWR STEAM GENERATOR

Case 4 Class: cross-flow-induced vibration Date: < 1971

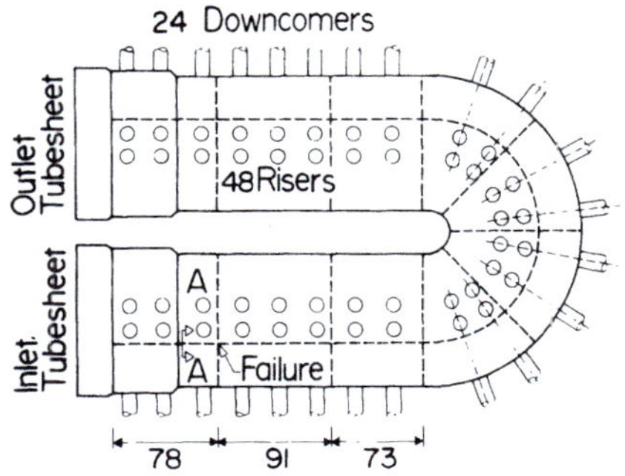

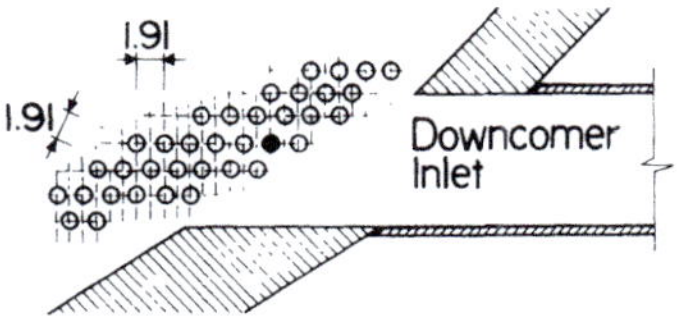

Sect A-A Showing Proximity of Failed Tube to Downcomer Ports

All Dimensions in cm

Problem: In this shell-and-tube heat exchanger, with a U-tube bundle in a U-shell, the recirculated liquid returns through 24 downcomers, entering the shell almost tangentially. One tube failed, several years after commissioning, near the downcomer inlet (second row in); it failed by being "cut" at first support plate, causing a leak; there was similar damage at the second support plate.

Data: tubing pattern: normal triangle, $p/D=1.5$; $U/f_nD=4.2$ for nominal velocity of 2.6 m/s of D_2O; $m/\rho D^2=4.4$.

Mechanism: Cross-flow-induced vibration, probably due to vortex shedding; cf. $U/f_nD=4.2$ with $1/S_{FH}=(1/0.26)=3.85$ according to Fitz-Hugh's charts. However, on Connors-type diagram, with $K=3.3$ and $\delta=0.1$ assumed, it lies in unstable region. Long life before failure and post-cure behaviour tends to support vortex shedding - in any case, a weaker fluid-elastic coupling than that for fluid-elastic instability.

Cure: The tube was plugged and nothing else done. No further damage was encountered.

References: 1. Hartlen, R.T., Paper 611, *International Symposium on Vibration Problems in Industry*, Keswick 1973.

2. Stevens-Guille, P.D., *Atomic Energy of Canada AECL-4449*, Apr 1973.

CHART III: (b) TUBING IN PWR STEAM GENERATOR

Case 5 Class: cross-flow-induced vibration Date: 1975

Problem: A leak developed during power testing in the tubing through abrasion, causing a three-month delay in commercial operation.

Mechanism: Flow-induced movements of a metal measuring tape against the tubing; the tape was inadvertently left in the steam generator during fabrication.

Cure: Plugging the leak and removing the measuring tape.

Reference: Hare, M.G., *Nuclear Safety*, Vol.18, 1977, pp.355-364.

CHART IV: TUBING IN PHWR STEAM GENERATOR

Case 6 Class: cross-flow-induced vibration Date: < 1971

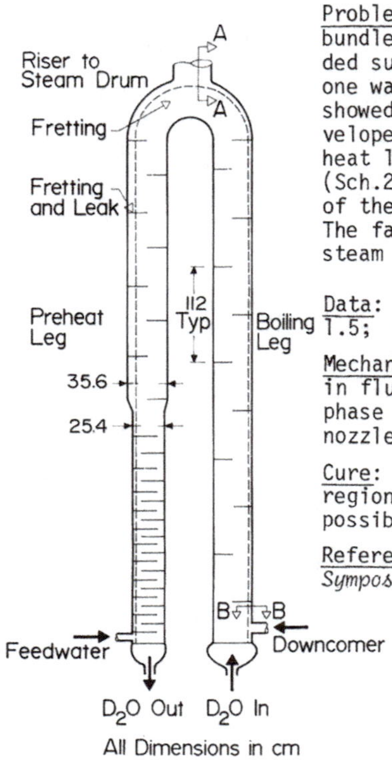

Problem: This steam generator consists of a U-tube bundle in a U-shell (Sch.1). In most other units, added support was provided in U-bend region, but this one was tested without. After 10,000 hrs, inspection showed no fretting, but after 13,000 hrs, a leak developed through a fretted hole in a tube in the preheat leg at the third support plate from the top (Sch.2); there was also extensive fretting at the top of the U-bend by contact with adjacent tubes (Sch.3). The failed tube was close to both downcomer inlet and steam outlet ports.

Data: Tubing pattern: normal triangular, $p/D=1.25$, 1.5; $U/f_nD=48$, $m/\rho D^2=18.9$, $\delta=0.38$.

Mechanism: Analysis shows operating conditions are in fluid-elastic instability range in the outlet two-phase region, exacerbated by tube proximity to inlet nozzle.

Cure: Provision of additional support in U-bend region to sufficiently increase f_n, and to reduce possibility of impact damage.

References: 1. Hartlen, R.T., Paper 611, Proc. *Int'l Symposium Vibration Problems in Industry*, Keswick 1973.

2. Pettigrew, M.J., *Atomic Energy of Canada AECL-5852*, Sep 1977.

3. Suurmann, S., *Canadian Nuclear Association Paper 71-CNA-203*, Jun 1971.

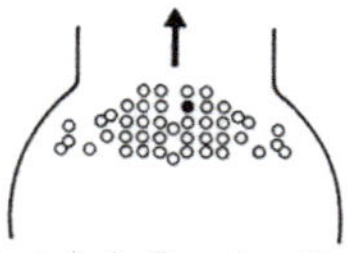

Sect A-A Showing Outlet Geometry and Position of Failed Tube

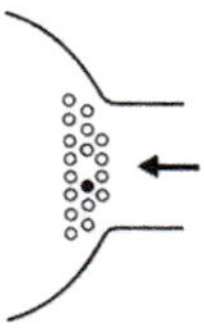

Sect B-B Showing Inlet Geometry and Position of Failed Tube

①

②

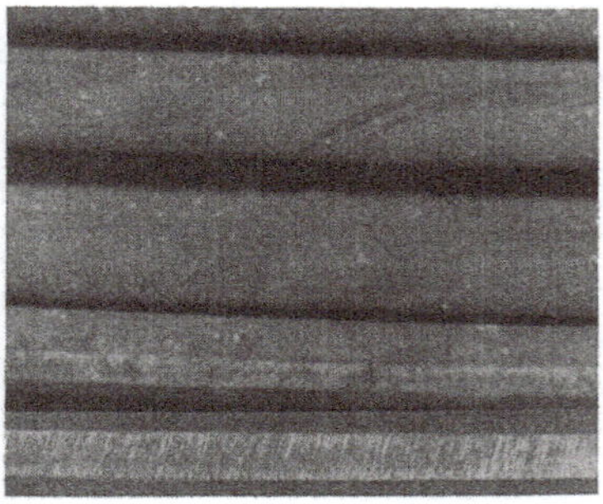

③

CHART V: TUBING IN LMFBR STEAM GENERATOR

Case 7 Class: cross-flow-induced vibration Date: 1962

Problem: During preliminary testing at 80% of design flow rates, after 13 days' operation, a tube failed in the steam generator, allowing water to enter the shell side and react with the liquid sodium. The steam generator had to remain operating to remove the reaction products while the tubes were kept filled with inert gas, resulting in numerous additional tube failures. In all, 45 tubes failed, mostly close to the inlet nozzles, by impact with one another.

Data: Full-scale tests using water with bundle from an identical steam generator indicated vibrations at f=30 Hz= f_n for the risers, most severely at outer ones. Maximum amplitude at design U was ~ 0.25 cm. Introduction of inlet baffle displaced the take-off point of amplitude-flow curve from 68% of design flow to 88%, but did not solve problem.

Mechanism: The constancy of the vibration at f_n over wide range of flows precludes vortex-shedding excitation. It was concluded (1964) that the mechanism must have been buffeting. However, the amplitude of vibration, A, depended on U - after the take-off point - as $A=kU^n$, with n=4.5 -12.9. Hence, this is not the likely mechanism, and one must suspect that it was fluid-elastic instability.

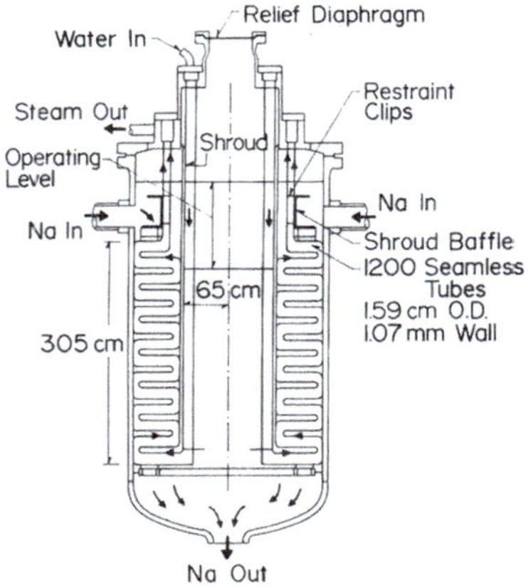

Cure: A shroud-type inlet baffle was introduced, together with lacing clip restraints. The maximum amplitude (in full scale tests) was reduced to 0.01 cm, and f_n increased to 55 Hz.

References: 1. Smith, J.O., McCarthy, J.H., & Boelter, W.A., *ASME Paper 64-WA/PWR-4*.

2. Shin, Y.S., & Wambsganss, M.W., *Nuclear Engng & Design*, Vol.40, 1977, pp 235-284.

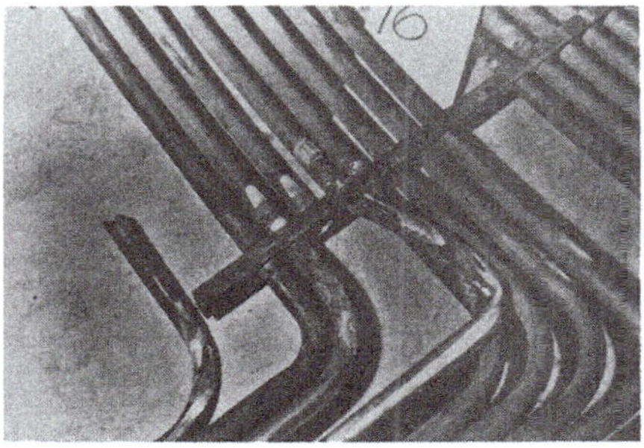

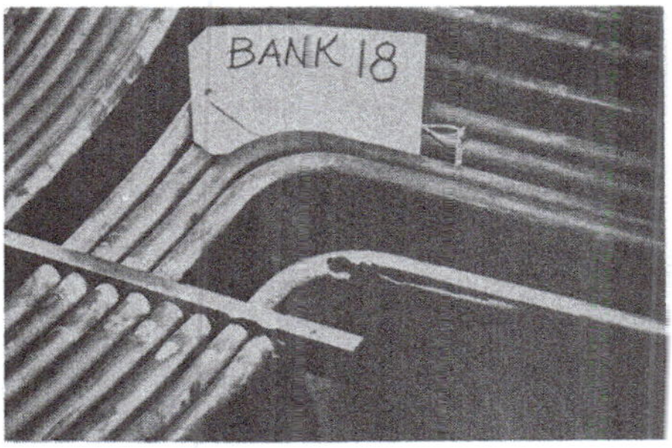

CHART VI: TUBING IN THERMOSIPHON REBOILER

Case 8 Class: cross-flow-induced vibration Date: ~ 1973

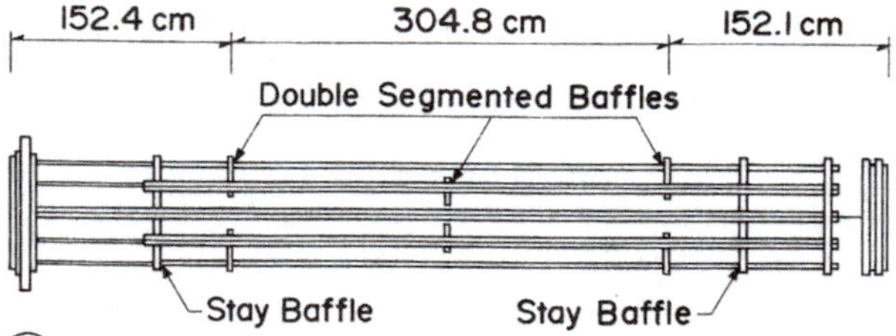

① ORIGINAL BUNDLE DESIGN (PLAN)

②

③

Problem: 36 hrs. after start-up the tube bundle of this shell and tube heat exchanger failed (tube side: boiling water; shell side: effluent gas). Inspection showed 20% of tubes leaked, mostly where unsupported tubes were worn by rubbing against edges of segmented baffles; tubes on periphery showed flat spots where they had pounded against one another. "Others had sawed themselves in two where they passed through stay-baffle holes" (Sch.2). Of the unfailed tubes many were loose and showed some wear.

Mechanism: Cross-flow-induced vibration. Authors suspect vortex shedding. The severity and extent of damage would suggest coincidence of f_{vs} with f_n, and probably with f_a as well, i.e. coincidence with both structural natural frequency and acoustic resonant modes.

Cure: Detuning was achieved by inserting bar stock between tube rows with zero clearance; the bars were not inserted at same transverse plane, but in a plane at an angle to the shell axis, so that the bar ends, cut to length, defined an ellipse. A second set of bars was installed but rotated at 90^0 to the first; the two ellipses criss-crossed (Sch.3). This gave more support points; different spans of adjacent tubes produced a detuning effect. This "solution" was retained in a redesigned bundle, where the three double-segmented baffles were completely replaced by the "rod baffles" described.

Reference: Eilers, J.F. & Small, W.M., *Chemical Engineering Progress*, Vol.69, 1973, pp.57-61.

CHART VII: TUBING IN POWER STATION FEED-WATER HEATER

Case 9 Class: cross-flow-induced vibration Date: < 1973

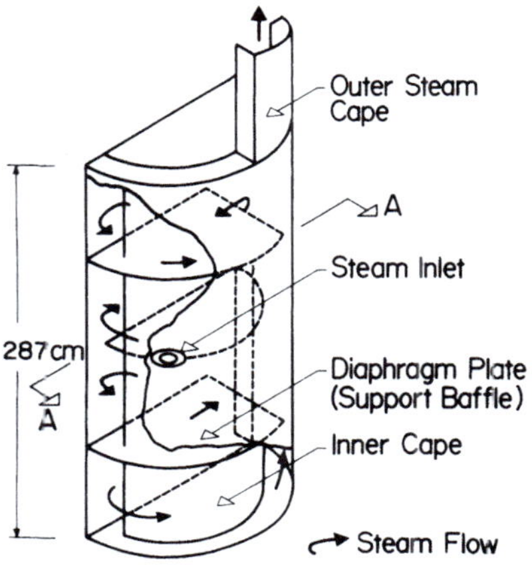

①

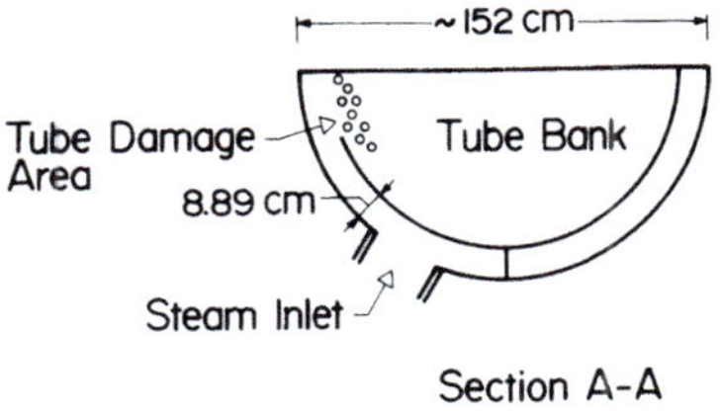

Section A-A

②

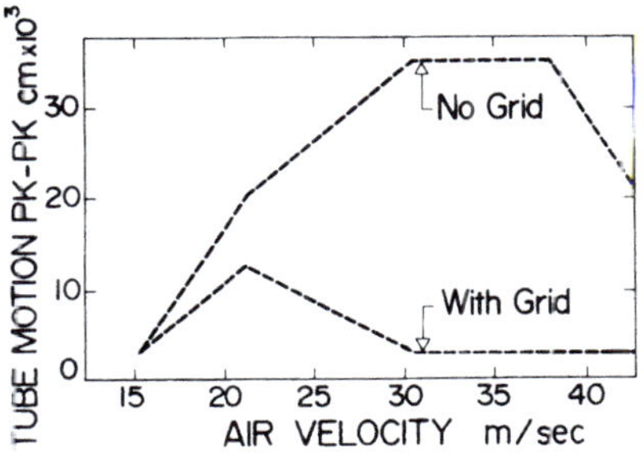

③

Problem: A number of individual tube failures were recorded in the high pressure feed-water heater of a modern power station, involving extremely heavy fretting of the tube wall against the baffle plates, with eventual wall penetration in a few tubes; "the subsequent escape of high pressure water caused severe local erosion eventually penetrating adjacent tubes". The failed tubes were located close to the edge of the steam inlet impingement baffle (Sch.2) - across the intake "windows".

Data: Tubing pattern: triangular, p/D=1.25; U=24.4 m/s of steam at 2070 kPa and 316°C; D=1.91 cm; $f_n \approx 89$-109 Hz†).

Mechanism: At the time of reporting, vortex shedding based on S=0.4-0.5 was eliminated as a possible cause. This is confirmed here; $S_{FH}=0.35$ and $S_C=0.85$ yield $f_{vs}>448$ Hz. Model tests using air indicated a slow rise of amplitude with flow, the amplitude eventually reaching a plateau, suggesting that fluidelastic instability is not the cause. (Insufficient information does not allow for a positive conclusion. One has $U/f_nD \simeq 12$ to 14; these values, although $m/\rho D^2$ and δ are unknown, are in the range of fluidelastic instability for steam flows). Presuming correct modelling in the air tests one must conclude that this was a case of buffeting, as concluded in [1].

Cure: A perforated grid was welded across the intake "windows". In the model tests this showed a reduction in turbulence level and an attendant threefold to sevenfold reduction in amplitude of vibration (Sch.3).

Reference: 1. Brown, R.D., Paper 423, *Int'l Symposium on Vibration Problems in Industry;* Keswick, 1973.

† Note that these frequencies differ from those of [1], where the total length of the tubes is used to calculate f_n, neglecting the support baffles, presumably because of the existing clearance in the baffle holes.

CHART VIII: TUBING & IMPINGEMENT BAFFLES of TURBINE-STEAM REHEATER

Case 10 Class: cross-flow-induced vibration Date: < 1973

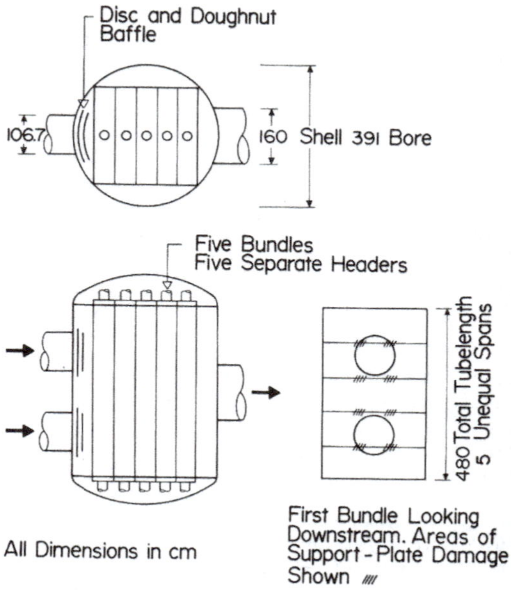

Problem: "After a relatively short time in service it was discovered that the impingement baffles had torn away completely, falling into the bottom of the vessel". Also, vibration of the tubes caused hole enlargement and breaking through of the ligaments in the support plates. In this steam to steam cross-flow heat exchanger there are five rectangular tube bundles, in series, in a cylindrical shell.

Data: Tubing pattern: normal triangular, $p/D=1.58$; $U/f_nD=83$ for nominal velocity of 26 m/s; $m/\rho D^2=1320$; $\delta=0.05$ in air. The inlet flow had a swirl component.

Mechanism: The baffle failures were attributed to fluctuating aerodynamic load due to swirl, coupled with inferior welding. Analysis and some field testing established clearly that the tube vibration was caused by fluid-elastic instability; also, the presence of mechanical vibration close to f_n probably contributed.

Cure: The baffles, originally installed for erosion protection which was found to be unnecessary, were eliminated. The effective tube-spans were reduced by driving rectangular, slightly oversized steel bars between the tubes, so that U/f_nD is below the threshold for fluid-elastic instability.

Reference: Hartlen, R.T., Paper 611, *Int'l Symposium on Vibration Problems in Industry;* Keswick, 1973.

CHART IX: (a) TUBING IN COMPONENT COOLING WATER HEAT EXCHANGERS

Case 11 | Class: cross-flow-induced vibration | Date: 1970-1971

Problem: Three leaking tubes were found in pre-operational testing in 1970, and three more in pre-criticality testing in 1971. The tubes were broken through fretting at the first baffle from the inlet-end tube sheet; they were in the vicinity of the impact-baffle edges.

Mechanism: The failures were thought to be "caused by excessive velocity over the tubes adjacent to the edges of the impact baffle at the shell side inlet nozzle". Probably this translates to buffeting.

Cure: After the second incident, in addition to the leaking tubes, another 40 in the vicinity of the failed ones were plugged.

References: 1. Verna, B.(ed.), *Nuclear Power Experience*, Section PWR-2, VIII.B, Item 1.
2. Anonymous, *Power Reactor Docket Information*, Docket 50255-57, Nov 1971.

Case 12 | Class: cross-flow-induced vibration | Date: 1972

Problem: Tube leaks occurred during filling of one of these units. Several tubes had broken adjacent to the tubesheet, nearest to the shell outlet nozzle.

Mechanism: Cross-flow-induced vibration due to excessive tube-bundle penetration velocities at inlet and exit nozzles, associated with insufficient annular clearance between outer tube limit and the inside of the shell. Thus, although outlet nozzle velocity was 4.45 m/s, the penetration velocity was 11.6 m/s.

Cure: The heat exchanger was disassembled, redesigned and rebuilt, incorporating flow distribution channels with large flow passages all around the tube bundle. The outlet penetration velocity was reduced to 1.1 m/s.

References: 1. Verna, B.(ed.), *op. cit.*, Section PWR-2, VIII.B, Item 5.
2. Jurgensen, R.W., private communication, 21 May 1979.

CHART IX: (b) TUBING IN RESIDUAL HEAT REMOVAL HEAT EXCHANGERS

Case 13 | Class: cross-flow-induced vibration | Date: 1974

Problem: A total of 22 tubes had to be plugged in this shell-and-tube heat exchanger after inspection, following recurrent tube failures. This heat exchanger is used in shutdown operations. The cause was identified as wear between tubes and support bars in the U-bend region. There were clearances because of faulty welding sequence and poor installation of some tubes

Mechanism: Without the support of the support bars, the tubes are probably susceptible to fluid-elastic instability by cross-flow at the U-bend region (cf. [2]). Buffeting alone might be sufficient to cause the damage.

Cure: Plugging of all tubes not tightly against the U-bend support bars.

References: 1. Verna, B.(ed.), *op. cit.*, Section BWR-2, VII.D, Item 52.
2. Lee, B., Jr., *Power Reactor Docket Information*, Docket 50295-68, Oct 1972.

Case 14 | Class: cross-flow-induced vibration | Date: 1975

Problem: A leaking tube in this heat exchanger caused a release of radioactivity.

Mechanism: The failure was caused by high inlet water impingement.

Cure: Plugging of this and another three tubes in high velocity inlet area. Later, stiffeners on these tubes were also installed.

References: 1. Verna, B.(ed.), *op. cit.*, Section BWR-2, VII.D, Item 68.
2. Anonymous, *Power Reactor Docket Information*, Docket 50277-613, Feb 1976.

CHART X: TUBING IN LIQUID-LIQUID CROSS-FLOW HEAT EXCHANGERS

Case 15 Class: cross-flow-induced vibration Date: < 1973

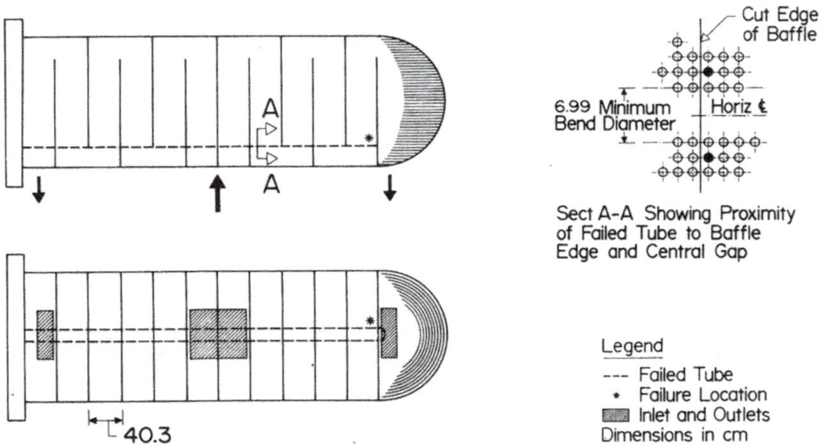

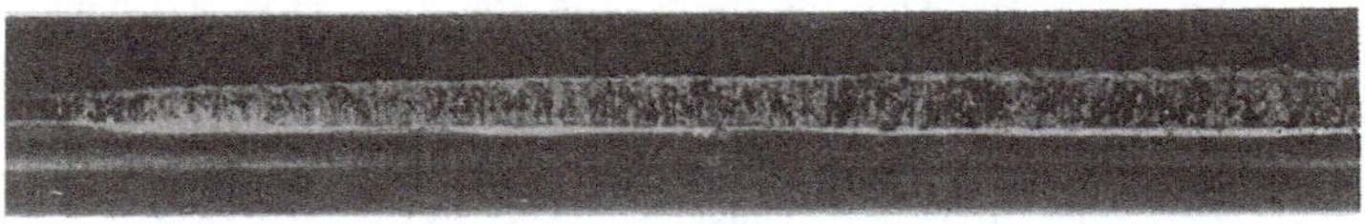

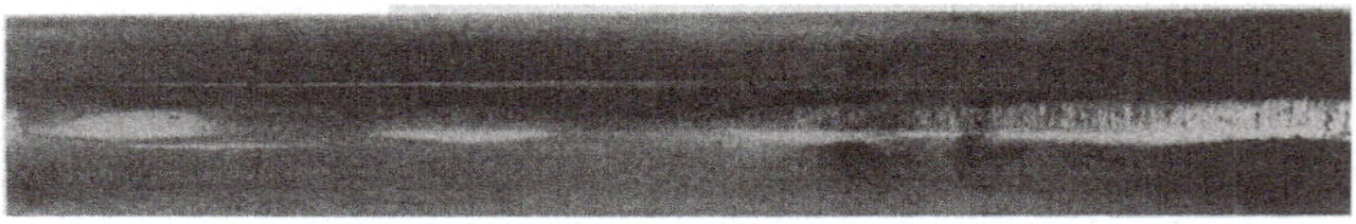

Problem: A tube of this shell-and-tube heat exchanger failed near the U-bend region, close to one of the shell-side exit ports. The tube was just beyond the baffle cut, so that its tube-span was twice as long as for most others. After audibly rattling since commissioning, eventually a leak developed, which was identified to be caused by fatigue where the tube passed through baffle plate - with wear patterns showing at mid-span (as shown in photographs).

Data: Tubing pattern: normal square, $p/D=1.5$; $U/f_nD=4.5$ for nominal velocity of 1.2 m/s of D_2O; $m/\rho D^2=4.0$.

Mechanism: Cross-flow-induced vibration, most probably due to fluid-elastic instability. On Connors-type diagram with K=3.3, assuming $\delta=0.1$, this is in unstable region.

Cure: All tubes with similar "double" span were constrained by introduction of transverse metal rods at mid-span. In later design all such tubes were eliminated.

Reference: Hartlen, R.T., Paper 611, *Int'l Symposium on Vibration Problems in Industry*; Keswick, 1973.

CHART XI: TUBING IN HEAT EXCHANGERS OF MOLTEN SALT REACTOR EXPERIMENT

Case 16 Class: cross-flow-induced vibration Date: ~ 1964

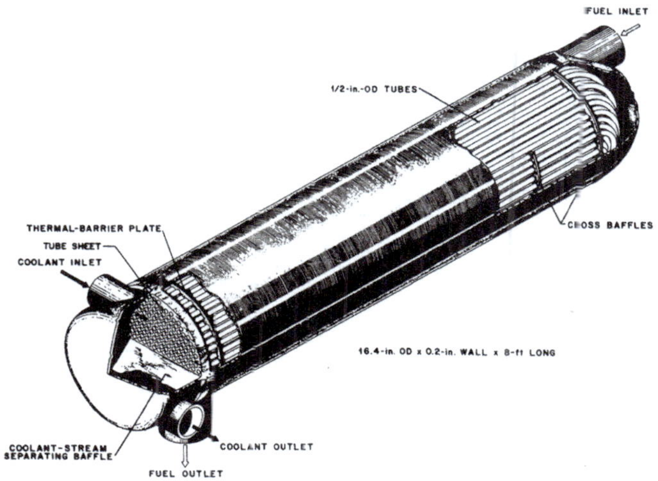

Problem: In the primary shell-and-tube heat exchanger of this experimental reactor the two fluids were molten salt (various fluorides). Some pre-commissioning water tests indicated widespread large amplitude vibrations, especially in the U-bend region (≃ 6.5 mm), with two tubes vibrating against the shell. The shell-side fluid enters in the U-bend region.

Data: Tubing pattern: triangular, p/D=1.55; assuming δ=0.1 and water flow, $m\delta/\rho D^2 \simeq 0.35$; taking U=1.85 m/s, the baffle-edge velocity, and f_n=28-78 Hz for various positions, $U/f_nD \simeq 1.9$ to 5.2, the higher value applying conservatively to the U-bend region.

Mechanism: When originally reported (1970) the vibration was thought to be due to vortex shedding [2]. Taking K=3.3 in Connors-type relationship, one obtains $(U/f_nD)_c \simeq 1.9$. Hence, most likely this was a case of fluid-elastic instability. The same holds for salt design-flow conditions.

Cure: (i) Removal of the four outermost tubes where U is high; (ii) wire lacing between tubes at each baffle level and in U-bend, to increase f_n; (iii) installation of an impingement baffle at inlet to reduce direct impingement and buffeting; (iv) enlargement of shell-side outlet to reduce U near the outlet (from 5.85 to 3.17 m/s in nozzle). The system has subsequently performed satisfactorily for at least 5 years.

References: 1. Nelms, H.A., *Flow-Induced Vibration in Heat Exchangers*, ASME, New York, 1971; pp.8-17.

2. Nelms, H.A., & Segaser, C.L., *Oak Ridge National Laboratory ORNL-4399*, Apr 1969.

CHART XII: TUBING IN HEAT EXCHANGERS OF HIGH FLUX ISOTOPE REACTORS

Case 17 | Class: cross-flow-induced vibration | Date: 1966-1968

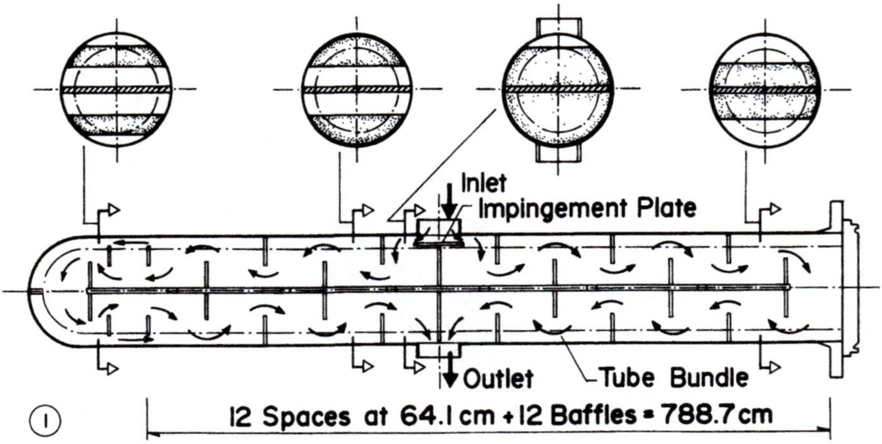

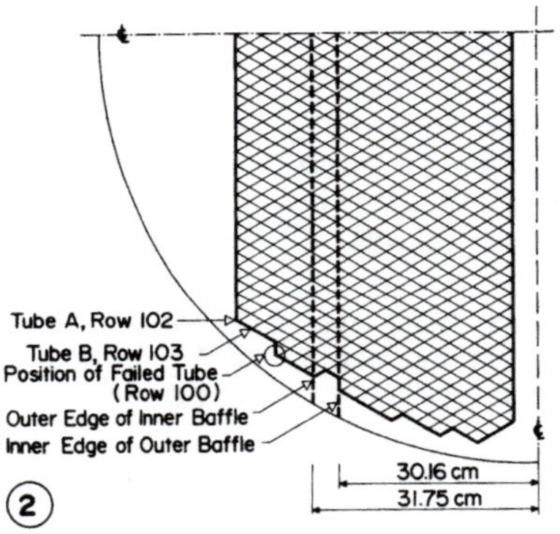

Problem: In each of three units of these water-water shell-and-tube heat exchangers, there was one ruptured tube, in the same position - namely, near the shell-side outlet, in the outer perimeter of the tube bundle. Metallurgical analysis showed this to be a fatigue failure.

Data: Tubing pattern: triangular, p/D=1.3; based on U=0.84 m/s [3], and on measured f_n= 22 Hz, $U/f_nD \simeq 2.4$; assuming $\delta=0.1$, $m\delta/\rho D^2 \simeq 0.28$. The units operated at times at 115 to 120% of rated flows.

Mechanism: When originally reported (1970) the cause was thought to be "vibrational stresses enhanced by defects". Testing for fluid-elastic instability, one obtains $(U/f_nD)_c=1.76$, based on K=3.3. Hence, instability would be indicated; yet the specificity of damage does not support this. Testing for vortex shedding, in this case S_C= 0.85 and S_{FH}=0.35; the latter gives $f_{VS} \simeq 18.4$, close to f_n; at 115 and 120% of rated flow, f_{VS} increases to 21 or 22 Hz. Moreover, experiments indicated the existence of an amplitude peak close to rated flow. Hence, vortex shedding is the more likely mechanism.

Cure: (i) Plugging of the tubes in locations where failures occurred; (ii) limiting velocities to below the rated flows (recommended operation at 85% of rated flow).

References: 1. Nelms, H.A., *Flow-Induced Vibration in Heat Exchangers*, ASME, New York, 1970, pp.8-17.

2. Nelms, H.A., & Segaser, C.L., *Oak Ridge National Laboratory ORNL-4399*, Apr 1969.

3. Fraas, A.P., *Oak Ridge National Laboratory ORNL-TM-2467*, Mar 1969.

CHART XIII: (a) TUBING IN WATER-WATER HEAT EXCHANGERS

Case 18 | Class: cross-flow-induced vibration | Date: < 1969

Problem: Tube failures occurred in this heat exchanger, attributed to bursting after excessive thinning of tube wall by wear.

Data: Tubing pattern: triangular, p/D=1.44; assuming $\delta=0.10$, $m\delta/\rho D^2=0.315$; $U/f_nD=4.84$.

Mechanism: Cross-flow-induced vibration. The information available at time of reporting would indicate that vortex shedding might be most probable cause. Testing for fluid-elastic instability with Connors-type relationship, taking K=3.3, $(U/f_nD)_c=1.85$, showing that this was the likely cause.

Cure: (i) Reduction of tube span from 152.4 cm to 96.5 cm; (ii) reduction of average velocity from 1.84 m/s to 0.99 m/s. Based on these parameters, $U/f_nD=0.65-0.74$, which is below the threshold for fluid-elastic instability. Checking for vortex shedding, the observed $f_n=70-80$ Hz, whereas $f_{vs}\simeq 50$ according to Y.N. Chen and $f_{vs}\simeq 18$ Hz according to Fitz-Hugh - which shows that the new design is safe in this respect also.

Reference: Nelms, H.A., & Segaser, C.L., *Oak Ridge National Laboratory* ORNL-4399 Apr 1969.

Case 19 | Class: cross-flow-induced vibration | Date: < 1969

Problem: Tube failures attributed to fatigue, close to tube sheet, were recorded, as well as heavy wear at baffle-plate holes.

Data: Tubing pattern: rectangular, p/D=1.25; taking $\delta=0.10$, $m\delta/\rho D^2=0.35$; $U/f_nD=5.95$ and 2.66, based on $f_n=31.3$ Hz and $f_n=71$ Hz, for simply supported and clamped tubes, respectively.

Mechanism: At the time of reporting, based on S=0.17, $f_{vs}=31.6$ Hz was calculated, and it was concluded [1] that vortex-shedding was the cause. Based on $S_C\simeq 0.84$ and $S_{FH}=0.35$, one obtains $f_{vs}=156$ and 65 Hz, respectively. The latter is close to $f_n=71$ Hz, assuming clamped supports. Testing for fluid-elastic instability with K=3.3, $(U/f_nD)_c=1.95$. Hence, fluid-elastic instability is the likely mechanism, taking into account conservativeness of δ assumed. Even if $\delta>0.1$ (!) and tubes initially clamped, with baffle-hole wear the supports would eventually become more like simple supports, allowing the system to become fluid-elastically unstable.

Cure: No recorded solution. Reduction of maximum flow velocity and possible lacing of tubes would have been indicated.

Reference: Nelms, H.A., & Segaser, C.L., *op. cit.*

CHART XIII: (b) TUBING IN Na-Na HEAT EXCHANGER

Case 20 | Class: cross-flow-induced vibration | Date: < 1969

Problem: Failure of tubes occurred and was attributed to wear at the baffle plates.

Data: Tubing pattern: rotated square, p/D=1.3; assuming $\delta=0.1$, $m\delta/\rho D^2\simeq 0.26$; $U/f_nD=3.66-5.65$.

Mechanism: Based on S=0.31, at time of reporting it was thought that this was a vortex-excited vibration. However, utilizing $S_{FH}=0.97$ and $S_C=1.24$ one obtains $f_{vs}=128$ and 163 Hz, far in excess of $f_n=35-37$ Hz. The fluid-elastic instability threshold is defined by $(U/f_nD)_c\simeq 1.7$, which suggests this to be the likely mechanism involved.

Cure: No recorded solution in [1].

Reference: 1. Nelms, H.A., & Segaser, C.L., *op. cit.*

CHART XIV: (a) TUBING IN GCR STEAM RAISING UNIT

Case 21 Class: cross-flow-induced vibration Date: < 1973

(1)

(2)

Problem: Fatigue failure of several tubes (Sch.1) in the steam raising unit of a CO_2, gas-cooled reactor. The heat exchanger is essentially a boiler with CO_2 on the shell side and water turning to steam in the horizontal tubes.

Data: The mechanism involved was established with the aid of model tests (Sch.2), modelling $m\delta/\rho D^2$=3.14, tubing pattern parameters† s_t=1.67 to 2.74 and s_ℓ=1.33-2.00, in various combinations, and U/f_nD. In the prototype δ=0.001, the tubes being welded to rigid supports at either end.

Mechanism: In the model tests both cases were obtained of (i) coincidence of f_n and f_{vs}, and (ii) coincidence of f_{vs} and f_a, corresponding to acoustic standing waves in the duct (normal both to tube axes and flow direction). The mechanism involved in the prototype failure was coincidence of f_{vs} and f_n, exacerbated by the very low value of δ involved [2].

Cure: Stiffening the assembly, thereby increasing f_n. This was achieved by lacing sprocket chains through the tube bank, tensioning them by weights hung at their ends (Sch.3).

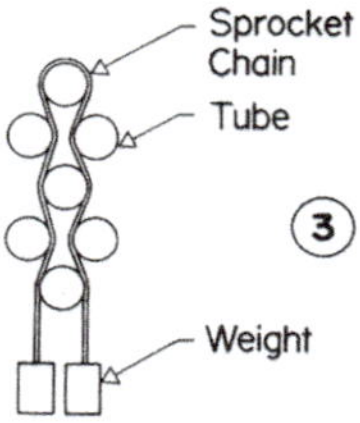

(3)

Reference: 1. Funakawa, M., Paper 428, *Int'l Symposium on Vibration Problems in Industry*, Keswick, 1973.
2. Funakawa, M., *Private communication*, 27 Jul 1979.

† Transverse pitch=T/D and longitudinal pitch=L/D, T being the distance between columns and L between rows. Note that s_ℓ=½(L/D) is used in [1].

CHART XIV: (b) TUBING IN H_2O-D_2O HEAT EXCHANGER

Case 22 Class: cross-flow-induced vibration Date: < 1969

Problem: Early in plant operations there were indications of excessive vibration, but no tube failures occurred. Nevertheless, modifications were made to avoid possibility of failures.

Data: Tubing pattern: triangular, p/D=1.25; assuming δ=0.1, $m\delta/\rho D^2$=0.34; for the straight tube spans, U/f_nD=0.53, based on observed f_n=140 Hz. In U-bend region, observed f_n=30-40 Hz and 5-15 Hz, for in-plane and out-plane motions, but the flow velocities there are not given.

Mechanism: Cross-flow-induced vibration. At time of reporting the vortex shedding frequencies were found not to coincide with any f_n's. Using S_{FH}=0.38, a frequency of f_{vs}=29 Hz is obtained. Vortex excitation at the last straight leg before the U-bend, so close to f_n=30, could be reasoned to be the cause of the observed vibration. Checking for fluid-elastic instability, for K=3.3 one obtains $(U/f_nD)_c \simeq 1.92$, so that this is not the mechanism in the straight legs; it may be, however, in the U-bend region, where although U is less, f_n is up to 28 times less than for the straight legs. Also, general buffeting could be the mechanism.

Cure: Changes in the flow pattern, reducing the flow velocities by a factor of 2.

Reference: Nelms, H.A., & Segaser, C.L., *Oak Ridge National Laboratory ORNL-4399*, Apr 1969.

CHART XV: DUCTS AND WALLS OF RADIANT-REHEAT BOILER

Case 23 Class: cross-flow-induced vibration Date: 1953

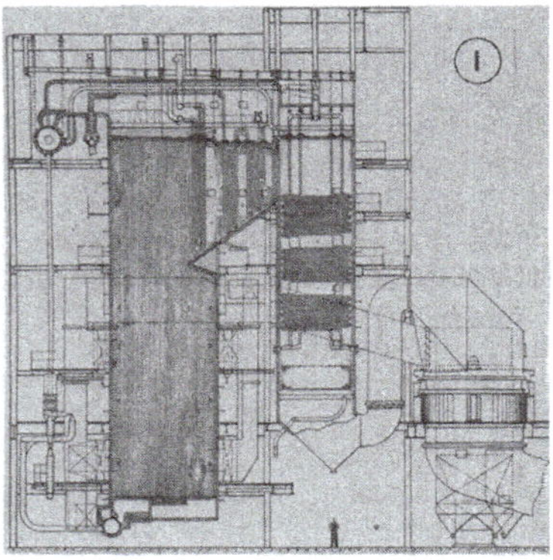

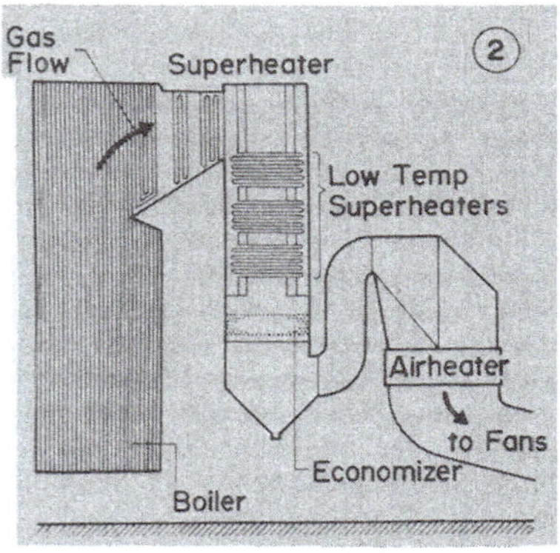

Problem: Severe vibration developed in the forced-draft duct, gas duct and superheater-economizer sections of this 135 MW boiler at steam flows associated with 85 to 105 MW. Angle-iron bracing had to be welded on wall panels, because high pulsating pressure levels ($\simeq$ 3 lb/in^2 or 20 kPa) damaged the refractory material and produced hot spots on the economizer and superheater duct walls (Sch.3). Yet this did not stop the problem. Also, severe sound levels of 115 db were detrimental to health of personnel.

Data: The acoustical frequency, $f_a \simeq 45$ Hz, indicated a full standing wave perpendicular to tube and flow axes. "It was indeed fortunate that none of the natural frequencies of the duct or exchanger tubes coincided with the standing wave frequency", for then "... the structure would not have lasted long enough for a cure to be effected". Tubing pattern is in-line with $s_\ell \simeq 2$, $s_t \simeq 2$.

Mechanism: Taking f_{vs}=40-50 Hz and utilizing S=0.18-0.27, the flow velocities obtained were found to be in the range of those actually present. Using S_{FH}=0.19-0.26 confirms that the problem was due to coincidence of f_{vs} and f_a. The dependence of vibration on power level reflects changes in the speed of sound, c, with temperature; similarly, the localization in the superheater-economizer section is dependent on the local c, the gas cooling down as it progresses through the low temperature superheaters and the economizer.

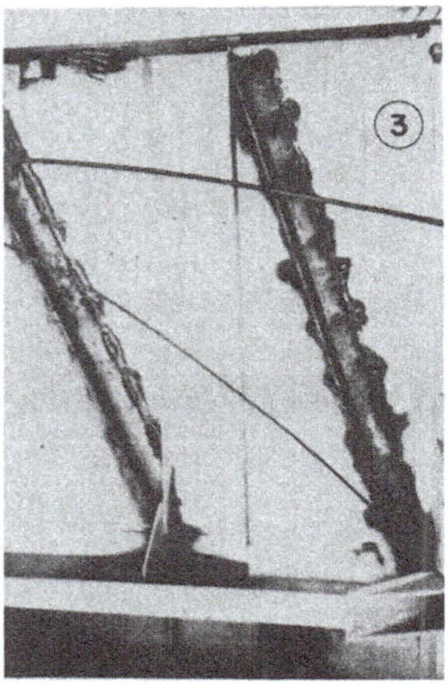

Cure: Installation of two deresonating baffles at quarter points from the walls extending right through, and short baffles at ½ point just between tube banks. This would theoretically allow half-wave formation in the unbaffled parts of middle sections, but was sufficient to detune the system. Significantly, the quarter-point baffles alone could not eliminate the problem [2]. More modern solution [1] would place baffles at unequal intervals (e.g. at 25%, 18.75% and 14.06% of width).

References: 1. Cohan, L.J., & Deane, W.J., *Trans.ASME, J.Engng Power*, Vol.87, 1965, pp.223-228.

2. Baird, R.C., *Combustion*, Vol.25, 1954, pp.38-44.

3. Rogers, J.D. & Penterson, C.A., *C-E Report TIS-5154*, 1977.

CHART XVI: TUBING IN GCR BOILER

Case 24 Class: cross-flow-induced vibration Date: 1972-1976

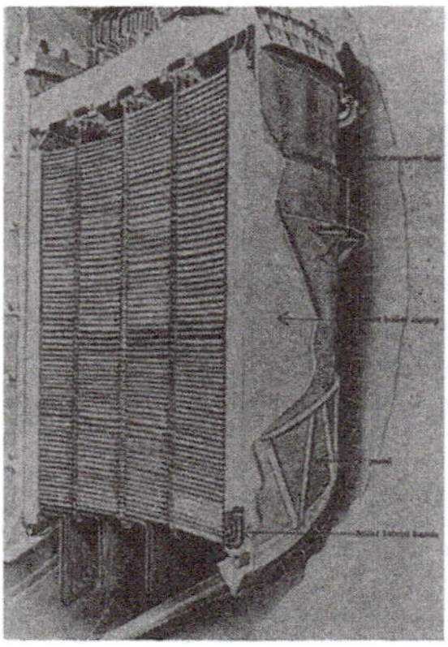

①

Problem: The boiler of this CO_2-cooled reactor is annular, surrounding the reactor itself. Its tubes, in groups of three, are arranged in vertical platens, and 16 platens go together to form a "tube packet". 62 of these tube packets form the entire boiler; they are disposed in the shape of an iris, i.e. not radially in the annulus, but spirally (Sch.1). The packets are meant to fit closely together, but some flexibility is allowed for thermal expansion. Boiler tube leaks began appearing in 1972 and continued through 1974, when their frequency was one every few days. The reactor was shut down and inspection showed severe fretting of the tubes where they were supported by clips; thinning of the tubes led eventually to the leaks (Sch.2). Also, tube packets had separated and moved into permanently distorted arrays. Vertical mismatch of the order of ½D and horizontal gaps of the order of D were discovered, the latter allowing gas to stream down between the packets.

②

Data: s_ℓ=2, 2.5 and 3, s_ℓ=2.17, 2.60 and 3.25, corresponding to different D's for different vertical locations in the boiler.

Mechanism: The mechanism is one of cross-flow-induced resonant vibrations of a type "never before seen in a nuclear boiler". The problem seems to be associated with high gas flows streaming between inter-packet gaps. In isothermal tests it was found that, whereas the inner tubes in a packet vibrated at ~39 Hz, responding to the expected local f_{vs}, for edge-of-packet tubes, with inter-packet gaps there was strong excitation at 80-120 Hz, which is in the range of the tube f_n. Moreover, 80-120 Hz corresponds to *twice* the "fundamental" f_{vs} in these higher flow regions! (Unfortunately, the values of S and U are not given). Moreover, there was a form of lock-in over a range of flows, once resonant vibration was established. Hence, this is a case of vibration excited by coincidence of f_n and f_{vs}, or the first harmonic of f_{vs} - apparently without acoustic coupling.

Cure: It took seven months, working 120 hrs per week, for crews of 40 men in protective clothing to (i) repair the tubes, (ii) shift the packets back into position, and (iii) fit restraining plates so that the packets could not move apart again. The reactor was down 1½ yr.

References: 1. Crimmin, W.R.C., Gray, I., Richards, D.J.W., & Dell, J.D. *Central Electricity Generating Board G823*, Sep 1977.

2. Whitton, P.N., & Durrans, R.F., Paper 7.5, *Int'l Conference on Vibration in Nuclear Plant*, Keswick, 1978.

CHART XVII: (a) TUBING IN LARGE SURFACE CONDENSER

Case 25 Class: cross-flow-induced vibration Date: < 1966

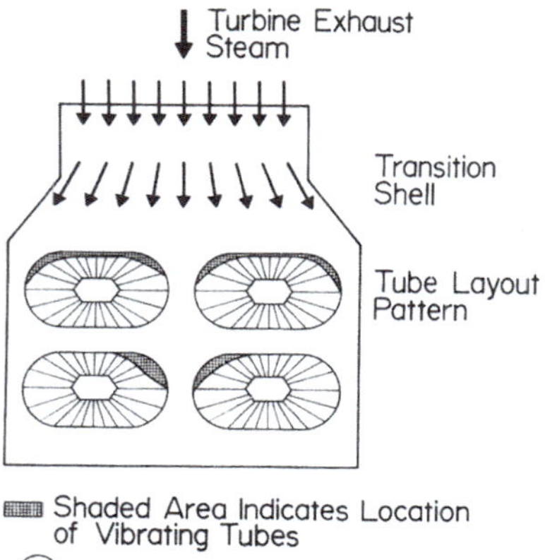

(1)

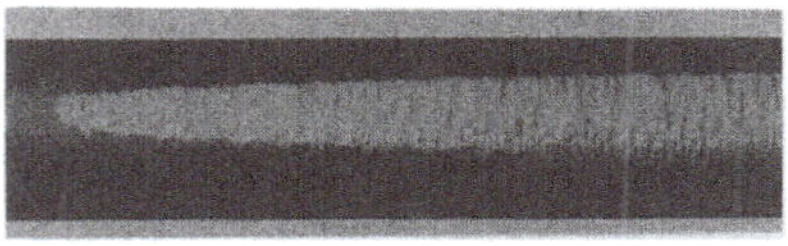

(2)

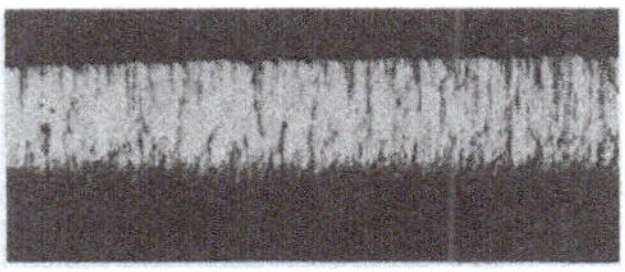

Problem: In this surface condenser, turbine exhaust steam enters from the top and passes through a 2x2 array of clusters of water-cooled tubes (Sch.1). In this unit, involving 2.54 cm Cu-Ni tubes with a span of 1.52m, the tube failure occurred four months after installation. Diamond-shaped wear pattern on the tubes indicated a whirling motion (Sch.2), as they hit and rubbed against one another. The failed tubes were located in the upper part of the two top tube clusters, but also at corner locations of the clusters below.

Mechanism: From data given, one estimates $m\delta/\rho D^2 \simeq 2.3 \times 10^3$ and $U/f_nD \simeq 4.7 \times 10^2$; this and the observations of wear pattern clearly indicate that the system was subject to fluid-elastic instability, the threshold of which is found to be $(U/f_nD)_c = 1.6 \times 10^2$, based on K=3.3. (The authors [1] developed an equivalent "critical severity factor", in this case exceeded, which is discussed in Section 3.)

Cure: Intermediate partial tube supports were installed reducing the tube-span to one half the original. This gives $U/f_nD \simeq 1.2 \times 10^2$, which with Connors' constant of K=3.3 brings the system below the instability threshold.

Reference: Coit, R.L., Peake, C.C., & Lohmeier, A., Proc. *American Power Conference* Vol.28, 1966, pp.469-483.

CHART XVII: (b) TUBING IN NUCLEAR PLANT CONDENSER

Case 26 Class: cross-flow-induced vibration Date: 1974-1975

Problem: 230 tubes in one unit and 250 tubes in the other had to be plugged in this major condenser tube damage. The tubes failed by (a) longitudinal thinning and fretting caused by inter-cylinder impact, and (b) transverse fatigue fractures. Damaged tubes were located mostly in the inboard peripheral side and upper corner tubes.

Mechanism: Cross-flow-induced vibration; most probably fluidelastic instability. Cf. Case 28. This case was analysed in terms of "severity factor" concepts [2] and found to have been nonconservatively designed; as shown in Section 3, this implies fluid-elastic instability. An ancillary mechanism in this case was "gross damage from foreign objects. These included pieces of wood scaffolding left in the condenser..."

Cure: Anti-vibration clips were installed in all areas where vibration has been determined to exist.

References: 1. Verna, B.(ed.), *op. cit.*, Section PWR-2, VI.C, Item 19.

2. Peake, C.C., Gestenkorn, G.R., & Arnold, T.R., Proc. *American Power Conference*, Vol.37, 1975, pp.562-574.

CHART XVIII: TUBING IN NUCLEAR PLANT CONDENSERS

Case 27 Class: cross-flow-induced vibration Date: 1972-1974

Problem: Leaks in main condenser tubing began after ~2 months' operation, necessitating several shutdowns to plug leaking tubes. In 1974, 1100 tubes were plugged. Well defined, fatigue failures with a sharp circumferential break mid-point between supports were found, mostly in peripheral tubes, but some as far as 18 tubes in from periphery.

Mechanism: Cross-flow-induced vibration, "resonant tube vibration"; probably fluid-elastic instability.

Cure: Installation of (i) more than 3,000 "stakes" (flat carbon steel bars) mid-point between support plates to reduce vibration, (ii) baffles to reduce steam impingement, and (iii) polyethylene inserts to prevent inlet end erosion. However, failures continued to occur in outer peripheral tubes after stake installation.

References: 1. Verna, B.(ed.), *Nuclear Power Experience*, Section PWR-2, VI.C, Item 7.
2. Anonymous, *Report on Nuclear Plants*, Consumers Power Co., Oct 1973.
3. Anonymous, *Power Reactor Docket Information*, Docket 50255-98, Aug 1972.

Case 28 Class: cross-flow-induced vibration Date: 1971-1974

Problem: Twelve tubes were found to be either cracked or broken in a 1971 inspection (and others leaked later). Eleven of the twelve were in the top three rows of the tube bundles; some had broken straight through and some had worn thin by contact with adjacent tubes.

Mechanism: Cross-flow-induced vibration, probably caused by fluid-elastic instability. Damage localization could be explained by reduction of U with condensing action, faster than $(1/\rho)^{\frac{1}{2}}$, in the Connors-type chart. *Alternatively*, lacking data, one could assume a vortex-shedding mechanism, which has been found to occur predominantly in the upstream rows of tubes [2] in the absence of acoustic reinforcement; moreover, deeper in tube bank, steam becomes progressively wetter, and regular vortex shedding has not been observed in two-phase mixtures. The published explanation: "vibration was caused by the impingement of steam on the tubes, particularly near the steam dump inlets".

Cure: Phenolic tube supports were installed to support four top tube rows in each bundle.

References: 1. Verna, B.(ed.), *op. cit.*, Section PWR-2, VI.C, Item 10.
2. Weaver, D.S., & Grover, L.K., *ASME Paper 78-PVP-25*, Jun 1978.

Case 29 Class: cross-flow-induced vibration Date: 1973-1977

Problem: Several, recurring cases of tube leakage in condensers of at least four different nuclear plants. In some cases at least, it is clear that the damage is vibration induced, where fatigue failures of corrosion-resistant materials (e.g. titanium) were involved.

Mechanism: In two cases it was identified as due to steam impingement associated with defective impingement plates; in the others it is unknown.

Cure: Installation of improved impingement plates. In other cases, only plugging of the tubes was reported.

Reference: Verna, B.(ed.), *op. cit.*, Section BWR-2, VI.C, Items 30,33,34,40.

CHART XIX: "EVACUATION TUBE" IN INLET PIPE OF EXPERIMENTAL LMFBR

Case 30 Class: cross-flow-induced vibration Date: 1970

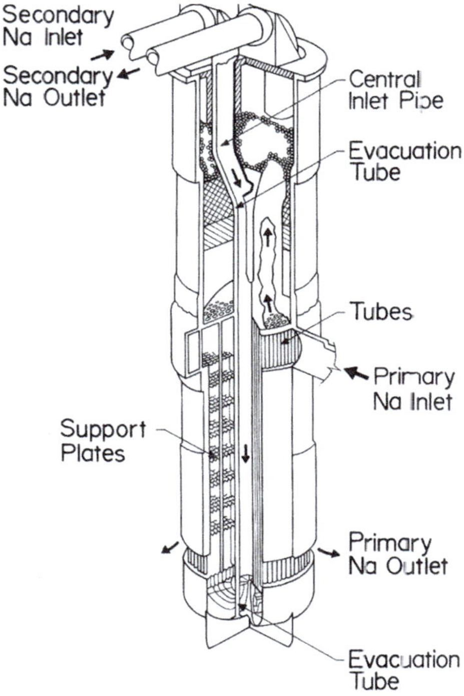

①

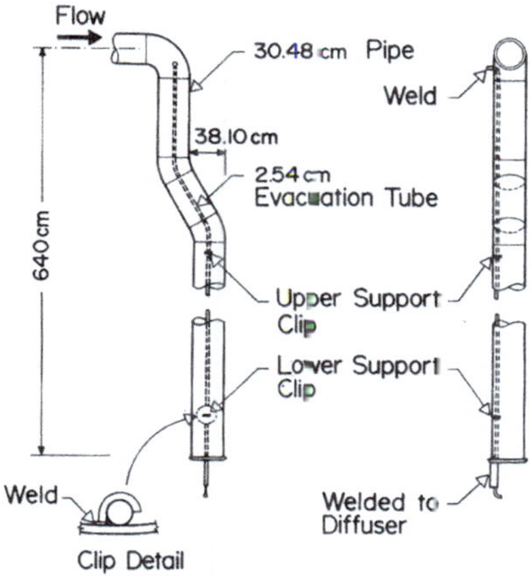

②

Problem: The 2.54 cm evacuation tube is clipped on inside wall of 30.48 cm secondary fluid inlet pipe, running down entire length of this sodium-sodium tube-and-shell intermediate heat exchanger (Sch.1). The pipe has a 90° elbow on top and an S-section with two 45° elbows, but is otherwise straight (Sch.2). The tube is welded at top entry point inside 90° elbow and at bottom exit point and is also supported by two clips along its 6.4 m length. Loud banging in Nov 1970 and inspection revealed that the top clip was sprung open and lower clip was missing. Wear on pipe and tube was found.

Mechanism: Steady drag force of cross-flow in 90° bend and secondary flows produced by bends; buffeting. (Mean flow velocity ≃ 4.7 m/s). Also, so-called "Bernoulli force", because of velocity profile near the pipe wall.

Cure: Removal of 2.54 cm tube altogether.

Reference: Buschman, H.W., Cerutti, B.C., & Clark, A.F., *Argonne National Laboratory* ANL-7834, Aug 1971.

CHART XX: FUEL ROD VIBRATION IN PWR CORNER FUEL ASSEMBLIES

Case 31 Class: cross-flow-induced vibration Date: 1971-1975

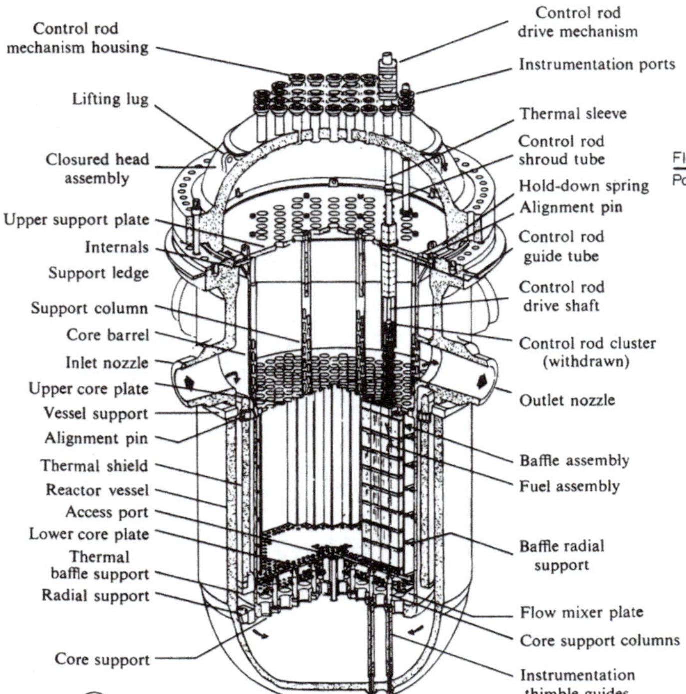

(1)

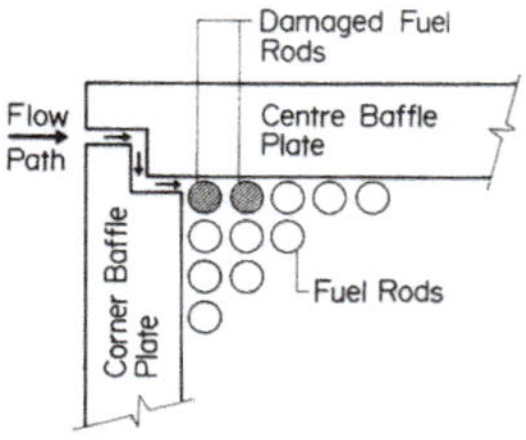

(2)

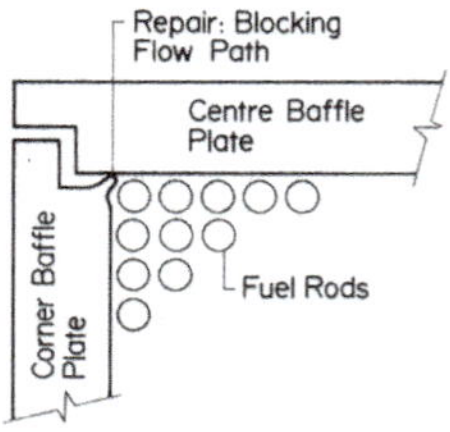

(3)

Problem: Several rods in the corner fuel assemblies of several reactors suffered damage at different times. Fuel integrity was lost in some cases, and cladding and fuel fragments were found in strange places; a portion of a fuel rod was entirely missing in one case. The flow is nominally axial in the fuel assembly.

Mechanism: High velocity cross-flow impingement leaking through gaps in the corner joints of the core baffle (Sch.2). The cross-flow jetted across a slight gap in the baffle ($\simeq$ 0.5 mm) causing buffeting of the corner rods and eventual fretting through the cladding in the grid support areas.

Cure: The baffle joints were peened to close the gap (Sch.3), and return to axial flow conditions was achieved.

References: 1. Verna, B.(ed.), *Nuclear Power Experience*, Section PWR-2, I, Items 11, 28.

2. Painter, W.B., *Westinghouse Nuclear Service Division NSD-TB-76-1*, Jan 1976.

3. Burstein, S., *Power Reactor Docket Information*, Docket 50266-362, Dec 1975.

4. Wellhofer, F.C., *Private communication*, 20 June 1979.

CHART XXI: SURVEILLANCE SPECIMEN HOLDER TUBES (SSHT) IN SEVERAL PWR's

Case 32 Class: cross-flow-induced vibration Date: 1976

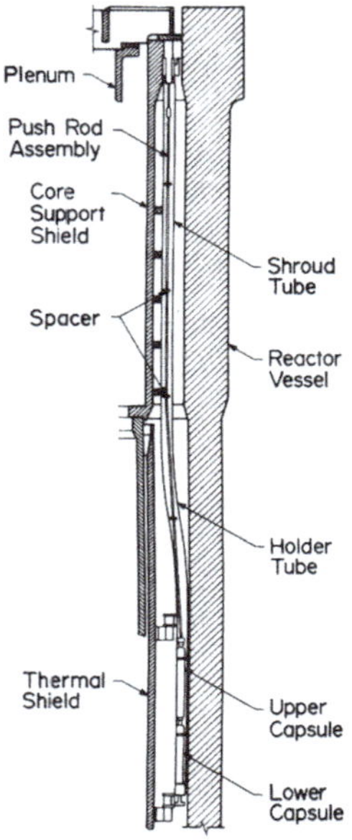

SURVEILLANCE SPECIMEN HOLDER TUBE

(1)

Problem: These are tubes located between the thermal shield and the pressure vessel, into which metallic samples are placed to be examined later for irradiation effects. The specimens are lowered into lowest point of the tube, which is perforated for cooling, by a pushrod (Sch.1). Because of offset of lower part of the SSHT and "dog-leg" transition tube shape, the pushrod has cylindrical spacers to guide it down and keep it from resting on SSHT. In 1976 it was discovered that some SSHT's had disintegrated and débris was carried by coolant into fuel assemblies and elsewhere.

Data: The upper, well-supported "shroud" is partly in cross-flow with U≈23 m/s max., and lower down originally calculated U=0.6-3.7 m/s, later revised to 3.7-13.7 m/s (Sch.2). There is also axial flow.

Mechanism: Flow through holes of perforated lower part caused capsule tubes containing specimens to "flutter". These capsule tubes, flexibly hung by pushrod, could move from one side of the SSHT to the other, as holes were blocked and unblocked - a basically unstable situation. There was also buffeting of the SSHT by cross-flow. Vibration induced by vortex-shedding is also possible in centre dog-leg portion. According to original max.U of 4.94 m/s, f_{vs}=32.5-67.9 Hz for S=0.45, or 21.7-45.3 Hz using S=0.3; this compares with f_n=56.4 Hz which excluded the effect of hydrodynamic added mass. Yet, in-service strain measurements indicated very low stress levels (200 psi rms) [4]. The motion of the capsule tubes transmitted to pushrod, and relative motion between it and vibrating SSHT, caused the pushrod cylindrical spacers to act as pipe cutters, cutting through SSHT.

Cure: The dog-leg portion of SSHT was eliminated (the offset also) to avoid vortex-shedding and buffeting excitation. The new SSHT (Sch.3, bottom) consists of two unconnected parts: the upper guide tube and the lowermost perforated capsule holder. The pushrod is retracted after placement of capsule tubes, eliminating cutting action. Also, the capsule tubes were redesigned to be wedged against each other, and a top closure cap puts capsules in compression and lower SSHT in tension, reducing possibility of rattling between the two due to flow through perforations.

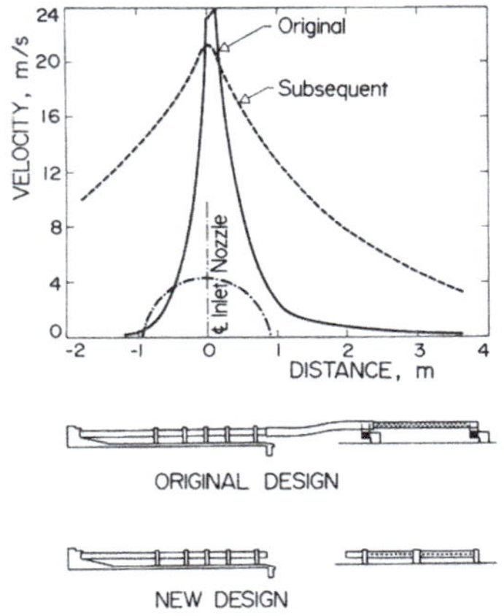

(2) CROSSFLOW VELOCITY ON SSHT

References: 1. Verna, B.(ed.), *Nuclear Power Experience*, Section PWR-2, Items 21,28.
2. Pryor, C.W. *et al.*, *Babcock & Wilcox BAW-10051 (Supp.1)*, Aug 1976.
3. Anonymous, *Arkansas Power & Light Co.* SSHT Repair Report.
4. Thoren, D.E., & Fortune, H.J., *Babcock & Wilcox BAW-10039*, Apr 1973.
5. Thoren, D.E., *et al.*, *Babcock & Wilcox BAW-10039 (Supp.1)*, Aug 1977.
6. Hooker, E.O., *Babcock & Wilcox BAW-10051, Rev.1*, Nov 1972.

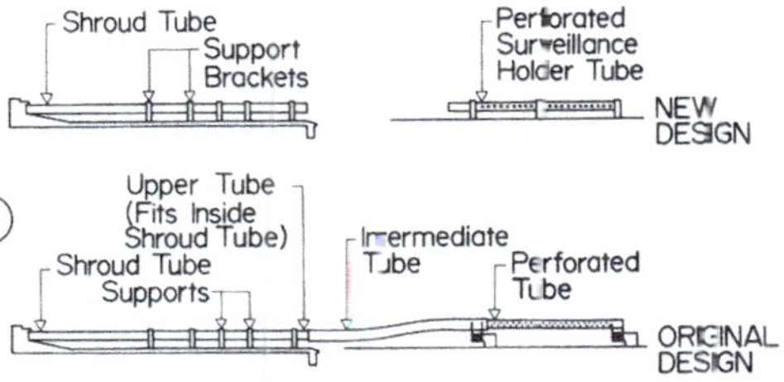

(3)

CHART XXII: IN-CORE INSTRUMENT NOZZLES & GUIDE TUBES OF PWR

Case 33 Class: cross-flow-induced vibration Date: 1972

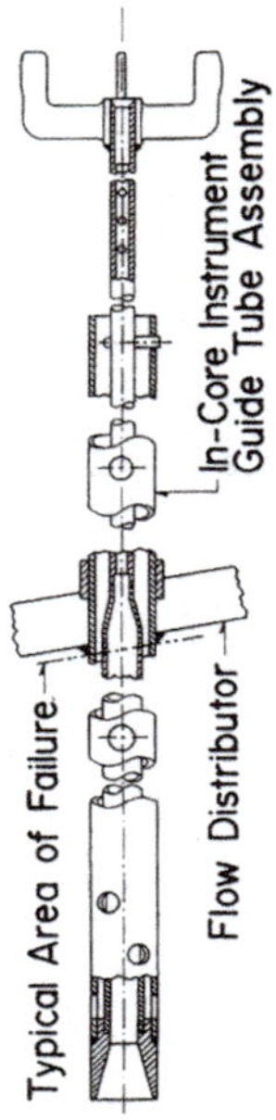

GUIDE TUBE

①

Problem: The 52 in-core instrument nozzles are essentially cantilevered pipes at the bottom of the reactor pressure vessel (Sch.2&3). They mate with the guide tubes (Sch.1), which fit over the end of the nozzles and extend inwards, towards the fuel assemblies; they are partly attached to the dish-shaped flow distributor and partly to the lower grid. The purpose of the nozzles and guide tubes is to guide the in-core neutron monitors into the fuel assemblies above.

In 1972, unusual noises from one of the heat exchangers prompted a shut-down, and inspection revealed that pieces of the nozzles had been carried by the coolant through the core to the steam generator head, causing extensive damage. 21 of the 52 in-core instrument nozzles had broken off just above the weld on the inner surface of the reactor pressure vessel and 14 others had cracked, by fatigue. Also, four guide tubes had broken off and fell to the bottom of the vessel (Sch.4 & 5).

Mechanism: *Post factum* analysis proposes that the mechanism was excessive cross-flow-induced vibration by vortex shedding. This seems to be related to high velocity areas in the lower head of the reactor (maximum/average velocity factor of 1.3) due to the location of the reactor inlet nozzles. Basing calculations on the higher-than-calculated U=10.7 m/s (35 ft/s) and D=2.54 and 6.03 cm for the nozzles and guide tubes, respectively, one finds that both are in the critical range of Re for vortex shedding. Taking then S=0.45 one obtains f_{VS}=190 Hz for the nozzles, while f_n=200-215 Hz; for the guide tubes, $f_{VS}\simeq$80 Hz and f_n=85-200 Hz, the wide range for the latter reflecting the differing lengths involved. Another proposed possible cause of failure is excitation due to random frequencies existing within the system as a result of turbulence and other sources such as the reactor coolant pumps.

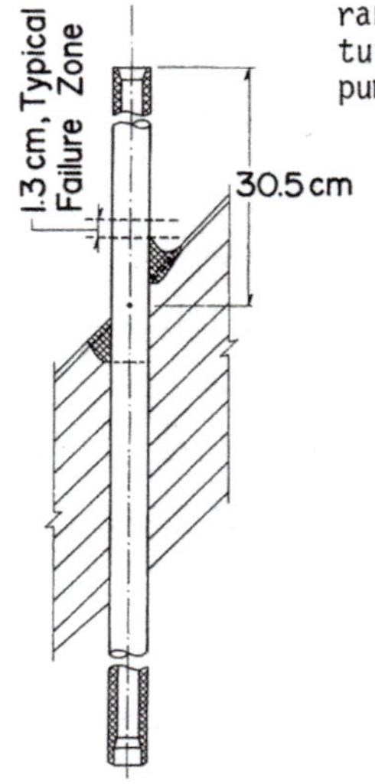

IN-CORE INSTRUMENT NOZZLE

②

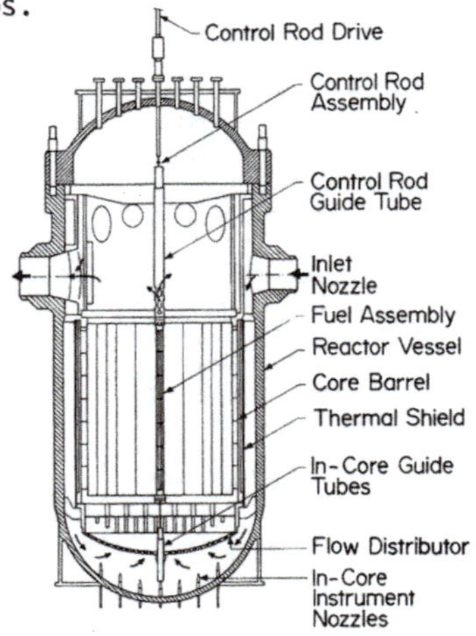

REACTOR VESSEL AND INTERNALS ARRANGEMENT ③

The vortex shedding mechanism is tenuous because of the large U taken uniformly over the length of the components - which measurements show to be overly conservative [4]. As shown by Savkar and So [7], on the other hand, the fluctuating lift forces in the vicinity of f_{VS} are appreciable, even in the critical and super-critical Re régime.

Cure: The nozzles and tubes were redesigned into more massive structures (Sch.6&7) such that f_{VS} (based on S=0.45) would be four times less than f_n. The alternating stresses to a given force amplitude were much reduced; also, the method of

Continued on next page

CHART XXIII Case 33 (cont) Class: cross-flow-induced vibration

attachment of both to their supports was beefed up; both measures aiming to eliminate the possibility of fatigue failures (e.g. by using gussets). The repairs took 10 months to effect.

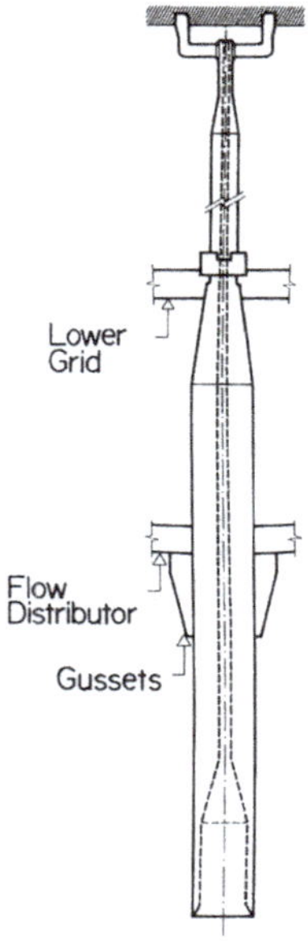

NEW DESIGN OF GUIDE TUBE
⑥

④

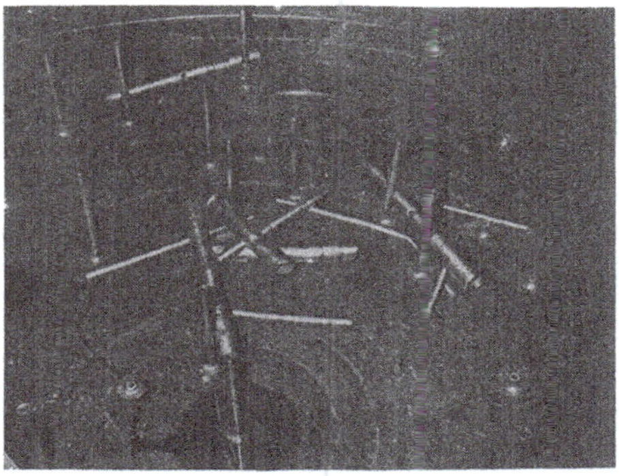

⑤

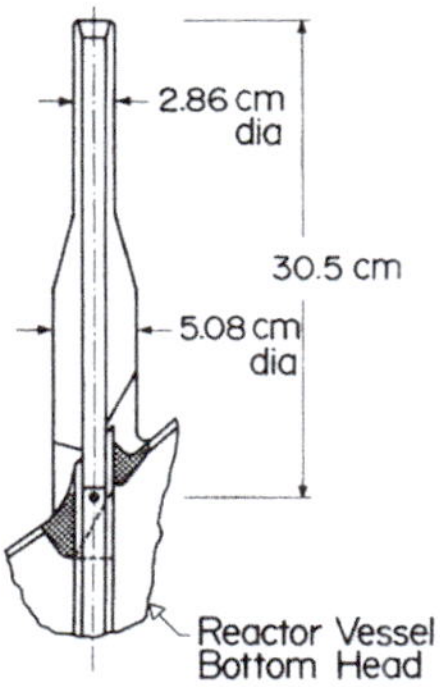

NEW DESIGN OF IN-CORE INSTRUMENT NOZZLE
⑦

References: 1. Anonymous, *Babcock & Wilcox BAW-10050, Rev.1*, Nov 1972.

2. Verna, B.(ed.), *Nuclear Power Experience*, Section PWR-2, II, Item 5.

3. Thoren, D.E., & Harris, R.J., *Babcock & Wilcox BAW-10038*, Sep 1972.

4. Thoren, D.E., & Fortune, H.J., *Babcock & Wilcox BAW-10039*, Apr 1973.

5. Hooker, E.O., *Babcock & Wilcox BAW-10051, Rev.1*, Nov 1972.

6. Thoren, D.E., & Harris, R.J., *Babcock & Wilcox BAW-10038, Rev.1*, Nov 1977.

7. Savkar, S.D., & So, R.M.C., *General Electric GEAP-24149*, Sep 1978.

CHART XXIV: BELLOWS SHROUDS IN AGR FUEL CHANNELS

Case 34 Class: internal axial-flow-induced vibration Date: 1975

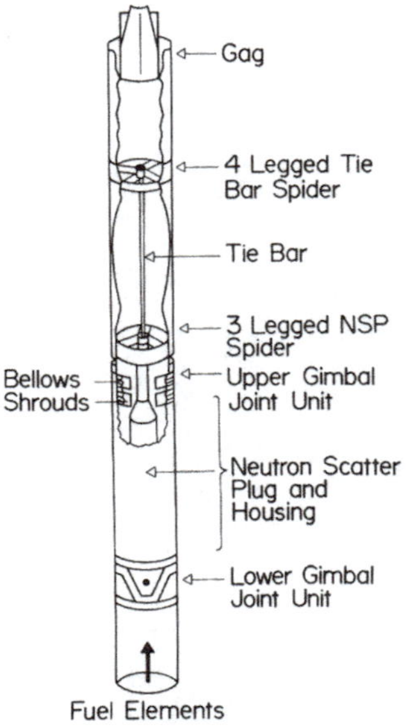

(1)

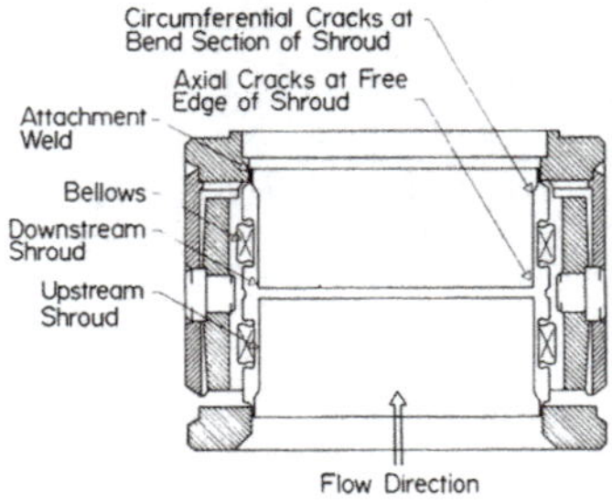

(2)

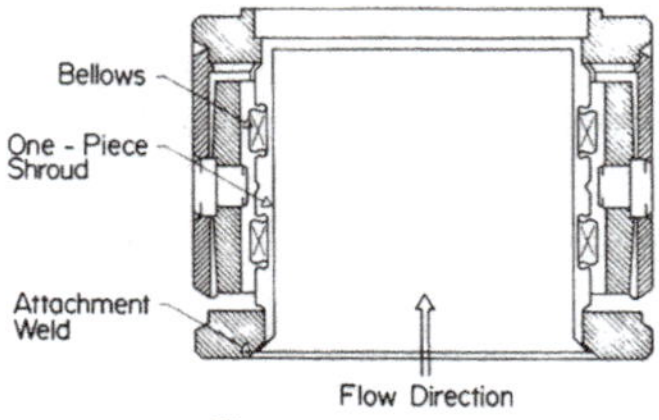

(3)

Problem: When the reactor was shut down during commissioning tests, to investigate another problem (Case 52), it was discovered that the protective shrouds fitted inside the bellows of the lower articulating gimbal joints of several fuel channels had cracked (Sch.1&2). In all cases it was the downstream shroud that had cracked (Sch.2). The purpose of the gimbal joints is to allow for thermal movements and to take up misalignments between the fuel channels in the core and the corresponding concrete penetrations. The flow rate in each fuel channel is designed to be variable, with velocities ranging from 16 to 40 m/s of CO_2 of 25 kg/m^3 density.

Mechanism: After very extensive testing in several laboratories it was concluded that the most likely mechanism was acoustically excited, resonant vibration of the shroud in one or more of its shell modes. One possible sound source is flow induced, generated by coincidence of (i) the vortex-shedding frequency from the spider legs of the so-called "tie-bar and neutron-scatter-plug spiders" and (ii) the frequency of travelling or standing acoustic non-planar waves (acoustic "cross modes") originating in the annular space between the spider tie-bar and the channel wall - of the type discussed in [2]; there appeared to be lock-in of f_{VS} to the acoustical wave,and reinforcement, resulting in loud whistles in the range of 1,200 - 2,700 Hz in the tests, their appearance and disappearance varying with flow rate. The proposed mechanism cannot be said to have been unequivocally established, having not explained all the test observations, nor why only the downstream halves of the split shroud were affected.

Cure: (i) The neutron-scatter-plug spider was reversed, so that the former leading edge became the trailing edge, resulting in a thicker wake and low enough vortex-shedding frequency (even at max. flow) to preclude coincidence with the acoustic frequencies; (ii) special spoilers were welded onto the tie-bar spiders for the same purpose; (iii) since no failures had occurred on the upstream half of the split shroud, the split shroud was replaced by a one-piece shroud, still cantilevered but facing downstream (Sch.3). Endurance tests showed that the cure was successful.

References: 1. Sansom, K., & Hopkins, I.H.G., Paper 5.3, *Int'l Conf. Vibration in Nuclear Plant*, Keswick, 1978.

2. Parker, R., & Pryce, D.C., *J.Sound & Vibration*, Vol.37, 1974, pp.247-261.

CHART XXV: PIPING OF GCR NUCLEAR PLANT

Case 35 Class: internal axial-flow-induced vibration Date: 1970-1974

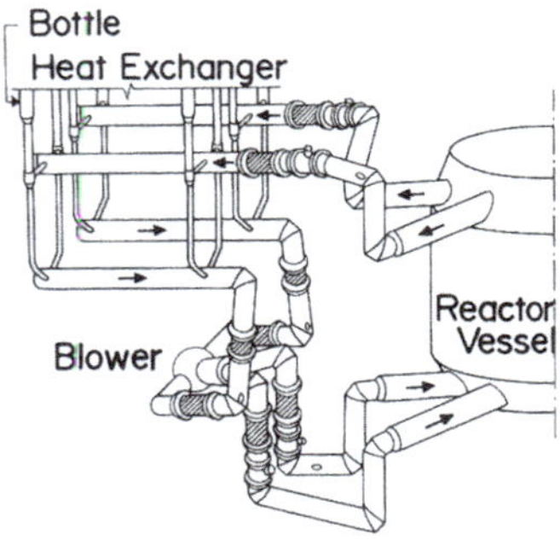

(1)

Problem: The complex piping system connecting this CO_2-cooled reactor to the heat exchanger (Sch.1,2) exhibited severe acoustically induced vibration. The sound could be heard 3 km away.

Data: Evaporator tubing: in-line, s_t=1.51, s_ℓ=1.13; length of heat exchanger, L=3.39 m (equivalent "simple pipe" length according to 2-D analysis [3,4]: L"=3.804 m); equivalent cross-sectional area, S"=0.326 m^2. Economizer tubing: in-line, s_t=1.51, s_ℓ=1.11; L=0.676 m (L"=0.759 m); S"=0.322 m^2. Amplitude of pressure pulsations in heat exchangers *versus* blower speed: Sch.3 .

Mechanism: The source of noise was in the heat exchangers, due to coincidence of vortex-shedding and acoustical frequencies. Calculated frequencies [3,4] for resonance: 96, 111, 113 and 145 Hz (cf. Sch.3). Moreover, one of these frequencies (112 Hz) coincided with the first transverse acoustic frequency of the cold manifold. This was responsible for the intense sound and was capable of exciting mechanical vibration of the whole piping.

Cure: The source of excitation could only be removed by redesigning and replacing the tubing pattern, or whole, of heat exchangers. Hence, it was decided to *attenuate* the noise source by installing Helmholtz resonators at the points of maximum pressure amplitude in heat exchangers (Sch.4) - i.e. in the evaporators (Sch.5) [4]. The sound levels were attenuated dramatically.

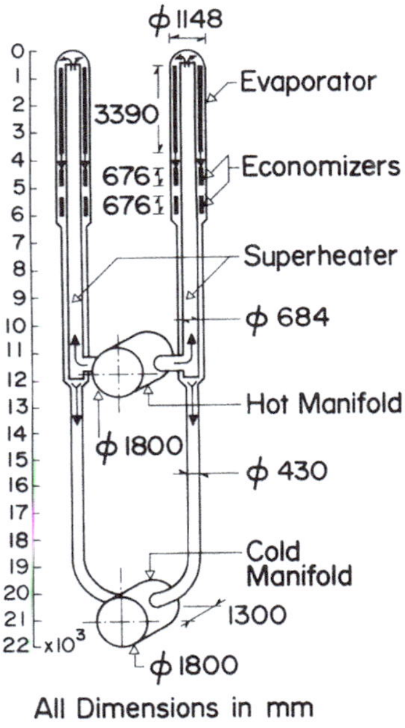

(2)

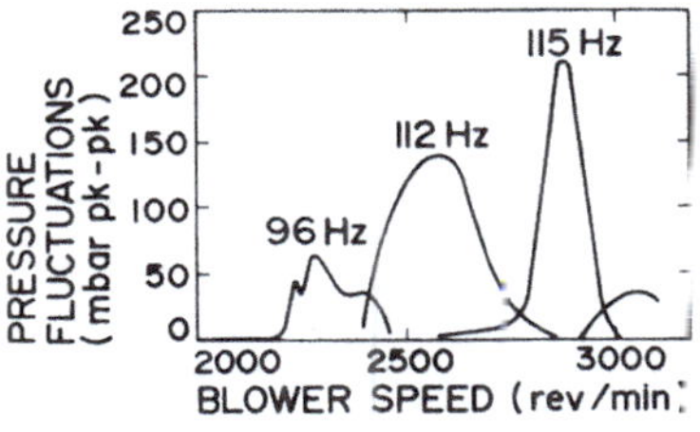

(3)

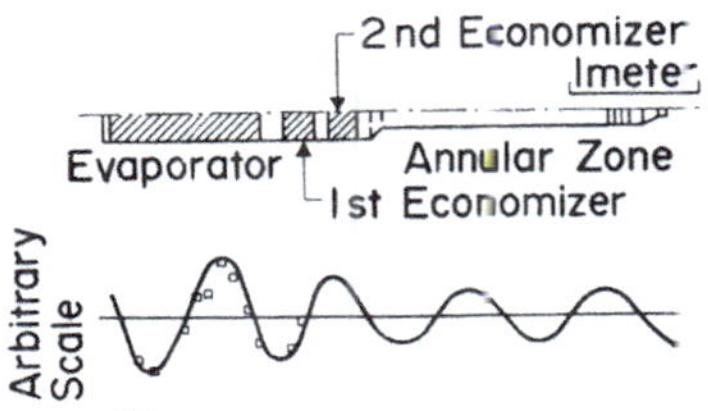

(4)

Continued on next page (bottom)

CHART XXVI: (a) FEED-WATER PIPING IN NUCLEAR PLANT

Case 36 Class: internal axial-flow-induced vibration Date: 1974-1975

Problem: Several failures, at different times, occurred in the piping of the low-flow feed-water lines, causing feed-water leakage. The failures were related to flow-induced piping vibrations at specific openings of the main feed-water regulating valve.

Mechanism: Probably valve oscillations of the main feed-water regulating valve, with attendant acoustical impedance matching at specific flow rates, or simply flow-induced noise, caused the piping vibrations.

Cure: (i) Installation of a more appropriate regulating valve; (ii) redesign of the piping for a less tortuous flow path; (iii) added bracing of the piping.

Reference: Operating Experiences; Reactor Safety, U.S. Atomic Energy Commission, 1974/75.

CHART XXVI: (b) WATER HAMMER IN VARIOUS SERVICE PIPING

Case 37 Class: internal axial-flow-induced vibration Date: 1971,1975

Problem: Several cases of supports being pulled out from walls, end caps in piping being blown off; mainly relatively minor problems. Reference [1] gives two examples.

Mechanism: Water hammer. This, properly speaking, is not a flow-induced vibration problem, but it does involve flow and piping vibration, and is included for completeness.

Cure: Replace supports, enlarge drain holes,if steam flows are involved, revise operating procedures.
Reference: 1. Verna, B.(ed.), *Nuclear Power Experience*, Section BWR-2, VIII.D, Item 71, and VI.C, Item 7.

CHART XXVI: (c) Case 35 (cont) Class: cross-flow-induced vibration

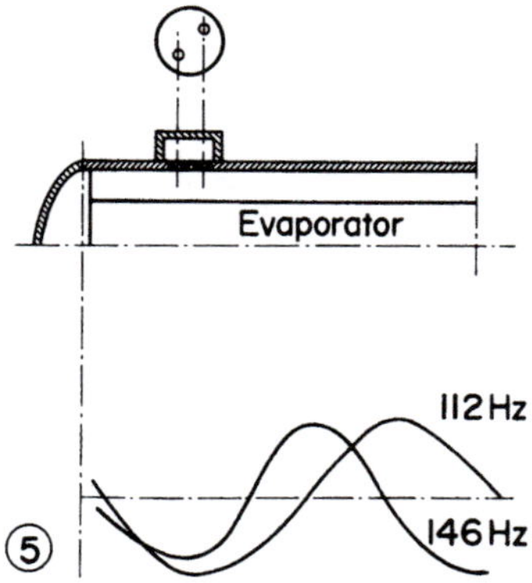

Later, it was discovered that in pre-cure operations the turning vanes in the elbows had cracked, due to poor aerodynamic design. At this point it was decided to redesign and replace the complete piping. To avoid recurrence of problems, thorough analytical and model experimental studies were undertaken [1-4] on generation, transmission and amplification of sound in complex circuits.

References: 1. Baylac, G., Planchard, J., Grégoire, J.P., Pailly, C., *ASME Paper 75-DET-64*, 1975.

2. Baylac, G., & Copin, A., Paper F4/3, *3rd Int'l Conf. Structural Mechanics in Reactor Technology*, London, 1975.

3. Baylac, G., Gibert, R.J., Livolant, M., *La Houille Blanche*, Vol.26, 1971, pp.407-428.

4. Baylac, G., Grégoire, J.P., *Journal of Sound & Vibration*, Vol.42, 1975, pp.31-48.

CHART XXVII: PIPING ASSOCIATED WITH FEED-WATER PUMP

Case 38 Class: internal axial-flow-induced vibration Date: < 1975

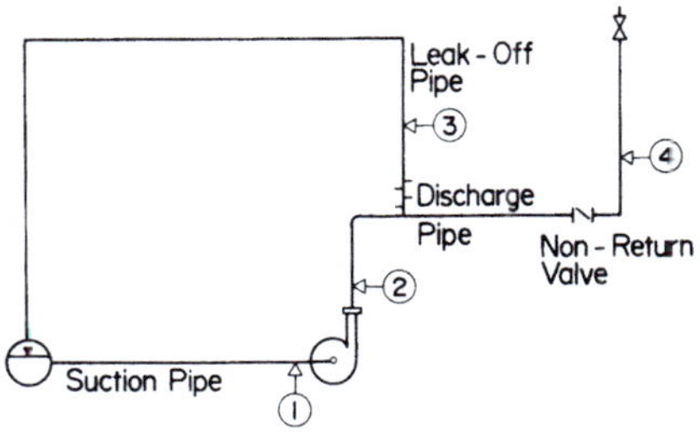

(1)

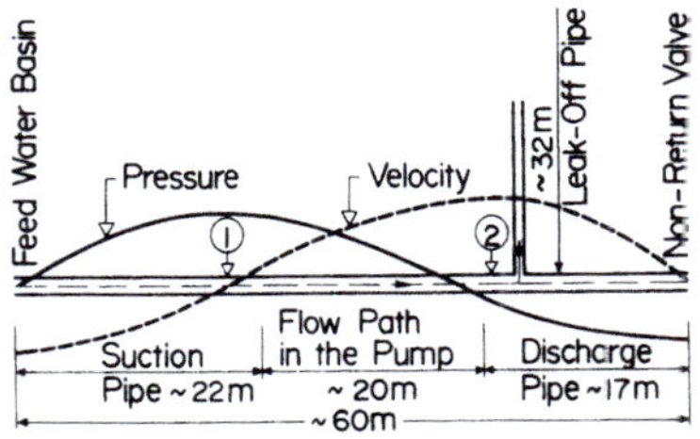

(2)

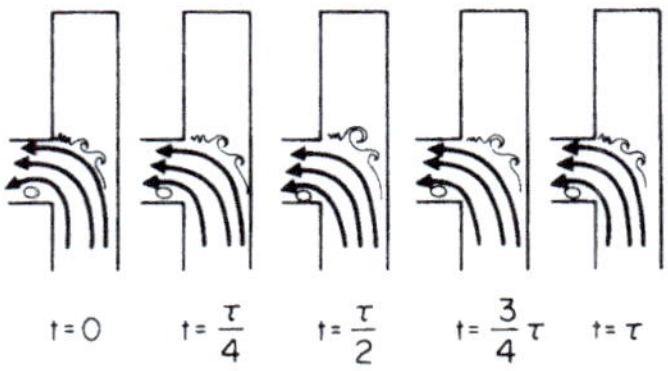

(3)

Problem: A feed-water pump exhibited strong vibration at constant f=15-16 Hz for pump speeds of 2,000-3,000 rpm, while the system was in leak-off mode of operation. In this mode of operation, the flow goes through the suction pipe and pump, and turns through 90° at a tee (Sch.1); beyond the straight-through side of the tee, the non-return valve is shut off. The vibration was strongest at ~2,700 rpm operation, i.e. at third harmonic of 15 Hz. A full scale test was mounted to determine the mode of excitation. (Main pipe: D_1=28.1 cm, U_1=0.23 m/s; leak-off pipe: D_2=5.9 cm, U_2=5.1 m/s).

Mechanism: An acoustical 3/4 standing wave was found in the main pipe, with a pressure node at the suction-pipe inlet and an antinode at the non-return valve (Sch.2). The tee was close to a velocity antinode. Taking L≃60 m and c=1,200 m/s, one obtains the observed f≃15 Hz. The excitation source proposed was the oscillation of vortices on the free shear layer of the flow turning in the tee (Sch.3). For the characteristic frequency, two approaches are taken. (i) One relates to the instability frequency of the shear layer by Tollmien-Schlichting wave generation - assuming the boundary layer is still laminar at the tee. Taking $S^*=f^*\delta^*/U_1$=0.01 - 0.04 for the formation of initial transition waves in the free shear layer, spanning the range for jets and flow over steps, and δ^*≃4.9 mm for the boundary-layer displacement thickness corresponding to Re_x=3.4 x 10^5, one obtains f^*=4.7 - 18.8 Hz, which brackets the observed one of ~15 Hz. (ii) The second relates to the frequency of vortex oscillation arising from the influence of the "dead-water zone", based on $S=fh/U_1$=0.08 - 0.32, where h=characteristic depth of that zone. This again brackets the calculated S=0.17, taking $h=D_2$.

It is suggested that the mechanism is one of feedback oscillation between the strength of vortices in the tee and the acoustic wave; this would be best served if the vortices were close to an acoustic velocity antinode, as is the case. No direct evidence is given that this indeed is the mechanism involved.

Cure: Reducing the distance between tee and non-return valve, which places the presumed vortices close to a velocity node, weakening the coupling between vortex oscillation and acoustic wave. The efficacy of this remedy was confirmed.

Reference: Chen, Y.N., & Florjancic, D., Paper C109/75, *I Mech.E. Conf. Vibrations and Noise in Pump, Fan, and Compressor Installations*, Southampton, 1975, pp.79-86.

CHART XXVIII: PIPE WITH SWIRLING FLOW AND ANNULAR HYDRAULIC JUMP

Case 39 Class: internal axial-flow-induced vibration Date: 1967

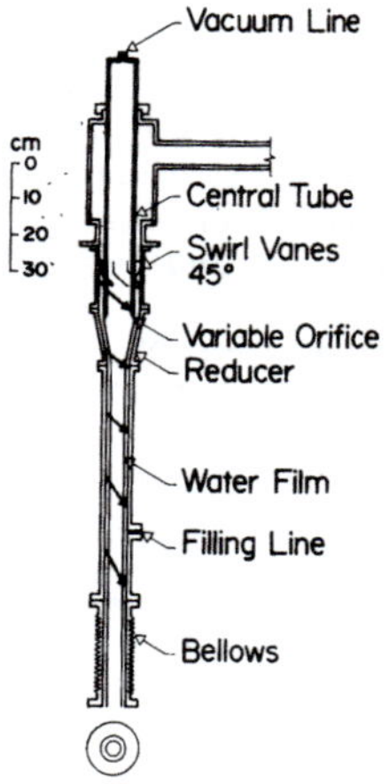

VANED SWIRL HEAD

Problem: This case relates to hydraulic simulation experiments modelling flow in a proposed nuclear apparatus (an intense neutron generator) in which swirling annular flow would go down a tube, with a gaseous core; somewhere below the top of the tube there would be a hydraulic jump, changing the annular flow to pipe flow - i.e. filling the tube. In the tests, water was used as the fluid, and air under vacuum as the gaseous core. Three different Al tubes of D=5.08, 7.62 and 10.16 cm, all 3.35 m long and 1.24 mm thick, were used. Violent vibrations were recorded. For the 5.08 and 7.62 cm tubes, flexural whirling vibrations in their first and, at higher flow, in their second modes developed, with max. amplitude of 2.5 cm; third-mode vibrations were also observed for the smallest tube. The 5.08 cm tube cracked at the lower support by fatigue. Vibration started in the first mode (frequency f_n^1) increasing in amplitude with flow, reaching a plateau, and then decreasing, before the next mode would start at $f=f_n^2$, etc. For the 10.16 cm tube, shell-mode vibrations in the second circumferential mode (n=2) were observed, involving several axial wavelengths, of 1.5 mm max. amplitude; n=3 vibrations were also obtained. In the apparatus used, swirl increased with increasing flow rate.

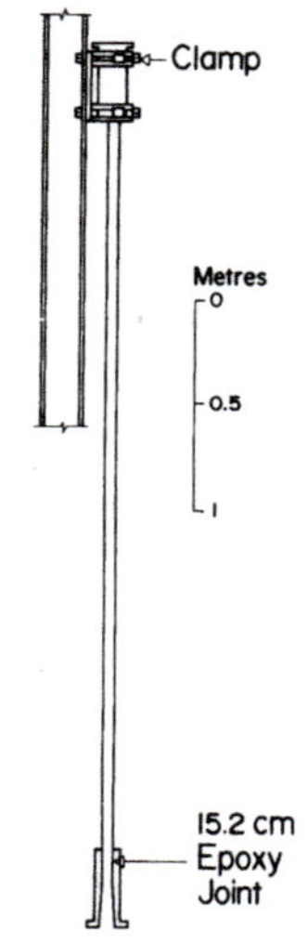

5.08cm TEST TUBE

Mechanism: The mechanism for the 5.08 and 7.62 cm tube vibration is akin to whirling of shafts. For a given flow rate, there exists in the tube a band of swirl frequencies, decreasing down the tube, as the swirl velocity decreases by friction. As flow increases, so does the swirl velocity, and eventually at the top of the tube it coincides with f_n^1, precipitating whirling. As flow increases further, the swirl at the top increases further and the resonant driving swirl area moves down, until maximum amplitude is produced when it reaches the tube midpoint. This explains the very broad resonance peaks. At higher flow, and higher swirl, second-mode resonance is reached, *et seq.* E.g. for the 5.08 cm tube, max. swirl velocity at the wall was 12.80 m/s, giving ω=503.9 rad/s or f=80 Hz, which may excite, in turn, each of the three lowest beam modes of 13, 36 and 71 Hz (calculated taking into account the internal fluid mass distribution), but not the fourth at f_n^4=118 Hz.

The mechanism for the 10.16 cm tube remains obscure, especially why flexural (beam) modes were not excited. Perhaps it is related to the collapse of annular flow or the existence of pressure pulses from the pump, some of which were in the range 150-200 Hz. Using simplified analysis, taking the tube to be simply supported, the second circumferential mode frequencies were found to be $f_{2,1}$=119 Hz, $f_{2,2}$=120 Hz,....., $f_{2,5}$=142 Hz,....., $f_{10,1}$=312 Hz, i.e. very closely spaced. The observed vibrations were in the range of 150-200 Hz, depending on the location of the hydraulic jump. Recently, flutter of shells subjected to either internal or external swirling flow has been treated theoretically [3,4].

Cure: Proposed cures were the introduction of stiffeners and the increase of damping to reduce vibration amplitude. Intermediate supports and thicker piping were precluded by other design parameters.

References: 1. Paidoussis, M.P., *Atomic Energy of Canada* Memorandum to R.I. Hodge, 25 Aug 1967.
2. Skiba,E.L.J., *Atomic Energy of Canada* Unpublished Report, Sep 1967.
3. Chen,T.L.C., & Bert,C.W., *J.Applied Mechanics*, Vol.99, 1977, pp.112-116.
4. Dowell,E.H. *et al.*, *AIAA Journal*, Vol.12, 1974, pp.1644-1651.

CHART XXIX: IN-CORE INSTRUMENT TUBES & FUEL CHANNELS IN SEVERAL BWR's

Case 40 Class: external axial-flow-induced vibration Date: 1974-1975

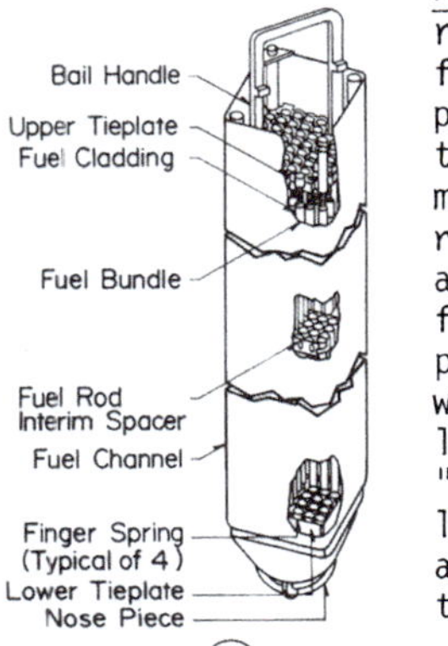

(1)

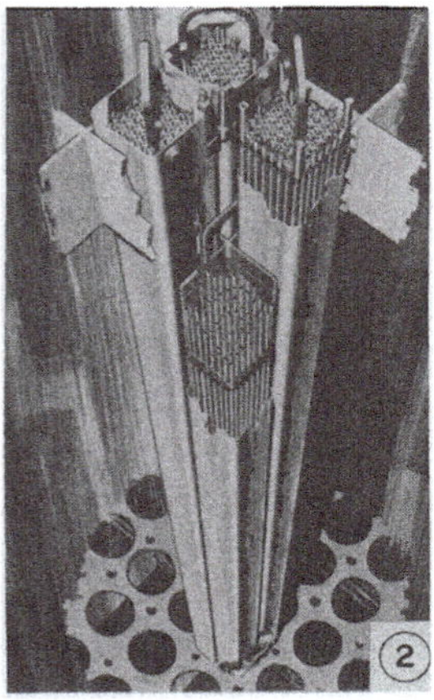

(2)

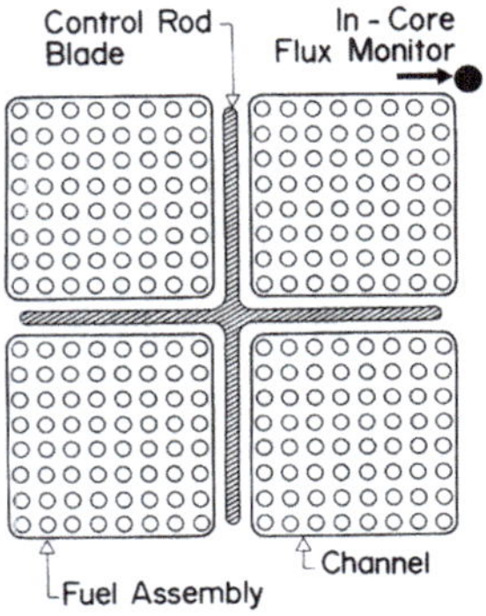

(3a)

Problem: In BWR's fuel assemblies are fitted in 2 mm thick rectangular "channel boxes" (Sch.1), placed in honeycomb fashion in the reactor core (Sch.2). Most of the flow passes axially through the channel boxes. In some locations, between the channel boxes, there are thin instrument tubes of various types, e.g. LPRM's (local power range monitors), SRM's (Sch.3). Each LPRM contains miniature fission chambers for measurement of local neutron flux, and a calibration tube for a traversing in-core probe (TIP). They are placed in a guide tube, D~1.9 cm, with evenly spaced holes along its length for water circulation to cool the detectors. The LPRM's are installed in "thimbles" permanently mounted at the bottom of the core lattice, penetrating the reactor vessel bottom, and are also supported at the top, but are unconstrained through the rest of the core region (L≃4 m) [2].

In BWR/4-type reactors, as opposed to the earlier BWR/3's, 2.54 cm "by-pass" holes were drilled in the lower core support plate, to improve cooling of LPRM's. The TIP signals tipped operators that something was wrong, when an unusual peak in 1-4 Hz range appeared in the neutron flux p.s.d. This occurred in one reactor first; later, at least twelve BWR/4's were involved. It was discovered that the signal was due to vibration and impact of the instrument tubes on the corners of the channel boxes, in some cases fracturing them and producing holes as large as 8.9x12.7 cm - a serious matter because coolant flow is thereby diverted from the fuel in the channel, also causing cross-flow in the fuel rods; missing pieces of channel box, carried away by the coolant, created added worry. In most cases power was reduced to 40% until the problem was solved. In one reactor, of the 192 channels inspected, 65% were considered rejects; four had been perforated [6]. In another, which also had "poison curtains" (for neutron absorption) in the channel box interstices, the curtains were found to vibrate and impact on the channels; upon removing them the problem was not solved, because then the in-core instrument tubes were found to impact on the channels, a problem that "was hidden 'behind the curtains' for the first two years" [1].

Mechanism: The problem was connected with the 2.54 cm by-pass holes in BWR/4's, which allowed highly turbulent, high velocity axial water flow, emerging in jet form from the lower support plate, to excite the slender and flexible instrument tubes to sufficiently high amplitude to hit the channel boxes. For the LPRM's f_n≃2.5 Hz, and in full-scale cold hydraulic tests impact levels of 4g-8g were recorded. Unfortunately no geometric or flow data have been released. A fundamental program investigating the effect of turbulence [5] in annular flows showed that, although the vibration of a cylinder in very narrow annular flows is

continued next page

CHART XXX Case 40 (cont) Class: external axial-flow-induced vibration

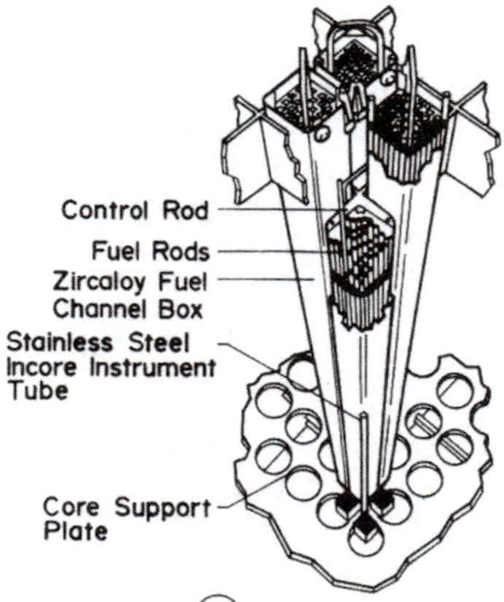

(3b)

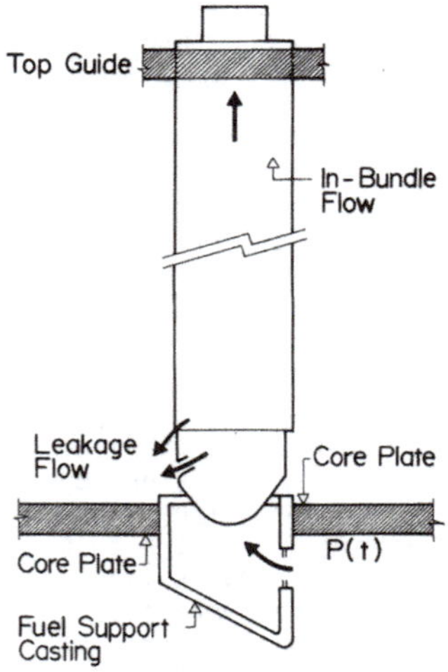

(4)

little affected by upstream turbulence (D_h/D= 2.85, D_h being the hydraulic diameter), in wider annuli (D_h/D=4.71) the effect is very strong.

Cure: (i) Plugging all by-pass holes in the core bottom support plate; (ii) drilling new holes in the tie-plate of the channel boxes, which restores the requisite by-pass flow (Sch.4); by directing this flow downwards, towards the core support plate, prior to mixing with the total by-pass flow, which is upwards, a more uniform flow is produced [4]. No further problems have been reported.

References: 1. Verna, B., (ed.), *Nuclear Power Experience*, Section BWR-2, II, Items 26,35,47.

2. Behringer, K., Kostič,Lj., & Seifritz, W., *Eidg. Institut für Reaktorforschung Würenlingen, Schweiz*, EIR-Bericht Nr. 328.

3. Anonymous, *Power Reactor Docket Information*, Docket 50259.

4. Anonymous, *U.S.NRC Safety Evaluation Report*, Supp. No.8, Jul 1976.

5. Mulcahy, T.M., Yeh, T.T., & Miskevics, A.J., *General Electric GEAP-24173*, Jan 1979.

6. Lessor, L.C., *Power Reactor Docket Information*, Docket 50298-465, Nov 1975.

7. Whitner, C.R., *Power Reactor Docket Information*, Docket 50321-417, Dec 1975.

CHART XXXI: SHIELD PLUGS OF PHWR FUEL CHANNELS

Case 41 Class: external axial-flow-induced vibration Date: 1976

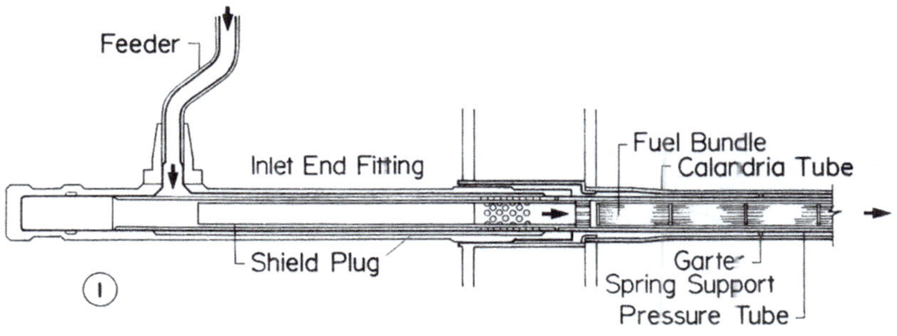

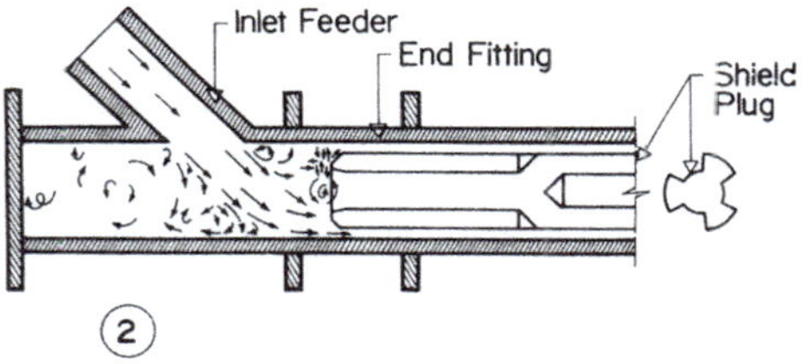

Problem: In PHWR's the fuel bundles rest on the horizontal pressure tube containing the coolant flow passing axially through the fuel rods in the bundles (Sch.1). At each end of the fuel-bundle train a shield plug, resting on an extension of the pressure tube, stops γ-rays and neutrons from escaping the reactor core. The shield plugs have an axially staggered, notched shape, which allows flow to pass through, but blocks radiation and particles (Sch.2). At the downstream end of the upstream shield plug there is a change in form from notched cylindrical to notched conical, just before the first fuel bundle. Experiments in a full-scale hydraulic facility indicated that amplitudes of vibration of 135 μm could be reached at 160% of design flow; this was deemed excessive, though not serious.

Mechanism: Flow visualization experiments (Sch.2) indicated poor flow conditions and turbulence generated by the sharp leading and trailing edges of the staggered notches and protrusions. Also, in the diffusing flow past the conical end there were large eddies, an added excitation source. Hence, the vibration was excited by turbulence.

Cure: (i) Streamlining the inlet section and leading and trailing edges of the protrusions, by rounding all sharp edges and blending them into smooth curved shapes, thereby reducing local turbulence generation; (ii) streamlining the abrupt conical downstream end; (iii) adding a bearing sleeve at the downstream end of the plug, to fit snuggly into the pressure tube inlet; the very narrow fluid annulus provides effective viscous damping at this end. The maximum vibration was thus reduced from 135 to 20 μm at 160% of design flow (Sch.3). As a bonus, the pressure drop over the shield plug dropped from 63.6 kPa to 52.2 kPa at design flow - a 22% gain.

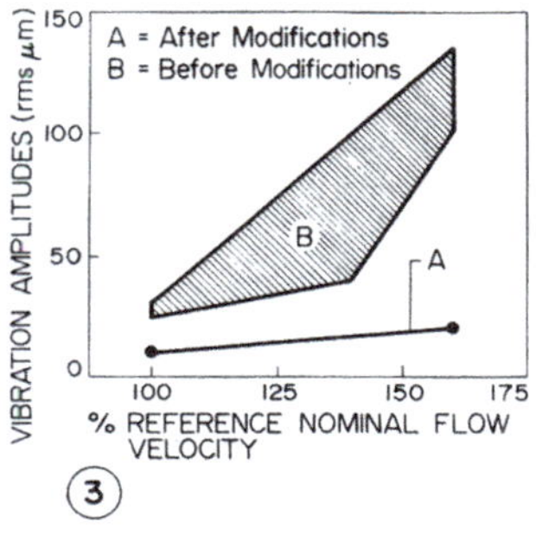

References: 1. Oldaker, I.E., *et al.*, *Whiteshell Nuclear Research Establishment* (AECL), *unpublished documents*, Jun 1976-May 1978.

2. Lane, A.D., *et al.*, Paper 1.1, *Int'l Conf. Vibration in Nuclear Plant*, Keswick, 1978.

CHART XXXII: FUEL ELEMENTS IN BLWR's & LWCHWR's

Case 42 Class: external axial-flow-induced vibration Date: 1974

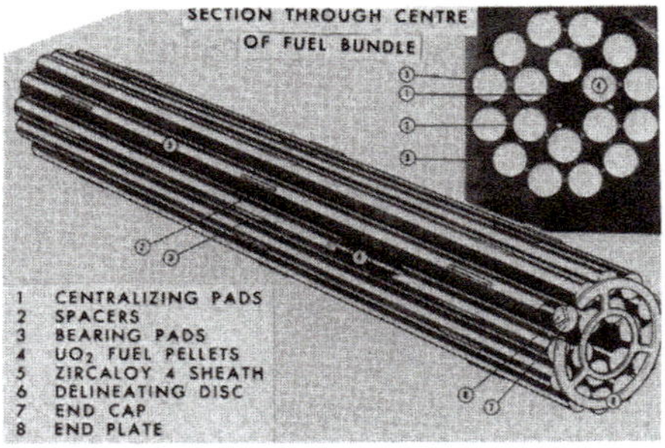

(1)

Problem: In BLWR's and LWCHWR's the fuel channels are vertical. The fuel bundles (Sch.1) - clusters of fuel elements or rods about 50 cm long - are stacked one on top of the other, and held together by a central support tube running through the centre of the whole fuel-bundle "string" (Sch.2). Each fuel string is located inside a pressure tube which contains the axially and upwardly flowing coolant. The string is flexibly supported at the bottom and is either free or simply supported at the top, where it is terminated by a flux suppressor. The great flexibility of the fuel string allows touching the pressure tube - the clearance being of the order of 2 mm - so that, inevitably, some flow-induced fretting wear takes place. Each fuel element consists of UO_2 encapsulated in a tube; on the latter are welded small "wear pads", which are sacrificially worn to protect the tube (Sch.1). Thus, a certain amount of wear is designed for, taking account of the design lives of the components [1].

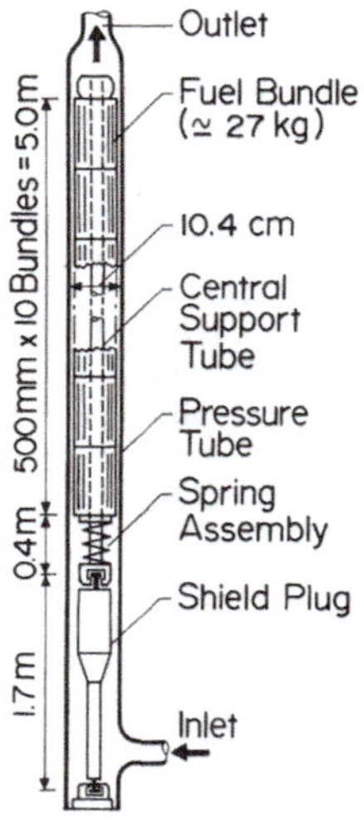

(2)

In one case, however, involving strings with a free top, it was found that accelerated fretting had taken place in the top fuel bundles of 40% of the high-flow fuel channels (Sch. 3). The wear was preferentially to one side [2].

Mechanism: It was determined that this was caused by excessive axial-flow-induced vibration and repeated impacting on the pressure tube. It was theorized that the relatively flexible string had become subject to axial-flow-induced flutter instability or buckling - in the latter case, the uppermost part touching the pressure tube and chattering against it. Extensive analytical modelling [3] of the dynamics of the fuel string in axial flow showed that the system, as designed, was free of flow-induced instabilities. The real problem was more mundane: due to manufacturing error, some of the flux suppressors at the top of the string were assembled slightly eccentrically; the steady flow thus forced the string to bend to one side and chatter against the pressure tube.

Cure: (i) Ensuring proper concentric assembly of the fuel strings; (ii) increasing the fuel string flexural rigidity to reduce vibration amplitude.

(3)

References: 1. degli Espinosa, P., Lelli, G., Possa, G., Vanoli,G., *Energia Nucleare*, Vol.25, 1978, pp.406-424.

2. Pettigrew, M.J., *Atomic Energy of Canada* AECL-5852, Oct 1976.

3. Païdoussis, M.P., *Annals of Nuclear Energy*, Vol.3, 1976, pp.19-30.

CHART XXXIII: CONTROL ROD BLADES OF PWR

Case 43 | Class: annular-flow-induced vibration | Date: < 1965

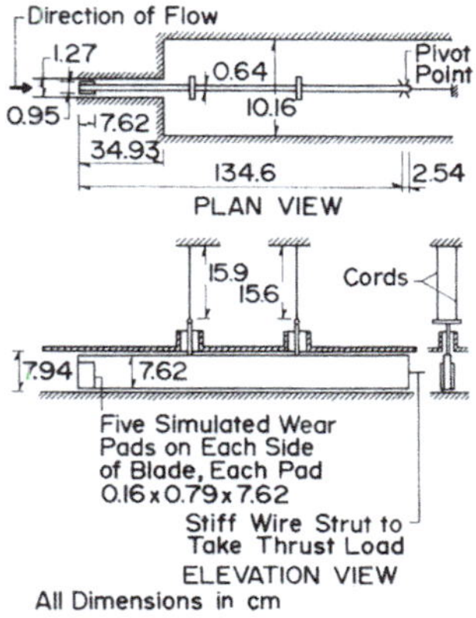

1

Problem: In this early PWR there was "evidence of hydraulic flutter of the control-rod blades and stationary shim rods". There is insufficient information on the original problem, geometry, etc.; some idea may be obtained, however, from model studies aiming to provide physical understanding of the problem. A rectangular blade was placed in flow in a water table, partly in a narrow upstream channel, and partly in a wider downstream one; the blade was flexibly supported at two points and pivoted at the downstream end (Sch.1). The wear pads at the upstream end gave an added local restriction. It was found that the blade began oscillating at ~1 Hz, at a flow velocity of 1.22 m/s.

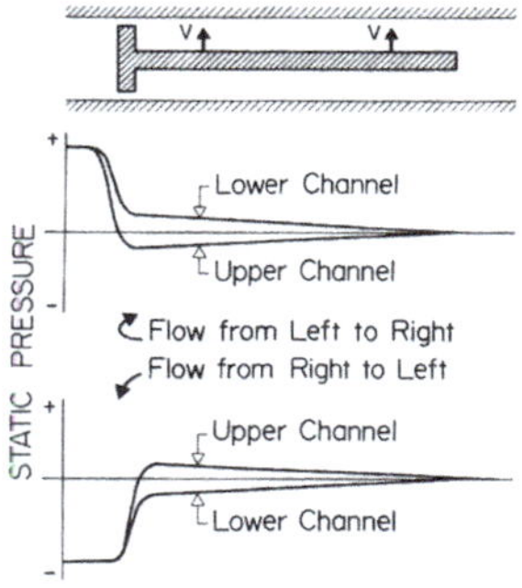

2

Mechanism: The mechanism advanced is one of flutter associated with negative damping. If the blade has an instantaneous velocity to one side of the channel, the flow rate in the sub-channel on that side will decrease with time; that means that an axial deceleration of the flow must take place, with an attendant depression of the static pressure along the sub-channel, to below what it was originally (Sch.2, top graph). The opposite occurs in the other sub-channel. The effect is magnified by the upstream constriction and the wear pad. Hence, a net force proportional and in the direction of the blade velocity is produced; i.e. a negative damping. When this overcomes the existing positive damping in the system, flutter ensues. Leakage around the blade diminishes, but does not eliminate the effect.

Cure: Changing the design so that there is a constriction at the downstream end of the narrow passage; then, an instantaneous velocity of the blade to one side produces decreasing outflow from the sub-channel and the inverse of the pressure effects described above, generating a positive damping, i.e. a stabilizing effect (Sch.2, bottom graph).

References: 1. Miller, D.R., & Kennison, R.G., *ASME Paper 66-WA/NE-1*, 1966.

2. Reed, G.A., Supplement *Trans. American Nuclear Society*, Vol.8, 1965, p.28.

3. Miller, D.R., *Argonne National Laboratory ANL-7685*, May 1970, pp.304-307.

CHART XXXIV: NUCLEAR ROCKET INTERNALS

Case 44 | Class: annular-flow-induced vibration | Date: 1962

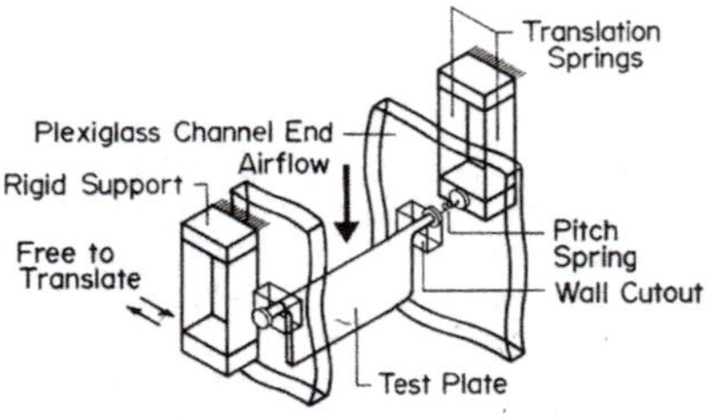

SCHEMATIC, NOT SHOWING SIDE WALLS

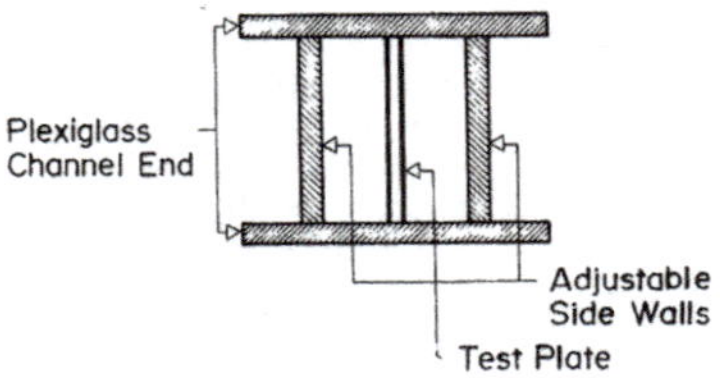

VIEW FROM UPSTREAM

Problem: The problem is described as "a dynamic instability that caused a severe vibration or 'flapping' in the reactor core" [1], where "the vibrations were so severe that the reactor core was seriously damaged. Subsequent cold-flow tests showed that the destructive vibrations were a result of leakage flow of the hydrogen gas (the coolant) through the very narrow passages between the reactor fuel elements". Indeed, it was the fuel elements which vibrated or "flapped", causing the core failure.

The elements were extruded graphite rods which were hexagonal in cross-section and contained enriched uranium. These rather long rods were supported in clusters of seven (six about a central support rod), and contained a large number of small passages along the interior of the rods to carry the propellent flow. The geometry of the hexagon shape was such that the clusters then fit tightly together throughout the whole engine core, rather like a large honeycomb [3]. The hydrogen coolant flow was supposed to be confined to the internal passages of the rods.

Mechanism: The mechanism involved was clearly one of leakage flow of the hydrogen gas *between* and parallel to the hexagonal rods, instead of being wholly confined to the internal passages. This evidently led to a leakage-flow instability, which destroyed the core.

Model tests to obtain a physical understanding of the phenomenon involved a blade suspended flexibly in a rectangular channel of variable width and subjected to airflow (see schematic). The blade was capable of torsional and lateral motions, restrained by springs. This model displayed flutter instability, similar to that of airplane wings, but with the flutter threshold affected by the width of the flow channel. As the ratio of channel width to plate length decreased from 1 to 0.1, both experiment and inviscid-flow theory predicted reductions of the critical flow velocity for flutter of the order of 200 to 400%. Instability in both the pitch (rotational) mode and the translational mode (lateral motion) were obtained.

Cure: The model tests indicated that stiffening the upstream support raises the flutter threshold. In the prototype, stronger "bundling forces" were applied, to keep the rods tightly pressed together and to effectively stop the leakage flow.

References: 1. Spence, R.W., & Durham, F.P., *Astronautics & Aeronautics*, Vol.3, 1965, pp.42-46.

2. Bland, S.R., Rhyne, R.H., & Pierce, H.B., Trans ASME, *J.Engng for Industry*, Vol.89, 1967, pp.824-830.

3. Rhyne, R.H., *Private communication*, 24 Jul 1979.

CHART XXXV: CONTROL RODS AND GUIDE TUBES IN SEVERAL PWR's

Case 45 Class: annular-flow-induced vibration Date: 1977

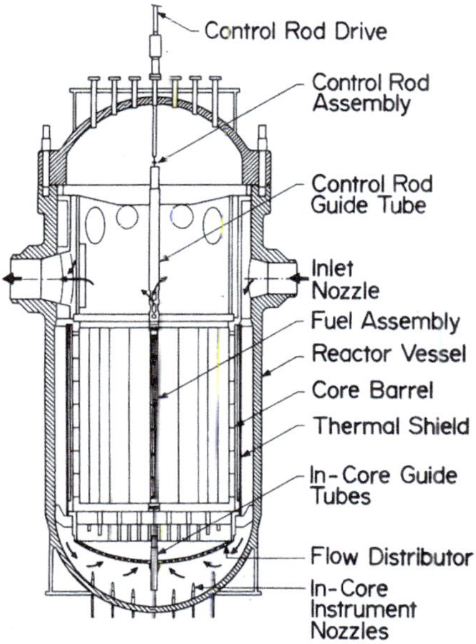

REACTOR VESSEL AND INTERNALS ARRANGEMENT

Problem: The guide tubes in PWR's serve a dual purpose. Being located on top of the fuel assemblies (square arrays of many fuel rods, supported at intervals), they serve as their principal structural members; they also serve as the guide channels for movement of the control rods, which control the reactor power output. Structural integrity is essential, under both normal and accident conditions. In Dec 1977, several holes (wear penetrations) were discovered during refuelling of a nuclear plant. Similar indications were later found in other plants of the same design. The wear occurs at the top of the guide tubes, adjacent to the tips of the control rods in their fully withdrawn position, where they normally remain most of the time. These reactors were operated from there on with the control rods 7.5 cm further into the core to improve emergency shut-down capability.

Mechanism: There is insufficient information on the geometry of the control-rod/guide-tube assembly and the hydrodynamics involved. It is clearly a case of flow-induced vibration of the control rods against the guide tubes (cf. Case 43).

Cure: Stainless steel sleeves were installed to reinforce the guide tubes.

References: 1. Anonymous, *U.S. Nuclear Regulatory Commission NUREG-0090*, Vol.1, No.2, Sep 1978.

2. Hennig, K., & Grunwald, G., Paper 4.9, *Int'l Conf. Vibration in Nuclear Plant*, Keswick, 1978.

CHART XXXVI: CORE BARREL & THERMAL SHIELD OF SEVERAL PWR's

Case 46 Class: annular-flow-induced vibration Date: 1967-1970

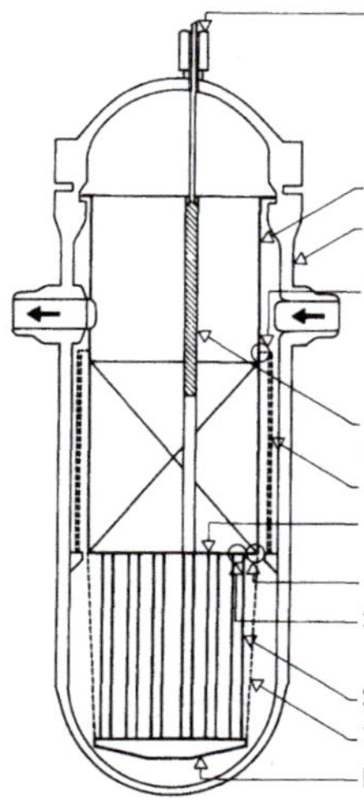

Problem: The cylindrical core barrel assembly contains and supports the fuel; it is supported at the top, on the pressure vessel. Over its lower part it is surrounded concentrically by a cylindrical thermal (neutron) shield, which, in these reactors, is supported at its bottom on the pressure vessel. Serious failures arising from vibration of the thermal shield and the core barrel occurred in various reactors around the world. In one reactor (i) the pins connecting the three segments, of which the thermal shield is made up, failed, allowing them to oscillate and knock against the core barrel and vessel; (ii) 1/3 of the bolts joining upper and lower parts of the barrel failed, débris finding its way to the steam generators; (iii) 2/3 of the tie rods connecting lower casting to barrel failed. The shut-down for clean-up and repairs took three years. In another reactor the whole shield dropped to the bottom when its supports failed. Typical core-barrel recorded amplitudes: 0.2 mm, and frequencies for pendular motion: 5-10 Hz. Typical annular flow velocities are 5-10 m/s.

Mechanism: One mechanism proposed is that small pendular oscillation of the core barrel induced large amplitude oscillation of the thermal shield, probably in an inverted pendular mode, or as a shell; or *vice versa*. The shield vibration generated swirling flow at the bottom of the annulus which caused the tie rods to fail by cross-flow-induced vibration. It was calculated that core barrel oscillation of 1 mm amplitude could generate unsteady pressures in the annulus of 0.1-0.2 atm. and loads on the thermal shield of 20-30 tons. The excitation causing pendular motion may be found in the unsteadiness of the multi-entry flow into the annular passage and its division by the thermal shield, pump pulsations, etc. It was calculated that gross velocity fluctuations in the annulus of a few percent were sufficient to produce the recorded 0.2 mm core barrel amplitudes [4]. In a 1/10 model study [7], a global flow instability was found, with large eddies forming at the top of the thermal shield, below the flow inlet, capable of inducing alternating loads of the order of 2 tons.

Cure: In most cases the thermal shield was eliminated altogether; some of the outermost fuel assemblies had to be removed, to protect the pressure vessel from undue radiation. The core barrel bolts and the tie-rod arrangement were redesigned. In one case the shield was retained with additional "stronger" supports.

References: 1. Verna, B.(ed.), *Nuclear Power Experience*, Section PWR-2,II,Item 2.
2. Calcagno, M., *et al.*, Paper at *Meeting on Vibration Phenomena inside Light Water Reactor Vessels*, Mol, Dec 1970.
3. Marchal,A., IAEA Symposium *Experience from Operating and Fuelling Nuclear Power Reactors*, Vienna, 1974, pp.243-258.
4. Iansiti,E., Lazzeri,L.,& Messore,G., *IAEA-SM-127/37*, 1970.
5. Bastl,W., *et al.*, *Atomwirtschaft*, May 1972, pp.263-264.
6. Lazzeri,L., & Dini, D., *Ingegneria Nucleare*, 1969, pp.48-57.
7. Daubert,A., Pugnet,L., & Boulot,F., *Flow-Induced Structural Vibrations* (ed. E. Naudascher); Springer-Verlag, Berlin, 1974; pp.75-80.

CHART XXXVII: CORE BARREL OF PWR

Case 47 Class: annular-flow-induced vibration Date: 1973

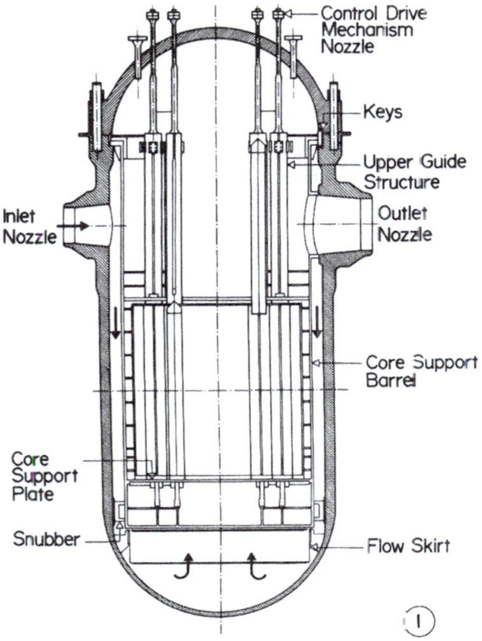

(1)

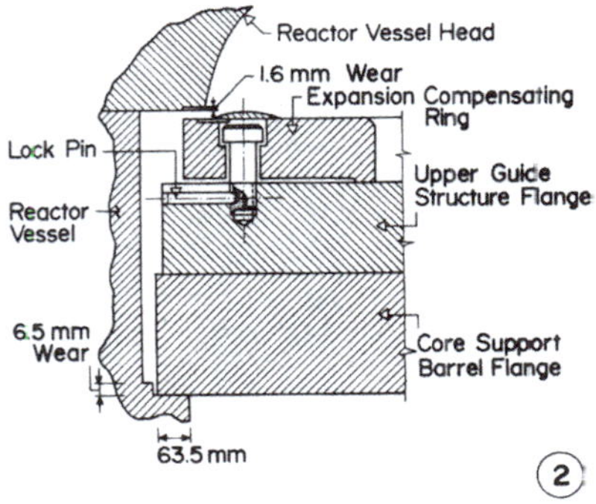

(2)

Problem: The core barrel contains and supports the fuel assemblies in the reactor. It is basically a cantilevered cylindrical vessel, clamped at the upper support flange of the pressure vessel; possible lateral vibrations are constrained by snubbers at the bottom (Sch.1). In this particular PWR design there is no thermal shield. After several indications of core movements (~0-6 Hz) by ex-core flux detectors, inspection of the reactor in Oct 1973 revealed that (i) all expansion compensating ring bolts were broken, (ii) the core support barrel flange had worn a 6.5 mm ledge in the vessel flange - effectively allowing 6.5 mm vertical movements of the barrel; (iii) a groove 1.6 mm deep and 6.5 mm wide was worn into the reactor vessel head (Sch.2).

Mechanism: Flow-induced rigid-body motion of the core barrel was established as the cause - based on the wear pattern and flux-detector signals. The hydraulic loads involved, together with insufficient "hold-down" load at the upper clamped support induced motions which began wearing the support, thereby allowing larger amplitudes of motion and more wear. The rigid-body motion was a complex combination of rocking, translation and nutation. The snubbers at the bottom constrained the motion into a quasi-elliptical path. No specific flow-induced vibration mechanism was advanced. However, the impingement of the multi-inlet flow, and the fundamental unsteadiness and high turbulence level of the flow in the narrow annulus, may be proposed here as the excitation mechanism, which can couple to the motions of the barrel.

Cure: A new hold-down device was installed to prevent the loss of clamped condition at the top, involving stacks of Belleville washers around the circumference of the support region; the hold-down load was thus increased from 15 x 10^3 to 33 x 10^6N.

References: 1. Anonymous, *Combustion Engineering CEN-5(P)*, Rev.01, Apr 1974.

2. Verna, B.(ed.), *Nuclear Power Experience*, Section PWR-2, II, Item 12.

3. Beckins, W.J., *Power Reactor Docket Information*, Docket 50255-182, Jul 1973.

CHART XXXVIII: THERMAL SHIELD OF A BWR

Case 48 | Class: annular-flow-induced vibration | Date: 1964-1965

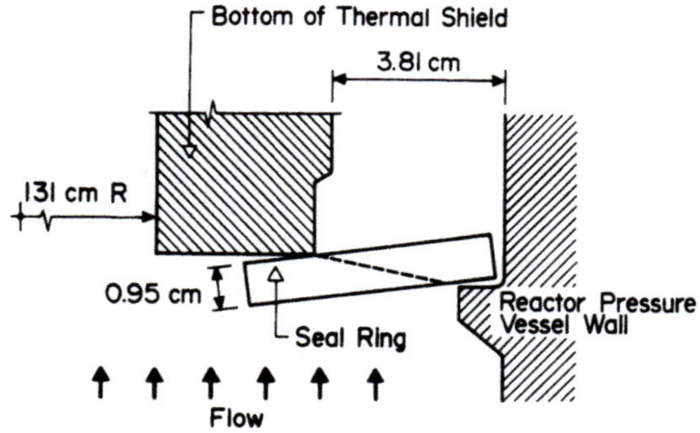

①

Problem: The thermal shield of this reactor is cylindrical, 261 cm in diameter and 234 cm long. It is supported at its bottom flexibly on the reactor pressure vessel, via a 0.95x5.56 cm Belleville washer (Sch.1). In 1964, several thermal shield retaining bolts were found to be broken. The support structure was stiffened and tests conducted in 1965. It was found that with increasing flow the shield first displayed "low level nervousness" at f≈4 Hz, and then suddenly developed rigid-body vibration "of intolerably large amplitude" of 0.6 mm, such that its centre described an elongated ellipse.

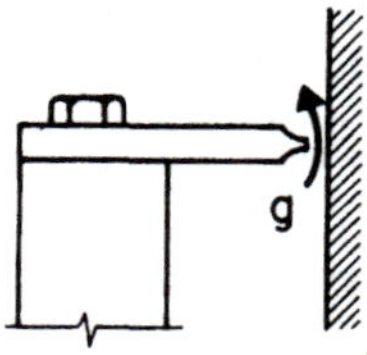

②

Mechanism: The mechanism proposed is that of a self-excited oscillation associated with annular flow leaking through the Belleville washer once a given pressure differential across it is established. The high pressure drop of the leakage flow at the inlet of the annulus gives a negative damping characteristic, capable of leading to self excitation in a lateral mode (cf.Case 43). Taking the leakage to vary locally, the mechanism proposed couples the motion to the flexibility of the Belleville washer seal.

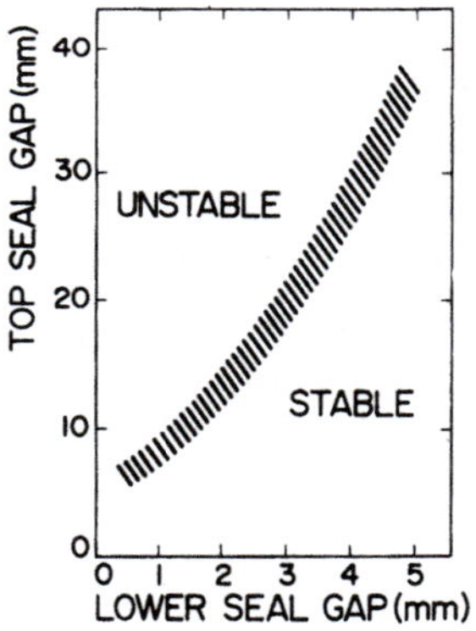

③

Cure: Immobilizing the Belleville washer by wedges, so that leakage would be equal on all sides; (ii) installing a top seal (Sch.2) which has a stabilizing effect (positive damping characteristic); (iii) inserting hold-down weights in the annular gap, resting on the Belleville washer, to immobilize the shield and disrupt the annular flow. Any one of these was considered capable of solving the problem; in any event, the three together did.

Reference: Corr, J.E., *Argonne National Laboratory ANL-7685*, May 1970, pp.272-289.

CHART XXXIX: THERMAL SHIELD OF PWR's

Case 49 Class: annular-flow-induced vibration Date: 1969-1970

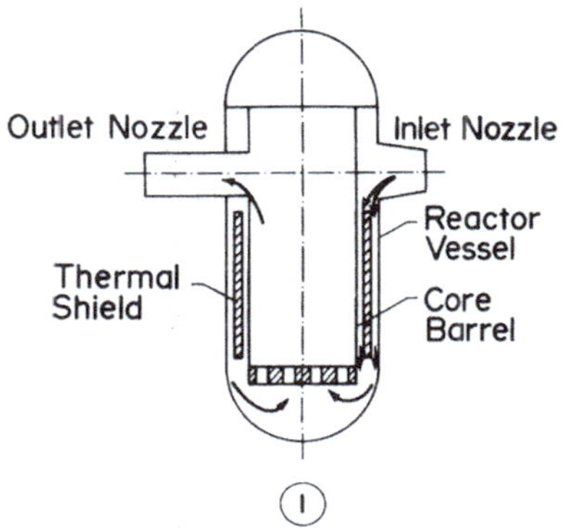

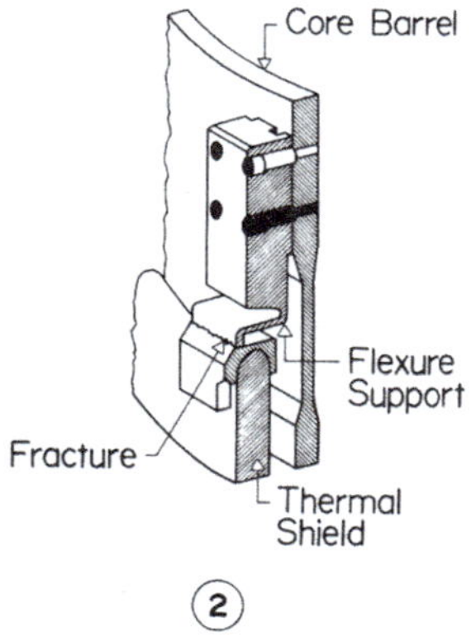

Problem: In this type of PWR, the thermal shield is attached by a number of supports around its circumference, top and bottom, on the core barrel itself. Some early cracking of welds in upper supports prompted the installation of additional flexible ("flexure") supports (Sch.2). This solved the problem in this reactor. Another similar reactor installed the same added supports; however, upon inspection it was found that all the added supports had failed by low-stress high-cycle fatigue.

Mechanism: In both cases the cause was diagnosed as flow-induced oscillation of the thermal shield in one of its shell modes.

Cure: The added flexible supports were removed from the second reactor. No further problems were encountered. The question remains why a successful solution in one case creates problems in another.

References: 1. Verna, B.(ed.), *Nuclear Power Experience*, Section PWR-2, II, Item 2.
2. Kufel, J.M., private communication, 2 Apr 1979.
3. Anonymous, *Power Reactor Docket Information*, Docket 50213-46, Jun 1970.

CHART XL: THERMAL SHIELD OF PWR

Case 50 Class: annular-flow-induced vibration Date: 1972.

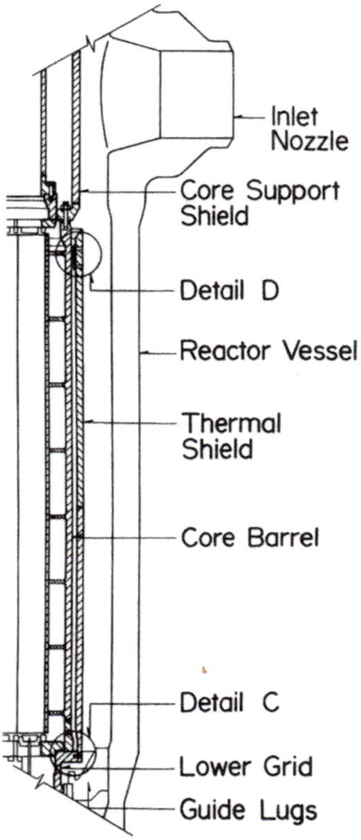

Problem: The thermal shield of this reactor is supported at the bottom, on the lower grid, as shown. At its upper end, it is supported and positioned radially by spacer blocks, at intervals around the circumference, installed with an interference fit (Detail D); the aim is to force the vibration of the thermal shield into the higher circumferential modes. In 1972 it was found that the retaining welds on the dowels at the bottom support had failed, and some deformation of the mating surfaces had taken place. The dowel holes had been worked oversize and clearances at the top spacer blocks were 0.2-0.5 mm, instead of interference fits.

Mechanism: It was proposed that "inbalanced pressure-forces" deformed the shield, creating clearances. Then, flow-induced vibration could take place, which fretted the spacer blocks, allowing larger amplitudes to develop. The vibration is thought to have been mostly in the second circumferential mode (at 31 Hz), and perhaps in the third (49 Hz).

Cure: (i) Improving and strengthening the lower support, and providing positive attachment to the core barrel at the top; (ii) reducing the peak downcomer velocities adjacent to the thermal shield.

References: 1. Anonymous, *Babcock & Wilcox BAW-10050, Rev.1*, Nov 1972

2. Verna, B.(ed.), *op. cit.*, Section PWR-2, II, Item 5.

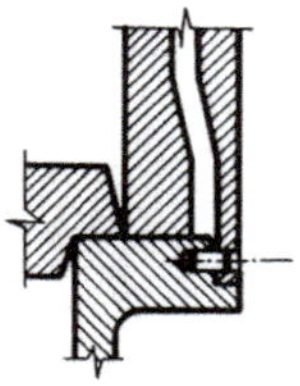

Detail C

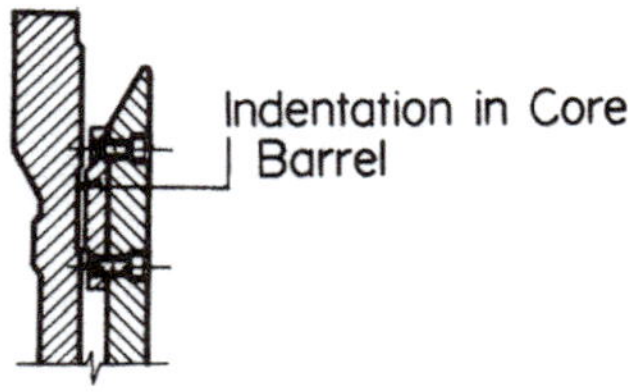

Detail D

CHART XLI: FEEDWATER SPARGERS IN BWR's

Case 51 Class: annular-flow-induced vibration Date: 1972-1977

Problem: The feedwater spargers are a spray system used to distribute the reactor feed water (Sch.1); they are sometimes a single pipe all around the interior of the pressure vessel (Sch.2), and sometimes independent curved segments on either side of a junction or tee-box, each fed individually by separate nozzles (Sch.3). Holes 1.27 to 3.81 cm in diameter all along the "header" pipe (typically 15.2 cm in diameter) discharge the water radially into the flow coming down from the steam separators towards the jet pumps. Wedges bearing against the pressure vessel were used to preload the spargers, so as to prevent vibrations arising from both internal and external flow excitations. Deep cracks (up to ~1.6 mm) were found in the spargers of one reactor after another, in some cases threatening the structural integrity of the system - in one case one crack extended 200° around the pipe. The cracks were mostly in the conjunctions of tee-box to header pipe, or of tee-box to header sleeve. Events will be described in terms of one reactor, although at least 16 reactors were implicated. Initially the high-cycle fatigue cracks were thought to have been caused by inadequate, or loss of, preload. "Design 2" spargers were installed, which involved heavier pipe (schedule 80 instead of 40) and larger preload. After ~6 weeks' operation, new fatigue failures occurred. This led to "Design 3" spargers, involving an entirely different support bracket, attached to the vessel wall near the feed-water nozzle. Yet, instrumented "Design 3" spargers indicated excessive vibration when the reactor operated at > 85% power. The frequencies ~47 Hz compared to one of the calculated natural frequencies of 60 Hz; strain amplitudes of as much as 100 με were obtained, sufficient to cause fatigue failure.

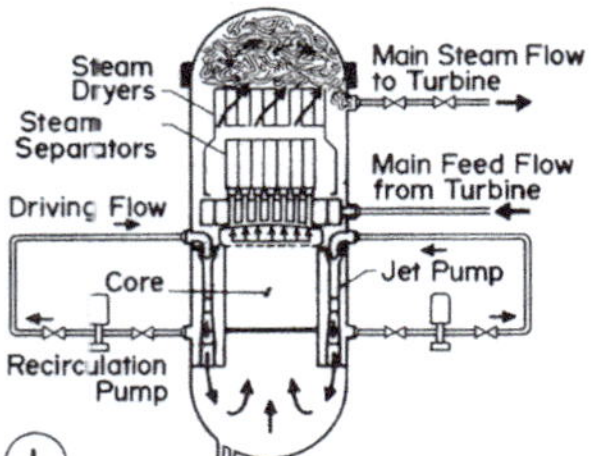

①

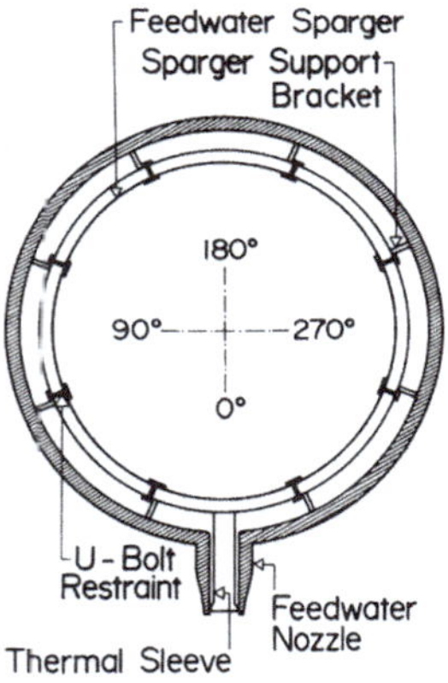

② SPARGER PLAN VIEW

Mechanism: Extensive full-scale tests eventually determined that the cause was a flow-induced, self-excited vibration involving annular ("leakage") flow between the feed-water nozzle and the thermal sleeve. Without preload, this occurred for radial gaps larger than 0.028 cm, at sparger flow rates of 284 l/s.

Cure: The "Design 4" spargers incorporate (i) interference fit between thermal sleeve and nozzle, to reduce annular leakage flow and to increase fluid damping in the gap, (ii) a reduced pressure difference between sparger inlet and discharge, (iii) stress-concentration reduction features.

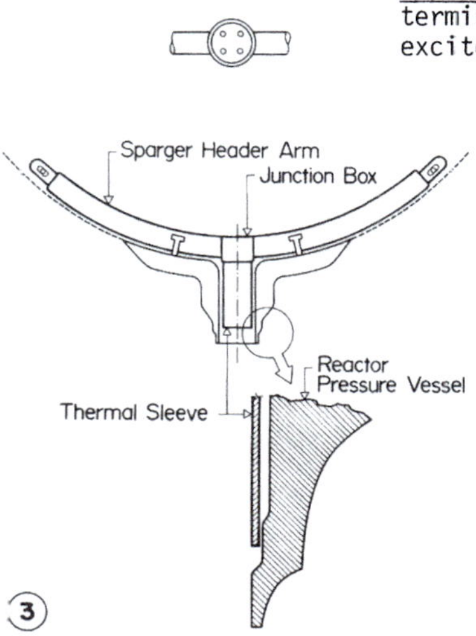

③

References: 1. Verna, B.(ed.), *Nuclear Power Experience*, Section BWR-2, II, Items 11,13,16 17,20-23,25,29,30,32,38,40,42,46,48.
2. *Power Reactor Docket Information*, Dockets 50245-294,Sep 1973; 50245-333,Jan 1974; 50245-401,Aug 1974; 50133-212,Nov 1974; 50133-227,Jan 1975; 50237-739, Jan 1975.

CHART XLII: FLOW CONTROL DEVICE OF AGR

Case 52 Class: annular-flow-induced vibration Date: 1975

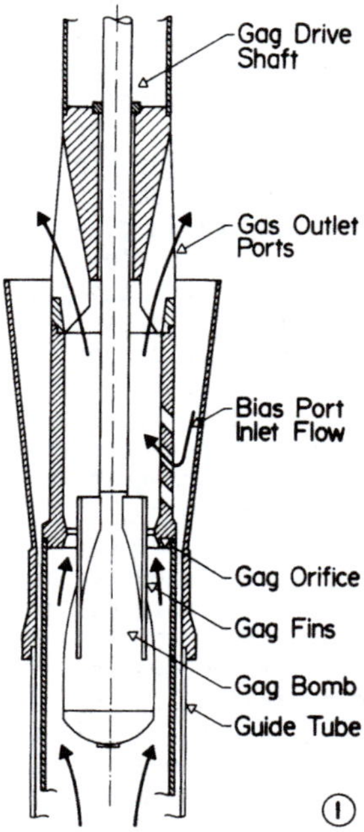

①

Problem: In an AGR the channel heat output varies both with position in the core and fuel age. This necessitates matching channel flow to channel heat output by means of an adjustable gag at each channel outlet (Sch.1). During engineering tests, with the AGR operating at design gas flow and the gags almost fully closed, noises were heard to emanate from the fuel stand pipes. Examination showed that a number of gag units had suffered mechanical damage: hammering and erosion of the gag plug locating fins, fracture of the gag plug and, in some cases, fracture of the webs of the gag outlet ports.

Mechanism: In the original design it had been realized that if the gag fins touched the orifice wall then there would be a net pressure force away from the wall, causing the gag to move to the other side of the orifice; this process would be repeated, so producing an instability. In an attempt to eliminate this, bias port holes were provided such that, due to the pressure recovery, there is an inward flow through the bias holes destroying diffusion and so increasing the side-force and causing the gag to remain on the wall opposite the bias holes (Sch.2). However, not only did the side force decrease as the gag moved away from the holes but the force was negated by the presence of small flow disturbances, such as irregular weld beads on the fin-centrebody corners, on the side opposite the bias ports (Sch.3). Once the gag is released from the wall, measurements made on a three-dimensional model forced to oscillate in a lateral manner, indicate that the pressure difference across the gag is in phase with the gag velocity so producing energy input to the oscillation (Sch.4).

Cure: The main defect of the original design is that the bias holes produce a large disturbance to the flow in a region where velocities are very low and thus the rate of diffusion small; their effect may easily be overcome by much smaller disturbances acting at the throat,where the velocities are high and the rate of diffusion large. Hence, holes were drilled, on the same side as the bias ports, right through the orifice, allowing gas from the surrounding plenum chamber to be sucked into the throat region. Also, providing vent holes in the gag fins, through to the hollow centre body which is vented downstream, enabled flow to be drawn into the throat area from the fins that were not in contact with the wall. The jets of gas act as trip wires, spoiling the diffusion and maintaining a side force. Both of the two above measures serve to provide an aerodynamic force, maintaining the gag on the wall opposite the original bias holes.

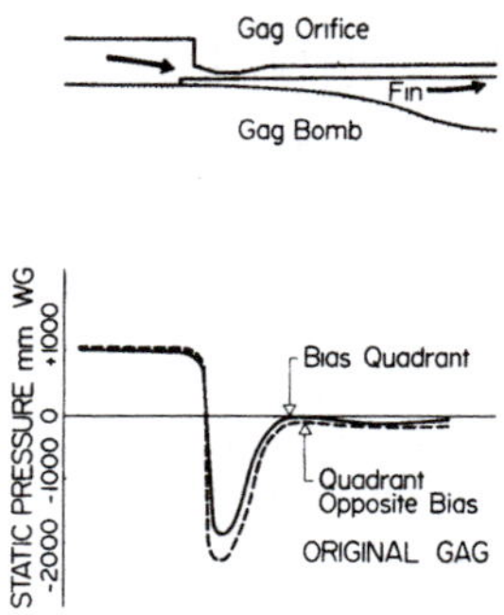

② PRESSURE DISTRIBUTION THROUGH THE GAG

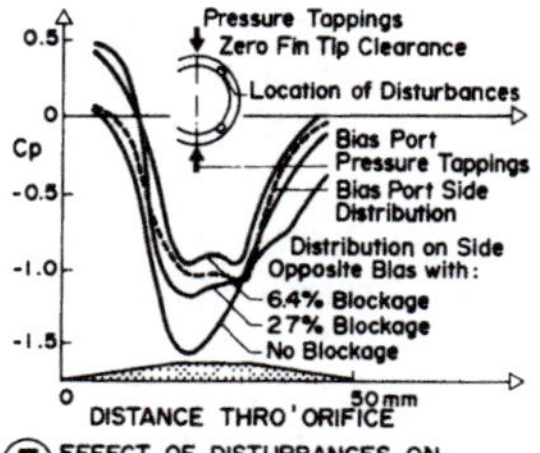

③ EFFECT OF DISTURBANCES ON ORIFICE PRESSURE DISTRIBUTION

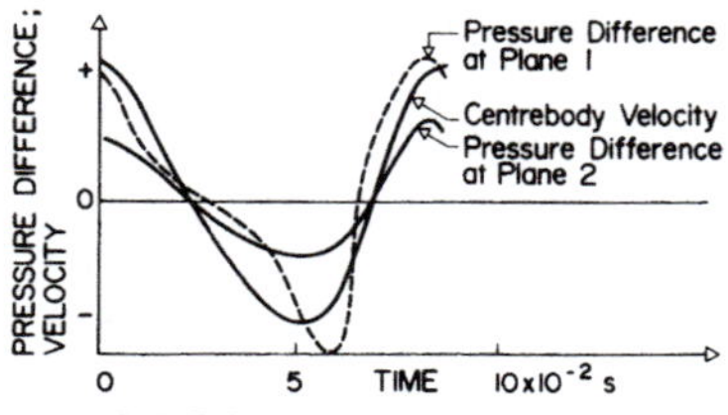

④ CENTRE BODY VELOCITY AND PRESSURE DIFFERENCE

References: 1. Joss, J.O., Emsley, G.W., & Hardingham, R.P., Paper 5.1. *Int'l Conf. Vibration in Nuclear Plant*, Keswick 1978.

2. Bolton, A.D., & Furber, B.N., Paper 5.2, *op. cit.*

3. Denton, J.D., Hutton, M.F., & Parkin, M.W., Paper 5.5, *op. cit.*

4. Fricke, F.R., *Nucl. Science & Engng.*, Vol.48, 1972, pp.87-102.

Discussion: FLOW-INDUCED VIBRATIONS IN NUCLEAR REACTOR AND HEAT EXCHANGERS (Paidoussis)

P. A. Kolkman
Delft Hydraulics Laboratory, Netherlands

This discussion focuses on the use of Connor's criterion in fluids other than air. Instead of using dimensionless parameters, I would like to consider the balance of forces, e.g., mechanical damping $c\partial y/\partial t$ versus hydrodynamic (positive or negative) damping forces due to flow $F = -\rho DV(\partial y/\partial t)*f(S)$, where $2\pi S = \omega_{f\ell}D/V$, and $\omega_{f\ell}$ = angular vibration frequency in flow.

Whether hydrodynamic damping is positive or negative depends on S. Instability occurs when the negative damping dominates the mechanical damping.

The ratio of mechanical damping to hydrodynamic damping determines whether the system is stable, and the stability criterion becomes:

$$\frac{c\dot{y}}{\rho DV\dot{y}} = \frac{c}{\rho \cdot DV} = f(\frac{\omega_{f\ell}D}{V}) \tag{1}$$

The damping factor c is related to the dimensionless damping δ as follows:

$$c = 2\delta_{dry}\, m\, \omega_{dry} \tag{2}$$

Theoretically, δ_{dry} is measured in vacuum conditions. We can show, from (1) and (2).

$$\frac{\delta_{dry}\, m\, \omega_{dry}}{\rho DV} \cdot \frac{V}{\omega_{f\ell}\, D} = f'(\frac{\omega_{f\ell}\, D}{V}) \tag{3}$$

f' is now a new function of the Strouhal number in the flow. Only in the case $\omega_{dry} = \omega_{flow}$ (which is more or less true in air) does the left hand term correspond to the Connor's expression of $\delta m/\rho D^2$. In other fluids, eq. (3) should be used in the data analysis, and I have the impression that this is neglected by several authors.

Reply to Discussion of: FLOW-INDUCED VIBRATIONS IN NUCLEAR REACTOR AND HEAT EXCHANGERS

M. Paidoussis

As I understand it, Dr. Kolkman suggests that in the analysis of any particular system for fluidelastic instability by means of a Connors-type relationship, the actual frequency of vibration of the cylinders should be used, rather than the one measured in air or vacuum. If the discusser means that one should use the frequency in the medium in which the cylinder array is immersed, then I must say that this is actually what most workers in the field do. In any case, if he means that the frequency under flow conditions should be used, this is not done. Differences in frequency between still and flowing fluid should be accounted for by the theory, if properly done, as part of the aerodynamic stiffness. Existing theory does not do this and assumes no change in frequency with flow, relying on experimental evidence. Experiments show that the frequency of oscillation during fluidelastic instability is very close to the natural frequency in the still fluid medium.

A2 HEAT EXCHANGER TUBE VIBRATION: COMPARISON BETWEEN OPERATING EXPERIENCES AND VIBRATION ANALYSES

M. J. Pettigrew, Head
A. O. Campagna, Research Engineer

Engineering Research Branch
Atomic Energy of Canada Limited
Chalk River Nuclear Laboratories
Chalk River, KOJ 1JO, Ontario, Canada

SUMMARY

Tube failures due to excessive flow-induced vibration must be avoided to assure the reliable performance of heat exchangers. Such components must be thoroughly analysed for vibration at the design stage. Several flow-induced vibration mechanisms are possible. In this paper, particular attention is given to fluidelastic instability of tube bundles subjected to liquid and two-phase cross-flow. The performance of ten operating heat exchanger components is reviewed to validate our recommended vibration analysis guidelines. Both tube failure histories and satisfactory performances are considered. The results show that a fluidelastic instability constant K=3.3 is a reasonable design criterion.

I. INTRODUCTION

The satisfactory performance of heat exchangers and steam generators is crucial to the reliability of nuclear power stations. Process type heat exchangers are similarly important to the reliability of industrial plants. Excessive flow-induced vibration may cause tube failures by fatigue or by fretting-wear. Such failures often require plant shutdowns for repairs which are very undesirable in terms of lost production and maintenance cost. To avoid these problems, we require thorough vibration analyses at the design stage. Our techniques and guidelines to analyse tube-in-shell heat exchange components for flow-induced vibration are outlined in [1].

It is essential to validate analysis techniques against the performance of operating heat exchangers, hence the purpose of this paper. Case histories of ten operating components are presented. These include four steam generators and six heat exchangers subjected to both liquid and two-phase cross-flow. Many of these components experienced failures; others are operating satisfactorily. The performance histories are compared against our design criterion for fluidelastic instability. The failures were mostly due to fretting-wear although two cases of fatigue failures were also encountered.

To our knowledge, very few case histories of heat exchanger failures have been described in detail in the literature. Hartlen described four cases in 1973 [2]. There are a few others. However a comprehensive review is beyond the scope of this paper. There are two main reasons for the general lack of sound failure data. First, designers, manufacturers and owners do not generally like to discuss their failures. Second, it is often difficult to get reliable field data such as flow rates, vibration amplitudes, damping values and the exact nature of the failures. The present paper should contribute new field data and should give credibility to our design criterion.

2. ANALYTICAL CONSIDERATIONS

In general, both axial and cross-flow exist in heat exchangers. However, most of the problems we have encountered so far were related to cross-flow. Therefore we shall limit ourselves to the latter in this paper. The tubes on the shell side of heat exchangers may be subjected to liquid or gas flow. In other components, such as steam generators the flow may be two-phase. The operating experiences outlined in this paper comprise liquid and two-phase flow.

Generally, in cross-flow, we consider three basic flow-induced vibration excitation mechanisms: 1) fluidelastic instability, 2) random excitation due to flow turbulence and 3) periodic wake shedding. We have observed the first two mechanisms in both liquid and two-phase cross-flow [2]. Periodic wake shedding resonance is possible in liquid flow, but has not been observed in two-phase flow. Either it does not exist or it is dominated by the response to random flow turbulence. We do not consider it in two-phase flow.

We have generally observed fluidelastic instability to be the most severe vibration excitation mechanism in basic studies on tube bundles in cross-flow [3,4,5]. Hence we first start by correlating failure histories against this mechanism. Other mechanisms should be kept in mind and will be considered in further studies.

2.1 Fluidelastic Instability

Fluidelastic instabilities are possible in a tube bundle subjected to cross-flow when the interaction between the motions of individual tubes is such that it results in fluid force components that are both proportional to tube displacements and in-phase with tube velocities. Instability occurs when during one vibration cycle the energy absorbed from the fluid forces exceeds the energy dissipated by damping. It is shown in [1] that, for a tube bundle subjected to the non-uniform flow velocity $V_r(x) = V_r\psi(x)$, the non-dimensional critical velocity $V_{rc}/f_i d$ at which instability occurs in the ith mode may be expressed by:

$$V_{rc}/f_i d = K\left\{ c/[2f_i d^2 m\rho \int_0^{\ell} \psi^2(x)\ \phi_i^2(x)\ dx] \right\}^{\frac{1}{2}} \tag{1}$$

In the above expression f_i and $\phi_i(x)$ are respectively the tube frequency and mode shape in the ith natural mode of vibration, d is tube diameter, ℓ is tube length, m is tube mass per unit length including the hydrodynamic mass, c is the viscous coefficient of damping, ρ is the fluid density and $\psi(x)$ is the flow velocity distribution function. The instability factor K is determined experimentally. To allow for a realistic safety margin, we recommend a K factor of 3.3 as a guideline for vibration analysis [1]. If the actual $V_r/f_i d$ is less than $V_{rc}/f_i d$, fluidelastic instability should not be a problem. Note that $V_{rc}/f_i d$ in Equation (1) is effectively a normalized dimensionless velocity so that, distributed uniformly along the entire length of the tube, it has the same effect as the non-uniform flow velocity.

For tubes subjected to uniform flow over their entire length, Equation (1) reduces to Connor's formulation for fluidelastic instability:

$$V_{rc}/f_i d = K \left(m\delta/\rho d^2\right)^{\frac{1}{2}} \tag{2}$$

in which the logarithmic decrement $\delta = c/(2mf_i)$.

The dynamics of multispan tubes and the formulation for fluidelastic instability and other vibration excitation mechanisms form the basis of a computer program called "PIPEAU" to predict the vibration response of heat exchanger tubes. The program first calculates the mode shapes $\phi_i(x)$ and the natural frequencies f_i(i.e., eigenvalue solution). Then the vibration response and the critical velocities for fluidelastic instabilities are calculated. We used this program to analyse the case histories described here.

2.2 Flow Velocities

The first step in a flow-induced vibration analysis is to evaluate operating conditions and flow velocities. We use the so-called "reference gap velocity" V_r throughout this paper. Regardless of tube bundle configuration the reference gap velocity is defined as:

$$V_r = V_\alpha p/(p-d) \tag{3}$$

where p is the pitch between tubes and V_α is the free stream velocity, i.e., the flow velocity with the tubes removed.

In heat exchangers, the flow velocities are generally calculated simply from the flow areas. Between two baffle-supports, the flow is assumed to be uniform. When the flow is divided between several parallel flow paths, the flow calculations are based on equal pressure drop between regions common to all flow paths. This situation exists in some heat exchangers where the flow may go through the tube bundle and between the tube bundle and the shell.

Special difficulties arise near inlets and outlets. For instance, the flow may be jetting directly towards the tube bundle from the inlet nozzle. Then we take the reference gap velocity V_r between the tubes to be the jet velocity or, the velocity based on the total flow area if it is larger. The total flow area at the inlet comprises the area around the inlet nozzle, between the tube bundle and the shell, and the projected area (between the tubes) of the inlet nozzle on the tube bundle.

For steam generators, comprehensive thermal-hydraulic analyses were done as outlined in [6] to determine the flow velocity and steam quality distribution. Homogeneous flow velocities based on the average density of the steam-water mixture are used in two-phase flow.

2.3 Damping

Damping is expressed in the form of a normalized viscous damping coefficient $c_n = c/d$. Measurements taken in water in real heat exchangers suggest that c_n is roughly constant with frequency up to at least 140 Hz [1]. At higher frequencies structural damping probably becomes significant and c_n would increase with frequency as the damping ratio ζ would tend towards a constant value. For the cases where damping measurements were taken, we use the actual damping values in the analysis. In other cases we use $c_n = 400$ for water and $c_n = 1400\ kg\cdot s^{-1}\cdot m^{-2}$ for two-phase mixtures. The latter was obtained on a steam generator model under realistic flow conditions.

3. HEAT EXCHANGER PERFORMANCE: CASE HISTORIES

3.1 Case NPSG: Fretting-Wear of Nuclear Steam Generator

This component is a natural circulation type horizontal "U" tube steam generator (Chart 1) which has been in operation since 1962 in a small prototype nuclear station. Steam-water flows from the top of the tube bundle to the steam drum through 48 risers. The steam is separated and the remaining water is recirculated in 24 downcomers. The flow velocity in the downcomer is roughly 3 m/s and directly impinges on the tube bundle. This caused excessive tube vibration and eventually tube failures by fretting-wear (Chart 1) in the downcomer inlet region. The first failure occurred in 1969 and the second in 1978.

The vibration analysis NPSG-1 indicates fluidelastic instability which explains the failures. The ratio of actual velocity/critical velocity is 1.36. The downcomer flow velocity is taken as the reference gap velocity V_r between the tubes in the inlet region. Tube response to random forces due to downcomer jetting turbulence would probably have also explained the failure. On the other hand, the vibration analysis for the outlet tubes subjected to two-phase cross-flow near the risers shows no problem (NPSG-2, Table 1). There are no damaged tubes in the outlet region.

The failed tubes were plugged. As a preventive measure, the partially worn tubes in the inlet region were also plugged. It is interesting to note that the first tube failed after 7 years and the second after 16 years of operation. Fretting-wear in this case seems to be a slow and continuous process. Fretting-wear appears not to be self-limiting as the contact area between tube and support increases with wear.

TABLE 1: Outline of Fluidelastic Instability Analyses of Operating Heat Exchanger Components

CASE No	COMPONENT TYPE	REGION	p/d	d (mm)	MODE ORDER	f (Hz)	c $\left(\frac{kg}{m \cdot s}\right)$	$\frac{m\delta}{\rho c^2}$	$\frac{V_r}{V_r CRIT}$
NPSG-1	Steam Generator	Inlet	1.50	12.7	1st	42.5	5.1	0.44	1.36
NPSG-2	" "	Outlet	1.50	12.7	1st	48.9	17.8	11.75	0.85
DPSG-1	" "	U-tube	1.38	12.7	1st	11.4	5.3	16.2	2.27
							17.8	53.9	1.24
DPSG-2	" "	U-tube	1.38	12.7	1st	30.2	17.8	20.4	0.11
PASG	" "	U-tube	1.50	12.7	1st	17.0	8.0	10.4	1.30
							7.8	23.2	0.87
BASG	" "	U-tube	1.57	13.0	1st	35.5	8.1	16.2	0.49
DPBC-1	Cooler	Outlet	1.5	12.7	1st	103.	5.1	0.156	0.77
DPBC-2	"	Inlet	1.5	12.7	1st	87.1	5.1	0.181	0.75
DPBC-3	"	Interior	1.5	12.7	1st	104.	5.1	0.152	0.76
PASC-1	"	Outlet	1.26	12.5	1st	62.1	5.6	0.256	1.73
PASC-2	"	Outlet	1.26	12.5	3rd	209.	5.0	0.076	1.74
PAMX-1	Heat Exchanger	Interior	1.50	12.7	1st	37.1	5.1	0.43	2.04
PAMX-2	" "	Interior	1.50	12.7	1st	134.	5.1	0.12	1.07
GBHX-1	" "	Inlet	1.25	19.1	1st	17.7	9.6	0.751	0.70
GBHX-2	" "	Inlet	1.25	19.1	8th	42.7	9.6	0.311	1.37
GBHX-3	" "	Inlet	1.25	19.1	8th	69.2	9.6	0.192	0.35
GBCX-1	Cooler	Inlet	1.25	19.1	1st	18.15	7.6	0.581	1.71
PHRC-1	"	Inlet	1.33	19.1	15th	177.44	7.6	0.050	1.55

3.2 Case GBHX: Fretting-Wear of Process Heat Exchanger

This case concerns a triple segmental liquid-liquid process heat exchanger operating in a heavy water plant (Chart 2). Excessive tube vibration amplitudes were measured near the inlet during commissioning of the heat exchanger system. The measurements were taken with an accelerometer probe inserted in the tubes from the tubesheet. The flow in the inlet was relatively high and impinged directly on the tube bundle although a perforated distribution plate in the inlet nozzle made the flow uniform. The relatively long tube spans (1.45 m) also contributed to the problem.

Vibration initially caused fretting-wear of the tubes at midspan between supports (see photo in Chart 2). Lacing-strips were installed to support the tubes near the inlet. Unfortunately they were excessively loose. Fretting-wear occurred between the tube and the lacing-strips. Tubes and strips wore through. The problem was aggravated by the partial blockage of the perforated distributor plate due to debris which resulted in much higher than expected flow velocities in some areas. Eventually proper baffle-supports were installed. No further problem occurred.

Vibration analysis GBHX-2 shows that this heat exchanger was prone to fluidelastic instability in the eighth mode. This is explained by the shape of the eighth mode which is largest where high inlet flow velocity exists. Vibration analysis GBHX-3 shows that the addition of baffle-supports eliminate the problem. The ratio of actual velocity/critical velocity is much less than one.

3.3 Case DPBC: Fretting-Wear Damage in Cooler Heat Exchanger

This cooler is a hairpin type "U" tube heat exchanger (Chart 3). Cooling water flows on the shell side between segmental baffle-supports. Heavy water flows in the tubes. There are two hairpins positioned one on top of the other. Originally the shell sides were connected in series. A few years later they were rearranged in parallel to increase heat transfer. Thus, for the bottom hairpin, the inlet became the outlet. The impact plate at the inlet became ineffective at the outlet.

Extensive fretting-wear damage was observed throughout the bottom hairpin as shown on Chart 3. Some fretting-wear also occurred on the outer tubes of the top hairpin. Severely damaged tubes were plugged.

Failures such as outlet tube DPBC-1 of the top hairpin (Chart 4), inlet tube DPBC-2 of the bottom hairpin and tube DPBC-3 (Chart 3) could not be explained by fluidelastic instability. Another mechanism was responsible.

We found that the piping system leading to the heat exchanger was extremely noisy. The noise was caused partly by flashing due to sudden expansion of the high pressure primary side water through bleed valves and partly by the subsequent collapse in the mixer with the tempering flow. The hydraulic noise induced severe vibrations of the system piping. These were structurally transmitted to the heat exchangers contributing to the fretting-wear damage. Piping vibration measurements were taken under various operating conditions. The bleed system is now being operated under the optimum conditions for minimum structural vibration. The whole system including the heat exchangers is being completely redesigned.

3.4 Case PASC: Liquid-Liquid Cooler Heat Exchanger

This cooler is a short "U" tube heat exchanger with water flow on the shell side and heavy water inside the tubes (Chart 4). It had operated satisfactorily for several years in a nuclear station until it was analysed for flow-induced vibration as required in the specifications for reordering additional units for a new nuclear station. The analyses PASC-1 in Table 1 and PASC-2 in Chart 4 indicate fluidelastic instability at 100% flow. An investigation at the station revealed that these coolers were never operated beyond 40% flow. In fact, at much higher flows, severe vibrations of the tubes were detected. The coolers are now limited to 60% flow which corresponds roughly to our design guidelines for fluidelastic instability.

3.5 Case PAMX: Fatigue of Liquid-Liquid Heat Exchanger

This case has been discussed previously [2]. The component in question is a "U" tube cross-flow heat exchanger with segmental baffles. The tubes are arranged in a square lattice of p/d=1.5. Cooling water flows on the shell side and heavy water in the tubes. Severe vibrations were observed during commissioning. Fatigue and tube-to-tube fretting resulted in a tube near the outlet. This tube was located beyond the baffle cut and hence only supported every second baffle.

Vibration analysis PAMX-1 (Chart 4) indicated fluidelastic instability. Lacing rods were installed to support the tubes beyond the baffle cut. Vibration analysis PAMX-2 (Table 1) shows this modification to be adequate to eliminate fluidelastic instability.

3.6 Case GBCX: Fatigue of Process Cooler

This component is a liquid-liquid triple segmental heat exchanger operating in a heavy water plant with cooling water on the shell side. The inlet nozzle is a wide angle cone which probably does not expand the flow uniformly. High flow velocities exist at the inlet resulting in direct impingement on the tube bundle. Tube failures by fatigue near the tubesheet have occurred several times since commissioning of the plant. Tube-to-tube fretting was also observed at midspan in the inlet region. However, no tube-to-support fretting was found. This is not surprising since the tubes were essentially jammed at the tube supports by corrosion products.

Vibration analysis GBCX-1 (Chart 4) indicates fluidelastic instability i.e., $V_r/V_{rc}=1.71>1$. The inlet nozzle was modified to distribute the flow more uniformly pending complete redesign of the tube bundle.

3.7 Case PHRC: Fretting of Process Heat Exchanger

This case concerns liquid-liquid cross-flow heat exchangers operating in a heavy water plant. They are coupled in sets of three installed one on top of the other. The tube bundles are designed to be interchangeable although inlets and outlets are not in the same location in the three exchangers. The tubes are arranged in a square lattice of p/d=1.33. Occasionally the heat exchangers were operated at 130% design flow.

Tube failures occurred due to fretting-wear at the baffle-supports. Vibration analyses indicate fluidelastic instability at the inlet (PHRC-1) at 130% flow. The tube bundles were redesigned with additional baffle-supports at the inlet. The inlet was modified to reduce flow velocities.

3.8 Case DPSG: Fretting-Wear in Steam Generator

The steam generators of the Douglas Point nuclear station consist of 80 vertical "U" tube hairpin type units grouped in 8 banks of 10 hairpin units each. The feedwater enters the preheater leg and the water being recirculated from the steam drum enters the boiling leg at the bottom. The steam-water outlet is at the top of the hairpin. The "U" bend tubes near the outlet are subjected to high velocity two-phase cross-flow as shown on Figure 1. In the 10 hairpins of one bank the "U" bend tubes were not supported at the top and vibrated with sufficient amplitude to cause tube-to-tube and tube-to-support fretting. A first tube failure occurred in 1971 and a second in 1977 ten years after the initial start-up of the station. The hairpins containing the failed tubes were removed.

Vibration analysis DPSG-1 indicates the likelihood of fluidelastic instability. The range of damping values assumed explains the range in V_r/V_{rc} in Figure 1. In the 70 other hairpins the "U" bend tubes were supported at the top and no tube failure occurred as expected from vibration analysis DPSG-2 in Table 1.

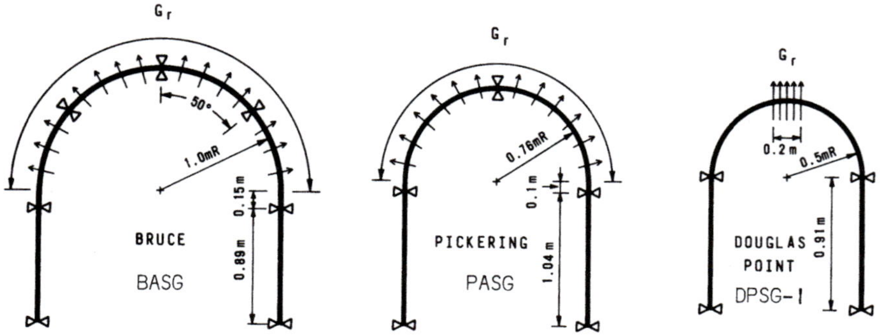

	G_r (kg $m^{-2}s^{-1}$)	X (%)	d (mm)	f_1 (H_z)	c (kg $m^{-1}s^{-1}$)	$\frac{m\delta}{\rho d^2}$	$\frac{V_r}{V_r CRIT}$
DOUGLAS POINT (+ top support)	853	21.2	12.7	11.4	5.4-17.9 17.9	16.3-54.2	2.26-1.24 0.11
PICKERING	446	12.5	12.7	17.0	8.0-17.9	10.5-23.3	1.3-0.87
BRUCE	328	20	13.0	35.5	18.2	15.9	0.49

FIGURE 1: Vibration Analysis of Steam Generator "U" Bend Tube Region (Fluid-elastic Instability in Two-Phase Cross-Flow).

3.9 Case PASG & BASG: Two-Phase Cross-Flow in Steam Generators

Current steam generators such as those for Pickering or Bruce nuclear power stations are much larger than the steam generators described above. The tubes are subjected to two-phase cross-flow over the entire "U" bend region. Vibration analysis PASG on Figure 1 shows that, in Pickering, the steam generator "U" tubes are near the recommended instability criterion. No problems have been experienced in seven years of operation. No problems have occurred in Bruce either over a few years of operation as expected from vibration analysis PASG.

4. SUMMARY OF RESULTS AND DISCUSSION

The operating experiences discussed above have been analysed in terms of fluidelastic instability and are summarized in Table 1. It is convenient to present the instability results in terms of the dimensionless reference gap velocity V_r/fd and the dimensionless damping $m\delta/\rho d^2$ as shown on Figure 2. All the tube failures occurred for values of fluidelastic instability constant K larger than 4.4 which is the equivalent of an actual velocity/critical velocity ratio V_r/V_{rc}=1.33. Thus, our recommended guidelines of K=3.3 appears very realistic in practice as it gives a reasonable safety margin. The credibility of this criterion is further enhanced by the wide variety of components considered and by the range of operating conditions encountered (i.e., liquid and two-phase flow). We expect that some components operate satisfactorily at fluidelastic instability constant K larger than 3.3. The important point is that no failures occurred below 3.3.

Some of the tube failures could also be explained in terms of vibration excitation mechanisms other than fluidelastic instability. In most cases it was not possible to identify positively the excitation mechanism from the field data. Although in all failure cases severe vibrations were observed. In one case these were structurally transmitted from the piping system.

The formulation for fluidelastic instability relates the fluid forces to the flexural rigidity of the structure, i.e.: multispan tubes. There is a similarity between the latter and the formulation of vibration response to other excitation mechanisms [1]. Hence the above instability criterion may well take care of all flow-induced vibration excitation mechanisms in many cases. However the designer should not make this assumption until it is completely validated and until the range of conditions and components over which it may apply is well defined.

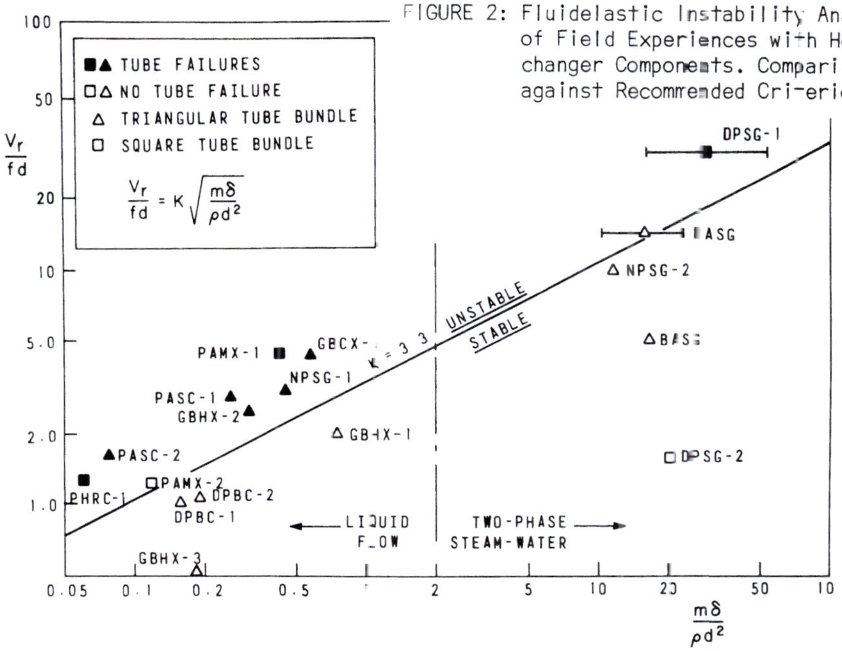

FIGURE 2: Fluidelastic Instability Analysis of Field Experiences with Heat Exchanger Components. Comparison against Recommended Criterion K=3.3

5. CONCLUDING REMARKS

The operating performance of ten heat exchanger components was reviewed from a fluidelastic instability point of view. Both failures and satisfactory performances were considered. The results show that a fluidelastic instability constant K=3.3 is a reasonable design criterion.

ACKNOWLEDGEMENT: The authors are grateful to the many staff members of Canadian nuclear stations and heavy water plants who have contributed valuable information on field experiences. The assistance of G.A. Wolgemuth, J.A. Aikin, J.L. Platten and P.L. Ko of the Chalk River Nuclear Laboratories and R.T. Hartlen and W. Jaster of Ontario Hydro Research with some of the measurements and analyses is gratefully acknowledged.

6. REFERENCES

1. M.J. Pettigrew, Y. Sylvestre and A.O. Campagna, "Vibration Analysis of Heat Exchanger and Steam Generator Designs", Nucl. Eng. Des. 48 (1978) 97-115.

2. R.T. Hartlen, "Recent Field Experience with Flow-induced Vibration of Heat Exchanger Tubes", Paper No. 611, Int. Symp. on Vibration, Keswick, U.K., 1973.

3. M.J. Pettigrew and D.J. Gorman, "Vibration of Heat Exchange Components in Liquid and Two-Phase Cross-Flow", Paper 2.3, Int. Conf. Vibration in Nuclear Plant, Keswick, U.K. 1978, also Atomic Energy of Canada Ltd., report AECL-6184.

4. D.J. Gorman, "Experimental Development of Design Criteria to Limit Liquid Cross-Flow Induced Vibration in Nuclear Reactor Heat Exchange Equipment", Nucl. Sci. Eng. 61, 324-336, (1977).

5. M.J. Pettigrew and D.J. Gorman, "Vibration of Small Tube Bundles in Liquid and Two-Phase Cross-Flow", Atomic Energy of Canada Ltd., report AECL-5804, (1977).

6. W.W.R. Inch and R.H. Shill, "Thermal-Hydraulic Analysis of Nuclear Steam Generators", Proc. 7th Canadian Congress of Applied Mechanics, Sherbrooke, Québec, Canada, May 1979, pp 667-668.

Chart I: CASE NPSG: FRETTING-WEAR OF NUCLEAR STEAM GENERATOR

TUBE FAILURE 1969 ↑

LOCATION OF TUBE FRETTING ↓

Problem: Excessive tube vibration caused fretting-wear and tube leakage at baffle - support near downcomer inlet in horizontal steam generator of prototype nuclear station. First failure occurred in 1969, second in 1978 after 16 years operation. Inspection revealed several other damaged tubes near the inlet.

Geometry: Multi-span "U" tubes of diameter d=12.7 mm assembled in triangular array of pitch p=19 mm. Tube ends clamped to tubesheet.

Flow Conditions: Natural circulation steam generator. Water flow velocity in downcomer roughly 3 m/s. Reference gap velocity V_r between tubes also taken as 3 m/s, allowing for two-thirds of flow in space between shell and tube bundle.

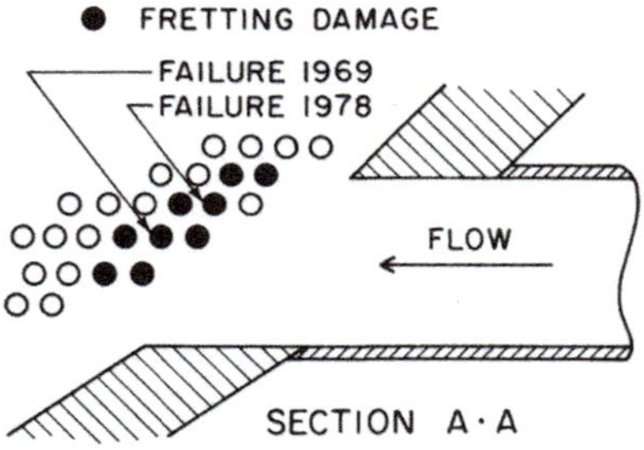

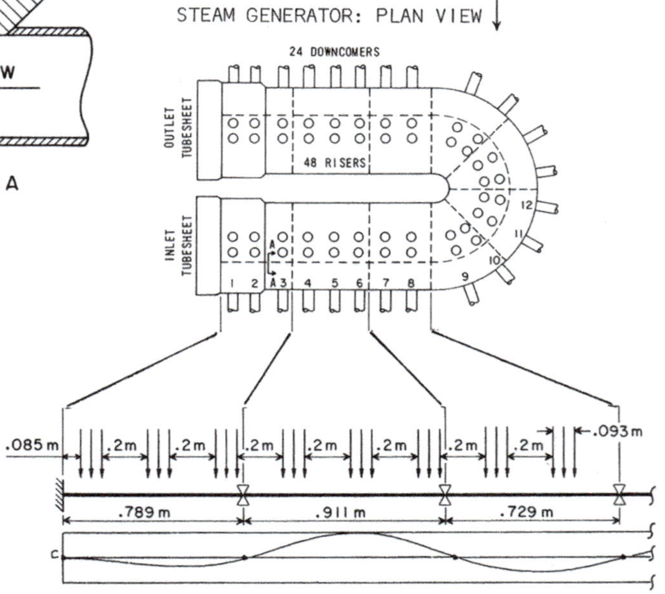

FLUIDELASTIC INSTABILITY ANALYSIS

Mode No.	f (Hz)	V_r / V_r CRIT.
1	42.5	1.36

Structural Dynamics: Mass per unit length m/ℓ = 0.55 kg/m; viscous damping coefficient c = 5.1 $kg \cdot s^{-1} \cdot m^{-1}$; fluid density ρ = 839 kg/m^3; flexural rigidity EI = 159. $N \cdot m^2$.

Vibration Analysis: Calculations indicate fluidelastic instability in fundamental mode at 42.5 Hz. Ratio of actual velocity/instability velocity is 1.36.

Solution: Plug tubes near inlet area.

Comment: Note that 1st failure occurred after 7 and 2nd failure after 16 years operation. It appears in this case that fretting-wear is a slow and continuous process.

Chart 2: CASE GBHX: FRETTING-WEAR OF PROCESS HEAT EXCHANGER

Problem: Fretting-wear of tubes in tube-in-shell heat exchanger. Excessive vibrations of tube measured near inlet region during commissioning of exchanger. This resulted in tube-to-tube fretting at midspan (GBHX-1&2).

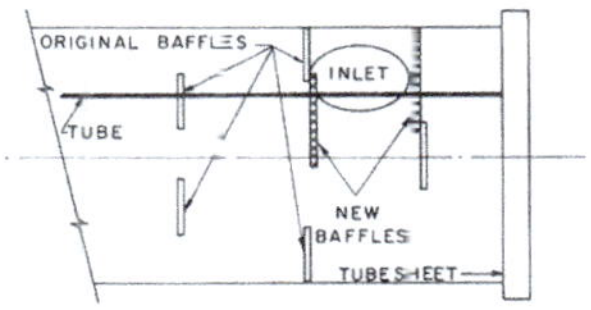

Geometry: Multi-span tube bundle clamped at both ends in tubesheet, outside tube diameter d=19 mm, triangular pitch p=23.8 mm.

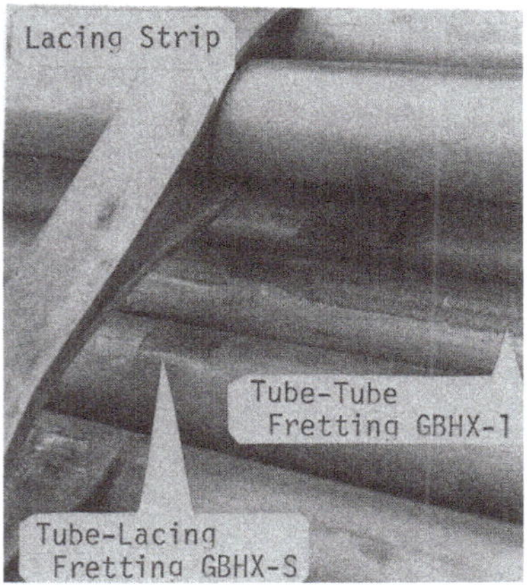

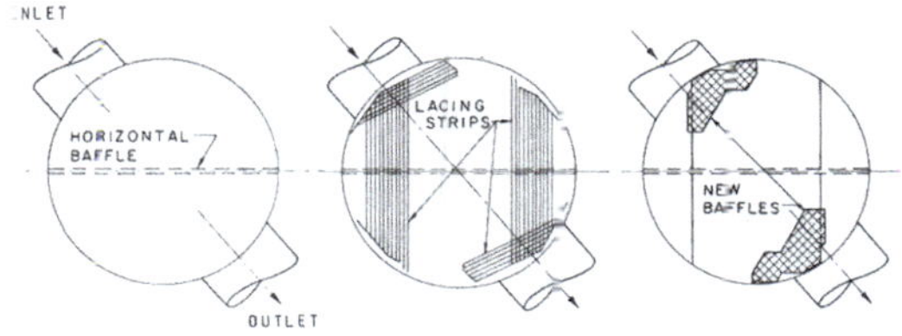

GBHX-1 GBHX-S GBHX-3

Flow Conditions: Water at 22.75 °C, non-uniform cross-flow, high flow velocity near the inlet nozzle (Sketch below). Water in tubes.

Structural Dynamics: Mass per unit length $m/\ell=1.05$ kg/m; measured viscous damping coefficient $c=9.6$ $kg \cdot s^{-1} \cdot m^{-1}$; fluid density $\rho=1000$ kg/m^3; flexural rigidity $EI=535$ $N \cdot m^2$.

Vibration Analysis: Analysis revealed fluid-elastic instability in the 8th mode of vibration at 42.7 Hz. Ratio of flow velocity/critical velocity for fluidelastic instability is 1.37. Note high amplitude in 8th mode corresponding to high inlet flow velocity

Solution: Initially lacing strips were installed to support the tubes near the inlet (GBHX-S). However the problem persisted due to the combination of excessively loose strips and partial blockage of the inlet which resulted in unexpectedly high velocity (See fretting-wear at the location of strips in Photo). Eventually, proper baffle-supports were installed (GBHX-3). No further problem occurred as expected from the vibration analysis (V_r/V_r CRIT<1).

FLUIDELASTIC INSTABILITY ANALYSIS

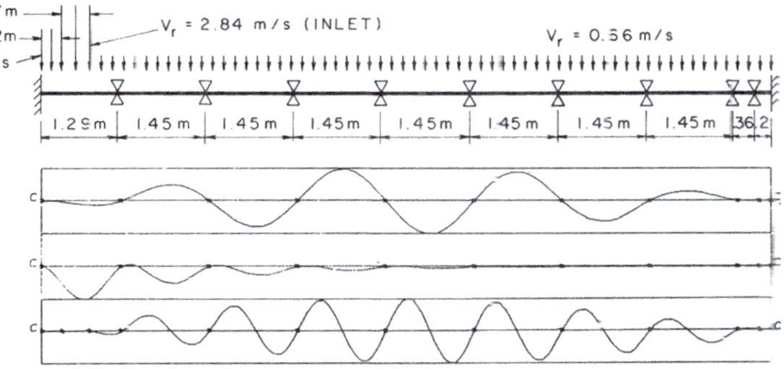

Case	Mode No	f (Hz)	$V_r / V_r CRIT$
GBHX 1	1	17.7	0.70
GBHX 2	8	42.7	1.37
GBHX 3	8	69.2	0.35

Chart 3: CASE DPBC: FRETTING-WEAR DAMAGE IN COOLER HEAT EXCHANGER

Problem: Fretting-wear and tube failures in many areas of hairpin type heat exchanger. Failures occurred several times during ten years operation. Inspection revealed many damaged tubes.

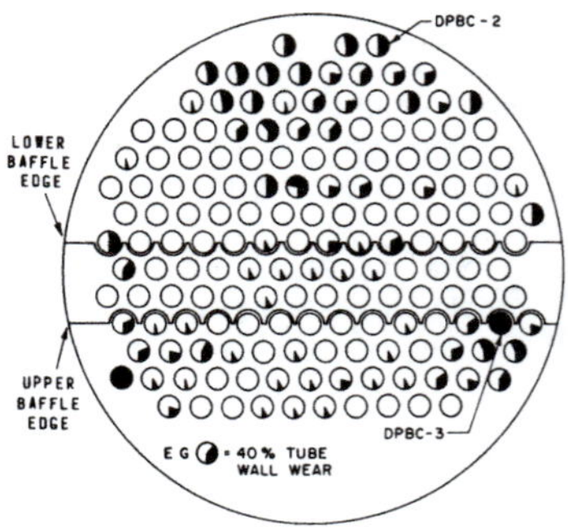

Geometry: Hairpin type tube-in-shell heat exchanger. Multispan "U" tubes clamped at tubesheets, diameter d=12.7 mm, triangular pitch p=19.1 mm.

Flow Conditions: Shell side water at ∿16°C, mostly cross-flow, inlet and outlet were reversed after a few years of operation as part of system modification. Tube side: heavy water at 121°C.

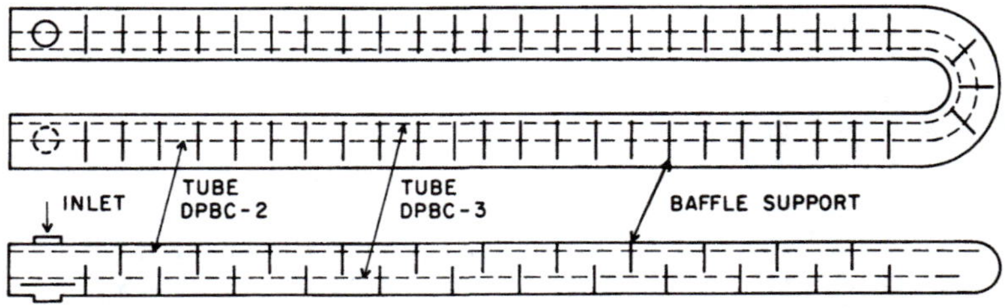

LOCATION OF TUBE FRETTING ↑

FLUIDELASTIC INSTABILITY ANALYSIS ↓

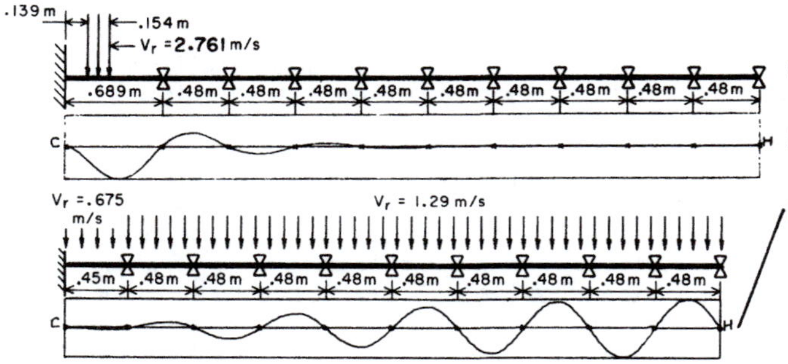

Case	Mode No.	f (Hz)	$\frac{V_r}{V_r\ \text{CRIT.}}$
DPBC 2	1st	87	0.75
DPBC 3	1st	104	0.76

HYDRAULIC SYSTEM ↓

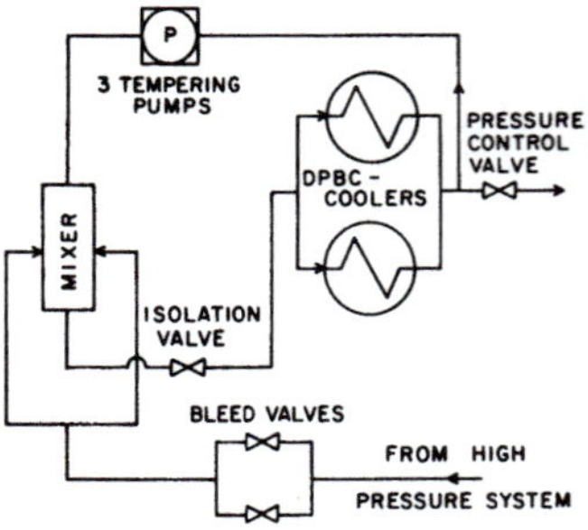

Structural Dynamics: Mass per unit length m/ℓ=0.606 kg/m, viscous damping coefficient, c=5.07 kg·m^{-1}·s^{-1}; shell side fluid density ρ=998 kg/m^3; flexural rigidity EI=133 N·m^2.

Vibration Analysis: Calculations of vibration response due to shell side flow show that fluidelastic instability could not explain failure of tube DPBC-2 or of tube DPBC-3. The ratio of actual velocity/instability velocity was 0.75 for the former and 0.76 for the latter.

Structural Vibration: Severe hydraulic noise in system leading to heat exchanger. Noise produced by flashing in high to low pressure bleed valve and subsequent collapse in mixer with tempering flow. Hydraulic noise induced vibration of system piping. Vibration structurally transmitted to heat exchanger contributing to fretting-wear damage.

Solution: Plug damaged tubes and optimize operation of bleed system for minimum hydraulic noise. Long term solution: complete redesign of system.

Chart 4: <u>SUMMARY OF FIVE LIQUID-LIQUID CROSS-FLOW HEAT EXCHANGER FAILURES</u>

<u>Case DPBC-1</u>: PROBLEM: Outer tube fretting at baffles; SOLUTION: Plug tubes for now and redesign system; FLUIDELASTIC INSTABILITY: No; $V_r/V_{rc}=0.77<1$; 1st mode, 103 Hz.

d = 12.7 mm
p/d = 1.5
m/ℓ = 0.61 kg/m
EI = 133 N·m^2
ρ = 971 kg/m^3
c = 5.07 kg·s^{-1}·m^{-1}

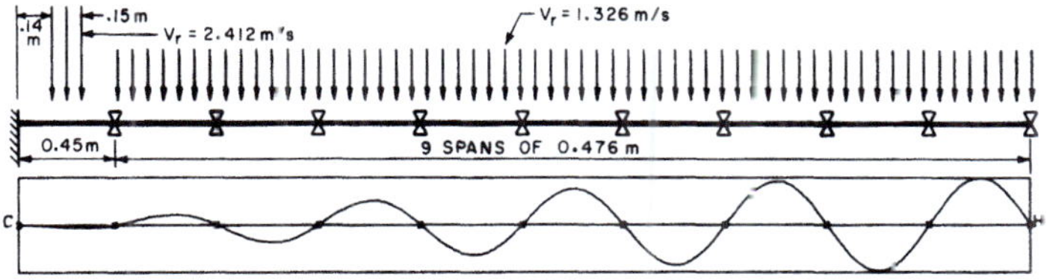

<u>Case PASC-2</u>: PROBLEM: Excessive vibration at 100% flow; SOLUTION: Operate below 60% flow; FLUIDELASTIC INSTABILITY: Yes; $V_r/V_{rc}=1.74>1$; 3rd mode, 209 Hz.

d = 12.6 mm
p/d = 1.26
m/ℓ = 0.61 kg/m
EI = 130 N·m^2
ρ = 998 kg/m^3
c = 5.04 kg·s^{-1}·m^{-1}

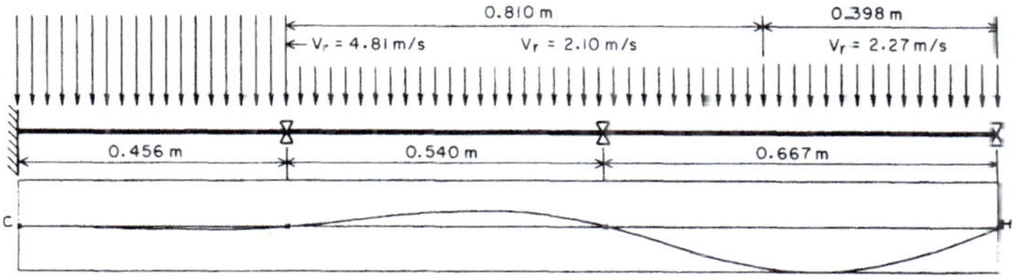

<u>Case PAMX-1</u>: PROBLEM: Tube fatigue & fretting near outlet; SOLUTION: Install lacing rod supports; FLUIDELASTIC INSTABILITY: Yes; $V_r/V_{rc}=2.04>1$; 1st mode, 37.1 Hz.

d = 12.7 mm
p/d = 1.5
m/ℓ = 0.67 kg/m
EI = 127 N·m^2
ρ = 997 kg/m^3
c = 5.07 kg·s^{-1}·m^{-1}

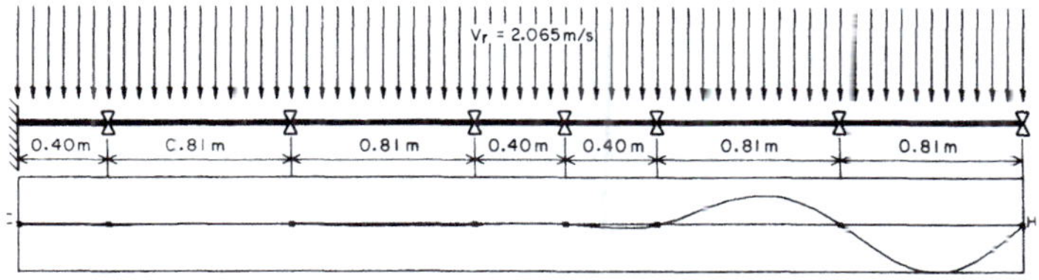

<u>Case GBCX-1</u>: PROBLEM: Tube fatigue at tubesheet near inlet, fretting; SOLUTION: Modify inlet now, redesign tube bundle; FLUIDELASTIC INSTABILITY: Yes; $V_r/V_{rc}=1.7$… 1st mode, 18.2 Hz.

d = 19.1 mm
p/d = 1.25
m/ℓ = 1.20 kg/m
EI = 665 N·m^2
ρ = 996 kg/m^3
c = 7.63 kg·s^{-1}·m^{-1}

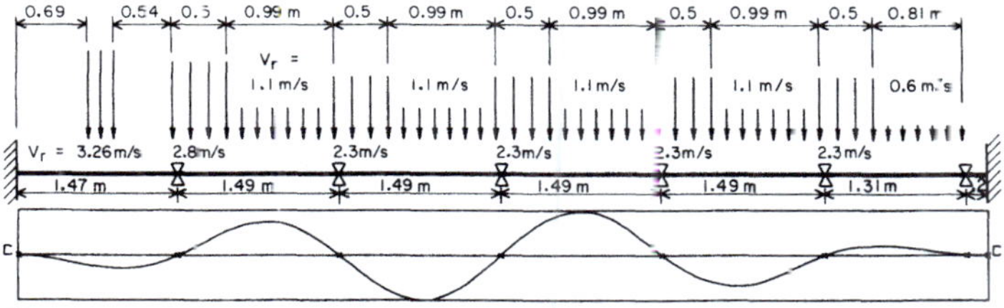

<u>Case PHRC-1</u>: PROBLEM: Tube fretting at supports; SOLUTION: Add supports, modify inlet; FLUIDELASTIC INSTABILITY: at inlet; $V_r/V_{rc}=1.37>1$; 15th mode, 177 Hz.

d = 19.1 mm
p/d = 1.33
m/ℓ = 1.25 kg/m
EI = 716 N·m^2
ρ = 993 kg/m^3
c = 7.63 kg·s^{-1}·m^{-1}

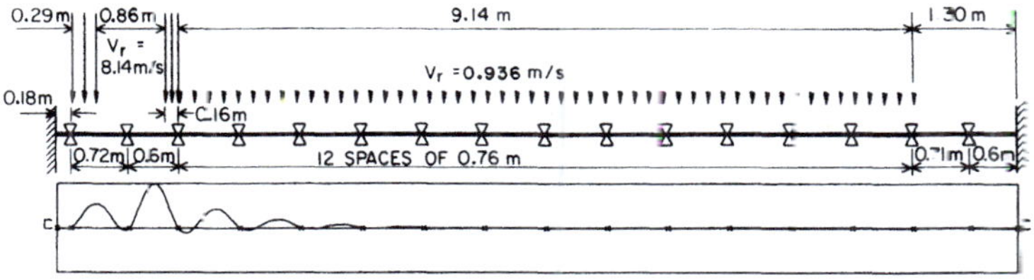

A3

TUBE VIBRATIONS IN LARGE CONDENSERS

Daniel Bai

Research Engineer

Electricite de France, Chatou, France

SUMMARY

This paper deals with an experimental study of tube vibrations in large condensers. The model is partial full scale, composed of about a hundred thin-walled titanium tubes, each having a length of 10.9 meter. A test rig has been built to recreate the operating conditions of a condenser characterized by low pressure steam flow at high velocities and condensation on the tubes. Fluid-elastic instability was observed, and the efficiency of antivibration strips was tested.

1. INTRODUCTION

With the introduction of larger power plants, both the size of components and system flow rates have increased. As a result, failures have begun to occur on the tubes of condensers, resulting in tube wear and, in some cases, fracture of tubes. Depending on the tube nest arrangement, tube failures can be located in small areas in the condenser, or be more widespread. The large amplitude vibrations occur generally, but not necessarily, when turbines are heavily loaded and/or when circulating water inlet temperature is very cold. Current design rules, based on vibration tests to avoid mechanical excitation by the turbine or by the alternator, do not take into account flow induced-vibration risks. This paper deals with the study of flow-induced vibrations on a model, with running conditions as close as possible to the actual system.

2. TEST RIG AND MODEL

In order to recreate the running conditions of a real condenser (high velocity, wet steam at low pressure) a test rig has been built (see Figures 1 and 2). This rig is a partially closed loop, with part of the steam being recirculated by four ejectors. The maximum steam flow of this rig is 11.7 kg/s. The rig has been specially designed by use of screens to have a uniform steam velocity profile over the entire length of the tubes for both dry and wet steam. The model is composed of about a hundred 10.9 meter long tubes. Water circulates in all the tubes, even in the instrumented ones, which allows simulation of condensation phenomena. The thickness of the tubes is 0.5 mm. Two types of bundles, as shown in Sketches 1 and 2, have been tested. For the type 1 bundle, the maximum gap velocity between tubes is 280 m/s at 50 mbar pressure, whereas for the type 2 bundle, the maximum velocity in the gap between leading tubes and the moveable panel is 270 m/s at a pressure of 70 mbar.

3. INSTRUMENTATION - MEASUREMENTS

Steam velocity and flow rate were determined for temperature and static pressure measurement. Visualization was carried out on a hydraulic model (scale 1/4) of the steam model to verify that the flow field upstream of the tube bundle remains steady.

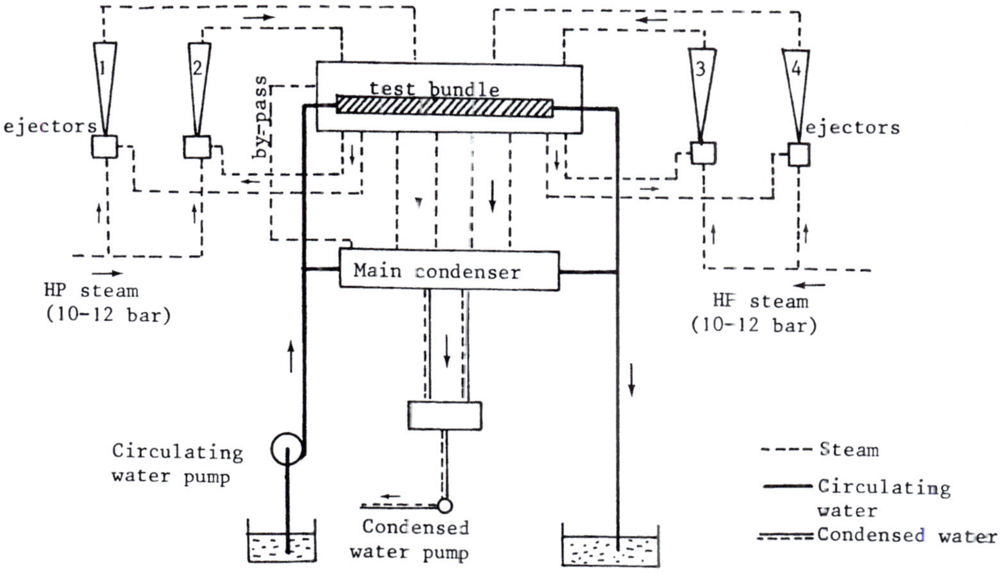

FIG. 1 SCHEMATIC DIAGRAM OF THE TEST RIG

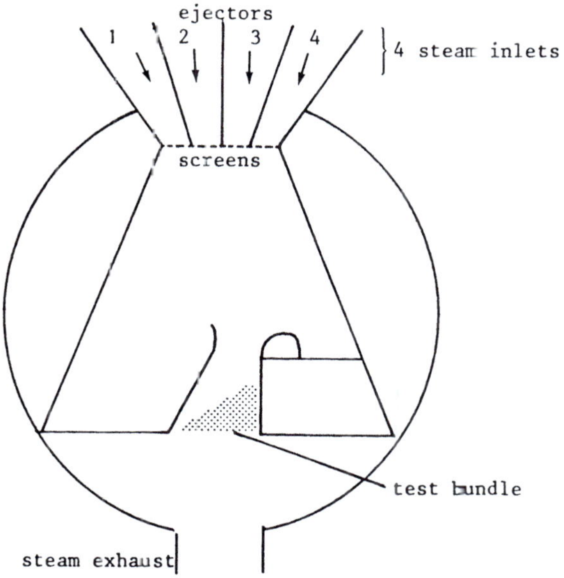

FIG. 2 SCHEMATIC SECTION OF THE RIG SHOWING THE POSITION OF THE TEST BUNDLE

The amplitude of tube vibration was found from accelerometers located inside the tubes (Columbia biaxial accelerometers HEVP 12) (Figure 3). For each type of bundle, the amplitude of tube vibration was measured as a function of steam flow rate. The influence of the dynamic head parameter (ρV^2) as well as the humidity ratio and variations of the near flow field were studied. The efficiency of anti-vibration strips set up the middle of all the larger spans of tubes was evaluated.

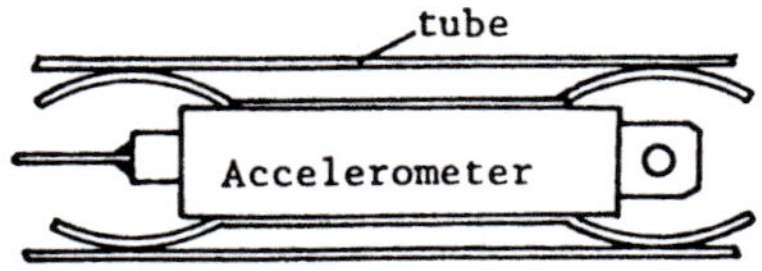

FIG. 3 MOUNTING OF COLUMBIA HEVP 12 ACCELEROMETER

4. RESULTS

As far as the first type of bundle (shown in Sketch 1) is concerned, large vibration amplitudes of the tubes without strips were observed. There appears to be a critical flow below which vibration amplitude remains very low and above which amplitude increases abruptly (Figure 4), involving whirling motion of the tubes. The "critical" ρV^2 value was realized with several different values of ρ and V.

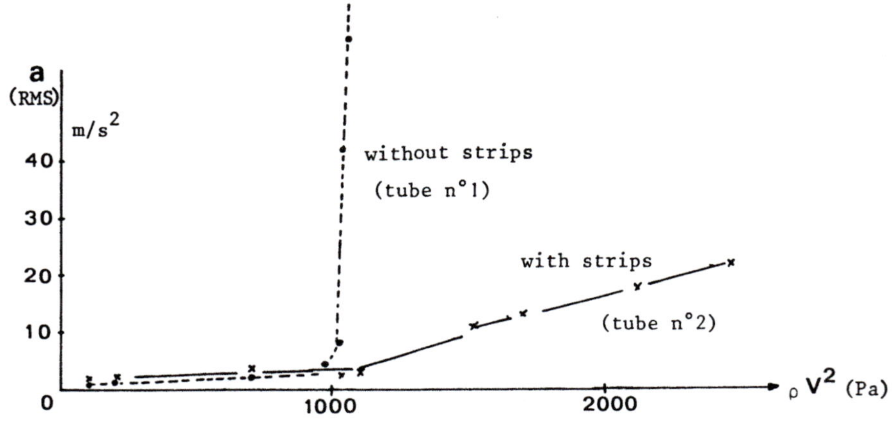

FIG. 4 VIBRATIONAL AMPLITUDE FOR TUBES WITH AND WITHOUT ANTIVIBRATION STRIPS TYPE 1 BUNDLE. (V = gap velocity between tubes)

For the tubes without strips, the critical ρV^2 value is equal to 1050 Pa (V = gap velocity between tubes). In the fluid-elastic formula $\frac{V}{fd} = K\sqrt{\frac{m\delta}{\rho d^2}}$, this leads to a $K\sqrt{\delta}$ value of 1.6. As it is difficult to accurately measure the damping (δ) of multi-spanned tubes, it is more appropriate to evaluate $K\sqrt{\delta}$ instead of K. Time dependent traces and spectra are different below and at the stability (Figure 5). Below instability, a range of frequencies is represented on the spectra. At the instability, most of the energy is concentrated at the lower frequency of the tubes.

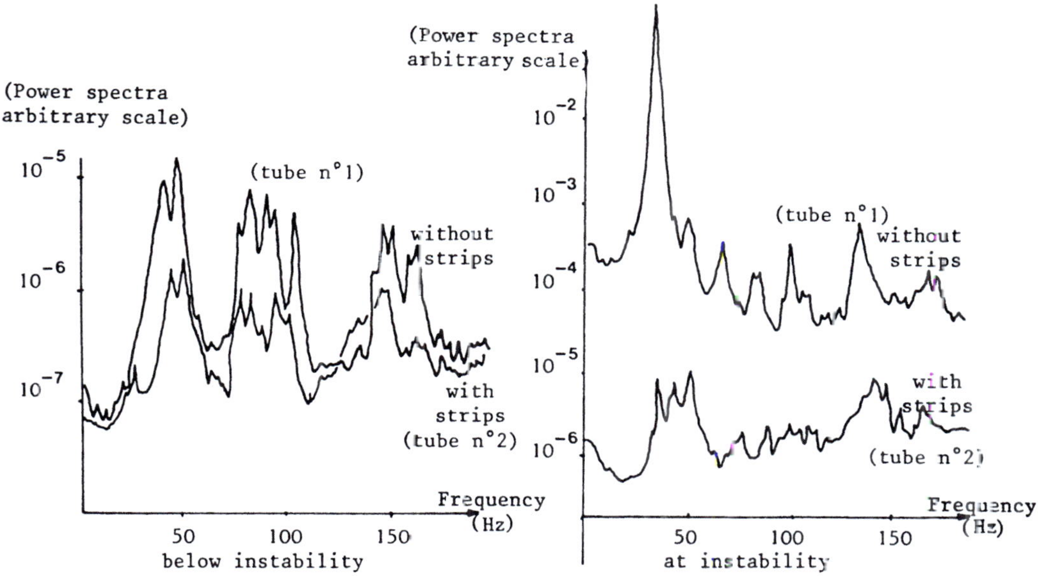

FIG. 5 ACCELERATION SPECTRA BELOW AND AT THE INSTABILITY, FOR TUBES WITH AND WITHOUT ANTI-VIBRATION STRIPS - (TYPE 1 BUNDLE)

For tubes with anti-vibration strips, the amplitude of vibration remains very low, even for high values of flow rate (Figure 4). In fact for the maximum steam flow rate allowable by the rig, which corresponds to $\rho V^2 = 2500$ Pa, no significant vibration of tubes with strips was observed, and spectral analysis shows no change of shape of the spectra as flow rate varies. No significant effect of the humidity ratio on the critical value ρV^2 was pointed out, and no excitation by vortex shedding was observed.

The second type of bundle (see Sketch 2) represents quite well half the top part of a tube nest where the near flow field is rather complex. It displayed the same instability phenomenon as for the first type of bundle. There exists a critical steam flow above which the amplitude of vibration increases abruptly with whirling motion of the tubes. As for the first type of bundle, the critical ρV^2 value was realized with different values of ρ and V.

5. CONCLUSIONS

These tests have pointed out the possibility for vibration of tubes in large condensers due to fluid-elastic instabilities. This instability can be avoided by setting anti-vibration strips at the middle of all the larger spans or by designing the tube nests so that no local velocity leads to a value of the local ρV^2 parameter greater than the critical one.

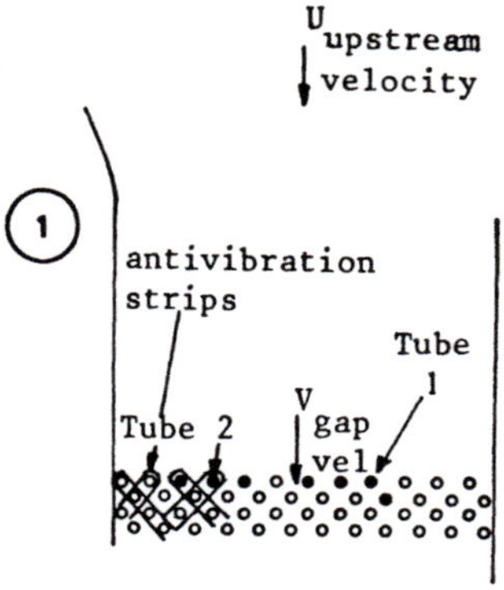

TUBE VIBRATION IN LARGE CONDENSERS

Problem : Vibration of tubes detected in large condensers. Tube wear and breakage.

Geometry : Study of the problem on scale model, with two arrangements (sketches 1,2) : length of tubes = 10.9 meter, O.D. = 19 mm; I.D. = 18 mm. Tubes made of titanium. Triangular arrangement. Pitch to diameter ratio = 1.37. Tubes expanded at each end (Dudgeon's system). 11 support plates. Tube hole clearance = 0.4 mm. Arrangement of support plates shown in sketch 3.

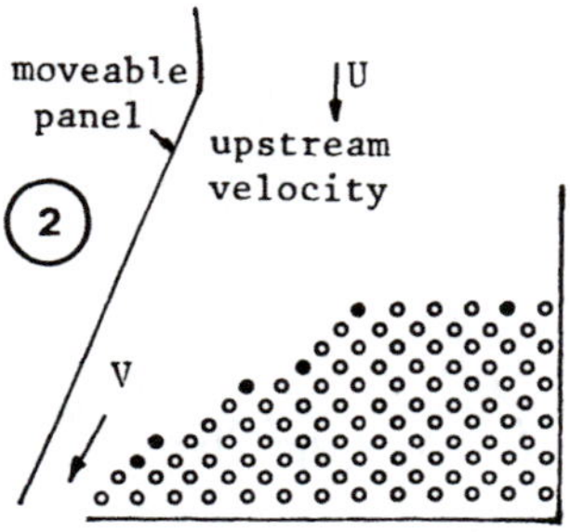

Flow conditions : Circulating water inside tubes and low pressure steam flow outside. Humidity of the steam adjusted by spraying water upstream of test bundle. Steam flow quite uniform over the whole length and width of the bundle, as verified by specific tests on a hydraulic model.

Structural dynamics : Frequencies are computed and validated by spectral analysis of accelerometer signals. The lower mode is the most important but all higher modes appear clearly on the spectra.

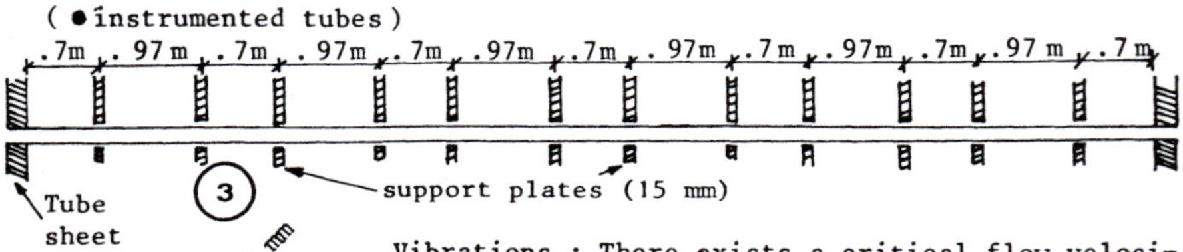

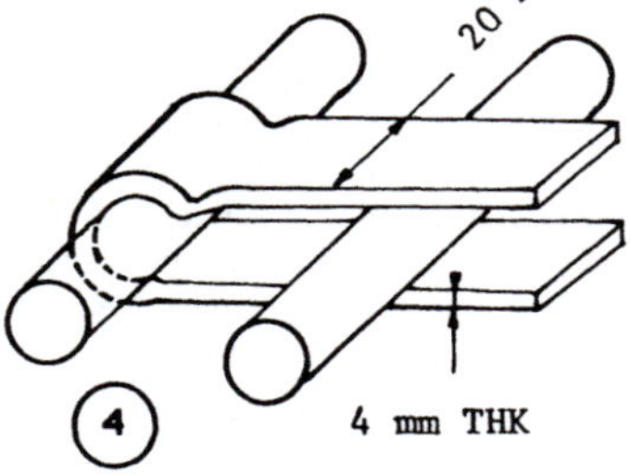

Vibrations : There exists a critical flow velocity for which amplitude increases abruptly : most of the energy is then concentrated at the lower resonant frequency of the tubes. The critical value of the corresponding ρV^2 parameter is about 1050 Pa ($V/f\,D \simeq 330$).

In the instability formula $V/fd = K \sqrt{\frac{m\,\delta}{\rho d^2}}$, it leads to value of $K\sqrt{\delta} = 1.6$ ·
No vortex shedding excitation was observed.

Cure : Anti-vibration strips (Sketch 4) made of nylon. Set up in the middle of all the larger spans. Strips clipped in two directions (sketch 5) Amplitude level decreased at least by a factor 20. No instability threshold detected even for high values of the ρV^2 parameter ($\rho V^2 \leq 2500$ Pa).

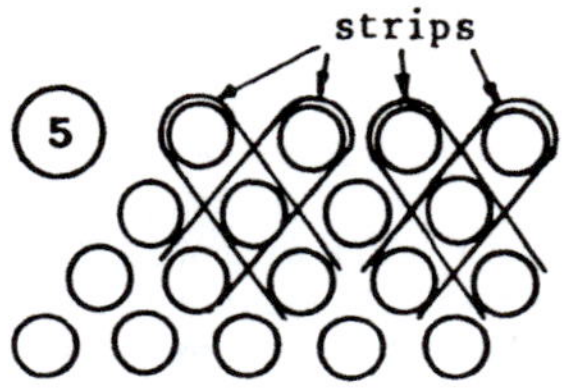

Reference : Bai,D. Unpublished work

A4

FLOW-INDUCED VIBRATION OF REACTOR SYSTEM COMPONENTS: HEAT EXCHANGER TUBES, FLOW COLLECTOR, AND THERMOWELL

Henry Halle
Principal Investigator

Thomas M. Mulcahy
Principal Investigator

Martin W. Wambsganss
Section Manager

Components Technology Division
Argonne National Laboratory
Argonne, Illinois 60439
USA

SUMMARY

Argonne National Laboratory (ANL) provides support to reactor designers in the design evaluation of specific Liquid Metal Fast Breeder Reactor (LMFBR) components from the standpoint of flow-induced vibrations (FIV). The performance of such evaluations involves both analyses and tests. In the course of testing, the potential for large amplitude flow induced vibration or instability was identified for three components: heat exchanger tubes in a sodium-to-air dump heat exchanger, a flow collector assembly in the upper plenum of a reactor vessel, and a thermowell. For each of the three components, this paper reports the results of the FIV design evaluation tests, assessment of the excitation mechanism, and suggested design fixes or methods for avoiding detrimental vibration or instability.

1. INTRODUCTION

The work scope of the flow-induced vibration program at Argonne National Laboratory includes the performance of analyses and tests to evaluate the design adequacy of specific breeder reactor system components. While state-of-the-art analyses have proven useful in providing guidance, it has often been found necessary to rely on testing for final design verification. Both scale-model and prototype testing are performed. In many cases, a sector out of a circular cross-sectioned unit, or a reduced number of components out of the full complement, is selected for testing. In general, component response is measured as a function of a flow parameter. The primary objective of the testing is to identify conditions which give rise to large amplitude motion; for example, a fluid-elastic instability or a resonance between a structural natural frequency and characteristic frequency associated with the flow. Once such a condition is identified, the next step is to attempt to develop an understanding and, if possible, a mathematical characterization of the excitation mechanism. As necessary, design fixes are developed and verified on a test rig.

Argonne is providing support in the area of component design evaluation for two LMFBR projects: the FFTF (Fast Flux Test Facility) and CRBR (Clinch River Breeder Reactor). The FFTF features a 400 MWt fast test reactor (FTR) which is scheduled to go critical in 1979. The primary purpose of the FTR is to provide a high-intensity, fast-neutron flux for irradiation testing of fuels and materials to be used in future fast breeder reactors. The CRBR is a 350 MWe LMFBR demonstration plant. The major objective of the CRBR project is to obtain data on the design, component fabrication, construction, and operation of a licensed LMFBR on a utility grid.

Three components that were tested, and found to exhibit a susceptibility to flow-induced vibration are: the U-tubes in the FFTF heat transport system dump heat exchanger; the CRBR flow collector which is in the reactor upper plenum; and an LMFBR thermowell assembly. The component design, test description, test results, assessment of excitation mechanism, and recommendations regarding avoidance of detrimental vibration are reported herein.

2. HEAT EXCHANGER TUBES

Sodium-to-air dump heat exchangers are employed to dissipate, to the atmosphere, the heat created by the Fast-Flux Test Facility. There are twelve heat exchanger modules, each with a design heat capacity of 33 thermal MW. The sodium inlet and outlet headers are connected by 66 rows of serpentine heat exchanger tubes in parallel. Each serpentine consists of a single run of tubing arranged in four parallel legs, each 10 m long, connected by three U-bend turns (Sketch 1 of Heat Exchanger Tubes Chart). Ambient air is forced over the tube bank [1].

To reduce tube stress levels caused by thermal expansion, the unit was designed with the tubes supported on only four tube support assemblies; this arrangement resulted in relatively long overhanging U-bend sections. Because of the flexibility (low natural frequency) of the U-bend overhangs, the possibility of flow-induced vibrations being excited by the external air flow passing over the tubes and the potential for damage were present. An investigation was deemed appropriate and subsequently performed [1].

The tests simulated, with five rows of single U-turn (double run) tubes (Sketch 2 of Heat Exchanger Tubes Chart), the 66 rows of three U-turn (four run) tubes of the DHX prototype design. Testing was performed with prototypic tubing and four different U-bend overhang lengths varying from 1.84 m (intermediate bend) to 2.27 m (large overhang). Tube to tube support hole diametral clearance was 2.8 mm. Prior to flow testing of any support configuration, the natural frequency and damping were determined in still air. Upon flow testing, each configuration was exposed to about 7 or 8 different flow velocities. The vibration amplitudes and frequencies of the ends of the three central U-tubes were monitored by means of an electro optical tracker through a transparent duct wall. The signals were tape recorded and processed on a spectrum analyzer.

The test results indicated the existence of a critical flow velocity above which large amplitude vibrations take place. Below the critical flow velocity the amplitudes were small, and considered acceptable. Typical test data are given in Sketch 3 (Heat Exchanger Tubes Chart) which, in this case, pertains to a 2.26 m long U-bend overhang with horizontal restraints. The effects of the air-flow drag approaching the tubes from below actually raised the U-bend overhangs (a steady-state displacement) to the extent that the support reactions and the vibration frequencies were influenced. The test results were correlated as indicated below.

Following Connors' analysis [2], which equates work done by the fluid on the tubes to energy dissipated, the critical mean gap velocity associated with fluid-elastic excitation can be characterized by the parameter grouping

$$\left(\frac{\rho U_{CR}^2}{\zeta m f_H f_V}\right) = K^2$$

where, ρ is air density, U_{CR} is the critical flow velocity, ζ is equivalent viscous damping factor (assumed to be the same in both horizontal and vertical directions), f_H and f_V are tube natural frequencies in the horizontal and vertical directions, respectively, m is virtual mass per unit length, and K is a constant for a given tube geometry and spacing.

With ρ = 1.20 kg/m^3, m = 5.00 kg/m, and ζ = 0.015, the test results provided an average value of about K = 38 for the subject tube bank. The comparable constant pertinent to Connors' equation is $K/\sqrt{2\pi}$ = 15, somewhat higher than the generally quoted values for tubes with equal natural frequencies in all directions.

Figure 1 shows details of the restraint device. The diametral clearance between the tube and the fingers is about 1.5 mm. The restraint device is located about 0.3 m from the U-bend end; one restraint for each run of 2.27 m long overhangs. It appeared that during the prototypic tests most of the tubes were subjected to side loads (transverse to the flow direction) against the restraint fingers as a consequence of inherent construction or thermally induced deformations.

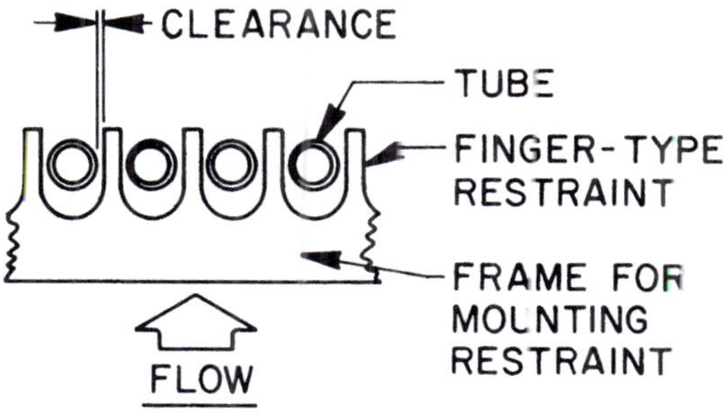

FIG. 1 FINGER-TYPE RESTRAINT DEVICE TO LIMIT TRANSVERSE-TO-FLOW MOTION OF TUBES

3. FLOW COLLECTOR

A massive Flow Collector hangs from the cover plate of the CRBR reactor vessel over the downstream end of the fuel assemblies, as shown in Sketch 1 of the Flow Collector Chart. The purpose of the Flow Collector is to channel the high temperature ($\sim$ 600°C) and velocity ($\sim$ 6 m/sec) coolant flow high into the upper plenum of the reactor to promote complete thermal mixing before the coolant enters the piping to the heat exchangers. Early in the design phase, there was concern that the relatively high velocity coolant flow from the fuel assemblies could force the essentially separate Flow Collector to vibrate. A flow-induced vibration mechanism discovered in scale model testing is reported, but to understand the mechanism a knowledge of the structural and flow geometry is required.

The coolant flow from groups of nine (hexagonally packed) fuel assemblies is collected by a single Outlet Module. Thirty-five Outlet Modules constitute the flow channels of the Flow Collector in the upper plenum. The flow exiting from each fuel assembly nozzle jets into corresponding individual holes in a grid plate attached to the upstream end of the Outlet Module, Sketch 2 of Flow Collector Chart. The holes through the grid plate are shaped like an inverted, elliptic funnel. The grid serves as a secondary constraint to vertical movement of the fuel assemblies. The jet flow is highly disturbed and widely dispersed by the nozzle internal geometry which is designed to enable removal of the fuel assembly from the reactor.

The structure of the Flow Collector is essentially an inverted frame: four columns ($\sim$ 6 m long) attached to the reactor cover plate at one end and fixed to two plates at the other end. The Outlet Modules are individually attached between the two plates. The fundamental vibration mode of the Flow Collector consists of a lateral movement across the top of the fuel assemblies at 0.9 Hz with less than 2% critical damping. Although the Flow Collector is massive ($\sim$ 5400 kg), if an alternating lateral fluid force exists, then large amplitude vibrations are possible which are limited only by the structural damping. The conditions for generating an alternating force appeared to be present. Because the fuel assemblies are relatively stationary, lateral movement of the grid plate would alter the locations where the jet flow from the fuel assembly nozzle impinges. If the lateral movement results in an increase in fluid force on the Flow Collector in the direction of motion, then the potential for self-excited vibrations is high. Since analytical prediction of the variation in fluid forces with lateral movement for such complicated geometry would be difficult, if not impossible, model testing was performed [3].

To assess the potential for self-excitation, a one-third scale model of a single, elastically-supported Outlet Module, Fig. 2, was tested. The mass of the partial scale model was determined by maintaining the prototypic ratio of contained fluid mass to structural mass. The structural stiffness of the model was determined by maintaining the prototypic ratio of structural potential energy to fluid kinetic energy: the reduced velocity was considered to be the most important parameter for flow-induced vibration simulation. The flow rates were made as large as possible to minimize distortion of Reynolds number to one-tenth that of the prototype. As such, the scale model actually represented a Flow Collector with every Outlet Module identical and subject to the same flow field. The model was considered conservative in that the excitation mechanism was more likely to occur in the model than in the prototype. Although only a single outlet module out of the full complement of thirty-five was tested, the prototype possessed enough spatial uniformity and symmetry of geometry and flow, that a design change would be deemed necessary if the mechanism was found in the model tests.

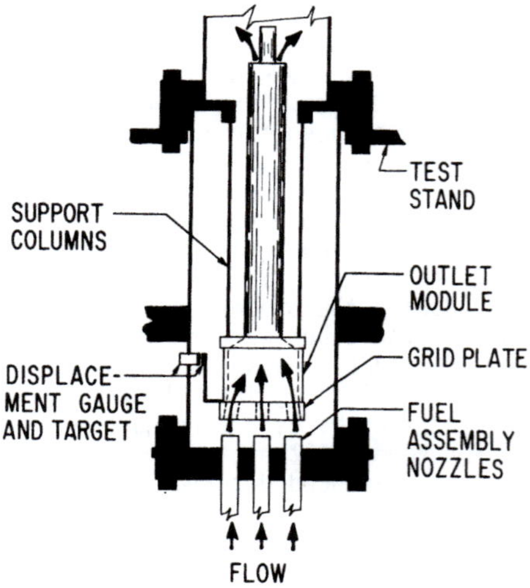

FIG. 2 SCALE MODEL OF FLOW COLLECTOR

The model was tested for several different parameters: vertical gap between the fuel assembly nozzles and grid plate (1.2 - 5 cm), lateral misalignment between the holes in the grid plate and the fuel assembly nozzles (0 - 1 cm), different flow distributions from nozzles, and total flow rates (0 - 120% of the normally operating flow rate of 19 x 10^6 kg/hr). Although modified by the different parameters, a fluidelastic excitation, typified by the results shown in Sketch 3 of Flow Collector Chart, was found to exist after a threshold flow rate was exceeded, conservatively estimated to be 70% of normal operating flow rate. Once excited, the motion eventually became large and always occurred in the direction corresponding to the short axis of the elliptic hole in the grid plate. Largest motion occurred for an intermediate vertical gap, maximum misalignment in the direction of motion, and for flow distributions with the largest flow velocities.

The existence of the self-excitation mechanism identified in the model test was eliminated from the prototype design when the bottom of the Flow Collector was keyed to the core support structure. Additional scale model tests indicated that the mechanism also could have been eliminated if the internal nozzle geometry was changed to that of a smooth pipe, as depicted in Sketch 3 of the Flow Collector Chart. This latter change was thought to concentrate the plume of the jet and greatly reduce interaction with the holes in the grid plates of the Outlet Modules.

4. THERMOWELL

The purpose of a thermowell is to install instrumentation, for instance a thermocouple, in a location that enables valid measurement of the flow temperature while providing protection from the environment encountered. A vibration test was performed to experimentally evaluate the potential for flow induced vibration of thermowell systems designed for liquid metal service [4].

Two thermowells of different designs were tested in a water loop to evaluate the potential of flow-induced vibrations. The cylindrical thermowell stems were cantilever mounted to protrude into the cross section of a water pipe and exposed to crossflow at various velocities. The test results and conclusions are presented further below for one of the thermowell designs.

The thermowell design is presented and described on the Thermowell Chart. Initial testing in air determined a natural stem frequency of 1245 Hz. After assembly, a small force generator was mounted to the water filled test section and energized to provide random excitation. Under the then present pressure and zero flow conditions, the natural frequency of the thermowell stem was about 1160 Hz and the natural frequency of the ovalling mode of the 102 mm I.D. piping was found to be approximately 950 Hz.

The test results and conclusions of the water flow tests are summarized as follows:

- The vibration amplitudes of the exposed thermowell stem were strongly influenced by the absence or presence of coupling with the connecting pipe structure. The structural changes required to effect or eliminate coupling were subtle.

- Considered as isolated components in the absence of such coupling, the thermowell responded to the water flow excitation with narrow-band random vibrations centered about the fundamental natural frequency of the stem in water. The vibration amplitudes were generally small. Under these conditions no flow-dependent resonant conditions or instabilities were observed.

- On the other hand, when vibration coupling provided interaction with the connecting pipe and/or support structure, the thermowell stem response was significantly changed and generated relatively large, almost sinusoidal vibration amplitudes in the drag (parallel-to-flow) direction. The coupling involved a shell type ovalling mode of the water carrying pipe as shown in Fig. 3 insert. When the flow velocity dependent resonance was established, both the pipe walls and the thermowell stem vibrated at the natural frequency of the thermowell stem at 1147 Hz.

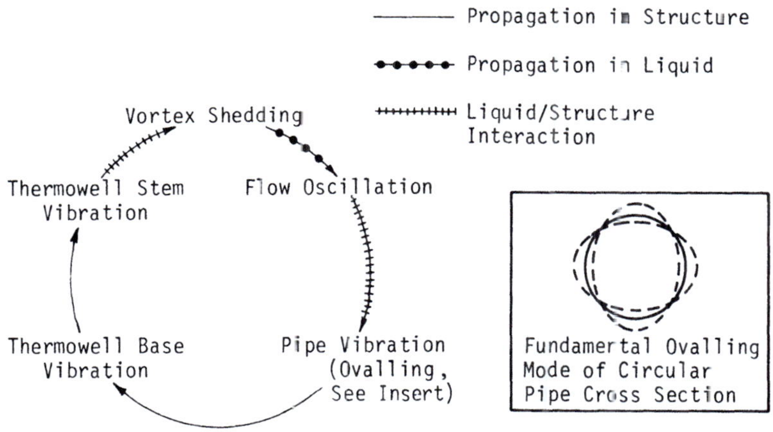

FIG. 3 UNPROVEN HYPOTHESIS OF THERMOWELL EXCITATION: FEEDBACK LOOP

- Tightening (or occasionally loosening) of the pipe support holders or application of a simple circumferential clamp around the test section pipe was effective to inhibit the thermowell resonance.

- Resonant vibrations of the thermowell were also dependent on the water pressure level.

- Because vortex shedding provides two pulses per cycle in the parallel-to-flow direction and because the resonant vibrations were generated in that direction, the coincidence of vortex shedding from the stem tip with a frequency half of the thermowell natural frequency was considered to be a contributing factor. Based on the measured thermowell/piping vibration frequency of 1147 Hz, the excitation could have been facilitated by vortex shedding at 573.5 Hz. Assuming a Strouhal number of 0.2, the corresponding calculated flow velocity based on the 4.1 mm tip diameter is 11.7 m/sec, which falls into the range of velocities where the resonant excitation of the thermowell was prominent. However, it was not possible to prove the existence of such vortex shedding experimentally with the available test equipment as explained below.

  To measure vortex shedding frequencies a hot film anemometer probe was introduced into the wake flow of thermowell through a penetration of the pipe. This fixed penetration permitted adjustment of the location of the anemometer probe in only one direction. During flow testing no distinct vortex shedding frequency could be observed from the available probe positions.

- A possible but unproven hypothesis to explain the resonance phenomenon as some type of feedback loop is presented in Fig. 3. Although the test work did not provide an extensive investigation of the overall problem area, the results suggest that coincidence of piping, pipe support, and thermowell frequencies be avoided by designers because of the potential for coupling with the flow.

5. CONCLUDING REMARKS

State-of-the-art prediction methods often are not sufficiently accurate, or applicable for reliable design evaluation of reactor system components from the standpoint of flow induced vibration. At best, they are useful in providing guidance in design evaluation and in identifying potential problem areas. With regard to the latter, it is also necessary to use engineering judgment based on experience, to identify potential problem areas. The FFTF heat exchanger tubes and CRBR flow collector are examples of components that were suspected to be susceptible to flow induced vibration by virtue of their low natural frequencies and exposure to high velocity flows. In the case of the FFTF dump heat exchanger tubes, a quantitative analysis using Connors stability diagram for tube rows exposed to crossflow [2] called attention to a potential problem. In these two cases, testing confirmed in the design stage that a flow-induced vibration problem existed and appropriate fixes were developed and evaluated. In the case of the thermowell, while vortex shedding was assumed to be the dominant excitation mechanism, no significant response was expected. However, testing revealed that the excitation mechanism was actually much more complicated, involving coupling among the piping, thermowell, and fluid.

The three components and associated FIV testing discussed herein serve to illustrate the importance of testing in the FIV evaluation of reactor system component designs. Additionally, the heat exchanger tube evaluation illustrated how the results of laboratory tests [2] can be used to identify potential problem areas; the flow collector testing demonstrated how rather subtle changes in geometry (grooved to circular nozzles) can effectively eliminate a fluidelastic instability; and the thermowell testing showed how interaction of a component, support, and fluid flow can result in an unexpected vibration problem.

6. REFERENCES

1. Halle, H., Boers, B. L., and Wambsganss, M. W., "Fluidelastic Tube Vibration in a Heat Exchanger Designed for Sodium-to-Air Operation," ASME J. Eng. for Power, 97(4), 561-568, (Oct. 1975).

2. Connors, H. J., Jr., "Fluidelastic Vibration of Tube Arrays Excited by Cross Flow," Proceedings of the Symposium on Flow-Induced Vibration in Heat Exchangers, ASME Winter Annual Meeting, New York, 42-56, (Dec. 1, 1970).

3. Mulcahy, T. M., Yeh, T. T., and Jendrzejczyk, J. A., "Flow-Induced Vibration Model Test of the CRBR Upper Plenum Structure," American Nuclear Society Transactions, 23, 447, (1976).

4. Halle, H., and Lawrence, W. P., "Waterflow-Induced Vibration Test of Liquid Metal Thermowells," Argonne National Laboratory Technical Memorandum ANL-CT-76-29 (March 1976).

HEAT EXCHANGER TUBES

Sodium-to-Air Dump Heat Exchanger of the Fast Flux Test Facility

H. Halle, B. L. Boers and M. W. Wambsganss, Argonne National Laboratory, USA

Problem: Serpentine heat exchanger tubes designed with long, overhanging U-bend sections to reduce stress levels caused by thermal expansion (Sketch 1). Potential for flow induced vibration studied in 5-tube mock-up with prototypic tubes (Sketch 2). Large-amplitude, fluidelastic vibration of tubes experienced at airflow rates above 120 percent of design conditions. Potential for impacting wear between adjacent U-bends or impacting/rubbing wear at support plate demonstrated.

Geometry: See Sketches 1 and 2. Finned tubes of stainless steel, type 304. Tube diameter: Across fins = 43.7 mm, Fin root and bare tube = 38.1 mm, Inside = 31.8 mm. Helical fin pitch = 2.3 mm. Transverse-to-flow tube spacing = 54.0 mm.

Boundary Conditions: U-tubes and supports prototypic. Tubes filled with stagnant water to simulate sodium. Five U-tubes in row. Duct sidewalls contoured to simulate tube-to-tube gaps.

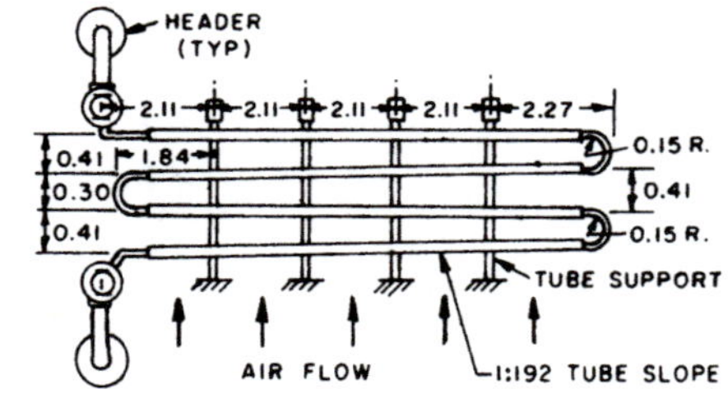

(1) PROTOTYPE DHX SERPENTINE TUBE
66 SERPENTINE TUBES IN ROW

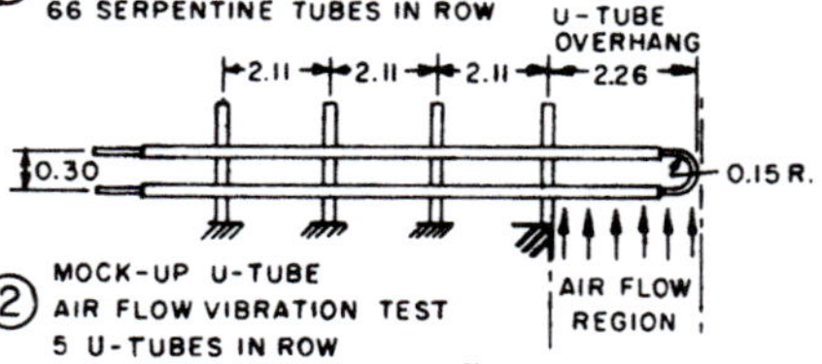

(2) MOCK-UP U-TUBE
AIR FLOW VIBRATION TEST
5 U-TUBES IN ROW

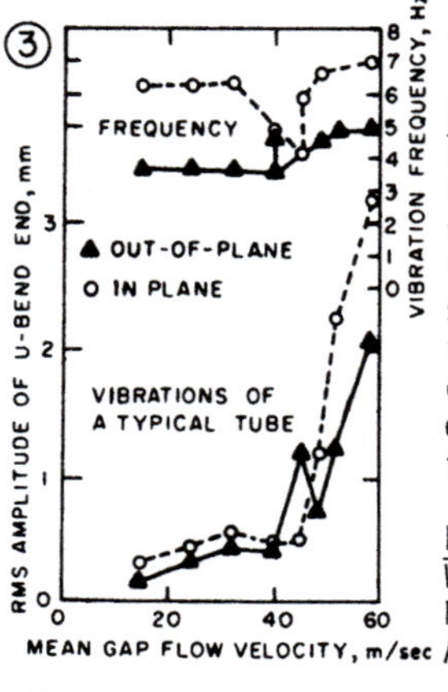

Flow Conditions: U-tube overhang exposed to ≈ 25°C air pulled through rectangular test duct (with converging entrance) with downstream blowers.

Operating Conditions: Airflow rate over U-tubes increased incrementally; approach velocity measured with a pitot tube and gap velocity computed.

Structural Dynamics: Fundamental tube natural frequencies of 3.6 and 6.3 Hz in horizontal (out-of-plane) and vertical (in-plane) directions. Damping factor $\zeta \approx 0.015$.

Vibrations: RMS displacement and frequency of the free end of the U-tube measured in horizontal and vertical directions and plotted as a function of mean gap flow velocity (Sketch 3). A critical gap flow velocity experienced above which large amplitude tube motions result. Decreasing length of U-tube overhang increased critical flow velocity, characteristic of a Connors-type stability relationship.

Cures: A restraint system with fingers between tubes developed to suppress horizontal motion and eliminate tube impacting; however, unless a side load exists, vertical motion is essentially unaffected. Prototype tested, with restraints, and sodium flow to 500°C. Resulting tube vibration was at higher than fundamental frequencies, or random, but generally of low level, considered acceptable. Leakage permitted around sides of tube bank can increase flow velocities and vibration potential of tubes in extreme outside positions.

Reference: Halle, H., Boers, B. L., and Wambsganss, M. W., "Fluidelastic Tube Vibration in a Heat Exchanger Designed for Sodium-to-Air Operation," ASME J. Eng. for Power 97(4), 561-568 (Oct. 1975).

CRBR REACTOR FLOW COLLECTOR

T. M. Mulcahy, T. T. Yeh and J. Jendrzejczyk, Argonne National Laboratory, USA

Problem: Flow from fuel assembly exit nozzles may excite lateral vibration of flow collector, Sketch 1, via fluidelastic (galloping) mechanism.

Boundary Conditions: The flow collector is a four column frame cantilevered from the reactor vessel cover plate. The fuel assemblies are stationary.

Flow Conditions: Complex geometry of fuel assembly exit nozzle, Sketch 2, creates wide, turbulent jet of flow which impinges on inverted funnel shaped holes in grid at upstream end of outlet modules.

Structural Dynamics: Although the flow collector is massive (∿ 5400 kg), large lateral motion is possible in the fundamental vibration mode at 0.9 Hz having less than 2% of critical damping.

Vibration: Large lateral movement of flow collector significantly alters locations on the outlet module grid where each fuel assembly jet impinges, creating alternating lateral fluid forces. Threshold for self-excitation determined, Sketch 3, to be 70% of the normal operating flow rate (19 x 10^6 kg/hr) in a one-third scale model consisting of a single, elastically mounted outlet module. This corresponds to a reduced velocity of 30, based on nozzle diameter.

Cures: Eliminate lateral vibration mode by fixing bottom of flow collector to core support structure, or modify nozzle geometry to reduce area of flow impingement on outlet module grid. Concentrated jet flow from smooth nozzle did not lead to self-excitation, Sketch 3.

References: Mulcahy, T. M., Yeh, T. T., and Jendrzejczyk, J. A., "Flow-Induced Vibration Model Test of the CRBR Upper Plenum Structure," American Nuclear Society Transactions 23, p. 447 (1976).

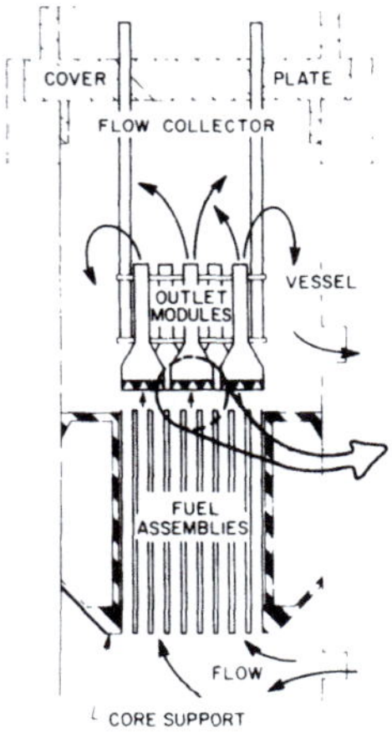

①

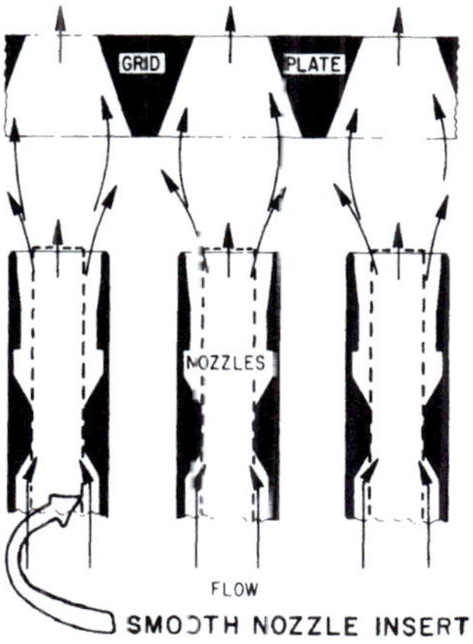

②

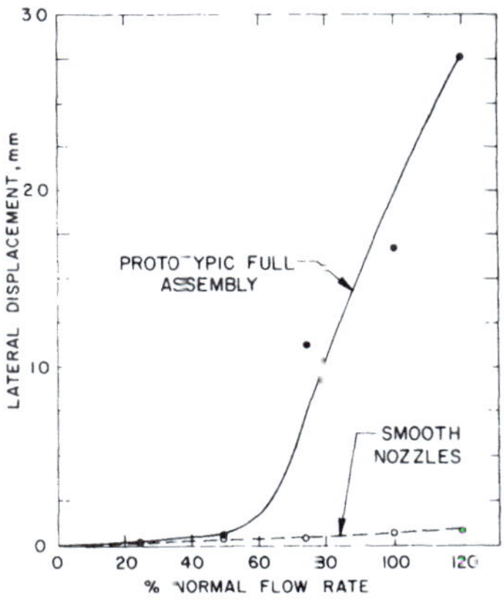

③

THERMOWELL

H. Halle and W. P. Lawrence, Argonne National Laboratory, Argonne, IL USA

Problem: Thermowell, designed for liquid metal service, with stem protruding into cross section of pipe. Potential of flow-induced vibration examined during waterflow testing. Excitation of strong vibrations possible depending on support conditions of connecting piping. Resonance readily eliminated, but required accessibility to pipe structure which may not be available in actual service.

Geometry: See Sketch 1 of section through thermowell and pipe. Thermowell and piping of stainless steel.

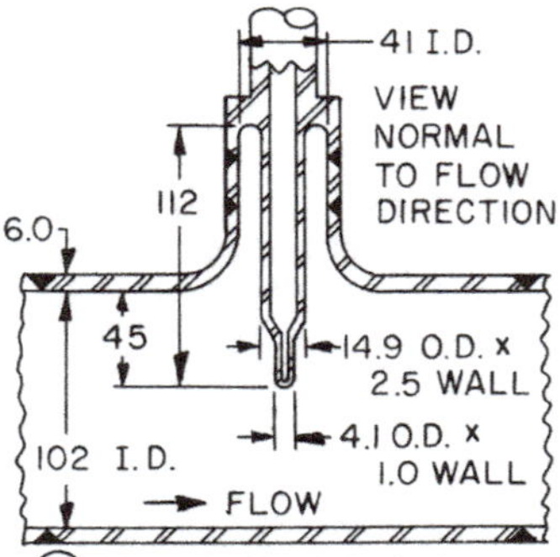

① DIMENSIONS IN mm

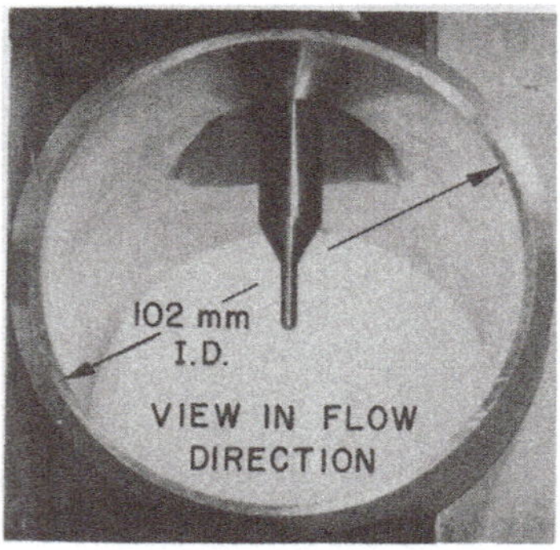

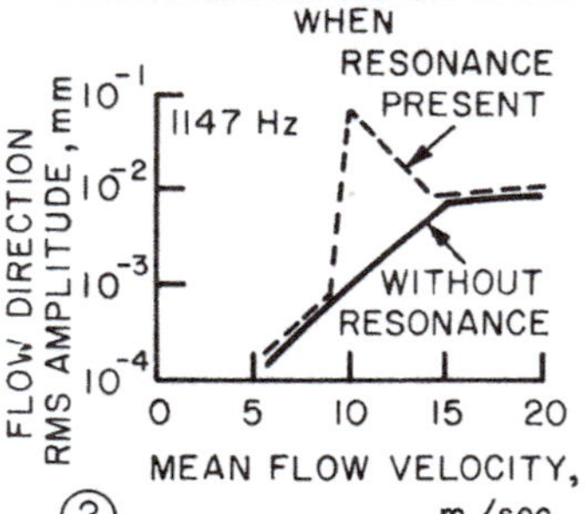

②

Boundary Conditions: Thermowell stem penetrating 45 mm into pipe cross section (See Photo taken during installation). Pipe clamped to support structure.

Flow Conditions: Uniform and straight approach for 20 diameters. Room temperature water, pressure from 210 to 380 kPa. Flow rate measured by turbine flowmeter, mean velocity calculated.

Operating Conditions: Waterflow across thermowell stem incrementally increased beyond rates causing resonance (Sketch 2).

Structural Dynamics: Natural stem frequency: 1147 Hz in water (1245 Hz in air). Equivalent viscous damping factor averaging about 0.007 in water.

Vibrations: Two orthogonal accelerometers mounted inside stem, 31 mm from tip end. Depending on pipe support/clamping conditions, flow velocity dependent resonant interaction between the stem and the pipe could be excited at 1147 Hz. In that case, the stem vibration amplitudes (Sketch 2) were considered significant for fatigue considerations and the 102 mm I.D. water pipe vibrated in a shell-type ovalling mode. Vortex shedding from 4.1 mm diameter tip a possible but not a proven factor.

Cures: Adjustment of pipe supports or placing a clamp around ovalling pipe.

Note: Test approach conservative, because 45 mm penetration in 102 mm diameter larger than specified by pertinent U.S. standards.

Reference: H. Halle and W. P. Lawrence, "Water-flow-Induced Vibration Test of Liquid Metal Thermowells," Argonne National Laboratory Technical Memorandum ANL-CT-76-29 (March 1976).

A5 EXCITATION, AMPLIFICATION AND SUPPRESSION OF FLOW-INDUCED VIBRATIONS IN HEAT EXCHANGERS

Momchilo M Zdravkovich
Reader
University of Salford, U.K.

John E. Namork
Research Engineer
Det Norske Veritas, Oslo, Norway

SUMMARY

Flow-induced vibrations in a six-row deep and triangularly staggered tube bank were investigated experimentally and analysed by separating the mechanism of excitation from the mechanism of amplification of vibration. The development of vibration was examined in two stages:

(i) Excitation of low-amplitude random vibration was estimated by measuring the distribution of pressure fluctuations around a stationary tube positioned sequentially in all rows. The measurements revealed that the distribution of the fluctuating pressure was distinctly different in first two rows from the rest of the bank. The peak of the fluctuating pressure was highest at the sides of the tube in the second row. The aeroelastic tests showed that the tubes in the second row always started to vibrate first with irregular small amplitudes. There was a definite correlation between the position of the peak in the fluctuating pressure distribution and the dominant direction of vibrational response.

(ii) Amplification of vibration by fluid-elastic forces led to large amplitude vibration at the natural frequency of the tubes. The time-averaged drag and lift forces were evaluated for the case where the tube in the second row was displaced in various positions relative to the first and third row and in various degrees of stagger with respect to the in-line position. The field of these fluid-dynamic forces exerted on the tube shows that considerable energy can be gained from the fluid along some orbits. This, in turn, either amplifies or maintains large-amplitude vibration.

The susceptibility of tubes in the second row to flow-induced vibration can be suppressed by fitting detached wires around the tubes in the first row. The resulting flow structure behind the first row then resembles that found around the second and subsequent rows. Similar arrangement of detached wires fitted on the tubes in the second row suppressed the amplification of fluid-elastic vibration of tubes in the second row. The device for suppressing flow-induced vibrations has an attractive feature in that it enhances the heat transfer rate in the first row, which is normally below that in the other rows.

1. INTRODUCTION

Closely packed tubes in modern heat exchangers are subjected to significant fluid-dynamic forces, subsequent elastic displacement, and mutual interaction between the two. The vibration of tubes can be excited by sufficiently high pressure fluctuations around initially stationary tubes. The amplitude of vibration, however, is likely to be amplified by the time-averaged fluid-elastic forces produced by a considerable variation of the interstitial spacing and the associated constriction of the fluid flow.

It was believed, at first, that vortex shedding or some modified version of it was the primary cause of flow-induced vibration in tube banks (1). This belief was held despite the fact that the vortex shedding became irregular and weak in closely spaced tube banks.

The Strouhal number, which was always constant at about 0.2 for a single cylinder, reached values an order of magnitude higher, and varied widely with the arrangement of tubes. The vibration of tubes was not limited to the resonant range when the tube natural frequency was near the frequency of vortex shedding, as observed for a single elastic circular cylinder.

This prompted Owen [2] to turn to buffeting as a primary mechanism of excitation of closely spaced tubes. High intensity of turbulent fluctuations generated within the tube bank produced an equally random fluctuating pressure distribution around each tube. The latter responded only to those fluctuations which were near to the natural frequency. Some kind of a selective resonance near the natural frequency of the tube was assumed to have taken place. The essential feature of excitation was generation of sufficient turbulent energy having a frequency content around the natural frequency of the tubes.

The third, distinctly different mechanism of tube vibration was found by Connors [3]. The displacement of one tube in the bank from its equilibrium position altered the flow field, thereby upsetting the force balance on the neighbouring tubes and causing them to also change their positions in a vibratory manner. If, during a cycle of vibration, the energy extracted from the flowing fluid by the tubes exceeded the energy dissipating by structural damping, a fluid-elastic vibration will be established. The fluid-elastic displacement mechanism, however, requires initial displacement of tubes in order to commence the build-up of large-amplitude vibration of tubes along oval orbits.

In summing up the variety of mechanisms involved in a flow induced vibration of tubes, it might be helpful to divide the vibration in two stages: excitation - low amplitude, and amplification - large amplitude. In the case of a single tube the vortex shedding is responsible for excitation of vibration and its subsequent modification by synchronisation - locking in leads to the amplification of vibration. The vibration of tubes in banks seems to be excited by buffeting but it is unlikely that the same mechanism can lead to large-amplitude vibrations. The fluid-elastic displacement, jet-switch, wake-switch and other related phenomena can build-up and maintain large amplitude vibration but cannot excite vibrations per se. This classification suggests that the pressure fluctuation around stationary tubes excites vibration and that the time-averaged force on vibrating, i.e. variously displaced tubes, amplifies and maintains large amplitude vibration.

The systematic experimental research at Salford was directed towards detailed measurements of pressure fluctuations around tubes placed in various rows and evaluation of time-averaged fluid-elastic forces exerted on a tube in various displacement from the neutral position. Some means of suppressing flow-induced vibrations in closely spaced tube banks were attempted. The resultant 'cures' will be described in this paper.

2. EXCITATION OF VIBRATION BY FLUCTUATING PRESSURE AROUND A TUBE SURFACE

Systematic measurements have been carried out involving fluctuating pressure distributions around a monitored tube located sequentially in various rows in a triangularly staggered tube bank. Fig.1a shows the actual arrangement of tubes having transverse to diameter ratio of 1.375 and longitudinal to diameter ratio of 1.1875. The six-row deep bank was tested first in a stationary mode when all tubes were mounted on two end plates made of perspex. The mean gap velocity was 69 m/s which corresponded to Re_g = $1.1x10^5$. The measuring tube was fitted with a pressure transducer mounted inside and connected via a small cavity with a single pressure tap at the midspan of the tube. The pressure transducer measured the difference between the pressure detected by the tap and a static pressure tap in the upper wall located five tube diameters upstream. The wind tunnel and instrumentation are described elsewhere [4].

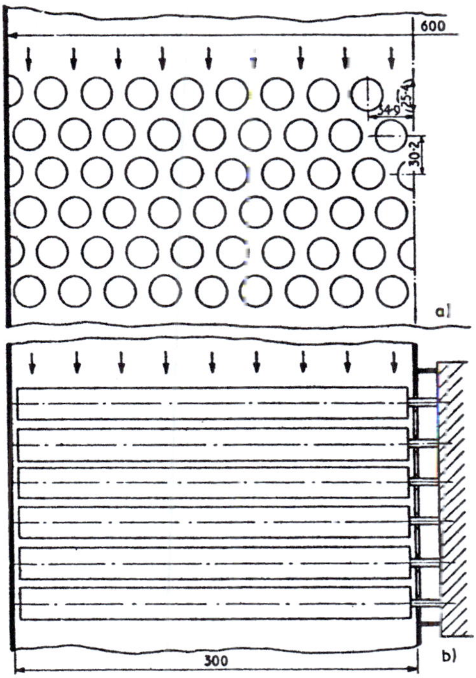

FIG.1. TUBE ARRANGEMENT; a) RIGID b) FLEXIBLY MOUNTED

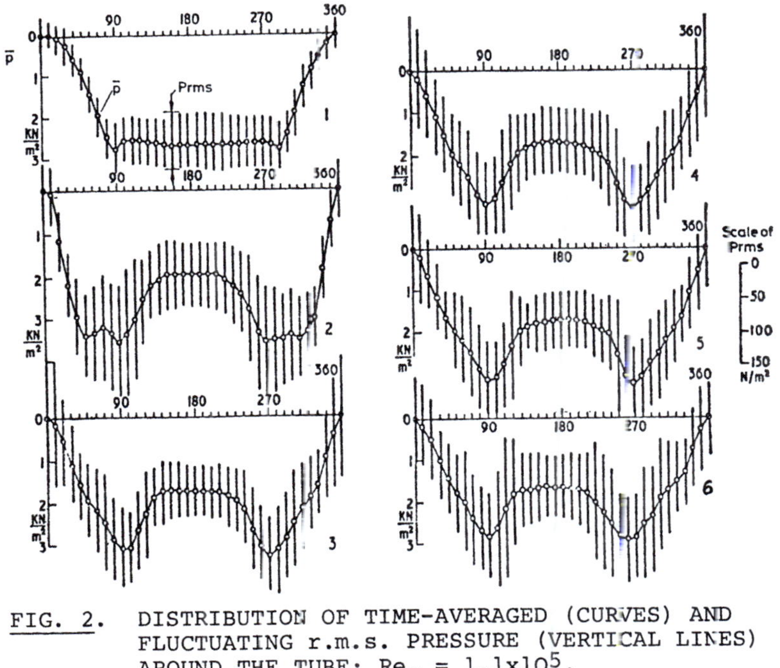

FIG. 2. DISTRIBUTION OF TIME-AVERAGED (CURVES) AND FLUCTUATING r.m.s. PRESSURE (VERTICAL LINES) AROUND THE TUBE; Re_g = $1.1x10^5$.

Fig.2 shows the measured distribution of root-mean-square of the fluctuating pressure (vertical lines), superimposed on the time-averaged pressure distribution (open circles). It is evident that both pressure distributions in the first two rows are distinctly different from the rest of the bank. The differences in pressure distribution of the first three rows can be explained by the following changes of the structure of the interstitial flow:

(i) The tube in the first row is subjected to interference by the adjacent side tubes and low-turbulence free-stream flow. A steep pressure gradient around the upstream part of the tube is accompanied by low-intensity pressure fluctuations.

(ii) The tube in the second row is affected not only by the adjacent side tubes but also by the wide wakes produced behind the first row of tubes. The result is a steeper pressure gradient than in the first row and a huge increase of pressure fluctuations around both sides of the tube. The wake behind the tubes in the second row is narrow, as can be deduced from Fig.2, by noting the reduced extent of the constant base pressure around the tube.

(iii) The tubes in the third and subsequent rows are exposed to a highly turbulent interstitial flow which produces a sharp increase in pressure fluctuations around the stagnation point, as seen in Fig.2. The pressure gradient is less steep than in the second row, due to the narrower wake behind the second row. The pressure fluctuations around the sides of the tube are also markedly reduced. The latter, however, gradually increase and reach a pronounced peak around the separation point in the last row, Fig.2.

It was expected that the susceptibility to flow-induced vibration of tubes in various rows could be correlated to the measured distribution of pressure fluctuations. Random small-amplitude vibrations should be excited at the natural frequency of the tube, their direction being predominantly in-line with the position of peaks in the distribution of pressure fluctuations. The tubes in the second row are obviously prone to start vibrations in a transverse direction before the rest, due to the huge peak and high intensity of pressure fluctuations around the side of the tube. The tubes in the third row are likely to follow with vibrations mostly in the streamwise direction while the remaining rows should respond along an oval orbit due to a series of peaks distributed around the tube circumference. The tubes in the first row should be least prone to vibrations having an amplitude predominantly in a streamwise direction.

The same tube bank was mounted on flexible cantilevers as shown in Fig.1(b). The tests on such an aero-elastic tube bank showed a qualitative agreement with the predictions based on the fluctuating pressure distribution. It was observed that always one of the tubes located in the second row started to vibrate first in the transverse direction at V_g = 21 m/s*. The other tubes in the second row joined in with a rapid increase in amplitude. The tubes in the third and other rows started to vibrate at V_g = 26 m/s when the tubes in the second row had already attained large amplitudes of vibration. The fluid-elastic displacement mechanism took over at these velocities and large-amplitude vibrations were rapidly built-up. Clashing of some tubes started at V_g = 33.5 m/s,

*V_g was a mean gap velocity evaluated by dividing the flow rate through the bank by the area of all transverse gaps between the tubes.

and a cine-film of these severe vibrations was made. Four typical frames from that film are shown in Fig.3.

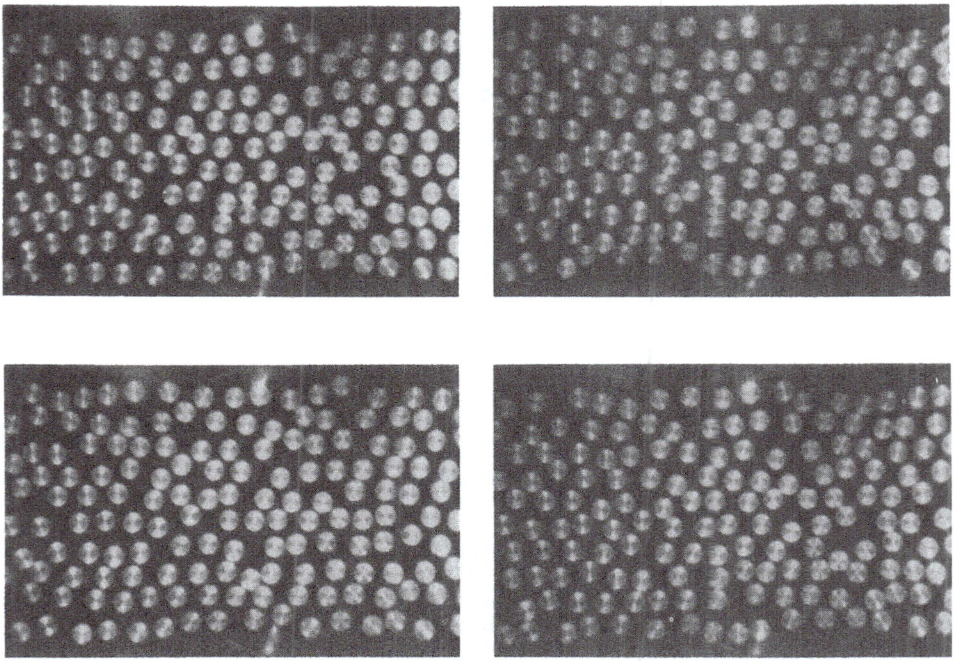

FIG. 3. FOUR FRAMES FROM THE FAST CINE FILM OF THE 10-ROW DEEP FULLY AEROELASTIC TUBE BANK

3. AMPLIFICATION OF VIBRATION BY FLUID-ELASTIC FORCES

The fluid-dynamic forces exerted on a tube displaced in various positions were calculated from the time-averaged pressure distribution [5]. The tube bank consisted of only three rows arranged as shown in Fig.4. The transverse spacing of tubes was kept constant at T/D = 1.66 while the second row as a whole assumed longitudinal spacings in relation to the first row of L/D = 1.2, 2.0 and 2.5. The second row was also displaced in the transverse direction from the fully staggered symmetric arrangement to 2/3, 1/3 and 0 stagger, the last being an in-line arrangement. The description of the experimental arrangement and actual pressure distribution were reported elsewhere [5].

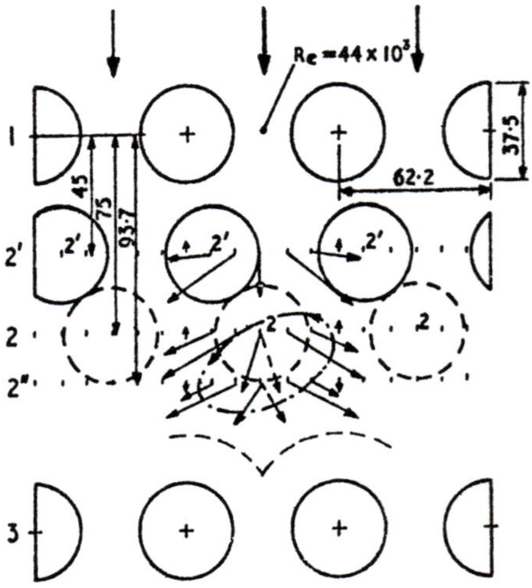

FIG. 4. FLUID-ELASTIC FORCES EXERTED ON TUBE IN ROW 2. (ONE DISPLACED POSITION IS SHOWN AS 2').

Fig.4. is a graphical display of resultant fluid-dynamic forces exerted on the monitored tube at 12 positions. The monitored tube was displaced with the second row as a whole. The resulting force field obtained in this manner is believed to be a better approximation of the observed, mostly in-phase vibrations of tubes in the second row than the displacement of a single tube in a fixed second row, as done by Connors [3]. It was assumed that the quasi-steady force field may be taken as a first approximation of the actual force field produced by vibrating flexible tubes.

The interference produced by the displaced second row of tubes caused a drastic change in the interstitial flow pattern accompanied with large fluid dynamic forces, see Fig.4. The force was mostly in the transverse direction as it was found to be for only two tubes in staggered arrangement [6].

The observed vibrational orbit produced by the centre of the tube in the second row is superimposed as a chain-dot line in Fig.4. It is evident that the fluid-elastic forces exerted on the tube along the upstream part of the orbit are in-line with the tube movement and the tube gained energy from the fluid flow. Some energy is lost along the downstream part of the orbit but considerably less than during the previous part of the orbit. The total energy gained along the orbit can build up or maintain large-amplitude vibration.

This was a simplified description of the fluid-elastic vibration. The fully aeroelastic tube bank produced a wide variety of orbits which changed with time. The basic feature of the phenomenon was that at any instant there always existed a force field produced by the instantaneous displacement of the tubes, which governed the subsequent vibrational motion of tubes-see Fig.3.

4. SUPPRESSION OF EXCITATION AND AMPLIFICATION OF TUBE VIBRATION

Once the salient features of the mechanisms of tube vibrations were deduced, it was possible to devise a means of suppressing the vibrations. The early excitation of tubes in the second row was caused by the wide wake behind the tubes in the first row. The usual tripping-wire device attached to the tube is not feasible for heat-exchanger tubes because it is likely to cause hot spots within the locally separated flow behind the tripping wire. A modification of the tripping wire technique was tried by placing the wire one diameter away from the tube surface. This modificatic is likely to enhance the heat transfer in the first row which is known to be below the average heat-transfer in the other rows. After a series of tests an optimum position for the four detached wires was found, as described in detail elsewhere [7], and shown in Fig.5.

The detached wires fitted on the tubes in the first row stabilised the tubes in the second row and postponed the onset of small-amplitude vibration but did not suppress tube vibration at higher speeds. The latter were fully stabilised only when the detached wires were fitted also to the tubes in the second row. Fig.5 shows the detached wire arrangement which suppressed excitation and amplification of tube vibration in the second row. The positioning and attachment of detached wires may be particularly straightforward for tubes with fins. The optimum position of detached wires depends slightly on the tube arrangement and may be accomplished by disk spacers arranged evenly along the tubes [7].

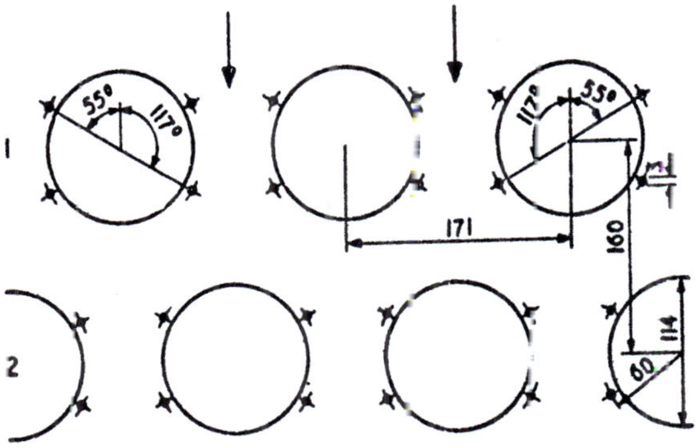

FIG. 5. MEANS FOR SUPPRESSING VIBRATION IN ROW 2, (TESTED WITHIN $8.6 \times 10^4 < Re_g < 2.1 \times 10^5$).

5. CONCLUSION

Systematic tests on inelastic and aeroelastic tube banks have shown that it is possible to separate the excitation mechanism of tube vibration from the subsequent amplification mechanism.

The distribution of pressure fluctuations around the tubes was found to be different in each of the first three rows, the worst being for the tubes in the second row. The cause was the wide wake behind the tubes in the first row. A novel means of suppressing vibrations consisted of four detached wires fitted on the tubes in the first row stabilised the tubes in the second row.

The force field produced by the tube displacement from the initial arrangement was measured for the second row of tubes. It was found that the displaced tube was subjected to a large fluid-dynamic force directed mostly in the transverse direction. The force was caused by the switch of the interstitial flow towards a smaller gap. This switch was suppressed and considerably reduced with the detached wires fixed around the tubes in the second row. The aeroelastic tubes, fitted with that device, were stabilised and did not develop large-amplitude vibration even beyond the critical speed.

6. REFERENCES

1. Livesey, J. L. and Dye, R. C. F., "Vortex excited vibration of a heat exchanger tube row", Journal of Mechanical Engineering Science, 4, 349-352, (1962).
2. Owen, P. R., "Buffeting excitation of boiler tubes vibration", Journal of Mechanical Engineering Sciences, 7, 431-439, (1965).
3. Connors, H. J., "Fluid-elastic vibration of tube arrays excited by cross flow", Proceedings on Flow Induced Vibration in Heat Exchangers, Winter Annual Meeting of American Society of Mechanical Engineers, 42-56, (1970).
4. Zdravkovich, M. M. and Namork, J. E., "Flow structure within both stationary and vibrating tube banks with triangular pitch", Proceedings of Vibration in Nuclear Power Plant, Keswick, U.K. paper 2:5, (1978).
5. Zdravkovich, M. M., Singh, S., Nuttall, J. A., and Causon, D. M., "Flow-induced vibrations in staggered tube banks", Proceedings 6th Thermo- and Fluid Mechanics Convention, Durham, 237-243, (1976).
6. Zdravkovich, M. M., "Review of flow interference between two circular cylinders in various arrangements", Trans. A.S.M.E. Journal Fluids Engineering, 99, Series 1, 618-633, (1977).
7. Zdravkovich, M. M., "Improvements in and relating to flow stabilisation", U.K. Patent Appl. 40768/71, (1971).

EXCITATION, AMPLIFICATION AND SUPRESSION OF FLOW-INDUCED VIBRATIONS OF TUBE BANK

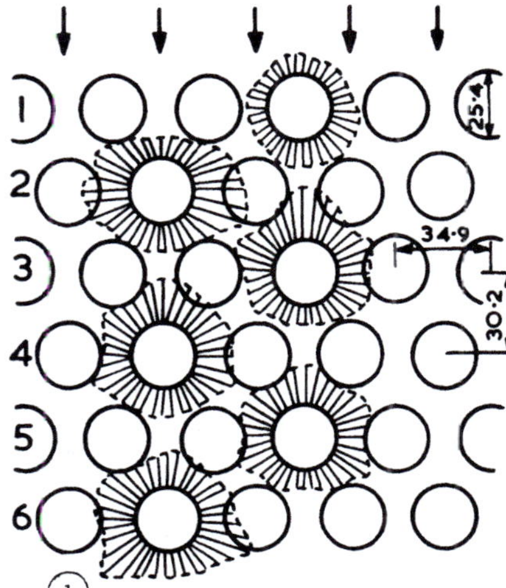

(1) Tube arrangement and fluctuating pressure distribution; $Re_g = 1.1 \times 10^5$ and Vg = 69 m/s.

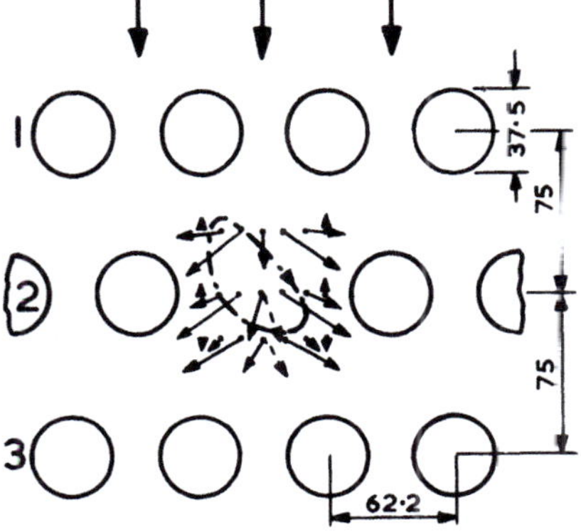

(2) Fluid-elastic forces and vibrational orbit.

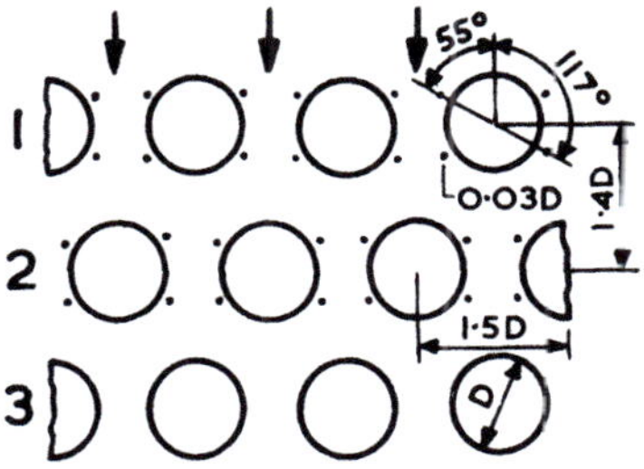

(3) Means for suppressing vibration.

Problem: Heat exchanger tube vibrations. The development of vibration can be separated in two stages:

a) excitation of random small amplitudes
b) amplification to large amplitude orbits.

Geometry: Skh.1 shows tube arrangement (dimensions in mm). The tubes were mounted either rigidly or flexibly on cantilever rods with caps at both ends.

Structural dynamics: Natural frequency of flexible tubes 75 Hz ± 3%: logarithmic decrement 0.013 ± 60%; critical gap velocity for buffeting in second row 21.2 m/s, reduced gap velocity $\frac{V_g}{fD} = 10.9$; tube clashing starts at $V_g = 39.7$ m/s.

Mechanism of excitation: Fluctuating pressure distribution (Prms) around tubes as shown in Skh.1. Note pronounced side peaks in row 2, in which tubes always start to vibrate first with random small amplitude.

Mechanism of amplification: Fluid elastic displacement brings tubes in various positions relative to each other. Time average forces acting on the monitored tube in row 2 are shown in Skh.1. Orbital vibration (chain-dot oval) is fed by energy gain during each cycle.

Cures: Optimum arrangement of detached wires is shown in Skh.3. The wires are less than 3% of the tube diameter and one wire diameter away from the tube surface. Those around tubes in row 1 delay the onset of excitation of vibration of tubes in row 2. The wires fitted around tubes in row 2 suppress large-amplitude vibrations by stabilising gap flow and drastically reducing transverse component of fluid-elastic forces. Note that the detached wires around tubes in row 1 enhance heat transfer.

References: Zdravkovich, M. M., and Namork, J. E. "Flow structure within both stationary and vibrating tube banks with triangular pitch," Proc. Vibration in Nuclear Plant, Keswick, U.K. 1978, paper 2:5.

Zdravkovich, M. M. Improvements in and relating to flow stabilisation. U.K. Patent Appl. 40768/71, 1971.

A6 RESPONSE OF A TUBE BANK TO CROSSFLOW INDUCED EXCITATION AT TRANS- AND SUPERCRITICAL REYNOLDS NUMBERS

B. T. Lubin
H. N. Guerrero
R. C. Kantayya

C-E Power Systems
Combustion Engineering, Inc.
Windsor, Connecticut 06095, U.S.A.

SUMMARY

Tests on a tube bank section subjected to crossflow resulted in amplitude response spectra that, while primarily random, did show evidence of periodic response. The Reynolds number range for these tests was from 5×10^5 to 5×10^6 (trans- to supercritical). Turbulent buffeting was predominant at low flows. As flow increased above a critical velocity, large amplitude periodic response was noted at the tube natural frequency. Incremental increases in flow caused reduction in amplitude and the response was once again random. With further increases, some tubes returned to a periodic response. The turbulent buffeting Strouhal number was ~ 0.18, while for the periodic response a value of ~ 0.45 was calculated.

1. INTRODUCTION

Several experimenters, e.g. [1-7], have run flow-induced vibration tests on tube banks. Almost exclusively, these tests have consisted of subjecting the tube banks to crossflow at low Reynolds numbers ($Re \leq 10^5$). Results from the tests have shown that, in general, once the fluid velocity through the tube bank exceeds a critical value, a fluid-elastic instability can occur [4]. The objective of the tests reported here was to investigate the dynamic response of tubes in a tube bank with $Re > 10^6$ and to identify possible mechanisms for the observed flow-induced vibrations.

2. EXPERIMENTAL SYSTEM

The tests were conducted in a high pressure, high temperature water test facility. A schematic of the test loop is shown in Figure 1A. The tube bank test model consisted of 201 tubes on a square pitch array mounted on the quarter circle shape tube sheets, Figure 1B. The tubes are subjected to axially directed flow, especially near the far end center of the tube bank; then to a diagonally directed flow midway towards the outlet pipe; and predominantly to crossflow on the tubes nearest the outlet pipe. Four strain gages per tube, aligned axially with the tube, were welded onto the inner surface of certain tubes, chosen because they were in an area that had primarily crossflow velocities with different up- and downstream conditions (Figure 1B).

Outlet velocity was measured by means of an orifice plate. Values of intertube velocity were obtained from direct static and dynamic pressure measurements on a similar model designed to obtain only pressure loss and flow distribution data. Though pressure instrumentation was not present on any tube, wall pressure transducers behind the last tube row gave spectra consistent with homogenous turbulence. Based on these data, the RMS turbulent intensity is approximately 1%.

3. NO-FLOW TESTS

Pre-test measurements were taken to establish the natural frequencies and the damping for the instrumented tubes. These characteristics were determined by means of impact tests carried out in air and water. Natural

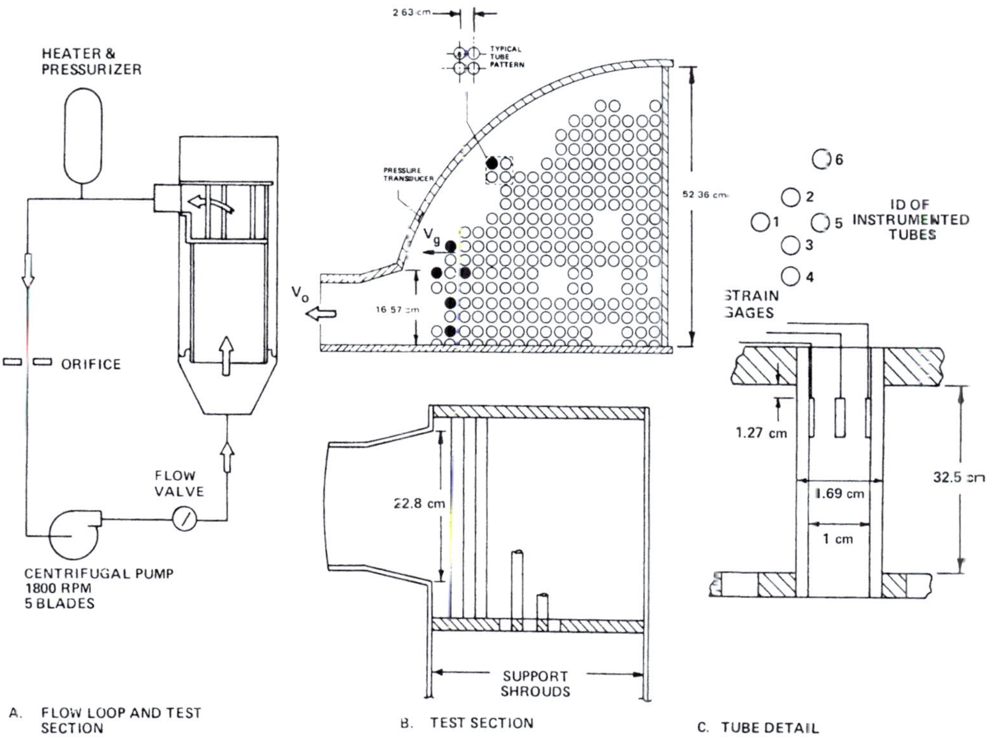

FIG. 1 EXPERIMENTAL SET-UP

frequencies and damping values are as follows: air, 770-797 Hz, 0.0043-0.0089%; water (21 C), 705-734 Hz, 0.035-0.058%. Dynamic tests were run on each tube to calibrate each set of strain gages. Data showed a linear relationship of load vs. strain, with a slope of 22.8μ strain/mil, providing means of relating strain to midspan deflection.

4. EXPERIMENTAL CONDITIONS AND PROCEDURES

Tests were run over the following range of dimensionless parameters:

$$1.2 \leq \text{Strouhal no., } St_o \leq 0.35;\ 0.007 \leq \text{damping parameter, } \left(\frac{2\pi\xi M}{\rho D^2}\right) \leq 0.013;$$

$$5 \times 10^5 \leq \text{Reynolds no., } Re_o \leq 5 \times 10^6.$$

A typical test procedure involved setting the system pressure, $2.06 \leq P(\text{MPa}) \leq 16$, and temperature $165 \leq T(^oC) \leq 326$. Then flow rate was increased until the outlet nozzle coolant velocity, $12 \leq V_o$ (m/s) ≤ 30, reached the desired test set point. Velocities were below calculated values at which cavitation would occur. Data from strain gage tests were recorded for periods up to 30 minutes for later analysis, by methods of random data reduction [8]. After completing a given test point, the flow rate was increased to set the next outlet nozzle velocity. When moving on to the next (higher) velocity point, the strain gage output was monitored with a real time analyzer, anticipating large responses with the possible onset of fluid-elastic instability.

Sample points were also taken as the flow rate was decreased. There was no discernable sensitivity in the magnitude of tube response to the direction of approaching a test point.

5. EXPERIMENTAL

5.1. Flow Related Mechanisms for Flow Induced Response

Flow-induced excitation of tube in an array is usually attributed to one of three, what may be described as, near field mechanisms: vortex shedding [3], fluid-elastic instability [4], or turbulent buffeting [2]. In addition, there are far-field effects, present in most tube banks, the result of pressure fluctuations due to a variety of periodic (e.g., pump rotor and blade passing frequencies, acoustic pulsations) and random (e.g., flow separation from structures, upstream turbulence) sources [7]. Amplitude response to periodic vortex shedding is characterized by a narrow band forced response at the frequency of the vortex shedding plus response at the tube's natural frequency due to random near and/or far-field effects.

Spectral response to the mechanism of fluid-elastic instability is normally characterized by a narrow band large amplitude peak at the tube natural frequency. A plot of RMS amplitude vs. flow shows, as for a true instability, a large increase in amplitude for a small change in velocity.

Turbulent buffeting response spectra are characterized by wide band random motion at the characteristic frequencies of the turbulent pressure fluctuations on the tube (near field effects), plus a narrow band response at the tube natural frequency, due to near and/or far field effects.

5.2 Observation Based on Present Experiments

Response spectra from tube locations representative of different crossflow/tube conditions are shown in Figures 2A to F. These plots show amplitude (strain) spectra (PSD) and amplitude probability distributions (APD) at each flow rate for each of the selected tubes. Since there is a great degree of qualitative similarity in these spectra at different temperatures, only the spectra at 326°C will be reviewed. Based on a review of these data, the following observations can be made, which are applicable to all of these representative tubes.

Review of the APDs show that some tubes have evidence of distinct periodicity, while all have a combination of periodic and random response. The APDs that are Gaussian in nature, but with a flattened peak will be defined as evidence of incipient periodicity. In general the PSDs are similar for all tubes: a broad, low magnitude random response peak at about 200-300 Hz in addition to a number of higher amplitude narrow band peaks within the frequency. The difference may, in part, be due to flow effects on hydrodynamic mass.

At the lower flow rates, the Gaussian nature of the APDs and the multi-peak character of the PSDs near the natural frequency indicate that the response is due primarily to random turbulent buffeting. This buffeting is due to near field effects, quite evident in the "band pass" nature of the spectra, e.g., tube 3Y at 200-300 Hz, and far field effects that cause the response about the natural frequency. Since the tube is lightly damped, the transfer function has a small half power band width at fn; the transfer function is almost constant until f ≈ fn. This means that the response spectrum is a good qualitative representation of the turbulent buffeting spectrum. The dominant peak of this spectrum can be reduced to a value of Strouhal number $(f_tD/V_g)^+$ and plotted vs. Re_g as shown in Figure 3B. For these test conditions, the turbulent buffeting Strouhal number is about 0.18 independent of Re_g. This value is close to that of 0.2 observed by Weaver [5], and 0.16 predicted from equation (5) of Owen [2].

$^+$refer to nomenclature

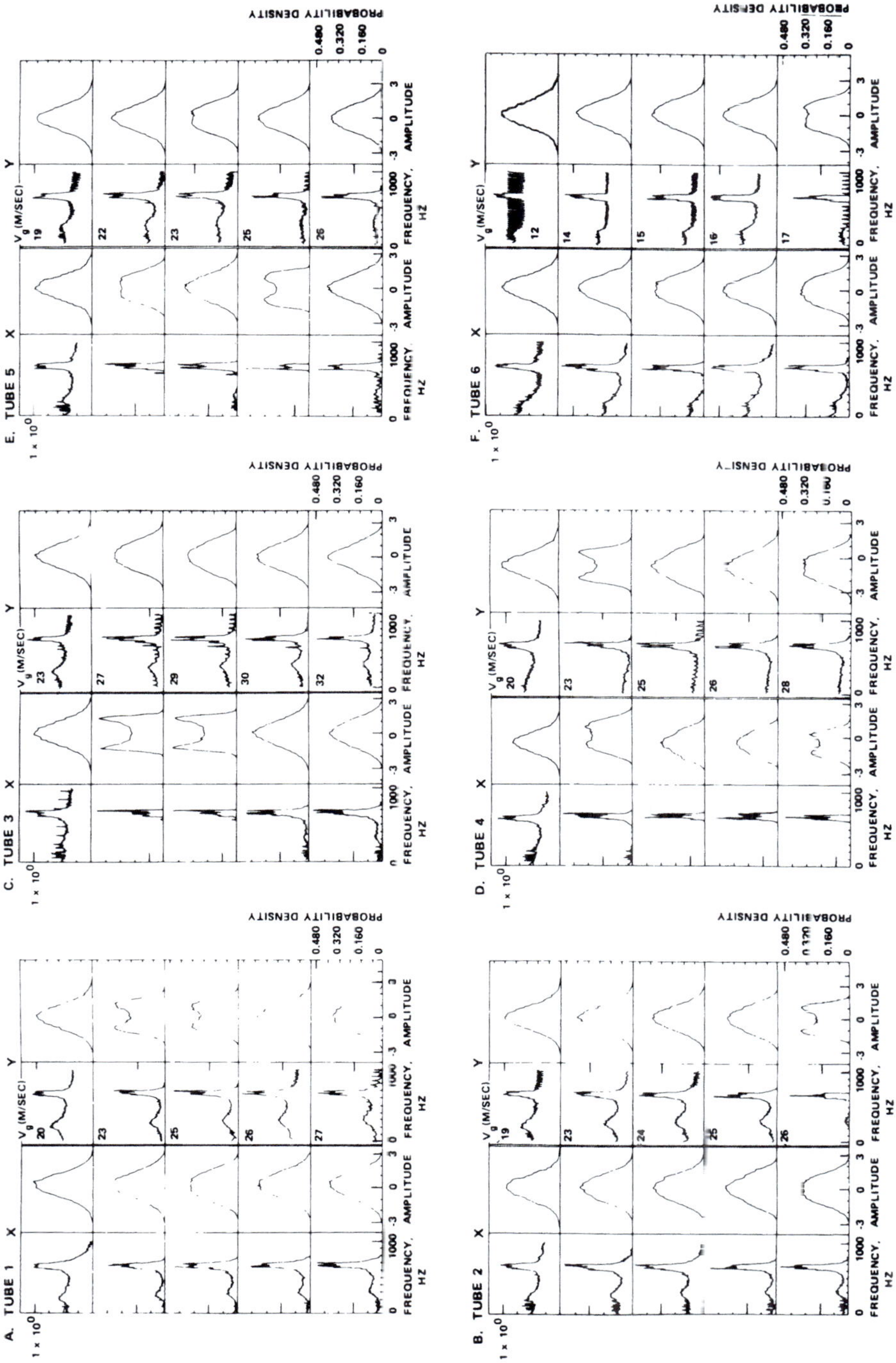

FIG. 2 RESPONSE SPECTRA vs INTERTUBE VELOCITY

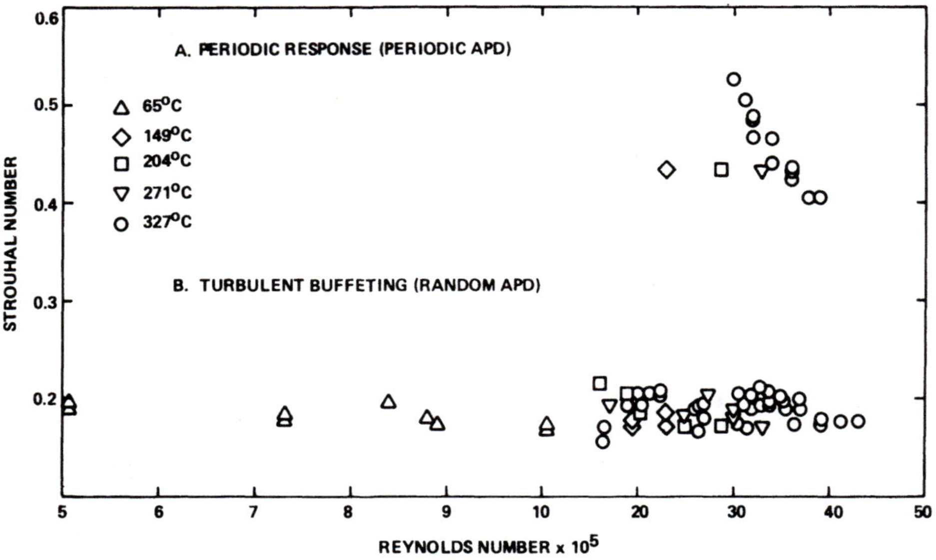

FIG. 3 STROUHAL vs REYNOLDS NUMBER

As the flow rate is increased, the APDs are less Gaussian in nature, growing wider and flatter. Some attain a shape characteristic of an APD for mixed random and periodic response, similar to those observed for single tubes [8]. As flow is further increased, some tubes (e.g., 3X, 5X) display a definite periodic response at their natural frequency. A Strouhal number for this periodic response (f_pD/V_g) can be calculated from these data and is summarized versus Re_g in Figure 3A. The nominal value of about 0.45 falls within the range observed by Pettigrew [6] for vortex shedding in tube banks at $10^4 < Re_g < 10^5$. However, it is 40% higher than the Chen [3] value of 0.31 for vortex shedding Strouhal number.

This vortex shedding cannot be classified as truly periodic. Indeed repeat of the same test point often showed the motion to be random. Its nature may thus be classified, per Jones [9], as quasi-periodic, or intermittent, in that the coherent flow pattern required for periodicity appears to break down, causing the vortex shedding to appear and disappear alternately.

Fluid-elastic response could also exhibit this large amplitude periodic behavior. Use of Weaver's [5] equation for prediction of fluid elastic motion results in a critical Strouhal number of about 0.40, though Connor's [4] results predict a value of 4. Closeness of the first value to that computed from these data suggest that these large amplitude periodic responses may be related to a fluid-elastic instability. However, rms amplitude would continue to increase rapidly with small increase in flow, whereas here the amplitude decreased as flow increased (Figure 4). If the motion is due to a coupled fluid/tube interaction, then this decrease could be explained as follows. Studies, e.g., Chen [10], have shown that, due to hydrodynamic coupling, both hydrodynamic mass and damping are functions of the vibration response pattern of the coupled structures. Thus it may be possible that an increase in velocity changes the response pattern between adjacent tubes, resulting in variations in both hydrodynamic mass (or frequency) and damping of the tube being moni-

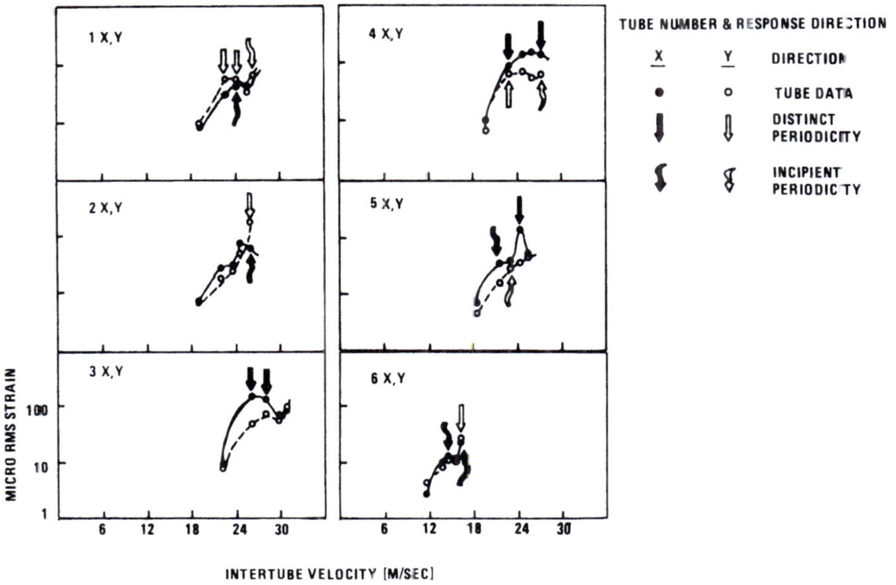

FIG. 4 RMS AMPLITUDE vs INTERTUBE VELOCITY

tored. This shift in response frequency with flow is generally evident in the plots of Fig. 2. These changes may explain the decrease in amplitude with flow contrary to what has previously been observed with tests in air where any hydrodynamic coupling is small. The subsequent increase of amplitude with further increase in flow suggests a different coupled mode of motion which then succumbs to instability at higher flows.

6. CONCLUSIONS

A series of flow tests with water were done on a section of a tube bank, in which the flow pattern consisted of both axial and crossflow. Maximum velocities were found to occur at the tubes adjacent to the exit nozzle where the flow was primarily crossflow. The intertube velocity range for this crossflow was 12 to 30 m/sec, at temperatures from 66 C to 326 C. The corresponding Reynolds number range for these conditions varied from $5x10^5$ to $5x10^5$ (trans - to supercritical).

Dynamic response data showed that at the lower flow rates, the tubes had a broad band response to turbulent buffeting over a frequency range below that of the first mode natural frequency, plus narrow band response centered about the first mode natural frequency. The rms value of amplitude increased in proportion to the velocity, until at certain values of velocity, large increases in rms amplitude were noted. Examination of the spectral data led to the conclusion that neither intermittent vortex shedding nor fluid-elastic induced instability could be identified as the primary excitation mechanism. Strouhal numbers, based on tube diameter and intertube velocity, were computed from the data, the values being $\sim$0.18 for the random response at lower frequency and $\sim$0.45 for the periodic response at the natural frequency.

These experimental results suggest that within the trans - to supercritical Reynolds number range, tubes within a tube bank, particularly the exit rows, can be subjected to flow-induced excitation consisting of near field turbulent buffeting and a not clearly identified mechanism that produced periodic response.

7. NOMENCLATURE

f_i	frequency
n	natural (first axial mode)
t	turbulent buffeting, random response
v	vortex shedding
p	periodic response
V_i	fluid velocity
o	outlet (average)
g	intertube
c	critical
D	tube diameter
M	mass
ρ	mass density
ξ	damping ratio
ν	kinematic viscosity

Re_i	$= VD/\nu$	Reynolds number
g	V_gD/ν	Intertube velocity
o	V_oD/ν	Outlet velocity
St_i	$= f\ D/V$	Strouhal number
c	f_nD/V_c	critical velocity, fluid elastic motion
t	f_tD/V_g	turbulent buffeting
v	f_vD/V_g	vortex shedding
p	f_pD/V_g	periodic response
o	f_nD/V_o	outlet velocity

8. REFERENCES

1. Savkar, S. D., "A Survey of Flow Induced Vibrations of Cylindrical Arrays in Cross Flow," ASME-76-NO/FE-21.

2. Owen, P. R., "Buffeting Excitation of Boiler Tube Vibration," J. Mech. Science, 7, 4, 891-914, (1965).

3. Chen, Y. N., "Flow-Induced Vibration and Noise in Tube-Bank Heat Exchangers Due to Von Karman Streets," J. Engr. for Industry, 90, 134-146, (1968).

4. Connors, H. J., "Fluid-Elastic Vibration of Tube Arrays Excited by Cross Flow," Proc. Symp on Flow-Induced Vibrations in Heat Exchangers, ASME, New York, 42-56, (Dec. 1970).

5. Weaver, D. W., Grover, L. K., "Cross Flow Induced Vibrations in a Tube Bank," J. Sound and Vibration, 59, 263-294, (1978).

6. Pettigrew, M., "Vibration of Heat Exchanger Components in Liquid & Two Phase Cross-Flow," Vibration in Nuclear Plant, Keswick, U.K., May 1978.

7. Lubin, B. T., "Response of a Tube Bank to Turbulent Crossflow Induced Excitation," ASME 77-DET-142.

8. Bendat, J. S., Piersol, A. G., Random Data: Analysis and Measurement Procedures, Wiley Interscience, New York, (1971).

9. Jones, G. W., et al., "Aerodynamic Forces on Stationary and Oscillating Circular Cylinders at High Reynolds Numbers," NASA-TR-R-300.

10. Chen, S. S., "A Mathematical Model for Cross Flow Induced Vibrations of Tube Rows," ANL-CT-77-4.

TUBE BANK

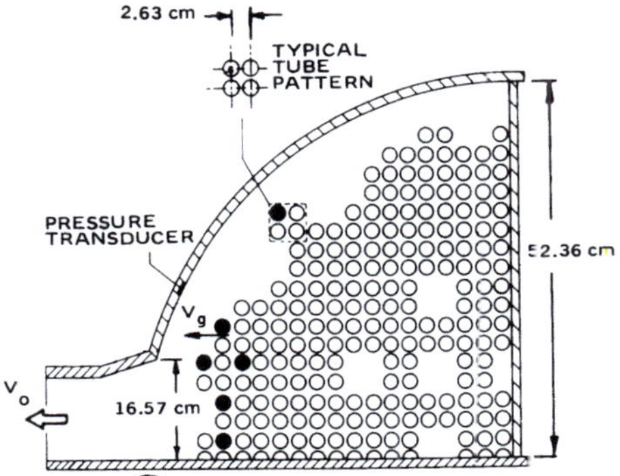

① EXPERIMENTAL SET-UP

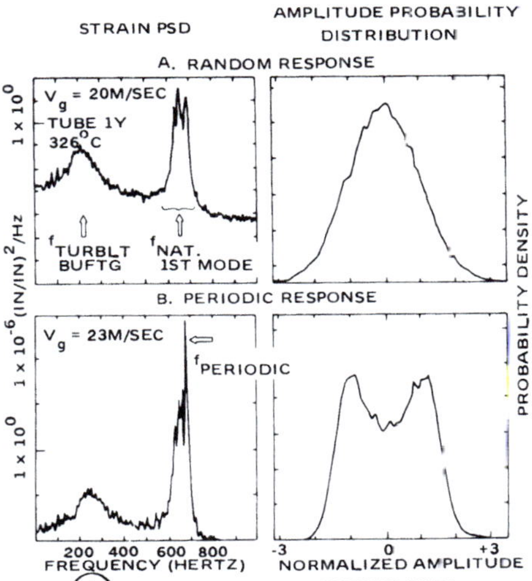

② REPRESENTATIVE RESPONSE DATA

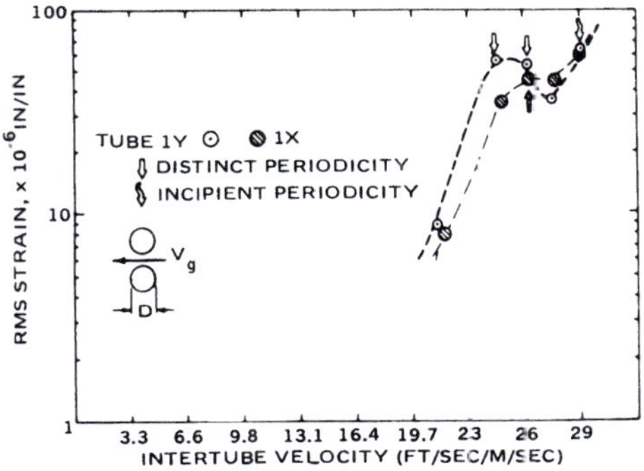

③. REPRESENTATIVE RMS AMPLITUDE RESPONSE vs INTERTUBE VELOCITY

Problem: Determine dynamic response of tubes in a tube bank subject to crossflow of water at Re $> 10^6$. Identify mechanism for flow-induced excitation.

Geometry: 201 stainless steel tubes, square pitch array mounted on quarter-circle tube sheets (Sketch 1).

Dynamic Characteristics: No flow tests, frequency and damping; air, 770-797 Hz, .0043-.0089%; water 21°C, 705-732 Hz, .035-.058%.

Flow conditions: Flow enters axially at bottom, becoming predominantly crossflow to tubes near exit. Turbulence levels ~ 1%.

Operating conditions: $2.06 \le P(MPa) \le 16$

$65^{\circ} \le T(^{\circ}C) \le 326^{\circ}$, $12 \le V_o$ (m/s) ≤ 30

Measurements: Velocity; outlet (V_o) orifice plate, intertube (V_g) based on ratio of outlet to local from geometrically similar test, response; strain gages (4) on id of selected tubes.

Data acquisition & reduction: Response recorded as time histories. Treated as random data reduced to PSDs and APDs (Sketch 2).

Observations: At lower flow rates, tubes have broad-band response to turbulent buffeting over a frequency range below that of the first mode natural frequency, plus narrow-band response centered about first mode natural frequency (Sketch 2A). Rms amplitude (Sketch 3) increases proportional to velocity until at certain velocities large increases were noted. Response was then periodic (Sketch 2B). Review of spectral data led to conclusion that neither intermittent vortex shedding nor fluid-elastic induced instability could be identified as primary excitation mechanism. Strouhal numbers ($f_i D/V_g$), computed from these data were .18 for the random response (f_t) and .45 for the periodic response (f_p).

A7 FLOW INDUCED VIBRATION PROBLEMS IN GAS COOLED REACTORS

Michael Parkin
Engineer

UKAEA, Windscale, England

SUMMARY

This paper describes practical experience with flow-induced vibration of fuel and other components in the fuel channels of two gas cooled reactors. The Dungeness 'B' examples are related to refuelling with the reactor operating, whilst the Windscale problems are related to "in-core" conditions. All but one of the problems are associated with annular flow, and taken together, they present an indication of the range of instability mechanisms likely to be found in annular flow situations.

1. INTRODUCTION

In common with other reactor systems, the Advanced Gas-cooled Reactors (AGR) now coming into operation in the United Kingdom contain equipment which is normally inaccessible, and which may be subject to flow-induced vibration. The remote nature of reactor equipment means that it is possible for a vibration problem to remain undetected until there is some degree of damage. Furthermore, investigations into the cause of the problem and subsequent repairs or modifications would be difficult and expensive. For these reasons "in-the-field" experience is sought in full-scale engineering demonstration facilities prior to reactor operation.

By adopting a policy of testing fully-developed production components, the final demonstration comes at an advanced stage of the power station construction. Thus the solution of problems found during the demonstration tests tends to fall into a remedial rather than developmental category, as would be the case "in-the-field".

This paper illustrates the benefits of this approach by describing the experience obtained in testing the Dungeness 'B' AGR fuel and channel components when subjected to the extreme and variable flows during refuelling, which is carried out with the reactor at full operating conditions. The test facility used in this work is described elsewhere by Robson [1]. Also offered are two examples of vibration problems which occurred, and were solved, in the fully operational Windscale AGR. These problems have the common feature of being flow-induced vibrations in annular flow passages.

2. DUNGENESS 'B' - VIBRATION DURING REFUELLING OPERATIONS

2.1 General description of reactor

The Dungeness 'B' AGR combines biological shielding and pressure containment in a concrete pressure vessel. The reactor core, heat exchangers and gas circulators are all situated within this vessel as shown in Figure 1. The basic gas flow path linking the three components is controlled by the gas baffle which separates the relatively cool core inlet gas from the hot outlet gas. Sub-divisions of the core inlet gas flow path maintain the gas baffle, core and fuel channel components at inlet gas temperature.

The reactors were designed so that refuelling could take place at full power operating conditions. To give access, there is a penetration through the roof of the concrete pressure vessel for each fuel channel, forming the standpipe. A guide tube provides continuity of the refuelling route between the standpipe and associated fuel channel. The guide tube is supported by the gas baffle, and accommodates misalignment of the fuel channel relative to the standpipe due to thermal and irradiation-induced distortion of the core. There is also provision for discharge of the hot gas from the channel at the guide tube/standpipe intersection. These features are shown in more detail in Figure 2.

To allow refuelling to be carried out in a single operation, the fuel is built up into a single assembly - a fuel stringer. This includes the fuel elements, neutron, gamma and thermal shields, a flow control valve and finally a combined pressure seal and hold down mechanism - the latch. This fuel stringer is illustrated in Figure 3. From the neutron shield upwards the components form a linked articulated unit. The fuel elements are supported from the bottom, the bottom support being suspended from a central tie bar which is connected to the non-fuel components just below the neutron shield.

A refuelling machine completes the requirement. This is essentially a hoist mechanism enclosed in a pressure vessel which is moved over and sealed to the channel requiring refuelling. The refuelling machine is indicated in Figure 1. It is capable of removing a spent fuel stringer and inserting a new fuel stringer in a single continuous operation.

2.2 Flow conditions during refuelling

There are more than 300 fuel channels in the reactor. Removal of a fuel stringer from one channel does not significantly alter the overall core flow resistance so the flow in that channel increases as the fuel stringer is withdrawn. The various flow paths into the channel are shown in Figure 2. Each flow serves a specific cooling function when the fuel stringer is fully inserted in the channel, but during refuelling they join to form a bulk upward flow referred to as empty channel flow. The way this empty channel flow varies in magnitude with depth of insertion of the fuel stringer is illustrated in Figure 4.

When the empty channel flow meets the fuel stringer, part flows up the annulus between the fuel stringer and the channel, and the balance flows up inside the fuel stringer, out of the gas ports and down the fuel stringer-channel annulus. These two annular flows meet and leave the channel at the channel gas discharge region. These flow paths are illustrated in Figure 2, and some idea of the ratio of the two annular flows may be obtained from Figure 4.

The work described in this section is concerned with the vibrational effects of these flows upon both the fuel stringer and channel components.

2.3 Fuel stringer vibration

Whilst there were no problems with fuel stringer vibration, the topic is reviewed here because of its relationship to channel component vibration.

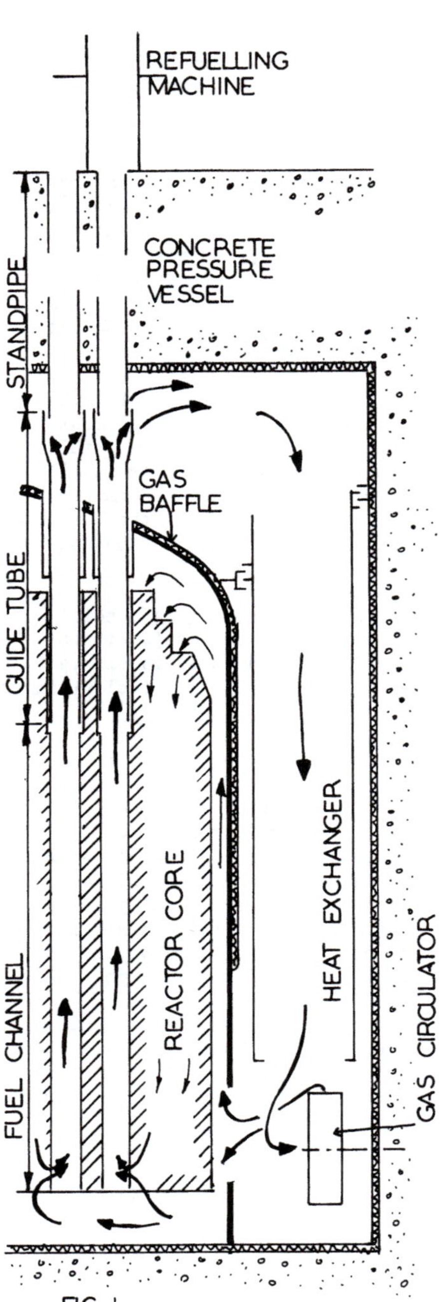

FIG 1
DUNGENESS "B" REACTOR

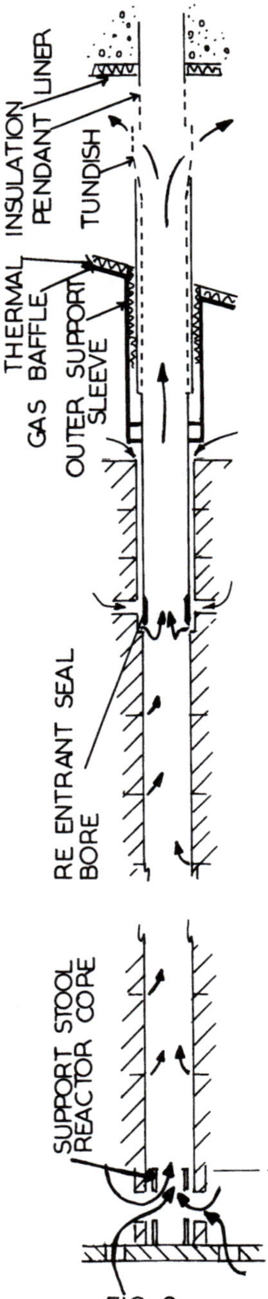

FIG 2
TYPICAL CHANNEL

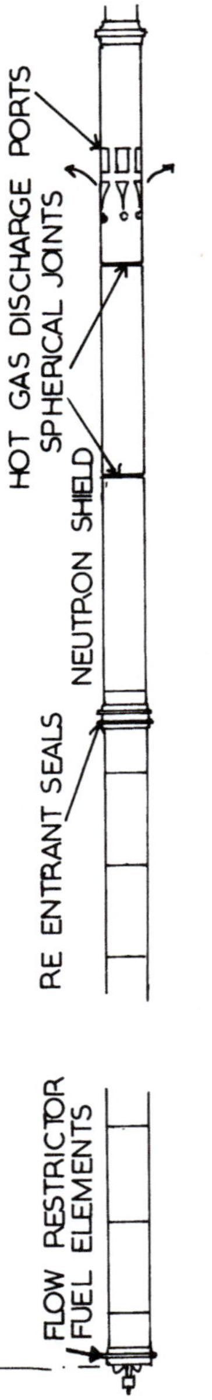

FIG 3
FUEL
STRINGER

During the refuelling process, the fuel stringer is freely suspended from the refuelling machine hoist chain. The flow energy dissipated in the fuel stringer and channel varies from 680 KW just before the fuel stringer enters the guide tube to 90 KW when fully inserted in the core. There are several mechanisms providing the force to vibrate the fuel stringer.

a. Consideration of the aerodynamic forces around the moving fuel stringer suggests that negative aerodynamic damping provides means of supplying energy from flow into vibration. This is discussed by Denton [2] with reference to a valve plug suspended in a channel, another reactor problem.

b. Fricke [3] found a vortex bursting phenomena in an annular diffuser with an eccentric centre body. This is a geometry likely to occur in a reactor channel.

c. Horne [4] describes an instability caused by opposing flows discharging from a common peripheral opening, like that in the guide tube hot gas discharge port. Here a stagnation region in the lower velocity passage outlet (when $\frac{v1}{v2} \leqslant 0.3$) restricts that outlet and leads to peripheral flow switching.

Under these circumstances it is extremely difficult to eliminate vibration of the fuel stringer, so the channel design was evolved to minimise vibration to a level where there is no risk of damaging the graphite outer sleeves of the fuel elements. Definition of the limiting criterion for graphite damage is the velocity of approach on impact. A practical limiting value for these fuel elements has been determined to be 0.45 m/s. Measurement of velocity of vibration is made by capacitive displacement transducers devised for this test work by Walton [5], with subsequent differentiation to provide velocity information.

Initial test work had shown that of the several planes of transducers tried, those shown at plane xx, Figure 4, proved to be at the most convenient location, and measured the highest velocities. This observation also fitted the position and simple mode shape of the fuel stringer during the most significant periods of vibration.

On conclusion of the demonstration tests, the worst combination of fuel stringer and channel variables were found to subject the fuel stringer to scalar velocities which vary with fuel stringer insertion as shown in Figure 5. An inspection of the vibrational behaviour of the fuel stringer revealed that whilst the velocities recorded in Figure 5 were in some cases higher than the limiting velocity of approach there was in fact a substantial safety margin.

2.4 Channel component vibration

As described in the previous section, production channel components were assembled in the test rig for demonstration tests of the refuelling cycle. This presented an ideal opportunity to check the channel components themselves. Ironically, the greatest cause for concern proved to be the channel inlet and channel outlet components which were not subject to measurements as was the guide tube. The behaviour of the guide tube will be outlined below and the channel inlet and channel outlet components are each the subject of a Chart.

2.4.1 Guide tube

Vibration of the guide tube was of interest from three points of view, one, the integrity of the guide tube; two, the possibility that a vibrating guide tube

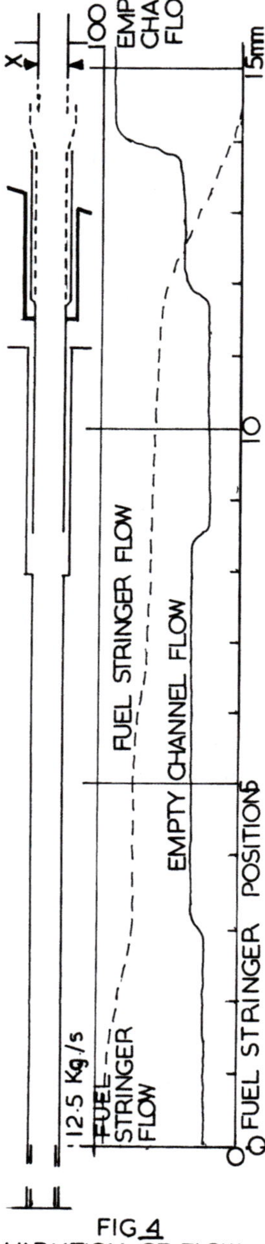

FIG 4
VARIATION OF FLOW DURING REFUELLING

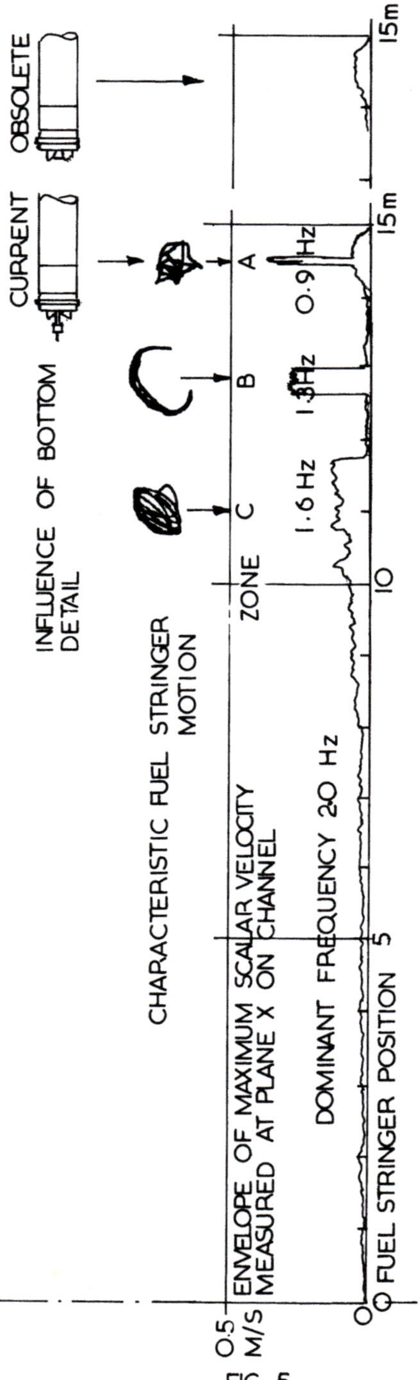

FIG 5
FUEL STRINGER VIBRATION

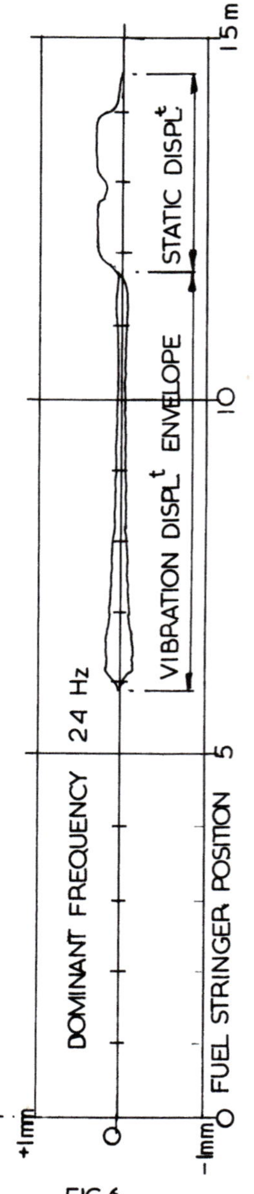

FIG 6
GUIDE TUBE VIBRATION

could damage gas baffle insulation shown in Figure 2; and three, that a vibrating guide tube could aggravate vibration of the fuel stringer.

The characteristic vibration of the guide tube revealed two features of importance which are identified in Figure 6, which is a graph of typical guide tube displacement plotted against fuel stringer insertion.

a. When the fuel stringer enters the upper part of the guide tube it is deflected radially by the gas forces, presses against the guide tube and produces the static deflection shown in Figure 6. This deflection becomes zero again when region C vibration of the fuel stringer commences as shown in Figure 5.

b. At greater fuel stringer insertion there is significant charge tube vibration coincident with the increase in annular flow shown in Figure 4.

Both phenomena were found to be acceptable.

2.4.2 Channel inlet component - Chart

This illustrates the value of full scale demonstration tests with production components for it was during the course of these tests that cracking was found in this component. This component had not been considered to be at risk and no measurements of its vibrational behaviour were available.

The support stool which is protected by the sleeve in question is located at the bottom of the fuel channel as shown in Figure 2. This takes the weight of the fuel stringer when the latter is in the fully loaded position. The fuel stringer coolant flow passes through ports in the support stool and the function of the sleeve is to shroud the support stool from this gas and thereby prevent corrosion of the support stool. The bottom edge of the sleeve is shaped to fit round the gas ports in the support stool. The cracking found could lead possibly to a portion of the sleeve becoming detached with consequent risk of flow restriction or damage in the gas circuit and especially within the fuel elements.

Investigation of the problem included answering the question "Is it a reactor phenomenon or an effect peculiar to the test rig, for example a consequence of the noise generated by the gas circulators?" The problem did prove to be a reactor problem, caused by flow through the ports and of significance only at the high empty channel mass flow occurring with the fuel stringer at or above 12 metres of lift, see Figure 4.

Excitation appeared to be an edge effect but this also provided damping reducing the magnitude of the problem. A possible solution proved to be that of ducting the flow through the ports. This was not adopted since the reactor core was complete and the extremely difficult access to gas ports threw doubts upon the subsequent reliability of any possible method of applying this modification.

Tests were carried out defining flows and durations necessary to cause cracking, and measuring propagation rates. From these data, and some flow endurance tests yet to come, a flow level will be confirmed which will give a life to crack initiation with a safety factor of 3 over the reactor requirement.

Measurements of strain at the crack site were also made to provide data for comparison with supporting mechanical fatigue tests. During these measurements it was observed that the inner/outer surface strain ratio was of the order of 5 even though the material was 1 mm thick. This was due to the adjacent stiffening flute affecting the position of the neutral axis. It explained the initial inability to justify fatigue on the basis of external surface strain measurements only.

2.4.3 Channel outlet component - Chart

The perforated pendant liner located at the lower end of the standpipe, together with the perforated tundish at the top end of the guide tube (Figure 2) provide a stable solution to vibration caused at the channel gas discharge region.

The high vibration levels discovered in the pendant liner were first observed in a full scale test facility at the Marchwood Engineering Laboratories (MEL) of the CEGB. Tests were being carried out on a fuel stringer and a prototype pendant liner had been fitted as part of the simulated channel geometry. Cracks were observed in the ligaments between perforations in this component. Whilst it was recognised that this example did not fully represent the production unit (for example, it was constructed of thinner material) an investigation was mounted to clarify the situation. Unlike the previous example, the pendant liners were manufactured but not then installed, thus some degree of modification was a possibility. The test work described was carried out jointly by MEL and the UKAEA at Springfields and Windscale.

Initial test work confirmed that this was a reactor problem associated with flow between the pendant liner and the fuel stringer and important only with the fuel stringer at 1.8 to 10.4 metres of lift. The duration of excitation was 30 minutes per refuelling cycle, that is 180 minutes during the reactor lifetime. The modes of vibration were found to be predominantly the 4th and 5th ovalling modes with some vibration in the 3rd and 6th modes; the associated frequencies were 500, 930, 1400 and 1960 Hz. The structure proved to be very lightly damped, damping factors were typically 0.03% - 0.1%.

The mechanism of vibration was not positively identified. Several possibilities exist, vortex shedding, an aerodynamic instability based on the lift forces acting on the ligaments between the perforations, or an instability in the flow diffusing through the perforations. Each would be associated with the elastic component vibration. It is also possible that the basic vibration mechainsm is amplified by a cascading effect across the 21 rows of holes. Whichever mechanism it was, there was every appearance of "locking-on" between the excitation mechanism and the structural response.

Two lines of modification were examined in the limited time available. Destroying the mechanism was tried by modifying the axial cross section of lines of holes. For example, chamfering the holes on the inside surface of the liner appeared effective but the limited test results were not entirely consistent. Damping was the alternative tried, but though it proved most effective it was abandoned because of the difficulties of applying it, without proper development, in a way which would be truly reliable at 650°C for 30 years.

Careful measurements of strain level were made, making correction for the damping effect of fitting strain gauges (often a factor as large as 2), and stress concentration factors relating measured to peak strains. From these data assessment of the fatigue life of the pendant liner indicated as adequate safety factor. This was supported by fatigue tests on real samoles. No modification was therefore made. Finally, it should be noted that once installed, this component is easily inspected.

3. WINDSCALE AGR VIBRATION

3.1 General

Experience with the prototype, Windscale AGR (Parkin [7]) indicated the need for careful development and proving of refuelling under operating conditions for the subsequent AGR's. The work illustrated in the first part of the paper is typical of this approach. There were in addition two channel vibration problems peculiar to the Windscale AGR and these are described in the two final charts.

Whilst there are many detail differences in design of the Windscale AGR compared with the later versions, the essential features of the reactors are common as are the flow paths, so Figure 2 applies in principle, and nomenclature used is the same for both reactors.

3.2 Reactor control mechanism - Chart

Unlike most of the later AGR's the control mechanisms are located in channels identical to fuel channels. The sleeve (Chart Sketch 1) is therefore provided to act as a channel flow restrictor, limiting the channel flow to the modest amount necessary to provide control rod cooling. The sleeve forms part of the control mechanism assembly which, like a fuel stringer can be taken out of the channel in a single operation. It is suspended from the winding mechanism assembly, the connection between the two being a spherical joint. There is another spherical joint further up the winding mechanism assembly.

The vibration problem was first observed with the original multicomponent sleeve which shook itself to pieces. This was replaced by an all welded design with a modified flow path, shown in the chart, with no attempt to tackle the vibration problem at source. Following serious damage to a high proportion of these modified sleeves an investigation was instituted with two objectives, one to find a solution and two to keep the reactor operating in the meantime.

The solution was sought in a 0.5 scale atmospheric pressure air rig first by reproducing the problem. This was concluded to be a fluid-impact excitation of random nature, consequent upon flow through an annular passage formed by one fixed (the sleeve) and one loosely located component (the guide tube). Solutions avoiding annular flow, or retaining it in a rigid passage were found and applied in the reactor. The first, an easy modification, was used as a short term expediant (Chart Figure 3). The second (Chart Figure 4), the permanent solution, avoided flow between the sleeve and the control rod, - a desirable reactor safety feature.

Meanwhile, continued reactor operation was achieved and protected by regular external vibration monitoring. As shown in the chart, a very characteristic difference in frequency spectra exists between healthy and the faulty conditions. The discrete frequencies are the gas circulator blade passing frequency and harmonics.

It is worth noting that, in the interests of speed, the air rig modelled only scale, buoyancy and dynamic flow forces. Emphasis on modelling the seals was placed on flow restriction, with only nominal reproduction of spring forces. This limited modelling proved quite adequate. Another example of investigation of a reactor problem, which illustrates the value of limited modelling, is described by Bolton [8].

3.3 Reactor core pressure tube - Chart

The Windscale AGR incorporates a number of experimental facilities. To provide independent control of the gas composition and flow in these facilities, each is contained in a pressure tube. The pressure tube passes through a standard fuel channel, access at the top is the same as for a normal fuel channel, and a gas circulator and associated plant is connected to the bottom. Part of the cooling is provided by the reactor gas flow passing up the annulus between the pressure tube and the core. It was this flow which caused vibration of the pressure tube seen at the top access point during initial commissioning.

The amplitude and mode shape of the vibrating pressure tube were measured by suspending a non-contacting displacement transducer within the tube. It was concluded that this was a fluid-elastic vibration of a general nature, rather than the consequence of a specific feature, as discussed in section 2.3. The preferred solution was that of fitting axial fins which would both locate the tube within the fuel channel and restrict peripheral flow movements, but this was rejected because of doubts about the practicality of fitting such a tube into a distorted channel, and also the need to minimise neutron absorption.

Instead, restraints were fitted at 5 alternate positions. This method was first checked by simulating the highest restraint only, the results were promising so all 5 were fitted in the form of pads attached to the tube and each giving a small positive radial deflection to the pressure tube when in position. These deflections were checked by an alignment survey and the loads associated with them estimated. This information was used in a wear assessment once final vibration measurements were available and it was concluded to be acceptable.

After 5 years of operation, there is no evidence of any deterioration.

4. CONCLUSIONS

With the present limited state of understanding of flow induced-vibration in annular flow passages, together with the many variables that have to be taken into account, it is clear from the experience illustrated in this paper that full scale testing of production components for this type of reactor application is highly desirable. These examples illustrate the need for a careful assessment of all components in contact with the flow, the assessment being based on vibration measurement supported by thorough inspection after testing. The understanding gained from this testing defines components which require further surveillance, during commissioning and into the early years of reactor operation. It is argued that the cost of testing in this manner is fully justified when set against the financial penalty of breskdown in an operational reactor, to say nothing of public concern for reliability and safety.

There is clearly a need for research into the basic mechanisms of flow induced vibration in an annular passage, especially the general case of a plain annulus with one bounding surface free or simply supported. There are also more specific problems of interest, for example the case of vibration induced by flow through and across one perforated boundary of the annulus. It is hoped that in the future, with a better understanding of such problems, or closer approach to a satisfactory solution may be achieved at the design stage.

5. ACKNOWLEDGEMENTS

The author would like to thank the Nuclear Power Company and the UKAEA for permission to publish this paper.

This has been very much a team effort and so I would like to express my special gratitude to my Windscale colleagues, and those throughout the industry, who have contributed so much.

6. REFERENCES

1. Robson J A, "Stability of reactor fuel elements during on-power loading." Paper 513 Symposium - Vibration Problems in Industry. NPL/UKAEA Keswick, (1973).

2. Denton J D et al, "Vibration mechanisms associated with annular flow through a flow control device," Paper 5.5 Symposium - Vibration in Nuclear Power Plant. BNES/UKAEA Keswick, (1978).

3. Fricke, F R et al, "Periodically in annular diffuser flow." Paper A6 Symposium - Flow Induced Structural Vibration, IUTAM/IAHR Karlsruhe, (1972).

4. Horne J L, "The attenuation of the unsteady flow state of two directly impinging co-axial annular flows," Paper 517 Symposium - Vibration Problems in Industry, NPL/UKAEA Keswick, (1975).

5. Walton H, "Capacitance transducers in severe environments." Conference - Transducer 77, London, (1977).

6. Jones C H, "Out-of-reactor endorsement of the on-load refuelling of MK 2 gas cooled reactors," Paper 516 Symposium - Vibration Problems in Industry - NPL/UKAEA Keswick, (1973).

7. Parkin M W, "A review of basic studies and modelling applied to annular flow," Paper 628 Symposium - Vibration Problems in Industry - NPL/UKAEA Keswick, (1973).

8. Bolton A D et al, "Evolution of the modifications designed to supress the damaging vibration of the Hinkley/Hunterston AGR gags," Paper 5.2 Symposium - Vibration in Nuclear Power Plant BNES/UKAEA Keswick, (1978).

REACTOR CHANNEL INLET COMPONENT

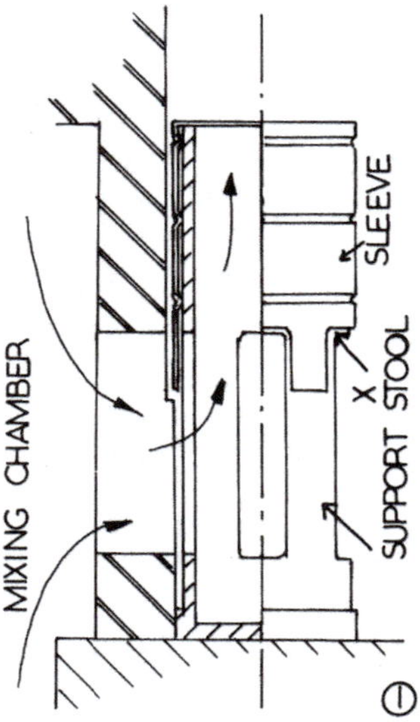

①

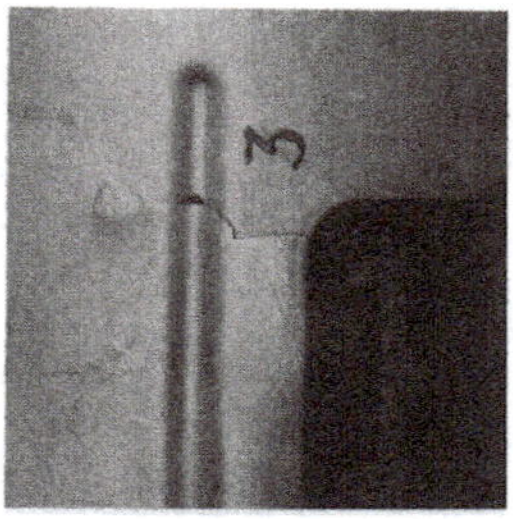

② SLEEVE DAMAGE AT X

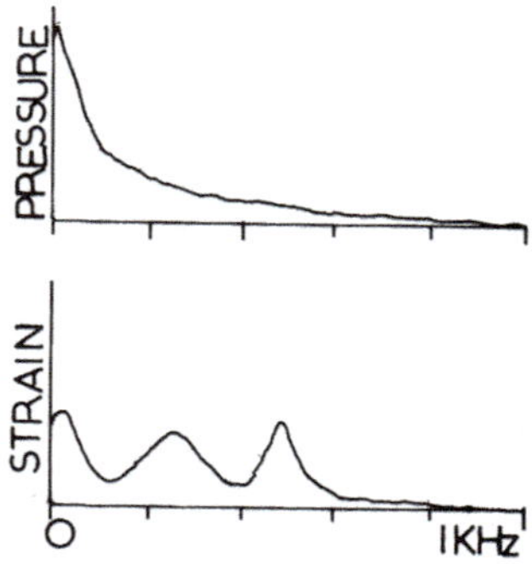

③

Problem: Stainless steel sleeves fitted to fuel support stools at channel inlet. These sleeves are required to shroud the support stools from hot gas thus preventing corrosion of the stools. The lower section of each sleeve cut away to fit round gas inlet ports - Sketch 1. Enhanced flow through ports, during refuelling, caused cracking in this region - Sketch 2.

Geometry: Sketch 1. Note, 4 equi-spaced ports fed from separate mixing chambers.

Boundary conditions: Defined by geometry.

Flow conditions: CO_2 at a density of 30.3 Kg/m^3, temperature 320°C. Temperature of investigation ~ 100°C. Maximum mass flowrate through 4 ports 59 Kg/s. Average velocity per port 33 m/s.

Structural dynamics: Damping factor, typically 0.2% at 230 Hz, 0.06% at 480 Hz, the dominant response modes - Sketch 3 - with mechanical excitation; but approximately 2% in flow situation.

Vibration: Vibration of the sleeve proportional to dynamic head. Maximum stress measure 80 N/mm^2 total, rms at the inner surface adjacent to point of crack initiation. Inner : outer stress ratio ≃ 5:1. Significant cracking predicted within reactor life.

Cures:
1. Ducting the flow through the ports reduced strain by 68%. Insertion of this modification proved impractical.
2. Reduction of maximum mass flow to 75% design value by reduced power refuelling gives adequate life. (Note each sleeve subject to only 10 hours refuelling conditions in reactor life).
3. Remotely operated sleeve removal machine developed. Muff removal in lower temperature peripheral channels may be acceptable.

References: Parkin, M W, UKAEA(WNL), Sansom K A, UKAEA(SNL) - Unpublished work.

REACTOR CHANNEL OUTLET COMPONENT

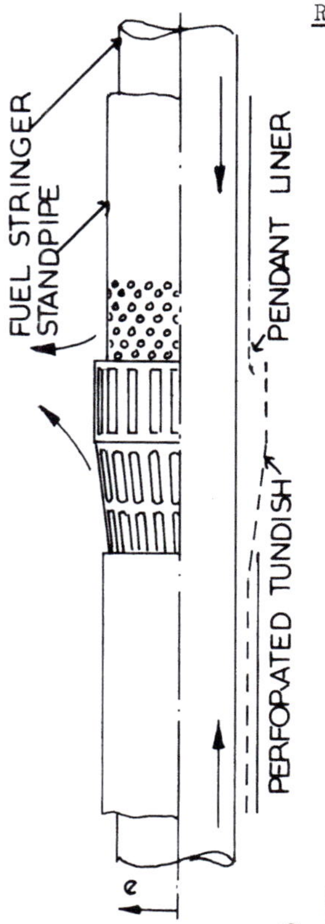

①

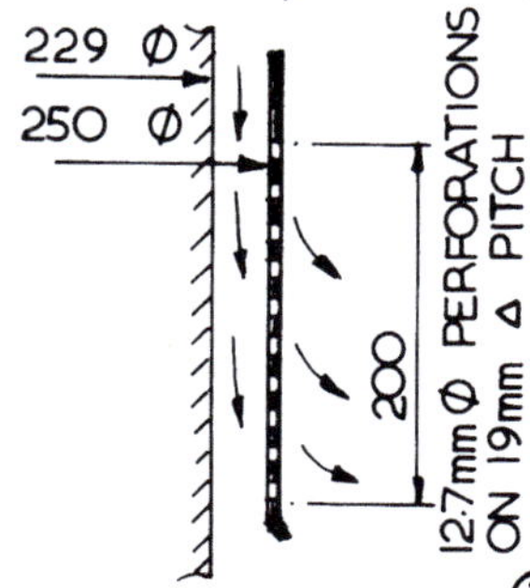

②

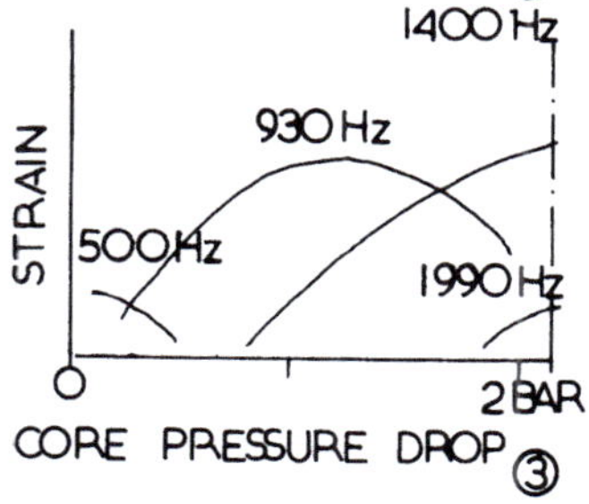

③

Problem: The perforated assembly in Sketch 1 forms the hot gas discharge section of a fuel channel. It is designed to minimize fuel stringer vibration during refuelling at full reactor power. At certain stages of refuelling the upper part of the assembly (the pendant liner) vibrates, with a risk of cracking between perforations.

Geometry: The significant geometry is formed by the pendant liner and the fuel stringer, Sketch 2. The eccentricity (e in Sketch 1) can be from zero to full annular clearance but this is unimportant providing there is no fuel/pendant liner contact. There is some variation in fuel stringer profile during pendant liner vibration.

Boundary conditions: Defined by geometry.

Flow conditions: Annular flow A in Sketch 2. Flow detail in and around perforations not known. Inlet conditions varaible but always $L/D \geqslant 80$. Approximate Re No 5×10^5.

Operating conditions: CO_2 at density of 20.8 Kg/m^3 and temperature 650°C. Temperature of investigations ~ 100°C. Estimated gas velocities in annulus in Fig 2 42 m/s maximum.

Structural dynamics: Pendant liner excited predominantly in 4th and 5th ring modes. Frequencies 928 and 1396 Hz. Damping factor 0.08% and 0.035%.

Vibration: Variation of flow indicated a "locking-on" effect, Sketch 3. Vibration levels saturated out at maximum strain of 1300×10^{-6} pk/pk at 1396 Hz. Strain concentration effects at the perforations require a further factor of 1.27.

Cures:

1. Damping very effective but no means found of applying this to 650°C reactor conditions.

2. Elimination of conjectured vortex shedding from the perforations plus cascading effect tried by spoiling techniques. Inconsistent results.

3. Life assessment supported by fatigue testing of perforated sections showed adequate life, though at strains greater than normally accepted by design. (NB. refuelling requires life of 10 hours per channel per reactor life).

References: Sansom, K UKAEA(SNL), Thompson, J G CEGB(MEL), Parkin M W, UKAEA(WNL) - Unpublished work.

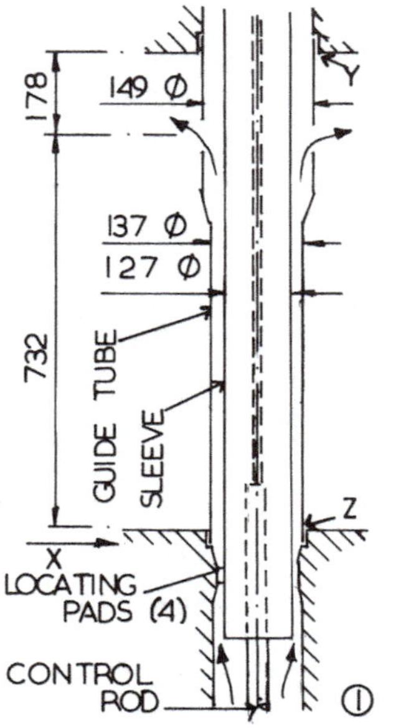

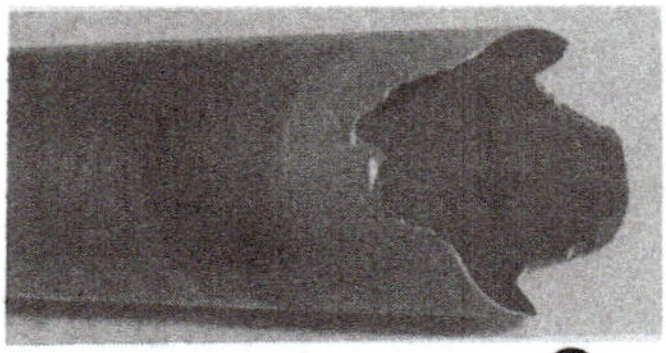

SLEEVE DAMAGE AT X ②

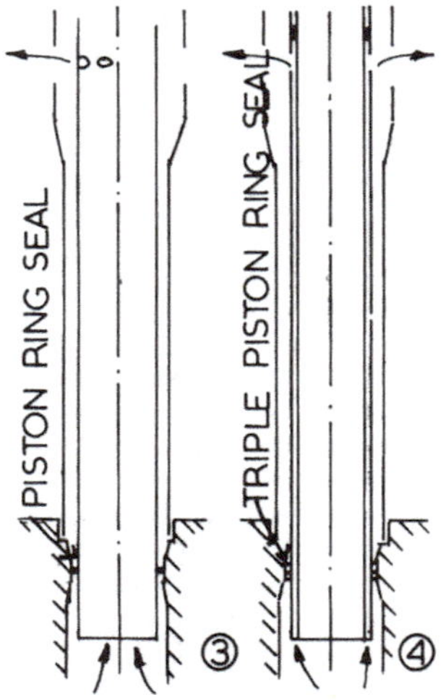

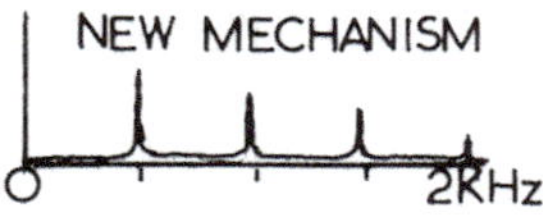

REACTOR CONTROL MECHANISM

Problem: Severe fretting damage of control mechanism sleeve and associated guide tube. Worst sleeve damage Sketch 2. There was guide tube damage at y and z (Sketch 1).

Geometry: The sleeve is the bottom section of a 7.45 m long unit fixed at the top, articulated at the one third lengths and located at the bottom by graphite pads. The guide tube is located, spigot and socket, at the top and bottom. See Sketch 1.

Boundary conditions: Axial flow as in Sketch 1. Sleeve has some radial freedom. Guide tube has radial freedom within tolerances of spigots and sockets (0.8 mm radial).

Flow conditions: CO_2 density 21 Kg/m^3, temperature 300°C. Estimated gas velocity in 5 mm annulus of 22 m/s. Pressure drop across section in Sketch 1. 145 kN/m^2. Upstream length 24m. Approximate Re No 9×10^4.

Operating conditions: Damage occurred during steady operation at the above flow conditions.

Structural dynamics: Not known.

Vibrations: Forced, random vibration of components with small radial freedom causing impact/wear.

Cures: Actual components inaccessible under operating conditions. Problem dealt with in 3 stages:-

(a) Remote vibration monitoring at the accessible top end of the mechanism provided operational safeguard, Sketch 5.

(b) Meanwhile, air model testing identified sleeve/guide tube annular flow as source of vibration and defined a temporary modification, Sketch 3.

(c) Replacement sleeve consistent with other reactor constraints designed, air rig tested and manufactured, see Sketch 4. 10 years trouble-free operation since fitting the replacements.

Reference: Parkin, M W. Paper 628. Symposium - Vibration problems in industry, UKAEA(NPL, Keswick, April 1973, Paper 628.

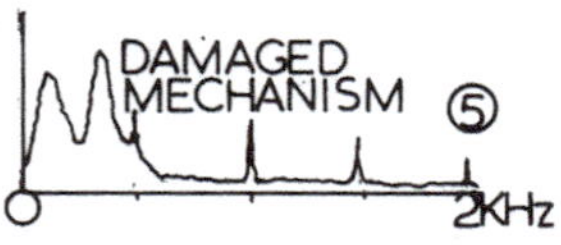

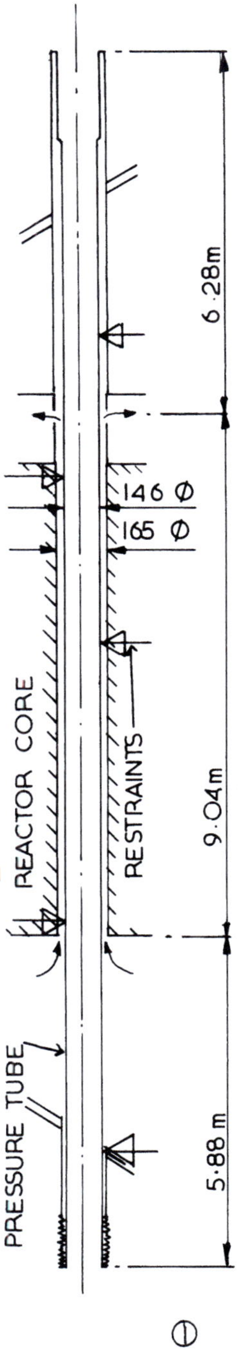

REACTOR CORE PRESSURE TUBE

Problem: The pressure tube provides for coolant control of the enclosed irradiation experiment separate from the main reactor coolant. The severity of beam-like vibration of the tube risked perforation of the pressure tube, damage to core components and damage to the enclosed irradiation experiment.

Geometry: The tube was welded to the outer ends of the top and bottom standpipes, (see Sketch 1). Spring loading, tensioning the tube when cold, avoided compressive collapse at operating conditions.

Boundary conditions: The central section of the tube was subject to annular flow and inlet and outlet conditions are equivalent to plenum chambers, see Sketch 1.

Flow conditions: CO_2 at density 21 Kg/m^2 and temperature 275°C. Estimated gas velocity in the annulus 11.8 m/s. Pressure drop across the annulus section 145 kN/m^2.

Operating conditions: Vibration observed on commissioning, testing carried out over range of conditions the maximum noted above.

Structural dynamics: Estimates of first beam mode indicates a frequency of from 1.3 - 3.9 Hz depending upon assumptions.

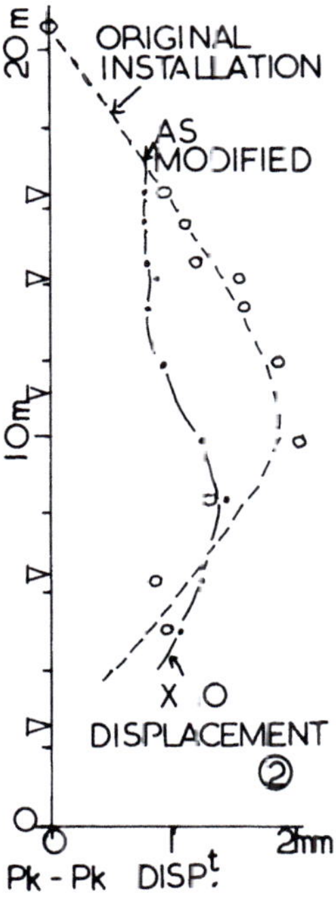

Vibration: Measurements made along the length of the tube showed a dominant frequency of 3.5 Hz. Maximum vibration amplitude 2.03 mm pk/pk. Vibration envelope as in Sketch 2.

Cure: 5 restraints fitted, marked ▷, in Sketch 1. Dominant frequency now shifted to 18 Hz. Maximum vibration amplitude 0.06 mm pk/pk, and vibration envelope as in Sketch 2. Wear assessments for the restraints indicated at worst 0.5 mm/year. Five years operation shows no increase in vibration, or damage to experiments.

Reference: Parkin M W. UKAEA(WNL) - Unpublished work.

A8 TREATMENT OF FLOW-INDUCED PENDULUM OSCILLATIONS

K. Hennig
E. Platen

Central Institute for Mathematics and Mechanics

Academy of Science GDR

G. Grunwald

Central Institute for Nuclear Research

Academy of Science GDR

SUMMARY

Flow-induced oscillations of pressurized water reactor control elements were experimentally examined. The investigation concentrated on determining the characteristics of the rod elements as a function of the water flow rate. With increasing flow rate, there is an increase in the frequency of pendulum-type oscillations, and the number of collision events between the rods and the channel walls.

1. EXPERIMENTAL SYSTEMS

The model of a control element, consisting of two coupled rods, was suspended in a hexagonal vertical channel (Fig. 1). Inductive transducers were installed to measure the rod displacement x. The acceleration b of the channel-walls and the pressure fluctuations p_u and p_o in the flow were measured with the help of piezoelectric-type transducers. An orifice plate was used to monitor the flow rate. As shown in Fig. 1, the channel was 2 m long and the mean gap S had values of 1, 2 and 3 mm. In the lower part of the channel, which had a diameter of 40 mm, the water flow reached a maximum velocity of 2 m/s, corresponding to a Reynolds number of 8×10^4.

2. EXPERIMENTAL RESULTS

The experimental results show that with increasing mass flow rate $\dot{m}$, the pendulum frequency and the number of collision processes per unit time between the control element and the channel walls increases. Fig. 2 shows a comparison of the experimental results employing water and air as flowing media (further results for air are to be seen in [1]). At a value of mass flow rate equal to $\dot{m}_A$, the friction force is equal to the weight (reduced by the lift) of the upper rod. For low flow rates, f increases more slowly with $\dot{m}$ than for high flow rates. This is typical for both flow media. There are two pendulum oscillation modes. In the case of low flow rates, the fundamental frequency is dominant, and for higher rates the harmonic oscillation mode, corresponding to the second frequency of a two-body pendulum system, becomes important.

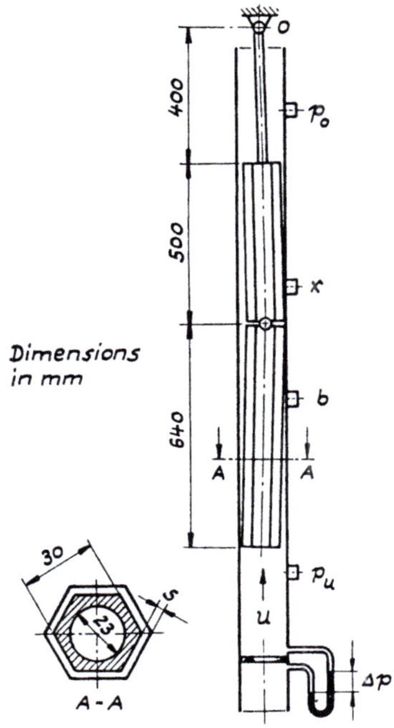

FIG. 1 SCHEMATIC OF EXPERIMENTAL SETUP

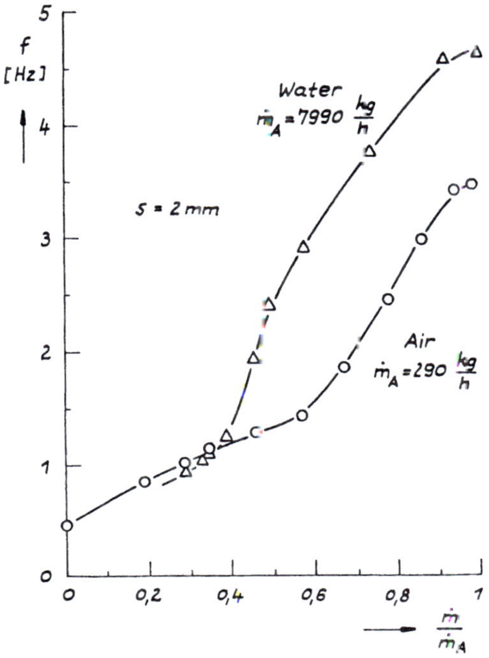

FIG. 2 PENDULUM FREQUENCY AS A FUNCTION OF FLOW RATE. EXPERIMENTAL RESULTS FOR WATER AND AIR AS FLOWING MEDIUM

The influence of the mean gaps is shown in Fig. 3. Curve 1 corresponds to S = 3 mm and to strongly coupled control elements. Therefore only pendulum oscillations with the fundamental frequency exist.

Further investigations in connection with the averaged trajectories of the control element motion and the sound level from the collision processes between the rods and the channel-walls are given in [2].

3. THEORETICAL MODEL

In order to describe the pendulum oscillations theoretically, a model was established. The arrangement and the coupling of the rods in a vertical flow channel, and the collision processes between the rods and the channel walls, were considered.

By neglecting phenomenona such as boundary-layer formation and separation of flow, nonlinear equations of motion were obtained in the following form:

$$\underline{M}\,\frac{d^2\vec{\phi}(t)}{dt^2} - \underline{D}(\dot{m})\frac{d\vec{\phi}(t)}{dt} + \underline{C}(\dot{m}^2)\vec{\phi}(t) + \vec{G}\left(\frac{d\vec{\phi}(t)}{dt}, \vec{\phi}(t), t\right) = 0$$

where $\vec{\phi}(t)$ is the vector of angular displacement of the rods. $\underline{M}$ is the mass-matrix; $\underline{D}$ the matrix of self-exciting forces; $\underline{C}$ the spring-matrix and $\vec{G}$ is a vector

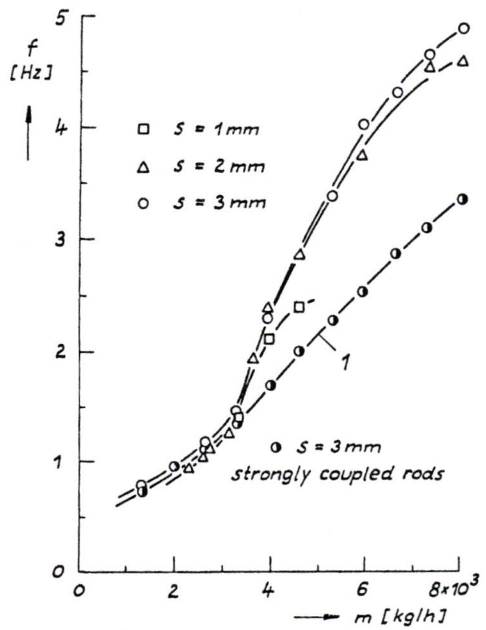

FIG. 3 PENDULUM FREQUENCY AS A FUNCTION OF FLOW RATE. EXPERIMENTAL RESULTS FOR WATER AS FLOWING MEDIUM

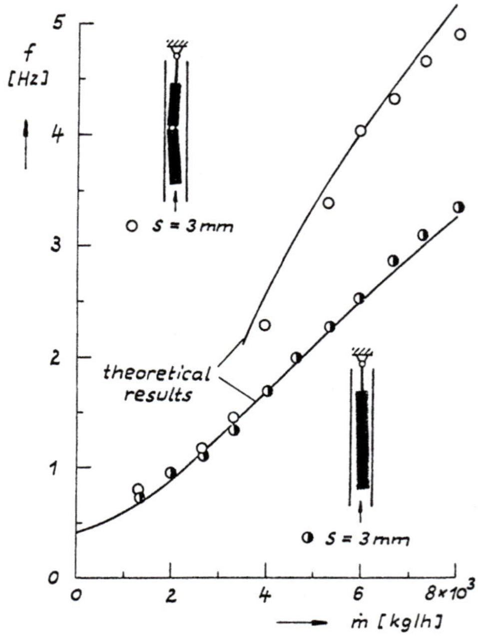

FIG. 4 PENDULUM FREQUENCY AS A FUNCTION OF FLOW RATE. COMPARISON BETWEEN THEORETICAL AND EXPERIMENTAL RESULTS

function of the angular displacement. The flow rate $\dot{m}$ and the fluid velocity U are connected by the linear equation

$$\dot{m} = A\ \rho_F\ U$$

where A is the inside area of the channel and ρ_F the fluid density.

If there are no collision events, the pendulum oscillation can be described by these equations according to the given initial values $\dot{\phi}(t_0)$. The geometry of the rods and of the channel determine the pendulum position where collisions are possible.

There are 3 types of collision: collision at the coupling point of the rods; at the lower end of the rod; and at both locations simultaneously.

For numerical solution of the oscillation equations under given initial and collision conditions, a Monte-Carlo method was developed. This method permits determination of the important parameters in a relatively short time. Calculations show that the theoretical results are in good agreement with the experiments (Fig.4

4. REFERENCES

1. Hennig, K., Grunwald, G., "Treatment of Flow-Induced Pendulum Oscillations", Int. Conf. Vibration in Nuclear Plant, Keswick, U.K.,(1978).

2. Grunwald, G., Liewers, P., Schumann, P., Weiss, F.-P., "Experimental Investigation of Flow-Induced Control-Element Movements by Noise Analysis", IAEA-SM-226/15, (1978).

PENDULUM OSCILLATIONS

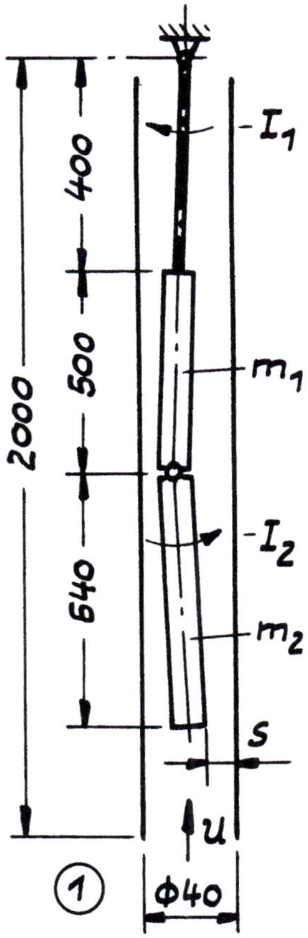

Problem: Flow-induced oscillations of reactor control elements (two coupled rods in a vertical channel) in model experiments. Dependence of vibration behaviour on flow rate.

Geometry: See sketch 1. Dimensions in mm.

Flow conditions: Closed water flow circuit, 20 °C, max. flow velocity U in lower part of the channel 2 m/s, giving Re = 8 x 10^4. Strong turbulence.

Structural dynamics: Mass and moment of inertia of rods (without added-mass effect): m_1 = 2.03 kg, m_2 = 2.75 kg, I_1 = 0.774 kg-m^2, I_2 = 0.354 kg-m^2. Two pendulum oscillation modes were observed. For low flow rates, the fundamental frequency, for high rates the harmonic oscillation mode.

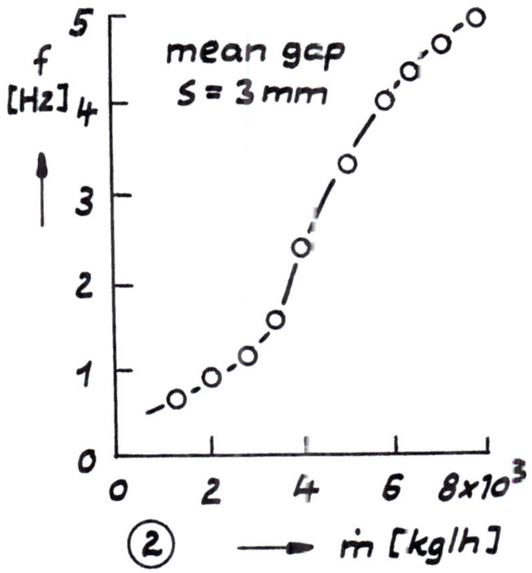

Vibration: See sketch 2. With increasing flow rate ṁ the frequency f of the pendulum oscillations and the number of collision events between the rods and the channel-walls increase.

References: Hennig, K., Grunwald, G., Treatment of Flow-Induced Pendulum Oscillations. Contrib. at Int. Conf. Vibration in Nuclear Plant, Keswick, U.K., May 1978.

A9 MAIN STEAM PIPING VIBRATION DRIVEN BY FLOW-ACOUSTIC EXCITATION

R.T. Hartlen
Supervising Engineer

W. Jaster
Engineer

Ontario Hydro Research Division
Toronto, Canada

SUMMARY

Low-frequency vibration was experienced in the main steam piping of 791 MW nuclear generating units. The severity of the vibration necessitated reduction of steam flow until additional piping supports were implemented.

A program of field measurements was conducted including vibration, dynamic strain, and fluctuating pressures. It was concluded that the excitation originated from a broadband flow noise source; from further reading and a small-scale laboratory model test, it was inferred that the source lay in the main control valve chest. The field data also showed that the noise source was acting through the long-wavelength axial acoustic modes of the piping circuit; that is to say the major frequency of excitation at a given piping elbow depends upon the elbow location with respect to the axial acoustic mode shapes.

On one particular line only there was a 'discrete tone' source. Its frequency increased linearly with flow velocity, and matched the acoustic-mode frequencies sequentially. This classical resonance effect caused large pressure fluctuations, and a more severe reduction of the generating unit load.

In keeping with the objectives of this Symposium, the main emphasis is given to the description and interpretation of the excitation mechanism; presumably this aspect of the field experience will contribute most toward the development of design methods. However for perspective and completeness, brief descriptions are given of the physical system, the vibration problem and the 'solution' by added mechanical restraints.

1. DESCRIPTION OF THE SYSTEM

Sketch 1 of the Chart is a schematic diagram that identifies some of the major components of the system. These components and their functions are described below.

1.1 The Loop

Pipes carry the steam from the governor-valve chests into the HP turbine. They are essentially U-shaped runs, lying mainly in the vertical plane.

Each of the governor-valve chests includes two valves, both essentially globe valves. On the upstream side is the emergency-stop valve, which is fully-open during normal operation. On the downstream side is the governor valve whose lift, of course, increases with unit load. The flow passage connecting the two valves is an integral part of the valve chest.

The upstream piping carries the steam from the steam drums to the governor-valve chest, the part between the balance header and the valves being of particular interest. These runs make a right angle turn in the horizontal plane at the balance header level, and then rise vertically to the valve level; note that this geometry gives four different line lengths between balance header and valve chest.

The balance header provides a collection point common to all four lines, and serves several functions: it is the takeoff point for the "dump-to-condenser" lines; it feeds the "live-steam-to-reheater" line; it tends to balance the flows to the four valves; and it facilitates running on three valves when required. Note that the balance header is a relatively small vessel, only 1.36 times the pipe diameter.

The boiler-stop valves just downstream of the balance header serve to isolate the downstream sections when necessary; they are fully open during normal operation. These valves include a venturi section, wherein the flow velocity increases to 1.70 times that in the piping.

The steam drums are relatively large vessels, representing the beginning of the main steam piping and the source of the steam. From the drums to the balance header, there are long, three-dimensional piping runs with no unusual flow discontinuities.

2. BRIEF DESCRIPTION AND HISTORY OF THE PROBLEM:

The full story of this field problem is very complex because of its physical nature, because more than one generating unit was affected, and because the investigation and the restraint system developed in stages. It appears neither practical nor desirable to attempt to describe it in detail. Rather, with reference to Figure 1, the main stages are summarized as follows.

The dominant problem in the original system was obvious; a predominantly vertical, low frequency vibration of the vertical piping runs rising through the turbine floor. Unit 1, with roughly equal amplitudes on all four lines, was limited to 725 MW. Unit 2 was more severely limited, to 625 MW, by one bad line (L4) with the other three lines much lower in amplitude. One of the early attempts at attenuation by partial restraint utilized a simple strut designed to brace the piping to the steel under the turbine floor. This provided much improved, if not optimum, control over the original problem, and allowed running at 750 MW on both units. With this strut in place the new operating limit was imposed by excessive axial vibration of the loop pipes; in retrospect, it was apparent that these loop pipe modes had been present from the beginning.

A perforated-shroud type distributor was installed to smooth the flow through the governor valve. These devices were first tried on Unit 3 which was just coming into service. Initial indications were that these devices had reduced the excitation. However, subsequent more rigorous testing on Units 1 and 2 showed that they had only a marginal effect on the low-frequency pressure fluctuations. Finally a much more extensive restraint system was designed and implemented, with emphasis on the loop piping. With this system the vibration and dynamic strain are well below allowable limits permitting continuous operation at full load of 791 MW.

3. INITIAL FIELD MEASUREMENTS, MODEL STUDY AND PROBLEM DIAGNOSIS

In the early stages, it was tempting to conclude that the only really serious problem was unique to Unit 2/Line 4; and that the other lines could be cured by minor modifications. However, the initial detailed field measurements disproved this; the general pattern of vibration and pressure fluctuations were remarkably similar from unit to unit; this signalled that we should expect a general, persistent problem.

Also at an early stage, we concluded that the vibration was of a rather complex "flow-induced" nature, with the following main characteristics:

(i) the root source is some generator of flow noise, ie, a periodic instability or random turbulence, creating pressure fluctuations.

(ii) the basic flow noise is selectively shaped and amplified by the one-dimensional axial acoustics of the circuit (in simpler terms, the 'organ pipe' effects).

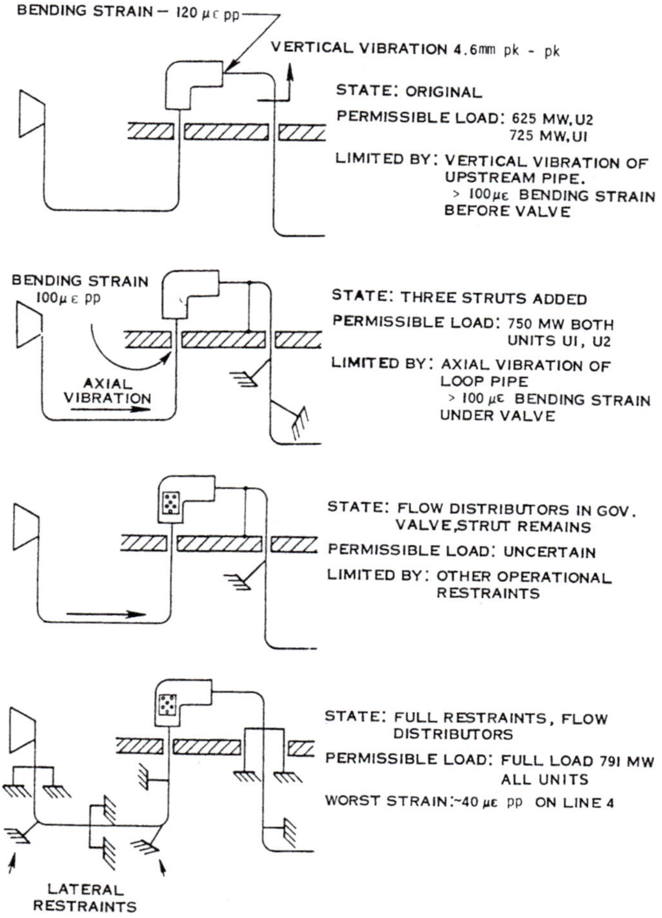

FIG. 1 SUMMARY OF HISTORY OF THE VIBRATION PROBLEM

(iii) the resulting mechanical vibration can be aggravated by unfortunate matching between acoustic and mechanical resonances.

Sketch 3 of the Chart illustrates the basis for the initial general diagnosis, i.e. basic flow noise modified by the individual line acoustics. Other more detailed measurements identified a discrete flow-dependent forcing frequency on one particular line. Unfortunate mechanical/acoustical coupling definitely played a role as well, where an acoustic mode could focus fluctuating pressure excitation on an elbow receptive for a given mechanical mode.

Early measurements were also made for a few pressure taps located near the valve chest, around the balance header, and also upstream of the balance header. The upstream section was quiet, significant amplitudes appeared downstream of the balance header, with the largest amplitudes near the valve chest. But one could not conclude immediately that the valve chest was the main source; it was conceivable that a "low-impedance source" (i.e. small pressure fluctuations, large velocity fluctuations) at the balance header could cause larger pressure fluctuations at the valve; indeed there was a large body of opinion, ours included, that the balance header was a likely source of excitation. At this point a small-scale air model study was conducted.

Although the 1/30 scale air model study lies well outside the scope of this paper, and will be treated elsewhere, a brief reference seems in order. The steam drums were represented by a large plenum connected to the air supply; the layout and internal geometry of the piping from steam drums to valve chests was modelled precisely, including the balance header details; the valve chests were represented approximately by simple globe valves exhausting directly to room ambient, i.e. no loop pipes nor turbine included. Fluctuating pressure measurements were obtained at various points along the lines. Although the overall symptoms of the excitation were successfully simulated, it was shown that the fluctuating pressure spectra were virtually dependent upon the particular valve in place (Figure 2). On the other hand, ad hoc modifications to the model balance header, based upon various flow instability ideas, were notably unsuccessful (Figure 3). Based largely upon these tests, the balance header was vindicated and it was concluded that the valve chest was the dominant noise source in the full-scale field system as well.

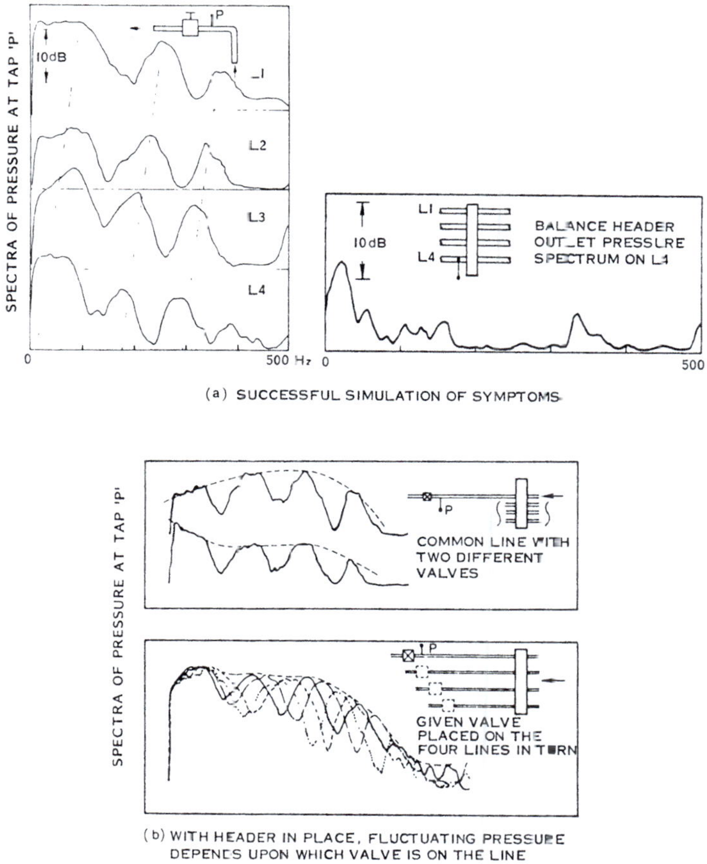

FIG. 2 SUMMARY OF MODEL RESULTS

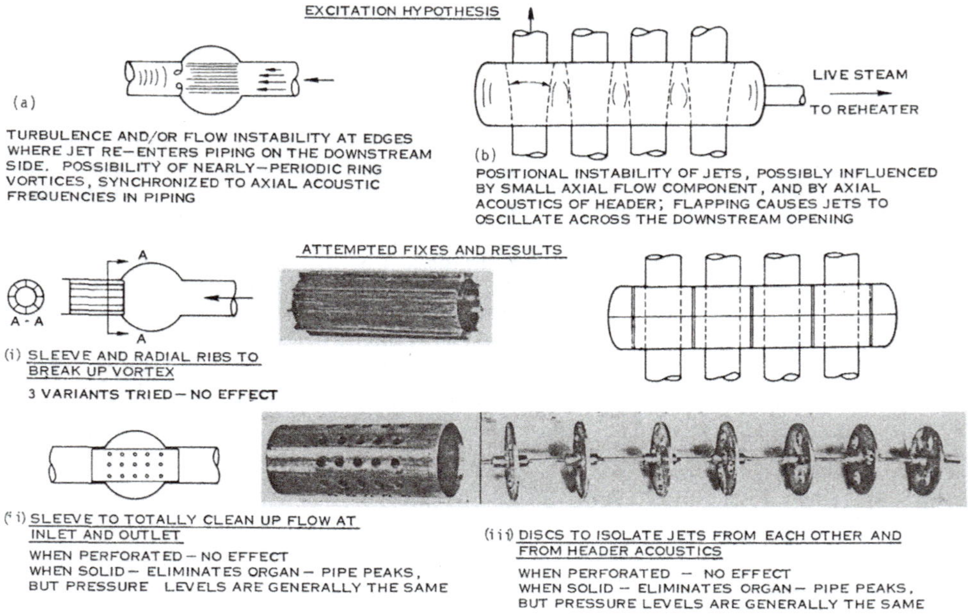

FIG. 3 HYPOTHESES OF FLOW EXCITATION IN BALANCE HEADER AND VARIOUS UNSUCCESSFUL AD HOC MODIFICATIONS TO MODEL

4. DETAILED MEASUREMENTS OF FLUCTUATING PRESSURES

The initial description and diagnosis of this problem was based upon a limited number of measurements. Later there was an opportunity to obtain more comprehensive data on the local pressure fluctuations within the pipe. In keeping with the objectives of this Symposium, some of the resulting data are presented here in considerable detail.

Figure 4 presents a description of a pressure wave form in various formats. The time waveform is continuous, but modulated. The r.m.s. value is 0.66 times the stream dynamic pressure. The maximum peak-to-peak magnitude, 48.3 kPa, is 4.64 times the r.m.s. The 'peak-hold' spectrum ordinates are about double those obtained in the averaging mode.

Previous sections of this paper have referred to the importance of the one-dimensional axial acoustic modes. This is quite clear from Figure 5 which shows pressure spectra from nine locations on the same line at full flow. For reference the top sketch shows the expected shape of a 3/4-wavelength mode between valve chest and balance header; the nominal expected frequency is about 8 Hz. The data fit the pattern well: the dominant frequency is 7.5 Hz; the relative amplitudes of P5, P6, P7 and P8 are as expected; further, the instantaneous traces of fluctuating pressure confirm the expected anti-phase relationship between P5 and P7.

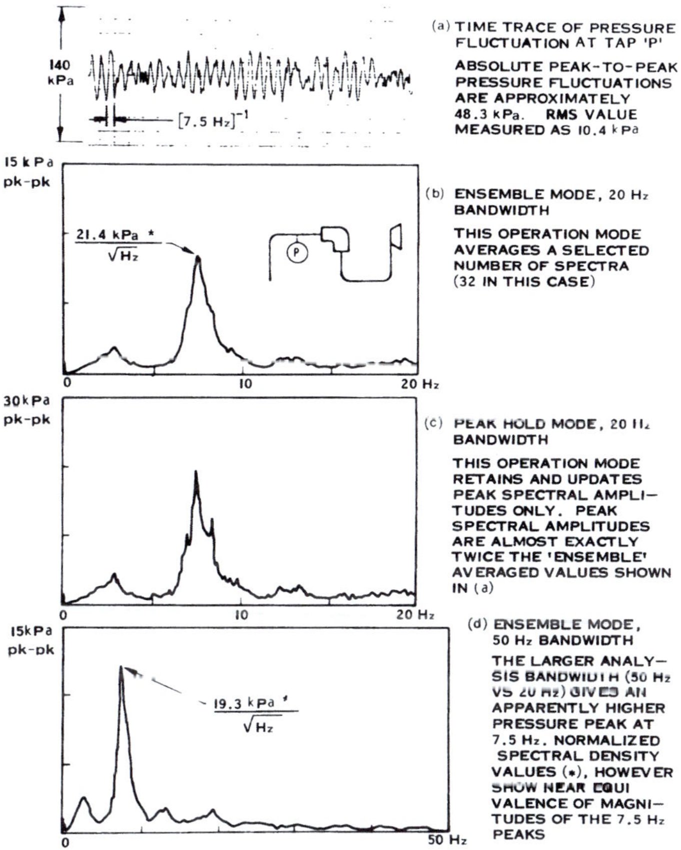

FIG. 4 PRESSURE FLUCTUATIONS IN VARIOUS FORMATS (FROM UNIT 2, LINE 4, AT 791 MW LOAD)

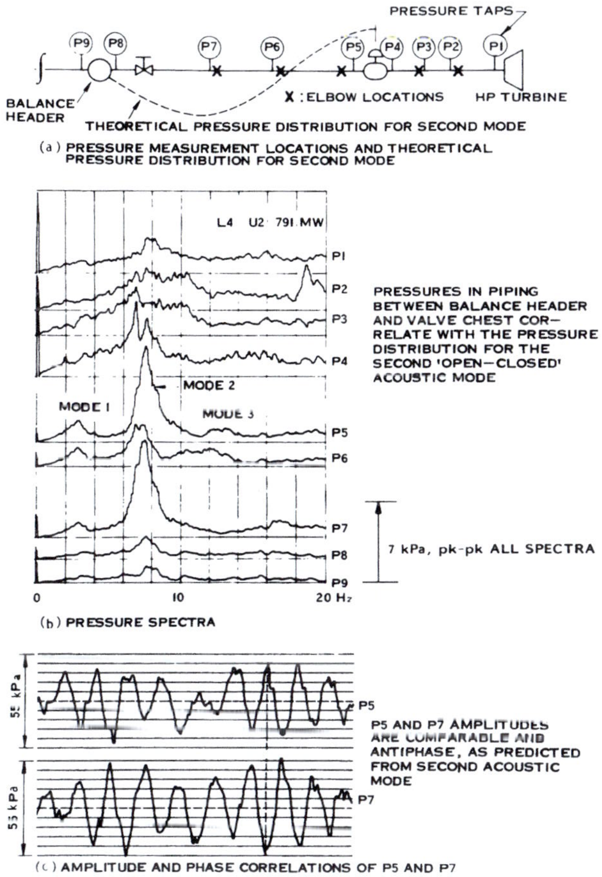

FIG. 5 VARIATION OF PRESSURE FLUCTUATIONS ALONG LINE 4 AT 791 MW LOAD

Previous sections of this paper have also referred to a broadband flow noise acting through the passive circuit acoustics. This is quite clear from Figure 6,

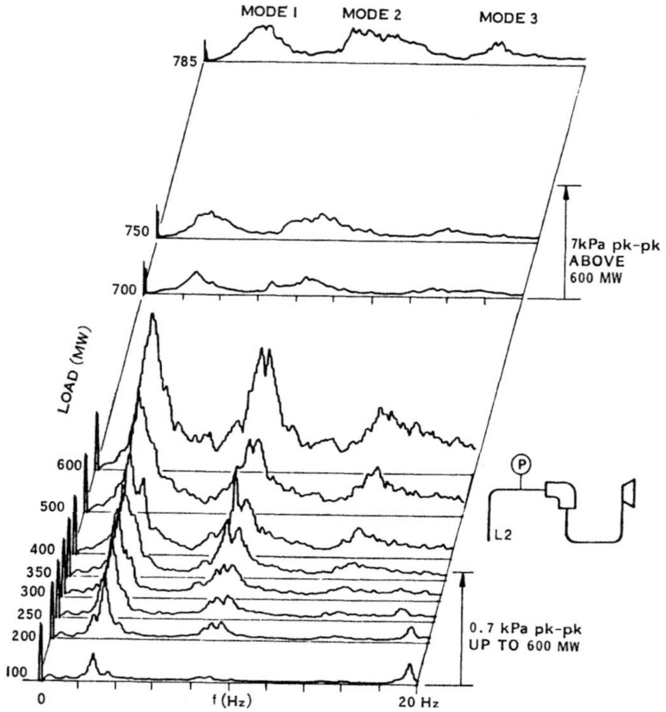

ON LINE 2, THE PRESSURE SPECTRA SHOW A GREATER RELATIVE INCREASE OF HIGH FREQUENCIES WITH INCREASING LOAD, AS THOUGH DRIVEN BY A BROAD-BAND FLOW-DEPENDENT SOURCE

FIG. 6 VARIATION OF PRESSURE SPECTRA WITH FLOW, FOR A TYPICAL LINE

which shows the variation of pressure spectra with increasing flow. All spectra are dominated by three fixed acoustic modes; the absolute amplitudes increase with flow rate; and the relative amplitudes change, progressively favouring the higher frequencies with increasing flow. This behaviour is considered typical of all lines but one in the plant.

Figure 7 shows the behaviour of the one atypical line; in addition to the normal broadband excitation there is very clear evidence of a discrete tone whose frequency scales with flow velocity. This tone first encounters the first acoustic mode which peaks out at about 250 MW. With further increase in flow, it can be seen shifting toward the second mode, apparently reaching it at around 600 MW, and then remaining locked in. By comparison with Figure 6 for a typical line, it can be seen that this tone results in a dominant second acoustic mode and much larger pressure amplitude at full load. This tone was clearly the cause of the larger vibration and greater load restriction on one of the generating units. To date its source has not been identified.

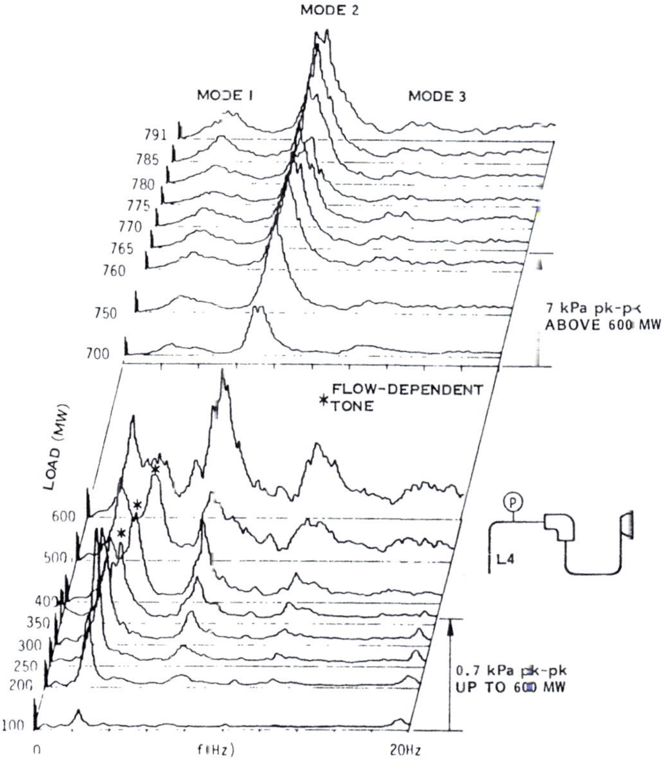

ON LINE 4, A DISCRETE FLOW–DEPENDENT TONE PROGRESSIVELY EXCITES THE PIPING ACOUSTIC MODES WITH INCREASING LOAD. ABOVE 600 MW LOAD, THE 'SHEDDING' TONE APPEARS TO 'LOCK–IN' ON THE SECOND ACOUSTIC MODE. OTHERWISE, THE SPECTRA SHOW A GENERALLY GREATER RELATIVE INCREASE OF HIGH FREQUENCIES WITH LOAD, SIMILAR TO LINE 2, (FIG 6)

FIG. 7 VARIATION OF PRESSURE SPECTRA WITH FLOW FOR THE LINE WITH A DISCRETE TONE

5. CONCLUDING REMARKS

The flow-induced piping vibration described here was a major obstacle during commissioning of a large nuclear power station.

The main source of excitation was a broadband flow noise associated, by inference, with the control valve chests. However a discrete flow-dependent tone markedly aggravated the situation on one particular line. The passive acoustic response of the piping circuit played a major role in both cases. It selectively filtered the broadband noise and thereby determined the amplitude and frequency of excitation at each piping elbow; and it amplified the flow-dependent tone at particular frequencies. Hopefully the detailed field data presented here will alert designers of future plants to the possible form and intensity of flow excitation.

Based upon this experience the authors would put forward two suggestions for further research. Firstly, the development of a predictive method whereby the intensity of pressure fluctuations can be estimated in advance, including any effects of source coupling with the acoustic circuit. Secondly, the development of experimental-data analysis procedures for rigourously identifying the nature and location of noise sources in lightly-damped acoustic circuits.

MAIN STEAM LINE PIPING VIBRATION

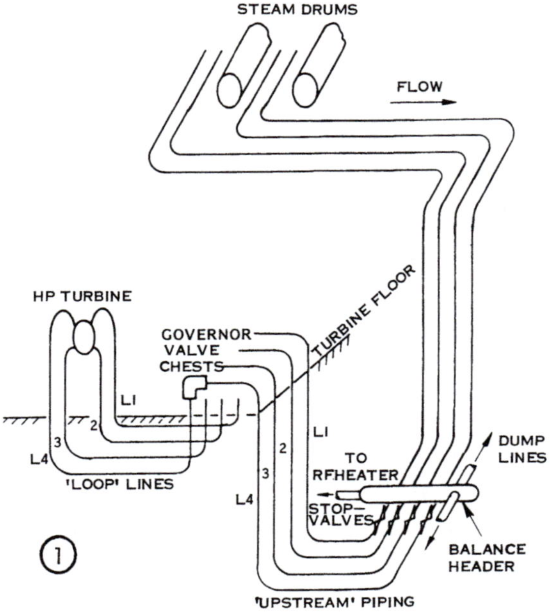

MAIN DIMENSIONS

UPSTREAM PIPING :	731.5 mm ID, 17.45 mm t, 321.8 kg/m
LOOP PIPING :	574.7 mm ID, 17.45 mm t, 254.8 kg/m
BALANCE HEADER :	990.6 mm ID, 9.45 m AXIAL LENGTH

NOMINAL LENGTHS (m)	L1	L2	L3	L4
DRUM TO HEADER	126.2	132.0	105.5	111.2
HEADER TO CHEST	34.4	39.0	43.3	47.9
CHEST TO TURBINE	25.3	20.7	20.7	25.3
TOTAL	185.9	191.7	169.5	184.4

STEAM CONDITIONS	: 4.24 MPa, 254°C
FLOW VELOCITY	: 38.4 m/s NOMINAL IN UPSTREAM PIPE AT 791 MW
SONIC VELOCITY	: 507.8 m/s
DYNAMIC PRESSURE	: $1/2\ \rho V^2 = 15.7$ kPa AT 791 MW

PROBLEM; LOW FREQUENCY BENDING VIBRATION, BOTH UPSTREAM AND DOWNSTREAM OF THE VALVE CHEST. NO DAMAGE, BUT REQUIRED LOAD RESTRICTION FOR SEVERAL MONTHS.

GEOMETRY; COMPLEX THREE-DIMENSIONAL LAYOUT. FOUR PARALLEL LINES OF SOMEWHAT DIFFERENT LENGTHS.

BOUNDARY CONDITIONS; MAJOR RELEVANT PIPING CIRCUIT ELEMENTS ARE COMMON BALANCE HEADER AND TURBINE, AND VARIOUS VALVES IN EACH LINE.

FLOW CONDITIONS; FLOW FROM STEAM DRUMS OF NUCLEAR STEAM GENERATORS TOWARD TURBINE, AS CONTROLLED BY GOVERNOR VALVES. NO KNOWN UNUSUAL FLOW CONDITIONS.

OPERATING CONDITIONS; VIBRATION FIRST BECAME SIGNIFICANT AT ABOUT 70 PER CENT LOAD AND GENERALLY INCREASED WITH FLOW.

STRUCTURAL DYNAMICS; VIBRATION PREDOMINANTLY IN A VERTICAL PLANE, INCLUDING HORIZONTAL MOTION OF THE LOOP LINES, VERTICAL MOTION OF THE UPSTREAM RISERS AND ROCKING OF THE VALVE CHEST. THUS BENDING STRAIN NEAR PIPE/CHEST JUNCTIONS.

VIBRATIONS; MODULATED, LOW FREQUENCIES IN THE 5 TO 10 Hz RANGE. ORIGINAL LOAD RESTRICTION WHEN 4.6 mm, pk-pk AT 7.5 Hz GAVE 120 MICROSTRAIN.

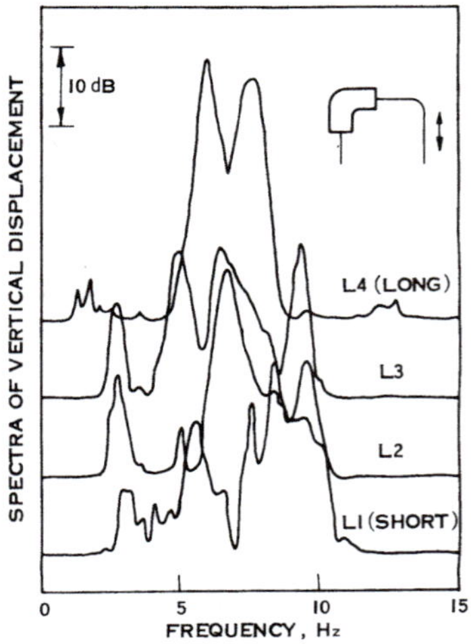

2

PRESSURE FLUCTUATIONS: CLEAR EVIDENCE OF THE BROAD FLOW NOISE SOURCE ACTING THROUGH THE PASSIVE CIRCUIT ACOUSTICS IN THE 0–20 Hz RANGE. ONE LINE HAD A 'DISCRETE TONE' IN RESONANCE WITH THE ACOUSTIC MODES; FLUCTUATING PRESSURE REACHED 55 kPa, PEAK-TO-PEAK AT 7.5 Hz.

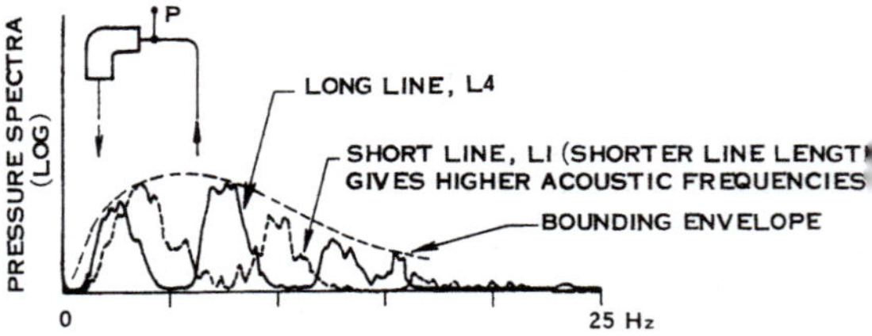

NOTE THE ODD- MULTIPLE MODES OF EACH INDIVIDUAL LINE 3

CURES: THE VIBRATION HAS SIMPLY BEEN 'CONTAINED' BY ADDITIONAL PIPE SUPPORTS; NO ELEGANT CURE.

A10

Discussion: PIPE LINE VIBRATION DRIVEN BY FLOW-ACOUSTIC EXCITATION (Chen, Florjancic, Hartlen, Hassouneh, Jaster, Paidoussis)

A. Schachenmann* and D. Rockwell
Department of Mechanical Engineering and Mechanics
Lehigh University
Bethlehem, PA 18015, USA

The authors' careful documentation of pipeline vibration is indeed valuable to the designer, and provides a stimulus for research in this area. Their observations seem to suggest that self-sustained oscillations of a fully turbulent flow are possible through coupling between the flow in the pipeline and the separated or vortical, flow in valves, valve header regions, or other flow regions tending to exhibit periodicity.

Our recent investigations show that it is possible to have highly organized oscillations due to coupling between the resonant modes of a pipe and the separated flow region in a cavity. The experimental system, involving a long smooth pipe terminated with an axisymmetric cavity is illustrated in Figure 1. The Reynolds number based on pipe diameter was 26,600<Re<82,900. The pipe diameter

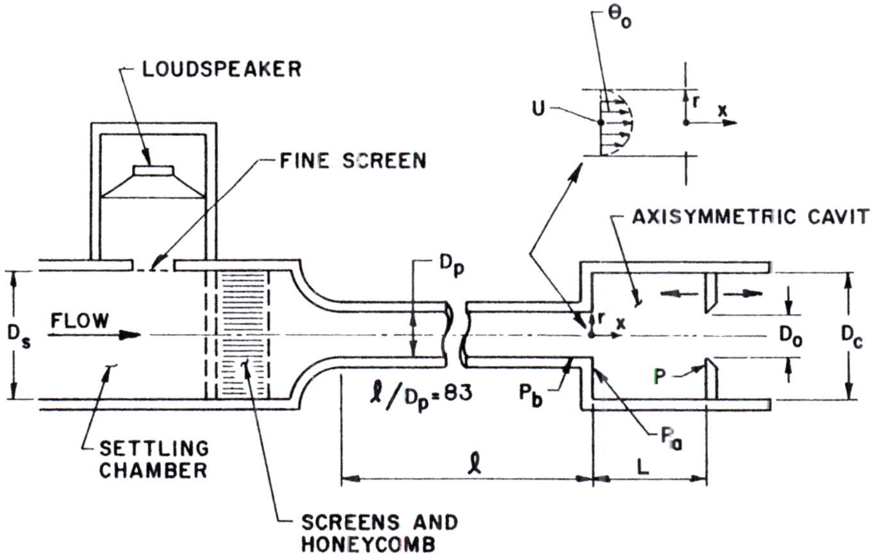

FIG. 1 EXPERIMENTAL SYSTEM

*Sulzer Bros., Inc., Winterthur, Switzerland, on leave from position of Research Engineer.

was 0.455 m, the cavity diameter 0.152 m, and the cavity length could be varied from 0.010 m to 0.150 m. The long approach pipe (ratio pipe length/diameter = 83) insured a fully-developed pipe flow velocity profile at the separation edge. No flow resonance of the system (with loudspeaker excitation) showed the presence of organ pipe modes. The measured fundamental frequency was 46 Hz. With flow through the system, strong, self-sustaining oscillations could be generated. The frequencies of these oscillations correspond closely to the pipe modes of the system. Depending on the flow velocity and cavity length, it was possible to excite the third through seventh pipe modes.

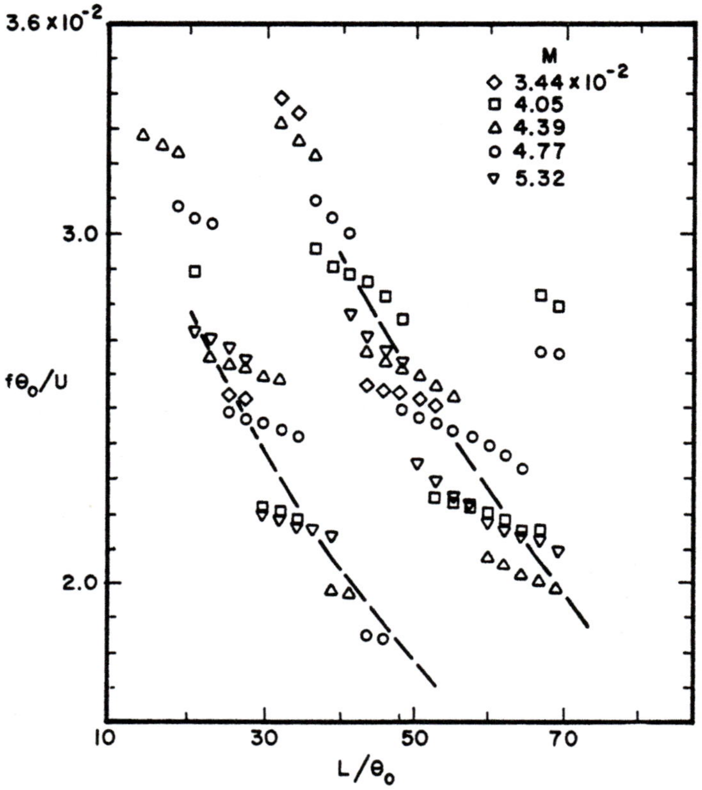

FIG. 2 DIMENSIONLESS FREQUENCY OF OSCILLATION VERSUS DIMENSIONLESS CAVITY LENGTH

Figure 2 shows a typical variation of dimensionless frequency of oscillation versus dimensionless cavity length at a constant value of maximum velocity measured at the exit of the pipe. As the cavity length is increased, there are both upward and downward jumps of the primary frequency of oscillation. Comparison of the dimensionless frequencies of oscillation with those predicted using inviscid stability theory, shows that the frequencies fall within the range of disturbances experiencing the largest amplification in the shear layer along the mouth of the cavity. That is, one may view the separated flow along the cavity as an "unstable turbulent" flow, which triggers the pipe modes. The resonant condition of the pipe at a given frequency in turn enhances the cavity oscillation and so on.

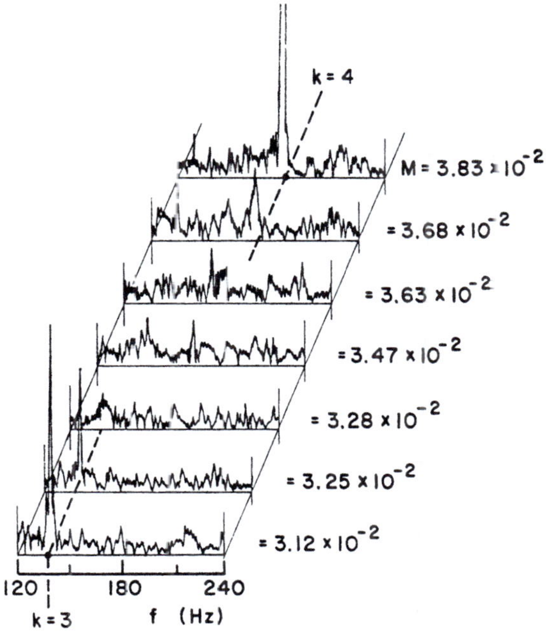

FIG. 3 FREQUENCY SPECTRA OF PRESSURE AT IMPINGEMENT

Figure 3 shows a typical variation of frequency spectra of the fluctuating pressure measured at impingement (Figure 1). Although two kinds of frequency jump were observed, and will be reported in a forthcoming paper, only one type of jump is illustrated here. It is evident that the process of a frequency "jump" actually involves a gradual transition from a predominant oscillation at one frequency to another frequency, each of which corresponds to a pipe mode.

Our recently completed work includes the effects of variations in flow velocity (and Reynolds number) on the frequency of oscillation, information on the frequency spectra for two basic types of frequency jumps, amplitude variation of the predominant oscillation frequency, and effects of external excitation (using a loudspeaker) on the unsteady flow features. An investigation is currently underway to characterize the unsteady velocity and pressure fields within the cavity, in order to better understand the resonant coupling pheromenon. This will involve synthesizing the oscillation into compressible and incompressible contributions.

A11 VIBRATIONS IN PIPE LINE SYSTEMS

Mohamed A. Hassouneh
Mechanical Project Engineer
Kuwait National Petroleum Company
Shuaiba Refinery, Shuaiba, Kuwait

SUMMARY

The potential destructiveness of flow-induced vibrations in high temperature, high pressure pipe lines is discussed, and our experiences with flow-induced vibrations in a pre-flash tower feed line and a crude heater outlet line are reviewed.

1. INTRODUCTION

Pipe line vibrations may be caused by internal flow in pipes and by plant machinery such as compressors, pumps, motors, etc. Of these sources, those due to flow pulsations are usually more common and more difficult to predict. While mechanically-induced vibrations are normally limited to the running speed of the compressor (or pump) and its lower order multiples caused by machine unbalance and misalignment, flow-induced vibrations are due to a kind of self-excited instability, and prediction of the frequency of instability is quite involved. A review of our experience with flow-induced vibrations in the pre-flash tower feed line and crude heater outlet line is given to aid in establishing design guidelines which will minimize the occurrence of objectionable or damaging pipe line vibrations. These vibrations may undergo resonance; steps to control or eliminate excessive vibration when it develops in practice are discussed.

2. PRE-FLASH TOWER FEED LINE

The problem represents a case where the designer has numerous considerations in laying out a pipe line system other than vibrations of the line. It is not suggested here that an elaborate vibration analysis of all standard pipe line systems be carried out. Nevertheless, the engineer will usually be justified in spending the time needed to insure that the fundamental natural frequency of a piping system bearing pulsating flow will not be in the neighbourhood of the frequency of pulsation. Proper choice and spacing of supports and braces (guides and damping devices) may be accounted for in the original design at little cost; these measures are much less expensive than correcting problems encountered in the field.

Sketch 1 on the attached Chart gives the layout of the last portion of a pipe line. Due to requirements of thermal expansion, this part of the line was left with no other anchor except the tower inlet nozzle, although a few stops and guides are located in order to allow thermal expansion in a preferential direction. The line dead weight is taken care of by spring hangers at joints 8 and 14. This line is used to transfer hot crude feed from the effluent heat exchanger to the pre-flash tower. Average flow velocity is 35 m/sec as calculated for the 400 mm section. Calculated exciting frequencies range from 2.6 to 25 Hz. The layout given has many sources capable of generating significant pulsation energy, which can couple with the acoustic and mechanical responses of the system:

a) Control valve: Typically, the requirements are to operate with a relatively severe pressure drop across the control valve. The high flow velocities achieved through the valve cause the generation of significant pulsation energy. Besides generating low frequency turbulence, flashing is expected to occur in the low pressure region of the valve. Although the generated pulsation spectrum is relatively broad band, it is characteristically centered around a peak, or frequency, corresponding to a Strouhal number of approximately 0.2 based on pipe diameter. It is felt that the amplitude of vibration could be minimized if the pipe could be anchored very close to the control valve. However, the elevation of the control valve (approximately 18 m from ground level), its distance from the vessel wall, and inadequate flexibility of the piping downstream of the control valve, would not allow such an anchor location. A vibration damper was used at this location, but it could not prevent the line from vibrating vigorously.

b) Piping Bends: The shaking force results from the change in momentum due to the change in direction of flow and is equal to 2pA cos θ/2, where p is the maximum pressure pulsation amplitude in the piping run, A is the area of the pipe and θ is the angle between the legs of the bend. For reasons associated with the process, the presence of the control valve and the two phase flow cannot be avoided. As such, the only way to remove the source of vibration is to modify the layout by using a smaller number of bends and locating the unavoidable bends close to the anchors, as shown in sketch 3. The modification in layout also gives us an opportunity to anchor and support the downstream piping in the most beneficial way.

3. CRUDE HEATER OUTLET LINE

Unlike the previous case, the piping layout has no unnecessary bends; in addition, clamps have been added as near as possible to the bends. However, in this case, the header represents another mechanism of acoustical energy coupling to enhance the vibration process. In such elements, the acoustical forces present at each end, and at each baffle or restriction (if any), must be considered to determine the total resultant unbalanced force. Thus not only differences in pulsation amplitude but also the phase differences and the areas on which the pulsations act must be accounted for in obtaining the vector sum for each frequency component. In this case, the pulsations are generated inside the heater tubes due to pulsations in the fuel gas, combustion air etc. These pulsations are normally small but due to phase differences, the excitation force produced is nearly doubled. To eliminate this problem, an internal baffle plate was added to break up standing waves and vortices.

To minimize vibration problems in oil refinery piping, the mechanical natural frequency of individual piping spans and piping components should not be resonant to any of the major excitation frequencies; the clamp and support stiffness should be adequate to restrain the shaking force in the piping to acceptable limits; and all unnecessary bends should be eliminated. If bends must be used, clamps should be located as near as possible to each side of the bend; pipe supports should be installed near all large masses and piping discontinuities including valves; flow velocities should be maintained as low as possible in flow and pressure control valves; finally, small auxiliary piping connections should be designed such that the masses can be braced to the main pipe to eliminate relative vibrations.

REFERENCES

1. Harris, C. M. and Crede, C. E., Shock Vibration Handbook, 2d ed., N.Y., McGraw-Hill, 1976.

2. M. W. Kellog Co., Design of Piping Systems, 2d ed., N.Y., John Wiley & Sons Inc., 1956.

3. Den Hartog, J. P., Mechanical Vibrations, 4th ed., N.Y., McGraw-Hill, 1956.

4. Wachel, J. C. and Bates C. L., "Escape Piping Vibrations While Designing," Hydrocarbon Processing, vol. 55, no. 10, p. 152-156, October 1976.

5. Chen, Y.N., "Flow-Induced Vibrations in Tube Bundle Heat Exchangers with Cross and Parallel Flow," Proceedings of the Symposium on Flow-Induced Vibration in Heat Exchangers (ASME Winter Annual Meetings, New York, 12/1/70), N.Y., ASME, 1970, p. 59-74.

PREFLASH TOWER FEED LINE

Problem: Vigorous pipe vibration caused failure of two pipe supports, resulting in crack propagation into vessel wall.

Geometry: See sketch 1. Outside vessel diameter 4880 mm.

Flow Conditions:

Flow medium	Crude Oil
Total flow	1153 t/hr
Liquid flow	1018 t/hr
Vapor flow	135 t/hr
Liquid density	750 kg/m^3
Vapor density	9.2 kg/m^3
Liquid viscosity	0.511 cp
Vapor viscosity	0.013 cp
Flowing temp.	325 °C
Pressure	3.85 kg/cm^2

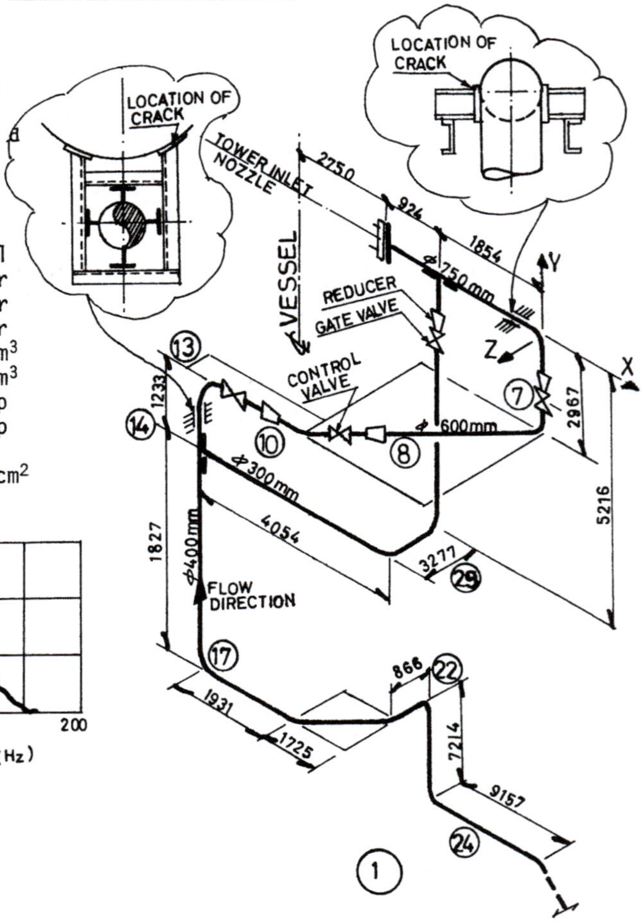

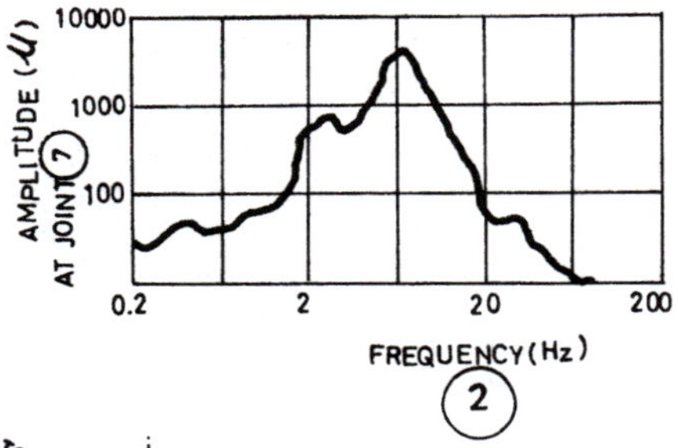

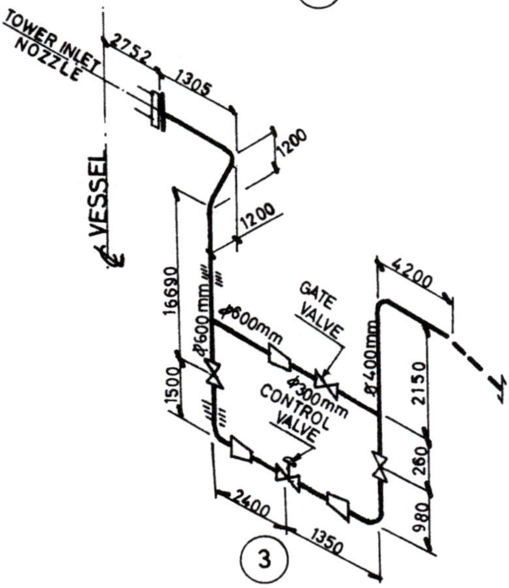

Structural Dynamics:
300 mm pipe, W = 119 kg/m, I = 7991 cm^4
400 mm pipe, W = 179 kg/m, I = 19730 cm
600 mm pipe, W = 358 kg/m, I = 54942 cm

Vibration:
Broad band random vibration with maximum aplitude of 3.5 m measured at joint 7 (corresponds to Strouhal number of 0.18).

Cures:
Change pipe geometry to eliminate most of the bends - sketch 3. Additional support and vibration snubber at joint 8 were unsuccessful.

Reference:
Design of Piping Systems, MW Kellog Company, 1965.

CRUDE HEATER DISCHARGE HEADER

Problem: Excessive vibration in crude heater discharge header

Geometry: See sketch -1. Nominal diameter of header 600 mm, nominal diameter of branches 200 mm.

Boundary Conditions: Branches and headers are rigidly clamped at positions shown.

Flow Conditions:

Flow medium	Atmospheric tower feed.
Total flow	680.7 t/hr.
Liquid flow	447 t/hr.
Vapour flow	233.7 t/hr.
Liquid density	800 kg./m^3
Vapour density	9.6 kg./m^3
Temperature	360 °C.
Pressure	2.1 kg./cm^2

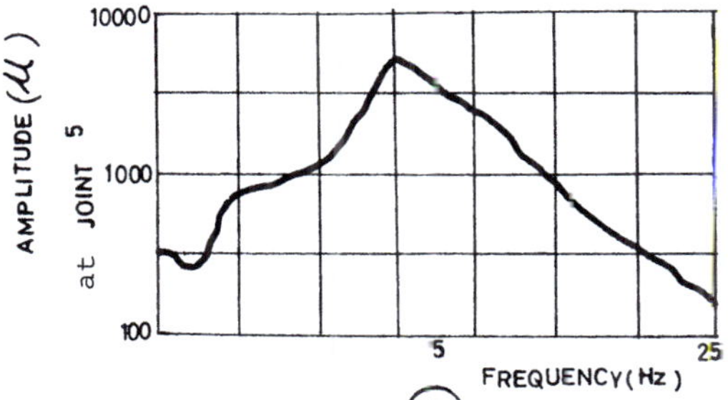

2

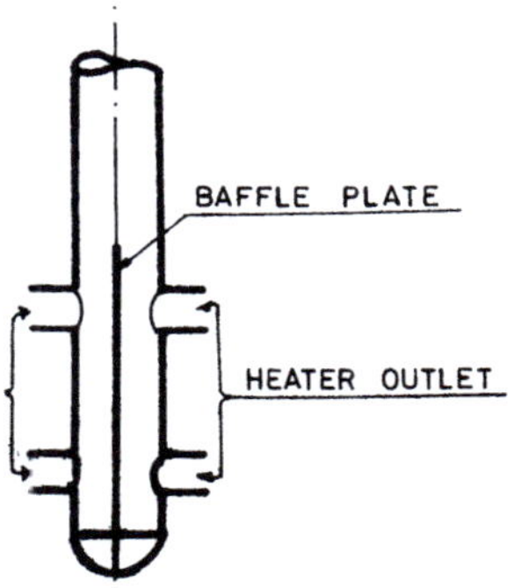

3

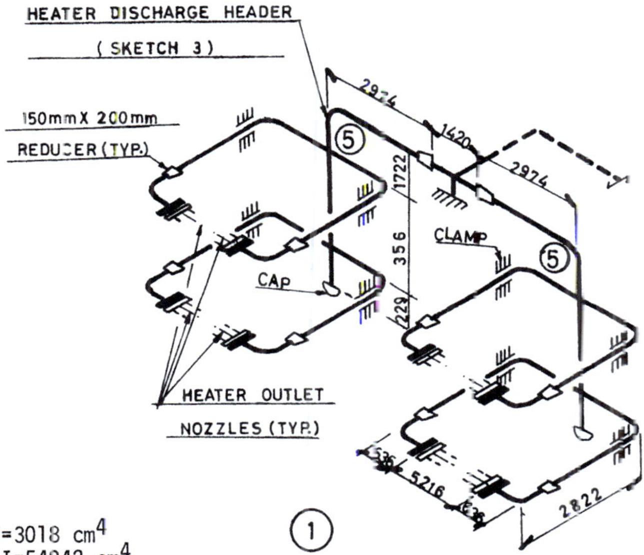

1

Structural Dynamics:

200 mm pipe W = 73 kg./m, I=3018 cm^4
600 mm pipe W = 358 kg./m, I=54942 cm^4
750 mm pipe W = 448 kg./m, I-133610 cm^4
Estimated natural frequency = 5.72 Hz.

Vibration: Frequency 4-6 Hz, maximum amplitude 4 mm at joint 5. (Corresponds to Strouhal Number 0.18-0.27)

Cures: Addition of baffle plate reduced amplitude of vibration to less than 1 mm.

Reference: Design of Piping Systems, MW Kellog Company, 1965

A12 FLUTTER OF PLATES AND SHELLS IN PRACTICE

Earl H. Dowell
Professor

Princeton University, Princeton, N.J., USA

SUMMARY

Several case histories of the flutter of plates and shells are recounted. These will provide the reader with a perspective on the variety of applications which are encountered in practice. They will also illustrate how the advances in our fundamental understanding which have been made in the last twenty years are being used to solve real life problems.

1. INTRODUCTION

One of the most significant success stories in aeroelasticity in the last twenty years has been the emergence of a sound fundamental understanding of the flutter of plates and shells. Not only is there a body of theory based upon first principles of solid and fluid mechanics which includes structural non-linearities and the effect of the fluid boundary layer, but there is also a comprehensive set of experimental data which largely confirms the best available theory, Ref. [1]. There are, to be sure, some important research questions which remain. For example, how quantitatively accurate are the limit cycle amplitudes predicted by the theory and can they be used with confidence to estimate fatigue life of fluttering plates and shells? Nevertheless, the transition of the subject from one notorious for its difficulty and poor agreement between theory and experiment to one which is on a firm fundamental basis of understanding is, on the whole, remarkable.

Indeed for conventional, metallic skin panels typical of airframe construction, design guides are available which make flutter of such panels an infrequent event, Refs. [2, 3, 4]. To achieve such prevention, however, a substantial theoretical and experimental effort may be required for a specific structural design. As one might expect, the flutter problems which most frequently arise after the design is complete are those for which flutter was not anticipated to be a problem; e.g., because the plates or shells were made of unusual materials or geometries or were placed in unusual flow environments.

In the present paper, several case histories of flutter problems are recounted. These will provide the reader with a perspective on the variety of applications which are encountered in practice and also illustrate how the advances in our fundamental understanding are being used to solve real life problems.

A word about the questions which the designer poses to the flutter expert is in order here. Before the structure goes into service, he asks under what flow conditions (e.g., at what flow velocity) will flutter occur? If flutter is predicted under anticipated operating flow conditions, he then asks for a range of options for design modifications which will suppress the flutter. After the structure is in service, if flutter should still occur despite everyone's best efforts, the first question changes to, how can we be sure it is a flutter induced oscillation? If it is determined to be flutter, the second question remains; how can flutter be suppressed? Now, however, the range of options for doing so which are acceptable to the designer is likely to be reduced. These questions are dealt with further in the text.

2. CASE HISTORIES

2.1 Case I: Shell Flutter of the Main Shaft of a Jet Engine

A few years ago when the Pratt and Whitney JT9D engine, which powers many of the current wide body jet aircraft, was being developed, there was a series of breakages of the main shaft due to excessive vibration. See Chart A. The shaft is a cylindrical shell of dimensions, 2.44 m long, 20 cm in diameter and 0.635-1.27 cm thick. Moreover, the temperature of the surrounding air is about 800°F and the mean pressure 15 atmospheres. See Sketch 1 for an overview of the engine. As is usual in such a case, conventional explanations were sought not involving flutter. These included shaft whirl instability, forced resonance due to a variety of possible mechanical sources, and acoustic excitation. After study, none of these explained the measured evidence. This included a traveling wave oscillation around the shell circumference of the shaft and a dramatic increase in vibration amplitude as the RPM of the concentric shell surrounding the inner vibrating shell was increased above a certain value. Moreover, a continuing increase in amplitude and eventual failure occurred as the RPM was further increased.

An extensive theoretical and experimental program was conducted by Pratt and Whitney and Princeton University which led to quantitative understanding of the phenomenon and several options for design changes to suppress flutter. This work is documented in Refs. [5-7] and the highlights are discussed here.

A dynamically scaled model of the shaft and its flow environment was constructed to study the phenomenon. The experimental apparatus is shown in Sketch 2. A photograph is shown in Fig. 1a. Two concentric shells are present and the outer one rotates. In the engine itself, the inner shaft also rotates, but this is not qualitatively important and in this particular application is of minor quantitative importance. The outer shell has a number of holes around the circumference at one axial station. Their practical use is to provide cooling air which is drawn through the holes by a vacuum. This cooling air combines with the outer shell rotation to form a circumferential vortex flow about the inner cylinder. If the vortex flow achieves a sufficiently large velocity, the inner shell will flutter in the form of a circumferential traveling wave.

The experimental procedure was representative of that used in all work involving flutter of a plate or shell, Ref. [1]. First the natural modes and frequencies of the inner shell were measured and compared with theory, Ref. [6], with satisfactory results, thus demonstrating that the structural aspects of the phenomenon were understood. The experimental procedure was to use a loudspeaker as an exciting force and a non-contacting magnetic or inductance transducer to detect shell resonant modes, as shown in Fig. 1b.

Next the flutter experiment itself was conducted. For a given outer cylinder rotational speed (RPM), the mass flow through the holes was varied by controlling the vacuum pump. At low RPM, no flutter occurred, but at higher RPM flutter existed for a range of mass flows. The maximum shell response which occurs at some mass flow is displayed in Fig. 2 as a function of RPM. Near 2,000 RPM the dramatic increase in shell response signifies the onset of flutter. Associated with this is a change from random shell oscillations at RPM below 2,000, where the shell is responding to turbulent pressure fluctuations due to the mass flow through the cooling holes, to sinusoidal shell motion above 2,000 RPM, which is

FIG. 1a STANDARD EXPERIMENTAL MODEL

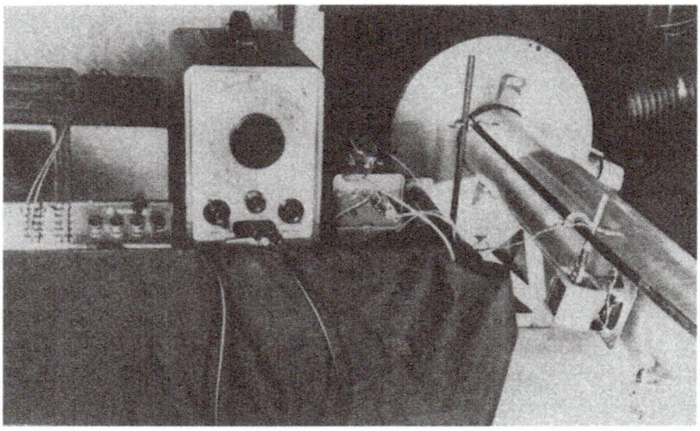

FIG. 1b EXPERIMENTAL APPARATUS FOR MEASURING INNER CYLINDER NATURAL MODES AND FREQUENCIES

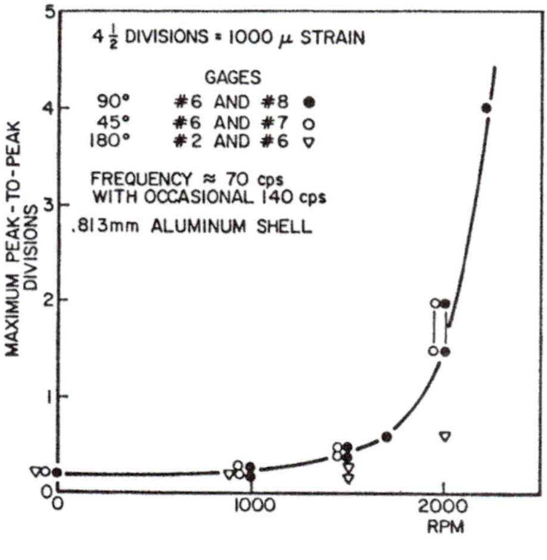

FIG. 2 MAXIMUM INNER CYLINDER STRAIN vs OUTER CYLINDER RPM

characteristic of flutter. Flutter can lead to violent oscillations and Fig. 3 displays a close-up photograph of an inner shell permanently plastically deformed due to flutter. Although difficult to see in the photograph tears in the shell skin were detectable by eye.

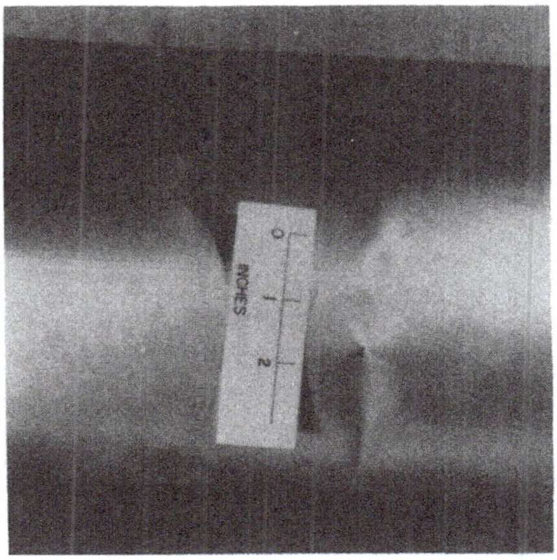

FIG. 3 PHOTOGRAPH OF FLUTTER DAMAGED .508 mm CYLINDER, #1

The flutter oscillations can be suppressed by stiffening the inner shell or reducing the severity of the flow environment. Usually the former is done, however in this case the latter was accomplished by reducing the strength of the vortex flow. To do this, the hole geometry was modified by beveling or the addition of tubes internal to the annulus between shells, Ref. [7]. Thus flutter was eliminated with only a modes weight, peformance and cost penalty.

2.2 Case II: Flutter of the Heat Shield in the Rocket Engine of the Solid Rocket Booster of the Space Shuttle

In 1976, a small scale (3%) model of the heat shield was tested to evaluate its thermal characteristics and also what was thought to be its favorable effect on reducing acoustic loads on the rocket engine nozzle, Ref. [8]. The heat shield covers an annulus between the rim of the conical rocket engine nozzle and a concentric conical fairing on the aft end of the Solid Rocket Booster. See Chart B. It is an annular, flat impregnated fabric which sags under gravity load and is pressurized to provide a membrane tension stiffness. See Sketch 1. The wind tunnel test revealed that large vibrations of the shield could occur and these were identified from motion pictures as traveling waves. The waves were absent when the Mach number over the heat shield was subsonic, but were present at supersonic Mach numbers. Subsequent mathematical analysis for flutter showed that the frequency and wavelength of the vibrations at Mach number conditions observed experimentally could be explained on a theoretical basis, Ref. [9].

Currently, additional analysis is being carried out to identify ways of suppressing flutter at supersonic conditions. Also a full scale wind tunnel test of a portion of the heat shield is being planned. The current and planned efforts on this problem, conducted by the NASA and Bolt, Beranek and Newman, Inc. are reviewed here.

In Sketch 2, the flow geometry is shown. The incoming supersonic flow goes through an oblique shock and is reduced to a lower supersonic Mach number (for sufficiently large angles, θ, the flow may become subsonic). This supersonic flow over a flexible surface creates the possibility of panel flutter, Ref. [1].

To make the problem tractable to analysis, the annular geometry was straightened conceptually into a long rectangular plate, Sketch 3. Traveling wave, low supersonic Mach number flutter analysis and high Mach number standing wave flutter analysis were invoked (Refs. [1 and 9]) to predict the heat shield flutter characteristics including wavespeed, wavelength and frequency. The results were compared to the measured values and the comparison is shown in Fig. 4. The calculated and measured results were in qualitative agreement, which is perhaps the most that can be expected in view of the complex structural characteristics of the heat shield. It is a nylon impregnated with rubber in the model scale flutter tests (and a fibre impregnated quartz material in the full scale heat shield).

	MEASURED	FLUTTER THEORY
WAVESPEED	152-178 cm/s	150-760 cm/s
WAVELENGTH	4-12 cm	7-37 cm
FREQUENCY	120-250 Hz	50-800 Hz
AMPLITUDE	1.2 cm (APPROXIMATELY)	---

FIG. 4 TABLE OF MODEL HEAT SHIELD OSCILLATIONS

The amplitude of the flutter oscillation (see Fig. 4) was the order of the sag distance of the heat shield (see section A-A of Sketch 3), which is what nonlinear flutter theory predicts, Ref. [1].

A typical analytical flutter boundary is shown in Fig. 5 in terms of flow velocity vs. Mach number for various wind tunnel stagnation pressures, P_o, and pressure differentials across the heat shield, ΔP. The wind tunnel trajectory which was followed during the flutter experiments is also shown. As can be seen, at the larger P_o, flutter is predicted at low supersonic Mach numbers for $\Delta P = 7\,kN/m^2$, and at virtually all supersonic Mach numbers for an unpressurized heat shield, $\Delta P = 0$.

To date, the optimum flutter cure for this configuration has not been identified. To stiffen the heat shield, thereby preventing flutter, is undesirable because this will constrain the desired movement of the inner rocket nozzle to which the heat shield is attached. A better alternative may be to build an aerodynamic "fence" from the edge of the outer rocket nozzle which will protect the heat shield by reducing the local velocities and Mach numbers over its surface.

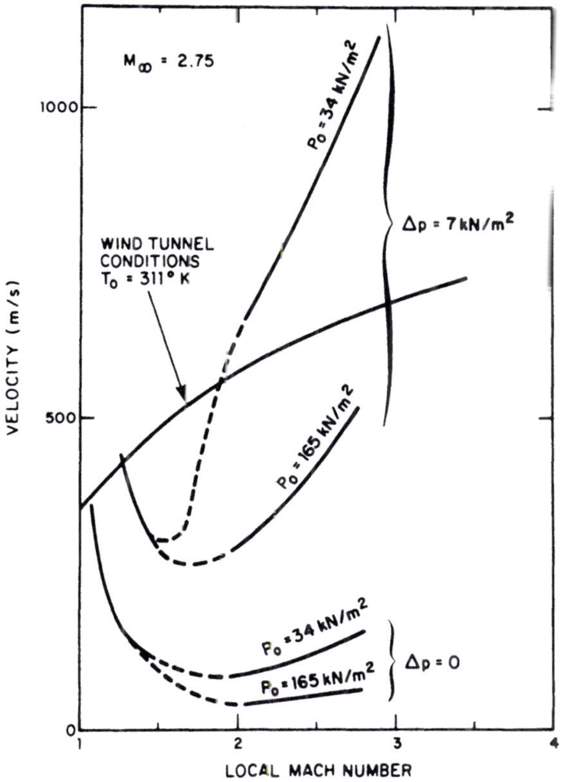

FIG. 5 COMPOSITE STABILITY BOUNDARIED FOR 3% SCALE HEAT SHIELDS, $M_\infty = 2.75$

2.3 Case III: Flutter of a Wind Tunnel Wall

This case (see Chart C) is rather straightforward as compared to the first two, except in one respect. The flat plate which flutters is a portion of a transonic wind tunnel wall and is quite large and massive, indeed it weighs 5448 kg! It is a flat steel plate of length and width, 5.18 m x 3.35 m, and with a tapered thickness from 3.18 cm to 5.08 cm. See Sketch 1 for a schematic of the wind tunnel and Sketch 2 for the wall dimensions. The plate is a portion of the NASA Ames Research Center Unitary Wind Tunnel and has been in place for many years. Recently, as part of the refurbishment of the wind tunnel, the rubber seals along the side edges of this plate were replaced. The next wind tunnel test at supersonic speeds led to flutter of the plate with an amplitude estimated to be 22.9 cm. Needless to say, this was a dramatic event.

The Mach number at which flutter occurred was 1.15 and the flutter frequency was at nearly the fundamental natural frequency of the plate, 7 Hz. Conventional flutter analysis (Ref. [1]) agreed quite closely with the measured results and suggested that the flutter was caused by a loss of damping when the old rubber seals were replaced by new (less inelastic) ones. Hence, additional shims between seal and wall, as well as a hydraulic damper, were added to the plate to restore an adequate amount of damping and thereby suppress flutter. The wind tunnel wall is now operating again successfully at supersonic Mach numbers. The diagnosis and cure of this flutter problem by the NASA (Ref. [10]) will be discusse

In Ref. [10], a fascinating, detailed account of this flutter case is given including a description of the wind tunnel apparatus. Erickson has described the wind tunnel geometry as follows. "A planform view of the wind tunnel nozzle is shown in Sketch 1. The two flex-walls are attached to the movable thrust blocks by a double row of bolts in the 45.7-cm (18.0 in.) length just upstream of station 00 which corresponds to the nozzle throat. The flex-walls are 3.35 m (11.0 ft) high and extend downstream to station 239 where the test section begins. (Station numbers give length in inches.) The downstream ends of the flex-walls are welded to a box beam which is bolted to a massive girder that circumscribes the nozzle between stations 211 and 239. Stiffening ribs between stations 188 and 211 are also welded to the walls and box beam. With the walls flat, the nozzle area downstream of station 00 is about the same as in the test section. Note that the flex-wall thickness varies in the streamwise direction. This variation provides the proper nozzle contour when the wall is in a flexed position for supersonic flow." The stabilizer prevents wall twisting, but does not constrain the wall from bending, as illustrated in Sketch 1.

These flex walls underwent flutter during two separate incidents. After the first flutter incident, accelerometers were mounted on the flexible walls. A time history of the wall motion as flutter began in the second incident is shown in Sketch 3. Scuff marks were left on the wind tunnel floor by the walls as they fluttered giving a graphic display of the flutter mode shape, as illustrated in Fig. 6. The flutter frequency ranged from 7.1 to 7.7 Hz, increasing with increasing flutter amplitude as expected from theory, Ref. [1]. The measured fundamental bending frequency of the wall is 7.3 - 7.5 Hz indicating, along with the flutter mode shape of Fig. 6, that flutter occurred in the first bending mode. The calculated fundamental bending frequency, using a NASTRAN finite element code, was 7.9 Hz and the second bending mode frequency was 21 Hz.

The theoretical flutter boundary for fundamental bending mode (single-degree-of-freedom) flutter is shown in Fig. 7 from Ref. [1]. The boundary shown is for zero mechanical damping in the wall, and is given in terms of a non-dimensional flow dynamic pressure, λ^*, vs. Mach number.

$$\lambda^* \equiv 2\, q\, a^3/D$$

where

$q \equiv$ dynamic pressure
$a \equiv$ plate (wall length)
$D \equiv$ bending stiffness of an equivalent wall of uniform thickness

The value of λ^* at the time of the first wall flutter incident is also shown in Fig. 7. As can be seen, this falls well above the theoretical flutter boundary, indicating that flutter would occur if the wall had no mechanical damping.

This result led to the hypothesis that originally the wall had sufficient mechanical damping to prevent flutter, but recent changes to the seals between wind tunnel wall and floor had led to loss of mechanical damping. Theory suggests that mechanical damping corresponding to a critical damping ratio of the order of 0.02 would increase the λ^* required to achieve flutter by a factor of 2-4 (Ref. [1]) over that for zero mechanical damping.

In point of fact, the seals and shims between the wind tunnel wall and floor had been changed shortly before the flutter incidents occurred and this apparently caused a loss of mechanical damping. Subsequent detailed measurements with various shim and seal configurations allowed measurement of wall critical damping

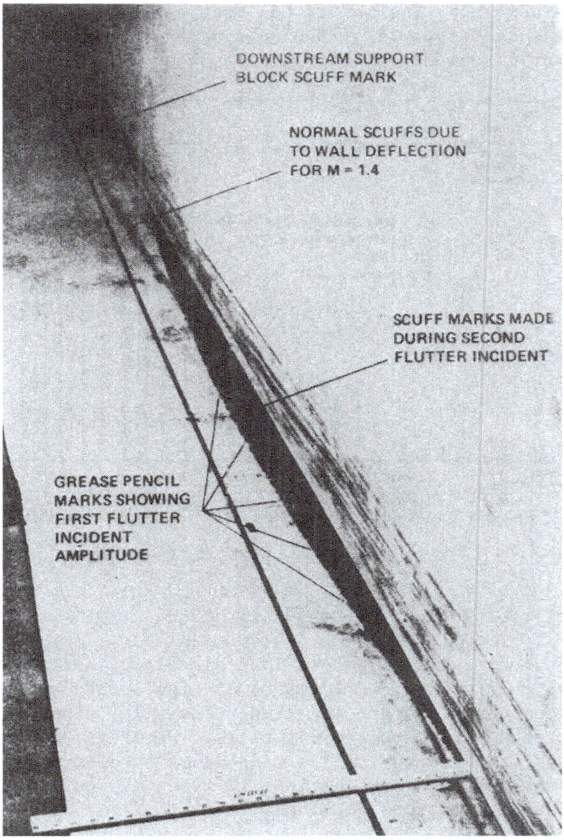

FIG. 6 SOUTH FLEX-WALL SCUFF MARKS

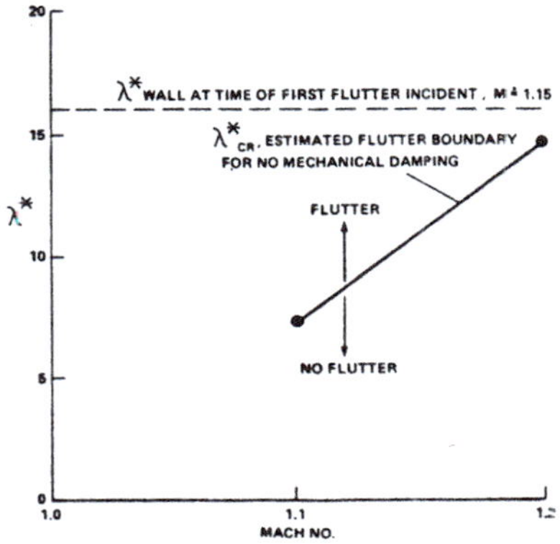

FIG. 7 ESTIMATED FLUTTER BOUNDARY OF FLEX-WALL

ratios. In addition, a hydraulic damper was attached to each wind tunnel wall to provide yet more damping. The result has been a flutter free wind tunnel wall since its damping was restored. Fig. 8 displays the damping achieved by various seal, shim and hydraulic damper configurations.

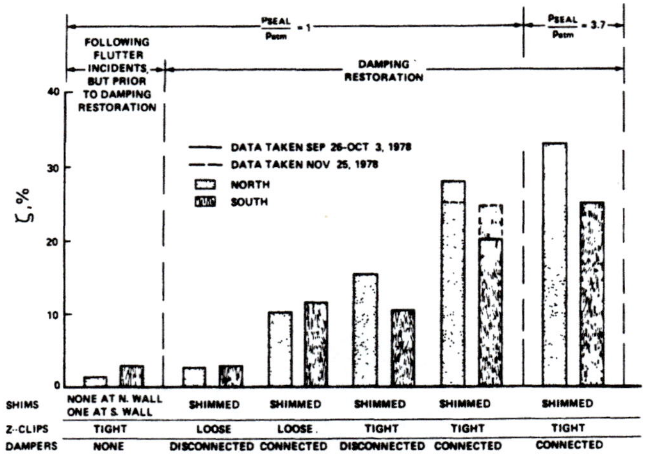

FIG. 8 RESTORATION OF FLEX-WALL DAMPING

2.4 Case IV: Flutter of Re-Entry Missile Exit Cone

On a number of re-entry missiles, it is desirable to add an extension to the exit cone to improve their thrust performance at high speeds and high altitudes. These extensions are of light weight and highly flexible. Some are rigidized cloth and others are made of composite materials. See Fig. 9. Flutter of such flexible cones has occurred though only a modes published literature is available, Ref. [11]. The essence of this problem is briefly discussed and possible flutter suppression design options outlined.

There is a theoretical literature on conical shell flutter which indicates that the qualitative features of the problem are similar to those of cylindrical shell flutter (Refs. [1,12]). For a small length to radius ratio, the conical shell may be treated even more simply as a flat surface. If most of the stiffness of the shell derives from tension forces rather than bending stiffness, the shell may be further simplified and considered as a membrane.

For these re-entry cones, according to Ref. [11], reasonable correlation with flight experiments has been obtained using the flat, membrane theory of Spriggs, Messiter and Anderson (Refs. [1,12]). See Fig. 10. Unfortunately, there is no published correlation between theoretical and experimental data. Apparently the flutter has been suppressed by using re-entry cones of greater stiffness.

2.5 Case V: Flutter of a Toroidal Membrane at Low Flow Velocities

A pressurized toroidal membrane (oblong dough-nut) has been suggested as an air cushion device suitable for various ground vehicles and landing gears of aircraft operating over rough terrain. A cross-section of such a device is shown in Fig. 11. For a given trunk pressure, P_t, as the cushion pressure, P_c, is increased, flutter of the trunk (membrane) occurs. This has been observed on both small scale experimental models and full scale, operational landing gear systems, Ref. [14 and 15]. The flow speeds beneath the trunk are of the order

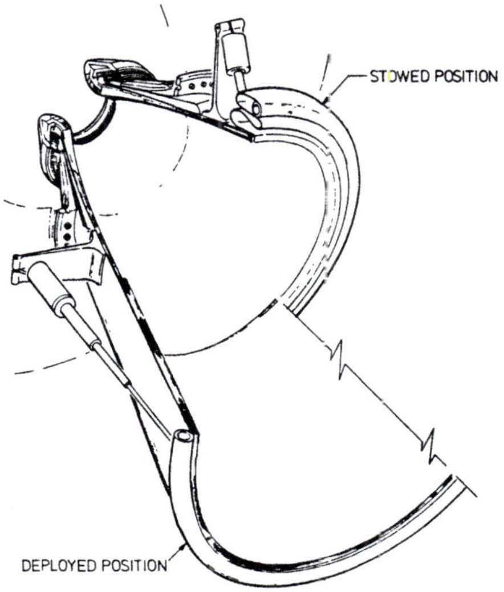

FIG. 9 RADIATION COOLED CLOTH EXTENDIBLE EXIT CONE CONCEPT

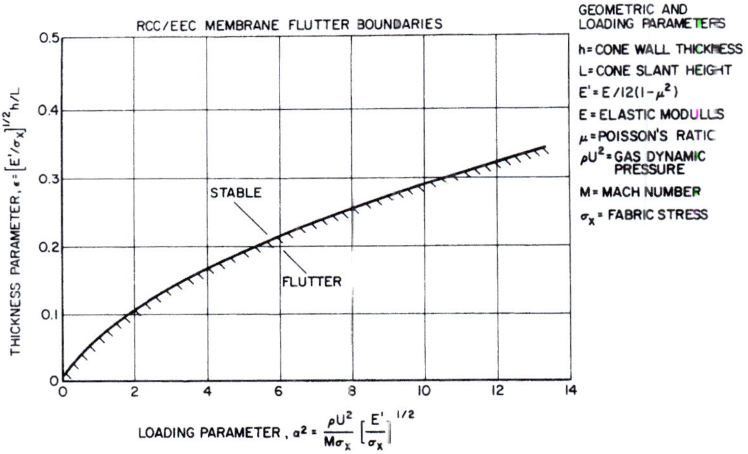

FIG. 10 RCC/EEC MEMBRANE FLUTTER BOUNDARIES

of 30.5 m/s. At present the Foster-Miller Corporation is engaged in a combined experimental-theoretical study to explain this phenomenon.

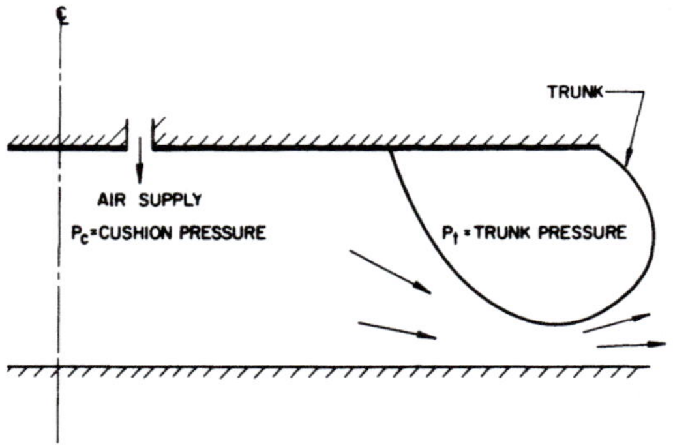

FIG. 11 SCHEMATIC OF AIR CUSHION LANDING SYSTEM

To date, there are no published discussions of this problem, although its existence is well-known to air cushion designers. At present, the essential aspects of the flutter mechanism are undetermined, both with respect to what constitutes an adequate fluid mechanical model, as well as with respect to what is required to model the air cushion structure. A simple one-dimensional model of the fluid (incompressible channel flow) and a one-degree-of-freedom spring-supported plate model of the air bag predicts a static instability, Ref. [14]. Experiment, however, suggests the instability is dynamic. A two-dimensional fluid mechanical model and multi-degree-of-freedom structural model are being formulated in an effort to resolve this issue. More systematic experiments would also be helpful.

3. CONCLUDING REMARKS

It is well to emphasize once again that, when flutter has occurred for a plate or shell in practice, it becomes important to verify that a flutter mechanism is indeed the source of the oscillation. It is most advantageous if a theory is available which predicts the nature of the motion (frequency, mode, etc.). Lacking such, one must rely on experiment and the characteristic (if not universal) rapid change in amplitude and character (from random to periodic) of the structural motion with some flow parameter (e.g., velocity) to identify the oscillation as flutter induced. As for cures, changes in structural mass, stiffness or damping are most often used. Again, at least a qualitatively correct theoretical model will be most helpful in selecting the best of these. Occasionally a modification to the flow field to suppress flutter will be possible and, if so, it is sometimes the method of choice for reasons of economy or simplicity.

In each of the case histories discussed, flutter was not anticipated and prior to first testing or use of the design, no flutter analysis by theoretical or experimental means was performed. In Cases I and II, significant, if not fundamental, advances of the available knowledge were required to understand and suppress the flutter vibrations. These advances were subsequently made. In Cases III and IV, available methods were probably sufficient to identify flutter as a potential problem had they been used. These four cases taken together suggest that it is important to use our available knowledge for design evaluation of flutter of plates and shells in a flow on a timely basis. Even so, it will continue to be necessary from time to time to advance our knowledge to account for new structure and flow geometries as well as environmental factors. Case V is an example where advances are currently needed.

4. REFERENCES

1. Dowell, E. H., "Aeroelasticity of Plates and Shells", Noordhoff International Publishing, Leyden, THE NETHERLANDS, (1975).

2. Dowell, E. H. (principal author), "Panel Flutter, NASA Space Vehicle Design Criteria", NASA SP-8004, (1972)

3. Lemley, C. E., "Design Criteria for the Prediction of Panel Flutter", Air Force Flight Dynamics Laboratory, TR 67-140, I, II, August 1968.

4. Laurenson, R. M. and McPherson, J. I., "Design Procedures for Flutter-Free Surface Panels", NASA CR-2801, (1977).

5. Srinivasan, A. V., "Flutter Analysis of Rotating Cylindrical Shells Immersed in a Circular Helical Flowfield of Air", AIAA Journal, 9, 3, 394-400, (1971).

6. Dowell, E. H., Srinivasan, A. V., McLean, J. D. and Ambrose, J., 'Aeroelastic Stability of Cylindrical Shells Subjected to a Rotating Flow", AIAA Journal, 12, 12, 1644-1651, (1974).

7. McLean, J. D. and Dowell, E. H., "Swirling Flows Between Coaxial Cylinders with Injection by Radial Jets", AIAA Journal, 13, 11, 1435-1440, (1975).

8. Porter, J. H., Jr., "An Aero-Acoustic Test of a 3-Percent Scale SRB with Various Flexible Heat Shields and Aero-Fixes", AEDC-DR-76-11, Von Karman Gas Dynamics Facility, Arnold Engineering Development Center, February 1976.

9. Ventres, C. S. and Dowell, E. H., "The SRB Heat Shield: Aeroelastic Stability During Reentry", Bolt, Beranek and Newman, Inc., Report No. 3532, September 1977.

10. Erickson, L. L., Kassner, D. L., Guist, L. R. and Chargin, M. K., "Investigation of Flexible Nozzle Wall-Flutter Incidents in the NASA Ames Research Center 11 by 11 foot Transonic Wind Tunnel", AIAA Paper 79-0797, presented at the AIAA/ASME/ASCE/AHS 20th Structures, Structural Dynamics, and Materials Conference, St. Louis, MO., April 1979.

11. Inman, F. S. and Garield, J. R., "Development of a Radiation Cooled Cloth Extendible Exit Cone (RCC/EEC)", Phase II, Final Report AFRPL-TR-77-41, Thiokol/Wasatch Division, Brigham City, Utah, September 1977.

12. Dixon, S. C. and Hudson, M. L., "Flutter, Vibration, and Buckling of Truncated Orthotropic Conical Shells with Generalized Elastic Edge Restraint", NASA TN D-5759, (1970).

13. Spriggs, J. H., Messiter, A. F. and Anderson, W. J., "Membrane Flutter Paradox - An Explanation by Singular Perturbation Methods", AIAA Journal, 7, 9, 1704-1709, (1969).

14. Erickson, A., Foster Miller Corporation, private communication, (1978).

15. Wormley, D., Massachusetts Institute of Technology, private communication, (1978).

CHART A: SHELL FLUTTER OF THE MAIN SHAFT OF A JET ENGINE

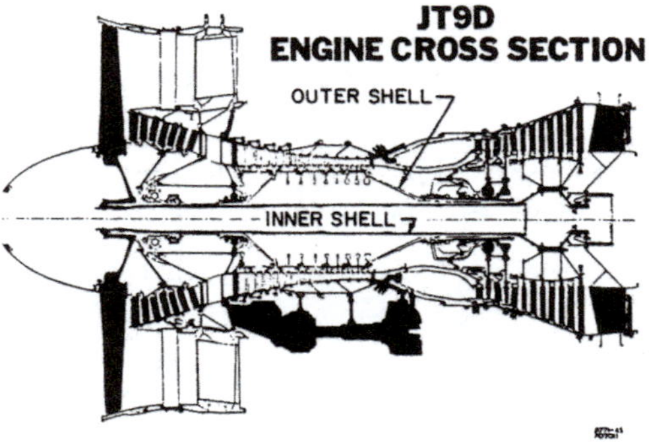

(1)

Problem: Inner shell of two concentri shells undergoes flutter as outer shel RPM is increased. Shell breaks.

Geometry: See Sketches 1 and 2.

Boundary Conditions: Inner shell is clamped.

Flow Conditions: A circumferential vo flow is generated by outer shell rotat and the cooling air which is drawn thr holes in outer shell.

Operating Conditions: For the laborat model, outer shell rotation up to 2000 and cooling air mass flow to 1 lbm (.4 sec at ambient conditions. For the je engine, the corresponding numbers are 11,000 RPM and 10 lbm (4.54 kg)/sec at and 15 atmospheres.

Structural Dynamics: The aluminum inn shell natural modes and frequencies we well predicted by theory and ranged fr 90 Hz upwards. Typical dimensions were 2.44 m long, 20 cm diameter. Thicknes were .635-1.27 cm for full scale engin and .508-.813 mm for dynamically scale models.

Vibration: A circumferential travelin wave with a frequency of 135 Hz, wavesp of 23.2 m/sec and wavelength of 17.1 cm was typical of the flutter motion.

Cure: Cooling hole geometry was modi to decrease the severity of the vortex flow.

References: Srinivasan, A. V., Flutter Analysis of Rotating Cylindrical Shells Immersed in a Circular Helical Flowfield of Air. AIAA Journal, Vol. 9, No. 3, pp. 394-400, March 1971.

Dowell, E. H., Srinivasan, A. V., McLean, J. D. and Ambrose, J., Aeroelastic Stability of Cylindrical Shells Subjected to a Rotating Flow. AIAA Journal, Vol. 12, No. 12, pp. 1644-1651, December 1974.

McLean, J. D. and Dowell, E. H., Swirling Flows Between Coaxial Cylinders with Injection by Radial Jets. AIAA Journal, Vol. 12, No. 11, pp. 1435-1440, November 1975.

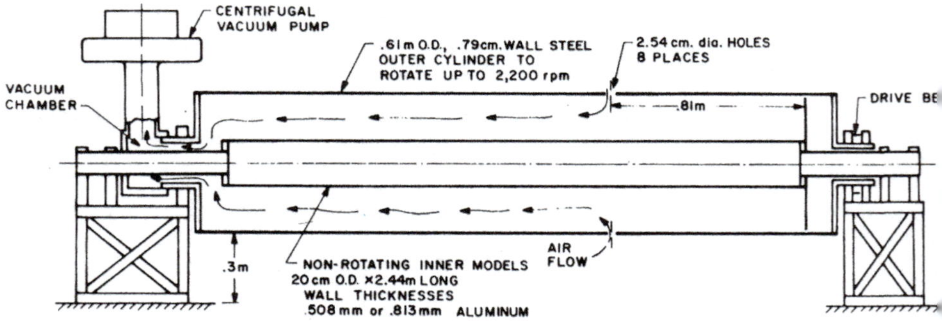

(2) STANDARD EXPERIMENTAL MODEL

CHART B: HEAT SHIELD FLUTTER

Problem: Heat shield in aft end of Solid Rocket Booster of Space Shuttle flutters according to 3% scale model wind tunnel tests.

Geometry: A curved annular membrane. See Sketches 1 and 3.

Boundary Conditions: Membrane is pinned on side edges.

Flow Conditions: Membrane is embedded in a rocket exit cone. Flow over cone face is supersonic and includes an oblique shock wave. See Fig. 2a.

Operating Conditions: Supersonic Mach numbers, 2.5-3.5, high temperatures $\sim$ 1000-2000°F and turbulent flow in full scale device. Wind tunnel conditions only simulated Mach number and flow dynamic pressures.

Structural Dynamics: Stiffness of membrane poorly defined. Wind tunnel model has a mass of .20 k /m^2, width of 2.54 cm, sag height of 1.27 cm, thickness of .015 cm and modulus times thickness of 2 x 10^4 N/m with an overall mean diameter of 15 cm.

Vibrations: A traveling wave with a frequency of 120-250 Hz, wavespeed of 152-178 cm/sec and wavelength of 4-12 cm was typical of the flutter motion.

Cure: Not yet determined. Possibilities include stiffening heat shield and modification of flow environment.

References: Ventres, C. S. and Dowell, E. H., The SRB Heat Shield: Aeroelastic Stability During Reentry. Bolt, Beranek and Newman, Inc. Report No. 3532, September 1977.

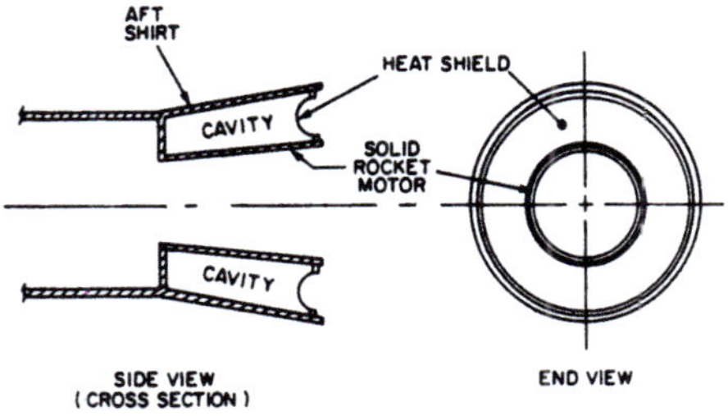

(1) SCHEMATIC OF AFT END OF SRB.

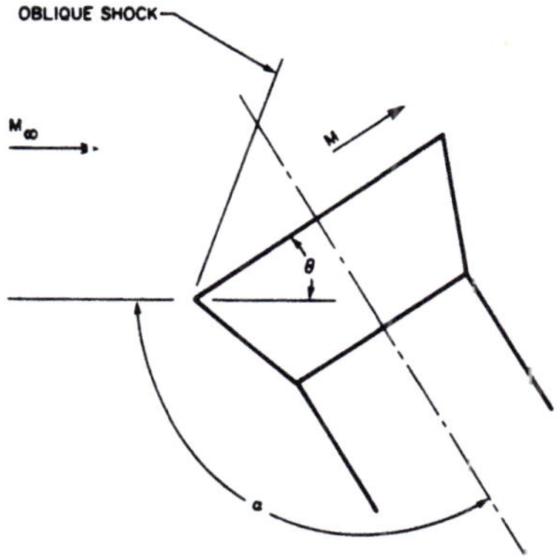

(2) LOCAL FLOW GEOMETRY

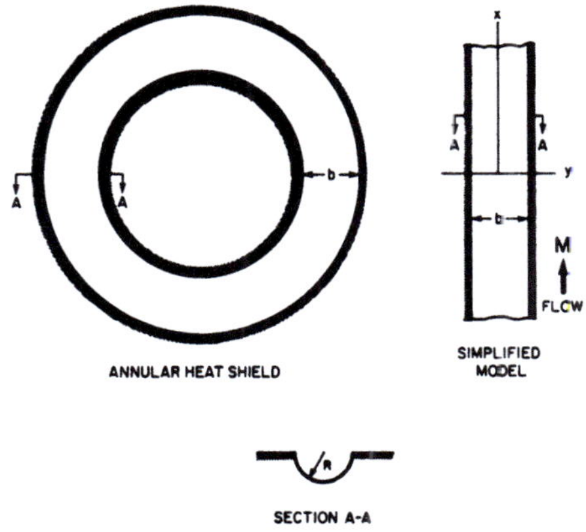

(3) SCHEMATIC DRAWING OF ANNULAR HEAT SHIELD AND THE SIMPLIFIED GEOMETRICAL MODEL

CHART C: WIND TUNNEL WALL FLUTTER

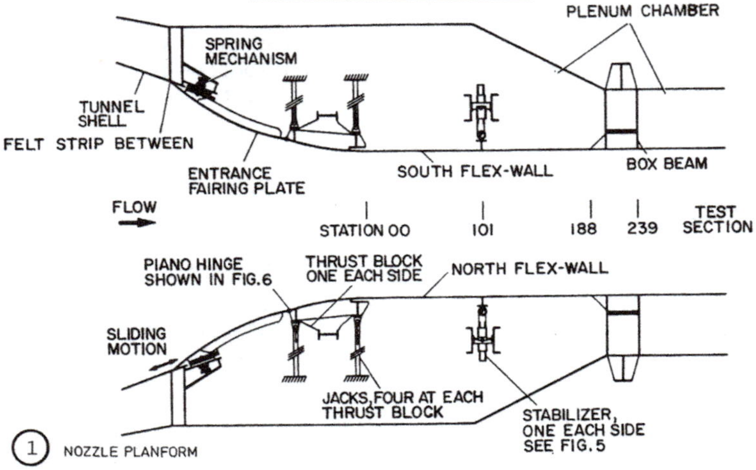

(1) NOZZLE PLANFORM

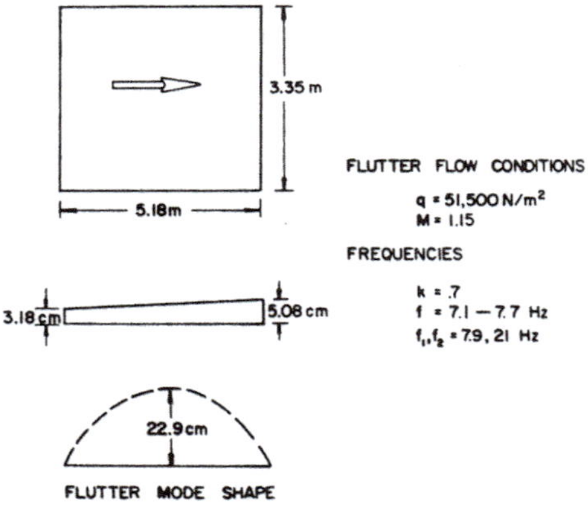

(2) WALL FLUTTER

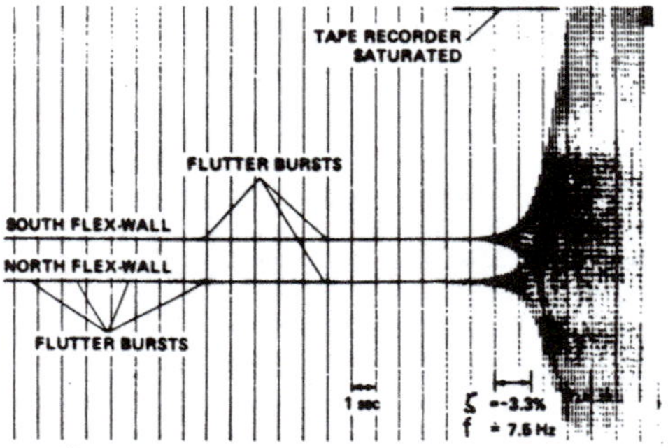

(3) GROWTH OF FLUTTER AMPLITUDE

Problem: A 3.35 m x 5.18 m, 5448 kg wall of a wind tunnel flutters.

Geometry: See Sketches 1 and 2.

Boundary Conditions: Wall is clamped on leading and trailing edges and free on side edges.

Flow Conditions: Low supersonic Mach numbers and smooth attached flow.

Operating Conditions: M = 1.15 with dynamic pressure of order 5 x 10^4 N/m^2.

Structural Dynamics: A 5448 kg clamped plate, 3.35 m x 5.18 m long and thickness from 3.18-5.08 cm, with free side edges (beam-like motion) and a fundamental resonant frequency of the order of 7 Hz.

Vibrations: A standing wave in fundamental beam-like mode with a frequency of 7 Hz and a peak to peak amplitude of 22.9 cm was typical of flutter motion.

Cure: Restore mechanical damping to wall. The damping had been, inadvertently lost during refurbishment and routine maintainence of wind tunnel wall.

References: Erickson, L. L., Kassner, D. Guist, L. R. and Chargin, M. K., Investigation of Flexible Nozzle Wall-Flutter Incidents in the NASA-Ames Research Center 11 by 11 foot Transonic Wind Tunnel. AIAA Paper 79-0797, Presented at the AIAA/ASME/ASCE/AHS 20th Structures, Structural Dynam and Materials Conference, St. Louis, MO, April 1979.

A13

THE FLOW INDUCED RESPONSE OF REACTOR PRESSURE VESSEL THERMAL INSULATION

Martin E. Drake

A. Graham Watson

Peter N. Whitton

Research Officers

Central Electricity Generating Board, Berkeley Nuclear Laboratories, Berkeley England.

SUMMARY

This paper reports the results of laboratory and in-reactor experimental work, along with theoretical studies related to the mechanical vibration behaviour of a reactor pressure vessel thermal insulation. On-load measurements show that the major in-reactor response is non-resonant and due to turbulent buffeting. The important flow and mechanical parameters are identified.

1. INTRODUCTION

The Central Electricity Generating Board operates a number of nuclear reactors cooled by pressurised carbon dioxide. The pressure containment in later stations is constructed from prestressed concrete which must be maintained at a low temperature. This is achieved by insulating the pressure containment from the high temperature coolant gas using a metallic thermal insulation. A flexible structure is necessary in order to minimise thermal stresses in the retention system securing the insulation to the cool pressure vessel or pressure vessel liner. The consequence of this flexibility is that the behaviour of the system in the reactor environment, which includes both flow and acoustic excitation, is of concern.

As part of an investigation of the stresses in the retention system of the insulation in the boiler inlet region of one of the CEGB's magnox reactors, in-reactor measurements have been made to establish the dynamic behaviour during power operation. The measurements have been supported by theoretical studies of the dynamics of the insulation system and further experimental studies on a laboratory specimen. The work provides an understanding of the factors influencing the in-reactor vibration behaviour of this type of insulation.

2. REACTOR GEOMETRY

2.1 Gas Circuit

Pressurised carbon dioxide at 27×10^5 N/m^2 is circulated through the reactor core where it is heated to a temperature of 360°C. The hot gas then passes through gas/steam heat exchangers. Sketch 1 is a schematic of the boiler inlet region, showing the insulated surfaces and the flow paths. The main flow path involves two 90° bends before the gas emerges from an annulus into the insulated region. The radial outward flow has a mean velocity through this annulus of $\sim$16.5 m/s and impinges directly on the pressure vessel thermal insulation. The height of this annulus (450 mm) is maintained by 25 mm thick support webs periodically arranged around its circumference.

2.2 Thermal Insulation Design

As shown in Sketch 2, the insulant consists of 28 alternate layers of 0.1 mm thick stainless steel foil separated by a 1.22 mm thick mesh of wires on a 50 mm square pitch. The insulant is covered by two 0.71 mm thick stainless steel sealing skins and retained by an array of 6.35 mm thick mild steel coverplates of nominal dimensions 600 mm x 450 mm. These coverplates are exposed to the gas at 360°C whilst the steel pressure vessel liner, in contact with the concrete, operates at <90°C. The coverplate is held in contact with the insulant by a 12.7 mm diameter stud welded to the pressure vessel liner. A single stud is used to minimise thermal stresses in the retention arrangement. It is the stud dynamic stresses which are of prime concern in the vibration behaviour. To guard against the possibility of failure of the retention stud, due to manufacturing defects or other causes, a secondary retention system is provided consisting of tags which overlap adjacent coverplates ensuring the plate is retained. Normally the tags do not contact the adjacent plates.

3. REACTOR ON-LOAD MEASUREMENTS

3.1 Dynamic Pressure

Limited data are available from pressure transducers fitted at selected positions in the insulated region. A typical pressure power spectral density, recorded at close to full reactor volumetric flow, is shown in Figure 1. 60 per cent of the total mean square pressure is below 25 Hz and the spectrum is dominated by peaks in the region 5-10 Hz. Very little energy is associated with the pressure signal above 100 Hz. Consideration of the geometry upstream of the insulated region suggests that the radial outward flow from the annulus may be important. Using the annulus height of 450 mm to relate to the scale of turbulence in the sheared annular flow and the estimated flow velocity in this region gives frequencies of ∿7 Hz. Similar length scales and flow velocities are associated with other parts of the upstream geometry and these may also be of significance.

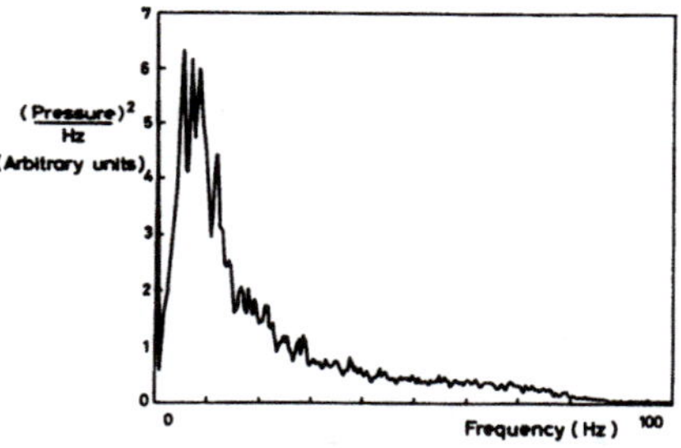

FIG. 1. PRESSURE POWER SPECTRAL DENSITY (REPRODUCED BY PERMISSION OF NPC (W) LTD.)

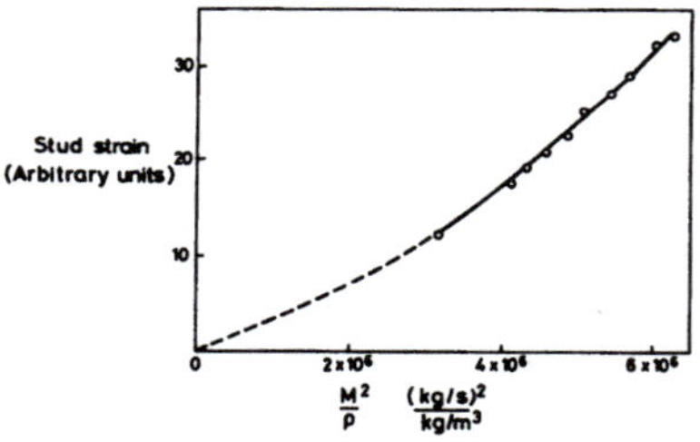

FIG. 2. VARIATION OF STUD DYNAMIC STRAIN WITH DYNAMIC HEAD

3.2 Insulation Dynamic Strain

Strain in the coverplate retention stud is the main parameter of interest. It has been investigated by fitting weldable strain gauges to selected studs. Sketch 3 presents the strain power spectral density at two reactor gas mass flows. The major response is associated with bending of the stud. 98% of the total mean square strain is below 25 Hz. The frequency of the response peaks increase linearly with flow velocity and relate to those observed from pressure transducer measurements made during a separate test. Prior to the on-load measurements the mechanical vibration characteristics of these systems were established using mechanical excitation. The lowest natural frequency for the stud/coverplate system, whose strain is presented in Sketch 3, was 26 Hz and a heavily damped response can be seen at this frequency in the spectrum. Thus non-resonant flow excitation dominates the response.

Figure 2 gives the variation of the root sum of squares strain, from two gauges spaced at 90° around the stud, with (mass flow)^2/density, (M^2/ρ) which is proportional to gas dynamic head. It can be seen that stud strain increases with M^2/ρ to a power greater than one. This is consistent with the increase in frequency of the excitation with mass flow and the increase in the system compliance as the frequency of the excitation approaches that of the lowest resonance. It may be shown theoretically that for a single degree of freedom system excited by a band limited force of constant spectral density well below resonance, the dependence on flow velocity V is

$$\text{response} \propto V^2.V^{\frac{1}{2}}\,[1 + K_1V^2 + K_2V^4]$$

where K_1 and K_2 are constants dependent on the relation between the highest excitation forcing frequency and the resonant frequency.

4. INSULATION DYNAMICS

The dynamics of the insulation system have been studied in the laboratory using a single coverplate specimen and supported by calculations using a finite element idealisation of the structure. In-reactor measurements have been made to confirm the mechanical characteristics of the installed insulation.

The dynamics of the stud-coverplate system are strongly influenced by interactions with the foil and mesh insulant. Static stiffness measurements on the laboratory sample showed that the insulation stiffness increases with compression. The assembly procedure for this type of insulation involves applying a pre-load to the coverplate and subsequently locking the plate into position using a nut on the threaded retention stud (see Sketch 2). Thus it may be expected that the installation procedure will influence the mechanical behaviour.

The finite element model represented the stud by beam elements and divided the coverplate into 16 rectangular plate elements. The insulant stiffness was included as lumped stiffnesses at the plate element nodes and the additional, calculated, mass associated with the insulant as a distributed increase in density of the coverplate. This model confirmed the interpretation of the in-reactor strain measurements described in Section 3 showing that the lowest two resonances of the structure are in the range 25-40 Hz and involve rocking of an essentially rigid coverplate on the retention stud. Higher modes involve flexural vibration of the plate but do not produce a concentration of strain energy in the stud.

Approximate expressions for the resonant frequencies (f_{ob} and f_{oa}) of the two lower modes can be derived by solving the eigenvalue equation obtained by equating stiffness and inertia forces assuming a rigid coverplate rocking on a central flexible stud:

$$f_{ob} = \frac{1}{2\pi}\left[\frac{\dfrac{3E\pi D^4}{16\ell} + Ka^3b}{\rho_p a^3bc + \rho_I\dfrac{\ell a^3b}{2}}\right]^{\frac{1}{2}}$$

f_{ob} applies for rocking about an axis through the stud parallel to the side of length b. f_{oa} is obtained by interchanging a and b. a and b are the length and width of the coverplate and c its thickness. E Young's modulus, ρ_P the density of the coverplate, D the diameter of the stud and ℓ its length. ρ_I and K are the density the stiffness/unit area of the insulant. The expressions clearly show the relative importance of stud and insulant stiffness. At the upper end of the range of preloads relevant to the reactor, the bending stiffness of the stud is small compared to the stiffness associated with the insulant.

Mechanical tests were performed on a 3x3 coverplate matrix of the installed insulation within the reactor. Excitation was provided by a standard electro-dynamic shaker and applied through a force transducer to the central plate. Measurements of the vibration transmitted through the insulant to adjacent coverplates showed that the energy transmission was less than 2 per cent, confirming that valid conclusions could be drawn using experimental or theoretical models based on a single coverplate unit.

The 3x3 matrix was successively compressed and the compliance of the central plate measured as a function of the applied pre-load (500 N to 2000 N) to confirm the effect of the underlying insulant. Sketch 4 shows the variation of compliance with frequency at two different pre-loads. It can be seen that at frequencies below 20 Hz where the majority of the flow energy is concentrated, the compliance decreases with increasing pre-load. The natural frequencies and associated modal damping increase with pre-load.

Measured mode shapes, obtained by curve fitting the transfer function between the applied point force and acceleration measurements on the coverplate, are shown in Figure 3 for the lowest two modes. These modes basically involve rocking about the major and minor axes of the plate, but with the nodal lines rotated from these axes because of the asymmetry introduced by the dynamic mass on the tagging system. The frequencies and mode shapes were in agreement with those predicted using the finite element model and the lower modes could also be adequately determined by simple theory.

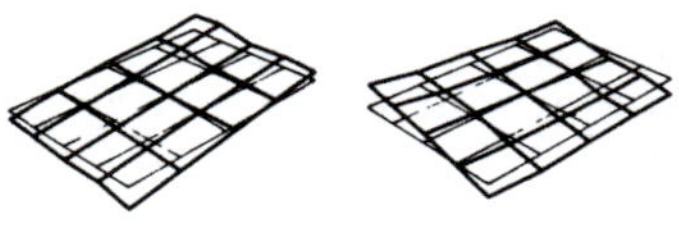

FIG. 3. MODE SHAPES OF THE LOWEST TWO MODES OF THE INSULATION SYSTEM

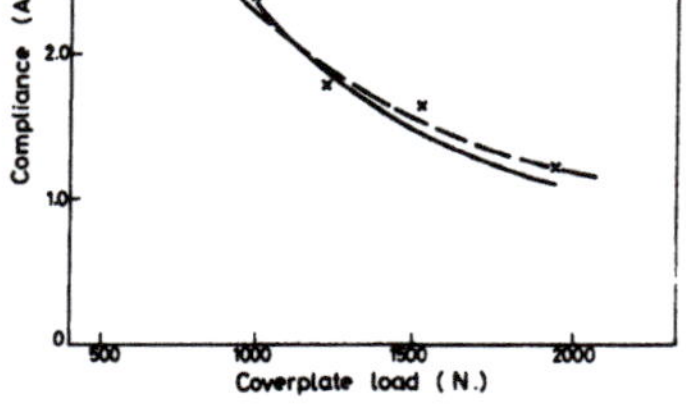

FIG. 4. VARIATION OF COMPLIANCE WITH COVERPLATE LOAD-REACTOR INSULATION.

Figure 4 presents the variation of compliance with coverplate installation pre-load over the range of pre-loads in the frequency range 5-10 Hz associated with the major excitation forces (see Figure 1). The theoretical prediction based on the finite element model is also shown and is in good agreement. Over this pre-load range the damping loss factor associated with the lowest mode increased from 0.13 to 0.5. Thus the system damping is high and associated with dissipation in the insulant. The loss factor is approximately proportional to the applied pre-load, supporting a view that the damping is due to friction between the coverplate and foil and mesh layers.

5. DISCUSSION

The significant dynamic response of the insulation is due to non-resonant turbulent buffeting. The turbulent energy is concentrated in the range 5-10 Hz and is probably associated with the geometry of the flow path upstream of the insulated region. The higher modes of the mechanical system do not produce appreciable strains in the retention stud, thus confirming that this low frequency excitation is the dominant constraint on the high endurance behaviour of the studs.

The response of the retention system of this type of insulation is strongly influenced by its interaction with the insulant, and in turn this is controlled by the installation procedure. Increasing the assembly pre-load applied to the coverplate results in reduced response both below the lowest resonance, due to an increase in stiffness, and at resonance, due to an increase in damping. Similar insulation is fitted to all U.K. magnox reactors and this behaviour may be expected to apply.

The retention stud strains are not directly proportional to dynamic head but increase with gas flow velocity at a higher rate. This dependence on flow velocity is consistent with the increase in compliance of the system with frequency up to the lowest resonance, and the increased frequency of the flow force spectrum with gas velocity. Thus small changes in reactor operating conditions may be used to minimise the vibration response to flow buffeting. This behaviour can be expected for any similar mechanical system exicted below resonance.

ACKNOWLEDGEMENTS

This paper is published by permission of the Central Electricity Generating Board.

REACTOR THERMAL INSULATION RETENTION SYSTEM

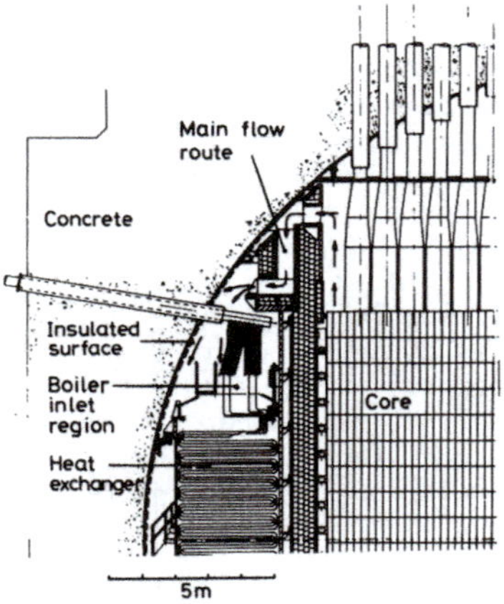

① SCHEMATIC OF BOILER INLET REGION

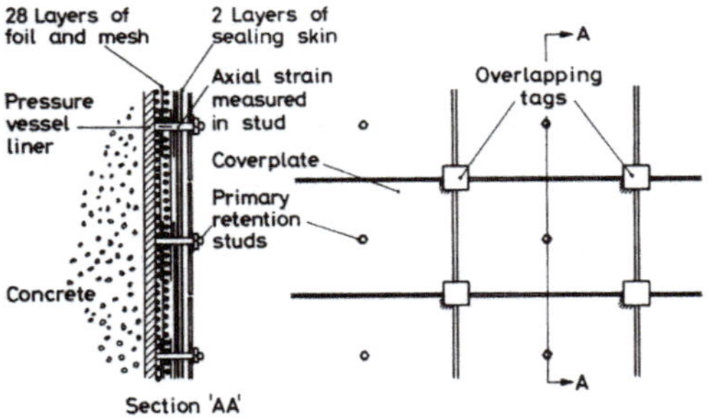

② SCHEMATIC OF FOIL & MESH THERMAL INSULATION

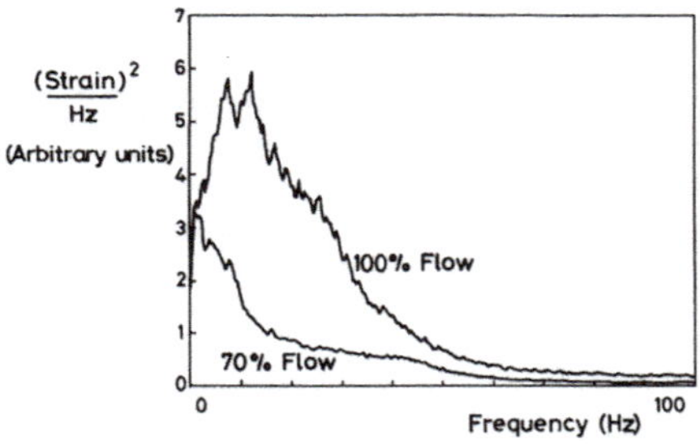

③ STRAIN POWER SPECTRAL DENSITY AT TWO REACTOR GAS MASS FLOWS

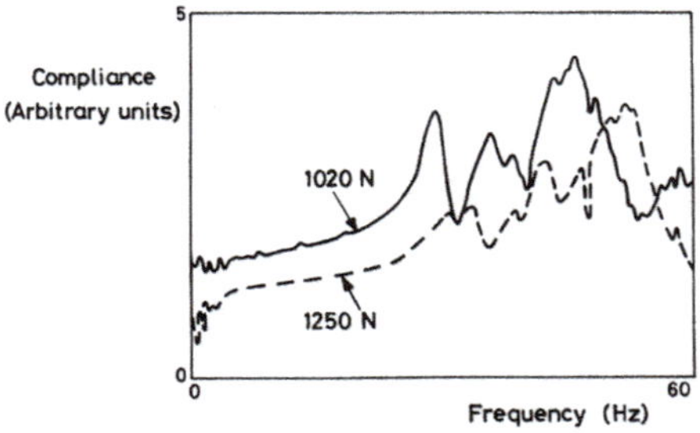

④ VARIATION OF COMPLIANCE WITH COVERPLATE PRE-LOAD

Problem: Non-resonant response of retention system to turbulent buffeting. Fatigue damage of stud possible.

Geometry: See sketch 2.

Boundary Conditions: See sketch 1. Radial flow from Gas Outlet Annulus impinges directly onto insulation.

Flow Conditions: See sketch 1. Maximum mean gas velocity through annulus $\sim$16.5 m/s. Carbon dioxide at 27×10^5 N/m^2 and 360°C. Pressure power spectrum peaks 5-10 Hz. Peaks relate to upstream flow geometry scale.

Operating Conditions: Vibration increases with flow velocity, V, but at rate greater than dynamic head, ρV^2.

Structural Dynamics: See sketch 4. Lower two modes involve rigid coverplate with bending retention stud. Insulant compression important, increase in compression gives: lower compliance below resonance; increase in natural frequency; and increase in system damping (in proportion to pre-load).

Vibrations: See sketch 3. Major response below 10 Hz.

Cures: Stud strain levels can be reduced by: increasing installation compression due to interaction with insulant; reduction in reactor gas volumetric flow.

A14 CORE BARREL OSCILLATIONS DURING BLOWDOWN OF A PRESSURIZED WATER REACTOR

U. Schumann

Kernforschungszentrum Karlsruhe, Institut für Reaktorentwicklung
Projekt Nukleare Sicherheit, Karlsruhe, Germany

SUMMARY

In this paper, a survey of an experimental and theoretical program on fluid-structure interactions during blowdown of a nuclear reactor is given. In particular, the dynamics of the core barrel - a thin cylindrical shell inside the pressure vessel - are discussed. Oscillations are induced by transient spatial pressure differences. In the initial period, while most of the fluid is in a liquid state, the motions are strongly influenced by fluid-structure coupling. Without this coupling, the computed response is misleading because of the possibility of unphysical resonance. The virtual fluid mass results in a considerable reduction of the frequencies of motion. The effective speed of sound is reduced by the structural motion and is finite even in case of an incompressible fluid. Evaporating water inside the core barrel results in quasi-steady and large outflow velocities which in turn cause large local pressure changes according to Bernoulli's law and large core barrel deformations near the blowdown nozzle.

1. INTRODUCTION

The core barrel in a pressurized water reactor (PWR), under normal operating conditions, controls the flow path from the inlet nozzle of the reactor vessel through the downcomer and the reactor core into the outlets. In case of a blowdown, initiated by a complete and sudden break of the inlet pipe, a depressurization wave migrates through the pipe and the downcomer into the inner region. The resultant pressure differences cause oscillations of the core barrel. The core barrel is designed so that it should resist these loadings. In the past, a decoupled analysis was used for this purpose [1] in which the pressure load was computed assuming rigid walls and the structural response to this load was determined subsequently without accounting for fluid-structure interactions. As will be shown, the coupling has a strong effect on the structural dynamics. In order to understand the basic physical effects and to evaluate the actual safety margins, an extensive experimental and theoretical program [1] has been initiated.

For theoretical analysis, several computer codes have been developed and are still being refined. They should permit a "best estimate" analysis. Such codes account for the fluid-structure interactions. Overviews of such code developments are given in [2]. The validity of the codes has to be assessed by comparison with large scale experiments. As yet, relevant large-scale experimental data are available only from the "RS16B experiments" at Battelle [3]. The interpretation of these experiments is limited by the rather slim geometry of the vessel, see chart, and insufficient instrumentation. Also, the rather complicated internal structures make it difficult to discriminate between errors induced by non-proper modelling of the structure and those caused by deficiencies in the fluid-structure interaction models. Therefore, a new set of experiments is under preparation in which the former "HDR reactor" near Karlstein/Main is used as a test facility. This setup is rather close to the actual size and shape of a typical German PWR. The details of the reactor core, which would complicate the analysis considerably, are excluded from the model. Only the mass of the core is approximately represented by the mass ring at the lower edge of the core barrel. The core barrel is a simple cylindrical shell of constant thickness with clear clamping conditions which should be well represented by quasi-analytical shell models [2, 4]. Subsequently, the principal effects of fluid-structure coupling will be explained by reference to code predictions in case of the HDR and a typical PWR and by comparison with the RS16B experiments.

2. THE COMPUTATIONAL MODEL FLUX2

Parametric studies of the fluid-structure interaction have been performed with the code FLUX2 [2, 5, 6]. The code simulates the blowdown in the early period while most of the fluid is still in the subcooled state. Two-phase effects at the pipe exit can be roughly accounted for by prescribing the saturation pressure as a boundary value. The fluid part of FLUX2 is based on a three-dimensional compressible potential flow model with constant speed of sound. As a limiting case, incompressible fluid can be assumed. In contrast to purely acoustical models, large flow velocities and the kinetic energy, as it appears in Euler's momentum equation for irrotational flow, are accounted for. The structural part can be any linear-elastic model. At present, the model CYLDY2 [4] is used for the core barrel and the pressure vessel is assumed to behave rigidly. At the structure-fluid interface normal velocities and accelerations are set equal for both parts at the undeformed wall. The numerical scheme employs finite differences in space and a second order implicit time integration scheme [6].

The validity of the code has been demonstrated by comparison to one-dimensional analytical solutions [7] and by comparison to the RS16B experiments. Qualitatively, the motions in this experiment as computed with FLUX2 are depicted in Fig. 1. The lower two thirds of the core barrel are much thicker than the upper and have been modeled as a rigid cylinder, therefore, in these simulations. Comparisons between measured and computed pressure signals at several positions show good agreement [2, 5].

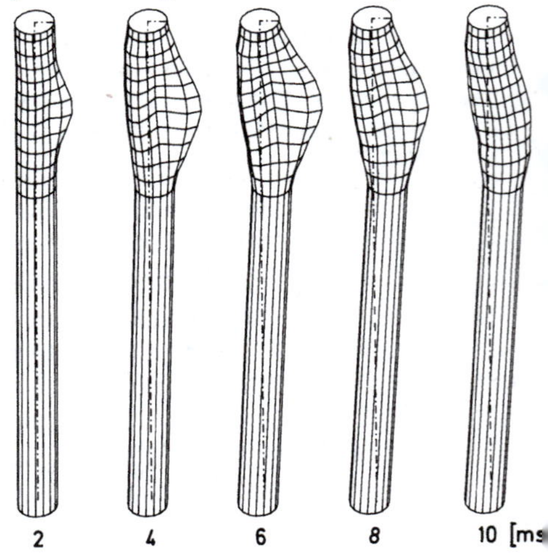

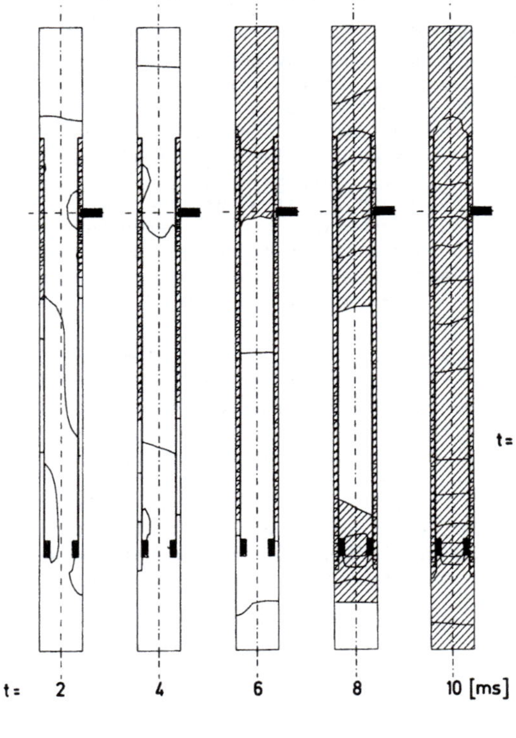

FIG. 1 ABOVE: CORE BARREL DEFLECTIONS (300 TIMES ENLARGED). LEFT: PRESSURE ISOBARS AT 1 BAR DISTANCE. IN THE SHADED REGION THE PRESSURE HAS DROPPED BY 2 BAR BELOW ITS INITIAL VALUE (142 BAR).

3. FLUID-STRUCTURE INTERACTION EFFECTS

3.1 Virtual Fluid Mass

For the case of incompressible fluid, the eigen-oscillations of the core barrel have been computed, see Fig. 2. The large reduction of eigenfrequencies is due to the effect of the added virtual fluid mass. It is particularly large for the case of the breathing mode in which the core barrel motion requires the fluid in the downcomer to flow through a long and narrow flow path around the lower edge. Because of the relatively large eigenfrequency, the tipping motion of the second beam mode dominates at early times of the blowdown, as shown in Fig. 1. The computed eigenfrequency of the ovaling mode has been clearly seen in the experimental results [2, 5]. The enlarged effective mass of the core barrel makes its dynamic response to the blowdown loading slower and smoother, and reduces the maximum amplitudes by a factor of the order 0.5.

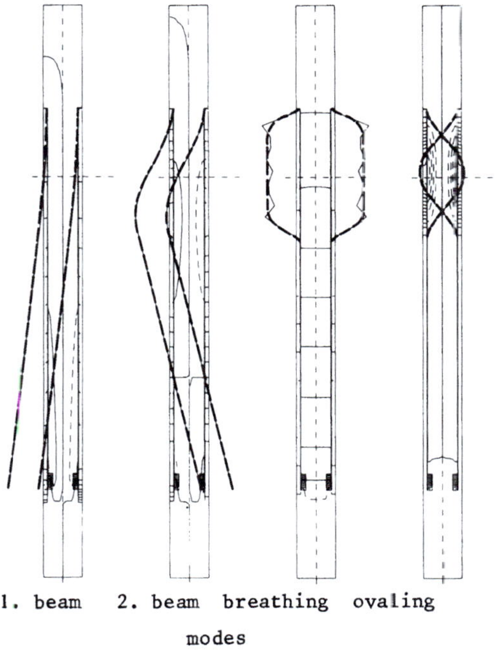

FIG. 2 COMPUTED EIGENSOLUTIONS OF THE CORE BARREL OF THE RS16B EXPERIMENT FOR THE LOWEST EIGENFREQUENCIES IN WATER. THE FREQUENCIES ARE 4.1 (21.4), 44.2 (92.8), 56.5 (2086), 73.9 (182) HZ IN WATER (VACOU). THE ISOBARS REPRESENT THE RELATED PRESSURE FIELD.

3.2 Reduced Effective Speed of Sound

Changes of the downcomer cross-sectional area due to the core barrel flexibility result in a smaller effective speed of sound a,

$$a = a_o \left(1 + \rho a_o^2 R^2/E h b\right)^{-1/2}$$

(a_o = fluid speed of sound, ρ = fluid density, R = core barrel radius, E = Young's modulus, h = wall thickness, b = downcomer width). The ratio a/a_o is 0.6 in case of the HDR and the resultant delayed spreading of the depressurization can be detected from Fig. 3. The effective speed does not change drastically if a_o is replaced by infinity, so an incompressible fluid model gives nearly the same results as a compressible one, see Fig. 4. Thus pressure waves are controlled by the structural motions and this is an important effect of fluid-structure interactions.

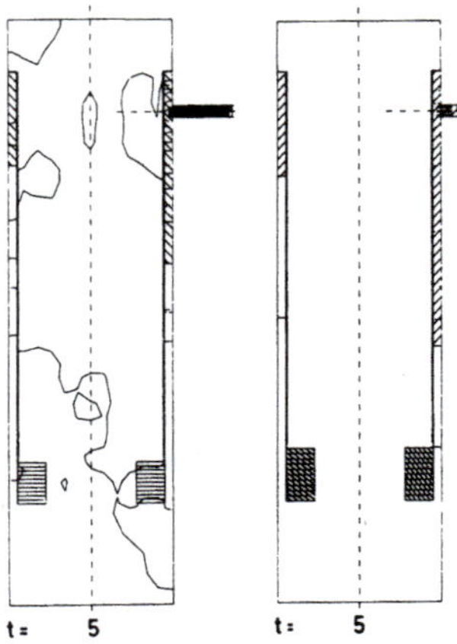

FIG. 3

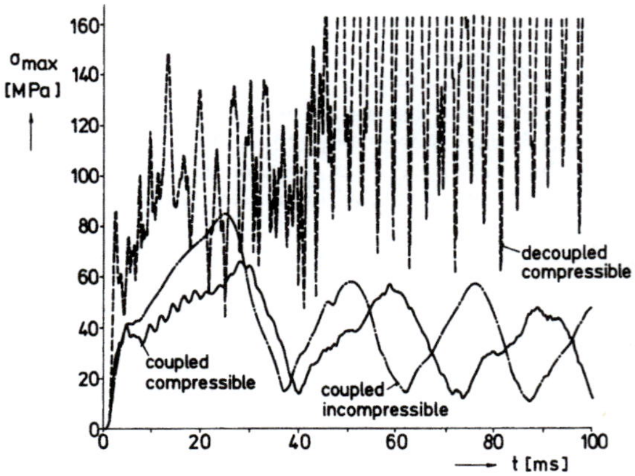

FIG. 4

Fig. 3: Computed depressurization in the HDR at t = 5 ms with (left) and without (right) fluid-structure coupling. Isobars are plotted at 1 bar distance. In the shaded region the pressure has dropped by 1 bar below the initial value of 110 bar.

Fig. 4: Maximum stress in the core barrel of the HDR versus time for different fluid models and coupling conditions.

3.3 Pressure Reduction Inside the Core Barrel

The pressure difference across the downcomer wall causes an increase of the inner domain which in turn reduces the inside pressure. Thus the inside water adds its stiffness to the core barrel. The internal pressure drop at early times is of the order 2 bar or 10 % of the pressure difference across the wall. This effect can be seen from the isobars plotted in Figs. 1 and 3.

3.4 Energy Exchange in a Decoupled Analysis

Fig. 4 shows that a decoupled analysis overestimates the stresses initially by a factor 2.5, which is a consequence of the above effects. Later, the decoupled analysis drastically overestimates the structural response. The growing oscillations are a consequence of imposing an oscillating (fluid) force to a (structural) dynamical system. In case of resonance between the oscillating pressure loads and one of the many structural eigenfrequencies an unlimited amount of energy can be supplied to the structure while the energy loss in the fluid is not accounted for. Thus a decoupled analysis can result in a physically meaningless overestimation of the structural response and this clearly indicates the need for coupled analysis tools.

3.5 Transmission of Internal Fluid Energy

The pipe break essentially supplies no forcing or energy to the system. Rather the energy exists initially in terms of internal energy in the fluid. This internal energy would be zero for an incompressible fluid and $1/2\ V\ p_o^2/(\rho\ a_o^2)$ (V = vessel volume) in case of an acoustic medium. The internal energy is much larger if we account for the evaporation heat which comes free in a two-phase flow. If a mechanism exists by which a considerable amount of this internal energy can be transferred into energy associated with the structural motion, then excessive structural vibrations are possible.

Such a mechanism has been found to exist in case of a simple linear one-dimensional acoustical model as shown in Fig. 5. If the pressure wave frequency in the tube is close to one of the eigenfrequencies of the piston-fluid-system then a large amount of energy is transmitted into the piston's motion. Of course, this model cannot be directly compared with the reactor case where we have a larger number of degrees of freedom and where the pressure cannot drop below the saturation value. Nevertheless, this effect deserves further investigations.

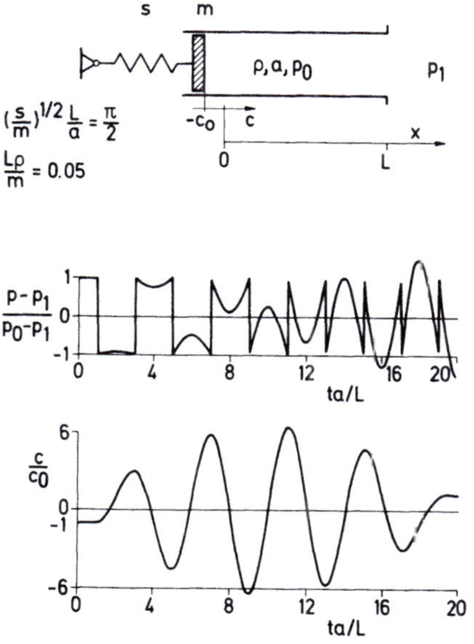

FIG. 5 A MODEL PROBLEM WITH QUASI-ANALYTICAL SOLUTION [1] SHOWING LARGE STRUCTURAL AMPLITUDES DUE TO ENERGY TRANSFER FROM THE FLUID. S = STIFFNESS AND m = MASS OF THE PISTON PER UNIT AREA, L = LENGTH OF PIPE, ρ = DENSITY, a = SPEED OF SOUND, P_0 = INITIAL PRESSURE, P_1 = OUTSIDE PRESSURE, C_0 = INITIAL DEFLECTION, C = DEFLECTION, P = PRESSURE AT THE PISTON WALL, t = TIME.

3.6 Large Local Pressure Differences Near the Nozzle under Two-Phase Conditions

In order to check the existence of the above mechanism, two-phase flow inside the core barrel has been roughly simulated using a much smaller value a_2 for the speed of sound in this region while the value for subcooled water a_1 is prescribed for the remaining fluid part. By this means the pressure is kept essentially constant in the inner domain for a long time and a quasi-steady outflow velocity in the nozzle evolves. For the PWR situation, Fig. 6 shows that this "two-phase" model results in much larger stresses than the "one-phase model" where a_1 applies to all fluid parts. Careful investigation of this situation has shown, however, that this is not the effect of energy transfer described in 3.5. In particular, the enlarged stresses disappear if the nonlinear kinetic fluid energy is neglected in the computer model. The enlarged stresses are rather due to the high velocity change near the blowdown nozzle and related pressure change of roughly 33 bar over a distance of the order of 40 cm (the radius of the nozzle). The peak pressure arises at the stagnation point lying on the nozzle axis at the core barrel wall. The explanation is confirmed by the resultant core barrel deflections, see Fig. 7. Whether the computed effect is much reduced by friction or choking at the break is an open question. In the subcooled case this effect does not appear because the vessel empties long before the maximum velocity is reached. Also, this effect has not been found in case of the HDR because of its much smaller nozzle diameter.

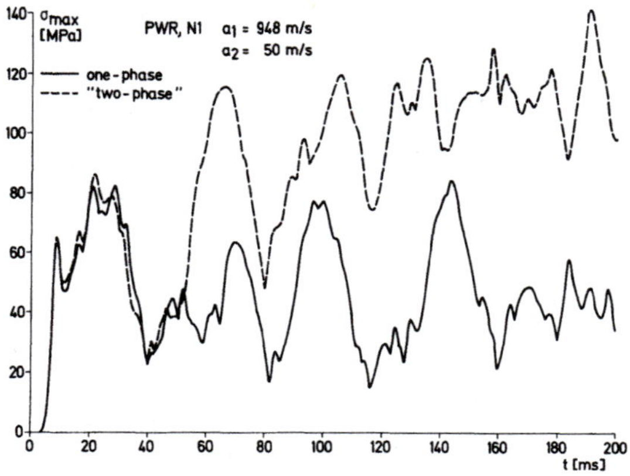

FIG. 6 MAXIMUM STRESS IN THE CORE BARREL OF A PWR VERSUS TIME FOR A SINGLE-PHASE MODEL AND A ROUGH TWO-PHASE MODEL. (THE ABSOLUTE STRESS VALUES ARE AFFECTED BY DISCETIZATION ERRORS.)

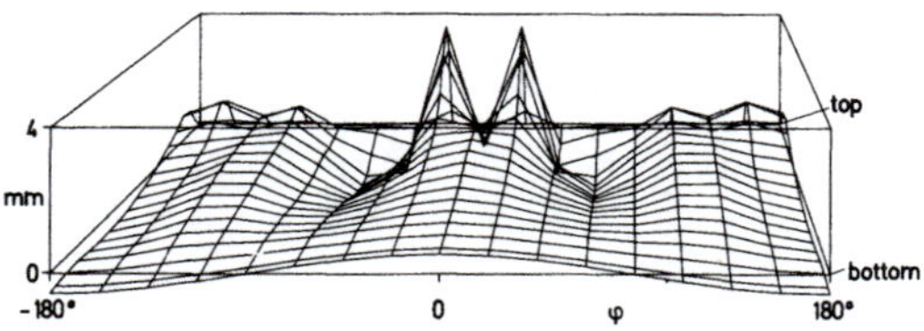

FIG. 7 RADIAL PWR CORE BARREL DEFLECTION IN THE UNWRAPPED PLANE AT T = 50 MS SHOWING LARGE DEFORMATIONS IN THE VICINITY OF THE BLOWDOWN NOZZLE.

4. REFERENCES

1. R. Krieg, E. G. Schlechtendahl, K. H. Scholl, "Design of the HDR experimental program of blowdown loading and dynamic response of PWR-vessel internals," Nucl. Eng. Design 43, 419-435, (1977).

2. Proc. 5th Int. Conf. on Struct. Mech. in Reactor Technology, Berlin, Germany, 13-17 Aug. 1979, Session B6, (1979).

3. Battelle-Institut e.V. Frankfurt, "Ergebnisse der ersten DRW-Versuche mit Einbauten (DWR1-DWR5)," Report BF-RS 0016B-10-1, (1977).

4. A. Ludwig and R. Krieg, "Dynamic Response of a Clamped/Ring-Stiffened Circular Cylindrical Shell under Non-axisymmetric loading," Nucl. Eng. Design, 43, 437-453, (1977).

5. U. Schumann, "Effektive Berechnung dreidimensionaler Fluid-Struktur-Wechselwirkung beim Kuhlmittelverluststörfall eines Druckwasserreaktors - FLUX," Kernforschungszentrum Karsruhe, Report KFK 2645, (1979).

6. U. Schumann, "Fast elliptic solvers and three-dimensional fluid-structure interactions in a pressurized water reactor," J. Comp. Phys., accepted for publication, (1979).

7. U. Schumann, "Fluid-structure interactions in one-dimensional linear cases," Kernforschungszentrum Karlsruhe, Report KFK 2723B, (1979).

CORE BARREL IN A PRESSURIZED WATER REACTOR

Problem: Depressurization waves and fluid flow (see Fig. 1) resulting from a break of the coolant inlet pipe of a pressurized water reactor ("blowdown") cause large loadings on and oscillations of the core barrel. How large are the safety margins?

Geometry: See Fig. 2. In the HDR experiment, the mass of the reactor core only is represented by a mass ring (≃ 12000 kg) at the lower barrel end.

Initial State and Boundary Conditions: Initial pressure 160 bar, temperature ≈ 290°C (subcooled water). Sudden complete break postulated. At break, pressure drops to saturation pressure ≈ 74 bar.

Characteristic Times: Break time 1 - 15 ms, time of two phase non-equilibrium 0.1 - 3 ms, wave spreading 1 - 20 ms, acceleration of fluid 10 - 30 ms, time until water in vessel evaporates 40 - 100 ms.

Characteristic Frequencies: Beam mode 20 Hz, breathing mode 28 Hz (in vacuo: 346 Hz). Many shell modes between 30 and 200 Hz.

Vibrations: Induced by pressure load. Strongly dependent on flow velocity. Internal resonance effects cause concern.

Research: (1) Development of best estimate codes to compute safety margin. (2) Comparison with experiments RS16B and HDR). It is expected that such "best estimate" analysis will show that the design of existing reactors is "conservative" so that no cures are necessary.

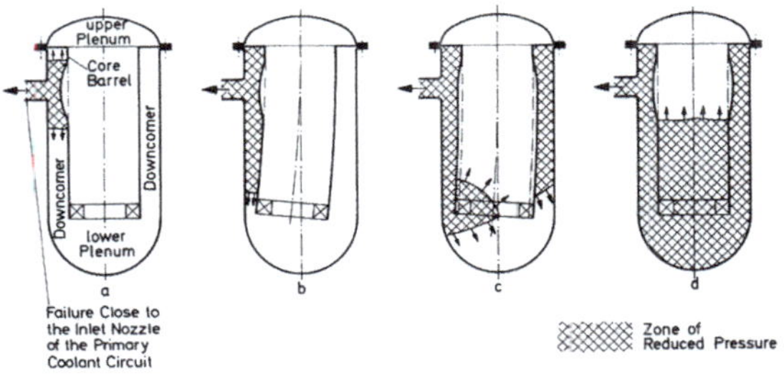

Schematic Sequence of Depressurization and Core Barrel Deflection

①

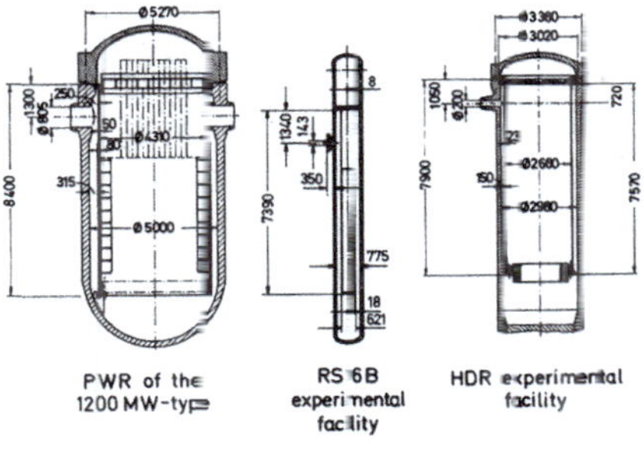

②

Reference: Krieg, R. Schlechtendahl, E.G., Scholl, K.-H., "Design of the HDR Experimental Program on Blowdown Loading and Dynamic Response of PWR-Vessel Internals", Nucl. Eng. Design, 43, 419-435, (1977).

A15 FLOW-INDUCED VIBRATION IN HEAT-EXCHANGER TUBE BANKS

Algirdas Zukauskas

Vladislovas Katinas

Institute of Physical and Technical Problems of Energetics
Lithuanian Academy of Sciences, Kaunas, USSR

SUMMARY

We report here experimental results on flow-induced vibration, involving forces and flow dynamics in staggered banks of equidistant triangular arrangement with pitch ratios $S_1/d = S_2'/d = 2.68$, 2.0, 1.61, 1.34 and 1.15, as well as in banks of in-line arrangement with pitch ratios $S_1/d = S_2/d = 1.61$, 1.34 and 1.15. The vibration characteristics were studied using resiliently mounted tubes in water flows. To study the unsteady flow dynamics, the same banks were mounted rigidly in water and air flows. The range of Reynolds number was from 10^4 to 2×10^5.

Three different mechanisms of excitation were observed: i) pressure fluctuations in a turbulent flow; ii) vortex shedding; and iii) hydroelastic tube-fluid interaction. Relations were derived for practical determination of the vortex shedding frequency, and the critical velocity of hydroelastic interaction as well. Suggestions are made for prevention of vibration in compact heat exchangers during the design stage.

1. INTRODUCTION

Reliable operation of heat exchangers is often limited by the stability of their heat transfer surfaces. Failure of tubes in banks due to the unsteady forces can lead to considerable damage and economic loss in nuclear reactors and other apparatus. The problems of failure are studied internationally, and a close relation with the unsteady flow dynamics is acknowledged (see, for example, Refs. [1] through [13]).

At the Institute of Physical and Technical Problems of Energetics, Lithuanian Academy of Sciences, experiments have been performed in extensive studies of flow-induced vibration. Experiments involved large-scale models of tube banks in water. The same bank models, but mounted rigidly, were used in unsteady flow studies of tube banks immersed in air and water. The experiments covered Reynolds numbers (Re) from 10^4 to 2×10^5, where Re = $U_{max}d/\nu$, and ν = kinematic fluid viscosity.

2. EXPERIMENTAL METHODS AND EQUIPMENT

A closed-loop water channel was used for the studies. The water was pumped through screens and a contraction to a test section. Flow velocity profiles were determined by Pitot-Prandtl tubes upstream of the bank.

Flow temperature was constant within 0.01°C. Nearly isotropic small scale turbulence was maintained at 1%. Bank arrangements most common in tube-and-shell heat exchangers, Sketch 1, were studied. In staggered banks, the tubes were arranged in equidistant triangles, with pitch ratios $S_1/d = S'_2/d = 2.68$, 2.0, 1.61 and 1.15. In banks of in-line arrangement, pitch ratios $S_1/d = S_2/d = 1.61$, 1.34 and 1.15 were studied. Vibration was studied in each row of the seven row banks.

In the test section of 300 x 129 x 800 mm, Figure 1, rigid tubes 300 mm long, and 16 mm diameter, were mounted vertically on resilient supports of 200 mm. By varying the material and the diameter of the supports, the natural frequency of

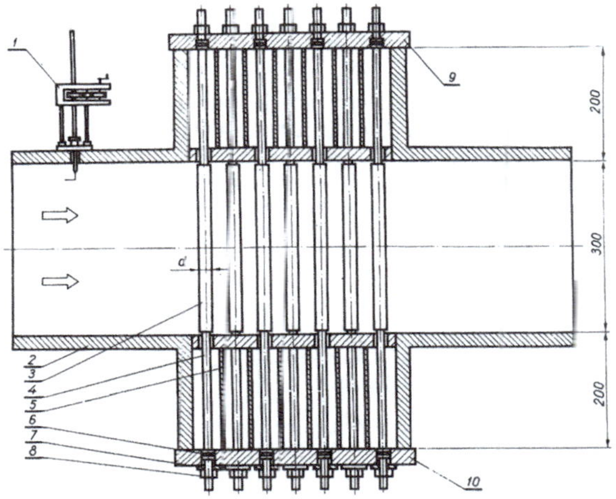

FIG. 1 TEST SECTION (DIMENSIONS IN MM). 1 - TRAVERSING DEVICE AND PITOT TUBE, 2 - TEST CHANNEL, 3 - RIGID TUBE, 4 - RESILIENT SUPPORT, 5 - PLATE BETWEEN TRANSVERSAL ROWS OF RESILIENT SUPPORTS, 6 - RUBBER SEAL, 7,8 - WASHER AND NUT, 9,10 - UPPER AND LOWER TUBE-FIXING PLATES

the tube vibration could be controlled within 20 to 99 Hz. Tubes on resilient supports were mounted in special plates, their ends fixed rigidly, with a certain degree of longitudinal freedom, to exclude axial forces from the vibration. The test tubes were under the effects of flow, while their resilient supports were located in a chamber of static liquid.

Mechanical vibration induced by the pump was insignificant compared to the flow-induced vibration level, and never exceded the experimental error of 5%. In preliminary tests, vibration level and dominant frequency of the loop and of the test section were determined. Three-dimensional measurements revealed a nearly homogeneous mechanical vibration with a dominant frequency of 150 Hz, which is 1.5 times as high as the highest natural frequency of the test tubes.

Tensometers, piezoelectric sensors and Brüel and Kjaer - instrumentation were used in the measurements. The data were interpreted by a the Fourier technique and computer analyzed. Flow structure in the banks was studied with DISA anemometers. Signals were recorded on a magnetic tape and transferred to a computer for analysis.

3. RESULTS AND DISCUSSION

Vibration of resiliently mounted tubes in crossflow is associated with fluid-dynamic forces acting on the tubes. The mean square value of the amplitude A, Figure 2, is a function of the flow velocity. In this study, flow velocity through the minimum area between adjacent tubes in same row was used as reference. The reference velocity is defined as

$$U_{max} = U_o \frac{S_1/d}{S_1/d-1} \qquad (1)$$

where U_o is flow velocity approaching the bank.

A continuous increase of the vibration amplitude with flow velocity was only noted in the range of flow velocity $U_{max}/f_n d << 1.3$, which is the range of excitation by the pressure fluctuations in the turbulent flow and in the near wake.

At a certain frequency of vortex shedding, maximum vibration amplitudes are induced by its coincidence with the natural frequency of tube vibration. The frequency of vortex shedding is a function of the pitch ratio and of the flow velocity. In the amplitude-velocity curve for flow-induced vibration of a bank, Figure 2, resonance peaks can be easily distinguished at the frequency of vortex shedding. In our studies, the peaks occur in the velocity range of $U_{max}/f_n d$ from 1.3 to 3.

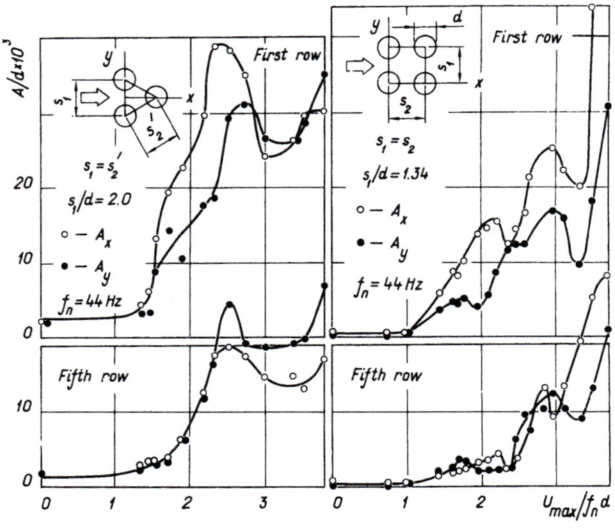

FIG. 2 AMPLITUDE VARIATION AS A FUNCTION OF FLOW VELOCITY FOR LONGITUDINAL A_x AND TRANSVERSE A_y VIBRATION IN STAGGERED AND IN-LINE BANKS

A sharp increase of the vibration amplitude at $U_{max}/f_n d > 3.2$ is a reflection of the hydroelastic tube-fluid interaction. This sort of excitation is described by the degree of asymmetry of the flow past a tube, which is associated with the tube deflection. This gives different passages between the neighboring tubes in a row, and consequently an asymmetrical pressure distribution around a tube in a bank. The fluctuating dynamic force acting on the tube increases sharply with either flow velocity or vibration amplitude. A minor increase of flow velocity can lead to a considerable change in the amplitude. When the flow excitation force exceeds that of tube resilience and damping, rapid failure ensues because of the dynamic stress.

Studies of parallel banks of tubes suggest three possible mechanisms of flow-induced vibration (Sketch 2): i) pressure fluctuations in the incident turbulent flow and in the near wake, ii) vortex shedding, and iii) hydroelastic tube-fluid interaction (fluid-elastic excitation in Connors [1]). These three mechanisms of excitation may act simultaneously and result in flow-induced vibration. But in certain velocity ranges, only one of the three mechanisms prevails, contributions of the other two being at negligible levels.

Specific causes of vibration are determined by the flow structure inside a bank, which is related to the nature of the pressure fluctuations. Anemometer measurements (Figure 3) yielded spectral density of the local velocity fluctuations in air at a location of 0.7 mm from tubes in rigid staggered banks. For tubes in the first rows, two amplitude peaks were observed for $S_1/d \geq 2$, compared to one for $S_1/d < 2$. They were related to vortex shedding. In the amplitude versus velocity curves, the peaks were observed at the frequency of vortex shedding, and at double this frequency. In these cases, excitation due to vortex shedding exceeds considerably the energy of the other two sources of excitation.

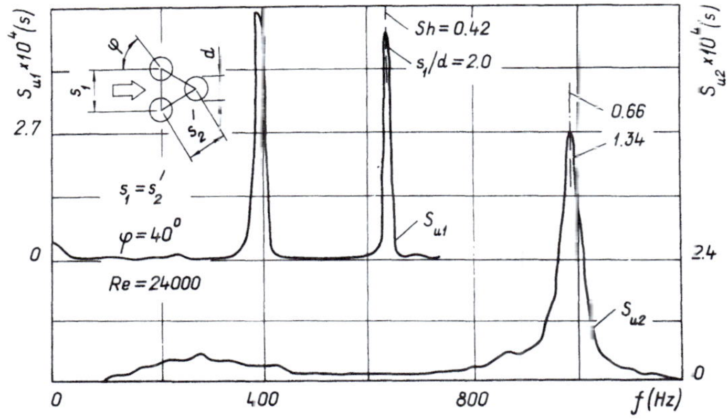

FIG. 3 SPECTRAL DENSITY S_u OF LOCAL VELOCITY FLUCTUATION AT LOCATION 0.7 mm FROM A TUBE. $\psi = 40°$ FROM FRONT STAGNATION POINT. DATA TAKEN IN LEADING ROWS OF STAGGERED BANKS WITH $S_1/d = 2.0$, 1.34

Tubes in the first rows are subjected to the low turbulence levels and uniform velocity profiles of the main flow. In the inner rows, high turbulence and non-uniform velocity distributions are observed. Flow turbulence is usually about 40%, the exact value depending on the bank geometry. Turbulence is higher inside banks of lower pitch ratios. Inner tubes are mainly excited by turbulent pressure fluctuations. Excitation by vortex shedding and hydroelastic tube-fluid interaction is negligible. This is supported by the spectral analysis of the local velocity fluctuations in air. In the frequency range of interest, the power spectrum of the fluctuations is nearly flat. The energy of vortex shedding is less pronounced than in the first rows.

Because of the flow excitation, tubes perform regular movements with well-defined orbits (Sketch 3), which vary with time because of variable vibration amplitudes. In most cases, amplitudes of the vibrations vary with time, in both longitudinal and transverse directions. For tubes located either within the bank or in the leading rows, the amplitude distributions are approximated by a normal law. This applies to most of the banks studied and to different velocities. A significant deviation from the normal distribution was noted only in the range of flow velocity related to the resonance of vortex shedding and natural tube vibration. Here, an approximately bimodal amplitude distribution was found, because tube vibration was due to comparable excitation forces of two different mechanisms - turbulent pressure fluctuations and vortex shedding. A normal amplitude distribution implies a prevailing effect of the turbulent pressure fluctuations.

Statistical analysis of the vibration, Figure 4, suggested frequencies close to the natural vibrations of the tubes. A deviation of about 20% from the natural frequency range was noted only in the leading rows. The leading rows are excited by pressure fluctuations having narrow spectra, the energy of excitation being concentrated in a narrow frequency range. In the inner rows, the excitation comes from wide-spectrum pressure fluctuations, and the deviation of the frequency from the natural frequency range was less than 10%. Only first mode vibration was noted in our studies, but second mode vibrations cannot be excluded.

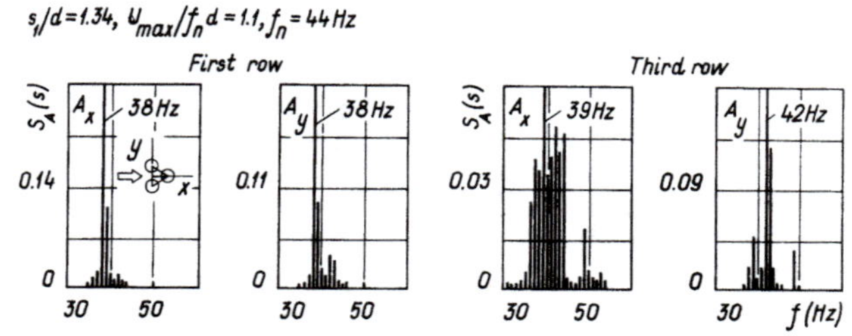

FIG. 4 NORMAL SPECTRAL DENSITY FUNCTION S_A FOR LONGITUDINAL A_X AND TRANSVERSE A_y TUBE VIBRATION IN DIFFERENT ROWS OF BANKS FOR VARIABLE FLOW VELOCITIES AND NATURAL FREQUENCY OF TUBE VIBRATION f_n = = 44 Hz

Fluid-elastic instability of tubes in banks due to the fluid-tube interaction is a function of the damping parameter $m_o\delta_o/\rho d^2$. The critical flow velocity for the hydroelastic interaction, Figure 5, can be written in a dimensionless form

$$\frac{U_{max}}{f_n d} = k \frac{m_o \delta_o}{\rho d^2} , \tag{2}$$

where f_n is the natural frequency of tube vibration, m_o is mass of tube plus added mass of displaced fluid per unit length, ρ is fluid density, and δ_o is log decrement.

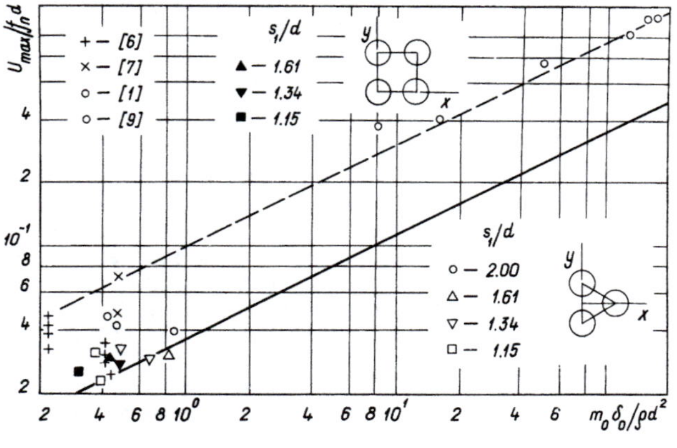

FIG. 5 CRITICAL FLOW VELOCITY IN DIMENSIONLESS FORM. HYDROELASTIC TUBE-FLUID INTERACTION IN STAGGERED AND IN-LINE BANKS AS A FUNCTION OF DAMPING PARAMETER.

In determining the velocity at which hydroelastic tube-fluid interaction is initiated, the value of k = 3.7 is recommended in eq. (2). A similar value of k was found by Gorman [6] for banks of tubes in water, though the value is different in Connors [1] for a single row of tubes in air crossflow. The reason must lie in different boundary conditions for a bank and for a single row, and in different fluid densities. Eq. (2) gives a satisfactory description of the hydroelastic tube-fluid interaction for staggered and in-line banks. The fluid-elastic force must be generated mainly by the transverse vibration of the tubes. In eq. (1) for U_{max}, an account of the transversal pitch effect was also given. In our studies, pitch ratios were considered constant, and effects of the longitudinal relative pitch were neglected in determining the fluid-elastic force.

The results suggest the presence of all three excitation mechanisms in the leading rows. The most probable excitation in the inner rows comes from turbulent pressure fluctuations and from the hydorelastic tube-fluid interaction. If this is so, the amplitude of tube vibration must increase with flow velocity up to failure.

In engineering practice, the means of control for flow-induced vibration is a matter of utmost importance. The first approach is to vary structural parameters included in eq. (2). We have also studied other ways of vibration control, one of them consisted of combining tubes with different natural vibration frequencies in a single bank. By mounting tubes with lower natural frequencies of f_n among tubes with higher f_n, a 10% increase of stability can be achieved. In engineering practice, to allow increase of flow velocities in heat exchangers, tubes with higher natural vibration frequencies can be mounted in the leading rows of banks. A possible hazard lies in the application of tubes with multiple natural vibration frequencies. In such banks, peaks of vibration amplitudes could be generated in the multiple frequency ranges.

4. PRACTICAL SUGGESTIONS

As a result of the above-described studies and of the experimental information gained on full-scale thermal apparatus, some practical suggestions can be made for eliminating tube vibration.

(1.) Vibration of tube banks presents no serious hazard, if it is excited by turbulent pressure fluctuations in the main flow and in the near wake, Figure 2. It is observed only with low flow velocities. Amplitude peaks of tube vibration due to vortex shedding can be avoided by choosing tubes with natural vibration frequencies different from that of vortex shedding. Multiple frequency ranges are to be avoided as well. Dynamic stability of banks of tubes is a function of the critical velocity of the hydroelastic tube-fluid interaction eq. (2).

(2.) The frequency of vortex shedding in parallel banks (Figure 6) can be described by

$$Sh = 0.2 + \exp\left\{ - \frac{(S_1/d)^{1.8}}{2.3} \right\} \tag{3}$$

for staggered banks of $S_1/d = S_2'/d \geq 1.15$, and by

$$Sh = 0.2 + \exp\left\{ - \frac{(S_1/d)^{1.83}}{0.88} \right\} \tag{4}$$

for in-line banks with $S_1/d = S_2/d \geq 1.15$.

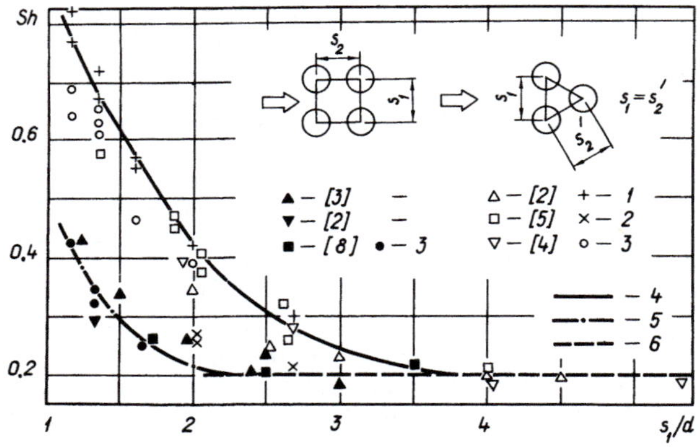

FIG. 6 VORTEX SHEDDING FROM LEADING ROWS OF STAGGERED AND IN-LINE BANKS IN CROSSFLOW. DATA FROM SPECTRAL ANALYSIS OF VELOCITY FLUCTUATIONS IN AIR (FIGURE 3): 1 - FREQUENCY OF VORTEX SHEDDING, 2 - FREQUENCY OF THE SECOND PEAK IN THE SPECTRAL DENSITY FUNCTION FOR STAGGERED BANKS WITH $S_1/d \geq 2$ (DATA ON AMPLITUDE PEAKS FOR WATER, FIGURE 2), 3 - FREQUENCY OF VORTEX SHEDDING. ANALYTICAL RELATIONS: 4 - EQ. (3) STAGGERED BANKS, 5 - EQ. (4) IN-LINE BANKS, 6 - EQ. FOR VORTEX SHEDDING FOR A SINGLE CIRCULAR CYLINDER

The expression for $Sh = fd/U_{max}$ (f is frequency of vortex shedding) from eqs. (3) and (4) gives a reasonably accurate approximation for describing the measured amplitude peaks at resonance and for the results of other authors as well.

(3.) Highest amplitude occurs in the leading rows of both in-line and staggered banks, with excitation due to vortex shedding. The vibration can be avoided by eliminating regular vortex shedding.

(4.) With damping, when the log decrement is under 0.3, tube vibration is close to the natural frequency of the tubes. Maximum difference of the two frequency ranges, within 20%, was observed in the leading rows.

(5.) To avoid flow-induced vibration, design calculations must be based on the control of structural parameters included in eq. (2). The effect is most pronounced when the natural frequency of the tube vibration is increased. An increase of damping δ_0 has only a moderate effect, and is by far more difficult to carry out.

(6.) To increase dynamic stability of a bank, tubes in the leading rows must have 20 to 30% higher natural vibration frequencies, as compared to those in the inner rows. Multiple frequencies must be avoided, to exclude amplitude peaks in these ranges.

5. ACKNOWLEDGEMENTS

The authors are indebted to A. Markevicius and V. Suksteris for their help in performing the experiments and to V. Ziugzda for important advice.

6. REFERENCES

1. Connors, H.J., Jr., "Fluidelastic Vibration of Tube Arrays Excited by Cross Flow", In: ASME Winter Annual Meeting. Proc. of the Symposium of Flow-Induced Vibration in Heat Exchangers. New York, Dec. 1, 1970, New York, 42-56, (1970).

2. Chen, Y.N., "Flow-Induced Vibration and Noise in Tube - Bank Heat Exchangers Due to von Karman Streets", Trans. ASME, Ser. B, J. of Engineering for Industry, 90, No. 1, 134-146, (1968).

3. Grotz, B.J., Arnold, F.R., "Flow-Induced Vibration in Heat Exchargers", Techn. Rept. No. 31 to Office of Noval Research from Stanford University. AD 104568. Stanford, California, (1956).

4. Borges, A.R.J., "Vortex Shedding Frequencies of the Flow Through Two-Row Banks of Tubes", J. of Mechanical Engineering Science, 11, No. 5, 498-502, (1969).

5. Heinecke, E., "Stationary and Instationary Flow Phenomena in and Behind Staggered and In-line Tube Banks", In: International Symposium on Vibration Problems in Industry, Keswick, England. Paper No. 412, Keswick, 1-12, (1973).

6. Gorman, D.J., "Experimental Studies of Liquid Cross Flow Induced Vibration of Heat Exchanger Tube Bundles", In: Seminar on Vibration of Reactor Fuel and Steam Generator Tubes. Moscow, September 9-12, 1975. Paper No. 3a, Moscow, 1-18, (1975).

7. Pettigrew, M.J., Gorman, D.J., "Experimental Studies on Tube Bundle Vibration in Liquid and Two-Phase Cross-Flow", In: Seminar on Vibration of Reactor Fuel and Steam Generator Tubes. Moscow, September 9-12, 1975. Paper No. 3b, 1-13, (1975).

8. Clasen, P., Gregorig, R., "Ein Schwingungskriterium eines quer angeströmten Rohres. T. 4, Schwingversuche in einem fluchtenden Rohrbündel", Chemie Ingenieur Technik, 43 Jahrgang, Heft 17, 982-985, (1971).

9. Hartlen R.T., "Recent Field Experience with Flow-Induced Vibration of Heat Exchanger Tubes", In: International Symposium on Vibration Problems in Industry. Keswick, England. Paper No. 611, 1-21, (1973).

10. Zukauskas, A.A., Katinas, V.J., Suksteris, V.S., Sobolev, V.A., "Flow and Vibrations of Staggered Banks of Tubes of Heat Exchangers in Crossflow of Liquid", Lietuvos TSR Mokslu akademijos darbai, B serija, 4(101), 59-71, (1977) (In Russian).

11. Zukauskas, A.A., Katinas, V.J., Perednis, E.E., Sobolev, V.A., "Flow and Vibration of the Radial Banks of Tubes of Heat Exchangers in Crossflow of Fluid", Lietuvos TSR Mokslu akademijos darbai, B serija, 6(109), 53-62, (1978) (In Russian).

12. Zukauskas, A.A., Katinas, V.J., Markevicius, A.A., "Flow Pattern an Hydrodynamic Vibration of Banks of Tubes in Crossflow of Air", Lietuvos TSR Mokslu akademijos darbai, B serija, 5(102), 89-101, (1977) (In Russian).

13. Blevins, R.D., Flow-Induced Vibration, Van Nostrand Reinhold Company, New York, 1-363, (1977).

HEAT EXCHANGER TUBE BANKS

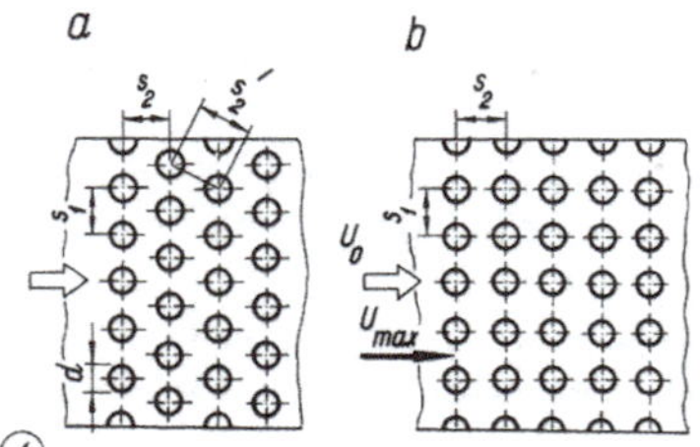

①

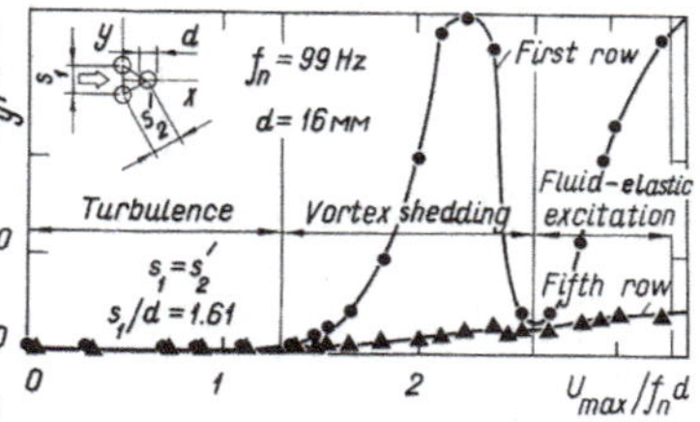

②

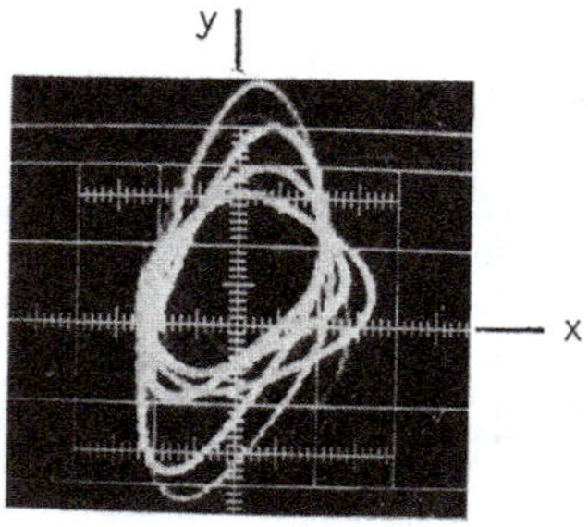

③

Vibration: Orbital movement of tubes was observed in the first row of banks under the flow excitation. Sketch 3, for $S_1/d = 1.61$ bank in crossflow of water with $U_{max}/f_n d = 2.8$. Natural frequency 99 Hz. Three mechanisms of excitation were stated: i) turbulent pressure fluctuations in the flow and in the near wake, ii) vortex shedding, iii) hydroelastic tube-fluid interaction (fluid-elastic excitation by Connors), Sketch 2. Each of the three mechanisms prevails in specific velocity ranges.

Problem: Vibration of heat exchanger tubes can occur with certain combinations of bank arrangement and nominal flow velocity. Flow-induced vibration leads to rapid failure of the tubes because of dynamic stress.

Geometry: Seven-row banks of circular tubes, 16 mm diam. were studied in crossflow, Sketch 1. Staggered arrangements of $S_1/d = S'_2/d$ = 2.68, 2.0, 1.61, 1.34, 1.15, and in-line arrangements of $S_1/d = S_2/d$ - 1.61, 1.34, 1.15 were covered.

Boundary conditions: Rigid tubes were mounted on resilient supports in the test channel. Natural frequency of tube vibration was controlled from 20 to 99 Hz.

Flow conditions: Flow-induced vibration was studied in water: flow dynamics in water and air. Flow velocity in the least free cross section of a bank U_{max} was from 0 to 8 m/s in water, and from 0 to 30 m/s in air, with the incident flow turbulence intensity of 1%.

Operating conditions: Tube failure occurs at the critical dimensionless value of the flow velocity $U_{max}/f_n d = 3.7\sqrt{m_o\delta_o/\rho d^2}$, where m_o is mass of tube plus added mass of fluid per unit length, δ_o is log decrement, ρ is fluid density. At $U_{max}/f_n d$ from 1.3 to 2.6, maximum vibration amplitude due to vortex shedding is observed.

Structural dynamics: First mode vibration was observed. Maximum acceptable flow velocity without failure was determined by frequency f_n, diameter, damping, tube mass and density.

Cures: A bank operates safely until the flow velocity reaches its critical value, which is determined by the parameters of vibration (see above). To achieve higher values of U_{max}, suitable choice of f_n or application of tubes with a higher f_n in the leading rows is recommended. Frequencies of vortex shedding coincident with or multiple of the value of f_n, must be avoided.

A16 VIBRATION EXPERIENCES WITH A FUEL PIN MODEL SYSTEM IN A PARALLEL TWO-PHASE FLOW

Fumio Hara
Associate Professor

Tokyo University of Science, Tokyo, Japan

SUMMARY

This paper reports on experiments with vibrations induced by a parallel two-phase (air and water) flow in a fuel pin system and describes their attenuation. The fuel pin model consists of two flat stainless steel rods fixed in a vertical circular channel conveying the two-phase fluid. Experimental information is presented on vibrational strain magnitude and psd, and relations between vibration magnitude, flow velocity, flow pattern and pressure fluctuation intensity. Vibration can be attenuated either by changing the flow pattern from slug to bubble or by reducing flow velocity.

1. INTRODUCTION

To ensure safety of fuel pin systems in pressure tube nuclear reactors, intensive research has been done on vibrations of fuel pin assemblies in parallel two-phase flows. It has been difficult, however, to collect information on such vibrations, especially those occurring under other than normal conditions: partial choking of coolant flow in a reactor due to heavily regulated operations, for example. This paper reports such information for the case of vibrations, and their attenuation, induced by a parallel two-phase slug flow in a fuel pin model.

2 GEOMETRY AND MEASUREMENT

The fuel pin model consisted of two pieces of flat stainless steel rods 10 mm wide, 2 mm thick and 836 mm long; both ends of the pin were simply supported, as shown in Sketch 1 of the Chart. The model represents the simplest configuration for uni-directional vibration, although it differs from an actual fuel pin having a circular cross section. The model was fixed at the center of a pyrex tube (inner diameter 35 mm) and the two-phase fluid flowed parallel and vertical to the model. The distance between the two fuel pin models, shown by d in Sketch 1 of the Chart, was 5 mm and 10 mm. The following variables were measured under conditions specified by water flow rate ($10 l/m \leq Q_w \leq 60 l/m$) and air flow rate ($10 l/m \leq Q \leq 70 l/m$), or, in other words, for a two-phase flow void fraction range $0.1 \leq F_a \leq 0.7$:

(1) vibrational strains of the two rods 200 mm from the downstream end of the model;
(2) pressure fluctuations on the inner surface of the channel at a point 435 mm from the upstream end;
(3) void signals expressing two-phase flow states in 0-1 (0 = water and 1 = air) valued data at 435 mm and 585 mm from the upstream end.

Power spectral densities (psd's) and variances (sampling period: 0.02 sec; length: 2000) were calculated experimentally, clarifying excitation and attenuation characteristics of the model's vibration.

3. FLOW CONDITIONS

Air was injected into the channel through 20 holes (diameter: 2 mm) located at 90° intervals around the tube type injector's circumference (see Sketch 1 of the Chart). Water flowed from a surge tank into the channel, producing the air-water mixture, which traveled through a flow development section 793 mm long. The test loop operated under atmospheric pressure.

Two-phase flow velocity, ranging from 1 m/s to 3.5 m/s, was evaluated by $0.15/\tau_d$ (m/s), where τ_d means time lag of cross-correlation function in void signals. Under these experimental conditions, two-phase flow patterns were observed (see Table 1) for bubble (B), bubble and slug (B+S), and slug (S) flows, illustrated in Sketch 1. The appearance of these flow patterns depends roughly on void fraction F $(=Q_a/(Q_a + Q_w))$: bubble, $F \leq 0.3$; bubble and slug, $0.3 \leq F \leq 0.5$; and slug, $F \geq 0.5$.

Table 1 Two-phase flow pattern diagram; S = slug flow, B = bubble flow, S+B = slug and bubble mixed flow

Air flow rate Qa l/min (columns); Water flow rate Qw l/min (rows)

Qw \ Qa	10	15	20	25	30	40	50	60	70
10	S	S	S						
15	S	S	S	S	S	S			
20	S+B	S+B	S	S	S				
30	B	B+S	B+S	B+S	S				
40	B	B+S	B+S	B+S	S	S	S	S	S
50	B	B	B+S	B+S	B+S	S	S	S	
60	B	B	B	B+S	B+S	B+S	S		

4. STRUCTURAL DYNAMICS OF THE FUEL PIN MODEL

The lowest natural frequency of each rod was measured when the model was wholly inserted in still water and in air, giving Table 2. Added mass reduced the rod's natural frequency by about 2 to 2.3 Hz less than that in air. The damping ratio ζ was evaluated by the impulse response technique for both these cases, resulting in ζ = 0.1 (in water) and 0.02 (in air). Using two small test pieces (20 mm long, 10 mm wide, 2 mm thick) located in still water with a distance (d in Sketch 1) of 5 mm or 10 mm corresponding to the system's configuration, pressure p acting on a fixed piece and acceleration α_r of another piece oscillating with a small amplitude were measured, giving

$$p \propto - \alpha_r \qquad (1)$$

Induced pressure has a 180° phase difference from acceleration.

Table 2 The lowest natural frequency in air and in water(Hz)

	rod no.	d=5mm	d=10mm
in water	1	5.2	4.5
	2	4.5	4.5
in air	1	7.5	6.8
	2	6.5	6.7

Other parameters concerning the dynamics of the model: Bending stiffness $EI = 1.33 \times 10^6$ N mm^2 ; added mass in still water per unit rod length = 0.78 g/cm; mass per unit rod length = 1.6 g/cm.

5. VIBRATION

A typical strain wave form of strong vibrations produced in the rod is shown in Sketch 2 of the Chart. It describes relatively random oscillations. The lowest natural vibration and its higher modes are recognizable in this Figure.

Vibrational strain magnitude (in variance σ_s^2) had a clearcut relation to flow parameter $\widehat{V}$, defined as $V + \alpha(Q_{wo} - Q_w) / Q_{wo}$, where V = two-phase flow velocity evaluated through $\lambda/\tau d$ (λ = distance between two void probes 0.15 m, τd = time lag of void signal cross-correlation function); Q_w = water flow rate in l/min; Q_{wo} = a specified water flow rate of 60 l/min; α = constant (=1.8) for d = 5 mm and (= 2.1) for d = 10 mm.

Table 3 Parameters in Equation (2) for vibration magnitude(x10^{-12})

d	F	σ_{so}^2	a	b
5 mm	F<0.4	30	9.8	2.0
	F>0.5	5	7.2	1.3
10 mm	F<0.4	120	14.0	4.4
	F>0.5	36	6.8	1.4

An empirical formula was obtained for vibrational strain magnitude (σ_s^2) in respect to void fraction F:

$$\sigma_s^2 = \sigma_{so}^2 + (aF - b)\, Q_w / (1 - F) \qquad (2)$$

Parameters in Equation 2 are given in Table 3.

Figure 1 shows the relation between σ_s^2 and two-phase flow pressure fluctuation (in variance). In a range of pressure fluctuation less than 0.5 $(N/cm^2)^2$, vibration magnitude σ_s^2 depends strongly on water flow rate Q_w, or, in other words, an increase in Q_w, changing the flow pattern from slug to bubble, markedly reduces vibration. However, over the range of pressure fluctuations greater than 0.5 $(N/cm^2)^2$, σ_s^2 correlates very well with pressure variance.

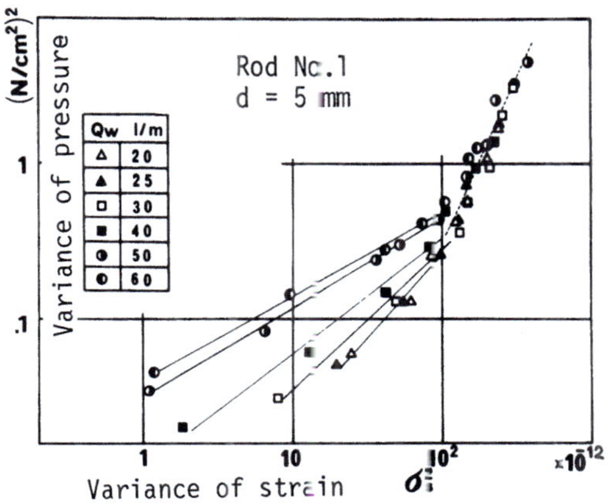

FIG.1 Relation between pressure and vibrational strain variances

According to the information above described, the vibration magnitude of the pin depends strongly on two-phase flow pattern and pressure fluctuation intensity; that is, although vibration is very weak in bubble flow, slug flow produces strong vibration in the model; two-phase flow pressure fluctuation might affect vibration as an external force.

Figures 2 and 3 show correlation coefficients of vibrational strains between the two rods (see Sketch 1) for d = 5 mm and 10 mm, respectively. In a close model configuration, each model vibrates rather in-phase regardless of flow patterns. More distant spacing produces out-of-phase vibration in the system but the degree of out-of-phaseness depends on flow pattern.

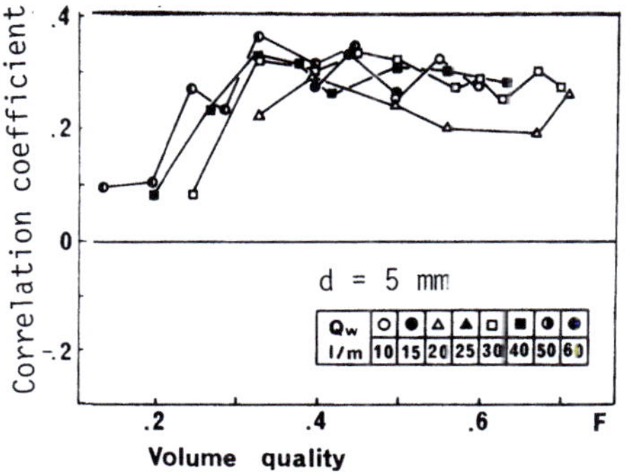

FIG.2 Correlation coefficient of Rod No. 1 and 2 strains against void fraction for d = 5 mm

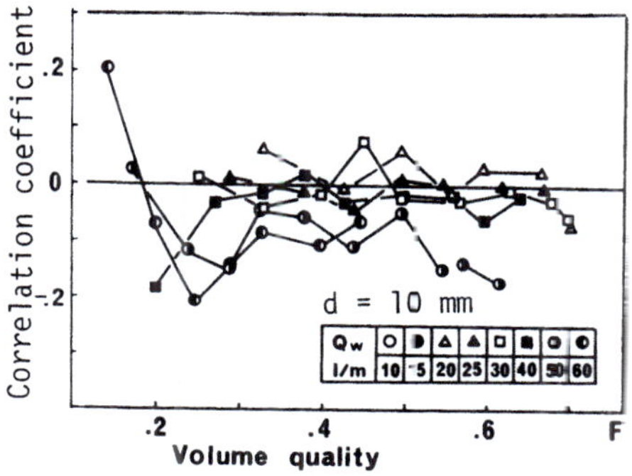

FIG.3 Correlation coefficient of Rod No. 1 and 2 strains against void fraction for d = 10 mm

6. ATTENUATION

Sketch 4 in the Chart shows vibration dependency on two-phase flow velocity, void fraction, and flow pattern. Changing the flow pattern from slug to bubble greatly reduces vibration (by 90% in σ_s^2) even at the same flow velocity (see arrow in Sketch 4). Under the same void fraction, vibration can be attenuated through reducing flow velocity (see dotted lines in Sketch 4).

In conclusion, information presented here must, as yet, remain largely preliminary. However, knowledge obtained in these experiments should be valuable in attenuation of vibrations causing fretting wear in the fuel pin assembly of a pressure tube reactor.

FUEL PIN MODEL IN PARALLEL TWO-PHASE FLOW

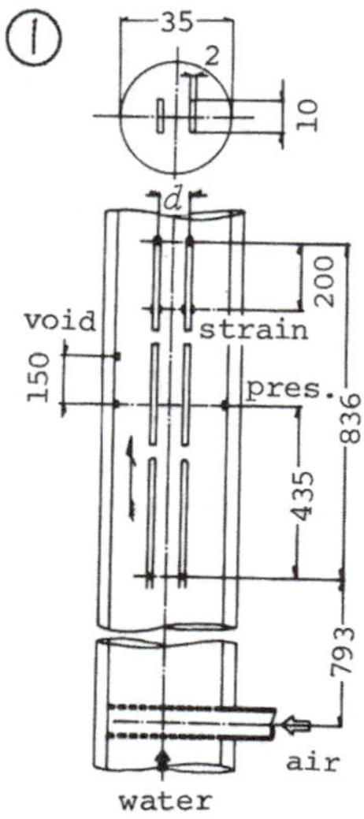

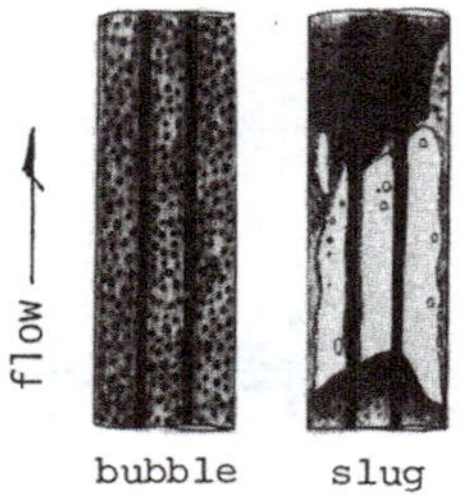

Problem: Strong vibrations of model pin induced by parallel two-phase slug flow. Expected damage in form of vibratory fretting erosion between pins or pin and spacer.

Geometry: See Sketch 1. Simplest configuration for one dimensional vibration. d = 5 mm, 10 mm.

Boundary conditions: Fluidically coupled two-pin model, both ends simply supported, in a vertical circular channel conveying two-phase fluid.

Flow conditions: Air and water two-phase flow, mixed 793 mm before the test section, under atmospheric pressure. Flow velocity (V) measured by cross-correlation technique. V = 1 to 3.5 m/s. Flow patterns: bubble, bubble and slug, and slug flow (see Sketch).

Operating conditions: Two-phase flow void fraction $F = Q_a/(Q_a + Q_w)$: 0.1 to 0.7. Strong vibrations were observed over the velocity parameter $\hat{V} = V + 1.8\,(Q_{w0} - Q_w)/Q_{w0}$ = 2.5 m/s (Q_a = air flow rate; Q_w = water flow rate; Q_{w0}=60 l/min).

Structural dynamics: Pin model's lowest natural frequency (including added mass effect, dependent on void fraction) = 4.0 to 5.5 Hz. Bending stiffness EI = 1.33 x 10^6 N mm^2, damping ratio ζ = 0.1 at F = 0.0, and 0.02 at F = 1.0.

Vibration: Dominant frequency F_o = 4.0 to 5.5 Hz, depending on void fraction, relatively random (see Sketch 2). Magnitude of vibrational strain in variance $\sigma_s{}^2$: see Sketch 3.

Attenuation: Changing flow pattern from slug to bubble reduces vibrational strain magnitude markedly (by 90% in $\sigma_s{}^2$); flow velocity reduction can also attenuate vibration, even in slug flow (see Sketch 4).

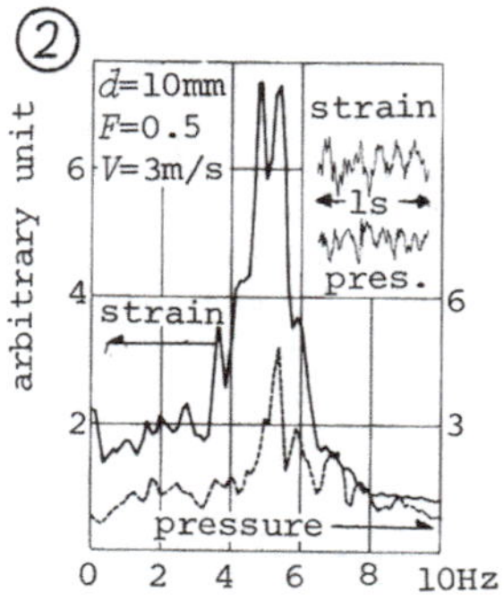

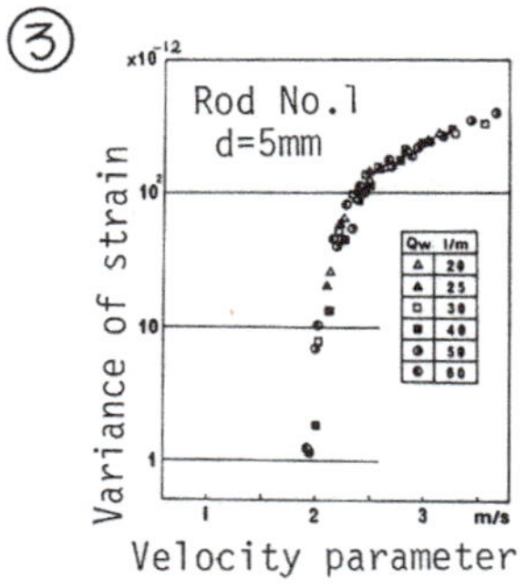

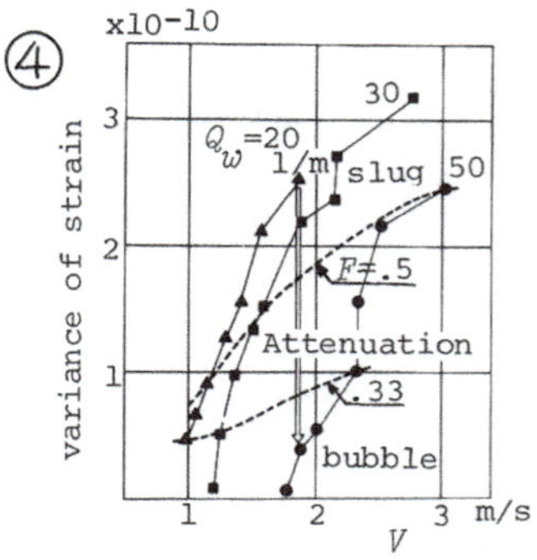

REFERENCE: Uehara, K., Hara, F., and Yamashita, T.: Parallel Two-Phase Flow Induced Vibrations in a 2-Elastic-Rod System, Trans. JSME 45: 495 (to appear) 1979.

A17 FLOW-INDUCED VIBRATIONS OF PLATES, SINGLE CYLINDERS, AND TUBE BUNDLES

Y. N. Chen

Scientific Consultant
Sulzer Brothers Limited, Winterthur, Switzerland

SUMMARY

The dynamic forces associated with Karman vortex streets shed by single cylinders of various shapes, and by plates with various trailing edges, can be correlated using the reaction theory of the Karman vortex street and the length of the vortex critical formation zone. Furthermore, the measures used for damping the Karman vortices are briefly discussed. Two excitation sources in tube bundles, namely the Karman vortex shedding and the turbulence buffeting, can be clearly separated into two different tube spacing regions. It will finally be shown that the fluid-elastic instability of tube bundles can only occur in narrowly spaced tube rows owing to susceptibility of the gap flows (jets) to the deflection and swing movements.

1. LIFT COEFFICIENTS OF SINGLE CYLINDERS OF VARIOUS CROSS-SECTIONAL SHAPES

The Karman vortex street of a cylinder will exhibit a swing movement of its near wake in accordance with the periodic reaction force exerted on the cylinder by the vortex street. The intensity of the swing depends on the effectiveness of the cylinder rear surface in guiding the separated free-shear layer [1]. Thus the shape of the afterbody of the cylinder, as expressed by the ratio s/d (s = distance of the flow separation point from the base of the cylinder, and d = its width), is a characterizing parameter. The lift force exerted by the Karman vortex will be inversely proportional to this swing movement, contrary to conventional expectation [2]. This concept, which is confirmed by the recent measurements carried out by Kramer, Gerhardt & Jansen [3], gives a good correlation between the lift coefficient C_L and s/d as shown in Fig. 1 [4].

The Karman vortex will be less coherent if there is a mean transverse flow component imposed on it. For example, the Karman vortex street shed by a strongly yawed circular cylinder will no longer exhibit the pattern of two orderly vortex columns, but will rather degenerate into a number of curved vortex tubes with alternate senses of rotation along the cylinder axis [5]. The lift coefficient will decrease with the increase of angle of yaw Λ, as shown by curve 1 in Fig. 2, namely

$$C_L \sim \text{ctg}\ \Lambda$$

The effect of helical fins on a stack is based on the generation of streamwise vorticity by the fin grooves with opposite direction on the two shoulders of the stack, making the Karman vortex very unstable and thus considerably reducing the dynamic force [5]. However, the

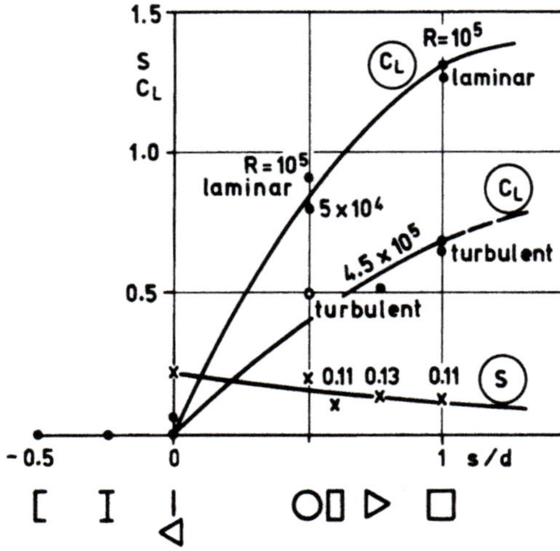

FIG. 1 LIFT COEFFICIENT C_L VERSUS s/d (s = DISTANCE OF THE FLOW SEPARATION POINT FROM THE CYLINDER BASE, d = CYLINDER WIDTH) [4]

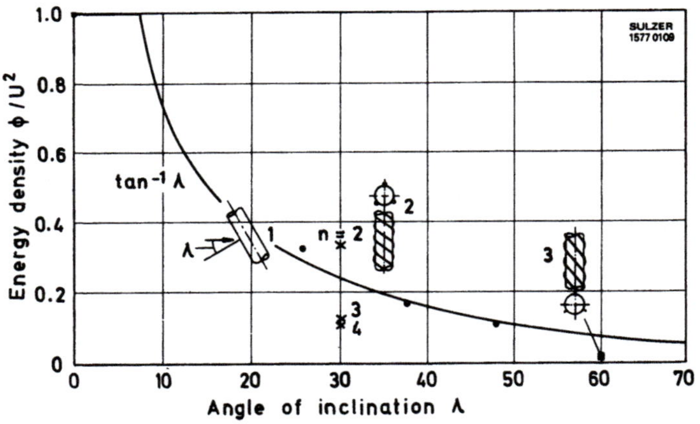

- *Point group 1:* Measured energy density Φ/U^2 of the vortex , normalized by the energy density for $\Lambda = 0$
- × *Point group 2:* Experimental result of the vibration amplitude obtained on circular cylinder models (d = 37 mm) with equally spaced helical wires (h = 4 mm, i.e. h/d = 0.11) with ascending angle of 30° (n = number of helical wires)
- □ *Point 3:* Experimental result on stack model with three equally spaced helical fins of height 0.09 d and ascending angle of 60°

FIG. 2 NORMALIZED LIFT VERSUS ANGLE OF YAW [6]

effect depends on the shape of the grooves formed by the fins. The rectangular fins are more effective than the circular ones (compare point 3 and point group 2 in Fig. 2 [6]).

2. HIGHER HARMONICS OF THE KARMAN VORTEX FORCE

The dynamic force due to Karman vortices shed from a circular cylinder may have, in addition to the fundamental frequency, a second harmonic in the streamwise direction (C_D) and a third harmonic with components in both the lateral (C_L) and streamwise (C_D) directions, as shown in the following table for a Reynolds number of 6.4×10^4 at the critical flow velocity [7]:

Fundamental frequency:	$C_L = 0.7$
Second harmonic:	$C_D = 0.14$
Third harmonic:	$C_L = 0.03$, $C_D = 0.05$

Therefore, a pile in a water flow may be excited to resonance not only at the critical flow rate, but also at half (in streamwise direction) and one-third (in an elliptical orbit) of it [8].

The field of the fluctuating velocity is shown in Fig. 3: Curves 1, 2, 3, 4, 7 and 9 correspond to the fundamental frequency with maximum values on the frontal surface: curves 5, 8 and 10 to the second harmonic with a peak on the base, and curves 6 and 11 to the third harmonic with two symmetrical peaks in the wake [7]. The second harmonic continues to exist far downstream along the wake axis.

3. KARMAN VORTEX STREETS SHED BY PLATES (GUIDE VANES, IMPELLER BLADES, AIRFOILS, ETC.)

The strength of the Karman vortex shed by a plate depends on the shape of its trailing edge, as shown in Fig. 4, with s/h as parameter [10]. The strength is largest if the trailing edge has the shape of a semi-circle (s/h = 0.5), followed by symmetrical wedges of 90° and 60° (s/h = 0.866) and the blunt edge (s/h = 0). The more slender the wedge than 60° (s/h > 1), the less the strength will be.

Moreover, an asymmetrical trailing edge will make the Karman vortices weaker than a symmetrical one. If, however, the wedge is very slender, for example in the case of an airfoil (s/h > 4), the strength can increase again. Finally, a trailing edge with cavity will make the plate practically free of vibration due to trapping of the vortex in it.

The reason for the influence of the trailing edge shape on the strength of the Karman vortex can be traced to the degree of the obstruction of the critical formation zone x = (0.8 - 1.3) h, which is affected by the trailing edge [10]. The Strouhal number will be determined both by the thickness of the plate h and a characteristic boundary layer thickness such as the sum of the displacement thicknesses of the two free-shear layers at separation δ^*, as well as their ratio. As the effect of this ratio is not yet known, it is

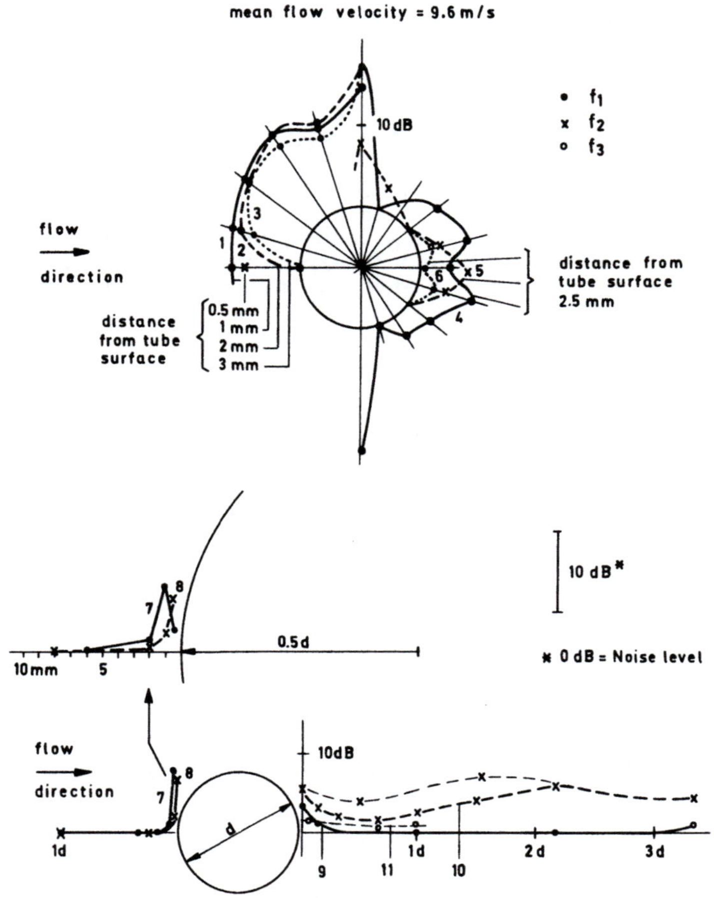

FIG. 3 DISTRIBUTION OF FLUCTUATING VELOCITIES AROUND A CIRCULAR CYLINDER [7]

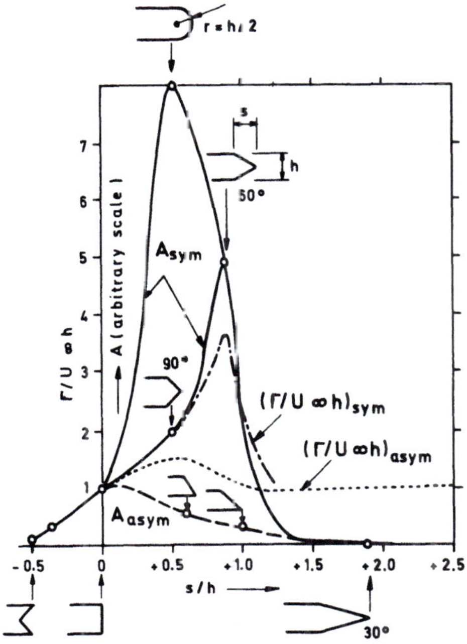

FIG. 4 DIMENSIONLESS CIRCULATION $\Gamma/U_\infty h$ OF KARMAN VORTEX AND NORMALIZED VIBRATION AMPLITUDE A VERSUS s/h [10]

obvious that the Strouhal S* based on h + 2δ* cannot yield a good correlation (Fig. 5) [9]. Another possibility may involve consideration of the momentum thickness (θ) at separation.

The Strouhal number of the Karman vortex street shed by plates or airfoils with slender trailing edges, behind which the two separated shear layers merge to form the wake without a dead-air zone, will be determined by the instability of the upstream boundary layer, characterized by Tollmien-Schlichting waves and the instability of the separated free-shear layer [10].

If the flow velocity is small, no Tollmien-Schlichting wave can be formed. Then the instability will eventually arise in the free-shear layer alone. A Karman vortex street will be formed further downstream (Fig. 6b). For large flow velocities, however, an interaction between the Tollmien-Schlichting wave and the free-shear layer instability will influence the generation of the Karman vortex street just behind the trailing edge (Fig. 6a) and a jump of its Strouhal number (Fig. 7) [11].

4. TURBULENCE AS EXCITATION SOURCE IN TUBE BUNDLES

The Karman vortex shedding as an excitation source applies only to tube bundles with wide tube spacings. The corresponding lift coef-

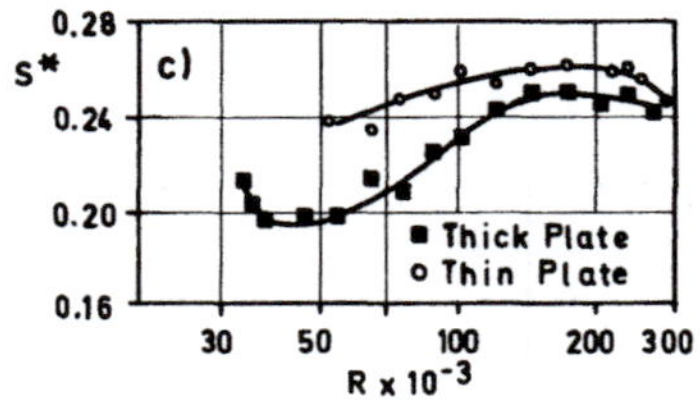

FIG. 5 STROUHAL NUMBER $S^* = f(h+2\delta^*)/U_\infty$ VERSUS REYNOLDS NUMBER [9]

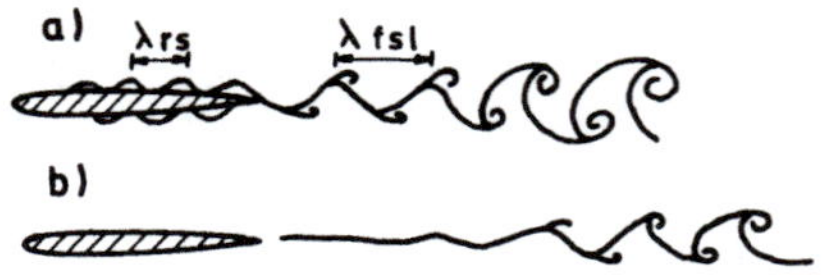

FIG. 6 KARMAN VORTEX STREET SHED BY A SLENDER PLATE [10] (TS = TOLLMIEN-SCHLICHTING WAVE, fsl = FREE SHEAR LAYER

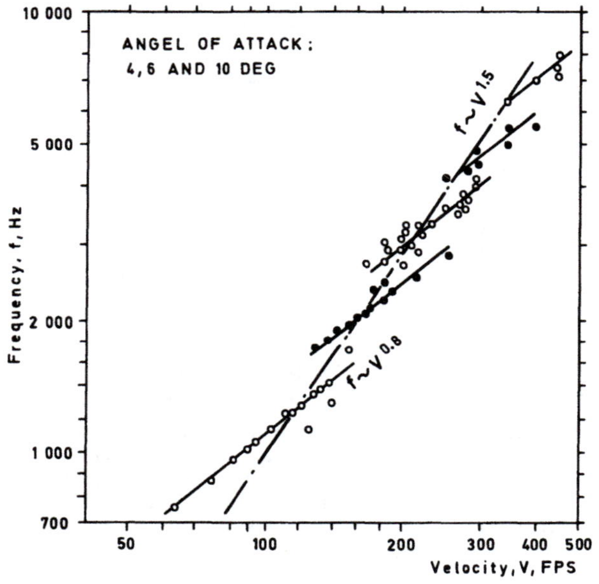

FIG. 7 FREQUENCY JUMP OF KARMAN VORTEX STREET SHED BY AN AIRFOIL [11]

ficients can be as high as 0.5 [12]. The turbulence associated with the vortex shedding is very weak [13]. No problem of turbulence buffeting will arise in this case. If, however, the tube spacings are narrow, other sources, such as wake swing, jet switch, jet instability, and vortex pairing may be the main excitation mechanisms [14, 15]. In this regime, the corresponding lift coefficients are very small ($C < 0.1$). The turbulence associated with these sources can be important for excitation of the mechanical vibration of the tube if the flow rate is high. The turbulence energy increases with the flow travelling downstream through the tube bundle [13].

Fig. 8 shows the fields of these two different categories of the excitation sources for staggered tube bundles. The turbulence can

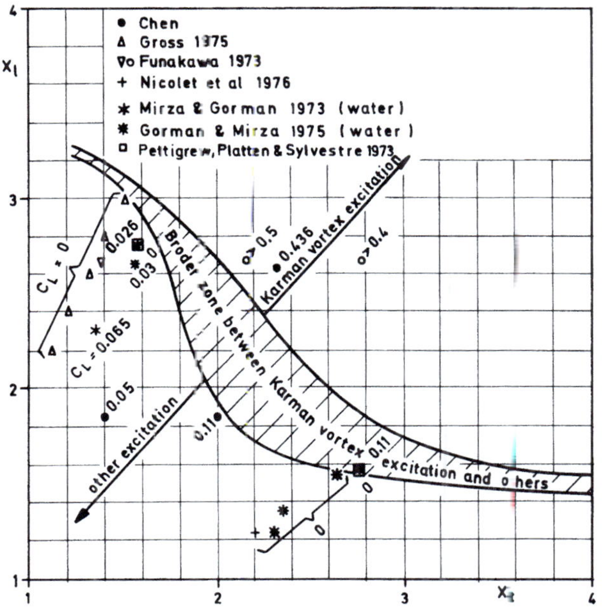

FIG. 8 LIFT COEFFICIENT FOR STAGGERED TUBE BUNDLE AS A FUNCTION OF TUBE SPACING RATIOS $x_t = T/d$ and $x_\ell = L/d$ [12]

reach an intensity corresponding to $C_L = 0.04$ to 0.07 in the lower field for excitation sources other than the Karman vortex [13]. An example for the fluctuating velocity in the tranverse gas column and the mechanical vibration of the tube for $x_t/x_\ell = 1.66/2.8$ is given in Fig. 9.

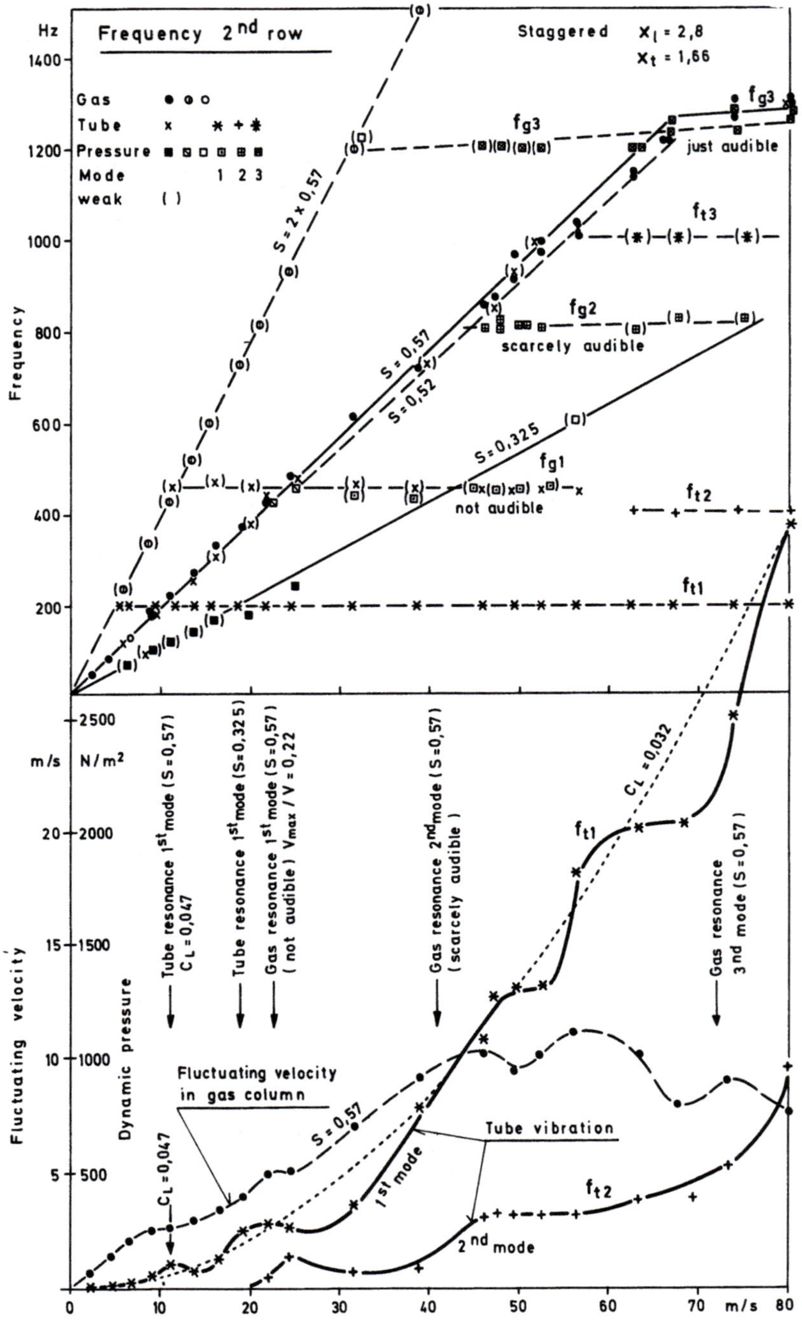

FIG. 9 FREQUENCY, FLUCTUATING VELOCITY OF FLOW BEHIND GAP, AND TUBE VIBRATION AMPLITUDE (IN TERMS OF DYNAMIC PRESSURE LOADING WHEN AMPLIFICATION FACTOR IS ASSUMED TO BE UNITY) VERSUS VELOCITY OF FLOW IN GAP FOR STAGGERED TUBE BUNDLE WITH $x_t/x_\ell = 1.66/2.8$

5. JET DEFLECTION AND SWING AS GOVERNING FACTOR FOR FLUID-ELASTIC INSTABILITY OF TUBE BUNLES

Fluid-elastic instability is caused by feedback between tube vibration at its fundamental mode and the flow path variation as a result of tube vibration. This flow path variation appears to originate from the capability of the jet to perform deflection or swing movement. The experimental results obtained by Ishigai et al. on the behavior of flow in tube banks reveal that in regions of small lateral tube spacing, there is susceptibility to jet deflection and swing [16,17]. If, in addition, the longitudinal tube spacings are also small, then the susceptibility is greater.

In reference [18], two charts show the experimentally determined instability factor $\beta = V_{cr}/f_o d\ (\delta m/\rho_g d^2)^{-1/2}$ for in-line and staggered tube bundles with tube spacings x_t = T/d and x_ℓ = L/d as coordinates. Two tentative curve groups are drawn through the points in each, one for the results obtained at ideal conditions in laboratory tests, and the other for measurements on site at industrial flow conditions with non-uniform mean flow distributions. The centre of greatest susceptibility to the fluid-elastic instability coincides well with the region of the most violent jet deflection as found by Ishigai et al. Thus, the theory of the jet deflection, or swing, as a factor for the fluid-elastic instability, would seem to be substantiated.

5.1 Remedy

1. The region around the spacing ratios of x_ℓ/x_t = 1.7/1.7 for the staggered spacing and that around x_ℓ/x_t = 1.5/1.5 for the in-line spacing are the most sensitive ones to the fluid elastic instability for industrial cases. The instability factor β can be as low as 3 - 4 for the in-line case and 3 for the staggered case. If possible, the spacings should be made larger.
2. Prevention of the wandering of the flow separation points on the tube surface, and thus hindering of the jet deflection and swing can apparently cure the problem of fluid-elastic instability. The effect of 4 parallel wires attached axially to the tube surface as suggested by Zdravkovich would appear to be a means of suppressing the jet deflection [19], thereby preventing fluid-elastic instability. Axial grooves in the surface of the tube should be as effective as the wires, because their edges would stabilize the flow separation points as well.

6. REFERENCES

1. Chen, Y.N., "Wake Swing and Vortex Shedding in a Cross Flow Past a Single Circular Cylinder", Flow-Induced Structural Vibrations, edited by E. Naudascher, Springer-Verlag Berlin/Heidelberg/New York, 143-157, (1974).

2. Iwan, W.D. and Blevins, R.D., "A Model for Vortex Induced Oscillation of Structure", Trans. ASME, 41, Series E., Journal of Applied Mechanics, 581-586, (1974).

3. Gerhardt, H.J., Kramer, C. and Jansen, H., "Untersuchungen zur Windbelastung von Stahlbaukonstruktionen, Bauwerke unter aerodynamischer Belastung", Mitteilungen aus dem Institut für Massivbau der Darmstadt, Heft 26, 1977.

Gerhardt, H.J., Kramer, C. and Jansen, H., "Wind Loads on Slender Prismatic Structures", Proceedings of the 3rd Colloquium on Industrial Aerodynamics, Aachen June 41-16, Buildings Aerodynamics Part 2, Fachhochschule Aachen, edited by C. Kramer and H.J. Gerhardt, 91-105, (1978).

4. Chen, Y.N., "The Vortex Shedding", Proceedings of the 2nd Colloquium on Industrial Aerodynamics, Fachhochschule Aachen, edited by C. Kramer and H.-J. Gerhardt, 107-133, (1976).

5. Thomson, K.D. and Morrison, D.F., "The Spacing Postion and Strength of Vortices in the Wake of Slender Cylinder Bodies at Large Incidence", Journal of Fluid Mechanics, 50, 751-783, (1971)

6. Chen, Y.N., "Damping of Karman-Vortex-Induced Vibrations by Axial Flow", Sulzer Technical Review, Research Number 1978, 27-31.

7. Chen, Y.N., "Experimental Investigation of Harmonics of the Dynamic Force Induced by the Karman Vortex", Paper presented at the IEA Conference, at Argonne USA, (February 20-21, 1979).

8. Wootton, L.R., Warmer, M.H. and Cooper, D.H., "Some Aspects of the Oscillations of Full-Scale Piles", Flow-Induced Structural Vibrations, ed. E. Naudascher, Springer-Verlag, 587-601, (1974).

9. Bauer, A.B., "Vortex Shedding from Thin Flat Plates", Journal of Aerospace Sciences, 341, (1961).

10. Chen, Y.N., "Behavior of Karman Vortex Streets Shed by Plates", Proceedings of the 3rd Colloguium on Industrial Aerodynamics, Buildings Aerodynamics, Part 2, Fachhochschule Aachen, edited by C. Kramer and H. J. Gerhardt, 1-17, (1978).

11. Paterson, W., Vogt, P. K., Fink, M.R. and Much, C.L., "Vortex Noise of Isolated Airfoils", Journal of Aircraft, 10, 296-302, (1973).

12. Chen, Y.N., "The Sensitive Tube Spacing Region of Tube Bank Heat Exchangers for Fluid-Elastic Coupling in Cross Flow", Fluid Structure Interaction Phenomena in Pressure Vessel and Piping Systems, ASME, edited by M.K. Au-Yand and S.J. Brown, Jr., 1-18, (1977).

13. Chen, Y.N., "Investigation of the Effect of Turbulence Associated with Vortex Shedding and Jet Instabilities on the Resonances Tube and Gas in Staggered Tube Bundle Heat Exchangers", Paper Presented at the IEA Conference, February 20-21, 1979 at Argonne, USA.

14. Chen, Y.N., "Excitation Sources of the Flow-Induced Vibrations and Noise in Tube Bank Heat-Exchangers", Noise and Fluid Engineering, ASME, edited by R. Hickling, 239-246, (1977).

15. Chen, Y.N., "Rauchgasseitige Schwingungen in Dampferzeugern - Verbrennungsinstabilität und Wärmeaustauscherschwingungen", VGB Kraftwerktechnik, 59, 420-433, (1979).

16. Ishigai, S. and Nishikawa, E., "Experimental Study of Structure of Gas Flow in Tube Banks with Tube Axes Normal to Flow, Part II: On the Structure of Gas Flow in Single Column, Single Row and Double Row Tube Bank", Bulletin of the JSME, 18, 528-535, (1975).

17. Ishigai, S., Nishikawa, E. and Yagi, E., "Structure of Gas Flow and Vibration in Tube Banks with Tube Axes Normal to Flow", International Symposium on Marine Engineering, Tokyo 1973, Marine Engineering Society of Japan, pp. 1-5-23 to 1-5-33.

18. Chen, Y.N., "Fluid-Elastic Instability in Tube Bank Heat Exchangers", Paper presented at the IEA Conference, April 11-12, 1978, Harwell, England.

19. Zdravkovich, M.M. and Namork, J.E., "Excitation, Amplification and Suppression of Flow-Induced Vibrations in Heat Exchangers", IAHR-IUTAM Symposium on Practical Experiences with Flow-Induced Vibrations, Karlsruhe, Session A, 135-142, September 3-6, 1979.

A18 INSTABILITY IN THE VESSEL OF A PRESSURIZED WATER REACTOR

F. Boulot, L. Pugnet

Electricite de France - Laboratoire National d'Hydraulique
6 Quai Watier - 78 400 - Chatou

SUMMARY

In 1968, after several months of problem-free operation of the SENA reactor, some bolts which joined the two segments of its reactor core-barrel were found broken, and it was observed that the thermal shield, initially resting on supports, was displaced by the flow of water.

With the aim of obtaining a better knowledge of the phenomena involved and thus avoiding these accidents in projected plants, Electricite de France started an experimental study on an hydraulic model of scale 1/10. The influence of certain parameters, such as the number of loops, geometry, Reynolds number, etc. has been investigated.

Visual observation of the Plexiglas model, under different working conditions, has shown the possible existence of an hydraulic instability phenomenon in the external annular space, which is between the thermal shield and the vessel. This instability occurs only when there is a thermal shield. During the instability, two large eddies were observed under the inlet nozzles. These generally unstable eddies were formed near the upper rim of the shield, where they broke away more or less regularly, causing an oscillation in the flow, and bringing about substantial alternating forces.

A19 ACOUSTIC PHENOMENA IN A STEAM GENERATOR

G. Baylac, J. P. Gregoire, J. Planchard

Electricite de France - Direction de l' Equipement
Tour EDF-GDF - 92080 Paris la Defense

Electricite de France - Direction des Etudes et
Recherches - 1, Avenue du Général de Gaulle
92140 Clamart

SUMMARY

With the increased capacity of modern power plants, the study of flow-induced vibration in heat exchangers becomes an important problem. In fact, tube-banks are sources of noise as well. This noise can be amplified by the acoustic resonances of the circuit, resulting in severe stresses in the affected structures. Such a problem was examined in the heat exchanger circuit of the CO_2-graphite nuclear reactor, CHINON 3. In order to reduce the pressure fluctuation level observed at resonance, it was decided to build a muffler into the circuit. The characteristics of the muffler were determined by means of computer codes accounting for the acoustic behavior of the flow circuit.

INSTABILITY IN THE VESSEL OF A PRESSURIZED WATER REACTOR

Problem: Serious damage occurred on the internal structure of Chooz PWR (325 MW); the thermal shield (resting on only 10 supports) was moved; 33 bolts out of 80 joining the two parts of the core barrel were found broken.

Flow conditions: Water jets from the 4 inlet nozzles enter the annular space just above the rim of the thermal shield. No instability without shield.

Operating conditions (results of scale model tests): A hydraulic instability occurs when a loop stops working. This phenomenon increases when more than 1 or 2 loops stop.

Vibrations: Scale model tests (see photo): in the external annular space, under every working inlet nozzle, 2 large eddies are alternately formed near the rim of the shield from which they are released more or less regularly, causing an oscillation in the flow and an oscillating force on the core barrel and the shield. No coupling expected between flow and structure.

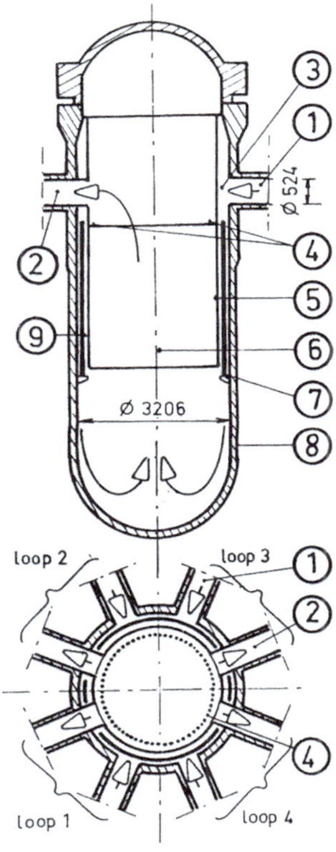

1 - inlet nozzle
2 - outlet nozzle
3 - annular space
4 - bolts
5 - core barrel
6 - core
7 - supports
8 - vessel
9 - thermal shield

Cures: Removal of the thermal shield.

References:

Daubert, A. et al,"Analyse des principales instabilités hydrauliques et des problèmes hydroélastiques dans les réacteurs à eau pressurisée", La Houille Blanche n°5 (1971).

Daubert, A. et al,"On a Global Hydraulic Instability in the Vessel of a Pressurized Water Reactor", IUTAM/IAHR Symposium, Karlsruhe (1972).

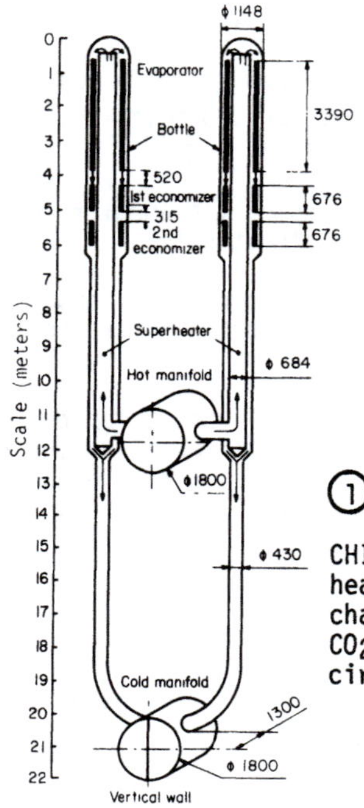

(1) CHINON 3 heat-exchangers CO_2 circuit

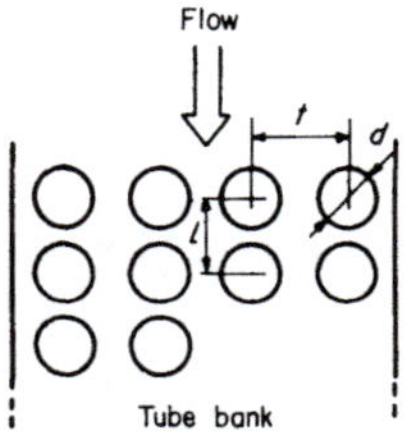

(2) Geometry of the tube bank

	$X_t = t/d$	$X_l = l/d$
Evaporator	1.51	1.13
Ist economizer	1.51	1.11
2nd economizer	1.51	1.11

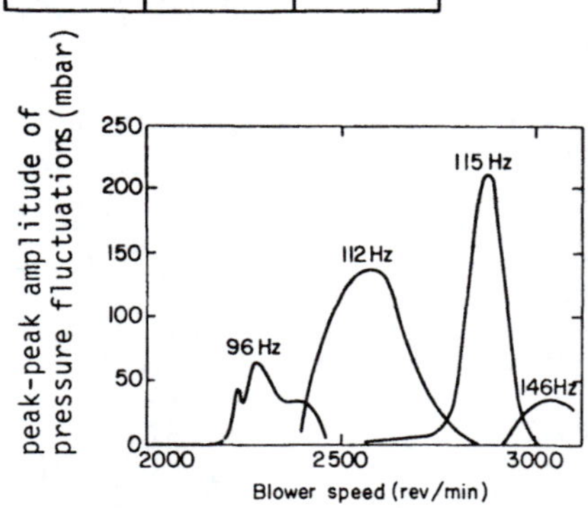

(3) Magnitude of pressure pulsation versus blower speed

ACOUSTIC PHENOMENA IN A STEAM GENERATOR

Problem: During commissioning test of the graphite-gas CHINON 3 reactor, a very stron noise (170 db) was observed in the heat exchanger circuit.

Geometry and boundary conditions: Heat exchanger CO_2-circuit consists of 24 identical elementary circuits represented in sketch 1.

Operating conditions: At blower speed of 1900 rev/min, 96 Hz vibration appeared in the CO_2 circuit, and was superseded by a stronger vibration (112-115 Hz) at 2300 rev/min. This intense vibration disappeare at 2900 rev/min, replaced by a 146 Hz vibration in the bottle-shaped heat exchangers (see sketch 3).

Structural dynamics: Most severe vibration (112 Hz) was at the first transverse acoustic vibration of the cold manifold, with pressure fluctuations too high for the structure.

Vibrations: It was concluded that the various acoustic resonances of the elementary circuit represented in sketch 1 were excited successively by narrow-band noise induced by tube-banks (evaporator and economizer, sketch 2).

Cures: In order to reduce pressure fluctuations, it was decided to modify acoustic response of circuit because modification of tube banks was impossible. A muffler was installed on each bottle, its characteristics being determined by computation and experiments. Code ACOUSTI was used to find an equivalent pipe for each tube bank (sketch 2) and code VIBRAPHONE was used to find the overall response of circuit.
With this muffler, the residual pressure level was below 150 db.

References: Baylac, G., Gregoire, J.P., "Acoustic phenomena in a steam generating unit", Journal of Sound and Vibration, Vol. 42, 31-48 (1975).
Gibert, R.J., Kivolant, M., "Sources et amplification des bruits dans un circuit", Programme VIBRAPHONE - CEA report SPM 1120 (1970).
Baylac, G., Planchard, J., Gregoire, J.P., Pailly, C., "Acoustical resonances in irreg lar cavities with application to noise-induced stress in expansion joints", Journa of Engineering for Industry, 101-108, (1977

A20 SOUND INDUCED VIBRATION IN CORRUGATED PIPES

A.M. Petrie
Department of Mechanical Engineering
I.D. Huntley
Department of Mathematics & Computing
Paisley College of Technology, Paisley, Scotland

SUMMARY

This paper is concerned with the noise produced by a steady flow of air through a duct which is internally corrugated. Thus, it is of relevance to situations such as heat exchanger tube bundles, dissipation-type silencers and the open undercarriage holds of an aircraft. Of particular interest is a commercial application where low velocity air travelling along a flexible corrugated duct (of the configuration used on vacuum cleaners) can result in acoustic outputs at the duct outlet in the region of 140 dB. Experiments are described which investigate the phenomenon, and several methods of attenuation are discussed.

1. INTRODUCTION

This project was initiated by an enquiry from a local manufacturer of flexible corrugated pipes of the type used on certain domestic appliances. The manufacturers were concerned about a whistle which was produced, for particular flow speeds and corrugation sizes, when air was passed down the pipe. A preliminary examination under controlled laboratory conditions (mainly with the tube lying flat on a bench) produced a whistle of up to 140 dB intensity using a duct of nominal diameter 30 mm and velocities in the range 8 to 60 m/s. Initially, the frequency of the whistle was observed to increase linearly with air velocity, and the whistle was therefore assumed to be associated with a vortex shedding process and a characteristic Strouhal number. Later experiments showed this to be substantially correct: step changes in frequency, corresponding closely with the axial duct harmonics, were observed as the air velocity through the duct was changed (see Figure 1), but the overall effect exhibited a linear relationship.

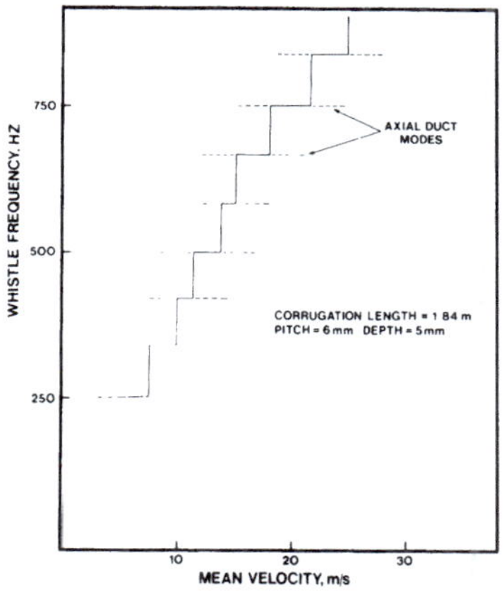

FIGURE 1

2. CURING

Several possibilities for avoiding this situation were available. Firstly, it was observed that by reducing the number of corrugations (in this case shortening the tube), the velocity range for which a whistle would occur was reduced. This, of course, altered the fundamental axial frequency, and so a higher mode was excited at the same velocity. Secondly, placing an obstruction in the duct inlet (in this case a piece of cardboard across about 30% of the area) created turbulence which was swept downstream and seemed to destroy the periodic excitation forces. Thirdly, the introduction of small wrinkles on the surface of the corrugations (of approximately 5% of the corrugation dimension as shown in Sketch 1 of the Chart) resulted in a curing of the whistle. This last preventative mechanism proved most convenient, since the tube walls were made from plastic and automatically wrinkled when the wall thickness was decreased during manufacture.

3. RELATED LITERATURE

This is by no means the first time that this phenomenon has been observed (see references). In fact, 70 years ago a study was completed using a tube with a single corrugation, which could be made to whistle. In recent years, however, the phenomenon has been observed in many situations of practical importance. For example, in references 2 & 3 it was observed that a steady airflow through a dissipation-type silencer containing the usual perforated lining could generate sound at certain frequencies. In this case the frequencies were related to the radial duct modes and the cavity natural frequencies. In references [4,6,7 and 8], a steady airflow across the open undercarriage hold of an aircraft was observed to have an apparently similar effect. More recently, in a private communication, analogous results have appeared with a heat exchanger tube which is carrying a liquid internally and whose internal surface has machined corrugations of depth less than 10% of the tube diameter.

4. PRESENT EXPERIMENTS

It is clear that this type of problem is becoming increasingly important, particularly in situations where environmental or structural considerations can result in expensive retrospective solutions. It was, therefore, the purpose of the present experimental investigation to establish whether there was a simple relationship between the geometry of the system and the flow characteristics. The apparently 'obvious' parameters to examine in such a study were the pitch, depth and number of corrugations, the duct length, flow velocity and turbulence characteristics.

A simple rig, shown schematically in Sketch 1 of the Chart, was constructed in which the capability existed to vary the velocity from 8 to 60 m/s through a range of ducts from 15 to 40 mm nominal internal diameter. The pitch and depth were varied in two ways: by stretching the flexible tubes used or, in the case of a perspex duct constructed from alternate large bore and small bore washers, by changing the number of washers used. The pitch and depth were thus varied in the ranges 2.5 to 8.0 mm and 2.5 to 7.5 mm respectively. The number of corrugations was varied between 1 and 340, and several lengths of smooth duct were used to increase the total length as required. The settling chamber (which was lined with foam and had internal baffles) was used to smooth the flow from the fan and reduce the fan noise transmitted to the duct. The constant-diameter section L_1 was used rather than a bell mouth (which would have given a much smoother flow) in order to define clearly the duct length for calculation of the duct natural frequencies. Initially air was blown through the duct as shown in Sketch 1. Later experiments required the air direction to be reversed, so that air was sucked through the corrugated duct, when it was found that turbulence generated within the smooth inlet length L_1 had the effect of destroying the whistle.

5. OBSERVATIONS

Three important observations were made. Firstly, the vortex shedding process always excited a duct axial mode. Thus the frequency did not change continuously but jumped in discrete steps as shown in Figure 1. This occurred for all the corrugations examined, and also for all the various geometries cited in the references. However, the number of modes which it was possible to excite did vary, particularly with duct length, as also did the bandwidth of each mode. Secondly, it was uncertain whether the whistle would always occur. For example in Sketch 2 the short length of duct exhibited a 'gap in the regular steps of at least 12 modes. The definition of the occurrence of a whistle was at first done aurally, but examination of the sound spectra showed that peaks were present at the 'gap' but were less than 5 dB above the background noise level compared with at least 20 dB when the whistle was clearly present. The phenomenon of 'gaps' cannot be explained. A preliminary calculation, however showed that it was not due to an interaction with radial or swirling modes of the duct. Thirdly, it was anticipated that the sound intensity would be dependent upon the number of corrugations. This was generally true, although initial tests produced no whistles with 50 corrugations even when other researchers had produced whistles with a single corrugation. This led to an important method of avoidance, since it was deduced that the boundary layer growth in the upstream length L_1 was responsible. Thus, by simply sucking air through the corrugated pipe (with $L_3 = 0$), whistles could be produced with only a single corrugation.

Additional experiments were completed using the duct constructed from perspex washers, which enabled the effects of pitch, depth and shape to be investigated independently. The sound intensity was independent of depth, except for shallow cavities (less than 2 mm) and for values of the pitch greater than 8 mm. This was the expected form of result, since for small depths or large pitches the vortex formation and/or separation of the shear layer is difficult. It also gave some indication of the size of the eddies involved. No difference was observed between semicircular, square and triangular shaped edges to the corrugations. There was no apparent simple relationship between intensity and the number of corrugations as can be seen by comparing Sketch 2 and Figure 2. It appears that the intensity is greatest when the higher modes are excited, and this must be related to the boundary layer growth along the tube. The data in Sketch 2, however, indicate that large duct lengths (and hence a large number of corrugations) are associated with high sound intensities, while the data in Figure 2 seem to contradict this.

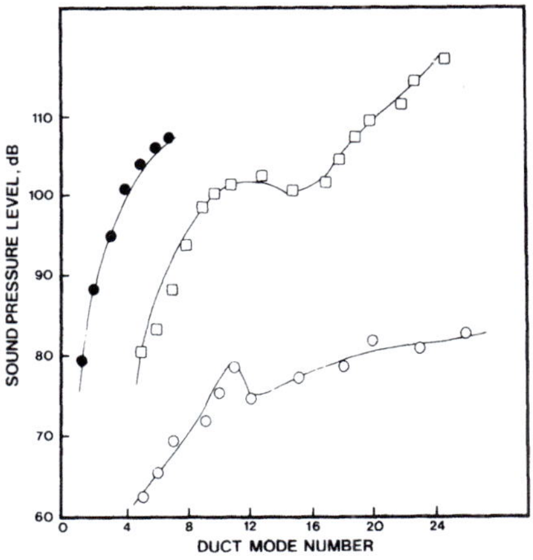

FIGURE 2

	NUMBER OF CORRUGATIONS	PITCH, mm	L_1, m	L_2, m	L_3, m
●	2	6	0·63	0·012	0
○	2	6	1·84	0·012	0
□	340	6·5	0·41	2·21	0

6. EFFECT OF TURBULENCE

Hot wire studies were expected to highlight some of the factors which would destroy the whistles. A comparison between the original flexible tubes whose only difference was the smooth or wrinkled corrugations, however, showed no changes in turbulence intensity across the duct section. Measurements close to the wall were difficult because of the small dimensions. The very small differences in corrugation shape which completely eliminate the whistle (see Sketches 1 and 3), do suggest that it is the changes in the turbulence spectrum close to the separation point which control the growth of the excitation mechanism.

Experiments with a single corrugation (using a duct length 0.63 m. and a corrugation of pitch 6 mm and depth 5 mm) showed that the introduction of only a few cm (L_3) of smooth duct upstream of the corrugation could also eliminate the whistle. However, for 188 corrugations (length 1.33 m) an upstream length of 1.88 m only reduced the intensity by 10 dB, and the whistle was still audible, as shown in Figure 3. Some experiments are now in progress, using a duct of larger section. These aim to examine boundary layer details, and it already seems that it is the turbulence spectrum close to the wall which is primarily responsible for the excitation. It may be, therefore, that wall obstructions or irregularities on the corrugations are sufficient to destroy the excitation mechanism (see the survey paper, reference 9).

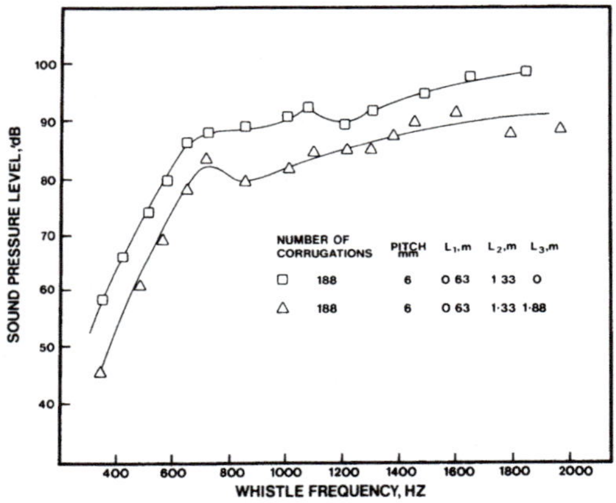

FIGURE 3

7. CONCLUSIONS

The so called 'self generation' of sound or vibration occurs in several practical situations, in some of which the lack of a practical solution can present serious difficulties. This investigation, which concentrates on only one of these problems, has attempted to identify the controlling parameters. In doing so, it has only met with limited success. It has shown that an axial mode is always excited, but the occurrence or intensity of the signal cannot be predicted with any accuracy and several contradictory results have been observed.

The avoidance of the problem must centre around a more detailed investigation of the turbulence structure in the system. Increasing the upstream length has provided a convenient cure, but it is suspected that this could be done more efficiently if obstructions can be introduced close to the flow separation region (as in the case of the wrinkled tube).

8. REFERENCES

1. Crawford, F.S., "Singing Corrugated Pipes", Am. J. Physics, 42, 278-288, (1974).

2. Tsui, C.Y., Flandro, G.A., "Self Induced Sound Generation by Flow Over Perforated Duct Liners", J. Sound Vib., pt. 3, 50, 315-331, (1977).

3. Bauer, A.B., Chapkis, R.L., "Noise Generated by Boundary Layer Interaction with Perforated Acoustic Liners", J. Aircraft, pt. 2, 14, 157-160, (1977).

4. Sarohia, V., Massier, P.F., "Control of Cavity Noise", J. Aircraft, pt. 9, 14, 833-837, (1977).

5. Wilson, T.A., et al, "Experiments on the Fluid Mechanics of Whistling", J. Ac. Soc. Am., pt. 1, 50, 366-372, (1971).

6. Heller, H.H., et al, "Flow Induced Pressure Oscillations in Shallow Cavities", J. Sound Vib., pt. 4, 18, 545-553, (1971).

7. Heller, H.H., Bliss, D.B., "The Physical Mechanism for Flow Induced Pressure Fluctuations in Cavities and Concepts for their Suppression", AIAA Paper 75-491, (1975).

8. Sarohia, V., "Experimental Investigation of Oscillation in Flows over Shallow Cavities", AIAA Paper 76-182, (1976).

9. Rockwell, D., Naudascher, E., "Self-sustained Oscillations of Impinging Free Shear Layers", Ann. Rev. Fluid Mech., 11, 67-94, (1979).

SOUND INDUCED VIBRATION IN CORRUGATED PIPES

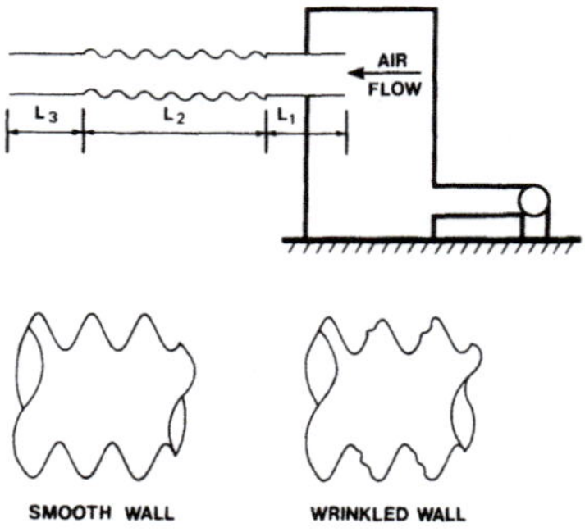

SKETCH 1

Problem: Attenuation of high intensity sound generated when a steady airflow passes through a corrugated duct. Possible acoustic fatigue.

Geometry: See Sketch 1. Range of ducts investigated as follows.

Pitch 2.5 - 8.0 mm
Depth 2.5 - 7.5 mm
Internal diameter 15 - 40 mm
Number of corrugations 1 - 340
Total duct length 0.64 - 3.84 m
Flow velocity 8 - 60 m/s

Boundary conditions: Outlet of pipe to free atmosphere. Pipe wall rigid.

Vibrations: Duct axial modes excited. Sound intensity can reach 140 dB at outlet. Occurrence of whistle not predictable. 'Gaps' in harmonic range possible, which may depend on duct length (see Sketch 2).

Primary cure: Small wrinkles in corrugation profile (see Sketch 1) can be introduced during duct manufacture and eliminate whistle (see Sketch 3).

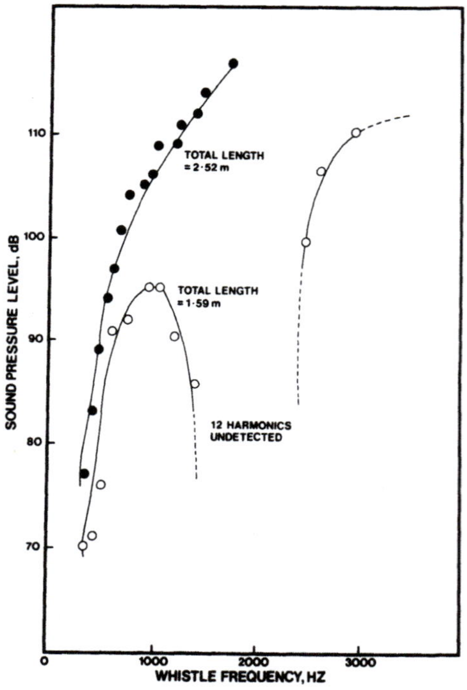

SKETCH 2

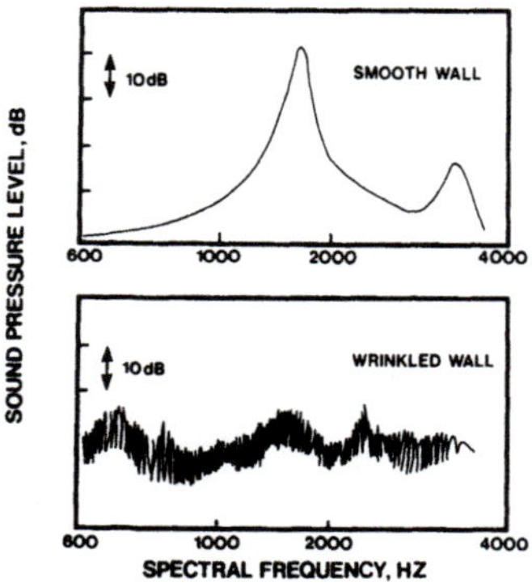

SKETCH 3

Secondary cures:

(i) Smooth upstream length L_1 attenuates whistl

(ii) Blockage at inlet increases turbulence and sometimes attenuates whistle.

(iii) Shorter corrugated lengths (and so less corrugations) may also attenuate whistle, depending on upstream conditions.

A 21

VIBRATIONS OF TUBE BUNDLES

Oskar Mahrenholtz
Prof. Dr.-Ing.

Herbert Tangemann
Dr.-Ing.

Institut für Mechanik, Universität Hannover
FR Germany

SUMMARY

Fluid-elastic problems which are a result of the interaction between the fluid flow and the tubes are the reason for damage in cross flow heat exchangers. This work should be considered as a contribution to the identification of excitation mechanisms. Experiments were carried out on a sectional model in a closed wind tunnel, starting with a single tube. Further, a pair of identical and small adjacent tubes were investigated and eventually staggered and in-line tube bundles as well.

1. INTRODUCTION

The fluid forces acting on a tube in a cross flow tube bundle depend on several important parameters. These are

- the fluid velocity
- the relative distances of the tubes
- the number of tube rows in front of the row with the reference tube
- the relative amplitude of the reference tube normal and that in the flow direction.

The test program was aimed at studying the effects of these parameters. In addition, many velocity profiles in the smallest gaps of the tube bundles were measured as well as pressure distributions on the tube surfaces.

2. VELOCITY PROFILES

In the row with the reference tube there are three test sections A, B, C, where velocity profiles have been measured. The tube length is $Y_{max} = 790$ mm . In a test section, the x-coordinate was always zero, the profile was measured at a particular ratio y/Y_{max}, and the position of the hot wire anemometer was changed in the z-direction (Figure 1).

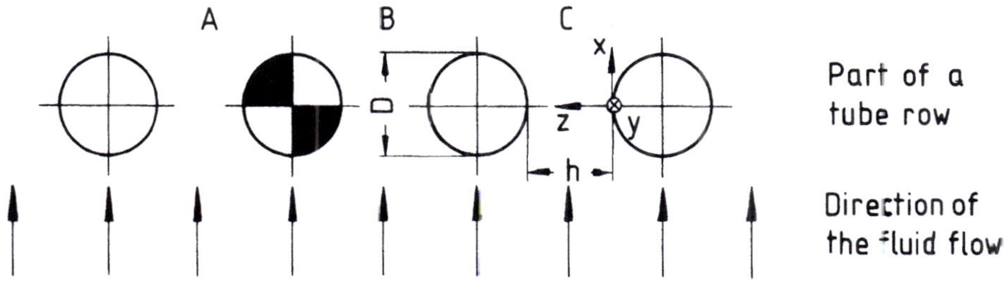

FIG 1 TEST SECTIONS A, B, C IN THE ROW WITH THE REFERENCE TUBE

To compare the velocity profiles, the following relative coordinates and velocities have been introduced:

$$HN = \frac{z}{h} \tag{2-1}$$

h is the height of the smallest gap and

$$0 \leq z \leq h$$

$$UN = \frac{\bar{u}(z)}{\bar{u}_{HN=0.5}} ; \tag{2-2}$$

$\bar{u}_{HN=0.5}$ is the velocity in the middle of the gap, and

$$Tu = \frac{u'(z)}{\bar{u}(z)} \tag{2-3}$$

Figure 2 shows the velocity profiles for an in-line bundle with a = 1.5. There exist two steady profile curves. Not only the number of tube rows (Figures 2.1, 2.3 and 2.4), but also the change of velocity in the gaps can cause a change from one to the other steady velocity profile (Figures 2.1, 2.2). This abrupt change in the profiles is due to the change in location of the separation points. This phenomenon occured only with small pitch/diameter ratios (a = 1.25 and 1.5).

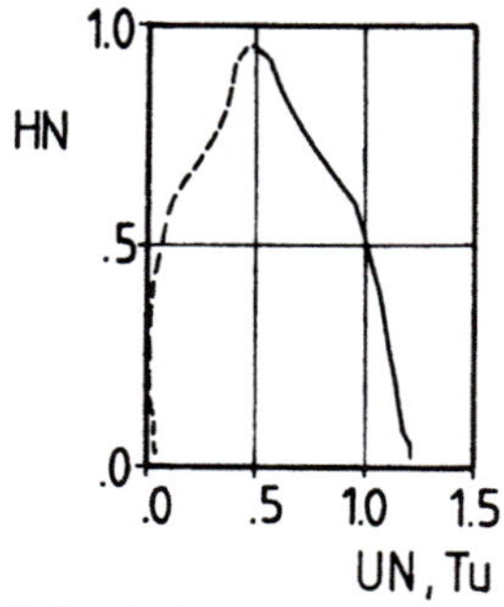

2.1 2nd row

$\bar{u}_{0.5}$ = 11.65 m/s

Re = 61720

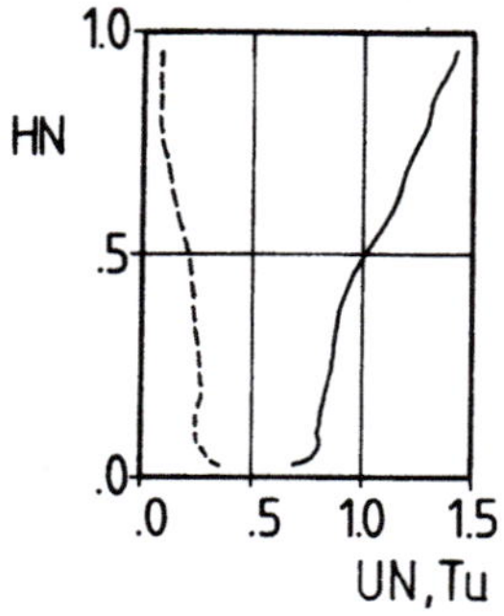

2.3 3rd row

$\bar{u}_{0.5}$ = 9.98 m/s

Re = 52850

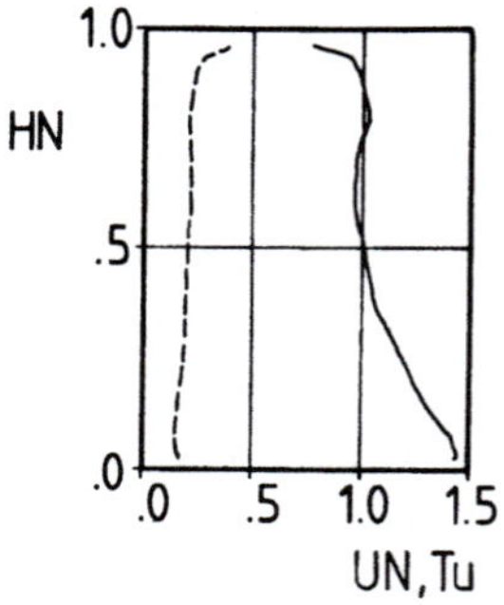

2.4 4th row

$\bar{u}_{0.5}$ = 8.4 m/s

Re = 44500

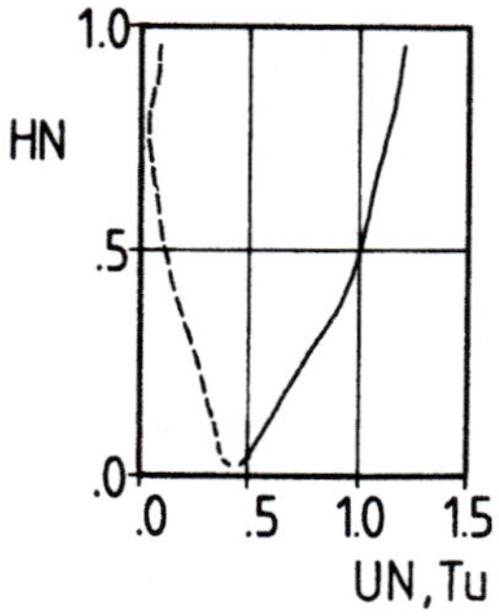

2.2 2nd row

$\bar{u}_{0.5}$ = 25.6 m/s

Re = 125630

——— UN, ------ Tu

bundle in-line, a = 1.5

test section C

y/Y_{max} = 0.375

FIG 2 VELOCITY PROFILES

3. STEADY STATE FLUID FORCES

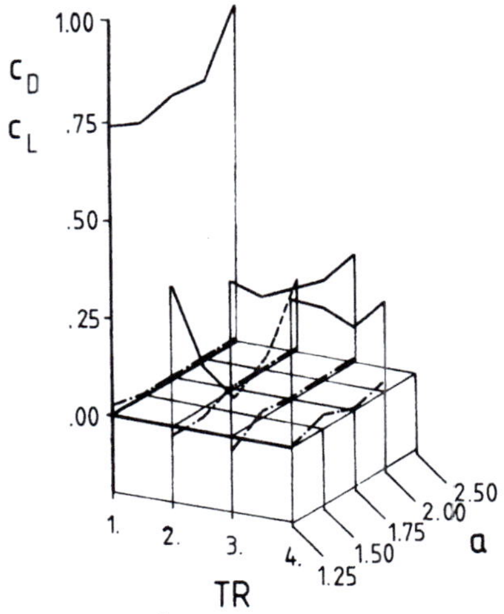

3.1 in-line bundle

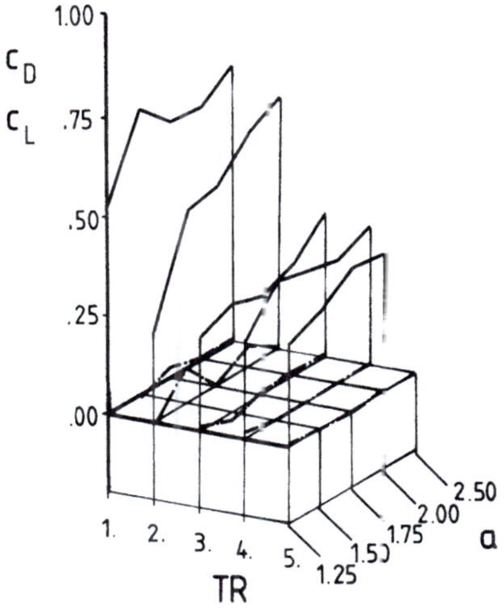

3.2 staggered bundle

———————— c_D , —·—·—·— c_L

no static displacement of the reference tube

$Re = 1.1 \cdot 10^5$

a = pitch / diameter ratio

TR = tube rows in front of the reference tube.

FIG 3 STEADY STATE FLUID FORCES

With staggered and in-line tube bundles, the steady state fluid forces acting on the tubes are dependent on the number of tubes in front of the reference tube. There are only small differences in the c_D and c_L coefficients when there are three or more rows in front of the row with the reference tube (Figure 3).

4. UNSTEADY FLUID FORCES

The main focus of this work was to study the unsteady fluid forces. Every tube of a bundle can be seen as a mechanical system with two degrees of freedom and small damping (Figure 4). Assuming that the tubes are oscillating mainly normal to or in the direction of the flow we can describe the motion of this reference tube as

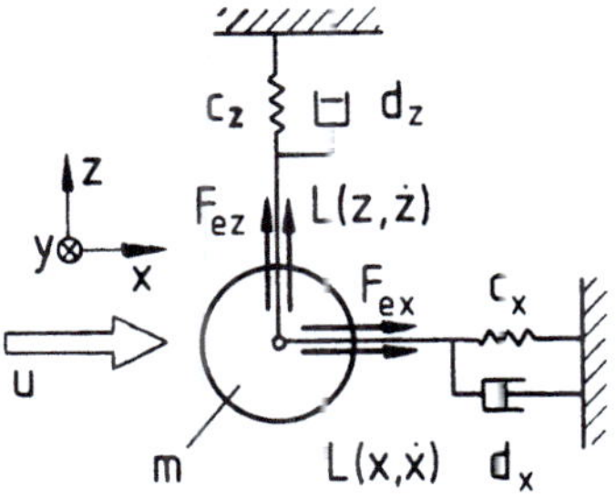

FIG 4 TUBE MODEL OF UNIT LENGTH

$$m\ddot{q} + d_q\dot{q} + c_q q = L(q, \dot{q}) + F_{eq}(t) \tag{4-1}$$

with $q = x, z$

The fluid forces according to Scruton [1] are

$$L(q,\dot{q}) = H_q q + K_q \dot{q} \tag{4-2}$$

The external exciter force is

$$F_{eq}(t) = \hat{F}_{eq} \cos\omega t \tag{4-3}$$

Only by making the aforementioned assumption can these investigations be carried out with a reasonable effort. The values H_q and K_q are expressed in the form of dimensionless coefficients h_q and k_q:

$$h_q = \frac{2m}{\rho D^2} 2\pi^2 \left(\frac{f_o^2}{f^2} - 1\right) \quad , \tag{4-4}$$

$$k_q = \frac{2m}{\rho D^2} 2\pi \vartheta_q \tag{4-5}$$

By using the phase resonance method, the tests showed that h_q could be neglected because the exciter frequency f was approximately equal to the natural frequency f_o. The k_D (x-coordinate) and the k_L (z-coordinate) coefficients are plotted vs. the reduced velocity

$$V_r = \frac{\bar{u}}{f \cdot D} \tag{4-6}$$

as stability diagrams. Sketches 2 and 3 are two examples of these stability diagrams. Sketch 3 represents a type of unsteady fluid force which is independent of the relative vibration amplitude. Changes within the range $0.025 \leq \hat{z}/D \leq 0.25$ did not affect $k_L(V_r)$. In contrast, in sketch 2, coefficients are dependent on the relative amplitude. With increasing $\hat{z}/D$ the energy transfer changes from damping to excitation. In many cases the k_D, k_L coefficients showed that flow-induced vibrations are possible. The coefficients are relatively high in the second tube row. This means the tubes of this row are highly susceptible to damage.

5. RESONANCE WAVE PHENOMENA

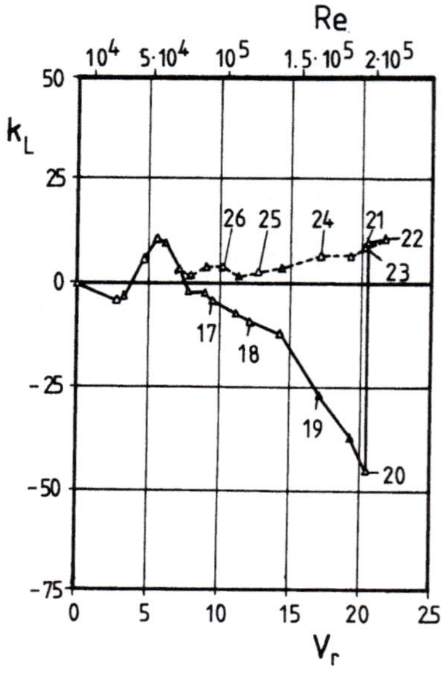

— Δ — $\bar{u}$ increasing

-- Δ --- $\bar{u}$ decreasing

bundle in-line

$a = 1.5$; $\hat{z}/D = 0.05$

$f = 22.4$ Hz; 2nd row

FIG 5 ABRUPT CHANGE IN THE $k_L(V_r)$ CURVE

Resonance wave phenomena are the reason for jumps in the $k_L(V_r)$ curves. To investigate these jumps, hot wire anemometers were used (Figure 6). Until a critical velocity u_{cr}, the flow around the reference tube was asymmetric (Figure 7). The corresponding k_L coefficients

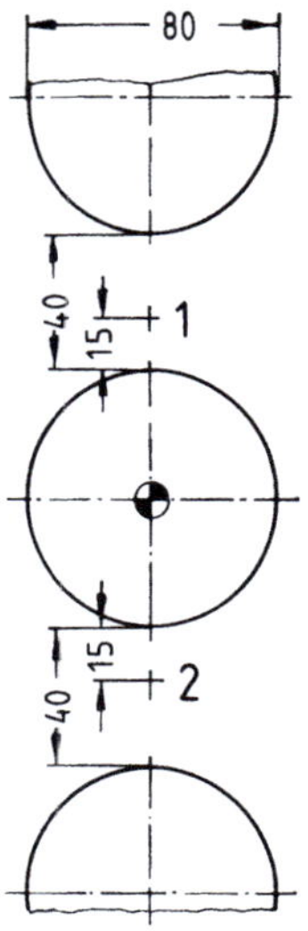

FIG 6 POSITIONS OF THE HOT WIRE ANEMOMETERS

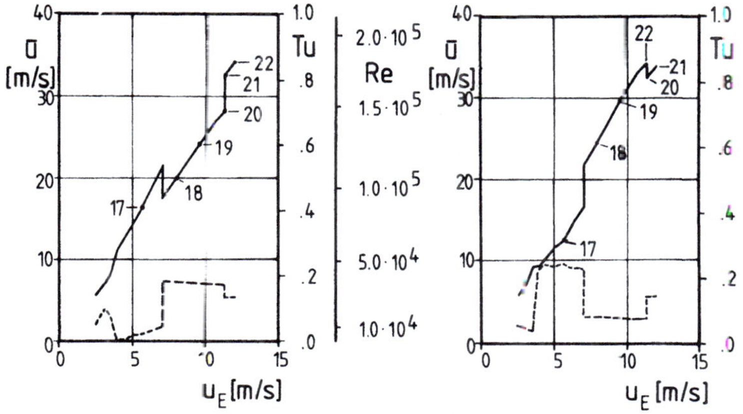

7.1 anemometer 1

7.2 anemometer 2

—— ū , --------- Tu

FIG 7 FLUID VELOCITIES NEAR THE REFERENCE TUBE

(point 17 til 20, Figure 5) were negative. From 20 to 21, the flow became symmetric and the k_L coefficients jumped to positive values. At this point 21, resonance wave phenomena could be analyzed in the anemometer signals (Figure 8). The resonance frequencies are linearly dependent on the fluid velocity $\bar{u}$. This leads to a constant number

$$S_{fn} = \frac{f_n D}{\bar{u}} \tag{5-1}$$

with n = 1,2,3,4.

The number n refers to the Fourier component. From 22 to 25, the velocity was decreased, the flow remained symmetric. The resonance wave phenomena continued as well. The k_L-coefficients gave the information that this phenomenon was responsible for the energy transfer from the fluid to the tubes, which led to this spectacular jump in the $k_L(V_r)$ curve at the critical velocity. Tripping wires on the surfaces of the upstream tube row led to a much lower critical velocity with the effect that the k_L-coefficients did not jump into the excitation region.

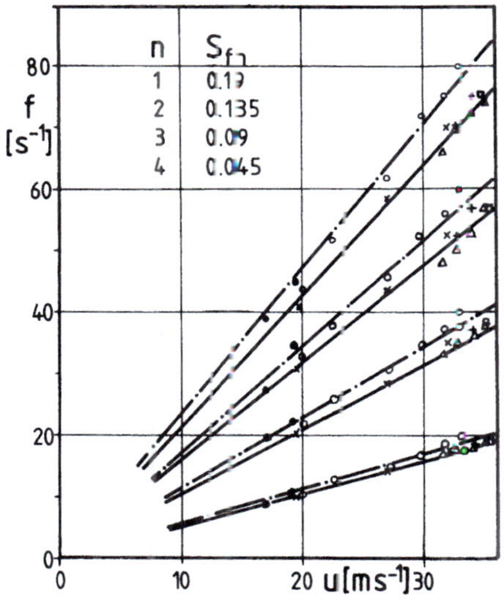

—— without } tripping wires
—·— with

Test	1	2	3	4
ū increasing	Δ	□	+	o
ū decreasing			x	●

FIG 8 RESONANCE WAVE FREQUENCIES

6. CONCLUSIONS

The unsteady fluid forces were determined for an elastically mounted tube within a tube bundle by the method of forced oscillations [2]. The influence of vibrating tubes in the proximity of the reference tube could not be investigated. However, first steps in this direction were made by Gross [3]. He showed, at a sectional model of a tube bundle with free oscillations, that tubes in one of the first three rows had the highest vibration amplitudes. This finding coincides with the results of this work. Moreover, these results can be used to quantitatively assess whether the fluid forces lead to damped or increasing vibrations of the tubes, and predictions of the fluidelastic stability and the critical flow velocities can be carried out.

7. REFERENCES

1. Scruton, C., On the Wind-Excited Oscillations of Stacks, Towers and Masts. Prof. 1st Conf. Wind Effects on Buildings and Structures, Teddington, (1963).

2. Tangemann, H., Fluidelastische Schwingungen von Einzelrohren in querangeströmten Rohrbündeln. Dissertation, Universität Hannover, (1979).

3. Gross, H.G., Untersuchung aeroelastischer Schwingungsmechanismen und deren Berücksichtigung bei der Auslegung von Rohrbündelwärmetauschern. Dissertation, Universität Hannover, (1975).

TUBE BUNDLES

Problem: Determination of critical flow velocities and of steady and unsteady fluid forces acting on a tube of a tube bundle in a cross flow leading to fluid-elastic vibrations of the tubes.

Geometry: 'Sectional model' in a closed wind tunnel, tube length 0.80 m. One tube - the reference tube - was prepared for the measurements. Removable tube rows in front of the row 1 with the reference tube, to check their influences on the fluid forces of this tube.

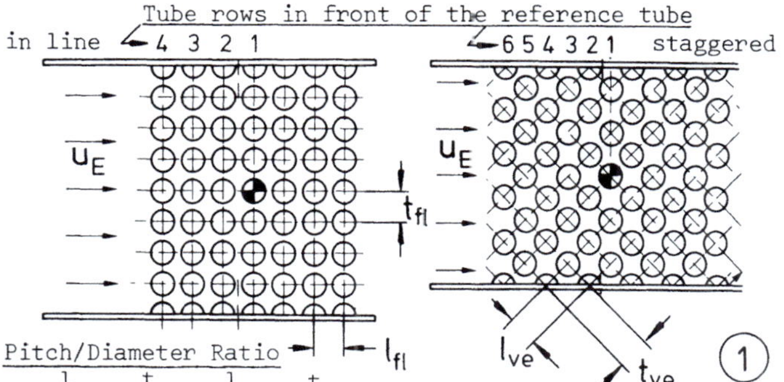

1

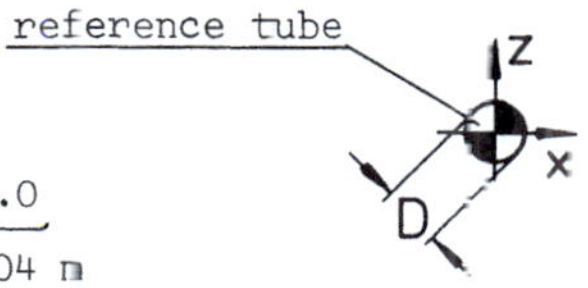

Pitch/Diameter Ratio

$$a = \frac{l_{fl}}{D} = \frac{t_{fl}}{D} = \frac{l_{ve}}{D} = \frac{t_{ve}}{D} = \underbrace{1.25,\ 1.5,\ 1.75,\ 2.0,}_{D\,=\,0.08\ \mathrm{m}}\ \ \underbrace{2.5,\ 3.0}_{D\,=\,0.04\ \mathrm{m}}$$

Flow conditions: The flow far upstream (u_E) is steady and non turbulent. Flow velocity u in the gaps up to 45 m/s and turbulent. Single phase flow: air.

Operating conditions: Externally forced oscillation of the reference tube controlled by phase resonance method. Constant relative amplitudes: $0.025 \leq (\hat{x}/D,\ \hat{z}/D) \leq 0.25$.

Structural dynamics: Natural frequency $f_o \approx 23.5$ Hz.

Vibrations: Forced vibrations; horizontal (x) or vertical (z). This research work leads to stability diagrams. They indicate if the fluid flow transfers energy to the tubes or not, resulting in excitation or damping of the tubes. Resonating wave phenomena in the fluid field are a possibility, which may transfer additional energy to the tubes.

Reference: Tangemann, H., Fluidelastische Schwingungen von Einzelrohren in queranqeströmten Rohrbündeln. Dissertation, Universität Hannover, 1979.

Stability diagrams:

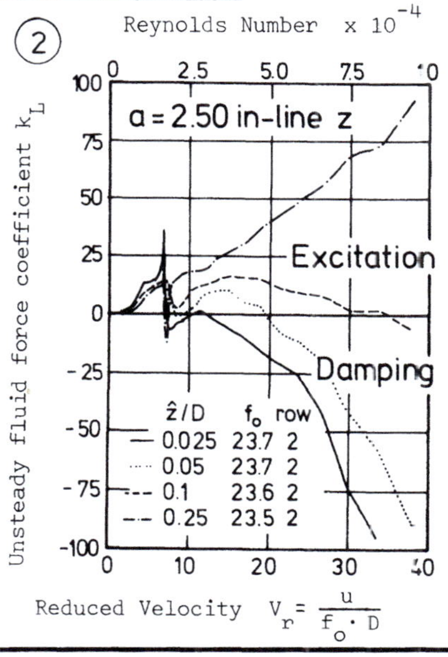

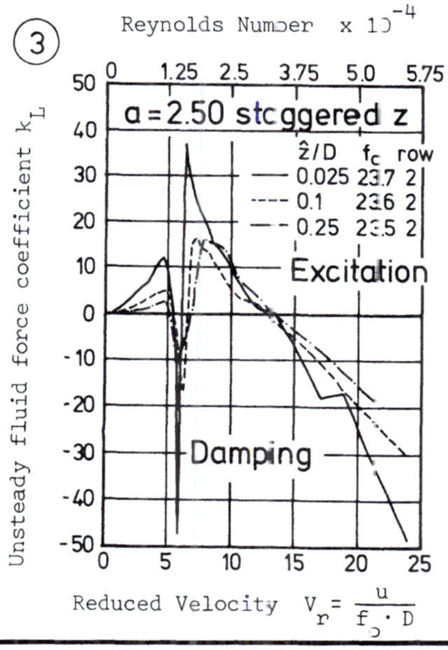

B1 SOME TVA EXPERIENCES WITH FLOW-INDUCED VIBRATIONS

Patrick A. March
Research Engineer

Svein Vigander
Supervisor, Research and Testing Section

Tennessee Valley Authority
Engineering Laboratory
Norris, Tennessee USA

1. INTRODUCTION

A variety of examples of flow-induced vibrations has been encountered within the TVA System. Three areas in which sufficient information is available to be of general interest have been selected for presentation. These areas include severe shaft vibrations in a fixed-blade hydraulic turbine, vibrations in a bulkhead gate slot-filler due to shear layer disturbances from a reversed tainter valve, and flow-induced vibrations in trashracks for a pumped storage project. No attempt has been made to explain specific design details, which can be obtained from the references.

2. VIBRATIONS IN A FIXED-BLADE HYDRAULIC TURBINE

Severe turbine shaft vibrations and structural vibrations occurred at intermediate and fully open wicket gate positions in three hydroelectric generating units at TVA's Wheeler Project [3,6]. Because these units are operated according to power demand, smooth operation at all gate openings is desirable.

The turbines have four fixed blades and are coupled to an umbrella-type generator with a 10.4-m vertical shaft. The rotating unit is supported by a single thrust bearing near the generator and is aligned by two guide bearings, with one guide bearing near the generator and one near the turbine, as shown in Sketch 1 on Chart A. Instrumentation locations for tests conducted on Wheeler Unit #11 are also shown in Sketch 1.

Tests were conducted at five percent intervals of wicket gate openings from 40 percent to 100 percent at synchronous speed (1.3 Hz) and at no load for 12 machine speeds ranging from 30 percent to 130 percent of synchronous speed. Lateral and vertical shaft displacements, bending and torsional strains in the shaft, and draft tube pressures were measured and recorded on magnetic tape. Flow velocities depended on wicket gate openings and ranged from five m/s to 12 m/s for wicket gate openings from 40 percent to 100 percent, respectively. Natural frequencies for the unit were computed analytically.

The ranges of computed natural frequencies and corresponding mode shapes, shown in Figure 1, were used to interpret the vibration test results. During vibration testing, shaft deflections up to one mm were observed. Relative amplitudes are shown as a function of wicket gate opening in Figure 2. The locations of amplitude maxima and minimia correspond well with plant operator experience. Spectra for torsional shaft strain under synchronous speed/variable load conditions are shown in Sketch 2 of Chart A. Spectra for lateral turbine shaft displacements are shown in Sketch 3, and spectra for vertical shaft displacements are shown in Sketch 4 of Chart A.

Vibrations at partial load and frequencies from 0.3 to 0.5 cycles per revolution were identified with an unsteady vortex core in the draft tube.

DIRECTION OF VIBRATION	MODE NO.	NATURAL FREQUENCY RANGE ORDER	NATURAL FREQUENCY RANGE Hz	APPROXIMATE DESCRIPTION
VERTICAL	1	10±	12.5-13.5	ENTIRE ROTATING UNIT BOUNCING ON THRUST BEARING AND BRACKET
	2	29 1 30.7	38 40	GENERATOR RIM BOUNCING ON SPIDER ARMS RELATIVE TO THE HUB
TORSIONAL	1	3.8 4 6	5.5 6	GENERATOR AND TURBINE ROTATING IN OPPOSITE DIRECTIONS
	2	23.0 38.4	30 50	GENERATOR ROTATING RELATIVE TO THE HUB
SHAFT LATERAL	1	6.1 9.2	8 12	SHAFT ALONE IN TRANSVERSE VIBRATIONS
UNIT LATERAL	1	3.8 5 4	5 7	GENERATOR, SHAFT, AND TURBINE ALL IN TRANSVERSE VIBRATIONS WITH GUIDE BEARING CONSTRAINTS
	2	6.1 7.7	8 10	AS ABOVE, WITH ONE NODE

MODE SHAPE SKETCH

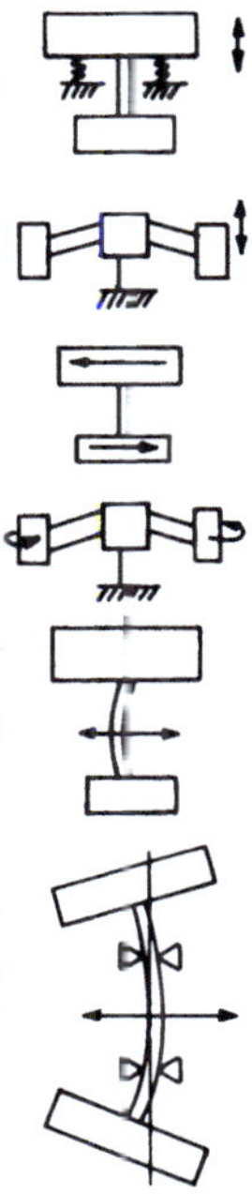

FIG. 1 SUMMARY OF COMPUTED NATURAL FREQUENCIES

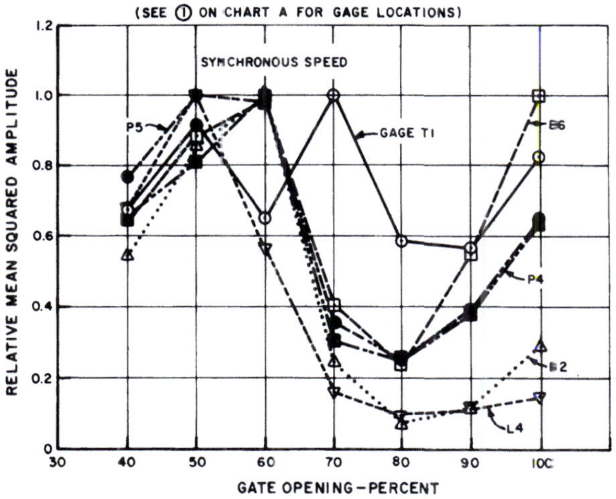

FIG. 2 RELATIVE AMPLITUDE AS A FUNCTION OF WICKET GATE OPENING

Vibrations near six Hz were attributed to a near resonance of flow disturbances caused by the four-bladed propeller with the torsional mode of shaft vibrations. Structural vibrations were primarily excited by the vertical mode resonance near 13 Hz. Excitation of this vertical mode was due to broad-band turbulence coupled through the angled turbine blades.

Three design modifications by the turbine manufacturer, including headcover baffles and inflow edge modifications, draft tube baffles, and turbine blade trailing edge pins, were tested. None of the modifications was considered as an acceptable solution to the vibration problem, and each modification was removed from the unit. The operational range of Wheeler Unit #11 is presently restricted to wicket gate openings between 65 percent and 85 percent. The use of signature analysis of selected measurements as an aid to preventive maintenance has been proposed.

3. VIBRATIONS IN A SLOT FILLER

Reversed tainter valves are used to control flow in the lock culverts at TVA's Wilson Lock, as shown in Sketch 1 of Chart B, and slot fillers are used in the bulkhead gate slots to provide a smooth culvert surface and prevent cavitation damage. Unacceptably large motions in the downstream slot fillers have resulted in damage to the supporting concrete.

Valve opening, vertical pressures, lateral pressures, and vertical accelerations were measured at the locations shown in Sketch 1 of Chart B. Velocities at the valve ranged from 0 to 26 m/s, depending on valve opening.

Vertical accelerations in the slot filler up to 0.8 g (peak-to-peak) were observed for expansion ratios (b/B, see Sketch 1) ranging from 0.3 to 0.6 [4]. These accelerations were forced vibrations, or buffeting, due to turbulence created in the shear layer downstream from the partially opened valve. Strouhal numbers ($S=f_s b/V_j$ where f is the frequency of vortex shedding, b is the valve opening, and V_j is the normal jet velocity at the valve) were computed by analyzing frequencies from sidewall pressure measurements (see Sketch 2 of Chart B). Sidewall RMS pressure coefficients ($([\Delta P^2]^{\frac{1}{2}})/\frac{1}{2}\rho V_j^2$ where $[\Delta P^2]^{\frac{1}{2}}$ is the RMS sidewall pressure and ρ is the fluid density) were also computed from sidewall pressure measurements (see Sketch 3 of Chart B). Measured Strouhal numbers were close to those observed in model tests for sudden expansions and sharp-edged gates [5,7]. Sidewall RMS pressure coefficients also agree reasonably well with model test data for sharp-edged gates and abrupt expansions [5,7]. The location of the maximum pressure coefficient at about the half-open valve position may, in part, result from the placement of the pressure transducers.

A satisfactory solution to the vibration problem was achieved by adding concrete weights to the slot fillers to increase the mass.

4. VIBRATIONS IN TRASHRACK (CIRCULAR BAR DESIGN)

Trashracks have been provided in the lower pool inlet/outlet structure for TVA's Raccoon Mountain Pumped Storage Plant. Due to conditions of flow reversal and relatively high velocities in the prototype and accounts of trashrack failures in other pumped storage systems, a half-scale model study of a typical trashrack unit was conducted [8].

The originally proposed design for an individual trashrack unit is shown in Sketch 1 of Chart C. The dimensions given in Sketch 1, as well as other values and results for the trashracks discussed in this paper, are for the half-scale model. The prototype trashrack units are stacked and clamped into slots in concrete support piers. The half-scale model was placed in a steel support frame and clamped at the appropriate locations. Mode shapes, frequencies, and damping values were measured in air using accelerometers and an electromagnetic shaker. Modal information with the rack vibrating transverse to the flow (heave) is shown in Sketch 2 of Chart C. Modal information with the rack vibrating parallel to the flow (plunge) is shown in Sketch 3.

A lock-filling culvert was used as a water tunnel for hydrodynamic tests with the half-scale trashrack. Approach velocities in the lock culvert ranged from 0 to 4.3 m/s. The velocity distribution at the trashrack section was measured with propeller-type current meters and determined to be within ± five percent of the mean throughout the cross-section.

Results of the hydrodynamic tests are summarized in Sketch 4 of Chart C. "Locked-in" vibrations were observed at 112 Hz for velocities from 1.7 m/s to 2.6 m/s, at 120 Hz for velocities from 2.6 m/s to 3.0 m/s, and at 185 Hz for velocities from 3.0 m/s to 4.3 m/s. Fractures occurred in 17 of the short rods in the central portion of the rack (see Sketch 1) after less than 30 minutes of operation at 4.3 m/s. The locked-in vibration for velocities from 1.7 m/s to 2.6 m/s was identified as the (0-3) plunge mode (see Sketch 3). For velocities from 2.6 m/s to 3.0 m/s, the locked-in vibration was identified as the (0-4) plunge mode (see Sketch 3). For velocities from 3.0 m/s to 4.3 m/s, the locked-in vibration was identified as the (0-5) plunge mode (see Sketch 3). In all cases, vibrations were in line with the flow, in modes with very low damping, and at frequencies corresponding to second harmonic excitation from the drag component of the unsteady hydrodynamic force [1]. Note the effect of vibrations in reducing the effective Strouhal number (see Sketch 4 of Chart C) for the trashrack bars [2].

5. VIBRATIONS IN TRASHRACK (RECTANGULAR BAR DESIGN)

The circular-bar trashrack design was modified to reduce or eliminate the destructive flow-induced vibrations described above. Rectangular sections with a minimum aspect ratio of 1:4, as shown in Sketch 1 of Chart D, were used to stiffen the rack in the plunge direction and lower the energy transfer from the water to the rack. A half-scale model of the modified design was clamped into the steel support frame. Mode shapes, frequencies, and damping values were measured in air using accelerometers and an electromagnetic shaker [8]. Modal information with the rack vibrating transverse to the flow (heave) is shown in Sketch 2 of Chart D. Modal information with the rack vibrating parallel to the flow (plunge) is shown in Sketch 3 of Chart D.

The modified rack was tested in the same lock-filling culvert under similar flow conditions. Approach velocities ranged from 0 to 5.5 m/s. The test results are summarized in Sketch 4 of Chart D. For much of the velocity range, the rack response corresponded to low-level random buffeting. Locked-in vibrations were observed at 30 Hz for velocities from 1.5 m/s to 1.9 m/s; at 116 Hz for velocities from 3.1 m/s to 3.7 m/s; and at 267 Hz for velocities from 4.6 m/s to 5.5 m/s. The maximum stress level determined from the test results was about one percent of the typical endurance limit for the prototype construction material, stainless steel. For velocities from 1.6 m/s to 1.9 m/s, the locked-in vibration was identified as the (0-0) heave mode (see Sketch 2). For velocities from 3.1 m/s to 3.7 m/s, the locked-in vibration was identified as the (0-2) plunge mode (see Sketch 3). The locked-in vibration for velocities from 4.6 m/s to 5.5 m/s was identified as the (0-3) plunge mode (see Sketch 3). The (0-0) heave mode responded to excitation from the lift component of the unsteady hydrodynamic force; the (0-2) plunge mode responded to the drag component; and the (0-3) plunge mode responded to the second harmonic of the drag component [1-8]. Note the effect of vibrations in reducing the effective Strouhal number (see Sketch 4) for the trashrack bars [2].

Neoprene pads placed between the trashrack and the adjacent racks eliminated lock-in throughout the velocity range tested. A baffle plate, which measured 15 cm x 30 cm and was mounted in the center of the trashrack, also eliminated lock-in. The recommended trashrack design includes both Neoprene damping pads and baffle plates.

6. SUMMARY

This paper presents background information and test results for four examples of flow-induced vibrations which we have encountered within the TVA system. The turbine vibration example illustrates a broad-band vibration response due to instability of the draft tube vortex, resonance of a turbine natural mode with pulsations at the blade-passage frequency, and structural vibrations related to flow-induced resonance of a vertical turbine mode. The slot-filler example provides information on frequency and magnitude of fluctuating pressures due to a shear layer instability. The trashrack examples illustrate the destructive effects from and the successful resolution of self-controlled vibrations occurring in-line with the flow. We hope that these practical examples will assist others in their efforts to understand and control flow-induced vibrations.

7. REFERENCES

1. Crandall, S., S. Vigander, and P. March, "Destructive Vibration of Trashracks due to Fluid-Structure Interaction", ASME Journal of Engineering for Industry, November 1975, pp. 1359-1365.

2. Eagleson, P., and J. Daily, "The Effect of Boundary Layer Thickness and Vibrational Amplitude on the Strouhal Numbers for Flat Plates ", Proceedings, IAHR Congress, London, England, 1963, pp. 73-80.

3. Elder, R., J. Dale, and S. Vigander, "Vibrations in a Fixed-Blade Hydraulic Tubine Unit", Proceedings of IAHR Symposium on Hydraulic Machinery, Equipment, and Cavitation, Lausanne, Switzerland, October 1968, pp. G2/1-G2/10.

4. Kirkpatrick, K., "Forces on Bulkhead Slot Filler, Wilson Main Lock", TVA-WSDB Report No. WM28-2-1-100, June 1978.

5. Naudascher, E., "On the Role of Eddies in Flow-Induced Vibrations" , Proceedings, IAHR Congress, London, England, 1963, pp. 61-70.

6. "Vibraticn Studies: A Summary of the Vibration Problem, Wheeler Units 9-11 ", TVA Engineering Laboratory Report No. 3-509, February 1970.

7. Vigander, S., "An Experimental Study of Wall-Pressure Fluctuations in a Cavitating Turbulent Shear Flow", Studies in Engineering Mechanics, Report No. 21, The University of Kansas, Lawrence, Kansas, 1965.

8. Vigander, S., and P. March, "Trashrack Vibration Studies, Raccoon Mountain Pumped Storage Project", TVA Engineering Laboratory Report No. 43-47, October 1974 (Revised March 1979).

Chart A : FIXED-BLADE HYDRAULIC TURBINE

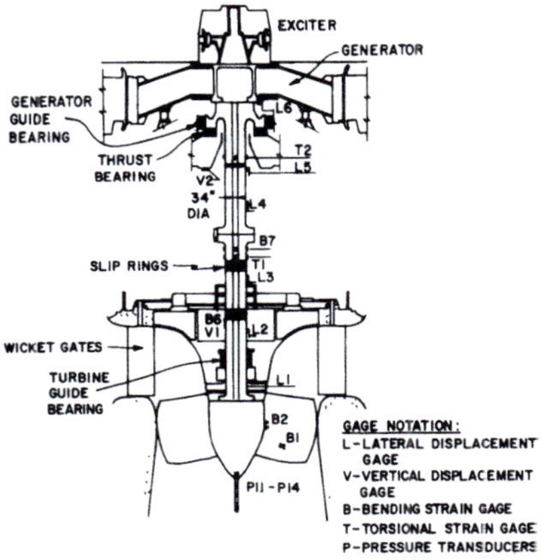

① Fixed Blade Turbine

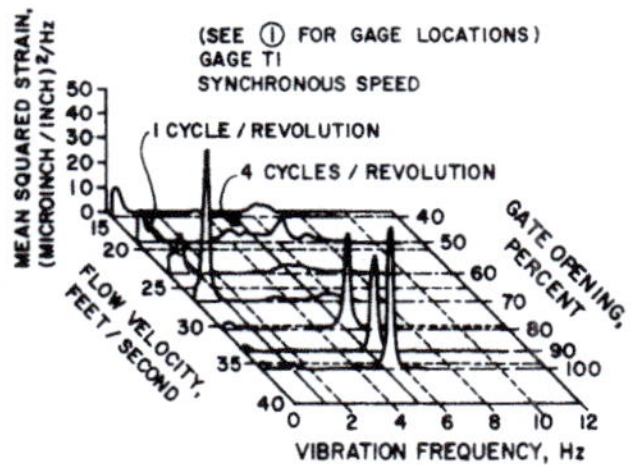

② Power Spectra for Shaft Torsion Under Variable Load Conditions

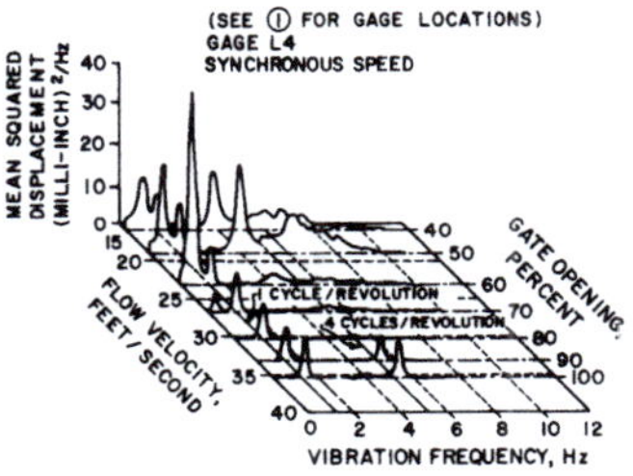

③ Power Spectra for Shaft Displacement Under Variable Load Conditions

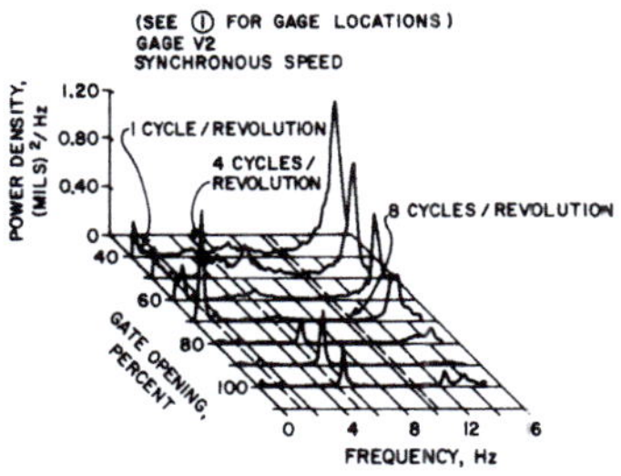

④ Power Spectra for Vertical Displacement at the Generator Thrust Bearing Bracket, Synchronous Speed Conditions

Problem: Severe turbine shaft vibrations and structural vibrations occurred at intermediate and fully open wicket gate positions.

Geometry: See ①.

Boundary conditions: Guide bearing and thrust bearing locations are shown in ①.

Flow conditions: Flow velocities ranged from 5 m/s to 12 m/s for wicket gate opening from 40% to 100%, respectively.

Operating conditions: Tests were conducted at 5% intervals of wicket gate openings from 40% to 100% at synchronous speed (1.3 Hz) and at no load for 12 machine speeds ranging from 30% to 130% of synchronous speed. Lateral and vertical displacements, bending and torsional strains, and draft tube pressures were measured and recorded.

Structural dynamics: Natural frequencies, computed analytically, are summarized in figure 1 (see text).

Vibrations: Relative amplitudes shown in figure 2 (see text). Vibrations at 0.3 to 0.5 cycles per revolution were (see ②, ③, ④) identified with unsteady vortex core in draft tube; vibrations near 6 Hz (see ②) were attributed to interaction between four-bladed propeller and torsional mode; structural vibrations were identified with 10th order vertical mode near 13 Hz (see ④).

Cures: Operational range of unit is restricted to wicket gate openings of 65% to 85%.

References (see text): 3. Elder, R., Dale, J., and Vigander, S., Vibrations in a Fixed-Blade Hydraulic Turbine Unit. Proceedings of IAHR Symposium on Hydraulic Machinery, Equipment, and Cavitation, Lausanne, Switzerland, October 1968, pp. G2/1-G2/10.

6. Vibration Studies: A Summary of the Vibration Problem, Wheeler Units 9-11, TVA Engineering Laboratory Report No. 3-509, February 1970.

Chart B : SLOT FILLER VIBRATIONS

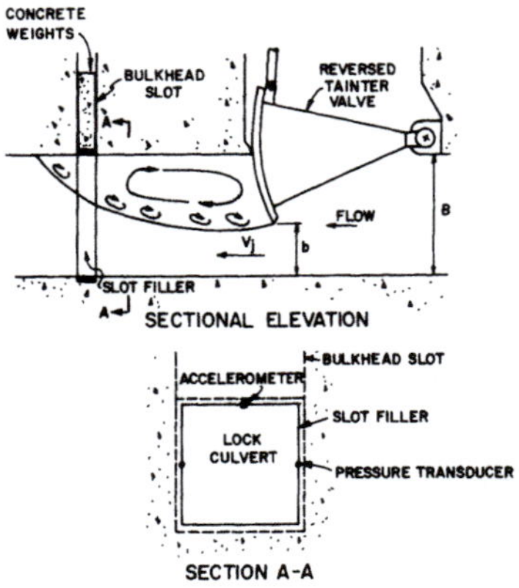

(1) Geometry of Lock Filling Valve

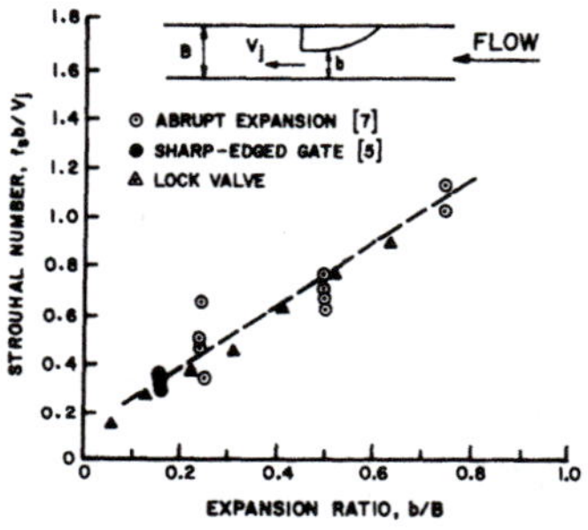

(2) Variation of Strouhal Number with Expansion Ratio

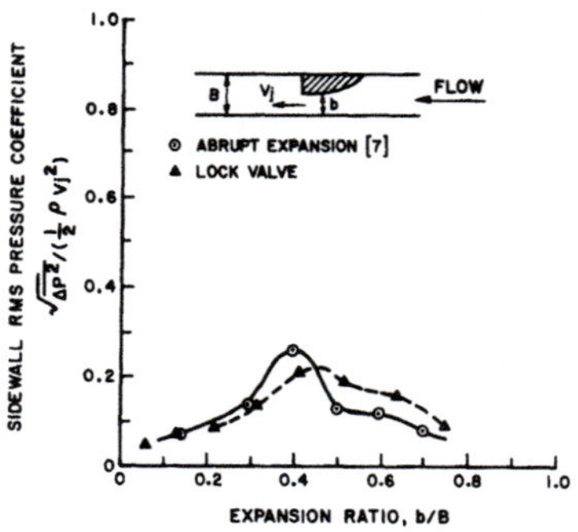

(3) Sidewall RMS Pressure Coefficient as a Function of Expansion Ratio

Problem: Unacceptable motion in Bulkhead gate slot fillers downstream from a reversed tainter valve resulted in damage to supporting concrete.

Geometry: See (1).

Boundary conditions: Velocity distribution varied with valve opening.

Operating conditions: Velocities at valve ranged from 0 to 26 m/s. Valve opening, vertical pressures, lateral pressures, and vertical accelerations were measured (see (1)).

Vibrations: Vertical accelerations up to 0.8 g (peak-to-peak) were observed for expansion ratios (b/B, see (1)) ranging from 0.3 to 0.6 [4]. Measured Strouhal numbers were close to those observed in model tests for sudden expansions and sharp-edged gates [5,7]. Sidewall RMS pressure coefficients also agree reasonably well with model test data for sharp-edged gates and abrupt expansions [5,7].

Cures: Concrete weights were added to slot filler to increase mass.

References (see text): 4. Kirkpatrick, K., Forces on Bulkhead Slot Filler, Wilson Main Lock. TVA-WSDB Rpt. No. WM28-2-1-100, June 1978.

5. Naudascher, E., On the Role of Eddies in Flow-Induced Vibrations. Proceedings, IAHR Congress, London, England, 1963, pp. 61-70.

7. Vigander, S., An Experimental Study of Wall-Pressure Fluctuations in a Cavitating Turbulent Shear Flow. Studies in Eng. Mechanics, Rpt. No. 21, The University of Kansas, Lawrence, Kansas, 1965.

Chart C : CIRCULAR BAR TRASHRACK

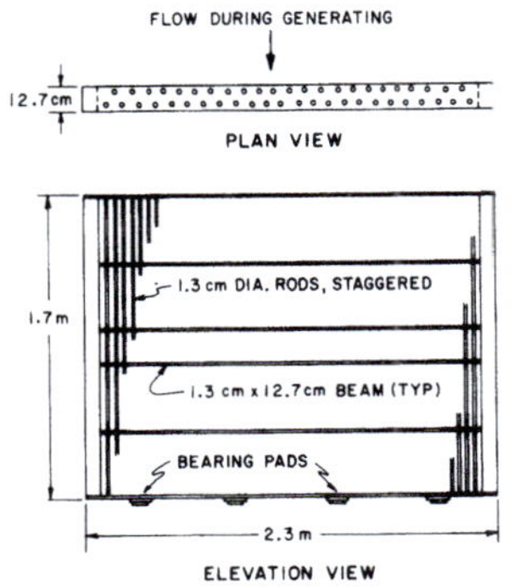

(1) Half-Scale Model of Circular Bar Trashrack

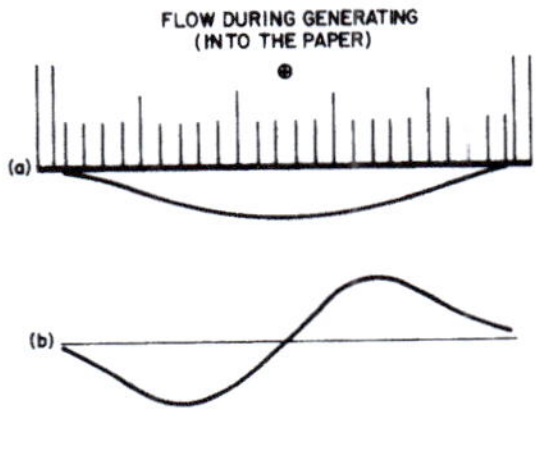

(2) Mode Shapes for Horizontal Bars of Test Rack ("Heave," Transverse to Flow)

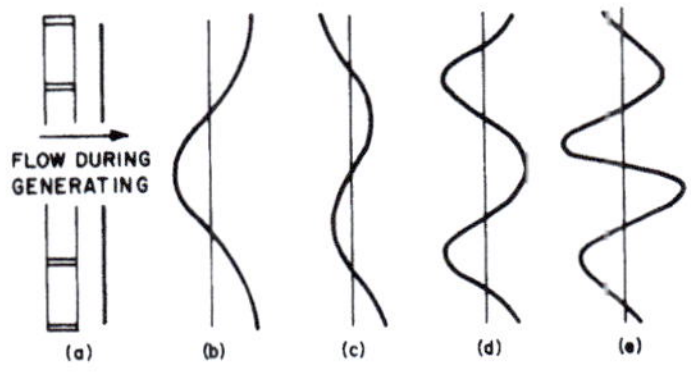

(3) Mode Shapes for Horizontal Bars of Test Rack ("Plunge", Parallel to Flow)

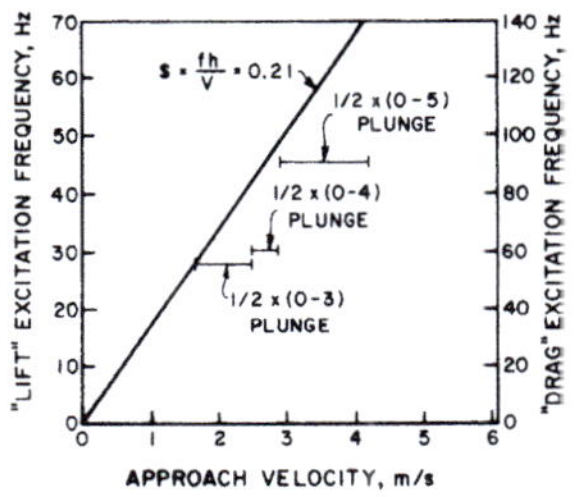

(4) Summary of Test Results, Circular Bar Trashrack

Problem: Flow-induced vibrations led to rapid fracturing of trashrack bars.

Geometry : See (1).

Boundary conditions: Trashrack clamped in slot at 3 points on each side; rack pressed against other racks at 4 points on top and bottom (see (1)).

Flow conditions: Trashrack was tested in lock filling culvert; velocity distribution within ± 5% of mean.

Operating conditions: Approach velocities from 0 to 4.3 m/s; "Lock-In" was observed at 112 Hz for velocities from 1.7 m/s to 2.6 m/s, at 120 Hz for velocities from 2.6 m/s to 3.0 m/s, and at 185 Hz for velocities from 3.0 m/s to 4.3 m/s. Fractures occurred in 17 of the short rods in central portion of rack (see (1)) after less than 30 minutes of operation at 4.3 m/s.

Structural dynamics: Heave modes in air shown in (2) [(a)(0-0) mode, f_n = 28 Hz, Q = 83; (b)(I-0) mode, f_n = 63 Hz, Q = 39]. Plunge modes in air shown in (3) [(a)(0-0) mode, f_n = 50 Hz, Q = 21; (b) (0-2) mode, f_n = 87 Hz, Q = 250; (c) (0-3) mode, f_n =113 Hz, Q = 940; (d) (0-4) mode, f_n =122 Hz, Q = 635; (e) (0-5) mode, f_n =193 Hz, Q = 1115].

Vibrations: Results are summarized in (4).

Cures: Problem solved by redesign. See accompanying Chart D, "Rectangular Bar Trashrack".

References (see text): 1. Crandall, S., Vigander, S., and March, P., Destructive Vibration of Trashracks due to Fluid-Structure interaction. ASME Journal of Eng. for Industry, Nov. 1975, pp. 1359-1365.

8. Vigander, S., and March, P., Trashrack Vibration Studies, Raccoon Mountain Pumped Storage Project. TVA Eng. Lab. Rpt. No. 43-47, Oct. 1974 (Revised March 1979).

Chart D : RECTANGULAR BAR TRASHRACK

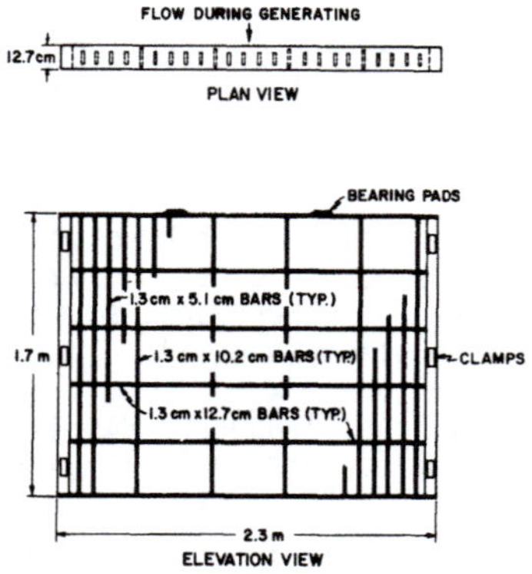

① Half-Scale Model of Rectangular Bar Trashrack

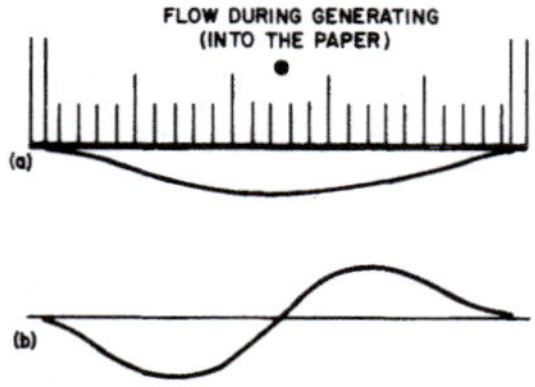

② Mode Shapes for Horizontal Bars of Test Rack ("Heave," Transverse to Flow)

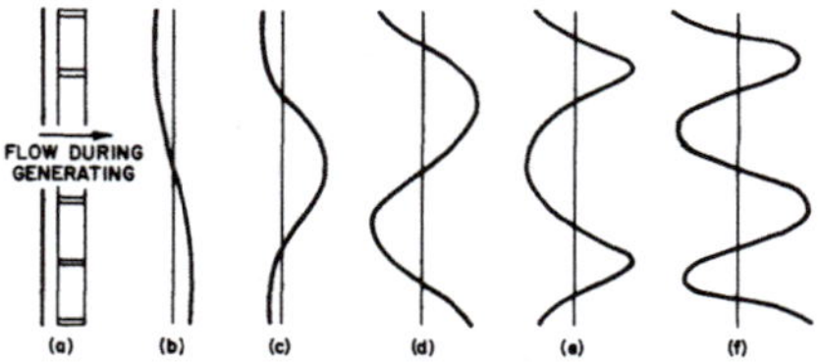

③ Mode Shapes for Horizontal Bars of Test Rack ("Plunge", Parallel to Flow)

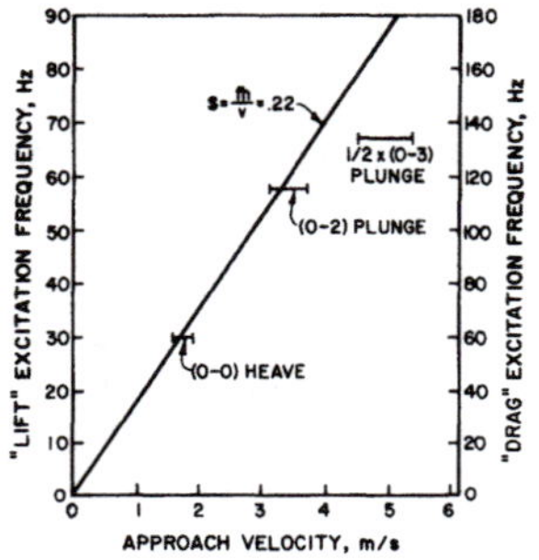

④ Summary of Test Results, Rectangular Bar Trashrack

Problem: Reduce or eliminate destructive Trashrack vibrations encountered with circular bar trashrack (see Chart C).

Geometry: See ①.

Boundary conditions: Trashrack clamped in slot at 3 points on each side; rack pressed against other racks at 2 points on top and bottom (see ①).

Flow conditions: Trashrack tested in lock filling culvert; velocity distribution within ± 5% of mean.

Operating conditions: Approach velocities from 0 to 5.5 m/s; "Lock In" observed at 30 Hz for velocities from 1.5 m/s to 1.9 m/s; at 116 Hz for velocities from 3.1 m/s to 3.7 m/s; and at 267 Hz for velocities from 4.6 m/s to 5.5 m/s. Maximum measured stress was 6.2 MN/m² (peak-to-peak).

Structural dynamics: Heave modes shown in ② [(a)(0-0) mode, f_n = 33 Hz, Q = 13; (b)(I-0) mode, f_n = 80 Hz, Q = 20]. Plunge modes shown in ③ [(a)(0-0) mode, f_n = 41 Hz, Q = 19; (b)(0-I) mode, f_n = 56 Hz, Q = 24; (c)(0-2) mode, f_n = 119 Hz, Q = 411; (d)(0-3) mode, f_n = 275 Hz, Q = 1042; (e)(0-4) mode, f_n = 502 Hz, Q = 1240; (f)(0-5) mode, f_n = 770 Hz, Q = 1500].

Vibrations: Results are summarized in ④.

Cures: Neoprene bearing pads eliminated lock-in. Baffle plate (15 cm x 30 cm) mounted in center of rack also eliminated lock-in. Recommended design includes both neoprene damping pads and baffle plates.

References (see text: 1. Crandall, S., Vigander, S., and March, P., Destructive Vibration of Trashracks due to Fluid-Structure Interaction". ASME Journal of Eng. for Industry, Nov. 1975, pp.1359-1365.

Vigander, S., and March, P., Trashrack Vibration Studies, Raccoon Mountain Pumped Storage Project. TVA Eng. Lab. Rpt. No.43-47, Oct. 1974 (Revised March 1979).

B2

HYDROTURBINE ROTOR VIBRATION

D.N. Gorelov
Chief of the Laboratory

Institute of hydrodynamics, Siberian Branch of the
USSR Academy of Sciences, Novosibirsk

V.A. Kovalenko
Research Manager

V.M. Malyshev
Deputy Chief Designer

S.N. Yavits
Research Engineer

Turbine Building Association
Leningradskij Metallitsheskij Zavod
USSR

SUMMARY

The aim of the present paper is to investigate causes of the high level vibrations of a shaft and damage of the bearings of a vertical hydroturbine designed for operation at relatively high heads. These vibrations were observed only within a narrow range of operating conditions of the hydroturbine. The vibration frequency was near the frequency of the first mode of the transverse oscillations of the rotor shaft. Data analysis has shown that the most probable cause of these vibrations is self-excited oscillation, which is due to the interaction of the turbine blades with the flow. Two hypotheses of the self-excited oscillations were investigated. From calculation of one of them, it was shown that rotor flutter is possible, and its velocity and frequency are near the natural ones.

1. INTRODUCTION

At one operating power station designed for relatively high heads, the high level vibration of a unit with bearing damage was observed. These vibrations appeared only within a narrow range of operating conditions of the units. The units are vertical and the runners are of a Kaplan-type, with four pairs of blades. Each pair of blades is fixed at a general flange. The unit arrangement and the main geometrical dimensions are presented in Sketch 1. The masses of the main rotor elements are as follows: runner = 41.5 t, unit rotor = 242 t, shaft = 18.1 t. The rotor is mounted with a segment type thrust bearing; the radial bearings are also of the segment type with oil lubrication. The gate is provided with 24 guide vanes. The units have been designed for operation at 60-80 m head of water, and a rotational speed of 250 rpm. The rated output of the unit is 68.3 Mw. Water supply to the spiral casing is through a conduit having a total length of about 260 m, and a diameter of 5.6 m. The turbine performance was published in a handbook [1].

2. OBSERVATION DATA

The operating conditions of the hydroturbines were in agreement with design conditions. The units performed quietly under all loads except in the region from 54 to 60 Mw. In this region, use of the unit is impossible because of high level vibrations of the shaft and radial bearings, especially in the turbine part of the unit. Even operation for a short time at a high vibration level causes damage of a turbine bearing.

To investigate the vibrational character of units at various operating conditions, special natural condition tests were performed. Some results of these tests are presented in Sketches 2 and 3. In Sketch 2, the levels of radial vibration of three bearings, and of the shaft near the turbine bearing at (a) stable and (b) unstable operating conditions, are presented. In Sketch 3, the locus of the shaft center near the turbine bearing (a), and the mode of the shaft transverse vibrations (b), at the same operating conditions are presented. These data show that the shaft vibration frequency coincided with the rotational frequency at the stable operating condition, and at unstable operating condition, it was equal to 7.1 to 8.3 Hertz. In this case, intense vibrations were observed near the turbine bearing.

Natural condition tests of the non-rotational rotor frequency were also performed. They showed that the natural transverse frequencies of the shaft are from 7 to 10 Hertz for the first mode, and from 15 to 16 Hertz for the second mode The frequency of the first mode of torsional oscillation of the shaft is 19.3 Hertz.

3. SOURCE OF VIBRATION

Data analysis has shown that rotor vibrations of the hydroturbine at an unstable operating condition are self-excited oscillations. Two hypotheses of rotor flutter were considered.

The first hypothesis was advanced by Thomas [2]. In accordance with this hypothesis, transverse oscillations of the rotor change the clearances between the blades and the discharge ring. Owing to this, the blades are affected by variable hydrodynamic forces which are the source of the shaft vibration. From a preliminary analysis, it was shown that such self-excited oscillations of the hydroturbine rotor are, in principle, possible. However, additional experimental data are necessary to clarify the actual conditions for onset of these oscillations.

The second hypothesis involved self-excited oscillations of the turbine rotor due to transverse oscillations of the shaft in two mutually perpendicular planes, at the same frequency in both planes. In this case, the unsteady hydrodynamic forces acting on the blades depend on the deflections of the shaft and the angles of the shaft slope at the point B1. (See Sketch 3). Theoretical analysis of such self-excited oscillations of the turbine rotor was carried out in another study [3]. Two parameters were found to have the strongest influence on the instability of oscillations: angle of shaft slope, which is due to its deformation; and the stagger angle of blades in a runner. In particular, the decrease of angle of the shaft slope at the point B1 may lead to elimination of rotor flutter.

4. CURES

From these two hypotheses, it follows that the self-excited oscillations can be eliminated by increasing the flexural stiffness of the shaft. It may be achieved, for example, by mounting the additional bearing behind the runner or between the turbine bearing and the runner, and also by moving the turbine bearing to the runner.

5. REFERENCES

1. Hydroturbine Designer Handbook, Mashgiz, L, 78, (1971).

2. Thomas, H.J. "Instabile Eigenschwingungen Turbinenlaufern angefacht durch die Spaltstromungen in Stopfbuchsen und Bescgaufelung", Bull. de IAIM, Nr. 11/12, 1039-1064, (1958).

3. Gorelov, D.N. "Theoretical investigation of flutter of runner of axial hydroturbine", Energomashinostroenie, No. 12, 16-19, (1977).

HYDROTURBINE ROTOR VIBRATION

Problem: Intense rotor vibration with damage of bearings within narrow range of operating regimes of the hydroturbine.
Geometry: See Sketch 1.
Boundary conditions: Flow supplied to the runner through spiral case by a conduit (260 m leng., 5.6 m dia.). Gate provided with 24 blades.
Flow conditions: Cavitation is practically absent.
Operating conditions: Generator rated output is 68.3 Mw, rotational speed is 250 rpm. Intense rotor vibration appears at operating conditions from 54 to 60 Mw.
Structural dynamics: Mass of the runner = 41.5 t, mass of shaft = 18.1 t, mass of the generator rotor = 242 t.
Vibration: At operating conditions, frequency of shaft vibration coincided with rotational frequency, and at the unstable operating conditions it was 1.8 times as high as a rotational frequency. Maximum vibration amplitude observed near the lower bearing B1; minimum vibration amplitude near the upper bearing B3 (Sketch 2). Vibration character as function of generator output shown in Sketches 2 and 3.
Cures: Intense vibration probably caused by self-excited oscillation of the rotor, which is due to the interaction of the turbine blades with flow. A theoretical analysis has shown that this vibration may be eliminated by increasing the flexural stiffness near the turbine bearing. It may be achieved, for example, by mounting the additional bearing behind the runner or between the turbine bearing and the runner.
References: D.N. Gorelov. "Theoretical investigation of flutter of runner of axial hydroturbine", Energomashinostroenie, No. 12, 1977.

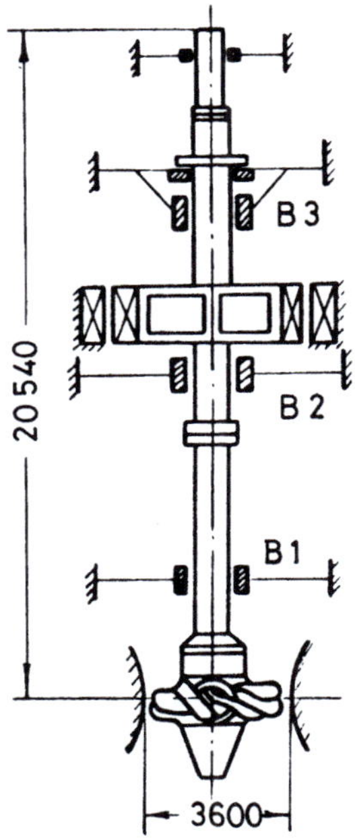

(1) SKETCH OF THE UNIT. DIMENSIONS IN mm

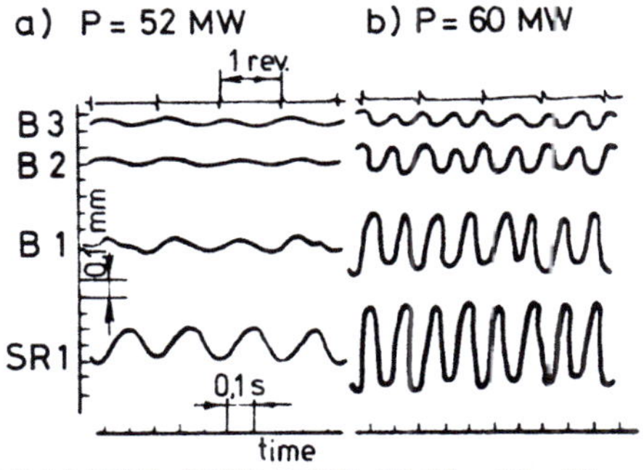

(2) RADIAL VIBRATIONS OF B1, B2, B3 BEARINGS AND SHAFT RUNOUT (SRI) AT B1 BEARING UNDER STABLE (a) AND UNSTABLE (b) OPERATING CONDITIONS

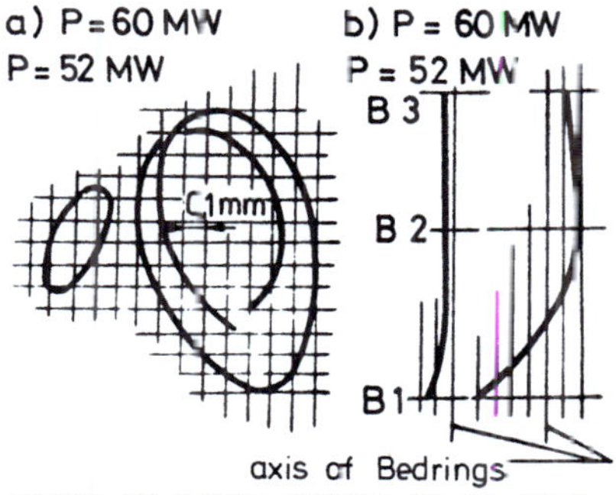

(3) LOCUS OF SHAFT CENTER AT B1 BEARING (a) AND MODE OF SHAFT TRANSVERSE VIBRATIONS (b) UNDER STABLE (52MW) AND UNSTABLE (60 MW) OPERATING CONDITIONS

B3 VIBRATION PHENOMENA ON HYDRAULIC AXIAL TURBINES

Otto Eichler

Research Engineer, Hydro-Turbine Division

J. M. VOITH GMBH, Heidenheim, Federal Republic of Germany

SUMMARY

In case of heavy swirl in the flow passage between the distributor and runner of hydraulic axial turbines, separation zones combined with vortex formation may develop due to small guide vane openings. This may result in machine vibrations, which is described by the example of vibrations occurring in a bulb turbine with fixed guide vanes and in a Kaplan turbine under special operating conditions. Measures to be taken towards successful elimination of the vibration phenomena are indicated.

1. INTRODUCTION

Because of its importance for the vibration behaviour of the machines, the swirl flow occurring in hydraulic turbines, particularly in radial-flow machines, (downstream of the runner in the draft tube and accompanied by the formation of rotating free vortices) has been investigated extensively. On the other hand, separation zones resulting from swirl flow and developing upstream of the runner may occur in axial-flow machines (Kaplan turbines, bulb turbines) under particular operating conditions. As they do not normally affect or jeopardize machine operation, not so much has been reported on this subject. Such separation zones develop due to heavy swirl in the so-called vaneless space between distributor and runner, proceeding axisymmetrically from the inner contour of the machine (headcover, water guide shield, hub). Depending on the magnitude of the swirl, they extend up to a well-defined radius in the flow passage.

Axial-turbine vibration phenomena related to these separation zones will be treated in the following, using examples taken from plants. However, it is advisable to first make some remarks on such separation zones in swirl flows.

2. SWIRL FLOWS

In the event of *swirl flows occurring in pipelines* and at sufficiently high peripheral flow velocities, zones of separating or recirculatory flow proceeding axisymmetrically from the axis may develop. Their dimensions – i.e. the radial extent of this deadwater – depend on the ratio of the axial velocity c_{ax} to the peripheral velocity c_u for the respective radius. In this connection, the ratio of the discharge Q $[m^3/s^{-1}]$, ($\sim c_{ax}$) to the product formed from circulation $\Gamma [m^2/s^{-1}]$ ($\sim c_u \cdot r$) and the local radius r [m] may be taken as the characteristic hydraulic quantity:

These swirl flows with deadwater core have been investigated theoretically and experimentally, Ref. [2], [3], and [4]. The form of the swirl flow, particularly the boundary of the separation zone, can be determined according to the theory of the kinematic equilibrium, as described in [3] and [4] (Fig. 1). According to

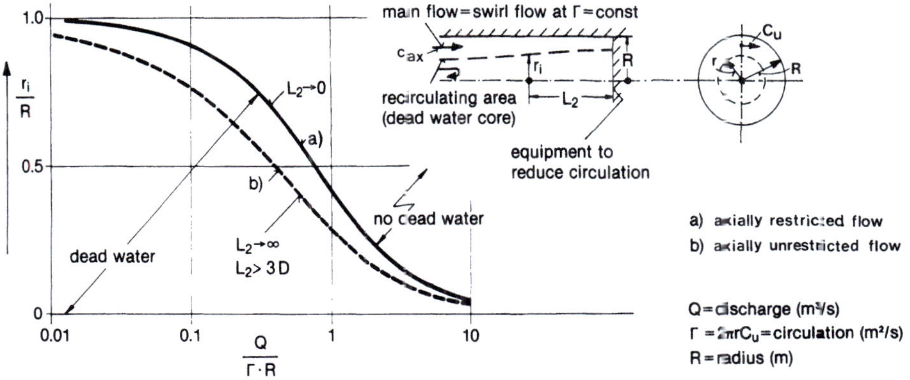

FIG. 1 EXTENT OF THE DEADWATER CORE OF A SWIRL FLOW IN A CYLINDRICAL PIPE

the calculation, at friction-free swirl flow, a kink develops in the velocity curve between separation zone and sound flow and, thus, the "deadwater core" and the "sound flow" (i.e. the main flow) are sharply separated by a free boundary (Fig. 2).

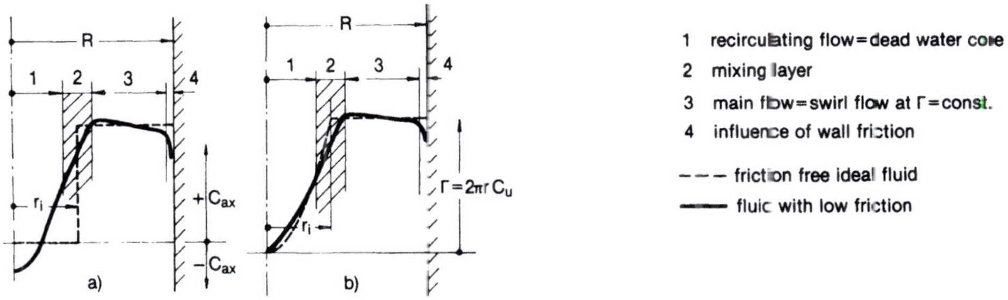

FIG. 2 CHARACTERISTICS OF AXIAL FLOW VELOCITY C_{ax} -(a)- AND OF CIRCULATION Γ-(b)- OF A SWIRL FLOW IN A CYLINDRICAL PIPE

However, in the case of the actual frictional flow, a mixing layer evolves, in which v o r t i c e s develop.

For conditions prevailing in axial-flow turbines, at small guide vane opening, such a high circulation can be generated that a deadwater zone develops, provided the radius of the outer boundary of the deadwater (mixing layer) is greater than the respective radius of the given shape of hub.

3. VIBRATION PHENOMENA ON A BULB TURBINE WITH FIXED GUIDE VANES (Chart 1)

3.1 Brief description of the plant

On the occasion of the modernization of a channel power station approximately 20 years ago, 5 Kaplan bulb turbines (head H = 9 to 10 m; maximum turbine output $P_T \approx 2.1$ MW; runner diameter D = 2.41 m) instead of Francis twin turbines had been installed in the existing powerhouse (sketch 1), [1] . In this connection, the drive shaft is brought out through the S-shaped draft tube to the external generator. The turbine is designed as a turbine with fixed guide vanes: these fixed guide vanes support the bulb containing the upstream bearing. Also accommodated in this bulb is the oil supply to the runner adjustment of the movable runner blades (sketch 2). From the curb ring (point B in sketch 1) to the first draft elbow portion, the machine is exposed, i.e. not embedded in concrete.

Sketch 3 shows the typical efficiency curve of this design with fixed guide vanes and movable runner blades in comparison to the double-regulated Kaplan turbines.

The shape of the intake is unusual for a bulb turbine; it leads from the headwater race via a fall shaft to the guide vane assembly, the outer shell of which is conical. This shape was determined in a special model test. The bulb turbine itself was tested in a turbine model facility.

3.2 Observed vibration phenomena

At the time of commissioning the first machine, vibrations occurred over wide operating ranges, particularly on the external housing part of the bulb turbine. During detailed measurements, the following was found:

3.2.1 Mechanical vibrations:

Vibrations occurred on the external steel plates (sketch 2), particularly on the curb ring (point B in sketch 1) with a range of fluctuations (peak-to-peak value) up to 100 µm at a frequency of approximately 140 Hz and 176 Hz, depending on the machine load. The measurements were conducted with the help of an inductive pick-up. On the concrete bases (point A in sketch 1) of the throat ring support, a fluctuation range of 20 µm, for a short time a range of 60 µm, was measured (fig. 3). The form of vibration is partially periodic and partially irregular.

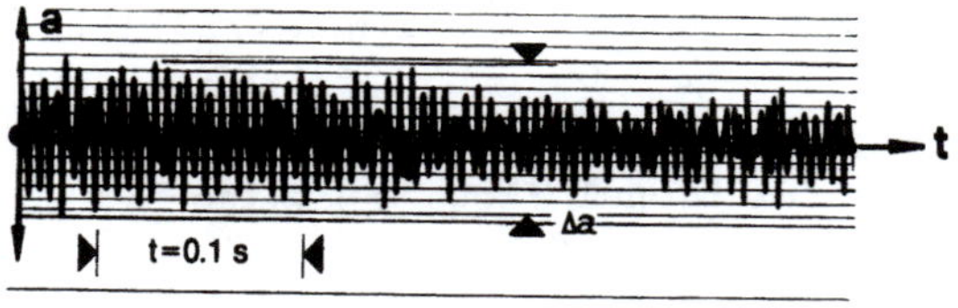

FIG. 3 MECHANICAL VIBRATION ON CONCRETE BASE A

3.2.2 Pressure fluctuations measured mainly on the outer shell of the guide vane assembly yielded the same characteristic vibration pattern as in the case of mechanical vibrations that occurred on the steel plates: part of them uniformly periodic and irregularly periodic with beats (fig. 4 and sketch 4). At maximum output, local fluctuations at 176 Hz with a fluctuation range of $3 \cdot 10^4$ Nm^{-2} max., which is approximately 30 % of the head, were measured.

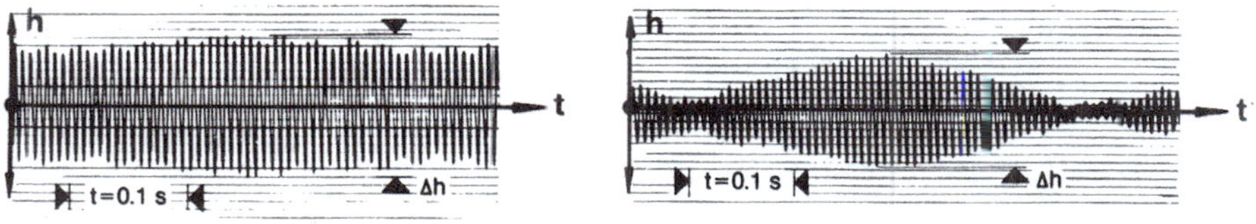

FIG. 4 PRESSURE FLUCTUATION AT THE OUTER SHELL OF GUIDE VANE ASSEMBLY

3.2.3 In the immediate vicinity of the machine, a sound pressure level of 96 dBA, up to a maximum of 103 dBA, was measured. In this connection, the noise was subjectively perceived as a "drone", which can also be explained by the vibration pattern with beats, ascertained during the pressure fluctuation measurements (sketch 4).

In general, the vibrations, except the vibrations occurring on the concrete bases, were not considered detrimental or dangerous.

3.3 Further observations and considerations - preliminary measures

No resonances associated with hydraulic or mechanical excitations by the turbine were observed: possible resonances could occur e.g. at the rotational frequency $f_n = 3.\overline{3}$ Hz, the rotational frequency times number of runner blades $f_n \cdot z_2 = 3.\overline{3} \cdot 5 = 16.\overline{6}$ Hz or the rotational frequency times number of runner blades times number of guide vanes $f_n \cdot z_2 \cdot z_o = 3.\overline{3} \cdot 5 \cdot 12 = 200$ Hz, or at frequencies associated with Karman vortices, shed from the trailing edge of the guide vane profiles. Furthermore, in the manufacturer's workshop natural frequencies in the range of 140 - 180 Hz and 220 - 240 Hz were measured on a guide vane assembly.

Initially, various measures taken on the machine could not reduce or markedly change the vibration phenomena.

- By air admission to various points, e.g. to the profile nose of the guide vanes via vent pipes, a certain reduction of the fluctuation range was achieved; pronounced vibrations of 33 Hz ($= 2 \cdot f_n \cdot z_2$) occurred in the draft tube. Without air admission, these vibrations could not be observed. On the whole, attempts to use air admission in order to exert a lasting influence on the noise level were unsuccessful.

- In order to clarify whether modifications to the supply section could yield any change in the vibration behaviour, the headwater gates were lowered down to 1 m immersion depth (both gate parts simultaneously or the left and the right half separately). On the whole, no consistent effect resulted: sometimes a slight increase, at other times slight reduction of the pressure fluctuations.

On the basis of all these measurements and observations, it was assumed that the vibrations resulted from flow disturbances ahead of, or in, the guide apparatus. After counterchecking, the intake shape, including the fall shaft, was not regarded as the cause of the vibrations. Finally, considerations were concentrated exclusively on the guide vanes.

3.4 Modifications to the guide vanes: remedy

3.4.1 Hydraulic and design conditions prevailing on the guide vanes

The guide vane assembly is a welded structure. The guide vanes are made of steel plates bent circular-cylindrical (see sketch 5). They are profiled only at the entrance and exit.

Due to the hydraulic layout of the turbine for a relatively low specific speed, the guide vanes, when viewed in the tangential direction, were set at relatively small exit vane opening and small exit angles (sketch 2). Therefore, it was natural to examine the swirl conditions prevailing at the guide vanes near the hub. A rough calculation showed that it was quite possible that, due to the strong swirl flow, produced by the vane geometry, a deadwater zone (so-called hub deadwater), may develop in the range of the hub (see section 2). A rough indication for this was the swirl flow condition inside straight pipes (fig. 1). This hub deadwater was assumed to influence the vibration behaviour.

When checking the shape of the manufactured guide vanes, it was also observed that, within the manufacturing tolerance, the exit vane opening and the exit angle of the guide vanes tended towards smaller values. This was also obvious from the measured characteristics of the turbine performance (sketch 3).

3.4.2 Modification to the guide vanes and its effect

Due to the reasons advanced in 3.4.1, it was necessary to design for: a) avoidance of hub deadwater and b) a desired slight correction of the performance characteristic towards a higher discharge. The result was a guide vane correction giving a wider exit angle and a greater vane opening at the exit particularly on the profiles near the hub. This was accomplished by cutting back the guide vanes at the profile exit (sketch 5).

This measure was very successful:

1) The vibrations were reduced quite considerably, i.e. to approximately 1/4 of the previous fluctuation ranges. This

represents a normal extent of vibration, in view of the large area of the exposed steel plate structure. As regards the steel plate vibrations, a frequency of approximately 190 - 250 Hz, presumably corresponding to steel plate natural frequencies, was measured after the corrections. At the concrete base, the maximum fluctuation range of 60 µm was reduced to 15 µm and the mean fluctuation range of 20 µm to 4 µm.

The noise level was reduced considerably: from 100 dBA to approximately 80 dBA, which is a reduction of the sound pressure to approximately 10 % of the previous value. The noise was no longer considered disturbing.

2) As a side effect, the desired performance characteristics were completely attained (sketch 3).

3.4.3 Interpretation of the vibration phenomena and of the effect of the modification to the guide vanes

As the cutting back of the guide vanes by a small amount has already resulted in a sufficient reduction of the observed vibrations, it is presumed that the existing hub deadwater was relatively small in its radial extent and not very stable (sketches 2 and 6). Presumably, this is why the outer boundary of the deadwater - the mixing layer between main flow and hub dead-water - was not very stable either.

Furthermore, as consequence of the quite small axial-flow velocities occurring in the deadwater zone, the flow towards the inner runner blade profiles near the hub was very unfavourable: so-called "negative" flow towards the profile involving separation zones at the pressure side of the profile nose (sketch 5). Most likely, these separation zones also fluctuated. Vortices originating from the mixing layer of the separation zone may possibly have added to the excitation of the vibrations as well.

In summary the measures taken met the objectives: the cutting back of the guide vanes eliminated, or at least appreciably reduced an existing hub deadwater so that no disturbing vibration phenomena occurred.

4. HUB DEADWATER IN KAPLAN TURBINES (Chart 2)

The vibration problem, which was dealt with in section 3 and which resulted from the hub deadwater at a bulb turbine with fixed guide vanes is a special case.

Under well defined operating conditions prevailing in double-regulated Kaplan turbines at small wicket gate openings, bangs and vibrations are due to vortices of the mixing layer of a dead-water zone above the runner, resulting from the high circulation at small wicket gate opening (see section 2).

4.1 Operation along propeller curves

Ref. [5] reports on such vortices, which occur during operation at a well defined wicket gate opening along the propeller curve during off-cam operation, i.e. during operation with non-optimal runner/wicket gate relationship (sketch 2). In the core of such vortices, the evaporation pressure of the water is frequently reached; thus, these vortices become visible as free vortices filled with cavitating bubbles (sketch 3).

4.2 On-cam operation at heavy part load

Such vortices from separation zones may also occur at heavy part load and correspondingly small wicket gate opening during on-cam operation at optimal runner/wicket gate relationship. On a large-size Kaplan turbine with 75 MW turbine maximum output, acoustically strong, individual bangs were perceived during part-load operations at 25 MW (runner blade opening $\varphi/\varphi_{opt} = 0.2$). The intensity of these bangs exceeded - subjectively - the noise at the cavitation limit or overload (sketches 1 and 2).

By a slight increase of the wicket gate opening $\Delta\gamma$ from $\Delta\gamma/\Delta\gamma_{opt} = 0.53$ to $\Delta\gamma/\Delta\gamma_{opt} = 0.59$, in which case a slight reduction of efficiency to 0.2 % was accepted, the bangs could be eliminated. Part of the cavitation-filled free vortices, of the same type as those shown in sketch 3, pass between the runner blades; part of them impinge the runner blade profile, which then produces irregular hard bangs. However, these vortices do not pose any problem for proper operation of the machine, not even with respect to cavitation pitting.

4.3 No-load operation

In the case of a bulb turbine operating at no-load, an intensive noise development could be observed (setting: runner blade opening $\varphi/\Delta\varphi \approx 0.1$, where $\Delta\varphi$ = overall adjusting angle, and wicket gate opening $\Delta\gamma/\Delta\gamma_{max} = 0.05$). By increasing the wicket gate opening to $\Delta\gamma/\Delta\gamma_{max} = 0.1$, the machine immediately became very quiet. However, to accomplish this, the runner blade opening had to be increased to approximately $\varphi/\Delta\varphi \approx 0.15$ in order to maintain the synchronous no-load speed, which involved a slight increase of the no-load discharge.

The measures taken in the above-mentioned cases 4.2 and 4.3 can be applied generally: In the operating range with necessary small wicket gate openings, a possible hub deadwater with acoustically annoying vortex formation can be eliminated, or at least reduced appreciably, by a slight increase of the wicket gate opening.

5. SYMBOLS

Q	Discharge	$[m^3s^{-1}]$
Γ	Circulation = $2 \cdot \pi \cdot r \cdot c_u$	$[m^2s^{-1}]$
r	Radius	$[m]$
R	Radius of pipe	$[m]$
φ	Runner blade opening	$[^o]$
$\Delta\gamma$	Wicket gate opening	$[^o]$
c_{ax}	Axial-flow velocity	$[ms^{-1}]$
c_u	Peripheral-flow velocity	$[ms^{-1}]$
w	Relative flow velocity	$[ms^{-1}]$
u	Peripheral velocity of runner	$[ms^{-1}]$
f_n	Rotational frequency of turbine	$[s^{-1}]$
z_2	Number of runner blades	
z_o	Number of guide vanes	

6. REFERENCES

1. Schiechtl,H., de Lorenzo,J., Hillenbrand,G., "Sanierung und Modernisierung der Wasserkraftanlage Gersthofen am Lech". Mitteilung Nr. 33 der Versuchsanstalt für Wasserbau, EHT Zürich.

2. Meldau,E., "Drehströmungen im Drehhohlraum". Dissertation, Hannover 1935.

3. Strscheletzky,M., "Gleichgewichtsformen der rotationssymmetrischen Strömungen mit konstantem Drall in geraden, zylindrischen Rotationshohlräumen". Voith Forschung und Konstruktion, Vol.5 (1959), Repr.No.1502.

4. Strscheletzky,M., "Gleichgewichtsformen einer aus zwei Teilströmungen bestehenden inkompressiblen Drallströmung in einem geraden, langen zylindrischen Rotationshohlraum". Voith Forschung und Konstruktion, Vol.9 (1962), Repr.No.1633 and Vol.12 (1964), Repr.No.1684.

5. Thuß,W., "Betriebszustände von Kaplanturbinen". Voith Forschung und Konstruktion, Vol.4 (1958), Repr.No.1463

VIBRATION ON A BULB TURBINE WITH FIXED GUIDE VANES

Problem: Vibrations of the external housing and acoustic noise.

Geometry: Sketches (1) and (2); width of inlet and outlet = 7 m.

Boundary conditions: parts of turbine are exposed from B to C (1), i.e. not embedded in concrete.

Flow cond.: see (1), (2) and (5); discharge Q_{rat} = 25 m^3/s, head H ≈ 10 m. Inflow from head race channel (1) via fallshaft to guide vanes. Performance: (3).

Structural dynamics: Pressure fluctuations and mechanical vibrations. No resonance with hydraulic or mech.excitation.

Vibration: Vibr.over the whole operating range. Pressure fluctuation Δh_{max} ≈ 30 % of turbine head; mech. vibration: fluctuation range Δa_{max} = 100 µm on steel plates and 60 µm on concrete base (A in (1)). Frequencies 140 and 176 Hz. Generally: uniformly and irregularly periodic vibrations, partly surging, e.g. (4).

Cure: Shortening of guide vanes at exit (5); reduction of circulation: sufficient reduction of vibration; also favourable shifting of performance.

Interpretation: Because of high circulation downstream of guide vanes a separation zone formed.
From this, fluctuations were induced (e.g. (4)). By reduction of circulation (6): elimination/reduction of separation zone.

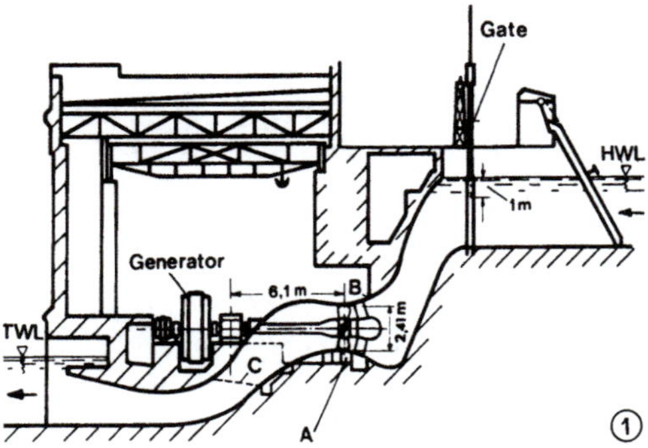

(1)

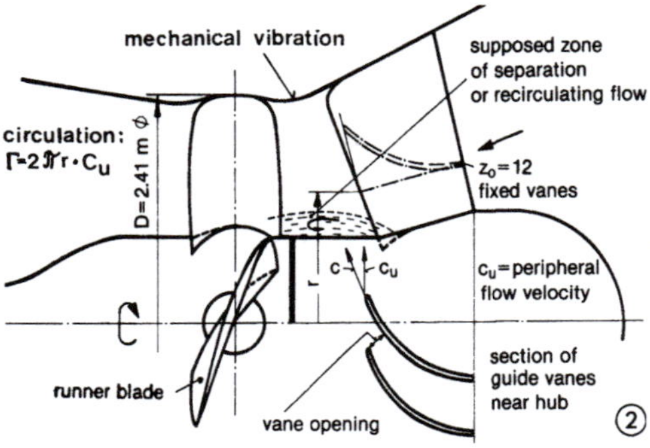

(2)

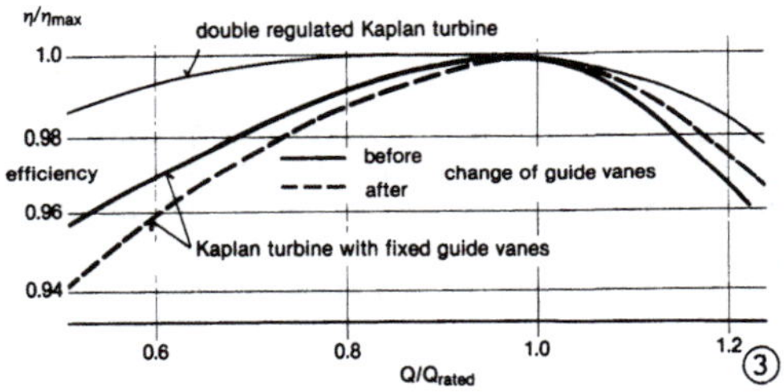

(3)

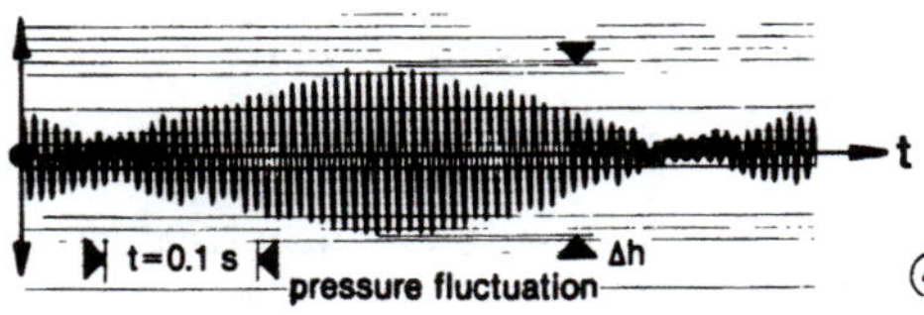

(4)

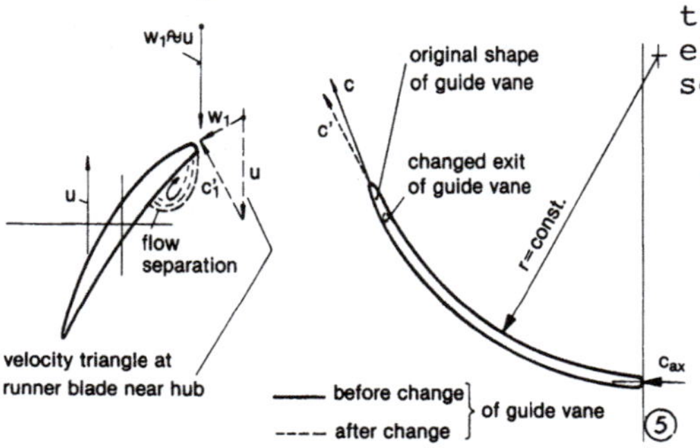

(5)

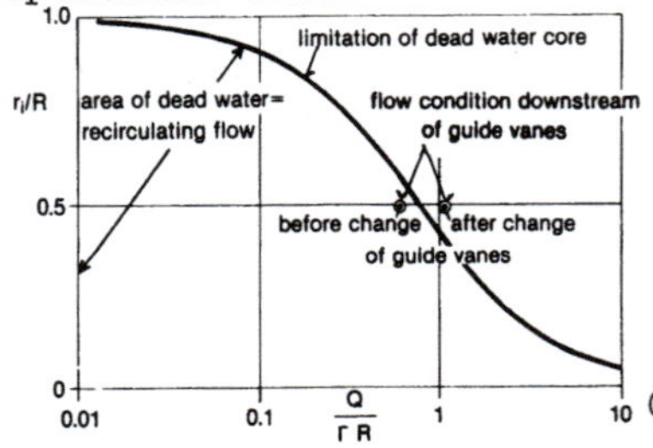

(6)

SWIRL FLOW UPSTREAM OF THE RUNNER OF KAPLAN TURBINES

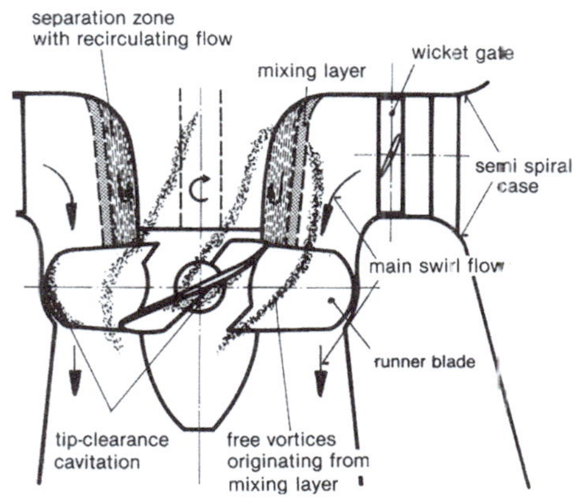

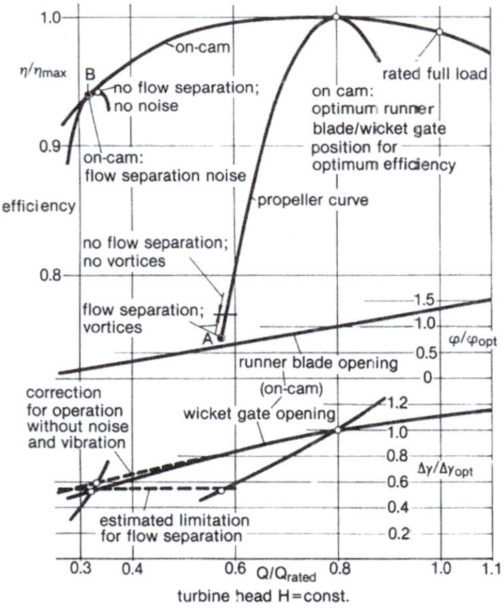

Problem: Vibration, esp. acoustic noise occurring at off-cam operation (case A, sketch ②) or at heavy part load during on-cam operation (case B).

Geometry: Double-regulated Kaplan turbine. Passage way see ①.

Boundary condition: Swirl flow as a function of wicket gate position: small angle $\Delta\gamma$ of wicket gate → high circulation.

Flow cond.: Normal inflow conditions: swirl flow, coming from spiral case via cascade of stay vanes and wicket gates.

Operating cond.: Vibrations occurring at a certain small wicket gate position and below.

Structural dynamics: At operating points A and B can arise individual bangs (blows), producing a hard metallic sound and single beats on the structure.

Vibrations: No measurements of vibrations were made. At operating point B at heavy part load the noise exceeds subjectively the noise at full load.

Cure: In case B the bangs could be eliminated by slight increase of the wicket gate opening ②.

Interpretation: The swirl flow in the space between wicket gates and runner can at a certain high circulation originate a separation zone near the hub. In the mixing layer are generated single free vortices ① and ③, which can impinge the runner blade profile; this produces the noise. For information see section 2 and 4 of paper.

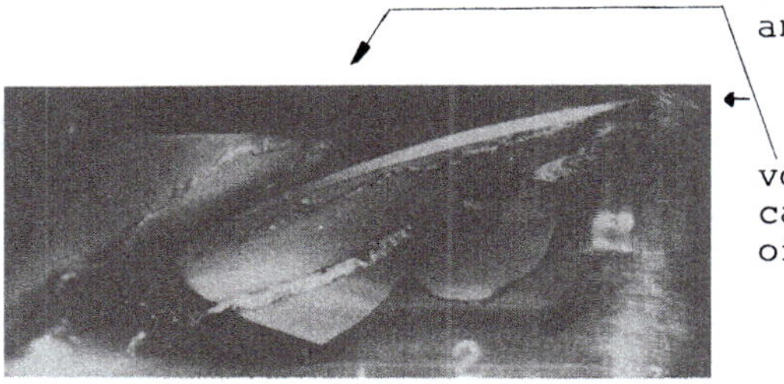

vortices, visible by cavitation in the core of vortex

B4 DYNAMIC STRESSES ON STAY VANES OF VERTICAL-SHAFT LARGE-SIZE FRANCIS TURBINES

by

Dipl.-Ing. Helmut Kirchner
Head of the Turbine Design Department
J.M. Voith GMBH, Heidenheim, Federal Republic of Germany

SUMMARY

This is a report on fatigue phenomena occurring in stay vanes of Francis turbines. Part of the fatigue results from vibration phenomena, although vibrations do not play the dominant role. Investigation into and remedy of the damage are reported.

1. INTRODUCTION

Hydraulic turbine engineering is characterized by a trend towards increasingly higher outputs of the machines. Particularly in the case of Francis turbines, but also for Kaplan turbines operating under low heads, this leads to slender structures. It is well known that many fractures have already occurred on the stay vanes of such machines. We had to deal with this phenomenon in a plant with 8 large-size machines. Each machine developes 200 MW at a speed of 100 rpm, under a head of 65 m.

It may be added that, in the case of the present design, for hydraulic reasons the last three stay vanes ahead of the nose are manufactured shorter than others. In the plan view, the end stay vane is the shortest (Sketch 1). All stay vanes are welded into the upper and lower part of the box-type stay ring (see Sketch 2).

2. VIBRATION OF VANES

Damage was observed on stay vanes after a short period of operation, beginning after about 2,000 hours. Chiefly affected was the end stay vane ahead of the nose. On some machines, cracking occurred only at this stay vane. However, in the time that followed, damage was also observed to be irregularly distributed over all stay vanes. The jeopardized area was always the transition from the stay vanes to the stay ring. The incipient cracks extended not only from the trailing edge, but also from the leading edge. In addition, cracks were found in the middle of the stay vane, due to flame-cutting. The machines exhibited different susceptibility to cracking, one of the 8 machines being completely spared from it.

On the basis of the preliminary assessment, special attention was devoted to establishment of the extent of the dynamic stresses on the stay vanes. Therefore, the dynamic stresses occurring in the stay vanes were measured on one of the machines during essential operating conditions, including shutdown. The strain gauge measurements were conducted in measuring planes a-a and b-b (Sketch 2). During measurements, special attention was devoted to acquisition of the respective operating conditions. In contrast to expectations, only very low dynamic stresses were measured in the stay vanes. The stresses measured were smaller than $\pm$ 4 N/mm^2. It was only upon overload that an increase of these stresses to about $\pm$ 8 N/mm^2 was evidenced (Sketch 3).

Bending vibrations of the stay vanes occur at their natural frequency. The natural frequency was ascertained as being 83 Hz, the covibrating water mass effecting a reduction coefficient of 0.7. The observed frequency was 80 - 84 Hz.

Dependence of the vibration frequency on the output of the machine, i.e. water velocity, could not be observed, which is attributed to low damping. At a maximum discharge of $Q = 350\ m^3/s$, the velocity between the stay vanes was $V = 11\ m/s$.

Possible forms of vibrations with higher natural frequency, as would be expected in the event of stay vanes being subjected to torsional stresses, were not observed.

Excitation by the Karman street-type of trailing vortices at the trailing edge cannot be ruled out. On the basis of the water velocity, and of the stay vane dimensions, the possibility of an excitation can be evidenced in the formula for the Karman street-type of trailing vortices. It is pointed out that in application of this formula

$$f_{Karman} = S \frac{c}{d}$$

no exact values can be inserted for the effective stay vane thickness and for the Strouhal number.

To be on the safe side, we had already introduced various modifications to the trailing edges on some stay vanes during the determination of the dynamic stresses. These modifications yielded further reductions of the low level of the alternating stresses in the full-load and overload ranges. The modification with a tip welded onto the trailing edge, as shown in Sketch 4, was considered optimal. Other modifications, involving attachment of flow tripping arrangements to the 15° inclined face of the trailing edge, also exerted a favorable effect.

In order to obtain more information on the flow conditions, the flow past the stay vanes was observed in a model turbine. These tests confirmed the expectation that the flow impinging on the very last stay vane does so at a very small angle of attack. This suggests that the last stay vanes are most susceptible to vibrations excited by flow past the stay vanes.

Because of the low level of dynamic stresses, the vibrations induced by the flow cannot be regarded as the sole cause of the damage. Therefore, further investigations had to be conducted. The static stresses were measured, and were found to be at, or partly below, the design values, due to the supporting action of the concrete. Furthermore, it was found that high internal stresses were present in the structure, some of which approached the yeild point.

In summary, careful assessment of various results of investigations on these stay vanes yielded the following explanation for the cause of the damage: the static and dynamic operating stresses are not considered to be the sole cause of damage; nor are the minor welding errors, from which the individual cracks originate; only simultaneous inclusion of the very unequal, partly high, inherent stresses yields an explanation for irregularity of the observed damage.

3. CURE

In light of these insights, repair was carried out on the basis of a carefully compiled welding schedule to reduce, as much as possible, inherent stresses in the stay vanes. Modifications to the basic design were kept minimal. The trailing edges were modified according to Sketch 4. Furthermore, at the time of welding, care was exercised in welding the transition radii Since the repair, the machines have been in operation over an extended period, i.e. up to approximately 15,000 hours. No appreciable damage has occurred on the repaired stay vanes.

STAY VANES OF FRANCIS TURBINES WITH VERTICAL SHAFT

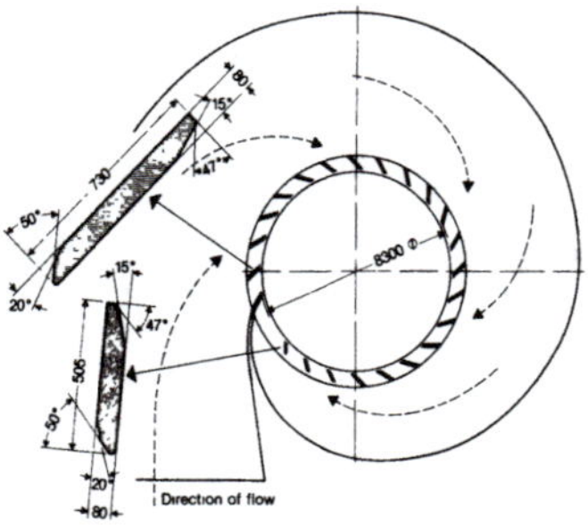

(1)

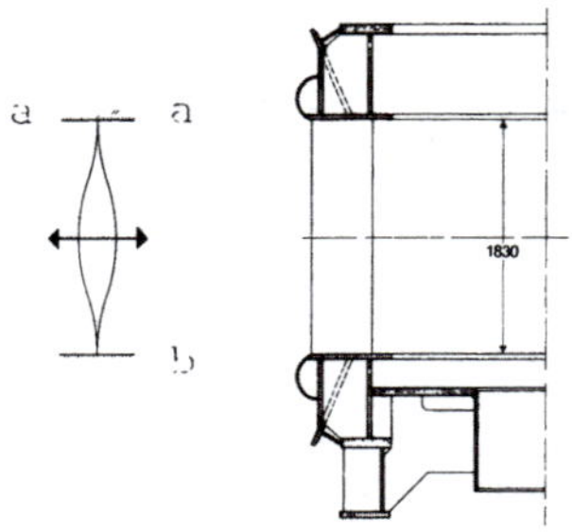

(2)

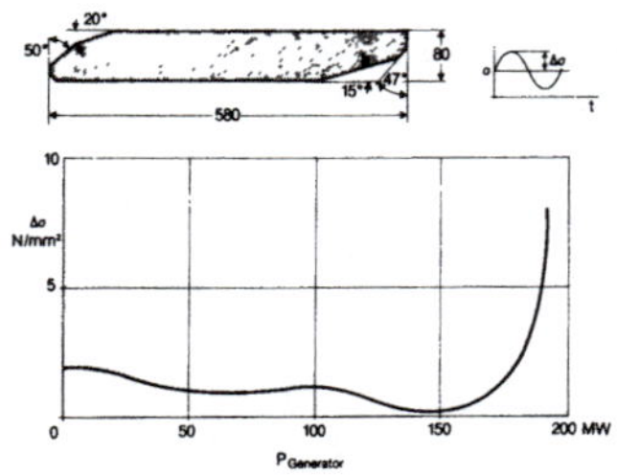

(3)

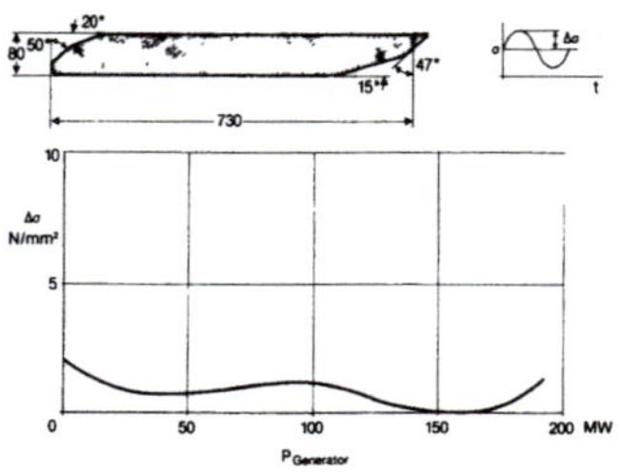

(4)

Problem: Dynamic vibration of stay vanes. Damage due to fatigue.

Geometry: see Sketch 1.

Boundary conditions: Stay vanes fixed to upper and lower stay ring parts with box-type section. See Sketch 2.

Flow conditions: Water flow is not exactly parallel to the stay vanes. Water is deflected between the wicket gates. See Sketch 1.

Operating conditions: Different discharge "Q" for part load and full load. Maximum $Q = 350\ m^3/s$ corresponds to $V = 11\ m/s$.

Structural dynamics: Frequency in water = 83 Hz for pure bending. See Sketch 2.

Vibration: Frequency = 80 to 85 Hz. Bending stresses measured in plan a-a and b-b. Increasing, but still small amplitudes for maximum load. See Sketch 3.

Cures: Change of trailing edge gives profile which results in smaller amplitudes. Apparently excitation by Karman vortex street was attenuated. See Sketch 4.

Remark: Besides the change of the trailing edge, other means were introduced (see main text): minimizing of internal stress during welding, grinding of surface of welds.

B5 FLOW-INDUCED CRACKS ON STAY VANES

Peter Ulith
Diplom-Ingenieur

J. N. Voith GmBH, Heidenheim
Federal Repuplic of Germany

SUMMARY

With this contribution to the problem of stay vane vibration, an attempt is made to clarify the cause of damage to stay vanes on the basis of model tests. The influences of angle of attack and of hydraulic load are addressed.

1. UNSTEADY LOADING OF STAY VANES

The occurrence of stay vane fracture in full spirals of Francis turbines, discussed in a related paper by Mr. Kirchner [1], has stimulated an investigation of the unsteady flow patterns around stay vanes. In the case cited by Mr. Kirchner, field measurements have shown that stay vanes vibrate at their natural frequency of 80 Hz. This was observed not only for stay vanes at the end of the spiral, which showed indications of fracture, but also for other stay vanes, such as those near the intake.

The possibility that stay vane vibrations at a frequency of 80 Hz may have been generated by a Karman vortex street follows from consideration of the dimensionless ratio for the Strouhal number, S.

$$f_{Karman} = S \frac{c}{d_{eff}}$$

For the case of a Reynolds number range (based on the effective thickness of the plate) of 3×10^5 to 5×10^5, a Strouhal number of S = 0.35 can be employed (see Figure 2 of Chen [2]). Here, the effective thickness of the vane (or plate) is defined as one half the thickness of the actual stay vane, and the velocity c incident upon the stay vane is computed at full load of the turbine. Of course the accuracy of predicting the frequency of vibration on the basis of this simple relationship is highly sensitive to the assumed Strouhal number, effective thickness, and incident velocity. There is large uncertainty in the value of Strouhal number, as essentially all data acquired to date have been for plates having zero angle of attack. In addition, the form of the trailing edge of the stay vane is also expected to influence the dimensionless Strouhal number. Both of these effects could be correlated in the form of "effective" plate thickness, but reliable information is lacking. Moreover, in practice, the velocity c of the incident flow can vary within wide limits depending on the turbine load.

For a given spiral arrangement of stay vanes, the angle of attack of the incident flow varies with plate location within the spiral. This effect of angle of attack, which has been studied in a preliminary sense for flow past an inclined plate by Abernathy [3], varies from one quadrant to another within a typical spiral case (see Sketch 1). In the first quadrant (I), the flow approaches the first few stay vanes at a moderately large angle of attack; in the second quadrant (II) the angle of attack rises to its maximum value; and in the third (III) and fourth quadrants, the angle of attack again decreases. It is important to note that the most significant cases of stay vane fracture occurred primarily in the fourth quadrant of the spiral shown in Sketch 1. These trends are based on model tests of a turbine having a moderate specific speed. Films taken by the author clearly show that in the last (fourth) quadrant, the angle of incidence of the flow upon the vane is approximately zero.

2. RECOMMENDATIONS FOR RESEARCH

On the basis of this investigation, it is apparent that there is a need for theoretical and experimental studies of flow-induced excitation of stationary plates at various angles of attack with respect to the incident free stream; furthermore, a cascade of plates should be considered in order to clearly define the interference effects due to proximity of stay vanes.

3. REFERENCES

1. Kirchner, H., Dynamic Stresses on Stay Vanes of Vertical Shaft Large Size Francis Turbine, IAHR/IUTAM Symposium, Karlsruhe, 1979.

2. Chen, Y. N., "60 Jahre Forschung über die Kármánschen Wirbelstrassen", Schweizer Bauzeitung, H. 44/1973.

3. Abernathy, F. H., Flow over an Inclined Plate, Transactions of ASME, September 1962.

Discussion: FLOW-INDUCED CRACKS ON STAY VANES (Ulith)

F.M. Neilson, USAE Waterways Experiment Station, Vicksburg, USA.

The purpose of this discussion is to mention some of the trailing-edge modifications that have been used in practice. Flow-induced vibrations have caused cracking of buckets, blades and runners at several Corps of Engineers power plants--the cure of these has been to modify the trailing-edge geometry. This cure is comparable to the cures for the stay vanes described in the subject paper. The importance of trailing-edge geometry has been emphasized in previous studies (1,2) of flow-induced vibration of vanes; in particular Ippen, et al, (2) show that a blade with a reentrant trailing edge exhibits less vibration than other non-streamlined trailing edge geometries. Three examples of remedies for vibration of prototype buckets, blades, and runners are shown in Figures 1, 2, and 3, respectively. The thinning of the Francis turbine runner bucket, 1, was required to increase turbine vent area as well as to solve the vibration problem.

The turbine runner modification, 3, is the least satisfactory of the three examples. These units are rated at 80 MW at 45.6m net head. The wicket gate opening was limited to 92% of full open because of excessive vibrations in a free standing 39m long by 7.3m diameter penstock. Tests (1) in air, using a flange mounted piston-type vibrator indicated that the natural frequency of the unmodified runner bucket in air was 165 Hz; severe vibration amplification began at about 118 Hz and stopped at about 229 Hz. The natural frequency of the runner bucket in water was about 122 Hz. Vibration with the 45° trailing edge was not significantly less than the vibration of a square-edged blade; the blade thinning, 3, was not 100% successful in eliminating the vibration and additional studies are being made.

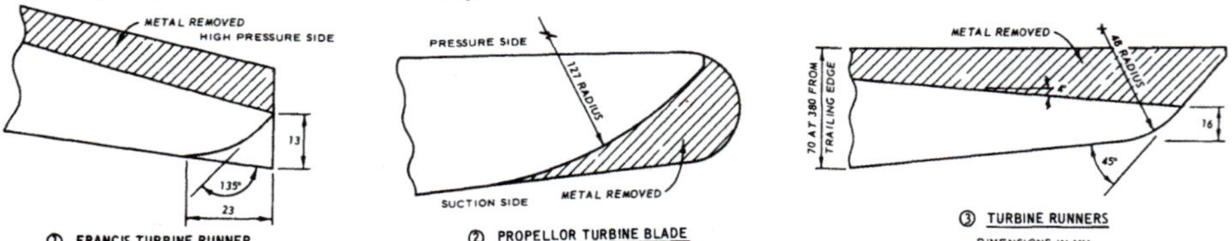

Trailing-Edge Modifications

References: 1. Heskestad, Gunnar, and P. D. Olberts, "Influence of Trailing Edge Geometry on Hydraulic Turbine Blade Vibration Resulting from Vortex Excitation, ASME Transactions, April 1960. 2. Ippen, A. T., G. H. Toebes, and P. S. Eagleson, "The Hydroelastic Behavior of Flat Plates as Influenced by Trailing-Edge Geometries," M.I.T., T.R. No. 36, April 1960.

FLOW-INDUCED CRACKS ON STAY VANES

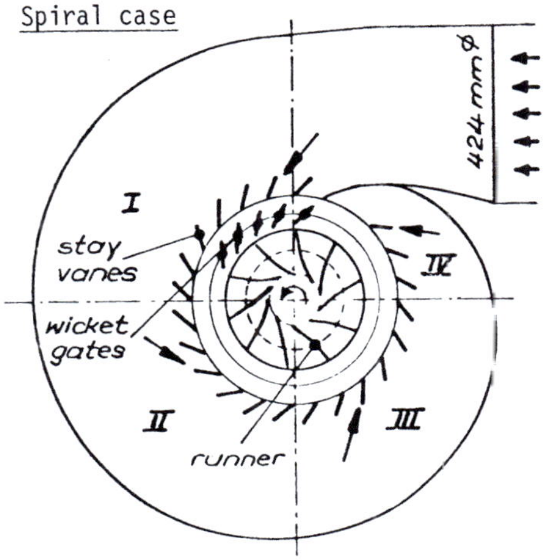

① ANGLE OF ATTACK OF INCIDENT FLOW IN THE FOUR QUADRANTS OF A SPIRAL

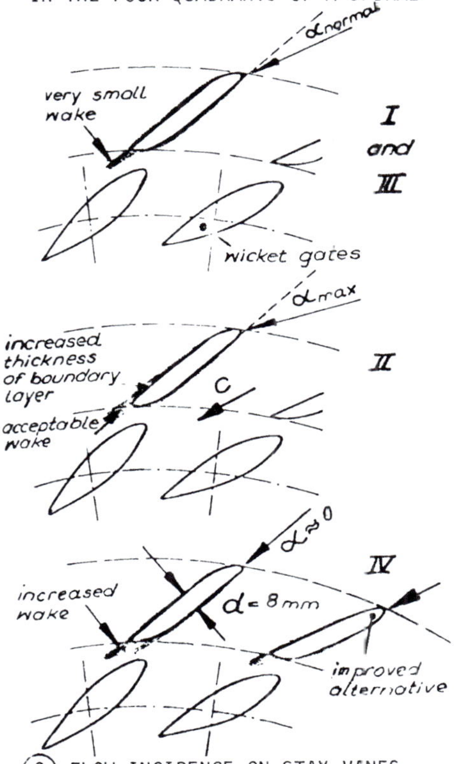

② FLOW INCIDENCE ON STAY VANES

Problem: Stay vane cracks, most often in quadrant IV of a spiral case.

Geometry: All stay vanes are identically dimensioned and positioned.

Modelling conditions:

Re = cd/ν

Prototype Re = 2.2 x 10^6

Model Re = 2.0 x 10^4

Flow condition: Angle of attack varies due to nonconstant circulation.

Operating condition: Flow rate changes between 20% and 110% of rating.

Structural dynamics: Amplitudes of stress fluctuations with magnitude of ± 2 to 8 N/m^2 measured on the prototype. No measurements on the model.

Vibrations: All stay vanes of prototype vibrating with natural frequency of about 80 Hz, which would correspond to

$$f_{Karman} \cong (0.35)\left(\frac{c}{0.5d}\right)$$

Cure: In the prototype, improvement of stay vane trailing profile. For the model, improvement of the stay vane leading edge profile as a result of flow visualization.

B6 AERODYNAMIC BLADE EXCITATION IN TURBOCOMPRESSORS

by

R.J. Jenny, U. Bolleter

SULZER Brothers Ltd., Winterthur, Switzerland

SUMMARY

Very reliable and sensitive experimental techniques for measuring and analyzing blade vibration are now widely used. This paper demonstrates how classical dynamic strain gauge measurements on stationary and rotating blades may be used to identify different excitation mechanisms and to extract some general excitation data.

The test blade is dynamically calibrated by experimental modal analysis making use of a viscoelastic damping model and a generalized force. The resonance frequencies and the generalized force of the test blade are experimentally determined by impact excitation. The generalized force is related to an assumed distribution of instationary aerodynamic loads.

Experimental results for blade excitations are presented for two types of flow. Firstly, the lower "engine order" excitations due to distorted oncoming flow are determined for both radial and axial compressor inlets of conventional design. Secondly, the blade excitations due to rotating stall are found to be a strong source of blade vibration. Stimuli for the periodic and the stochastic excitation are extracted. The averaged data (RMS-values) are complemented by the time history of the stress signal.

1. INTRODUCTION

Blade vibration in axial compressors and turbines is still a main concern of all manufacturers of turbomachinery equipment. Extensive blade vibration testing has been carried out all over the world in the past 25 years. The primary purpose of these tests has been to confirm that, for a specific design, no excessive vibrations are encountered within the range of compressor operation [1].

The present investigation also involves classical dynamic strain gauge measurements. Here, however, they are analyzed in order to provide basic excitation data which may enable the designer to evaluate vibrations at a very early stage of design. In this approach, the stress measured on the blade is not used to judge the actual state of stress but rather to find the aerodynamic instationary forces (stimuli) which are the origin of the vibration. This becomes possible if the test blade is calibrated dynamically by means of experimental modal analysis which correlates the instationary blade loading and the blade response.

2. CALIBRATION OF TEST BLADE, TEST INSTALLATION

The test blade equipped with strain gauges according to Fig. 1 may be looked upon as a dynamic "strain gauge balance". In Ref. [2], a simple method of calibration for this balance was used. The test blade was assumed to be excitated by a dynamic load equally distributed over the entire blade surface. By assuming the structural response of the blade to be limited to the first bending mode only, the measured stress signal could be related to the excitation force acting on the blade. The present paper shows a refinement of this method by taking into account a sufficient number of mode shapes of the test blade and allowing for more general load distributions.

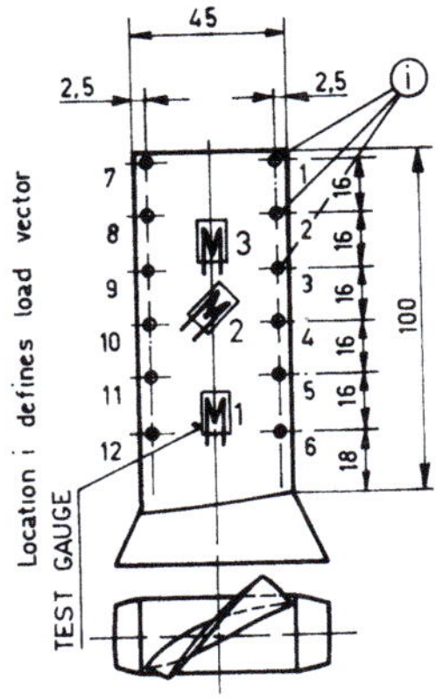

FIG. 1 TEST BLADE

Any forced vibrational pattern may be approximated by superimposing a sufficient number of modes with different modal amplitudes. An arbitrarily placed strain gauge measures, for each mode r, a stress σ_r which is exactly related to each modal amplitude. If m different modes are contributing to the overall motion, the overall stress becomes

$$\sigma = \left| \sum_{r=1}^{m} \sigma_r \right| \tag{1}$$

Using a viscoelastic damping model, the contribution of each mode σ_r can be written as:

$$\sigma_r = \frac{\{\Psi_{ri}\}^T \{f_i\} \; F}{\left| 1-\left(\frac{\omega}{\omega_r}\right)^2 + j\frac{\delta_r}{\pi}\frac{\omega}{\omega_r} \right|} \; ; \text{ with } \Sigma f_i = 1 \tag{2}$$

In this equation, Ψ_{ri} are the elements of the modal vector that give the influence of a normalized load, f_i, at a specific blade location i, on the stress for the different modes. F is a scale factor representing the actual magnitude of the load distribution, ω_r is the resonance frequency, and δ_r the logarithmic decrement of the modal damping. These data are determined for the test blade experimentally by impact excitation [3]. Introducing the generalized force

$$N_r = \{\Psi_{ri}\}^T \{f_i\} \tag{3}$$

and considering the first three modes only, equations (1), (2) and (3) give

$$\sigma = \left| \sum_{r=1}^{3} \frac{N_r}{1-\left(\frac{\omega}{\omega_r}\right)^2 + j\,\frac{\delta_r}{\pi}\,\frac{\omega}{\omega_r}} \right| \cdot F \tag{4}$$

Equation (4) relates directly the stress σ measured at the location of the strain gauge to the magnitude F representing the scale factor of the load distribution.

The generalized force N_r can be determined if the distribution of f_i, e.g., the aerodynamic load distribution, is approximately known. Fortunately, it seems that this load distribution does not significantly influence the generalized force N, thus allowing the use of an approximation. Several investigations, e.g. Ref. [4], seem to indicate that for many blade profiles and reduced frequencies of disturbances, the strongest variation of the pressure difference across the blade occurs near the leading edge. For the present calibration it is therefore assumed that the resulting instationary load attacks the blade profile at a distance of approximately 6% from the leading edge. To account for the variation in flow velocity W along the blade height, a reduction proportional to $(W/W_R)^2$ is used. The load vector f_i, and the corresponding normalized force N_r, are shown in Fig. 2 and compared with two other load distributions.

Measurements of aerodynamic excitation are made in an 8-stage axial compressor having a rotor of constant diameter D_R = 0.4 m. The compressor is installed in

LOAD DISTRIBUTION				
f_i	1	0.083	0	0
	2	0.083	0	0
	3	0.083	0	0
	4	0.083	0	0
	5	0.083	0	0
	6	0.083	0	0
$\sum_{i=1}^{12}=1$	7	0.083	0.167	0.245
	8	0.083	0.167	0.210
	9	0.083	0.167	0.178
	10	0.083	0.167	0.149
	11	0.083	0.167	0.122
	12	0.083	0.167	0.098
N_r		0,485	0.576	0.629
		0,033	- 0,059	- 0,095
		-0,0002	- 0,084	- 0,091

FIG. 2 NORMALIZED FORCE

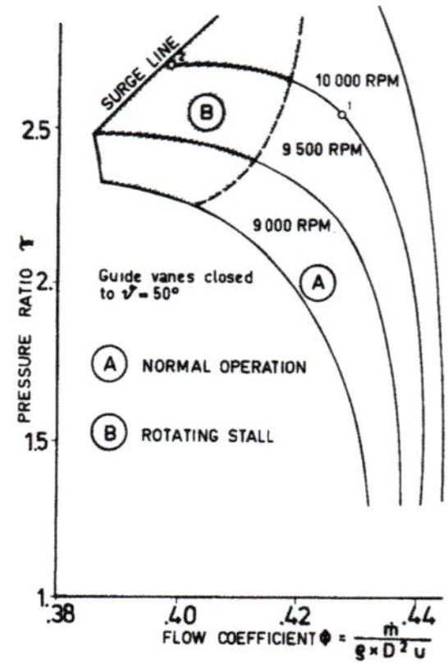

FIG. 3 COMPRESSOR MAP GIVING OPERATING POINT 1,2

an open loop with the main throttling valve on the suction side. The rotor blades have slightly modified NACA 65 sections. The diameter ratio D_T/D_R is 1.5 in the first stage.

3. BLADE EXCITATION DUE TO DISTORTED INLET FLOW

The distribution of velocity and flow angles of the flow entering the rotating blades may be far from uniform over the entire annulus. The blade passing through this distorted flow senses a varying force, which repeats after each revolution. If the total stress signal is analyzed and averaged (Fig. 4), it becomes evident that the strain gauge signal contains different frequencies, which are integer multiples of the rotational frequency. The maximum peak in Fig. 4, marked by Δ, is the natural frequency of the first bending mode excited by the broadband random excitation due to stochastic flow variation and turbulence. Each stress peak in the spectrum, with exception of the one marked by Δ, is converted according to the reduced form of formula (3),

$$\sigma_h = \left| \sum_{r=1}^{3} \frac{N_r}{1 - (\frac{\omega}{\omega_r})^2} \right| \cdot F_h \tag{5}$$

into a force, F_h, which corresponds to the harmonic of the excitation force. Defining a reference force by $F^* = \rho/2\ W^2 A_0$, the results may be expressed by a stimulus S_h

$$S_h = F_h/F^* \tag{6}$$

Fig. 5 shows the results for a rotational speed of 1000 rpm. At a constant speed, the results are practically independent of the flow coefficient. However, comparing other speeds, and other guide vane settings, there are considerable changes in the spectra. Some harmonics will always be present and reasonably constant, while others set in or disappear, or experience a considerable change in

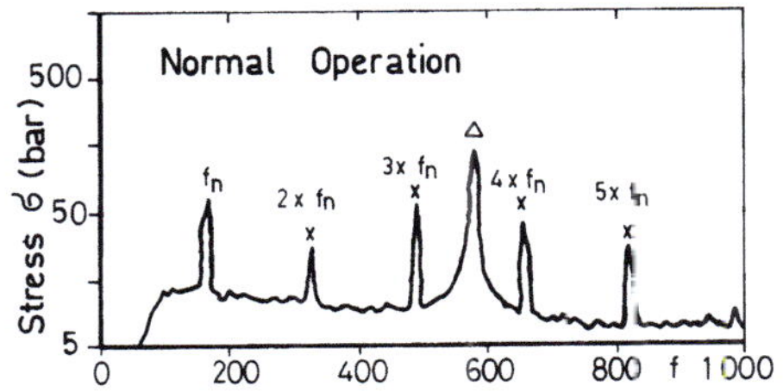

FIG. 4 SPECTRUM OF STRESS SIGNAL (NORMAL OPERATION)

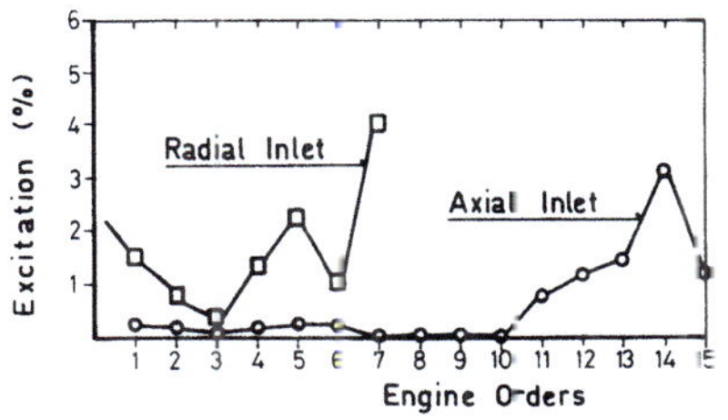

FIG. 5 INLET EXCITATION STIMULI

magnitude. This indicates that the flow distortion is modified depending on these parameters. The square symbols in the same Figure show the results corrected to the same load distribution obtained from earlier tests with a 17-stage axial compressor [2] having a radial inlet. The stimuli for both inlets are relatively small up to the 6th harmonic. Only in application of high density gases, and high relative velocity W, could the radial inlet lead to sufficiently high excitation which might become dangerous for the blade.

The harmonics corresponding to about half the blade passing frequency are relatively strong. They may be due to slightly irregular distances between the wakes of the different stator vanes.

The data of Fig. 5 may be used to estimate stresses in different blades. The generalized force and the modal frequencies are determined by modal analysis. The damping term in equation (4) has to be replaced. A safe assumption involves neglecting effects of blade fixation and considers only damping in the material. Damping logarithmic decrements measured during flexural vibration (Ref. [5]) show a strong dependence on the stress level, which can be accounted for by improving a first estimation for δ using successive approximation.

4. BLADE VIBRATION DUE TO ROTATING STALL

It is well known that rotating stall in the front stages of an axial compressor may cause quite severe blade vibration. Depending on the layout and the actual operating point of the compressor, some stages may operate beyond their individual stall limit. Some blade channels will incidentally stall and modify the flow in such a way that the stalled region will propagate in the circumferential direction with about half the speed of the rotating blades. For established rotating stall, this flow pattern does vary significantly with time.

The rotating disturbance pattern can be looked at in a similar way to the disturbance caused by inlet flow distortions. The difference is that the periodicity is not related to the rotational speed but to the relative speed (frequency)

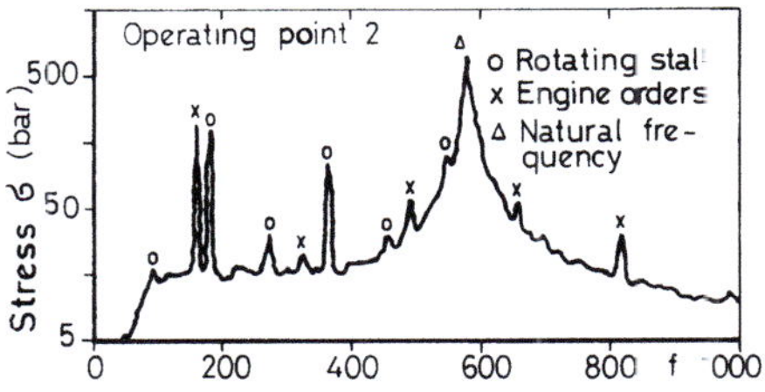

FIG. 6 SPECTRUM OF STRESS SIGNAL (ROTATING STALL)

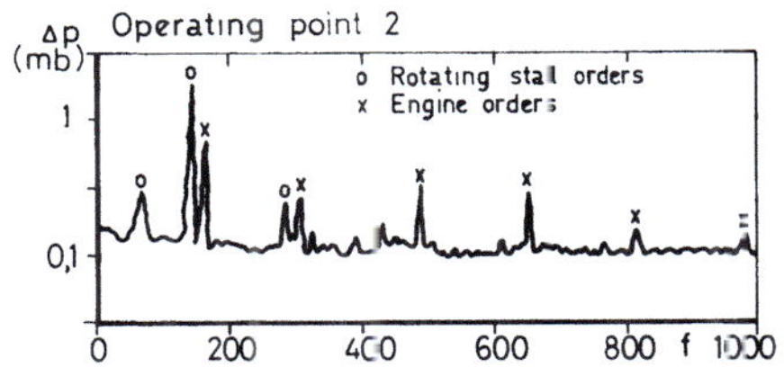

FIG. 7 SPECTRUM OF PRESSURE TRANSDUCER SIGNAL (ROTATING STALL)

between test blade and stall pattern. The Fourier spectrum of the stress signal of the blade operating in rotating stall is shown in Fig. 6. A new set of peaks (0) has appeared which are now integers of the relative speed.

$$f = f_n - f_s \tag{7}$$

The simplest way to identify rotating stall is to use stationary pressure transducers, preferably upstream of the stage being considered. If only light rotating stall is present, it is difficult to recognize the periodicity in the pressure signal. In this case, the frequency spectrum (Fig. 7) should be compared to the one with no stall present. If a new set of pressure peaks (0) appears, one may conclude that this is due to rotating stall. The lowest appearing frequency f_s is the rotational speed of the rotating stall pattern with respect to the stationary reference frame. The frequencies corresponding to the other pressure peaks are higher harmonics and are integer multiples of f_s. However, certain multiples of f_s may not be present in the spectrum, therefore attention must be given to whether the lowest appearing frequency f_s is really the rotational speed of the pattern. This is checked by comparing the harmonics of the stress signal and the pressure signal. Knowing the frequencies in the stationary frame, the corresponding frequency in the rotating frame must correspond exactly to f calculated from equation (7).

Once the different stress peaks, σ_h, due to rotating stall are identified, the excitation stimuli S are calculated from equation (5), (6). The results of Figs. 6 and 8 show that besides the periodic excitation stimuli, a strong broadband stochastic excitation is present which is attributed to large random flow disturbances of the partially stalled flow.

The results presented up to now are based on time averaged RMS values; they should be complemented by the time-history of the stress signals. Especially in the case of rotating stall, the stress fluctuates strongly with time. To illustrate this effect, the vibrational signals for the different operating points investigated are analyzed statistically for a time period of about one minute. A comparison of the maximum stress peak encountered during the one minute observation time, with the RMS averaged stress values in the same time period, show a ratio $\hat{\sigma}_{max}/\hat{\sigma}_{RMS}$ between 3 and 5. The maximum is reached at an intermediate operating point, e.g. prior to full development of the phenomenon. This means that near surge, although the average value is only about 500 bars, single events with much higher stresses have to be expected over a longer observation period.

Curve (a) of Fig. 9 shows the statistics of a stress envelope obtained for fully developed rotating stall. It is compared to a Rayleigh-distribution having a probability density defined by:

$$p(x) = x/c^2 \cdot e^{-x^2/2c^2} \tag{8}$$

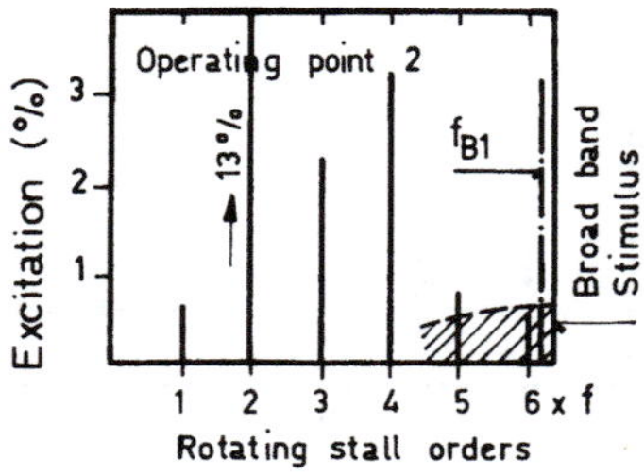

FIG. 8 PERIODIC AND STOCHASTIC STIMULI (ROTATING STALL)

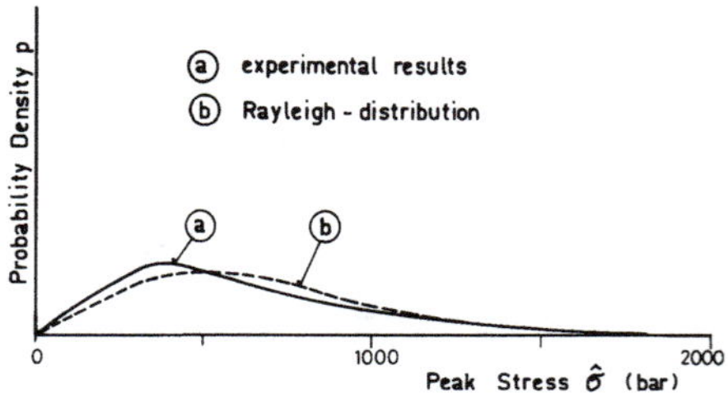

FIG. 9 AMPLITUDE DISTRIBUTION OF STRESS ENVELOPE

If the constant is chosen to be $c = 4.1$, and if $x = 1$ corresponds to $\sigma = 125$ bar, then the curve (b) fits well the experimental curve in the region of high stresses. Although this equation may be used to estimate high stress cycle accumulation, any extrapolation to very high stresses must be made with great care, as no data for lang observation times are yet available.

5. CONCLUSION

A classical dynamic strain gauge technique has been used to measure aerodynamic excitation forces on blades in turbocompressors. The results obtained may be useful to the designer to evaluate vibrations at a very early stage in the design of a compressor.

In this method, periodic and broad band random excitation data are calculated from the forced stress response of the test blade. The calibration of the test blade uses normalized forces based on "realistic" load vectors to approximate the aerodynamic load distribution of the blade. Although it is believed that the load vector employed is properly chosen for the frequency range considered, similar investigations in other compressors and with other test blades would be desirable.

6. LIST OF SYMBOLS

A_0	blade area (height x chord)
f_i	normalized load vector
f	frequency
F	scale factor of load distribution
F*	reference force
N_r	generalized force
S	excitation stimulus
W	relative flow velocity
π	pressure ratio
Φ	volume flow coefficient $(\dot{V}/R^2U_R)$
Ψ_{ri}	modal vector
ω	angular velocity $(2\pi f)$
σ	stress
δ	damping log. decrement
ρ	density of gas

7. SUBSCRIPTS

h	number of harmonics
i	number defining position on test blade
n	rotational
r	numbers of modes
R	rotor
T	tip
S	stall

8. REFERENCES

1. Strub, R.A.,"Field Measurements of Blade Stresses on Industrial Turbomachines' Sulzer Technical Review 2 (1975).
2. Aicher, W., Jenny, R., Roduner, H.,"Investigation in Turbocompressors: Rotor Vibration, Blade Vibration, Tests with Heavy Gases" Sulzer Technical Review (1978).
3. Bolleter, U., Eberl, J., Buhlmann, E., Modal Analysis of Compressor Blades by Means of Impulse Excitations. AGARD Proceedings, AGARD-CCP 248, Cleveland, Ohio, 23. - 27. Oct. 1978.
4. Gallus, H.E., Kummel, W., Wallmann, T.,"Messung zur Wechselwirkung von Laufrad- und Leitrad-Beschaufelung in einer Axialverdichterstufe" VDI-Bericht 264, (1976).
5. Schabtach, C., Fehr, R.O.,"Measurements of the Damping of Engineering Materials During Flexural Vibration at Elevated Temperatures' Trans. ASME, Journal of Applied Mechanics (June 1944).

Discussion: AERODYNAMIC BLADE EXCITATION IN TURBOCOMPRESSORS
(R.J. Jenny and U. Bolleter)

M.E. Ginniff, UKAEA, Seascale, England

Would Mr. R. J. Jenny please explain how he would have dealt with the failure of the strain gauge blade. In my experience it was necessary to have 2 or 3 strain gauge blades in case one failed. To obtain 2 or 3 identical specimen blades there is a need to mechanically tune the blades and also to use strain sensitive varnish (lacquer) so that the strain gauges are positioned in high strain areas where the strain pattern is relatively constant and hence strain gauge positioning is not over critical.

Reply to Discussion of: AERODYNAMIC BLADE EXCITATION IN TURBOCOMPRESSORS

R. Jenny

We have had no problems with blade failure as our experimental procedure is designed to avoid high stresses in the blades. The signal from the test blade, Figs. 4 and 6, indicates that stresses are too low to initiate blade failure. If, however, during the test, these stresses had appeared to be greater, indicating blade resonance, then we would not have measured at this point. We would have either untuned the blade or measured at a slightly different operating point, where the blade was no longer in resonance.

We are able to do this, because the purpose of our test was not to measure the actual stress field of the blade, but to find the aerodynamic excitation force, using the blade as a strain gauge balance. The stresses can afterwards be calculated from the knowledge of the aerodynamic excitation force.

Nevertheless, failure of strain gauges may be a problem, when testing high speed rotating machinery. In fact, during the four weeks of testing, two out of twelve gauges failed. In order to cope with this problem, each test blade was equipped with three strain gauges (Fig. 1) and each gauge calibrated seperately as indicated in the paper. The calibration equation being

$$F = \frac{1}{\left| \sum_{r=1}^{r} \frac{\gamma_{pr} \cdot N_r}{1 - \left(\frac{\omega}{\omega_r}\right)^2 + j \frac{\delta}{\pi} \frac{\omega}{\omega_r}} \right|} \cdot \sigma_p$$

where the matrix γ_{pr} accounts for the gauge location and is obtained during the calibration procedure.

During the tests, the stress signal σ_p of each strain gauge can be used to determine the aerodynamic excitation force F acting on the blade. The position of the strain gauge affects both the stress and the calibration factor in such a way as to give rise to an identical excitation force.

For example, the excitation force at 185 Hz for rotating stall (Fig. 6) has been found to be 0.694, 0.666 and 0.430 Newton respectively. This good agreement indicates that the measurements are consistent and when one gauge fails, the two remaining would have been sufficient.

BLADE VIBRATION DUE TO INLET FLOW DISTORTION

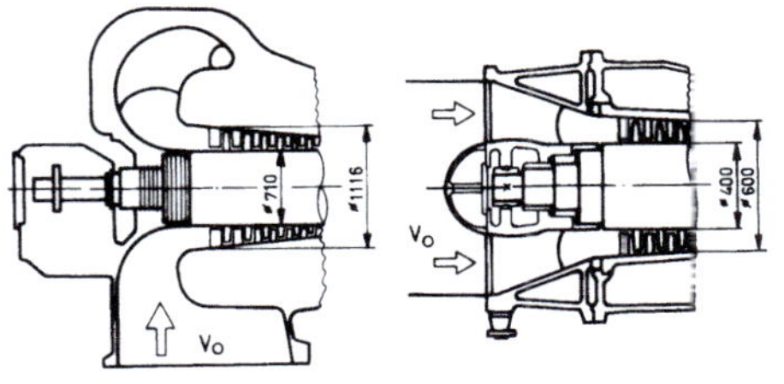

(a) (b)

①

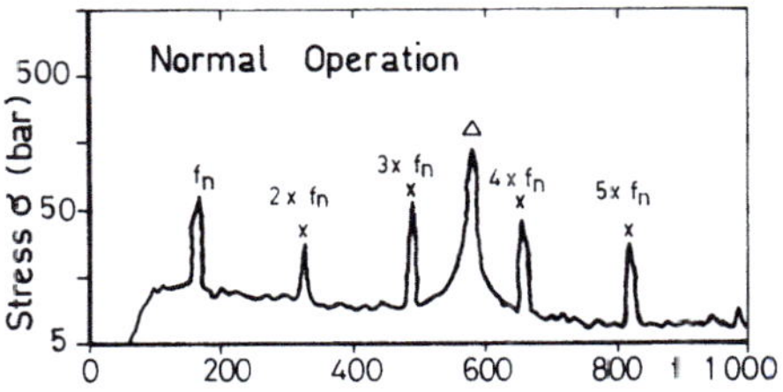

②

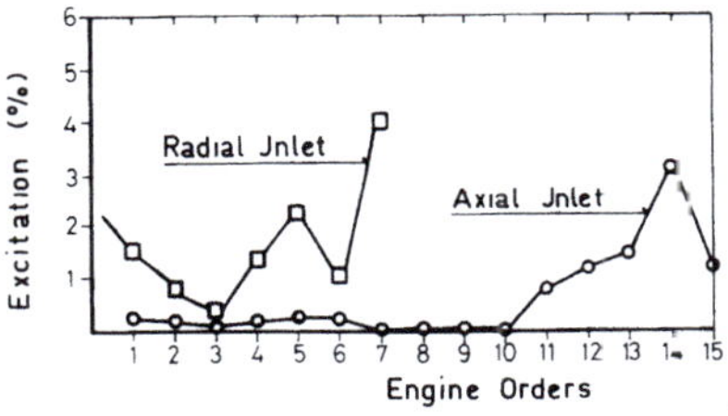

③

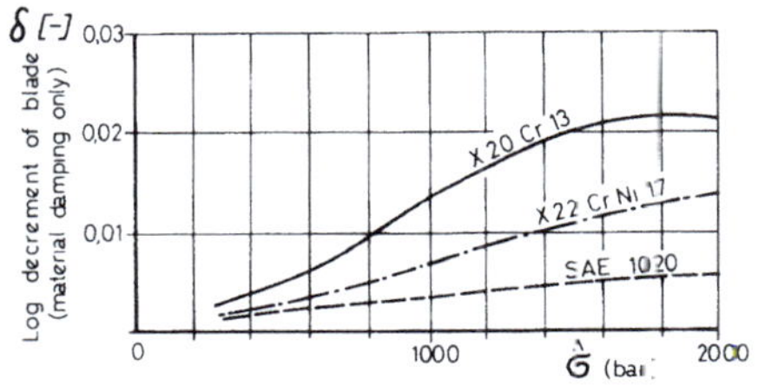

④

Problem: Due to irregularity of meridional flow fields, rotating blades are submitted to cyclic loads. Near resonant excitation leads to high cyclic stresses causing fatigue failure of the blade.

Geometry: Cross section of compressor inlet: a) radial, b) axial. Suction nozzle to annulus area ratio 3.8 and 5.3, respectively. Guide vanes closed to a stagger angle of $\theta = 50^{0}$ with respect to flow direction. Number of guide vanes, 26 and 36, respectively.

Flow conditions: Uniform flow in the suction nozzle. Velocity $V_0 = 16$ m/sec, circumferential speed of rotor $Un = 180$ m/sec and 210 m/sec respectively.

Operating conditions Operating point near optimum compressor performance. No rotating stall present.

Structural dynamic of blades: Test blade is identical with rotor blading (NACA 65 slightly modified). Blade structure described by modal analysis:
generalized force: $N_1 = 0.63$, $N_2 = 0.095$, $N_3 = 0.091$; resonance frequencies $\omega_1 = 490$ Hz, $\omega_2 = 1780$ Hz, $\omega_3 = 2170$ Hz, for bench tests; $\omega_1 = 580$ Hz, $\omega_2 = 1790$ Hz, $\omega_3 = 2175$ Hz, at operating speed.

Vibrations: Forced vibration with low stress amplitudes. Fourier spectrum (sketch 2) shows that test blade operates in off resonance condition. High excitation for the engine orders, corresponding to about half of stator blade passing frequency (sketch 3).

Cures: Increasing area ratio of radial inlet gives reduced circumferential flow irregularities. Axial inlet practically eliminates low order excitations up to about $10 \cdot f_n$. Higher orders are due to unequal spacing of stator blade wakes. Modifying the resonance frequency of the blades by modifying aspect ratio avoids resonances.

Thickening the blade and using material with high damping, such as 13 % Cr steel (sketch 4), reduces stresses.

BLADE VIBRATION DUE TO ROTATING STALL

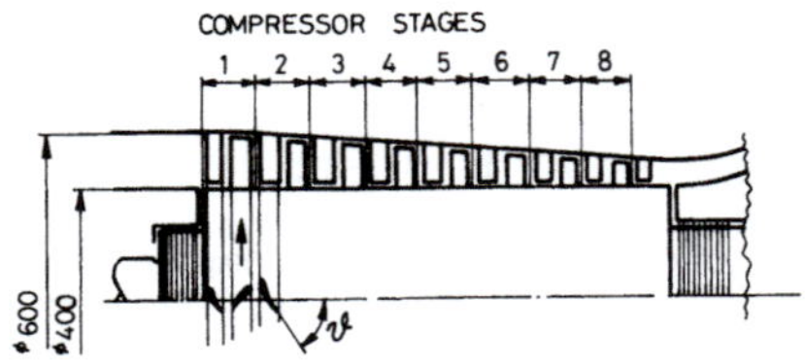

(1)

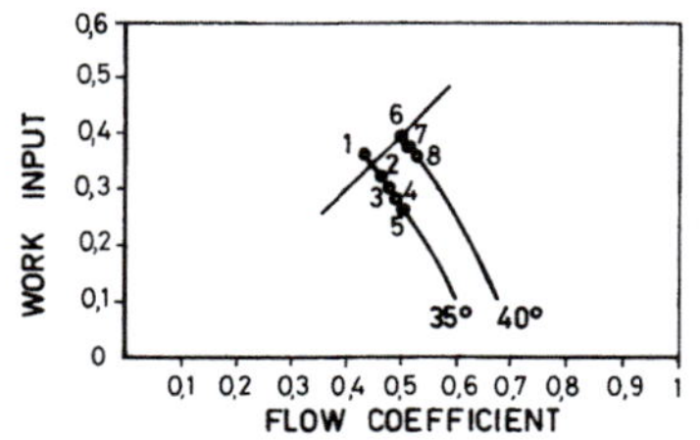

(2)

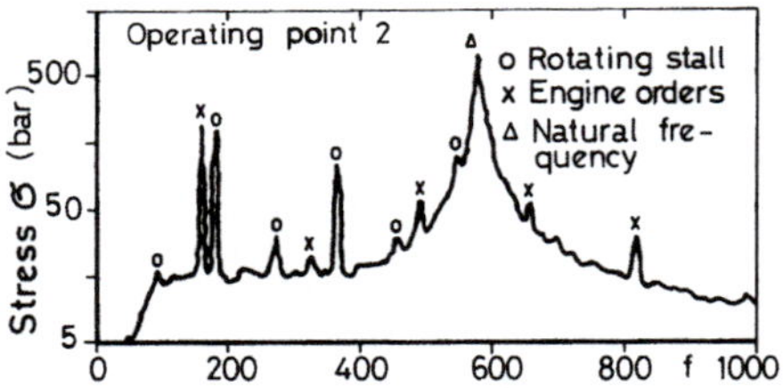

(3)

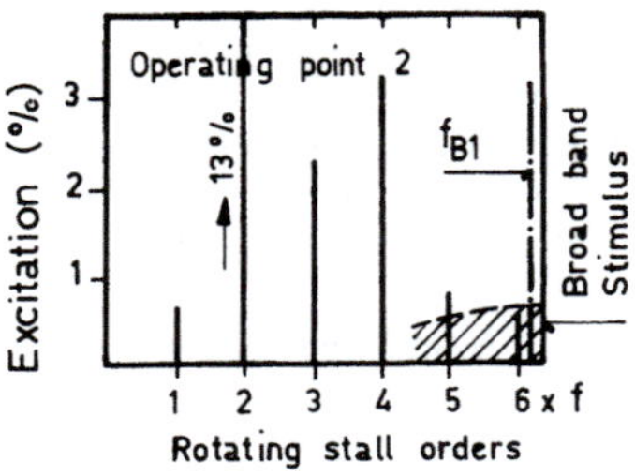

(4)

Problem: Off design operation of axial compressors favours rotating stall in the front stages of the compressor. The highly instationary and turbulent flow excites the blade not only periodically, but also stochastically. Accumulation of high stresses over longer time periods causes blade fatigue.

Geometry: Channel geometry and stator vane angles of 8-stage test compressor, see sketch 1.

Flow conditions: Uniform flow in the suction nozzle. Velocity $V_0 \cong 15$ m/sec, blade velocity $U_n = 210$ m/sec.

Structural dynamic of blades: Test blade is identical with rotor blading (NACA 65 slightly modified). Blade structure described by modal analysis:
generalized force: $N_1 = 0{,}63$, $N_2 = 0{,}095$, $N_3 = 0{,}091$; resonance frequencies: $\omega_1 =$ 490 Hz, $\omega_2 = 1780$ Hz, $\omega_3 = 2170$ Hz for the bench tests; $\omega_1 = 580$ Hz, $\omega_2 = 1790$ Hz, $\omega_3 = 2175$ Hz at operating speed.

Vibration: Fourier spectrum of stress at blade root given in sketch 3. Strong stochastic excitation. No resonance with harmonic of rotating stall. Peak to RMS-ratio of vibration is shown on sketch 5 indicating that infrequent stress levels of 3 to 5 times the RMS-values are possible. RMS average of excitation stimuli shown in sketch 4 extracted from sketch 3.

Cures: Modification of stator vane setting angle. Stagger angle gradually decreasing along the compressor. This layout brings the rotating stall limit and the surge limit together. As the compressor is protected against surge, it is also protected against stall.

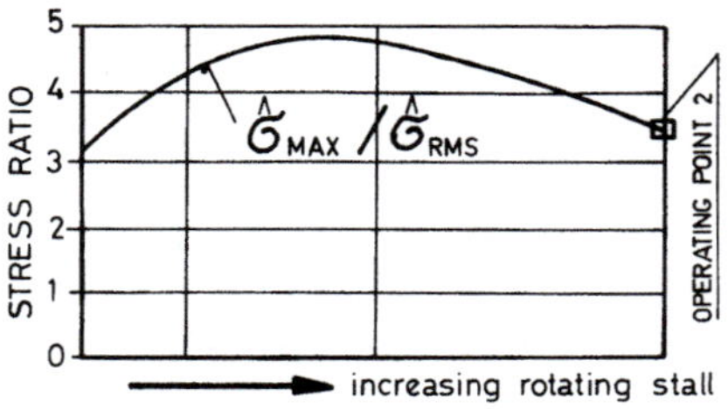

(5)

B7 EXPERIENCES WITH FLOW-INDUCED VIBRATIONS AT SULZER

Y. N. Chen
Scientific Consultant
Sulzer Brothers Limited, Winterthur, Switzerland

SUMMARY

In this paper, a number of flow-induced vibrations in industrial plants, with excitation sources involving vortices, turbulence, swirling flow and thermal relaxation, will be covered. The topics encompass piping systems for high pressure steam, turbomachinery inlet vortices, cooling of high temperature flows, cyclone separators, and combustion instability of swirl burners.

1. INTRODUCTION

Some vortex and flow-instability induced vibrations in piping systems of steam flow with high pressure and high temperature (reduction valve stations and pressure transmitters) will be summarized. A number of cases concerning swirling flow-induced vibrations will also be given: for example behind inlet swirling guide vanes of ventilators, at pump intakes, in combustion chambers, and in cyclone separators. Thermal relaxation vibrations in flows due to cooling water injection into flue gas and streams downstream of reduction valves form the additional areas of this paper.

Working charts for the determination of the properties of the excitation sources, and remedies for elimination of the flow-induced vibrations will be compiled as guidance for reaching vibration-free operation of the respective components.

2. SWIRLING FLOW-INDUCED VIBRATIONS

The instability of swirling flows shows a common nature: either in the form of the precession of the vortex core (with a low frequency) or in the generation of circumferential wave motions (with high frequencies). Some cases encountered in cyclone separators, inlet guide vanes of ventilators, and draft tubes of water turbines will be summarized.

A special class of swirling flow is the intake vortex of the pump, whereby the axial-flow-induced instability is the main source of vortex generation.

A still more complicated case is the toroidal ring vortex induced by swirl burners in combustion chambers. The flutter of the flames will be induced by the instability of this vortex owing either to the vortex filament formation itself or to the impulsive ignition of the turbulence cells caused by this formation.

3. THERMAL RELAXATION-INDUCED VIBRATIONS

The thermal relaxation of jets in a flow can produce vibrations, if the heat release and the pressure fluctuations fulfill the Rayleigh criterion. Some of the causes lie either in the uniform distribution of the heat release (positive or negative), or in the unfavorable delay of the heat release due to unfavorable injection of the jet.

4. FLOW-INDUCED VIBRATIONS IN PIPING SYSTEMS

A series of flow-induced vibrations have been encountered in piping systems of power generation plants with steam flows of high pressures up to 209 bars, and high temperatures up to 535°C. The excitation source is the turbulence transported by flow in the duct due to flow separation upstream. The instability of the boundary layer can further strengthen the excitation by means of vortex shedding over the apertures of the branch tubes. The vortices can then be coupled between the neighboring tubes, resulting in reinforcement of the vibration. The spectrum of such a type of turbulence will be shown. Some cases dealing with reduction valve stations and pressure transmitters will also be discussed.

5. REFERENCES

1. Chen, Y. N., "Aerodynamically Excited Vibrations and Noise in Cyclone Separators Due to the Reverse of the Spiral Flow", Fifth International Congress on Acoustics at Lege, Paper F 57, (1965).

2. Syred, N., Beér, J. M., "Combustion in Swirling Flows: A Review. Combustion and Flame", 23 143-201, (1974).

3. Coles, D., "Transition in Circular Couette Flow", Journal Fluid Mechanics, 21, 385-425, (1965).

4. Chanaud, R. C., "Observation of Oscillatory Motion in Certain Swirling Flows", Journal Fluid Mechanics, 21, 111-127, (1965).

5. Chanaud, R. C., "Experiments Concerning the Vortex Whistle", Journal ASA, 35, 953-960, (1963).

6. Bolton, A. N., "Pressure Pulsations and Rotating Stall in Centrifugal Fans", Proceedings of Conference on Vibrations and Noise in Pump, Fan and Compressor Installations, Southampton, pp. 175-181, Instn. Mech. Engrs. (16-18 September 1975).

7. Chen, Y. N., Comment on the previous paper.

8. Gilkey, C. H., Rogers, J. D., Wyrough, D. J., Pettersson, C. H. and Howitt, I., "Centrifugal Fan/Duct Systems Vibrations", Combustion, 15-21, (March 1975).

Gilkey, C. H. and Rubin, A. N., "Recent Experience with Vibration Problems Associated wtih Fossil Fuel Stean Generating Equipment", American Power Conference (1970).

9. Jacobs, J. and Günther, R. , "Periodische Störungen in Drallströmungen", Forschung-Ingenieurwesen, 41, 85-91, (1975).

10. Chen, Y. N., "Studies into Swirling Flow in Furnace with a View to Flame Stability", Gaswärme International, 33-50, (1979).

11. Chen, Y. N., "From Bath-Tub Vortex to Pump-Intake Vortex Axial-Flow-Induced Instability as Their Formation Mechanism", to be published.

12. Daggett, L. L. and Keulegan, G. H., "Similitude in Free-Surface Vortex Formations", Proceedings ASCE, Journal Hydraulics Div., 100, No. HY11, 1565-1581, (November 1974).

13. Brown, C. E., "Aerodynamics of Wake Vortices", AIAA Journal, 11, 531-536, (1973).

14. Prosser, M. J., "The Hydraulic Design of Pump Sumps and Intakes", British Hydromechanics Reserach Association (July 1977).

15. Chen, Y. N. and Bolleter, U., "Thermo-Driven Self-Sustained Vibration and Noise of a Su½ercritical Reduction Valve", INTER-NOISE 79, 103-122, Warsaw, (1979).

16. Hay, J. A., "Problems of Cabin Noise Estimation for Supersonic Transports", Journal of Sound and Vibration, 1, No. 2. 113-126, (1964).

17. Schlichting, H., "Grenzschicht-Theorie, 555-556, Verlag G. Braun, (1965).

18. Motzfeld, H., "Frequenzanalyse Turbulenter Schwankungen. Zeitschrift für Angewandte Mathematic und Mechanik, 18, 362-365, (1938).

19. Chen, Y. N. and Stürchler, R., "Flow-Induced Vibraticns and Noise in the Pipe System with Blind Branches Due to Coupling of Vortex Shedding", Proceedings of International Conference on Noise Control Engineering INTER-NOISE 77, B 189-203, Zürich, (March 1-3, 1977).

20. Chen, Y. N., "Calculation of Gas Vibrations Due to Simultaneous Excitations in Reciprocating Compressor Piping Systems with Allowance for Frictional Effect and Temperature Change in the Flow", Journal of Sound and Vibration, London, 5, No. 2, 215-256, (1967).

21. Dziallas, R., "Francisturbinen bei Teil-und Ueberlast", VDI-Berichte Nr. 75, Hydraulische Strömungsmaschinen, 53-64, (1964).

CHART A: SWIRLING FLOW-INDUCED VIBRATIONS IN CYCLONE SEPARATOR [1-5]

Problem 1:

The swirling flow in a cyclone separator can be unstable with its core performing a precession movement (Sketch 1). The precessing frequency is generally in the range of

$f = (0,2 \text{ to } 0,4) f_o$ (f_o = angular velocity in RPS of the swirling flow).

This precession will cause a low-frequency vibration.
A further case is shown in Sketch 2. The flow pattern changes periodically from position n to o causing strong pressure fluctuation.

Remedy:

Guidance of the swirl flow either by means of an extension of the central tube (Sketch 1) or a conical column as shown in Sketch 3.

Problem 2:

A high-frequency instability can arise from the free-shear layer mixing between the two axial components of the incoming outer spiral flow and the inner return flow (Sketch 4) with frequencies shown in Sketch 5. The threshold of the instability (reciprocal Rossby number $\xi = w/fD$) increases with the length-diameter ratio L/D. (Sketch 6), where w is the axial velocity.

Remedy: Make L/D as large as possible.

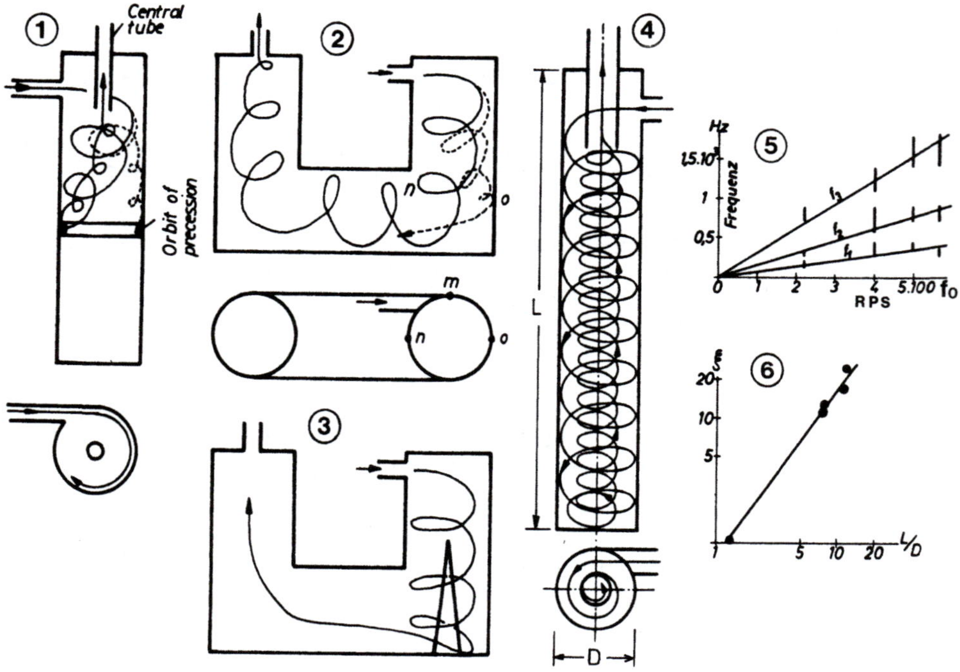

CHART B: SWIRLING FLOW-INDUCED VIBRATIONS DUE TO SWIRL INLET VANES IN VENTILATOR OR SWIRL FLOW IN DRAFT TUBE OF WATER TURBINE

Problem 1: The swirl inlet vanes of a ventilator produce a swirling flow through the entrance of the annular space of the impeller (Sketch 1). The core can undergo precession [6, 7]. Due to variation of the mass flow along the impeller circumference, a periodic pressure fluctuation can be produced according to the precessing frequency of

$f = (0.2 \text{ to } 0.4) f_o$ (f_o = R.P.S. of the swirling flow)

Remedy: Installation of a shaft with radial fins for damping the swirl (Sketch 1) [8].

Problem 2: The precessing swirl flow in the draft tube of a water turbine will have the same effect on producing periodic pressure fluctuations, as shown in Sketch 2 [21].

Remedy: Installation of a conical column as shown in Sketch 3 in order to centre the swirl flow; however, a sacrifice of the efficiency must be taken into consideration.

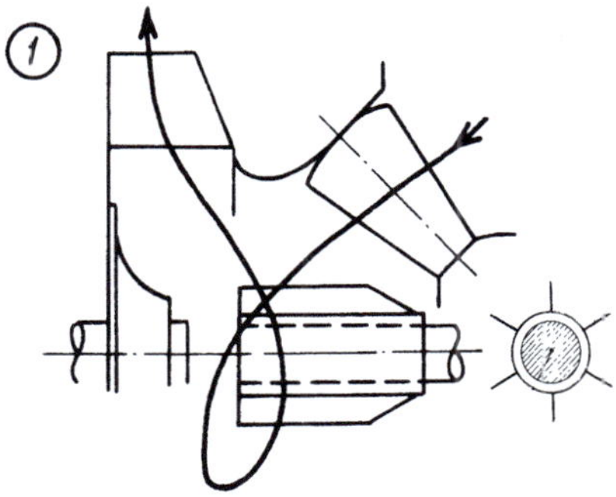

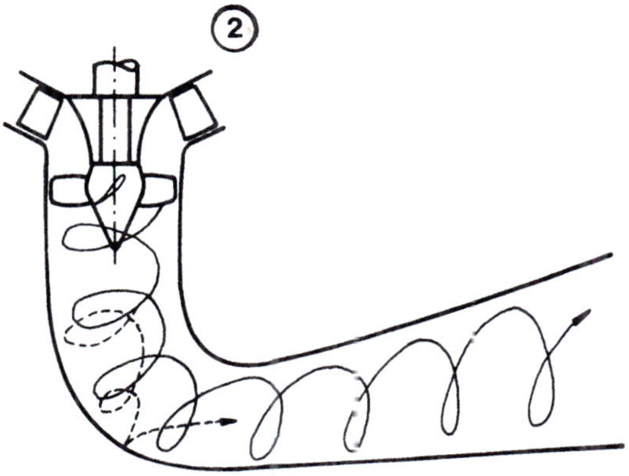

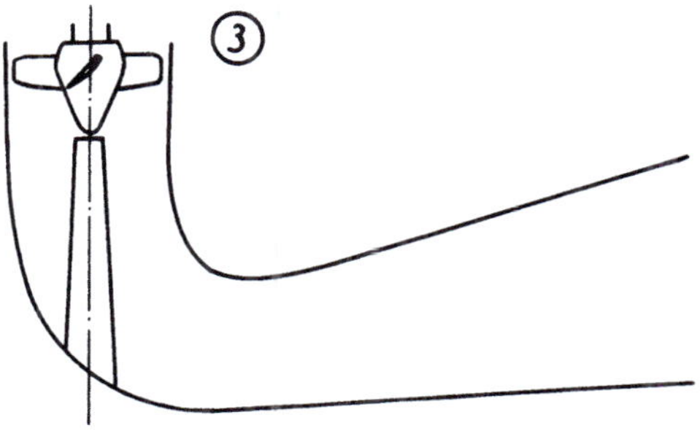

CHART C: SWIRLING FLOW-INDUCED VIBRATIONS - COMBUSTION INSTABILITY OF SWIRL BURNER [9,10]

Problem:

The toroidal ring vortex formed in a combustion chamber owing to induction of the swirl burner flow has the following behaviour (Sketch 1):

1) Reversal flow zone along the axis carrying high-temperature flue gas due to turbulence activity in the wake. This heat should be applied for the generation of a pilot flame.
2) Precessing vortex filament on the shell exciting a strong pressure fluctuation in the combustion chamber in the case of a large swirl number.
3) Turbulence zone behind this vortex filament causes impulsive ignition of fuel and thus vibration.

The 8 gas lances arranged in a circle with injection holes entirely directed to the front of the toroidal ring vortex caused strong vibration due to ignition in the turbulence zone: flutter and precession of the flame.

Cure:

The gas lances were modified to produce inclined and axial fuel jets (Sketch 2) with a random pattern (Sketch 3), so that the pilot flame in the reversal flow can ignite the fuel along the front of the toroidal ring-vortex, causing the flame to hang firmly on the burner. A very stable combustion could be achieved with the result of vibration-free operation of the combustion chamber up to 620 MW output (with 18 burners in operation).

Design criteria:

The fuel jets should be well distributed in three directions in order to enable fresh air to reach their surfaces. The momentum flux of the fuel jets should be capable of strengthening the swirling movement of the toroidal ring vortex.

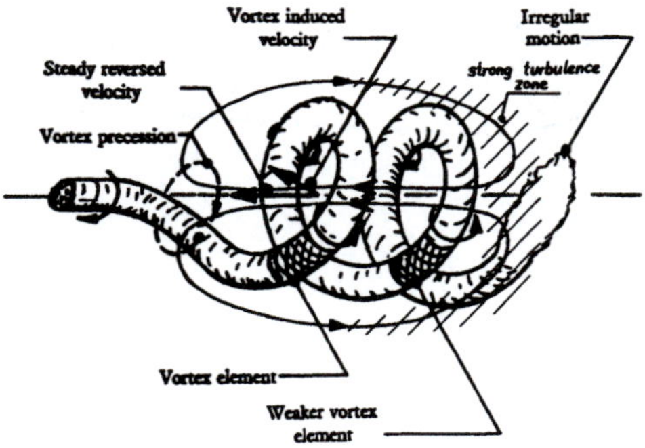

(1) A toroidal ring vortex formed in the core of a swirling flow expanding from a swirl burner into a combustion chamber

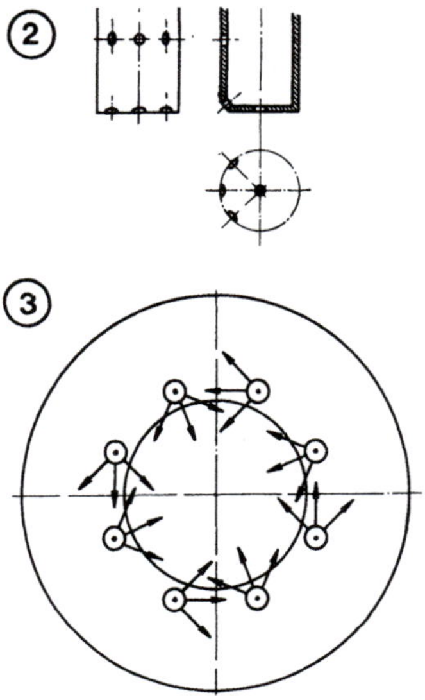

CHART D: SWIRL FLOW-INDUCED VIBRATIONS - INTAKE VORTEX OF PUMP [11-14]

Problem:

The intake vortex of a pump can produce low-frequency pressure fluctuations, impairing the pump operation.

Theory:

The intake vortex of a pump is quite similar to the bathtub-vortex, based on the transfer of circulation from the outer potential field to the inner viscous core, due to axial flow imposed by the suction in the former case, and by the drain in the latter case (Sketch 1). The result of this transfer is the increase of the swirl speed of the inner core, until a breakdown stage is reached at which a deep air core is formed, impairing the pump impeller.
If the suction pipe is vertical, the intake vortex will be formed behind the pipe (Sketch 2), initiated by the two vortices originating from the flow separation. One of the vortex pair will be strengthened by the prerotation of the inflow or/and the preswirl of the impeller, with the result that it develops into an intake vortex.

Suppression of the intake vortex:

All the measures which minimize the prerotation in the inflow and the preswirl of the pump impeller, and which reduce the flow separation behind the vertical suction pipe, will lead to prevention of the intake vortex, for example (Sketch 3):
- Small clearance of the suction pipe against the sump walls and bottom,
- Cross vanes inside or in front of the intake,
- Elimination of the free surface of the sump by means of an adequately inclined roof,
- A raft behind the suction tube to damp the vortex formation.

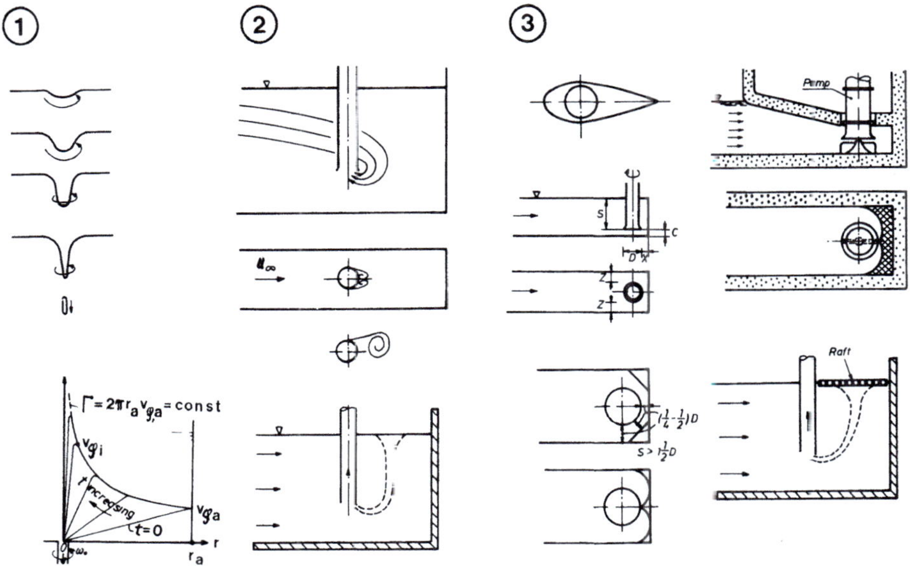

CHART E: THERMAL RELAXATION INDUCED VIBRATIONS - COOLING WATER INJECTION INTO FLUE GAS FLOW

Problem: Cooling water injection into a flue gas by means of a tube ring with holes equally spaced on its inner wall caused strong vibration of modes with one and two diameter nodal lines, see Sketch 1. The spectra of the pressure fluctuation showed peaks in the Strouhal number range of the cooling water jets of $S = fd/V = 0.2$ to 0.8 corresponding well to the turbulence energy distribution of the jet.

Theory: The excitation mechanism was found to lie in the thermal relaxation of the cooling water. As soon as the gas flow is disturbed by the turbulence of the water jets so that a natural vibration, e.g. of a fundamental mode with a diameter nodal-line, is excited, however weak it might be, the water jets situated, for example at, the side of the negative dynamic pressure will be accelerated.
The result of this is that an additional water quantity is injected into the gas flow on this side. If the preparing time for this excess of water to release its cooling effect in the turbulence cells during the travelling downstream is such that the Rayleigh criterion:

$\oint p\, h\, dt > 0$ (p = dynamic pressure, h = heat release)

is fulfilled, then a self-excited vibration arises. The coupling component is the uniform distribution of the cooling jets along the circumference of the duct. Only then can the pressure fluctuation of the gas column be amplified by the well distributed heat withdrawal of the cooling jets.

Cure: Modification of the jet distribution to a random pattern shown in Sketch 2. The water jets were injected from a three-dimensional tube structure in quite random directions.

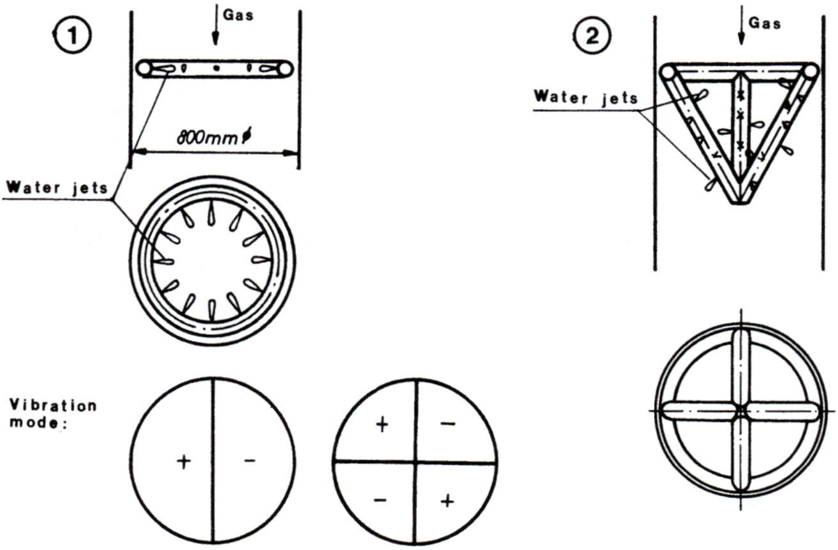

CHART F: THERMAL RELAXATION INDUCED VIBRATIONS - COOLING WATER INJECTION INTO STEAM FLOW

Problem: Thermal relaxation vibration was found in steam flow behind a reduction valve when cooling water was injected uniformly through the circumference of two nozzles (Sketch 1). The natural frequencies were calculated to have the following values

Vibration mode	Before nozzle	Behind nozzle
⦶	540 Hz	360 Hz
⊕	940 Hz	640 Hz
◎	1200 Hz	800 Hz

All these natural vibrations were excited, but the fundamental mode with a diameter nodal line and measured values of 565/345 Hz was especially strong (Sketch 3), causing failure of the duct. The pressure in the water feedline fluctuated heavily with ± 1 bar at the first mode frequency. The coupling component in this thermal relaxation-induced vibration was again found to lie in the uniform distribution of the water injection along the nozzle circumference (Sketch 1).

Cure: Modification of the uniform distribution of the water jets to a strongly asymmetrical pattern, whereby the steam jets were deflected inwards by this asymmetrical cooling: one directed upwards, and the other directed downwards to avoid impact against each other (Sketch 2). The noise spectra measured at a distance of 50 mm outside the duct before and after modification (corresponding to the upper and lower curves in Sketch 3) showed a reduction of the peak at 345 Hz by 6 dB and the peak at 565 Hz by 8 dB, corresponding to pressure reduction by a factor of 2 and 2.5 respectively. The other peaks of the spectra, corresponding to turbulence and higher modes, no longer caused concern for the present problem of duct failure.

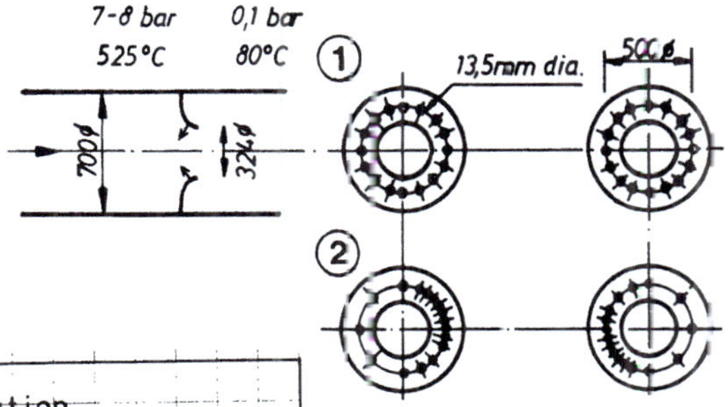

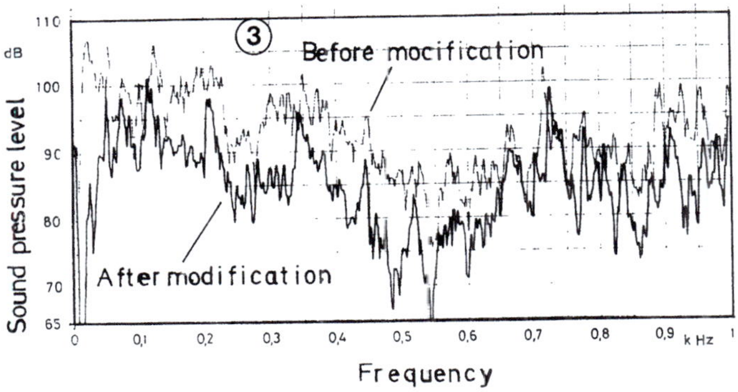

CHART G: THERMAL RELAXATION INDUCED VIBRATIONS - COOLING WATER INJECTION BEHIND THE STEAM REDUCTION VALVE [15]

Problem:

A reduction valve station with supercritical pressure reduction from 264 bar (455°C) to 54 bar (295°C), involving cooling water injection through a center hole of the valve, exhibited a strong vibration according to the fundamental mode of 750-780 Hz. The vibration had a diameter nodal line (0.45 m duct diameter). It was not possible to damp this vibration by means of perforated plates installed downstream of the valve. Thus, the excitation source did not directly originate from the feedback of the acoustic wave to the origin of the shock wave pattern at the nozzle exit similar to the mechanism of a supercritical free jet (Sketch 1), in which one wavelength roughly embodies two shock cells with a universal Strouhal number of

$$S_L = fL/U_\infty = 0.33.$$

Theory:

A thermal relaxation induced vibration, due to coupling between the injected cooling water jet and the pressure fluctuations in the duct, was the prevailing cause of the present problem. The concentrated water jet produced such large droplets that heating and evaporation time fulfilled the Rayleigh criterion. The water jet appeared to be deflected to a curved path by the pressure field at the positive pressure side. The part of the jet on this side was apparently decelerated and the other part on the negative pressure side was apparently accelerated. In this manner, a kind of thermal relaxation of the cooling water jet could be established.

Cure:

Dividing the concentrated water jet (Sketch 2) into a number of smaller axial ones (Sketch 3), which could be very well disintegrated by the large number of radial, well distributed steam jets (Sketch 3). In this way, the evaporation time of the fine water droplets could be shortened such that no coupling between the thermal relaxation and the pressure fluctuation could occur.

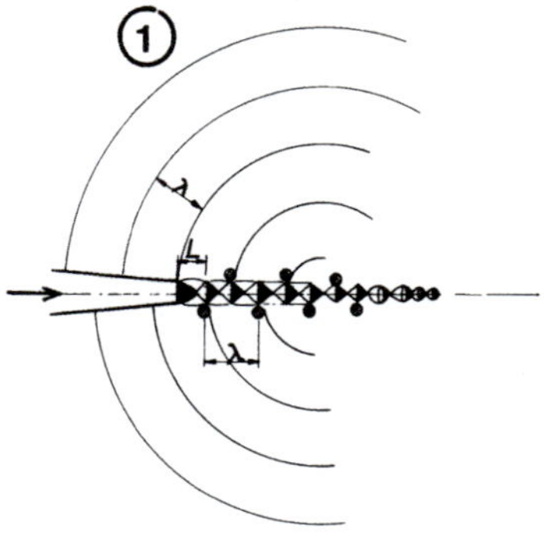

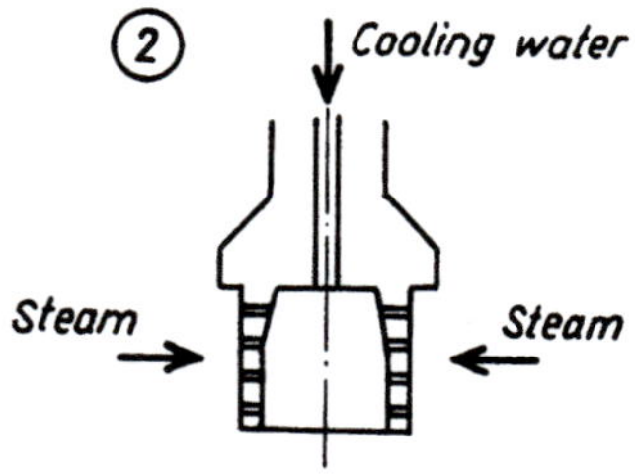

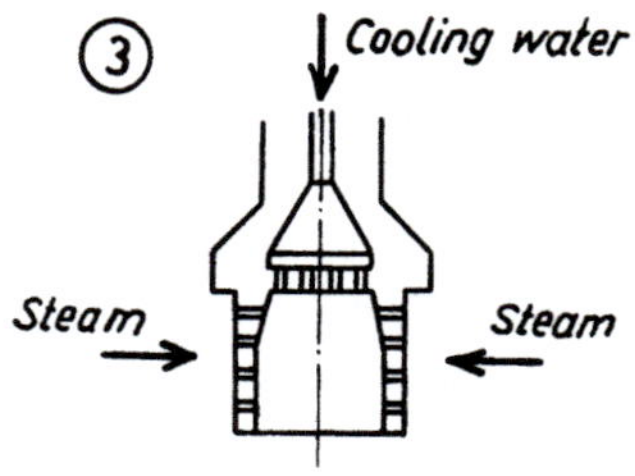

CHART H: FLOW-INDUCED VIBRATIONS IN PIPING SYSTEMS - FLOW-SEPARATION-CAUSED TURBULENCE AS EXCITATION SOURCES [16]-[19]

The excitation sources of flow-induced vibration in a pipe are generally the turbulence and vortices generated by flow separation behind bends, etc.
From the behaviour of a smooth flow along a flat plate [16], it can be expected that a very long duct without bends and discontinuities will exhibit fluctuating pressure spectra with a flat plateau of maximum energy density in the low Strouhal number region extending from $S_d = fd/U_\infty = 0.1$ to about 0.5 to 3 [17]. Bends, joints and discontinuities in actual ducts, however, will cause peaks in the spectra within the corresponding Strouhal number range. This is shown in Sketch 2 by curve 1 for a laboratory test arrangement [18], curve 2 for a transmitter station (see Sketch 1a), and curve 3 for a reduction valve station (measured at location 1 as shown in Sketch 1b, with the corresponding spectra given in Sketch 3 [19]). The spectra at location 8 further downstream (see Sketch 1b) are much narrower and weaker than those at location 1 (see Sketch 3), because location 1 is just behind a bend.

Working criterion:

The turbulence of a duct flow has maximum energy densities in the range of $S_d = fd/U_\infty = 0.1$ to 1.5, originating from separation of flow upstream. The intensity and broadness of this range decreases as the distance from the flow separation increases. The Strouhal number at which the peak occurs depends on the characteristics of the flow separation region and the resonant characteristics of the piping.

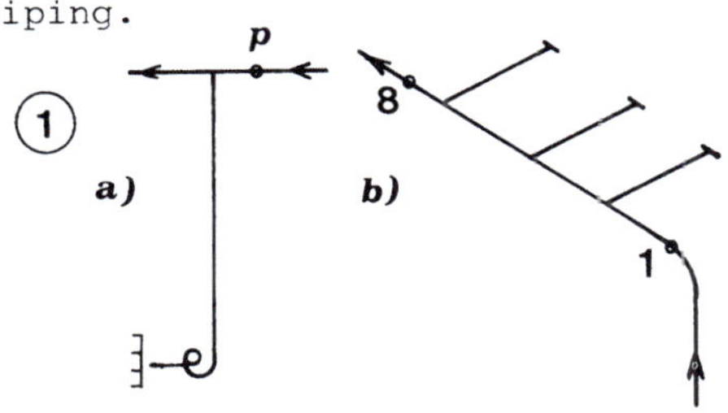

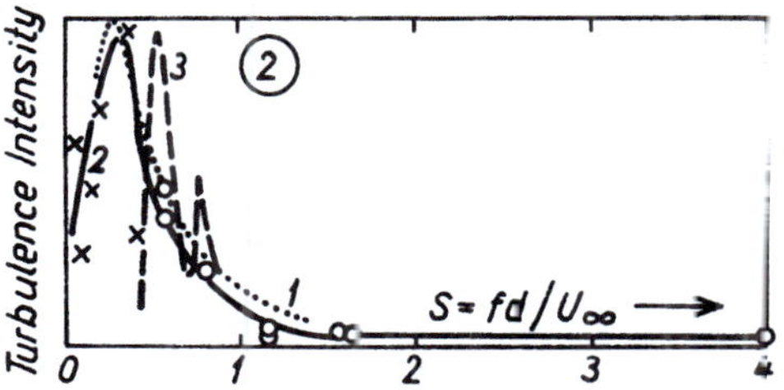

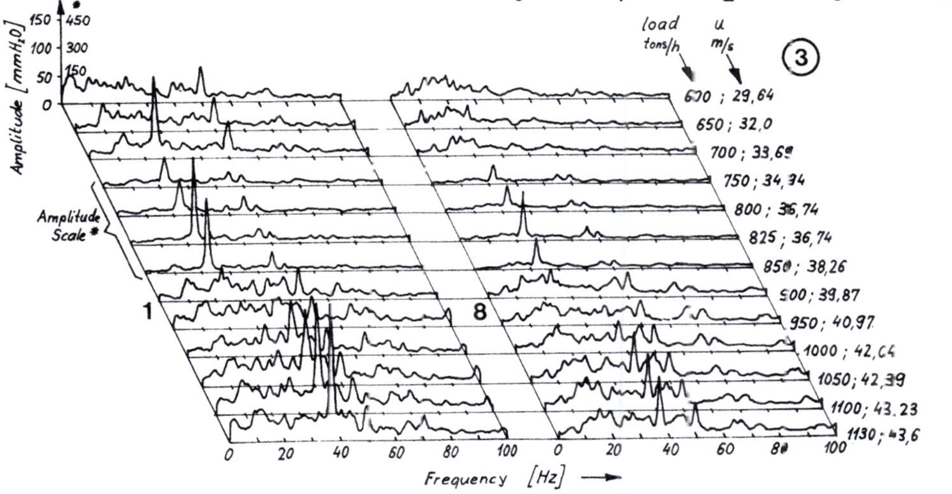

CHART I: FLOW-INDUCED VIBRATIONS IN PIPING SYSTEMS - TRANSMITTER VIBRATIONS

Problem: The resonance in the transmitter tube (Sketch 1) caused by the turbulence in the main duct, as shown previously, will excite the pressure transducer to vibrate, so that the spring spiral of the transducer can be damaged. Such damage would cause an interruption of the operation for several hours: a very large financial loss for the power plant. It was found that the strong pressure fluctuation of the low frequency resonance (on the order of about 100 Hz) was accompanied by a very weak one of the high natural frequency of the spring spiral (on the order of about 500 Hz). A number of cases of the corresponding failure occurred before the investigations had been carried out.

Remedy: The remedy was to install a vibration damper of the expansion chamber type, as shown in Sketch 2. The tube extends into the chamber by a length of 1/2 ℓ and 1/4 ℓ in order to make the frequency range of the damping very broad [20].

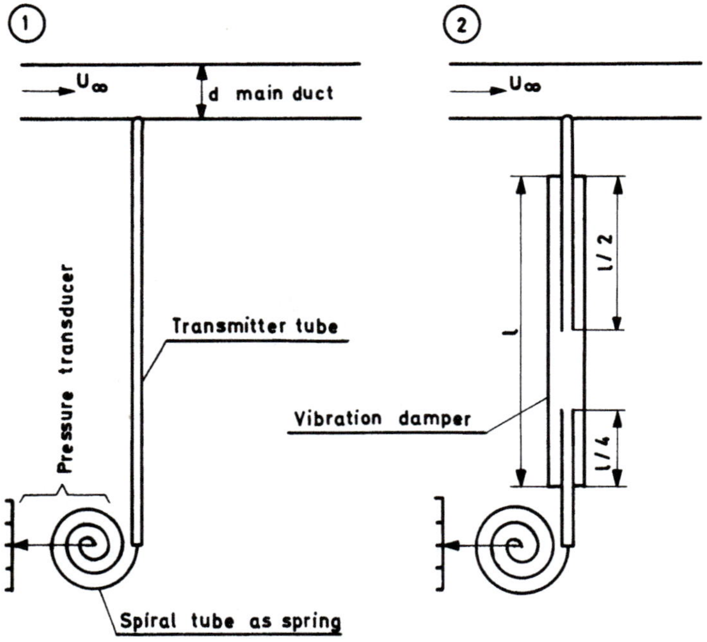

CHART J: FLOW-INDUCED VIBRATIONS IN PIPING SYSTEMS - REDUCTION VALVE STATION WITH IN-LINE ARRANGEMENT [19]

Problem:

A reduction valve station with 3 valves arranged in an in-line pattern showed (in their closed condition) strong vibrations at the second and third modes of 23 and 38 Hz, corresponding to the Strouhal number range of maximum energy density of the turbulence in the flow of the main duct carrying steam from turbine to boiler (Sketch 1). In addition, this main excition source was accompanied by the coupling with vortices shed at the valve tube apertures of two neighbouring valves. Therefore, the pressure fluctuations in the neighbouring valve tubes were always out of phase.

Theory:

A period of the vortex generation in rhythm with the pressure fluctuation of the valve tube is given in Sketch 2. The convection velocity of the vortex along the main duct is determined to be

$$U_c = (0.71 \text{ to } 0.75) U_\infty$$

which is slightly less than the convection velocity of 0.8 U_∞ of large eddies in a boundary layer, this due to movement with the vibrating steam column of the tube.
The vortex coupling will take place at a frequency f which must fulfill the following equation:

$$s = m\lambda$$

where s is the distance of the tubes, $\lambda = U_c/f$ is the wavelength and m is an integer. For the present case, m = 2 for 23 Hz and m = 3 for 38 Hz.

Cure:

As the reduction valve station was situated directly behind a bend of the main duct, the strong turbulence originating from the separated flow was the primary excitation source, also inducing strong vortex coupling.
The bend was moved far upstream to a distance of about 21 d from the first valve tube, and the vibration was practically eliminated.

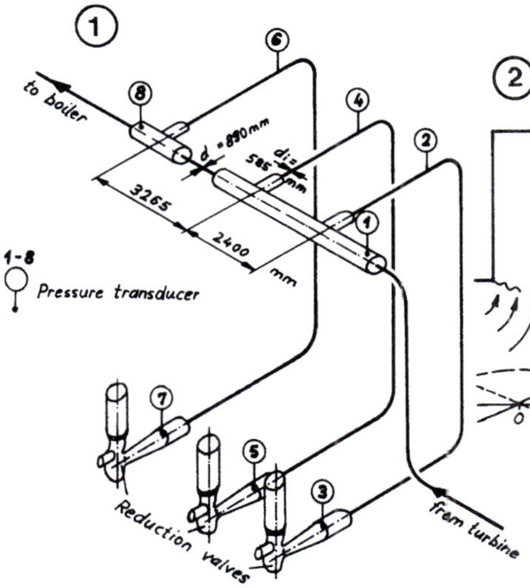

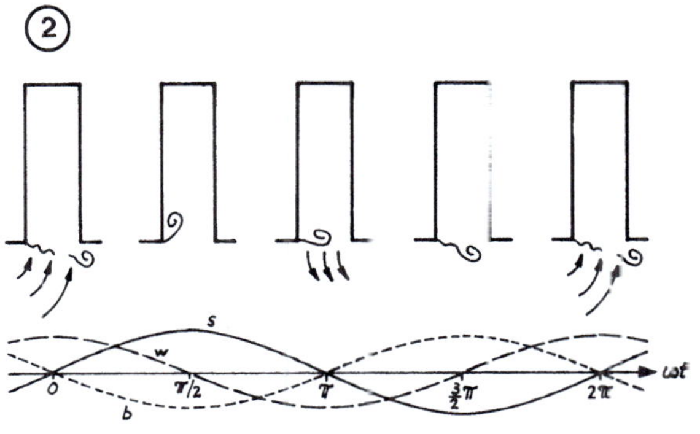

CHART K: FLOW-INDUCED VIBRATIONS IN PIPING SYSTEMS - REDUCTION VALVE STATION WITH H-ARRANGEMENT [19]

Problem:

A reduction valve station with four valves arranged in an H-position of equally long tubes of 1750 mm (Sketch 1) showed strong vibration primarily at the fundamental frequency of 91 Hz during normal operation, with the valves being kept closed (Sketch 2, curve 1). But only the second mode of 182 Hz propagated into the main duct, due to its nearly monopole nature, whilst the third mode of 273 Hz remained within the tube region owing to its dipole property (Sketch 4). These vibrations caused failure in the armatures fixed to the main duct.

Theory:

As shown in Sketch 2, the pressure fluctuations in the tubes arranged opposite to each other were coupled. In addition, the pressure fluctuations in the tubes situated in series were also coupled via the vortex interaction, because the tube distance of 1.3 m fulfilled the following equation:

$$s = m\lambda = 3\ (0.71 \text{ to } 0.75)\ (50 \text{ to } 60)/91 = 1.2 \text{ to } 1.4 \text{ m}$$

where U_∞ = 50 to 60 m/s, m = 3.

Cure:

To get rid of both the couplings along the flow and across it, a modification of the equal length of the four valve tubes to the different lengths of 2100, 1750, 2450 and 2815 mm was carried out, as shown in Sketch 3. Afterwards no resonance appeared, despite occurrence of different lower modes in different tubes. The amplitudes of vibrations detected at the armatures were reduced by a factor of 20.

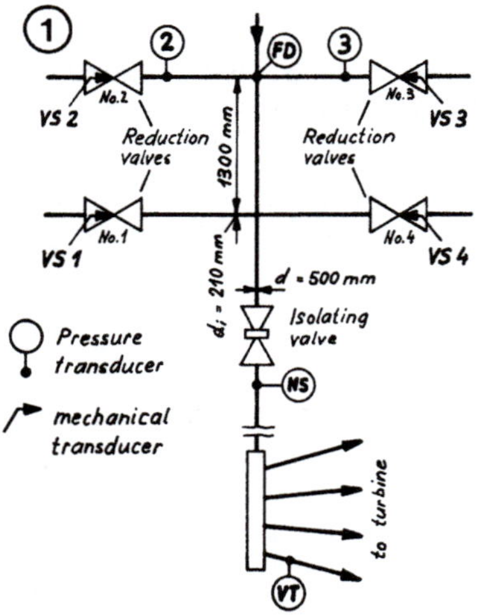

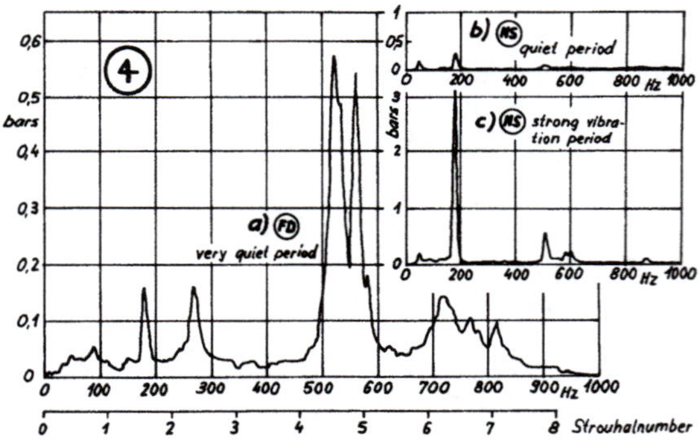

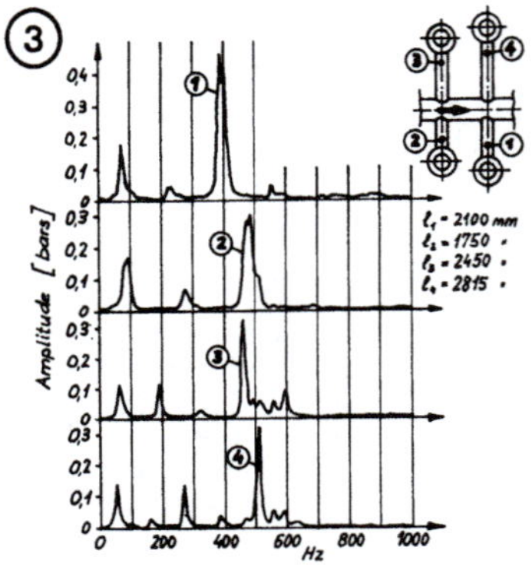

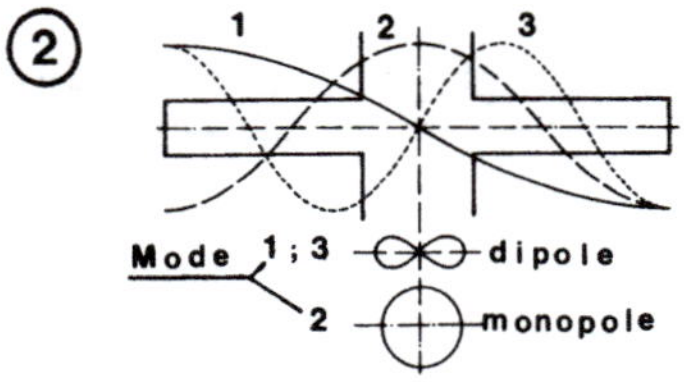

B8 PRESSURE SURGE IN THE DRAFT TUBE OF FRANCIS TURBINE

Takashi Kubota
Chief Hydraulic Engineer

Hiroyuki Aoki
Senior Hydraulic Engineer

FUJI ELECTRIC CO., LTD., Kawasaki Factory,
Tanabe-Shinden 1-1, Kawasaki, 210, Japan

SUMMARY

Pressure surge in draft tubes was experimentally investigated with a model Francis hydroturbine under partial load operation. The flow in a draft tube with inlet cone angle of nine degrees was observed under a velocity field having strong swirl. When the forced vortex core in the annular free-vortex flow grows extensively along the axis of the draft tube, cavitates under the low pressure flow field, is turned to the dead-water core, and breaks down at the elbow end of the draft tube, the resultant violent pressure pulsations produce vibration of the draft tube.

The inlet part of the draft tube was modified to the increased cone angle of fifteen degrees in order to attenuate the pressure surge. The high pressure recovery along the axis of the 15-degree cone suppressed the growth of the cavitating core under the strong swirl flow, resulting in successful cure of the draft tube surge.

1. INTRODUCTION

As is well known, air supply into the draft tube is very effective in suppressing pressure surges of a Francis turbine under partial load operation. The recent trend in using higher speeds of hydroturbines, however, results in deeper submergence of the draft tube, which generates difficulty in natural aeration. As forced aeration is expensive, the development of a new draft tube is required to minimize the pressure surge without aeration.

Palde [1] investigated the influence of draft tube shape on pressure surge. In his extensive experimental study, conical diffusers having large cone angle show smaller pressure pulsations, the reason remaining unclarified. On the other hand, Senoo et al. [2] investigated the swirling flow in straight conical diffusers in detail. They show that static pressure is very low at the center of the diffuser inlet under the swirling flow; however, it recovers rapidly downstream along the axis of diffuser, this tendency becoming larger with increasing cone angle. Both investigations [1,2] were performed in a flow field without a rotating runner (that is, without a total pressure gradient of the flow across the diffuser inlet) using air as a working fluid. Consequently, cavitation effects were absent.

Pressure surge in the draft tube of a hydroturbine is strongly influenced by the cavitation parameter. It is reported [3,4] that pressure pulsations increase at a critical value of cavitation parameter. Therefore, it can be expected that cone angle of the draft tube inlet has a deciding influence on the dependence of pressure surge on the cavitation parameter. This paper discusses experimental investigations of the influence of cone angle of the draft tube inlet on the pressure surge of a model Francis turbine in water, incorporating the above mentioned considerations, described in detail in reference [3].

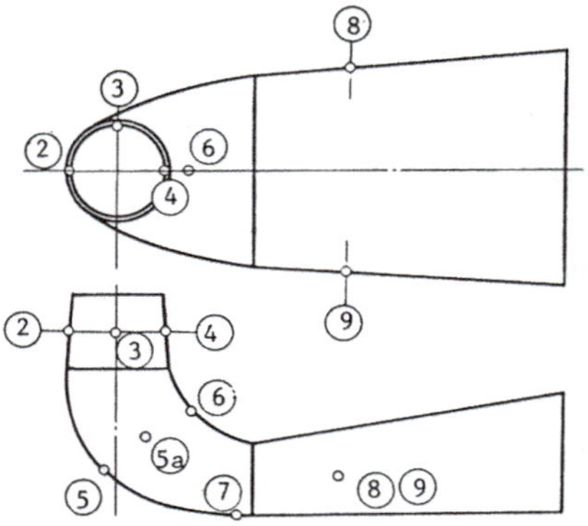

FIG. 1 MEASURING POSITIONS OF PRESSURE PULSATIONS

FIG. 2 MODEL FRANCIS TURBINE

2. GEOMETRY OF ORIGINAL DRAFT TUBE

The shape and the size of the original draft tube are shown in Sketches 1 and 2. The inlet cone angle (2θ) was selected as nine degrees which is the normal value for a straight conical diffuser. The measuring locations of pressure pulsations are shown in Fig. 1.

3. MODEL TURBINE AND TEST CONDITIONS

The draft tube was attached to a model Francis turbine shown in Fig. 2. The specific speed of the turbine is 46.1 (rpm-m^3/s-m) at the optimum efficiency point. The runner has an outlet diameter of 380 mm and thirteen buckets. Measurements of pressure pulsations were made under a test head of 40 m with changing suction head. The rated output of the model turbine was 264 kW under a rotating speed of 999 rpm.

4. FLOW AND OPERATING CONDITIONS

When the Francis type hydroturbine operates at partial load (and in some cases at extreme overload), the velocity of flow discharging from the runner with fixed blades has a swirl component which generates the Rankine-vortex flow pattern (annular free-vortex flow including forced-vortex core) at the draft tube inlet. The typical flow pattern measured using a pitot probe is shown in Fig. 3. As the static pressure in the forced-vortex core, in general, is very low at the draft tube inlet, the vortex core cavitates and becomes a dead water core, which can be visually observed, as shown in Fig. 4.

When the turbine discharge decreases at partial load operation, the axial component of flow velocity decreases and the swirl component increases at the runner outlet. The momentum ratio MR_2 of swirl to axial flow is defined as;

$$MR_2 = \int_0^{R_2} Vu \cdot Vz \cdot R^2 \cdot dR \,/\, (R_2 \cdot \int_0^{R_2} Vz^2 \cdot R \cdot dR)$$

It increases with decreasing discharge shown in Sketch 3. The pressure surge in the draft tube is closely related to the magnitude of momentum ratio [1,2]. The measured surge region is shown in Sketch 3.

5. PRESSURE SURGE

When the static pressure in the flow field becomes low at reduced suction pressure, and reaches the critical value of cavitation parameter [3,4], the cavitating core grows extensively downstream along the axis of the draft tube and breaks down violently at the near elbow end (Sketch 1 and Fig. 5). Strong pressure surge occurs as shown in Sketch 4 with a frequency of 4.2 Hz, which corresponds to 0.25 times the rotating speed. The axial and radial thrust on the runner, and the turbine output, oscillate with the same frequency. Figure 6 shows a typical oscillograph of the pressure pulsations at various measuring locations given in Fig. 1. There is no difference in phase among the pressure pulsations in the same cross-section (points 2, 3 and 4 in Fig. 1) in spite of eccentric rotation of the cavitating core.

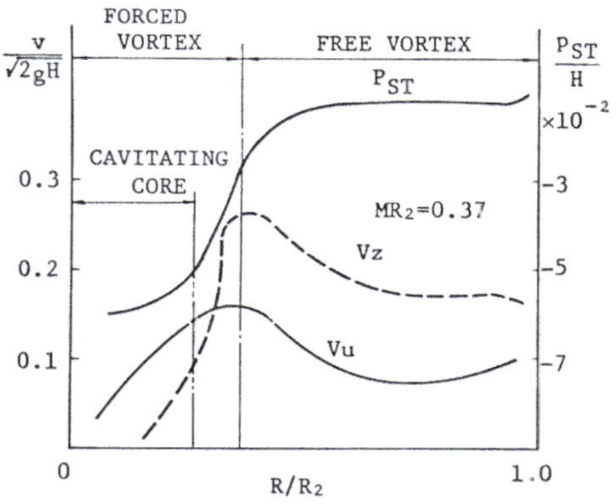

FIG. 3 FLOW DISTRIBUTION AT THE DRAFT TUBE INLET

FIG. 4 CAVITATING CORE IN THE CENTER OF DRAFT TUBE INLET

6. CURE

The inlet cone angle of the draft tube can be expected to play the decisive role in determining surge characteristics, as seen in the INTRODUCTION. When the swirl component is very small in the inlet flow, the use of a large cone angle is at the risk of flow separation. The possibility of flow separation, however, will reduce with increase of swirl. Therefore, the pressure recovery at the center will increase along the axis of the diffuser so rapidly under the strong swirl flow that backflow is apt to occur at the center of the diffuser. Consequently, it is expected that use of an inlet cone having a large cone angle and a shorter length could suppress the growth of the cavitating core, which is related to the pressure surge.

The inlet cone angle (2θ) of the draft tube was increased from nine to fifteen degrees. Figure 7 shows the modified draft tube with the 15-degree cone. The amplitude of pressure surge decreased remarkably in the 15-degree tube

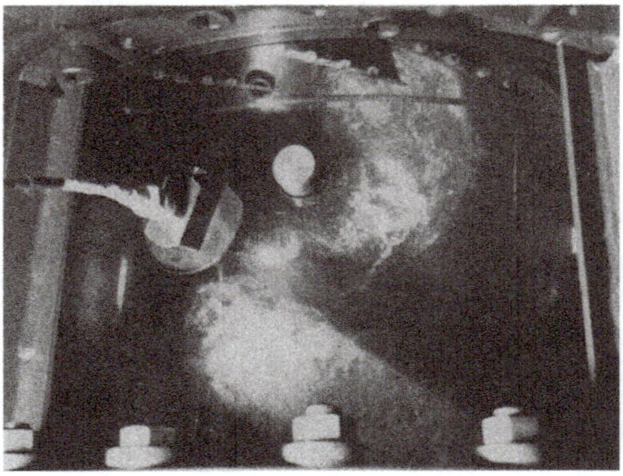

FIG. 5 CAVITATING CORE WITH STRONG SURGE (9-DEG. CONE)

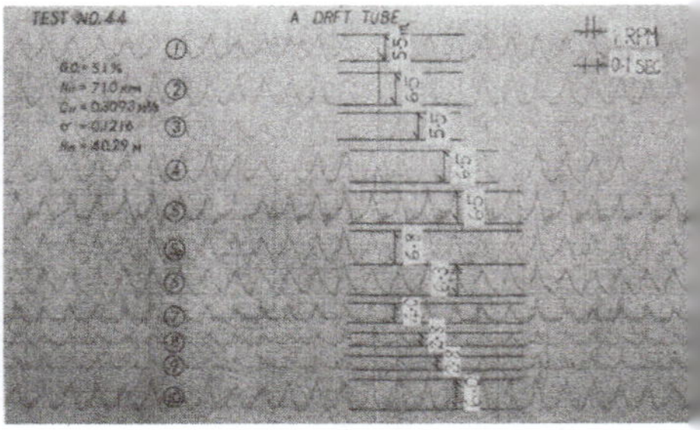

FIG. 6 OSCILLOGRAPH OF PRESSURE PULSATIONS (9-DEG. CONE)

compared with the 9-degree tube, as shown in Sketch 4 and Fig. 8. However, the frequency did not change. The observed cavitating core at the inlet did not grow downstream along the axis of the tube, and disappeared just below the entrance of the elbow of the tube (as shown in Figs. 7 and 9), because of high pressure recovery.

Measured turbine efficiencies of the model turbine having the advanced 15-degree tube were almost the same values as those of the original 9-degree tube, within the range from the optimum efficiency point to full load. At partial load with strong swirl, however, the 15-degree tube showed better diffuser performance (approx. 0.5 point higher turbine efficiency) compared with the 9-degree tube.

7. CONCLUSION

An advanced draft tube design having a 15-degree short cone proved to be the successful cure for pressure surge encountered in a Francis-type hydroturbine under partial load operation. It was especially effective regarding the dependence of pressure surge on the critical cavitation parameter.

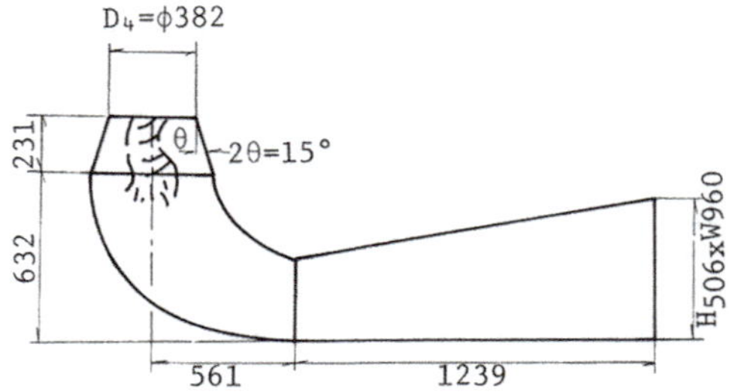

FIG. 7 ADVANCED DRAFT TUBE HAVING SHORT 15-DEG. CONE

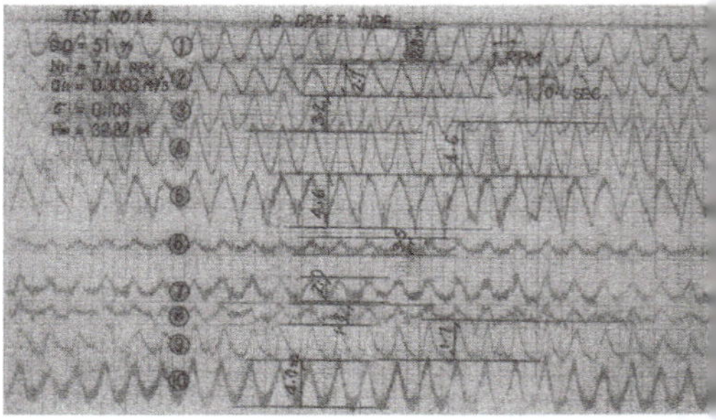

FIG. 8 OSCILLOGRAPH OF PRESSURE PULSATIONS (15-DEG. CONE)

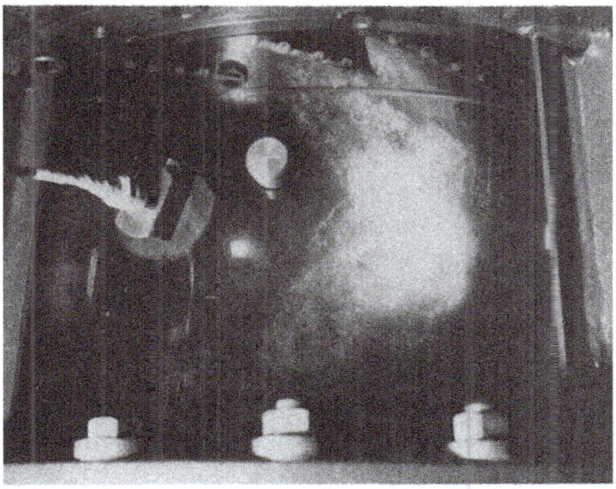

FIG. 9 CAVITATING CORE WITH 15-DEG. CONE

8. REFERENCES

1. Palde, U.J., "Influence of Draft Tube Shape on Surging Characteristics of Reaction Turbines", Bureau of Reclamation, REC-ERC-72-24, (July 1972).

2. Senoo, Y. et al., "Swirl Flow in Conical Diffusers", Bulletin of Japan Soc. Mech. Engrs., 21-151, 112-119, (1978).

3. Kubota, T. et al, "Cavitation Characteristics of Forced Vortex Core in the Flow of the Francis Turbine", Fuji Electric Review, Vol. 18, No. 3(1972).

4. Ulith, P. et al, "Contribution to Clarifying the Inception of Nonstationary Flow Phenomena in the Draft Tube of High Specific Speed Francis Turbines Operating at Part Load", Trans. IAHR 7th Symposium in Vienna,(1974).

DRAFT TUBE OF HYDRO-TURBINE

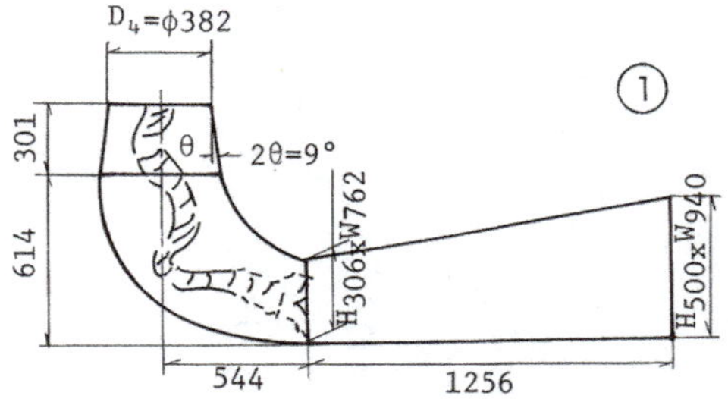

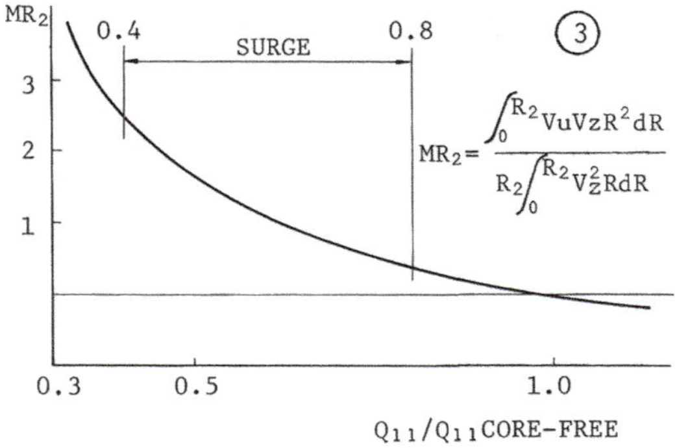

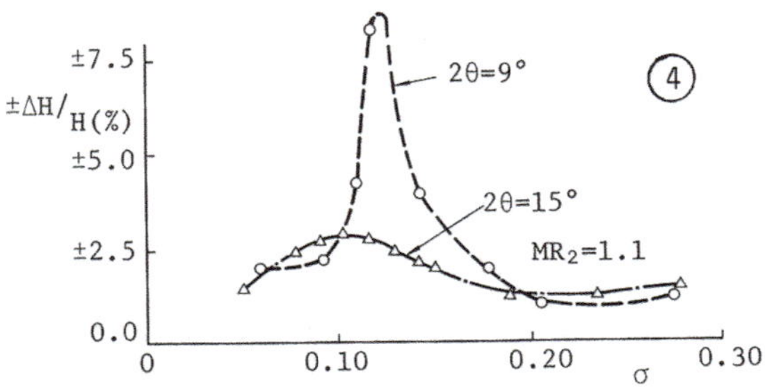

Problem: Rough operation of hydro-turbines (power swing, noise and vibration of machine, trembling of power house).

Geometry: See Sketches 1 and 2.

Boundary Condition: Inlet cone angle 9 degrees (normal value for the conical diffuser).

Flow conditions: Flow pattern at draft tube inlet consists of annular free-vortex flow field with inner forced-vortex core due to strong swirl component of flow discharging from fixed-bladed runner. Cavitating vortex core breaks down at end of elbow. (Sketch 1).

Operating Conditions: The ratio MR_2 of swirl flow momentum to axial flow momentum at inlet given in Sketch 3.

Vibration: Frequency f = 4.2 Hz (0.25 times runner rotating speed). Amplitude of pressure pulsation relative to test head of 40m at inlet given in Sketch 4.

Cure: Inlet cone angle increased to 15 degrees. Static pressure in the vortex core increased downstream along the axis of draft tube so rapidly that cavitating core broke down easily and pressure pulsation was suppressed remarkably (see Sketch 4)

Reference: Palde, U.J., Influence of Draft Tube Shape on Surging Characteristics of Reaction Turbines. U.S. Bureau of Reclamation REC-ERC-72-24, July 1972.

B9 VIBRATIONAL PROBLEMS WITH MIXED-FLOW PUMPS

Rex A. Elder
Engineering Manager
Hydraulics/Hydrology
Bechtel, Inc.
San Francisco, California

It is well known that air entraining surface vortices will cause severe vibrations and often structural failure in this type of pump. What has only lately become apparent is that subsurface vortices can develop. To my knowledge no definitive tests have been run to establish the conditions under which these develop. However, from our experience, the back wall vortices are reduced in intensity as the pump is moved away from the back wall. Undoubtedly, the side wall ones will respond similarly. We do know, however, that as the pump is moved away from the back wall, the side wall vortices increase in intensity. While two counter rotation vortices can develop on the floor or back wall, usually the flow approach conditions cause one to dominate over the other. The more uniform the flow, the more likely that both will appear.

These internal or subsurface vortices, if strong enough, will cause a structural vibration at the rotating blade frequency i.e. — machine speed times the number of blades. In some cases, the vortex effect appears to be strong enough to persist through the impeller, resulting in a frequency equal to machine speed times the number of blades times the number of diffuser vanes.

Small pumps have sufficient rigidity so that the internal vortex problem is essentially unnoticed. In very large pumps of the type now used for condenser water systems, the structural rigidity tends to be much lower, the hydraulic forces larger, and the resonant frequencies of all parts lower. Therefore, there is a possibility of the internal vortices exciting some part of the large pumps at the resonant frequency.

We have found the possibility that flow conditions approaching a wet well pump result in some amount of pre-rotation of the flow entering the pump. This pre-rotating flow by itself does not appear to cause vibrations directly. Instead, it may create cavitation which then can feed high frequency vibrations into the machine. In addition, with progressive loss of metal due to cavitation erosion, mechanical imbalance can result, along with its associated vibration. The high frequency vibrations, if they occur, can feed into any area of the machine, since the pump structure exhibits essentially no damping for high frequencies.

Actual vibrational damage to the large mixed flow pumps can be spectacular, i.e. thirty-three out of thirty-six, 1 1/2 inch, bolts failing due to fatigue; and tearing of one half to three quarter inch plate in the main pump body cylinders. Stiffening vanes on the intake bells were snapped off or broken out of the parent metal to which they were welded. Some secondary failures also occurred after the first structural failure, due to the high power required to drive these pumps. The unleashed power can tear the motor mounts loose, rip the discharge elbows, etc.

Hydrodynamic solutions to each of the causitive forces have been developed on a case by case basis but no definitive basic designs have been developed. Floor splitters and cones under the bell work sometimes are effective. Splitters on the back wall seem to be effective for elimination of most back wall vortices. The side wall vortices are particularly difficult to control, as is pre-rotation.

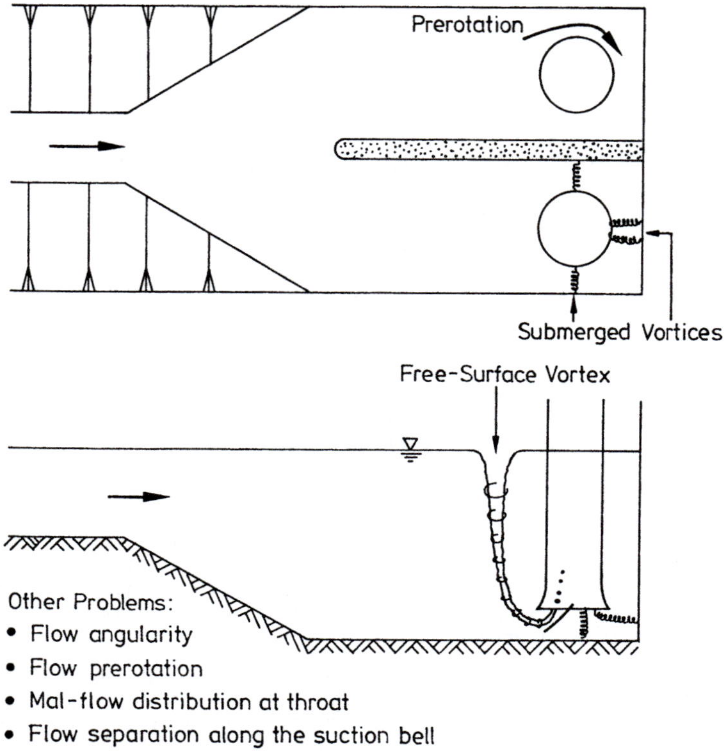

FIG. 1 HYDRAULIC PROBLEMS AT PUMP INTAKES

B 10

SWIRLING-FLOW INDUCED VIBRATIONS IN TURBOMACHINE EXIT CHAMBERS

Marcel Escudier

Leader, Fluid Mechanics Research Group
Brown Boveri Research Centre
Baden, Switzerland

SUMMARY

The paper describes a detailed model investigation into the cause and suppression of periodic pressure fluctuations in exit chambers of the type commonly employed in an axial-flow turbomachine to collect the throughflow from an annular passage (diffuser, blading, etc.) and divert it radially away from the machine's axis. That the oscillation frequency obeys a constant Strouhal number law (i.e. frequency proportional to flow rate) suggests that vortex-flow effects are at the root of the problem. This is shown to be the case, a strong vortex being created by tangential inflow into the torus-like main body of the exit chamber. A simple baffle in the radial exit pipe, to separate the two arms of the vortex which meet and interact there, suppresses the oscillation at the cost of increased pressure loss. Geometric changes aimed at weakening the vortex itself are effective in avoiding periodic pressure fluctuations whilst at the same time reducing the loss, which in large part also stems from the vortex flow.

1. INTRODUCTION

A problem inherent in the design of many axial flow turbomachines is that of collecting the axial throughflow and diverting it radially away from the machine's main axis. This is the case, for example, for a large turboset where steam flows into and out of several turbines (high, medium and low pressure) running on a common shaft and also for the valves which control the steam flow. A similar situation is often encountered in the interconnection of the component parts of industrial gas turbines: the compressor, the combustor and the turbine. A commonly encountered solution to this problem is the use of a ring-type exit chamber (or ring diffuser) with annular inflow and radial outflow, such as the model geometry shown schematically in Sketch 1 of the Chart. In practice, the ring cross section is usually non-circular, either for reasons of strength or to save space; and the axial inflow may be complicated by such upstream influences as a diffuser or blading stages, leading to a highly non-uniform, swirling turbulent flow. The proportions of the model shown are close to those encountered in practice: the ratio t/R (t being the width of the inlet annulus and R the radius of the ring cross section) may be varied through the use of inserts, but has a value of 0.31 unless stated otherwise. For this latter case, the inlet annulus, torus and outlet tube all have the same cross-sectional area. Practical exit chambers may have diameters ranging in size from less than a metre to several metres, and the fluid medium may be steam, hot compressed air or even water. It may be remarked in passing that the geometry of Sketch 1 is also close to that employed for the inlet chambers of axial-flow machines, the flow direction then being in the opposite direction to that shown. Here too vortex phenomena play an important role, as discussed by Merkli and Escudier [1].

The tangential inflow into a ring chamber generates a strong vortex which is unstable and leads to well-defined periodic fluctuations in pressure the understanding and avoidance of which are the subject of this paper. Previous workers, such as Deych and Zarayankin [2] and Suter and Girsberger [3], who were concerned primarily with the question of pressure loss, have apparently not appreciated the dominant role played by the vortex in determining the flow characteristics of ring type exit chambers (see also Escudier [4]).

2. OBSERVATIONS OF PERIODIC BEHAVIOUR

The periodic nature of the flow was first revealed by observations of water flow in the model described above, which is constructed largely of clear plexiglas to permit flow visualization and LDV measurements. At sufficiently high flow rates, the vortex core becomes visible due to gaseous cavitation in this low pressure zone, as seen in Figure 1. The two views of the vortex core shown, a plan view and a composite side view (to reduce optical distortion), reveal its spatial structure to be that of a helix wrapped around the ring axis. Also shown is the circumferential distribution of static pressure which evidently "mirrors" the vortex core. Unless suppressed, as described below, the vortex is observed to oscillate periodically about the central axis of the torus.

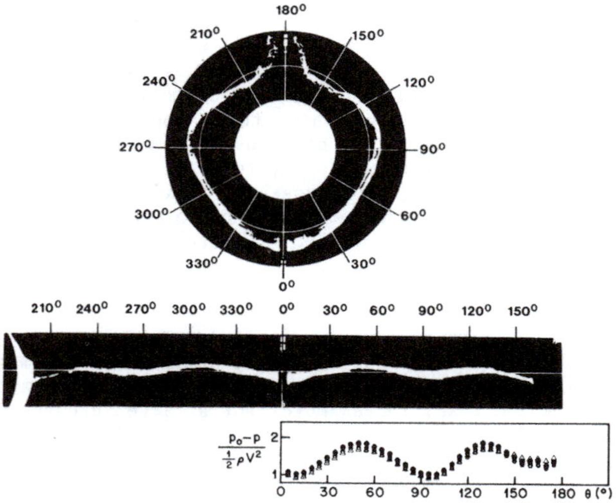

FIGURE 1 APPEARANCE OF VORTEX CORE IN WATER FLOW AND CORRESPONDING DISTRIBUTION OF STATIC PRESSURE

Measurements of the surface pressure in airflow through a similar model to that used for water flow confirm that the fluctuations are strongly periodic, as seen from the frequency spectra of Sketch 2. The frequency f of the main peaks in this case ($t/R = 0.31$) turns out to be given by $f\rho D^3/\dot{m} = 1$, ρ being the fluid density, $\dot{m}$ the mass flow rate and D the mean ring diameter. However, a more general scaling is found to be $fR/V = 0.026$, wherein V is the flow velocity in the annular inlet. The Strouhal number defined in this way is plotted against the Reynolds number $Re = 2\rho Vt/\mu$ in Sketch 3 for various values of the inlet slit width t. The data shown represent the fundamental and harmonics of the oscillation up to sixth order, as well as some data between the second and third harmonics which do not fit the main pattern. The harmonic order is indicated by the number adjacent to each set of data points.

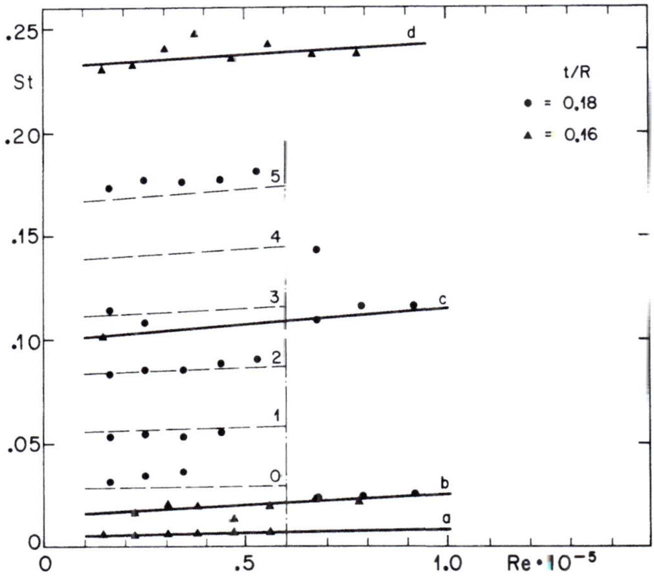

FIGURE 2 STROUHAL NUMBER VS REYNOLDS NUMBER FOR NARROW INLET ANNULI

The additional data for very narrow inlet annuli, shown in Figure 2, whilst still appearing to obey a constant Strouhal number law, do not follow precisely the same trends as those of Sketch 3.

When the frequency of the vortex fluctuations matches an acoustic eigenfrequency of the ring-chamber cavity, an interaction occurs resulting in the bi-stable behaviour shown by Figure 3. Although true frequency locking does not occur, presumably because phase matching is not possible, there is a random variation with time of the rms fluctuating pressure p' between two well-defined levels: a lower normal level, and an upper enhanced level.

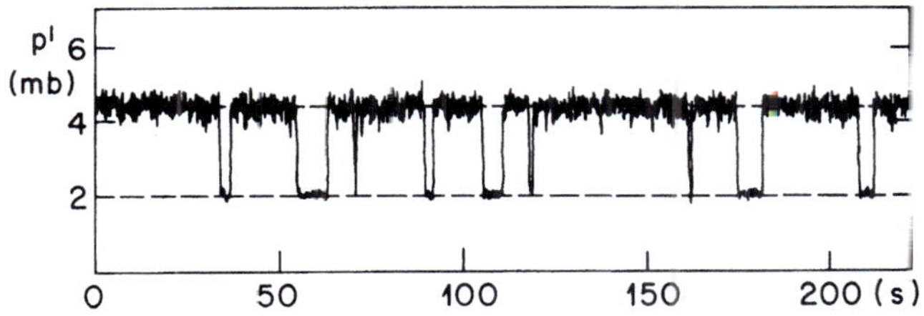

FIGURE 3 BI-STABLE BEHAVIOUR OF FLUCTUATING WALL PRESSURE FOR $t/R = 0.16$, $Re = 4 \cdot 10^4$

3. SUPPRESSION OF PERIODIC BEHAVIOUR

A flow divider installed at $\theta = 0°$ (see Sketch 1) has the effect of slightly changing the Strouhal number of the vortex oscillation, but is ineffective in reducing its amplitude. A splitter plate in the exit tube, on the other hand, is much more successful as is seen from the frequency spectra shown in Figure 4. The principal effect of the splitter plate is to separate the two arms of the vortex where they meet at the ring exit, and it is therefore concluded that it is a vortex interaction in the exit tube which is responsible for the oscillation. Other measurements reveal that the vortex structure in the ring is unchanged by the installation of the splitter plate.

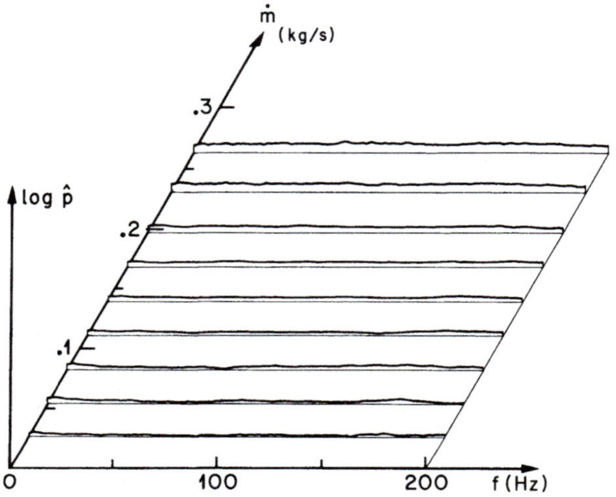

FIGURE 4 FREQUENCY SPECTRA OF FLUCTUATING PRESSURE WITH EXIT-TUBE BAFFLE INSTALLED

As discussed by Escudier [4], a large amount of the available flow energy is imparted to the flow as swirl and the high pressure loss of ring-type exit chambers is a direct consequence of the dissipation of this swirling kinetic energy. It is unfortunately the case that the exit-tube splitter plate has the effect of increasing the loss coefficient $\xi \equiv 2\Delta p_o/\rho V^2$ by about 25%, from an already high value of about 3.0 in its absence. By contrast, simple geometrical changes can be made to the ring cross section which have the direct influence of weakening the vortex by increasing the frictional constraints on the vortex but at the same time reducing the overall losses. Figure 5 shows three alternative ring cross sections which have been investigated, together with the corresponding distributions of swirl velocity (LDV measurements) and loss coefficient ξ. A significant reduction in ξ with reduction of the vortex strength (in each case the flow rate is the same) is clearly seen and is accompanied by complete elimination of periodic pressure fluctuations in each case. The important difference between suppression, with a splitter plate, and elimination at source is clearly evident. It is also found that the angular distribution (i.e. with θ) of the mean static pressure becomes much more uniform as the vortex strength is reduced. Further details of the flow in ring-type exit chambers are discussed by Escudier and Merkli [5].

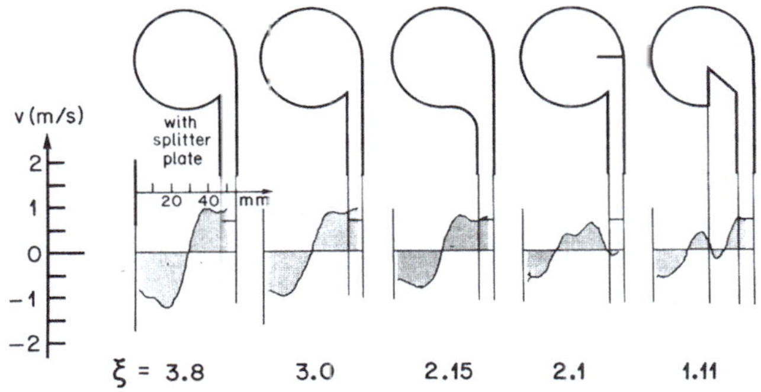

FIGURE 5 SWIRL VELOCITY DISTRIBUTIONS FOR VARIOUS RING CROSS SECTION AND THE ASSOCIATED VALUES OF ξ

4. ACKNOWLEDGEMENT

The author gratefully acknowledges the contributions to the work on which this paper is based of J. Bornstein, P. Merkli, W. Kunz and A. Biland.

5. REFERENCES

1. Merkli, P. and Escudier, M.P., "Observations of flow in a ring inlet chamber", Transactions of the ASME, Journal of Fluids Engineering, 101, no. 1, 135-142, (1979).
2. Deych, M.Y. and Zarayankin, A.Y., "Gas dynamics of diffusers and exhaust ducts of turbomachines", FTD-MT-24-1450-71, (1972) (Translated from Russian)
3. Suter, P. and Girsberger, R., "Strömungstechnische Gestaltung des Austrittsstutzens von Axialmaschinen", W. Traupel Festschrift, Junis-Verlag, Zürich, (1974)
4. Escudier, M.P., "Estimation of pressure loss in ring-type exit chambers". BBC Report KLR 79-08 C (1979) (accepted for publication, Transactions of the ASME Journal of Fluids Engineering)
5. Escudier, M.P. and Merkli, P., "Observations of the oscillatory behaviour of a confined ring vortex", AIAA Journal, 17, pt. 3, 253-260, (1979).

TURBOMACHINE EXIT CHAMBERS

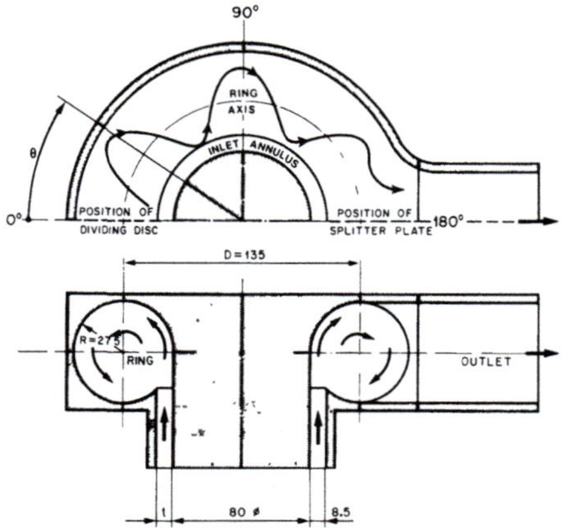

① Schematic diagram of model ring chamber (dimensions in mm)

Problem: Vortex generated by tangentia inflow leads to periodic pressure fluc tuations.

Geometry: See Sketch 1. Model geometr with D = 135 mm. Practical geometries ma reach several metres in diameter.

Flow conditions: Upstream flow may be swirling, non uniform, turbulent. Flui may be steam or hot gas.

Oscillation: Oscillations with $f \sim \dot{m}$ occu for most flow conditions - see Sketches 2,3 Particularly severe when acoustic inter action at cavity eigenfrequency occurs

Cures: Splitter plate at θ = 180° effectiv (divider at θ = 0° ineffective) but increase pressure loss. Better approach to weaken vor tex directly by modifications to ring cros section - oscillation eliminated and losse reduced.

Reference: Escudier, M.P. and Merkli, P. Observations of the oscillatory behaviou of a confined ring vortex. AIAA Journal 17, pt. 3, 253-260, 1979.

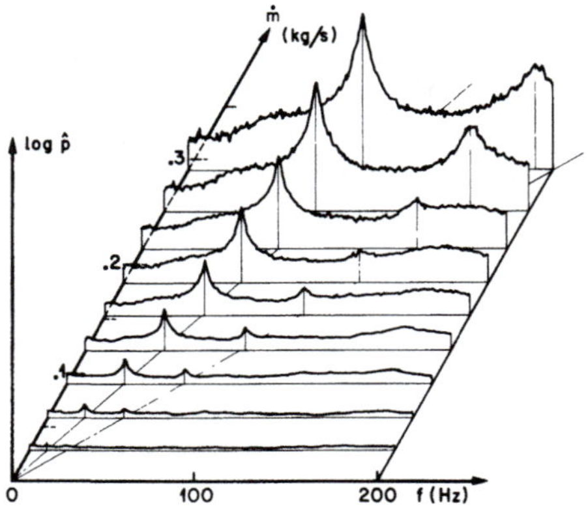

② Frequency spectra of fluctuating pressure

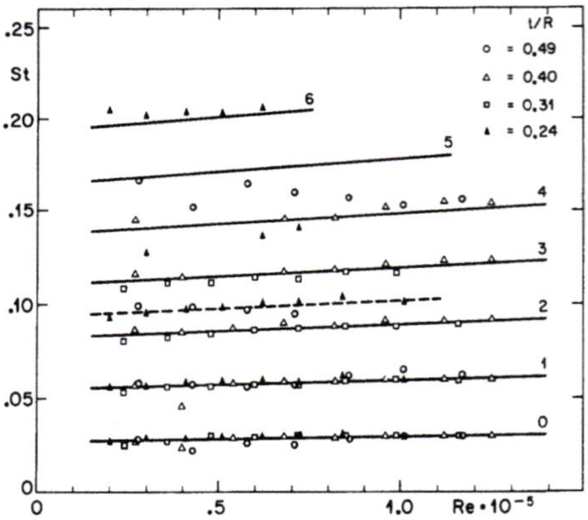

③ Strouhal number vs Reynolds number. St ≡ fR/V, Re ≡ 2Vt/υ.

B 11 SELF-EXCITED OSCILLATIONS IN A HYDROELECTRIC UNIT

Adolf Hermann Glattfelder
Herbert Grein
Ludwig Huser

ESCHER WYSS LIMITED, Zuerich, Switzerland

SUMMARY

In a power plant with three identical Francis turbines, severe pressure pulsations, not consistent with previous experience, appeared at low partial load. A cure had to be found, as any prolonged operation in the critical load range would considerably reduce the life expectation of some system components. The paper describes the main effects of the phenomenon, proposes an excitation mechanism, and gives a cure.

1. OBSERVED VIBRATION

A series of tests were performed on the prototype (sketch 1). Analysis of the results showed that the pressure pulsations (sketch 2) always appeared in the 16% to 25% range of nominal generator power for slow loading, as well as unloading. However if a normal loading, or unloading, change of speed (> 3%/s) was applied, the pulsation intensity remained below generally accepted limits (5% of head). The pulsation appeared on each of the three units at the same power level, independent of the power level of the other machines. Furthermore the pulsation intensity was found to increase with the head [1].

The main spectral component was 1.8 ± 0.2 Hz (or 0.8 f_n with f_n being the runner rotational frequency) for all three units. This does not coincide with well known excitations by f_n, or by the partial load draft tube vortex for Francis turbines, which was clearly visible at 0.25 f_n and at 25% to 40% of nominal generator power. The second harmonic (2 f_n) appeared only for strong pressure pulsations. Moreover there was no clear dependence of frequency on head, in contrast to work reported in [1].

The pressure amplitude at the spiral casing was always lower than in the turbine. No phase shifts of the predominant spectral component of pressure were observed between the spiral casing, turbine and draft tube entry. However the generator output oscillation (sketch 2) showed a phase shift of approximately -270°; -90° of which is due to the generator subsystem being near resonance. This leads to a phase shift of ca. 180° between pressure and turbine power (or turbine angular momentum). No oscillations of turbine speed were observed, which is consistent with calculations, which predict very small variations.

In the critical range, the 1.8 Hz component was clearly visible on the axial bearing pressure recording, but not on the recording of the radial position of the rotor. However, the latter recording clearly showed the partial load vortex.

Finally, the guide vane position was clearly not influenced by the pressure pulsation.

This list of observations, together with those given on the Chart, along with other details, provided a good basis for screening possible excitation mechanisms. The following one was consistent with present observations and those reported in references [2] and [3].

2. SOURCE OF VIBRATION

The system (see sketch 1) consists of four main subsystems: penstock, turbine, draft tube and generator. These subsystems clearly are stable if they are isolated from the other subsystems. However their interactions may produce an unstable system, as, for instance, in a feedback control loop: pressure variations in the penstock and draft tube change the flow pattern in the turbine, and thus the momentum or/and torque in such a way that the initial pressure fluctuations are amplified.

The turbine flow pattern change is explained in Fig. 1. Consider the pressure between guide vanes and runner to be lower than its steady state value. Consequently the velocity at the guide vane exit increases (Fig. 1a). After a finite delay, C_T at the turbine blade will increase accordingly (Fig. 1b), changing the angle of attack [4]. If then the flow separates from the low pressure side of the blade, the momentum at the runner exit will switch according to Fig. 1c. The higher axial component increases the pressure in the turbine, which reduces the velocities and finally the flow will switch back again, but with a hysteresis [2]. The impulse at the exit of the runner also acts on the draft tube subsystem. Its response will be a pressure oscillation, with low damping, transmitted back into the turbine. If the draft tube resonance frequency is compatible with the finite delay time at flow conditions where the hysteresis effect is also present, then a self-induced oscillation can result.

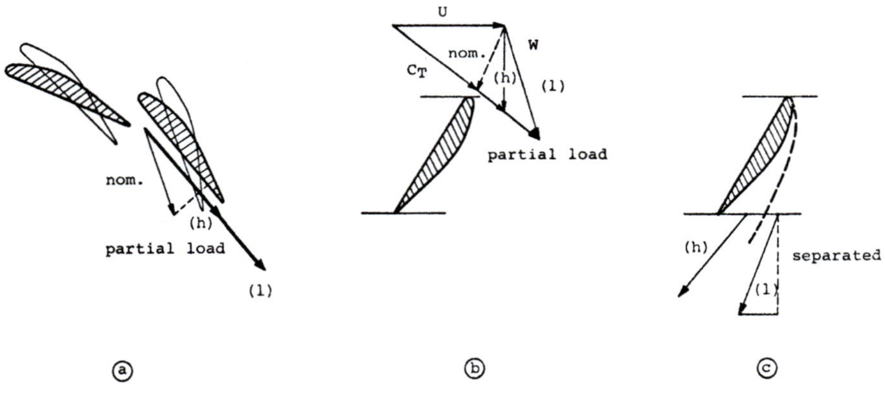

FIG. 1 TURBINE FLOW PATTERN AT NOMINAL AND PARTIAL LOAD CONDITIONS WITH TURBINE PRESSURE LOWER (l) AND HIGHER (h) THAN STEADY STATE AT THE GUIDE VANE EXIT (a), RUNNER ENTRY (b) AND RUNNER EXIT (c).

This instability was investigated further using a mathematical model, Fig. 2. It is valid for small deviations of variables from their steady state values in the critical flow range, and only around the main spectral component observed. The mathematical models of the subsystems are therefore linear and of the lumped parameter type [5], describable by rational transfer functions, except for the turbine with the hysteresis effect. The factor β defines its weight relative to the normal linear transfer of momentum.

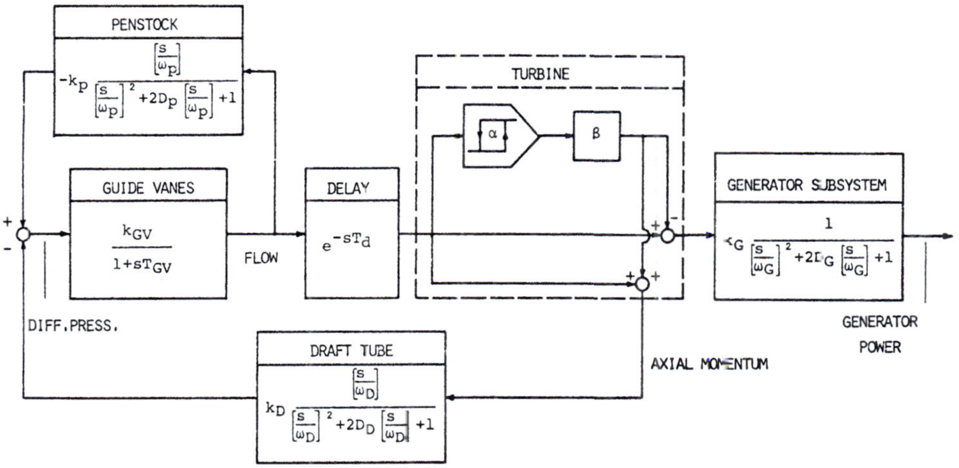

FIG. 2 THE SYSTEM MODEL IN BLOCK DIAGRAM FORM, VALID FOR SMALL DEVIATIONS OF THE CRITICAL POWER REGION AROUND THE MAIN SPECTRAL COMPONENT.

The system stability is examined by the "describing function method" often used in control theory [5] for this class of nonlinear system. Fig. 3 indeed predicts a limit cycle if the hysteresis effect exceeds a certain limit, which is determined by the coefficients of the linear part of the system (Fig. 2).

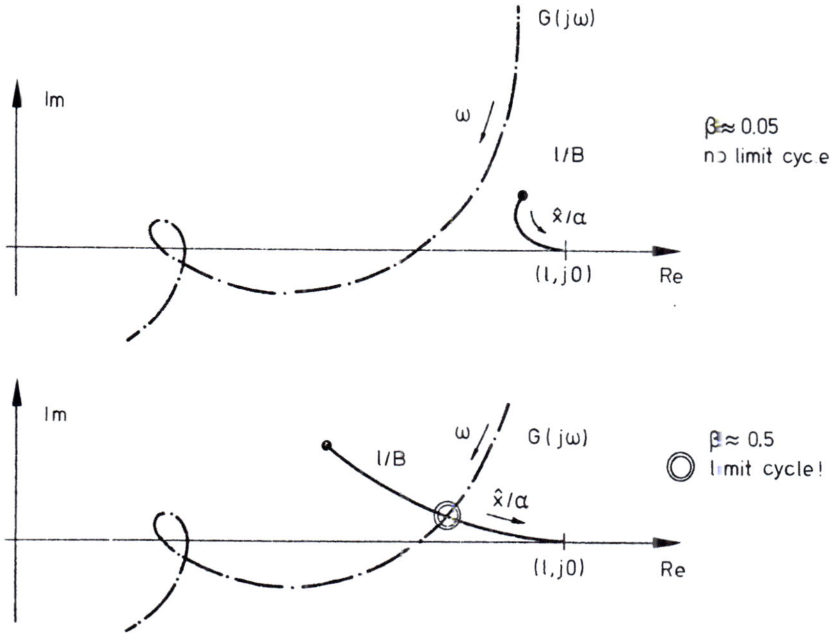

FIG. 3 STABILITY ANALYSIS OF THE SYSTEM MODEL USING THE DESCRIBING FUNCTION TECHNIQUE.

3. THE CURE

Load swings on Francis turbine systems can be attenuated by several well known methods [1]. In this case, however, the analysis suggests another solution. The basic idea is to avoid the intermittent flow separation by reducing C_T in Fig. 1b. This is achieved by holding every other guide vane in the shutdown position through the critical load range (Fig. 4); that is, by use of a deliberate guide vane asynchronism. It was easy to implement in this case, as each guide vane has its own positioner feedback loop. The test results on the prototype are not quite complete. For low heads the pressure amplitudes have been reduced to about one third their original value, with no appreciable negative side effects.

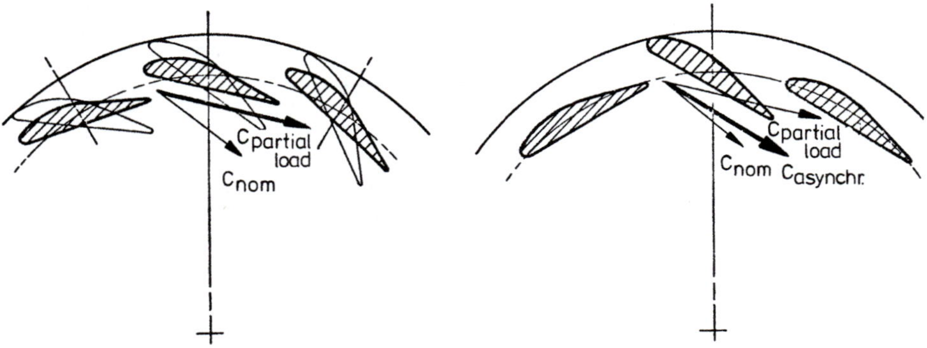

FIG. 4 GUIDE VANE EXIT VELOCITIES FOR SYNCHRONOUS (LEFT) AND ASYNCHRONOUS (RIGHT) GUIDE VANE POSITIONING AT PARTIAL LOAD.

4. CONCLUSION

The identification of such a complex excitation mechanism requires a systematic approach. Extensive measurements of many system variables on the prototype, not only of peak to peak values, but also of frequencies and phase relations as well, are necessary for subsequent analysis. On this basis, possible cures can be tested prior to their implementation on the prototype. Lately, this approach has been used to solve another case of overload power swings on a Kaplan turbine system.

5. REFERENCES

1. Purdy, C.C., "Reducing power swings of Tarbela's turbines", Water power & Dam construction, April 79, 23-27.

2. Yamabe, M., "Hysteresis characteristics of Francis pump turbines when operated as turbines", Trans. ASME, J. Basis Engng., March 1971, 80-84.

3. Brennan, C., "The unsteady dynamic characterization of hydraulic systems with emphasis on cavitation and turbomachines", Proc. ASCE-IAHR-ASME Joint Symposium, Ft.Collins, 1978, Vol. 1, 90.

4. Jansen, W., "Flow analysis in Francis water turbines", Trans. ASME, J. Engng. Power, July 1967, 445.

5. Truxal, J.G., "Automatic control system synthesis", Mc Graw Hill, 1955.

HYDROELECTRIC 3-UNIT PLANT WITH FRANCIS TURBINES

Problem: Pressure and power pulsations at very low partial load.

Geometry:

①

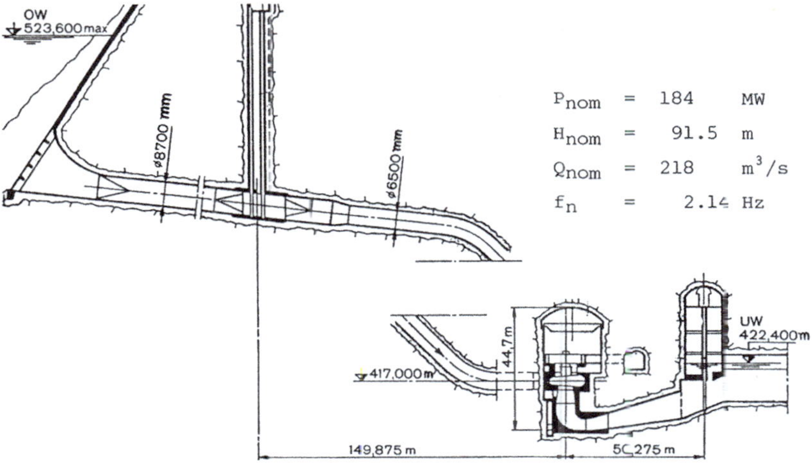

P_{nom} = 184 MW

H_{nom} = 91.5 m

Q_{nom} = 218 m^3/s

f_n = 2.14 Hz

Operating conditions: Pulsation at 30 to 50 MW (or 16% to 25% of nominal power) and heads of 90 to 110 m, independent of load conditions of other units.

Structural dynamics: Lowest resonance of penstock (observed at no-flow): 1.67 Hz. Draft tube resonances at 1.6 to 2.0 Hz and 3.2 to 3.7 Hz (observed at critical flow). Damping ζ below 0.10. Generator subsystem resonance at 1.8 Hz, ζ = 0.15 at 50 MW.

Pulsation: Main spectral component at 1.8 ± 0.2 Hz.

②

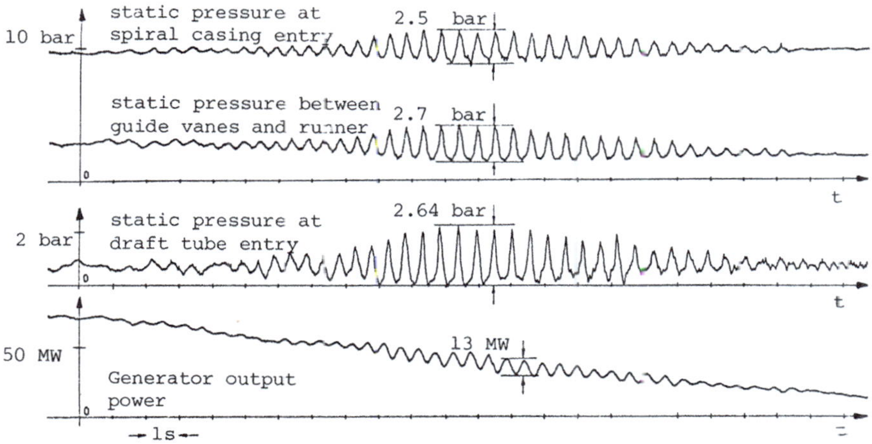

Cure: "Asynchronous" guide vane opening from 0 to 60 MW proposed in order to change flow conditions through runner.

B12 REDUCTION OF VIBRATIONS IN A CENTRIFUGAL PUMP HYDRAULIC SYSTEM

Y.N. Chen, Dr.-Ing., D. Florjancic, Dr.-Ing., R. Stürchler, Ing.
SULZER, Winterthur, Switzerland

SUMMARY

Strong, self-excited pressure fluctuations in the hydraulic system of a feed pump occurred at a small load, with a low frequency corresponding to the second resonant mode of the system. The cause was found to be unsteady flow separation in the non-return valve. This strong second mode of pressure fluctuation gave way to a weak third mode, when the feedback loop was disconnected. The damping capability of the pump against such a self-excited vibration is discussed.

1. PROBLEM

Self-excited vibrations with a frequency of about 11.9 Hz occurred in the piping system of a centrifugal pump unit during the combined operation of leak-off and main discharge flow. This frequency was much lower than the running speeds (24.9 Hz for the booster pump, and 114.2 Hz for the main feed pump). The mechanical vibrations of the piping system reached amplitudes as large as 2.3 mm (peak to peak).

2. ARRANGEMENT OF THE UNIT

See Fig. 1 for location of stations for measuring pressure fluctuations (P) and mechanical vibrations (M). The unit consists of a booster pump, motor, gear box, main feed pump, leak-off valve, non-return valve, control valve, shut-down valve, suction pipe, cross-over pipe, discharge pipe and leak-off pipe. Piezoelectric transducers (Kistler) were used for the measurement of the pressure fluctuations.

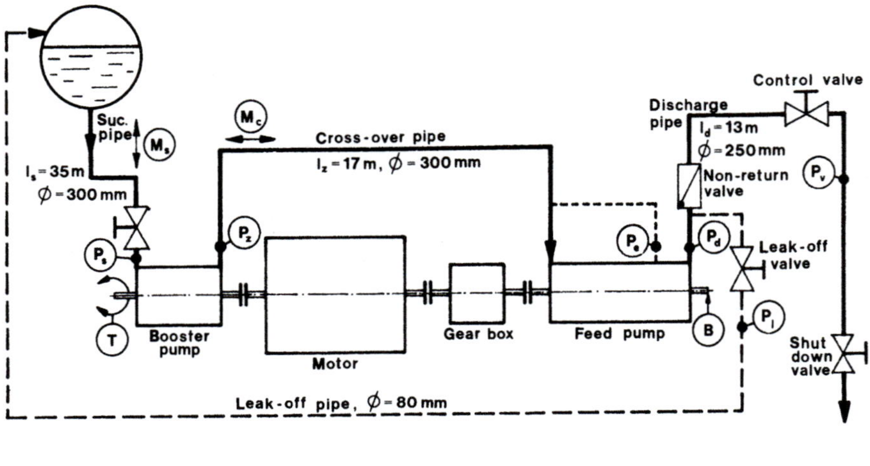

FIG. 1 SKETCH OF THE PUMP UNIT

3. VIBRATION

The vibration pattern in the hydraulic column extending from the suction pipe to the control valve, as measured by means of pressure transcucers (Fig. 1), revealed a natural vibration in the second mode (about 11.9 Hz), in which a pressure nodal point appeared within the main feed pump region (see Sketch 1a).

4. EXCITATION SOURCE

Periodic vortices were formed in the free shear layer between the jet and the dead water zone behind the non-return valve due to shear layer instability (see Figs. 2 and 3).

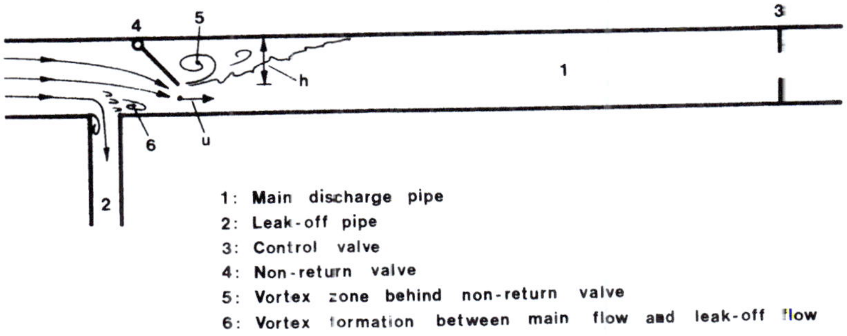

FIG. 2 VORTEX ZONES BEHIND NON-RETURN VALVE AND BETWEEN MAIN FLOW AND LEAK-OFF FLOW

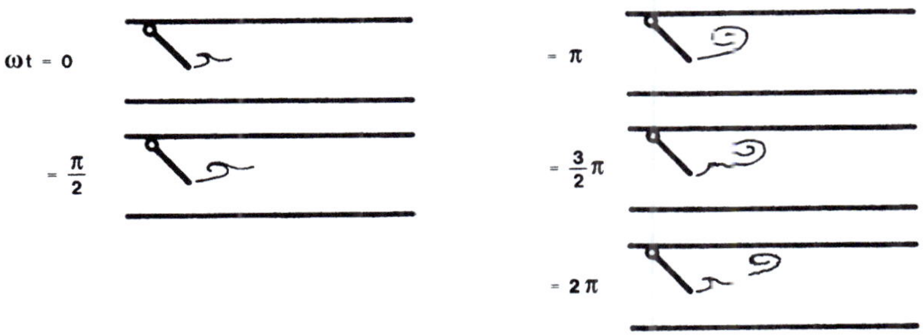

FIG. 3 VORTEX EXCITATION MECHANISM

The Strouhal number of the vortex formation possessed the following values

$$S = f\,h/U = 0.08 \text{ to } 0.32 \qquad (1)$$

according to the evaluation given in reference [1], where U is the flow velocity through the opening of the valve and h is the characteristic length corresponding to the depth of the dead water zone. For the present case, we have U = 5.5 m/s and h = 170 mm for a flow rate of $\dot{m}$ = 50 kg/s during the period of strong vibrations. The Strouhal number for the vibration of 11.9 Hz was found to be

$$S = 11.9 \times 0.170/5.5 = 0.37$$

a value which corresponds very well to eqn. (1). The vortex growth generates a field of velocity and pressure fluctuations, which excite the water column to resonance. This feeds back energy to the vortex, strengthening its intensity. In this manner, a coupling cycle for the flow-induced, self-excited vibration is completed. The periodicity of the vortex development is then controlled by the frequency of the water column resonance [4]. Any fluid-elastic coupling between the non-return valve and the fluid column must be excluded, because no torsional vibration of the shaft of the butterfly plate of the valve could be measured during the test.

5. CURE

In order to avoid the coupling mentioned, the control valve was moved from the original position of 13 m behind the non-return valve to a position of 2 m behind it. This was the minimum distance which could be achieved. This measure reduced the vibration amplitudes to considerably lower levels. A comparison of the vibration before and after the modification is given in Fig. 4 for the mechanical vibration of the suction pipe, and the pressure fluctuations in the suction and discharge pipe. The effectiveness of the measure is obvious.

6. DAMPING CAPABILITY OF PUMP AGAINST SELF-EXCITED VIBRATIONS

Fig. 4 reveals a very important feature: the vibrations were not only reduced to a great extent by the replacement of the control valve, but at the same time their frequency was raised from about 11.9 Hz (corresponding to the second mode of the hydraulic column) to about 25 Hz. As shown in Sketch 1b, this new frequency corresponded to the third mode of the hydraulic column extending from the suction pipe to the replaced control valve. A pressure nodal point appeared again within the main feed pump. It is thus clear that the self-excited vibration described here will arise in a pump-hydraulic system with such an order that a pressure nodal point lies within the hydraulic path of the pump itself. This behaviour must be connected with the lack of damping capability of the pump.

7. STABILITY REGION OF PUMP OPERATION

For the purpose of studying the pump stability behaviour, the hydraulic system of the pump unit is reduced to a simplified model, with lumped mass m in the discharge pipe having a short length ℓ and a cross-section A and elasticity, represented by the air cushion in the surge tank with a mean volume of V_N, as shown in Fig. 5. Fluid discharges from the surge tank through a short pipe to the discharge valve (control valve), and returns over the reservoir directly to the pump having a suction tube of negligible length [3].

Fig. 6 shows the pump stability region in a coordinate system with axes X and Y:

$$X = \frac{dP_P}{dQ_P}\,\frac{Q_N}{P_N} \qquad \text{pump characteristic} \quad (P : \text{pump})$$

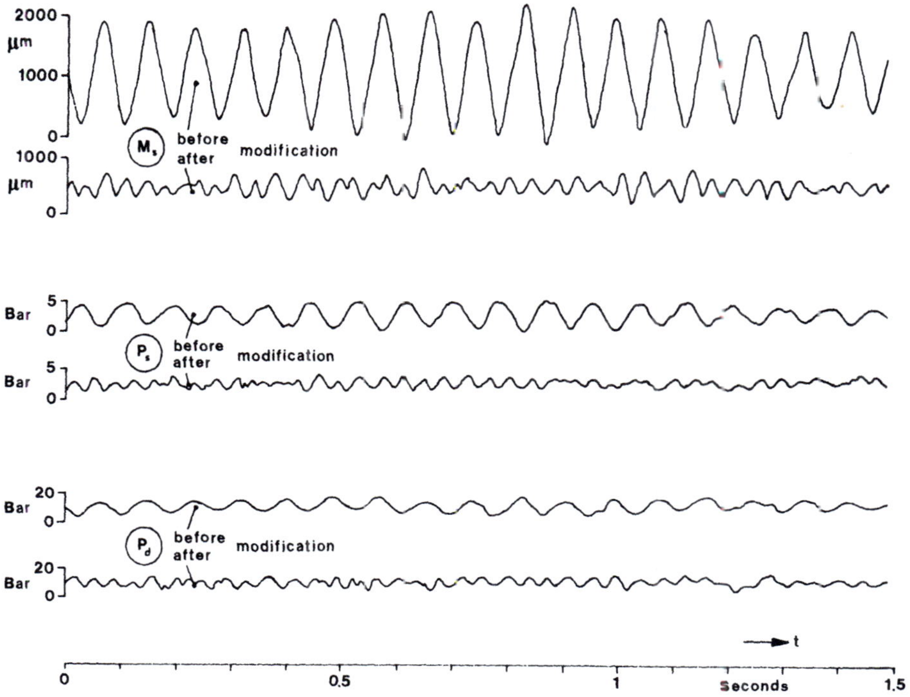

FIG. 4 VIBRATION DIAGRAMS BEFORE AND AFTER MODIFICATION

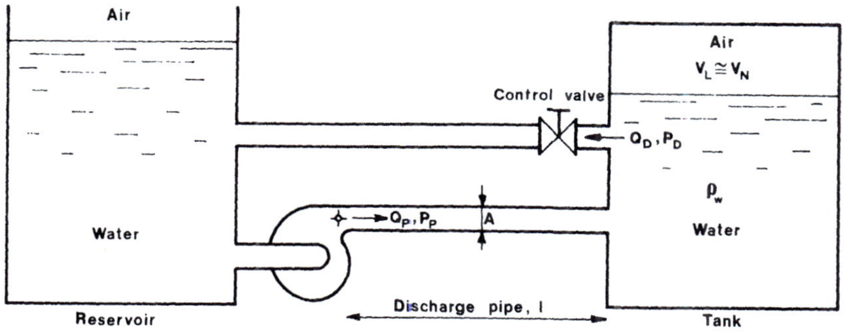

FIG. 5 MODEL

$$Y = \frac{dP_D}{dQ_D} \quad \frac{Q_N}{P_N} \qquad \text{discharge valve characteristic} \quad (D\ ;\ \text{valve})$$

and following time constants:

$$F = Q_N/\rho_w\ V_N \qquad (N = \text{operating point, } w = \text{water})$$

$$T = \ell Q_N/P_N A$$

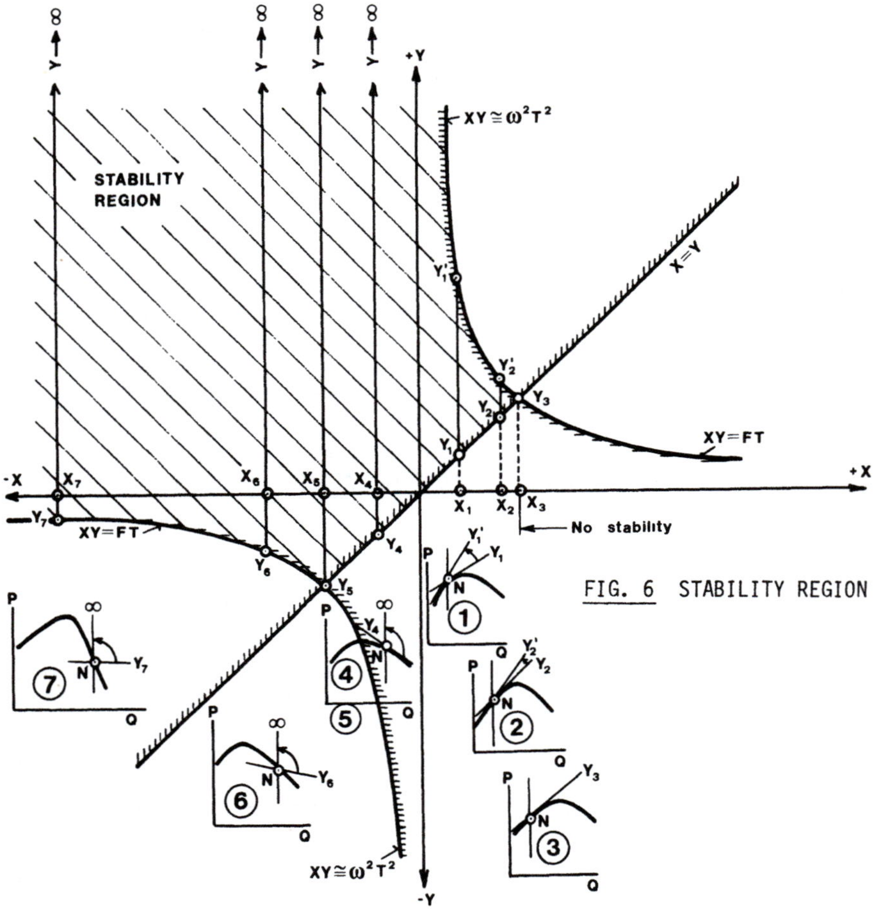

FIG. 6 STABILITY REGION

The threshold of the stability region is determined by $X = Y$ (static stability condition) and $XY = FT \cong \omega^2 T^2$ (dynamic stability condition) [2,3]. Only in the region which fulfills both these conditions is the pump operation stable. It corresponds to the hatched area shown in Fig. 6. The higher the angular natural frequency of the hydraulic system ω, the broader the stability region. The stable zone for operation with a system characteristic Y can then be found as shown by arrows on the P-Q diagram in Figure 6, corresponding to seven conditions (1 to 7). The angle α between the lower limit of this stable zone and the system characteristic is then a measure of the damping capability of the pump against self-excited vibration (Fig. 7).

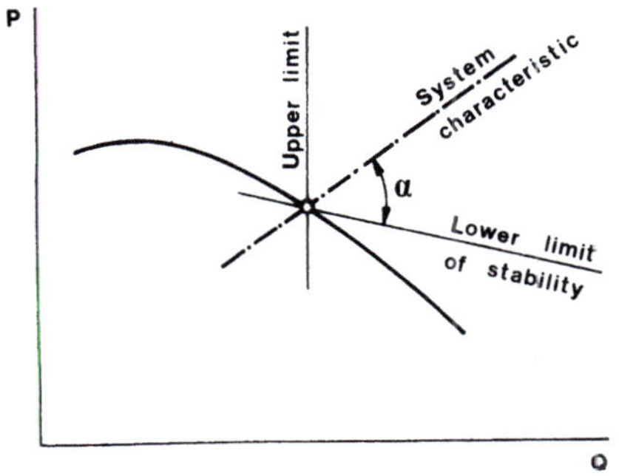

FIG. 7 ANGLE α AS A MEASURE OF THE DAMPING CAPABILITY OF A PUMP

For operation on the negative branch, the upper limit of the stable zone is $Y = \infty$, meaning that the pump is working against a very stiff hydraulic system; the operation will always be very stable. For the most part, the lower limit of this zone is practically a horizontal line, representing a softer (more elastic) hydraulic system.

For operation on the positive branch, however, the stable zone is very limited (see condition Nos. 1, 2 and 3). Beyond (X_3, Y_3) no stable operation is available. It is thus obvious that stable operation on the positive branch is scarcely possible.

8. SYSTEM CHARACTERISTIC OF A VIBRATING HYDRAULIC SYSTEM

If the vibration pattern of the hydarulic system shows a pressure nodal point within the hydraulic path of the pump, the system characteristic will be a horizontal line owing to $\partial P/\partial Q = 0$ or $Y = 0$. It will then be impossible to operate on the positive branch, because Y will be outside the stable zone. A strong vibration will be the result. Even for operation on the negative branch, Y will only be slightly above the lower limit of the stable zone. The pump can no longer offer resistance to this vibration; a self-excited vibration can arise.

If, however, the vibration would have a pressure anti-nodal point within the pump, the system characteristic would be $\partial P/\partial Q = \infty$ or $Y = \infty$. The pump could then bring about a strong resistance against the oscillation and it would be damped out at once.

The phenomenon of changing-over of the mode number from 2 to 3, after replaceing the control valve near to the non-return valve, can thus be very well explained by the model developed here.

9. REFERENCES

1. Chen, Y. N. and Florjancic, D., "Vortex-Induced Resonance in a Pipe System due to Branching," Proceedings of the Conference on Vibrations and Noise in Pump, Fan, and Compressor Installations, 16.-18, The Institution of Mechanical Engineers, Southampton, 79-86, (September 1975).

2. Taylor, E. S., "The Centrifugal Compressor," Aerodynamics of Turbines and Compressors, High Speed Aerodynamics and Jet Propulsion, Vol. X, Princeton University Press, 573-580, (1964).

3. Rothe, P. H. and Rundstadler, Jr., "First Order Pump Surge Behavior," ASME Paper 77-WA/FE-12, (1977).

4. Chen, Y. N., Florjancic, D., and Stürchler, R., "Instability of the Hydraulic System of a Centrifugal Pump Group Due to Feedback of the Flow in the Discharge Pipe," Pumpentagung 78, Karlsruhe 1978 Paper of Section K 3, and 3 R International, Vol. 18, No. 2, 136-143, (1979).

REDUCTION OF VIBRATIONS IN A CENTRIFUGAL PUMP HYDRAULIC SYSTEM

Problem: Strong, self-excited hydraulic vibrations with a frequency of about 11.9 Hz occurred in the piping system of a centrifugal pump unit during the combined operation of leak-off and main discharge flow. This frequency was much lower than the running speeds of 24.9 and 114.2 Hz for the booster pump/motor and the main feed pump, respectively. The mechanical vibrations of the piping system reached values of as large as 2.3 mm peak to peak (sketch 1a).

Cure: In order to avoid the coupling mentioned, the control valve was moved from the original position of 13 m behind the non-return valve to a position of 2 m behind it. This was the minimum distance which could be achieved.

This measure reduced the vibration amplitudes to considerably lower levels. A comparison of the vibration of the suction pipe before and after the modification is given in Sketch 2. The effectiveness of the measure is obvious. At the same time, the frequency was raised from about 11.9 Hz (corresponding to the second mode of the hydraulic column) to about 25 Hz (corresponding to its third mode), see sketch 1b. In both cases, a pressure nodal point appeared within the main feed pump. This behaviour is connected with the lack of damping capability of the pump.

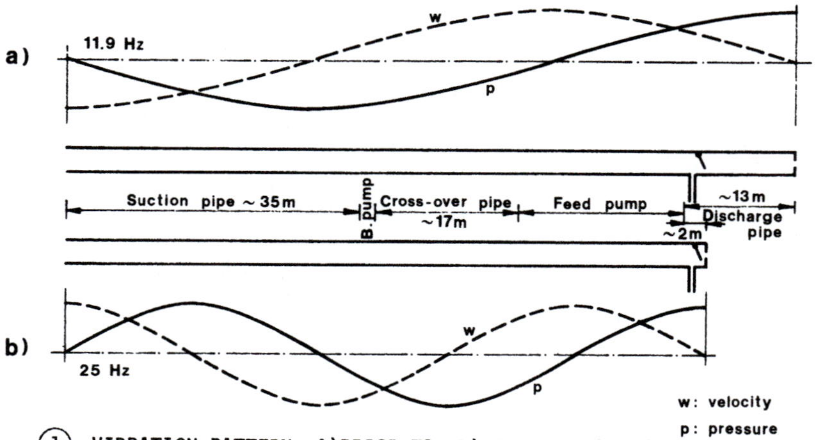

(1) VIBRATION PATTERN A) PRIOR TO B) AFTER MODIFICATION

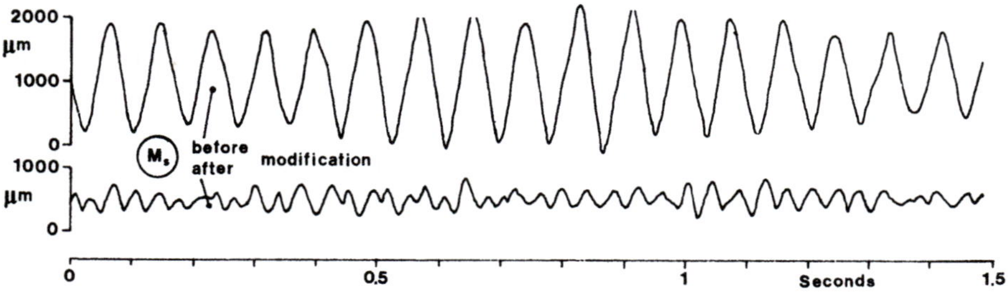

(2) VIBRATION DIAGRAMS BEFORE AND AFTER MODIFICATION

B 13 FLOW INDUCED VIBRATIONS IN VALVES OPERATING AT SMALL OPENINGS

D.S. Weaver
Associate Professor

Dept. of Mechanical Engineering
McMaster University,
Hamilton, Ontario, Canada.

Visiting Principal Engineer,
BHRA Fluid Engineering, U.K.

SUMMARY

This paper considers the flow-induced vibrations of a variety of valves operating at small openings. These are categorized in terms of a jet flow-inertia mechanism, turbulence and acoustic excitation. In each case, problems are described, and the mechanisms elucidated. Where possible, predictors are suggested and guidance is given for design against vibration problems.

1. INTRODUCTION

Flow-induced vibrations are a common problem in many types of valves when they are operating at small openings. Most people will be familiar with such vibrations in household plumbing such as bathtub or sink stoppers, toilets and hot water heating systems. More severe problems are encountered in commercial applications, primarily because of the much larger size of the hardware, and higher pressures and mass flows involved.

In spite of the pervasiveness of these problems, the literature is rather scant. This is probably because of difficulty in studying the vibrations and the relative rarity of catastrophic failure. In most cases, unacceptable noise levels and excessive wear of the valve trim are the primary results. However, fatigue failures do often occur and in many countries, more strict noise regulations are being enacted. With regard to the difficulties in studying these vibration phenomena, the vibrating components are internal and not easy to observe or instrument. In addition, the internal valve geometries are often complex and the unsteady, three dimensional, separated flows are difficult to analyze.

It would appear that the flow induced vibrations in valves operating at small openings can generally be divided into three categories:

1. jet flow-inertia mechanism
2. turbulence
3. acoustic resonance

In this paper, each of these mechanisms will be treated, in turn, with a description of the phenomenon as well as the measures which may be effective in eliminating the vibration problems.

Specifically excluded are vibrations in butterfly and hollow cone valves. It would appear that the mechanism of excitation of these phenomena are not well understood and, as the vibration problems arise at large openings, they are outside the scope of this paper. Suffice it to say that some spectacular failures have been associated with hollow cone valves and that a small reduced velocity parameter has been developed as a safety indicator using the failure and safety records of valves in service, [1], [2].

2. JET FLOW-INERTIA MECHANISM

This phenomena is so-called because a necessary condition for its occurrence is the existence of a high velocity coherent flow through the valve orifice at small openings. When the valve moves, the jet pulses, thereby altering the forces on the valve. The fluid inertia creates a hysteresis in these forces which is instrumental in feeding energy into the structure to perpetuate its motion. The two charts appended to this paper outline examples of this mechanism causing vibrations in swing check and globe valves. Note that in the case of the swing check valve the mean flow is tending to force the valve shut whereas the flow is tending to open the globe valve.

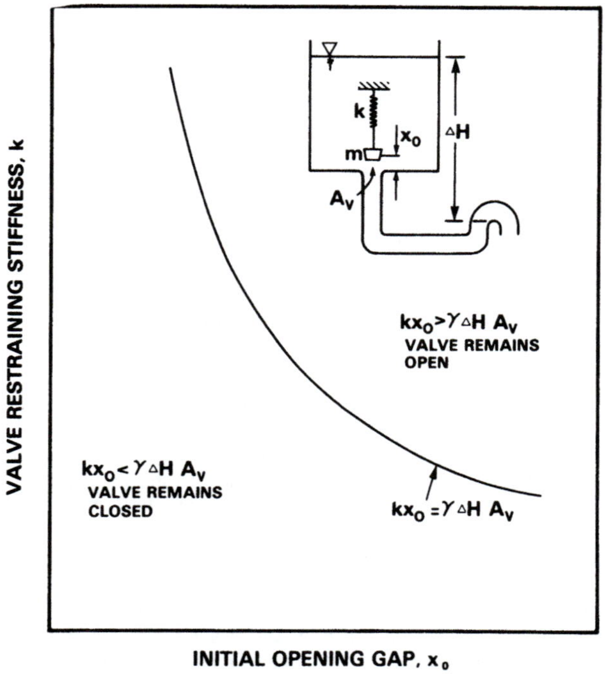

FIG. 1 STATIC PLUG VALVE CHARACTERISTIC.

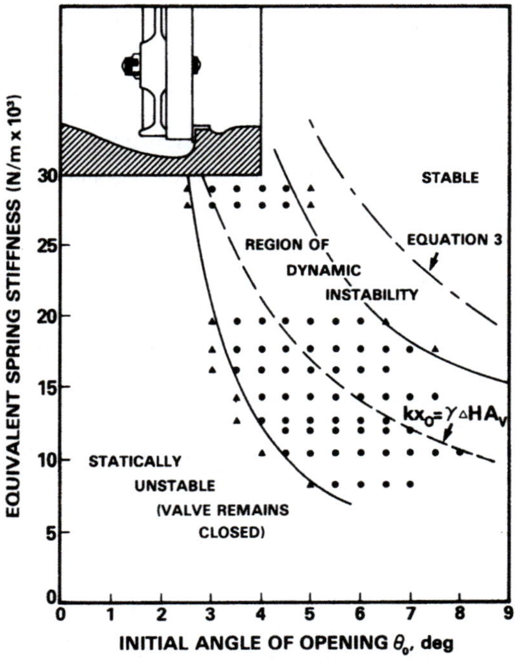

FIG. 2 SWING CHECK VALVE STABILITY BEHAVIOUR. Experimental points: ● dynamically unstable, ▲ statically unstable or stable (adapted from reference [7]).

Interestingly, this phenomenon has been known for some time and was used by Whitehurst in 1772 in his invention of the hydraulic ram [3]. More recently several authors have analyzed the performance of the hydraulic ram [4], [5] but, significantly, its development is limited by the lack of experimental data on the drive valve dynamics.

2.1 Flow-Close Valve Vibrations

Vibrations of check, poppet or plug valves where the flow tends to close the valve are the easiest to understand. These vibrations are also the most violent - one large improperly seated check valve was reported to be heard thumping more than two kilometers away [6].

It is instructive to examine the static pressure-displacement characteristic of such values as shown in figure 1. Given the available hydrostatic head across the valve, ΔH the effective cross-sectional area of the valve, A_v, and the fluid density γ, the static closing force is $\gamma\ \Delta H\ A_v$. For any given value of restraining stiffness, k, an initial zero load valve opening, x_o, exists such that the restraining force just balances the static fluid closing force.

$$k\ x_o = \gamma\ \Delta H\ A_v \tag{1}$$

For any combination of $k\ x_o$ below that in equation (1), the valve will be held closed against the seat under static conditions, whereas larger values of $k\ x_o$ will not permit valve closure under static conditions. However, under conditions of flow, the head difference across the valve is dynamic because of the high velocity jet past the plug and the flow acceleration and deceleration resulting from plug motion. Thus a region of dynamic instability develops about the static characteristic as shown in figure 2 [6]. In the unstable region above the static characteristic, the fluid inertia associated with the reduction of discharge creates sufficient dynamic head to close the valve. When the fluid comes to rest, the dynamic head reduces and the spring can open the valve (although a low pressure wave may assist). In the dynamically unstable region below the static characteristic, the available static head is sufficient to hold the valve closed. However, the sudden closure causes pressure waves which are significant in liquid flows and these cause the valve to move off the seat enough to re-establish the flow and the valve continues to open. In the statically unstable region, the valve's elastic restraining force is insufficient to hold the valve open against the flow and the valve closes and remains closed. Fuller details are given in references [6] and [7]. These studies also show that water hammer and vortex shedding are not significant factors in the vibration mechanism.

The effect of fluid inertia is to delay the reestablishment of flow while the valve is opening and to maintain a high rate of flow while the valve is closing. Thus fluid inertia is responsible for the hysteresis in the fluid force on the valve over a vibration cycle and hence the net energy transfer from the fluid to the structure to perpetuate its motion. This is illustrated in figure 3 which shows the variation in flow velocity just upstream of the valve plate obtained from flow visualization films [7]. The flow slowly increases as the valve opens and does not reach a maximum until the valve is well on its way shut again. There is then a very rapid rate of change of discharge over the last 2° of closure which gives rise to the large dynamic head.

Recently, Weaver and Ziada [8] developed a theoretical model for this problem. The valve opening was modelled as an orifice of time varying area and the fluid inertia and flow losses were included. The coupled nonlinear differential equations were solved numerically and the amplitude and frequency of the check valve vibrations in chart 1 were predicted within 20%. A typical result for the predicted discharge variation over a vibration cycle is shown in figure 4. The curve for large fluid inertia shows the same type of discharge variation as was obtained in the model experiments (figure 3). As expected, if the fluid inertia is reduced considerably, the maximum discharge occurs at about the point of maximum displacement and both are considerably reduced. The "small fluid inertia" case in figure 4 is about one tenth of the "large fluid inertia" case, all other parameters being held constant.

When the fluid inertia is significant, the fluid forces on the valve are obviously quite different between the opening and closing portions of the cycle and it is not surprising that the motion is not simple harmonic. On the other hand, when the fluid inertia is small, the oscillations are more nearly simple harmonic and the frequency is considerably increased. The theoretically predicted displacement-time curves corresponding to the cases shown in figure 4 are given in figure 5. These curves show that the opening times are about the same, being dominated by the elastic valve restraint. On the other hand, the closing portion is dominated by the fluid forces which accounts for the large variation in frequency observed experimentally, independent of valve natural frequency in quiescent

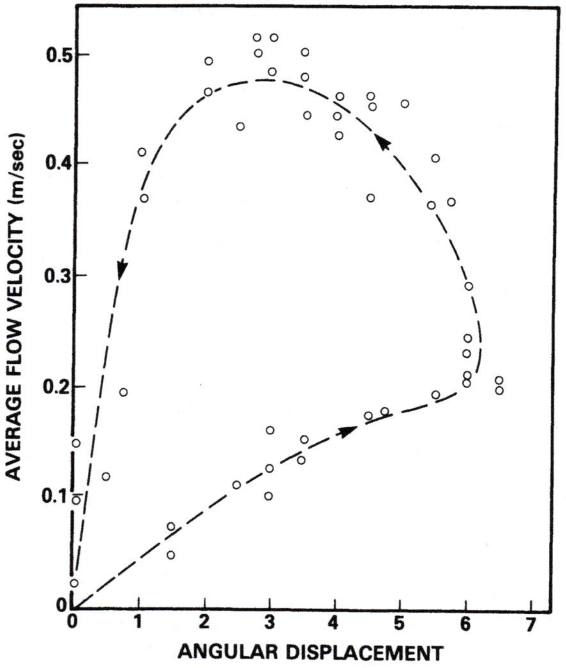

FIG. 3 FLOW VELOCITY-DISPLACEMENT BEHAVIOUR FROM FLOW VISUALIZATION TESTS (adapted from reference [8]).

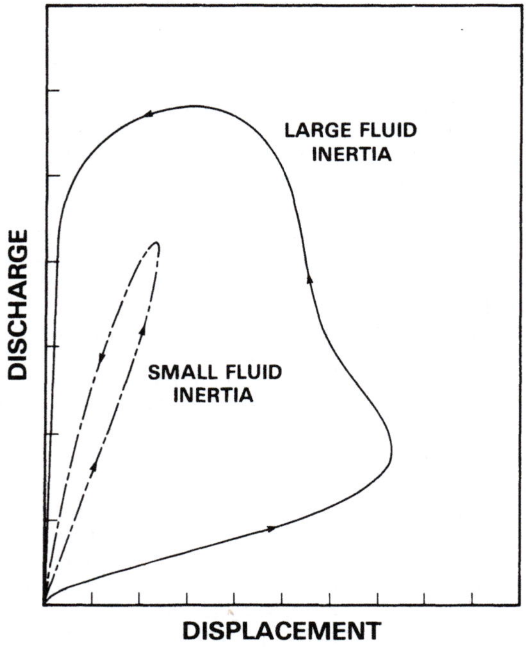

FIG. 4 THEORETICAL PREDICTIONS FOR DISCHARGE-DISPLACEMENT BEHAVIOUR (adapted from reference [9]).

fluid. This also explains why an increase in valve restraint stiffness can result in a decrease in vibration frequency. While the opening time will be reduced, the closing time can be significantly increased because of the relative increase in elastic restraining force compared with the dynamic fluid closing force.

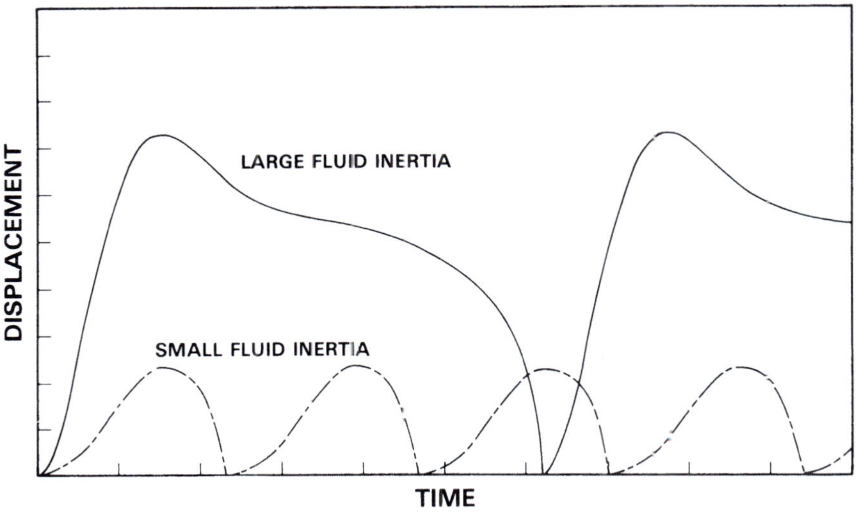

FIG. 5 THEORETICAL PREDICTIONS FOR DISPLACEMENT-TIME BEHAVIOUR (adapted from reference [9]).

Vibration amplitude and frequency are also strongly dependent on the initial (zero load) opening displacement. This is illustrated in figure 6 which gives experimental displacement-time traces for a simple plug valve such as that shown schematically in figure 1. The natural frequency of the valve in quiescent water was 5 Hz for all four cases, the only differences being pipe length, L, (fluid inertia) and initial displacement, x_o. Clearly, the motion is generally not of a simple harmonic nature and the effects of increasing fluid inertia and initial displacement are to decrease the frequency and increase the amplitude of oscillation.

This jet flow-inertia mechanism is fundamentally the same as that causing vibrations in hydraulic gates and seals operating at small openings. An important contribution to our ability to predict this phenomenon has been made by Kolkman [10]. He assumed small displacements, simple harmonic motion and a constant discharge coefficient. This permitted the linearized equations to be solved analytically and a simple stability predictor to be established. If the discharge varies linearly with valve displacement at small openings (which is quite a reasonable assumption in many cases of practical interest), the stability threshold is given by:

$$k\,x_o = 2\,\gamma\,\Delta H\,A_v \left(\frac{m}{\rho\,AL} + 1 \right) \qquad (2)$$

where m is the mass of the valve including added mass and ρ is the mass of fluid in the pipe. If the latter is large, as in the case of the check valve vibrations shown in figure 2, the stability threshold is given by:

$$k\,x_o = 2\,\gamma\,\Delta H\,A_v \qquad (3)$$

Note that the right hand side of this equation is simply twice the maximum available static closing load indicated in equation (1). Equation (3) is plotted in figure 2 as the chain line just above the unstable region and clearly gives a reasonable conservative estimate of the stability threshold.

As Kolkman [10] gave little experimental support for his analysis, the author has conducted additional parametric tests with a simple plug valve, examples of which were given in figure 6. In all applicable cases, it was found that Kolkman's approach gave conservative predictions of the stability threshold. Of course, the linear theory could not be expected to give reasonable predictions of vibration frequency and amplitude.

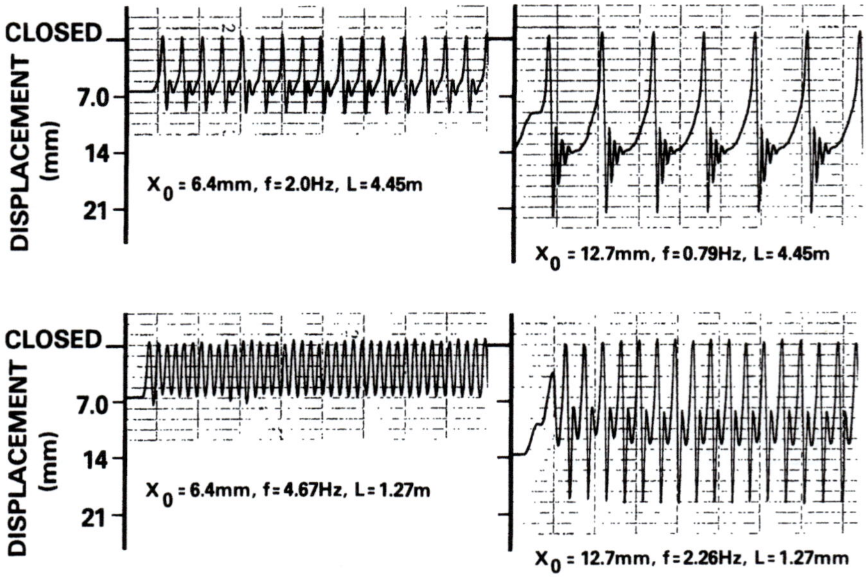

FIG. 6 EXPERIMENTAL PLUG VALVE DISPLACEMENT TRACES, (vertical time lines are 1 second apart).

Several further observations regarding this stability indicator should be made. Firstly, in most valves, there will be a maximum initial displacement for which the assumption regarding variation of flow area with displacement is valid. Secondly, the indicator suggests that there will be at least a small region of instability for any practical level of restraint stiffness, i.e., an increase in stiffness is unlikely to eliminate the problem entirely. This in fact has been the author's experience. Finally, the assumption of constant discharge coefficient is well justified in view of our lack of knowledge of dynamic discharge coefficients but represents a significant limitation of the theory. More will be said of this below.

2.2 Annular Flow Valve Vibrations

The vibrations of valves with an annular flow which does not necessarily tend to close the valve are caused by a similar mechanism as described above although the flow characteristics are usually somewhat more complex. The globe valve vibrations of chart 2 are an example. In this case, the valve plug chatters from side to side and while these oscillations are less violent than those described above, high noise levels and damage to the plug, seat and guide may occur.

The centre location of the plug in the orifice is actually unstable. Any disturbance from this position will create an increased flow velocity on the side of the smaller gap and a decreased velocity on the other side. These fluid forces tend to increase the eccentric displacement. When the gap is closed, the jet flow on this side is destroyed and a combination of elastic and positive fluid dynamic stiffness will tend to restore the plug to its central location. As discussed above, the rapid rate of change of velocity with gap width and the phase lag due to fluid inertia create a hysteresis in the fluid force displacement cycle and hence a net energy transfer from the fluid to the valve.

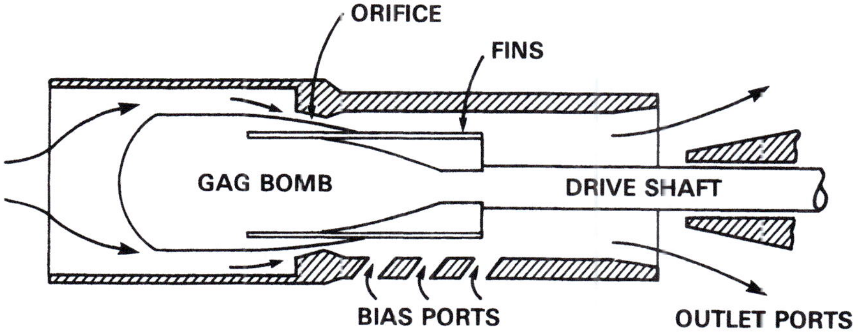

FIG. 7 SCHEMATIC OF CO_2 FLOW CONTROL GAG.

An interesting example of this phenomenon is the damaging vibrations of the Hinkley Point 'B' and Hunterston 'B' gags [11], [12]. These gags, shown schematically in figure 7, have a minimum diametral fin tip to orifice clearance of 0.7 mm to 1.1 mm when closed and a maximum CO_2 gas velocity of 160 m/sec. It was recognized that a vibration problem could develop and hence the bias ports were introduced to force the vertically hanging gags to one side of the orifice. However, asymmetries due to manufacturing tolerances coupled with ineffective biasing led to serious vibration problems on some of the gags. The biasing was ineffective because it was too far downstream from the point of highest flow velocity in the orifice (where the largest fluid dynamic forces are developed) and the bias force was too small compared to buffeting forces when the gag was against the orifice opposite the bias ports. As with the check valve vibrations described above, the cure was to modify the orifice flow. In particular, at the expense of an increase in pressure drop when fully open, vents were used to spoil the jet flow through the minimum gap when the gag was closed.

2.3 Methods of Alleviation of Jet Flow-Inertia Type Vibrations

The following guidelines should be followed to avoid vibrations induced by this mechanism.

(a) Whenever possible, avoid valve configurations in which the flow creates a large closing force at small openings.

(b) When valve configurations susceptible to the jet flow-inertia mechanism cannot be avoided, such as in many check valve designs or annular flows through plug valves, the best approach is to use high loss trim. In this way, the rate of change of discharge with valve displacement at small openings is greatly reduced and thus the fluid-dynamic loading becomes much less sensitive to valve motion. Disturbance of the valve by turbulent buffeting is difficult to avoid but the design goal should be to eliminate as much as possible any significant discharge variations resulting from small motions of the valve.

(c) Make the valve restraint stiffness as great as possible. While this mechanism is extremely persistent and may not be eliminated by an increase in stiffness only, the range of openings over which instability occurs becomes small.

It should be noted that in cases of severe vibrations, the displacements and discharge variations are generally not small and are highly nonlinear. In such cases, models of the phenomenon which reduce to simple harmonic oscillators with negative damping are not valid. More importantly, the addition of sufficient positive damping will not eliminate the vibration problem, contrary to the predictions of such a model.

2.4 Research Required

The mathematical models discussed above, [9], [10], both assume a constant discharge coefficient. In the absence of empirical data this is a reasonable assumption and largely accounts for the difference between experimental results and theoretical predictions. In fact, the contraction and discharge coefficients will depend on flow acceleration and valve motion in a way which is not understood at this time and therefore not predictable. Daily et al [13] have shown that the discharge coefficient of a square edged orifice increases when the flow accelerates and decreases when the flow decelerates. McCloy [14] demonstrated that the mean discharge through an orifice decreases as the pulsation frequency of orifice area increases. Alpay [15] argued that the mean dynamic discharge coefficient depends on Reynolds number, Strouhal number (frequency, initial opening and pressure drop) and the ratio of vibration amplitude to initial opening. His experiments indicated that fluid inertia could actually increase the dynamic discharge coefficient while vibration frequency had the opposite effect.

In hydraulic structures and devices where vibration amplitudes are significant, the dynamic discharge coefficient will undoubtedly undergo large non-linear variations. Considerably more research is necessary in this area before our theoretical predictions of these flow induced vibrations can be improved substantially.

3. TURBULENCE INDUCED VIBRATIONS

In valves used for flow or process control such as the diaphragm actuated single port control valve in figure 8, the flow path is tortuous and there is significant energy dissipation. Turbulent pressure fluctuations as well as direct flow impingement on the internal valve components may cause damaging mechanical vibrations. Schuder [16] states that the buffeting force spectrum is relatively flat up to 100 Hz and in certain valve designs can have peak-to-peak forces in excess of 4.45 kN. Thus, the plug in a control valve can be forced into large amplitude vibrations. More common are the vibrations associated with lateral plug motions between the guide or sealing surfaces. These are usually less than 1500 Hz and sound like metallic rattling [17]. The result may be valve stem fatigue and destruction of the seat, guides and packing.

While there seems to be little published experimental data, the broad band random pressure fluctuations must have significant energy levels up to quite high frequencies. Allen [17] reports that flexible valve components such as hollow skirt plugs and ball valve seal rings can be excited to resonate at their natural frequencies. The high pitched tones emitted are from 3000 to 7000 Hz and may result in fatigue failure.

It should be noted that the turbulence induced mechanical noise is only a part of the noise in control valves. Cavitation, flashing and random velocity fluctuations are significant acoustic sources which may produce intolerable noise levels. However, given empirical data from the valve manufacturers, the aero- or hydrodynamic sound pressure levels from a given valve may be predicted [17], [18]. The mechanical noise is a useful indicator that conditions exist which may produce valve failure.

3.1 Alleviation of Turbulence Induced Vibrations

(a) Since these vibrations are forced by broad band random pressure or impinging velocity fluctuations with the greatest energy levels in the lower frequency range, the response may be reduced by an increase in stiffness and damping. In plug valves, especially if the pressure drop is high, the plug stem should be heavily guided above and below the port(s). Flexible trim should be avoided. While low friction packing materials make valve operation easier, the reduced damping aggravates turbulence response problems.

(b) Since turbulence related problems are often difficult to predict, care should be taken to select the proper valve for a given job. For example, gate valves have a tendency to vibrate when operating at small openings. The work of Spanton and Ashton [19] suggests that we can expect a well defined peak in the turbulence spectrum downstream of a partially closed gate valve. The corresponding Strouhal number based on vena contracta velocity is about 0.01. Therefore, significant low frequency excitation may exist at small openings. Gate valves should not be used for throttling purposes but rather in services where they are either fully open or fully closed.

In addition to selecting the correct type of valve, it is important to select the proper size and not exceed the manufacturer's specified pressure drop for a given valve. The buffeting force and frequency, sound pressure level and tendency towards cavitation and flashing are all proportional to the pressure drop.

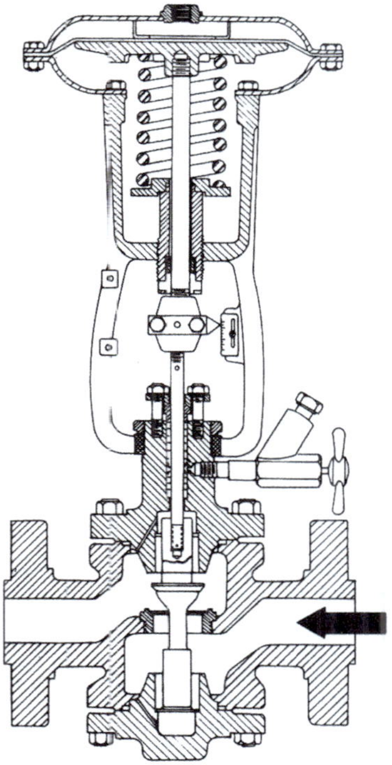

FIG. 8 DIAPHRAGM ACTUATED CONTROL VALVE.

It should be noted that improper actuator selection for automatic control valves can lead to potentially disastrous large amplitude valve oscillations as a result of feedback control loop instability. Such considerations are beyond the scope of this paper.

4. ACOUSTIC RESONANCE IN VALVE SYSTEMS

In section 3, mention has been made of the noise sources associated with highly turbulent flows through control valves. Generally, these cause a random buffeting of the valve components and, in some cases of broad band excitation, mechanical resonance. Amplification of the pressure disturbances will occur if acoustic resonance occurs in the associated piping system. In most cases of practical interest, the fundamental frequency or the lowest harmonics are the problem source. Therefore, the wave length is very long and the disturbances propagate nearly as plane waves up and down the pipes. These can couple with pipe bending modes and lead to dangerous vibrations of the piping system. Such problems have recently attracted attention because of serious vibrations in nuclear reactor steam mains where the acoustic excitation source is apparently the governor valves (see, for example, Gibert et al [20]). However, there would appear to be no related valve vibration.

Our concern here is restricted to valve vibrations driven by acoustic resonance in the piping. Basically, this has been found to occur in plug and spool valves which oscillate and act as wave amplifying boundaries of the pipe. Examples are shown schematically in figures 9 and 10.

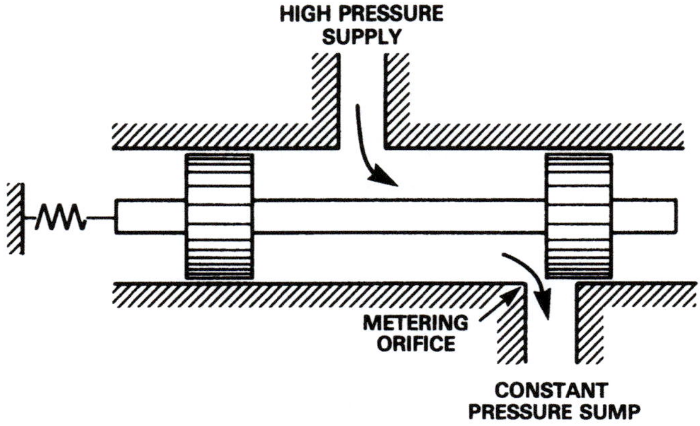

FIG. 9 SCHEMATIC OF A SPOOL VALVE.

In the case of poppet valves, it is interesting to note that acoustic resonance may occur for flow in either direction. Thomann's theoretical and experimental results [21] indicate that the frequency of oscillation may be slightly higher or lower than the natural frequency of the valve in quiescent fluid depending on whether the flow tends to open or close the valve respectively. This is probably the result of a change in sign of the aerodynamic component of stiffness. In contrast, Kolkman [10] has shown that the jet flow-inertia mechanism for plug motion parallel to the stem axis is only possible when the flow tends to close the valve (flow direction opposite to that shown in figure 10).

In spool valves, the peculiar flow characteristics make the analysis rather more difficult [22]. Such valves have a backward curving pressure-discharge characteristic such that, above a critical value of pressure depending on the ratio of valve restraint to hydraulic stiffness, an increase in pressure will reduce the discharge through the valve. Thus, an unstable mechanism is established similar to that in plug valves with a closing flow. This is aggravated by acoustic resonance. While these valves may be extremely unstable if the wrong flow direction is used, Ainsworth [23] and Ezekial [24] have shown that acoustic resonance can cause vibration even when the flow direction is proper.

In the case of both plug and spool valves, vibrations are amplified when the acoustic frequency of the piping is close to the natural frequency of the valve. The system is extremely unstable when these frequencies are the same, i.e., when:

$$f_n = \frac{nc}{4\ell} \tag{4}$$

where f_n is the natural frequency of the valve, n is an integer, c is the speed of wave propagation in the pipe and ℓ is the effective length of the piping system.

4.1 Alleviation of Acoustic Resonance Vibrations

(a) The most obvious and effective cure for acoustically excited valve vibrations is to design the system with the widest possible separation between the valve natural frequency and the acoustic frequencies of the pipe.

(b) The addition of a viscous damper will reduce the resonant amplitudes as well as the range of frequencies and pipe lengths over which resonance can occur.

(c) It is generally desirable to make the effective pipe length as short as possible so that its acoustic natural frequency is high. It has also been found effective to use a composite upstream pipe where the diameter is suddenly reduced some distance from the valve [24]. However given the plane wave nature of the pressure disturbance, the difference in cross-sectional area would have to be significant, and in the case cited above, the area reduction probably behaved like a partially closed fluid column.

(d) In self-actuated poppet valves, it is desirable to have the valve fast acting so that the valve remains in the small opening range for as short a time as possible. If the valve lingers in the low lift range, a short burst of oscillations may occur.

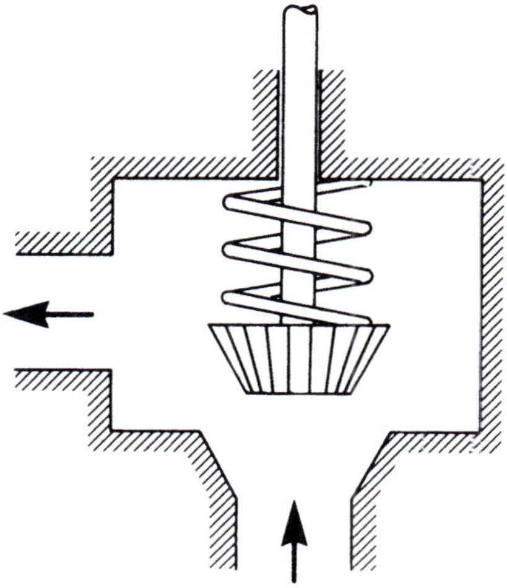

FIG. 10 SCHEMATIC OF A PLUG VALVE.

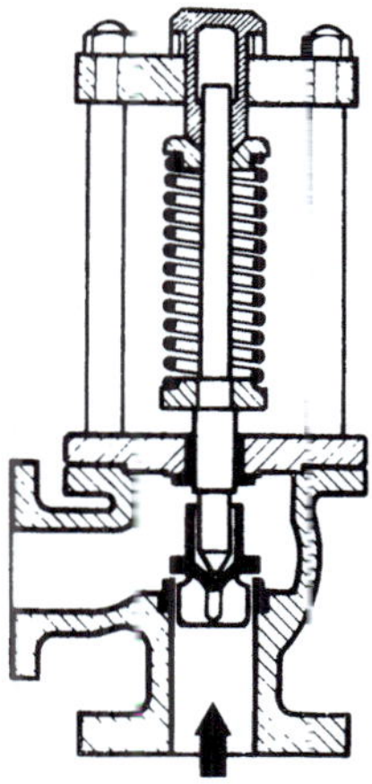

FIG. 11 RELIEF VALVE.

It should be noted that acoustic resonance may be a problem in standard safety or relief valves, such as that shown in figure 11, as these are designed to operate at small openings. Typically, the lift is 1/24th of the valve port diameter. This is especially true if the valve is oversized.

(e) The use of a chamber or accumulator just upstream of the valve may also be an effective cure for acoustic oscillations. However, for automatic control valves this may have an adverse effect on control system dynamics.

5. CONCLUSIONS

Valve vibrations have been characterized as being the result of a jet flow-inertia mechanism, turbulent pressure fluctuations, or acoustic resonance. In each case a number of methods of vibration alleviation have been suggested. While complete avoidance of flow induced vibrations is sometimes as much an art as a science, careful adherence to the guidelines presented will prove effective in most cases.

6. REFERENCES

1. Mercer, A.G., "Vane Failures of Hollow-Cone Valves", IAHR Symposium Proceedings, Part 1, G4, Stockholm, Sweden, 1971.
2. Wang, C., "Analysis of Vibration of Hollow-Cone Valves", ASCE Vol. 99, No. EM6, pp. 1147-1163, 1973.

3. Conway, H.G., "Fluid Pressure Mechanisms", Sir Isaac Pitman & Sons, Ltd., London, 2nd ed., 1958.
4. Iverson, H.W., "An Analysis of the Hydraulic Ram", ASME Jo. of Fluids Engineering, pp. 191-196, June, 1975.
5. Krol, J., "The Automatic Hydraulic Ram: Its Theory and Design", ASME paper 76-DE-17, Design Engineering Conf., Chicago, 1976.
6. Griffiths, P.T.A., "Large Gates and Valves: Vibration", Proc. 3rd Colloquium for Hydraulics Lab. Staff., Australian Water Resources Council, Cooma, NSW, pp. 40-75, 1969.
7. Weaver, D.S., Adubi, F.A., and Kouwen, N., "Flow Induced Vibrations of a Hydraulic Valve and Their Elimination", ASME Jo. of Fluids Engineering, Vol. 100, pp. 239-245, June, 1978.
8. Weaver, D.S. and Adubi, F.A., "Flow Visualization Studies of Check Valve Vibrations", Proc. 4th Biennial Symp. on Turbulence in Liquids, University of Missouri-Rolla, J.L. Zakin, G.K. Patterson editors, Science Press, Princeton, N.J., 1975.
9. Weaver, D.S. and Ziada, S., "A Theoretical Model for Self-Excited Vibrations in Hydraulic Gates, Valves and Seals", ASME PVP Conf., San Francisco, Ca., June, 1979.
10. Kolkman, P.A., "Flow Induced Gate Vibrations", Ph.D dissertation, Delft University of Technology publication 164, 1976.
11. Denton, J.D., Hutton, M.F. and Parkin, M.W., "Vibration Mechanisms Associated with Annular Flow Through a Flow Control Device", British Nuclear Energy Society Conf., Vibration in Nuclear Plant, Keswick, U.K., paper 5:5, 1978.
12. Bolton, A.D. and Furber, N.B., "Evolution of the Modifications Designed to Suppress the Damaging Vibration of the Hinkley Point 'B' and Hunterston 'B' Gags", British Nuclear Energy Society Conf., Vibration in Nuclear Plant, Keswick, U.K., paper 5:2, 1978.
13. Daily, J.W., Hankey, W.L., Olive, R.W. and Jordan, J.M., "Resistance Coefficients for Accelerated and Decelerated Flows Through Smooth Tubes and Orifices", ASME Trans., Vol. 78, no. 5, p. 1071, 1956.
14. McCloy, D., "Effects of Fluid Inertia and Compressibility on the Performance of Valves and Flow Meters Operating Under Unsteady Conditions", Jo. Mech. Eng. Sci., Vol. 8, no. 1, pp. 52-61, 1966.
15. Alpay, S.A., "The Flow Coefficient of a Hydraulic Control Valve Under Dynamic Conditions", Flow; Its Measurement and Control in Science and Industry, Vol. 1, part 3, W.E. Vannah and H. Wayland editors, Instrument Society of America, Pittsburgh, pp. 1227-1234, 1974.
16. Schuder, C.B., "Understanding Fluid Forces in Control Valves", Instrumentation Technology, pp. 48-52, May, 1971.
17. Allen, E.E., "Prediction and Abatement of Control Valve Noise", Flow; Its Measurement and Control in Science and Industry, Vol. 1, part 3, W.E. Vannah and H. Wayland editors, Instrument Society of America, Pittsburgh, pp. 1167-1173, 1974.
18. Seebold, J.G., "Valve Noise and Piping System Design", Flow; Its Measurement and Control in Science and Industry, Vol. 1, part 3, W.E. Vannah and H. Wayland editors, Instrument Society of America, Pittsburgh, pp. 1151-1159, 1974.
19. Spanton, J.H. and Ashton, M.W., "Measurements of Fluctuating Forces Generated by Internal Flow in Pipeline Components", British Nuclear Energy Society Conf., Vibration in Nuclear Plant, Keswick, U.K., paper 6:3, 1978.
20. Gibert, R.J., Axisa, F. and Villard, B., "Flow Induced Vibrations of Piping System (Vibration Sources - Mechanical Response of the Pipes)", British Energy Society Conf., Vibration in Nuclear Plant, Keswick, U.K., paper 6:2, 1978.
21. Thomann, H., "Oscillations of a Simple Valve Connected to a Pipe", Jo. Applied Math. Phys. (ZAMP), Vol. 27, pp. 23-40, 1976, and "Part II Experiments", Jo. Applied Math. Phys. (ZAMP), Vol. 29, pp. 75-84, 1978.
22. Ito, T., Muto, T., Suzuki, K. and Tsukahara, K., "Study of the Self-Sustained Oscillation of a Piston-Type Valving System 1st Report", Bulletin JSME, Vol. 10, no. 41, pp. 793-807, 1967, "2nd Report", Bulletin JSME, Vol. 11, no. 45, pp. 487-495, 1968.
23. Ainsworth, F.W., "The Effect of Oil-Column Acoustic Resonance on Hydraulic Valve Squeal", Trans. ASME, Vol. 78, pp. 773-778, 1956.
24. Ezekiel, F.D., "The Effect of Conduit Dynamics on Control-Valve Stability", Trans. ASME, Vol. 80, pp. 904-908, 1958.

Discussion: FLOW-INDUCED VIBRATIONS IN VALVES OPERATING AT SMALL OPENINGS (Weaver)

M.J. Kenn
Civil Engineering Department,
Imperial College, London SW7 2BU, UK

An effective and inexpensive way to overcome the slamming of a "Swing check valve" is to limit the travel of the flap and to dispense with any damper. The valve is thus able to close quickly and quietly before flow reversal occurs. Properly designed non-return valves, of course, already have moving parts with small inertias and small movements, suitably housed in appropriately streamlined water passages.

Reference

Kenn, M. J. Free - acting reflux valves can be silenced. J.I.W.E., 1966, Vol. 20, No. 1, February, p. 78.

Reply to Discussion of: FLOW INDUCED VIBRATIONS IN VALVES OPERATING AT SMALL OPENINGS

D. Weaver

Mr. Kenn is quite correct in his general suggestions for eliminating the slamming of reflux valves. However, such solutions were not possible in the case of the valve cited in my paper. That valve had to pass a test in which a sphere, the same size as the pipe upstream and downstream of the valve, passed through the valve unobstructed. Hence no travel limiting stop could be used to reduce the motion of the flap. It is worth noting as well that such a stop would also increase the valve loss coefficient considerably, thereby eliminating another important advantage of swing check valves.

SWING CHECK VALVE

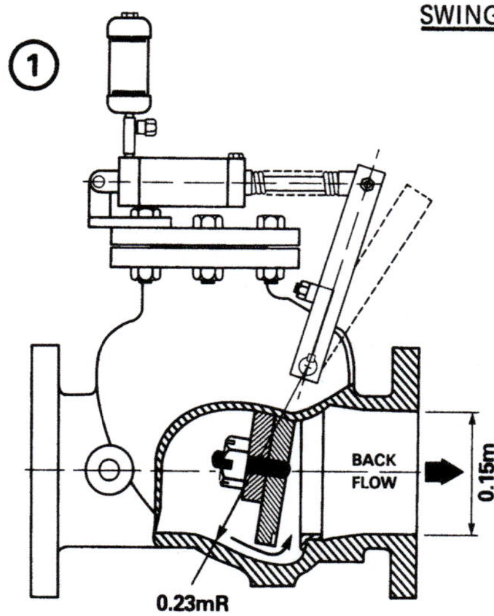

Problem: A damper affixed to a swing check valve to prevent slamming produced multiple slams and violent oscillations for sufficiently high damping. Excessive noise and shearing of clapper trunion pins resulted.

Geometry: See sketch 1. Tests were conducted with standard 6 inch and 12 inch valves of the same design as well as a two-dimensional full scale model of the 6 inch valve [1], [2].

Boundary conditions: The pipe and valve were stiffly supported but the slamming and subsequent water hammer moved the whole structure. Some cavitation was produced at closure. The water hammer waves attenuated before the next closure. See sketch 2.

Flow conditions: The backwards flow causing closure was smooth with no upstream flow disturbances. There were no large scale eddies approaching the valve and cavitation and vortices in the wake played no significant role.

Operating conditions: The vibrations were extremely persistent and occurred over a large range of initial openings, damper, clapper restraint structural stiffness, and pressure drop ($4m < \Delta H < 40m$)

Structural dynamics: Inaccessibility prevented measurements of valve natural frequency and damping. However, the motion was not simple harmonic as seen in sketch 2. It is estimated that the oscillations observed were generally at frequencies less than 20% of the valve's natural frequency in quiescent fluid. Parameter values in model tests are given in reference 1.

Vibrations: Typical results for the 6 inch valve are shown in sketch 2. Pressure measurements were taken just upstream and downstream of the valve.

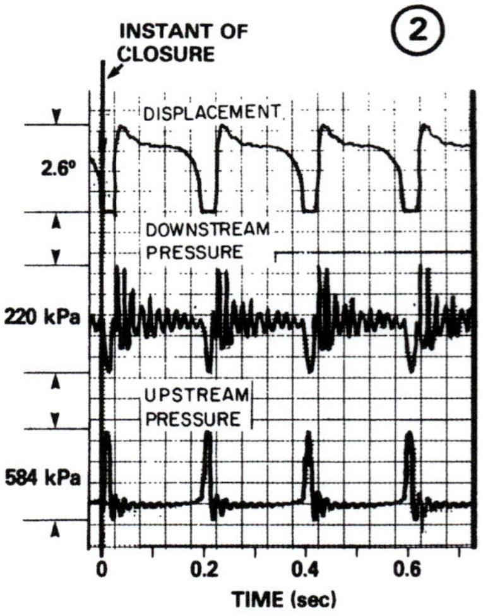

Cure: The valve geometry was changed by filling in area between valve plate and body and adding a lip to the seat so that the rate of change of discharge was considerably reduced over the final 25° of closure (reference 1). (See sketch 3)

References:
1. Weaver, D.S., Adubi, F.A. and Kouwen, N., "Flow Induced Vibrations of a Hydraulic Valve and Their Elimination", ASME Jo. of Fluids Engineering, Vol. 100, pp. 239-245, 1978.
2. Weaver, D.S. and Adubi, F.A., "Flow Visualization Studies of Check Valve Vibrations", Proc. 4th Biennial Symposium on Turbulence in Liquids, Science Press, Princeton, N.J., pp. 201-208, 1975.

GLOBE VALVE

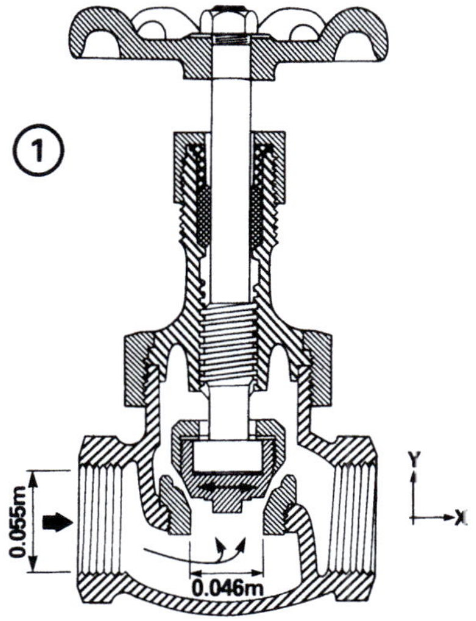

Problem: The valve vibrated severely at small openings both during opening and closing. No structural damage resulted but the noise level (94 dBA) was excessive and the valve could not be continually operated at small openings.

Geometry: Standard 2 inch globe valve with a stainless steel plug and seat as shown in sketch 1.

Boundary conditions: The piping upstream and downstream of the valve was rigidly supported. The valve vibration did not excite resonances in the piping system.

Flow conditions: The mean flow of water through the valve was measured using a rotameter downstream of the valve. There were no significant flow disturbances immediately upstream or downstream of the valve.

Operating conditions: The vibrations were observed for a range of pressure drop across the valve 275 kpa $<\Delta p<$ 400 kpa. The maximum mean flow rate was 0.8 litres/sec. The exact valve lift is unknown but estimated to be between 1 and 3 mm.

Structural dynamics: The stiffness of the valve stem in the body cannot be estimated. However, it would appear that the motion was primarily a rocking mode of the valve plug on the stem where there is considerable free play to allow for proper centering and sealing of the valve. Most of the effective plug stiffness and damping would be hydrodynamic in nature. The noise is generated by the plug chattering from side to side of the seat.

Vibrations: The vibrations were relatively steady at frequencies between 380 Hz and 470 Hz depending on valve opening. The peak to peak amplitudes were estimated to be up to 2 mm. Typical accelerometer records from the valve body are shown in sketch 2. The relatively large, regular x and z traces indicate lateral motion of the plug. Vertical motion would accompany cyclic closure of the valve for flow in the opposite direction and would be indicated by larger y-direction accelerations.

Cures: Some attenuation was achieved by tightening the packing nut which would have the effect of increasing stem stiffness. The stem is actually quite a loose fit in the lifting threads. The problem was eliminated by greatly reducing the play between the plug and the end of the stem.

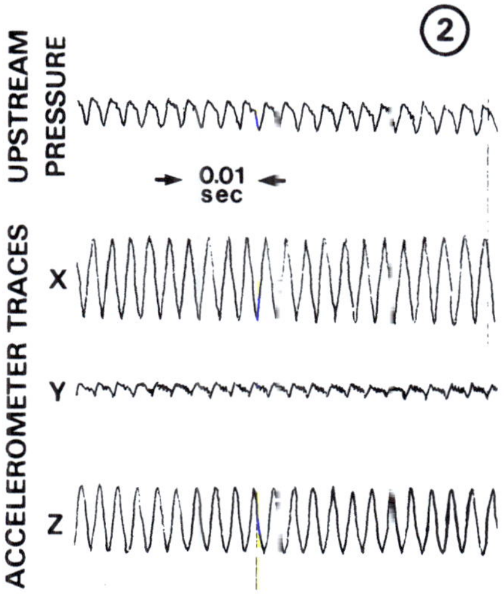

B14

GOVERNING VALVE VIBRATIONS IN A LARGE STEAM TURBINE

Karl-Erik Widell

Manager, Turbine Engineering
Product Development Division

STAL-LAVAL TURBIN AB
S-612 20 FINSPONG, Sweden

SUMMARY

This paper describes vibrations in the governing valve of a large steam turbine caused by an acoustically induced valve-pipe instability in one of the nine steam pipes from the valve. The vibrations, which were causing a number of servo spindle failures, could have been eliminated by changing the length of the steam pipe after the valve, but for practical reasons this was not possible; therefore the cure was to remove the actual valve and plug the corresponding steam pipe, which solved the problem completely.

1. VALVE SYSTEM

During the initial operation of a 460 MW steam turbine plant, vibrations of the governing valve occurred within a limited load range around 300 MW. The vibrations caused a number of failures of the spindle and its couplings. The governing valve is of the bar-lift type with nine mushroom valves which are lifted one by one using a single hydraulic servomotor connected to two hinged beams as shown in sketch 1. The live steam enters the valve casing through two emergency stop valves at the ends of the casing. The pipes from the valves, which are of different lengths, lead to two steam chests, one on each side of the high pressure turbine (of the contra-rotating radial flow type).

2. VIBRATIONS OF VALVE

The vibrations occurred spontaneously, but sometimes erratically, at turbine loads over 270 MW, which closely coincided with the opening of valve no. 5 (the valves are numbered in order of opening). It was observed that the valve vibrated so violently that the springs of the servo were seen in a blur. Subsequent measurements with accelerometers and strain-gauges on the valve casing, servo, and pipes, revealed that the dominant frequency was around 63 Hz on all parts. A typical amplitude-frequency plot from an accelerometer on the servo spindle is shown in sketch 2. The maximum amplitudes measured were typically of the order of 100-400 μm on the servo spindle and the casing. Dynamic pressure measurements in the steam pipes were not possible, but it was observed that pipe no. 5 was vibrating more than the other pipes.

The good correlation between all the external vibrations indicated a single source within the valve itself. Of the possible mechanisms that could cause the observed vibration, the most probable was self-sustained oscillations of the valve-bar system as a result of the coupling between its mechanical vibration modes and the acoustic oscillation modes of a steam pipe. In this type of acoustically induced oscillations the mechanical vibrations of the valve plug will produce a corresponding fluctuation of the mass flow rate through the valve. This fluctuation will induce an acoustic pressure response in the steam pipe which in turn closes the feed-back loop by producing a reaction force on the valve plug which can promote the initial vibration. The frequency of the coupled valve-pipe oscillations is mainly determined by the mechanical eigen-frequency of the valve-bar system, even if the acoustic reaction force tends to lower it slightly. The feed-back amplification and phase shift is, on the other hand, mainly controlled by the flow characteristic of the valve and the acoustic impedance of the steam pipe, i.e., of its length and termination.

The problem was treated theoretically by formulating the equations of motion of the valve system including the flow coupling terms and the acoustic reaction forces for a vibration with a harmonic time dependence of the type: $x = x_o \exp(-i(\omega_r + i\,\omega_i)t)$. Solving for $\omega_i = 0$, i.e., neutral stability, one can obtain the stability limits of this phenomenon, e.g., in the form of a stability diagram as shown in figure 3. A more detailed presentation of the theory can be found in references [1] and [2].

The theoretical considerations were backed up by a model test with air on a scaled valve with an open-ended pipe. The test confirmed all the essential aspects of the theory and demonstrated most convincingly the existence of the assumed instability mechanism. Calculations of the mechanical vibration modes and -frequencies of the valve beam system revealed an S-shaped mode with a frequency close to 64 Hz - see sketch 1. This was later confirmed through a vibrator test on the valve beams, after the valve had been dismantled to introduce the modifications described below. The acoustic impedances of the steam pipes could be calculated from the pipe lengths and estimated values of the terminal impedances at the turbine end. These were also checked through direct measurements on the pipes.

Using the information gathered in this way to analyse the behaviour of the valve, it was found that valve no. 5 was definitely unstable, and that all other valves were stable although some were close to the stability limit. Acting on this conclusion, the logical cure of the problem would have been to lengthen pipe no. 5 to bring it into a stable region in the same way as the other valves - see sketch 3. This would have been easy at the design stage, but for practical reasons it was impossible to do in the existing installation. Instead valve no. 5 was removed and the pipe plugged. This eliminated the main vibration source completely and could be done with only negligible losses in turbine performance because the valve originally had some unused margins in valve area. Furthermore the valve beams were stiffened somewhat to bring some of the other valves further into the stable region. This cure completely solved the problem and the valve has now been operating for more than seven years without trouble.

It should be added that several hundreds of valves of this type have been operating for many years in smaller turbines without any sign of this type of instability. The reason for this is that the absolute size of the valve enters into the stability criteria for this phenomenon, and that the actual valve was larger than any previous built.

3. REFERENCES

1. Ingard, Uno, "Valve Noise and Vibration". Report prepared for Stiftelsen för Värmeteknisk Forskning, Sweden. (1977).

2. Singhal, Vijay Kumar, "Acoustics of ducts with flow and its relation to acoustically induced valve-pipe instability". Thesis, Aeronautics Dept., MIT, (Dec. 1975).

GOVERNING VALVE IN LARGE STEAM TURBINE

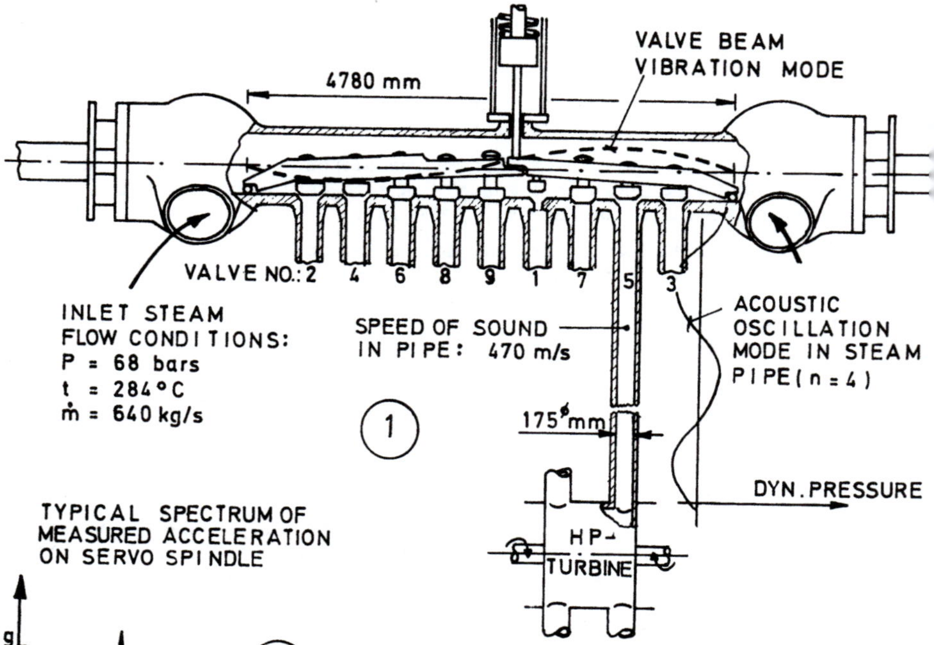

TYPICAL SPECTRUM OF MEASURED ACCELERATION ON SERVO SPINDLE

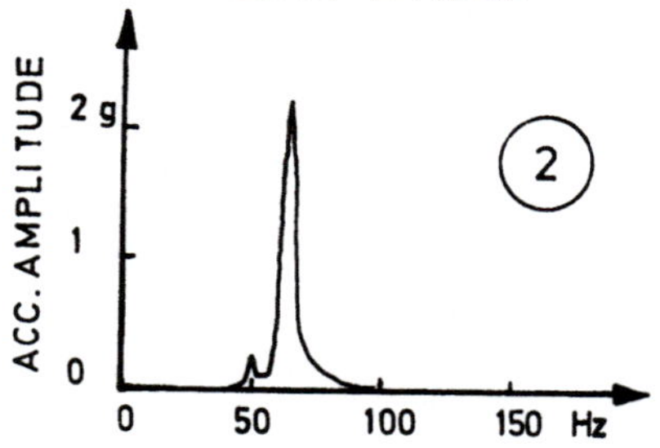

STABILITY DIAGRAM (ref.1)

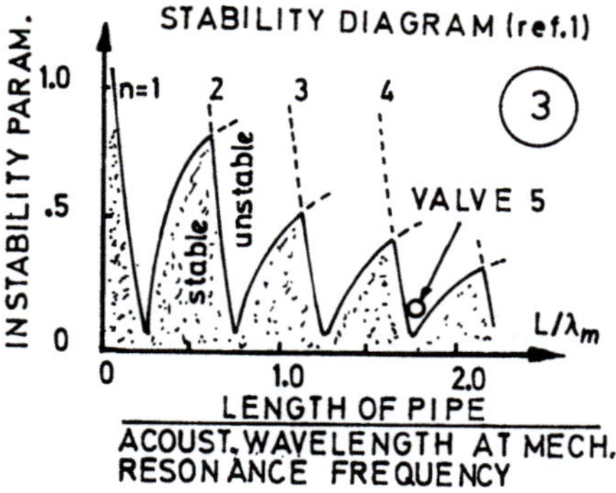

LENGTH OF PIPE / ACOUST. WAVELENGTH AT MECH. RESONANCE FREQUENCY

Ref. 1: Ingard, U., Valve Noise and Vibration. Report to Stiftelsen för Värmeteknisk Forskning, Sweden. 1977.

Problem: Valve vibrations caused by acoustically induced, coupled oscillations of the valve-pipe system, valve no.5. Failures of servo spindle and couplings.

Mechanism:

Initial disturbance → Valve vibration → Fluctuations of flow through valve → Induced acoustic pressure in pipe → Fluctuating reaction force on valve → Valve vibration

Operating conditions: The vibrations occurred in the operating range of valve no. 5 (ca. 270-320 MW turbine power).

Vibration: Distinct peak in frequency at ca. 63 Hz - see sketch 2. Typical amplitudes: 100-400 μm on valve casing and servo spindle.

Structural data: Mechanical eigen-freq. of valve beams in S-shaped mode: 64 Hz. Effective acoustic length of pipe no. 5: 13.0

Cure: Removal of valve no. 5 and plugging of corresponding pipe to eliminate coincidence between mechanical and acoustic resonance solved problem completely. Could also have been achieved by changing length of pipe no. 5 (cf. sketch 3).

B 15 FLOW INDUCED VIBRATIONS OF AN ANCHOR AGITATOR

by

Roger King, Principal Engineer,
BHRA Fluid Engineering,
Cranfield, Bedford, U.K.

Richard Jones, Division Engineer,
I.C.I. Organics Ltd.,
Manchester, U.K.

SUMMARY

This paper describes a programme of tests carried out to investigate the cause of vibration of an anchor agitator, used in the process and chemical industries. A hydroelastic scale model of 1/2.25 was used; bending and torsion measurements were made using strain gauges and a small transmitter mounted on the central shaft of the agitator.

The hydroelastic model tests established that the vibrations were sustained by interactions between vortex shedding and the anchor flexibility. The effect of attaching fins to the agitator prongs was studied to determine a method of suppressing these oscillations. A wide range of fluid viscosities was used in the tests to investigate oscillatory reponse, Power number and Reynolds number.

A step-by-step guide was developed as a basis for future designs of similar agitators.

NOTATION

a	Length of Anchor Prong	m_a	Added Mass of Water/Unit Length
C_d	Drag Coefficient	m_s	Structural Mass/Unit Length
D_v	Vessel Diameter	n	Natural Frequency of Anchor
D	Outer Diameter of Anchor Prong	R	Radius of Anchor
d	Inner Diameter of Anchor Prong	v	Fluid Velocity
E	Young's Modulus of Elasticity	V	Slip Velocity
G	Gap between Anchor and Vessel Wall	V_r	Reduced Velocity = $V/_{nD}$
H	Depth of Fluid in Vessel	x	Rotational Speed
h	Space Dip from Bearing to Water Surface	δ	Logarithmic Decrement of Anchor Material
I	Moment of Inertia of Anchor Shaft	λ	Scale Factor
L	Length of Anchor Shaft	S	Strouhal Number
$m\delta$	Mass-damping Parameter		

1. INTRODUCTION

In many chemical processes where reactions are carried out in large cylindrical mixing vessels, it is essential that the reactants or components are brought into contact and distributed uniformly throughout the vessel, and for this purpose, mechanical mixers, or agitators are used. Typically, the vessels are about 7 m^3 in volume and the majority are each fitted with a central shaft, supported on a steel lid and driven from above through an external gearbox and motor (see Fig. 1).

In viscous media an agitator shaped like an anchor frequently is used; this sweeps past the walls of the vessel (which sometimes are heated) to prevent local accretion or hot spots. Normally anchor blades or arms would be oval in cross section, with the larger axis radial to the vessel wall and equal to one tenth of the vessel diameter.

In order to mix one particularly corrosive brew, the decision was made to construct an agitator from Hastelloy 'C' steel, because of its remarkably good corrosion resistance. However, difficulties were encountered in procuring this material in appropriate form and the agitator was fabricated from circular section tubing 114 mm outer diameter, as compared with the desired 180 mm to comply with the one tenth of vessel diameter specification.

This resulted in a non-standard agitator having arms and shaft that were less stiff than the norm, of smaller cross-section, and circular instead of oval.

In service, the agitator and drive assembly vibrated, flexing the vessel top (40 mm cast iron) on which they were supported, putting at risk the glass-lining of the vessel if the agitator should contact it. Initially, the speed of rotation was reduced from 48 r.p.m. to 39 r.p.m. and the gearbox braced to the edge of the vessel top. This stopped the conspicuous vibrations but the shaft still flexed in the guide bush at the gland (by up to ± 1 mm) causing rapid wear and eventually poor gland sealing. Running the agitator at even lower speds avoided excessively large deflections but did not give acceptable mixing.

If the cause of the vibrations was vortex shedding from the tubular arms of the agitator, no useful reference to this could be found in the literature. A series of tests was arranged to investigate the cause and cure of the vibrations in more detail and for this purpose the tank was filled with water and various vortex spoiling devices attached to the arms of the anchor.

A trailing fin (splitter plate) was attached to each arm to act as a vortex breaker, and in this configuration the vibrational amplitude was reduced to about the eccentricity of the shaft (± 0.15 mm at the gland). However, this was an unacceptable cure because it made the arm into a streamlined body and insufficient mixing and turbulence were generated. Side plates were then welded to the leading faces of the arms, giving various projected areas. In order to give a projection of one tenth of the vessel diameter, the fins had to be of different sizes (because of the small wall clearance on one side); these were entirely satisfactory from a mixing consideration and increased the power consumption by 10%. The shaft amplitude at the gland was not quite so small as with the splitter plate but was still a factor of four better than the originally recorded deflections. The side fins were used in the final design and the plant has operated satisfactorily for the past five years.

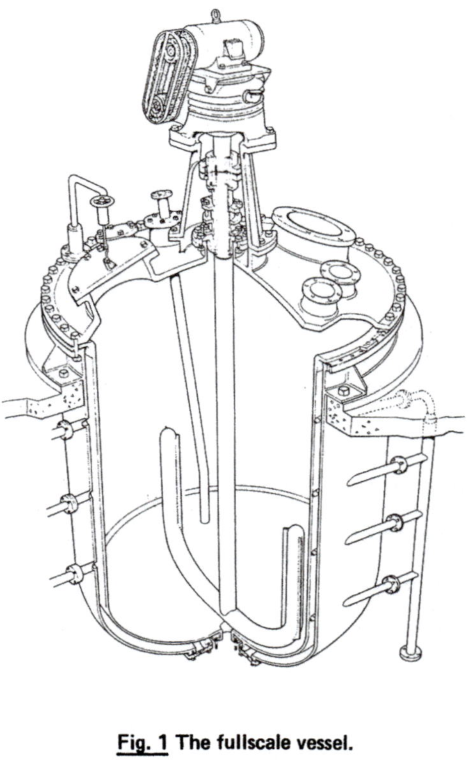

Fig. 1 The fullscale vessel.

Fig. 2. Model anchor and vessel

The operators (I.C.I.) wished to extend the sizes of anchor agitator and contracted BHRA to build a hydro-elastic model of the anchor with the following objectives:

(i) Positively identifying the cause of vibration.
(ii) Optimising the sizes and location of the fins.
(iii) Varying the gap between the anchor arms and the vessel wall to observe the resulting influence on vibration.
(iv) Estimating the power consumption for the various configurations.

These tests form the substance of this paper, and resulted in a step-by-step guide being defined for use in the design stages.

2. PREVIOUS WORK

There is evidently no documented previous work specifically relating to the hydrodynamics of an anchor agitator rotating in a mixing vessel. The work most closely associated with the anchor is that of a cylinder in a flowing fluid mounted parallel to a solid plane or wall although here major differences will arise due to the variation in boundary layer formation and relative motion. There are several reports dealing with this set up and the majority reach similar conclusions from tests with rigid cylinders. In the literature, Goktun (1) showed that two types of eddy shedding occurred; the frequency of shedding between the cylinder and the wall (gap frequency) was different to that from the top surface of the cylinder. He demonstrated that the frequency from the top of the cylinder was approximately 7% lower than the isolated cylinder value as the cylinder was moved away from the wall. There was a peak frequency at G = r, approximately 10% higher than the isolated cylinder frequency and the isolated cylinder condition was reached at G = 2r. Similarly the gap frequency varied from zero (zero gap) to the isolated cylinder value at G = 2r. A peak in the shedding frequency of the gap was recorded at G = r, similar to the result from the top of the cylinder. The tests were conducted at only one Reynolds number so the generality of the findings was not proven.

Buresti and Lanciotti (2) tested a smooth and a rough cylinder in the Reynolds number range $0.85 \times 10^5 < Re < 3 \times 10^5$. The hot wire probe was placed in the gap for part of the test and just above the cylinder for others; thus it is difficult to compare these results with those recorded by Goktun. Measurement of frequencies from the top surface of the cylinder showed that the isolated cylinder situation was approximated for all gaps greater than 0.8r. Regular vortex shedding disappeared at G = 0.6r.

Bearman and Zdravkovich (3) observed effects similar to those of Buresti and Lanciotti in that regular vortex shedding disappeared at G = 0.6r, although the tests were run in a slightly lower Reynolds number range $2.5 \times 10^4 < Re < 4.8 \times 10^4$.

Roshko and Steinolfson (4) recorded an abrupt reduction in drag coefficient at G = 0.6r as the gap was decreased from the isolated cylinder value. It is speculated that this may be associated with changes in the vortex shedding processes.

Sarpkaya (5) tested a cylinder near a wall in harmonically oscillating flow at high Reynolds numbers, demonstrating that regular vortex shedding ceases for $G \simeq r$. Sarpkaya attributes this to the proximity of the wall decoupling the frequency of oscillations in the top and bottom shear layers.

In what seems to be the only available paper on tests with a flexible cylinder near a rigid boundary Wilson and Cauldwell (6) showed that sustained vibrations can be excited by the flow for a wide range of cylinder-wall spacings. For each spacing, the Reduced Velocity V/nD at which vibrations were first observed was a function of Reynolds number. Reynolds number evidently had the greatest effect at the smallest spacing (G = 0.5r), became less important as the spacing was increased to G = 14r and eventually had no discernible influence for the isolated cylinder case.

However, as may be seen the literature was not in general agreement, and so far removed from the conditions applicable to the anchor agitator that it was felt the only satisfactory solution would be achieved through a series of hydroelastic model tests.

3. THE APPARATUS

3.1 The Hydroelastic Model

Fig. 2 shows a general view of the vessel used in the study. The gap between the anchor prongs and the side of the vessel could be varied by inserting wooden liners round the edge; these enabled the gap to be changed from 0.025 m to 0.011 m and 0.0046 m.

Three independent sets of strain gauges could be connected to the transmitter which measured the three modes of bending of the centre shaft. Signals from the transmitter were routed to an ultra-violet recorder where a permanent record could be made.

3.2 Selection of Scale and Material for the Model Anchor Agitator

For a model to be accurately representative of the full scale it must satisfy simultaneously a number of similarity criteria. These naturally put severe constraints on the choice of model materials and it is generally accepted that some relaxation of requirements is necessary, with only selected criteria being satisfied. The criteria considered necessary for modelling the anchor were as follows: (subscripts f and m refer to full scale and model respectively).

a) Froude Number V^2/gD

i.e. $(V^2/gD)_m = (V^2/gD)_f$

$g_m = g_f$ and D_f/D_m = the linear scale factor λ

$V_m = 1/\sqrt{\lambda} V_f$

b) Natural Frequency n

$n_m = \sqrt{\lambda} n_f$

c) Mass-damping Parameter $m\delta$

$(m\delta)_m = 1/\lambda^2 \ (m\delta)_f$

Dimensional analysis shows that at resonance the mass ratio and damping parameters can be combined to form the group $m\delta$, where m is the mass ratio (or density ratio) and δ is the logarithmic decrement of the material, indicating the internal damping and thus energy dissipation properties of the material.

d) Diameter of Prong D

$D_m = 1/\lambda D_f$

e) Gap between Prong and Vessel Wall G

$(G/D)_m = (G/D)_f$

The diameter of the vessel and the anchor length were considered secondary to the quantities listed above. It was decided to test the anchor in a 0.6 m diameter research vessel and this suggested a scale factor λ of about 3, since the fullscale vessel was 1.80 m diameter. Next it was argued that D_m should be made as large as possible and to use ABS plastic from which to manufacture the model: 0.05 m diameter tube was commercially available and this was decided upon, fixing $\lambda = 2.25$ (the fullscale anchor was 0.114 m diameter). Table 1 gives details of the fullscale and model anchors.

TABLE 1
Details of model and fullscale anchors

Property	Fullscale	Model
D_v (m)	1.800	0.600
D (m)	0.114	0.051
d (m)	0.102	0.044
EI (Nm^2)	5.98×10^5	4.52×10^2
G/D	0.500	0.500
n (Hz)	2.300	3.200
δ	0.100	0.065
R (m)	0.790	0.025
L (m)	3.420	1.060
m_s (kgm^{-1})	16.000	0.550

4. RESULTS

The first tests in the mixing vessel showed that the anchor did vibrate and could be made to hit the side of the vessel by increasing the speed sufficiently. However, before investigating the oscillation, it was thought necessary to obtain some basic results concerning the model test rig.

A miniature propeller meter was used to measure the water speed for various anchor speeds and distances from the centre of the vessel, and for various heights above the base of the vessel. The centre portion of the flow was very close to that of a forced vortex with a linear increase of speed with radius, but the outer portion showed that the maximum speed attained was equal to approximately half of the anchor speed. The velocity of the water did not vary noticeably with depth and therefore was taken to be constant.

It was found that while vibrating, the largest signal by far was obtained from the radial sway mode. The natural frequency of the anchor was measured in different depths of water so that the excited frequency could be compared with the natural frequency. The natural frequency in air was also measured so that the 'added mass' for the anchor could be calculated. The results showed that the added mass coefficient was approximately 1.5.

Having calibrated the model it was possible to do some quantitative tests on the amplitude and frequency of vibration. These were measured for different values of rotational speed x, and depths of water H. Fig. 3 shows the results obtained from these tests; there is a critical value of x below which there is little vibration and above which the amplitude increases rapidly.

It was possible to obtain the water speed at this critical speed and so calculate the corresponding Reduced Velocity V_r i.e.:

$$V = 2\pi Rx - v \quad \text{(Slip or Relative Velocity)}$$

$$V_r = V/nD \quad \text{(Reduced Velocity)}$$

The results are given below:

TABLE 2
Criteria for the onset of oscillations

H (m)	x (revs per sec)	v (cm per sec)	V (cm per sec)	n* (Hertz)	V_r
0.46	0.85	77	59	3.13	3.72
0.53	0.82	75	56	3.06	3.60
0.61	0.79	72	54	3.04	3.50
0.69	0.79	72	54	2.99	3.56

*n is the frequency of oscillations.

Problems were encountered in the full scale vessel when the rotational speed exceeded about x = 32 r.p.m. For the Froude similarity criterion to be satisfied the equivalent model conditions would be:

$$V_m = 1/\lambda^{1/2} V_f$$

$$R_m x_m = 1/\lambda^{1/2} R_f x_f$$

$$x_m = 1/\lambda^{1/2} R_f/R_m x_f$$

$$x_m = 1/\sqrt{2.25}\ 32/60\ 1.80/0.6$$

$$x_m \sim 1.0 \text{ rev/s}$$

Thus the amplitude of oscillations would be expected to increase noticeably at a model rotational speed of $x_m \sim 1.0$ rev/s. Fig. 3 confirmed this and demonstrated the existence of sustained oscillations at speeds below this value, although amplitudes were correspondingly smaller.

It was considered that the validity of the model had been established beyond reasonable doubt and the main tests were carried out to investigate the effect of varying the gap between the anchor and the wall of the mixing vessel, using wooden liners. During these tests, torsion, as well as bending was studied. Additionally, tests were conducted using fins of varying sizes to observe the influence of each on the resulting amplitude of oscillation. The viscosity of the fluid was varied in the model vessel using aqueous solutions of glycerol. Power and vibrational amplitudes were measured.

4.1 Vibration Tests with Variable Wall Clearance

Initial observations showed that the anchor vibrated with either of the sets of liners in position. Measurements of amplitude and frequency of oscillation were taken in the same way as before : values for V_r were calculated, and are tabulated below:

TABLE 3
The effect of wall clearance on the onset of oscillations

H (m)	x (revs per sec)	v (cm per sec)	V (cm per sec)	n* Hertz	V_r
0.46 no liners	0.85	77	59	3.13	3.72
0.46 thin liners	0.60	57	39	2.99	2.57
0.46 thick liners	0.42	42	25	2.92	1.69

*n is the frequency of oscillations.

Thus, from the table, the threshold reduced velocity decreases as the gap is made smaller. As the gap tends to zero, the oscillation should tend to some limiting value; evidently the frequency will be zero when the anchor arms actually come into comtact with the vessel walls. As the gap tends to infinity, $V_r \rightarrow 5$ which approximates to the value for an isolated cylinder.

4.2 Torque Measurements for Variable Wall Clearances

The results of these tests showed that decreasing the gap between the anchor and the side of the mixing vessel increased the torque. This was not unexpected, since as the gap was made smaller, there was a larger velocity gradient between the anchor and the vessel wall, and this increased the drag on the arms of the anchor. The addition of trailing fins, one diameter in length, did not appreciably alter the torque measured.

In the full scale tests investigating the effects of trailing fins, a 40% reduction in torque was recorded when trailing fins projecting for about one diameter were fitted. However, there is a possibility that this was due to the gearbox oil warming up and no definite conclusions should be drawn.

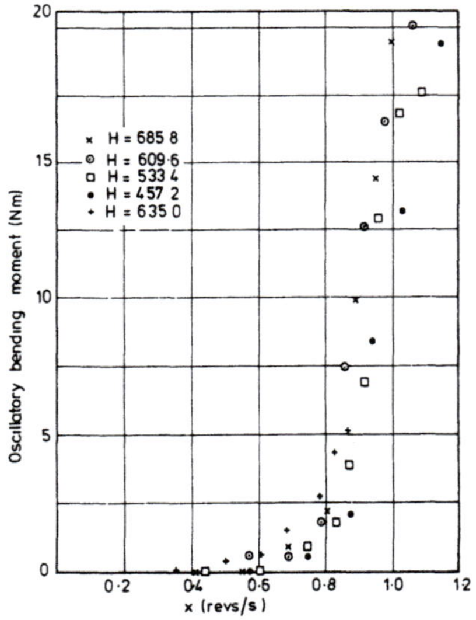

Fig. 3 Amplitude response in water.

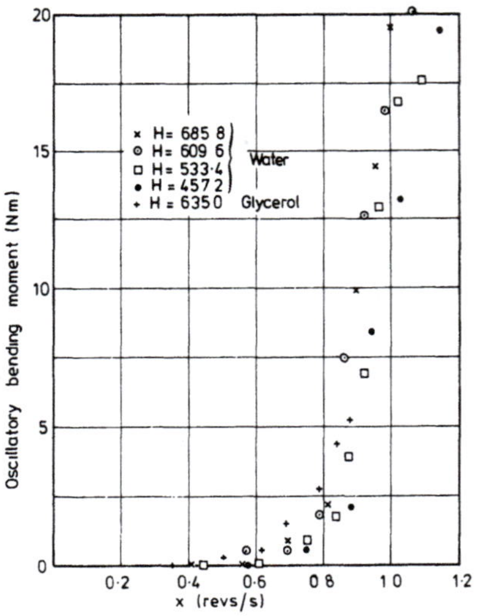

Fig. 4 Amplitude response in water and in glycerol solution.

4.3 The Effect of Varying Viscosity on the Onset of Oscillations

Some of the water used in the previous experiments was removed from the mixing vessel and pure glycerol was poured in until the concentration became 80% glycerol, 20% water, by volume. Fig. 4 shows the oscillatory amplitude of the anchor in this mixture. It will be seen that the velocity at onset of large amplitude oscillations is very close to that recorded in pure water, although the Reynolds numbers of the comparable cases are Re ~ 23,000 for the pure water test and ~ 400 for the glycerol test.

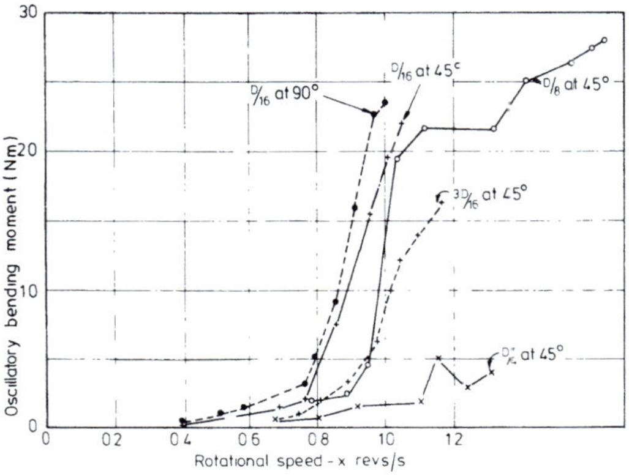

Fig. 5 **Amplitude response for different types of fin.**

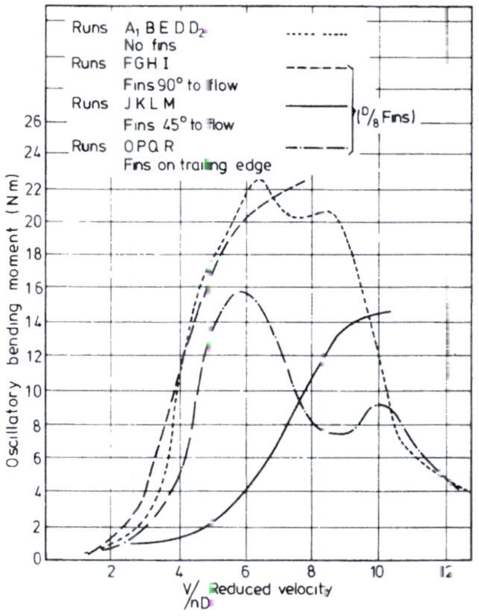

Fig. 6 **Amplitude response for different fin locations.**

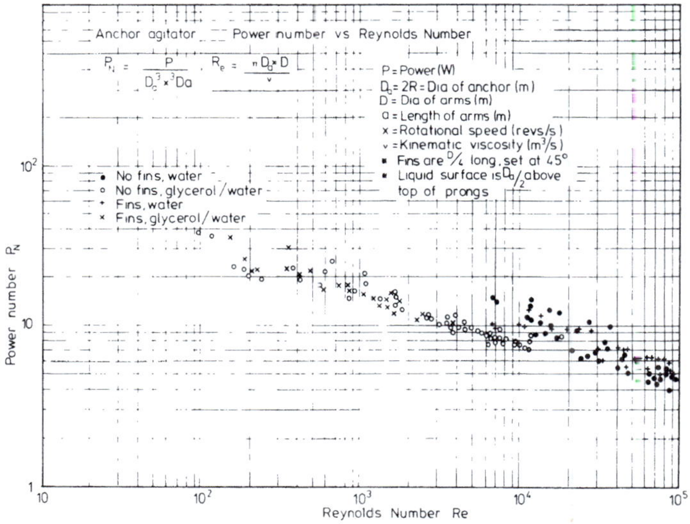

Fig. 7 **Power number versus Reynolds number - a collection of the overall test results.**

It is believed that vortex shedding was identified beyond all reasonable doubt as the cause of the oscillations recorded in the full scale tests. Additionally the present tests confirmed the selection of the modelling criteria and the assumption that Reynolds number was of secondary importance.

4.4 The Stabilising Effects of Adding Fins

In these tests, fins of 1/16, 1/8, 3/16 and 1/4 of a diameter were attached in sets of two to the leading faces of each of the anchor arms to determine the minimum size of fin to suppress excitation. In the majority of these tests, the fins were set at ± 45° from the front stagnation point whilst in one test the fins were also set at ± 90°. Following these tests, a comprehensive series of additional tests were run using fins 1/8 of a diameter to determine the influence of fin angle on the resulting amplitudes of oscillation.

Fig. 5 summarises the initial tests, showing amplitude of oscillation against rotational speed of the anchor. It will be noted that the amplitude of oscillation decreased as the fin sizes were increased; the D/16 fins gave results very similar to those recorded for the 'bare' agitator, the D/8 fins showed little improvement and the 3D/16 fins caused a considerable reduction both in the amplitude and rate of onset of oscillations. However, the D/4 fins gave yet more dramatic reduction in amplitude and it was concluded that for absolute safety, these fins should be fitted to the full scale installation. Fig. 6 is a collection of results from the series of tests with the 'bare' anchor, the anchor plus D/8 fins at ± 90° and for one fin at 180°. This graph emphasises the comparatively small influence of these fins at ± 90°, and demonstrates that ± 45° is a considerably more effective location. It should be pointed out that in Section 4.3 the D/4 fins set at 90° were shown to be completely adequate in suppressing the oscillatory response of the anchor; the comments in this present section apply to the D/8 fins and smaller. Surprisingly, the trailing fin is also as effective as the fins at ± 45° and indeed introduced increases in the velocity at which large amplitude oscillations occur; this would imply that it should be possible to run the anchor at considerably higher speeds before encountering the same oscillatory amplitude. However, this was interesting but unattractive as a means of suppressing the oscillations because the fin is believed to decrease the agitation efficiency.

4.5 Power Consumption

In this section the power consumption of the anchor, (with and without fins) was recorded for a total of three fluid depths and for a range of operational speeds and viscosities. In the initial tests the power was expressed in watts; however, in the overall analysis a composite non-dimensional graph of a power number (P_N) versus Reynolds number (Re) was assembled to make the results generally applicable (Fig. 7).

The large Re ranges were achieved by varying the viscosity using aqueous solutions of glycerol. By this device, the viscosity was varied from 1 centi-stoke to 50 centi-stoke.

5. CONCLUSIONS

The tests described in this paper have shown:

1. That vortices can be shed from a cylinder even when close to a plane boundary. The effect of the boundary is to increase both the drag coefficient and the apparent frequency with which the vortices are shed. The shedding of these vortices interacted with the anchor flexibility and resulted in sustained oscillations at the anchor's natural frequency.

   Two methods of eliminating these vibrations have been studied in this report. The first method simply required the anchor to be run below the critical excitation speed. This can be found using the values of V_r obtained for different gap sizes. The second method is the addition of the fins at least D/4 wide to the outer prongs of the anchor, and it would seem that the optimum position is ± 45° from the stagnation point, although their effect on the mixing process was not quantified.
2. The more general tests showed that reducing the clearance between the anchor prongs and the vessel wall reduced the speed at which vibrations were first excited and also increased the power consumption.
3. The tests using the solutions of glycerol have shown that the critical parameter governing the onset of vibrations is the reduced velocity and not Reynolds number.
4. The added mass associated with the anchor is dependent upon vessel wall clearance and to a first approximation can be about 50% higher than for an ideal, infinitely long isolated cylinder.
5. The power number vs. Reynolds number graph can be used for geometrically similar systems.

6. The oscillatory behaviour of the anchor varied with water depth, although this was not fully explored in these tests and the general conclusions represent conservative estimates of safety margins.

The fins recommended were fitted to the anchor agitator and have operated satisfactorily from both vibration and mixing considerations. It should be noted that this recommendation results in a frontal area of only $D_V/15$ compared with the originally desired $D_V/10$. However, when the drag coefficients of the two configurations are included in the expression for fluid resistance (and hence power) the respective power consumptions become approximately equal.

6. DESIGN GUIDE

Step 1

Estimate the stillwater frequency of the anchor (n) using the simplified equation noting that the added mass coefficient will probably be about 1.5.

$$n = \frac{1}{2\pi} \sqrt{\frac{3EI}{m_a L^4 \; [2a/L \; (1 + m_s/m_a) \quad 0.2427 \; (1 + m_s/m_a - (h/L)^4)]}}$$

Step 2

The threshold of vibrational instability is reached at $V/nD \sim 3$, although this will depend upon gap between vessel wall and anchor prongs. Assume 100% slip such that $V = 2\pi R$; the maximum permitted speed of rotation x (rps) to avoid vibrations is thus $x = 3nD/2\pi R$. If this is unacceptable some redesigning may be necessary to adjust n, or another value of slip (say 0.6) assumed, and the calculations repeated to qualify the probability of vibrations occurring. Alternatively put on D/4 fins at ± 45° from the stagnation point.

Step 3

Read off power number from Fig. 7, at the appropriate Reynolds number to estimate power consumption.

7. REFERENCES

1. Goktun, S. "The drag and lift characteristics of a cylinder placed near a plane surface". MSc Thesis, Naval Postgraduate School Monterey (1975).
2. Buresti, G. and Lanciotti, A. "Vortex shedding from smooth and roughened cylinders in cross-flow near a plane surface". Aeronautical Quarterly pp 305-321, (1979).
3. Bearman, P.W. and Zdravkovich, M.M. "Flow around a circular cylinder near a plane boundary". J. Fluid Mechanics, (1978).
4. Roshko, A., Steinolfson, A. and Chattoorgoon, V. "Flow forces on a cylinder near a wall or near another cylinder". Proc. 2nd US Conf. Wind Res. Eng. Fort Collins, Paper IV-15, (1975).
5. Sarpkaya, T. "In-line and transverse forces on cylinders near a wall in oscillatory flow at high Reynolds numbers". Offshore Technology Conference, U.S.A., Paper OTC 2898, (1977).
6. Wilson, J.F. and Cauldwell, H.M. "Force and stability measurements on model submerged pipelines". Offshore Technology Conference, U.S.A., Paper OTC 1224, (1970).
7. King, R. and Priddin, K.G. "Model testing of an anchor agitator". BHRA Report, RR 1456, (1978).

8. ACKNOWLEDGEMENTS

This programme of work was undertaken for I.C.I. Ltd., Organics Division. The authors acknowledge the co-operation of I.C.I. throughout this project; they would also like to thank the Director and Council of BHRA for permission to publish this paper.

ANCHOR AGITATOR

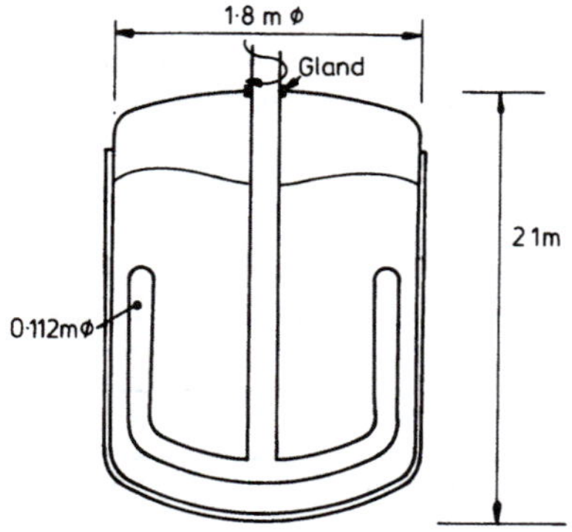

①

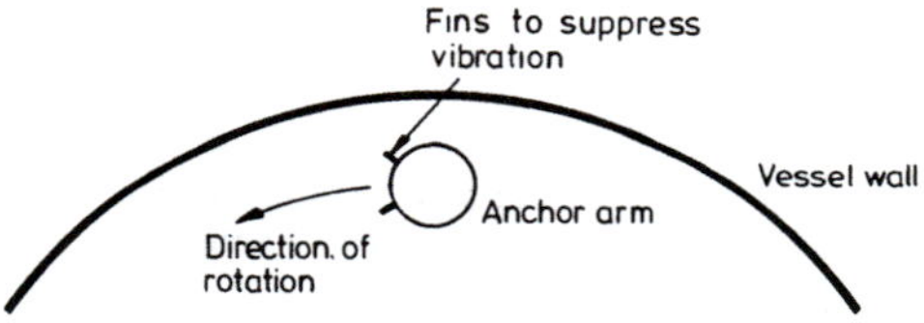

②

Problem : Agitator rotates about vertical axis to promote shear velocity field and enhanced mixing of fluids.
Agitator constructed from circular section steel. As arms sweep past walls of vessel, agitator becomes unstable and sustained oscillations ensue. Severity of oscillations is apparently a function of speed and agitator cannot be run at design speed (48 r.p.m.).
Vessel is glass lined: possibility of anchor hitting and destroying this lining. Additionally, large amplitude vibrations cause rapid wear of gland on shaft: regular re-packing is necessary.

Geometry : See Sketch 1.

Boundary conditions : Vessel walls containing water; typical of vessel is 6.8 m^3.

Flow conditions : Determined exclusively by speed of rotation of agitator. Very little vertical velocity. Turbulence not measured. Point velocities measured in model, with miniature propeller meter; not measured on site.

Operating conditions : Rotational speeds of 35.39 r.p.m.
Vibrational response at natural frequency of 2.5 Hz. Mode of vibration: response in radial direction.

Structural dynamics : Mass of anchor (excluding added mass effects) = 225 kg.

Vibration : Frequency ≈ 2.5 Hz (approximately equal to still-water natural frequency).
Maximum amplitude at gland ± 1.0 mm.
Maximum amplitude of arm ± 15 mm (estimated).

Cures : Model tests with hydroelastic model used to determine cause of oscillations. Vortex excitation confirmed. Various spoilers tested and radial fins of at least ¼ D disposed at ± 45° from front stagnation point found to be most effective. (see Sketch 2)
Amplitude at arm reduced to ± 3 mm (estimated).

B16 VIBRATIONS OF A PUMP COLUMN UNDER VORTEXING CONDITIONS

James C. Wang
Engineer

Wei-Yih Chow
Engineer

Alexander B. Rudavsky
President

Hydro Research Science, Santa Clara, California, USA

SUMMARY

Flow-induced vibrations of a pump column under vortexing conditions are discussed in this paper. Experimental tests were conducted in a flume and the following parameters were monitored during the test program: discharge pressure, flow, motor current, local velocity of inflow, water depth, acceleration at bottom of pump column, strain at top of pump column, and pressure at various locations inside the pump column. The test results showed that vortex flow reduced pumping efficiency; it also caused severe vibration which could possibly damage the pump column, shaft, and impeller. The solution involved improvement of flow conditions and the placement of geometric configurations that eliminated vortices.

1. DESCRIPTION OF THE PUMP AND PUMP BAY

As indicated in Figure 1, the model, supplemented by a small pump, a motor, and a valve to control flow, was set up in a 3 ft wide, 4 ft deep, 25 ft long water flume. The pump was an AC 7x5AX, driven by a 5 hp motor running at a speed of 3,495 rpm. Pump bays of various widths, from 2 to 10 times the bellmouth diameter, were tested. Figure 2 indicates the location of the pump. The Chart shows the various dimension parameters.

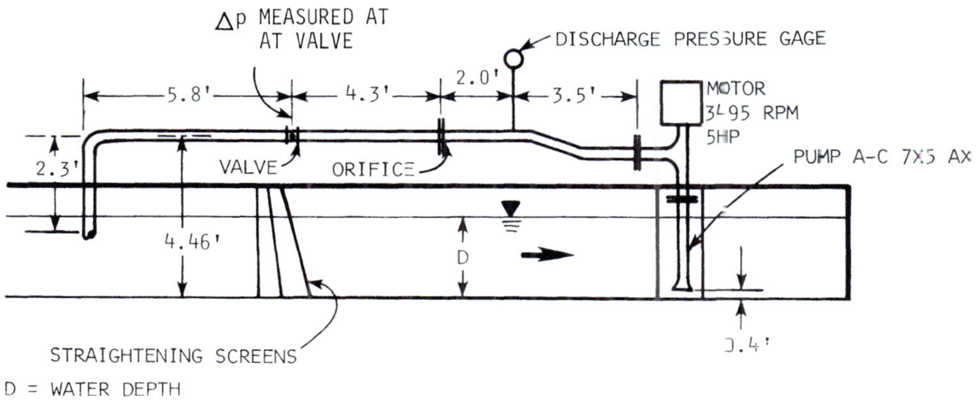

FIG. 1 MODEL SETUP (ELEVATION VIEW)

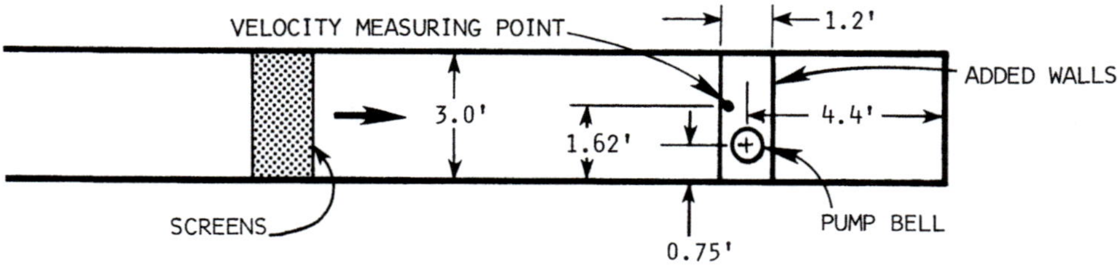

FIG. 2 LOCATION OF PUMP (PLAN VIEW)

2. INSTRUMENTATION

The phenomena measured during this study were the discharge pressure using a Bourdon gage, the flow using orifice and mercury manometers, the motor current using a Dayton Ampere probe, the velocity using a calibrated pinwheel-type meter, the acceleration using accelerometers, the strain using strain gages, and the column pressure using pressure transducers. The discharge pressure gage and orifice flow meter were located as shown in Figure 1. The amp probe was mounted on the motor input wiring. Water depth was measured at the suction bell. Accelerometers, strain gages, and pressure taps were located as shown in the chart.

3. TESTING PROGRAM

The testing program consisted of three phases. The first phase included experimentation at five water surface elevations and at five points on the pump operating curve for each elevation. These elevations are in terms of 1, 1.5, 2, 2.5, and 3 pump bell diameter submergences. The tests showed that low-submerged conditions (1, 1.5, and 2 bell diameters), resulting in a surface vortex and a subsurface vortex entering the pump, produced pump pitting and vibrations. For high submergence conditions, there was only a surface vortex; it did not enter the pump, as no severe vibrations were observed. The pump was modified for this phase of the experimentation by adding walls, indicated by the solid lines in Figure 2. Velocity measurements were taken at the location of the symbol • in Figure 2 and at a point 0.5 ft below the water surface. Only during this phase were velocity data assembled.

During the second phase of the study, the effect of a 0.5 in. thick splitter on shutoff discharge pressure was determined. The effect was investigated at three water surface elevations and at a discharge pressure of 11.5 psi (shutoff). The effect was also determined at the same three elevations for a discharge pressure of 5 psi, to determine any possible effect of the splitter on flows other than shutoff. The data collected during each of these runs are listed in Table 1.

TABLE 1

EFFECT OF SPLITTER

RUN	WATER DEPTH (ft)	SHUTOFF HEAD ON DISCHARGE VALVE WITHOUT SPLITTER (ft)	WITHOUT SPLITTER (ft)	INCREASE DUE TO SPLITTER (%)
# 3	1.75	27.4	34.2	25
# 2	1.46	27.4	33.4	22
# 1	1.17	26.9	33.8	26

The presence of a splitter not only reduced the occurrence of vortexing phenomena but also improved the flow conditions in the vicinity of the pump bell. During the third phase, experimentation on the effect of various pump conditions on the vibration and strain of the pipe column was conducted. The conditions were as follows:

(1) Smooth inflow to pumps; no surface or subsurface disturbances; no vortices; valve wide open; water depth of four bellmouth diameters; pump operating pressure of 2.5 psig;

(2) Same inflow conditions as case (1), but valve completely closed;

(3) Disturbed inflow to pumps; vortices present; data taken at water depths of 2, 1.5, and 1.0 bellmouth diameters.

The dynamic strain measurements for conditions (1) and (3) were all at about 5 microinch per inch whereas the shutoff test (condition 2) showed a maximum peak of 40 microinch per inch reading.

The dynamic strain with vortices present at different submerged conditions was 12μ, 22μ, and 29μ for 2, 1.5, and 1 bell diameters respectively.

4. CONCLUSIONS

(1) The size of the surface vortex core increases as the water level decreases, and the vibration and dynamic strain of the pump column increase as the size of the vortex core increases.

(2) The presence of a suction vane or splitter significantly increases the shutoff discharge head. However, the splitter does improve the flow patterns at the entrance of the pump so as to reduce the occurrence of vortexing phenomena.

(3) By comparing the data of shut-off experiments and the data of pump vibration due to vortexing in Phase 3, it may be concluded that even under the most severe conditions of vortex action, the maximum dynamic strains exerted upon the pump column are considerable; yet, they are small compared to the strains resulting from the pump shutoff operation.

VIBRATIONS OF PUMP COLUMNS UNDER VORTEXING CONDITIONS

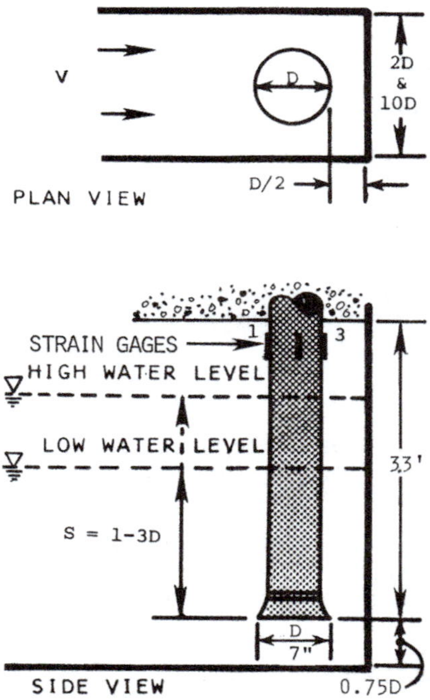

① PUMP COLUMN

Problem: Vortex-induced vibration of pump column during pumping process, reducing pumping efficiency and possibly damaging the impeller because of cavitation.

Geometry: See side view. Column length: 3.3 ft; bellmouth diameter: D = 7 in.

Boundary Conditions: Pump bell submergence: 7 in. to 21 in.; and Pumping rate: 385 gpm.

Flow Conditions: Uniform flow approach to the pump.

Operation Conditions: Flow rate: Q = 385 gpm. A vortex was generated under low submergence conditions. Excessive vibrations were observed during the generation of the vortex.

Structural Dynamics: Modulus of elasticity $E = 3 \times 10^7$ psi. Average moment of inertia of the pump shaft section (including added mass for lowest submergence condition) $= 11$ in.^{4}.

Vibration: Frequency varies with vortex formation. Maximum amplitude at the end of the column = 0.4 in.

Solutions: (1) Increase the depth of submergence; (2) change the geometric conditions at the approach area and pump forebay to eliminate vortices; and (3) place splitter plate underneath the bell to prevent the submerged vortices.

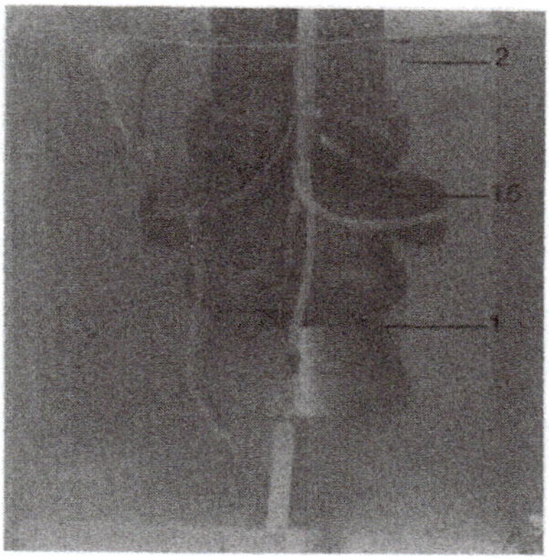

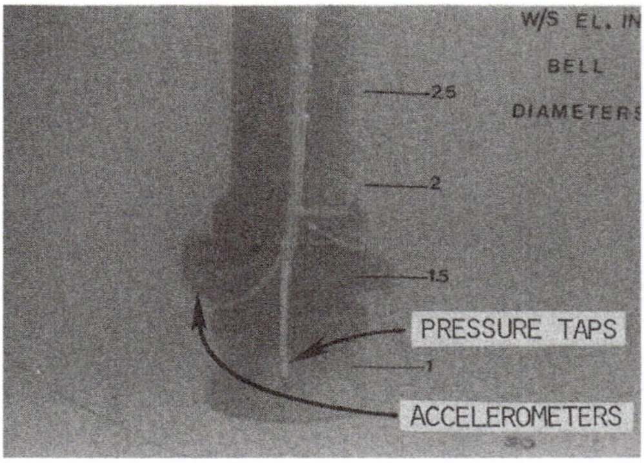

② VORTEXING PHENOMENA AT DIFFERENT SUBMERGENCE

B17 VIBRATIONS OF A CANTILEVERED PUMP COLUMN

Wei-Yih Chow
Engineer

Alexander B. Rudavsky
President

Hydro Research Science, Santa Clara, California, USA

SUMMARY

Flow-induced transverse vibrations of a cantilevered pump column during the backflushing process are discussed in this paper. The backflushing process involved reversing the flow through the pumping system in order to flush out sediments deposited within the pump bay. Severe column vibrations were observed during this process. The nature of the vibrations and the factors affecting them were experimentally investigated, and a method of reducing or eliminating the vibration was sought.

1. DESCRIPTION OF THE STRUCTURE AND OPERATION

Pump intake structures are, of course, primarily designed to intake water from a source area. As a result of sediment deposition or detritus blockage, backflushing (or reverse flow) of the water through the pump bays is necessary to wash out sediment or clear blockage of the system.

The pump intake structure discussed in this paper comprises four identical pump bays; within each pump bay is a river water pump capable of drawing 40,000 gpm during normal operation. The pump column, which is cantilevered 28.25 ft from an upper concrete deck, has a diameter of 3.0 ft. During the backflushing sequence considered in this study, a discharge of 60,000 gpm could be forced out of any one pump by gravity.

As shown in the Chart several flow-stabilizing structures are located within the pump bay: a 2.0 x 1.22 x 18 ft horizontal beam placed in front of the pump column across the entire width of the pump bay, a 4.6 ft high baffle wall, and 6 vertical guide vanes, 12 ft high, spaced equidistantly across the pump bay entrances. These structures were designed to stabilize the flow during the pumping (suction) mode. In addition, a wedge-shaped splitter, installed immediately beneath the bellmouth, guided the flow back to the source area during the backflushing mode and aided in the clearance of deposition from the bay floor.

2. VIBRATIONS

Vibration of the cantilevered pump column during the backflushing process is primarily induced by flow instability. Experimental investigations of the nature and severity of this vibration were conducted by means of the model illustrated in Figure 1. The model was constructed to an undistorted scale, model to prototype, of 1:12; accurate simulation of the stiffness and natural frequency of the prototype column were of primary importance.

The vibrations of the pump column were monitored by four strain gages. As indicated in the Chart, two gages were oriented in the X direction, and two in the Y direction. A Sanborn recorder registered the strain gage outputs. Calibration of the model involved the application of a known constant force to the end of the column and measurements of the resulting displacement and strain on the pump column. Figure 1 shows the calibration test setup. The fundamental natural frequency of the system during vibration was found by applying a periodic constant force over a range of frequencies to the structure and again monitoring the response of the structure. The results indicated that the natural frequency of the modeled system was 13.2 Hz. Since the Froudian model (1:12) was employed during this study, a natural frequency of 3.8 Hz in the prototype column was therefore obtained. This prototype column frequency was consistent with computational results.

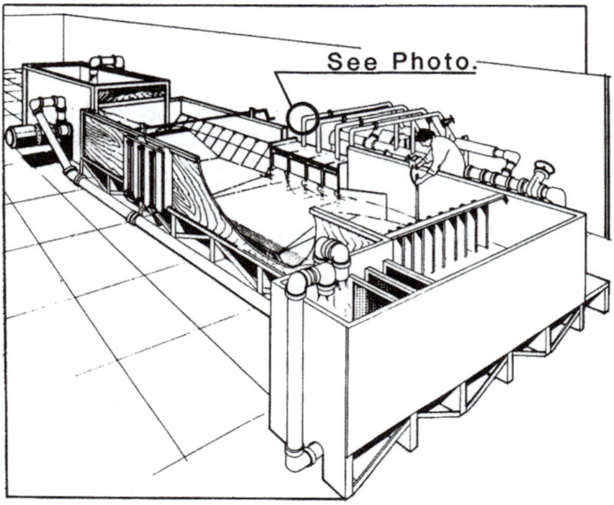

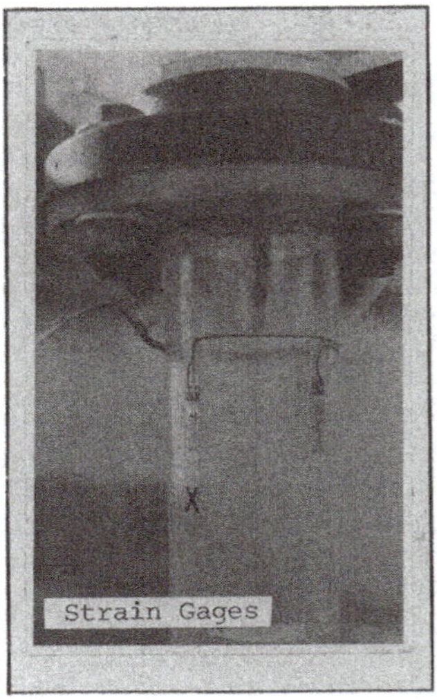

FIG. 1 MODEL OF RIVER FLUME AND PUMP INTAKE COMPLEX

Twenty-seven vibration tests were conducted at different submerged water surface elevations and different flow discharge rates. As indicated in the Chart, at a system flow rate of Q = 60,000 gpm and a submerged water surface elevation of 334.5 ft, a fundamental frequency of 3.5 Hz occurred during the backflush process. The resultant maximum displacement at the lip of the bellmouth during this resonant condition was 1.85 in.

Further tests involved raising the water surface elevation. The results showed that when reduced discharge rates were coupled with higher submergence elevations (Q = 40,000 gpm, water surface elevation at 342 ft), vibration levels decreased significantly. By increasing the submergence (i.e., raising the water surface elevation), the magnitude of vibration of the pump column could be reduced. The test results indicated, however, that this magnitude might still be critical to the system at a submerged water surface elevation of 336.5 ft or above. Reduction of the pump column to 1.5 times less than that produced by the 60,000 gpm condition; the magnitude of displacement at the lip of the bellmouth is correspondingly reduced to 0.8 in.

3. CONCLUSIONS

On the basis of this study, the following recommendations can be made: (1) since the vibration induced at discharges higher than 40,000 gpm and at water surface elevations lower than 342 ft may cause damage, backflushing processes should be conducted at rates lower than 40,000 gpm and at elevations above 342 ft; (2) lateral support of pump columns should be instituted if significant vibrations occur.

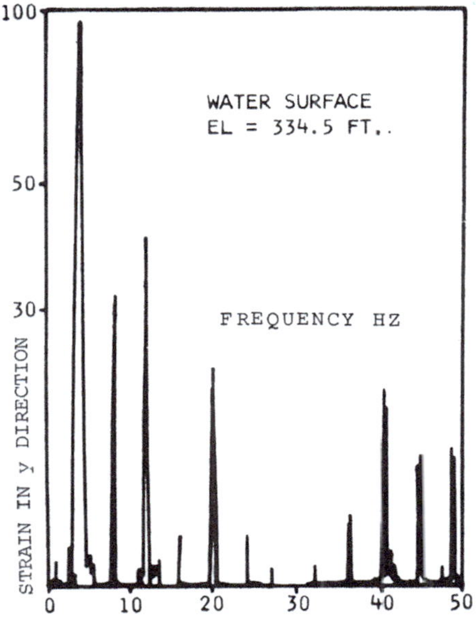

(1) FORCED VIBRATION SPECTRA DUE TO FLOW DISTURBANCE FREQUENCY = 3.5 HZ

VIBRATIONS OF A CANTILEVERED COLUMN

Problem: Flow-induced transverse vibration of a long cantilever pump column during the backflushing process caused damage.

Geometry: Diameter of the pipe = 3.0 ft, column length = 28.25 ft, wall thickness = 0.5 inch.

Flow Conditions: Discharge flow with turbulence.

Operating Conditions: Backflushing at a pump discharge rate of Q = 130 cfs. Water surface elevation ranges from El. 334.5 ft to El. 360.0 ft.

Structural Dynamics: Modulus of elasticity $E = 3.0 \times 10^7$ psi. Average moment of inertia of the pump column section (including added mass for submergence condition) = 9160 in^4.

Vibration: Frequency 3.5 hz. Maximum amplitude at the end of the column = 1.85 in.

Solutions: Backflushing process only during high submergence condition; change discharge rate .

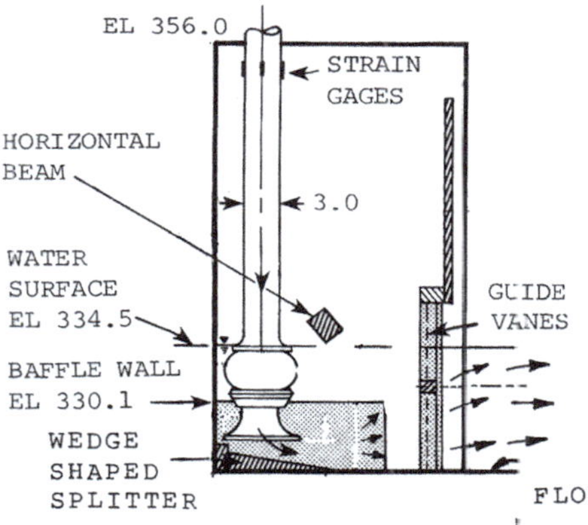

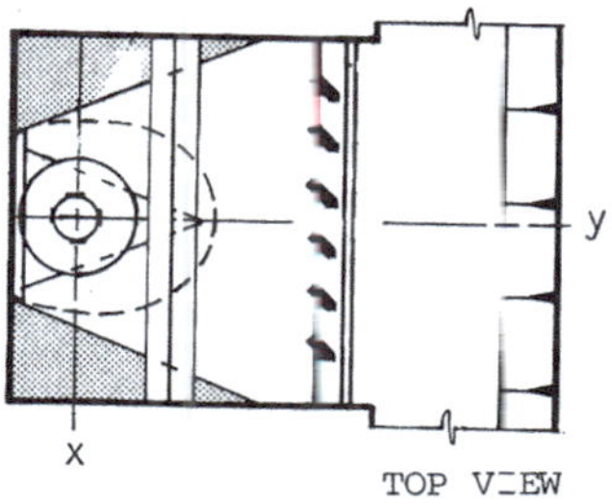

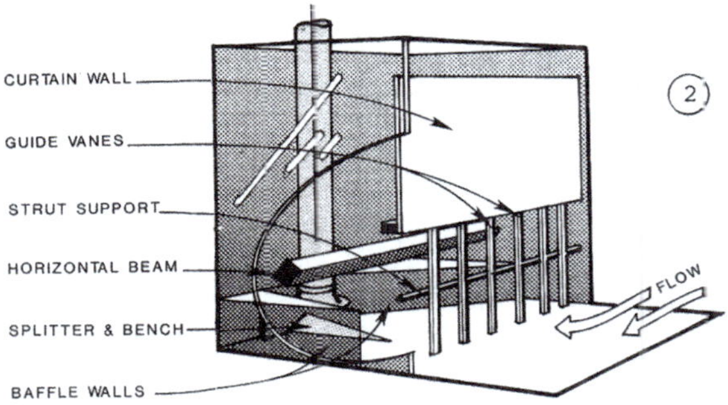

(2) MODEL DETAILS

B18 ROCK DRILLING TOOL WITH PULSED JETS

D. Milan
G. Berthollon
NEYRPIC Company, Grenoble, France

R. Toueille
S.M.F. Company, Tarbes, France

SUMMARY

This paper deals with development of a hydraulic device intended to generate pulsed jets around a drilling tool so as to facilitate chip removal. Its principle is based on a hydroelastic instability in a given range of flow rates.

1. INTRODUCTION

1.1 Purpose

General circulation of the drilling mud around a bit is shown in Sketch 1 of the Chart. The essential role of the fluid is to remove the chips resulting from rock destruction by percussion of the cutting teeth. For this purpose, mud jets are provided by nozzles at velocities of about 100 m/s. The time that chips reside near the tool is a significant factor conditioning the bit penetration rate. In fact, the recycling of chips and possible clogging of the tool directly affect its performance. It is therefore of interest to improve removal of chips around the rotating bit.

As a general rule, the transport of solid particles in a rising flow is improved by flow pulsation. In fact, the diffusivity is increased by pulsation [1]. As a result, the apparent sedimentation rate of the particles in suspension is reduced [2] and [3]. Experiments [4] illustrate this phenomenon in the case of erosion of a sand bed by continuous and pulsatory jets (Fig. 1). Furthermore, our preliminary experiments show that pulsating jets at 100 Hz can increase the particle removal rate by as much as 300%

1.2 Drilling requirements

Various requirements must be complied with: the pulsation source must be in the tool so that its action is independent of the bit position in the well; the flow rates and head losses of the nozzles should not be significantly altered, so as to retain the tool hydraulic power; and the pulsation characteristics must be selected to optimize the particle removal in the area of high turbulence around the bit.

1.3 Principle of the device

Sketch 2 on the Chart shows the principle of the selected device, based on a hydroelastic instability, and self-operated by the mud discharge. Fluid enters vertically from the top and leaves the device through two lateral arms fitted with nozzles at their ends. A ball is provided in the unit to plug each of the two lateral ports. This ball can move between both extreme positions, corresponding to the cutting-off of each discharge, and is brought back to the

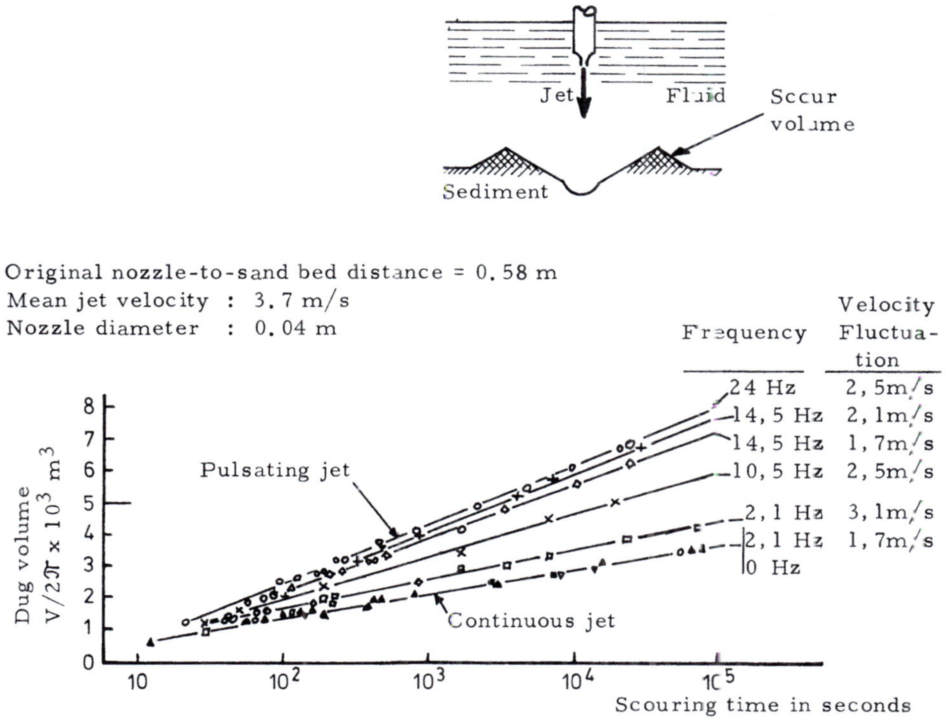

FIG. 1 WESTRICH'S [4] EXPERIMENTAL RESULTS ILLUSTRATING THE IMPROVED EROSION OF A SAND BED BY THE FLOW RATE PULSATION.

center position by a torsional shaft. When the ball is in the center position, the system is symmetrical and the flow rates are equal. When the ball is in the intermediate position between the center and the closing positions, the two flow rates are imbalanced. This induces hydrodynamic forces which act on the ball. The system is so dimensioned that it starts to vibrate for a determined flow rate range. When vibrating, the ball moves alternately between the two extreme positions, thus creating two alternately pulsating flow rates.

2. THEORY

The equation of ball motion involves inertia forces, damping forces, hydrodynamic forces and a spring constant. To obtain the first two terms, we consider the dynamic characteristics of the system in a still fluid; i.e. mainly the factors of added mass and the damping rate are considered. We therefore assume that these two parameters are independent of the fluid velocity. To determine the hydrodynamic forces applied to the ball, we write the momentum equation, considering the terms due to unsteadiness of the flow, and employing a control volume including the nozzles. For this purpose, we must accommodate the head loss factors dependent on the conduit arrangement. The corresponding equation of the ball, written in accordance with the above, allows determination of the range of unstable operation. The threshold flow rate is that value, at which, for small perturbations, the negative damping of the system equals its natural damping. The upper operating limit, or immobilization of the system, is reached when the hydrodynamic force, depending on the velocity, equals the return force.

3. EXPERIMENTS

The experiment was carried out on a model using water flow as shown in Fig. 2. The system was preliminarily tested both dynamically in presence of still fluid to obtain the factor of added mass and the damping rate, and hydrodynamically with the ball maintained at various openings by mechanical means to determine the quasi-steady values of head loss factors. We thus obtained the evolution of the head loss factors and of the forces applied to the ball; these forces can be compared to those resulting from the application of the momentum theorem in steady flow. The complete tests of the system with the ball in motion consisted of determining the operating range and the oscillation frequency. A typical comparison between experiment and calculations is shown on Fig. 3; it confirms that the theoretical description of the system operation is satisfactory.

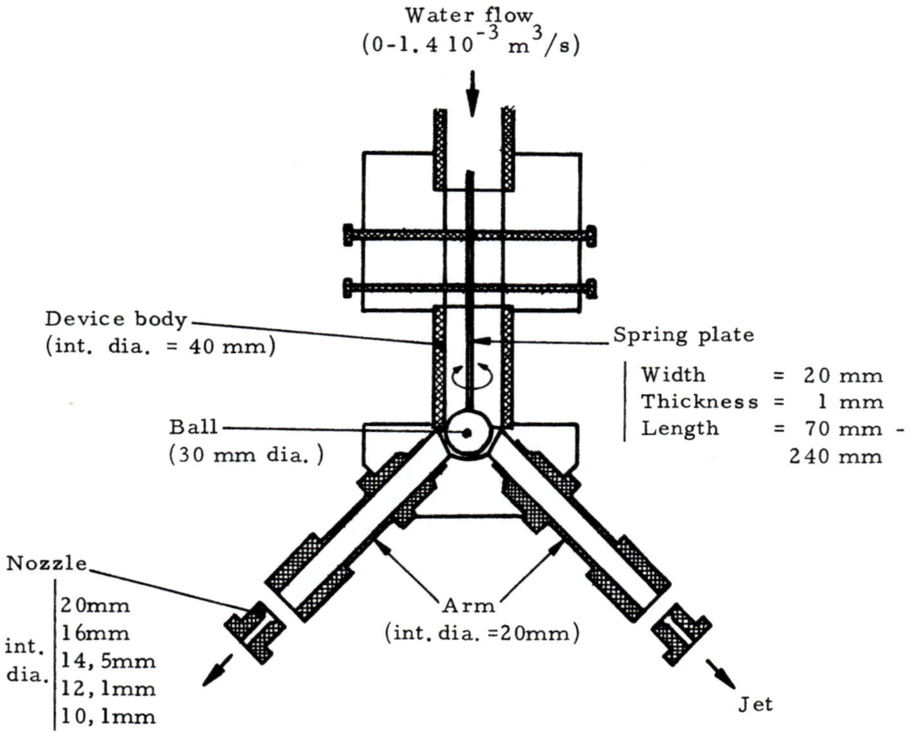

FIG. 2 EXPERIMENTAL MODEL

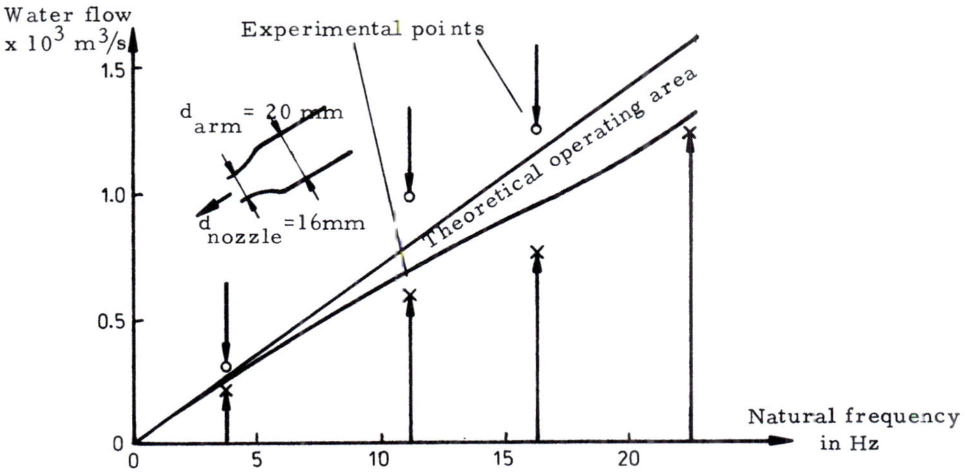

FIG. 3 COMPARISON OF TYPICAL THEORETICAL AND EXPERIMENTAL OPERATING RANGES OF THE MODEL PULSATING DEVICE

4. CONCLUSION

The above analysis allows dimensioning the prototype whose characteristics are given on the Chart. The prototype was tested on a drilling-bench for the given conditions. The results match those expected, which proves that the damping rate, as measured in still water, is valid as a first approximation for the muddy operation.

The subsequent phases of the study will consist of checking the effectiveness of the device in removing chips and to determine its action on the drilling efficiency.

This study can also assist in understanding and modelling the instabilities of other types of units, such as valves and cocks in hydraulic systems [5,6].

5. REFERENCES

1. Favre Marinet M., Structure des jets pulsants, These a l'Universite Scientifique et Medicale de Grenoble, Avril 1975.
2. Larras, Hydraulique et Granulats, Collection du Centre de Recherches de Chatou, 1972.
3. Molerus et Werther, Chemical Ing. Tech., 40-522, 1968.
4. Westrich, "Erosion of a Uniform Sand Bed by Continuous and Pulsating Jets", XVe Congres AIRH, Vol. 1, p. 91/98, Istanbul 1973.
5. Kasai, K., "On the Stability of a Poppet Valve with an Elastic Support", Bulletin of JSME, Vol. 11, n° 48, p. 1068/1083, 1968.
6. Maeda, T., Studies on the Dynamic Characteristic of a Poppet Valve, Bulletin of JSME, Vol. 13, n°56, p. 281-297, 1970.

ROCK DRILLING TOOL WITH PULSED JETS

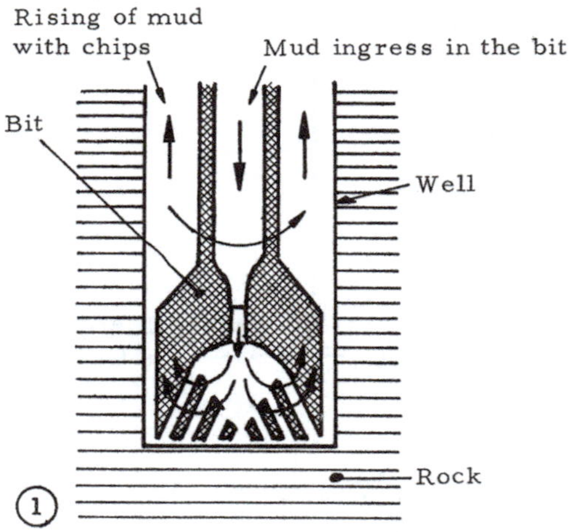

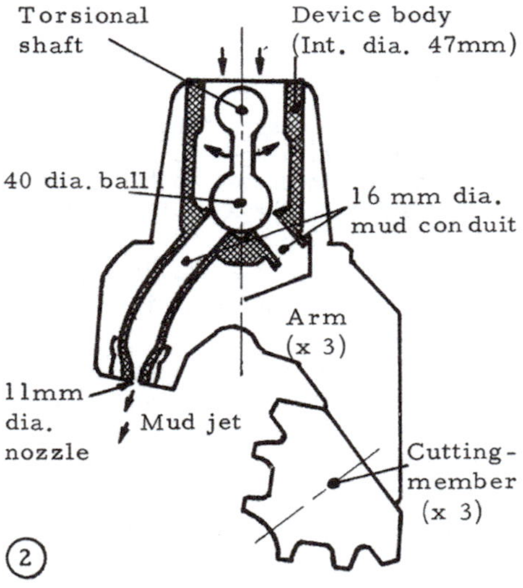

Problem: Definition of device producing a pulsed jet around a drilling bit to improve removal of rock chips.

Boundary conditions: See Sketch 1 showing the general configuration.

Geometry: See Sketch 2. The ball in rotational motion around the torsional shaft alternately plugs both mud conduits.

Discharge conditions: Flow rate of 0 to 8.5×10^{-3} m^3/s of drilling mud (ρ = 1200 kg/m^3, μ equivalent = 1.2×10^{-2} Poiseuille).

Dynamic characteristics: The ball describes a pendular motion around the torsional shaft. Torsional moment of inertia is 4×10^{-3} kg m^2. Torsional moment 75 Nm for a rotation of 1°. Natural damping in air: less than 1%. Natural frequency in air: 164 Hz. Maximum angular displacement $\pm$ 1.5°.

Vibration: Up to a flow rate of 2.7×10^{-3} m^3/s, no vibrations. From 2.7×10^{-3} m^3/s, vibrations at frequency of 125 Hz.

Calculation predicts threshold flow rate of 2.4×10^{-3} m^3/s and an immobilization flow rate of 10^{-2} m^3/s. Below threshold flow rate, ball is stable in center position. Above immobilization flow rate, hydraulic forces maintain the ball in position of closing a conduit. Between both flow rates, the ball vibrates at the operating frequency of 130 Hz.

References

Patent n° 4.114.705 dated 19/9/78 "Procédé de forage de roches et dispositif pour la mise en oeuvre de ce procédé" (Rock drilling procedure and device for the application of such procedure).

B19

FLUID WAVE-ROTOR MOTION RESONANCE IN A FLUID LINE-ROOTS FLOW METER SYSTEM

Takehiko Kawase
Professor

Tadahiro Machiyama
Professor

Waseda University, Tokyo, Japan

SUMMARY

This paper describes an experimental model study on the fluid wave-rotor motion resonance in a fluid line-roots flow meter system. A simple and practical pulsation absorber using a small air bag is proposed to attenuate the oscillating rotor speed and also the flow pulsation under the resonance. Explicit expressions which give dynamic quantities are presented in terms of system parameters and operating conditions, and a condition which determines the above resonance is given.

1. INTRODUCTION

Fluid wave resonance in fluid line-positive displacement flow meter systems has well been observed, and, in some cases, vibrations of the pipe and serious noise have been experienced. It is supposed that, in many such cases, the rotor motion of the flow meter also resonates with the fluid motion. This type of resonance, however, has not yet been reported anywhere except by the authors[2].

In the present paper, an experimental model study on the above wave-rotor motion resonance is reported, and a simple method to attenuate the flow pulsation, and also the oscillating rotor motion under such resonance, is proposed. Furthermore, some experimental results showing its effectiveness are presented.

2. THE SYSTEM DESCRIPTION AND THE ROOTS FLOW METER

In the experiment, the system was composed of a flow meter and a straight pipe together with a volute pump and a head tank as shown in Sketch 1, in order to model actual system configurations encountered in many process plants. To find the relation which determines the resonance condition, supplementary experiments were also performed for a system having simpler boundary conditions, by replacing the volute pump with a head tank.

Specifications and dimensions of the roots flow meter used in this investigation are listed in Table 1(see Fig.1 for nomenclature). The dimensionless fluid volume delivered per unit rotor angle is shown in Fig.2. The periodicity of this function $dV^*/d\theta$
where

V^* = dimensionless fluid volume = $V(\theta)/V_o$,

$V(\theta)$ = fluid volume delivered,

V_o = average fluid volume delivered = $V(2\pi)/2\pi$,

θ = rotor angle,

is the very cause of the pulsating fluid flow, and the nonlinearity of the function produces higher harmonic components of dynamic quantities as shown later.

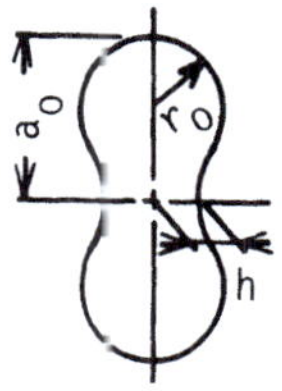

FIG. 1 ROTOR PROFILE

Table 1 SPESIFICATIONS AND DIMENSIONS

Spesifications	Type; Roots type, Nominal dia.; 50 mm, Capacity; 1,500-6,500 ℓ/h
Dimensions	r_0/a_0; 0.473, h/a_0; 0.667, ℓ/a_0; 2.333, c/a_0; 4/3

c : Distance between rotating axes.

ℓ : Rotor length

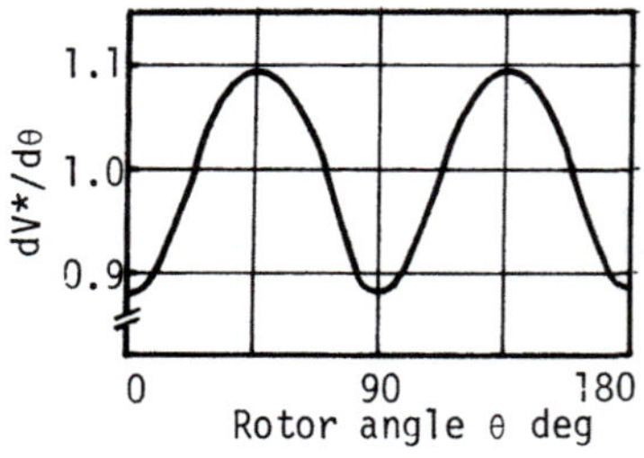

FIG. 2 dV*/dθ VS ROTOR ANGLE

3. EXPERIMENTAL RESULTS AND THE ESTIMATION OF THE RESONANCE CONDITION

In the experiments, angular velocity of the rotor and the fluid pressure at the outlet of the flow meter were measured(see Sketch 2). Observed amplitude of the angular velocity and that of pressure at the outlet of the flow meter are shown in Sketch 3. From the results, we can recognize the strong wave-rotor motion resonance at a rotor speed of approximately 9 rps.

Fig.3 shows the observed dimensionless amplitudes of the above dynamic quantities. These dimensionless quantities are defined using the average values as follows.

$$\omega^* = \hat{\omega}/\omega_0, \quad p_2^* = 2\hat{p}_2 V_0/T_0, \quad \text{and} \quad n^* = 8\ell n/c_0 \qquad (1)$$

where

$\hat{\omega}$ = amplitude of the angular velocity of the rotor

$\hat{p}_2$ = pressure amplitude

n = average rotor speed

ω_0 = average angular velocity
 = $2\pi n$

T_0 = average kinetic energy of the rotors
 = $J\omega_0^2/2$

J = moment of inertia of the rotors

c_0 = sound velocity

ℓ = pipe length.

The subscript 2 indicates the variables at the outlet of the flow meter as indicated in Sketch 2.

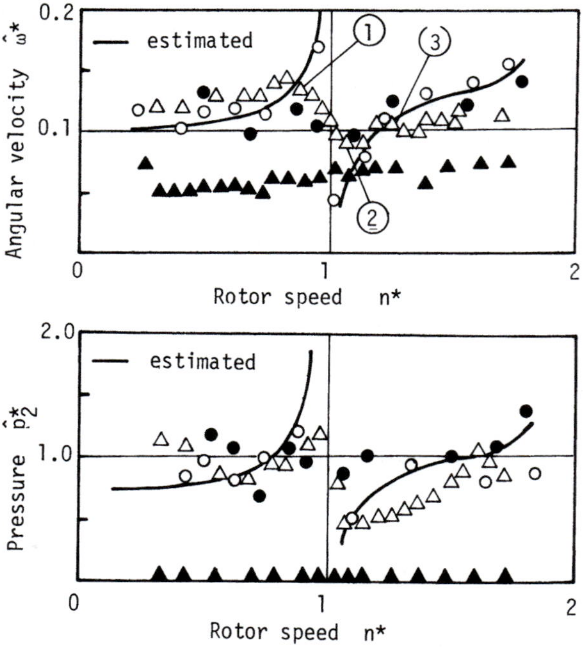

Experimental conditions:
● pipe length = 25.0 m, without absorber
○ 15.7 m, without absorber
▲ pipe length = 16.6 m, with absorber
△ 16.6 m, without absorber

FIG. 3 OBSERVED AMPLITUDES OF THE ANGULAR VELOCITY AND FLUID PRESSURE

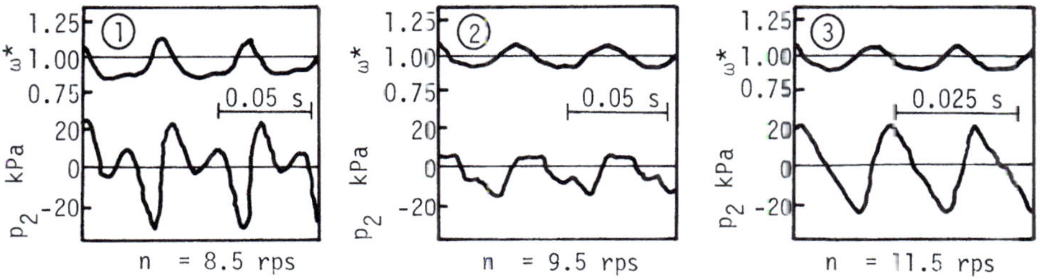

FIG. 4 OBSERVED RESPONSES

The symbols ○ and ● in Fig.3 indicate the amplitudes observed in the supplementary experiments performed for the system having a head tank at the far end of the upstream pipe instead of the volute pump. On the other hand, experimental results for the system shown in Sketch 1 are illustrated by symbols △ and ▲. Thick lines in Fig.3 indicate the estimated amplitudes from the analytical results to be described later. Fig.4 illustrates some of the observed responses at various operating points indicated in Fig.3. Numbers in circles correspond to those in Fig.3, and these responses are selected so that they show the change of response patterns across a resonance point. From these results, we can make the following points:

1) The angular velocity and the fluid pressure p_2 abruptly change their amplitudes and wave forms across the resonance point.
2) Their amplitudes generally increase in proportion to the square of the mean rotor speed, and hence the mean flow velocity.
3) The rotor speed at the resonance point corresponds nearly to that calculated from the frequency f_0 of the fluid wave resonance in the downstream pipe, that is, $f_0 = c_0/2\ell$, assuming the boundary conditions to be open ends.
4) As Fig.4 shows, observed responses include higher harmonic components, especially near the resonance points, which come from the nonlinearity of the function $dV^*/d\theta$[2]. Fig.3 and Fig.4 suggest the existence of the resonance of the second harmonic.

To obtain fundamental information for system design, we briefly describe the analytical results reported elsewhere by the authors[2]. Let us consider the system in which fluid is supplied from a head tank through a sufficiently short pipe. Taking note of the fact that the frequency range of concern is rather high(10-60 Hz), we use the linear wave equation for the fluid motion, and we neglect the energy loss due to the friction torque exerted on the rotor[1]. Thus we have the following energy equation for the rotor motion.

$$\frac{dT}{d\mu} = K_1[\,\psi(\mu)^2 - \psi(\mu + \mu_0)^2] \qquad (2)$$

where

T = kinetic energy of the rotors = $J\omega^2/2$

ψ = arbitrary function which has the unit of pressure and satisfies the wave equation

ω = angular velocity of the rotor

μ = dimensionless time = $2\pi t/t_0$

t_0 = fundamental period = $1/4n$

t = time

$K_1 = St_o/2\pi\rho_o c_o$

$\mu_o = 4\pi\ell/c_o t_o$

S = cross-sectional area of the pipe

ρ_o = average fluid density.

Note, however, that we used the boundary condition $p(\ell,\mu) = 0$ at the exit of the downstream pipe.

Assuming the system to be under steady oscillating condition, we employ the Fourier series expansion of $\psi(\mu)^2_5$ and in order to linearize the governing equations, we assume any oscillating component of the dynamic quantities to be sufficiently small compared with the corresponding steady state value. Thus, by using the continuity of the fluid flow at the outlet of the flow meter, we finally have the following equations in terms of dimensionless quantities, as the first order approximation.

$$\omega^* = 1 + (1/2)\sigma_1 \cos\mu \qquad (3)$$

$$q_2^* = 1 - \frac{1}{2\rho_o c_o}\frac{SJ}{V_o^2}\tan\left(\frac{\mu_o + \pi}{2}\right)\omega_o\sigma_1\cos\mu \qquad (4)$$

$$p_2^* = 4\sigma_1 \sin\mu \qquad (5)$$

where

q_2^* = dimensionless flow rate at the outlet of the flow meter = q_2/q_o

q_o = time mean flow rate = $\omega_o V_o$

and the coefficient σ_1 is given below.

$$\sigma_1 = \frac{2(V_m^{*\prime} - 1)}{V_m^{*\prime} - \frac{1}{\rho_o c_o}\frac{SJ}{V_o^2}\tan\left(\frac{\mu_o + \pi}{2}\right)\omega_o} \qquad (6)$$

In the above, $V_m^{*\prime}$ indicates the extremum value of $dV^*/d\theta$.

The resonance points are thus obtained by equating the denominator of eq.6 to zero. The rotor speeds obtained slightly differ from those estimated from the resonance frequency f_o mentioned previously. The above difference is less than 2 %. However, it is rather obvious that the relation of eq. 6 is necessary to quantitatively estimate the amplitude of any oscillating dynamic quantity. The solid lines in Fig.3 indicate the amplitudes estimated using eq.3 through eq.5. These estimated results well explain the observed. From Fig.3, we see that the replacement of the volute pump with the head tank does not produce any significant change in the resonance frequency and in the amplitudes of the quantities shown in Fig.3.

4. PULSATION ABSORBER

As is easily seen from eq.2, to attenuate the fluid wave-rotor motion resonance, it is desirable to place the pulsation absorber as close to the flow meter as possible to absorb the fluid power transmitted to the rotors by the approaching waves. To absorb such fluid power by inserting a high resistance for fluid flow or rotor motion is not advantageous because, in this case, it considerably increases the pressure loss, and may cause serious error in measuring flow rate. So we used an alternative method to reduce the flow pulsation. It involved a small air bag with elastic wall in conjunction with a flow resistor. The air bag and the resistor were placed in a strainer to be installed with the flow meter. The pulsation absorber proposed is shown in Sketch 5. The strainer has a wire mesh filter(60 mm dia. x 130 mm length), and the air bag and the resistor were placed in this filter.

The air volume and the pressure in the air bag used in the experiments were 150 cm^3 and 0.03 MPa respectively. As a resistor, a porous material was used. Its pressure loss characteristics are shown in Fig.5. The additional pressure loss induced by inserting the above resistor is not significant.

Observed amplitudes of the angular velocity and the fluid pressure are indicated in Sketch 3 and also in Fig.3. The effectiveness of the absorber is rather apparent; by the insertion of the absorber, the amplitudes of the angular velocity decreased to approximately 50 % of that in the case without it.

Absorbers having an air bag or a flow resistor only were also attempted. However, the results were not successful. In the case having an air bag only, the resonance amplitude was quite high at low rotating speed, and the case of a resistor alone considerably increased the pressure loss. Thus, by use of the proposed absorber, we can expect steady rotor motion, and a decreased pulsation amplitude of the fluid flow.

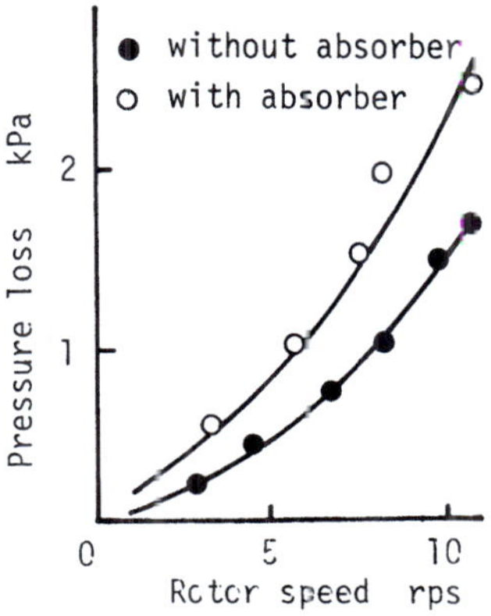

FIG. 5 PRESSURE LOSS INDUCED BY THE ABSORBER

5. CONCLUSION

From the foregoing discussion, we have the following conclusions concerning the fluid wave-rotor motion resonance:

1) Across the resonance point, the angular velocity of the rotor and the fluid pulsation greatly change their amplitudes within a very narrow range of the rotor speed.
2) The frequency of the above resonance can be approximated by the resonance frequency of the fluid waves in the pipe. However, to estimate the amplitudes of the dynamic quantities under such resonance, eq.6 is necessary.
3) The pulsation absorber proposed here is quite effective to attenuate fluid pulsation and oscillating rotor speed under resonance.

6. REFERENCES

1. Kawase, T. and Machiyama, T., Pulsative Fluid Flow caused by Volumetric Flow Meter, Trans. JSME, 44, 384, 2741-2750, (1978).
2. Kawase, T. and Machiyama, T., Experimental Study on the Oscillatory Fluid Flow caused by a Volumetric Flow Meter, Bull. Sci. Eng. Res. Lab., Waseda Univ., 82, 1-22, (1978).

ROTORS OF A FLOW METER IN A PIPE LINE

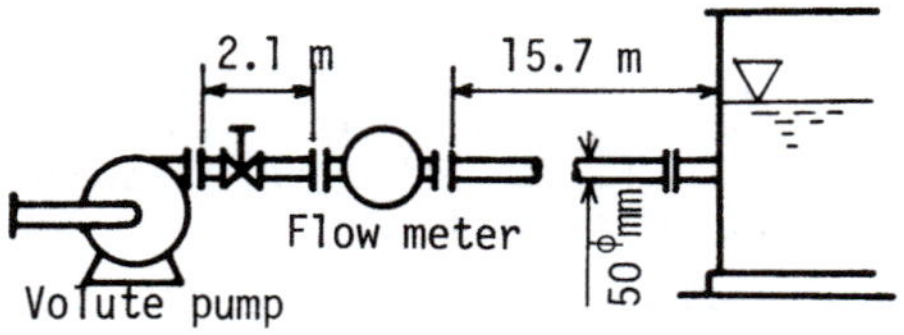

① FLOW METER-PIPE LINE SYSTEM

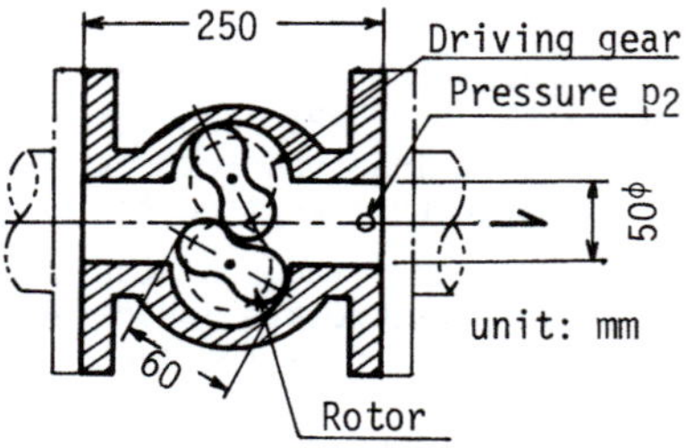

② FLOW METER

Problem: Wave-rotor motion resonance at certain flow and boundary conditions. Generation of vibrations of the pipe and considerable sound.

Geometry and boundary conditions: See Skh.1 and Skh.2.

Flow conditions: Fluid velocity = 0.3 to 1.5 m/s($Re = 10^4$ to 5×10^4). Pressure = 0.02 to 0.1 MPa. Supplied fluid flow contains small pulsation of 170 Hz.

Operating conditions: Resonance was observed at: flow velocity = 0.8 to 1.0 m/s and rotor speed = 8 to 10 rps. See Skh.3.

Structural dynamics: Moment of inertia of the rotors = 5.71×10^{-4} kgm^2. Wave-rotor motion interaction coefficient is a nonlinear periodic function.

Vibrations: Resonance frequency = 36 Hz. Amplitudes: 7 rad/s and 30 kPa(P_2). See Skh.3 and Skh.4. Resonance frequency and amplitudes can be estimated.

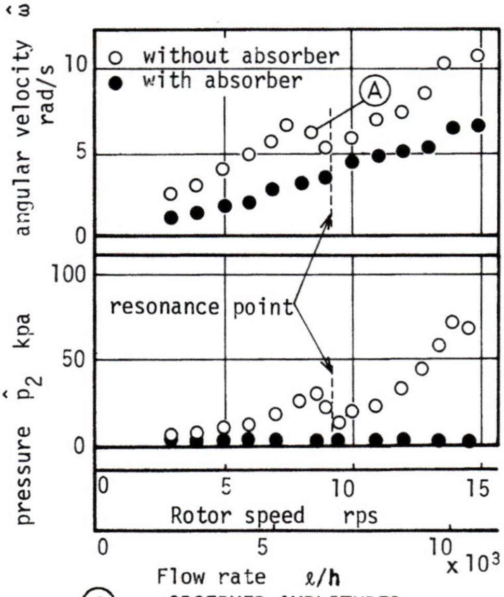

③ OBSERVED AMPLITUDES

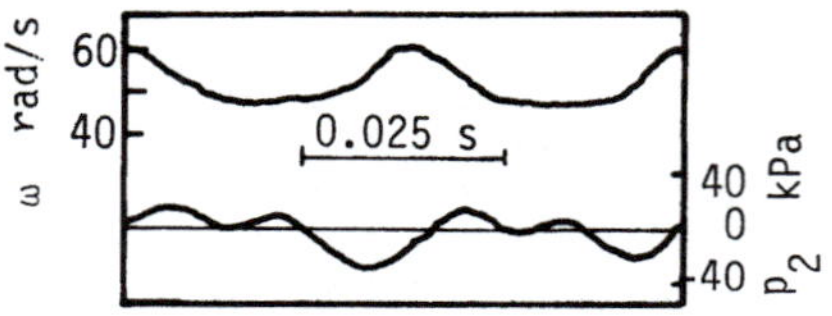

④ OSCILLATION AT THE POINT A

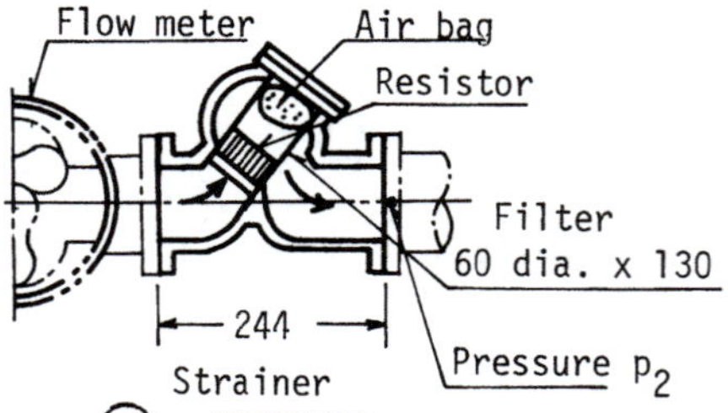

⑤ ABSORBER

Cures: absorber: air volume = 150 cm^3; resistance = 0.5×10^3 kPa/(m^3/s)^2. See Skh.3. Absorbers having an air bag or resistor only were not successful.

Reference: Kawase,T. and Machiyama, T., Experimental Study on the Oscillatory Fluid Flow caused by a Volumetric Flow Meter, Bull. Sci. Eng. Res. Lab., Waseda Univ., No.82, 1978, p.1-22.

C1 DEVELOPMENT OF VIBRATION-FREE GATE DESIGN: LEARNING FROM EXPERIENCE AND THEORY

P.A. Kolkman

Head of the Locks, Weirs and Sluices Branch
of the Delft Hydraulics Laboratory and
Senior Research Officer at the
Delft University of Technology
The Netherlands

SUMMARY

Gate vibrations experienced by the author can be classified into three types:

- Vibration modes with flow-gap width variation due to gate vibration.
- Vibration modes with constant flow-gap width; often high-frequency plate vibrations are observed.
- Vibrations occuring at gate positions where the flow is wavering between reattachment and full separation.

The first type can be analysed theoretically. In the other two types, self-exciting vibration is probably due to a mechanism involving a fluctuating discharge coefficient, induced by the "added mass flow" of the vibrating gate (i.e. velocities in phase with the vibratory movement).

Different experiences concerning dynamic phenomena due to flow instability, sometimes amplified by partial aeration, are presented. One example demonstrates how a dangerous case of flow separation instability could not be reliably reproduced in a hydraulic model.

Finally, criteria for vibration-free gate design are formulated. The paper is followed by 12 illustrative charts concerning vibrations encountered in prototypes and in hydroelastic models.

1. INTRODUCTION

This paper contains cases from our experience, both practical and theoretical, with flow-induced gate vibrations in prototypes and hydraulic models. This will be presented in such manner that some generally applicable design criteria can be deduced.

Limitation to our own experiences has, on the one hand, the disadvantage that they come in at random and do not afford a complete picture of what may happen. On the other hand, the personal approach and incompleteness are more likely to provoke discussion. I also hope to convey some of the pleasures of pioneering in the field of flow-induced vibrations, which lies between the specialized branches of applied structural mechanics and fluid mechanics.

2. CONCLUSIONS FROM THE APPLIED STRUCTURAL MECHANICS

The simplest form of a vibrating system is the single oscillator with linear components (Fig. 2.1):

$$m\ddot{y} + c\dot{y} + ky = F \qquad (2.1)$$

For a system in water, F represents all hydrodynamic forces. In F one can sometimes distinguish components which are proportional to y, $\dot{y}$ or $\ddot{y}$, in which case a convenient representation of Equation (2.1) is:

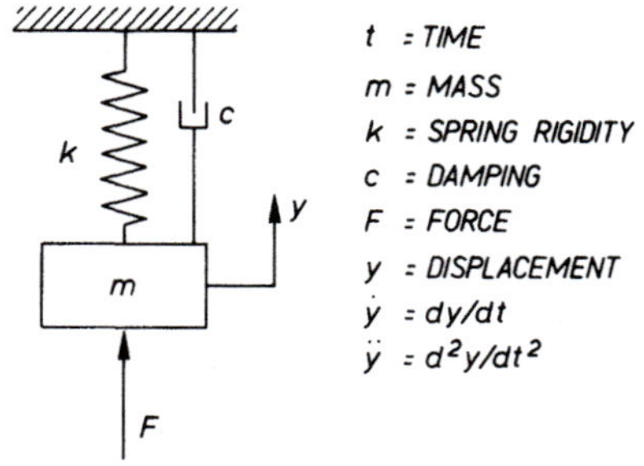

FIG. 2.1 THE SINGLE OSCILLATOR

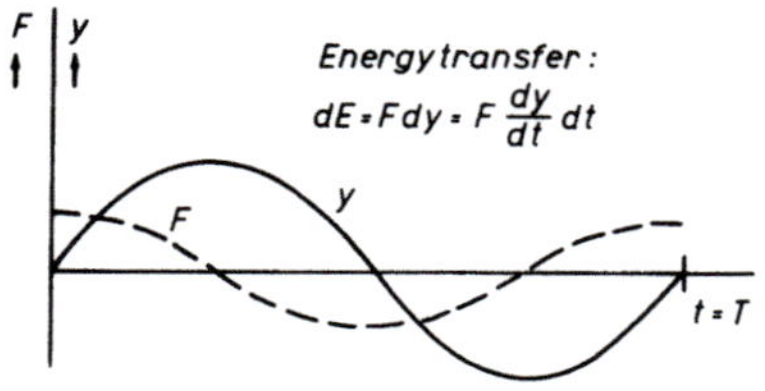

FIG. 2.2 ENERGY TRANSFER BETWEEN EXTERNAL FORCE AND VIBRATING BODY

$$(m+m_w)\,\ddot{y} + (c+c_w)\,\dot{y} + (k+k_w)y = F_w \tag{2.2}$$

F_w now still contains components which are independent of movement (uncoupled components), but also components coupled with the vibrational movement; only the linear terms have been brought to the left hand side. In Equation (2.2) the factors m_w, c_w and k_w are the hydrodynamic (or added) mass, damping and rigidity; it is not yet defined whether these hydrodynamic terms are positive or negative.

Whereas the total mass $(m+m_w)$ and the total rigidity $(k+k_w)$ together determine the resonance frequency of the system, the factor $(c+c_w)$ determines whether the system is

- stable (or positively damped) namely when

$$(c+c_w) > 0, \text{ or} \tag{2.3}$$

- unstable (or negatively damped) namely when

$$(c+c_w) < 0 \tag{2.4}$$

In the latter case, self-excitation will take place and one can conclude that the design is wrong. When design criteria are to be defined, the prevention of unstable vibration should be one of the first.

To see if negative hydrodynamic damping can occur, it is convenient to analyse the hydrodynamic forces induced by purely harmonic movement (Fig. 2.2):

$$y = Y_o \sin\omega t \tag{2.5}$$

As long as small vibration velocities $\dot{y}$ are involved (i.e. small compared to water velocities), one can reason that the hydrodynamic forces coupled to the vibrations are linearly related to y, $\dot{y}$ and $\ddot{y}$. So, F_w will be a harmonic force. When we define

$$F_w = F_1 \sin\omega t + F_2 \cos\omega t \tag{2.6}$$

the energy transferred from the flow to the vibrating body is

$$E = \int_o^T F_w \, dy = \int_o^T F_w \, \dot{y} \, dt =$$

$$= F_1 \, Y_o \int_o^T \sin\omega t \, \cos\omega t \, dt + F_2 \, Y_o \int_o^T \cos^2\omega t \, dt \qquad (2.7)$$

It is obvious that only the second term transmits energy, which means that only a force in phase with the vibrational velocity and working in the same direction will transfer energy. It is demonstrable that this is the case when in Equation (2.2) c_w is negative.

Conclusions:

From the general analysis of Equation (2.2), two evident design criteria can be derived:

- Negative hydrodynamic damping should not occur, or must at least be compensated by stronger mechanical damping.
- Dominant excitation frequencies should be well away from resonance frequencies, given by the equation

$$f_r = \frac{1}{2\pi} \sqrt{(k+k_w)/(m+m_w)} \qquad (2.8)$$

In the case of flow-induced vibrations, it is the Strouhal number $S = f\,L/V$ which is the main descriptive parameter of the excitation frequency (f = excitation frequency; L = a representative length of the flow geometry; and V = a representative flow velocity). When V is variable, the excitation frequency will also vary; this means that the resonance frequency f_r must be a factor higher than the excitation frequency f belonging to the maximum flow velocity. How great this factor should be is still under discussion, but to our opinion one should try to obtain at least a factor 3. In this case, the amplification factor results in an overshoot of 10% above the static deflection, while available data show that there is still no tendency of the excitation frequency to shift towards the resonance frequency.

3. CONCLUSIONS FROM THE APPLIED FLUID MECHANICS

The added, or hydrodynamic, mass in its simplest form is represented by a piston in a cylinder partially filled with water (See [1] and Fig. 3.1):

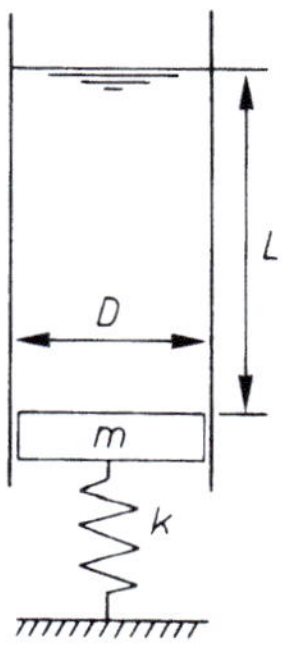

FIG. 3.1 ADDED MASS OF A PISTON

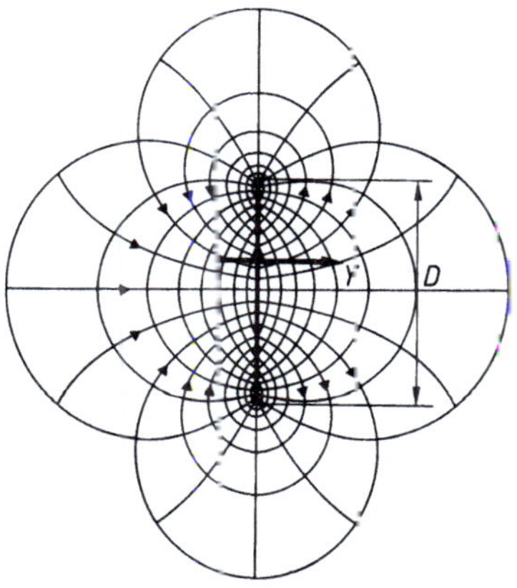

FIG. 3.2 ADDED MASS FLOW OF AN OSCILLATING STRIP

$$p = \rho L \ddot{y} \tag{3.1}$$

$$m_w = \rho L \frac{\pi}{4} D^2 \tag{3.2}$$

For vibrating bodies completely submerged in stagnant fluid, Lamb [2] already showed that m_w can be deduced from potential flow theory and he found an analytical solution for the added mass of slender cylinders of circular and elliptical cross-section, with as limiting case a flat strip (Fig. 3.2):

$$m_w = \rho L \frac{\pi}{4} D^2 \tag{3.3}$$

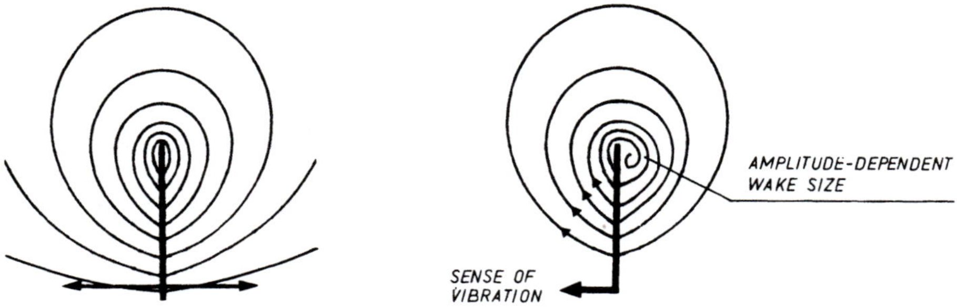

FIG. 3.3 ADDED MASS FLOW WITHOUT AND WITH FLOW SEPARATION

What is neglected is that flow velocities at sharp edges are infinitely high in potential-flow theory (Fig. 3.3 left), whereas in reality flow separation will appear (Fig. 3.3 right). In stationary flow, flow patterns with separation are considerably different; but in the case of added-mass flow, fluid is sucked around the edge because of a fluid deficiency at the momentary downstream face. Moreover, no vortex street will appear when locally-induced vortices are not transported by the flow.

When the vibrating body is positioned in non-stagnant fluid, the added mass remains unchanged as long as flow velocities are not too great. The determinative parameter is again the Strouhal number $S = f\,L/V$, but now f is the frequency of the vibrating body.
When, for instance, in the terms of the Navier-Stokes equation for the x-direction

$$\partial p/\partial x = \rho\, \partial v/\partial t + \rho v_x\, \partial v_x/\partial x \quad \text{etc.} \tag{3.4}$$

periodic disturbance terms of small amplitude are introduced, the additional terms (indicated by accents) give:

$$\partial p'/\partial x = \underbrace{\rho\, \partial v'/\partial t}_{A} + \underbrace{\rho v_x'\, \partial v_x/\partial x + \rho v_x\, \partial v_x'/\partial x}_{B} + \underbrace{\rho v_x'\, \partial v_x'/\partial x}_{\text{negligible}} \tag{3.5}$$

If v_x' is proportional to $\dot{y}$ of the vibrating body, $\partial v_x'/\partial x$ will be related to $\dot{y}/L$ (wherein L is a representative length of the flow geometry) and $\partial v_x'/\partial t$ is proportional to $\omega\dot{y}$.
It can be demonstrated that the B-terms can be neglected as long as the order of magnitude of S is greater than $(2\pi)^{-1}$, and then the velocity v does not affect m_w.

Knowing that potential flow problems can be solved with electrical analogues (if the vibration frequency is high enough to neglect wave radiation), or by means of finite-element methods (also applicable when wave radiation is not negligible), it is possible to determine the added mass for two-dimensional gate vibrations. In an electrical analogue, the free-liquid surface has to be represented by a zero-potential line, as has been shown by Zienkievicz and Nath [3].

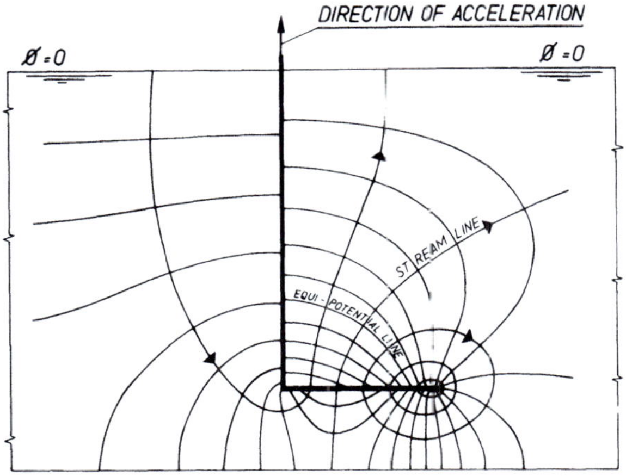

FIG. 3.4 ADDED MASS OF AN L-SHAPED GATE

A solution for a vertically vibrating gate (Fig. 3.4) shows potential differences across the vertical plate; and because potentials represent liquid pressures proportional to the acceleration $\ddot{y}$, there are found coupling forces which are due to the added-mass flow (vertical acceleration $\ddot{y}$ causes a horizontal force).

Figure 3.5 represents in gradual steps how added-mass flow has to be introduced in mechanical systems. It is the third case, which is in general representative of the hydrodynamic mass. This explains why, for instance, vibration modes can differ considerably depending on whether the gate is submerged to a greater or lesser degree in the water.

When resonance frequencies of a submerged structure are calculated and when the schematization into a lumped mass-spring system with n degrees of freedom is made, the added-mass effect results in coupling forces wherein the movement of each mass, and this in each of its degrees of freedom, causes resultant pressures on all the masses. The added mass, including the coupling terms, has to be expressed in an n x n matrix and can then be introduced into the eigen-value calculations of the lumped-mass system. In [9] this method is introduced for ships and the added-mass matrix calculation is performed by the finite-element method.

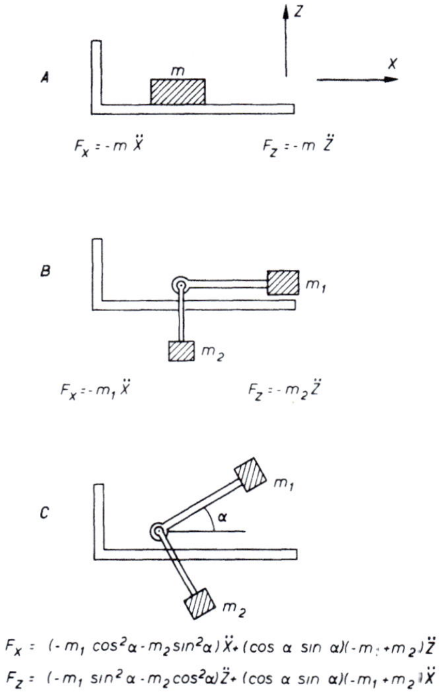

FIG. 3.5 COUPLING EFFECTS OF THE ADDED MASS

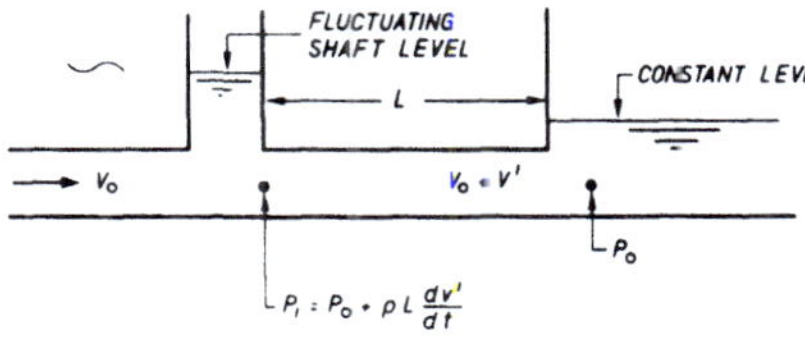

FIG. 3.6 FLOW INERTIA IN A CULVERT

Inertia of the water also plays a role under non-stationary flow conditions, namely when the water is accelerated and decelerated. This inertia effect is of a character similar to that of the added-mass flow. The parallel roles of flow inertia and added mass flow in gate vibration can be demonstrated by comparing Figure 3.6 with Figure 3.1.

Inertia of the water in the culvert causes a pressure rise (above p_o) in the shaft which is equal to

$$p' = \rho L \frac{\partial v}{\partial t} \tag{3.5}$$

This equation is similar to Equation (3.2). L can be seen as an "added mass", not of a vibrating body, but the mass of one of the components of an oscillating flow system.

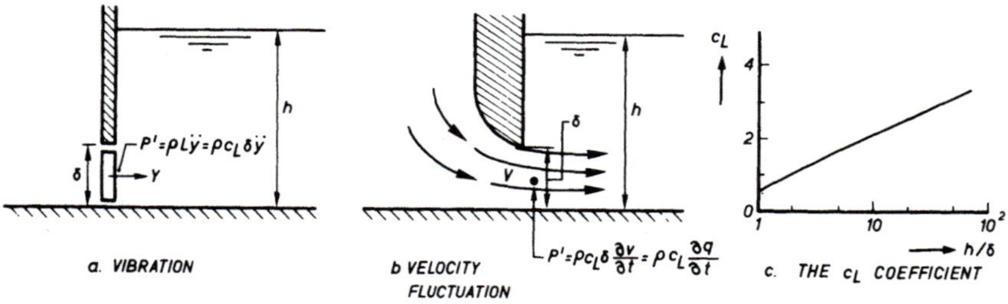

FIG. 3.7 FLOW INERTIA AT VIBRATING GATE AND AT FLUCTUATING DISCHARGE

In much the same way, there is a parallel in the two cases presented in Figure 3.7: the added mass of a gate and the inertia of a fluctuating flow. The factor c_L, introduced here, is derived from an imaginary tube with a length $L = c_L\ \delta$, which is assumed to represent the flow inertia.

In the first case, the added mass follows, as explained before, from the potential flow theory. In the second case, this gives only an approximation because

- the velocity v and its fluctuation dv/dt are not evenly distributed across the opening and the distribution pattern of dv/dt is unknown;
- the initial velocity should not be too great: when the velocity fluctuation is periodic, that is when $v' = \hat{v}\ \sin\omega t$, the Strouhal number $S = (\omega/2\pi)\ \delta/v$ should, again, not be below a certain minimum value, or else c_L will be affected.

4. GATE VIRBATIONS EXPERIENCED; PART I

The double-leaf gate of Chart J was rebuilt, making use of favourable experience with the old one over many years. There was a discussion on whether the seal between the two leaves should be asymmetric with a single seal beam, or symmetric with equal beams on both the upper and the lower leaf. The asymmetric seal, with the beam on the lower leaf, was chosen, because this construction relieves the load on the upper leaf when this is lowered (the condition wherein the water pressure would otherwise exert its maximum horizontal load). This solution had already been applied in the old gate, which had functioned correctly during a long period.

After installation of the new gate, strong vibrations occurred. Lowering the upper leaf more than 1.6 m was even unacceptably dangerous. Inspection of the old gate showed that the paint had been scratched over the entire overlap area. So, there had been no leak gap! Putting a plastic rain pipe on top of the lower leaf to seal the leak gap immediately stopped the vibrations. A similar permanent solution was then applied (Sketch 3 of Chart J).

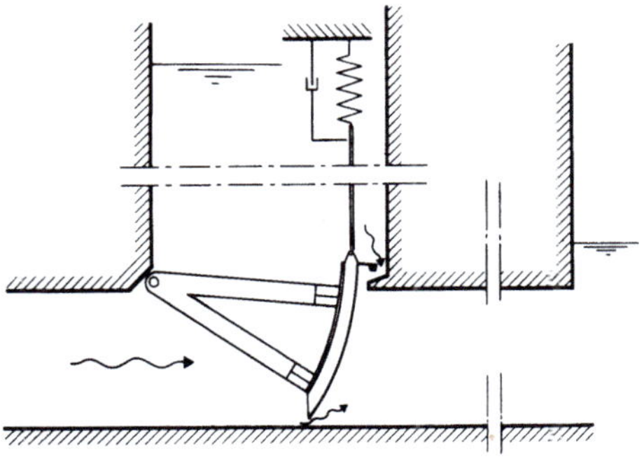

FIG. 4.1

VIBRATION OF REVERSE TAINTER VALVE DUE TO PROTRUDING LIP

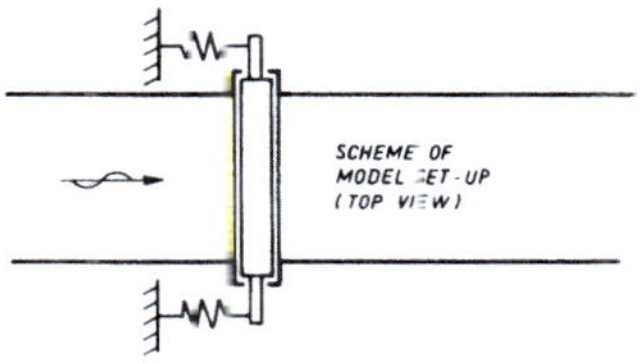

FIG. 4.2

SCHEME OF MODEL SETUP FOR HORIZONTAL VIBRATION STUDY OF VERTICAL-LIFT CULVERT GATE

The reversed Tainter valve of Figure 4.1 (see also Chart C) had to be free of vibration and was therefore designed with a sharp lower edge.
In the bottom of the culvert, a strip of hard rubber was installed to provide a tight seal. Wear of this bottom rubber and the resulting change of position of the closed gate demanded a softer, slidable top seal. Therefore, a top seal was proposed with a protruding beam (solution A), which was tested in the existing hydraulic model. At small openings, the tests showed very strong vibrations, with hoist forces of 1.5 to 2 times the weight of the gate; sometimes these vibrations were only limited because the gate hit the bottom.

The solution decided on (sol. B) was based on two principles:
- reduction of the upper gate lip, and
- reduction of the clearance between the gate and the shaft wall by means of a protruding steel plate to prevent pressure pulsations in the downstream culvert acting on the lip.

The principle of this solution was probably first published by Petrikat.

After all, a further improvement might have been found in a rubber profile clamped along both edges (sol. C of Chart C), as there still can be heard a weak vibration in the actual gate during a (very short) period after the gate has begun to rise. This probably occurs in the range where the initial gap is narrower than, or of the same order as, the gap between the skin plate of the gate and the protruding plate in the shaft.

Chart K presents a vertical-lift culvert gate for a navigation lock between a salt-water and a fresh-water basin wherein leakage has to be fully prevented in closed condition. This demand led to a gate with a circumferential seal on its downstream face, sealing against a seat recessed in the wall of the culvert. To prevent seal damage during movement of the gate, the gate was released from its seat by means of cranked wheel axles coupled to the hoisting bars.

In a hydraulic model, horizontal vibrations should be studied by providing the gate, instead of wheels, with four arms projecting outside the culvert and supported on vertical leaf springs which allowed horizontal and torsional gate vibrations (Fig. 4.2). It became apparent in the model that a strong horizontal vibration of the gate began the moment it came free from its seat, and stopped when the release distance was increased. The frequency of the gate/spring oscillator decreased when the gap was small, as hindrance of added-mass currents around the gate edges caused an important part of the culvert water mass to act as added mass.

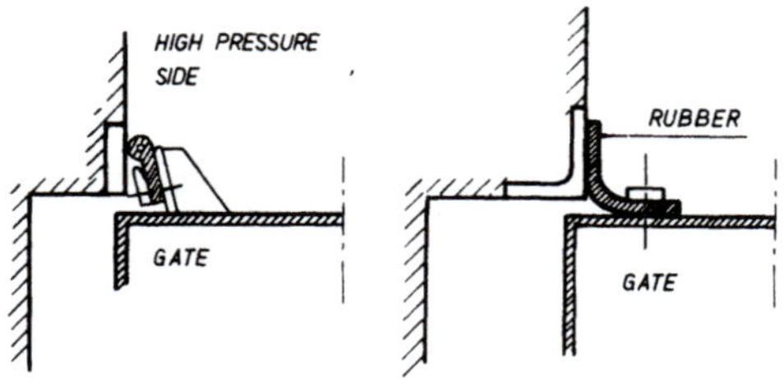

FIG. 4.3 MUSIC NOTE SEAL AND FLAP SEAL

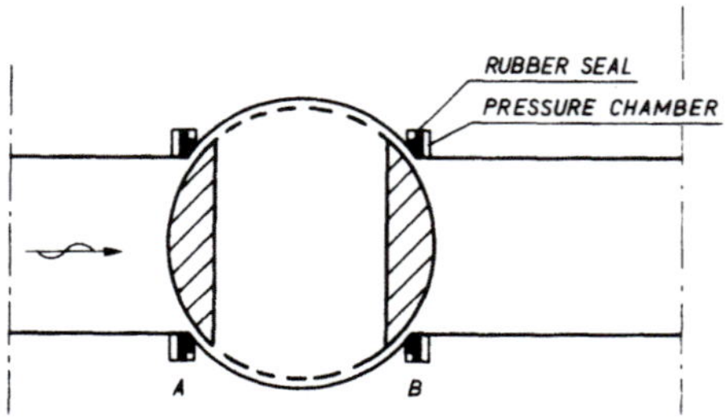

FIG. 4.4 INFLATABLE SEAL OF BALL VALVE

In all cases where an elastical seal (e.g. a music note seal) is tightened by the water pressure, a small leak gap can cause strong vibrations (Fig. 4.3). Such a seal was applied at a mitre gate in a medium-head navigation lock wherein a leak gap had developed due to damage caused, it was suggested, by a floating log that had deformed the seal. When the gate was under its full head of 8 m, no vibrations occurred; but at heads between 2 m and 6 m, vibration of the entire gate became so strong that extra damage to the gate pivots was suspected to be imminent.

At the lower end of a penstock of 500 m head, carrying water to a Pelton turbine, a ball-type shut-off valve and a needle-type regulating valve were provided. When the ball valve (Fig. 4.4) was closed, but the seals not yet inflated, a strong vibration developed with a period of 3 seconds and pressure pulses which damaged the seal operation system. The vibration period corresponded with the basic period of the standing waves that could occur in the penstock. A similar case is described by Streeter and Wylie [4], wherein water hammer calculations revealed that the water column/valve system could become unstable when the valve characteristics showed a decrease of discharge capacity caused by an increasing pressure head. This condition is a possibility when an elastical seal has a tendency to be closed by an increasing pressure, like for instance the music note seal described above.

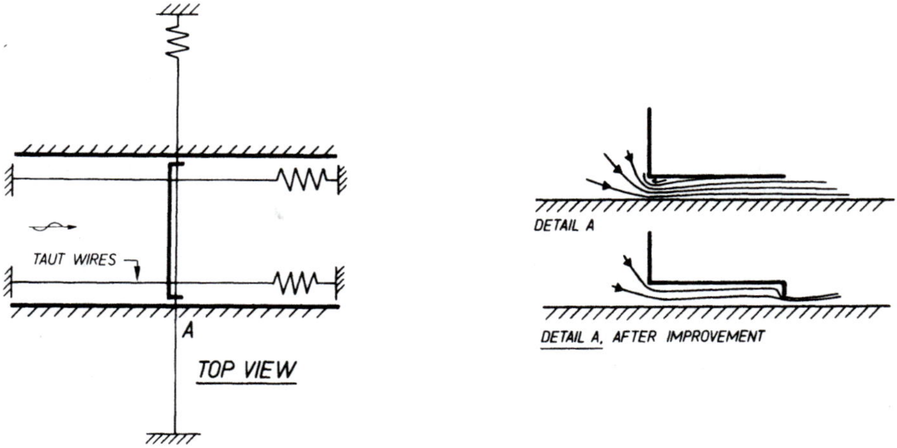

FIG. 4.5 TOP VIEW OF MODEL FOR STUDYING VERTICAL VIBRATION, BUT WHICH VIBRATED TRANSVERSELY

Figure 4.5 shows a hydraulic section model of an underflow-type gate; side leakage was accepted between gate and flume wall to permit frictionless vertical gate vibration. The gate was stabilized by a series of horizontal wires, attached at two levels (above the water) so as to permit only vertical vibration of the gate. It was suspended by a vertical wire from a leaf spring of adjustable rigidity. To our surprise the gate started to vibrate in transverse direction and, due to the relatively high rigidity in the horizontal wires, quite high frequencies of 10-15 c/s were involved. The depth of the water at the downstream side was on the average 30 cm and the head difference varied between 10 and 45 cm. The leak gap was about 4 mm. The remedy was found in decreasing the gap locally to 2 mm. The narrow part had to be at the downstream end of the gap to be effective.

During preliminary investigation of an underflow gate with 30 m span, a section model was built to see if it would be acceptable to protect the girders from wave attack by a second vertical plate. The horizontal girders were designed as an open frame; see further Chart A. The model was only suited to investigate vertical vibrations and had hardly any structural damping. In a later stage, when decisions about the general shape of the gate had been made, a three-dimensional elastic model was planned for an overall check of the design.

The investigation in the two-dimensional model revealed that condition I_0 (see Sketch 3 of Chart A) showed a critical gate position in which the (single) amplitude of the hoist force became about 25% of the gate weight.

The model showed that a stable gate design could be arrived at (except for very small openings, which were not further investigated at that moment) on the condition that the upstream gate lip was the one with the bottom seal. A further improvement could be obtained by extension of the upstream gate lip.

As further presented in Chart B, the same gate was subsequently reproduced in an elastic model (scale 1:25) to check the results of the first model and to establish the optimum seal shape for the prevention of vibrations at small openings (for details see Sketches 3 and 4 of Chart B). A problem with this model was that it was impossible to reproduce wheel friction. Although this friction was not very high, it hindered reproducibility in the vertical vibration tests. For these reasons the model gate supplied with wheels was not suited to study vertical vibrations, so a part of the tests were done with the gate supported by horizontal leaf springs which allowed it to move only vertically and without friction. In other tests wheels were used, while force meters between the wheels and the gate measured the horizontal vibration forces.

Various types of vibration arose in the model. Vibrations occurred at gate openings of the order of the seal width, independent of the value thereof. These vibrations were hardly acceptable; because of variable gate flexure due to tidal action, the provision of a leak gap at closed gate position was proposed. These vibrations occurred only in the setup for vertical vibration at head differences greater than 3 m prototype value.

After initial vertical vibration, horizontal vibration became dominant. During vertical vibration, waves radiated away from the gate in the upstream basin (Fig. 4.6a), whereas horizontal vibration resulted in a pattern of intersecting waves generated mainly at the corners (Fig. 4.6b).
Vibration stopped when the gate was touched with the hand at the abutments and also stopped when the leak flow in the recesses was reduced to nearly zero; a gap of less than 75 mm prototype value did already reduce the intensity.

The older gate of Chart L is the downstream gate of a 5 m head difference navigation lock. Problems arose after a revision in which, among other things, the wooden seal beam was replaced by a new one. At lifts between 0.13 and 0.35 m, the 14 m span wheel gate became subject to strong horizontal vibration of about

FIG. 4.6 WAVE RADIATION UPSTREAM AT VERTICAL AND AT HORIZONTAL GATE VIBRATION (VOLKERAK SLUICE GATE; SEE CHART B)

5 c/s and 5 mm amplitude (single). It started with vertical vibration of much smaller amplitude, whereupon horizontal vibration became dominant, generating a pattern of intersecting waves (Photo in Chart L).
A remedy for the vibrations was found in the provision of 1 m spaced, 0.4 m long pieces of steel angle bar as disturbance elements along the upstream side of the seal beam in order to prevent a complete synchronization of the vortex street. For further data see Chart L.

Conclusions:

- All the above vibrations occurred (or started) in the direction wherein the flow gap width or the leakage fluctuated due to the vibration.
- The excitation mechanism appeared to be strong and was probably much stronger than can be explained by local vortex shedding. It seems justified to say that the resulting discharge fluctuations - due to the inertia of the upstream and downstream water - give rise to pressure fluctuations which act on the gate even in zones remote from the seal. This explanation is confirmed by a recent publication of Miller of the Hydraulic Institute at Stuttgart [5], showing that pressure measurements at the skin plate of a gate at different distances from the gap revealed amplitudes that decreased with the distance; but no phase shift occurred.
- In all the examples it can be seen that - when one assumes a constant discharge - the buildup of pressure when the gap narrows results in forces on the gate and the seal in a direction tending to a further decrease of the gap (a sort of negative hydrodynamic rigidity). This empirical conclusion has long been used in the Delft Hydraulics Laboratory as an instability indicator [8]. Sometimes the pressure acts directly; sometimes indirectly, namely when a narrower gap results in an extra head difference, which results again in suction forces at the gate edge.
- The use of the instability indicator reveals that those gates where stationary suction forces do occur (forces tending to decrease the gap) are sensitive to vibration.

5. FLOW BEHAVIOUR WHEN THE GAP WIDTH FLUCTUATES

To illustrate flow phenomena that are related to vibrations due to fluctuating gap width, elements from my lecture held at Baden-Baden will be used [1]

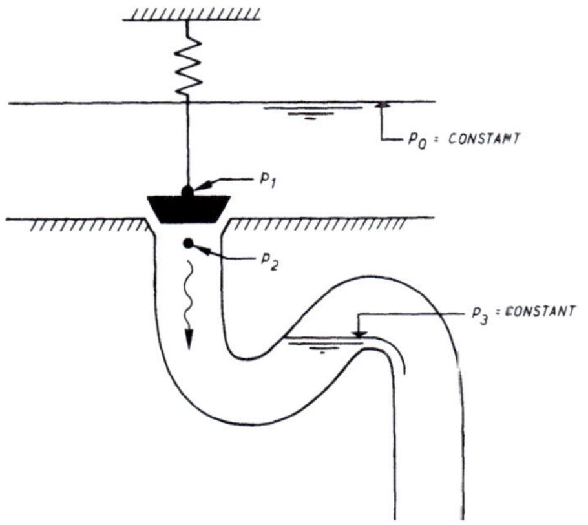

FIG. 5.1 THE VIBRATING BATH-TUB PLUG

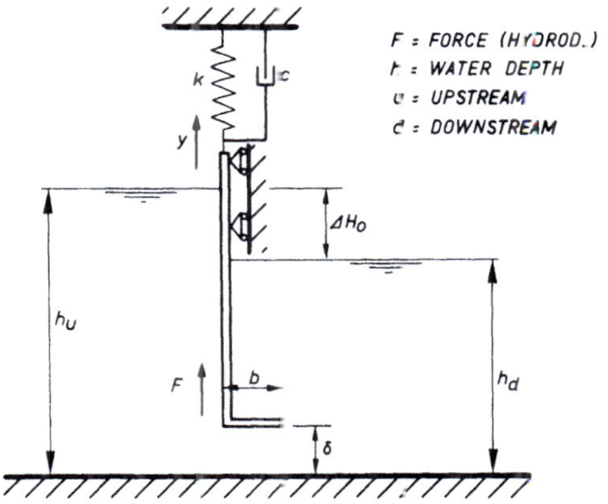

FIG. 5.2 SCHEME OF VERTICALLY VIBRATING GATE

The bath-tub plug of Figure 5.1 is a good illustration, because everyone knows from experience that when one closes the drain while the bath is emptying, a great force is felt to be exercised on the plug during the final closing motion. In a way, the plug is sucked in by the flow and, apparently, the inertia of the water in the drain pipe makes it difficult to stop the discharge immediately. Moreover, holding the plug so as to leave a very small opening always leads to a shaking movement of the plug.

We shall now examine what forces are induced by the water during harmonic plug vibration, and then analyse if there are forces in phase with the vibration velocity, which are indicative of self-excitation.

At low frequencies, the discharge will follow the momentary discharge capacity, which is related to the gap size: an upward plug movement y causes an extra discharge Q' (appr. Q'::y). The pressure difference (p_2-p_3) over the length of the pipe is related to $\partial Q/dt$ of the pipe flow, and, because p_3 is constant, p_2 increases proportionally to $\dot{Q}'$ and also proportionally to $\dot{y}$; this causes self-excitation (see Par. 2).

At high frequencies, the discharge remains constant due to pipe flow inertia. When, for instance, the gap increases by a value y, the head difference across the gap will decrease; this means that a negative hydrodynamic rigidity is found and the negative hydrodynamic damping is gone. So the hydrodynamic forces get a 90° phase shift compared to the low-frequency condition (this conclusion will be used in Par. 7), and it can be concluded that an increase of the vibration frequency (increase of rigidity) is a measure which will decrease negative hydrodynamic damping.

A similar mechanism can be found in the case of vertical vibration of an underflow gate (Fig. 5.2) when one assumes that the seal shape is such that a suction force is generated proportional to the local head difference across the gate. In the upstream and downstream water there will be inertia effects, such that an increase of discharge (under the gate) per unit of time ($\partial Q'/dt > 0$) causes a decrease of pressure upstream and an increase downstream. (One can, for example, use Fig. 3.7 to find the order of magnitude.) This means that with a positive value of $\partial Q'/dt$ there is a proportional decrease of head difference across the gate. Now a similar line of reasoning can be followed as was given for the bath-tub plug.

At low frequencies, when y is positive, again Q' becomes positive. The local head difference decreases proportionally to $\dot{Q}'$ and proportionally to $\dot{y}$, and the related suction force will also decrease proportionally to $\dot{y}$. This again, as in Paragraph 2, indicates self-excitation.

At high frequencies, when y is positive, and now assuming that Q remains constant due to the inertia of the flow, the gap height increases and, together with the constant discharge, the head difference across the gate will decrease, as will the suction force. Now the harmonic force is in phase with y and there is a negative hydrodynamic rigidity.

This theory has been evaluated quantitatively, and Figure 5.3 shows the ranges of the Strouhal number where self-excitation and where negative hydrodynamic rigidity can occur. The contraction coefficient μ depends on the seal shape, while the inertia coefficient c_i is obtained by adding the imaginary tube lengths upstream and downstream (to be found with the aid of Fig. 3.7) and the seal width. For practical purposes c_i varies between 4 and 10 for free-surface flow gates. Experimental verifications of the theory [6] showed a sensitive region, as indicated in Figure 5.3. This corresponds to a Strouhal number $S_\delta = f\,\delta/\sqrt{2g\,\Delta H}$ in the range of 0.025 to 0.1 (see the maximum of the "experiment" curve in Fig. 5.3 obtained with $\mu = 0.6$ and with c_i varying between 4 and 10).

Note that gate vibrations at a rectangular seal do not only occur at smaller gate openings, where reattachment of the flow against the seal causes a suction force. At openings where reattachment is absent, however, vibrations are weaker. A theoretical explanation of this phenomenon is also presented in [6]; it defines an imaginary suction casued by flow inertia in the downstream water only.

Conclusions:

From the foregoing one may deduce the so-called fluctuating-gap theory.

- The occurrence of a negative hydrodynamic rigidity at high frequencies coincides with the occurrence of self-excitation at low frequencies; this is the theoretical base of the use of the instability indicator.
- The instability indicator can also be used in a second way: when a slow closing of the gap is considered, the inertia of the water creates regions with an increase and with a decrease of pressure; when the resultant force on the gate is in closing direction, negative hydrodynamic damping occurs.
- The use of the indicator reveals that stationary suction forces (tending to close the gap) result in a tendency towards self-excitation.
- Extra safety agianst this type of vibration is obtained by increasing the rigidity of the gate so as to increase the resonance frequency, and consequently the S_δ-value; the theoretical damping curve of Figure 5.3 illustrates the decrease of self-excitation.
- For vertical vibrations rather low critical S_δ-values are found: between 0.025 and 0.1.
- From Figure 5.4 one can see that horizontal as well as vertical vibration can induce discharge variations. When, due to flow inertia in the culvert, these discharge variations, in their turn, cause pressure variations downstream of the gate, horizontal and vertical forces are induced. This is, in addition to the coupling by added-mass flow, the second coupling phenomenon of various vibration modes.

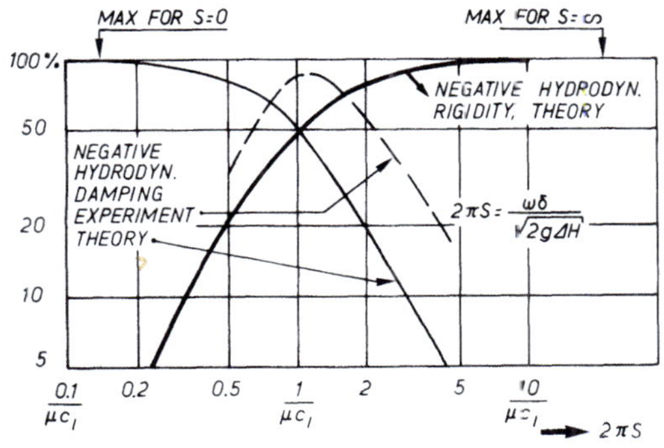

FIG. 5.3 NEGATIVE HYDRODYNAMIC ELASTICITY AND DAMPING OF RECTANGULAR SHAPED GATE (OR GATE LIP) AS FUNCTION OF STROUHAL NUMBER

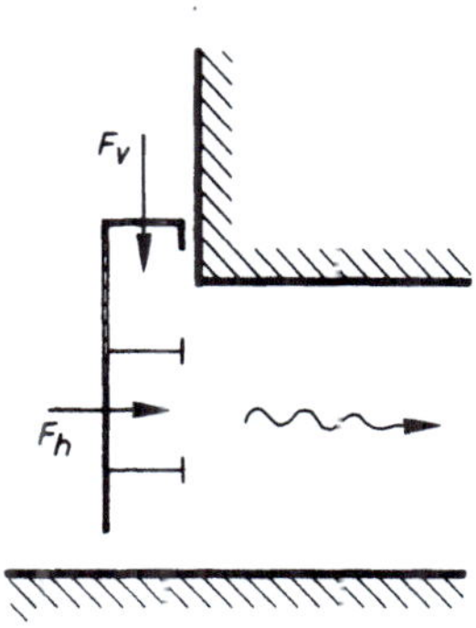

FIG. 5.4 SITUATION WHERE COUPLING EFFECTS BETWEEN HORIZONTAL AND VERTICAL VIBRATION MAY OCCUR

6. GATE VIBRATIONS EXPERIENCED; PART II

The visor gates of the Lower Rhine canalization in the Netherlands (see also Chart G) have already functioned for more than 15 years now. They have been used to check elastic similarity models; models which reproduce the lower-frequency vibration modes, but not local high-frequency plate vibrations.

The design of their lower edge (Sketch 3 of Chart G) was based on the argument that, in closed position, the gates would be adequately supported by the fender profile extending along their length and that, when partly raised, no adverse effects had been found at the four lower resonant vibration modes (up to about 6 c/s) in the hydro-elastic models. Theoretically, a negative aspect of the fender profile is that at a semicircular profile the flow separation point is not well defined and may be largely influenced by a small disturbance.

After construction, high-frequency plate vibrations occurred at the corners of the gate when the local gate lift there was of the order of 0.5 m. Strengthening and stiffening of the plates led to only partial success. The vibrations have now been accepted, though in critical circumstances a remedy is available in the form of an added lift of one gate and a reduced lift of the other, thus maintaining the same total discharge capacity.

In a hydraulic model, we intended to perform pressure measurements with an unfavourably shaped quarter-elliptical gate seal (Chart M). A 2 cm steel plate was wedged between the flume walls. The construction was strong and stiff enough for a maximum head difference of 50 cm water. Nevertheless, horizontal plate vibration became apparent, generating a cross-pattern of waves in the upstream water. Frequencies were not measured; analysis of the wave pattern (with $\lambda \approx 2$ cm) shown in the photograph of Chart M indicated a frequency of about 11 c/s, but personal impressions (from the rumbling noise) pointed to about 20 c/s, which frequency agreed with the calculated value of the first natural frequency of the submerged plate.

Chart D shows an emergency flap gate, arranged to lie down flat in normal circumstances. These gates are to provide safety, should a dike rupture locally, by preventing the entire canal system to empty through the gap. During closing movement, flow can occur in two directions; let us call them normal flow (as shown in Sketch 1 of Chart D) and reverse flow. To prevent nappe vibration during normal flow, the gate top had to have higher and lower sections of 2 m length, because splitter plates were not allowed, as ship's anchors might get fouled therein. For reverse flow, precautions had to be taken, such as a row of spaced blocks under the edge (found in Chart D), to keep the flow attached in all gate positions in order to prevent vibration.

During verification tests with normal flow in the hydraulic model, extremely heavy vibration occurred when the nappe was relatively thick and the air under the nappe was sucked off. This was one of the very few examples of gate vibration in a model, wherein increased hoist rigidity had an adverse effect. A little aeration was sufficient to suppress the vibration.
It was decided to install two splitter plates near the abutments. In this case, the great distance between the splitter plates (30 m) was considered acceptable, which, by the way, is not the case when they serve to prevent nappe oscillation. For further details of the conditions wherein vibrations occurred, see Chart D.

Conclusions:

Three empirical conclusions can be drawn from the foregoing:
- Edges at which unstable flow separation might occur cause danger of vibration, even vibration in the direction in which the gap size remains constant. Such vibrations can also occur when the gate lift is a few times greater than the seal width.
- These vibrations tend to have relatively high frequencies; the Strouhal numbers of the examples were:
  $S_\delta = f\ \delta/(2g\ \Delta H)^{1/2}$ of the order of 0.5 - 2 (δ = gate opening),
  $S_b = f\ b/(2g\ \Delta H)^{1/2}$ of the order of 0.2 - 1 (b = width of seal beam).
- In conditions with an overflow nappe where the air underneath can be sucked off by the nappe, resulting in a low-pressure zone, aeration should be applied.

7. FLOW CONTRACTION FLUCTUATIONS

The vibrations mentioned in Paragraph 6 were all in a direction in which the gap width or nappe thickness were not, at first impression, varied by gate vibration. Flow and force analysis, assuming a constant discharge coefficient of the gap flow, leads to the conclusion that only damping occurs; the gate acts in the same way as does a body in flow which vibrates in the flow direction.

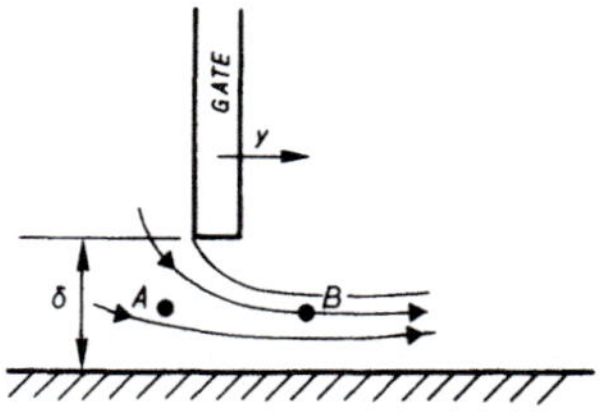

FIG. 7.1 CALCULATION SCHEME FOR HORIZONTAL VIBRATION (WITH CONSTANT DISCHARGE COEFFICIENT)

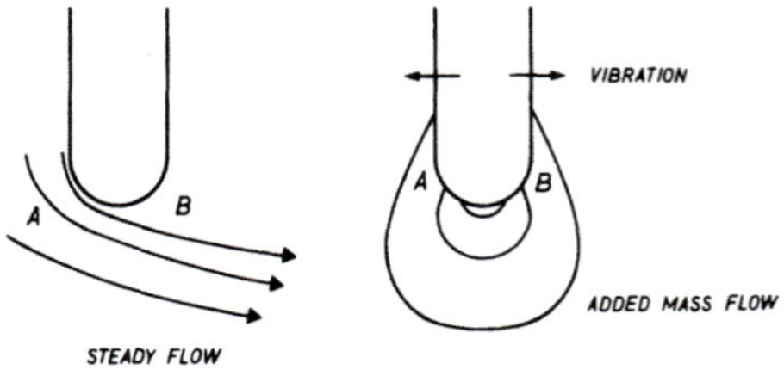

FIG. 7.2 THE ELEMENTS OF SIMULTANEOUSLY OCCURRING GAP FLOW AND ADDED-MASS FLOW

The detailed computation, presented in [8], is based on the following principles (see also Fig. 7.1):
- A horizontal gate oscillation is assumed; the displacement is y.
- A fluctuating additional discharge q' is assumed, which is superposed on the initial discharge q_o.
- Both gate vibration (related to added mass flow) and discharge fluctuation (related to flow inertia) cause an extra head difference across the gate

(ΔH_{gate}) and across the gap ($\Delta H'_{gap}$).

One can now express the equalities:

$$\Delta p'_{AB} = \rho g\ \Delta H'_{gap} = -c_L\ \rho\ \frac{\partial q}{\partial t} - c_{gate}\ \rho\delta\ \frac{\partial^2 y}{\partial t^2} \tag{7.1}$$

(points A and B are indicated in Fig. 7.1; c_L is related to c_L in Fig. 3.7), and

$$\rho g\ \Delta H'_{gate} = -c'_L\ \rho\ \frac{\partial q}{\partial t} - c'_{gate}\ \rho\delta\ \frac{\partial^2 y}{\partial t^2} \tag{7.2}$$

The discharge equation

$$q_o + q' = \mu\ \delta_o \sqrt{2g(\Delta H_o + \Delta H'_{gap})} \tag{7.3}$$

relates q' to $\Delta H'_{gap}$.

Of the four variables $\Delta H'_{gap}$, $\Delta H'_{gate}$, q' and y, one can, by using these three equations, find a relation between $\Delta H'_{gate}$ and y; and one can also find the phase difference. This results in an added-mass effect and a (high) damping term, but there is no self-excitation. This analysis is presented in Reference [8], Chapter 4.

When one has to assume that the flow contraction coefficient fluctuates during vibration, a possible explanation is that the added-mass flow - i.e. the flow wherein the velocities vary in phase with the vibration velocity of the gate edge - can have an effect on the flow contraction, especially, but not exclusively, in the case where a rounded edge already induces a very sensitive flow pattern (Fig. 7.2). This effect would operate in such a way that edge movement in the y-direction (to the right) causes a flow around the edge which tends to decrease the discharge coefficient.

When the contraction coefficient is also expressed in a constant part μ_o and a fluctuating part μ', it is of interest to know if μ' varies proportionally to the vibration displacement $-y$, to the vibration velocity $-\dot{y}$, or to the pressure gradient (related to $-\ddot{y}$). If the vibration is entirely parallel to the bottom, the relation of μ' to $-y$ is improbable. There remain:

possibility a: the flow is sucked around the edge by the right-hand part of the added-mass flow ($\mu'::-\dot{y}$); and

possibility b: the boundary layer becomes unstable (early flow separation) when there is a positive pressure gradient in the flow direction ($\mu'::-\ddot{y}$).

Similar reasoning as used in the gap-flow theory of Paragraph 5 leads to two possibilities: low-frequency and high-frequency vibration.
At low frequency, the discharge fluctuates proportionally to and in phase with μ', resulting in a pressure difference across the gate and the gap proportional to $-\partial q'/\partial t$, due to the inertia of the up- and downstream water.
At high frequency, the discharge remains constant due to flow intertia, and in this case the head across gate and gap varies proportionally to $-\mu'$.

Between these two cases there is again a 90° phase shift of the forces involved. In the fluctuating-gap theory, this results in negative hydrodynamic damping for low-frequency vibration and negative hydrodynamic rigidity for high-frequency vibration (Par. 5). Qualitative analysis of the phase shift between the head difference across the gate and the vibration displacement y reveals that negative hydrodynamic damping ($\Delta H'::\dot{y}$) occurs only at a combination of possibility a and high-frequency vibration.

That this combination indeed leads to self-excitation can be explained in the following way:
When at high-frequency vibration, where the discharge remains constant, a smaller

value of μ results in a greater head difference across the gate ($\Delta H' :: -\mu'$) and also $\mu :: -\dot{y}$ (possibility a), the head difference is in phase with and in the direction of the vibration velocity. These conditions being fulfilled, negative hydrodynamic damping will occur (when other components do not give too high a positive damping).

In the case of the flap gate with overflow (Chart D), the flow is probably stably attached at the upstream rounding of the crest. But at the moment when vibration lowers the gate, the added-mass flow will tend to influence the nappe path such that the water will fall farther away. This, however, is prevented by the constant volume of water under the nappe, resulting in a low pressure in phase with the vibration velocity.

8. GATE VIBRATIONS EXPERIENCED; PART III

In the same series of experiments wherein the quarter-elliptical gate seal was tested (Chart M), a rectangularly shaped gate edge was tested for suction forces, using a rigid model. The gate, including the edge, was again the 2 cm thick steel plate, wedged between the walls of a flume (Chart M). High-frequency bending vibration was observed (15-20 c/s) at gate lifts of 1.35 and 2 cm, but not at the other lifts of 1, 3 and 14 cm. The head differences were 0.2-0.4 m. Especially at a lift of 1.35 cm (and ΔH = 0.4 m), vibration was intense.

There is a tendency in the Netherlands to suppress seal beams and make sharp-edged gate lips whenever watertight closure is not needed. This holds particularly for storm-surge barriers, where the varying load at closed position can cause wear of the seals. However, problems can arise.

The 80 m span gate at Krimpen (Chart H) was tested at head differences of up to 2 m. Only in "closed" position, when a minimum gap width of officially 3.5 cm is available, a high-frequency plate vibration was felt. Above water (depth about 7 m), a spectrum of vibration frequencies was measured, with a clear peak at 88 c/s, which corresponds roughly to the natural frequency of the gate lip. When the gate was raised only a little, the vibration stopped.

In a complex of restored antique navigation locks (the Great Sea Locks at Muiden, near Amsterdam), one of the locks was adapted for sluicing. To hide the sluicing gates from sight, a steel frame with two butterfly gates was mounted under a bridge (Chart E). Here also, a leak gap at closed gate position was designed, and the gate plate (with reinforcement ribs at 65 cm intervals) ended with a sharp edge. In a 1:10 model, the rigidity was reproduced (but not the plate mass, because the added water mass was considered to be predominant). So, the plate thickness of the model was 1.25 cm (prototype value) instead of 2 cm. With a leak gap of 1 cm, strong plate vibration was felt between closed gate position ($\phi = 0^o$) and $\phi = 3^o$, and the torque meter at the central gate shaft indicated a frequency of 50 c/s (prototype value).

A remedy was found in mounting seats on the jambs of the surrounding frame, against which the plate was pressed in closed position. Moreover, the plate edge was stiffened; and the jamb edges were bevelled in order to obtain a rapid increase of the gap width at initial opening of the gate without causing a complete change of direction of the water jet. Further details can be found in Chart E.

Chart B was already presented in Paragraph 4; the thickness of the lower edge of the 1:25 elastic similarity model of the Volkerak gates (span 30 m) was varied in order to study the suppression of gate vibration at small openings, that is, at closed position of the gate. Whenever the gate lift was between 0.7 and 2 times the width of the gate edge, vibration occurred, always starting as vertical vibration (2.9 - 3.6 c/s prototype value), but continuing as horizontal vibration (2 - 2.5 c/s) mainly of gate-bending type. The advice was to make the edge sharp and to maintain a gap of a few cm in closed position of the gate.

Conclusions:

- Rectangular-section gate seals do not only induce vibration in the direction in which the gap width varies, but also in other directions.
- For a gap between a rectangular gate edge and a flat bottom, the most critical ratio between gap width and gate width is 0.7 - 2.
- Frequencies are relatively high; most of them correspond to a Strouhal number of $S_b = f\ b/\sqrt{2g\ \Delta H} = 0.1 - 0.25$.

9. GATE VIBRATIONS EXPERIENCED; PART IV

The butterfly gate of the restored old navigation locks with butterfly sluicing gates (Chart E) showed the same high-frequency plate vibration of about 50 c/s (prototype value) at gate opening angles of about $\phi = 50^o$. This angle corresponds roughly to the gate position at which local flow separation becomes complete flow separation at the downstream side; this can be concluded from the stationary torque measurements.

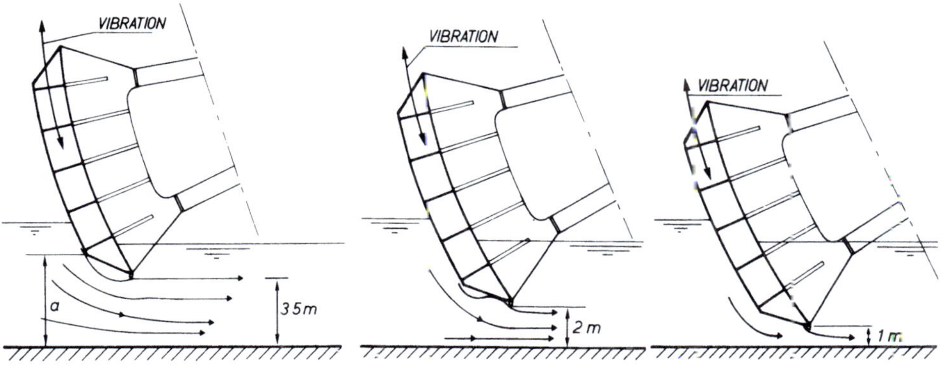

FIG. 9.1 TANGENTIAL VIBRATIONS OF SECTOR GATES OF HARINGVLIET SLUICES; ELEMENTS OF FLOW PATTERN WHICH CAN CAUSE VIBRATIONS AT DIFFERENT GATE POSITIONS (CHART F)

Figure 9.1 shows the Haringvliet sluices, controlling the discharge of Rhine water into the sea and closing off a fresh-water reservoir; they comprise a double row of sector gates, hinged to the seaside and to the landside of a bridge (see also Chart F). The landside gates have their cylindrically curved plate, and the seal, at the seaside to prevent tangential force components due to wave action.

At the time model studies on vibration were executed, attention was concentrated on the shape of the seal and therefore only gate lifts of up to 1 m were studied. In the prototype, however, vertical (or better: tangential) vibrations occurred at gate lifts of 1 - 4.5 m. The 500 ton gates with a span of about 60 m came into a regular vibration with a maximum (single) amplitude of 2 mm and a frequency of about 2.5 c/s. It must be remarked that, due to the inflatable side seals, mechanical damping at raised gate position is extremely low.

It is only at higher gate positions (condition A) that one can explain such vibrations with the fluctuating-gap theory, or that one can use the instability indicator in the following way: a lowering of the gate at constant discharge causes extra suction, because this suction is related to the increased head difference across the gate. (When the jet touches the seal, the contracted flow has already expanded again so that near the contraction zone there is a low-pressure zone with underpressures proportional to the head difference.)

At lower gate positions (condition C), a lowering of the gate at constant discharge causes, to the contrary, an increased pressure under the gate, because the gate edge, which regulates discharge, is situated at the downstream side of the gap. Near the upstream side there is a suction zone, but because of the wider flow section there, water velocities increase only slightly and therefore lowering of the gate does hardly create an extra suction force there either.

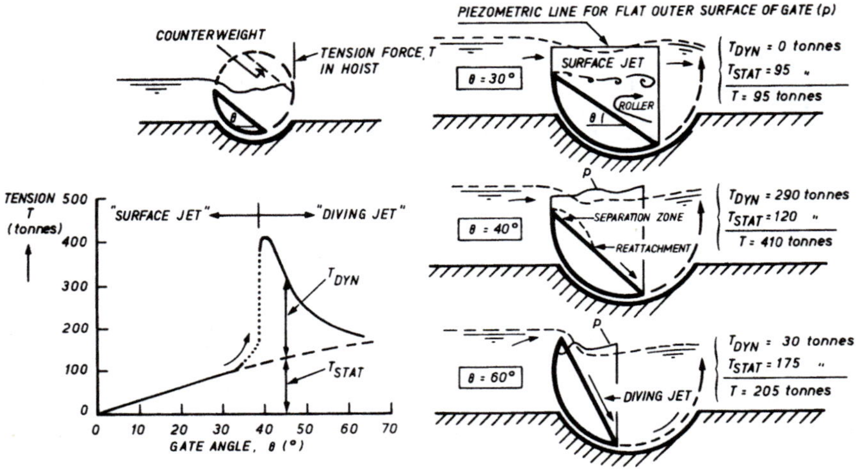

FIG. 9.2 TANGENTIAL VIBRATION OF THAMES BARRIER GATE; INSTABILITY OF FLOW REATTACHMENT (REF. [7])

The example of Figure 9.2 is included in this paper, although it is not from the Delft Hydraulics Laboratory but from the Imperial College in London. Dr. Hardwick showed me a small section model of the Thames Barrier and he demonstrated that, in conditions wherein the flow wavered between full flow separation and reattachment, vertical (or better: tangential) vibrations occurred. In his paper [7], from which he kindly permitted me to use Figure 9.2, he also demonstrated that by perforating the gate in the low-pressure zone vibrations were reduced. After all, this remedy was not necessary, because in reality the gate frequency was outside the frequency range of these self-exciting vibrations.

Conclusions: The examples of this series have in common that a local low-pressure zone between flow separation and reattachment is involved and reattachment is unstable.

10. CONSIDERATIONS ABOUT FLOW SEPARATION AND REATTACHMENT

In Paragraph 7 it has been suggested that the added-mass flow could influence the point of flow separation at a rounded edge, where the point of separation is unstable and undefined. It is conceivable that the same mechanism could, in the case of a stable point of separation, influence the flow contraction directly, but to a lesser degree.
As yet, however, no vibrations have been attributable to such a case (except for the case with an overflow nappe discussed in Par. 7); the rectangular edge of Chart M did not vibrate at wider openings than the seal width.
But vibrations can occur when there is a stable flow separation point followed by

reattachment, especially when reattachment of the jet is uncertain. These two conditions, full separation and reattachment, make a considerable difference in the discharge coefficient.

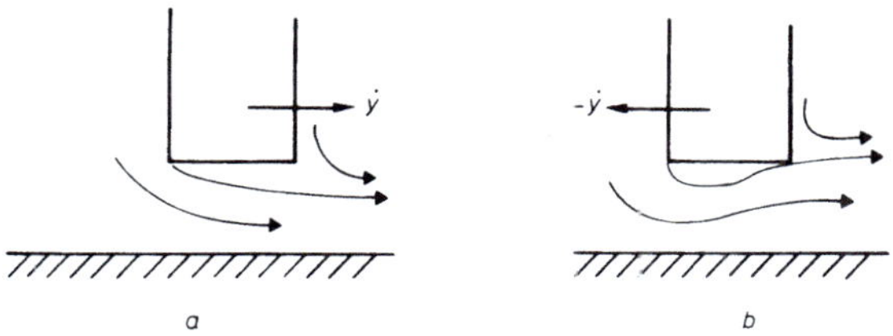

FIG. 10.1 FLOW PATTERN AT VIBRATION STAGES WITH VIBRATION VELOCITY IN FLOW AND COUNTER-FLOW DIRECTION

When we consider (Fig. 10.1) a high-frequency horizontal plate vibration wherein Q = constant (due to the high frequency) and the velocity = $-\dot{y}$ (directed to the left), the jet will be sucked upwards by the added-mass flow and reattachment is possible. This increases the discharge coefficient considerably, while the head difference across the gate will decrease accordingly, causing a force in phase with $\dot{y}$. With this mechanism all examples of Paragraph 8 can be explained.

When, downstream of the edge, the flow wavers between remaining attached to the face plate of the gate and getting separated, the added-mass flow can also have substantial effects. This is the case with the butterfly gates in the old navigation locks at Muiden (Chart E) and also with the gates in the Thames Barrier (Sketch 4 of Chart E and Fig. 9.2).

The vertical gate vibrations at the Haringvliet sluices (Fig. 9.1) form a special case, but here too the same line of reasoning can be followed. Due to the cylindrical curvature of the skin plates, vertical (tangential) gate vibrations only cause an added-mass flow by displacing water underneath the gate (like a heaving ship). The added-mass flow is more or less symmetric, and the flow contraction and the point of reattachment of the continuous flow are both affected. During downward movement, the contraction coefficient decreases, but at the same time the local low-pressure zone between the points of flow separation and reattachment increases in size, and consequently extra suction is created. Because this suction now again is in phase with the vibration velocity, self-excitation can occur.

11. DIVERSE GATE VIBRATIONS AND OTHER DYNAMIC EFFECTS

In the storm surge barrier in the Ooster Schelde, which is actually in the final stage of design and the first stage of construction, vertical lift (sliding) gates of 6 - 12 m height and 41 m span will be used. To permit independent placement of the gates and the sills and masks, the gates are positioned to the seaside of the sills. (Were they positioned on the landside, strong wave forces from below could be exercised on the masks.) A wide leak gap is accepted, which ranges between 6 cm under load and 15 cm when unloaded.

The mechanism of the rather strong dynamic flow behaviour found in an elastic similarity model (Fig. 11.1) cannot be explained in an exact way, but is most probably related to the flow capacity, which varies considerably with small changes of the upper (sea-)water level. When the flow separates at the upstream gate (girder) edge, the capacity is greater than when separation occurs at the lower (downstream) edge. Maximum capacity is reached when flow reattachment takes place exactly at the upstream edge of the mask bottom, just beyond the point of maximum flow contraction. So, at a rising upper level the discharge capacity increases and at a dropping upper level it decreases, and this occurs rather abruptly. Moreover, at greater water level oscillations there are still other effects. For instance, at a very low upper water level the mask is no longer touched by the flow and no contraction occurs. Theoretically, a mechanism of self-sustained water level osciilations is possible, similarly as has been formulated by Streeter and Wylie [4] for self-sustained shock waves (see Par. 4 and Fig. 4.4 of this paper): a self-sustained oscillation can theoretically occur when an increase of the upstream head causes a decrease of discharge.

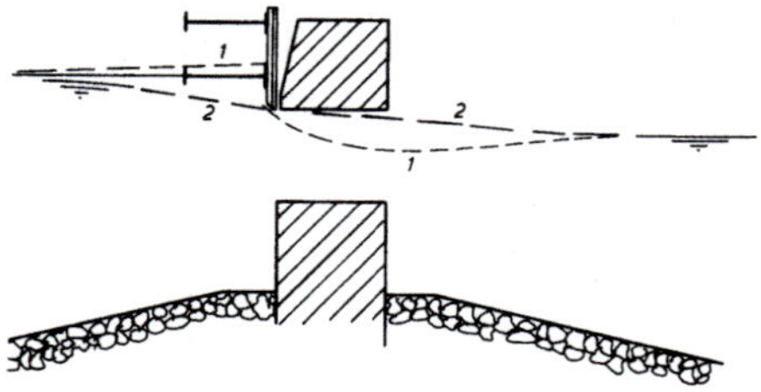

FIG. 11.1 FLOW INSTABILITY AT VERTICAL-LIFT GATE OF OOSTER SCHELDE BARRIER

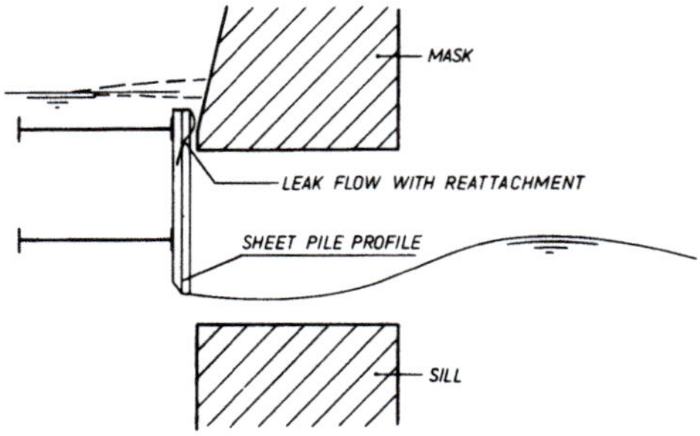

FIG. 11.2 OOSTER SCHELDE BARRIER: VIBRATIONS OF GATE, MASK AND SILL DUE TO FLUCTUATING GAP SIZE

When the same gate is nearly closed (see Fig. 11.2), intense vibrations (bending as well as torsion) of the gate and the mask have been found in the elastic similarity/hydraulic model. Probably, the point of maximum flow contraction was situated near the upstream end of the gap passage, a condition which according to the instability indicator is sensitive to vibration. (See the example of Fig. 4.6.) The solution was found in providing the concrete mask with a lip at the edge.

When the 30 m span wheel gate (initial design) described in Chart B and discussed in Paragraphs 4 and 8 was nearly closed and the water levels were such that the upper wheel was on the point of being unloaded, a strong rotational vibration occurred, centered on the lower wheel shaft. In the final design, the gate became a sliding gate and thus the problem was eliminated.

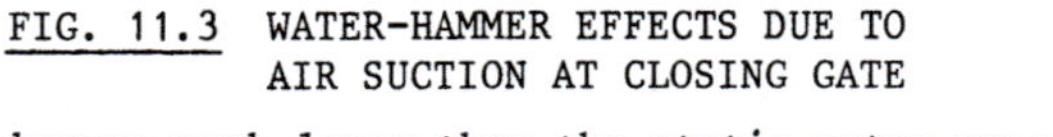

FIG. 11.3 WATER-HAMMER EFFECTS DUE TO AIR SUCTION AT CLOSING GATE

In a downstream lock culvert of a high-head lock, aeration tubes with screen were installed to suppress gate-induced cavitation (Fig. 11.3). A pressure cell was mounted in the ceiling of the culvert. When the culvert valve was closed, inertia kept the water flowing and air was collected near the gate. As the pressure in this elastic air cushion became much lower than the static water pressure in the downstream pool (even lower than the atmospheric pressure), the water flow was decelerated and then reversed. With an initial head difference of 20 m, a water-hammer shock of 110 m water column was recorded.

Conclusions:

The following design criteria can be deduced from the examples given.

- Play in pivots, hoist chains, articulation points, guide wheels, etc. must be preloaded in one definite direction.
- Free-surface flow can cause considerable changes in flow characteristics

when the surface is oscillating near girders, ceiling, etcetera. Whenever possible, gates should be shaped in such way that a well-defined flow pattern is obtained.
- In culverts, free-surface flow and total submersion are both acceptable, but not situations in between. Oscillating flow surfaces should never be able to hit structural surfaces parallel thereto; nor should a moving or oscillating water surface be allowed near dead corners, as this can cause water-hammer shocks. This holds true for free surfaces related to both air and cavitation pockets, and for whether the free surface oscillation is due to unstable hydraulic jump, wind waves, or water column separation.

12. DIFFICULTIES WITH VIBRATION MODELS DUE TO REDUCED REYNOLDS NUMBER

When cylindrical stop logs are used in flow, they show several disadvantages:
- Vortex shedding results in forces of extreme magnitude.
- When they are lowered, there is high pressure underneath (the upstream water is stagnant in the dead-water zones) and the pressure on top is low because flow separation takes place far beyond the point of lowest pressure (due to a regain of pressure head); as a result, the log can be lifted again by the flow.
- Cylinders can roll.

In practice, circular cylinders cannot be lowered in flow; they bump, dance, and even jump out of the water. Still, there is the economic advantage of using commercial tubes. Investigation in our laboratory showed that this is possible when a longitudinal fin is applied so as to give a downward hydrodynamic force; but then rotation of the log must be prevented by means of wheeled slot blocks. There remained a small instability when a log was lowered onto a preceding one (see also Fig. 12.1).

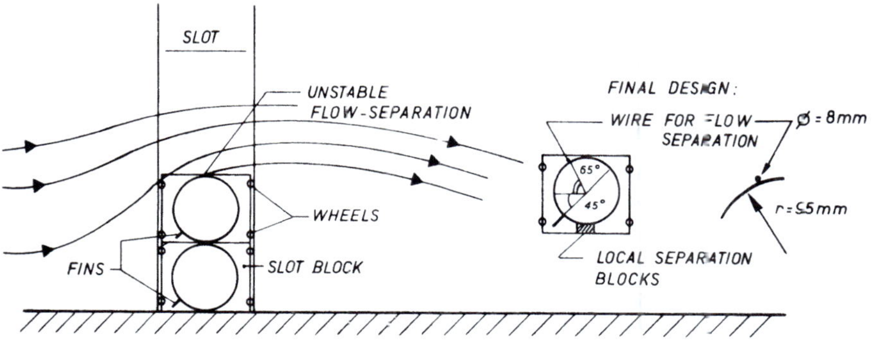

FIG. 12.1 VIBRATION OF CIRCULAR STOP LOGS BY RETARDED FLOW SEPARATION

A prototype experiment showed that the design did not work; so the model result proved invalid. The cause was found by looking into the width of the vortex zone at subcritical (model) and supercritical (prototype) Reynolds numbers: in the model, flow separation took place much farther upstream than in the prototype and low pressure on top of the log did barely occur. The solution for obtaining the same favourable condition in the prototype was found in welding a wire on the tube slightly upstream of the point of flow separation in the model. The optimum position of the wire was determined with a model placed in a closed high-velocity tunnel to obtain a relatively high Reynolds number.

Conclusions:

- Stop logs must be designed carefully, because vertical forces tend to be important compared to the weight of the log.
- Unstable flow separation should be prevented, not only because it makes hydraulic models less reliable, but especially because surface irregularities (varying from milimetre-size fabrication roughness to centrimetre-thick mussel growth in salt water) can strongly influence the flow pattern; initial turbulence can also strongly change the flow character.

13. DESIGN CRITERIA DERIVED FROM EXPERIENCE AND THEORY

Recollecting the conclusions of the foregoing, we can formulate the following design criteria for gates:

1. To prevent self-excitation
   a in cases of vibration modes in which the gap size fluctuates, suction forces (in general: forces tending to close the gap) should be prevented and the rigidity and/or structural damping should not be too low;
   b in all cases unstable flow separation and/or reattachment zones should be prevented;
   c low-pressure regions should preferably be vented by admitting air or water from a high-pressure region;
   d the width of leak gaps should be at least 1.5 times (preferably 2 times or more) the width of the gate edge;
   e when play is of influence, no change of force direction should be permitted; and
   f slender stop logs and gates should be carefully designed so that static and dynamic loads do not prevail over their weight.
2. Resonance frequencies of gate hoist systems and gate members should be well above the dominant excitation frequencies.
3. Water surfaces, unstable because of turbulence, waves, hydraulic jump, water column separation or cavitation, should have no opportunity to hit structural surfaces, especially when these are parallel to the water surface or have dead corners.

It is important to note that when dominant vibration modes are considered, coupling between vibration modes due to added-mass effects and to self-excitation mechanisms has to be taken into account. Moreover, shapes can be influenced by wear, growth of algae and mussels, etc.

14. FINAL REMARKS

From this paper have been excluded a few instances of nappe oscillation to which classical remedies, such as splitter plates, have been applied.

Actually, at the University of Karlsruhe, a systems approach is being developed to interrelate all types of aero- and hydro-elastic vibration mechanisms. I hope that this symposium will bring together a lot of information and also that this paper has contributed elements to this approach.

This paper is the result of many design-orientated studies and model investigations in the Delft Hydraulics Laboratory. Most of these projects have been done in close cooperation with the design offices of the Dutch Ministry of Public Works, who have made possible the development of advanced techniques, using expensive and complex combined hydraulic/vibration models (single-oscillator models and continuous elastic similarity models). A few years ago, a general research programme has been set up by this Ministry to improve gate design criteria, and largely thanks to this programme our theoretical base and understanding of vibration mechanisms has been expanded.

15. REFERENCES

1. Kolkman, P.A., "Self-Exciting Gate Vibrations", Invited lecture 17th Congress of Int. Ass. of Hydraulic Research, Aug. 1977, Baden-Baden; also Delft Hydraulics Laboratory Publ. 186.
2. Lamb, H., "Hydrodynamics", 6th Ed. Cambridge University Press (1932).
3. Zienkiewicz, O.C., Nath, B., "Analogue Procedure for Determination of Virtual Mass", Proc. of Am. Soc. of Civ. Eng.; Journal of Hydr. Div. Sept. 1964, HY5, pp. 69-81.
4. Streeter, V.L., Wylie, E.B., "Hydraulic Transients", McGraw-Hill Book CY (1967).
5. Miller, C., "Ein Beitrag zur Bestimmung der schwingungserregenden Kräfte an unterströmten Wehren", Inst. f. Wasserbau, Univ. Stuttgart, Publ. 42 (1977).
6. Kolkman, P.A., Vrijer, A., "Gate Edge Suction as a Cause of Self-Exciting Vertical Vibrations", 17th Congress of Int. Ass. of Hydr. Research, Aug. 1977, Baden-Baden, Paper C49; also Delft Hydr. Lab. Publ. 188.
7. Hardwick, J.D., "Hydraulic Model Studies of the Rising Sector Gate Conducted at Imperial College", Conf. on Thames Barrier Design, Oct. 1977, London; Pub. Inst. of Civ. Eng. Paper 12, p. 133-143.
8. Kolkman, P.A., "Flow-Induced Gate Vibrations", Thesis Delf Univ. of Technology; also Publ. 164 of Delft Hydr. Lab., 1976.
9. Meyers, P., "Numerical Hull Vibration Analysis of a Far-East Container Ship", Netherlands Ship Research Centre, TNO Report 195 S (1974).

CHART A

VERTICAL LIFT GATE

(Section model 1:30; dimensions etc. in proto-value)

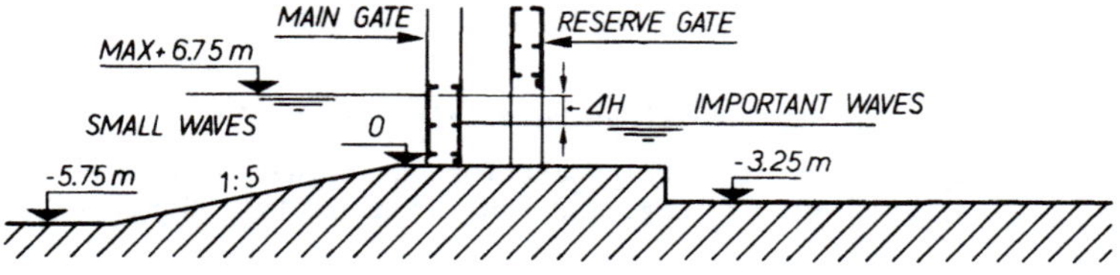

A1 PROTOTYPE SECTION

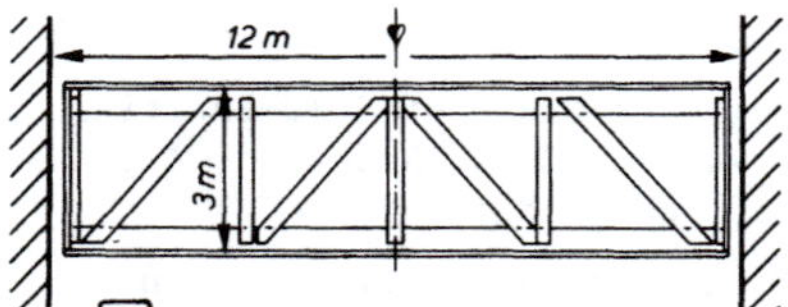

A2 TOP VIEW OF MODEL

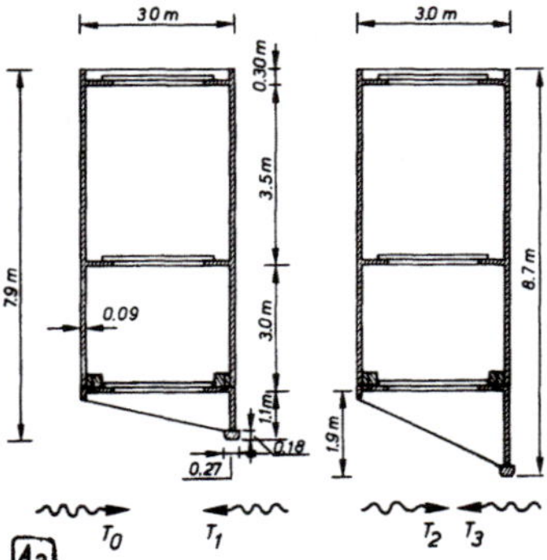

A3 FLOW DIRECTION

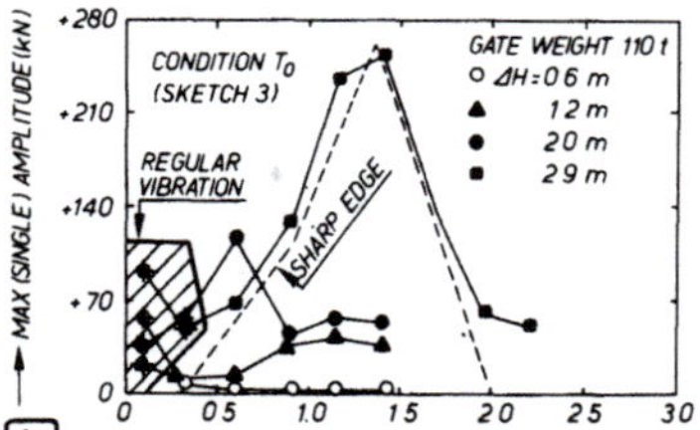

A4

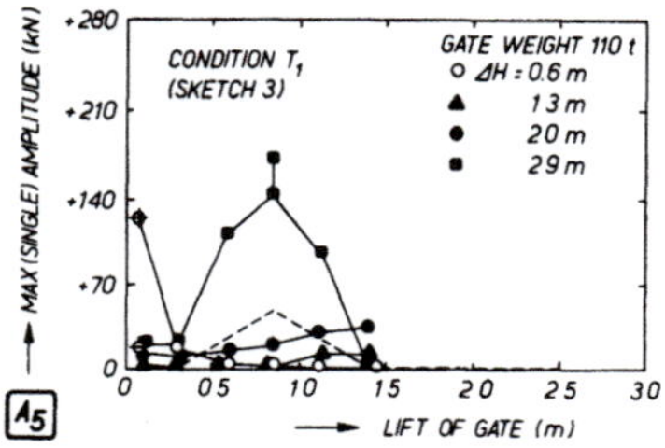

A5

Problem: In a single oscillator model (only vertical vibrations) of Volkerak sluice gates, strong vibrations occurred (max. force ampl. = 28 tonf). Total weight of 30 m span gate = 120 tonf (120 x 10^4 N).

Geometry: Sketches 1-3; the 11.9 m long midsection was reproduced in model.

Conditions: Upstream head 6.75 m (above bottom); ΔH = 0.6 - 2.9 m. Tests to decide if double-plate gate was acceptable. Vibrations at small openings (due to local edge geometry) were not considered, although they did occur. Great preference for smooth double-plate gate (wave forces low) with one plate extended at important wave side (T_o, T_2). Final check of design later in continuous elastic model (Chart B).

Structural dynamics: Model suspended in horizontal leaf springs; 11.9 m long gate section with m = 47 ton; elastic suspension K = 20.2×10^6 N/m (2020 tonf/m). Resonance freq. dry 3.3 c/s (in water 2.3 - 3 c/s); negligible damping.

Vibrations: Small openings (up to 0.2 m): regular vibration. Greater openings: T_o condition was worst (less regular vibration), see Sketch A4 (forces are for whole 30 m span gate). For T_1 see Sketch A5. T_2 also showed rather great amplitudes (210 kN at 1.2 m lift). T_3 had only 50 kN force ampl. (two peaks: at lift 0.8 m and 2.7 m). Mechanism: fluctuating gap theory (paper Par. 5) or unstable flow reattachment (paper Par. 10).

Decision: Double-plate solution with extended edge (Sketch A3, T_3) was chosen, although at high openings waves can slam against the lower girder.

References: See Paragraph 4 of this paper.

CHART B

VERTICAL LIFT GATE

(Hydraulic/continuous elastic model 1:25; all dimensions in proto-value)

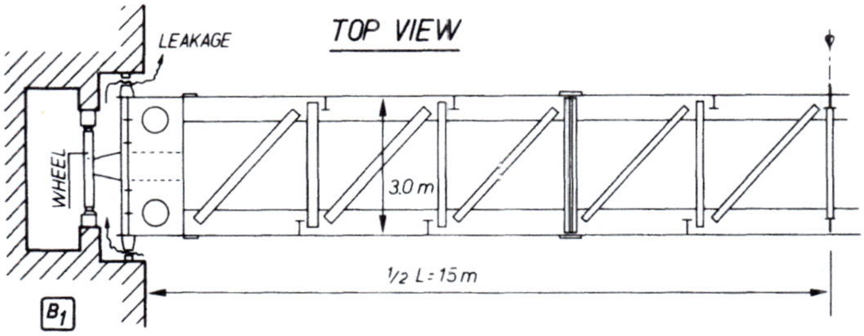

Problem: Check of final design of Volkerak sluice gates (see also Chart A) revealed vibrations at nearly closed position (the gate was designed with a leak gap in "closed" position).

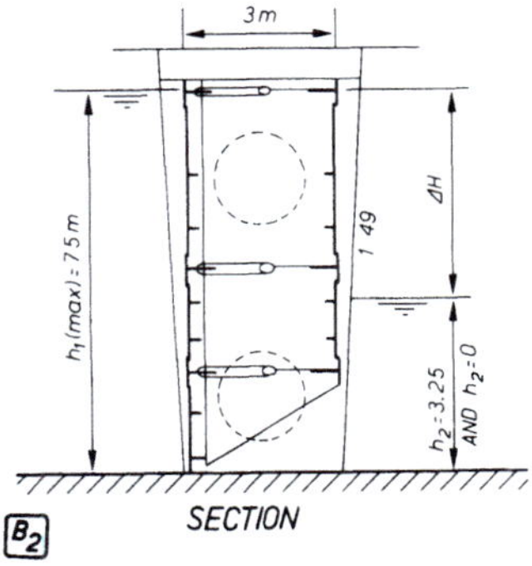

Geometry: Double-plate vertical lift gate, Sketches B1-B3; original design with wheels.

Conditions: Downstream water depth $h_2 = 0$ and $h_2 = 3.25$ m in most tests. Upstream water depth h_1 variable with max. 7.5 m. Original design as wheel gate and with edge as shown in B3.

Structural dynamics: Mass 120 tons, hoist rigidity $2 \times 2.5 \; 10^6$ N/m. Bending rigidity: 3 girders, each $E \times I = (2.1 \times 10^{11}) \times (0.124) = 26 \times 10^9$ N/m²; under dynamic conditions model showed 40% more rigidity. Tests without wheels (to suppress damping) and with a pair of leaf springs for only vertical freedom revealed vibrations of type I.

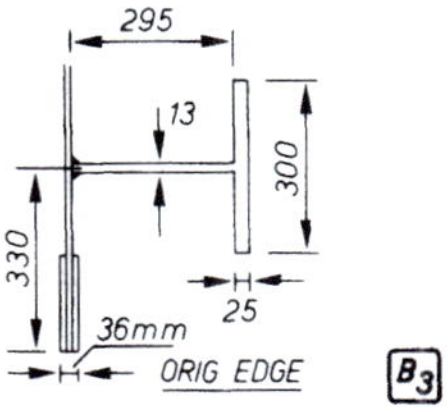

Vibrations, type I (edge, see B3): Regular vibrations starting vertically (2.9 - 3.5 c/s) and later sometimes (and intense) horizontally. Wave pattern parallel to gate (vert. vibr.) or waves generated at corners (hor. vibr.); see paper Figure 4.5. When $h_2 = 0$ and $\Delta H > 2.75$ m, vibrations started. Intense horizontal vibration at $\Delta H \sim 3.2$ m, disappearing at $\Delta H = 3.75$ m.
When $h_2 = 3.25$ m, vibrations starting at $\Delta H = 3.25$ m and increasing with ΔH. Gate edge thickness variations (see B4): d = 0, 25, 75 and 300 mm. Critical positions at gap width $\delta = 0.7 - 1.5$ d. When d = 300 mm, vibrations started already at $\Delta H = 1.25$ m. When d = 0, 25 or 75 mm and $h_2 = 3.25$ m no vibrations, but for $h_2 = 0$ and $\Delta H = 3.5$ m vibrations also started. Vibrations stopped when vertical friction at the abutments was introduced, or when leakage in slots was nearly blocked. Mechanism: probably unstable flow separation and/or reattachment (paper Par. 7 and 10).

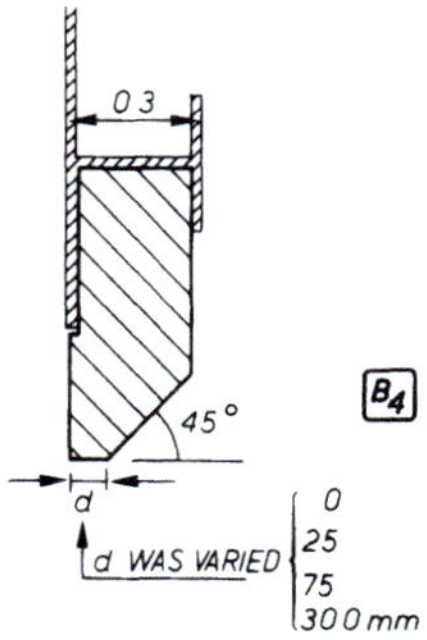

Vibrations, type II: When $h_2 = 1.75$m, $h_1 = 3.25$ m and upper wheel just unloaded (gate nearly closed), strong horizontal vibrations and waves parallel to the gate were observed.

Final gate design: Sharp edge (d = 10 mm); a minimum gap width of $\delta = 30$ mm at "closed" position; stiffening girder near edge moved to higher position. Wheels removed (slide gate).

References: See CHART A, and Paragraph 4 of this paper.

CHART C

REVERSED TAINTER VALVE IN LOCK CULVERT

(Hydraulic/vibration model 1:10; all dimensions etc. in proto-value)

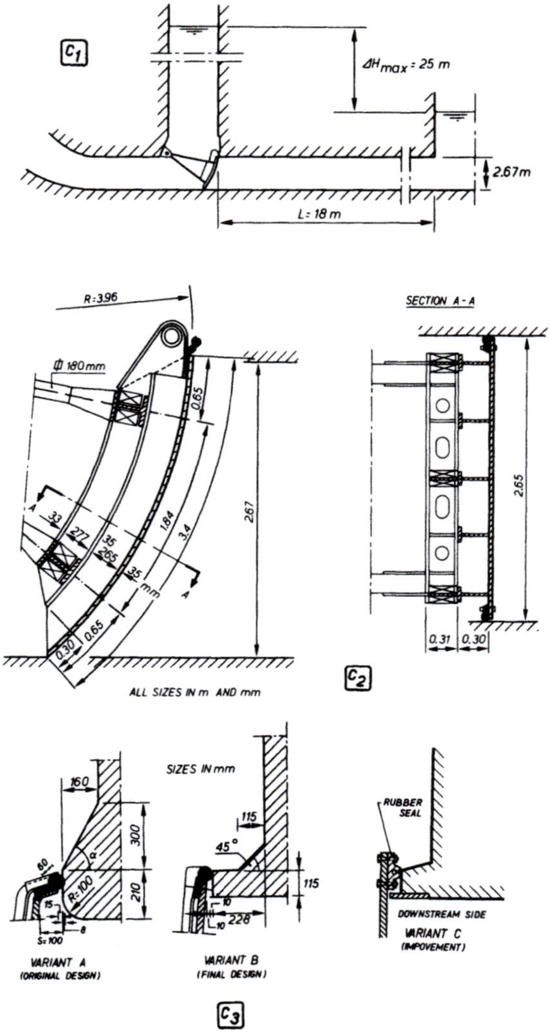

Problem: In hydraulic model of inlet gates of Kainji locks (with inlet culvert reproduced up to 18 m downstream of gate) very intense vibrati occurred at nearly closed position.

Geometry: Sketch C1-C3, sharp lower edge (with rubber in the gate floor)

Hydraulic conditions: When valve is nearly closed, ΔH = 25 m. Whole gate plus culvert in submerged condition.

Structural dynamics: Single oscillat model (rotation). Gate mass 9300 kg, no mechanical damping (side seal removed), hoist rigidity $k = 5 \times 10^6$, 10×10^6 and 18.5×10^6 N/m.

Vibrations: Regular; occurring at ga openings less than 5% of culvert height. Max. (single) force amplitu (at $k = 5 \times 10^6$ N/m) was 40×10^4 N (40 tonf) with displacement $y \sim 0.1$ m. A small openings gate touched bottom. Added mass of the order of 700 kg.

Cure: Variant B in Sketch C3 was applied; in prototype weak vibrations were still experienced when top seal was just released; a rubber seal clamped along both edges would have been preferable (Variant C).

Reference: See Paragraphs 4 and 5 an Figure 4.1 of this paper.
Kolkman, P.A., "Flow-induced Gate Vibrations", Publication 164, Delft Hydraulics Laboratory, 1976, Chapter I Appendix II.

CHART D

FLAP GATE

(Hydraulic/vibration model 1:15; dimensions etc. in proto-value)

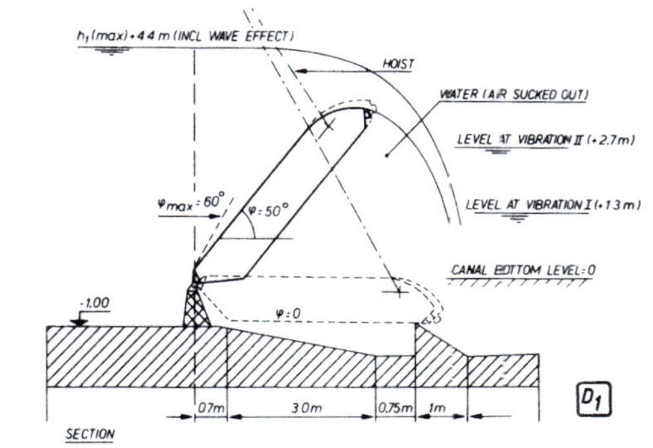

D1

Problem: Emergency gate, which must always be ready to be closed in case of dike rupture. Flow direction normal (as in Sketch D1) or reverse. Model tests showed extremely intense vibrations (rotational) when both downwater h_2 was rather high and nappe was thick, such that air under nappe was sucked away. Nappe spoilers preferably not to be used (ship's anchors).

Geometry: Sketches D1 and D2, span 36 or 50 m, hoist at one side. Rectangular-toothed crest to prevent nappe vibration at small nappe thickness; flow deflection blocks to keep flow attached at reversed flow direction (extra torque, but safety against vibration).

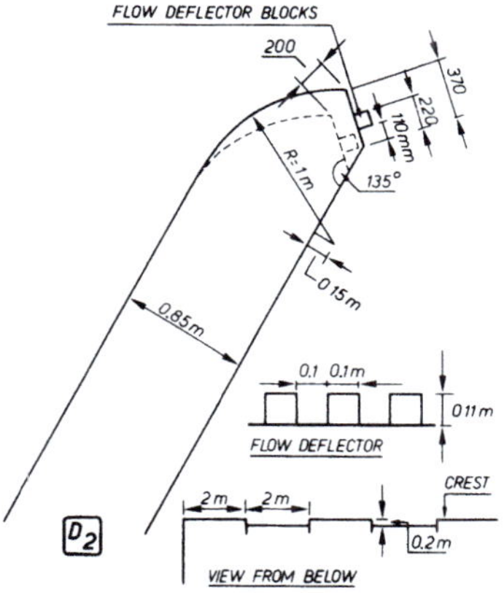

D2

Flow conditions: Normal canal level +3.7 m; but during closing lower, and when nearly closed higher due to translatory waves. Max. upstream level h_{max} = +4.4 m. During vibrations $h_1 = h_{max}$ and h_2 = +1.3 m (vibration I) and h_2 = +2.7 m (vibration II). Other h-values not checked.

Structural dynamics: In the model a 12.5 m gate section was tested; gate filled with water; polar mass of this section (dry) $I_p = 0.2 \times 10^6$ kgm^2. Elastic hoist with variable rigidity.

Vibrations:
Vibr. I, res. freq. dry 3.5 c/s, in flow 1.3 c/s
Vibr. II, res. freq. dry 5.2 c/s, in flow 1.7 c/s.
Both vibrations at $40° < \phi < 55°$ (max. at $\phi = 50°$). Max. amplitude of 12.5 m section was 1.25×10^6 Nm torque. Mechanism: flow-contraction fluctuation, see Paragraph 7 of this paper.

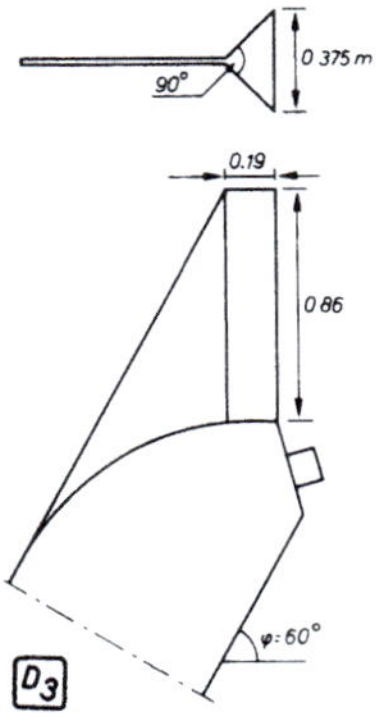

D3

Cures: Aeration pipes (one must always remain unsubmerged) or nappe spoilers. Second solution applied: one spoiler near each abutment (Fig. D3). Sufficient condition is that air remains under nappe; this in contrast to thin oscillating nappes, where more aeration is needed to prevent pressure oscillations.

References: See Paragraphs 6 and 7 of this paper.

CHART E

BUTTERFLY GATES

(Hydraulic vibration model 1:10;
dimensions etc. all in proto-value)

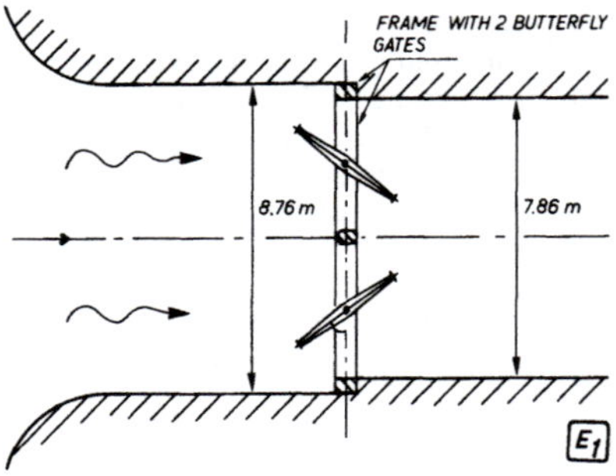

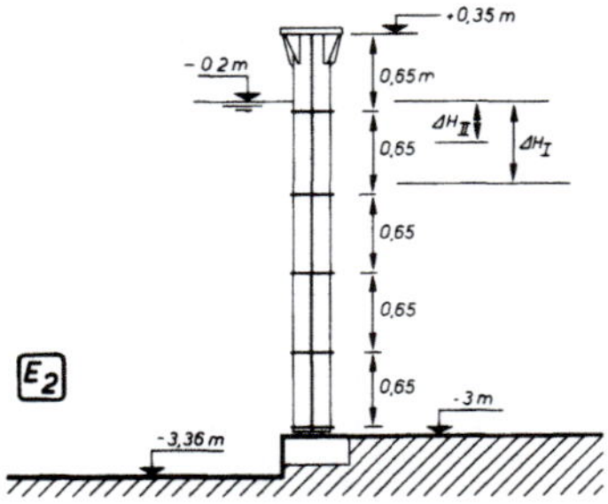

VERTICAL SECTION OF GATE

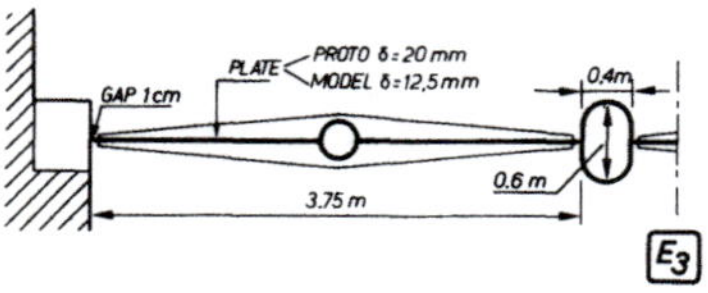

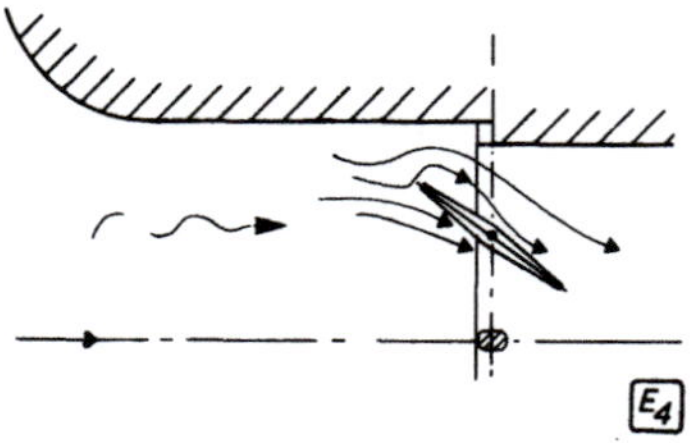

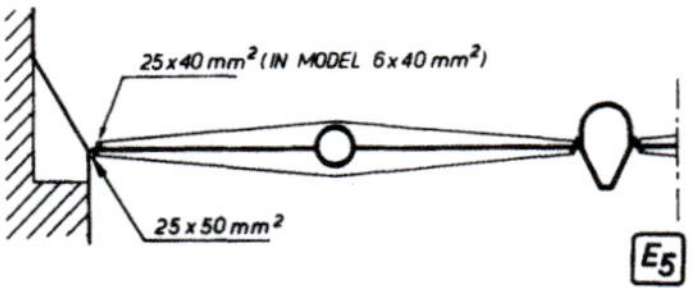

Problem: In a row of old navigation locks at Muiden one lock is adapted for sluicing. For monument conservation reasons, 2 butterfly sluicing gates in a steel frame are hidden under a fixed bridge. The gates consist of a plate with horizontal stiffening ribs (spacing 0.65 m). In "closed" position (ϕ = 0 - 3°) a leak gap was initially accepted. In this position there was strong plate vibration (vibration I), both audible and with wave radiation. The same occurred at ϕ = 50° (vibration II).

Remark: The model had two elastic elements scaled down: the torsion rigidity of the shaft and the bending rigidity EI of the plate. For this reason, the model thickness of the plate does not completely correspond to prototype values. Mass was not corrected because added mass was considered strongly dominant.

Geometry: Sketches 1-3.

Flow conditions wherein vibrations occurred: Upstr. water level about -0.2 m., head difference ΔH up to 2 m. Precise data on ΔH at which vibration started not available, but at 0.7 m vibration I was experienced; for vibration II this was 0.35 m.

Structural dynamics: Plate in prototype 20 mm steel; in model 12.5 mm (proto-value) brass.

Vibrations: Both vibrations were measurable with a torque meter; unknown magnitude. Vibrations were audible, with frequency 50 c/s (in the model 150 c/s). Vibration I is due to unstable flow separation; vibration II occurred just in the ϕ region where flow wavered at reattachment, see Sketch 4 and Paragraphs 8 and 9 of this paper.

Cures: See Sketch E5: a. stiffening ribs on the plate (thinner ribs were tried but were not effective); b. seats at the centre post and abutments to press the gate against in closed position; c. bevelling of the frame so that the gap widens quickly when opening the gate.

CHART F SECTOR GATES WITH UNDERFLOW

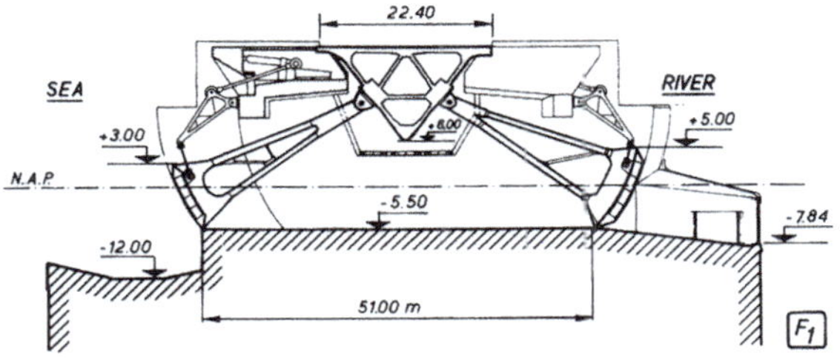

Problem: Vertical (= tangential) vibration of river gates of Haringvliet sluices when raised between 1 and 2 m, or between 3 and 4 m; flow from river, sea gates open.

Geometry: Sketches 1-3. Gate span 56.5 m. Struts supporting gates from bridge girder have 14.8 m spacing.

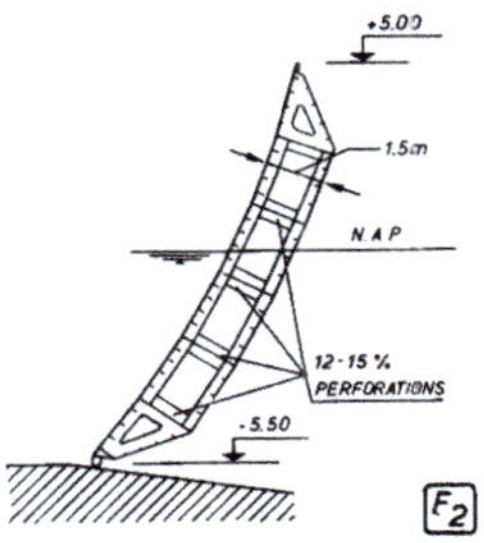

Flow conditions during vibrations: River level between NAP and NAP +0.5 m; sea levels 0.5-2 m lower.

Structural dynamics: Weight of gate (= hoist force in dry condition) 520 tonf (520×10^4 N). Hoist rigidity (of the 2 hoists together) 115×10^6 N/m in closed gate position and somewhat less when raised. Resonance frequency at closed position: dry, f_{res} = 2.3 c/s; at water level -4 m, f_{res} = 2.4 c/s; at -2 m, f_{res} = 2.1 c/s; at +0.3 m, f_{res} = 1.95 c/s.

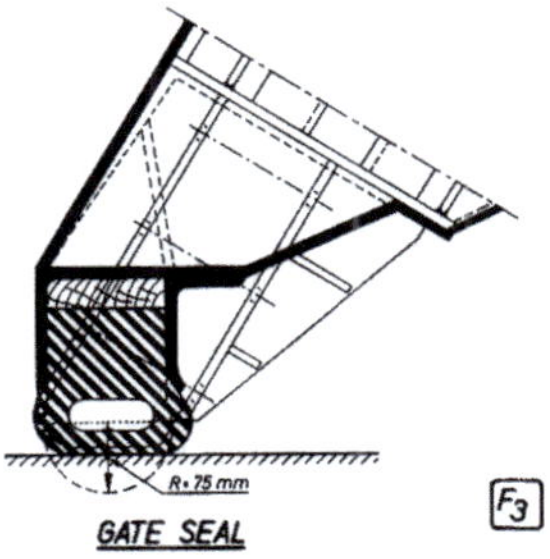

Vibration: Regular vibration, frequency between 2 and 2.5 c/s (first natural mode with roughly hoist = spring, gate = mass). Typical experience during one test period: Sketch F4. Max. acceleration amplitude $\hat{\ddot{y}} = 0.3$ m/s^2, both when gate was raised 2 m and raised between 3.5 and 4 m. Vibration amplitude $\hat{y} = \hat{\ddot{y}} : \omega^2$ = 1.8 mm for 2 m gate opening and 1.2 mm for 4 m opening. Vibration mechanism: fluctuating gap theory (see paper Par. 4) or unstable flow reattachment (see paper Par. 10).

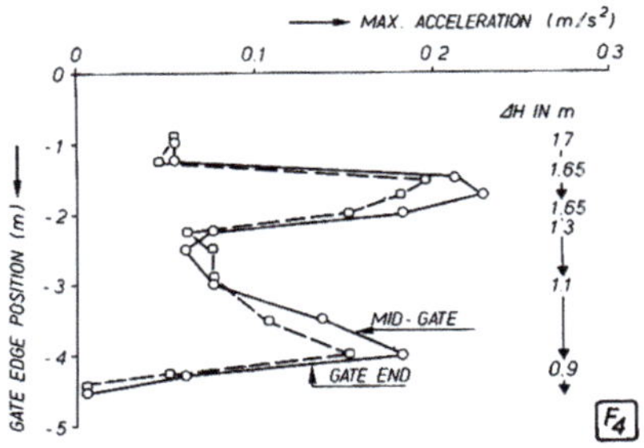

Cures: No cures tested.

References: See Paragraphs 9 (Fig. 9.1) and 10 of this paper. Structural description: "The Haringvliet Sluices", Rijkswaterstaat Communications Nr. 11, 1970 (Ministry of Public Works of The Netherlands).

CHART G

VISOR GATE

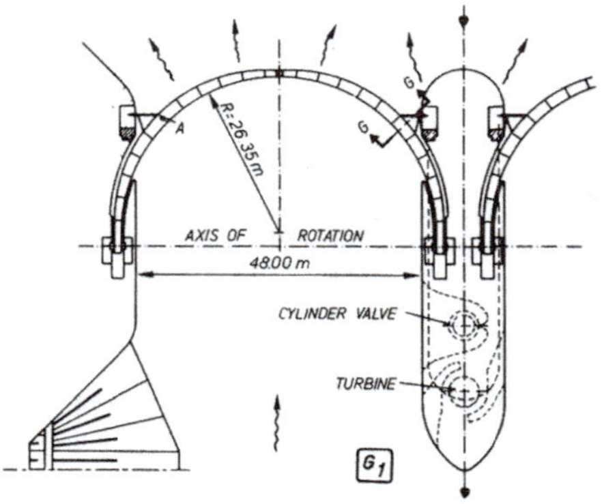

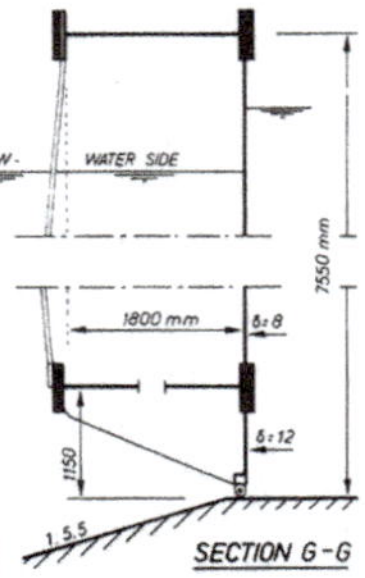

G2

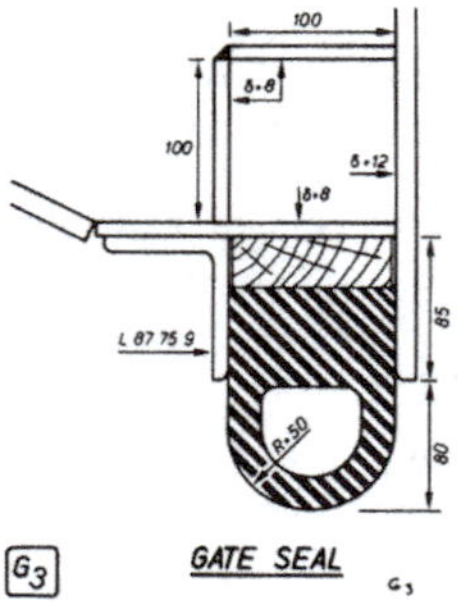

G3

Problem: At gate lift (midgate) of 0.8 - 1.1 m, local plate vibrations occur near section G. Wave radiation is observed locally.

Geometry: Sketches 1-3.

Conditions: Bottom -4.5 m, downstream level -0.5 to +2.0 m, upstream level +2 to +3 m. During operation, when ΔH is greater than about 1 m, vibrations occur. For flow pattern and geometry at critical opening see Sketch 4.

Structural dynamics: Local plate thickness δ = 12 mm in lower part of the gate. The beam in which the seal is clamped has I = 12×10^{-6} m^4. The seal beam (incl. added mass) between supports (3.7 m spacing) has an estimated resonance frequency of 10 - 15 c/s.

Vibrations: Exact data not available; frequencies about 12 - 20 c/s, local gate lift 0.45 - 0.55 m. Sometimes vibrations regular, sometimes "breathing". Mechanism: unstable flow separation. See also Paragraph 6 of this paper.

Cures: Normally, both gates are raised symmetrically; at critical gate openings one gate is somewhat lowered and the other raised. Sharper edge would be preferable.

References: See Paragraphs 6 and 7 of this paper.
Kolkman, P.A. "Flow-induced Gate Vibrations", Publication 164 of the Delft Hydraulics Laboratory, 1976.

CHART H

80 m SPAN VERTICAL LIFT GATE

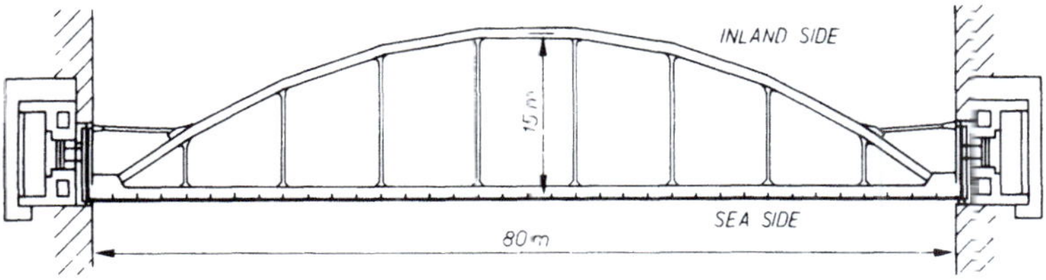

H1 TOP VIEW

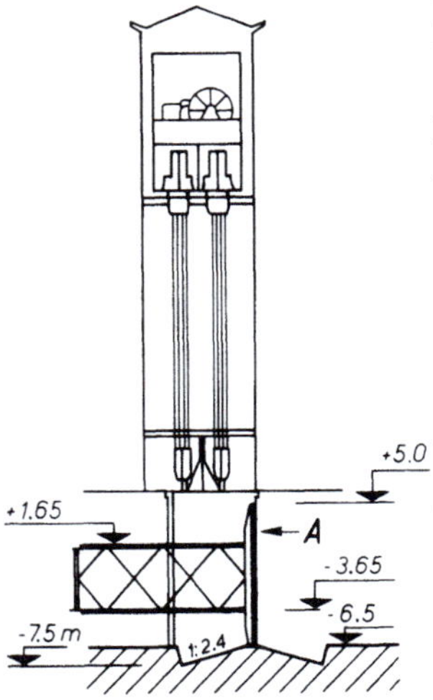

SECTION

(A: MEASURING POINT) H2

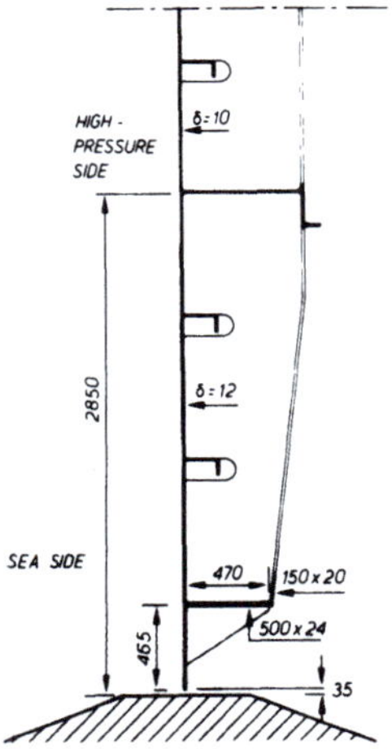

H3 GATE LIP

Problem: The vertical lift gate of the storm surge barrier at Krimpen closes at extremely high tide, but can also serve for regulation purposes (against salt penetration). At closed position, a gap of 35 mm is designed (real value not known) to prevent scraping due to gate deflection when water levels vary; to prevent vibrations, a sharp edge (δ = 12 mm) is provided. At closed position with said gap of a few cm, a buzzing sound was heard with a measured frequency (local plate vibrations above the water as indicated in Sketch H2) of 88 c/s, once even of 100 c/s, at flow in both directions.

Geometry: Sketches 1-3.

Conditions: During measurements, level at seaside was varying from -0.65 m to +1.2 m, at inland side from -0.4 m to +1 m; all sorts of level combinations were met. From ΔH = 0.8 m on, the high-frequency buzzing was heard and felt at currents in both directions.

Structural dynamics: Retaining plate thickness and edge are δ = 12 mm; see also Sketch H3. Support distance between vertical frames 9 m. Approximating calculations show that the 88 c/s frequency is probably due to bending of the lip below the lowest L-shaped stiffening beam.

Vibrations: Spectrum (Sketch H4) found in horizontal direction at point A (Sketch 2) at sea level +1 m, inland level -0.2 m. Probable mechanism: unstable flow separation and/ or reattachment at the 12 mm edge (see Par. 8 of this paper).

Cure: Lifting the gate a few cm.

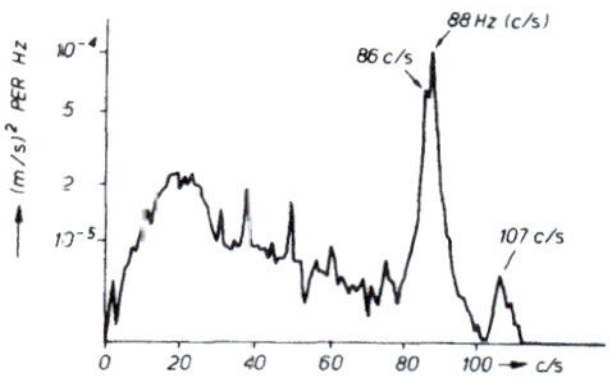

H4 "VIBRATION VELOCITY" SPECTRUM

CHART J

DOUBLE-LEAF (STONEY) GATE WITH SEAL BEAMS BETWEEN LEAVES

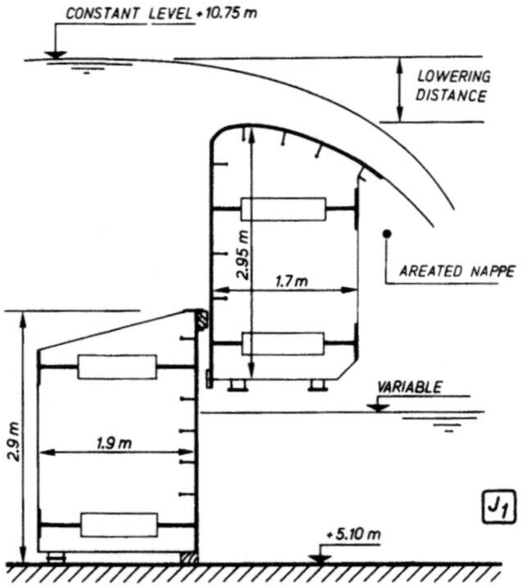

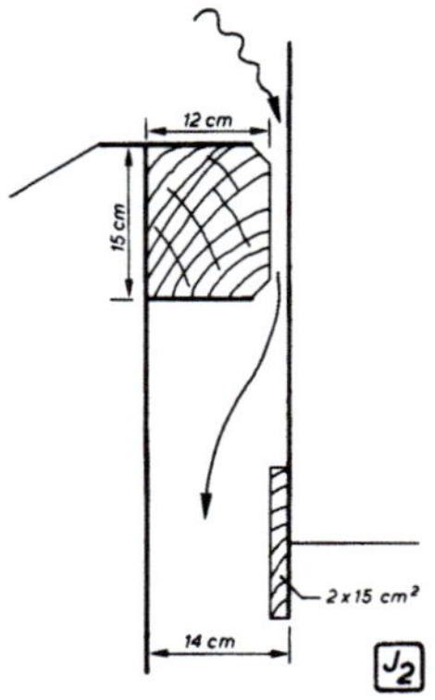

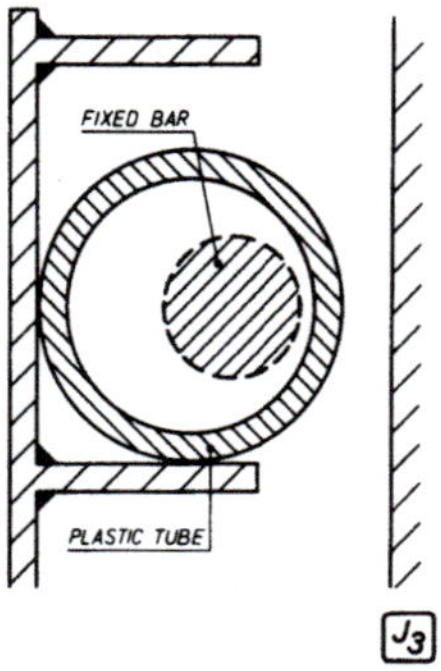

Problem: When the upper gate was lowered more than 0.6 m, combined horizontal and torsional vibration of both leaves started. Vibrations (like a heavy engine in a ship, about 3 c/s) were alarming. After a lowering of ΔH = 1.6 m nobody dared to go any further because of increased trembling.

Geometry: Sketches 1 and 2. Span between wheels of upper gate 17.6 m, and of lower gate 18.8 m. Width of flow section 17 m. Normally, the lower gate remains on the bottom, but at high river discharge both leaves are taken out.

Flow conditions: Sketch 1. Low turbulence upstream. Nappe aerated from both sides. Constant upstream water level.

Operating conditions: Vibrations observed when lowering of upper leaf from 0.6 to 1.6m; at latter value further lowering was forbidden. Design condition: variation between completely closed (raised) and completely open (lowered) position of the upper leaf.

Structural dynamics: Horizontal rigidity of each girder of lower gate I_{yy} = 0.03 m^4, and of upper gate 0.018 m^4 with an extra influence of the crest rigidity of 0.007 m^4. The torsional rigidity at the first harmonic vibration follows completely from these horizontal rigidities, but is not calculated. Weight of the upper leaf 22.5 tons and weight of the lower leaf 24 tons.

Vibration: Not measured, only felt with a stick; 3 c/s vibration; combined torsion and bending (in the first harmonic) of both leaves. (Single) amplitudes up to 2.5 mm at 1.6 m lowering. Hoist chain vibration amplitude a few cm.

Cures: Type of vibration follows from fluctuating gap width theory (Par. 5 of this paper). Vibration stopped when leakage flow was provisionally sealed with a plastic rain pipe. Final solution Sketch 3. Theoretical considerations and use of the instability indicator (see conclusions of Par. 4 of this paper) reveal that another solution is to exchange the two seal beams (and fix the thick one to the upper leaf).

Reference: See Paragraphs 4 and 5 of this paper.

CHART K VERTICAL-LIFT CULVERT GATE WITH RELEASE SYSTEM
(Hydraulic model 1:15; all dimensions etc. proto-value)

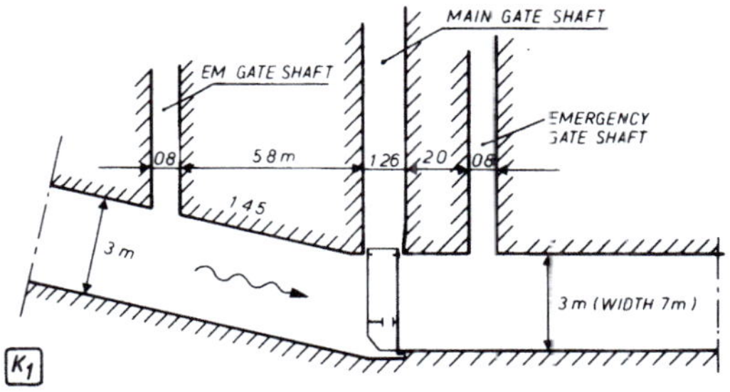

Problem: During model investigation it was found that, when the gate is closed but released from its seat, intense horizontal vibrations occurred; amplitudes were probably limited because the gate hit its seat.

Geometry: Sketches 1, 2 and 3. In the model, the wheels of the gate were replaced by lateral arms so that the gate could be supported by 4 vertical leaf springs to permit horizontal and rotational vibrations with low damping. Slot geometry was kept realistic; the leaf springs were submerged.

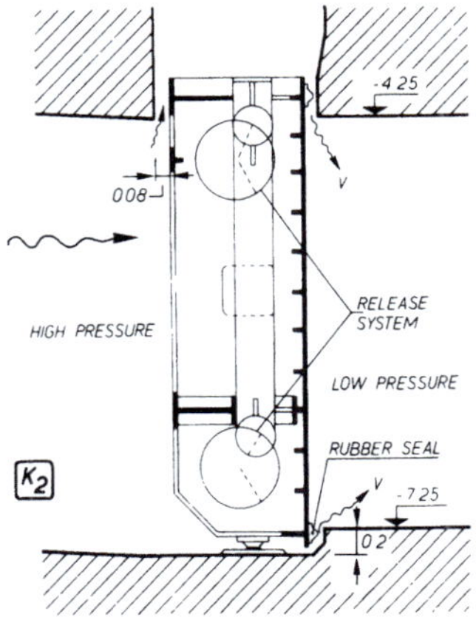

Flow conditions: Completely submerged; upstream level 9.3 m above culvert bottom.

Operating conditions: Front seal applied (with a release system) to guarantee absolute water tightness in closed position. Maximum head difference 5.5 m, which is the determining condition at the moment when the gate is just released and vibration commences.

Structural dynamics: Model gate designed as single oscillator. Horizontal rigidity 108×10^6 N/m; gate mass (incl. release construction, hoist bars, etc.) 18×10^3 kg. Horizontal bending vibration in main girders and wheel shaft is feared for in prototype; model was only schematically representative (Fig. 4.2 of this paper).

Vibration: When gate was released with a small opening: very intense horizontal vibrations. Vibrations started purely incremental, max. amplitude probably when gate touched its seat. Excitation mechanism: see bath-tub plug vibration in Paragraph 5 of this paper. Resonance frequency of horizontal gate vibration was in general 6 - 10 c/s, but at nearly closed position 2.5 c/s because part of the culvert volume then acted as added mass volume.

Cure: Release system should give minimum real gap width of 15 mm, taking into account reshaping of the squeezed rubber seal.

Reference: See Paragraphs 4 (Fig. 4.3) and 5 of this paper. For the total lock structure: Kolkman, P.A. and Slagter, J.C., "The Kreekrak Locks on the Scheldt-Rhine connection", Delft Hydraulics Laboratory, Publication 60 (1976).

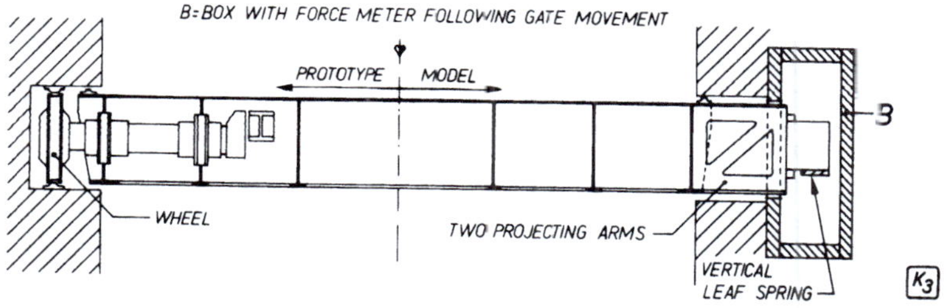

CHART L LIFT GATE IN DOWNSTREAM LOCK HEAD

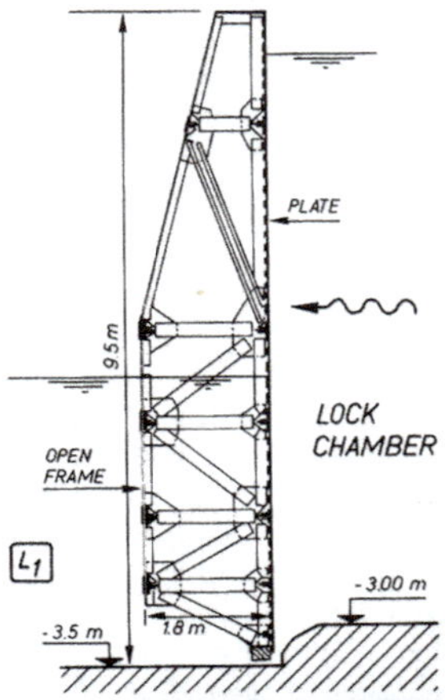

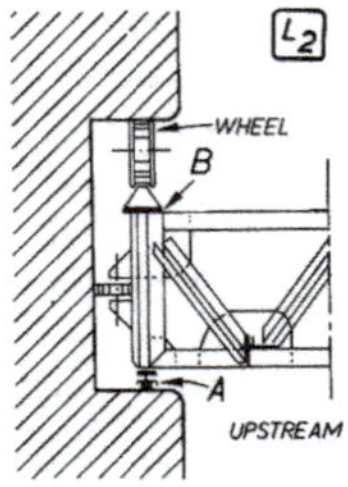

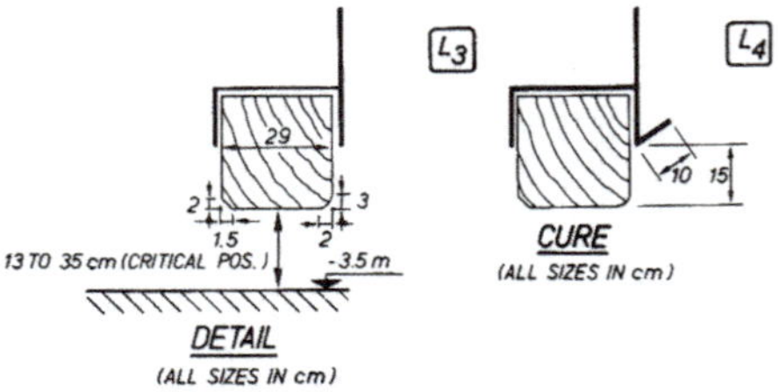

Problem: Lock at Lith; emptying by raising gate. Vibration after a revision: renewal of wooden seal beam and truing of main wheels (1 cm radius reduction). Although strip B (Sketch 2) was adjusted, seal beam A (normally compressed) started leaking because its rubber was worn off 2 cm; this probably reduced damping.

Geometry: Sketch 1. Gate span between walls 14 m. Retaining plate and side seal beams at upstream side. Wooden lower seal: Sketch 3. Details: Plates I and II.

Boundary conditions: Wheel gate; varying lock water level during emptying.

Operating conditions: Upstream level before gate raising: +4.5 m. Downstream level varying during tests between +1.05 and +1.4 m. Critical gate openings 0.13 - 0.35 m (most critical 0.25 m). When $\Delta H < 1.9$ m, vibration stopped.

Structural dynamics: Gate mass 90 tons. Total horizontal bending rigidity $EI = 250 \times 10^{12}$. Gate is cable-suspended ($\Sigma K = 1.3 \times 10^7$ N/m), but is also counterbalanced. Rigidity between gate and counterweight: $k = 1.2 \times 10^7$ N/m (both k-values estimated from cable data).

Vibrations: Starting vertically (wave crest parallel to gate plate), but then becoming intense as horizontal vibration (pattern of intersecting waves; f between 4.8 c/s and 5.2 c/s; only first harmonic vibr. involved. Max. (single) vertical ampl. 0.25 mm; horizontal mid-gate ampl. 5 mm. Vertical vibr.: fluctuating gap theory (Par. 5 of this paper). Horizontal vibr.: hypothesis of flow contraction fluctuation (Par. 7).

Cure: 1 m spaced, 0.4 m long pieces of steel angle bar (Sketch 4) to prevent: 1) complete synchronization of vortex street, and 2) a uniform critical gate position over the entire length. At the middle of gate, 2 pieces and at each side one piece were left out for construction reasons.

CHART M ELLIPTICAL AND RECTANGULAR PLATE EDGES

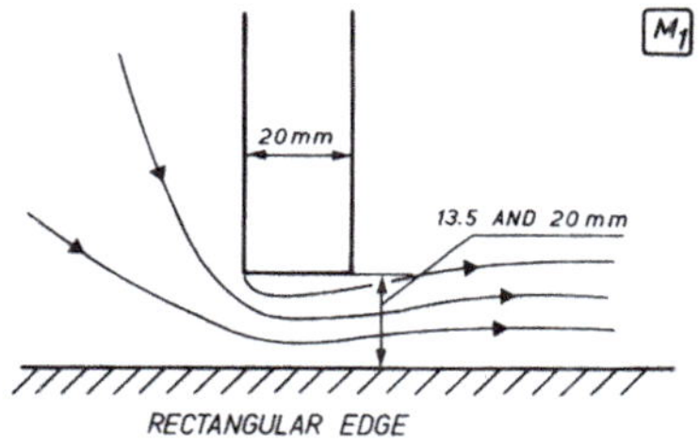

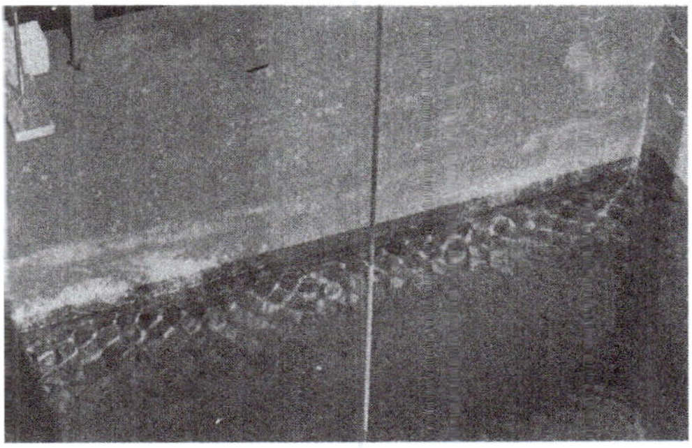

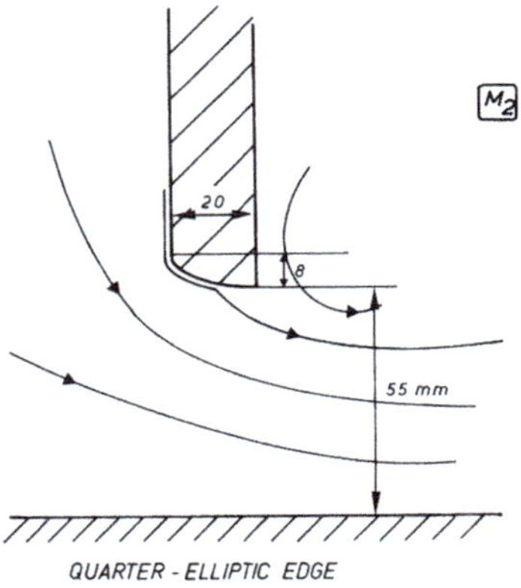

Problem: When pressure measurements at gate edge were performed in hydraulic model, strong bending vibrations of the plate occurred. This example is typical of high-frequency edge vibrations at gaps with thin-edged gates.

Geometry: Sketches 1 and 2. The 20 mm plate was about 1 m high and extended above the maximum water surface. Flume width 1 m.

Boundary conditions: Low-turbulence approach flow.

Flow conditions: Upstream depth 0.5 - 1 m, downstream depth 0.1 - 0.8 m, max. ΔH = 0.5 m.

Structural dynamics: Steel plate 2 cm thick, 1 m high, 1 m wide; secured by wedges which were fixed to the walls. Bottom edge (about 3 cm high) made of Trovidur.

Critical conditions: a) Rectangular edge: lifts of 13.5 and 20 mm were sensitive; 10, 30, 140 mm were not. b) Elliptical edge: only 55 mm lift was tested.

Rectangular edge: vibration started at $\Delta H > 0.1$ m. At the 13.5 mm lift the following combinations h_{up}/h_{down} were checked: 0.5 m/0.1 m, 0.5 m/0.3 m, 0.7 m/0.3 m, 0.9 m/0.3 m, 0.7 m/0.5 m, 0.9 m/0.5 m and 1.0 m/0.8 m. At all levels vibrations occurred, the strongest at 0.7 m/0.3 m. At the 20 mm lift the following tests were done: h_{up}/h_{down} = 0.5 m/0.3 m, 0.7 m/0.3 m, 0.9 m/0.3 m, 0.7 m/0.5 m, 0.9 m/0.5 m, 0.9 m/0.8 m and 1.0/0.8 m. At all levels vibration occurred, but at 0.9 m/0.3 m and 0.9 m/0.8 m they were only weak.

Elliptical edge: rather strong vibrations at every tested h_{up}/h_{down} combination: 1.0 m/0.6 m, 0.6 m/0.2 m, 0.8 m/0.2 m, 0.8 m/0.4 m and 1.0 m/0.4 m, all with a lift of 55 mm.

Frequencies: Not measured; wave length of the cross-wave pattern λ = 2 cm, which could indicate 11 c/s; but there was a rumbling noise of probably 20 c/s, which is the calculated first natural frequency of the sumberged plate.

C2 BUREAU OF RECLAMATION EXPERIENCE WITH FLOW-INDUCED VIBRATIONS

Henry T. Falvey
Hydraulic Research Engineer

U.S. Department of the Interior
Bureau of Reclamation

Denver, Colorado, USA

SUMMARY

The paper presents a summary of model and field investigations that have been conducted by the Bureau of Reclamation over the past 15 years. The studies are arranged according to the type of flow control mechanism that induces the vibration. In each case the type of remedial measures taken to reduce or eliminate the vibrations are discussed.

1. INTRODUCTION

The Bureau of Reclamation has been concerned with the problem of flow-induced vibrations for more than 40 years. The first research report dealing with vibrations was published in 1938 [1]. In this study, a model of a prototype slide gate was constructed to a scale of 1:30. Great care was exercised in maintaining similarity of mass and geometry in the model. However, it is interesting to note that the vibrations were evaluated by "lightly touching the gate with the fingers of the hand." It was reasoned that if no vibrations could be felt in the model, no serious problems from vibrations would develop in the field. Since that time, the analysis and measurement techniques have obviously improved greatly.

In 1964, Simmons [2] presented a summary of Bureau experiences with flow induced vibrations. The report cites examples in which vibrations created operational problems in several broad classes of hydraulic structures. These include radial gates, power and pumping plants, pipeline distribution systems, and spillways and spillway gates.

The purpose of this paper is to present a summary of the vibrational problems that have been encountered in the past 15 years. Rather than just citing a list of all the problems that have occurred, typical examples have been chosen which illustrate the basic control mechanisms of flow-induced structural vibrations. These flow mechanisms include:

- Fluid-dynamic excitation [3]
- Fluid-resonant excitation [3]
- Fluid-elastic excitation [3]
- Extraneous periodic or random excitation

2. FLUID-DYNAMIC EXCITATION

This type of vibration is characterized by two critical locations in the flow. One location is sensitive to disturbances and the other location is a site where new disturbances are self-generating. The new disturbances feed back through the fluid and affect the sensitive location. This type of mechanism can be observed with swirling flow in a tube, chart A. With sufficient angular momentum a reverse flow region forms. A single processing spiral-shaped vortex commonly forms at the boundary of the reverse flow region. The sensitive site is located where the vortex attaches to the upper flow boundary. The site of new disturbances is located at the stagnation point and the feedback path is formed by the reverse flow. When appropriate, these locations will be identified in the examples which are discussed below.

Three examples of the fluid-dynamic feedback have been studied by the Bureau. These are the draft tube surge, overflow nappe, and spillway-outlet junctions.

The draft tube surge has been found to be a function of the ratio of the angular to linear momentum of the flow entering the draft tube [4]. For a typical turbine, constant values of this ratio, called swirl, define the surging range of the unit, Chart A. All reaction turbines exhibit surging characteristics. The three most significant parameters affecting the swirl are the wicket gate geometry, the runner performance characteristics, and the draft tube geometry. Corrective measures to reduce the surging have included changing the runner characteristics; stabilizing the point of new disturbances by cones rising from the draft tube floor; eliminating the feedback path by filling the reverse flow region with air, water vapor, or solid cores; dissipating the angular momentum with fins on the draft tube walls; introducing counter rotating swirl [5]; and changing the draft tube shape.

The vibration problems of overflow nappes at small discharges are well known. Therefore, the Bureau investigates this type of vibration in every new hydraulic structure that contains an overflow nappe [6], chart B. For these vibrations, the sensitive site is located where the nappe leaves the structure. The site of the new disturbances is located where the nappe strikes the downstream pool. The disturbances are fed back through the air column beneath the nappe. Experiments have shown that the vibrations can be eliminated by disrupting the feedback path. One method of accomplishing this is to break the nappe into short segments through the use of spoilers.

The fall height of the nappe in the model was 560 mm. With this fall height, oscillations were observed with a discharge up to 1.23 (liters/s) per meter width of crest. Two modes of vibration were observed. The frequencies of these modes were approximately 3.4 and 6.6 Hz. The frequency was found to increase with discharge. The vibration was stopped by the addition of spoilers to the lip of the spillway. The spacing between the spoilers was 430 mm in the model. This solution has not been verified as water has never passed over the prototype spillway. A curious phenomenon was noted in the model. Quite accidently the model spillway surface was roughened by an overspray of paint droplets. This additional roughness apparently acted upon the location which is sensitive to disturbances and stopped the vibration.

Hydraulic model studies of spillway flow passing over a nonoperating outlet portal revealed an unsuspected flow vibration [7], chart C. The intersection between the crown of the outlet portal and the spillway face was the location of the sensitive site. When the pressure at this location is low, the spillway jet is deflected somewhat into the outlet portal. The point where it impacts is the source of the new disturbances. Water from the deflected spillway jet gradually fills the outlet conduit causing pressures at the sensitive site to increase. When the pressures at the sensitive site become sufficiently large, the spillway jet is deflected out of the outlet portal and the water in the outlet conduit quickly drains. This process decreases the pressure at the sensitive point and the cycle repeats. It may just be coincidental, but the largest pressure fluctuations in the structure coincided with the most heavily damaged areas in the prototype. The frequency of the oscillations was in the range of 1.5 to 2.5 cycles per minute (prototype). During the surging, heavy audible pulsations were noted in the model air vent. The solution of this problem was to eliminate the site of new disturbances by adding an eyebrow over the outlet. The effect of the eyebrows is to always keep the impact point of the spillway jet downstream of the outlet. The eyebrows also provided aeration to the sensitive area at all times thus preventing the formation of low pressures.

3. FLUID-RESONANCE EXCITATION

This type of vibration is identical with the fluid-dynamic feedback except that a resonant element is placed in the feedback path.

The Bureau experience with this type of vibration has been limited to open channel flow problems. One typical example is at bridge piers in canals.

The occurrence of waves traveling upstream of bridge piers in canals has been studied both in the field and in the laboratory [8]. In this case, the bridge pier consisted of five identical rectangular piles separated by short spacings, chart D. It has been shown that the cause of these waves is the alternate vortex shedding at the trailing end of the pier or pier elements [9]. If the shedding frequency matches the resonant frequency of the space between the row of piers and the canal bank, then large amplitude waves can be formed. In the case of rectangular cross sections, the resonant frequency can be predicted analytically. However, with trapezoidal sections, model studies must be used to predict the resonant conditions.

In the field, crest to trough amplitudes of about 600 mm were observed. The frequency of the waves was about 0.4 Hz for a mean flow velocity of 1.1 m/s and a flow depth of 4.8 m. The distance between wave crests was between 7.5 and 10.5 m. In addition to the waves, a beat frequency of about 0.07 Hz was observed in the wave trains.

The problem was cured by filling the spaces between the piers. The effect of this modification is to eliminate all but the most downstream source of disturbances at the pier.

4. FLUID-ELASTIC EXCITATION

This type of vibration is identical with the fluid-dynamic feedback except that the elastic properties of the structure influence the excitation. Two examples of this type of vibration are fixed cone valves and trashrack bars. Since the problem of trashrack vibrations has just begun in the Bureau, only the vibrations of hollow cone valves will be discussed.

A fixed cone valve is a flow-regulating device which consists of a cone supported by a series of vanes. Closure of the valve is accomplished by a sleeve which passes over both the vanes and the fixed cone. The vanes and the the fixed cone represent a spring-mass system within the flow which can react with fluctuations within the flow. This reaction causes new disturbances which then interact with the valve and the entire fluid-elastic system locks on to a specific frequency. Hydraulic model studies were made of two fixed cone valves discharging into a long circular tunnel [10], Chart E. In the model studies all geometric sclaes were similar, but no effort was made to scale the elasticity of the valve. Pressure traces taken in the valve chamber did not reveal any significant periodic disturbances.

Subsequent field tests indicated a very large pressure fluctuation occurred with a frequency of about 85 Hz, chart E. This frequency coincided very closely with that which can be predicted from the elastic vibration of the fixed cone in a torsional mode [11]. It is reasonable to expect that the vibration of the valve caused discharge fluctuations from the valve. These manifested themselves as pressure fluctuations in the valve chamber. The only solution to this problem would be a redesign of the valve to change its elastic properties. In this particular field installation severe vibrations of the operating mechanism were observed for valve openings exceeding 80 percent. Therefore, the valve travel was limited to this value.

5. EXTRANEOUS PERIODIC OR RANDOM EXCITATION

In this type of vibration, the affected structure is separate from the source of the excitation and neither contributes to nor attenuates the source. An example of periodic excitation is penstock resonance. Two examples of random excitation are center walls in hollow jet valve stilling basins and butterfly valves located downstream of a Y-branch.

The Fremont Canyon Powerplant in Wyoming consists of two units located at the end of a very long power conduit, chart F. Each unit could cause severe pressure fluctuations to occur in its respective penstock when operated at gate openings greater than 78 percent. This is normally a smooth operating range for a reaction turbine. However, field measurements eventually showed that the exit area of the turbines was about 14 percent too small for this particular installation. The original units exhibited draft tube surges with frequencies ranging between approximately 1.0 and 1.9 Hz.

When the draft tube surge frequency was between 1.4 and 1.9 Hz, the penstocks would resonate. The amplitude of the penstock pressure fluctuations was 16.9 m with a mean pressure head of 84.4 m.

A large number of remedies were tried to eliminate the resonance before it was discovered that the discharge area of the runner was too small. None of these remedies were successful. The cure for this problem was to trim the trailing edges fo the turbine blades, chart F. This essentially changed the runner characteristics so that it ran smoothly above 95 percent gate. Resonance still occurs for gate openings between 85 and 95 percent. However, this plant is operated in such a fashion that it can rapidly pass through this rough zone.

Hydraulic model studies were made of a hollow-jet valve stilling basin following extensive damage to the center wall [12], chart G. This basin contains two hollow-jet valves that discharge into a common structure. Since the valves may operate either singly or together, their respective flows were kept separate in the basin by a center wall. The annular flow from each valve is transformed into a solid jet before entering the stilling basin by wedges located on the walls just downstream of the valves. The wedges create separation zones on each side of the basin. Vertically oriented air-entraining vortices form within the separation zones. These vortices coupled with the horizontally oriented vortices in the hydraulic jump, which forms in the basin, create a very turbulent three-dimensional flow pattern. The spectrum of this turbulence can be characterized as being narrow-band low-frequency noise [13]. Fatigue cracks have been found at the bottom of the center walls of several basins. The natural frequency of the wall lies within the narrow-band turbulent pressure fluctuation spectrum. The solution at Navajo Dam was to eliminate both the center wall and the wedges. This solution eliminated the damage but now symmetrical operation of both valves is required.

The final example of extraneous control deals with the vibration of a butterfly valve located downstream from a Y-branch [14], chart H. The butterfly valve vibrated with a period of 2 to 3 seconds at a discharge of 165 m^3/s and a head of 37 m at the Y-branch. In addition, the valve tended to vibrate toward a closed position at these flow conditions. Laboratory studies showed that the source of the problem was nearly random excitation of the valve. The random excitations were created by separation of the flow in the Y-branch. The problem was eliminated by streamlining the Y-branch through the addition of filler blocks.

6. CONCLUSIONS

In all of these examples of flow-induced vibrations, it has been possible to identify the type of excitation which controlled the vibration. Identification of the excitation mechanism makes it possible to identify possible remedial actions that can be taken to eliminate or reduce the magnitude of the problem.

7. REFERENCES

[1] Streeter, V. L., Progress Report on Parker Dam Gate Tests, Vibration Studies, Parker Dam Project, Bureau of Reclamation, Hydraulics Branch Report HYD-38, 1938

[2] Simmons, W. P., Experiences of the Bureau of Reclamation with Flow-induced Vibrations in Hydraulic Structures, Bureau of Reclamation, Hydraulics Branch Report HYD-538, 1964

[3] Rockwell, D., and Naudascher, E., Review - Self-Sustaining Oscillations of Flow Past Cavities, Transactions of the American Society of Mechanical Engineers, Journal of Fluids Engineering, Vol. 100, pp. 152-165, 1978

[4] Falvey, H. T., and Cassidy, J. J., Frequency and Amplitude of Pressure Surges Generated by Swirling Flow, International Association for Hydraulic Research, Hydraulic Machinery Symposium, Stockholm, Sweden, 1970

[5] Seybert, T. A., Gearhart, W. S., and Falvey, H. T., Studies of a Method to Prevent Draft Tube Surge in Pump-Turbines, IAHR Hydraulic Machinery Symposium, Fort Collins, Colorado, 1978

[6] Isbester, T. J., Hydraulic Model Studies of the Pueblo Dam Spillway and Plunge Basin, Bureau of Reclamation, Hydraulic Branch Report REC-ERC-71-18, 1971

[7] Isbester, T. J., Hydraulic Model Studies of Folsom Spillway-Outlet Junction, Bureau of Reclamation, Hydraulics Branch Report REC-ERC-71-12, 1971

[8] Schuster, J. C., Canal Capacity Studies Wave Formation by Bridge Piers, Bureau of Reclamation, Hydraulics Branch Report HYD-485, 1967

[9] Toebes, G. H., Flow-induced Structural Vibrations, Journal of the Engineering Mechanics Division, American Society of Civil Engineers, Vol. 91, December 1965, pp. 39-66

[10] Beichley, G. L., Hydraulic Model Studies of Portage Mountain Development Low-level Outlet Works, British Columbia, Canada, Bureau of Reclamation, Hydraulics Branch Report HYD-562, 1966

[11] Mercer, A. G., Vane Failures of Hollow-Cone Valves, International Association for Hydraulic research, Hydraulic Machinery Symposium, Stockholm, Sweden, 1970

[12] King, D. L., Hydraulic Model Studies of the Modified Outlet Works Stilling Basin, Navajo Dam, Colorado River Storage Project, New Mexico, Bureau of Reclamation, Hydraulics Branch Report HYD-573, 1967

[13] Falvey, H. T., and King, D. L., Spectral Analysis, A Prototype Study, Paper presented at the Annual Hydraulics Division Conference, American Society of Civil Engineers, Logan, Utah, August 20-22, 1969

[14] Ball, J. W., Calibration of Hollow-jet Valves and Vibration Studies of Outlet Works Y-branch, Falcon Dam, Bureau of Reclamation, Hydraulics Branch Report HYD-474, 1960

CHART A DRAFT TUBE SURGE

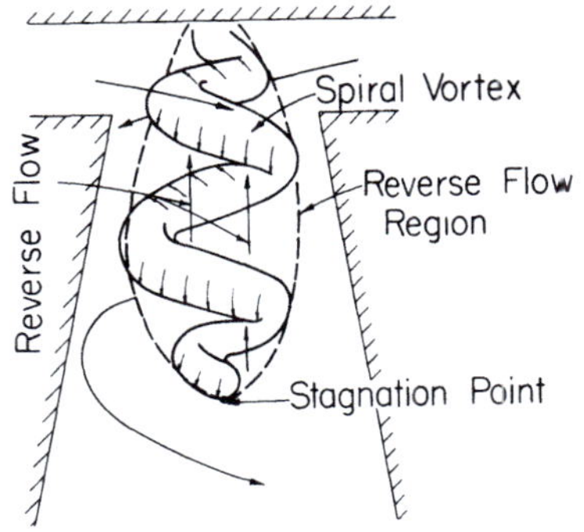

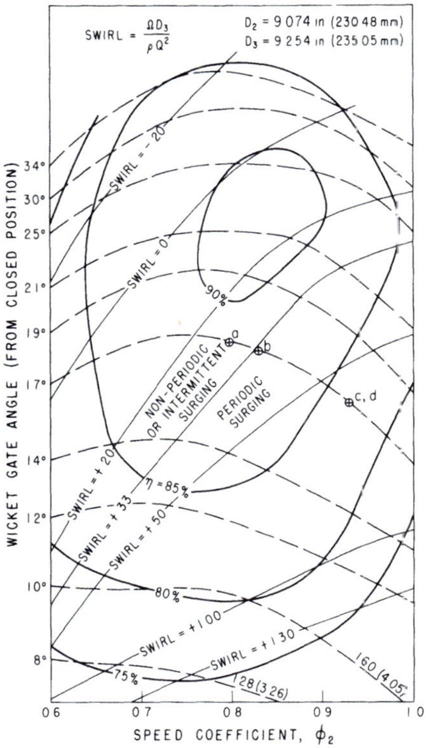

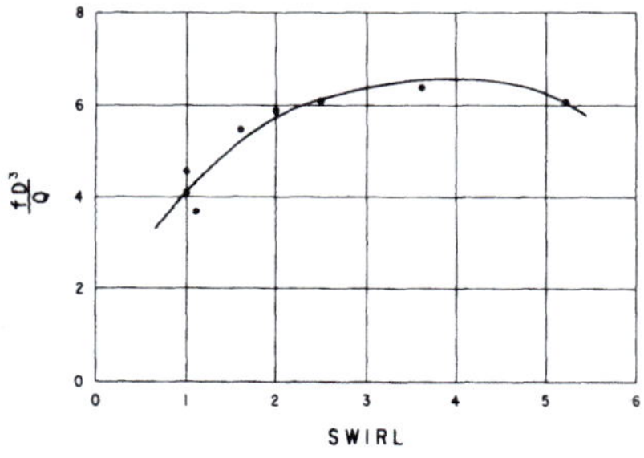

Problem: The draft tube surge can induce penstock resonance, power swings, and excessive motion of the turbine.

Geometry: An idealized sketch of the spiral vortex is shown in sketch.

Boundary conditions: In prototype installations the spiral vortex occurs in an elbow draft tube. The factors influencing its occurrence and magnitude with turbines are the wicket gate geometry, the turbine characteristics, and draft tube geometry.

Flow conditions: The spiral vortex occurs over a limited range of the ratio of angular momentum to linear momentum in the draft tube, (Swirl). See figure.

Operating conditions: For a given draft tube geometry the instability is a function of the momentum ratio.

Structural dynamics: None

Vibrations: Frequencies can be given as a function of the momentum ratio. See figure.

Cures: Alter wicket gate or draft tube geometry and change turbine characteristics by altering blade geometry. Introduce reverse swirl in draft tube by bleeding off some flow from the penstock and injecting it into the draft tube. Interrupt the feedback path by air injection.

References: Palde, U. J., Model and Prototype Turbine Draft Tube Surge Analysis by the Swirl Momentum Method, IAHR Symposium, Austria, 1974. Seybert, T. A., Gearhart, W. S., and H. T. Falvey, Studies of a Method to Prevent Draft Tube Surge in Pump-Turbines, IAHR Symposium, Fort Collins, Colorado, 1978

CHART B SPILLWAY NAPPE VIBRATION

Problem: Nappe vibration was observed in a hydraulic model. It was feared that a similar occurrence in the prototype would cause shaking of nearby structures.

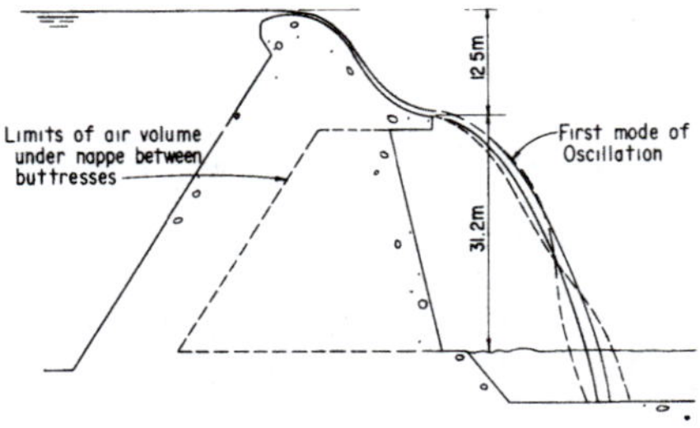

Geometry: See sketch.

Boundary conditions: The nappe is fully aerated on both sides. Approach flow was without disturbances.

Operating conditions: Fall height of nappe was 560 mm. Oscillations observed up to a discharge of 1.23 L/m·s.

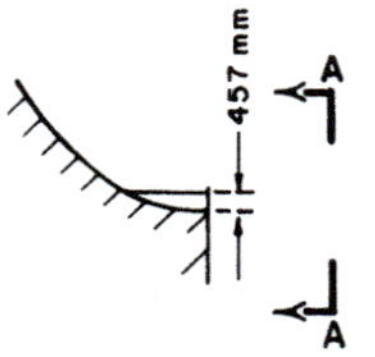

SECTION AT SPLITTER

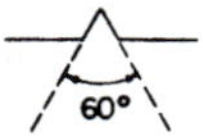

SECTION AA

Structural dynamics: The structural dynamics played no part in the oscillations.

Vibration: Two modes of vibration were observed. The frequencies of the modes were approximately 3,4 and 6.6 Hz. The frequency increased with discharge.

Cures: A very rough surface on the spillway crest stopped the vibration. Spoilers on the lip broke the nappe into 8 segments for flows less than 2.5 L/m·s. Spacing between model spoilers 430 mm. Prototype has not operated.

References: Isbester, T. J., Hydraulic Model Studies of the Pueblo Dam Spillway and Plunge Basin.

CHART C SPILLWAY-OUTLET JUNCTION

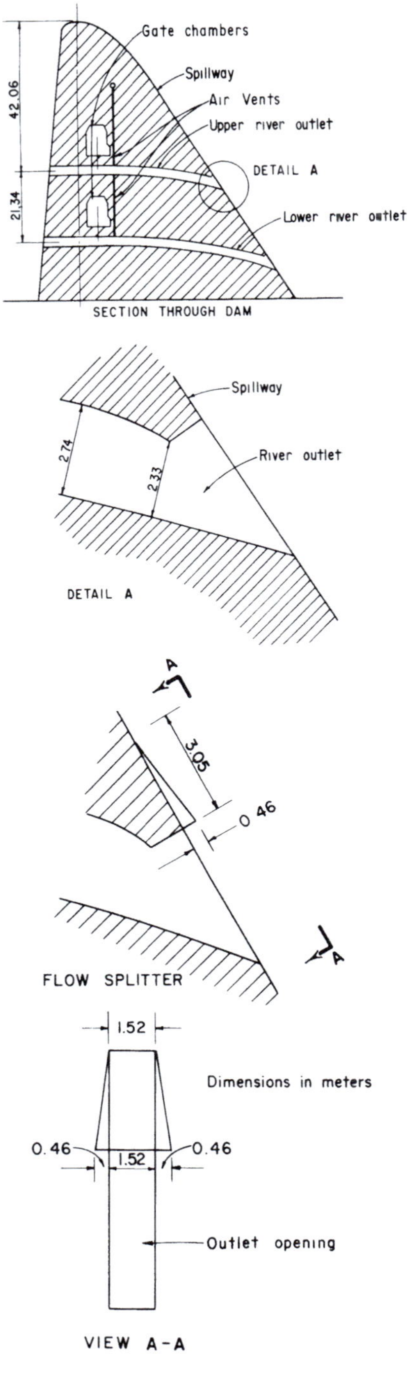

Problem: Cavitation damage in prototype. Heavy audible pulsations in model air intake.

Geometry: See sketch.

Boundary conditions: Air vent into outlet conduit at gates. No venting at intersection of outlet and spillway.

Flow conditions: Surging occurred for spillway flow only. Outlet gates closed. Outlets contained water only near intersection of outlet and spillway. A separation zone exists at crown of outlet tunnel.

Operating conditions: Surging occurred for unit spillway discharges of 16-, 33-, and 44 m^3/s per m width.

Structural dynamics: Not significant.

Vibrations: Frequencies of 0.025 to 0.042 Hz. The model pressure head fluctuations (scaled to prototype values) exceeded 25 m near the impact zone.

Cures: An eyebrow over the outlet works caused the jet to impinge downstream of the outlet invert and it kept the nappe at the crown of the outlet aerated.

References: Isbester, T. J., Hydraulic Model Studies of Folsom Spillway-Outlet Junction, Bureau of Reclamation, Hydraulics Branch Report REC-ERC-71-12, 1971.

CHART D WAVE FORMATION BY BRIDGE PIERS

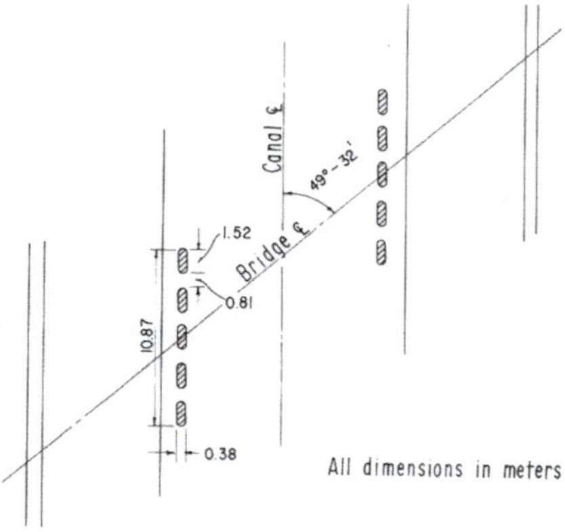

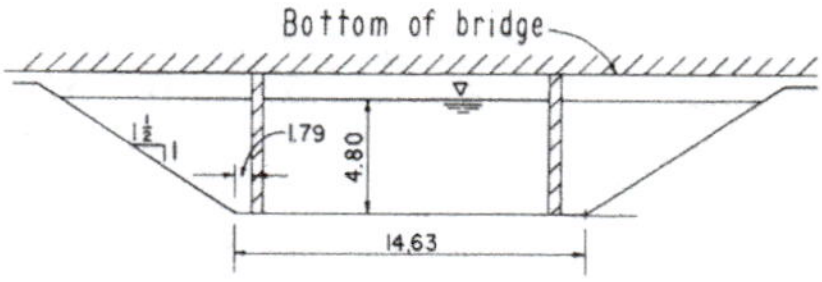

Wave at canal bank upstream of bridge

Waves progressing upstream of bridge for about 1 km

Problem: Overtopping a canal liner due to resonance induced waves at bridge piers. These waves travel upstream in a canal.

Geometry: See sketch.

Boundary conditions: Piers supported both ends.

Flow conditions: Mean flow velocity of 1.1 m/s with a depth of 4800 mm.

Operating conditions: Waves formed when flo rate in canal was about 96 m^3/s.

Structural dynamics: Movement of the piers does not contribute to generation of waves.

Vibrations: Trough to crest amplitude of 600 mm at a frequency of 0.38 Hz and a wavelength of 7.5 to 10.5 m. A beat frequency of 0.07 Hz was observed. Vortex shedding from piers not reported, but probably was present.

Cures: The pier geometry was altered by filling in the spaces between the piles.

References: Schuster, J. C., Canal Capacity Studies, Wave Formation by Bridge Piers, Bureau of Reclamation, Hydraulics Branch Report HYD-485, 1967.

CHART E FIXED-CONE VALVES

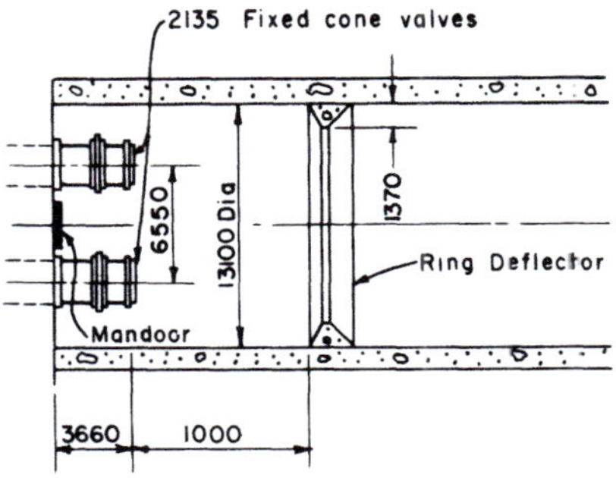

PLAN

Problem: Severe vibrations of the valves and their operating mechanism.

Geometry: See sketch.

Boundary conditions: Valves are aerated by a vent pipe from the downstream tunnel.

Operating conditions: Valves tested in the prototype through full range of positions.

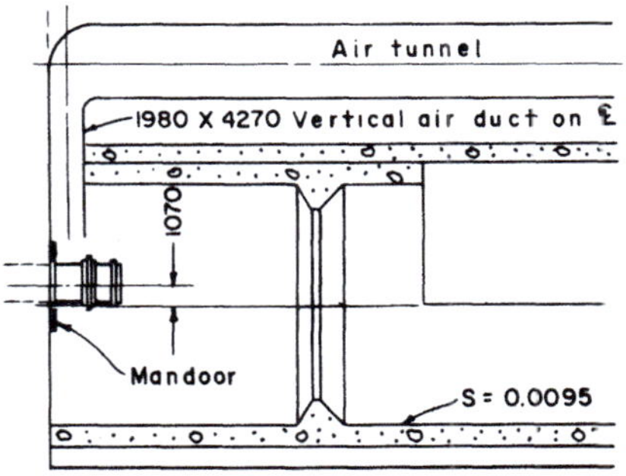

SECTION ON ℄

Structural dynamics: Unknown.

Vibration: Measurements made at steel man-door, (see sketch). The observed 85-Hz frequency correlates well with estimates of natural vibration based on paper by Mercer. This frequency not observed in model study. Elasticity of valve not simulated.

Cures: In prototype, valve opening restricted to 80 percent.

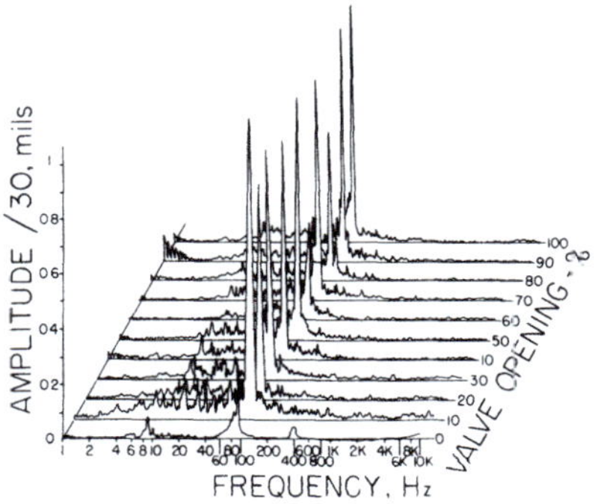

References: Mercer, A. G., "Vane Failures of Hollow-Cone Valves," IAHR Hydraulic Machinery Symposium, Sweden, 1970. Also, Falvey, H. T., Bureau of Reclamation, Denver, Colorado, U.S.A.

CHART F PENSTOCK RESONANCE

Problem: Severe pressure fluctuations in the penstock prevented turbine operation at more than 78 percent capacity.

Geometry: See sketch.

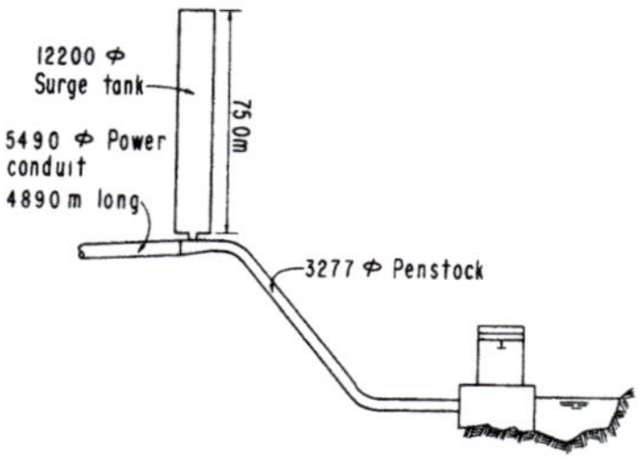

Boundary conditions: Surge tank essentially decouples inlet conduit from penstock. Two turbines whose penstocks branch immediately downstream of surge talk.

Flow conditions: The occurrence of the draft tube surge caused the penstock resonance.

Operating conditions: Each turbine is rated at 25 MW at a head of 88.4 m. The rotational speed is 257 r/min.

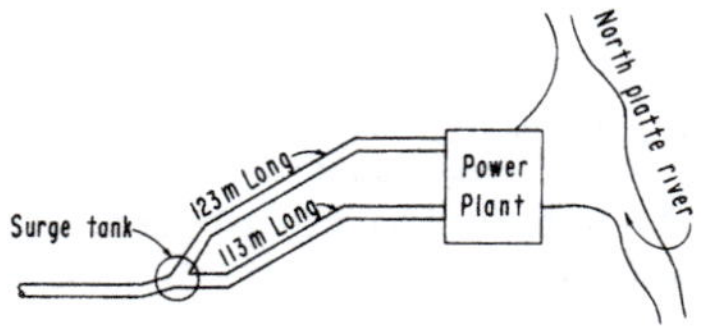

Structural dynamics: No significant structural considerations.

Vibrations: The resonance occurred for gate openings of 78 to 100 percent. The resonant frequency matched the draft tube surge frequency between 1.4 and 1.9 Hz. When the draft tube surge frequency dropped below about 1.4 Hz, the resonance dissappeared. At resonance, the amplitude of pressure head fluctuations in penstock was 16.9 m for a mean pressure head of 84.4 m.

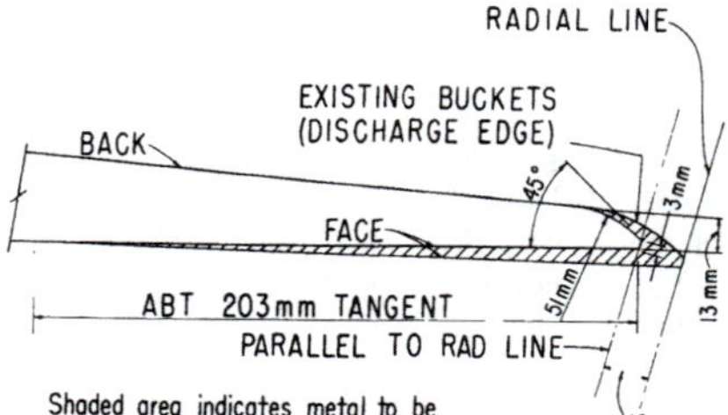

Shaded area indicates metal to be removed from existing buckets to increase vent openings approx. 12 %

Cures: Turbine characteristics changed by trimming 25 mm off trailing edge of each blade. Air admission and fins on draft tube throat were of small benefit.

References: For more information, contact Henry T. Falvey, Bureau of Reclamation, Denver, Colorado, U.S.A. No publications.

CHART G STILLING BASIN WALL VIBRATIONS

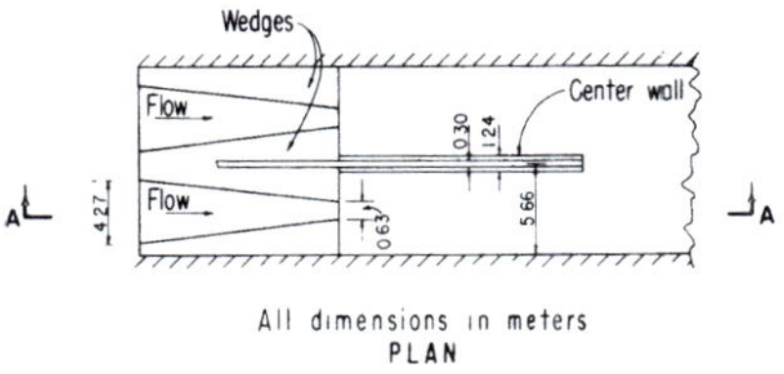

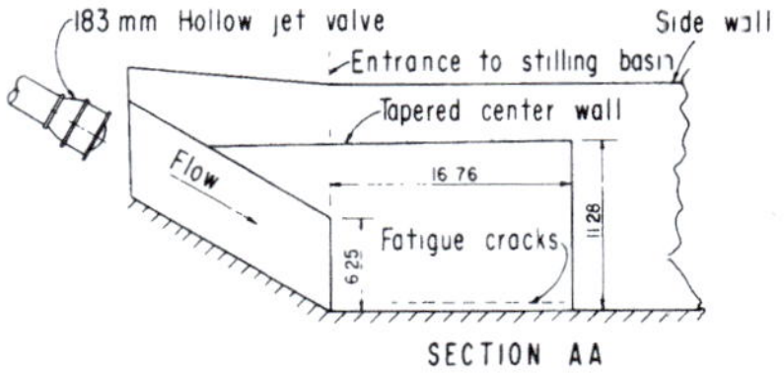

Problem: The center wall in a stilling basin had fatigue cracks at its base.

Geometry: See sketch.

Boundary conditions: Wall supported at base and one end.

Flow conditions: Separation zone and abrupt increase in section of entrance of basin. Head on valve is about 10 m. Highly turbulent flow and large surface boils in basin. Vortex action at wedges Air entrainment along center wall. No standing waves were observed in basin.

Operating conditions: Discharge of 50 m^3/s required to produce vibrations.

Structural dynamics: Natural frequency of wall matches frequency of pressure fluctuations. Movement of wall does not contribute to the flow induced pressure fluctuations.

Vibrations: Frequency and amplitude of pressure fluctuations a function of discharge and tailwater elevation. Frequency varied from 3.0 to 4.6 Hz.

Cures: Remove wedges causing abrupt section change of basin entrance. Remove wall.

References: King, D. L. Hydraulic Model Studies of the Modified Outlet Works Stilling Basin, Navajo Dam, Colorado River Storage Project, New Mexico, Bureau of Reclamation, Hydraulics Branch Report HYD-573, 1967.

CHART H Y-BRANCH AND DOWNSTREAM BUTTERFLY VALVE

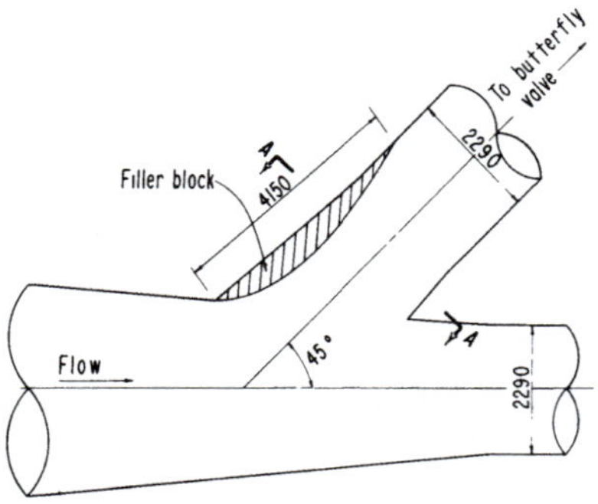

Problem: Vibration and noise in Y-branch and tencency for downstream butterfly valve to vibrate and move toward the closed position.

Geometry: See sketch.

Flow conditions: Approach flow to Y-branch was symmetrical. Cavitation sounds were reported in the field.

Operating conditions: 36.9-m head on Y-branch with a flow rate of 164.8 m^3/s.

Structural dynamics: Not known for butterfly valve.

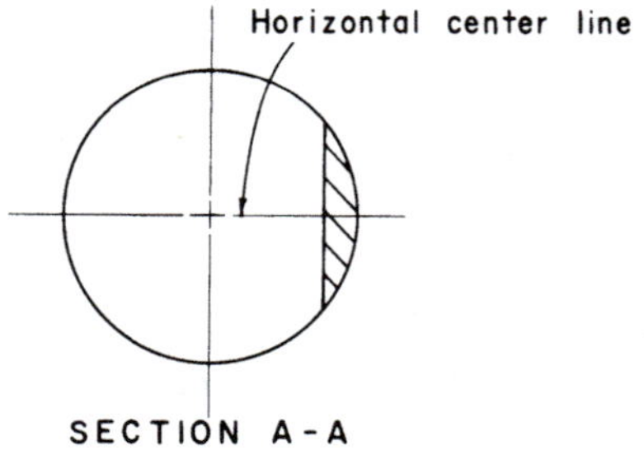

Vibrations: Period of butterfly valve motion was reported to be between 2 and 3 s. Oscillograph records of model studies show a predominant 0.4 s, low amplitude pressure head fluctuation and an intermittant, high amplitude fluctuation with 1- to 2-s periods. Spectral analysis not performed.

Cures: Streamlining the Y-branch eliminated the noise and vibration, see sketch.

References: Ball, J. W., Calibration of Hollow-Jet Valves and Vibration Studies of Outlet Works Y-Branch, Falcon, Dam, Bureau of Reclamation, Hydraulics Branch Report HYD-474, 1960.

C3 CORPS OF ENGINEERS EXPERIENCES WITH FLOW-INDUCED VIBRATIONS

Frank M. Neilson
Hydraulic Engineer

Ellis B. Pickett
Hydraulic Engineer

U. S. Army Engineer
Waterways Experiment Station
Vicksburg, Mississippi
United States of America

SUMMARY

Flow-induced vibrations of hydraulic structures, in some cases, have been the consequence of inadequate knowledge or oversight during the original design process; more commonly, however, an initially satisfactory design became unsatisfactory following an unforeseen change in operating conditions (increased head differential, for example) or following structural deterioration. Remedies, in some cases, involved detailed test programs that resulted in definite changes in design criteria; commonly, however, the problem was remedied by more rigorous maintenance requirements (and ad hoc modifications in the field to the structure) or by altering the operating conditions so as to avoid a particularly harmful flow condition.

1. INTRODUCTION

A survey of Corps of Engineers (CE) District and Division Offices concerning flow-induced vibrations indicated, first, that many circumstances involving flow-induced vibration have occurred in earlier designs and, second, that most of these problems would now be avoided by using current state-of-the-art practice. This paper is concerned with ordering and identifying the responses received from the field offices. No attempt is made to analyze any problem area in view of the wide diversity of the structures and flow conditions involved and the absence of detailed information in most cases.

2. EXAMPLES REPORTED

A summary of the responses is listed in Table 1. Multiple examples in particular problem categories were received (i.e., similar problems at separate projects); these are the common problems in the listing. The "status" denotation, Table 1, refers to the following four situations:

Status

1 Remedied by field maintenance or modification.

2 Remedied after detailed field and laboratory testing and analysis.

3 Existing problem but of only nominal concern; vibrations are tolerable under current operating conditions.

4 Existing problem area of major concern; field and laboratory testing and analysis are currently underway.

A status of 1 or 2 corresponds directly to problems that are considered solved and should not occur in new designs; a 3 would indicate additional studies probably should precede acceptance of designs that are similar to these projects; and a 4 indicates a current point of interest in Corps of Engineers design practice.

Vibration is normally first reported as a consequence of operator discomfort, or concern, due to either noise or movement. The most common flow-induced vibration problem, the J-seal, occurs as a result of poor adjustment, material deterioration, or fouling--and usually is easily corrected in the field. If, however, the vibration is not remedied, a major structure may suffer severe structural damage even for projects with low heads and low flow velocities. Despite the seemingly trivial nature of the seal problem from an academic point of view, it is emphasized here simply because it was, by far, the most common problem reported.

3. FIELD REMEDIES

Field remedies usually require that the exciting force be identified (not necessarily evaluated) and the structural members involved be located. Given this information, an adjustment to the structural members or an alternative design that does not vibrate, may be obvious--this is true of all problems in the status 1 category. Commonly, however, neither the exact nature of the exciting force or its point of application are known and field and laboratory testing are required. The field test may be used to locate the problem area and to place a level of severity on the vibration of the structure; a laboratory model test is then used to determine an effective and economical remedy; and a final field test may be made to evaluate the success of the remedy in the prototype situation. This general situation pertains to the resolved problems identified in the status 2 category as well as the yet unresolved problems in the status 4 category. Major study reports having to do with flow-induced vibration at CE projects are listed in the REFERENCES. Since time is an important consideration (the structure may have to be at least partially operable while the cure is being derived), each of the test programs may, in practice, be abbreviated as compared to a pure research study. When an acceptable, but limited, operational capability or procedure is vibration free, the search for more complete remedies may be delayed--these cases are status 3 items.

4. VIBRATION MEASUREMENT ANALYSES

The basic circumstances surrounding the solution of flow-induced vibration problems have changed substantially during the last decade. Improvements in measurement equipment (normally accelerometers, pressure transducers, and strain gages) and in recording equipment (normally producing oscillograph charts and analog magnetic tape recordings) have occurred. However, the major advance, particularly as far as prototype testing is concerned, is in the use of digital equipment. Measurements (as well as visual observations) in the prototype are usually not extensive enough and usually not in the most preferred locations. Digital time series analysis techniques are used to obtain far more information from limited data than was previously obtained. Modern high-speed digitizing, computing, and plotting equipment provides any combination of data playbacks, transforms (Fast Fourier Transforms, FFT's, for example), correlograms, filters, etc. to be rapidly evaluated. These analyses, when carefully applied and interpreted, provide substantially more insight into the vibration phenomena than the scaled time-history playback that previously was the main study method. Expertise on the part of the civil engineer in the area of analog electronic data reduction procedures is now only rarely needed. For example, the analog FFT's shown in Chart I, which were obtained in 1969, required substantial specialized electronic equipment and are of questionable reproducibility and accuracy; this information would now be evaluated (and checked) by digital means at substantially less cost.

5. OPERATIONAL MONITORING OF VIBRATION

The vibration phenomena are not always regarded in a negative sense. It is not uncommon for vibration measurements to be used as a detector in regard to other structural or flow conditions; two recent examples of using flow-induced vibration as a detector are as follow.

5.1 Structural Stability

During flood flows a major diversion structure (Neilson and Tool [8]) was undermined due to soil erosion. A vibration monitor program was initiated at the project in order to obtain a continuous measure of the dynamic stability of the structure. As the erosion continued down to the support piling, the dominant frequency of structural vibration decreased from about 12 Hz to about 2.5 Hz indicating severe weakening of the structure; subsequently as the scour hole was filled with stone the dominant frequency increased to about 11 Hz. Accelerometers were used as measuring devices; oscillograms were made and inspected several times each day; and magnetic tape records were analyzed in detail.

5.2 Setting Maximum Valve Openings

As a hollow cone valve is opened, the flow control may shift from the sleeve, point A in Figure 1, to the valve body, point B. A limit is usually placed on the sleeve travel so that this shift does not occur. The limit is established by continuously moving the sleeve (opening the valve) until an alternating control (corresponding to intense vibration) is observed. The sleeve travel is then set such that the valve openings do not extend into the range corresponding to intense vibration. This adjustment is made in the field; accelerometers or strain gages are used to sense the valve vibration.

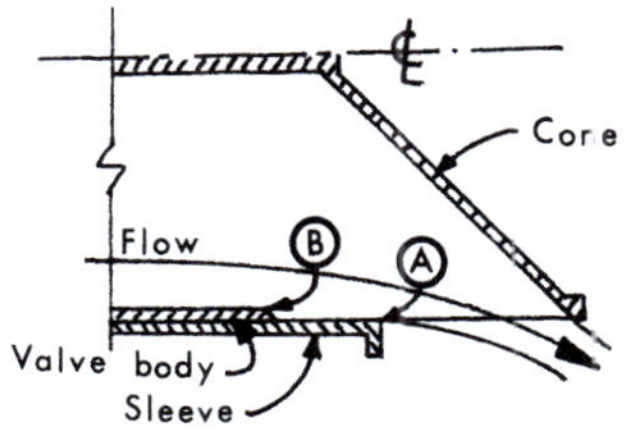

FIG. 1 FLOW-CONTROL POINTS IN HOLLOW CONE VALVE (see Ref. 6 for details on the operation of this type of valve.)

6. PROBLEM AREAS

Numerous examples of flow-induced vibration occur in hydraulic structures, many of which can be avoided using current design guidance and careful maintenance procedures. As far as current research is concerned, all items listed as status 4 are of practical current interest to the Corps of Engineers. Of course, any new design or operating condition could warrant testing and analysis with regard to applied vibration studies; and many aspects of fluid flow involving turbulence, vortex shedding, etc. in and around complex boundaries as well as related analytical techniques, remain to be studied as basic research in hydrodynamics facilities.

TABLE 1. FLOW-INDUCED VIBRATION - REPORTED EXPERIENCES

Chart	Device	Occurrences	Status*
A	Submergible tainter gate	1	1
B	Vertical lift gate	1	2
C	Hinged leaf gate	1	3
D	Tainter gate	Common	2
E	Vertical slide gate	Several	1
F	Trash rack support vanes	1	1
G	Trash-rack bars	Several	1
H	Tainter valve	1	3
I	Hollow-cone (Howell-Bunger) valve	2	4
J	Reverse tainter valve	Common	4
K	Tandem gate valves	1	3
**	Seal vibration	Common	1

* See text for description of status ratings.

** Charts regarding seal vibration and turbine blade cracking are not included; however, these problems have occurred at Corps of Engineers projects. In fact, seal vibration was the most prevalent example of flow-induced vibration received, and blade cracking is a topic of current concern (Status 4).

7. REFERENCES (U.S. Army Engineer Waterways Experiment Station Reports)

7.1 Prototype

1. Guyton, Benson, "Pressure and Air Demand Tests in Flood-Control Conduit, Denison Dam, Red River, Oklahoma & Texas," MP 2-31, USAEWES, (1953).
2. Guyton, Benson, "Vibration, Pressure and Air-Demand Tests in Flood-Control Sluice, Pine Flat Dam, Kings River, CA," MP 2-75, USAEWES (1954).
3. Murphy, T. E., "Model and Prototype Observations of Gate Oscillations," MP 2-563, USAEWES (1963).
4. Yates, R. A., "Frequency Spectrum Analysis of the Old River Control Structure Vibration," MP 2-774, USAEWES (1965).
5. Guyton, Benson and Pickett, E. B., "Vibration and Pressure-Cell Tests, Flood-Control Intake Gates, Ft. Randall Dam, Missouri River, South Dakota," TR 2-435, USAEWES (1956).
6. Neilson, F. M., "Howell-Bunger Valve Vibration, Summersville Dam Prototype Tests," TR H-71-6, USAEWES (1971).
7. Neilson, F. M., "Barkley Lock Prototype Tests, Cumberland River, Kentucky," TR H-75-11, USAEWES (1975).
8. Neilson, F. M. and Tool, A. R., "Prototype Tests, Old River Low-Sill Control Structure, April 1973-June 1975," TR H-76-15, USAEWES (1976).
9. Hart, E. D. and Tool, A. R., "Sluice Pressures, Gate Vibrations, and Stilling Basin Wall Pressures, Libby Dam, Kootenai River, Montana," TR H-76-17, USAEWES (1976).
10. Pugh, C. A., "Spillway Vibration, Pressure, and Velocity Measurements, Ozark Lock & Dam, Arkansas River, Arkansas," TR H-77-6, USAEWES (1977).
11. Hart, E. D.and Hite, J. E., "Prototype Gate Vibration Tests, Barkely Dam, Cumberland River, Kentucky," TR HL-79-8, USAEWES (1979).

7.2 Model

12. Pickering, G. A., "Spillway Gate Vibrations on Arkansas River Dams, Arkansas and Oklahoma," TR H-71-5, USAEWES (1971).

7.3 Other

13. Campbell, F. B., "Vibration Problems in Hydraulic Structures," MP 2-414, USAEWES (1960).
14. Locher, F. A., "Some Aspects of Flow-Induced Vibrations of Hydraulic Control Gates," CR H-69-1, USAEWES (1969).
15. Campbell, F. B., "Report on Spillway Vibration Studies," CR Unnumbered, USAEWES (1969).
16. Locher, F. A. "Some Characteristics of Pressure Fluctuations on Low-Ogee Crest Spillways Relevant to Flow-Induced Structural Vibrations," CR H-71-1, USAEWES (1971).

CHART A: SUBMERGIBLE TAINTER GATE

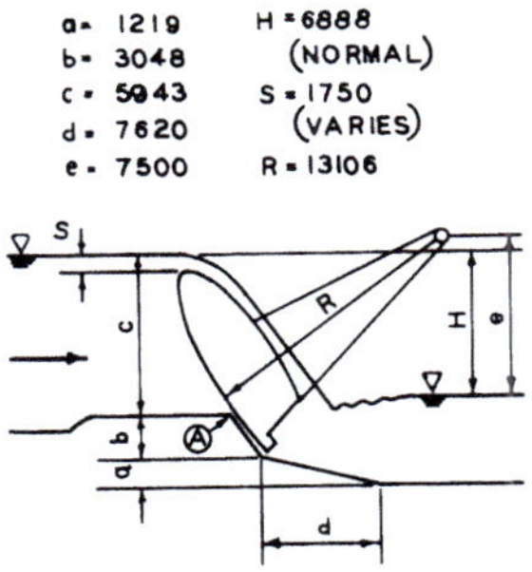

① SUBMERGIBLE TAINTER GATE (SECTION)

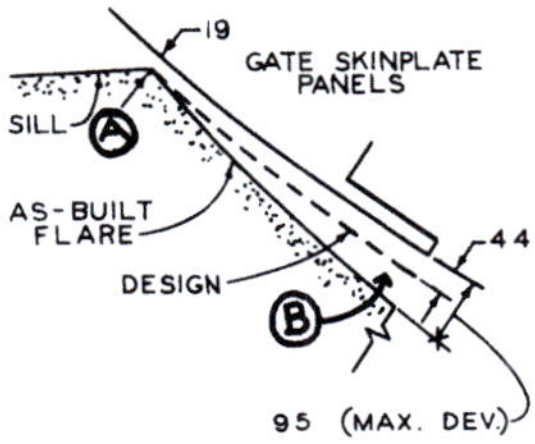

② DETAIL OF SILL DIMENSIONS IN mm

Problem: Serious vibration of gate panels; damage due to fatigue a definite possibility.

Geometry: See sketch 1 (gate) and sketch 2 (sill). The width of the gate is 33,500 mm. Note larger as-built than as-designed divergence between gate and sill below location A, sketch 2.

Flow conditions: See sketch 1. Gate vibrates during overflow over a broad range of submergence, s, values. No vibration during underflow (i.e., when control is by raising the gate).

Operating conditions: Most severe vibration occurred at a submergence of 1700 mm. Minor evidence of possible cavitation at location A in sketch 1.

Structural dynamics: Fundamental-mode natural frequency of individual 19 x 870 x 2640 mm panels of the gate skinplate expected to be intermediate between about 68 Hz and 144 Hz (unsubmerged). The overall unsubmerged gate weight is about 2028 KN.

Vibration: Panel frequency 100 Hz with 1700 mm submergence; predominant frequencies range from 30 to 110 Hz with displacements of 0.005 to 1.5 mm depending on location, submergence, etc.

Cures: (a) Partial. Build up the sill shape with epoxy to maintain design flare between sill and gate; i.e., fill region B in sketch 2. (b) Total. Do not operate gate in lowered position.

References: None on this vibration problem; submergible tainter gates: (1) U. S. Army Engineer Waterways Experiment Station, CE, Spillway for Markland Locks and Dam, Ohio River, Kentucky and Indiana, TR 2-566, Vicksburg, MS, May 1961. (2) U. S. Army Engineer Waterways Experiment Station, CE, Spillway for New Cumberland Locks and Dam (Final Design), Ohio River, West Virginia, TR 2-585, Vicksburg, MS, Nov 1961.

CHART B: VERTICAL LIFT GATE

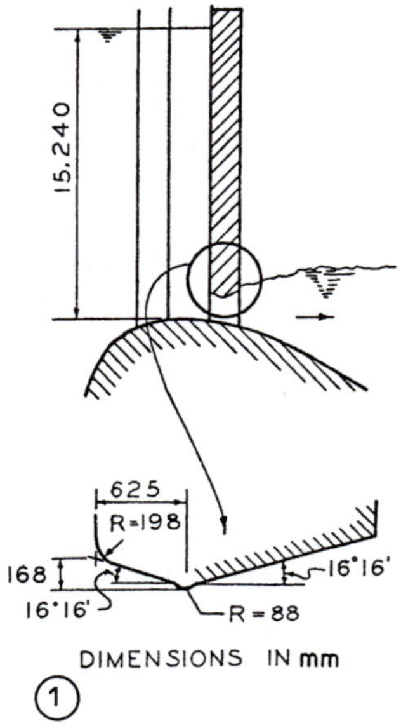

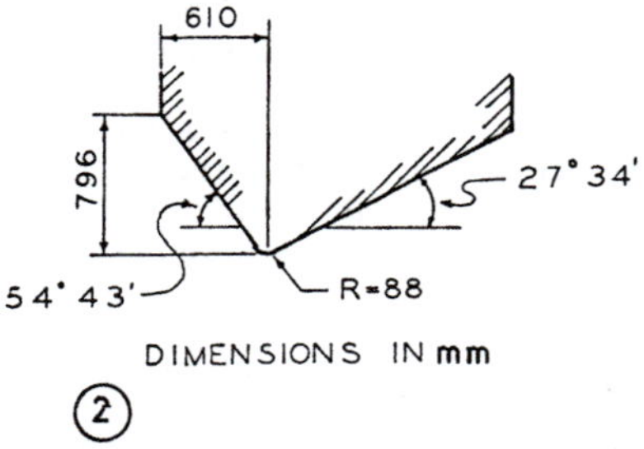

Problem: Vertical movement or bouncing of gate when bottom just submerged by tailwater. Suspension cables separated from sheaves making pulley block inoperative.

Geometry: See sketch 1.

Boundary conditions: Gate suspended from cables.

Flow conditions: Head from 6,400 to 13,700 mm. Tailwater alternately submerged and exposed bottom of gate. Flow control alternated from seal near center of bottom to upstream edge of bottom.

Operating conditions: Gate openings of 600 to 2400 mm.

Structural dynamics: Spring constant of cables 1.3×10^7 kg/m (4.5×10^6 kg/m in model).

Vibrations: Random.

Cures: Bottom reshaped to create positive flow control (with 1:25-scale model). See sketch 2.

References: Edmister, T.D., and P. M. Smith, "Spillway Gate Modification, Bonneville Dam, Columbia River, Oregon and Washington," Hydraulic Model Investigation, Division Hydraulic Laboratory, U. S. Army Engineer Division, North Pacific, CE, Bonneville, Oregon (May 1975).

CHART C: HINGED-LEAF GATE (NAPPE OSCILLATIONS)

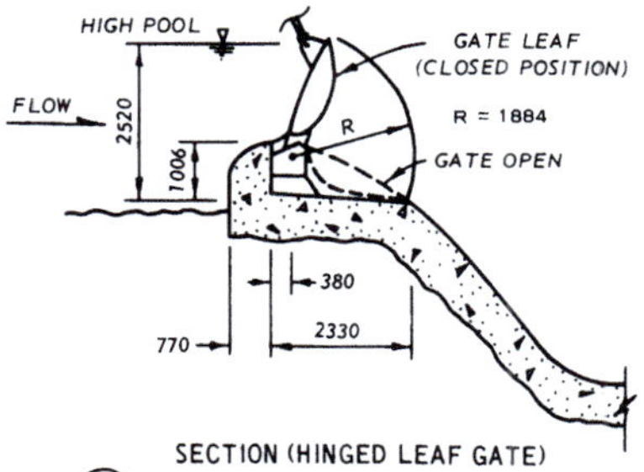

SECTION (HINGED LEAF GATE)

(1)

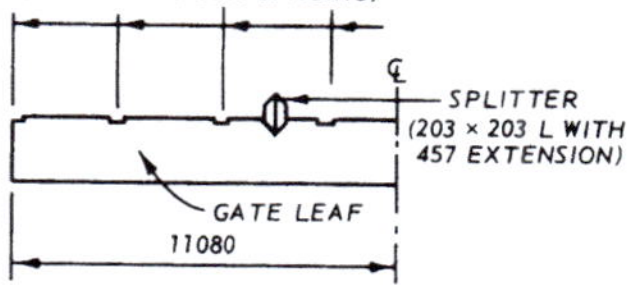

UNSUCCESSFUL MODIFICATIONS
(LOOKING DOWNSTREAM)

DIMENSIONS IN MM

(2)

Problem: Flow induced vibrations were observed in a hydraulic modified hinged-leaf gate during the first year of operation (1976).

Geometry: See sketch 1. The total width of the gate is 21900 mm. No air ducts were provided.

Boundary conditions: Mechanical opening devices (hydraulic cylinders) with pinned connections at the gate are used for gate operation; flow conditions are as shown in sketch 1.

Flow conditions: Nappe springs free from the gate. Tranquil flow without excessive turbulence or flow disturbances exists upstream. The gate is not submerged on the downstream side.

Operating conditions: The original design called for flow splitters at 3048 mm spacing; these were changed to notches as shown in sketch 2 in order to avoid debris and icing problems. The notches are entirely ineffective for preventing self-excited nappe oscillation and the resulting vibration. The nappe has a relatively small fall height (2520 mm) and variable thickness. Vibrations were observed at overflow heights of 146, 183, and 207 mm.

Structural dynamics: Unknown.

Cures: The problem has not been resolved. Two flow splitters (see sketch 2) were added to the gate; the splitters extend through the nappe but do not reduce the vibration.

References: None on this project; nappe oscillations have been discussed extensively in the literature. Schwartz, H. I., "Nappe Oscillation," Journal of the Hydraulics Division, ASCE, Vol 90, No. HY6, proc. paper 4138, Nov 1964, pp 129-143.

CHART D: TAINTER GATES

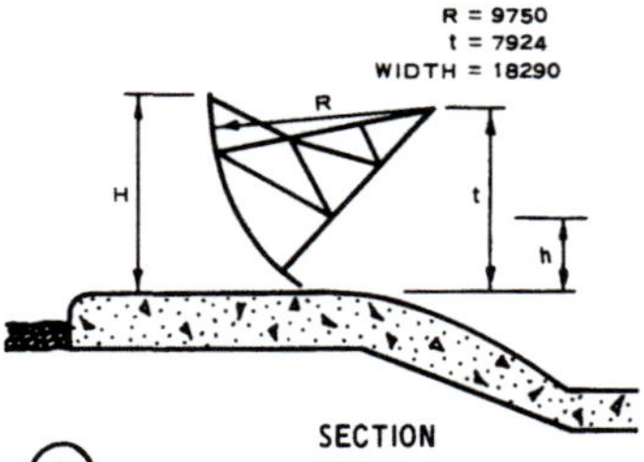

(1)

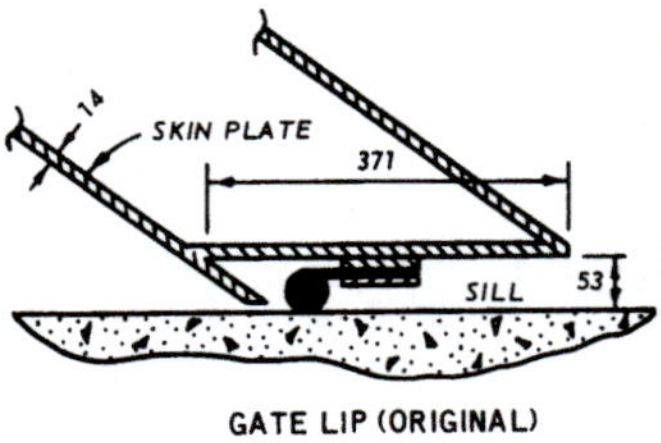

(2)

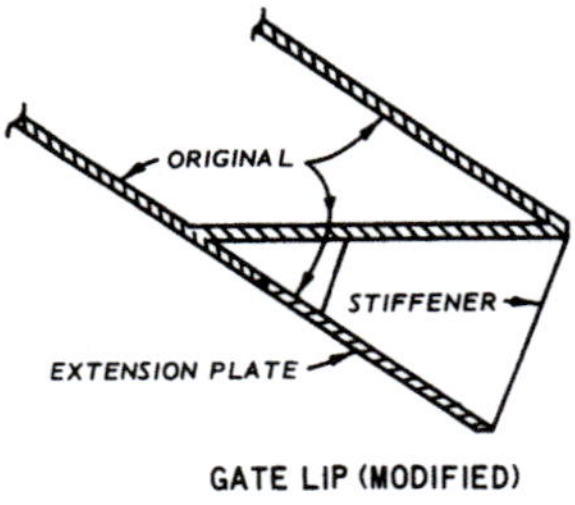

(3)

Note; This is one of several vibration-free designs; see references 1 and 2 as well as the Symposium presentation (K. Petrikat, Chart 6) concerning the details of design of gate seals.

Problem: These tainter gates are used to maintain navigation pools and to regulate discharges. Field investigations of the gates at six projects revealed cracks in structural members and in welded connections; these were determined to be fatigue failures resulting from severe vibration.

Geometry: The basic geometrical arrangement is shown in sketch 1, details of one original gate lip in sketch 2, and details of one modification in sketch 3. The gates are commonly 15,200 mm or 18,240 mm wide. A bottom plate is used (not shown in sketch 2) when leakage is not permissible.

Boundary conditions: The gates are suspended from cables. The problem was studied in the field and by means of hydraulic model tests.

Operating conditions: The vibrations typically occur at small gate openings (240-900 mm). A model having the prototype dimensions shown in sketches 1 and 2 showed the most severe vibration occurred at the following gate openings:

H(mm)	h(mm)	Gate Opening (mm)
7300	5470	180
7300	4860	210
7300	4260	240
7300	3650	490

Structural dynamics: Unknown.

Vibrations: Operating personnel could hear the fluttering of structural members as well as feel the vibration of the spillway walkway and see a ripple pattern upstream from the gates. The model reproduced the prototype with regard to size, shape, and weight, but did not reproduce the gate elastic properties; the vibration in the model (2) was measured by means of a load cell in the support wire and could be observed visually.

Cures: Modifications, such as shown in sketch 3, to the gate lips were undertaken and were succesful in eliminating the vibrations. Basically the revised shapes contain a sharp and clean flow breakoff point and, whenever leakage can be tolerated, do not use a rubber seal as shown in sketch 3.

References: 1. Schmidgall, Tasso, "Spillway Gate Vibrations on Arkansas River Dams," Journal of the Hydraulics Division, ASCE, Vol. 98, No. HY1, Jan 1972, page 219. 2. Pickering, G. A., "Spillway Gate Vibrations on Arkansas River Dams, Arkansas and Oklahoma," USAE Waterways Experiment Station, T.R. H-71-5, Jun 1971, Vicksburg, MS, USA.

CHART E: VERTICAL SLIDE GATE

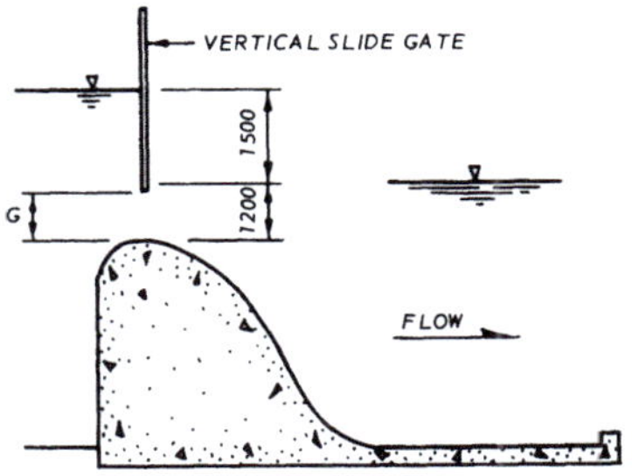

SECTION: GATE PARTIALLY OPEN

(1)

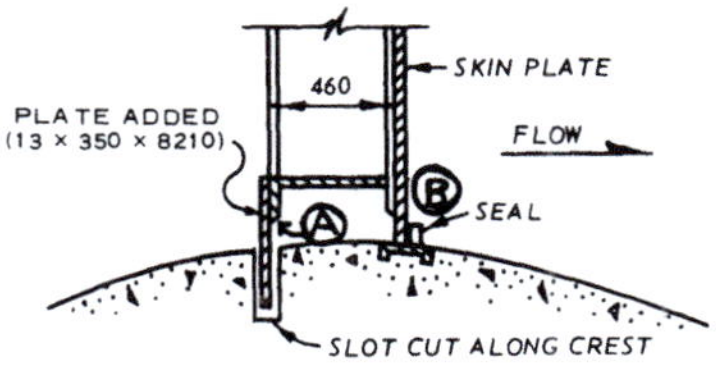

MODIFIED LIP IN CLOSED POSITION

(2)

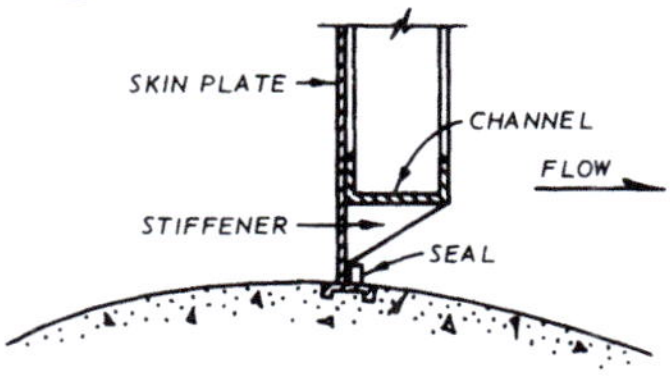

REVISED LIP DESIGN

(3)

Problem: Shifting of the control (separation) point appeared to cause the vibration of the gates. Structural damage could have been caused by fatigue if the vibration was allowed to continue.

Geometry: See sketch 1; the structure has two 8210 mm wide vertical slide gates. The "plate added" in sketch 2 was not present in the original design. The location of the seal (not to scale) is shown in sketches 2 and 3; the seal and keeper plate are bolted to the downstream face of the gate skin plate.

Boundary conditions: The weir conforms to WES standard shape (1960). With the original design, the upper nappe was in contact with either the upstream flange of the original cross beam (A in sketch 2) or the gate seal (B in sketch 2) when the gates were partially open.

Flow conditions: The approach flow is tranquil; local disturbances caused by the vibrating gate were observed.

Operating conditions: The structure is operated to maintain a pool about 1500 mm above tailwater. The gates vibrate at partial openings, G, up to about 900 mm. The maximum amplitude of vibration occurs at gate openings of about 450 mm.

Vibration: No measurements were made; it appeared that vibration frequency and amplitude increased with head.

Cures: A plate was welded to the upstream flange of the cross beam for the entire width of the gate (see sketch 2). A slot was cut into the top of the weir to accept the plate when the gate is closed. (This provided for continued use of the original seal.) With this modification vibration was eliminated; control is the lower edge of the added plate.

The gate lip configuration shown in sketch 3 is less likely to cause gate vibration. For this design, the channel beam flanges are turned upward. The skin plate is mounted on the upstream side of the flanges and extends below the channel beam not less than one-half the width of the beam. This design has been used for more recent projects with no reported vibration problems.

References: None.

CHART F: TRASHRACK (REAR SUPPORT VANES)

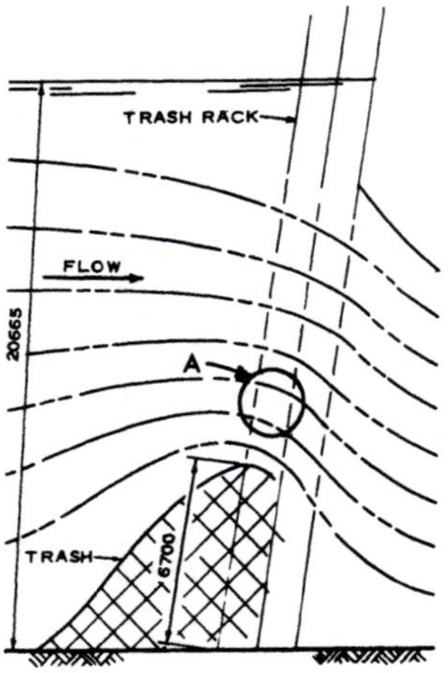

(1) FLOW STREAMLINES

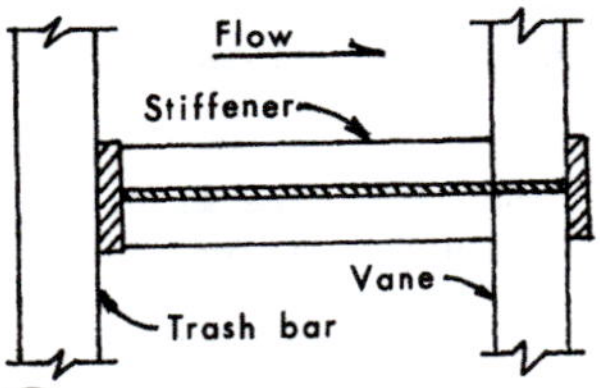

(2) DETAIL A (ORIGINAL)

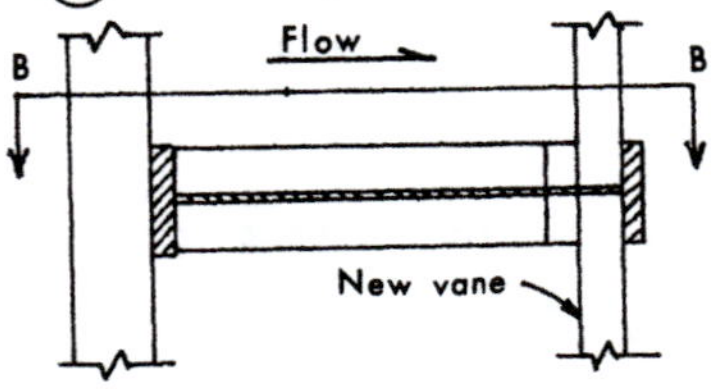

(3) DETAIL A (NEW)

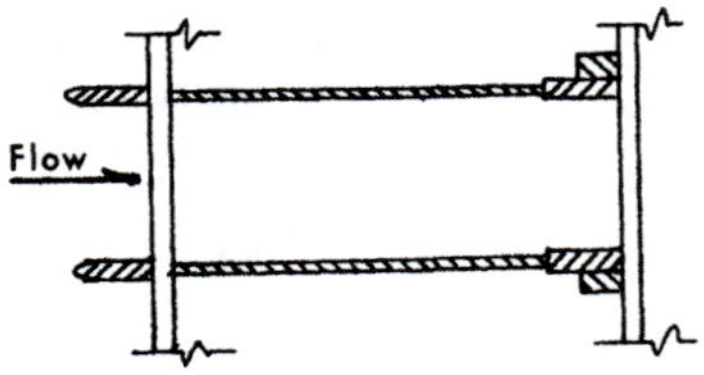

(4) SECTION B—B

Problem: Trash accumulation (sketch 1) causes increased velocities through the racks. The trashrack rear support vanes (see sketch 2) failed; examination of the fractures indicated a fatigue-type failure.

Geometry: See sketches 1, 2, and 3. The flow passage through the rack at the support vanes is 81 m^2 (5 trashrack sections). The maximum unsupported length of a rear support vane is 1565 mm.

Boundary conditions: The average flow velocity through the rack with maximum trash buildup as shown in sketch 1 is 2100 mm/sec.

Operating conditions: Trash accumulates as shown in sketch 1. Note that similar accumulation occurs at the site having the trashbar failure, Chart G.

Structural dynamics: The vibration (not measured) was assumed to be a transverse vibration of the rear support vanes; a simplified analysis (vane vibration independent of other rack members) indicated a natural frequency of about 24 Hz.

Vibrations: The Strouhal number for the vortex-shedding frequency of the vanes was estimated to be about 0.174 which corresponds to a forcing frequency of 19 Hz (see reference 1). The natural frequency and the forcing frequency are sufficiently close so as to suggest the possibility of resonance or self-excited vibration.

Cure: Since the trash accumulation cannot be eliminated, the support vanes were replaced by vanes nearly equal in cross-sectional area but of higher natural frequencies as shown in sketch 3. The new natural frequency is about 46 Hz.

References: 1. Sell, L. E., "Hydroelectric Power Plant Trashrack Design," Journal of the Power Division, ASCE, Vol. 97, No. PO1, January 1971, pp 115-121.

Dimensions (mm):
- Original vane = 19 x 102 x 1565 long
- New vane = 32 x 64 x 1565
- Trashbar = 19 x 116 x 765
- Stiffener = 9.5 x 140 x 533

CHART G: TRASH RACK BARS

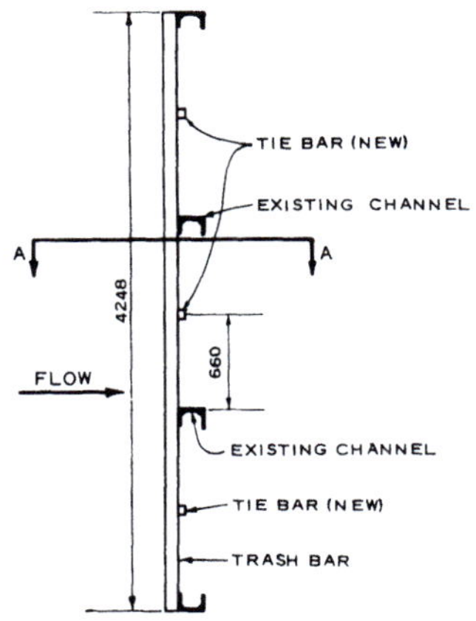

TRASH BAR AND SUPPORTS
VERTICAL SECTION

(1)

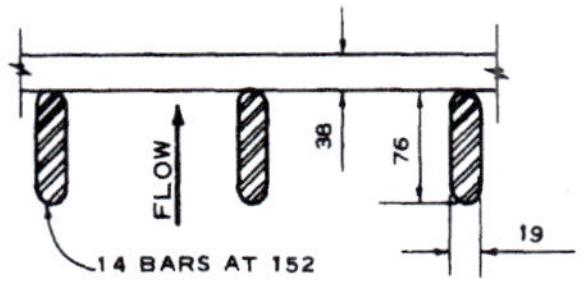

SECTION A-A

(2) DIMENSIONS IN MM

Problem: Broken steel bars downstream of a service gate were evidence of trashrack failure. The failing members were the trash bars rather than support members.

Geometry: The geometry is as shown in sketches 1 and 2. The net flow area of the water passage (clean trashrack) is 13.8 m^2. The trash bar dimensions are: Length = 1320 mm (longest bars; unsupported); thickness = 19 mm; width = 76 mm.

Boundary conditions: The trashrack is a welded structure with supports as shown in sketch 2. Trash is accumulated at the base of the bars as previously shown in Chart F.

Flow conditions: The average flow velocity (no accumulation) through the trashrack net flow area is 1570 mm/sec. With trash accumulation this is increased to 2510 mm/sec.

Operating conditions: Trash was allowed to accumulate upstream from the rack (see Chart F).

Structural dynamics: The natural frequency of the bars was calculated to be about 36 Hz (see reference 1).

Vibrations: The forcing frequencies were calculated to be about 22 Hz (clean) and 35 Hz (with accumulation). These values suggest a tendency toward a resonance condition as trash accumulation continues.

Cures: Two steps were taken. First, a tie bar was welded, as shown in sketch 1 to the mid point of the trash bars (increasing the natural frequency to about 147 Hz) to avoid the resonance condition. Second, although trash accumulations cannot be eliminated at this project, the benefit of minimizing the accumulation was emphasized.

References: 1. Sell, L. E., "Hydroelectric Power Plant Trashrack Design," Journal of the Power Division, ASCE, Vol. 97, No. PO1, January 1971, pp 115-121.

CHART H: TAINTER VALVE

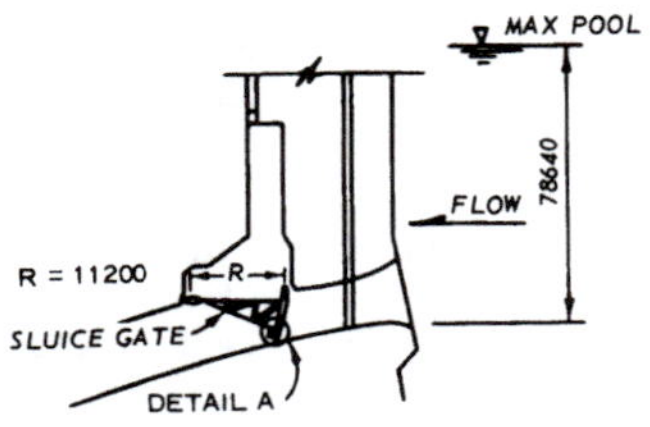

(1) SECTION OF SLUICE

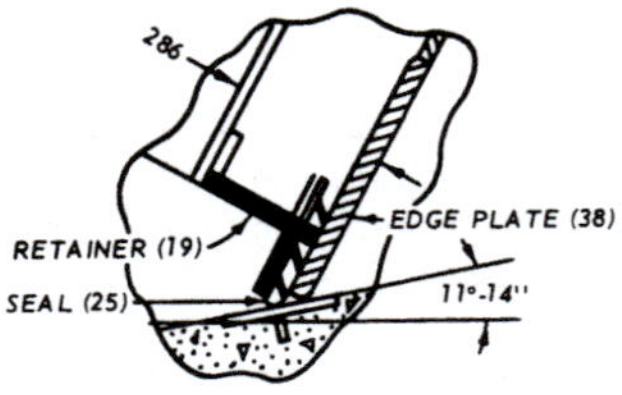

(2) DETAIL A (SEAL)

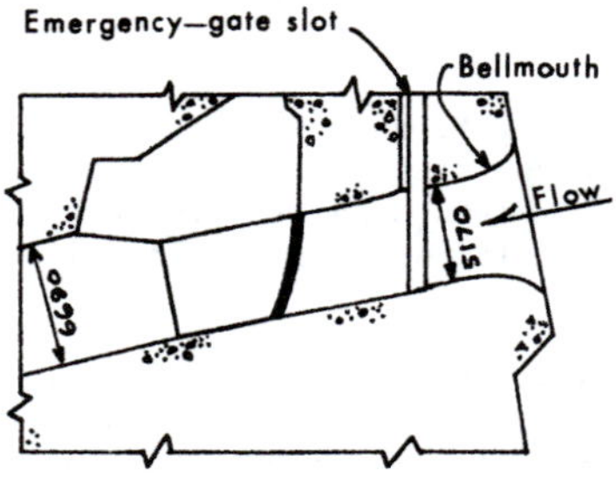

(3) INTAKE (VALVE NOT SHOWN)

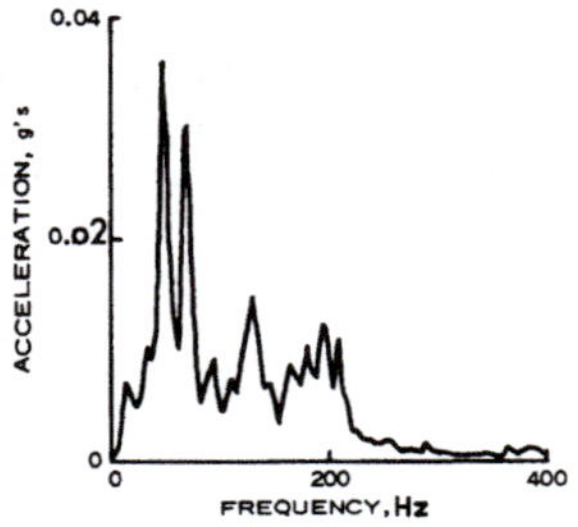

(4) RADIAL ACCELERATION

DIMENSIONS IN MM

Problem: The valve vibrates when near the full open position; the severity of the vibration is currently being assessed. Prototype tests were used to determine whether or not an unstable oscillation of control between bellmouth, gate slot, and gate lip occurs. No damage to the gate has been observed.

Geometry: The gate and lip geometries are shown in sketches 1 and 2, respectively. The conduit upstream from the valve is 5170 mm high and 3040 mm wide; free-surface flow occurs in the downstream conduit which is 6690 mm high.

Boundary conditions: The gate opening is controlled by means of a hydraulic mechanism; the natural frequencies associated with this mechanism are unknown.

Flow conditions: The maximum flow corresponding to the head shown in sketch 1 is about 1320 m^3/sec.

Operating conditions: The gate is used for flow control. For test purposes this gate was opened in 304 mm increments to determine zones of severe vibration.

Structural Dynamics: Unknown.

Vibrations: The frequency decomposition of the radial acceleration is shown in sketch 4. The most severe radial vibration occurs when the gate is 4600 mm open (as compared to 5170 mm when fully open); the dominant frequency for a sample record was 56 Hz and the maximum acceleration was 0.87 g's. The transverse and tangential accelerations of the gate were substantially less.

Cures: None at this time.

References: 1. Hart, E. D., "Sluice Pressures, Gate Vibrations, and Stilling Basin Wall Pressure, Libby Dam, Kootenai River Montana," USAE Waterways Experiment Station, TR H-76-17, Oct 1976, Vicksburg, MS, U.S.A.

CHART I: HOWELL-BUNGER VALVE

Problem: A major vane failure (fatigue type) and the need for a replacement required that a sturdier valve be used as a replacement. Prototype test data shown here are for the replacement valve which is currently operating satisfactorily.

Geometry: The geometry is shown in sketches 1, 2, and 3. Pertinent values are as follow:

Dimension Symbol	Valve (Original) (mm)	Valve (New) (mm)
a	15350	15350
b	3900	3900
c	1720	1720
d	3340	3340
e	2740	2740
f	912	912
g	9350	9350
h	25	59
i	51	289

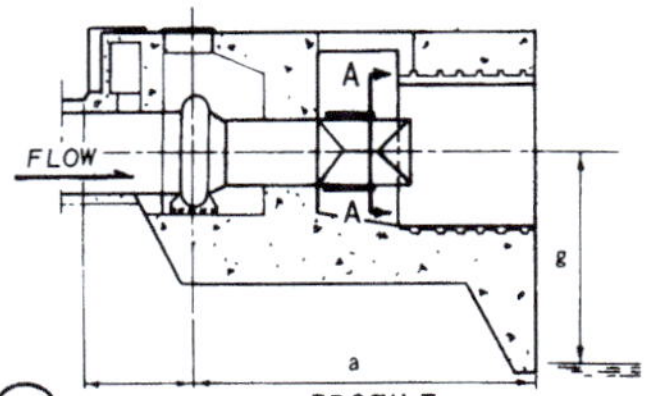

1 PROFILE

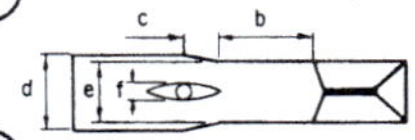

2 PLAN

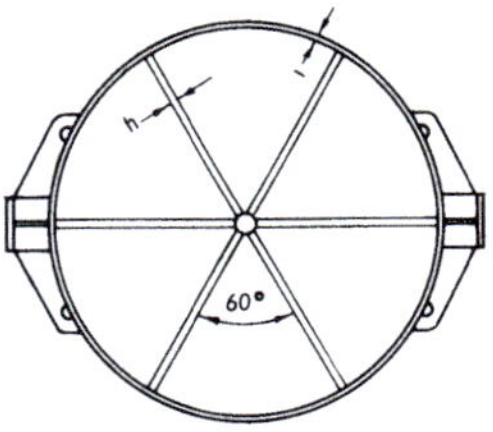

3 SECTION A-A

Boundary conditions: Interior fabrication is welded steel; the valve is attached to the conduit by means of bolted flanges.

Flow conditions: Test conditions were at a net head of 68400 mm and a flowrate of 166 m^3/sec (valve 90% open). The original valve failed during a period in which the head (and flow) varied. Pressure fluctuations (sketch 4) upstream from the valve are largest at a frequency less than 5 Hz.

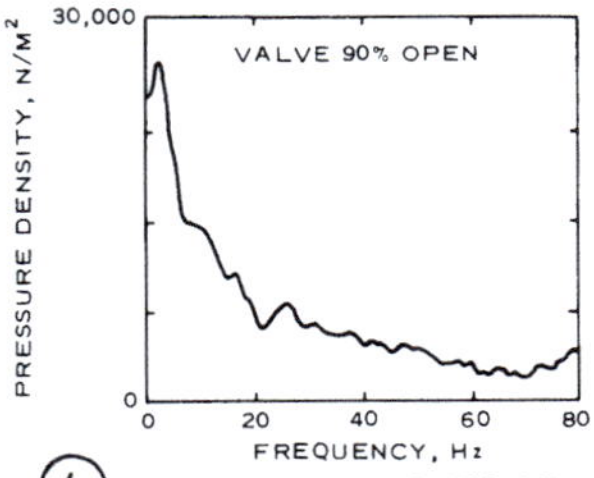

4 UPSTREAM PRESSURE

Operating conditions: The valve is used for control purposes; the upstream butterfly valve is for emergency closure.

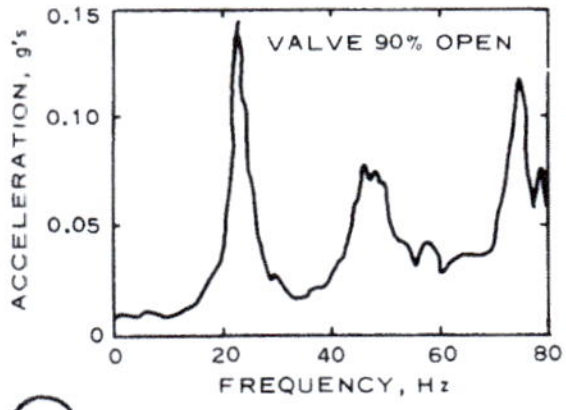

5 PERIPHERAL ACCELERATION

Structural dynamics and vibration: Acceleration recordings on the downstream cone face indicate the following dominant frequencies (sketch 5).

Frequency (Hz)	Probable Source
25	Torsional vibration of the cantilevered valve
46	Transverse vibration of the cantilevered valve
75	Transverse vibration of a vane

Cures: The sturdier replacement valves are operating satisfactorily and are considered, in this case, to be a cure. Recent hollow-jet cone valves are of designs that tend to minimize the exposed vane surface area and are less apt to suffer vane failure.

References: 1. Neilson, F. M., "Howell-Bunger Valve Vibration, Summersville Prototype Tests," USAE Waterways Experiment Station, TR No. H-71-6, Sep 1971, Vicksburg, MS, USA. 2. Mercer, A.G., "Vane Failures of Hollow-Cone Valves," IAHR Symposium, Paper G4, Stockholm, 1970.

CHART J: REVERSE TAINTER VALVE (HIGH-LIFT LOCK)

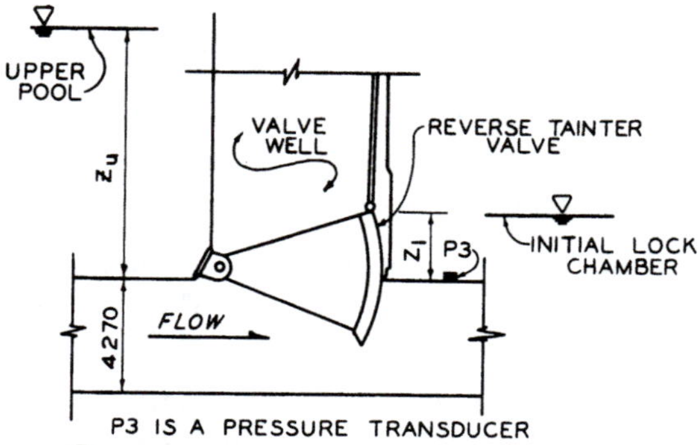

(1) SECTION OF VALVE IN CULVERT

Problem: During the valve opening period (from about 30 to 70 percent open) large "booming" sounds occur at irregular intervals; these sounds are accompanied by severe valve vibration. The probable cause is the collapse of large vapor cavities near the valve.

Geometry: See sketch 1; the values listed here pertain to a specific project; however, the problem also exists at several others. The conduit is rectangular in section, 3660 mm wide.

Boundary conditions: The valve is opened by means of a hoist rod and cylinder. Lock operation requires that the lock chamber water-surface rise to upper pool level.

Cures: At some projects adequate air venting downstream from the valve apparently eliminates the problem. At this specific project, the relatively high pressure (P3) on the culvert roof prevents venting. Therefore, a slower valving (i.e., open to 30%, hold valve at this position to allow pressure P3 to rise, then open completely) is used. Alternatives to this procedure are being studied.

References: A report of field measurements is being prepared for this project; similar conditions are described in the following prototype test reports. 1. U. S. Army Engineer Waterways Experiment Station, CE, McNary Lock, Columbia River, Washington, T. R. No. 2-552, Vicksburg, MS Jun 1960. 2. Tool, A.R., "Filling and Emptying System Measurements, New Bankhead Lock," USAE Waterways Experiment Station, Vicksburg, MS, 1980 (in preparation).

Operating conditions: Typical values are as follow (datum is the culvert roof elevation downstream from the valve):

Time (sec)	Upper Pool (mm)	Lock Chamber (mm)	Flowrate (m^3/sec)
0	41200	13100	0
171	41200	17100	140
248	41200	22900	258
598	Lock operation complete		

The pounding initiated 131 sec after start of the operation, the maximum pounding occurred at 171 sec, and pounding ceased at 203 sec. The instantaneous pressures at the roof of the culvert during maximum pounding ranged from a maximum of 30000 mm to a minimum of -2100 mm. Valve is full open at 248 sec.

Structural dynamics: Unknown.

Vibration: Nonstationary inasmuch as large pressure fluctuations with frequencies greater than 1000 Hz occur at irregular intervals; these large pressure fluctuations appear to initiate valve vibration which tends to decay thereafter.

CHART K: TANDEM LOW-FLOW SLUICE GATES

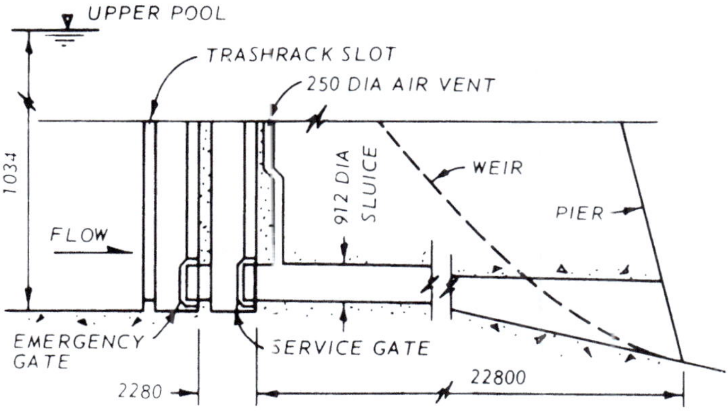

① VERTICAL SECTION

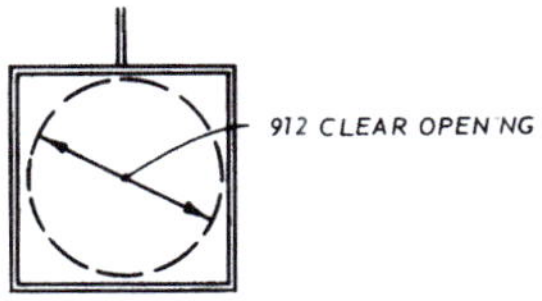

GATE CLOSED

②

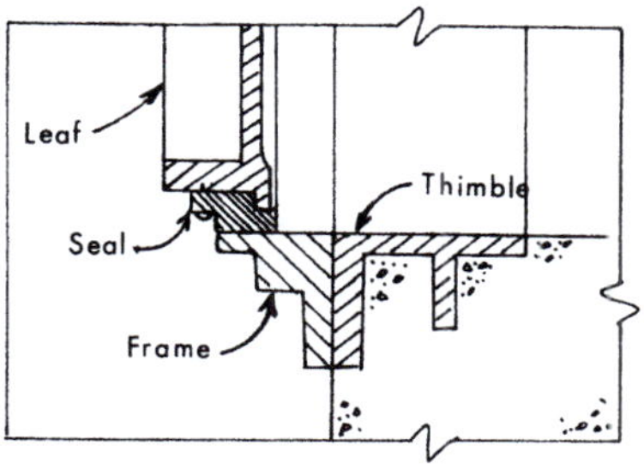

GATE SEAL

③

Problem: When the two identical gates (emergency gate and service gate in sketch 1) are in the full open position a large amount of noise and some vibration of the concrete occurs. The vibration may have contributed to the erosion of concrete near a construction joint 3040 mm downstream from the service gate.

Geometry: See sketches 1 and 2. Note that the gates are identical in size and geometry; they are flat-bottomed.

Flow conditions: Upper pool as shown in sketch 1.

Operating conditions: Vibrations occur when both gates are full open.

Structural dynamics: Unknown.

Vibrations: Low-frequency noise "chatter" disburbing to campers nearly 4 km from the site.

Cures: Use upstream gate in full open position but limit downstream gate to 745 mm maximum opening. Future designs probably will require the upstream gate to have a waterway area twice that of the lower gate; the gate guides will be extended to more securely support the service gates at the full open position.

References: None.

C4 HYDROELASTICITY EFFECTS IN HYDRAULIC ENGINEERING PRACTICE

V. M. Ljatkher, Prof., Dr. Sc. (Eng.)
"Hydroproject" Institute Research Centre, Moscow, USSR

M. F. Skladnev, Prof.
The B. E. Vedeneev VNIIG, Leningrad, USSR

I. S. Sheinin, Prof., Dr. Sc. (Eng.)
The B. E. Vedeneev VNIIG, Leningrad, USSR

SUMMARY

Effects of hydroelasticity are widely encountered in hydraulic engineering. They have to be taken into consideration in designing gates of outlet works, trash-racks of water intakes and fish-barriers, linings of gate chambers, high-pressure conduits, and protection and training walls in the structure tail-water, storage reservoirs slope protection, etc.

The results of studies of dynamic phenomena carried out in the USSR in connection with the design of these hydraulic structures are regularly published in "Trudy Koordinatsionnykh Soveshchanii po Gidrotekhnike" ("Proceedings of Coordination Meetings on Hydraulic Engineering"), e.g., No 28, 1966; No 52, 1969; No 54, 1970; No 64, 1972; No 116, 1977. Below some typical hydroelasticity problems are considered, which in many respects determine the integrity or damage of certain structural members, and representative examples from the hydraulic-engineering practice are given.

Hydrodynamic loads onto a rigid construction (river-bed protection in the tailrace, for instance) may be represented as the sum of the pulsating components. As far as the latter is concerned, usually in the first stage of research only the amplitude and the predominant period of pressure pulsation are determined. For example, regular measurements performed by G. A. Yuditskii [1,2] provide information on the amplitude of pulsation of forces acting on the downstream apron slab and the downstream river-bed protection at various relative dimensions of slabs and various ranges of flow parameters. A. S. Abelev et al. [3] have studied in detail the amplitudes of pulsating load on gates. Numereous studies of this type are generalized in many books [4,5] , and pertinent references may also be found in review articles [6,7,8] . However, in spite of the practical importance of these studies, the data on maximum amplitudes and on the predominant period of pulsation are still inadequate for prediction of the structural vibrations, and the results of the studies may be applied only in case that the computation model of the construction under design is similar to the computation model of the construction on which the experiments have been carried out.

The approach based on the analysis of the interrelation between the pressure pulsation and the kinematics of the flow is more promising. Yu. D. Gorodkov [9] was the first researcher who pointed out the important part which the size of the area of averaging plays and who has given a structural formula for the amplitude of the pressure pulsation on a uniform flow boundary. D. I. Koumin has drawn attention to the important part played by turbulent waves forming on the free surface of flow [10] . The rigorous treatment of the problem of interrelation between the pulsation of velocity and of pressure started by R. H. Kraichnan [11] as applied to hydraulic structures has been developed further in references [12,13]. These research findings deal in particular with the procedure of analysis of hydrodynamic load distortion under the effect of boundary vibration. The classification of the observed phenomena is given in [12,14] together with examples for the calculation of the effect of vibration on the flow-induced hydrodynamic load. These general methods can be used for analysis of different emergency situations brought about by the flow.

1. HIGH-HEAD GATES

In the thirties, in the USSR, V.G. Tishchenko, D.A. Harin [15] and others - and abroad Müller, Fischer, Seifert and others - investigated flow-induced structural vibrations as applied to gates, and the results were generalized by F.Linev and Z.F. Nichipurenko [16]. In November of 1956, a gate of one of the Volga navigation locks failed 17 after water flushed through the lock chamber for a short time (the upper gate was open, the lower-closed and the vertical-lift gate was lifted 1.5 to 1.0 m above the sill). The inspection revealed 135 mm deflection in the lower part of the gate, several guide rollers were torn off, there were breakages in girders and in welds between the plating and the horizontal beams etc. With the gate replaced (the change was performed by representatives of the Scientific Research Centre of the Hydroproject [18,19]), a detailed vibration study of the new gate was performed, along with studies of vibration of similar gates for some other locks, see Fig. 1. Special measuring instruments registered oscillations or displacements

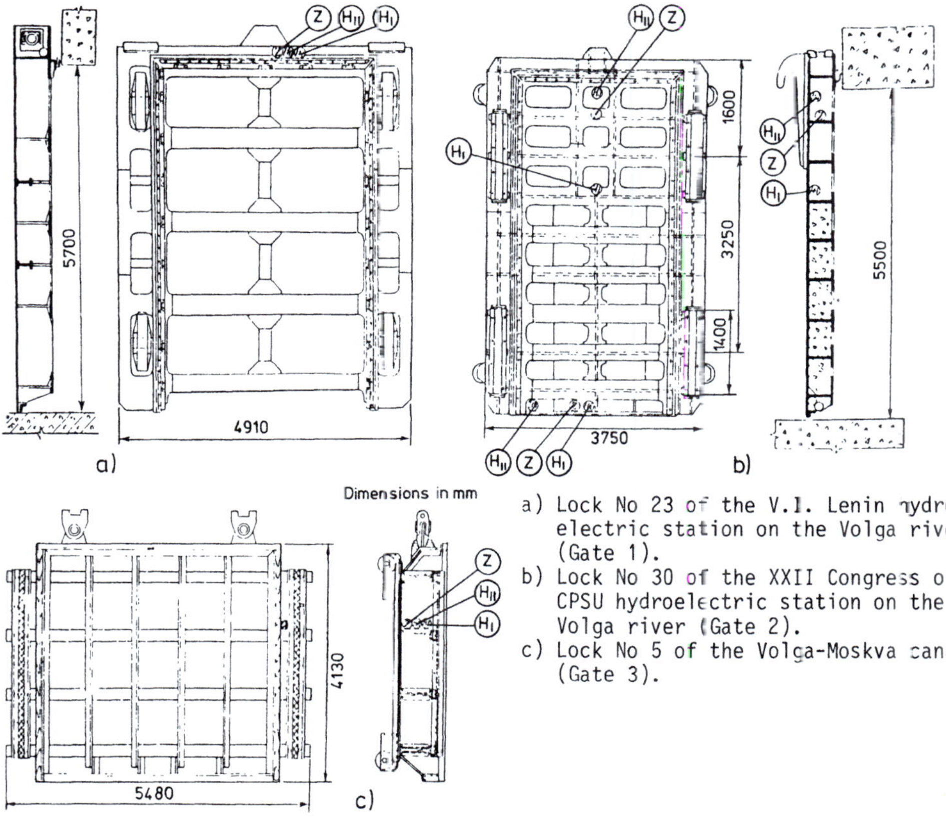

a) Lock No 23 of the V.I. Lenin hydroelectric station on the Volga river (Gate 1).
b) Lock No 30 of the XXII Congress of CPSU hydroelectric station on the Volga river (Gate 2).
c) Lock No 5 of the Volga-Moskva canal (Gate 3).

FIG. 1 GATES FOR CULVERS OF NAVIGATION LOCKS

in the points indicated in Fig. 1, where Z is vertical vibration component, H_{ii} and H_i are horizontal components along and across the flow, respectively. Vibrations of three main types were observed (Fig. 2):

I. Random Vibrations (Fig. 2a). These vibrations feature lack of apparent regularity and a wide variation of amplitudes and frequencies with some prevailance of high-frequency vibrations. Such vibrations appear from the moment, when the gate is detached from the sill, up to relative openings $Q_o \simeq 0.4$; they insignificantly vary in intensity and vanish only at full gate opening. The vibrations do not cease with the gate fixed. The random nature of the vibrations testifies about their origin; they seem to belong to the group of forced oscillations caused by pressure pulsation due to turbulent flow.

II. Self-Excited or Auto Oscillations in the System "Gate-Flow". Such oscillations are characterized by regularity and considerable intensity. They are generally observed at relative openings $Q_o > 0.5$ (depending on pressure and design features of the gate). They appear during gate lifting at constant head (Fig. 2,b) or at varying head but at constant gate opening. In the first case they do not vanish when the gate lifting stops (Fig.2,c). Time until the amplitude increases to the maximum values can be very small - the order of 5 to 10 periods of the natural gate vibrations (Fig. 2,b). Sometimes vibrations of two frequencies and modes are excited at once (Fig. 2,d).

III. Mechanical self-excited oscillations developed during the gate movement as a consequence of a negative slope in the initial portion of the relationship between the frictional force and rate of movement. It is characterized by an approximately constant amplitude and frequency of oscillations at constant rates of gate lifting (Fig. 2,e).

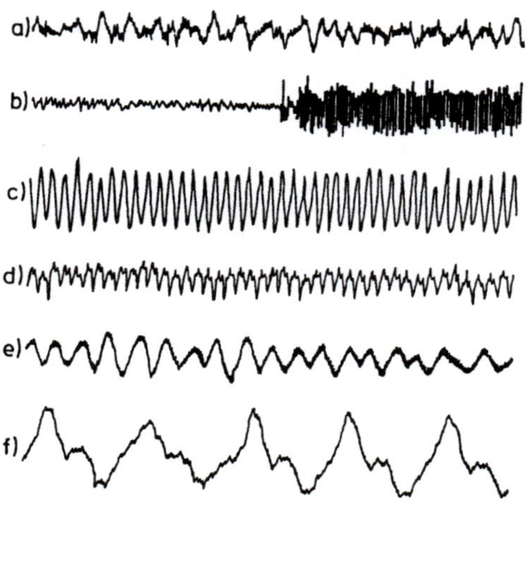

FIG. 2 MAIN TYPES OF GATE VIBRATIONS

(a) Vibrations of type 1 (random vibrations)

(b) Vibrations of type 2 (onset of flow-induced self-excited oscillations)

(c) Vibrations of type 2

(d) Vibrations of type 2 (superimposed bending vibrations are seen)

(e) Vibrations of type 3 (mechanical self-excited oscillations)

(f) Vibrations observed only at the gate of XXII-Congress-of-CPSU hydroelectric station (of the self-excited oscillation type)

Specific data about the test conditions and quantitative results are as follows:

Gate No. 1 is a vertical-lift roller gate. At the moment of testing, the head was about 13.5 m. The conditions of continuous gate lifting to the full height up to relative openings $Q_o = 1.0$ and the gate stopped at $Q_o = 0.17$; 0.35; 0.53 and 0.7 were studied. Vibrations of the types I and II were registered: their parameters are given in Table 1. Vibrations of the type 1 are noted at

Q_o < 0.53 both with the gate moving or stopping. At $Q_o \geq 0.53$ and heads 9.5 to 12.5 m, vibrations of the type II sharply come into action (Fig. 2b), irrespective of whether the gate moves upward or not. Vibrations of type II terminate at $Q_o \approx 1.0$.

Gate No. 2 is a vertical-lift gate with sliding supports (lignefol). The test head amounted to about 22.5 m. The study aims were: conditions of continuous gate lifting up to Q_o = 0.85 to 0.90 with lifting rates from 0.26 to 1.47 m/min, openings with the gate stopped at Q_o = 0.17; 0.34 and 0.69, and also some other regimes. Vibrations of all three types were noted, their parameters are summarized in Table 1. Vibrations of type I exist in all regimes and terminate only after the flow stops. Vibrations of type II appeared after the gate was lifted in excess of 3 m (Q_o = 0.54 - 0.66; head 12.5 - 15.0 m). Their amplitudes depend upon the gate lift rate (Table 2). They terminated approximately 40 sec after their appearance when the head decreased till 6.8 - 9.0 m and at relative openings Q_o = 0.54 - 0.66. We succeeded in making a synchronous record of vibrations in several gate points in the plane parallel to the flow (ZH.). The record reveals the fact that the gate experienced mainly rotational vibrations around the center located almost in the middle of the gate, and they are superimposed by bending vibrations with the frequency twice or thrice as high and the amplitude no greater than 15 μm (Fig. 2b). Vibrations at relative openings Q_o > 0.6 were observed only at this gate; they started with the gate stopping in the extreme upper position, or at the beginning of its lowering (Q_o = 0.85 - 0.90, head 1.8 - 3.5 m) and they terminated at Q_o = 0.6 - 0.75 and 0.8 - 3.2 m. Vertical amplitudes of these vibrations amounted to 1.5 mm, and horizontal ones to more than 0.3 mm; the frequencies were 2-3 Hz and 6-9 Hz at the beginning, and 4-5 Hz and 12-15 Hz at the end. Relatively high-amplitude stability and a narrow spectrum of vibration testify to the fact that in this case the vibrations seemed to be self-excited. Vibrations of type III were observed at any height but only with the gate moving. The amplitudes amounted to 1.3 - 1.5 mm at the moment when the gate is detached from the sill, and they decreased further along to 0.5 - 0.8 mm.

Vibration type	Component of vibration	Gate No 1		Gate No 2		Gate No 3	
		Freq., Hz	Accel. cm/s²	Freq. Hz	Displac. mm	Freq. Hz	Accel. m/s²
I. Aperiodic random vibrations	Z	15-23	15-18	*	-	30-45	-10
	H_I	15-23	18-20	8-14	0.05-0.06	**	5
	H_{II}	40	7-10	not registered			4
II. Auto-oscillations in "gate-flow" system	Z	***		-	-	not observed	
	H_I			25-28	0.08-0.10		
	H_{II}			not registered			
III. Mechanical auto-oscillats.	Z	Not observed		2-6	0.5-1.5	3.5-3.9	20
	H_I			-	-	4	2-3

TABLE 1 PARAMETERS OF GATE VIBRATION

* Vibrations of type 1 were not registered due to considerable submergence of vertical vibrographs

** Frequency decreases from 40-45 Hz to 20-25 Hz as gate is lifted

*** Smooth frequency decrease from 60-62 Hz to 20-25 Hz as gate is lifted

Rate of gate lifting m/min	0.26	0.64	0.960	1.170	1.380
Average vibration amplitude mm	0.04	0.05	0.075	0.110	0.150

TABLE 2 RELATIONSHIP BETWEEN INTENSITY OF VIBRATIONS OF TYPE II AND GATE LIFTING RATE

Gate No 3 is a vertical-lift gate with sliding supports (lignefol). The test head was about 8 m. Conditions of continuous gate lifting at different rates (0.2 - 0.5 m/min) and of the gate stopped at different openings were examined. The Authors [18] noted the existence of vibrations of types I and III (Table I); vibrations of type II apparently were not excited due to the relatively low heads.

2. LOW-HEAD STRUCTURES

Investigations of the low-head gate arranged at the end of the pressure outlet of the V.I. Lenin hydroelectric station [20] were performed by simultaneously measuring the gate vibrations, pressure pulsation in many points, and the stresses in the gate elements. Only random gate bending vibrations of type I were observed, i.e. vibrations as a plate supported at two sides and subject to pressure pulsation due to turbulent flow. A great contribution of the gate bending rigidity to the averaging of the pressure pulsation over the gate area was noted. The natural frequencies of the gate in water were shown to be half or less than in air.

Further investigations lead to the formulation in codes of the following requirements. The studies performed during the gate-design stage should ensure that no auto-oscillation can be generated under any operation conditions. Moreover, to forecast forced vibrations under the action of pressure pulsation due to turbulent flow, two aspects should be investigated: flow loads and inertia and damping effect of the medium of flow. Studies of the first aspect were performed by A.S. Abelev [3] , P.E. Lysenko [21] and others. As to the second aspect, a brief review of problems pertaining to different hydraulic structures is presented below.

Vibrations of the trash-racks in the Saratov hydroelectric station [22]is similar in a way to the case of low-level gate vibration due to flow. These vibrations caused numerous failures of lattice struts of the trash racks during the release of the first flood. The studies showed that at first the structural members underwent fatigue damage due to vibrations resulting from the Karman vortices at particular velocities and because the frequencies of the structural members were near the vortex frequencies [22] . After the trash-racks were reconstructed and the lattice members strengthened, the trash-rack geometry and support conditions were changed and unfavorable regimes were revealed. Recommendations for operating personnel were given accordingly which permitted satisfactory operation for more than 10 years.

A unique case of excitation of pressure pulsation and penstock vibrations due to disturbances propagating from the end of the penstock towards the dam-spillway face [23] was observed at the hydroproject Baine-Bashta. These disturbances were found to be amplified in the penstock like in a resonator [24].

In the last decade, the trend was to increase specific discharges through spillway structures with simultaneously lightening the structural parts of the spillways. This trend resulted sometimes in the development of severe vibrations of the spillway slabs and in cavitation damage of the spillway surface. All cases of cavitation observed were accompanied by vibration, but not vice versa. Analysis of one of the first damages of a spillway slab at the hydroelectric station of the integrated type is given in Ref. [25] . At this hydroelectric station piers are provided at the end portion of the pressure outlet which divide them into two branches: rectilinear and curvilinear. Here the outlet structure is 3.38 m high, each branch being 4.35 m wide the pier is 1.80 m wide. In front of the piers, the outlet structure is curved in plan; moreover, the upstream face of the piers was well-rounded. The roof and walls of the outlet structure and the pier were covered with facing slabs. The pier nosings had steel lining 16 mm thick. The lined portion was almost 1 m long.

When the construction period was nearing completion, the outlet structures were allowed to operate under nonsubmerged conditions with only a single outlet completely opened; the average velocity in this outlet end was almost 19 m/sec. As a result, the side portions of the piers failed. The failure started with the damage of the facing slabs due to cavitation. They were finally torn off, caused by the following process: Below the slabs, pit-holes formed which were in communication with the zones of higher pressure at the moment when a decreased hydrodynamic pressure acted on the facing slab externally. A further cause for the failure of the pier proper was vibration of the exposed steel rods. The vibration effect of flow on the steel rods was so pronounced that 45 mm-diameter rods, exposed over the length of 2.5-3 m (the ends being embedded in concrete), burst and the middle pieces were carried away with the flow; only rods of 30-35 cm length stuck out of the concrete. These remaining rod ends continued to vibrate, so that deep holes formed around them. In some cases, rods were broken at the depth of 70 cm. This prepared the next step of the rupture. Further damage to the pier-mass concrete continued after the facing slabs had been torn out. It was caused by the abrasive effect of suspended sediments, floating construction debris etc. Damage to the roof of the outlet structure, though less severe than at the pier and mainly of the cavitation type, was also observed.

Another case of vibration of an o u t l e t - s t r u c t u r e roof at the power plant integrated with a spillway was examined in Ref. [26,27] . Similar phenomena were also observed at the ski-jump end of the spillway and also at the apron and flow dissipators of hydroelectric stations [28] .

Marked vibrations of the integrated-type p o w e r p l a n t as a whole and of its separate structural members, may also develop in the process of spillway operation. The investigations of Kiev and Saratov hydroelectric stations [29], [30, 31] and those at Pavlovsk and Piyavinas [32] demonstrated that main vibrations of these stations resulted from pressure pulsation at the guarding constructions of spillways and also from the vibrations of the tailwater protection slabs transferred through the soil foundation. Vibrations of power-house buildings under the action of hydropower-units operation and pressure-pulsation effects on the guarding constructions of turbine water passages appeared to be markedly less severe than in case of spillway operation.

Typical examples of failure of g u i d e w a l l s (piers) in stilling basins and spillway tunnels are described in Ref. [12] . Analysis of dynamic loads appearing in similar situations is given in Ref. [33] . Here effects of averaging of the fluctuating load acting on the guide-pier surface were assessed on the basis of general methods developed in [12] with the use of data from testing the large-scale model of Sayano-Shushensk hydroelectric station. Values of dynamic response factor were established, which characterize the relation between the force moments at the pier footing in cases of static- and dynamic-load application. It was found that resonance phenomena can result in doubling the design moment at the guide-pier footing.

3. SLABS AT SPILLWAY TAILRACES AND LININGS OF CONDUITS

While designing the river-bed protection in the tailrace and the protection constructions of water-saturated slopes of dikes and dams, it is very important to take into consideration the hydrodynamical loads not only at the flow side but also at the back surface of the p r o t e c t i o n s l a b s , i.e., on the contact between the slab and the soil. These loads may arise due to penetration of pressure pulsations from the external side (through leaky joints and drainage well) or directly due to vibrations of slabs. Various models of the water-soil foundation system have been used in the calculation of pressure under the slabs: without consideration of the soil skeleton deformability in Refs. [31, 35] ; with integral consideration of the soil skeleton and fluid deformability in the form of the "piezo-conductivity" equations [12, 36, 37] ; and by way of solution of the dynamic equations of the two-phase medium with account of elastic or elasto-plastic deformations of the soil skeleton [13, 38, 39] . The statement of the boundary-value problem may be different for these soil models (with or without consideration of the gap between the slab and the foundation). The differences correspond to various criteria of the protection designs; it is assumed that under normal operation conditions the protection-slab vibrations should not cause irreversible deformations of soil and formation of a gap between the foundation and the protection slab; under emergency conditions when the normal gate-operation conditions are disturbed (e.g., one water outlet in operation with low tailwater level), the protection should not fail. However, no limitations are imposed on the soil conditions and on the soil-slab contact. Under such operation conditions the separation of the slab from the soil foundation for a value stipulated in the design specification is still admissible. The admissible value of the gap be prescribed to be, e.g., one half of the average soil or filter-grain diameter under the slab, this condition ensuring that no quick re-formations of soil under the slab will take place.

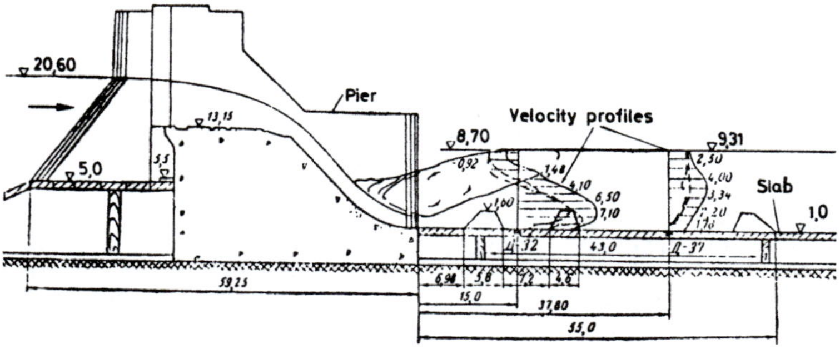

FIG. 3 TYPICAL SITUATION FOR THE LOADING OF SLABS AT SPILLWAY TAILRACES

The first computation model has been studied in great detail. Under conditions when no filter is made under the protection slabs, the problem amounts to the boundary-value dynamic problem of the theory of elasticity for the "equivalent" single-phase medium with modified Lame's parameters and [13, p. 37] :

$$\lambda = \lambda_o + K_f \quad , \quad \mu = \mu_o$$

where $K_f = E_f/n^{n-1}$; E_f is the elasticity modulus of the pore fluid (with proper account for gas inclusions); and n is the porosity.

The calculations yield the "total" dynamic stresses ard the stresses in the soil skeleton σ_{ij} [13] :

$$\sigma_{ij} = \tilde{\sigma}_{ij} \quad i \neq j \quad ; \quad \sigma_{ij} = \frac{\tilde{\sigma}_{ii}}{1 + K_f/(\lambda_o + 2\mu_o/3)}$$

In case the protection slabs are resting on a filter layer, the calculations are to be made using the equations of a two-phase medium [13, 37, 38] . The calculated dynamic stresses in the soil skeleton are added to the statical ones and are used then to check the soil strength conditions. The stability of slabs is ensured, if at any moment of time, and with an adequate degree of probability, compression stresses are acting in the soil skeleton on the surface of contact with the slab and the condition of the soil strength in all the points of the foundation is met.

At present, the methods of calculating processes of separation of slabs from the foundation are being developed intensively. For the first time these calculations have been applied in practice for the design of the tailwater-bottom protection constructions of the Kiev and Kanev hydroelectric stations on the Dnieper river [40] , which have been in successful operation for over 10 years.

Further developments can be found in Refs. [36, 37, 41] . It has been shown that the optimized design of the protection drainage enables appreciable decrease of the required slab thickness with increase of the admissible size of the gap[42]. Thus, the consideration of the hydroelastic effects in the process of the protection-slab uprise makes it possible in this case to lighten the protection construction.

Usually the protection slabs may be considered as rigid in the stability analysis. Deformability of the protection as a whole depends to a great extent on the deformability of the joints. Mutual correlation of vibration of individual slabs resting on a common elastic foundation may result in the amplification of the vibration of an individual slab, in an apparent decrease of the damping decrement, and in a severe increase with respect to vibrations with frequencies which are beyond the limits of the principal energy spectrum of action. Besides measurements of hydrodynamic loads and vibrations in the tailwater and in the power house of the Saratov hydroelectric power plant on the Volga river [30] , during the period preceding the construction-pit flooding measurements of parameters and the free vibration of the protection slabs were carried out. This made possible the comparison of the predicted and observed parameters of vibrations. It has been found that a satisfactory agreement is observed only for relatively low frequencies. At $\omega > 12$ Hz, the theoretical relationship for the spectral density of vibration of an individual slab gives considerably lower results than the relationship obtained by experimental studies. Thus, the mutual swinging of slabs must be taken into consideration for real situations. There are some data which allow one to anticipate the self-synchronization of vibrations of slabs, as well as that of vibrations of individual sections of the power houser or of the dam. Such a self-synchronization has taken place during the passing of a high spring-flood discharge over the spillway dam of the V.I. Lenin Volga hydroelectric power plant in May, 1979 and cased a perceptible vibration of houses situated at distances up to 3 km from the dam.

Figure 4 shows by way of example such self-synchronized vibrations, excited at uniform discharge distribution over the front of the dam of V.I. Lenin hydroelectric power plant. Examples of frequencies of vibrations at the foundation and the 8th floor of a house situated at the distance of 2.8 km from the dam are presented in Fig. 5.

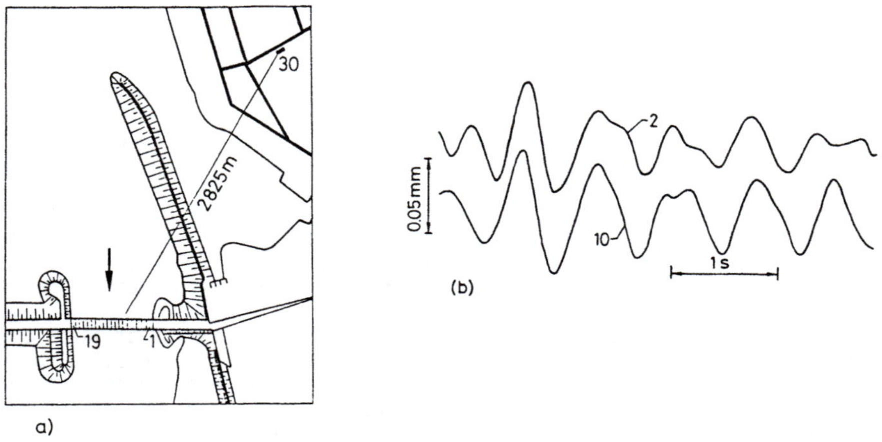

FIG. 4 SELF-SYNCHRONIZED VIBRATIONS OF SPILLWAY DAM

V.I. Lenin hydroelectric station; uniform distribution of discharge over width of the dam (in May 19, 1979) of 28 m³/sec per unit width and 36 800 m³/s in total.

(a) Scheme of the dam sections (1-19) and a domestic building (30), its records of vibrations are shown in Fig. 5

(b) Vertical displacements of dam sections 2 and 10.

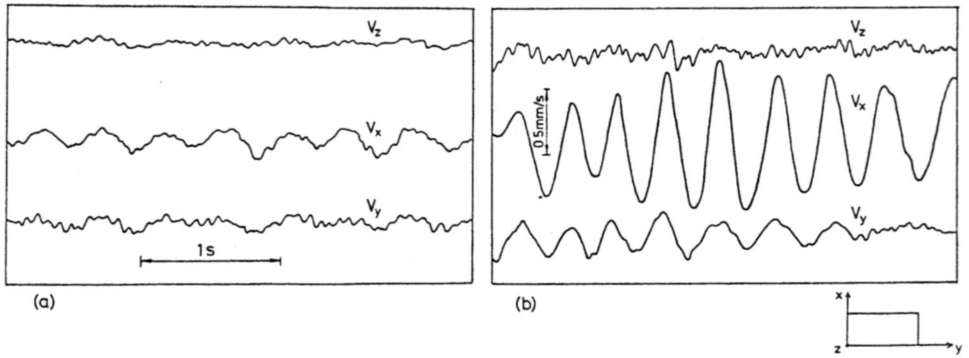

FIG. 5 EFFECT OF SPILLWAY VIBRATION ON A DISTANT HOUSE

Vibrations of the 9-story, 4-entrance panel domestic building shown in Fig. 4 on May 19, 1979 (axes x parallel to the short side of the building)

(a) Foundation of the building

(b) 9th story

Regularly repeated observations on a number of hydroelectric power plants revealed the variations of parameters of structural vibrations with time. For instance, the mean-square value z' of vertical vibrations of the spillway dam of the XXIInd Congress of the CPSU Volga hydroelectric power plant, the flood-discharge conditions reamining the same, has increased for about 30 per cent in two years at high specific discharges and more than twice at low specific discharges, see Fig. 6a [43] . In all these cases, the spectral density of vibrations featured a sharp maximum.The frequency corresponding to this maximum increased from 22 to 27 per cent during the above-mentioned time (Fig. 6b). Collection of data on this particular problem is being continued.

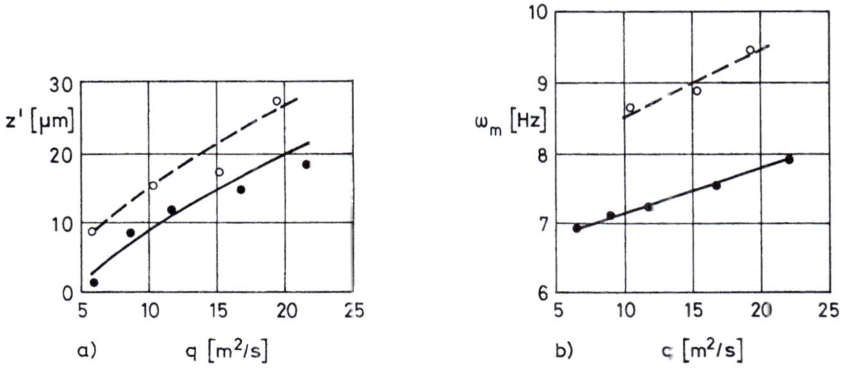

FIG. 6 VERTICAL VIBRATIONS OF A SPILLWAY DAM AS A FUNCTION OF SPECIFIC DISCHARGE on the Volga river; date of field measurements in 1963 (lower line) and in 1965 (upper line).

(a) Root-mean square of vibrations

(b) Cyclic frequency corresponding to maximum spectral density of vibrations

Hydroelastic effects are of great importance in designing c o n d u i t l i n i n g s . Appearance of local cavitation or flow aeration may be an important feature complicating this case (see Ref. 44-46). Stability and strength analyses of linings are also carried out taking into consideration the hydroelasticity effects, especially in the space between the lining and the tunnel walls [47, 48]. In principle, this problem is similar to the problem of calculation of bottom protections in the railrace of spillways, but it is appreciably complicated by the necessity of taking into consideration the relatively higher deformability of the lining.

It is important in all dynamic analyses to determine the parameters of the transfer function of the structure in water, i.e., the characteristics of the free vibration of the structure first of all.

The natural frequencies and forms of vibrations of structures oscillating in fluids are found by solving the boundary-value problems of the flow/body interaction. The analytical work initiated in the last few years [49, 50] has demonstrated that in most cases encountered in the practice of hydraulic engineering, a possibility exists to study the initial hydrodynamic field (neglecting the deformations and displacements of the elastic body and considering the hydrodynamic pressure on the surface of this body as a load) and the hydrodynamic field induced by the motion of the elastic body in the "fluid at rest" independent from each other. The simple superposition of these both fields yields an adequate picture. Only in the zones where the high-frequency velocity and pressure pulsations are playing an important part, and the frequencies of the elastic-body vibrations are far in excess of a certain limit frequency, the nonlinear nature of interaction of these two fields may manifest itself. For example, for protection slabs in the tailwater adjoining the spillway without lodges, this limit frequency (cyclic frequency) is $\omega = 8U_1/h_1$, where U_1 and h_1 are the velocity and the depth in the contracted section of flow at the beginning of the hydraulic jump. The velocities at which the allowance for the nonlinear nature of interaction would be required are not attained in hydraulic engineering under these conditions and, in particular, proceeding from conditions of cavitation. The same prevails for other hydraulic structures as well.

This circumstance permits an independent investigation into the hydrodynamics of flow with rigid boundaries, yielding as one of the results the characteristics of the hydrodynamical load on these boundaries, as it has been demonstrated in the first section of this paper. In this case, the problem of hydroelasticity is simplified due to the possibility of examining the vibrations of elastic constructions under the given loads in the "fluid at rest". The word at rest is set in quotation marks here because, actually, the velocity and pressure pulsations induced by the motion of the structure are not only taken into consideration-determination of these pulsations is one of the principal steps in solving the problem. At present, general solutions of this problem are beginning to come forth [51, 55] , and a number of papers have appeared considering the limits of applicability of assumptions usually adopted in solution of such problems.

REFERENCES

1. Recommendations on the determination of hydrodynamic loads acting on downstream apron slabs of spillway dams. P73-78 / VNIIG, Leningrad, 1979.
2. G.A.Yuditskii, "Experimental Studies of Pusation of Hydrodynamic Load on Protection Slabs in the Tailwater". Izvestia VNIIG, v. 58, 1958; v. 65, 1960.
3. A.S.Abelev, "Determination of Pulsations of the Total Hydrodynamic Load Acting on Vertical-Lift Bottom Gates", Izvestia VNIIG, v. 68, 1961; see also v. 58, 1958; v. 67, 1961; v. 69, 1962.
4. N.P.Rozanov, "Problems of Designing Water Outlet Structures Operating Under Conditions of Vacuum and at High Velocities of Flow". Gosenergoizdat Publishers, 1959.
5. F.G.Gun'ko (Edit.), "Hydraulic Design of Tunnel-Type and Tubular Water Outlets of Hydraulic Projects", Leningrad, 1974.
6. F.G.Gun'ko, M.F.Skladnev, V.M.Ljatkher, "Principal Trends and Results of Hydraulic Studies of Water Outlet Structures of High-Head Hydroprojects". Trudy Koordinatsionnykh Soveshchanii po Gidrotekhnike, No 52, Energia Publishers, Leningrad, 1969, pp. 9-33.
7. N.P.Rozanov, A.S.Abelev, "Studies in the Field of Hydroelasticity", Izvestia VNIIG, v. 81, 1966.
8. "Studies and Designing of Hydraulic Structures for the Action of Dynamic Loads", Trudy Gidroprojekta, v. 7, 1962; v. 10, 1963; v. 13, 1966.
9. Yu.D.Gorodkov, "On Stability of Facings Slabs in a Water Flow". Doklady AN SSSR, v. 28, No 9.
10. D.I.Kumin, "Pressure Pulsation in Pools Junction Arrangements and Its Relation to Velocity Pulsation", Izvestia VNIIG, v. 62, 1959.
11. R.H.Kraichnan, "Pressure Field Within Anisotropic Turbulence", J. Acoustical Soc. Amer., 1956, v. 28, p. 64, 378; 1957, v. 29, p. 65.
12. V.M.Ljatkher, "Turbulence in Hydraulic Structures", Moscow, Energia Publishers, 1968.

13. V.M.Ljatkher, Yu.S.Yakovlev (Edit.), "Dynamics of Continuous Media in Designing of Hydraulic Structures", Moscow, Energia Publishers, 1976.

14. V.M.Ljatkher, "Pulsation of Pressure on the Bottom of a Turbulent Flow Behind a Ledge". Trudy Koordinatsionnykh Soveshcharii po Gidrotekhnike , 1969, v. 52, pp. 191-208.

15. T.N.Krassilnikov (Mrs), "Calculation of Statistical Characteristics of Pulsations of Pressure on a Surface From the Second Moments of Longitudinal Velocity Pulsations in the Boundary Layer'. In the book "Turbulent Flows", Moscow, Nauka Publishers, 1970, pp. 44-49.

16. Yu.G.Blindze, O.N.Dokuchaiev, "Measurement of Pulsations of Velocities and Pressures in Turbulent Boundary Layers", Izvestia AN SSSR, Mekhanika Zhidkosti i Gaza", 1969, No 5.

17. S.Aki, "Field Measurements of Velocity and Pressure on Spillway Chute", Proc. 13 Congr. IAHR", Kyoto, 1969, v. 5-1, pp. 167-172.

18. I.Ya.Miniovich, A.D.Pernik, V. S.Petrovskii, "Hydrodynamic Sources of Noise", Leningrad, Sudostroienie Publishers, 1972, 592 pp.

19. F.M.Greshnikov, L.M.Liamshev, "On the Spectrum and Correlation of Near-Wall Pressure Pulsations at a Flow Past a Rough Wall", Akusticheskii Zhurnal, 1969, v. 15, No 1.

20. V.M.Ljatkher, L.V.Smirnov, "Pulsations of Pressure on the Turbulent Flow Boundary", In the book "Turbulent Flows", Moscow, Nauka, Publishers, 1970, pp. 73-78.

21. V.M.Ljatkher, "Hydrodynamic Effect of Turbulent Flow on Flow Boundaries", "Flow-Induced Structural Vibrations", Berlin, Springer-Verlag, 1974, pp. 203-223.

22. V.M.Ljatkher, N.V.Halturina, "Pulsation of Pressure Under the Downstream Floor Slab", Trudy Gidroprojekta", No 7, Moscow, 1962.

23. V.M.Ljatkher, N.V.Halturina, L.I.Bozhich, "Pulsation of Pressure Under the Downstream Floor Slabs of Spillway Dam", Moscow, 1963.

24. V.V.Bukhanov, "On the Problem of Stability of Protection Slabs in the Hydroelectric Power Plant Tailwater", Izvestia VNIIG, v. 100, 1972, pp. 96-112.

25. V.V.Bukhanov, "On Dynamic Interaction of Rigid Protection Slabs in the Tailwater with the Domain Under the Slab", Izvestia VNIIG, v. 105, 1974, pp. 268-280.

26. V.M.Ljatkher, B.I.Didoukh, "Unidimensional Boundary Value Problems of the Water-Saturated Soil Medium Dynamics", Trudy Gidroprojekta, 1971, v. 20, pp. 167-196.

27. V.M.Ljatkher, B.I.Didoukh, "Investigation of Motion of Water-Saturated Soil Medium Subjected to Earthquakes", Proc. Symp. on Behaviour of Earth Struct. Subject to Earthquakes", Roorkee, 1973, v. II.

28. V.M.Ljatkher, N.V.Halturina, "Downstream Floor of a New Design at the Kanev Hydroelectric Power Plant", Express-Informatsia, Ser.Stroitel'stvo GES, Ministry of Power and Electrification of the USSR, Moscow, 1968, No 203.

29. L.V.Moshkov, V.V.Bukhanov, "Characteristic Properties of Behaviour of a Rigid "Imponderable" Slab Vibrating on a Water-Saturated Soil Foundation", Trudy Koordinatsionnykh Soveshchanii po Gidrotekhnike, v. 116, 1977, pp. 138-141.

30. V.M.Ljatkher, O.N.Chernykh, "Estimation of Vibrations and Stability of the Downstream Floor Slabs", Izvestia vuz'ov, Ser. Stroitel'stvo i Arkhitektura, No 6, 1979.

31. I.I.Borzenko, V.M.Ljatkher, N.V.Halturina, "Field Dynamic Studies of the Saratov Hydroelectric Power Plant", Trudy Koordinatsionnykh Soveshchanii po Gidrotekhnike, v. 64, 1972, Leningrad, pp. 104-112.

32. L.A.Goncharov, "Variations of Structure Vibrations Parameters with Time", Trudy Koordinatsionnykh Soveshchanii po Gidrotekhnike, v. 54, Leningrad, Energia Publishers, 1970, pp. 373-377.

33. V.M.Ljatkher, L.V.Smirnov, "Hydrodynamic Loads on the Boundaries of a Pressure Flow in the Separation Zone", Trudy Koordinatsionnykh Soveshchanii po Gidrotekhnike, v. 54, Leningrad, Energia Publishers, 1970, pp. 33-41.

34. V.M.Ljatkher, L.V.Smirnov, "Hydrodynamic Loads on the Boundaries of a Turbulent Flow in the Separation Zone", Collection of Papers: "Turbulent Flows", Nauka Publishers, 1970.

35. V.M.Ljatkher, L.V.Smirnov, "Pressure Pulsations in a Pressure-Flow Spillway as Measured on the Prototype and on the Model", Izvestia VNIIG, v. 85, 1967.

36. L.V.Moshkov, "Some Problems of Dynamic Interaction of Decks (Linings) of Pressure-Flow Spillway with the Flow", Izvestia VNIIG, v. 94, 1970.

37. "Recommendations of Accounting for the Cavitation in Designs of Outlet Hydraulic Structures", P38-75 / VNIIG, Leningrad, 1976.

38. V.M.Ljatkher, "Hydroelasticity Effects in Hydraulic Structures", Proc. XI Congr. IAHR, Leningrad, 1965; "Gidrotekhnicheskoye Stroitel'stvo", 1965, No 8.

39. S.M.Levina, "Calculation of Hydrodynamic Pressure Distribution on the Surface of an Oscillating Elastic Structure in a Turbulent Flow", Izvestia VNIIG, v. 97, 1971, pp. 29-39.

40. I.S.Sheinin, "Vibrations of Hydraulic Structures in Fluid", Leningrad, Energia Publishers, 1967, 314 pp.

41. I.S.Sheinin, "Some Problems of Hydroelasticity in the Dynamic of Hydraulic Structures", Izvestia VNIIG, v. 84, 1967, Leningrad, pp. 277-298.

42. I.S.Sheinin, "On Multiparametric Problems of Shell and Plate Dynamic Hydroelasticity", Trudy IX Vsesoyuznoi Konferentsii po Teorii Obolochek i Plastin, 1975, pp. 103-109.

43. M.F.Skladnev, I.S.Sheinin, "On Flow-Induced Structural Oscillations", Flow-Induced Structural Vibrations", Berlin, Springer-Verlag, 1974, pp. 249-269.

44. I.S.Sheinin, G.V.Vilin, I.S.Kalitseva, "Method of Solution of a Two-Dimensional Problem on Structural Vibrations on the Boundary of Fluid Volume of Complicated Shape". In the Collection of Papers: Dynamics of Power Plant Structures", v. 5, Leningrad, 1976, pp. 3-17.

45. P.Bradshaw, D.H.Ferris, R.F.Johnson, "Turbulence in the Noise-Producing Region of a Circular Jet", Journal Fl. Mech., 1964, v. 19.

46. S.Kotsubo, "Dynamic Water Pressure on Dams During Earthquakes", Proc. 2nd World Conf. on Earthquakes Engineering, 1960, pp. 799-814.

47. I.S.Sheinin, L.D.Noskov, L.V.Moshkov, "Experimental Studies of Slab Vibrations at the Boundary with Fluid in a Hydraulic Test Flume", Izvestia VNIIG, v. 87, 1968, Leningrad, pp. 183-192.

48. I.S.Sheinin, V.A.Zabolotnaya, "Motion of Multiphase Fluid with Density Varying Over the Depth in Case of Vibrations of Confining Structural Members", Izvestia VNIIG, v. 91, 1969, Leningrad, pp. 113-121.

49. V.A.Zabolotnaya, "Wall Vibrations in a Fluid with Depth-Varying Density", Trudy Seminara: Dynamika Sooruzhenii GES, v. 2, Energia Publishers, 1972, pp. 7-13.

50. I.S.Sheinin, "On Account for Surface Waves Resulting From Motion of a Vertical Wall in Fluids", Izvestia VNIIG, v. 83, 1967, Leningrad, pp. 204-213.

51. I.S.Sheinin, M.P.Shames, R.M.Mukhutdinova, "On the Effect of Ice Cover on the Water Surface on Vibrations of Hydraulic Structures in Water During Earthquakes". In the Collection of Papers: "Sovershenstvovaniye Metodov Rascheta; Proyektirovaniya Gidrotekhnicheskikh Sooruzhenii Vozvodimykh v Seismicheskikh Rayonakh", Energia Publishers, 1976, pp. 162-165.

52. I.S.Sheinin, L.S.Voronova, "On Problems of Dynamic Hydroelasticity of the Fluid Volume Confined by Percolating and Non-Ideally Pliable Walls", Proc. V Int. Conf. on Non-Linear Vibrations, v. 3, published by the Institute of Mathematics of the Ukrainian Academy of Sciences, 1970, Kiev, pp. 709-714.

53. L.S.Leibenson, "On Natural Period of Vibrations of the Dam Across a River", Uchenyie Zapiski MGU, No 2, pp. 121-125.

54. M.T.Urazbaiev, "Seismic Stability of Elastic and Hydroelastic Systems", Tashkent, Fan Publishers, 1966, 254 pp.

55. I.S.Sheinin, S.G.Shulman, "On the Vibrations of an Elastic Beam in Fluid", XI Congr. IAHR, Leningrad, v. 4, 12, 1965.

56. P.P.Kulmach, "Hydrodynamics of Hydraulic Structures", edit. by AN SSSR, 1963, 189 pp.

C5

STABILITY OF VERTICALLY MOVABLE GATES

A. Vrijer
Project Engineer

Locks, Weirs and Sluices Branch
Delft Hydraulics Laboratory, the Netherlands

SUMMARY

This paper describes the theoretical and model-investigation of gates with lower edges which cause a stable or unstable behaviour. The dynamic behaviour of the gates in this investigation is determined primarily by the pressure fluctuations at the gate opening caused by discharge fluctuations. The pressure fluctuations exert a force on the gate lower edge, causing an extra vertical force on the gate. If the extra force acts in the direction opposite to the direction of movement of the gate the gate behaviour is stable; if the extra force acts in the same direction as the direction of movement of the gate the gate behaviour is unstable.

1. INTRODUCTION

Various publications have appeared on the prevention of vibrations of vertical lift gates, which vibrations are most probably induced by flow-separation at the lower edge. Sharp gate edges turn out to be efficient in suppressing the phenomena, except when the gate raise is of the same order as the edge thickness. Investigators have directed their attention in particular to the frequency range of turbulent pressure variations without vibration of the gate. The mechanism of feedback of vibrations is little investigated: to be mentioned are references [1] to [5] . Reference [5] describes the unstable behaviour of vertically movable L-shaped gates. This paper is a sequel of [5] and describes the search for shapes of lower edges of gates that are more stable.

2. THE MECHANISM

2.1 The situation

The investigation relates to vertically movable gates, that are suspended from a spring. The water flows under the gate.

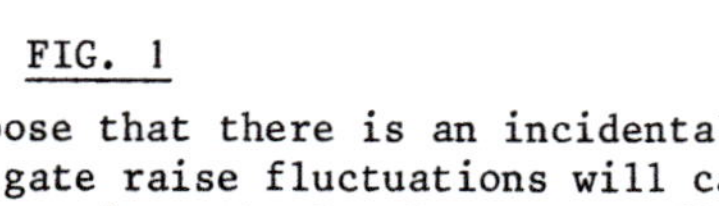

b = lip width
w = width of the gate
δ = gate raise
Q = discharge
ΔH = local head of the gate
F = hydrodynamic force
h_u = upstream waterdepth
h_d = downstream waterdepth

FIG. 1

Suppose that there is an incidental impact causing the gate to start vibrating. The gate raise fluctuations will cause discharge fluctuations. Due to discharge fluctuations the local pressure difference across the gate ΔH, and the vertical force, F, on the gate will fluctuate, too. With stationary vibration:

gate raise	$\delta = \delta_o + y\,e^{i\omega t}$
discharge	$Q = Q_o + Q'e^{i\omega t}$
pressure difference	$\Delta H = \Delta H_o + \Delta H'e^{i\omega t}$
hydrodynamic force	$F = F_o + F'e^{i\omega t}$

These fluctuations are not in phase with the gate raise due to the inertia of the water. The hydrodynamic force F at the gate includes the force due to the added mass of water and due to the discharge fluctuations.

2.2 Force due to added mass of water at lower edge of gate

Vertical displacement of the gate causes a vertical force F_m of the water on the gate lip (m_w = added mass of water)

$$F_m = m_w \ddot{y} \tag{1}$$

2.3 Force due to discharge fluctuations in the gate opening

2.3.1 Discharge adaption

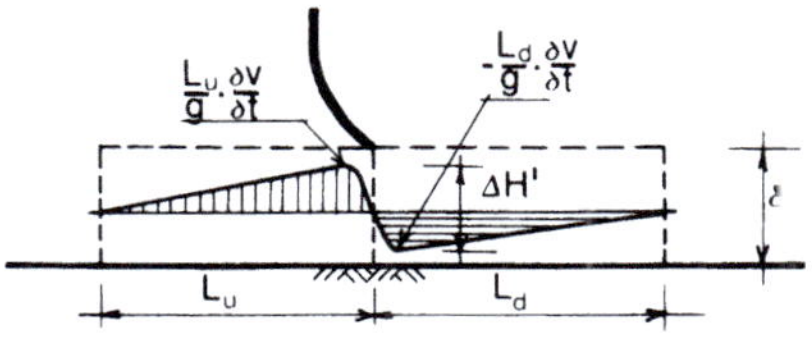

ρ = density of water
g = gravity acceleration
L_u = length block of water upstream, subjected to the extra pressures
L_d = length block of water downstream, subjected to the extra pressures
$L_u + L_d = c_i . \delta$

FIG. 2 EXTRA PRESSURES DUE TO DISCHARGE FLUCTUATIONS IN THE GATE OPENING

For some reason the gate may get a little vertical displacement downward in a short time; the discharge in the neighbourhood of the point of flow separation is therefore partly blocked. The section, approximately at that point of flow separation is decisive for the discharge; the discharge adapts to the instanteneous gate raise. The local (and instanteneous) pressure difference $\rho g \, \Delta H'$ across this section is then determined.
The pressure difference can be generally described by Navier-Stokes:

$$-\frac{\partial p}{\partial x} = \rho \frac{\partial v}{\partial t} + \rho v \frac{\partial v}{\partial x} + \ldots\ldots$$

$$-\Delta H' = \underbrace{\frac{1}{g}\frac{\partial v}{\partial t}\Delta x}_{\text{non-stat.}} + \underbrace{\frac{1}{g} v . \Delta v}_{\text{stationary}}$$

- Non-stationary term:
  The water, that is subject to the pressure fluctuations due to adaption of the discharge, is graphically represented as a rectangular block $(L_d + L_u).\delta$. These dimensions can be defined by an electric analogon: at the point where the discharge adapts is an imaginary vertical wall. This wall accelerates, horizontally, at a rate of $\partial v/\partial t$. The added mass of water of this wall equals exactly the block of water referred to above. In that case the formula below holds ($\Delta x = L_u + L_d = c_i . \delta$)

$$-\Delta H'_{\text{non-st.}} = \frac{L_u}{g} \cdot \frac{\partial v}{\partial t} + \frac{L_d}{g} \cdot \frac{\partial v}{\partial t} = \frac{c_i}{gw}\frac{\partial Q}{\partial t} \tag{2a}$$

- Stationary term:
  The discharge is fluctuating by $Q = Q_o + Q'e^{i\omega t}$. The contribution to the stationary term is:

$$-\Delta H'_{st.} = \frac{Q_o . \Delta Q'}{g\, w^2\, \delta^2} \tag{2b}$$

- Total pressure difference:

$$-\Delta H' = \frac{c_i}{gw} \cdot \frac{\partial Q}{\partial t} + \frac{Q_o . \Delta Q'}{g\, w^2\, \delta^2} \tag{2}$$

2.3.2 Effect on vertical force

A percentage of the pressure difference $\Delta H'$ acts on the lower edge of the gate, causing a vertical force F_f with a non-stationary and a stationary term:

$$F_f = c_F\ \rho gbw\ \Delta H' = \underbrace{\theta_n\ \rho gbw\ \Delta H'}_{\text{non-stat.term}} + \underbrace{c_s\ \rho gbw\ \Delta H'}_{\text{stat.term}} \qquad (3^a)$$

This force can also be written with an out of phase and an in-phase term (c_w = hydrodynamic damping and k_w = hydrodynamic rigidity):

$$F_f = -(c_w \dot{y} + k_w y) \qquad (3^b)$$

- Non-stationary term (as described by potential flow solution), see Fig. 3 :

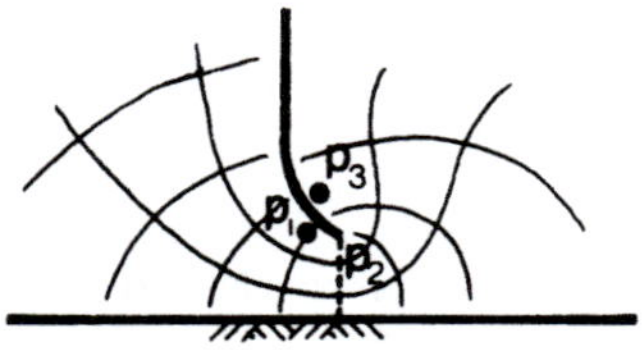

p = pressure

$$\theta_n = \frac{p_1 - p_3}{p_2}$$

FIG. 3 STREAM- AND POTENTIAL LINES IN THE CASE OF AN IMAGINARY, HORIZONTALLY OSCILLATING WALL IN THE OPENING

A pressure difference $p_2 = \rho g\ \Delta H'$ exists across the section of separation of the flow. Then, the pressures at the lower edge can be determined by means of an electric analogon. In this case the pressure difference across the lower edge is $(p_1 - p_3)$. The non-stationary vertical force F' is $(p_1 - p_3)\ bw = \theta_n\ \rho gbw\ \Delta H'$, where θ_n was determined in the analogon.

- Stationary term (c_s = suction/lift coefficient), see Fig. 1:
  In the stationary situation the vertical force on the gate has been measured in a model, in accordance with $F = c_s\ \rho gb\ \Delta H$ at waterlevels h_u and h_d. It is assumed that, when the pressure difference across the gate changes by $\Delta H'$, an extra (fluctuating) stationary force on the gate is produced of $F' = c_s\ \rho gB\ \Delta H'$.

2.4 Discharge fluctuations

The derivative of the discharge $Q = \mu\ \delta\ w\ \sqrt{2g\ \Delta H}$ (μ = discharge coefficient) is

$$\frac{\partial Q}{\partial t} = \mu\ \dot{y}\ w\ \sqrt{2g\ \Delta H} + \mu\ \delta\ w\ \tfrac{1}{2}\sqrt{\frac{2g}{\Delta H}}\ \frac{\partial\ \Delta H}{\partial t} \qquad (4)$$

2.5 Force equation

The force equation will be, with (1) and (3^b):

$$m\ddot{y} + c\dot{y} + ky = F_m + F_f = -(m_w \ddot{y} + c_w \dot{y} + k_w y) \qquad (5)$$

where
- m = gate mass; m_w = added mass of water
- c = imaginary damping; c_w = hydrodynamic damping
- k = suspension spring rigidity; k_w = hydrodynamic rigidity
- F_m = force exerted by added mass of water at lower edge of gate
- F_f = force exerted by discharge fluctuations in gate opening.

The force equation has been extended by the damping term $c\dot{y}$ to achieve a stationary vibration. Stationary vibration means $(c + c_w)\dot{y} = 0$ or $c = -c_w$.

2.6 Solving the set of equations

The set of equations (1), (2^a), (3^a), (3^b), (4) and (5) can be written as a third order differential equation in y by substitution. When the coefficients θ_n and c_i by means of an electric analogon and c_s and μ by means of a hydraulic model have been determined, then it is possible, upon solving the differential equation, to form an expression for the hydrodynamic damping c_w (see Ref. [5]):

$$c_w = c_F\ \rho bw\ \sqrt{2g\ \Delta H_o}\ \frac{c_i\ \mu}{1 + c_i^2\ \mu^2\ 4\pi^2\ S^2} \qquad (6)$$

where $c_F = \theta_n + c_s$
S = Strouhal-number = $\dfrac{f\ \delta}{\sqrt{2g\ \Delta H_o}}$
f = frequency
c_i = length coefficient
μ = discharge coefficient.

A factor of major importance is the sign of the hydrodynamic damping c_w. This sign determines the movement of the gate: damped or amplified according to an exponential curve. The sign of c_w is totally determined by $c_F = \theta_n + c_s$:

if $c_w > 0$ (or $c_F > 0$) damped gate movement
if $c_w < 0$ (or $c_F < 0$) amplified gate movement.

The damping can be expressed, dimensionless, as follows:

$$c_{SE} = \frac{c_w}{c_F\ \rho B\ \sqrt{2g\ \Delta H_o}} = \frac{c_i\ \mu}{1 + c_i^2\ \mu^2\ 4\pi^2\ S^2} \tag{7}$$

3. LOWER EDGE DESIGN

As said earlier, in 2.6, the sign of $c_F = \theta_n + c_s$ determines whether a specific gate movement is either damped or excited.
A similar interpretation is, if, with downward displacement of the gate, the sum of the non-stationary force ($\theta_n\ \rho gbw\ \Delta H'$) and the stationary force ($c_s\ \rho gbw\ \Delta H'$) is directed upward (Note: this is in the direction opposite to the direction of the gate movement) the movement is damped; if the sum is directed downward (Note: this is in the same direction as the direction of the gate movement) the movement is excited.

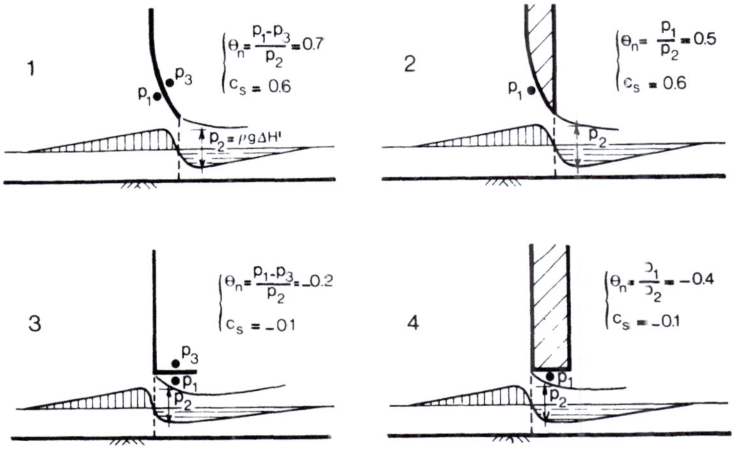

FIG. 4 FOUR SHAPES OF LOWER-EDGE DESIGNS WITH EXTRA PRESSURES DUE TO DISCHARGE FLUCTUATIONS IN THE GATE OPENING

Four examples are given of lower gate-edge design, see Fig. 4 :
- examples 1 and 2: the flow separates downstream
- examples 3 and 4: the flow separates upstream.

The values for θ_n and c_s have been rounded off; they have been determined by means of an electric analogon and a hydraulic model for a combination of waterlevels. The dynamic pressure change has been drawn for a downward gate movement. Successively, a distinction is made between downstream and upstream separation of the flow, and between open and massive gates.
- Downstream/upstream separation of the jet.
  The examples 1 and 2, with downstream separation of jet, show:
  - non-stationary with θ_n: the lower edge is located in an area of pressure increase, causing an upward force;

- stationary with c_s: the flow force is directed upward.

The examples 3 and 4, with upstream separation of jet, show:

- non-stationary with θ_n: the lower edge is located in an area of pressure decrease, causing a downward force;
- stationary with c_s: the flow force is very low, and directed downward.

- Open/massive gate.

Examples 1 and 3 show that, with the non-stationary term of the force, the top of the lower edge is located in an area of pressure decrease, causing an upward effect of the total force.

Examples 2 and 4, with massive gate do not show that effect.

All examples show that gate 1 is the most stable; gate 2 is less stable; gate 3 is unstable, and gate 4 shows the most unstable behaviour.

4. MODEL INVESTIGATION RESULTS

4.1 The model

Various lower edge designs were tested in a hydraulic model, as shown in Figure 4. The model consisted of a vertical gate in a flume. The gate was suspended from a set prestressed springs; only vertical displacement was possible, as the gate was tensioned by horizontal wires.

Model specifications:

Gate and flume widths	$w = 0.50$ m
Gate raises	$\delta = 0.01\text{-}0.10$ m
Lower edge width	$b = 0.01$ m
Suspension rigidities	$k = 1000\text{-}15000$ N/m
Gate mass	$m = 4\text{-}10$ kg
Frequencies	$f = 2\text{-}9$ Hz
Water levels	$h = 0.05\text{-}0.70$ m
Discharges	$Q = 3 - 100 \times 10^{-3}$ m^3/s.

4.2 Measurements

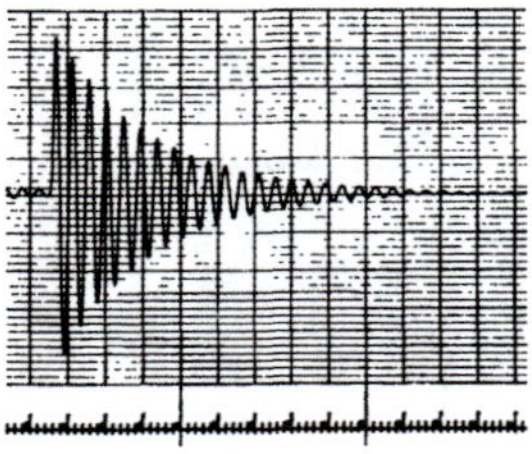

FIG. 5 EXAMPLE OF A DAMPING FORCE (TIME-REGISTRATION)

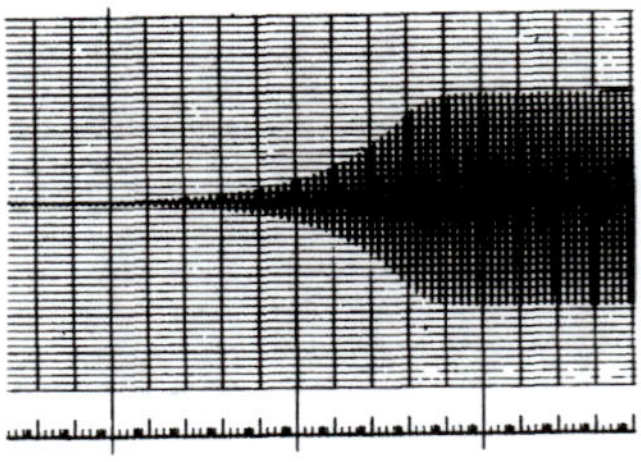

FIG. 6 EXAMPLE OF AN EXCITING FORCE (TIME-REGISTRATION)

The vertical forces acting on the gate were measured.

Gates 1 and 2 were pushed for various gate raises, suspension rigidities and water level combinations, the movement subsequently damping out in accordance with a power of e. An example is shown in Figure 5. The rate of decrease of the amplitudes was converted into the damping constant c_w. Figure 7 shows, for various gate raises, the minimum positive damping constants (in the dimensionless form $c_{SE} = c_w/(c_F \rho\, bw \sqrt{2g\, \Delta H_o})$, plotted against Strouhal's number $S = f\, \delta\sqrt{2g\, \Delta H_o}$.

Gates 3 and 4, for various gate raises, suspension rigidities and water level combinations, were exciting in accordance with a power of e, until a specific equilibrium had been reached (refer to Figure 6). The rate of increase of the amplitudes was converted into the negative damping constant c_w. Figure 8 shows, for various gate raises, the maximum negative damping constants (in the dimensionless form $c_{SE} = c_w/(c_F \rho\, bw \sqrt{2g\, \Delta H_o})$, plotted against Strouhal's number $S = f\, \delta/\sqrt{2g\, \Delta H_o}$.

Figure 7 and 8 show also the damping constants c_{SE}, theoretically computed in accordance with formula (7).

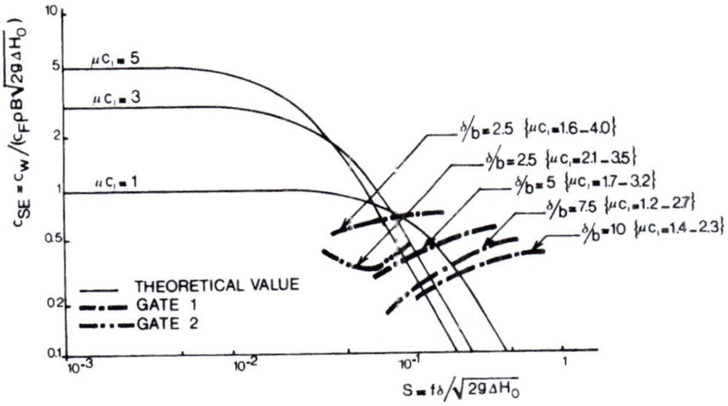

3 = b_w

FIG. 7 MINIMUM POSITIVE DAMPING AS FUNCTION OF STROUHAL NUMBER

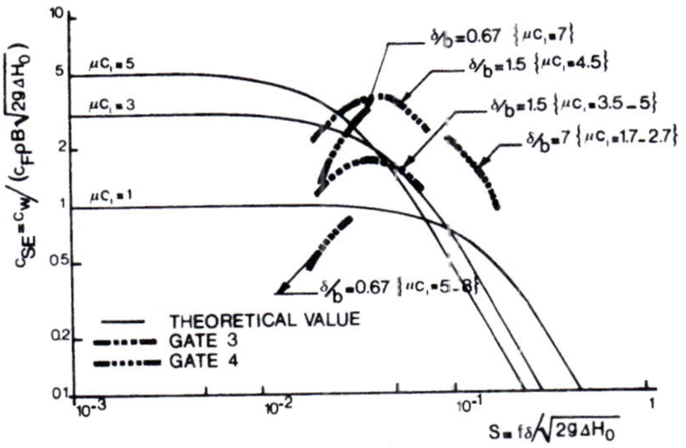

3 = b_w

FIG. 8 MAXIMUM NEGATIVE DAMPING AS FUNCTION OF STROUHAL NUMBER

5. CONCLUSIONS

The model investigation results have confirmed the theoretically forecasted behaviour of the gates:

- downstream separation of flow: for the greater part stable behaviour;
- upstream separation of flow : for the greater part unstable behaviour;
- open gate edge more stable than massive gate edge.

Figures 7 and 8 show, that theoretically forecasted values for the positive and negative damping do not fully match the values obtained through the model investigation by non-linear effects.

It should be noted that the examples do not deal with the situation where the flow separates and reattaches to the lower edge.

6. REFERENCES

1. Harding, D.A., "The stability of a Relief Valve connected to a Long Pipe", 10th IAHR Congress, London 1963, paper 3.18.
2. Abelev, A.S., Dolnikov, L.L., "Experimental Investigations of Self-Excited Vibrations of Submerged Vertical-Lift Hydraulic Gates", IUTAM/IAHR Symp. on Flow-Induced Vibrations, Karlsruhe 1972 (Springer-Verlag 1974).
3. Thomann, H., "Oscillations of a Simple Valve connected to a Pipe", Journ. of Applied Math. and Physics (ZAMP), Vol. 27, 1976, pp. 23-40.
4. Kolkman, P.A., "Flow-Induced Gate Vibrations", Publ. 164 of the Delft Hydraulics Laboratory, June 1976.
5. Kolkman, P.A. and Vrijer, A., "Gate Edge Suction as a Cause of Self-exciting Vertical Vibrations", 17th IAHR Congress, Baden-Baden, C-49, 1977.

VERTICAL VIBRATIONS of a GATE with VARIED EDGE SHAPE

(model investigation)

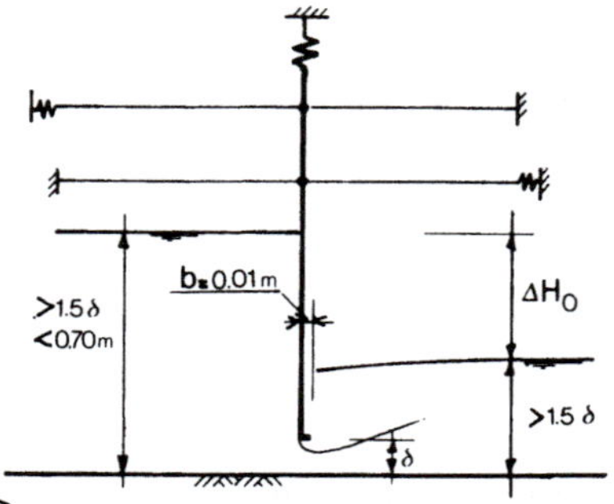

(1) HYDRAULIC MODEL OF GATE 3

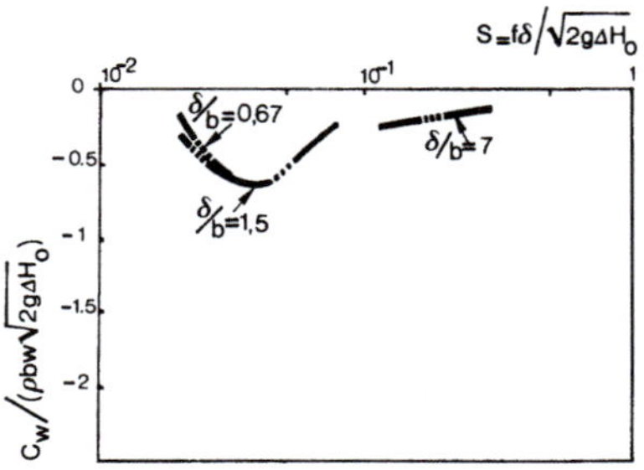

(2) NEGATIVE HYDRODYNAMIC DAMPING OF GATE 3

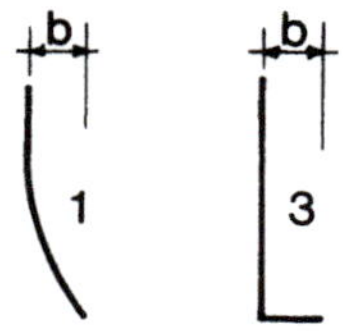

(3) SHAPE OF THE LOWER EDGES

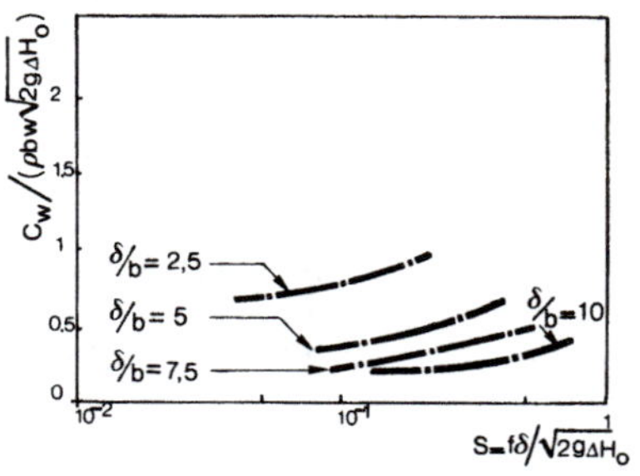

(4) POSITIVE HYDRODYNAMIC DAMPING OF GATE 1

Problem: Vertical vibration with a strictly exponential increase of the amplitude due to suction forces at the lower edge.

Geometry: See sketch 1 and 3/gate 3 as an example with so-called stable flow separation. (Remark: all shapes with unstable flow separation show also a similar problem.) Width of the gate w = 0.50 m.

Flow conditions: The critical range lies between S-values 0.012 and 0.2.
(S = Strouhal-number = $f\,\delta/\sqrt{2g\,\Delta H_O}$, where f = frequency and δ = gate raise)
Gate raise δ = 0.0067-0.07 m.

Gate 3 (sketch 3): separation of the flow at the upstream side of the gate. No reattachment of the flow.

Structural dynamics: Only a vertical vibration was allowed by means of horizontal tension wires. The gate was suspended from a spring (k = 1100-13000 N/m). No vertical mechanical damping. Frequencies 2-9 Hz. Mass of the gate m = 4-8 kg.

Vibrations: Strictly exponential increase of the amplitude in the eigen-frequency of the system until a maximum amplitude is reached. The amplitude reached a maximum due to non-linear effects of the model or due to hitting the bottom. See sketch 2, with maximum negative hydrodynamic damping c_w as function of Strouhal.
($A_n/A_{n+1} = e^{-c_w/(2mf)}$ with A = amplitude).

Mechanism: By some reason the gate moves, the discharge varies and by the inertia of the water there are extra pressure differences across the gate opening. This gives an extra vertical force in the same direction of the movement: self-excitation.

Cures: Vibration stopped with a shape of the lower edge with always downstream point of flow separation. (See sketch 3/gate 1). Each movement of the gate is then damped. (See sketch 4 with minimum positive hydrodynamic damping c_w).

C6

CAUSES AND REMEDY OF VIBRATIONS OF A HIGH-HEAD GATE

P.C. Saxena, Director

C.P. Venkataraman, Chief Research Engineer

V. Ramanathan, Research Officer

R.M. Sinnarkar, Research Assistant

Central Water and Power Research Station, Pune, India

SUMMARY

The paper highlights vibration problems encountered on a high-head gate in a tunnel and remedial measures for attenuating the vibrations. Horizontal girders supporting the hoist cylinder of the gate were found to experience severe vibrations when the gate was in its fully raised position. The tunnel had a 3.05 m long deflector along its roof just downstream of the gate grooves. Model studies following this field experience indicated that the sloping roof deflector caused rise of pressures in the gate housing. Transfer of energy from the turbulent flow in the tunnel to the water column in the gate housing caused pulsating pressures on the gate. Curtailing the length of the roof deflector from 3.05 m to 1.8 m reduced the amplitude of exciting forces by about 40 per cent.

1. INTRODUCTION

A fixed wheel gate of size 3.05 m x 6.4 m with upstream sealing and upstream skin plate was being used for closure of a diversion-cum-power tunnel (see Fig.1). The hydraulic hoist of the gate was supported on two horizontal steel girders placed on the top of the gate shaft in the body of the dam (see Fig. 2). A 3.05 m long deflector was provided on the roof of the tunnel just downstream of the gate grooves to avoid negative pressures in the transition from the circular tunnel to the rectangular section upstream of the gate grooves as well as in the vicinity of the gate grooves at full gate opening. A cross airvent of 250 mm diameter was provided for aerating the gate chamber.

At the fully open position of the gate, the girders supporting the hoist cylinder were reported to experience perceptible vibrations by the field engineers. Field observations under a head of 65 m of water revealed large-amplitude jerks followed by small amplitude vibrations in the frequency range of 18 and 20 Hz.

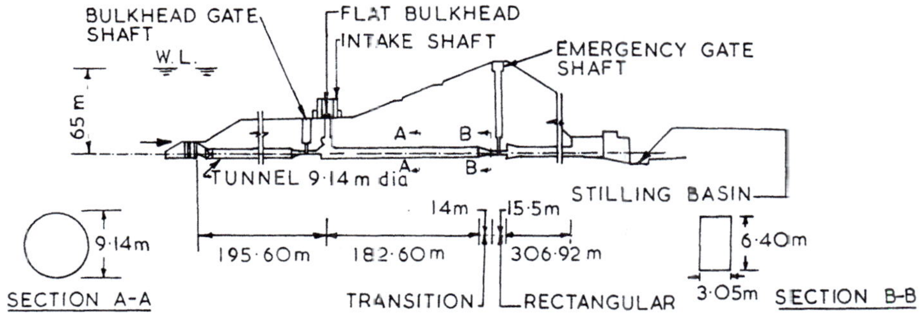

FIG. 1 SECTION OF THE DAM SHOWING THE TUNNEL

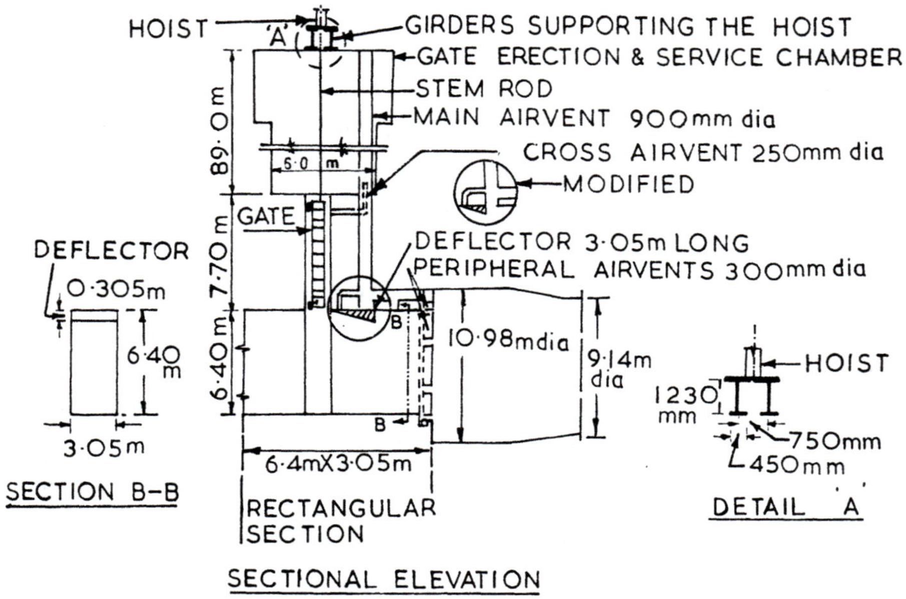

FIG. 2 DETAILS OF GATE IN SUSPENSION, AIRVENTS AND DEFLECTOR

2. MODEL STUDIES AND REMEDIAL MEASURES

Detailed studies were undertaken on a 1:30-scale hydraulic model to establish the causes of excitation and to find suitable remedial measures. The studies revealed that the roof deflector brought about pressurization of the gate shaft and the gate was fully submerged in the water column in its housing. The transfer of energy from the main turbulent stream in the tunnel to the standing water column in the gate shaft resulted in pulsating water pressures on the gate. The response of the gate to these pressure pulsations caused the vibrations which were transmitted to the supporting girders on the top of the shaft.

Pressure pulsations on the top and bottom girders of the gate were measured in order to characterize the hydrodynamic excitation force. The fluctuations of pressure on the top girder were very small in comparison with those on the bottom girder and hence were not considered. The amplitudes of exciting forces on the bottom girder were found to depend upon the length of the roof deflector. The amplitudes decreased with the decrease in the length of the deflector. Figure 3 shows the spectral densities as well as the root-mean-square amplitudes of the pressures on the girder of the gate for various lengths of the roof deflector. Curtailing the length of the roof deflector from 3.05 m to 1.80 m reduced the amplitudes of exciting forces by about 40 percent. The frequencies of the exciting forces did not undergo any appreciable change. Also, the intertial, damping, and elastic characteristics of the system were unaltered. Hence the "mechanical admittance function" [1] , which depended upon the frequencies and the system dynamics, was also unaltered. This implied that the amplitudes of vibrations would undergo changes proportional to the changes in the amplitudes of exciting forces, i.e., the amplitude of girder vibrations would be reduced by about 40 percent, thereby reducing the energy content of the vibration by about 65 percent. Further curtailment of the length of the deflector was not considered desirable as it was likely to affect unfavorably the pressures in the transition and in the vicinity of the gate grooves. Curtailing the length of the deflector from 3.05 m to 1.80 m also helped to open the airvent exit which was originally blocked by the deflector.

The studies also revealed that the water from the gate housing escaped through the cross airvent into the main airvent and choked it partially. The choking of the main airvent was stopped by extending the cross airvent independently above the maximum expected water level in the gate housing (see Fig. 2). This completely stopped the flow of water in the main airvent and created clear passage for air.

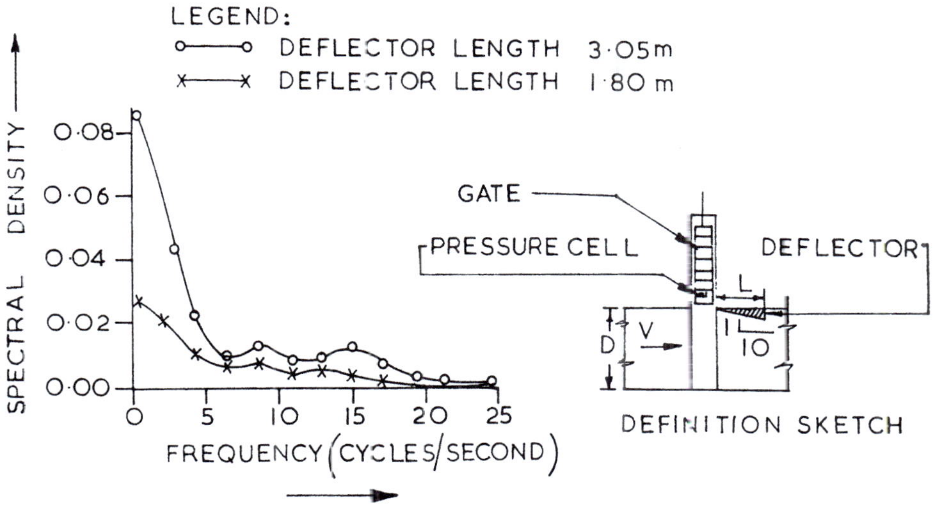

FIG. 3 EFFECT OF DEFLECTOR LENGTH ON EXCITING FORCES OF THE GATE

3. CONCLUSIONS

The experience gained in the prototype and subsequent model studies conclusively proved that curtailing the length of the roof deflector (thereby reducing the convergence) considerably reduced vibration of the emergency gate.

4. REFERENCE

1. Eduard Naudascher, "Flow-Induced Forces and Vibrations - Toward Safer Design in Hydraulic and Wind Engineering", Intensive Course held at CWPRS, Pune, India, Sept. 1978.

HIGH-HEAD GATE

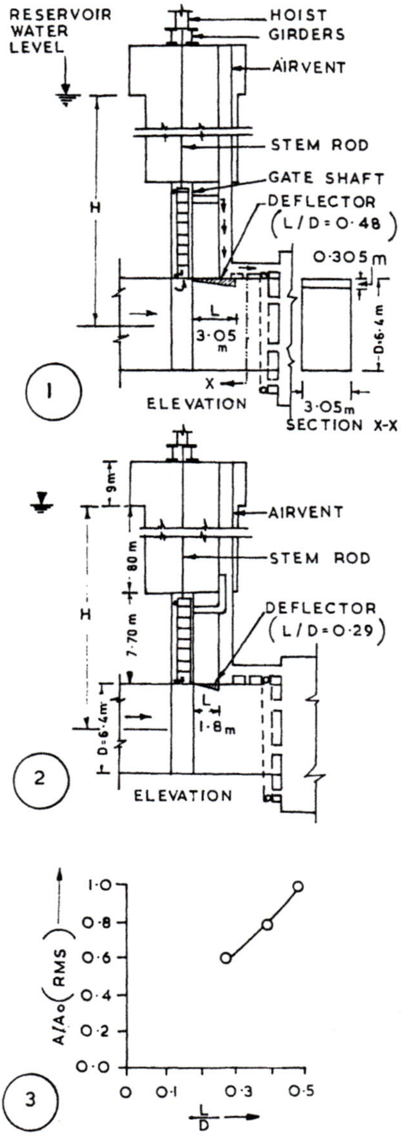

Problem: Vertical vibrations of horizontal girders supporting the hoist cylinder of a conduit gate. Gate in fully open position.

Geometry: See sketch 1.

Flow conditions: Flow in tunnel upstream of gate is turbulent.

Operating Conditions: H = 65.0 m with gate fully open.

Vibration: Roof deflector pressurizes gate shaft. Transfer of energy from the main turbulent flow to water column in gate shaft and consequent excitation of the gate. Girders supporting hoist experienced large amplitude jerks followed by small-amplitude vibrations in the frequency range of 18 to 20 Hz. Stron vibration intensity.

Structural dynamics: Natural frequency of the girders (vertical) = 9.0 Hz. Moment of inertia of the girders about horizontal axis = 0.016 m.

Cure: Reduce length of roof deflector from 3.05 m to 1.80 m (see sketch 2). Amplitudes of exciting forces reduced by 40 percent (see results of model study in sketch 3). A_o = Amplitude of exciting force for deflector length = 3.05 m. (L/D = 0.48). A = Amplitude of exciting force for reduced length of deflector.

C7 GATE VIBRATION AT EL CHOCON HYDRO-POWER SCHEME, ARGENTINA

J.D. Hardwick
Lecturer in Civil Engineering Hydraulics
M.J. Kenn
Senior Lecturer in Civil Engineering Hydraulics
Imperial College, London SW7, England
W.T. Mee
Consultant
Sir Alexander Gibb and Partners, Consulting Engineers,
Early House, Reading, England

SUMMARY

Emergency control gates in the intakes of a hydro-power scheme were observed to vibrate when fully open for a limited range of load on the turbines. There was a distinct possibility of damage to the supports of the gates if the magnitude of the vibrations was not reduced. A study based on field measurements and model investigations identified disturbances in the flow local to each gate and also water hammer pressure pulsations in each penstock and draft tube; it was suggested that a combination of these disturbances caused the excitation of the gates. A cure involving novel devices for providing increased damping was successfully applied without penalty to the operating characteristics of the gates.

THE SCHEME:

Description of the problem: The El Chocon Hydro-Power Scheme in North Patagonia, Argentina has six turbines, each with its associated intake, penstock and draft tube (see Chart). A 120 tonne vertical-lift gate in each intake is required to close rapidly in an emergency; during normal power generation the gates are designed to remain fully open and well clear of the mainstream flow (Detail A in Chart). On completion of the scheme it was observed that when a turbine was generating at part load, the fully open gate in the intake vibrated in a vertical direction with peak-to-peak amplitudes estimated to be as much as 4 mm. Concern for the safety of the gate-suspensions led to field measurements which showed that significant vibration was limited to a range of turbine load from 80-160 MW corresponding to a velocity range of 1.6 to 2.9 m/s near the gate. No vibration was observed at 180 and 200 MW. The vibration was irregular but with one clearly identifiable frequency component (Figs. 1 and 2). Pressure measurements in the scroll case (just upstream of the turbine) and in the draft tube (just downstream of the turbine) showed irregular pulsations over a wide range of frequency.

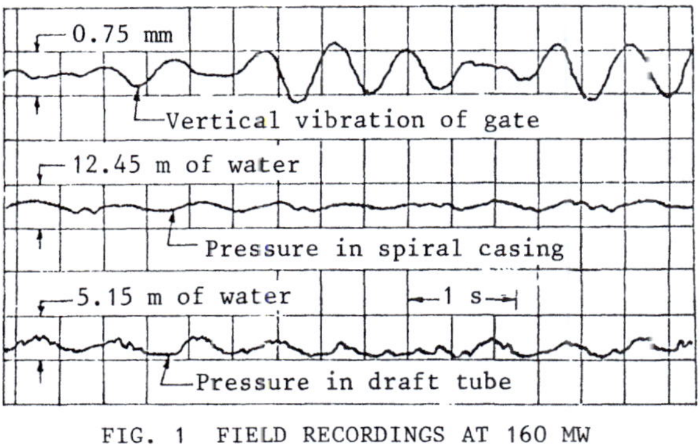

FIG. 1 FIELD RECORDINGS AT 160 MW

Preliminary investigations: An assessment of the field measurements and the characteristics of the gates showed the following.

(i) The most obvious component of recorded gate-vibration frequency varied from 1.4 Hz to 2.0 Hz with an average value of 1.7 Hz.

(ii) The effective mass of a gate was approximately 400 tonnes, comprising 120 tonnes of steel, 170 tonnes of water within the gate and 110 tonnes of added water mass external to the gate. The elasticity of the assembly was mostly in the light tension members connecting the gate to a hydraulic servomotor.

(iii) The natural frequency of the gates for vertical vibration based on the estimated mass and stiffness was calculated to be 1.8 Hz.

(iv) Each gate was designed with low-friction rollers and guides to facilitate emergency closure under the gate's own weight. The damping due to friction would be least when the gate was fully open and not transmitting any horizontal hydraulic thrust to its supports. It was not possible to ascertain the damping force with any certainty but a value was determined from an analysis of the gate-vibration shown in Fig. 2 where the trace has the appearance of a decaying transient following an impulse. The equivalent viscous damping relative to critical damping (c/c_c) appeared to be in the region of 0.17 for a fully open gate.

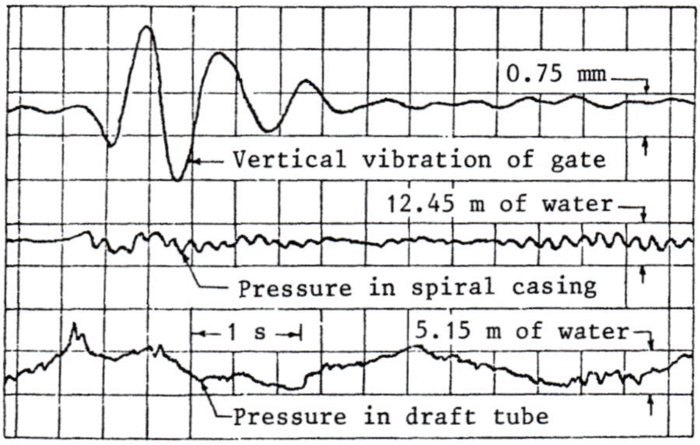

FIG. 2 FIELD RECORDINGS AT 100 MW

It was concluded from these studies that a gate and its supports could be represented as a lightly damped but heavy mass on a weak spring with a relatively low natural frequency[1]. The close proximity of the observed (1.7 Hz) and calculated (1.8 Hz) frequencies suggested that the vibration of a gate was either

(i) self-excited (or self-controlled)[2], or

(ii) forced by a fluctuating hydrodynamic load with a frequency close to the resonant frequency of the gate.

Model studies: Model studies were conducted with the aim of discovering the source of the exciting force causing gate-vibration. Initially two models at a scale of 1:100 were constructed to demonstrate qualitatively the accelerating flow patterns within the intake and close to the gate-bottom. For simplicity, an attempt was made to reproduce the accelerating flow by a *two-dimensional* contraction rather than with the *three-dimensional* contraction of the full-scale intake. In these models the gate was not free to vibrate and no influence of the penstock was incorporated.

The first model showed the separated flow and the free shear layer beneath the gate (Detail A, Skh.1) but there was no evidence of the fluctuating reattachment of the mainstream which can sometimes excite a virtually flat-bottomed gate [2.3]. The second model demonstrated the flow pattern in a plane at right angles to the first view (Section X-X, Skh. 1) and reproduced the three concrete piers supporting the superstructure of the intake. A periodic train of vertically-oriented eddies was shed from alternate sides of the long, central pier and this suggested the possibility of interaction involving these vertically-oriented eddies, the horizontally-oriented eddies of the free shear layer and the vibrations of the gate.

Accordingly a third model of the intake was constructed at a scale of 1:20 which included the central pier, an appropriate acceleration in two dimensions and a gate free to vibrate at a frequency determined by a Strouhal number criterion, i.e. the ratios of [gate frequency × gate width/mainstream velocity] were made the same for both model and prototype. The Reynolds number was one hundred times smaller than in the prototype but the boundary layers in the model were turbulent. Again, no attempt was made to model any features of the penstock. This study yielded the following results.

(i) The predicted range of frequency of single eddies shed from the full-scale central pier was 1 Hz to 2 Hz depending on the flow velocity. This range was regarded as approximate because errors in modelling the acceleration through the intake could lead to unrepresentative separation points on the tapering end of the pier.

(ii) Turbulent pressure pulsations in the separated flow beneath the model gate caused irregular vibration but there was no evidence of a strong interaction between the movement of the gate and the local flow pattern.

(iii) The *predicted* full-scale amplitude of vibration based on observations of model behaviour were only 20% of the *actual* amplitudes recorded for the prototype gate in the field.

It was concluded from these model studies that the vibration of the gate was not self-exicted in any local phenomenon. Nevertheless it was believed that the eddies shed from the central pier were significant because their frequency range (1 to 2 Hz) embraced the natural frequency of the gate (1.7 to 1.8 Hz). Also, since the *local* disturbances were evidently too weak to account for the large excitation of the full-scale gate, it was suggested that the pressure pulsations associated with the eddies had been amplified by some process downstream of the gate which was not reproduced in this 1:20 scale model.

Water hammer investigations: The field measurements shown in Fig. 1 indicated a correlation between vibration of the gate and pressure pulsations in the scroll case and draft tube. The correlation was less obvious in Fig. 2 but an analysis showed that if the gate had been exposed directly to the pressure fluctuation in the draft tube the vibration response of the gate would have been similar in form (but not amplitude) to that shown in Fig. 2. Water hammer pressure pulsations were the only feasible link between the gate and the draft tube and this led to a consideration of the frequencies at which the penstock could resonate.

The most likely resonant condition was that associated with waves traversing the distance between the intake and the draft tube exit; this mode (corresponding to that of an organ pipe open at each end) had a natural frequency of approximately 1.9 Hz given by the ratio [velocity of a pressure wave in conduit / (2 × length of conduit)]. Other modes involving reflections at the turbine or at the downward bend in the penstock gave frequencies outside the significant range of 1 to 2 Hz and could perhaps account in part for the wide range of frequency-components recorded in the draft tube (Figs. 1 and 2). The close similarity of the 1.9 Hz water hammer frequency to the eddy-shedding frequency range suggested an origin for the gate-vibration. Other sources of pressure fluctuations were recognized, e.g. from the turbines, but were not considered to be the primary source in this instance.

Origin of forces causing gate-vibration: Over the range of turbine load 80-160 MW, water hammer pulsations were generated in the penstock with a wide range of frequencies. Some of these pulsations were probably amplified and sustained over many cycles because they had frequencies close to a resonant frequency of the penstock (≈ 1.9 Hz). The origin of the pulsations in this narrow range of frequency was believed to be in the disturbances associated with eddies shed from the central pier in the intake. The frequency of these disturbances would vary with the velocity local to the central pier and thus be dependent on the turbine load; in a certain velocity range the shedding frequency would approach the resonant frequency of the penstock and amplification of pressure pulsations would occur. For turbine loads in excess of 160 MW the frequency of these eddies was probably too high to excite resonance.

The natural frequency of the gate appeared to be close to the resonant frequency of the penstock and when pulsations close to this frequency were present in the penstock the lightly-damped gate would vibrate with large amplitudes.

The *direct* influence of the eddies on the gate was shown in the model tests to be weak and the wake of the central pier could not have affected more than 10% of the gate-span. The *indirect* influence of the eddies became important when amplified by the resonating penstock and when extended over the full span of the gate-bottom by the pressure waves.

Suppression of gate-vibration: The vibration of a gate appeared to be *forced* rather than *self-excited* and had it been possible to remove the natural frequency of the gate from the 1 - 2 Hz range the amplitude of vibration would have been considerably reduced. However, it would not have been prudent either to increase the mass of the gate or to decrease the stiffness of the supports in any attempt to lower the natural frequency. Conversely, stiffening of the supports to raise the natural frequency would have created maintenance difficulties because of the confined space of the gate-well.

Some consideration was given to a method of modifying the eddy-shedding process but this was eventually rejected because it would have involved expensive alterations to the geometry of the central pier.

The solution finally recommended and eventually adopted was to increase the damping of the gate by the introduction of pads so designed as to generate vertical friction forces in opposition to the motion of the fully open gate. The pads (Skh. 2) would be engaged only when the gate was virtually fully open; during an emergency closure the pads would be ineffective and the advantage gained by using low-friction rollers and guides would not be lost.

A value for the constant friction force to be applied by the pads was deduced from the apparently transient vibration-record taken in the original field measurements. The equivalent constant friction force, F_o, associated with the decaying vibration shown in Fig.2 was estimated from

$$F_o = [\tfrac{\pi}{2} \times (\text{"average" amplitude of transient}) \times (\text{spring constant of gate}) \times (\text{relative damping})] \simeq 5\ \text{kN}.$$

It was recommended that the friction force should be of the order of 50 kN in order to suppress the vibration of the gate in the fully open position.

Eventually two pads were rigidly fixed to each gate at the ends of the gate-span. Each pad comprised a chamfered steel buffer plate backed by elastic packing of adjustable stiffness. When the gate was fully open the buffer made contact with a steel ramp fixed to the wall of the gate-well and the necessary normal force was provided by a compression of the elastic packing. The corresponding frictional force was designed to be 30 kN and so the total frictional force on a gate was approximately $(2 \times 30) + 5 = 65$ kN or $13\ F_o$.

Field tests conducted after installation of the pads showed that gate-vibration had been satisfactorily reduced to barely detectable levels.

Conclusion: Economic and operational considerations can usually be clearly defined at the design stage of a scheme and consequently they exert a strong influence on the design. In contrast, sources of fluid-dynamic excitation are more difficult to identify and inevitably tend to be given less weight and consideration. At El Chocon the prediction of the eddy-shedding frequency for the central piers would not have been straightforward, particularly as the pier is located in a region of strongly accelerating flow. In the event the solution adopted was to suppress the vibration rather than to remove the source.

Acknowledgements: The Authors are most grateful to Hidroelectrica Norpatagonica S.A. for their kind permission to publish the results of this investigation and also acknowledge the assistance of Boving & Co.Ltd. in obtaining field measurements.

REFERENCES:

1. Campbell, F.B., "Vibration problems in hydraulic structures", Journal of the Hydraulics Division, American Society of Civil Engineers, Vol. 87, No. HY2, pp. 61-77, March, 1961.

2. Naudascher, E., "On the role of eddies in flow-induced vibrations", Proceedings of the Tenth Congress of the International Association for Hydraulic Research, Vol. 3, pp. 61-72, London, 1963.

3. Hardwick, J.D., "Flow-induced vibration of a vertical-lift gate", Journal of the Hydraulics Division, American Society of Civil Engineers, Vol. 100, No. HY5, pp. 631-644, May, 1974.

DISCUSSION

In answer to various queries the authors wish to emphasize that the present investigations have shown that:

(a) eddies generated in one plane can be the indirect cause of gate vibration in an entirely different plane;

(b) flow-induced vibrations may be interlinked and naive attempts to "classify" all flow-induced vibrations into neat categories must not be permitted to obscure the vision of such interlinked problems in particular investigations;

(c) "water-hammer" in a large tunnel is accompanied by large reversals of water flow within the pipeline (except at any velocity nodes). The flow reversals are accompanied by (and associated with) substantial changes of water pressure, even near the tunnel intake (because of the mass and inertia of the adjacent water outside). It would therefore be erroneous to imagine pressure nodes to occur at the gate in the tunnel intake.

GATE VIBRATION AT EL CHOCON HYDRO-POWER SCHEME

Problem: Vibration of a fully open, vertical-lift gate. Possible damage to supports of gate if vibration allowed to persist.

Geometry: Skh. 1. Gate normally used in fully-open position (Detail A) and intended only for rapid emergency closure.

Boundary conditions: Gate (mass 120 tonnes) suspended on light tension members connected to a hydraulic servomotor. Low friction at rollers and guides.

Flow conditions: Alternate eddy-shedding from central pier just upstream of gate (Section XX) with free shear layer below gate (Detail A).

Operating conditions: Vibration significant for velocity range 1.6 to 2.9 m/s local to gate.

Structural dynamics: Calculated vertical natural frequency of gate (≈ 1.8 Hz) close to observed vibration frequency (≈ 1.7 Hz). Estimated frequency of eddies shed by pier 1 to 2 Hz. Estimate of one resonant frequency of penstock for water hammer pressure waves ≈ 1.9 Hz.

Vibration: Amplitudes of gate-vibration up to 4 mm (p.-p.) excited by water hammer pressure waves in penstock with frequency close to natural frequency of gate. These waves believed to be initiated by pressure pulsations associated with eddy-shedding from pier at frequency close to a resonant frequency of penstock.

Cure: Damping of gate increased by friction pads (Skh. 2) engaged only near highest region of gate-travel.

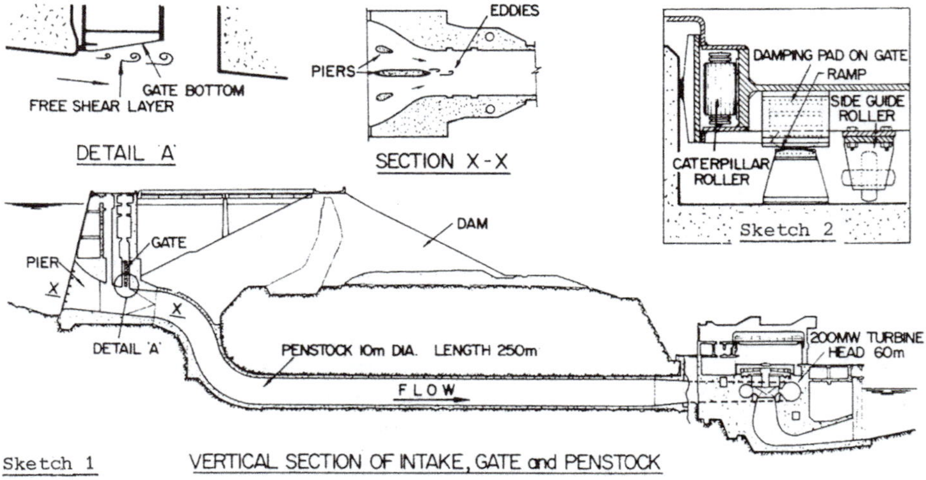

Sketch 2

Sketch 1 VERTICAL SECTION OF INTAKE, GATE and PENSTOCK

C8

EXCITATION AND VIBRATION OF A GRID GATE*

R.J. de Jong
Deputy head of the Locks, Weirs and Sluices Branch of the Delft Hydraulics Laboratory, the Netherlands

J.W.G. van Nunen**
Deputy Director of the German-Dutch Wind Tunnel, Emmeloord, the Netherlands

SUMMARY

This paper includes the results of an investigation relating to the static and dynamic loads and the response of a grid gate. Strong vibrations were found in open gate position, introduced by vortex generation and clearly reinforced by coupling effects. Vibrations also occur at 60% opening of gate, due to flow instability. Both types of vibration can be avoided. Reynolds effects, although present, were not predominant.

1. INTRODUCTION

Within the framework of studies for the storm surge barrier of the Eastern Scheldt in the Netherlands, in the years 1975-1976 a study was made of closure systems in big caissons. The object of the storm surge barrier is to let the tide pass (mean tide 2.30 m) into the Eastern Scheldt area (310 square km) but to close the barrier during extreme circumstances. This means that large openings (14.000 square m) in the barrier have to be closed. The whole study was performed in the preparation stage for changing the plans from closing the area and creating a stagnant fresh-water basin into the situation with the storm surge barrier. The decision to make a storm surge barrier was made in 1976. The whole work must be completed in 1985. As one of the alternative solutions the grid gate was suggested.
The gate consists of two frames located directly behind each other, in which beams are mounted in such a way that, in equal positions, a streamlined grid is formed while, after moving over one beam height, the closure is realized - see Sketch 1 of the Chart. The advantage of this gate type is that there is an optimum flow distribution over the cross-section in comparison to normal gates with the same discharge capacity and that the lifting system needs a short stroke. Later on in the project, foundation problems, however, led to the choice of another system in which this type of gate was no longer appropriate.

2. INVESTIGATIONS

An investigation scheme was drafted to become acquainted with Reynolds effects and to obtain static and dynamic loads for the design of the structure. Subsequently, response studies using elastic similarity models were planned.
At first, a simple model was used in a flume to study stream line patterns. Next to that, two models were built in a water tunnel - a two-beam pair model on a scale of 1:2.75 and a five-beam pair model on a scale of 1:6.9 to study static and dynamic loading with pressure cells for different gate-opening ratios.
After that a wind tunnel model study was done by the National Aerospace Laboratory (NLR) to look for Reynolds effects in a three-beam pair model on a scale of 1:1.9. In this test almost full-scale Reynolds numbers were attained. As all models dealt with different aspects of the prototype situation, much effort was

---

*The investigations were carried out for the State Public Works, Delta Works, The Hague

**Formerly with National Aerospace Laboratory, Amsterdam, the Netherlands placed in ensuring a good relationship between the various tests. Attention was also paid to wall effects of ceiling and bottom. The response study was done with the aid of an elastic similarity model on a scale of 1:23 in an open flume. The model consisted of twelve beam pairs and had a free water surface; one beam pair was instrumented with strain gauges and accelerometers. A more complete study on a 1:3 elastic model was cancelled because the design plans were changed at that time.

3. RESULTS

3.1 Reynolds effects

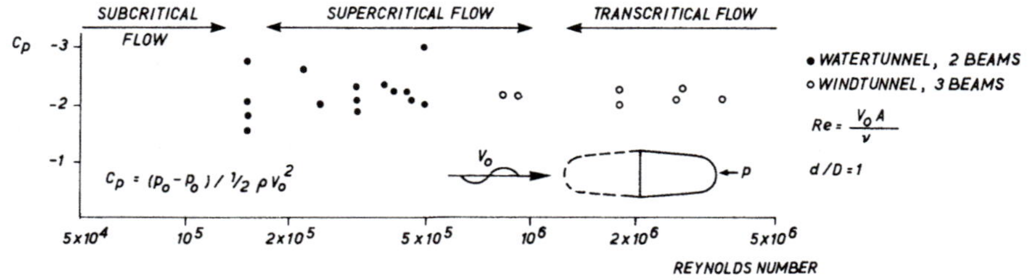

FIG. 1 AN EXAMPLE OF A C_p-COEFFICIENT AS A FUNCTION OF THE REYNOLDS NUMBER

Considering the Reynolds-number influence it appeared that, even at open-gate position, static pressures hardly vary; however, some scatter is found at supercritical Reynolds numbers - see Figure 1. The scatter of the fluctuating part of the local pressures is clearly greater, but spectra in the subcritical and transcritical Reynolds range exhibit a striking correspondence and moreover the same Strouhal number (about 0.5) was found - see Figure 2. In the area of supercritical Reynolds numbers, the energy was distributed like "noise" but still with some

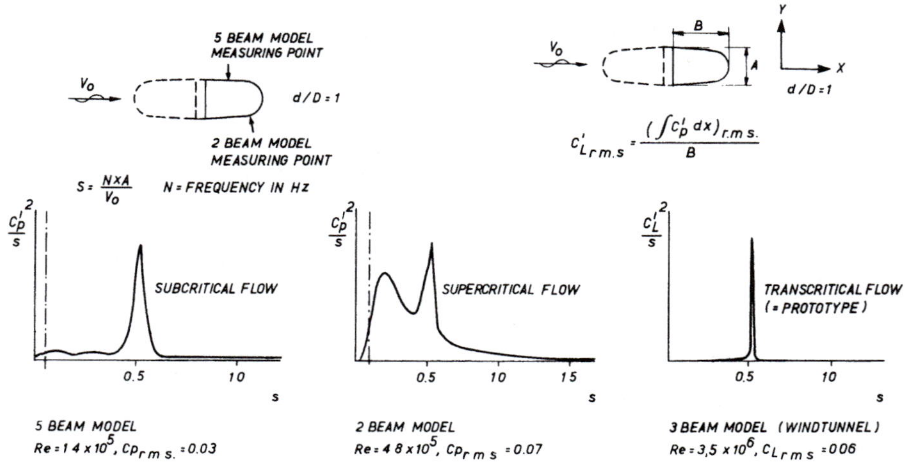

FIG. 2 SPECTRA OF RELATED RESULTS FOR DIFFERENT REYNOLDS NUMBERS

concentration of energy at the former Strouhal number. C_L- and C_D-values give less scatter than C_p-values. However, concerning C_D-values, there is an discrepancy in model results for an open-gate situation. The C_D-values from the 5-beam pair water tunnel model are 3 and -1.7 for the upstream and downstream beams, respectively 2 and -0.7 for the 3-beam pair wind tunnel model (values for the summation are the same). This is possibly due to effects of the slit between the upstream and downstream beam on the boundary layer or in other words, effects of the distance between a pair of beams. The Reynolds effects being small it could be concluded that the overall vibrations could be performed with the aid of a small elastic similarity model.

3.2 Static loads

The different tunnel models were also used to determine loads for different gate-opening ratios. In cases of partly closed situations, Reynolds effects are less probable but tunnel-wall effects become serious: the bottom and ceiling are a severe restriction for the development of a realistic flow pattern, so mainly the 5-beam pair model was used - see Figure 3. A remarkable result was obtained

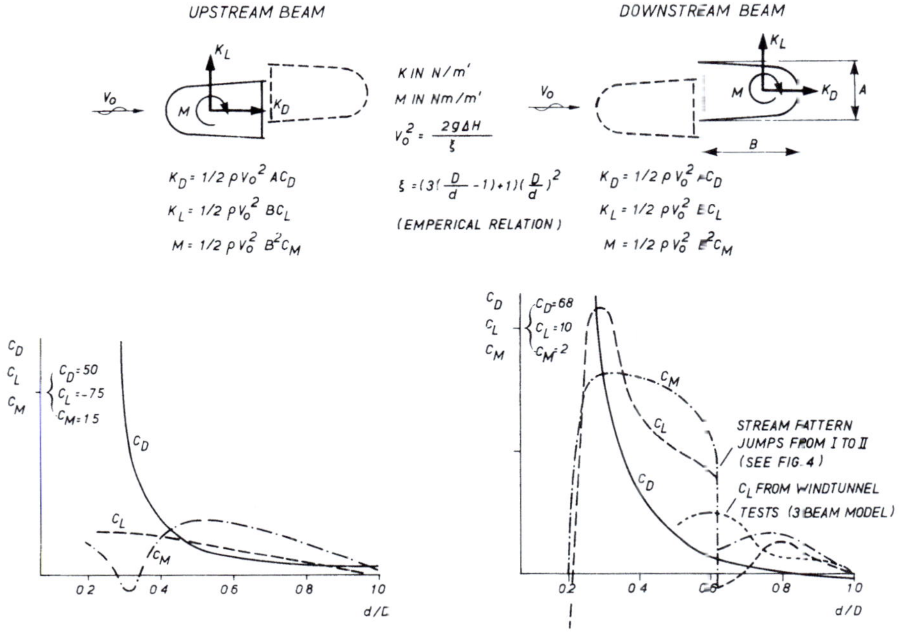

FIG. 3 DRAG, LIFT, AND TORSION COEFFICIENTS AS A FUNCTION OF THE OPENING RATIO d/D; RESULTS FROM THE 5-BEAM MODEL IN THE WATER TUNNEL

for the downstream beam at about 60% opening of the gate: C_L- and C_M- values change strongly. From a study of pressure distributions and with the aid of flow visualization it was found that the stream pattern I of Figure 4, normally valid for that opening, could jump to pattern II. Probably the pattern jumps only once and this will not cause dynamic excitation. However, if a steep but gradual change in forces occurs (how steep could not be measured), Figure 3 would indicate self-excitation following the instability criterion [1].
The wind-tunnel model results at 60% opening of the gate differ from the 5-beam pair model, because of great wall effects. However, in the five-beam pair model wall effects still can be expected.

3.3 Vibrations at 40-70% opening of gate

In the elastic model of the complete gate (12-beam pairs) it appeared that, at low-head differences, vibrations occurred in vertical direction at 40 till 70%

opening of the gate - see Figure 4. These vibrations were observed on both the upstream and downstream beam with about the same magnitude. In the horizontal direction the extra amplification could also be distinguished at the same frequency as in vertical direction, but less by about a factor of 2. Concerning the higher horizontal resonant frequencies, this means, that this is a quasi-static (forced) movement.

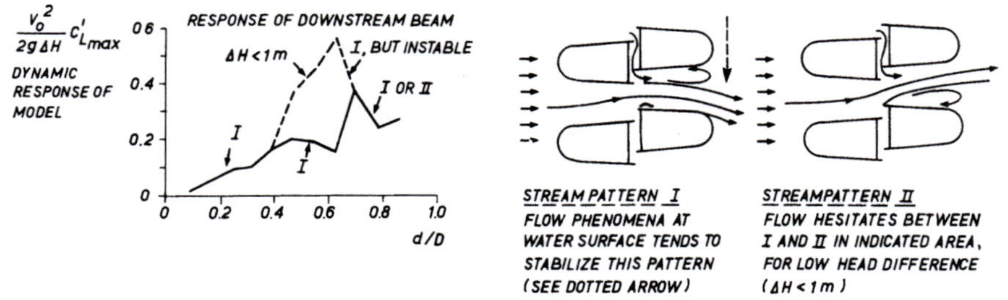

FIG. 4 RESPONSES AND STREAM PATTERNS, RESULTS FROM ELASTIC SIMILARITY MODEL

It was concluded that stream pattern I, which creates low pressures at the upper side of the downstream beam, causes these vibrations due to unstable reattachment of the flow.

At higher water-level differences in the elastic similarity model the outflow was pressed downward, thus enforcing stream pattern I and suppressing vibrations (see dotted arrow in Fig. 4).

Static vertical forces exhibit jumps similar to Figure 3, however, this happened at gate openings of 75 till 90%. The shift in opening ratio where the stream pattern jumps, was attributed to ceiling and bottom effects of the tunnel model.

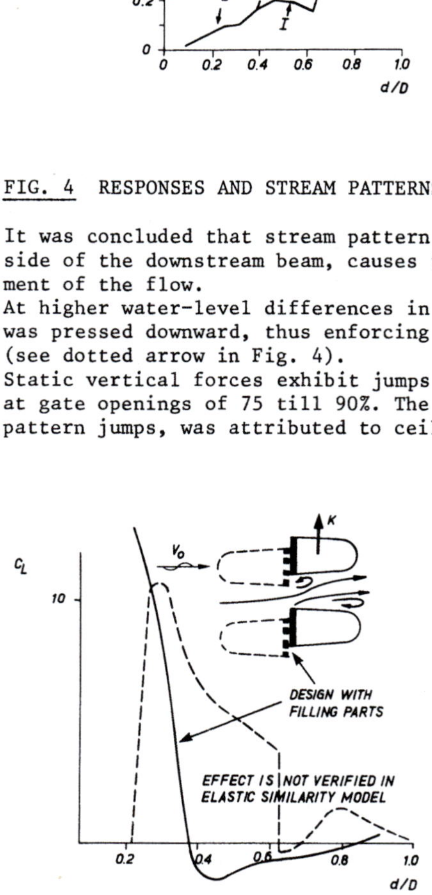

FIG. 5 EFFECT OF MODIFYING THE DESIGN TO PREVENT EXCITATION AT d/D = 0.4 TO 0.65 (RESULTS FROM 5-BEAM MODEL)

Despite the fact that vibrations were not very strong and disappeared at higher water-level differences, it is possible that higher gates are less affected by surface effects and vibrate more. For that reason, the influence of stopping the leakage flow between the beams, thus stabilizing stream pattern II, was studied in the 5-beam pair model. It seems - see Figure 5 - that this measure is effective. Still, there exists a possibility for a transition to stream pattern I, however, more gradual and for smaller gate openings, which means that forces are comparatively small, as can be seen on Sketch 2 of the Chart. (This sketch represents equal information as Fig. 5.) In what measure surface effects influence this behaviour was not checked further. (The observed effect holds for deep-laying beams; other forces were acting on the surface beams, but these were not decisive for the design.)

3.4 Vibrations at open-gate positions

On the basis of results on Strouhal number (about 0.52) and resonant frequency, it was expected that amplification would be obtained near v_o = 8.5 m/s.

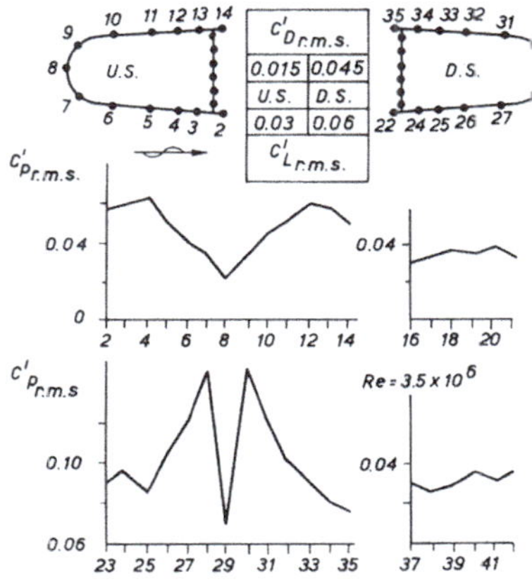

FIG. 6 EXCITATION OF BEAMS AT OPEN-GATE POSITIONS (RESULTS FROM WIND-TUNNEL MODEL)

From the information on dynamic loads - see Figures 2 and 6 - and the low total damping ratio (about 3%) an amplification up to 1 cm amplitude (prototype value) was calculated (all beam pairs were probably equally loaded as can be seen on Figure 7). However, the amplification occurred somewhat earlier and with an unexpected violence - see Sketch 3 of the Chart. As a result of the strong vibrations a deep-laying model beam broke out of the frame - see Figure 8. Both upstream and downstream beams vibrated with about the same magnitude (upstream 20% lower) but in counter-phase. At the same time, horizontal vibrations were rather low, whereas torsional vibrations were also violent and with the same frequency. The maximum torsion moment was 50% of the maximum vertical bending moment. In Figure 9 phase information is given on the vertical and torsional vibrations.

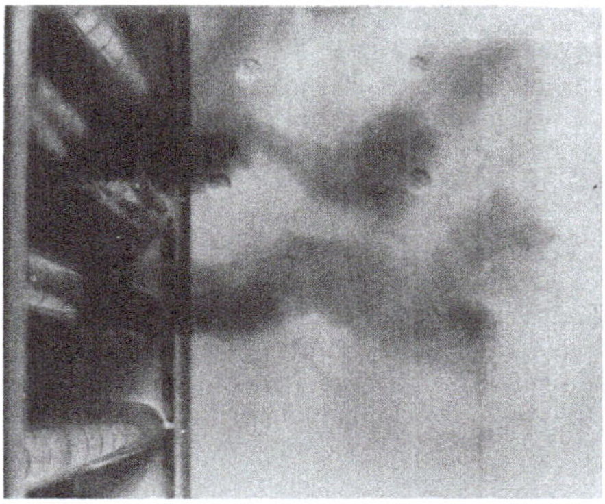

FIG. 7 VORTICES BEHIND THE GRID GATE IN OPEN POSITION (FROM ELASTIC SIMILARITY MODEL)

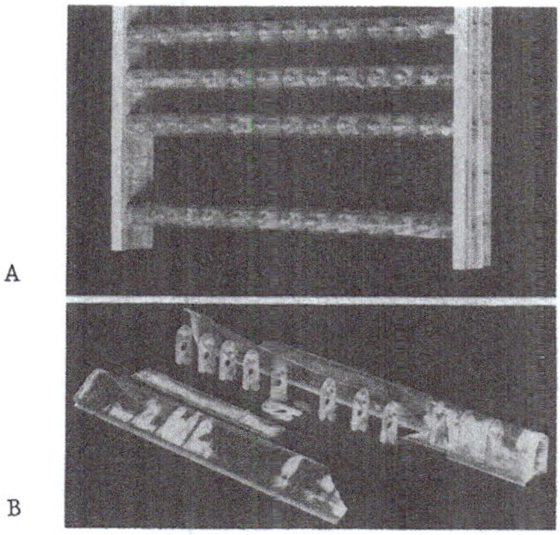

FIG. 8 PHOTOGRAPHS OF DESTROYED DOWNSTREAM BEAM FRAME (A) AND BEAM (B) (ELASTIC SIMILARITY MODEL)

The frequencies (6 Hz) were somewhat lower than the resonant frequency (about 7 Hz) of the beams in water, which is one reason why the vibrations occurred earlier. The shift in "amplification velocity" is probably also due to a local increase in velocity at the bottom of the gate (v_0 was defined as a mean velocity). At first, excitation factors were considered as given by Jones [2] for single circular cylinders, but on the basis of this alone the vibration amplitude could not be fully explained. The intense vibration can probably be better explained by interference phenomena and coupling effects. Connors [3] gives a criterion based on the displacement mechanism, separating stable and unstable situations of a row of circular cylinders exposed to cross flow. By defining such a criterion for this gate it will probably be possible to interpret the measurements more completely.

Studying the static forces, also an indication is obtained - see Sketch 2 of the Chart at d/D = nearly 1. Forces help to close the gate and grow with a growing displacement. This means that there exists a negative rigidity that counteracts

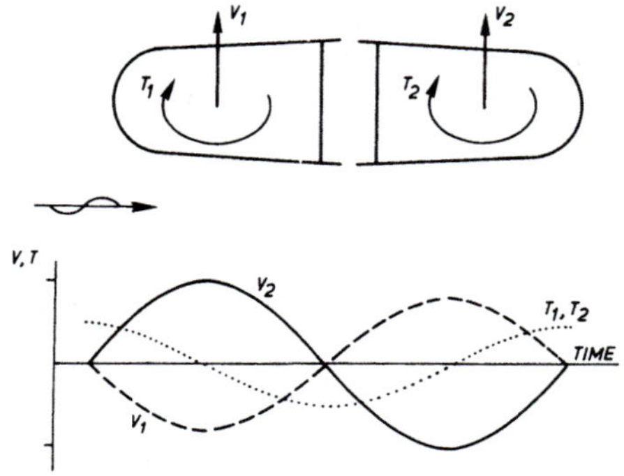

FIG. 9 TIME HISTORY OF VERTICAL BENDING MOTIONS AND TORSION MOTIONS OF UPSTREAM AND DOWNSTREAM BEAMS

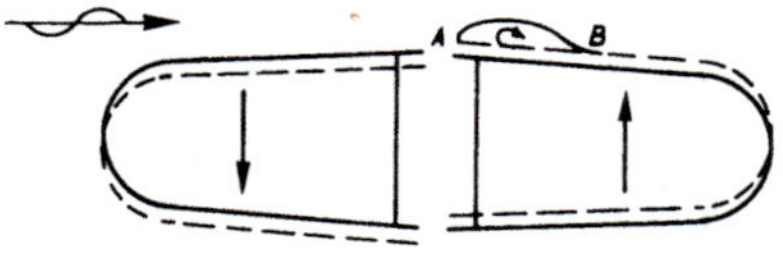

FIG. 10 STREAMLINE DURING UPWARD MOVEMENT OF DOWNSTREAM BEAM

the structural one. In case of vibration a phase shift can occur in these forces, which then results in an extra amplification. The next explanation is found.
When the downstream beam is in the upward position the streamline at A in Figure 10 separates and reattaches at B. Due to the movement in the upward direction potential flow effects enlarge the separation zone (see general lecture of Kolkman). This means an enlargement of vertical forces in the direction of the vibration velocity and energy transfers from the fluid to the gate structure, so heavy vibrations will occur. Stiffening of the beams in the vertical direction can suppress the vibrations.
Vibrations in the horizontal direction can occur at a relatively high frequency (19.5 Hz) due to the stiffness in that direction and the smaller added mass. The excitation spectra of the pressures at the downstream side of the downstream beam showed some energy in the double vortec frequency (S = 1). The high resonant frequency, however, prevents vibration in the horizontal direction at operating conditions.

3.5 Total loads

The forces, required to open or close the gate depend on water-level difference and stream direction. During part of the manipulation, hydraulic forces help to close; during another part they counteract gate closure. At open-gate position horizontal forces are high because of the low pressure between the beams.
To keep Coulomb friction forces low, slide materials must be chosen in prototype with low dynamic and static friction.

4. CONCLUSION AND RECOMMENDATIONS

In cases where it is important to keep a discharge well distributed over the height of a water course, a grid gate seems a good solution. Enough data are available to permit a reasonable design. However, more research would be recommendable concerning the required beam-skin stiffness. Depending on the circumstances one has to watch carefully for wave shocks and cavitation. The discharge capacity, moreover, can be greatly reduced by mussel growth.

5. REFERENCES

1 KOLKMAN, P.A., "Flow-Induced Gate Vibrations", Publ. 164 of Delft Hydraulics Laboratory.

2 JONES, G.W. Jr., CINCOTTA, J.J. and WALKER, R.W., "Aerodynamic forces on a stationary and oscillating circular cylinder at high Reynolds numbers", NASA TR-300, Washington D.C., 1969.

3 CONNORS, H.J. Jr., "Fluid elastic Vibration of Tube Arrays Excited by Cross Flow", Proceedings of the Symposium on Flow-Induced Vibration in Heat Exchangers, ASME Winter Annual Meeting, New York, Dec. 1, 1970, pp. 42-56.

GRID GATE

Model investigations (all sizes in prototype values)

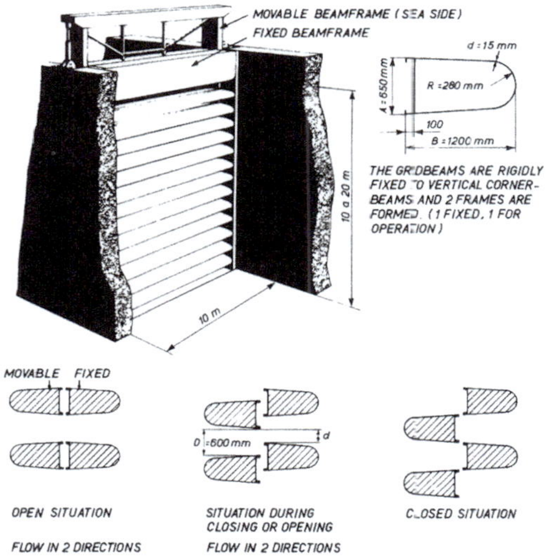

① Explanation of the principle of the grid gate

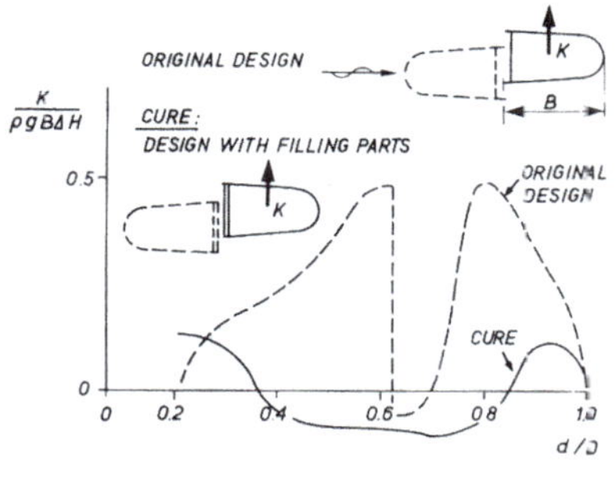

② Relative ratio of forces on downstream beam (from pressure measurements on a beam model)

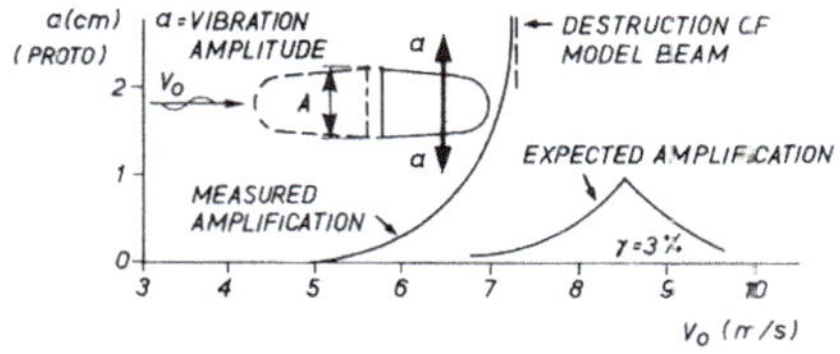

③ Measured and expected amplification of downstream beam (elastic similarity model)

Problem: Gate beam oscillations in mainly the vertical direction at 60% opening due to flow instability. Also vibrations in open position due to vortex excitation. In the latter case a model beam broke.

Geometry: See Sketch.

Flow conditions: in case of a head difference of ~1 m or more, a diving jet at the water surface influences the whole flow pattern. Flow in both directions possible, highest velocities at in-flow (in figures left to right). Approach flow without disturbance; almost no turbulence.

Operating conditions: Head differences up to 3.5 m for open gate, growing to 7 m for closed gate. Vibrations at 60% opening depend on head difference: at high ΔH, water-surface effects prevent instability of stream pattern, and dip in Sketch 2 disappears. Vibrations at open gate occur at high velocities, due to resonance between Strouhal number and natural frequency of beams.

Structural dynamics: Beam characteristics: $I_{vert.} = 0.0032\ m^4$, $I_{hor.} = 0.0073\ m^4$ $I_{torsion} = 0.0061\ m^4$.
Beams filled with water. Vertical moments at beam ends were 130% of moments in mid-sections. Structural damping ratio $\gamma = 2\%$.

Vibrations: Vibrations predominantly in vertical direction, especially in case of open gate. In general during heavy vibrations, both beams moved in counterphase.
Resonant frequency (vertical) = 7 Hz.
Excitation frequency in case of open gate = $S.v_0/A$ with $S = 0.52$.
Amplitude for 100% open gate see Sketch 3.
Upstream-beam vibration somewhat less.
Interference occurs.
Amplitude vibration at 60% opening and a head difference of 1 m is 0.5 mm for both upstream and downstream beam.

Cure: See Sketch 2 for 60% opening vibrations. To avoid vibrations at 100% opening higher rigidity is needed.

C9

DYNAMIC INSTABILITY OF TAINTER-GATES

Noriaki Ishii
Professor

Osaka Electro-Communication University
Neyegawa, Osaka 572, Japan

Kensaku Imaichi
Professor

Akio Hirose
Graduate Student

Osaka University, Toyonaka, Osaka 560, Japan

SUMMARY

One of the dynamic characteristics of the behavior of Tainter-gates is a flow-induced rotational vibration about the trunnion center. In this study, a model of a Tainter-gate is tested in order to clarify the dynamic properties of Tainter-gates. Furthermore, an actual collapse of a Tainter-gate in Japan, is investigated, and the possibility that the accident was induced by the dynamic instability discussed in this study is examined.

1. INTRODUCTION

Tainter-gates used as crest gates on dams are constructed to release large floods. Hence, the operation of the gates must be entirely free from danger. In 1967, a Tainter-gate of lighter design collapsed and was carried downstream suddenly during the test run [1]. The accident led to the question of whether the operation of Tainter-gates was reliable or not. Some studies [2,3,4] on the flow-induced dynamic instability of Tainter-gates point out that there is a possibility that dynamic factors indeed led to the accident.

As seen in Sketch 1 in Chart 2, the Tainter-gate is able to rotate about the trunnion center and is usually suspended by wire cables attached near the bottom of the weir plate. Since the wire cables generally act as an elastic support, the Tainter-gate and the wire cables compose an oscillatory system about the trunnion center. When the weir plate is slightly displaced from its position of equilibrium, a change in discharge rate is caused, wave motions are induced on the reservoir surface, and hence the hydraulic pressure on the weir plate fluctuates, corresponding to the rotational motion of the Tainter-gate. This fluctuating hydraulic pressure causes a self-excited vibration of the Tainter-gate under certain conditions since the hydraulic pressure is directly related to the motion of the Tainter-gate when the resultant of the hydraulic pressure does not pass through the trunnion center.

For a mathematical description of the above problem, the equation of motion of the Tainter-gate and the stability criteria have been derived [4]. However, the solution of this equation could not be obtained. Therefore, model tests were carried out in order to clarify the dynamic characteristics. The first model tests on small gate opening reveal the stability diagram, the instability, the frequency property and the flow-induced structural forces. The effects of the lightweight designs of Tainter-gates, the mechanism which causes the self-excited vibration and so on, are examined, on the basis of the results obtained. The second series of model tests on large gate openings confirm the above results. Furthermore, the field tests and remedies are shown and the possibility that the collapse of a Tainter-gate in Japan was caused by dynamic instability is examined.

2. MODEL TESTS

2.1 Model of Tainter-Gates

The model Tainter-gate is shown in Sketch 1 of Chart 1. The model gate consists of the weir plate with a radius R=135 mm and a width 40 mm, the two stays and the trunnion pin. The trunnion pin is supported by a small ball bearing. Since the weir plate can rotate on the pin A, the eccentricity of the trunnion pin is variable. The weir plate center of the model gate was always situated under the trunnion pin. The upper stay was set horizontal in the experiments. R_h is 86.5 mm, R_a 138.3 mm and hence Θ_0 39.2°. The model gate is lifted by a coiled spring attached to the upper stay. The position where the gate is lifted could be moved to the right and left, and the natural frequency of the model gate was hence adjustable. The moment of inertia I_0 per unit width of the model gate was 0.0575 Ns^2. The weir plate was fixed in the dam pier. But the side wall of the weir plate and the dam pier were not in contact. The depth of the gate opening d_0 and the gate opening b are variable since the base supporting the trunnion pin could be moved up and down.

2.2 Records on Vibrations

When a rotational displacement is initially given to the model gate which is at rest under the steady discharge, the model gate starts to vibrate. Some records on the vibrations are shown in Sketch 2 of Chart 1. The abscissa is the elapsed time and the ordinate the fluctuating gate opening. The maximum amplitude of the fluctuating gate opening is 10 mm. The records (a) and (b) show the intense self-excited vibrations. The model gate is so dynamically unstable that the vibrations increase very rapidly when only a very small disturbance is given to the gate. The record (c) shows a damped vibration; the model gate is so dynamically stable that the vibration decreases rapidly even for large initial disturbances. The instability of the model gate is clearly a function of the damping ratio ζ(=the logarithmic decrement/2π). The frequency property may be evaluated by $\omega'=\omega/\omega_0$.

2.3 Characteristic Factors

The characteristic factors determining the dynamic properties ζ and ω' of Tainter-gates are the five dimensionless quantities R_d', F_0', k', ζ_0, and b'. R_d' represents the water depth over the gate opening, F_0' the dimensionless natural frequency, k' a design characteristic of Tainter-gates and ζ_0 the damping effect. These four quantities are defined as follows[4]:

$$R_d'=d_0/R_h,\quad F_0'=\sqrt{d_0/g}\cdot\omega_0,\quad k'=2\sqrt{2}c_f c_g \rho R_a^3 \overline{R}_*/I_0\cdot\sin^2\Theta_0\cos\Theta_0,\quad \zeta_0=\gamma/(2\omega_0 I_0)$$

c_f and c_g for the model gate are, respectively, 0.97 and 1.04. k' becomes larger, as the gate is designed lighter in weight, or as the eccentricity of

Nomenclature

b=gate opening
$b'=b/d_0$
c_f=discharge coefficient
c_g=area coefficient
d_0=depth of gate opening
F_0=Froude number
F_0'=basic Froude number
g=gravity acceleration
h_0=water head
I_0=moment of inertia
k'=design factor
R=radius of weir plate
R_a=distance of trunnion center and gate opening
$R_d'=d_0/R_h$
R_h=height of trunnion pin
$\overline{R}_*$=eccentricity of trunnion pin
p_0'=water pressure
y=water depth
$y'=y/d_0$
β=retarded phase angle
γ=damping coefficient
δ=stroke of gate opening
$\delta'=\sqrt{2}c_f\delta$
ζ=damping ratio showing instability
ζ_0=damping ratio
Θ_0=angle between stays
ρ=specific mass of water
ω=actual frequency
ω_0=natural frequency
$\omega'=\omega/\omega_0$

the trunnion pin increases. b' represents the height of the gate opening and it is defined as $b'=b/d_0$. F_0 determining the fluctuating flow in the reservoir is given by $F_0=F_0'\cdot\omega'$.

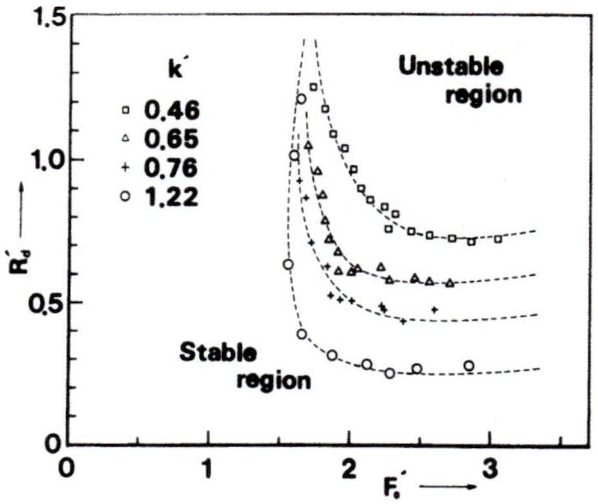

FIG. 1 STABILITY DIAGRAM (b'=0.06)

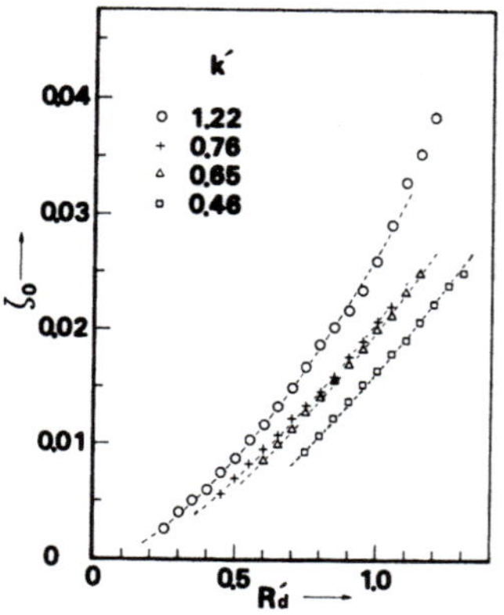

FIG. 2 DAMPING IN SYSTEM

2.4 Model Tests for a Small Gate Opening b'=0.06

Stability diagram in Figure 1 shows the effects of k' on the boundary between the stable and the unstable states. All boundary curves tend to ∞ near F_0'=1.6. Hence, when F_0' is smaller than about 1.6, self-excited vibrations do not occur. As F_0' increases, all boundary curves level off with values of R_d' which are almost inversely proportional to k'. It can hence be concluded that it is dangerous to design Tainter-gates with large k'. Terefore, careful consideration must be given to Tainter-gates which are designed lighter in weight.

The model gate was so designed as to minimize the damping effects. But, since the moment of inertia of the model gate is comparatively small, the damping effects cannot be neglected. They can be evaluated from the stability diagram shown in Figure 1. On the boundary curves, the energy supplied to the motion of the gate is equal to that dissipated by the damping. Hence, when the hydrodynamic forces exerted on the weir plate are obtained, the damping effects can be evaluated as ζ_0. The results obtained in this way are shown in Figure 2. ζ_0 becomes larger, as R_d' or k' increases.

The instability of the model gate for k'=1.22 is shown in Figure 3. When ζ is negative, there exists a negative resistance sufficient to overcome ζ_0 shown in Figure 2, and the absolute value $|\zeta|$ characterizes the instability. When R_d' is 0.2, ζ is positive for all values of F_0'. When R_d' is large, there exists a region in which ζ is negative, and the instability becomes larger, as R_d' increases. As F_0' increases, the instability becomes larger at first and then smaller. The value of F_0' for maximum instability gets nearer the stable/unstable boundary, as R_d' increases.

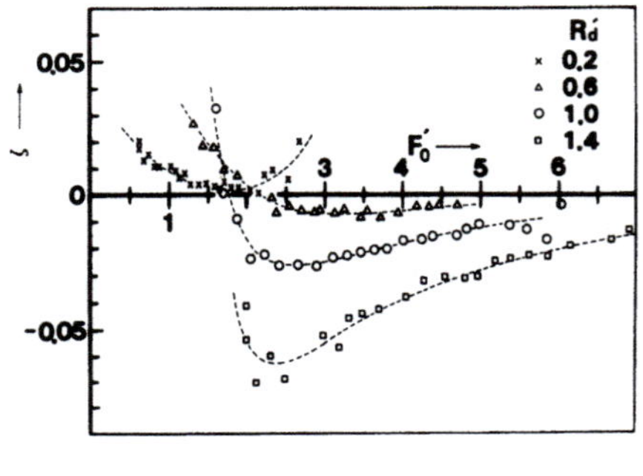

FIG. 3 INSTABILITY (k'=1.22, b'=0.06)

When R'_d is 1.4, the maximum instability value reaches 0.07. Since ζ_0 is about 0.05 in this case, it can be concluded that the negative damping due to the flow-induced forces corresponds to the considerably large damping ratio -0.12.

The frequency property for $k'=0.76$ is shown in Figure 4. Since the unstable region is the region of about $F'_0>1.6$, one may conclude that the frequency of the self-excited vibrations agrees with the natural one. But near the boundary, the former is slightly smaller than the latter. The frequency of the damped vibration is smaller than the natural one. This tendency becomes more and more pronounced, as R'_d increases.

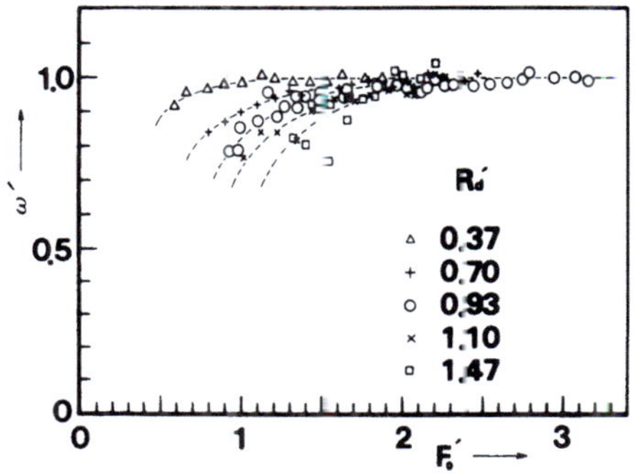

FIG. 4 FREQUENCY PROPERTY ($k'=0.76$, $b'=0.06$)

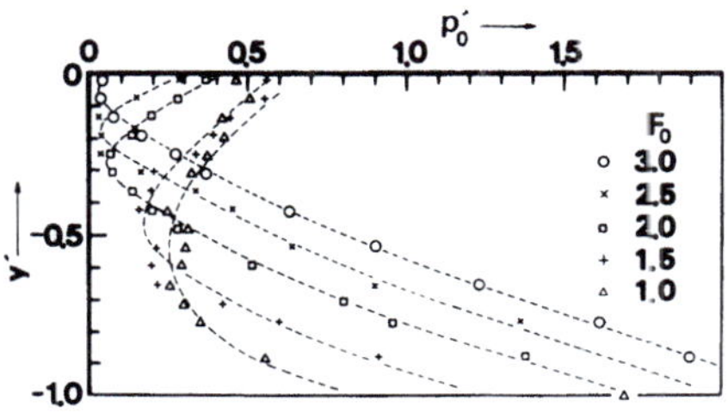

(a) AMPLITUDE

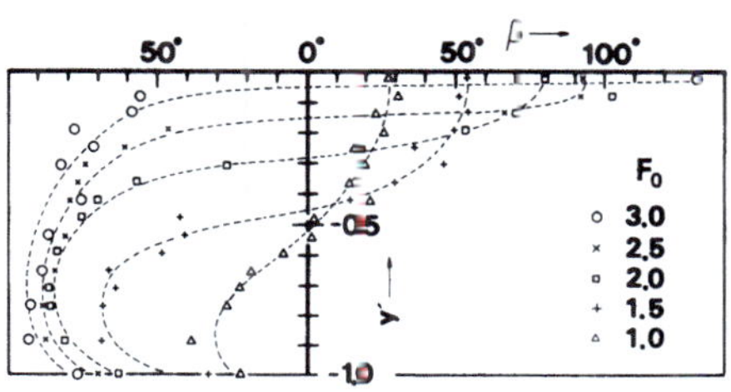

(b) PHASE

FIG. 5 FLUID-INDUCED STRUCTURAL PRESSURE ($R'_d=1.0$, $b'=0.06$)

In order to reveal the properties of the flow-induced structural forces, the model gate was forced to periodically vibrate by means of a motor, and the water pressure was measured at positions 5 mm apart on the surface of the weir plate. The measured results are shown in Figure 5 for $R'_d = 1.0$ and $b' = 0.06$. The ordinate represents the dimensionless water depth y', in which $y' = 0$ shows the position of the free surface and $y' = -1.0$ that of the gate opening. The abscissa p'_o is defined as h_o/δ'. The absicssa β is based on the downward motion of the gate. Hence, when the downward motion of the gate is defined as $\delta \cos\omega t$, the water pressure is represented by $p'_o \cos(\omega t-\beta)$. As the water depth increases, p'_o becomes smaller at first, and then it becomes larger. When p'_o decreases, β is positive. Reversely, when p'_o increases, β is negative. Hence, it can be concluded that the water pressure has reciprocal effects on the motion of the gate. In the case of the model gate, the energy which the water pressure supplies to the motion of the gate, is proportional to $-\int_1^0 p'_o \sin\beta \, dy'$. Hence, the water pressure with positive β dissipates the energy of the motion of the gate, whereas that with negative β supplies it to the gate. As F_o increases, the region of the water pressure with positive β on the free-surface side decreases, and vice versa. Hence, when F_o becomes larger than a certain value, the energy supplied to the gate overcomes that dissipated by the damping, and as a result, a self-excited vibration is generated.

2.5 Model Tests for Large Gate Openings

The effect of gate openings on the stability characteristics is shown in Figure 6, in which $k'=1.22$. When b' becomes larger, the same results are

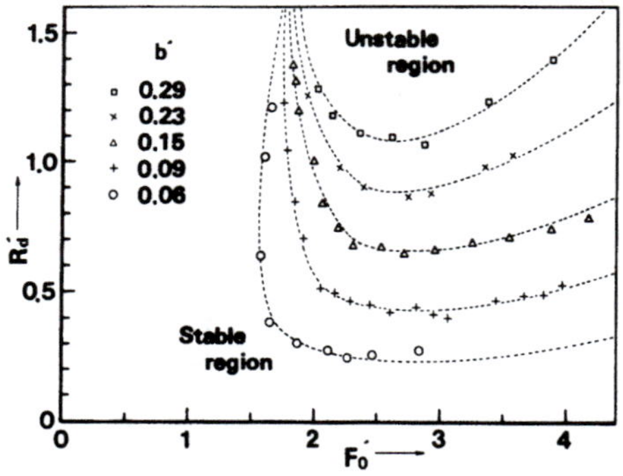

FIG. 6 STABILITY DIAGRAM (k´=1.22)

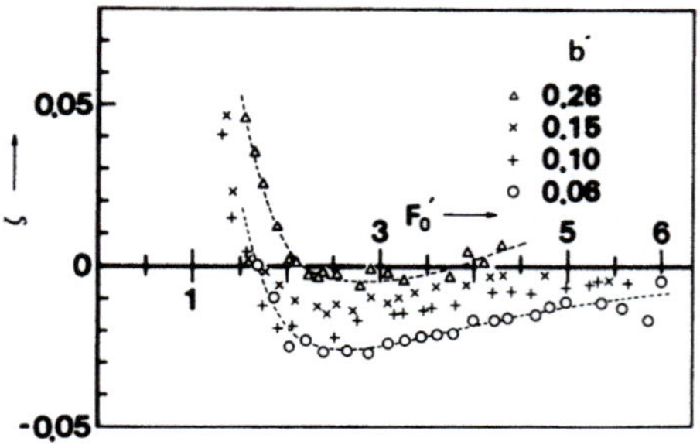

FIG. 7 INSTABILITY (R_d'=1.0, k´=1.22)

observed when k´ becomes smaller, namely, the unstable region decreases and the stable one increases. For example, as b´ increases, the maximum value of R_d' of the boundary curve becomes larger almost in proportion to b´. Hence, it can be concluded that with increasing gate opening the model gate becomes dynamically more stable.

The instability for R_d'=1.0 is shown in Figure 7, in which k´=1.22. As b´ increases, the data uniformly move toward larger ζ, that is, the instability decreases for all F_0'. As a result, the unstable region decreases as seen in Figure 6.

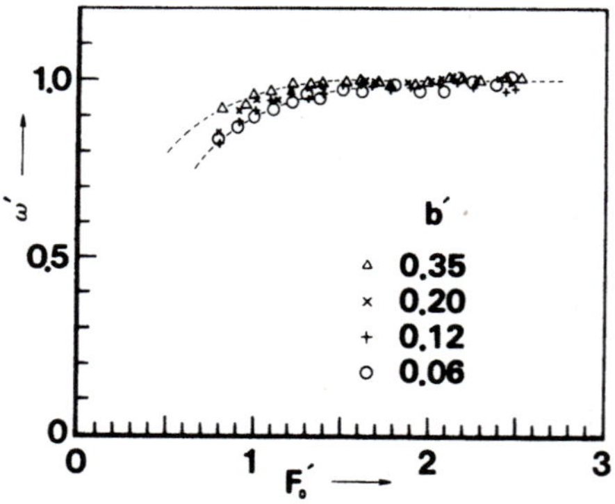

FIG. 8 FREQUENCY PROPERTY (R_d'=0.7, k´=0.76)

The frequency property for R_d'=0.7 is shown in Figure 8, in which k´=0.76. As b´ increases, all data approach 1.0. This tendency is most pronounced in the region of small F_0'. It can thus be concluded that the vibration frequency of the gate agrees more nearly with the natural one as the gate opening increases.

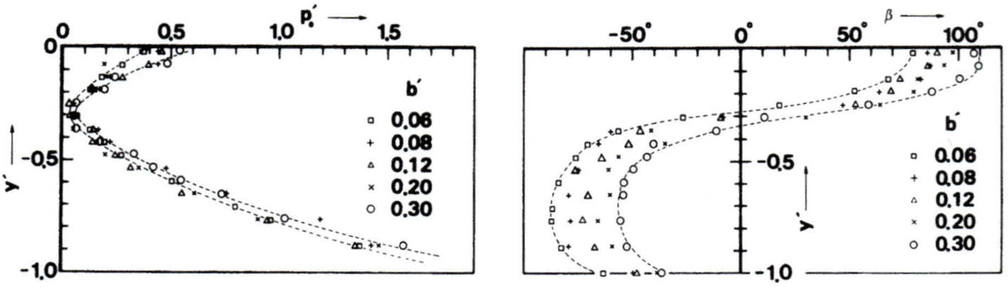

(a) AMPLITUDE (b) PHASE

FIG. 9 FLUID-INDUCED STRUCTURAL PRESSURE (R_d'=1.0, F_0=2.0)

The measured water pressure with the model gate forced to periodically vibrate is shown in Figure 9, in which $R_d'=1.0$ anf $F_0=2.0$. As b´ increases, p_0' becomes uniformly larger, but only slightly. On the other hand, the data of β move toward the positive direction of β quite considerably. Hence, the energy dissipated by the water pressure with positive β on the free-surface side increases, and that supplied by the water pressure with negative β on the gate-opening side decreases. As a result, the model gate tends to become dynamically stable.

3. COLLAPSE OF TAINTER-GATE IN JAPAN

Regarding the accident of a Tainter-gate in Japan [1], Prof. Yano reported [5] as follows. On July 2, 1967, a Tainter-gate suddenly collapsed, on the test run under a full load of $d_0=11.8$ m. The gate was one of four. The other three gate were perfectly closed when the accident happened. The size of the gate was 12 m high and 9 m wide, its radius was 13 m and the weight was 3.7×10^5 N. The construction of the gate is shown in Sketch 1 of Chart 2. The weir plate is supported by three stays with longitudinal members of two rows. The weight per unit area of the gate was 3.4×10^3 N/m^2, and it was lighter by 19 % than the previous, that is, the gate had been designed lighter in weight. The gate had been lifted by four wire cables with 37.5 mm diameter and 25 m length, and it had been operated with a small gate opening of 30 cm for 3 hours just before the accident. The gate had been closed down by about 30 cm per minute, and the accident happened at the same time that the gate was almost closed. Since the collapse occured suddenly, the situation cannot be reconstructed clearly, and the sliding down of the gate had to be estimated on the basis of scratches on the concrete pier as seen in Sketch 2 of Chart 2. The stays had been destroyed by buckling. The gate was swept downstream, as seen in Sketch 3 of Chart 2. After the accident, the static hydraulic pressure head of 12 m was exerted on the same Tainter-gate, and the stresses and deflections on the various elements of the gate were examined. The deflection curve of the stays, one of the results obtained, is shown in Figure 10. The maximum value y_1, y_2, y_3 of the deflection were about 20 to 24 mm downward. It was concluded that the accident was definitely caused by inadequate safety on the buckling of the stays. As a result, the stays were reinforced. A sketch of this "cure" is shown in Sketch 4 of Chart 2.

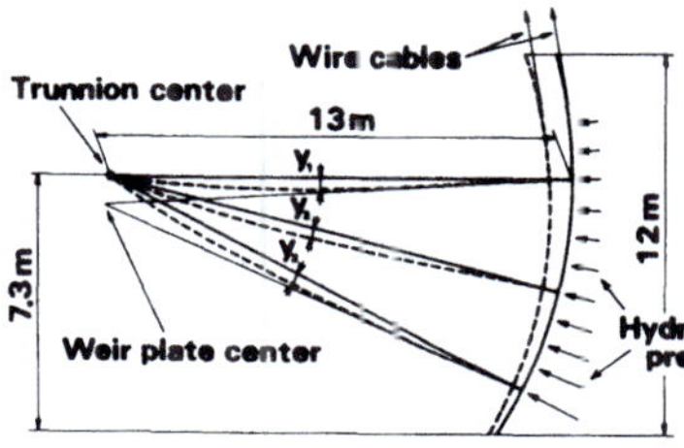

FIG. 10 DEFLECTION CURVES OF THE STAYS

As mentioned above, the cause of the accident was inadequate safety leading to the buckling of the stays, but since the only gate which collapsed was the one closing during a state of flow, it is supposed that dynamic factors contributed to the accident. Moreover, since the gate opening just before the accident was small, i.e., b´=0.04, it is concluded from the result reported above that the gate had been in an unstable situation. $\overline{R}_*$ of the collapsed gate is about 72 mm which is derived from the deflection curves of the stays shown in Figure 10, and the calculated value of I_0 is about 6.4×10^5 Ns^2. Hence, k´ is about 0.21. The damping in the system could not be obtained. However, when the damping ratio is estimated to be on the same level as that of the model gate, the stability diagram for k´= 0.21 can be evaluated as shown in Figure 11. When the accident happened, d_0 was 11.8 m and $R_d'=1.62$. Hence, the gate is unstable, when F_0' is larger than about 1.8, that is, when the natural frequency of the gate ω_0 is larger than about 0.27 Hz. It can be concluded from Figure 3 that the gate becomes most unstable when the natural frequency is about 0.35 Hz (i.e., $F_0'=2.4$).

A principal factor determining the natural frequency is the wire cables. Since the ratio of the load on the wire cable and the breaking load is about 0.1, the elasticity of the wire cable is estimated to be about 6.5× 10^5 N/m. Hence, the natural frequency of the gate is about 0.98 Hz when the wire cables on one side, that is, two of the four are effective, and it is about 1.39 Hz when all wire cables are effective. These calculated values seem to be slightly too large to make the gate dynamically unstable. But considering other factors in the system which make the elasticity decrease, the above can easily approach 0.35 Hz which makes the gate most unstable.

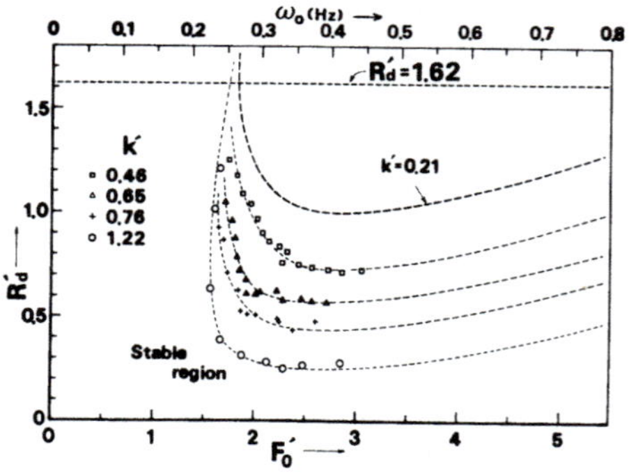

FIG. 11 STABILITY DIAGRAM OF SMALL GATE OPENING

4. CONCLUSIONS

Model tests on a small gate opening revealed that Tainter-gates of the same size become proportionally more unstable the lighter thev are made in weight. Hence, careful consideration has to be given to lightweight designs. When Tainter-gates are unstable, they vibrate at their own natural frequency. When they are stable, the vibration frequency is lower than the natural one. The instability is larger the greater the water depth of the gate opening. When the water depth is constant, the instability is highest near the stable/unstable boundary curve. The fluid-induced structural force exerted on the weir plate has reciprocal effects on the motion of Tainter-gates. That is, the force on the free-surface side dissipates the energy of the gate motion, and that on the gate-opening side supplies it to the gate, when the weir plate center is lower than the trunnion pin. Self-excited vibrations arise when the energy supplied exceeds that dissipated. Model tests with large gate openings revealed that as the gate opening increases, the negative effect of the fluid-induced force decreases, and the positive one increases. As a result, the instability decreases, and hence the unstable region becomes smaller. The vibration frequency agrees more with the natural one, as the gate opening increases. It can be concluded that one means for securing safe operation is to open the gate quickly as wide as possible.

Regarding the accident of a Tainter-gate in Japan, since the gate had been designed lighter in weight and it had been operated with a small gate opening, there is a good chance that the dynamic effects discussed in this paper contributed to the accident.

5. REFERENCES

1. ASAHI EVENING NEWS, July 3, 1967.
2. Ishii, N., & Imaichi, K., "Water Waves and Structural Loads Caused by a Periodic Change of Discharge from a Sluice-Gate", Bulletin of JSME, Vol. 20-146, 998-1007, (1977).
3. Imaichi, K., & Ishii, N., "Instability of an Idealized Tainter-Gate without Damping", Bulletin of JSME, Vol. 20-146, 963-970, (1977).
4. Ishii, N., & Imaichi, K., "Instability of Elastically Suspended Tainter-Gate System Caused by Surface Waves on a Dam", Transaction of ASME, Series I, Vol. 99, No. 4, 699-709, (1977).
5. Yano, K., "On the Event of the Gate Destruction of the Wachi-Dam", Disaster Prevention Research Institute Annuals of Kyoto University in Japan, 11-B, 1-17, (1968).

CHART 1: TAINTER - GATE (Model Tests)

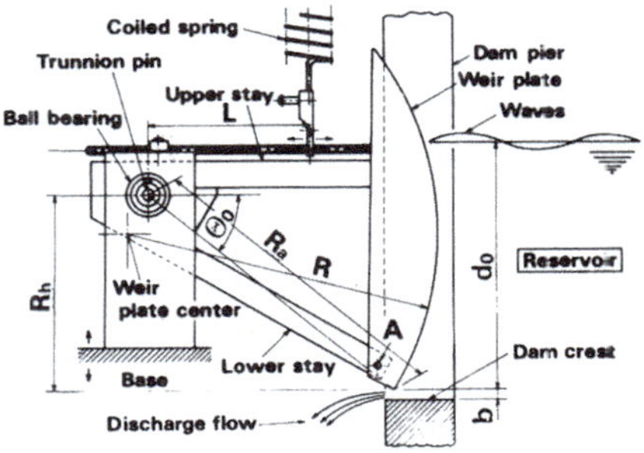

(1)

Problem: Self-excited rotational vibration about trunnion pin

Geometry: See Sketch 1. Radius of weir plate R=135 mm, distance between trunnion center and gate opening R_a=138.3 mm, height of trunnion pin R_h=86.5 mm and angle between stays Θ_0=39.2°. Width of weir plate 40 mm.

Boundary conditions: Gate vibration causes change of discharge. Reservoir with 40 mm is 230 mm deep from gate opening and 1500 mm long.

Flow conditions: See Sketch 1. Approach flow without disturbance and almost no turbulence

Operating conditions: Upper stay is set horizontally in experiments. Water depth over gate opening d_0=0 to 110 mm. Gate opening b=5 to 25 mm. Eccentricity of trunnion pin $\overline{R}_*$ 10 to 30 mm. Initial disturbance is given to gate under steady discharge.

Structural dynamics: Moment of inertia of gate per unit width I_0=0.057 Ns^2. Damping ratio ζ_0=0.003 to 0.05. Natural frequency ω_0=2 to 10 Hz. ζ is the damping ratio indicative of instability and $\omega'(=\omega/\omega_0)$ is the dimensionless frequency (see Sketch 2). Fluctuating gate opening was observed.

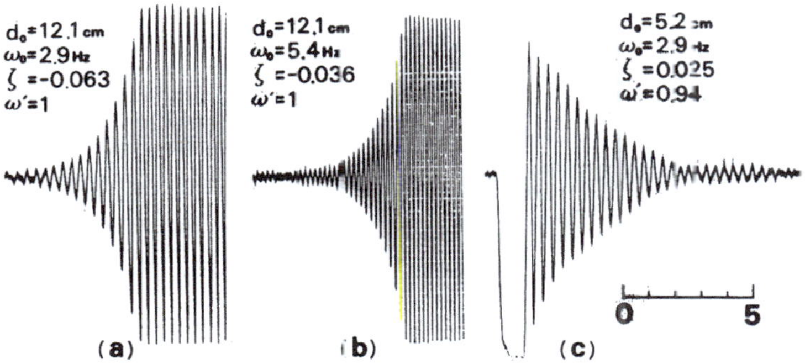

(2)

Records for R_*=30 mm, b=5 mm

Vibrations: See Sketch 2. Maximum amplitude of fluctuating gate opening 10 mm. (a) and (b) show intense self-excited vibrations, and (c) shows damped vibration.

Cures: Weir plate should be made exactly concentric to trunnion pin, and stays must be constructed rigidly, so as not to be deflected by static pressure giving rise of eccentricity. Safe operation of gate requires that the gate is opened rapidly as wide as possible, in order that the dynamically unstable situation at small gate openings is avoided.

References: Ishii, N, "Instability of Elastically Suspended Tainter-Gate System", Transaction of ASME, Series I, Vol.99, No.4, 699-709, (1977)

CHART 2: TAINTER-GATE (An Actual Collapse)

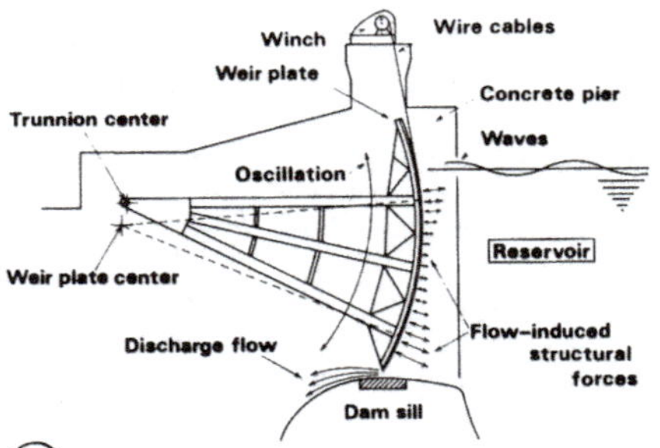

①

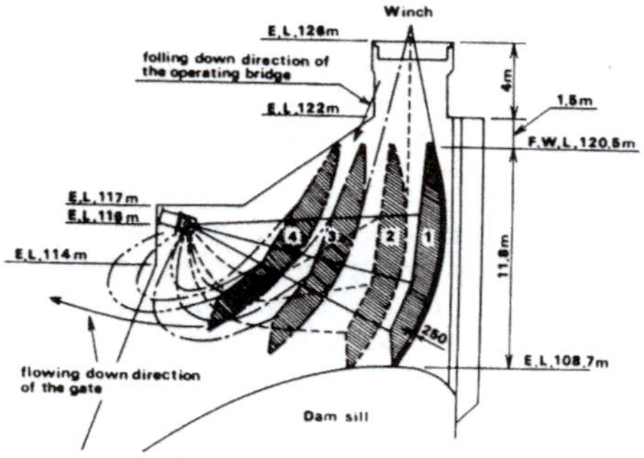

②

cited from Ref. 1

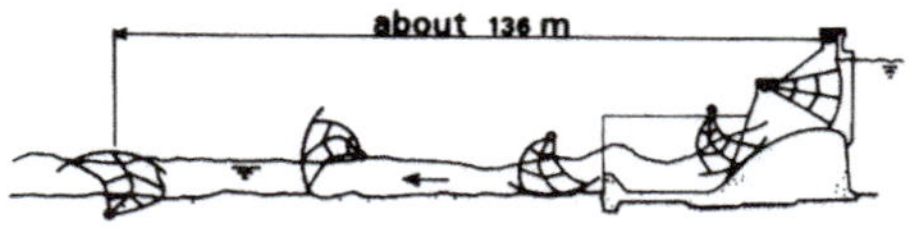

③

cited from Ref.1

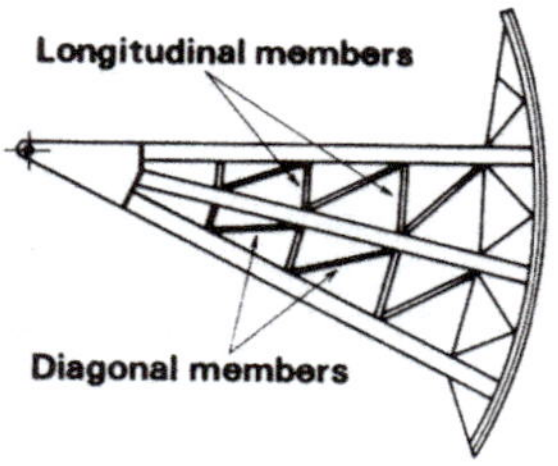

④

Problem: A Tainter-gate used as a crest gate of a dam collapsed and was swept downstream during a test run under full load.

Geometry: See Sketch 1. Radius of weir plate 13 m. Height and width of gate 12 m and 9 m.

Boudary conditions: Reservoir head is 11.7 m over the gate opening, and the upper reach is sufficiently wide.

Flow conditions: See Sketch 1. Approach flow without disturbance and almost no turbulence.

Operating conditions: Collapsed gate was one of four. Other were perfectly closed. Water depth over gate opening 11.8 m. Gate opening 30 cm. Gate had been lifted up by four wire cables with diameter 37.5 mm and length 25 m.

Structural dynamics: Moment of inertia of gate 5.76×10^6 Nms^2. Weight per unit area of gate was 3.4×10^3 N/m^2 which was lighter by 19 % than previously.

Collapse: See Sketches 2 & 3. Gate had been lowered by about 30 cm per minute from a small gate opening of 30 cm, and just when the gate was almost closed, the stays were destroyed by buckling. The gate then swept downstream about as shown in Sketch 2.

Cures: See Sketch 4. Three stays are rigidly joined by three rows of longitudinal members and three diagonal members, in order to reinforce stays against buckling.

References:

1. Yano, K., "On the Event of the Gate Destruction", Disaster Prevention Research Institute Annuals of Kyoto University in Japan, 11-B, 1-17, (1968).
2. Ishii, N. and Imaichi, K., "Instability of Elastically Suspended Tainter-Gate System Caused by Surface Waves on a Dam", Transaction of ASME, Series I, Vol. 99, No. 4, 699-709, (1977).

C10 VIBRATIONS DUE TO TWO-PHASE FLOW BELOW A TAINTER GATE

G. Rouvé
Professor, Hydraulic Eng. Resources

F.-J. Traut
Research Engineer

Technische Hochschule Aachen, Federal Republic of Germany

SUMMARY

This paper describes the results of model tests on a tainter gate, acting as an emergency gate in the flood tunnel of a reservoir. Special care was given to the fact, that there is flow of air and water under 5 bar pressure in site. Tests with flow of pure water led to routine results. Some minor vibrations of the structure were cured by changing small structural details. However, severe pressure fluctuations occured at two-phase flow with the gate 20 % to 40 % open. The flow-structure interaction is described in detail. No cure has been found to avoid these fluctuations, but it is believed that this phenomena is inherent to the system chosen.

As an additional provision to control the flood discharge through the emergency tunnel of a dam, a tainter gate is proposed for installation. This type of structure has proven to be adequate for flow with high velocities and relative high pressures. A general view of the site is shown in Fig. 1.

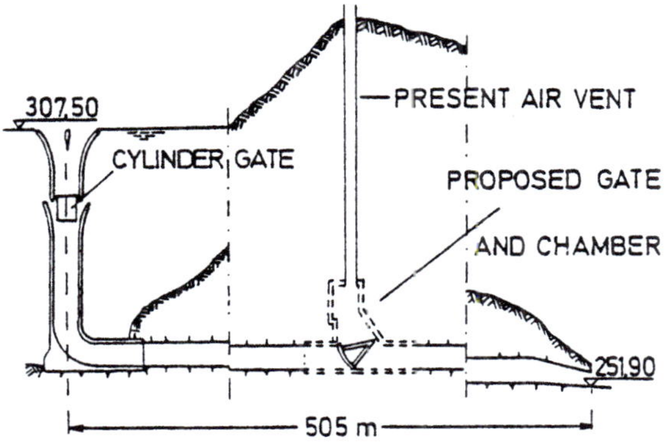

FIG. 1 GENERAL VIEW

Since there have been occasional difficulties to close the cylinder gate, it is expected that a mixture of air and water will discharge through the tunnel if in case of emergency the tainter gate has to be closed. This air-water mixture enters into the vertical shaft of the morning glory spillway and flows through the tunnel. At high velocities it can be expected that the distribution of air is almost uniform over the cross section upstream of the gate. The air contents of the mixture was roughly estimated in site to 30 % to 40 %. At minor velocities, which will occur when the proposed gate is partially closed, the distribution of air and water is not uniform, but the air will begin to rise to the ceiling of the tunnel.

Because of the unknown pressures and forces acting on the gate resulting from the flow of the air-water mixture, model tests were carried out. In the physical model (skale 1:25) the trunnion force, the lift force and the skin pressure were measured for flow conditions without and with air. The air entrainment of the prototype was simulated in the model by injecting air according to the observations in site. A sketch of the original design is shown in Fig. 2.

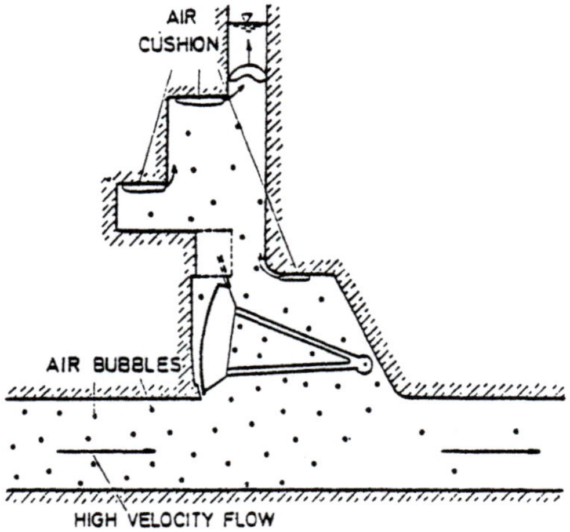

FIG. 2 ORIGINAL DESIGN

Two main conditions were observed: tailrace tunnel under pressure (large openings of the gate ≥ 70 % open) and free discharge in the tailrace tunnel (small openings of the gate ≤ 60 % open).

a) Tailrace tunnel under pressure

The experiments conducted with water only led to routine results. To minimize eddies in the gate chamber, the tailrace face of the gate was lined and the bearing arms were designed with closed square cross-sections.

At flow conditions with entrained air, bubbles rise in the gate chamber and form air cushions under the horizontal ceilings of the gate chamber. From time to time these cushions blow out into the air vent leading to minor pressure fluctuations in the chamber and behind the gate (see Fig. 2). This unwanted effect was eliminated by giving a slope to the ceilings. Now the bubbles rising in the chamber cannot accumulate and therefore discharge continously through the air vent (see Fig. 3).

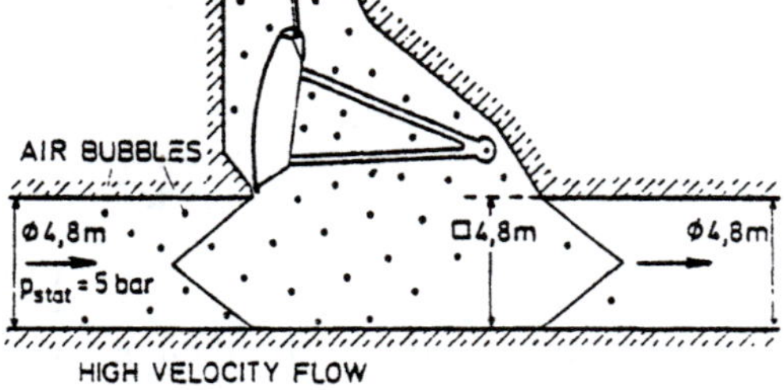

FIG. 3 FINAL DESIGN

After the modification of the design, no fluctuation in the trunnion forces of the gate was observed. The measured pressure at the skin plate due to the flow of the air-water mixture was constant and similar to the values measured for the flow of water without air.

b) Free discharge downstream of the gate

In the experiments with pure water, again all results were as expected. The forces and pressures at the gate were close to and never exceeded those calculated for static conditions. Because of the large air vent (diameter Ø 2,0 m) there is no problem to sufficiently vent the tailrace, regardless whether a hydraulic jump or spray flow occurs in the tunnel. Therefore, vibrations due to subatmospheric pressure or cavitation will not occur.

With air-water mixture, however, the results deviated markedly from those reported above. With the gate open between 20 % and 40 % a sort of water hammer generated by the unsteady, compressible flow lead to pressure peaks in front of the gate up to twice the static pressure (see Sketch 2 in the chart).

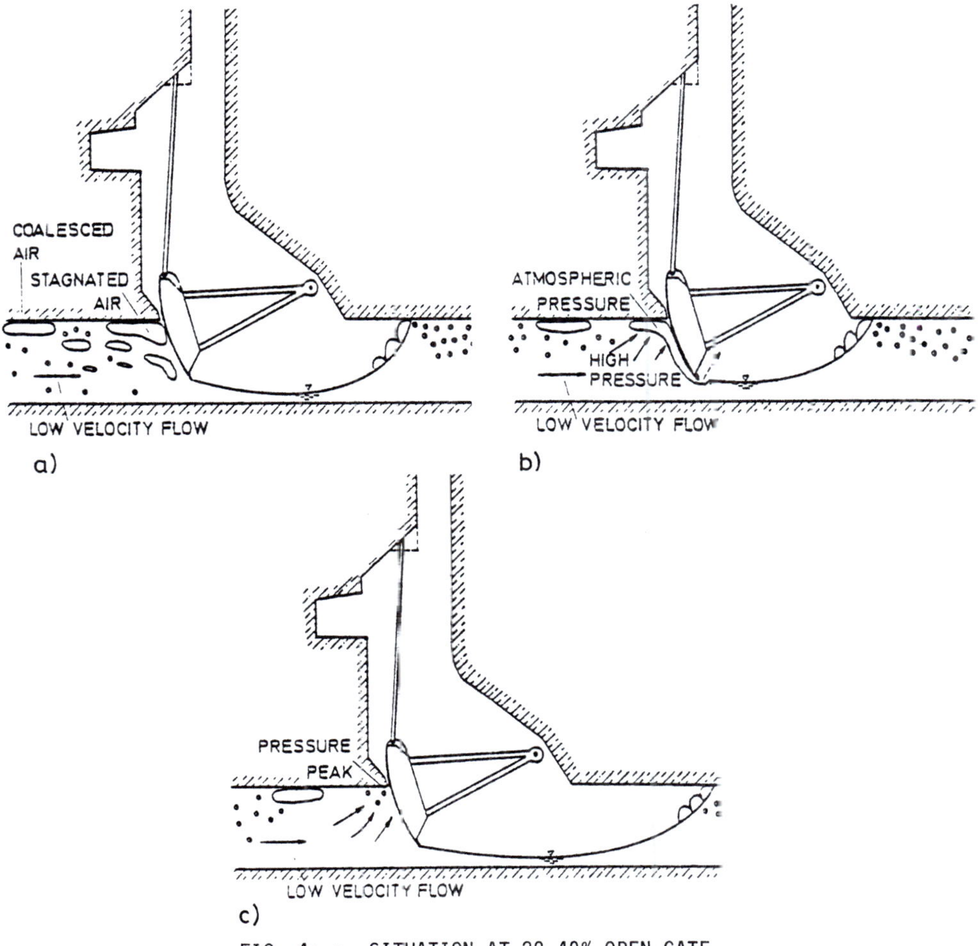

FIG. 4a-c SITUATION AT 20-40% OPEN GATE

The process by which these pressure fluctuations are generated is sketched in Fig. 4. The air, which is uniformly distributed at the entrance of the head race tunnel, accumulates to large bubbles due to the relatively low velocity of water and the long distance upstream from the gate.

The bubbles stagnate at the skin plate of the gate (Fig. 4a) until the air is partially drawn under the gate. Immediately it blows out into the free air behind the gate and for a moment the pressure at the skin plate sinks to nearly atmospheric value

(Fig. 4b). The flow accelerates into the low pressure space just in front of the gate, squeezing out the remaining air. When the water front reaches the plate a pressure peak occurs due to an abrupt clash (Fig. 4c). The next big air bubble reaches the gate and the cycle repeats.

In the model, the pressure peaks have a "frequency" between 0,2 and 0,7 Hz. A calculation of this frequency in the projected prototype is not possible, because of the unknown scale factor of this phenomena. But the frequency in site will not lead to resonance, for the frequency of the structure is estimated to lie above 1 Hz.

To prevent these pressure fluctuations, which can lead to severe vibrations and structural damage, no cure is known except for measures to let the air-bubbles along the ceiling escape before they can reach the gate. (One such measure would have consisted, e.g., in turning the tainter gate around; however, the gate chamber would then be under pressure at all times and a second independent air vent downstream from the gate has to be provided). It is believed that the process is inherent to every control structure, closing a pressurized tunnel with an air-water mixture from the ceiling. Therefore the structure was designed to withstand these pressure fluctuations. The gate chamber and the tunnel near the gate were steel-lined. Because the installation acts as an emergency gate, it is not meant to operate partially open in a steady position but will be closed continously. Since the gate is thus operated always from a completely open to the closed position, the unsteady loading will be limited to relatively small periods of time.

REFERENCES

1. Rouvé, G.; Traut, F.J.: Modellversuche für den Einbau eines Segmentverschlusses in den Hochwasserentlastungsstollen der Biggetalsperre. Versuchsbericht Institut für Wasserbau der RWTH Aachen, 1977 (unpublished)

2. Idel, K.-H.; Traut, F.J.: Untersuchungen für den Einbau eines Notverschlusses in einen Hochwasserentlastungsstollen. Mitteilungen Versuchsanstalt für Wasserbau, Hydrologie und Glaziologie, ETH Zürich Nr. 33, 1979

Discussion: VIBRATIONS DUE TO AIR-WATER FLOW BELOW A TAINTER GATE (Rouvé)

M.J. Kenn
Civil Engineering Department, Imperial College, London, England

Air-entraining water flows are notoriously difficult to model except perhaps when tested with full-scale velocities. In Professor Rouvé's model the air bubbles almost inevitably are relatively too large; they are necessarily inadequately sheared; and they are probably therefore insufficiently dispersed. Accordingly it is conceivable that any pressure fluctuations experienced with the full-sized structure (when it is built) may prove to be rather less severe than as suggested by the present model tests. Comparisons between the predicted and the ultimate behaviours of the full-sized tainter gate certainly will be of interest and value.

References:

1. Kenn, M.J. and K.J. Zanker. Aspects of similarity for air-entraining water flows.Nature, 1967, Vol. 213, No. 5071, pp. 59-60.

2. Kenn, M.J. Dynamical similarity for flow systems in which inertia effects are small. J.I.W.E., 1969, Vol. 23, No. 4, pp. 251-253.

3. Kenn, M.J., Further studies concerning the air control of siphonic water flow. J.I.W.E., 1965, Vol. 19, No. 3, pp. 231-238.

TWO-PHASE FLOW BELOW A TAINTER GATE

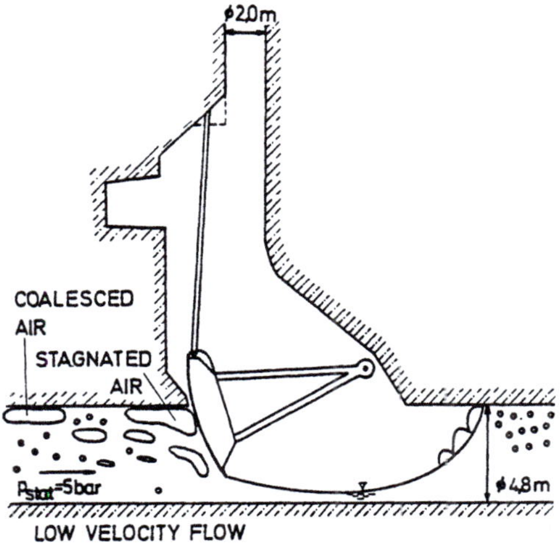

1

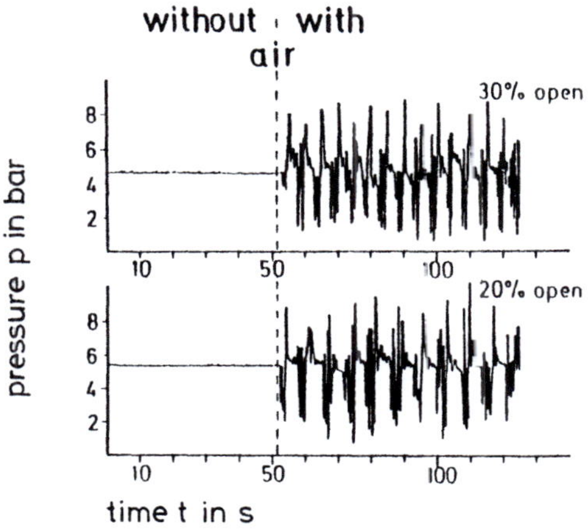

2 Pressures at skin plate

Problem: In model tests (scale 1:25) severe periodic pressure fluctuations (sort of water hammer) were observed up- and downstream the gate, the reason for some vibrations.

Geometry: See Sketch 1. Cross section of the tunnel: 4.80 m Ø, cross section at the gate: 4.80 x 4.80 m.

Boundary conditions: Gate chamber is aerated and deaerated by a vent of 2.00 m diameter.

Flow conditions: Approach flow under 5 bar pressure. Nearly uniformly distributed air bubbles (concentration approx 30-40 %) if the gate full open. Big air bubbles beneath the ceiling generate non-uniform flow (max. 40 % air) if the gate is 20-40 % open.

Operating conditions: Severe pressure fluctuations were observed in the model at a gate opening of 20-40 % at two-phase flow condition. No pressure fluctuations occured if only water flows (see Sketch 2).

Structural dynamics: No dynamic coefficients were measured in the model.

Vibrations: "Frequency" of the pressure fluctuations $n \sim 0.2 - 0.4$ Hz as obtained from the model. Prototype data unknown.

Cures: Due to the gate system (closing from the top) it seems to be impossible to cure the pressure fluctuations. Nevertheless, because of the low frequency of the fluctuations, resonance of the gate will not occur. Measuring equipment will be installed in the prototype to gain more information.

References

1. Idel, K.-H.; Traut, F.-J.: Untersuchungen für den Einbau eines Notverschlusses in einen Hochwasserentlastungsstollen. Mitteilungen, Versuchsanstalt für Wasserbau, Hydrologie und Glaziologie, ETH Zürich Nr. 33, 1979

C11

FLAP GATE OSCILLATION

Kunihiro Ogihara
Toyo University
Dept. of Civil Engineering
Kawagoe, Saitama, Japan

Sachihiko Ueda
Ishikawajima Harima Heavy Industries Co. Ltd.
Chiyodaku, Tokyo, Japan

SUMMARY

Results of field experiments on flap gate oscillations are reported in which regular periodic gate motion was observed with frequencies of about 4 Hz and amplitudes up to 1.8 mm. To prevent this oscillation, a new type of spoiler was developed, attached to the top of the flap gate.

1. INTRODUCTION

The gates of the Tone river project in Kanto plane, on which the field experiments were conducted, are as shown in Sketch 1 of the chart. The span of the gate is 40 m, the gate height is 3.5 m (1.1 m for the flap gate and 2.4 m for the main gate), and an operating velocity was used of v = 0.3 m/min. Measurements of oscillation were performed on the upper gate, most of them during operating conditions. In the only case in which regular periodic oscillations were observed, the gate was stopped. The points of measurement are shown in Fig. 1; the motion of the gate was measured at sections I and III and the pressure fluctuations on the gate surface at sections I and II. The load in the operating wire rope was measured by a tension gauge (Fig. 1). The motion of the gate was measured by accelerometers.

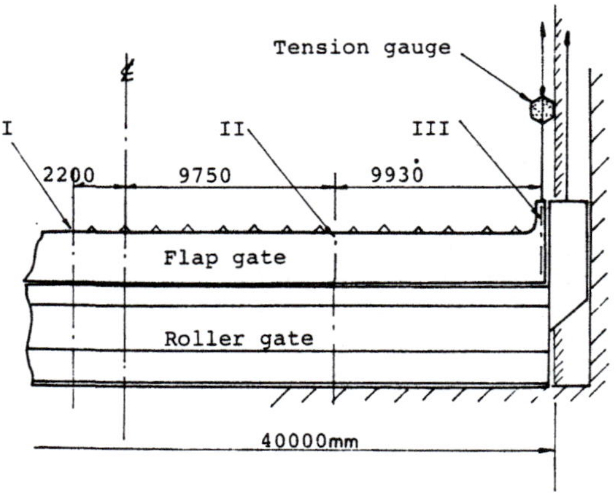

FIG. 1 POINTS WHERE METERS ARE ATTACHED

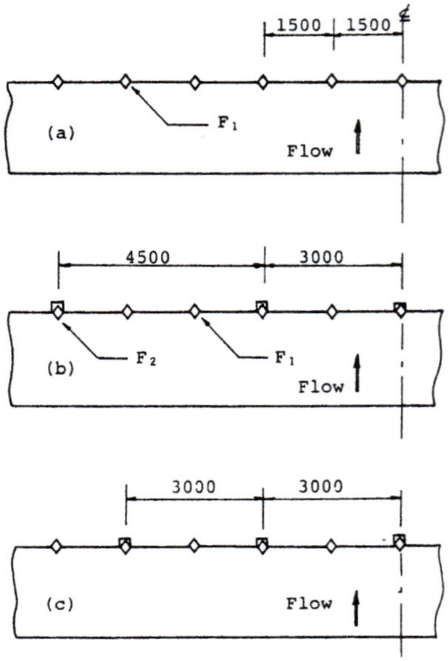

FIG. 2 ARRANGEMENTS OF SPOILERS

2. FIELD-TEST RESULTS

Two types of spoilers were used as shown in the Chart. The type F_2 is made by attaching a plate to the type F_1. With the type F_1 introduced in the original design, intense vibrations occurred. The type F_2 was then employed to stop this oscillation. Three arrangements were tested as shown in Fig 2: (a) type F_1 arrayed in equal distance, (b) type F_1 and F_2 arranged in irregular order and in equal distance pitch and (c) the same two spoilers arranged alternately in equal distance pitch.

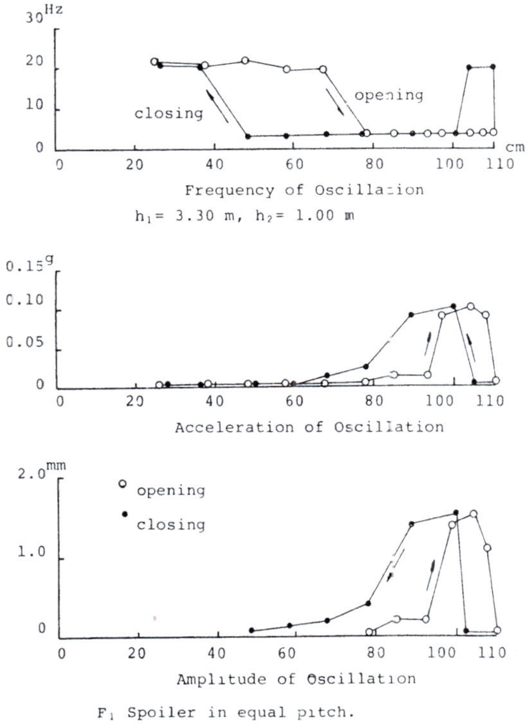

FIG. 3 TEST RESULTS FOR CASE (a)

In all these cases, the gate was operated continuously to the turning point where it was stopped for a while. After that the gate was raised again continuously. From the test results, shown in Fig. 3, two types of oscillation can be discerned: an irregular motion with mean frequency around 20 Hz and amplitudes below 0.01 mm; and a regular periodic vibration with 4 Hz frequency and amplitudes up to 1.5 mm. The second type of oscillations occurred when the gate opening was between 80 cm and 100 cm. They occurred over a wider range of gate openings in the closing stage than in the opening stage. This can be explained by the fact that separation of the nappe is hindered initially by surface tension. Another reason is connected to the fact that after the gate oscillation has been initiated, it continues for a while due to inertia.

The natural frequency of the gate system was calculated for three modes:

i) Gate motion considered as the rotational-oscillator motion of a rigid body,

ii) Gate motion controlled by elastic deflections of a beam supported at both ends,

iii) Gate motion controlled by both elastic-beam deformations and elongations of the operating wire rope.

In all these cases, the added mass on the gate surface was considered the same as for a plate. The frequencies derived from this calculation are as follows: i) f_n = 8.1 Hz, ii) f_n = 4.23 Hz, iii) f_n = 3.7 Hz. It is obvious, that the regular periodic oscillation described before corresponds to the second mode, i.e., the elastic-beam motion.

As revealed by a comparison of Fig. 4 with Fig. 3, regular periodic oscillations do no longer occur with F_2-spoilers added. Only the random oscillation with frequencies around 20 Hz and amplitudes of acceleration of 0.005 g were observed here. The arrangement of the spoilers F_1 and F_2, moreover, proved to be of little consequence.

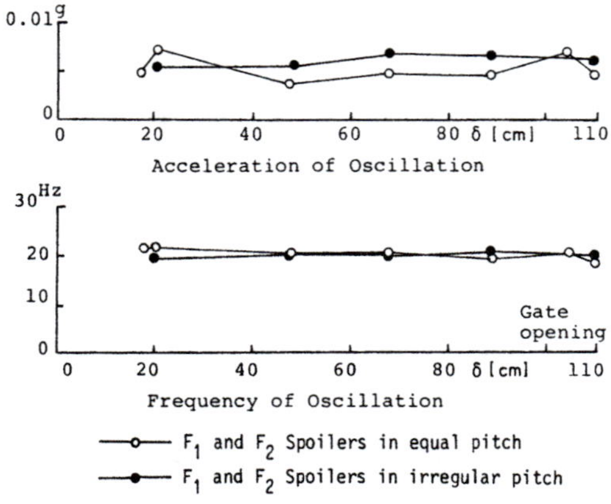

FIG. 4 TEST RESULTS
(h_1 = 3,45 m; h_2 = 1,00 m)

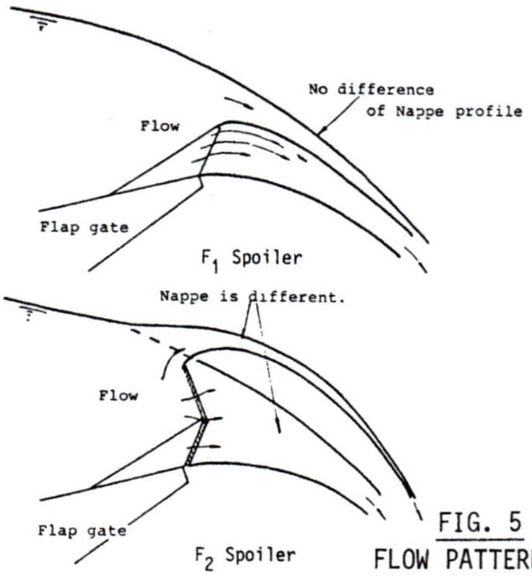

FIG. 5
FLOW PATTERNS

A record of oscillation as obtained from an oscillograph is shown in Sketch 3 in the chart. The first record showing a continuous sinusoidal signal corresponds to case (a). The second record with the substantially decreased amplitude indicates clearly that the spoiler type F_2 is indeed effective in the self-excited oscillation problem of flap gates. To understand how the spoilers act, the flow has been sketched in Fig. 5. In the case F_1, there is no difference in the upper profile of the nappe between the flow over the spoiler and the flow besides the spoiler and the difference in the lower nappe profiles extends over a small distance only. With the F_2 spoiler, on the other hand, the nappe profiles are changed significantly and the water sheet becomes separated and aerated up to a certain nappe thickness.

3. CONCLUSIONS

Spoilers of type F_2 act very satisfactorily in stopping self-excited flap-gate oscillations by cutting the most crucial link in the feedback loop - gate motion / nappe oscillation / volume change of air under nappe / pressure fluctuation - namely from pressure fluctuations to gate motion. A second reason for the effect of the spoilers is that the added mass over the spoiler is changed so that the period of natural oscillation no longer matches over the entire width of the gate.

4. REFERENCES

1. K. Petrikat, "Vibration test on weirs and bottom gates", Water power, May, 1958.
2. Schwartz H. Ivan, "Nappe Oscillation", Pr. of ASCE, vol.90, Hy. 6, July, 1963.
3. H.W. Partensky et al.,"Theoretical study on flap gate oscillation", Pr. IAHR, vol. 12, 1971.
4. B. Treiber, "Theoretical study on flap gate oscillation", IUTAM-IAHR Symp. Karlsruhe 1972, "Flow-Induced Structural Vibrations", edited by E. Naudascher
5. P.A. Kolkman, "Instability of a Vertical Water-Curtain Closing an Air-Chamber", IUTAM-IAHR Symposium Karlsruhe, 1977.
6. K. Ogihara and M. Homma, "Theoretical analysis of flap gate oscillation", Proc. of IAHR, C 51, 1977.
7. S. Ueda, "Study on a few types of gate oscillation under the water flow", Doctoral Thesis, (in Japanese, 1978.

FLAP GATE

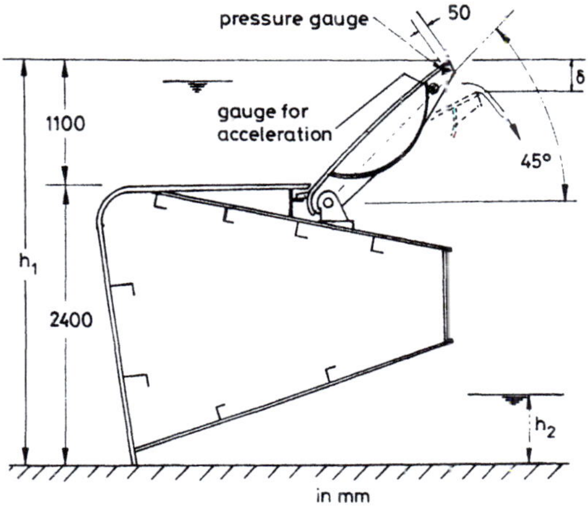

1

Problem: Regular periodic oscillation occurs with nappe undulation F_1 type spoiler.

Geometry: See Sketch 1. Width of gate is 40 m.

Boundary condition: Nappe is aerated through side opening and under the spoiler.

Flow and operating condition: Gate is operated continuously under a water depth of 3.5 m upstream and 1.0 m down-stream. Continuous oscillation occurs when the gate is lowered between 0.8 and 1.0 m.

Structural dynamics: Natural frequency is 3.7 to 4.23 Hz.

Vibration: Frequency 4 Hz. Maximum amplitude 1.8 mm at top of gate (see Sketch 3).

Cure: Spoilers of the F_2 type shown in Sketch 2 effectively stop the oscillation. Two kinds of spacings were used, 3 m in equal pitch and 3 m and 4.5 m in irregular pitch. The effect on the oscillations is clearly seen from Sketch 3. The main reason for stopping the self-excited flap-gate oscillation is sufficient flow disturbance and aeration by the spoiler.

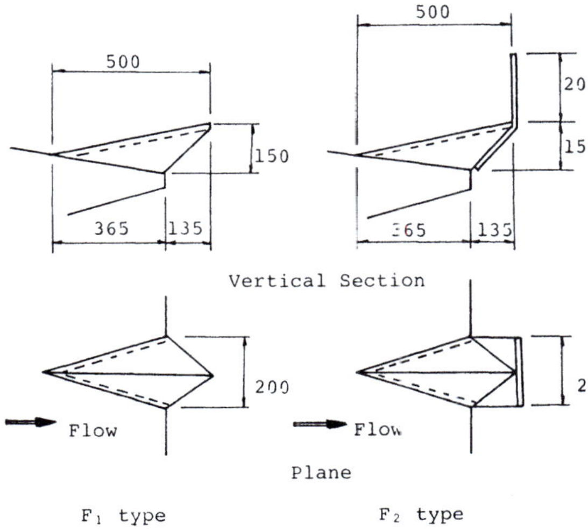

2

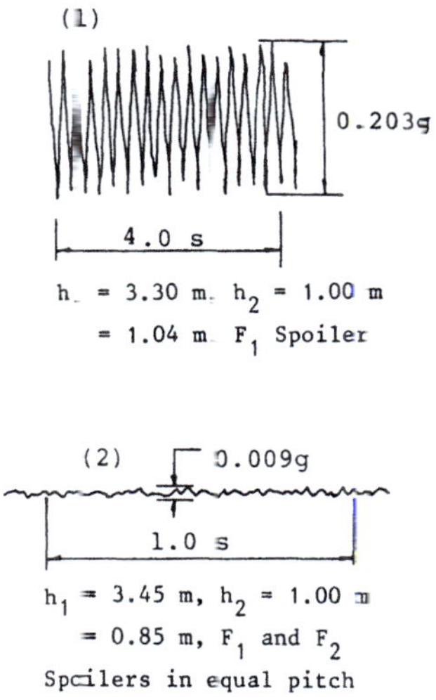

3

Discussion: FLAP GATE OSCILLATION (Ogihara and Ueda)

K. Petrikat, Universität Stuttgart, Germany

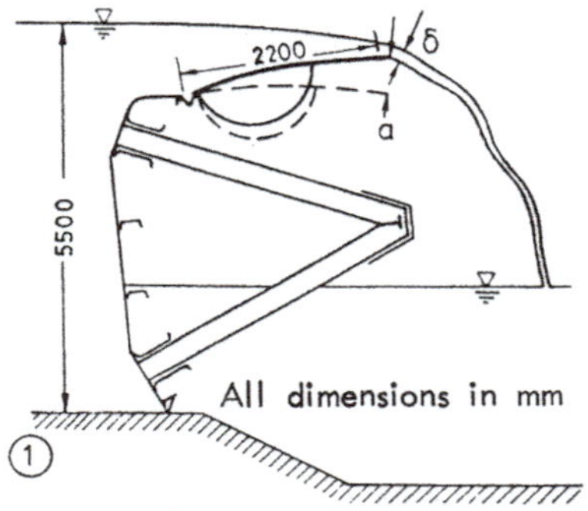

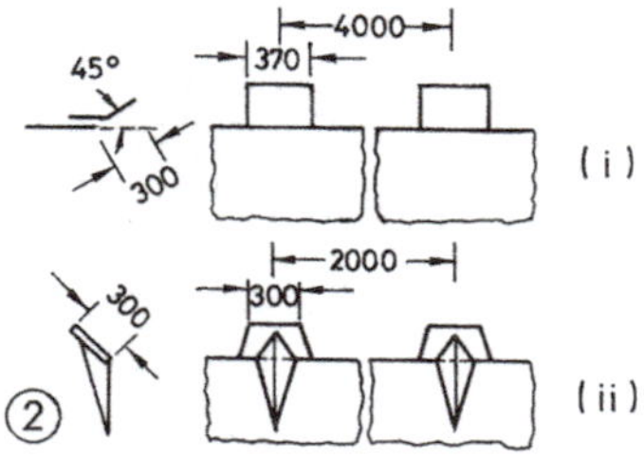

Cures: Nappe spoilers of proper geometry and spacing (ii) eliminate vibration through venting and 3D disturbances; original spoilers (i) were unsuccessful. If spoilers are too small to rip the nappe open, they should generate nappe portions with phase shift in oscillation of 180°.

Problem: Nappe oscillations at small depths of overflow excite vibration of flap. Damage due to fatigue possible. Trembling of windows excited over long distances.

Geometry: See sketch 1. Width of flap 30 m.

Boundary conditions: Nappe is aerated through side openings. Volume under the nappe 670 m^3.

Flow conditions: See sketch 1. Approach flow without disturbance and almost no turbulence.

Operating conditions: Overflow heights of 5 to 70 cm. Nappe oscillation and its mode with rigidly fixed flap depends on fall height. With elastically supported flap, fall height of 2.8 m, and nappe width at separation (δ) of 2.5 cm, vibration was observed as described below (see photo).

Structural dynamics: Moment of inertia of flap (including added-mass effect) acc. to vibration test: $I = 61800$ mNs^2; damping moment $- c d\varphi/dt$ with $c = 0.4 \times 8 \times 61800$ mNs; elastic restoring moment $-3.875 \times 10^7 \times \varphi$ mN; where φ is angular displacement of flap.

Vibration: Frequency $n = 4$ Hz; max. amplitude of tip of the flap: 2 mm. Amplitude (a) and airpressur record (Δp) relctively regular - see diagram 3. (There was no difference between pressure records in midsection and near end of flap.)

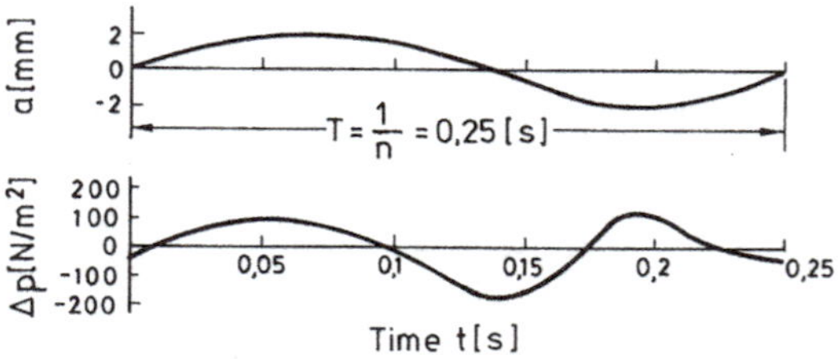

References: Petrikat, K., Schwingungsuntersuchungen an Stahlwasserbauten. DER STAHLBAU, Heft 9 u. 12, Sept/Dez 1955, p. 198-202, 272-289.

C12 SOME EXPERIENCES WITH CURING FLAP-GATE VIBRATION

Luděk Půlpitel
Research Engineer

ČKD Blansko, Czechoslovakia

SUMMARY

This paper describes some experience with flap-gate oscillations. The vibration appeared at four different flap positions and was eliminated by additional spoilers and closing the aerating pipes.

It is well known that the nappe-flap systems tend toward self-excited vibration [1] . The vibration appears usually at small depth of the overflow, and spoilers are the best cure.

During commissioning of a flap in the year 1975, nappe oscillations and vibration of the flap was observed. Since this flap is operated through a one-sided hydraulic servomotor, the torsional rigidity of the flap structure is relatively low. The overall arrangement and the fundamental dimensions of the structure are shown in the chart.

Some investigations were performed involving, among others, measurements of flap vibration (displacements A_1, A_2, A_3, frequency f) and of the pressure in the aeration pipes (mean value p and fluctuation amplitude Δp) [2] . At a steady river flow, the flap was very slowly lowered. The vibration appeared at four different flap positions. The corresponding positions are denoted by I, II, III, IV in the Table of the chart. In the position I, the nappe was separated into sections through the spoilers, whereas in the remaining positions II, III, IV the nappe was coherent over the whole length of the flap.

The effect of aeration was also investigated. The closing of the aerating pipes influenced the vibration in position III and IV only. In position III, the intensity of vibration increased, whereas vibration in position IV ceased completely.

The flap was further provided with additional spoilers and the measurements were repeated. The vibration in the positions I, II, III was eliminated. The vibration observed earlier in the position IV reappeared now at larger depths of the nappe and with an intensity lowered by about 70%. When closing the aerating pipes, the vibration ceased.

This observation can be explained as follows: In position IV, the flap is inclined close to the tailwater level. When closing the aerating pipes, the air underpressure under the flap increases from the value of about -400 Pa to the value of -1400 Pa, and thus the water level under the flap is lifted by about 100 mm. The flap is thus partially submerged and the corresponding damping supresses the vibration.

REFERENCES

1. Petrikat, K., "Vibration Tests on Weirs and Bottom Gates", Water Power, February 1958.

2. Půlpitel, L., "Unpublished report ČKD Blansko", Czechoslovakia, 1975.

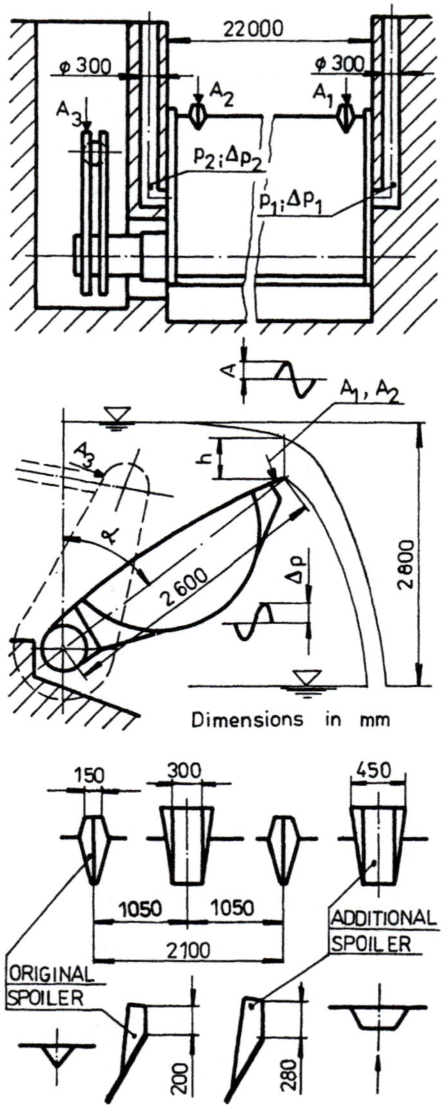

FLAP-GATE

Problem: Nappe-flap oscillations at four different positions.

Geometry: See figures.

Boundary conditions: See figures

Flow conditions: Approach flow without disturbance and with almost no turbulence.

Operating conditions: Overflow heights: h = 0 - 860 mm, flap position: $\alpha = 20^0 - 96^0$.

Structural dynamics: Dynamic characteristics of the flap were not measured.

Vibration: See Table.

Cures: Additional spoilers eliminated vibration in the flap positions I, II, III and reduced vibration in position IV (A_1 = 1.1 mm). However, in the latter case the overflow height was now h = 820 - 850 mm. The vibration in the position IV can always be eliminated by closing the aerating pipes.

Reference: Půlpitel L., Unpublished report ČKD Blansko, Czechoslovakia, 1975.

Position	h	α	f	A_1	A_2	A_3	p_1	Δp_1	p_2	Δp_2
	[mm]	[°]	[Hz]	[mm]	[mm]	[mm]	[Pa]	[Pa]	[Pa]	[Pa]
I	40 - 50	35 - 52	14,5-16,4	0,17	0,12	0,01	-31	62	-9	37
II	300	44 - 45	5	0,63	0,43	0,03	-15	38	-19	28
III	430 - 460	48 - 61	3,9-4,35	3,8	1,6	0,17	-92	276	-258	240
IV	700 - 720	61 - 63	2,75-3	2,8	1,12	0,09	-430	614	-369	462

C13

VIBRATION OF A WEIR WITH OVERFLOW AND UNDERFLOW

Alexander B. Rudavsky

Hydro Research Science, Santa Clara, California, USA

SUMMARY

Recent failures in ski-jump spillways and other overflow and underflow structures demonstrate the seriousness of flow instability and flow-induced vibrations encountered in existing hydraulic structures. HRS uses a conservative and exact experimental approach to develop solutions for this problem. The solutions have proven to be hydraulically stable and structurally sound. In this paper, the example of a weir with a strong flow-induced vibration is presented in a small drop structure with considerable overflow and underflow discharges. The vibration is caused by a combination of factors: the high velocity overflow and underflow; and the air cavity that is trapped between these two flows. This problem is solved by using a splitter to eliminate vibrations and to provide stable flow conditions.

1. INTRODUCTION

Recent model studies and prototype performance reviews indicate that overflow and underflow structures that have trapped-air cavities can create severe flow instability and structural vibrations. These adverse phenomena can be observed in ski-jump spillways, drops, overfall spillways, gate flows, and other merging jet flow structures. Figure 1 illustrates some examples of severe flow-induced vibration caused by trapped-air cavities that are due to overflow (Figure 1A), overflow and underflow (Figure 1B), and by a ski-jump jet (Figure 1C). Figure 2 shows an underflow and overflow structure which is being subjected to vibrations induced by junction flow. Figure 3A indicates an example of an induced stability: in this case, the jet does not merge with the tailwater, resulting in jet separation and causing an air pocket between the tailwater and the jet trajectory. Figure 3B shows a stable condition produced by increasing the tailwater elevation, which submerges the jet and eliminates the air pocket.

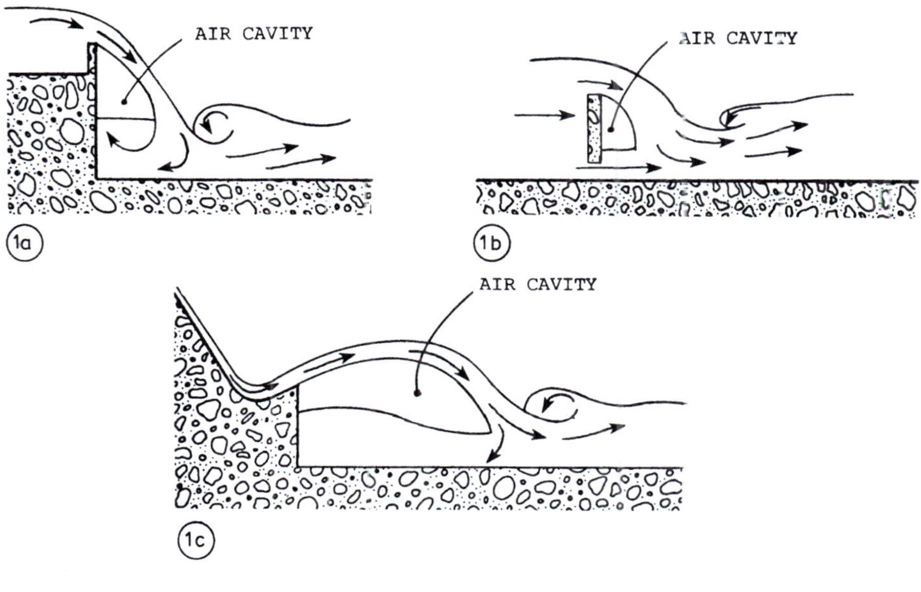

FIG. 1 STRUCTURES WHERE TRAPPED-AIR CAVITIES LEAD TO VIBRATIONS

2. OVERFLOW AND UNDERFLOW WEIRS

The particular weir considered in this paper is illustrated in Figure 2. A cross-section of this structure is shown in Figure 1B, as indicated, overflow and underflow conditions are present. As shown in Chart 1, the weir is 52 ft long and 4.5 ft high with a centered rectangular opening 3 ft wide and a 15-ft-long channel designed to discharge the underflow. The water surface level at 19,000 cfs is 18.7 ft; at 31,000 cfs the level is 23.3 ft. The 4.5-ft height of the weir allows for overflow when discharge is higher than the capacity of the underflow opening.

3. VIBRATIONS

In order to develop a solution to the problem of vibration that is induced in the weir by overflow and underflow, a hydraulic model of 1:40 scale was built based on Froude's Law.

Upon completion of model construction, discharges through the weir, ranging from 4,000 cfs to 31,000 cfs, were modeled. It was observed that oscillations of the overflow and the instability of the underflow, as well as the air pocket between them, caused vibration. The corresponding frequency of this vibration in the prototype in this case was 1.42 Hz.

It was found that when the weir height is relatively small, splitter blocks placed at the exit of the underflow will reduce vibration. As the weir height is increased, the solution to the vibration problem requires the introduction of air into the pocket under the overflow by means of a vent system.

Finally, model observations indicated that a solution to the flow-induced vibration was a splitter block that is 6 ft long by 6 ft wide by 3 ft high placed at the center of the underflow, immediately below the weir opening.

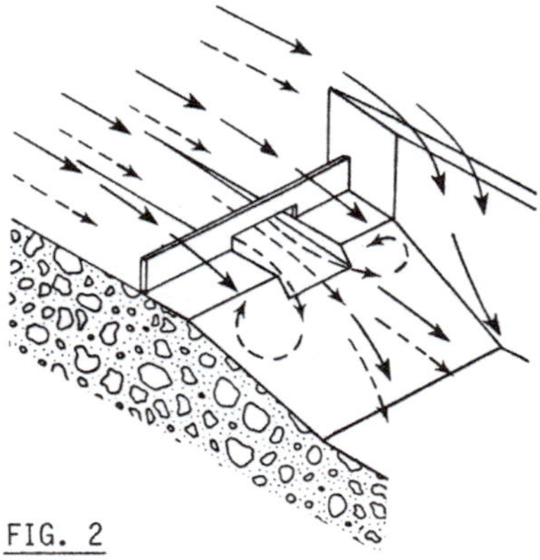

FIG. 2

(a)

(b)

FIG. 3

(a) SKIJUMP JET WITH THE AIR POCKET AND UNSTABLE FLOW SEPARATION

(b) STABLE CONDITIONS PRODUCED BY SUBMERGENCE OF THE JET

VIBRATION OF A WEIR WITH OVERFLOW AND UNDERFLOW

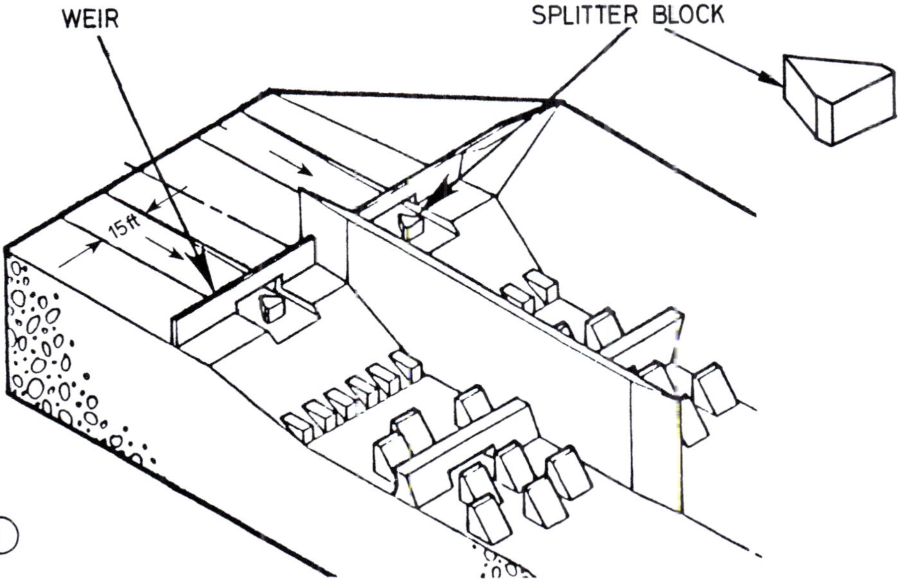

(1)

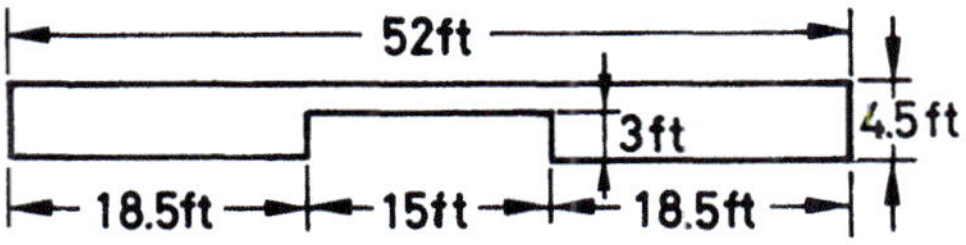

(2)

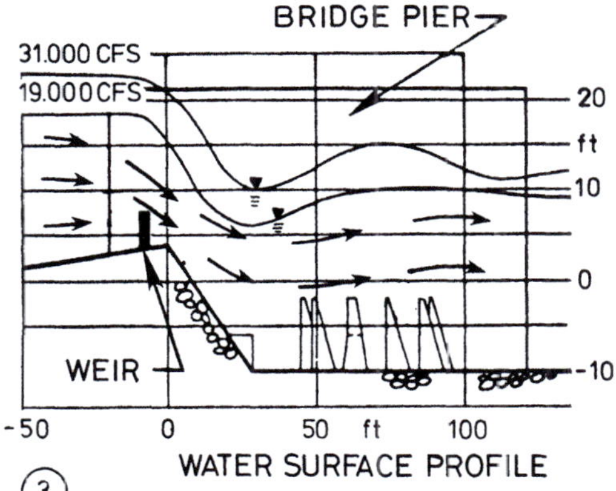

(3)

Problem: Nappe oscillations of the overflow caused vibration of the weir.

Geometry: see Sketch 1.

Boundary Conditions: The weir is 52 ft wide by 4.5 ft high, its bottom and two sides being fixed to the channel, with a 15 ft by 3 ft opening at bottom.

Flow conditions: Discharge varies from 5,000 cfs to 31,000 cfs. Drop of the head varies from 5 to 10 ft.

Structural Dynamics: Moment of inertia of the weir (including added mass effect) $I = 1,5000\ ft^4$
Modulus of elasticity
$E = 3.15 \times 10^6$ psi.

Vibration: Natural frequency (First Mode) of prototpye = 1.42 Hz.

Solutions: Two splitter blocks, with dimensions as shown in Sketch 1, placed in the channel eliminated the vibration.

C14 SEAL VIBRATION

Kurt Petrikat
Professor

Institut für Wasserbau
Universität Stuttgart
Federal Republic of Germany

SUMMARY

Elastic seals for sluice gates, lock gates, and bottom sluices are susceptible to vibration if, as a result of inaccurate construction, a gap remains between the sluice seal and the rigid reinforcement. The seepage through the gap and the resultant intermittent pressure cause the seal to vibrate. A gap between reinforcement and sluice is also formed, of course, when the sluice is opened. Thus, vibration is also possible during control movements, when the gates are opened or shortly before they are closed.

A slight deformability of the seal as a result of varying pressure in the gap can lead to changes in velocity in the gap and thus to an augmentation of the intermittent pressure. Thus, a process of self-excitation can be initiated, which leads to a considerable increase in seal vibration.

For this reason, seals with a high degree of rigidity frequently offer advantages, but they must be accompanied by effective damping. Broad seals on which the jet separates and reattaches (Fig. 1 A) are more susceptible to vibration than narrow seals of only a few millimeters of breadth. Unfortunately, a very hard and rigid seal often leaks because the reinforcement upon which the seal rests cannot be constructed and fitted accurately for cost reasons.

A further important influence on the extent of vibration is the shape of the seal. The various shapes are:
The old broad wooden sealing beam (Fig. 1 A), the rubber knife-edge seal (Fig. 1C), the music-note type, the hammer-head seal, the bulbous profile seal mounted on one or two sides, and the L-profile seal.
Thus, at the Hydraulic Institute of the University in Stuttgart and at the Laboratory of Voith in Heidenheim the intermittent pressure was determined with large, rigid models (not susceptible to vibration) at the scale of 1:1 (prototype) and 1:2 for flows through various gap widths. The width of the test canal and of the model was 172 mm.

The experiments showed that a very flexible and effective seal is susceptible to vibration, even if the reinforcement is not accurately constructed and installed. A very rigid seal which hardly makes allowances for unevenness in the concrete and reinforcement is less susceptible to vibration, especially if dampers are installed in addition. Effective sealing and freedom from vibration can only be obtained by carefully preparing the smooth seal surface of the reinforcement and by employing correspondingly hard rubber for the seal which has contact with a hydraulic friction damper.

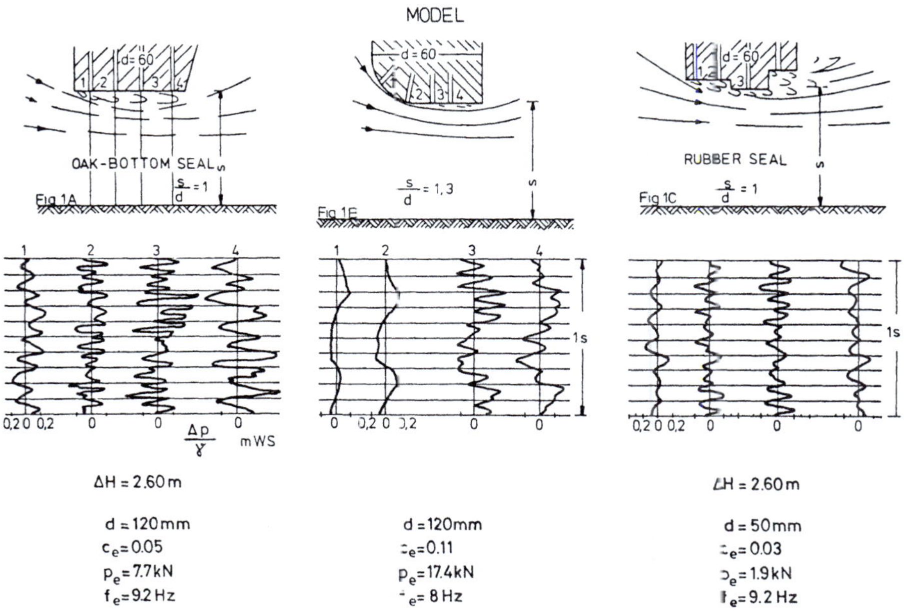

FIG. 1 SEAL-BEAM HORKHEIM; WIDTH: 24.6 m, HEIGHT: 5.2 m (see also Chart 1)

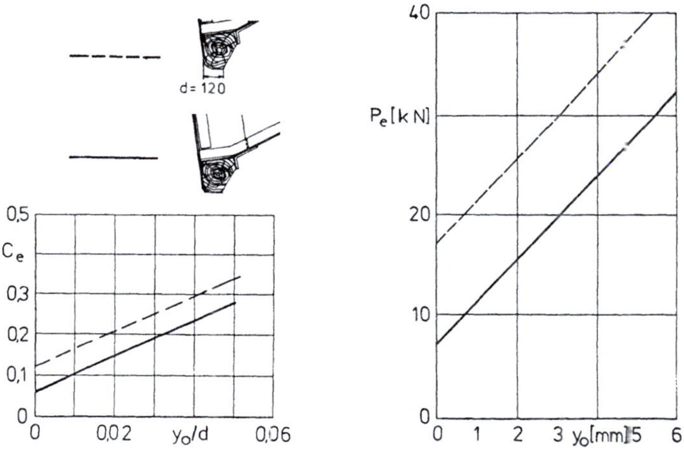

FIG. 2 EXCITATION FORCE P_{ϵ} AS FUNCTION OF THE AMPLITUDE y_o

1. BOTTOM SEALS FOR SLUICES

Wood was often chosen as a material for seals for sluice gates and also for turbine inlet sluices. If the sill was uneven, this form of seal could easily be shaped accordingly. The seal was cheap and easy to produce. Because of the minimal permissible unit pressure of the wood, the sealing beams had to be constructed from d = 7 cm to 30 cm in width, depending on the weight of the sluice. In the case of sluices fitted with these wooden beams, vibration frequently occurred at lifting heights between s = 2 cm and 50 cm.

In the zone of turbulence (Fig. 1 A) behind the sharp edge of the sluice lifted to s = 12 cm, the fluctuating pressure was measured at four points on the sealing beam with a thickness of d = 12 cm. The results are shown in the same figure as a function of time. In the first and last quarter of the bottom beam, the periodical pressure had a frequency of 7.5 Hz, while the pressure fluctuations in the middle section had a frequency of 13 Hz. For the bottom beam with a width of d = 12 cm and with a difference in water level of H = 5.2 m these readings showed a generating force of 7.7 kN with 9.2 Hz. Vibrations were not observed in this sluice.

On the sluice at Horkheim (Chart 1), through which the high flood was released over longer periods of time, the wooden bottom beam had become rounded at the front due to the floating debris, bottom ice and bed material (Fig. 1 B). The distribution of pressure on a rounded beam is quite different from that on a rectangular beam, and the excitation forces considerably larger, P_e = 17.4 kN, where $P_e = c_e \cdot A \cdot \Delta H \cdot \cos \omega t$ [kN] and c_e = exciting value[-], $A = d \cdot b$ = area [m²], ΔH = difference in water level [m]. The frequency of these vibrations almost coincides with the frequency of the sluice itself, and thus violent vibration was caused at the opening. The data is shown in Chart 1.

The vibrations were eliminated by replacing the wooden sealing beam by a rubber seal (Fig. 1 C). The rubber strip was secured between the skin-plate and a rubber clamping iron behind it. The vibration-generating forces amounted to no more than P_e = 3.2 kN, and the damping and the friction were sufficient to prevent vibrations.

In the case of the Augst-Whylen sluice (Chart 2), the vibration excitation was prevented by a jet dissipator. For about 30 years no vibration had been observed. Only after the water level was raised by 50 cm did vibrations begin to occur. By affixing jet spoilers (Chart 2, Sketch 3) to the face of the gate in the head water, a uniform vortex was prevented from forming under the beam as a result of the change in the distribution of pressure under the bottom beam, and thus no vibration was caused. A similar remedial measure was suggested by Dr.Otto Müller [2]. By means of alternating rounded and rectangular bottom beams the exciting forces can be eliminated (Chart 2, Sketch 4).

If the intermittent forces are large enough to cause a vibration of the sluice, then there occurs a change in the distribution of pressure. Intermittent changes in pressure leading to an excitation have been found to exist over the last third of the bottom beam, while in the first, upper quarter these intermittent forces were subjected to a damping effect. The influence of the dimensionless vibration amplitude y_o on the excitation coefficient C_e is shown in Fig. 2 on the left, while on the right the excitation force P_e is given for a sluice as a function of the amplitude.

The difficulty in determining the C_e value through the vibration equation lies in the calculation of the correct damping and the amplitude actually occurring. Both values are so important because they enter linearly into the calculation of C_e.

In Horkheim (Chart 1) the bottom seal with an overall width of 50 mm comprising 25 mm skin-plate, 15 mm rubber strip and 10 mm rubber clamping irons has proved to be vibrationfree.

One of these seals was installed in one of the bottom sluices (Chart 3) of the Feistritz Dam. In order to preclude excessive bending of the sluice under pressure from an increased imponding head a ledge was provided in the bottom. The overall width of the seal was only 45 mm hence differing from the Horkheim seal.

For slight sluice lifting heights up to approx. 20 mm, vibrations with an amplitude of 0.5 to 2 mm arose. In order to eliminate these vibrations, the width was reduced to 5 mm at the point of contact and the elastic seal placed into the bottom (Chart 3, Sketch 3).

Chart 4: If for the bottom sluices of reservoirs the emergency gate and control gate are constructed close one behind the other with a spacing as small as 80 cm, thereby reducing the cost (Sketch 1), unfavourable positioning of the gates leads to vibrations in the main control gate. These vibrations can be reduced by rubber jet dissipators (Chart 4, Sketch 3) so that operation of both sluices in all positions is possible without vibration.

2. MUSIC-NOTE TYPE SEALS

An even, rectangular sealing strip (Fig. 1 C) does not generally accomodate sufficiently the irregularities of a concrete bottom and the sealing performance of the gate may be unsatisfactory. For this reason, a bulbous rubber seal in the form of a music note was developed (Chart 5).

If the sluice is lifted slightly, these seals are susceptible to vibration. The measurements of distribution of pressure taken on a large model (scale 1:2), Fig. 3, show an intensive pressure drop over the rounded surface, which changes with a frequency of 16 Hz and which deforms the music-note type seal, especially if it is hollow, so that at frequent intervals a small gap is closed and then opened again. In striking the bottom, the seal rapidly disintegrates, and this can also lead to damage to the concrete bottom. If the gap becomes larger (Fig. 3 B), the intermittent pressure changes decrease.

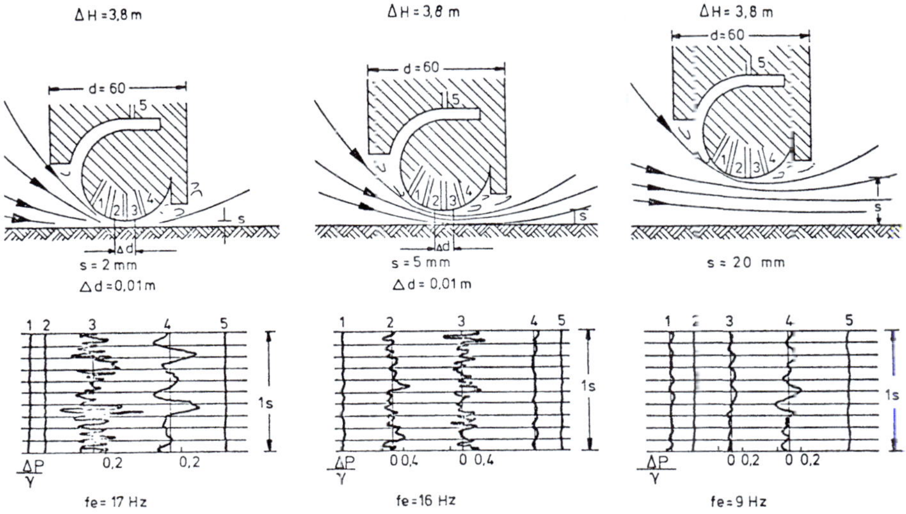

FIG. 3 PRESSURE FLUCTUATIONS AT DIFFERENT GAP OPENINGS

Chart 6: On the hook gate of the Horkheim lock, a bottom seal of the flap gate was installed of music-note shape (Sketch 2). When the gate was opened up to about 12 mm, vibrations occurred identical to those of the sluice in Chart 5. When the gate was opened only a few millimetres, the "note" profile was so distorted by lowered pressures acting on the rounded surface that it closed the gate. This initiated the cycle which caused the observed vibrations and disintegrated the seal. A remedy was found in replacing this music-note seal by a top seal as in Chart 1, Sketch 4 or Fig. 1 C.

Chart 7: Top seals on bottom sluices at the upper wall or ceiling of a bottom discharge tunnel are often constructed in music-note form. In the 9 bottom discharge sluices supplying the Bhavani River with water, strong vibrations at 35 to 50 Hz occurred at $\Delta H > 30$ m (Fig. 4, photo) at various opening positions. These originated at the top sealing.

The music-note profile was deformed as a result of the considerable pressure drop, so that the gap of 8 mm was reduced. Pressure pulsations were initiated and thereby the seal was caused to vibrate.

The five lateral canal sluices (Fig. 4), in which the difference in surface levels was only 25 m, were free from vibration although they were of the same construction and design as the river sluices. When the "note" profile was manufactured of full and very hard material, or reinforced with a steel core, no vibration occurred.

FIG. 4 VIBRATION RESEARCH ON THE TOP SEALS OF THE 9 RIVER SLUICES OF BHAVANI DAM

An elastic and well-fitting music-note profile seal is susceptible to vibrations. Hard and almost undeformable materials and shapes remain vibration-free when the difference in water levels is not excessive. Vibration is particularly apt to occur when the seal is affixed to the reinforcement (Chart 8, 1 and 2).

With designs by which the seal is positioned on the sluice (Chart 8.3 and 8.4), vibrations are only possible during the first few centimetres of lifting. In the case of bottom-discharge gates, this range can be passed through rapidly, so that this design can be regarded as almost free from vibration. A double fixing of the seal provides further reinforcement against vibrations (Chart 8.4).

Chart 9: If the water loss through bottom sluices in the closed position is to be reduced to a minimum, a fine seal must be fitted in addition to the coarse seal. Vibrations were observed directly in front of the closing unit when the wall of the radial sluice was not constructed uniformly enough and the coarse seal E did not sit tightly against it, the gap G thereby allowing water under pressure to penetrate into the Chamber F.

The fine seal was lifted for a short moment together with the sluice and the ensuing equalization of pressure caused the sluice to fall down. This cycle was repeated and gave rise to vibration observed on the sluice. When this seal (Sketch 2) was replaced by the L-shaped seal (Sketch 3), the vibration ceased.

3. ROLLER GATES WITH BULBOUS SEALS

Chart 10: Vibrations of seals occur likewise in the case of roller gates. On Sketch 1 the design of a roller gate with box sealing is represented. Under the influence of the sun's rays on the tailwater side, the gate becomes so deformed that a gap arises between sill and wooden sealing beam. The water pressure and the flow through the gap gives rise to variable forces on the wooden sealing beam and causes it to vibrate. An attempt was made to eliminate these vibrations by means of a fixed, adjustable spring pre-tensioning (see Chart 10, Sketch 2). With such a seal the exciting force in a gap of 20 mm is only 240 N/m. The frequency lies between 10 and 30 Hz (Fig. 5).

On the other hand, with a small gap of 2 mm, P_e becomes larger. The rigidity and friction in the springs must be sufficient to cancel out exciting forces.

Since the seal of this design has a large surface area, the intermittent forces are considerable if the gap is small. If the contact-pressure springs are not sufficiently pre-tensioned, or the hinges of the connecting rod are deflected, vibration is liable to occur, particularly if the friction damping of the springs was set too low.

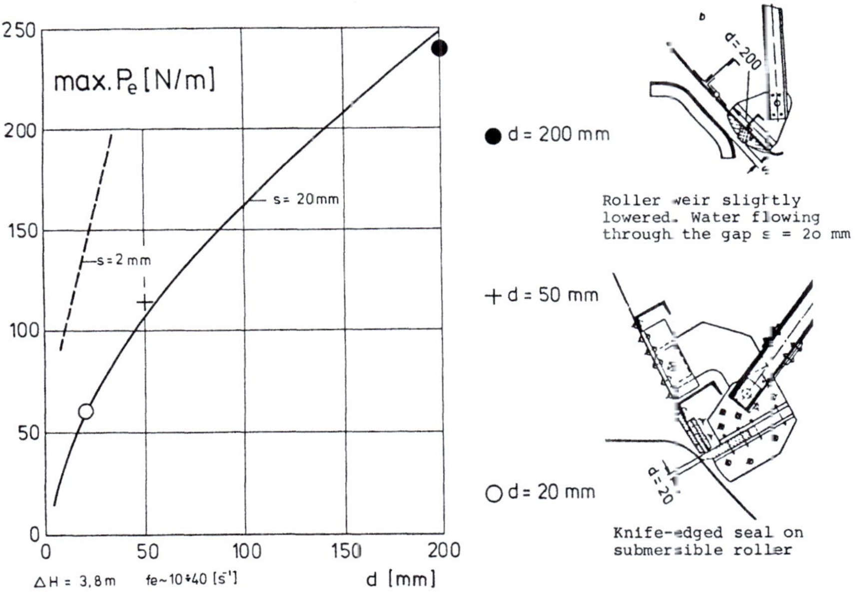

FIG. 5 EXCITATION AS A FUNCTION OF THE THICKNESS "d" OF THE SEAL

If the bulbous seal (Chart 10, Sketches 2,3) is replaced by a knife-edge seal (Chart 10, Sketch 4) and additional measures are taken for friction damping, a seal construction is achieved which is free from vibration.

Chart 11: shows this seal installed in the upper lock gate at Würzburg. In Sketch 2 this seal can be seen in detail.

4. VIBRATION OF BOTTOM SEALS ON FLAP GATES

Chart 12 shows the bottom seal of a flap gate. A steel plate is fastened to a small segment of the flap axis. This seals the flap gate against the tailwater.

If the river is carrying in bed material, inaccurate construction of the segment can lead to sand being washed in between the sealing plate and the segment. In this case, although the plate was of 5 mm thick steel (Chart 12, Sketch 1), it was lifted a few millimeters by coarse stones. The free end of approximately 5 m of the flap began to vibrate whereby the whole flap was shaken.

In this case there was no flow over the flap. When the steel plate was replaced by a rubber plate with high shore hardness (Chart 12, Sketch 3), there was no more vibration. It is important that the segment on the flap be accurately constructed and smoothly finished, so that there is no possibility of a gap remaining between the seal and the segment.

If the finishing is accurate, so that a rubber nose of lesser shore hardness has to be affixed to the front of the rubber seal, it is possible for the fine sand to draw in coarser particles with it; as a consequence, the seal can get loose and the free end can start to vibrate.

5. VIBRATIONS IN GATES WITH INCLINED SKIN-PLATES

On gates with an skin-plate inclined at angles $\alpha < 90^{0}$ towards the tailwater, a different type of self-excited vibration occurs at small gate openings of s = 8 to 25 mm, caused by pressure oscillations in the headwater region (Fig. 6) which act on the skin-plate. The vibration of the gate with amplitudes of $y_o = \pm 4$ mm changes the gate opening and hence the discharge periodically. The periodic velocity fluctuations cause corresponding periodic pressure fluctuations on the skin-plate, which lead to negative damping forces and hence to vibrations.

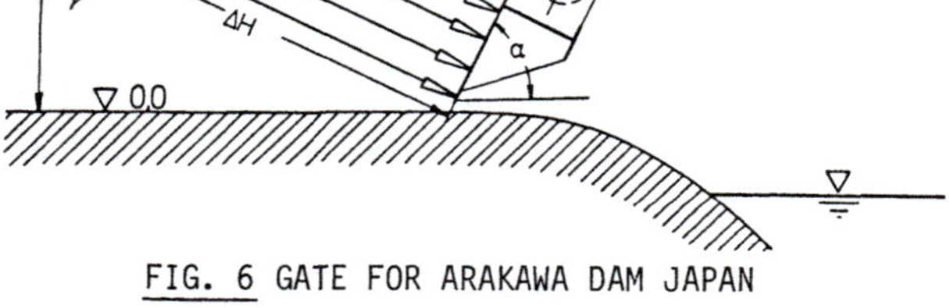

FIG. 6 GATE FOR ARAKAWA DAM JAPAN

Vibration: s 30 mm ; Seal: Chart 1, Sketch 4 or Chart 3, Sketch 3

Width: 28 m ; Amplitude y_o = 3 to 5 mm ; $f_o = f_e$ = 3 to 5 Hz

Reference: Dir. Shimoda, Sakai Iron Works 3 - 1 Dejima Nishimachi Sakei Osaka 590

6. VIBRATIONS OF LARGE BOTTOM GATES AND TURBINE OUTLET GATES

In the process of closing gates of this type, of which an example is shown in Figs. 7 and 8, vibrations may occur during the downward motion of the gate at openings of less than 2 m especially if the lower crossbeam is placed too close to the bottom edge of the gate.

In order to prevent these vibrations the lower beam should be arranged such that the angle between the bottom seal and the outer edge of that cross-beam is not less than 40°, as illustrated in Figure 7.

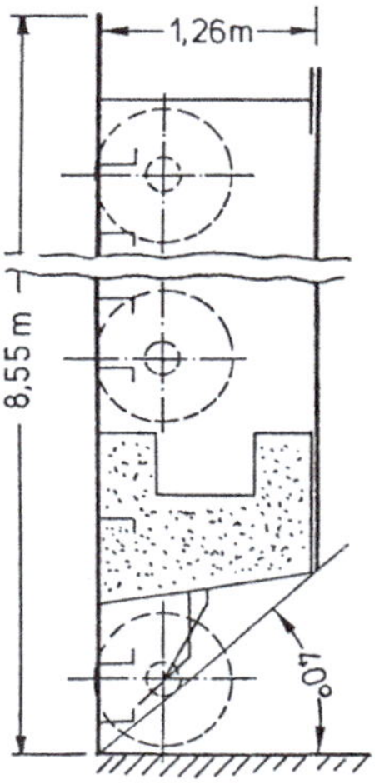

FIG. 7 DRAFT TUBE GATE IFFEZHEIM, DETAILS

FIG. 8 , PHOTO

Acknowledgment

I am greatly indebted to Dipl.-Ing. Dammel, MAN Gustavsburg and Director Dipl.-Ing. Erle, Dipl.-Ing. Fütterer, Badenwerk Karlsruhe for the comprehensive material supplied to me by them. I should also like to thank Dr. Schweikert, Dipl.-Ing. Hausenblass and Dipl.-Ing. Wurz, Voith-Heidenheim, for their cooperation and Dipl.-Ing. Kirchenbaur for carrying out the measurements on the seal models in Heidenheim and Stuttgart, and for his help with the design of the slides. Thanks are due to Dipl.-Ing. Schwarz of Zschokke-Wartmann Brukk for providing material on the gate of Schinznach.

Discussion: SEAL VIBRATION (Petrikat)

Neilson, F.M., USAE Waterways Experiment Station, Vicksburg, USA

The purpose of this discussion is to emphasize the predominance of seal-related vibration problems as far as prototype experiments with flow-induced vibration is concerned. The music-note type seal (J-seal) is used in a variety of geometrical configurations as noted in the subject paper and usually, due to material wear or fouling, undergoes flow-induced vibration at some time during its life. These vibrations, even when occurring under very low heads (often less than 3 m), can cause severe structural damage. Examples of this type of problem are the vibration of lock miter gates (the J-seal is between the gate and the sill) and the vibration of low-head tainter valves due to side-seal vibration. A field engineer must be able to recognize this type of problem; usually a design incorporates a means for making adjustments to the seal and for cleaning the seal and sealing surface. Numerous alternate bottom seal configurations for tainter gates were studied with regard to the Arkansas River Dams (Chart 6 of the subject paper) and other configurations could be construed. However, in most cases these vibrations were due more to the geometry of the lip than to the flexure of the seal material and were resolved by non-hydroelastically-similar model studies.

Chart 1: GATE HORKHEIM, GERMANY

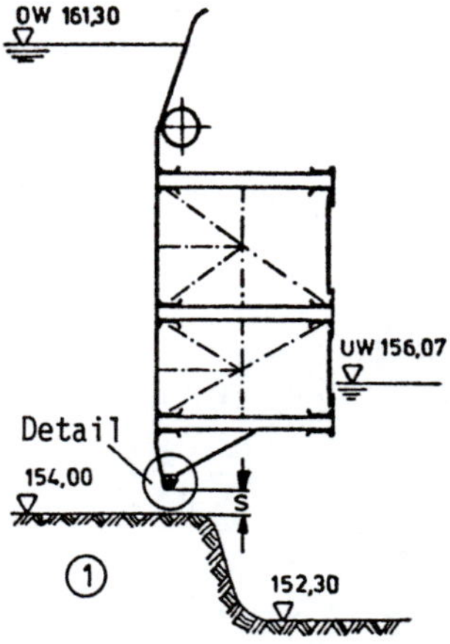

(1)

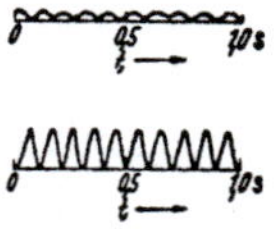

(4)

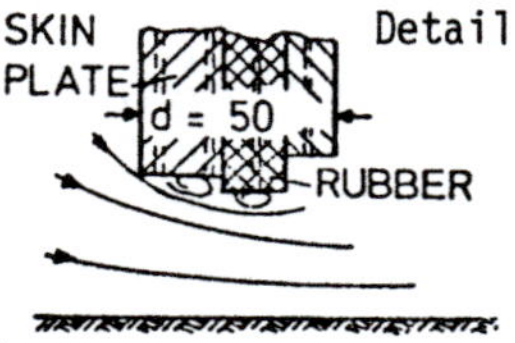

(5)

Problem: Gate oscillations at small openings s = 0,12 to 0,18 m.

Geometry: See Sketch 1. Width = 24.6 m; wooden seal beam, Sketch 2. By abrasion, the front edge was rounded, see Sketch 3. Breadth of the seal beam d = 0,12 m.

Flow conditions: See Sketch 5. Gate oscillations were observed only with rounded wooden seal-beam.

Structural dynamics: Mass = 111 t

Vibrations: n = 10 Hz, y_0 = 0,34 mm, see Sketch 4

Cures: Sketch 5 shows the new bottom seal which had no vibration; d was reduced from 120 to 50 mm.

References: Petrikat, K., Schwingungsuntersuchungen an Stahlwasserbauten. Der Stahlbau, Heft 9 und 12, Sept./ Dez. 1955, p. 198-202, 272-289.

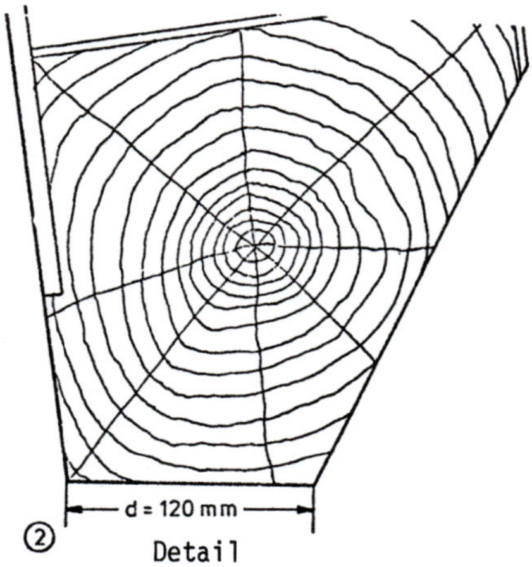

(2) Detail

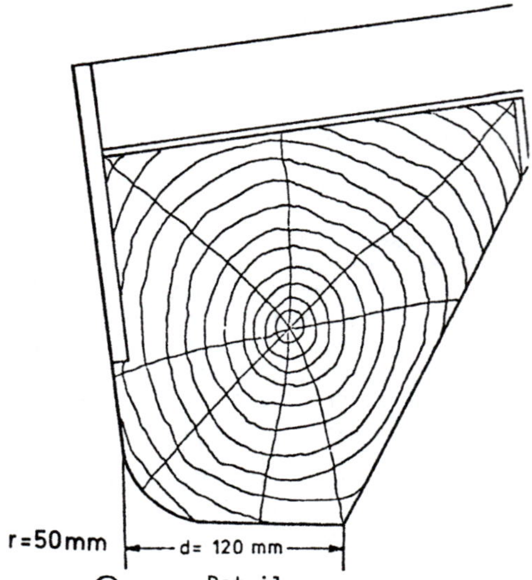

(3) Detail

Chart 2: SLUICE AT AUGST-WHYLEN, SWITZERLAND

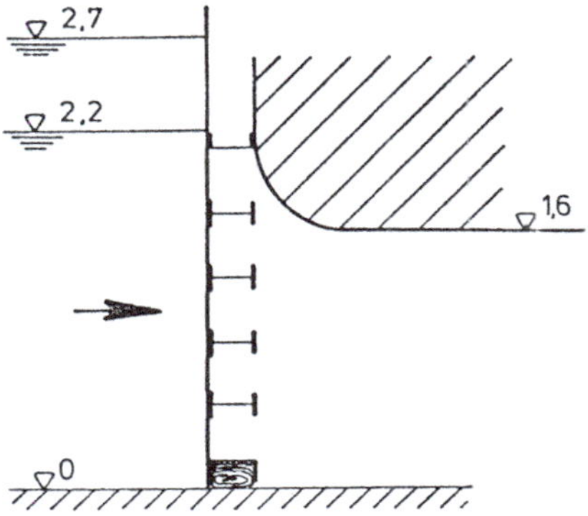

①

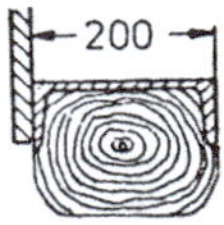

②

Problem: Vibration occurred in the sluice as soon as the headwater was raised by 0,5 m, from 2,2 to 2,7 m

Geometry: See Sketch 1. Width of the gate 2,1 m, sliding sluice.

Critical state: Wooden sealing beams (Sketch 2) d = 200 mm.

Vibration: Vibration occurred when the sluice was raised by s = 200 mm.

Cures: The sluice had been functioning vibration-free for 20 years at a surface level of 2,2 m. After the headwater had been increased by 0,5 m vibrations were observed when the sluice was raised by 200 mm. These vibrations were eliminated when spoilers as shown in Sketch 3 were installed in the headwater on the lower section of the upstream gate wall.

References: H. Ackermann, Würzburg, Schwingungen an unter- und überströmten Wehren und ihre Beseitigung, Der Bauingenieur 1962, Heft 9.
O. Müller, Neuere Schwingungsuntersuchungen an unterströmten Wehren, Die Bautechnik 1937, Heft 6.

③ Spoilers upstream of gate lip

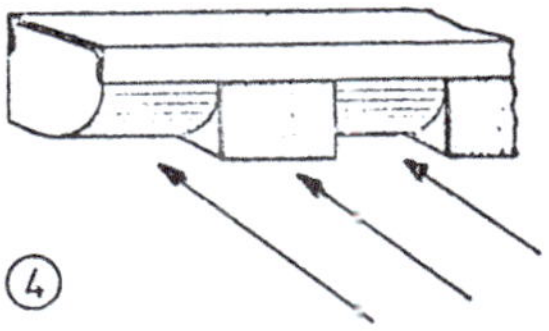

④

Elimination by means of alternating base beam sectional shapes.

Chart 3: OUTLET GATES FOR FEISTRITZ POWER PLANT, AUSTRIA

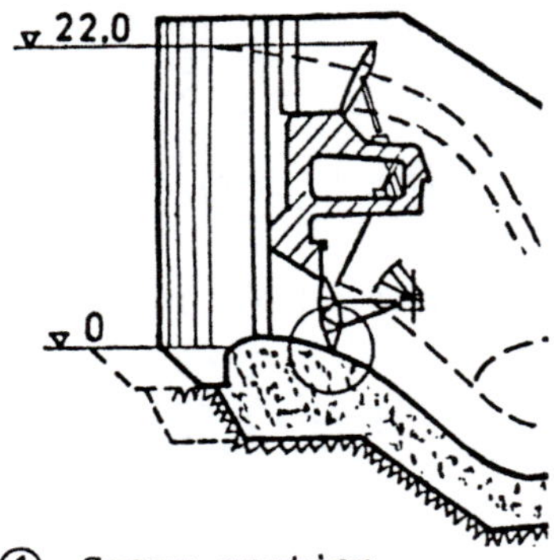

① Cross section

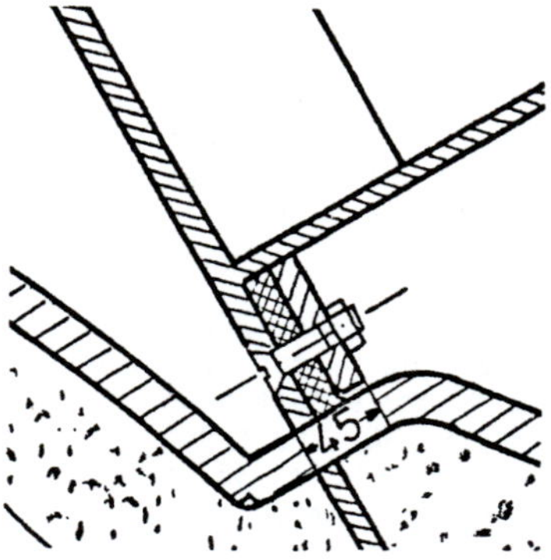

② Initial bottom sealing

Problem: Strong pulsations occurred at the bottom outlet gate which were caused by the shape of the sill beam.

Geometry: See Sketch 1. Width 15 m

Boundary conditions: The closing force of the outlet gate amounts to 10 kN/m.

Flow conditions: Sketch 2 shows the initial shape of the sill beam.

Structural dynamics: The gate has a dead weight of 45 tons.

Vibration: If the gate was lifted a few millimetres the gate vibrated. Amplitude 0,5 to 2 mm.

Cures: The vibrations of the radial gate at Feistritz Power Plant could be avoided by altering the shape of the sill as shown in Sketch 3. Care had to be taken that the gap causing danger of vibration (approx. a range of 20 cm) is automatically passed through during raising and lowering procedure.

References: Liebl A., High Pressure Sluicegates, Onzième Congrès des Grands Barrages, Madrid, 1973, Q 41 - R 42.

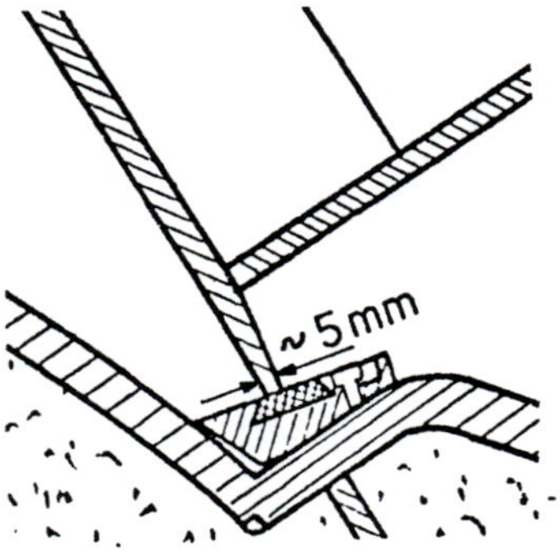

③ Final bottom sealing

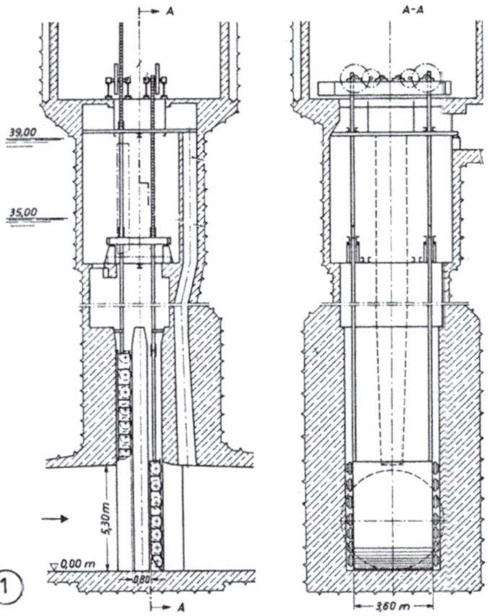

①

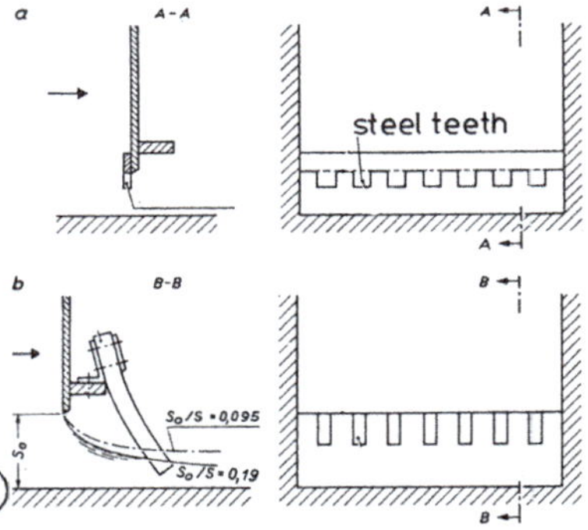

②

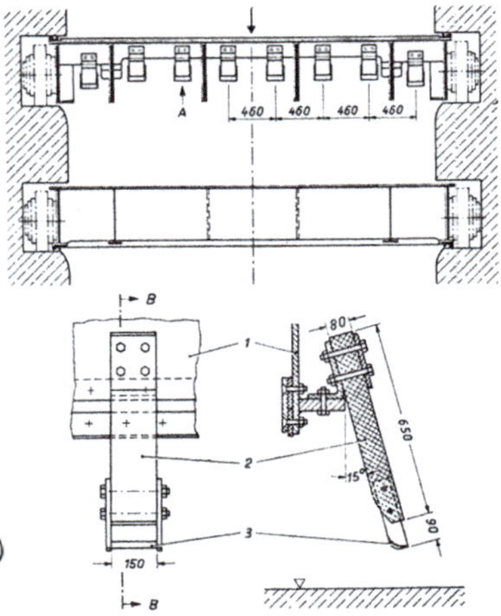

③

Chart 4: <u>TWO CONSECUTIVELY POSITIONED BOTTOM OUTLET GATES</u>

<u>Problem</u>: When two bottom outlet gates closely following one another are in operation, the main gate on the tailwater side is brought to vibrate.

<u>Geometry</u>: See Sketch 1. Longitudinal and cross-section of a bottom outlet gallery.

<u>Operating conditions</u>: When the emergency gate was opened at $s_o = 0{,}5$ m, the jet hit the main control sluice at $s_u = 0{,}35$ m, which with regard to the vibrations was in an unfavourable position. The jet gave rise to alternating forces in a vertical direction on the control sluice, which produced heavy blows on the winching gear.

<u>Vibration</u>: The maximal forces in the model were $P_N = 208$ kN and a frequency of 9,5 Hz was measured.

<u>Cures</u>: The installation of jet dissipators as shown in Sketches 2 and 3 proved to be an effective remedy for these vibrations. The dissipators break up the compact discharge jet, so that the changes in pressure and the violent roller in the zone between the two sluices were disrupted. In order to ensure that the jet dissipators are effective with openings greater than $s_0 = 1{,}32$ m, jet deflector plates were affixed to the rubber jet dissipators (Sketch 3) at lower end. Rubber was chosen as a material for the dissipators so that the dissipators can bend to ensure effective closing of the sluice.

<u>References</u>: Krummet, R., Untersuchung von Schwingungserscheinungen am unteren Verschluß zweier hintereinander angeordneter Tiefschützen, MAN-Forschungsheft No. 11, 1963/64.

① Longitudinal and cross-section of a bottom outlet gallery in the region of the sluice shaft with two consecutive bottom sluices.

② Schematic drawing of the dissipators tested on the upper sluice.

③ Positioning of the rubber jet dissipators on the final version.

1 Dam wall upper sluice, 2 Rubber jet dissipator, 3 Jet deflector plate

Chart 5: RADIAL GATE WITH MUSIC-NOTE TYPE BOTTOM SEAL

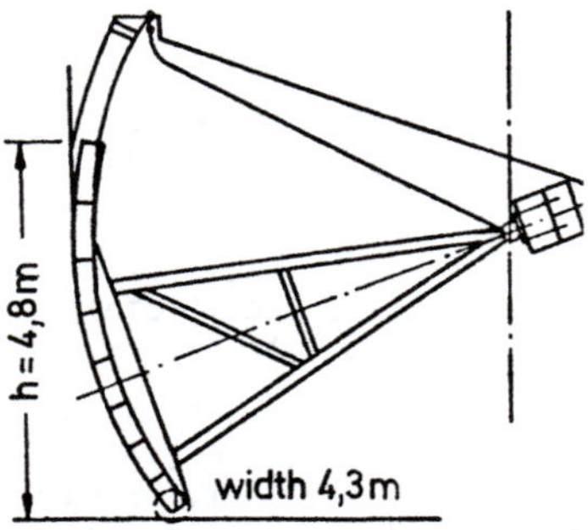

Detail

①

Problem: In order to obtain an effective sealing of the radial gate at the bottom (Sketch 1), a seal in the form of a music-note as in Sketch 2 was proposed. With openings of only a few centimeters, strong vibrations [1] occurred as a result of deformations (see Chart 6) caused by downward pressure (Sketch 3). Similarly, vibration occurred in the music-note type seals of the types A and B in the radial gates of the Arkansas River Dam (Sketch 3), [2].

Cures: The music-note type seal was replaced by a knife-edge seal of the type C or D. The vibration was thus eliminated (see Chart 1, Sketch 4). Type C does not offer as many advantages, since during the closing operation small stones can be washed through the last mm of the gap, and when the sluice is firmly pressed down the metal knife edge may be damaged, thus after a time causing it to leak.

References: 1. Bureau of Reclamation Denver, Colorado. ASCE, HY 5, July 1965, p. 188.
2. Douma, J.H.: Field Experiences with Hydraulic Structures, Symposium "Flow-Induced Structural Vibrations", Karlsruhe, 1972.

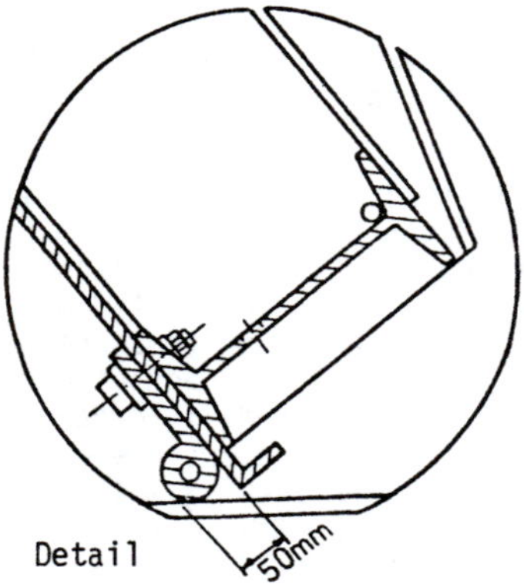

Detail

②

Type A und B , Vibrations

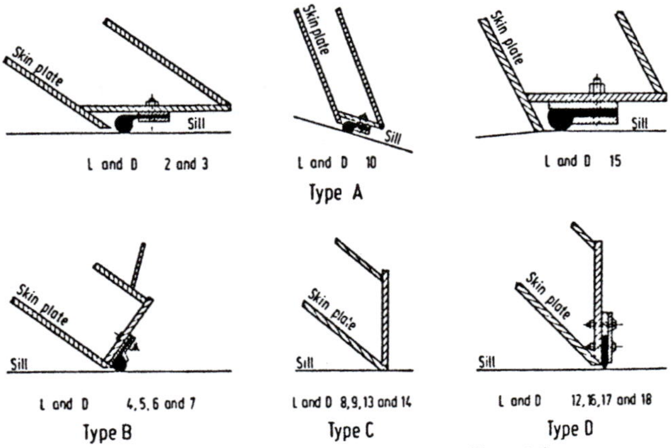

Type C und D, no vibrations

③ Tainter Gate Lips for Arkansas River Dams.

Chart 6: LOCK INLET GATE HORKHEIM, GERMANY

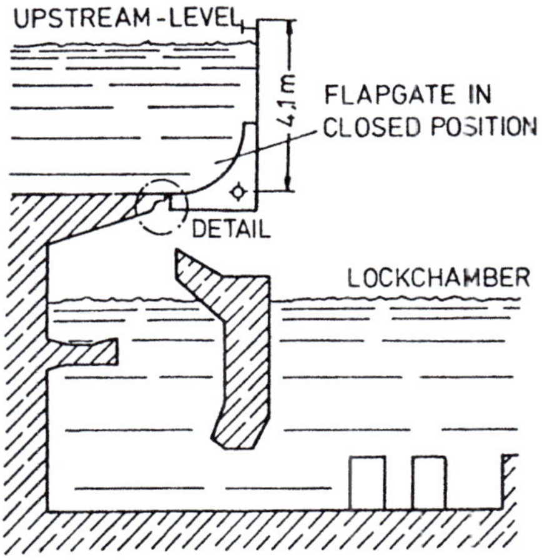

①

Problem: Vibration was observed on the hook gate of the Horkheim (Main) Lock (Sketch 1) at the beginning of the opening operation of the upper gate. The gate was sealed against the lock sill with a music-note type seal. For details see Sketch 2.

Geometry: For the filling procedure, the gate tilted towards the head water and the lock was filled through the gap. The width of the lock and gate was 12 m.

Boundary conditions: The door vibrated only in the first 15 s of the opening movement, over a gap distance of 15 mm.

Vibration: Frequency 10 Hz. The music-note shaped seal was so deformed by the lower pressure during the opening of the gate that the circular seal beam became oval. As a result of the elastic restoring force of the rubber the seal was subjected to severe vibrations. See Sketch 2.

Cures: The vibration was eliminated by replacing the music-note shaped seal by a narrow, rectangular rubber strip (Sketch 3).

References: Petrikat, K., Vibration Phenomena Observed on a Lock Inlet Flap Gate. IAHR Congress, Montreal 1959.

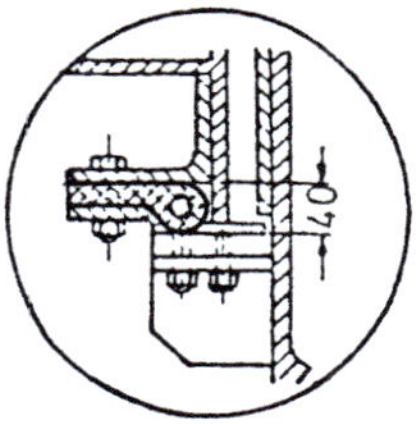

② FLAPGATE CLOSED

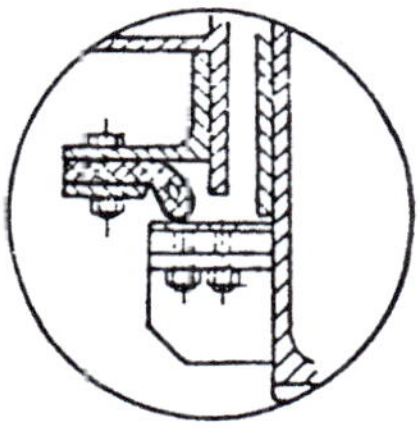

FLAPGATE 12 mm OPENED
seal vibrating

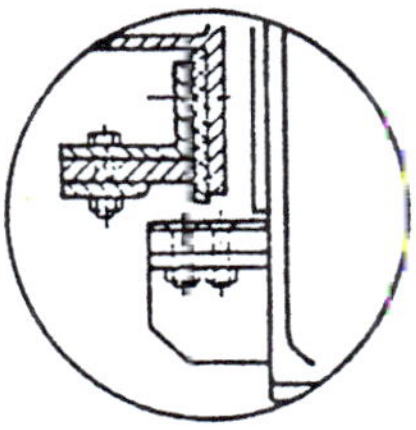

③

Chart 7: BOTTOM GATE BHAVANI DAM, INDIA

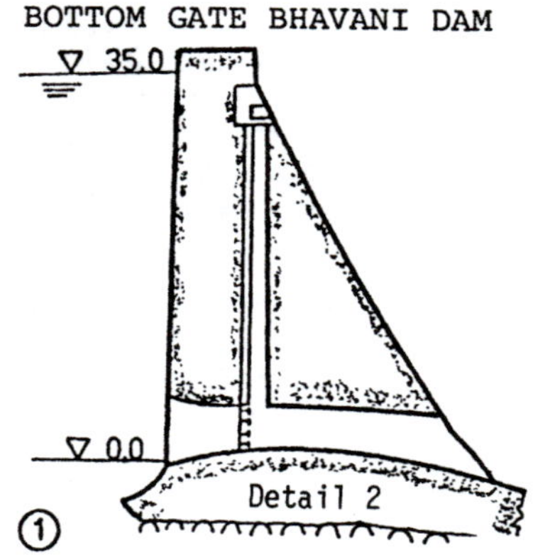

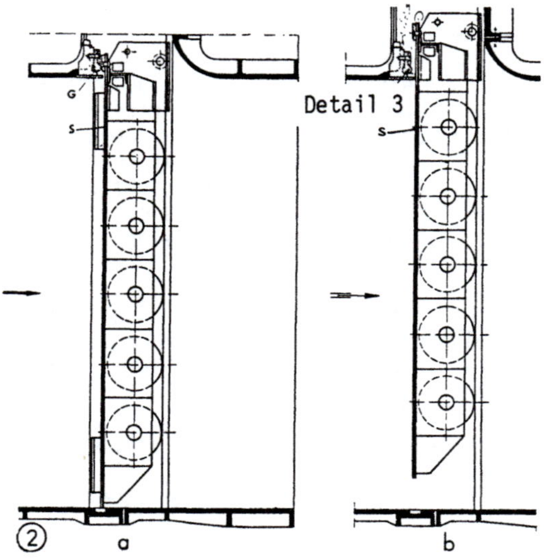

Detail 2

Problem: Top seals for bottom sluices near the upper wall, i.e. on the ceiling of a bottom discharge tunnel (Sketch 1) are ordinarily constructed in the form of a music-note for the purpose of obtaining an effective seal. The bulbous rubber with a diameter of approx. 4 cm has a 1cm hole in the middle so that the seal can accomodate any possible unevenness in the sluice. Thus, a bending of the sluice can also be accomodated. When the sluice is withdrawn (Sketch 2), the sealing strip on the sluice is raised from the seal affixed to the reinforcement and water shoots through the gap between the cover plate and the sluice-dam wall (Sketch 3). If the rubber has a lesser degree of shore scleroscope hardness (60) and the opening in the middle of the seal is large, the seal becomes deformed into an ellipse as in Sketch 3 and the rubber presses against the dam wall, thus preventing the water flow. This causes the downward pressure on the rubber to change into a return pressure and the rubber springs back. This cycle is repeated and produces a vibration of 35 to 50 Hz according to the hardness of the rubber and the difference in level between head- and tailwater.

Cures: Various forms of seals which are practically free from vibration are shown in Chart 8.

References: Petrikat, K., Vibration Test on Weirs and Bottom Gates, Water Power, 1958, p. 192.

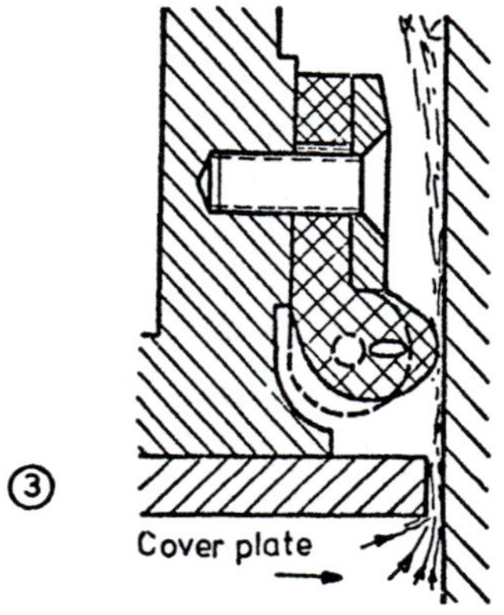

Detail 3

Chart 8: BOTTOM GATE WITH TOP SEAL

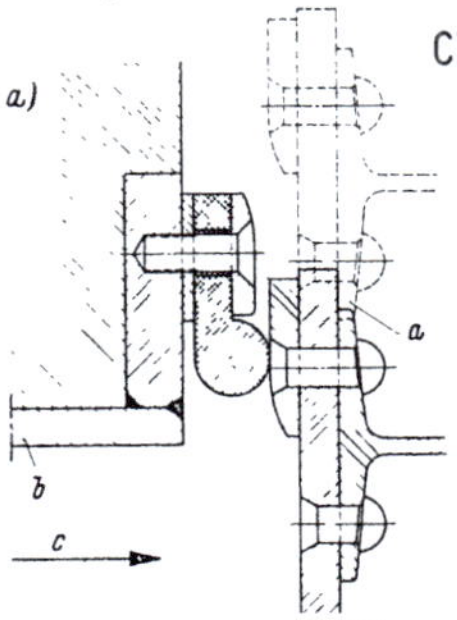

(1)

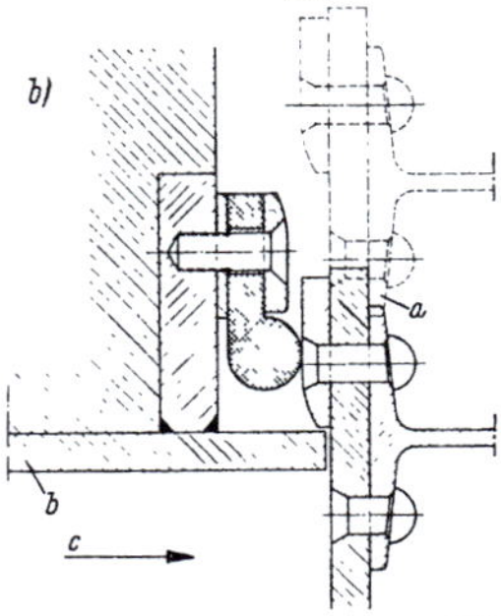

(2)

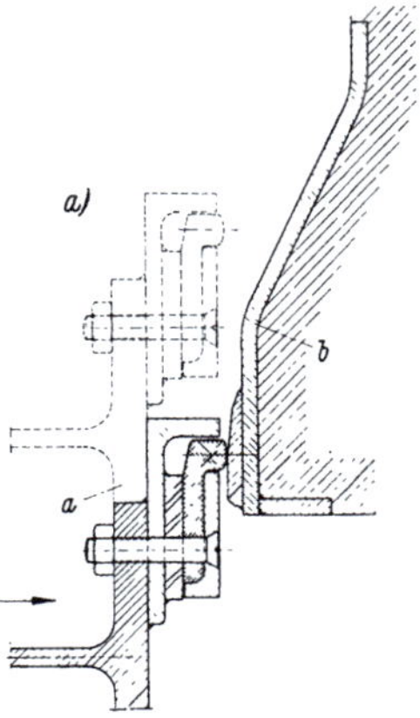

(3)

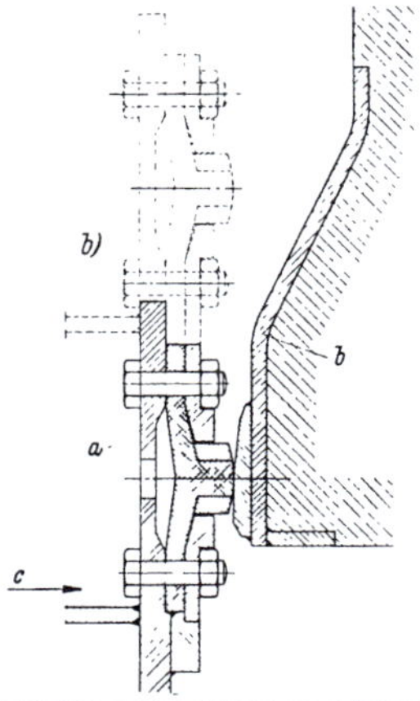

(4)

Problem: Top seals for discharge-tunnel gates are often affixed to the reinforcement of the tunnel in order to enable ease of replacement. When the sluice is partly raised, a jet shoots past the note profile. The underpressure exerted on the music-note and the pulsating pressure on the rounding (Sketch 1) causes vibration.

Flow conditions: The sluice is raised so that the sealing beam is lifted from the rubber sealing (broken line). The other drawings show the sluices in the closed positions.

Vibration: Frequency 35 to 50 Hz.

Cures: At smaller differences in pressure, $H < 25$ m, if a cover plate is fitted in front (see Sketch 2) the resulting shielding has the effect of preventing vibration. With greater differences in water level, $H > 30$ m, vibrations were also observed in the case illustrated in Sketch 2. The rubber sealing was quickly destroyed. If the music-note type of rubber seal was formed from very hard material and reinforced with a steel core, no vibrations arose.

When the top seal is secured to the sluice (Sketch 3), vibration is only possible in the first 10 cm of lift. This range can be easily avoided by rapid lifting. With the seal secured to the sluice, the replacement of a defective seal is considerably more difficult and time-consuming.

For high pressures, a double support of the top seal (Sketch 4) prevents any vibration even for high heads beyond a lift of 10 cm.

References: Krummet, R., Schwingungsverhalten von Verschlußorganen im Stahlwasserbau ..., Forschung im Ingenieurwesen, Bd. 31, 1965, Nr. 5.

Chart 9: TOP SEAL AT A BOTTOM OUTLET GATE

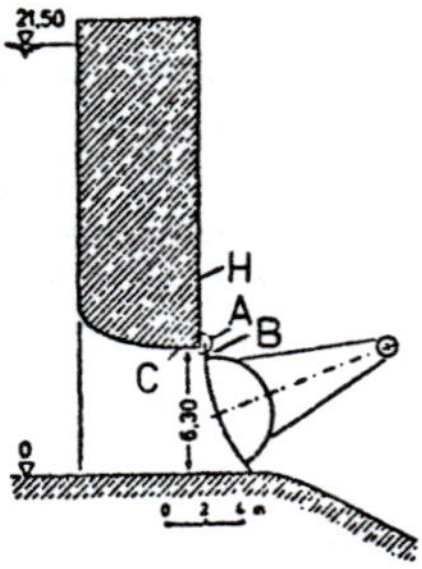

①

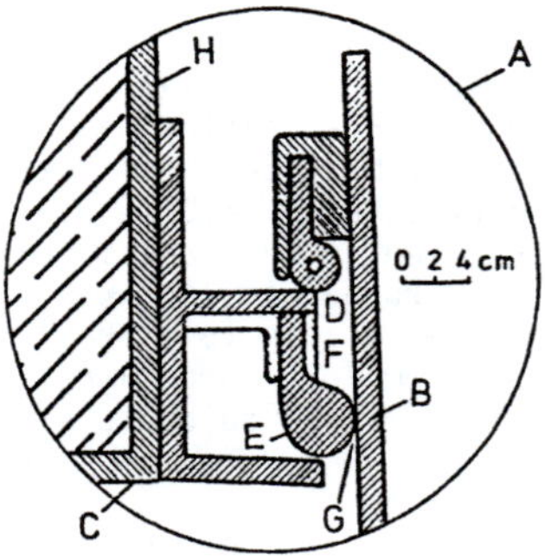

②

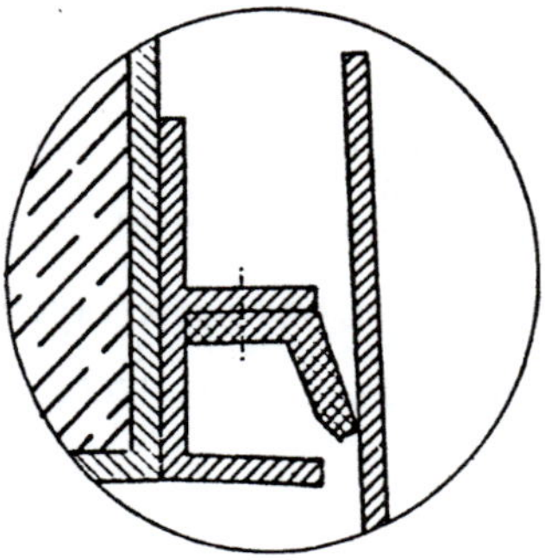

③

Problem: Vibrations of the bottom outlet gate were noticed which were caused by the top seal.

Geometry: See Sketch 1. Width: 11 m.

Vibration: See Sketch 2.
Vibrations of the segment gate were noticed with a gate opening between 5 mm and 10 mm. Frequency: 6,34 Hz; amplitude 0,2 mm.

These vibrations have resulted because of the top sealing arrangement (Sketch 2). The coarse sealing E does not provide sufficient sealing. With a small gap of the gate, the leaking water builds up pressure in the room F between the fine sealing D and the coarse sealing E. The pressure acts on the surface D and lifts the gate. Through the lifting at the gate, the fine-sealing is also lifted and, hence, the pressure in room F is reduced. The gate moves down in a zig-zag fashion and the fine-sealing closes the room F again so that once again pressure is built up in room F. This cycle repeats itself and results in a continuous vibration.

Cures: Vibrations are stopped if the music-note coarse sealing is replaced by an L-shape seal (Sketch 3).

References: Petrikat, K., Structure vibrations of segment gates. IAHR 8. Symposium,Section for Hydraulic Machinery Equipment and Cavitation, Sept. 6 - 9, 1976, Leningrad, USSR.

Chart 10: SUBMERSIBLE ROLLER WEIR

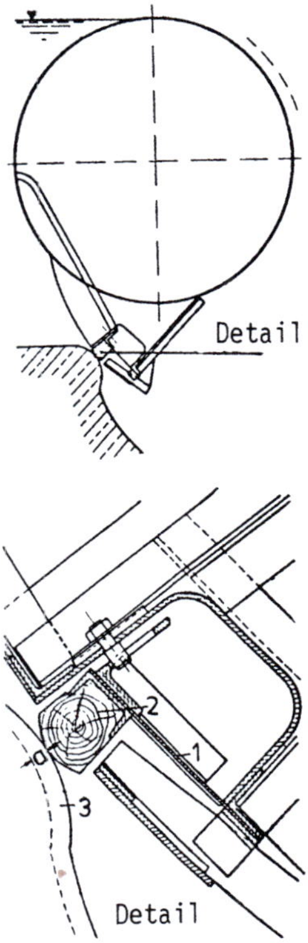

①

Problem: The effect of the sun's rays from the tail water side is to deform the roller, so that a space arises between the sill 3 and the sealing beam 2. The water pressure in chamber 1 is not sufficient to press the wooden seal 2 against the sill 3. The pulsating pressure on the wooden sealing sets the sealing beam in vibration, see Sketch 1.

Vibration: Frequency 6 to 10 Hz.

Cures: These vibrations were stopped by means of a strong spring pretensioning (Sketch 2, Pos. 3). If in this design a gap occurs due to the effect of the sun's rays or due to a lowering of the roller (Sketch 3), pulsating forces of P_e 1500 N/m with a frequency of 10 - 20 Hz act on the sealing through the resulting gap flow. These forces must be absorbed by friction in the springs 3 (Sketch 3). If the springs are not robust enough and the friction in the springs is insufficient, or the dynamic forces have dislocated the hinges of the connecting rod, vibration may also arise in this type of construction. If the seal 4 (Sketches 2 and 3) is replaced by a knife-edge seal (Sketch 4) and an adequate friction absorber installed, this construction will also be free from vibrations.

References: Petrikat, K., Vibration Tests on Weirs and Bottom Gates. Water Power, April 1958.

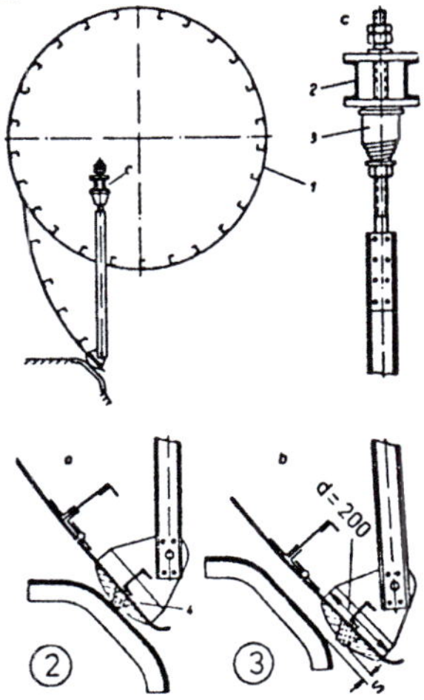

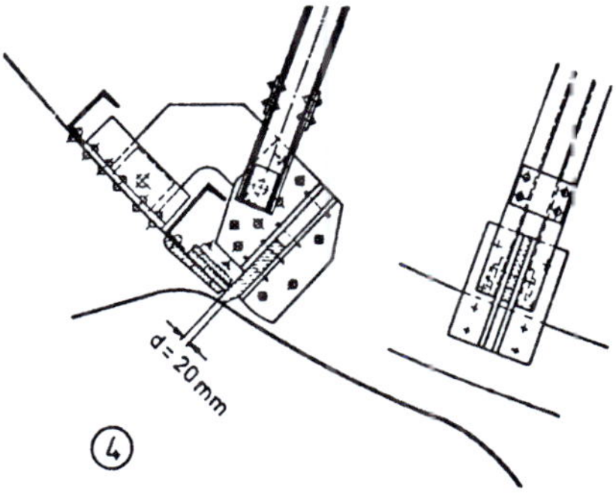

④

Chart 11: RADIAL LOCK AT WUERZBURG, GERMANY

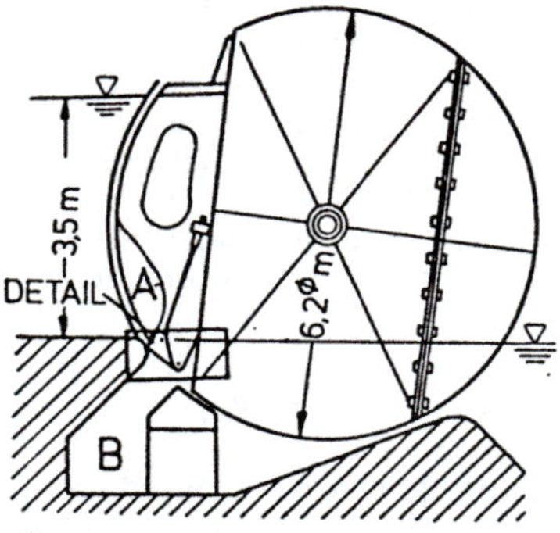

①

Problem: The lock gate in Wuerzburg is a turning radial gate (Sketch 1). For filling the lock, the upper gate is lowered by about one meter. The water flows over the whole lock width through the lowered filler shell A into the small power conversion chamber B and fills the lock chamber from the upper gate. At the same time the lock can be filled through bypass conduits from the upper gate.

Geometry: See Sketch 1. The lock gate and the lock are 12 m wide.

Operating conditions: The filling operation is begun with the lowering of the turning gate. The knife-edge sealing C in Sketch 2, constructed of steel, lifts up and in the first seconds only a small gap exists between sill D and seal C. Unsteady thrusts are exerted on the seal and can set these in vibration.

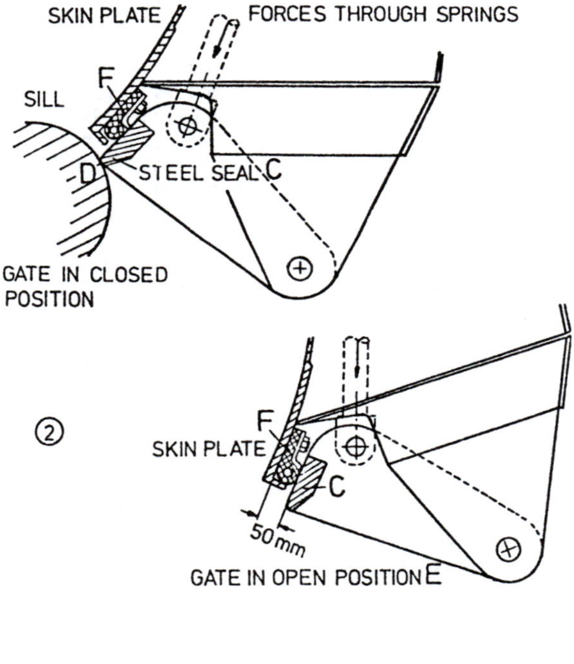

②

Cures: To prevent vibration, a music-note shaped seal F with a diameter of 50 mm is installed between the skin-plate and the knife-edge seal C. By means of water pressure and the appropriate prestressing, the absorption can be increased so as to stop the vibration. In Sketch 2 on the right-hand side, the gate is shown in the opened filling position. The knife-edge sealing C and the music-note type seal F are thereby shielded from the filling jet by the skin plate so that vibration-exciting forces have no effect on the sealing.

References: MAN-Gustavsburg, Germany, Prospekte.

Chart 12: HINGED SEAL OF THE FLAP GATE ON THE CONGO DAM

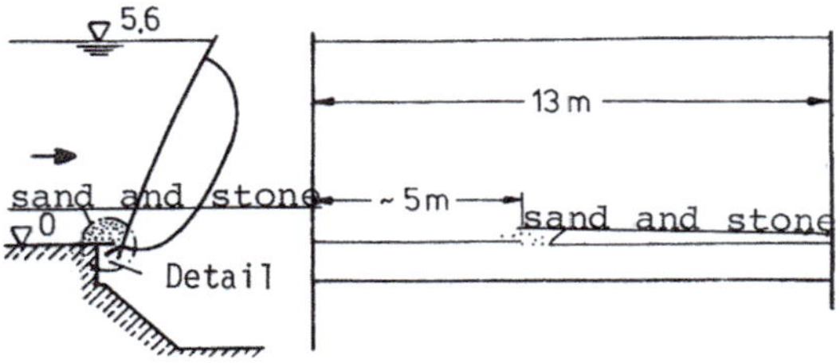

①

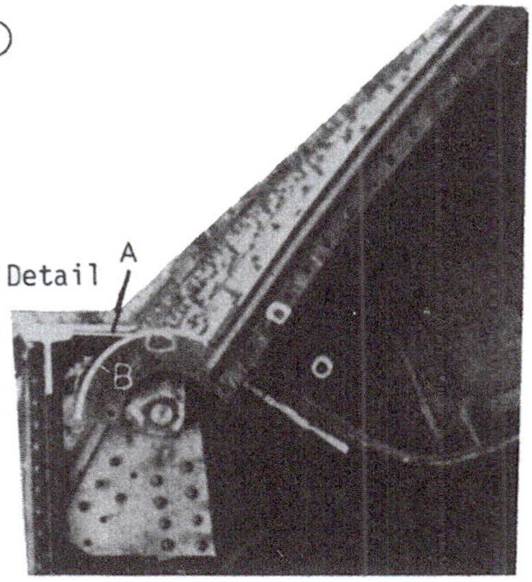

②

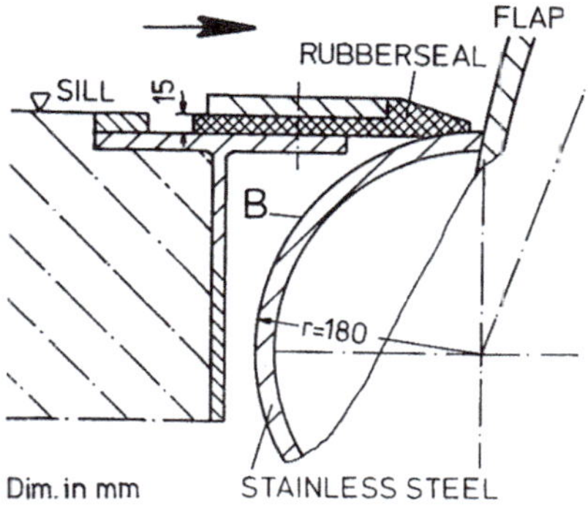

③

Problem: Vibrations occur in the bottom seals of the flaps (Sketch 1). When the river carries bed material, inaccurate construction of the segment can result in a gap between the steel plate A and segment B. First of all, fine sand and subsequently stones of up to 8 mm Ø are washed in (Sketch 2). The 5 mm thick steel plate installed as a seal is pushed up (e.g. in the centre of the weir 5 m from the piers, see Sketch 1).

Vibrations: The free end of the steel plate begins to oscillate, since the water flowing through the gap produces a vacuum and pushes the plate. When the gap is closed, the flow is stopped and thus also the vacuum, so that the plate springs back. The cycle then repeats itself. These vibrations shake the whole weir.

Cures: If the steel plate is replaced by a hard rubber seal (90 shore) as in Sketch 3, the seal will no longer be lifted. No further vibrations occur. It is essential to ensure that with a hard rubber seal the segment around the flap is even and smooth, so that there is no gap between seal and segment.

If the construction is not accurate enough, so that a soft rubber (30 shore) has to be chosen, this will result initially in the penetration of fine sand and later of coarser gravel, and the seal can lift and begin to vibrate.

Chart 13: SUBMERSIBLE GATE "KLEINE WESER", GERMANY

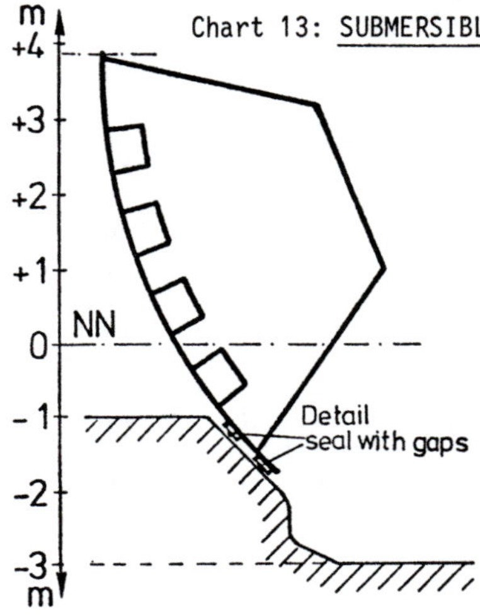

①

Problem: Inadequate sealing resulted in intermittent forces being exerted on the seal due to the passage of water through the gap. These forces were responsible for strong vibration in the seal and the whole weir.

Geometry: See Sketch 1; width: 24 m. The weirs were sealed by two rubber seals (Sketch 2) on the bottom.

Flow conditions: The water is impounded by two weirs. Regulation takes place by means of lateral discharge sluices. The weirs were subject to considerable undershooting at all water levels. The width of the gap at the point of leakage of the seal could not be measured.

Vibration: The strongest vibrations occurred where the throughflow of water was greatest in volume. Sketch 3 shows the results of the frequency analysis. At a water level H = 3,3 m the measurements of the vibration amplitudes y_o yielded 12,5 μm at 27 Hz and 55 Hz, and with a water level H = 3,5 m y_o = 15 μm at 20 Hz and y_o = 7,5 μm at 36 Hz.

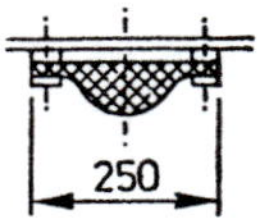

② Detail

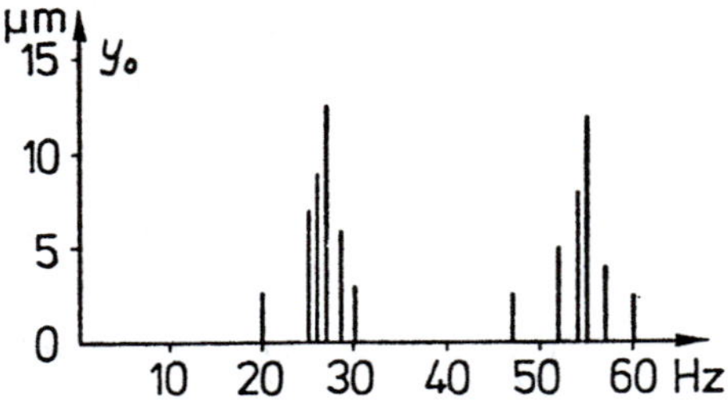

③

Cures: A modification of the seals to reduce the flow through the gap resulted in reduced vibration, but did not succeed in eliminating completely the leakage.

References: R. Zühlke, Neu Rautendorfer Str. 56, 2801 Grasberg.

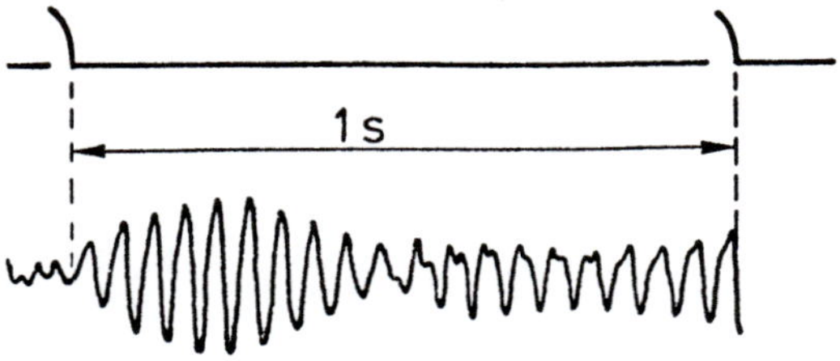

④

Chart 14: REVERSED TAINTER GATE SCHINZNACH, SWITZERLAND

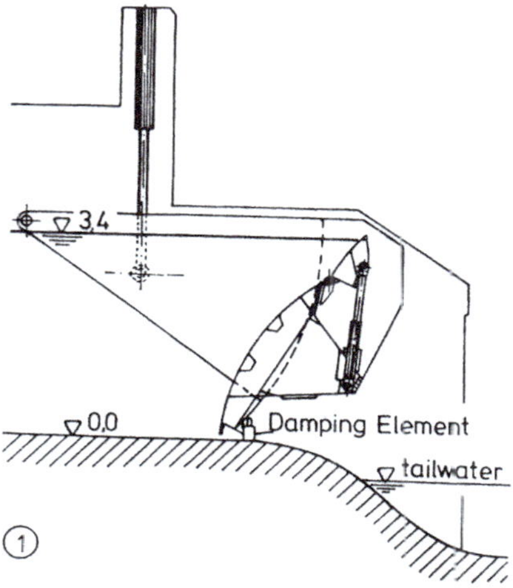

①

②

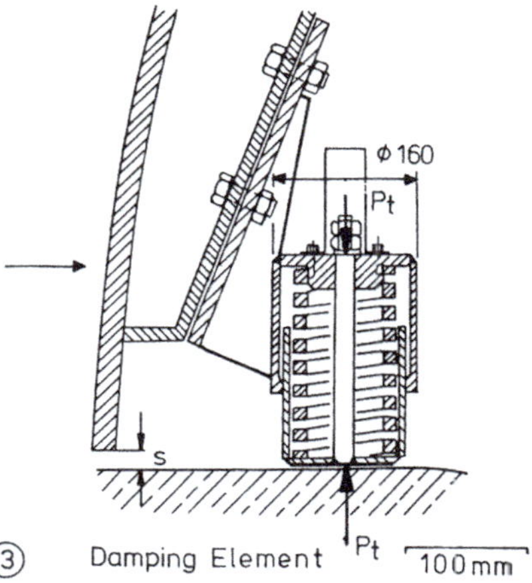

③

Problem: A self-excited gate vibration occurs for small gate opening s = 8 to 25 mm, caused by pressure-oscillations acting on the skin-plate in the headwater.

Geometry: See Sketch 1, width of gate: 22,5 m.

Flow conditions: The tailwater level is lower than the bottom of the gate and does not influence the bottom seal. The jet separates completely from the bottom seal.

Operating conditions: At a gate opening s = 25 mm on the sides, the opening in the middle of the gate is s = 8 mm.

Vibration: Frequency equal to natural frequency f_0 = 4,1 Hz (in a prototype test), logarithmic decrement: $\Lambda = 0{,}13$.

Cure: Damping cylinder (System Zschokke-Wartmann, Brugg, Switzerland) in the middle of the gate, see Sketch 3.
Model tests: scale 1:2 vertical oscillating gate.
P_t = 2 kN, where P_t = pretension.

References: Dr. C. Müller: Ein Beitrag zur Bestimmung der schwingungserregenden Kräfte an unterströmten Wehren. Mitteilungen Institut für Wasserbau, Universität Stuttgart, Heft 42, 1977.

C15

BARKLEY DAM GATE VIBRATIONS

E. Dale Hart
Research Hydraulic Engineer

John E. Hite, Jr.
Hydraulic Engineer

U.S. Army Corps of Engineers
Waterways Experiment Station
Vicksburg, Mississippi, USA

SUMMARY

This paper presents data obtained during prototype tests concerned with the vibration of the 16.8- by 15.2-m spillway tainter gates at Barkley Dam, KY. The gates vibrate during flow passage in which the gate lip is submerged by high tailwater. Field tests were conducted on two of the gates during high tailwater. Three pressure transducers were placed near the lip of the gate. In addition, fourteen accelerometers were attached at preselected locations on the gate. The second gate was instrumented with five accelerometers only for a limited data comparison. Data reduction indicated that many components vibrate at the same frequency as the gate-lip pressure fluctuations. Computations and ring tests indicate that this frequency approaches at least one component's natural frequency. The gate seal at the lip was removed and the test repeated under identical conditions. The measured vibrations were all dramatically reduced. By varying the opening of the gates, under identical conditions, it was found that vibration severity decreased with increasing tailwater submergence. Remedial measures in design and operation of the gates were recommended.

1. INTRODUCTION

The Barkley Dam tainter gates vibrate when discharging under certain lip submergence conditions caused by high tailwater.[1] Field observations and data analysis indicate that this submergence has two major effects on the gates. During submergence the upper streamlines of the flow are apparently altered so that this portion of the flow intermittently strikes the gate lip seal, causing a continuous shifting of the flow control point and creating pressure fluctuations. A second effect is apparently due to the depth of tailwater submergence, i.e., as the back pressure on the lip increases with submergence, the vibration intensity decreases.

2. SEAL EFFECT

To determine the degree of adverse effect (vibrations) caused by the gate lip seal, tests were made with and without the seal in place. Removal of the seal decreased the vibrations by a mean ratio of 15:1. This effect is illustrated by results from the time series analyis of the magnetic-tape data. Sketch 3 of the chart presents a comparison of accelerations for identical conditions except for seal removal. These data are from a transducer (designated CMP) which was mounted in the gate center near the skin plate and responded to movement tangent to the skin plate, thereby representing gate movement in the vertical plane.

3. SUBMERGENCE EFFECT

Head on the test gates was constant at approximately 6.2 metres. However, the tailwater lip submergence varied from 0.7 to 2.1 metres due primarily to the changes in gate opening. The intensity of gate vibration was found to decrease with increasing submergence. It was concluded that this was probably due to the increased back pressure at the lip which diminished the effect of the "feedback loop" phenomenon which will be discussed subsequently. Of course, changes in flow pattern close to the seal could have contributed to this effect as well.

4. FREQUENCIES

An estimated natural gate frequency was calculated to compare with the frequencies recorded in the test series. From this it could be determined if, under test conditions, the gates vibrated at or near their natural frequency. To make this calculation, the gate configuration was converted from a complicated arrangement of metal to a simple spring mass system (see Sketch 1). The natural frequency of the gate was found to be approximately 5 Hz using the equation:

$$fn = \frac{1}{2\pi}\left(\frac{L_1 + L_2}{L_1}\right)\sqrt{\frac{k}{m}}$$

A Fast Fourier Transform (FFT) of the magnetic tape data of the gate acceleration was accomplished to compare predominant test frequencies with the gate's natural frequency. Sketch 2 presents an FFT for transducer CMP during test 10 which was made with the lip seal in place. The lowest predominant frequency occurred at about 30 Hz, much higher than the computed natural frequency. This gate response frequency compares closely with the gate lip pressure fluctuations which will be discussed later.

5. FEEDBACK LOOP

Reference 2 states that in order to initiate flow-induced vibrations, some kind of fluid-dynamic "feedback" is necessary. This theory is also discussed by others [3,4,5]. A very brief description of the feedback loop, based on the references, follows:

a. A shear layer forms whenever flow separates from a boundary (as the upstream corner of the Barkley gate lips). Because the innermost portion of the layer moves much slower than the outermost portion, the layers roll up into discrete, swirling vortices (detail Sketch 1).

b. After these initial disturbances are amplified and convected downstream, they interact with the boundaries of the flow field (as, possibly, the gate seal) producing more disturbances. These, in turn, are partially transmitted back to the origin and trigger new disturbances which then proceed through the cycle. Because this is a selective amplification mechanism, the new disturbances result in fluctuations with a much narrower range of frequencies than the original perturbations. With subsequent cycles the fluctuations become more nearly periodic. Thus, interaction of the flow with its boundaries coupled with selective amplification forms what may be described as a "feedback loop." Naudascher[5] summarizes this phenomenon: "Although the fluctuating flow is the origin of the structural vibration, its pattern, periodicity, and intensity are ultimately dominated by the latter, and the vibration becomes sustained by a self-generated exciting force."

Sketch 2 shows that with the gate seal in place the gate vibrated and the lip pressures fluctuated at a discrete, common frequency of about 30 Hz. A cross-spectral density calculation of transducers CMP (gate acceleration) and PGL (gate lip pressure fluctuations) was also made and plotted. A very prominent spike occurred at 28.3 Hz, further implying a strong correlation at or near that frequency. On the other hand, with the seal removed, the oscillations were random and of small magnitude.

6. RECOMMENDATIONS

The following structural and operational alteration options were presented to the Nashville District (operating agency) for consideration:

6.1 Structural Alterations

a. Removal of the gate seal is an obvious alteration that will reduce vibrations. This procedure is recommended if the accompanying leakage can be tolerated.

b. If the leakage due to the removal of all gate seals cannot be tolerated, consideration should be given to removing the seal from a few gates which would be the primary operating gates (for example, the middle four gates).

c. Other possibilities include:

   (1) A narrower rubber seal.

   (2) Grinding the metal portion of the lip to a narrower width.

   (3) Placing the rubber seal in the gate sill bearing plate instead of the gate lip.

   (4) Installation of vibration absorbers. (A spring-mass vibration absorber has been designed and recommended. If approved it will be installed and tested during high tailwater in the spring of 1980.)

d. Vibrating support members should be braced or otherwise altered to change their natural frequency.

6.2 Operational Alterations

a. When the tailwater elevation is only a small distance above the sill, use fewer gates at higher openings so that their lips are not submerged.

b. When the tailwater elevation is high relative to the sill, use all gates with minimum openings to maximize the back pressure on the gate lips.

7. REFERENCES

1. U.S. Army Engineer District, Nashville, "Interim Report on the Tainter Gate Vibrations at Barkley Dam, Apr 1967, Nashville, TN.

2. Naudascher, E., and Locher, F. A., "Flow-Induced Forces on Protruding Walls," Journal, Hydraulics Division, American Society of Civil Engineers, No. HY2, Vol. 100, Feb 1974.

3. Blevins, R. D., Flow-Induced Vibration, Van Nostrand Reinhold Co., New York, 1977.

4. Goncharov, L. A. and Semenkov, V. M., "Field Investigations of Dam and Gate Vibrations," Proceedings, Flow-Induced Structural Vibrations Symposium, Karlshruhe, Germany, Aug 14-16, 1972.

5. Naudascher, E., "From Flow Instability to Flow-Induced Excitation," Journal, Hydraulics Division, American Society of Civil Engineers, No. HY4, Vol. 93, Jul 1967.

BARKLEY-DAM GATE VIBRATIONS

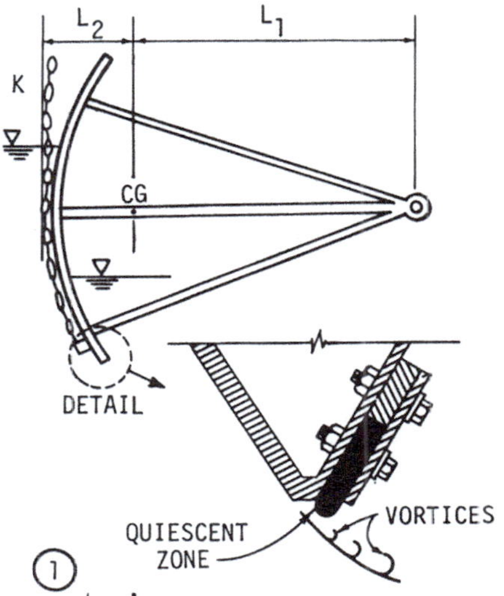

1

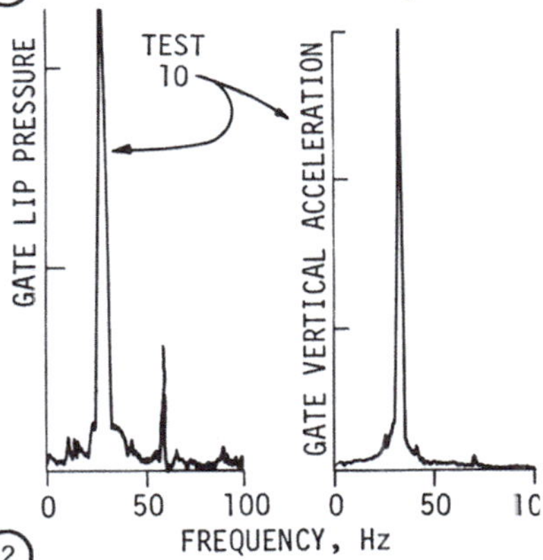

2

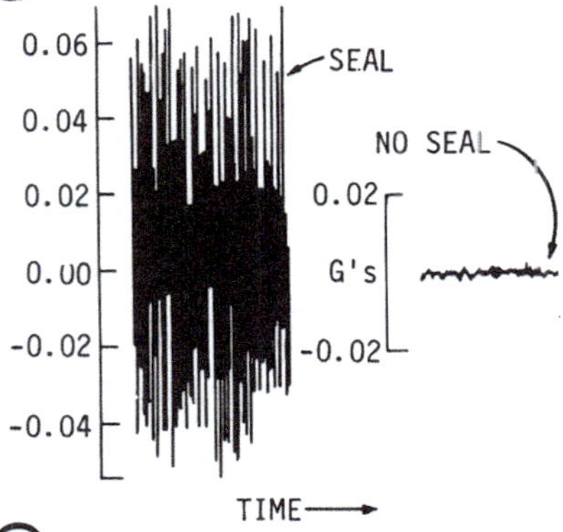

3

Problem: Gate lip submergence by the tail-water combined with shear-layer instability causes the gates to vibrate.

Geometry: Twelve 15.2- high by 16.8-m wide tainter gates control flow through the structure.

Flow conditions: See detail, Sketch 1. Shear layer forms at gate lip. Due to slower inner layer, the layers roll up into vortices. These disturbances are amplified and interact with protruding gate seal, producing a feedback loop with a narrow frequency range.

Operating conditions: Vibrations were detected between gate openings of 0.75 and 2.75 metres. The higher the tailwater submergence the smaller the vibration amplitudes, probably due to back-pressure effect. Vibration of some gate members was observed visually. Water ripples at the skin plate on the upstream side were observed.

Structural dynamics: Calculated natural frequency of gate = 5 Hz. Vortex shedding frequency calculated from Strouhal number = 30 Hz.

Vibration: See Sketch 2. With seal in place, the spectrum of gate-lip pressure (driving force) and gate vibration (response) each showed large spike at about 30 Hz. Without seal the spectra were random and small in amplitude.

Cures: See Sketch 3. Remove gate seal if resulting leakage will permit. Note acceleration amplitude change when seal removed. Proposed alternative: vibration absorber consisting of beam and weight to be attached on gate girder.

Reference: Hart, E. D., and Hite, J. E., Jr., "Gate Vibration Tests, Barkley Dam, Cumberland River, Kentucky," Technical Report HL-79-8, May 1979, U.S. Army Engineer Waterways Experiment Station, CE, Vicksburg, Miss.

C16 HYDRAULICALLY INDUCED VIBRATIONS IN A BEAR-TRAP WEIR

Theodor Merkle

J.M. Voith GmbH, Heidenheim, Federal Republic of Germany

SUMMARY

A brief description of the mode of operation of the bear-trap weir is followed by the chronological sequence from the observation of the first consequences of vibration, the model tests, the design considerations, the failure of the downstream seal during a flood, the rectification of the damage, as well as the conclusions for the planning of such installations.

1. INTRODUCTION

The bear-trap weir is a lowerable weir, essentially comprising two leaves: a curved downstream leaf with rounded crest and an upstream leaf, both leaves being pivoted at their bottom ends. Inserted between their trusses are the upstream side of the skin plates of wooden planks. Arranged at the end of each truss of the upstream leaf is a roller, through which the upstream leaf rests on the curved downstream leaf with rounded weir crest. The bear-trap chamber formed by the leaves is sealed and communicates with the headwater and tailwater through controllable canals. The control is fully hydraulic, use being made of the headwater pressure and, thus, of the varying water level in the bear-trap chamber. For raising and lowering the structure, the bear-trap weir requires no power source other than the static headwater pressure. Even with very small differences between head and tailwater of down to approximately 0.4 m, it is absolutely dependable in operation. On the basis of extensive tests, a bear-trap weir has been developed which sets up forces that tend to close the weir structure and ensures unfailing operation under any flow condition and for any position of the structure, as well as maintenance of the controlled headwater level under any operating condition by an automatically adjusted water level inside the chamber. The upstream leaf of the bear-trap weir in the range of the rotational axis can be sealed relatively easily by a wearing plate pressed on by the water pressure, and the downstream leaf by means of a rubber diaphragmn attached to the weir base and to the leaf.

What was decisive for providing the weir as bear-trap weir with an opening of 28 m clear width and 5.5 m head in the case reported, apart from the lower building costs, was that guarantee could be assumed that in all possible cases, particularly under severe icy conditions, the weir remains operable and that, in the event of sudden thawing setting in after an extended period of freezing weather, a flood can absolutely be prevented from occurring within the town area.

The design flood was established as being 640 m^3/s. The power station was developed up to 20.3 m^3/s at $H = 6.1$ m effective head. The maximum flood level that occurred between 1921 and 1965 was ascertained to be 400 m^3/s. Sketch 1 in the Chart shows a section through the weit plant with the powerhouse.

2. OBSERVATION OF DISTURBANCES AND MEASURES TAKEN TO ELIMINATE THEM

After a small flood with approximately 40 m^3/s, few months after commissioning, leakages were reported in the region of the diaphragm seal of the downstream leaf. The cause of the leakages were loosened screws, by means of which the holding device of the rubber diaphragm seal is attached to the downstream leaf (Chart, Sketch 2). Since bolts were also broken, these were replaced by bolts with higher strength, and the plate onto which the diaphragm seal was clamped was reinforced in the region of the leakages that occurred on the right weir side. After ocurrence of another flood, despite the measures taken, the same damage occurred, although not to the same extent. Therefore, pressure fluctuations and vibrations in the range of the sill seal were suspected as cause.

Preliminary investigations on the weir. After installation of a load cell in the region of the downstream leaf sill at M (Chart, Sketch 2), the measurements conducted confirmed that, in the case of a weir drop of approximately 25% corresponding to approximately 40 m³/s, pressure fluctuations occur due to the turbulence in the stilling basin, the magnitude of which urgently required more precise investigations and design changes in the region of the sill seal.

A comparison of the plants built and erected by Voith showed that at a head H higher than 3 m, either the downstream sill seal in the region of the lowerings of 20 to 30% is still above the tailwater level or that in the case of lower storage levels in the region H = 2 to 3 m the tailwater level is already considerably above the rotational axis of the downstream leaf. However, for the plant under consideration, at the discharge in question, the tailwater level stabilizes again in the region of the sill seal. Moreover, the storage level of 5.5 m provides a considerably higher flow velocity. Furthermore, the larger dimensions of the pivot bearings involve a larger surface of the diaphragm seal so that greater exciting forces act on a larger area.

On the basis of the preliminary investigations, Voith Heidenheim conducted model tests with the aim of changing the configuration in the range of the downstream sill seal and, thus, reducing the pressure fluctuations to an acceptable level. At the same time it was considered to replace the rubber diaphragm seal by a sliding seal.

Hydraulic model tests. For the systematic model tests, 7 different configurations of the range of the downstream sill seal were tested (Chart, Sketch 3):

I The rubber diaphragm seal was retained

- Test series 1 : existing configuration
- Test series 2 : rubber seal withdrawn
- Test series 3,4 : reduction in size of the space between downstream leaf and sill through steel girders
- Test series 5 : shortening of the concrete sill

II The seal was modified and the diaphragm seal replaced by a sliding seal.

- Test series 6 : sliding seal with rounded steel plate at the end of the downstream leaf
- Test series 7 : sliding seal with straight steel plate at the end of the downstream leaf with separation edge

During the model test series, the tailwater level was varied by 45 cm against the calculated value according to the run-off curve, and the pressure fluctuations were measured by means of load cell and recorded. The test results, a sample of which is shown in Sketch 3 of the Chart, were only regarded as measures for comparison, on the basis of which the method of cure was adopted.

It could be observed that the maximum aplitudes of the pressure fluctuations during the test series 1 to 4 occur at excessively high tailwater level when the surface roller in the very flat stilling basin pulsates and reaches up to the sealing section of the downstream leaf. A reduction in size of the space by means of fixtures in the region of the seal yields an improvement only in the event of minor drops of the bear trap. When the weir is lowered further, a suction effect by the water jet leaving the lower edge of the downstream leaf on the rubber diaphragm seal cannot be avoided by fixtures as in series 3 and 4 because of the resulting enlargement of the space between the steel angle and the lower edge of the bottom of the downstream leaf.

The design of a sliding seal was analyzed in test series 6 and 7. Test series 6 with a steel plate adapted to the radius of the sliding seal showed that, in the absence of a clear-cut separation edge, oscillating separations from the downstream leaf with strong pressure fluctuations occur. Using a straight steel plate with a slight upward kin in test series 7 , the discharging jet is better deflected into the stilling basin and the pressure fluctuations are minimized. Therefore, the design shown in Sketch 4 of the Chart was chosen.

For the modification, the fact had to be considered that there is only one weir opening and that the power station should remain in operation if possible. Therefore, the modification to the weir had to be carried out within the shortest possible time and during the low-water period under protection of stoplogs.

While the negotiations with the customer were still going on late in September after a heavy thunderstorm and with the river being flooded, the lower seal failed, the weir dropped and the water that had so far been held back drained off. Fortunately, the flood had not caused any further damage, and now it was necessary to repair the damage as quickly as possible, without waiting for clarification of the legal questions. After flood run-off, the new seal was installed in sections 2 m long in a 5 weeks day-and-night operation.

3. CONCLUSIONS

Early observations of disturbances, precautionary analysis, and tests combined with design considerations made it possible to repair the damages within a short time.

However, as a criticism of the planning it must be stated that a possibility of diverting the river up to a mean flood discharge of 40 to 50 m³/s, which, in the case under consideration, would have been possible during the construction without any appreciably higher expense for structural work, would have considerably reduced the risk taken at the time of repairing the damage.

REFERENCES

G. Wickert and G. Schmausser, "Stahlwasserbau", Springer Verlag Berlin - Heidelberg - New York, 1971.

P. Böss and H.P. Müller: Experimentelle und theoretische Untersuchungen zur Ermittlung einer neuen Dachwehrform, Bauplanung und Bautechnik Bd. 1, No. 2, Aug. 1947.

Druckschrift Voith No. 1047
Dachwehre mit selbsttätiger hydraulischer Steuerung, 1966

Voith Bericht No. 14297, Dachwehr-Modellversuche.

VIBRATIONS IN A BEAR-TRAP WEIR

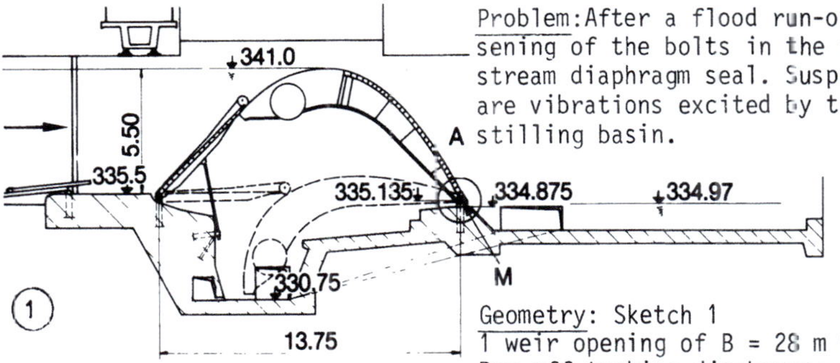

(1)

Problem: After a flood run-off, repeated loosening of the bolts in the range of the downstream diaphragm seal. Suspected as cause are vibrations excited by turbulence in the stilling basin.

Geometry: Sketch 1
1 weir opening of B = 28 m width;
Run-off turbine discharge: 20.3 m³/s;
mean discharge: Q = 20 m³/s;
mean flood: HQ = 400 m³/s;
design flood: HHQ = 640 m³/s.

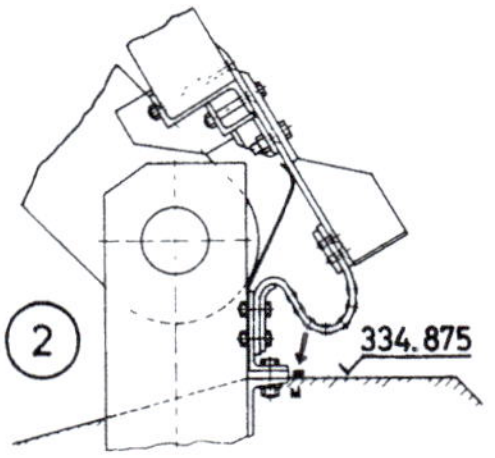

(2)

Determination of the cause for the pressure fluctuations at "M" through systematic model tests with 7 varieties of the design in the range of the downstream seals for discharges of Q = 40, 50, and up to 640 m³/s with 3 different tailwater levels.

Diaphragm seal, design 1: Sketch 2
Sliding seal , design 7: Sketch 5

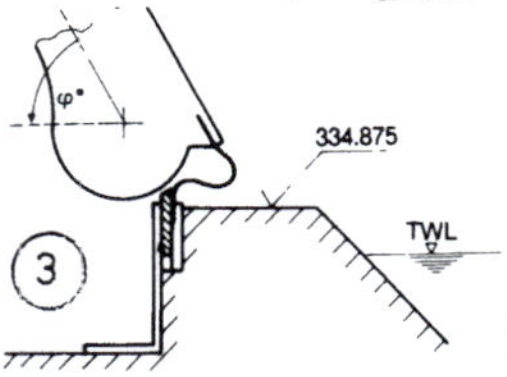

(3)

Weir position (°)	Weir crest (m0.NN)	Discharge (m³/s)	Tailwater level - TWL (m 0. NN)
55	340.58	40	335.31
			336.30
50	339.97	75	335.75
			335.20
			336.75
45	339.45	120	336.20

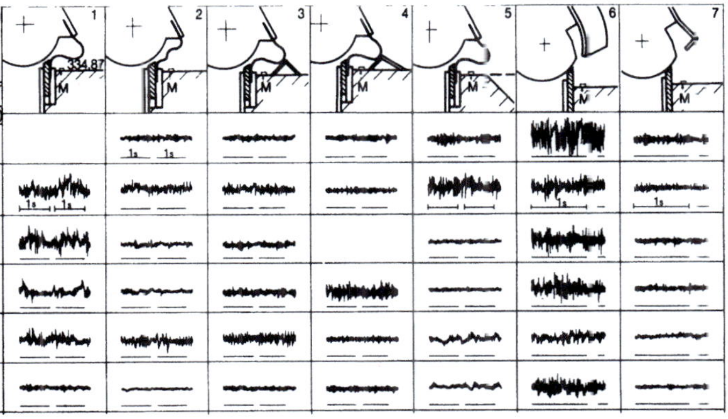

(4) Compilation of test results

Result: At a specific lowering of the weir crest and at a well-defined tailwater level, a maximum of the pressure fluctuations occurs in the range of the sill seal (see Sketch 3). By changing the shape of the flow boundary in the range of the bearing of the downstream flaps, the amplitudes of the pressure fluctuations acting on the seal can be influenced substantially.

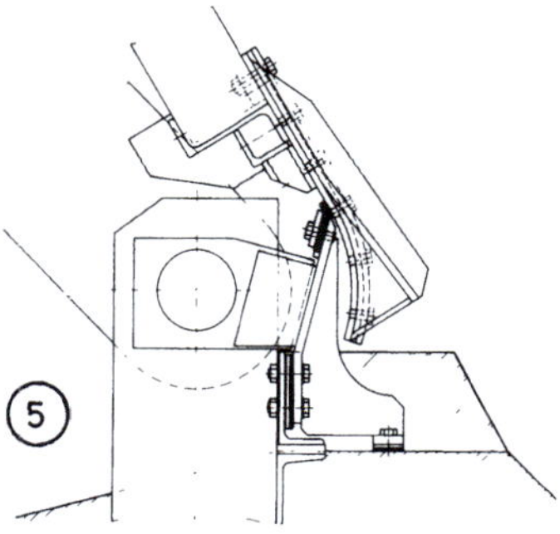

(5)

Cures: On the basis of the hydraulic investigations, besides a change in hydraulic contours, a sliding seal was chosen (Sketch 5) instead of the original diaphragm seal.

C17 VIBRATION DUE TO LEAKAGE THROUGH A REVERSE RADIAL GATE

W.R. Mitchell

Hydro-Electric Commission

of Tasmania

Hobart - Tasmania

SUMMARY

The Gordon Dam, situated in western Tasmania, has no outlet other than through the power station. In order to control the rate of filling, a conduit was constructed through the dam about half way up its height.

The conduit was sealed ba a bulkhead gate at its upstream end, and the rate of flow was controlled by a reverse-radial gate at the downstream end. The radial gate had no seals, so that with the bulkhead gate open and the radial gate closed some leakage flow existed through the small gap between the conduit end and the gate skinplate. It was found in operation that the leakage flow created auto-oscillation of the conduit system which was mainly at the seventh harmonic (171 Hz) of the natural frequency of the conduit system. The resonance would continue for up to about 2 radial degrees of opening of the gate but was suppressed at greater openings.

The auto oscillation was accompanied by a loud 'singing' sound showing an almost sinusoidal pressure trace. At other times a harsh discordant sound was heard which showed up as a rhythmical vibration with overtones of harmonics other than the seventh.

Considerable vibration of the downstream end of the conduit and the radial gate could be seen by observers.

It was considered that the pressure-discharge characteristics were such that the flow increased as the pressure decreased and vice versa, so creating the auto-oscillation condition. The reason would be a flexing of the radial gate skinplate and its supporting arms in the appropriate mode to create this condition.

The chart shows two samples of the pressure fluctuations taken by a transducer inserted in the side of the conduit. The transducer signal was modified and recorded on a cassette tape, because of the difficulty of taking more bulky equipment into the conduit area. The tape signal was then re-recorded by ultraviolet equipment to produce the samples shown. The photograph depicts the outlet at the time of oscillation. The radial gate is flexing, allowing the thickness of the leakage water to vary.

The cure was effected by opening an existing 10 cm drain offtake in the bottom of the conduit, which effectively changed the discharge characteristics of the system so that increased pressure resulted in increased discharge and vice-versa. The dam power station complex is now fully operational and the need for the controlled filling outlet no longer exists. The outlet has been dismantled and the conduit opening in the dam closed off with a concrete plug.

VIBRATION DUE TO LEAKAGE THROUGH A REVERSE RADIAL GATE

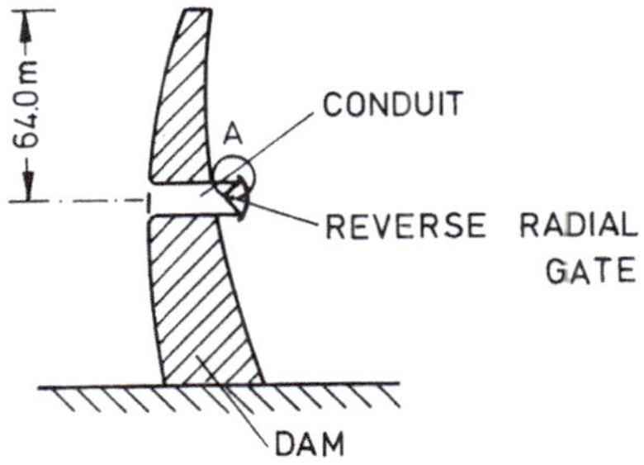

(1) Sketch of Gordon Dam and Conduit

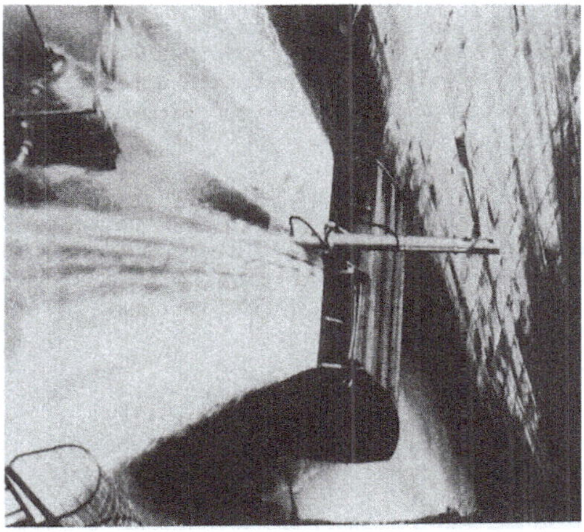

(2) View looking down face of dam. Note light and dark bands in water caused by variation in thickness caused by oscillation.

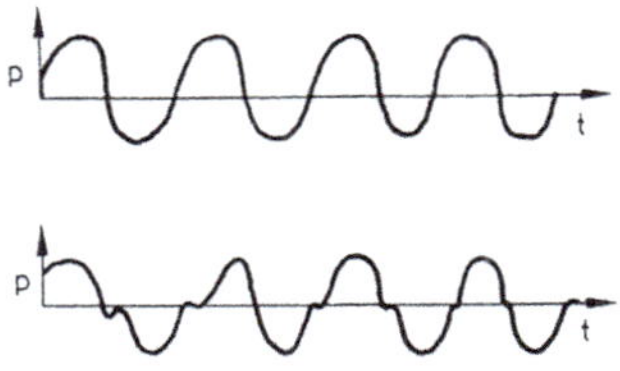

(3) Qualitative sketch of pressure variation in the conduit at two different times.

Problem. Auto-oscillation of a conduit for use as a controlled filling outlet in a concrete arch dam. Pressure fluctuations build-up which were observed as a vibration of the structure and a very loud resonant sound.

Geometry. The conduit is 12.8 m long and 2.6 m in diam. It is sealed by a bulkhead gate at its upstream end, and flow is regulated by a reverse-radial gate at the downstream end (Sketch 1).

Operating conditions. The maximum operating head is 64 m and the estimated maximum outflow is 240 m^3/s. To date the largest discharge through the system has been 70 m^3/s.

Auto-oscillation. Resonance occurred with the bulkhead gate open, the radial control gate closed, and leakage occurring between the end of the conduit and the skin plate of the radial gate (see photo). The radial gate reacted such that an increase in pressure caused a decrease in discharge, because of flexing of the gate.

The pressure fluctuations have been recorded by a pressure transducer inserted in the side of the conduit, and by photographs.

The amplitude of the pressure changes were not recorded. The frequency of the oscillations were measured at 171 Hz, which corresponds to the seventh harmonic of the natural frequency of the conduit system but other harmonics were found at times. A sketch of the recordings is shown.

Cure. A 0.13 m opening was constructed in the bottom of the conduit so that the net outflow from the system increased with increasing pressure.

C18

VIBRATIONS OF THE BOLARQUE DAM

M.J. Kenn
Senior Lecturer in Civil Engineering Hydraulics

A.C. Cassell
Reader in Structures

P. Grootenhuis
Professor of Mechanical Engineering Science

Imperial College of Science and Technology, London, England

SUMMARY:

Severe vibrations have been detected on the 30 m high concrete gravity dam at Bolarque, Spain, and in the 3.5 m diameter steel draft tubes, especially during part-load operation of the turbines and during shut-down. The draft tubes pass through the dam some way below the crest. The turbines can also be used as pumps and form part of a combined hydro-power and pumped-storage scheme. A finite element analysis of the motion of the dam and calculations of the hammer-blow frequency in a draft tube have shown a coincidence with a rocking mode of the dam. A considerable reduction in vibrations has been achieved by injecting a small quantity of air into the water being discharged from each turbine.

General layout: The concrete gravity dam was constructed in 1910 and was raised to its present height of 30 m in 1955. Further modifications have been made since 1971 to use the reservoir behind the dam as the supply for a pumped storage scheme or as the receiver during power generation in peak load periods. A sketch of the power house, a draft tube and the dam, is shown in Sketch 1. The total length of draft tube is about 81 m.

There are four reversible pump/turbine machines each with a maximum rated electrical output of 50 MW. The overall head available for power generation is about 260 m and the rated maximum flow for each turbine is 25 m^3/s. Each turbine has a runner with a diameter of 2.36 m and with 7 blades, an operational speed of 600 r.p.m. and there are 16 guide vanes.

Flow conditions: Under part-load conditions, and during turbine shut-down, considerable cavitation noise emanated from the turbines. Considerable noise and vibration emanated from the exposed draft tubes on the dam face and the sound resembled that of a series of heavy hammer-blows. This noise and vibration was believed to stem from cavitating vortices and associated 'water-hammer' within the draft tubes; phenomena which are not unusual with dual purpose machines, e.g. see references 1 and 2.

The calculated value of approximately 3.9 Hz for the water-hammer frequency of a draft tube was later verified from pressure measurements made by Escher Wyss, an example is given in Figure 1a. Fluctuating water pressures in the draft tube amounted to 6 m at 10 MW, 5 m at 40 MW (Figure 1a) and 15 m during turbine shut-down from 40 MW. Perceptible vibrations were observed on the dam crest.

Structural dynamics: A dynamic analysis of the dam, using a finite element idealization with 29 quadrilateral elements and 78 degrees of freedom, was compared with Fourier analyses of the velocities measured on the dam by Sociedad de Reconocimiento Geofisicos, and the lower frequencies and modes of the dam were studied. These showed (Fig. 2) that most of the vibration occurred in modes which were essentially of rigid-body type, with low stresses. One of the rigid-body rocking modes of the dam had a frequency of 3.91 Hz, which was very similar to the water-hammer frequency of 3.9 Hz calculated for the draft tube.

The maximum calculated direct stress is 0.04 N/mm² when one machine is running at 18 MW, which is the worst case. Vibration measurements of the draft tubes by Sociedad de Reconocimiento Geofisicos, showed unacceptably large amplitudes, with alternating stresses which could be as large as the nominal yield stress of the steel in between some of the stiffening rings.

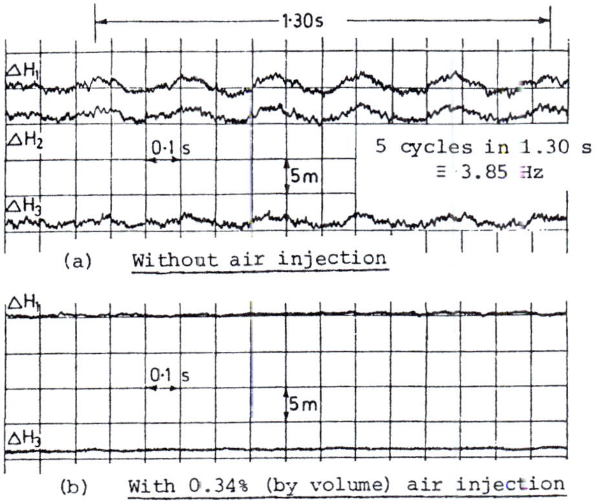

Figure 1 DRAFT TUBE PRESSURE FLUCTUATIONS AT 40 MW, WITHOUT AND WITH AIR INJECTION (PRESSURE TAPPINGS AS SHOWN IN SKETCH 1)

Cure: Attempts were successfully made to reduce the vibrations at their source by the injection of air into the zone of water deceleration below the turbine. The amounts of air injected were small, e.g. 0.3 - 0.4% by volume, measured as a percentage of the rated water flow of 25 m³/s for maximum power generation and measured at the pressure of the water at the injection point. The air was rapidly dispersed into small bubbles (see Plate 1) and hence cushioned the cavitation phenomena; changed the fluid elasticity; changed the natural water-hammer frequency of the draft tube; increased the system damping; suppressed the water-hammer and pressure-fluctuations in the draft tube, e.g. by a factor of 4 at 40 MW (Fig. 1a and b); and most markedly suppressed the amplitudes of vibration of the draft tube and of the dam.

With air injection the deflections of the pipes are at very low frequencies and there are no longer the sharp impacts which occur when no air is injected. It is therefore unlikely that any local deformation by the flexing of the pipe wall in between two stiffening rings will occur. Even allowing for this the maximum combined stress would not exceed the recommended limit of 20% of the yield stress.

Acknowledgement : The authors are greatly indebted to Sr M. Sendin, Inspector General, Union Electrica S.A., Madrid, for his kindness in granting them permission to publish the results of the present investigations.

REFERENCES:

1. Grein, H. and M. Jaquet. Measurements of pressure fluctuations in storage pumps and pump turbines. IAHR Symposium, Rome 1972, Transactions Part I. Reprinted in Escher Wyss News, 1974, No. 1.

2. Meier, W., J. Müller, H. Grein and M. Jaquet. Pump turbines and storage pumps. Escher Wyss News, 1971, No. 2.

Plate 1 AERATED TURBINE DISCHARGE CLEARLY SHOWING THE FINELY DISPERSED AIR BUBBLES (AS PREDICTED)

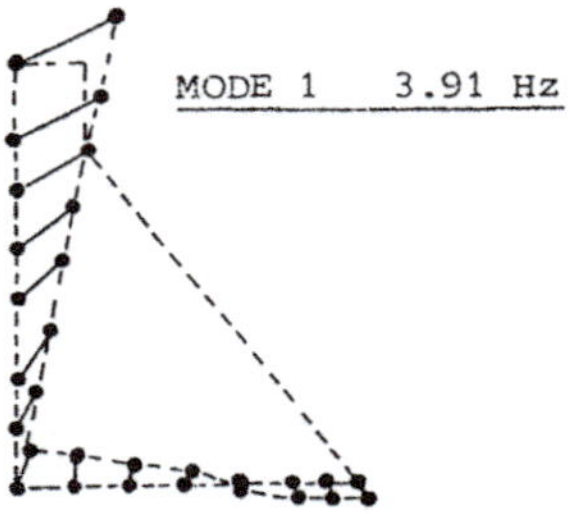

Predominantly a rocking motion; a rigid body motion with low stresses.

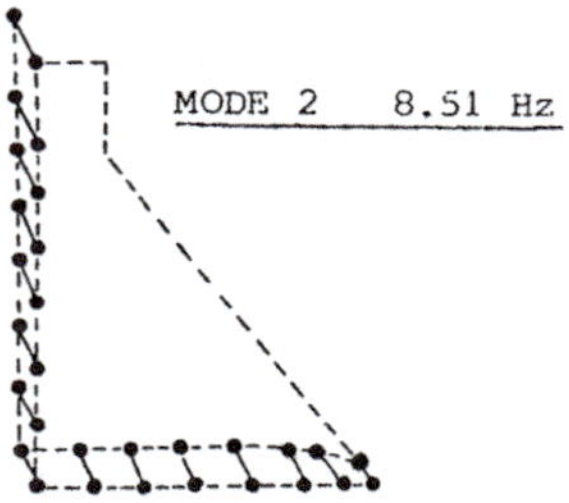

Predominantly a vertical motion; a rigid body motion with low stresses.

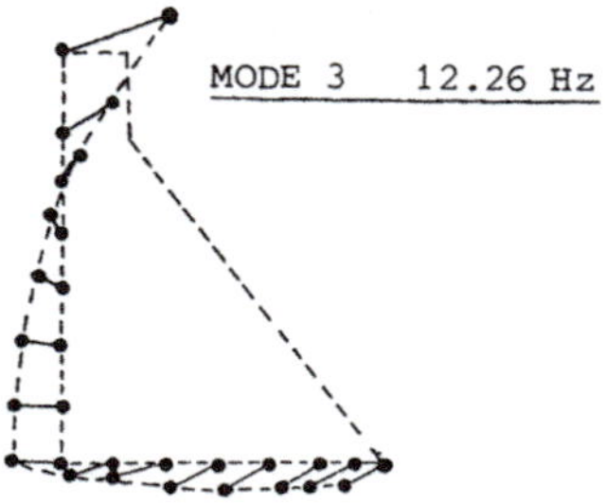

Predominantly a horizontal motion; some flexure is present, but stresses are still low.

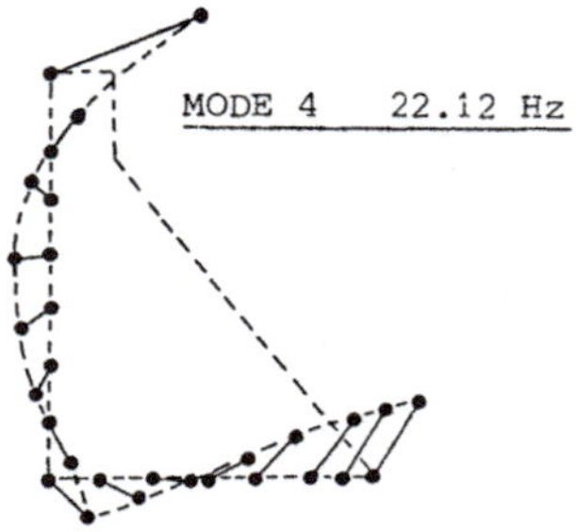

This is the first mode in which flexure predominates; stresses associated with this and subsequent modes are higher.

Figure 2 MODES OF DAM VIBRATION AT THE LOWER FREQUENCIES DETERMINED BY D. HICHINGS' PROGRAM FINEL AT IMPERIAL COLLEGE LONDON

VIBRATIONS OF BOLARQUE DAM

Problem: Severe vibrations (especially at part loads and during turbine shut down) of the 30 m high, concrete, gravity dam and of the four 3.5 m diameter, steel, draft tubes on the exposed dam face. Necessary to assess the likely effects of the vibrations of the dam and of the draft tubes.

Geometry: (see Sketch 1) 50 MW/turbine; 25 m^3/s/turbine; operational head 260 m approx.

Flow conditions: Considerable cavitation noise from turbines. "Hammer-blows" and vibrations from exposed 3.5 m diameter draft tubes. Perceptible vibrations at the dam crest.

Structural dynamics: Comparison of a dynamic, finite element analysis for the dam, with Fourier analyses of the velocities measured on the dam, showed that most of the vibration of the dam occurred in modes which were essentially of rigid-body type, with low stresses. One rigid-body vibration mode for the dam had a frequency of 3.91 Hz which was very similar to the draft-tube water-hammer frequency 3.9 Hz (calculated), or 3.85 Hz (measured). Worst maximum direct vertical stress for the dam appears to be 0.04 N/mm^2 (with one turbine running at the part-load of 18 MW). Vibration measurements of the draft tubes showed unacceptably large amplitudes.

Cure: Attempts were made to reduce the vibrations at source by the injection of air below the turbine. Small quantities of air suppressed the water-hammer and the pressure fluctuations in the draft tube (e.g. by a factor of 4 at 40 MW) and suppressed the amplitudes of vibration of the draft tube and of the dam. With air injection, the maximum combined stress for the draft tubes will not now exceed the recommended limit of 20% of the yield stress.

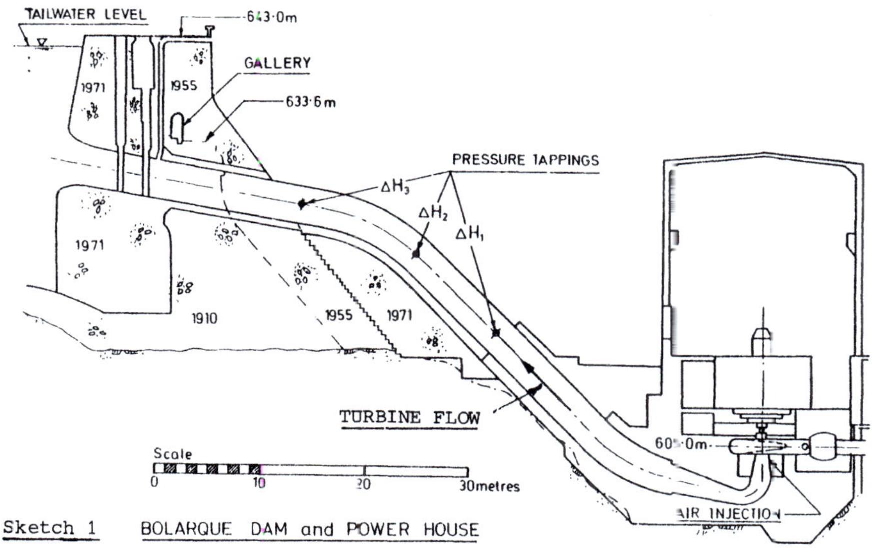

Sketch 1 BOLARQUE DAM and POWER HOUSE

C 19 ELIMINATING PRESSURE FLUCTUATIONS AT THE FLOATING BULKHEADS OF THE GURI DAM PROJECT

Konstantin Zagustin, Ph.D.

Pedro Solana

Instituto de Mecânica de Fluĩdos
Universidad Central de Venezuela
Caracas - Venezuela

SUMMARY

A procedure to eliminate pressure fluctuation occurring during the construction period of the spillway of the Guri dam project is presented in this paper. Model test showed that by placing a flow deflector at the end of the bulkhead, it was possible to avoid the large flow-induced fluctuations in the lateral side of the bulkhead by the strong flow separation. The widening of the pier nose, to coincide with the deflector plate, provided a better seat for the large bulkhead, without reducing the flow discharge through the spillway.

The Guri dam, located in the Caronĩ river in Venezuela, which at present is approximately 100 m heigh and produces 2.065 Megawatts, will be increased in its height up to 152 m bringing the total output to 9.000 Megawatts. The increase in the height of the dam will be accomplished with a simultaneous raising of the spillway which consist of three ski-jump-type channels, each provided with three radial sluices 15 m wide and 20 m high.

In order to keep the spillway in operation, two of the channels will be always in operation while the third one is under construction. The channel which is being worked on will be closed by three floating steel bulkheads, each 20 m wide and 35 m high, which will be sealed against the piers by hydrostatic pressure. Models of 1/50 and 1/30 scale were constructed in the Institute of Fluid Mechanics of the Central University of Venezuela in order to test the performance of the flow when the bulkhead would be in place. The dynamic pressure measurements using pressure transducers showed that due to the flow separation present at the end of the bulkhead (see Sketch 1 on Chart), very strong pressure fluctuations occurred in the neighborhood of the seal between the bulkhead and the pier. The magnitude of the peak-to-peak pressure was in the order of 8.000 kp/m^2 in prototype, which would produce severe vibrations in the bulkhead structure.

In order to eliminate this strong pressure fluctuation, it was decided to add to the side of the bulkhead a flow deflector, whose function was to guide the flow and reduce the magnitude of the pressure fluctuation.

After testing three different kinds of deflectors, it was decided to adopt a wedge-type deflector along with an increase in the width of the front of the pier as shown in the Chart (Sketch 3). This solution totally eliminated the pressure fluctuation in the bulkhead as can be seen from the pressure record in Sketch 2 of the Chart, where the first record corresponds to the case without the deflector while the second record corresponds to the same piezometric hole but with the flow deflector in place.

PRESSURE FLUCTUATIONS AT FLOATING BULKHEADS, GURI DAM

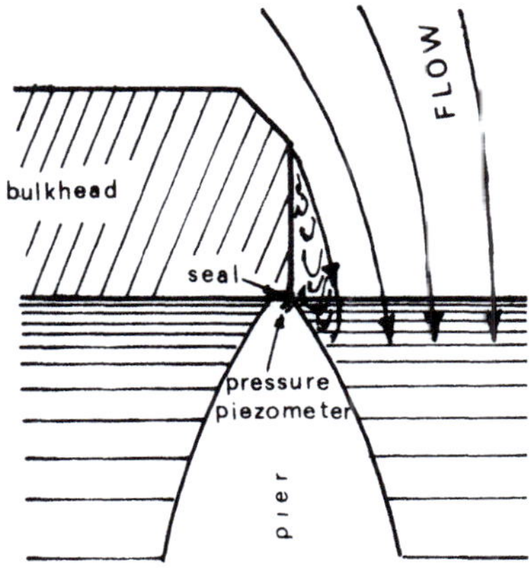

(1) Top view of original design

Problem: Flow separation at the end of a bulkhead placed between the piers of a spillway, produced large flow-induced fluctuation in the region of the seal (see Sketch 1).

Operating conditions: During the construction stage of the Guri dam spillway, one of the shutes has to be closed to the flow by means of a floating bulkhead. The prevailing conditions were tested in hydraulic models of 1/50 and 1/30 scale.

Pressure fluctuations: The magnitude of the peak to peak pressure recorded is in the order of 8000 kp/m² for the prototype when the spillway head is 25 m.

Cure: A supplement in the form of a deflector plate located at the end of the bulkhead, along with a pier nose widening eliminated the pressure fluctuations (see Sketch 3).

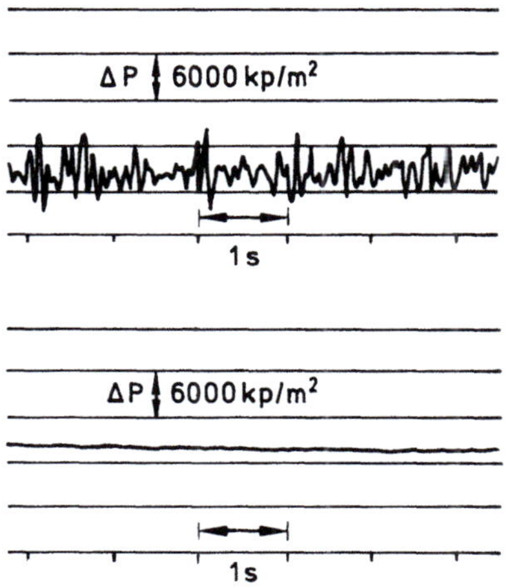

(2) Pressure fluctuations without and with supplement. H = 25 m.

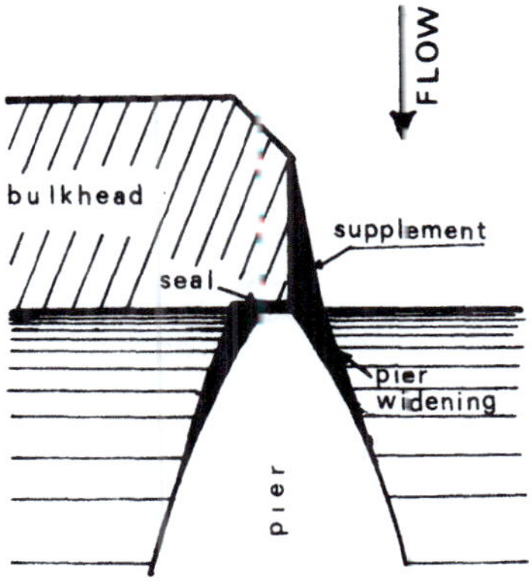

(3) Top view of the final design

C 20 SELF-EXCITED OSCILLATORY SURFACE WAVES AROUND CYLINDERS

F.G.Rohde
Professor, Hydropower Division

G.Rouvé
Professor, Hydraulic Eng. Resources

E.Pasche
Research Assistant

Technische Hochschule Aachen, Federal Republic of Germany

SUMMARY

This paper describes experience with self-excited oscillatory waves which arise around cylindrical obstacles placed centrally in a steady, uniform open-channel flow, provided certain geometric and hydraulic conditions are fulfilled. The excited wave represents a standing wave in the transverse direction and a progressive wave in the longitudinal direction, travelling always upstream. The mechanism of generation is explained by an interaction of vortex shedding and free surface oscillation involving fluid-dynamic and fluid-resonant feedback.

1. INTRODUCTION

Cylindrical obstacles, like bridge piers, placed centrally and with the axis vertical in a steady, uniform, subcritical open-channel flow give rise to violent, oscillating surface waves, provided that certain geometric and hydraulic conditions are fulfilled. The primary flow is steady but the secondary flow which is set up in the vicinity of the pier is unsteady, and provides an interesting example of a self-excited fluid oscillation. The wave height is maximum at the walls of the flume near the obstacles, and it decreases both up- and downstream. The wave length in the x-direction is approximately constant (Chart, Sketch 1).

2. THE MECHANISM OF SELF-EXCITATION

Self-excitation which in the present case leads to the formation of surface waves is caused by two mechanisms: Vortex shedding and fluid-resonant feedback. Sketch 2a in the Chart shows a possible streamline pattern due to vortex shedding. On account of the first vortex, the flow section is contracted. This gives rise to a surface disturbance in the form of a positive surge which will travel upstream and cause the flow in the section immediately upstream of the cylinder to become asymmetric with respect to the centre line such that more liquid will flow through the left sluiceway. In case of resonance between vortex shedding and wave (Sketch 2b), alternating surface waves as observed in the experiments develop in the vicinity of the obstacle.

Sketch 2b in the Chart demonstrates a strong asymmetry of the flow even some distance upstream of the cylinder. Due to this asymmetry, a further fluid-resonant feedback may become effective between any two points located symmetrically on the channel walls. The fact that in the experiment three dimensional waves are observed leads to the suggestion that two-dimensional transverse waves in a running stream are unstable. In order to test this hypothesis experimentally, an attempt was made to artificially excite two-dimensional transverse waves with an oscillating flap downstream of the obstacle. This was possible for standing water. With flow, however, the waves became immediately three-dimensional, which fact supports the above hypothesis.

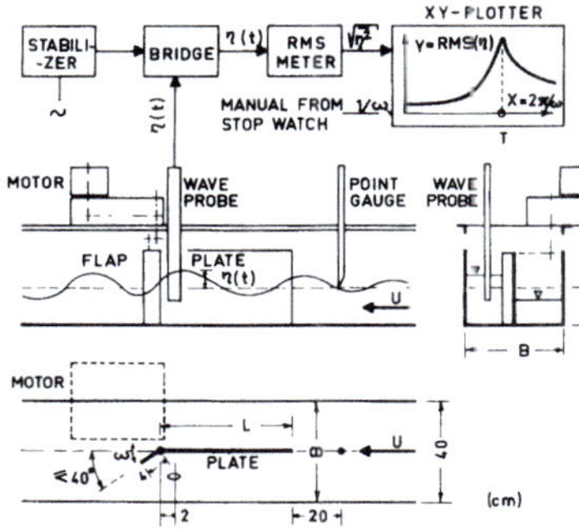

WAVE PROBE
Carbon Electrode Sensitivity
8mA/cm Linearity<3%, 0-10 cps
BRIDGE
25 V, Differential Amplifier
VOLTAGE STABILIZER
Magnetic Type, Accuracy 1%
RMS-METER
0.1-5 Kcs, 0.3 cps Low Frequency
cut-off for time constant 30 sec
output root-mean-square value
XY-PLOTTER
0-1.5 cps, sensitivity 0.2 V/cm
Linearity < ± 0.1%

FIG. 1 EXPERIMENTAL SET-UP 1 AND INSTRUMENTATION

3. EXPERIMENTAL STUDY

Experiments were carried out with a view to verify and eventually correct the expressions for the natural wave frequency f_o obtained theoretically [1] . For this purpose, the cylinder was replaced by a thin plate of length L with a rudder-like flap at the downstream end, Fig. 1. By oscillating the flap with a critical frequency, surface waves are created which are very similar to the self-excited waves observed.

The frequency which coinsides with the maximum wave amplitude - all other parameters kept constant - is taken as the natural wave frequency f_o and is evaluated in terms of the dimensionless natural wave period

$$S = \frac{T \cdot \sqrt{gh}}{2B} \quad , \quad \text{where } T = 2\pi/f_o \quad .$$

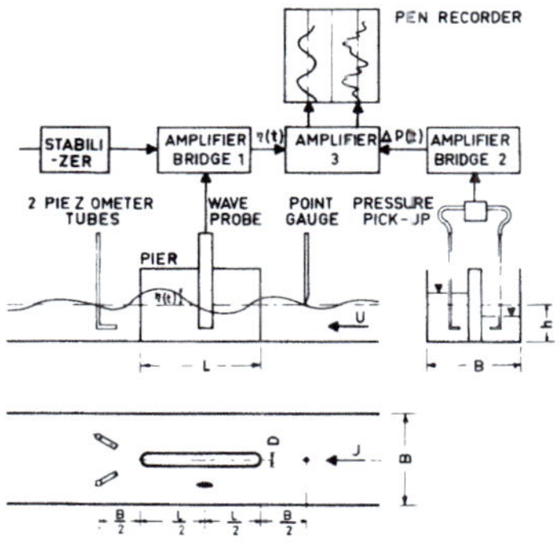

WAVE PROBE
Carbon Electrode Sensitivity
8mA/cm Linearity<3%, 0-10 cps
BRIDGE 1
25 V, Differential Amplifier
VOLTAGE STABILIZER
Magnetic Type, Accuracy 1%
PRESSURE PICK-UP
Inductive Type, 0-0.01 Kp/cm^2 <450cps
Carrier Frequency 5 Kcs,
Linearity < 1%
BRIDGE 2
Carrier Frequency Amplifier
Bridge 5 Kcs 0-1300 cps,
AMPLIFIER 3
Input 0.03-1V, Output 0.5-1.5 V,
0-10 Kcs Differential Amplifier
PEN RECORDER
Thermoscript, 0-50 cps, 6 cm/V, max.
Amplitude 3 cm, Paper Speed 50 mm/sec

FIG. 2 EXPERIMENTAL SET-UP 2 AND INSTRUMENTATION

The experimental set-up used and technical data of the instruments and a flow chart is shown in Fig. 2. For an exact description of the experiments see [1] . The evaluation of the experimental results permits the following conclusions (see Figs. 3a and 3b):

a) The typical spectra exhibit a pronounced peak at a particular frequency, corresponding to the natural wave period T.

b) For constant lengths of plate L, T decreases with h and increases with discharge Q. For larger water depths, i.e., $Fr \rightarrow 0$, T tends to a value which depends only on L.

c) Curves, described by $S = f\{h/B, Q\}$ approximately behave like the curve for $Fr = 0$, if the Froude number is small ($Fr < 0{,}25$).

d) For $L/2B = 0$, the dimensionless natural wave period S behaves like

$\sqrt{\frac{\pi h}{B} \operatorname{ctanh} \frac{\pi h}{B}}$ which corresponds to the period of a standing transverse wave. Based upon this, the experimental points for $Fr = 0$ can be fitted to curves described by

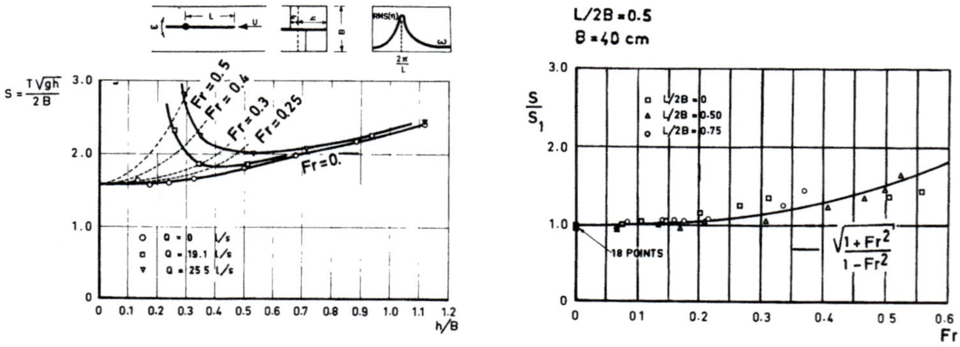

(a) (b)

FIG. 3 NATURAL PERIOD OF SURFACE WAVE

$$S_1 = \sqrt{3.0\left(\frac{L}{2B}\right)^3} + \sqrt{\frac{\pi h}{B} \operatorname{ctanh} \frac{\pi h}{B}} \tag{1}$$

with a deviation smaller than 10 %.

In order to obtain a correction of ep.(1) for $Fr \neq 0$, the ratio S/S_1 was plotted versus Fr in Fig. 3b. The points can be fitted to the curve

$$\frac{S}{S_1} = \frac{1 + Fr^2}{1 - Fr^2} \tag{2}$$

with a deviation of less than 10 %. From (1) and (2) a semi-empirical relation for the dimensionless natural period of waves S is obtained as follows:

$$S = \frac{\sqrt{1 + Fr^2}}{1 - Fr^2} \; 2 \left[\sqrt{3\left(\frac{L}{2B}\right)^3} + \sqrt{\frac{\pi h}{B} \operatorname{ctanh} \frac{\pi h}{B}} \right] . \tag{3}$$

Frequency measurements were conducted in two flumes of rectangular section in the wake of two cylinders of non-circular section and three cylinders of circular section (Fig. 2). Regular pressure fluctuations in the wake are obtained only within the range $0.02 \leq Fr \leq 0.20$ with optimal regularity in the presence of self-excited surface waves. For larger Froude numbers, the fluctuations become irregular; for smaller Froude numbers they become too weak to be detected by the measuring device.

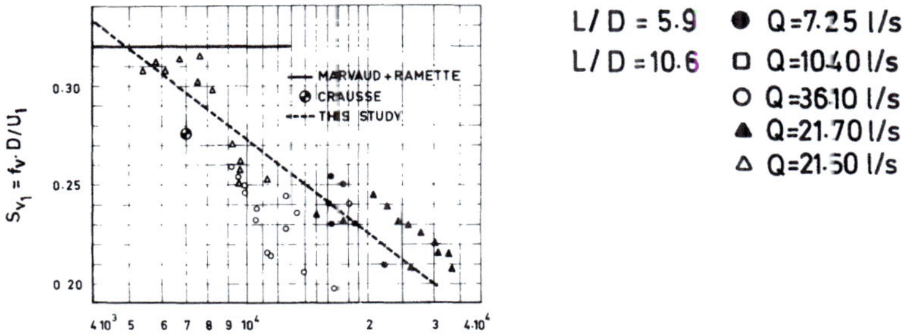

FIG. 4 NATURAL STROUHAL NUMBER S_{v1} FOR NON-CIRCULAR CYLINDERS

The natural frequency of vortex shedding determined in this way is plotted in Fig. 4 in terms of a Strouhal number S_{v1} based on the velocity $U_1 = Q/B(h-D)$ in the constricted section. Since the present results, obtained for different discharges and hence different Froude numbers, agree so well with the results from wind-tunnel measurements, it is concluded that the influence of Froude number on the frequency of shedding is small. In the plot S_{v1} versus Re_1, the Strouhal number shows a linear trend with $S_{v1} = 0.32$ at $Re_1 = 5.10^4$ and $S_{v1} = 0.20$ at $Re_1 = 2.10^5$. The experimental result of $S_v = 0.275$ at $Re = 7.10^4$ reported by CRAUSSE [2] for cylinders of oblong section falls well into the above range. Strouhal numbers as high as 0.32, as reported by MARVAUD and RAMETTE [3] , represent the upper limit in the present results.

Concerning the self-excited motion, the following results were obtained (Fig. 5):

a) For a given geometric configuration of channel and cylinder (L/2B, L/D) and a constant discharge (Q), waves are observed only within a narrow range of water depths. In this range the wave amplitude shows a maximum at a particular depth $h = h_{crit}$ and the frequency of the wave remains approximately constant.

b) The plot Q, T, a versus h_{crit} shows that the relation $Q(h_{crit})$ is nearly linear [1] . This indicates that self-excitation occurs around a critical velocity which is almost constant for a given geometry of cylinder and channel. The maximum wave amplitude $a = \eta_{max}$ varies with h_{crit} and has a maximum at a certain depth. The range of depth around h_{crit} where waves occur, increases with h_{crit} and the wave period T decreases with increasing depth.

The self-excited waves occur due to resonance between vortex shedding and free-surface oscillation. As observed in the experiments, the waves occur with a frequency which differs slightly from the natural shedding frequency, $f_v = \kappa \cdot f_o$, where κ was found to be approximately constant and equal to 1.09.

With $\kappa = 1.09$, $f_v = U \cdot S_v/D$ and $f_o = \sqrt{gh/2BS}$, where $S_v = 0.25$ and S is given by eq.(3), the resonance condition takes the form

$$\frac{D}{2B} \cdot \frac{1 - Fr^2}{Fr\sqrt{1 + Fr^2}\left[\sqrt{3\left(\frac{L}{2B}\right)^3} + \sqrt{\frac{\pi h}{B}\,\mathrm{ctanh}\,\frac{\pi h}{B}}\right]} = 0.23 \qquad (4)$$

with the introduction of a 'Strouhal number of the cylinder' [4], $S_o = f_o \cdot D/U$, eq.(3) is equivalent to $S_o = 0.23$.

This condition for self-excitation has been verified experimentally. With the critical values of B,D,L,h,Fr,Re measured at a state of resonance, S_o is plotted in Fig. 6. Based upon these results, a sufficient condition for non-occurence of self-excited surface waves is given in the form

$$S_o \neq 0.20 \text{ to } 0.25 \qquad (5)$$

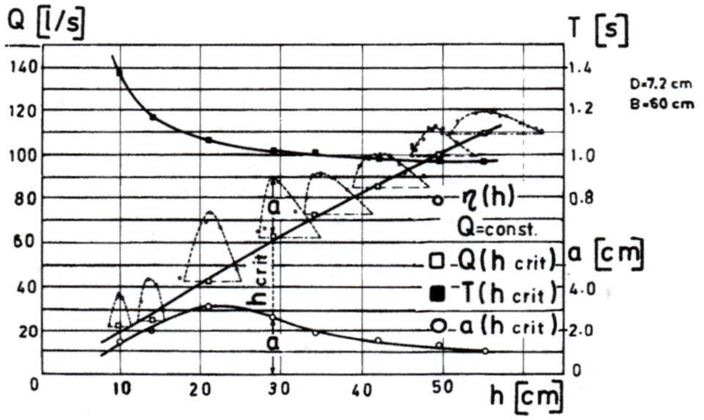

FIG. 5 RESONANCE STATE

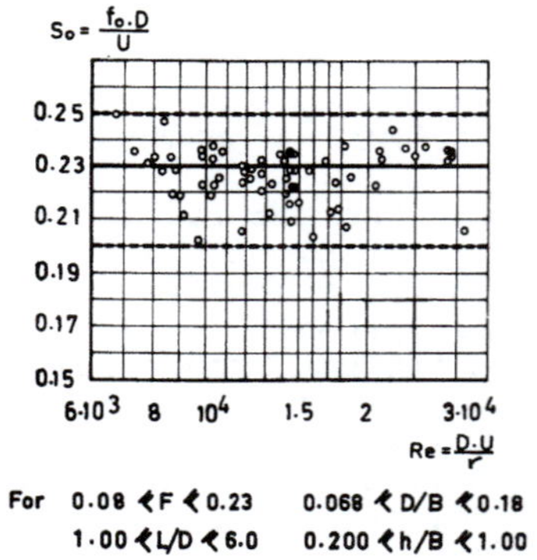

FIG. 6 RESONANCE CONDITION

As a further measure of preventing or mitigating the formation of self-excited surface waves, one may consider modification of the upstream and/or downstream end of the cylindrical obstacle. When using eq.(5) it is important to keep in mind that resonance is possible also with higher modes of wave oscillations, which were not investigated in the present study.

4. REFERENCES

1. Rohde, F.G.: Self-Excited Surface Waves Around Cylinders, Indian Institute of Technology, Madras 1970

2. Crausse, E.: Comptes-rendus de l'Academie de Sciences, Paris, Tome 209, No.4, 1939

3. Marvaud and Ramette: Comptes-rendus 8ièmes, Journées de l'hydraulique, Tome I, Paris, 1964

4. Tison, G.: Journal of the Hydraulic Division, Proceeding of ASCE, Vol.93, No. HY6, 1967

SELF-EXCITED OSCILLATORY SURFACE WAVES AROUND CYLINDERS

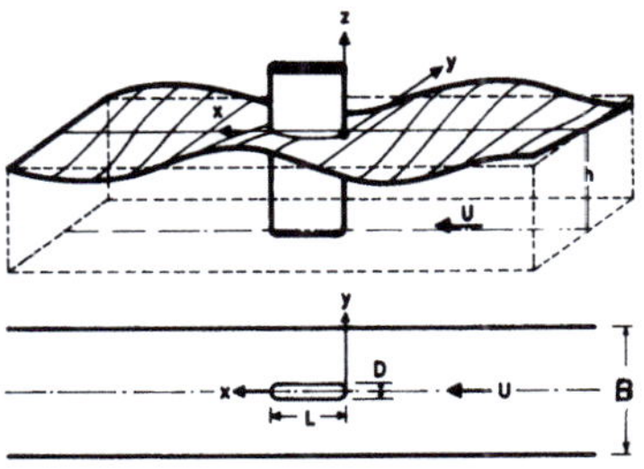

① SELF-INDUCED SURFACE WAVE

Problem: Cylindrical obstacles, like bridge piers, placed centrally and with the axis vertical in a steady uniform, subcritical open-channel flow give rise to violent oscillating surface waves. Damage because of fatigues failures is possible.

Geometry: See Sketch 1). B = 40/60 cm L/D = 1.0/5.9/8.1/10.6

Boundary conditions: a) The perturbations vanish far up- and downstream. b) The pressure on the free surface is the same as the pressure at infinity. c) The free surface is an important flow boundary.

Operating condition: Waves were observed for $0.08 \leq Fr \leq 0.23$; $0.068 \leq D/B \leq 0.18$; $1.00 \leq L/D \leq 6.00$; $0.200 \leq h/B \leq 1.00$

Structural dynamics: Not investigated.

Vibrations: A 'Strouhal number of the cylinder' was definded as

$$S_o = \frac{f_o \cdot D}{U}$$

where $f_o = f(B,D,L.h,Fr,Re)$ represents the frequency of the waves. Self-excited surface waves do not occur outside the range $0.20 < f_o < 0.25$. Maximum amplitudes of 15 % of the water depths were observed. For Froude number of $0.02 < Fr < 0.20$, periodic pressure fluctuation could be measured. The resulting regular vortex shedding interacts with the free surface oscillation. Two control-mechanisms are involved: a fluid-resonant and a fluid-dynamic feedback.

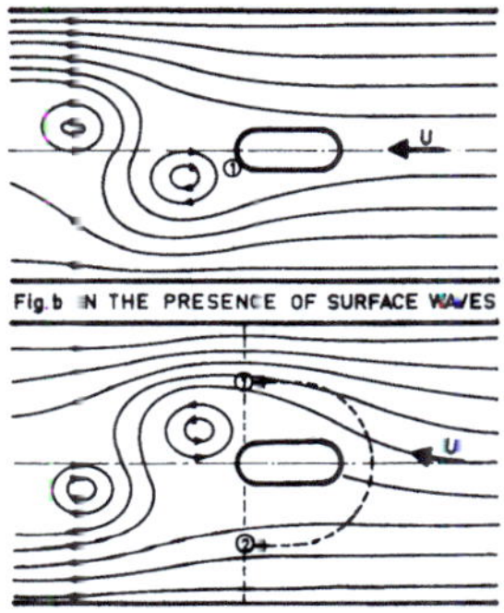

② STREAMLINE PATTERN

Cures: a) Modification of the parameters L,D,B,h,Fr such that the condition for fluid-resonant feedback and, thus, self-excited wave oscillations is avoided.

b) Form change of the upstream and/or downstream ends of the obstacle mitigates the formation of a vortex street.

References:

1. "Self-Excited Surface Waves Around Cylinders" by F.G.Rohde, Indian Institute of Technology, Madras 1970

C 21 ON IDENTIFICATION AND PRELIMINARY ASSESSMENT OF SOURCES OF FLOW-INDUCED VIBRATIONS

Eduard Naudascher

Institute of Hydromechanics
Karlsruhe University
Karlsruhe, Germany

After three days of deliberation on the subject of flow-induced vibrations, we have learned a great deal on a great variety of practical problems. And yet a practicing engineer might still be rather confused as to how to apply all this knowledge to his design task. What makes his task so difficult is the fact that even well understood vibration problems often come in disguise. Hardly any two gates, valves, or seals are alike in geometry and dynamic behaviour, so that application of practical experience becomes difficult; when different structures are involved, such application seems to be outright impossible even in cases in which the vibrations are induced by the same mechanism. Clearly the designer needs at least a scheme for identifying the common excitation mechanisms and vibration-inducing flow configurations in order to be in a position to diagnose a system.

To give you a glimpse of the classification scheme that Dr. Rockwell and I wish to suggest, and to provide us with an opportunity to obtain your reaction and criticism let me present briefly the main features of this scheme even though it still requires substantial improvement.

As a first step in the process of identifying sources of flow-induced structural vibrations, we wish to suggest the distinction between three basic types of excitation mechanism (Fig. 1). Extraneously induced excitation (EIE) is caused by a pulsation in flow or pressure which is not an intrinsic part of the vibrating system. Examples are the cylindrical structure and the sluice gate in Fig. 1 excited by the turbulence of the oncoming flow and the submerged surface roller, respectively. Instability-induced excitation (IIE) is brought about by an instability in the flow. In most cases, this instability is inseparably connected with the same structure which is endangered by vibration. Examples for this situation are the alternating vortex shedding from a cylindrical structure and the alternating shear-layer reattachment underneath a gate depicted in Fig. 1. Whereas the flow oscillation in these cases can be regarded as "self-excited", the structural vibration it induces is forced for the fluid-dynamic variety and resonator-controlled for the fluid-resonant and body-resonant variety of IIE. Movement-induced excitation (MIE), is due to exciting forces which arise from the movements of the vibrating structure. When the square cylinder and the rubber seal depicted in Fig. 1, e.g., are given a transverse movement, the flow will induce a force which tends to enhance this movement. Consequently, the structural vibration is clearly self-excited in this case.

Any one of these three basic excitation mechanisms can be further subdivided as indicated for the IIE in Fig. 1. In the fluid-dynamic cases, the exciting force is a function of the flow conditions alone; in the fluid-resonant and body-resonant (or fluid-elastic) cases, they depend on the dynamics of the flow as well as on the dynamics of a resonator within the system: a fluid resonator in the former and a body resonator in the latter case. In the examples of Fig. 1, a standing gravity wave in an open channel and a water column with an enclosed air pocket on top were chosen as illustrations of the fluid-resonator. Flow-induced vibrations can also be excited, finally, by any combination of the excitation categories described in Fig. 1. Identifying these possibilities is one of the tasks when assessing the danger for vibrations in a given system.

		EXTRANEOUSLY INDUCED EXCITATION (EIE)	INSTABILITY INDUCED EXCITATION (IIE): Excitation unaffected by oscillators FLUID-DYNAMIC	INSTABILITY INDUCED EXCITATION (IIE): Excitation affected by fluid oscillator FLUID-RESONANT	INSTABILITY INDUCED EXCITATION (IIE): Excitation affected by body oscillator BODY-RESONANT	MOVEMENT INDUCED EXCITATION (MIE)
FEATURES	Excitation due to	Flow pulsations not intrinsic to system	Flow instability	Flow instability controlled by fluid-resonator	Flow instability controlled by body-resonator	Self-excited movements of structure
	Excitation function of	Flow conditions alone (not affected by structural dynamics)		Flow conditions plus dynamics of fluid oscillator	Flow conditions plus dynamics of body oscillator	Flow cond. plus struct. dynamics
EXAMPLES	Cylindrical Structure					
	Conduit Gate					

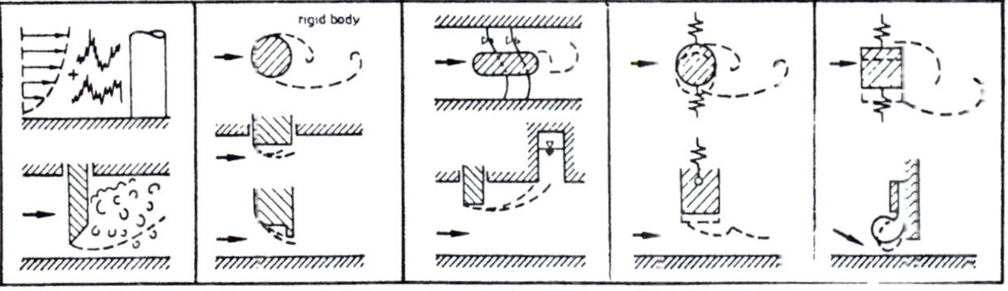

FIG. 1 BASIC EXCITATION MECHANISMS

A second task regarding such assessment is the search for (a) all possible sources of extraneously induced excitation (EIE) and control and (b) all flow configurations (or flow modules) in the given system which may lead to instability- or movement-induced excitation (IIE or MIE). And the final task is to analyse the excitations and structural vibrations to which all fluid and body resonators (or oscillators) of the system may give rise in conjunction with the identified sources for excitation (a) to (b). In addition to the scheme of Fig. 1, therefore, it is essential to provide a scheme for the sources of EIE, IIE, and MIE.

One possible form for such a scheme is shown in Fig. 2 for the category of instability-induced excitations (IIE). Rather than showing the basic flow configurations which act as the sources of excitation here in an abstract or generalized fashion, they are depicted as they have been found in hydraulic-engineering applications of IIE. To be sure, similar matrix representations can also be deduced for mechanical- and wind-engineering applications - a fact which offers the additional advantage that knowledge on methods of prediction or on practical solutions and cures can be transferred from one field to another with the aid of these charts. Perhaps more important for the designer, however: this matrix representation will guide him to center his attention on the basic sources of flow-induced vibration and help him not to overlook an element of the structural or fluid-flow system which may - either isolated or in combination with other elements - lead to a dangerous excitation under certain operating conditions. In addition, of course, this representation offers both a logical and convenient framework for listing the problems of flow-induced vibrations which have arisen in a certain engineering field in the past, as well as referring to the measures which have been found to be successful in curing these problems. With this checklist for identification and diagnosis, we can train ourselves to look out for the basic mechanisms of excitation, on the one hand, and the basic excitation-producing flow modules, on the other hand. And, as indicated already, we obtain a system herewith for the transfer of knowledge and experience from one case to another as well as from one field of application to another.

I am the first to admit, of course, that much can and must be improved in this classification scheme. Regarding the fluid-resonant excitations, moreover, it would undoubtedly be helpful to derive a similar "checklist" displaying all types of resonators which may play a role in amplifying flow-induced vibrations.

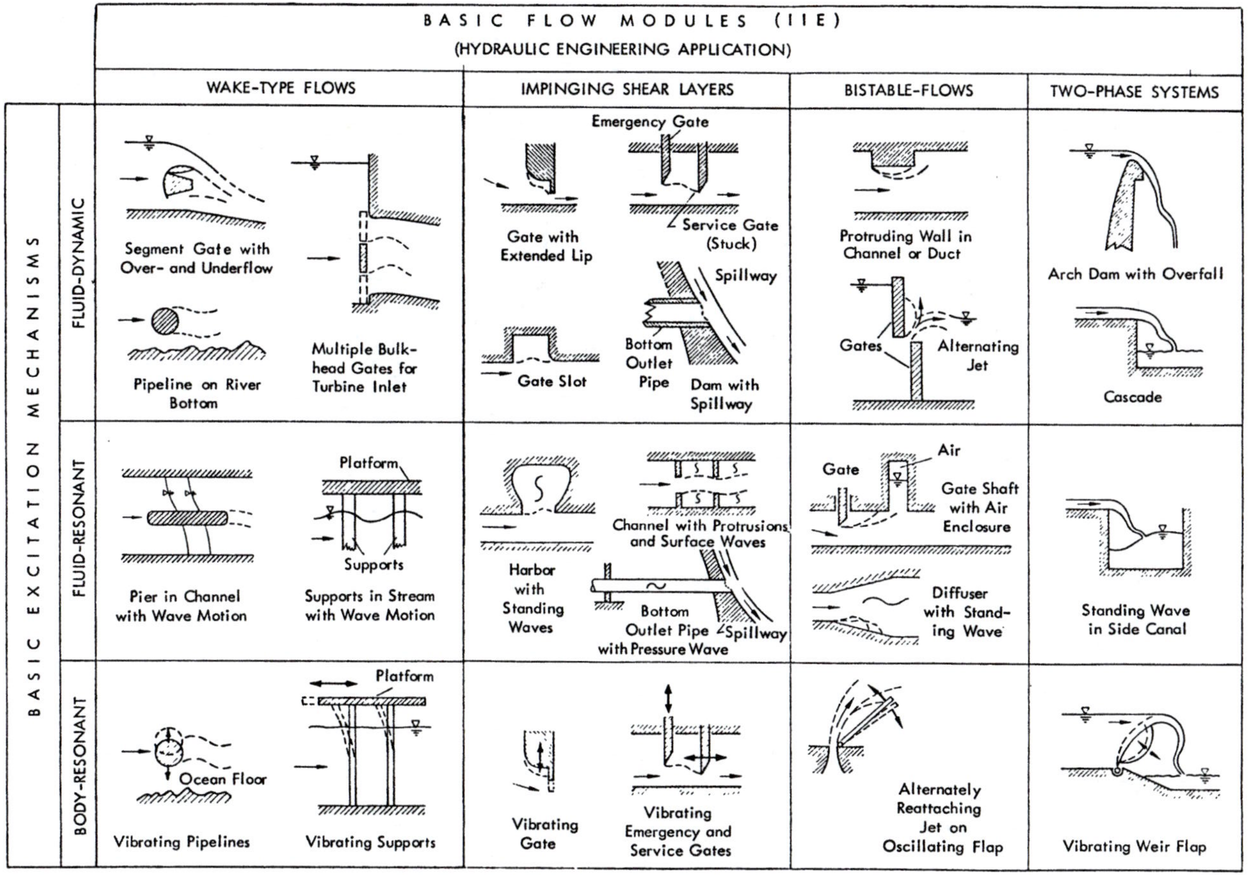

FIG. 2 "CHECKLIST" FOR BASIC EXCITATION MECHANISMS AND BASIC FLOW MODULES LEADING TO INSTABILITY-INDUCED EXCITATION (IIE) IN THE FIELD OF HYDRAULIC ENGINEERING

D1 REVIEW OF THEORETICAL AND EXPERIMENTAL PROCEDURES FOR PREDICTION OF SHIP-PROPELLER INDUCED EXCITATIONS

J. P. Breslin

Davidson Laboratory, Stevens Institute of Technology
Hoboken, New Jersey, USA

1. HISTORICAL SURVEY

An informative and highly readable account of the history of ship vibration from 1884 to 1961 is to be found in Dr. F. H. Todd's "Ship Hull Vibration" [1]. There we learn that early interest in ship vibration which dates back to the latter part of the last century was centered on the local and global response of the ship structure to excitations produced by engines and out-of-balance propellers and shafting. Studies were focused on the prediction of the natural frequencies of the lower hull modes. Propeller-induced excitation arising from the changing flow relative to the blades was not considered until after 1931 when the Society of Naval Architects and Marine Engineers (SNAME) set up a three(!) member committee to study propeller-generated vibration and engaged F. M. Lewis, Professor of Engineering at Webb Institute of Naval Architecture to develop experimental procedures. In the first of his landmark papers in 1935 [2], Lewis classified excitations by frequency. The first class are those at shaft frequency arising from out-of-balance of the masses of the propeller, shafting, shafting alignment or to errors in manufacture of the propeller or damage to one of the blades. All of these effects can be eliminated by sufficient care in construction and maintenance. The second class is characterized by frequencies given by the product of the number of blades and the shaft revolutions per unit time, referred to as blade frequency (bf) and integer multiples thereof, referred to as bf harmonics. Lewis further identified the bf excitations as those arising from the bearing reactions to the propeller forces and couples, called bearing forces and those induced on the hull and appendage surfaces by the near-field "fanning action" of the blades. These last were referred to as surface forces.

Lewis recognized that three approaches to the mitigation of ship vibration may be pursued: a) study of the hydrodynamic conditions at the propeller responsible for induction of bf forces with a view to design changes of the propeller and the hull configuration so as to minimize these forces; b) study of the hull elastic structure characteristics to avoid resonances; c) application of special devices such as anti-synchronization, dynamic vibration absorbers, dampers and rotating weights. He considered the accurate prediction of hull resonance in the blade rate range to be well-nigh impossible and argued that reduction of the excitation was a much more promising approach to the minimization of hull vibration. Consequently, he primarily chose to pursue experimental studies to define the significance of various propeller-hull parameters such as axial clearances from hull, bossings, rudders and blade-tip-to-hull clearances, bossings versus struts, blade number, pitch, diameter and blade skew.

Lewis devised an experimental procedure for determining what he called the total horizontal and vertical vibratory forces and applied that to a 6.1 m model of the President Hoover, a twin-screw ship (192.7 m) of 29,000 tons, reporting results in SNAME papers in 1935 and 1936 [3]. He used a tuned, resonating pickup to measure the model vibratory response with a model self-propelled and, with propellers removed, drove a set of unbalanced weights adjusted to give the same response at the pickup location as produced by the propellers. Since the rotating weight, its rotation arm and speed were known, the calibration-excitation force is readily calculated. (It should be pointed out that, in these and subsequent measurements made by Lewis, he did not separate the excitation of the propeller into a force and a couple, but rather determined "effective" forces which, when applied to a single fixed point, produced the same model response as the actual forces and couples produced at the single pickup point.) He found by this definition a ship-scale, single amplitude, vertical-exciting force of 24,000 lbs. from the synchronized twin screws which was 12.5 percent of the mean thrust and concluded that this was principally due to the surface forces. His further work on this model showed that the excitation could be materially reduced by exchanging the bossings for struts and that the forces fall off with axial distance or clearance from the bossings.

No further research was conducted until the early 1950's and, in 1954, Lewis and Tachmindji [4] gave results of similar measurements on a model of the Mariner class of cargo ships. Professor Lewis continued his work with rigid models of 3.06 m of a destroyer escort and a variety of propeller-Series 60 hull configurations, giving a summary of these measurements in a SNAME paper in 1969 [5]. More basic measurements of the pressure field around propeller models in a water tunnel have been reported by Tachmindji [6] and by Denny [7]. Theoretical field pressure theory using lifting line representations were initiated by Ramsay [8] in England, Pohl [9] and Breslin [10], who first noticed that the blade thickness distribution contributed significantly to the near-field pressures for propellers in uniform flow, and in 1961, observed the highly important contributions to the field pressure emanating from several of the blade-loading orders corresponding to the spatial harmonics of the hull wake.

Since 1958, methods for claculating unsteady propeller forces have proceeded in the U.S. by Tsakonas et al. [11], by Van Gent [12] and others in The Netherlands, and by Hanaoka [13] and others in Japan. Very recently, Kerwin has developed a time-stepping procedure [14]. Experimental methods for measuring these forces have also been advanced as given in some detail later. During the last decade, attention has been focused on intermittent blade cavitation (in the 12 o'clock, high wake region) which became a commonly occurring phenomena with the large rise in blade loading attending the rapid rise in ship size. The impact of intermittent blade cavitation on ship vibration is very great, giving rise as it does to large hull surface forces which are approximately proportional to the acceleration of the cavity volume. Approximate methods have been advanced [15, 16] for prediction of these radiating pressures and, currently, a three-dimensional computer method is being completed by Profs. Leehey and Kerwin at MIT. Methods for computing hull surface forces have been developed by Vorus [17,18] and by Breslin et al. [19]. We may now turn to a discussion of physical mechanisms, experimental results and some correlation with theory and assessment of future requirements.

2. FLUID MECHANICAL ASPECTS

2.1 Blade and Shaft Forces

Ship propellers are invariably located (for reasons of propulsive efficiency) at the stern where the flow pattern is temporally variable and spatially distributed. The temporal variations are due to turbulence and boundary layer intermittency of as yet undefined correlation scales. Measurements of shaft forces on models and ships show modulations of the amplitudes of the bf forces due presumably to temporal inconstancy of the flow. Thus far, no mathematical modeling has embraced this aspect. We may visualize any section of a propeller blade as a hydrofoil orbiting through a gust pattern which is fixed relative to the ship and idealized as independent of time and is, hence, spatially cyclic in blade position angle. The spatial nonuniformity of the hull wake flow in the plane of the propeller is defined by measurement of the longitudinal $u(r,\theta)$, tangential $v(r,\theta)$ and radial $w(r,\theta)$ velocity components by use of a multi-hole spherical pitot tube which measures the temporal mean of the flow at many points, generally around four concentric circles in the plane of the propeller. Typical curves of the longitudinal and tangential wake components for a single-screw ship at two radii are shown in Figure 1 where we can see that the total longitudinal components in fraction of the ship speed varies sharply from a maximum value of 0.92 to

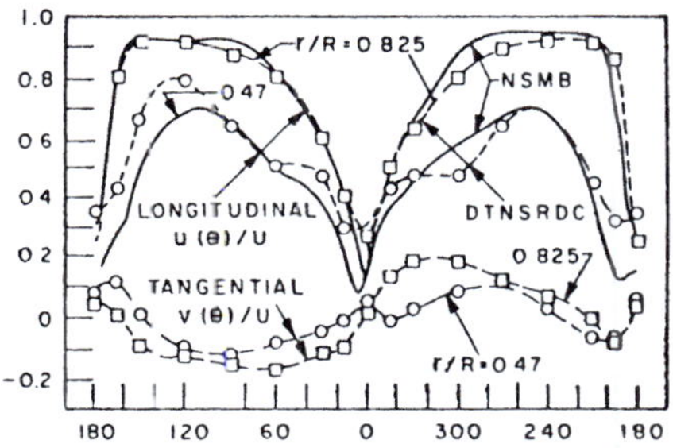

BLADE POSITION ANGLE, θ
TYPICAL MODEL WAKE VELOCITY VARIATIONS
ALONG TWO CIRCLES IN PROPELLER PLANE

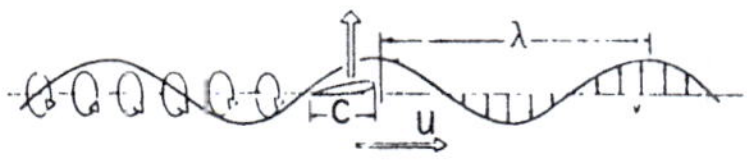

DOWNWASH DUE TO VORTEX SHED τ SECONDS EARLIER

FIG. 1 CIRCUMFERENTIAL VARIATIONS OF LONGITUDINAL AND TANGENTIAL COMPONENTS OF WAKE VELOCITY AND SCHEMATIC OF A TWO-DIMENSIONAL SECTION TRAVERSING A FIXED GUST PATTERN

a minimum value of ~ 0.15 at 12 o'clock. As with all cyclic functions, one can decompose $u(\theta)$ and $v(\theta)$ into harmonics varying as $\cos\nu\theta$ and $\sin\nu\theta$ as displayed in Figure 2. The advantage of this is that the thrust and torque transmitted to the shaft arise only from the angle of attack variations produced by the wake harmonic orders $\nu = qN$ and the horizontal and vertical forces, as well as the moments about the horizontal and vertical axes are produced by the angle of attack variations

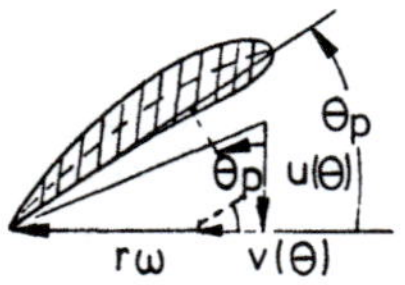

$$v_{normal} = v_n = u(r,\theta)\cos\theta_p - v(r,\theta)\sin\theta_p$$

$$v_n = \sum_{\nu=o}^{N+1} (u_\nu(r)\cos\nu\theta\ \cos\theta_p - u_\nu(r)\sin\nu\theta\ \sin\theta_p)$$

where

$$u_\nu(r) = \frac{\varepsilon_\nu}{\pi} \int_o^{2\pi} u(r,\theta')\cos\nu\theta' d\theta' ; \quad v_\nu(r) = \frac{1}{\pi} \int_o^{2\pi} v(r,\theta')\sin\nu\theta' d\theta'$$

where

$$\varepsilon_\nu = \frac{1}{2}, \quad \nu = 0$$
$$= 1, \quad \nu > 0$$

$u(r,\theta)$, $v(r,\theta)$ are obtained from measured nominal model wake in the plane of the propeller. (Propeller absent.)

Above formulation is restricted to single screw hulls.

FIG. 2 RESOLUTION OF RELATIVE VELOCITIES NORMAL TO A BLADE SECTION AND DECOMPOSITION INTO HARMONIC AMPLITUDES FOR EACH WAKE ORDER, ν

produced by the wake harmonic orders $\nu = qN\pm1$ where $q = 0,1,2...$ the harmonic orders of blade frequency. Figure 3 displays the harmonic components of the wake of a single screw ship around a circle of 0.7, the blade radius. Here we can observe that the flow seen by each of four blades is the same at any instant in the 4th harmonic of the wake.

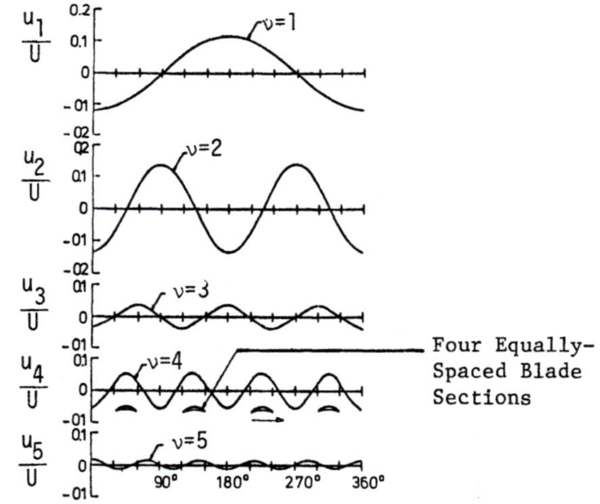

FIG. 3 HARMONIC COMPONENTS OF SPATIAL WAKE FOR A SINGLE-SCREW SHIP AT $r/b = 0.7$ IN FRACTION OF SHIP SPEED (MEAN WAKE VELOCITY $\bar{u} = 0.82U$)

As the incidence angle at all blade sections varies continuously, the flow is unsteady yielding lift components in phase and in quadrature to the incidence angle variations arising from downwash inductions feeding back from vorticity shed on to the helical wakes of the blades. Thus, the flow about the blades is neither two-dimensional nor is it steady.

2.2 Pressures Induced at Field Points

The pressures (and velocities), as would be sensed by an observer, at any point fixed in space in the neighborhood of a propeller operating in a *uniform infolw* varies harmonically at integer multiples of blade position angle because the distance of points on each blade very cyclically with respect to the fixed point. The magnitude of the pressure depends only on the nonvarying loading due to the distributed blade camber and angle of attack, as well as the distributed blade thickness. For a propeller in a *spatially non-uniform* inflow (as in a temporally invariant hull wake), the pressures (and velocities) sensed by a fixed observer, while at integer multiples of the harmonics of blade position angle, is now due to the joint or simultaneous action produced by the harmonically changing loadings and the harmonically changing distance of the field point from the blade elements.

2.3 Reaction of the Hull

The ship hull and appendages (rudder, bossings, duct) present barriers to the flow induced by the propeller. Consequently, there are developed on these surfaces blade-rate "reaction" pressures which depend on the magnitude and phase of the imposed flow from the propeller and its image in the water surface and also depend on the shape of these boundaries. For a hull whose surface is relatively flat and *deeply submerged*, the reaction pressures inboard of the beam are equal in magnitude and phase to those imposed by the propeller and, hence, the total pressure is twice that produced by the propeller in a boundless fluid. However, for hulls having curved surfaces, the hull reaction pressures are different in magnitude and phase from those of the propeller in a boundless fluid. For skeg and rudder surfaces, the reaction pressures can be three and four times that of the propeller alone.

3. THEORETICAL ASPECTS

Theoretical analyses of the fluid dynamical aspects of propellers in wakes yielding blade and shaft forces and field effects are based on the use of the linearized, unsteady Euler equations for irrotational motion of an inviscid incompressible fluid. The linearized pressure equation referred to fixed axes in a stream of longitudinal velocity U is frequently employed in the form

$$\frac{p}{\rho} + \frac{U\partial\phi}{\partial x} + \frac{\partial\phi}{\partial t} = 0 \qquad (1)$$

where

- p is the pressure change relative to ambient
- ρ is the fluid mass density
- U is the speed of the stream (or relative advance speed)
- ϕ is the velocity potential

x is the longitudinal streamwise coordinate

t is the time

The velocity potential must satisfy Laplace's equation

$$\left(\frac{\partial^2}{\partial x^2} + \frac{\partial^2}{\partial r^2} + \frac{1}{r}\frac{\partial \phi}{\partial r} + \frac{1}{r^2}\frac{\partial^2}{\partial \gamma^2}\right) \phi = 0 \tag{2}$$

(in cylindrical coordinates (x,r,γ)) and must be constructed to meet the boundary condition of zero relative flow normal to the blades or body boundary.

A basic solution of Laplace's equation in a boundless fluid is

$$\phi_s(x,r,\gamma,\xi,s,\theta) = \frac{M}{4\pi[(x-\xi)^2+r^2+s^2-2rs\cos(\gamma-\theta)]^{\frac{1}{2}}} = -\frac{M}{4\pi R(x,r,\gamma;\xi,s,\theta)} \tag{3}$$

which is identified as a point source of strength M located at cylindrical coordinates $x=\xi$, $r=s$, $\gamma=\theta$. It is seen to be a reciprocal function only of the straight line distance R from the source to the field point.

Derivatives of the fundamental solution are also solutions of the Laplace equation and, as is well known, solutions for the flow about bodies and propellers can be built up by surface distributions of such functions having unknown strength densities which are found by solving integral equations arising from the condition of zero normal flow relative to the particular surface.

For propeller problems, it appears attractive to me to exploit the fact that the potential of a rotating source to a field point can be expressed in a Fourier series, i.e.,

$$\frac{1}{R(x,r,\gamma;\xi,s,\theta)} = \sum_{m=0}^{\infty} A_m(x,r;\xi,s)\cos m(\gamma-\theta) \tag{4}$$

where the amplitude function has the alternate forms

$$A_m(x,r;\xi,s) = \frac{1}{\pi\sqrt{rs}} Q_{m-\frac{1}{2}}\left\{\frac{(x-\xi)^2+r^2+s^2}{2rs}\right\} \tag{5a}$$

$$= \varepsilon_m \int_0^\infty J_m(kr)J_m(ks)e^{-k|x-\xi|}dk \tag{5b}$$

$$\varepsilon_m = 1,\ m = 0; \quad = 2,\ m > 0$$

$$= \frac{\varepsilon_m}{\pi}\int_0^\infty I_m(kr)K_m(ks)e^{ik(x-\xi)}dk \ ; \quad r<s, \text{ for } r>s \text{ interchange arguments of } I_m, K_m \tag{5c}$$

where

Q_μ is the associated Legendre function of half-integer order

J_m is the Bessel function of the first kind

I_m, K_m are the modified Bessel functions of the first and second kinds respectively

The advantage of the Fourier expansion of R^{-1} is that the angular space and dummy angles are separated from the other coordinates, thus explicitly placing in evidence the harmonic variation of the source potential with blade angular position. Forms 5(b) and 5(c) also provide a quasi-separation of the axial variables x and ξ from the radial coordinates r and s.

The unsteadiness of the flow is incorporated in Equation (1) which can be solved for the velocity potential in the form

$$\phi(x,r,\gamma,t) = -\frac{1}{\rho U}\int_{\xi'=-\infty}^{x} p(\xi',r,\gamma,t-\frac{x-\xi'}{U})d\xi' \tag{6}$$

where the integral over ξ' accounts for the prior history of the flow.

The field pressure p induced by pressure jumps Δp at all dummy points on a single blade are given by the following integral over the blade surface S:

$$p(x,r,\gamma,t) = \frac{1}{4\pi}\iint \Delta p(x',s,t)\frac{\partial}{\partial n'}\frac{1}{R}\,dS \tag{7}$$

where n' is the local normal to the blade surface and R is the distance from any dummy point on the blade to any field point.

Now we can see the insertion of (7) to (6) and the use of (4) with (5b) or (5c) will enable the effect of all blades to be summed to give a potential, the normal derivative of which, when equated to the negative of the known onset flow provided by the hull wake yields an integral equation for the various unknown pressure jump amplitudes of the form

$$\iint_S \Delta\tilde{p}_\nu\, K\, dS = -\{u_\nu \cos\theta_p - v_\nu \sin\theta_p\} \tag{8}$$

where K is the kernel function which embodies the blade geometry and the unsteady feed back effects from all blades and their helical wakes.

Once the blade pressure distributions are found, the forces and moments applied to the shaft are determined by integration. Pressures at any field point can be found from (7) which, while at blade frequency, are composed of many terms involving the product of the loadings Δp with the kernel function at frequencies ν-qN and ν+qN, as well elucidated in our paper later today.

A completely different propeller force procedure has been recently developed by Prof. J. Kerwin [14] who represents the propeller with bound and shed vortex elements and calculates the evolution of the blade loading as the blades are time-stepped in rotation through the given wake. After about three revolutions, the amplitudes and phases of the blade loadings are found to plateau.

We may now turn to a look at some selected examples of measurements and correlation with theory where available.

4. EXPERIMENTAL RESULTS

4.1 Propeller Forces

Measurements of vibratory forces on propellers have been accomplished at David Taylor Naval Research and Development Center (DTNSRDC), Netherlands Ship Model Basin (NSMB), Hamburg Model Basin (HSVA), Berlin Tank (VSW), Japanese Shipbuilding Research Institute (SRI) and Mitsubishi Heavy Industries. Generally, the balances needed for these measurements are of the strain-gauge type and must be sufficiently stiff to secure natural frequencies sufficiently high to enable the use of static calibrations. Results of measurements at DTNSRDC and calculations at Davidson Laboratory (DL) on a set of 3-bladed propellers exhibit quite acceptable agreement in Figures 4 and 5.

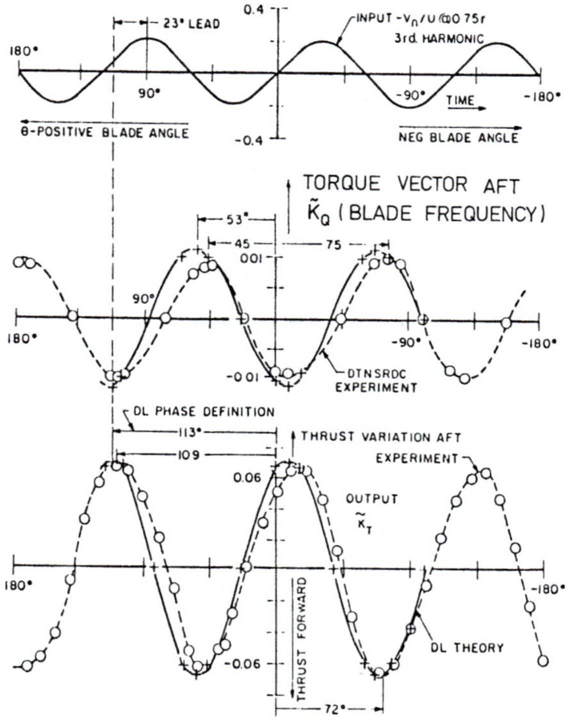

FIG. 4 MEASURED AND CALCULATED BLADE FREQUENCY THRUST AND TORQUE COEFFICIENTS FOR DTNSRDC PROPELLER 4118, EAR=0.6, USING WAKE VALUES OF TABLE V FROM 3-CYCLE SCREEN

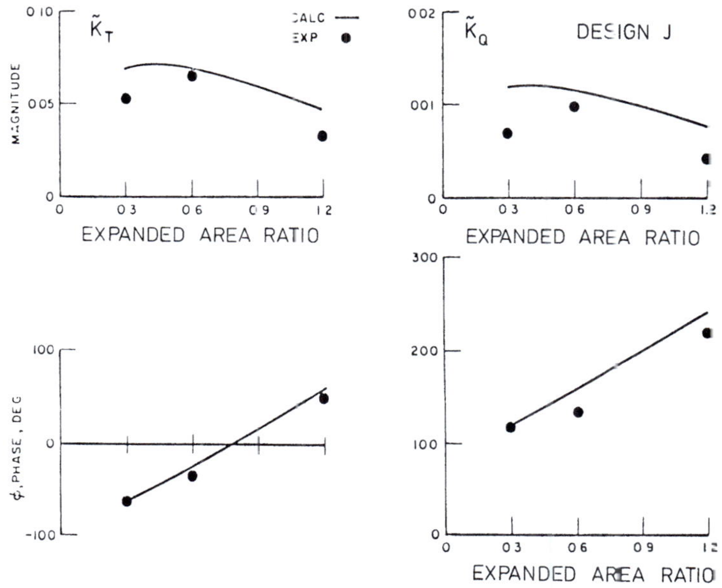

FIG. 5 BLADE-FREQUENCY THRUST AND TORQUE OF DTNSRDC 3-BLADED PROPELLERS (IN 3-CYCLE WAKE)

Measurements of forces on a series of skewed propellers presented by Nelka [20] of DTNSRDC are shown in Figure 6. Calculations made at DTNSRDC with the DL force program have shown close agreement with measurements of forces on skewed propellers. Results of the 3-bladed DTNSRDC data by Van Gent [12] appear to agree with these data and values obtained at DL, but unfortunately, the figures in Van Gent's paper are incorrect. Kerwin's [14] time-stepping method yields results which compare very favorably with data as shown in Figure 7 and with DL calculations for a propeller with skewed blades.

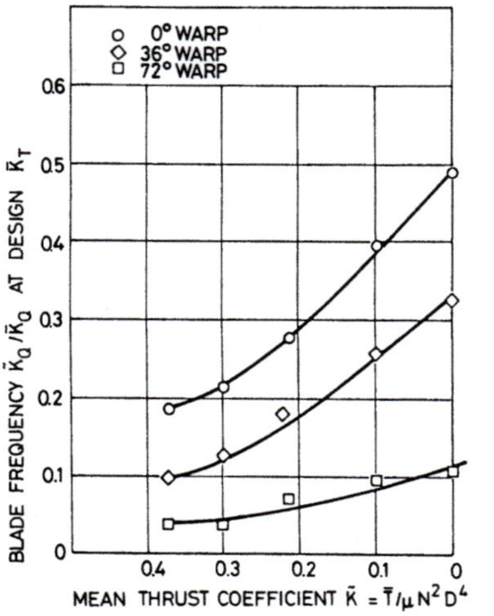

FIG. 6 BLADE FREQUENCY THRUST CO-EFFICIENT IN FRACTION OF THE MEAN DESIGN VALUE FOR THREE DESIGNS OF VARYING BLADE SKEW

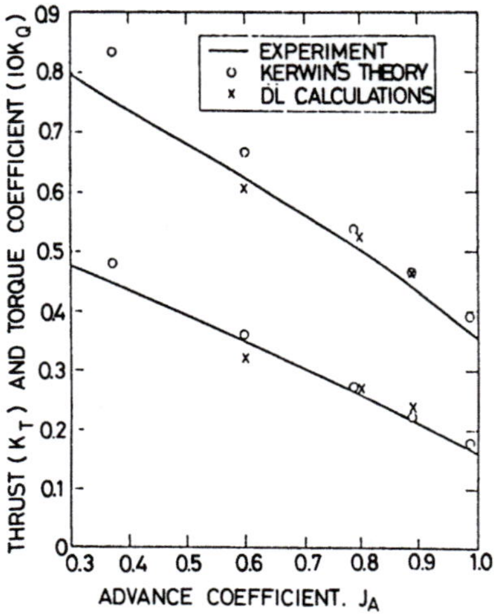

FIG. 7 COMPARISON OF KERWIN'S THEORY WITH MEASUREMENTS AND DL CALCULATIONS

4.2 Pressure Measurements

Pressure measurements in the neighborhood of model propellers, generally in water tunnels and with uniform inflow, have been made at DTNSRDC, HSVA, NSMB and at the National Physical Laboratory (NPL) (England). Excellent agreement of amplitude and phase between DTNSRDC and HSVA data and DL calculations are shown in Figures 8 and 9. The HSVA data was obtained with a wake screen upstream of the

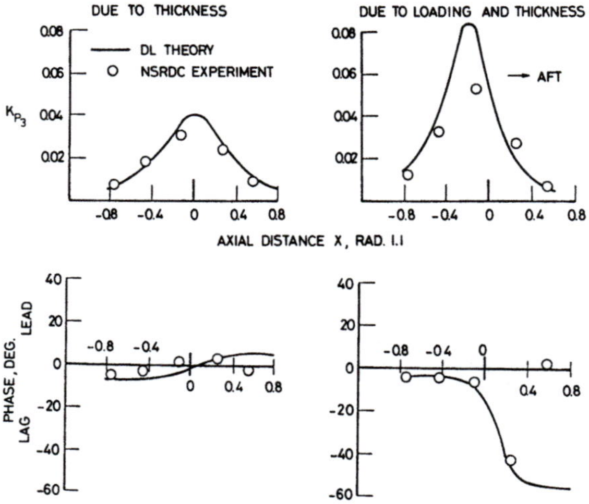

B-F PRESSURE AMPLITUDES AT POINTS ON A FLAT PLATE AT r=1.1 RADIUS AND ϕ=0, DUE TO 3-BLADED PROPELLOR 4118 IN UNIFORM INFLOW (J=0.83)

FIG. 8

CORRESPONDING B-F (BLADE FREQUENCY) PRESSURE PHASES (MECHANICAL)

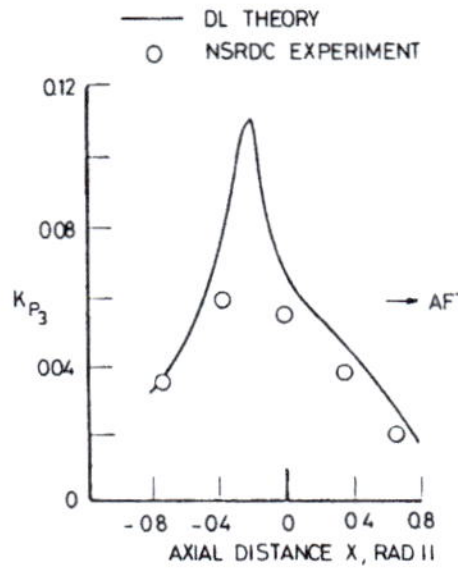

FIG. 9 TOTAL B-F PRESSURE AMPLITUDES AT POINTS ON A FLAT PLATE AT r=1.1 RADIUS AND ϕ=0, DUE TO LOADING AND THICKNESS OF 3-BLADED PROPELLER 4118 IN THE NONUNIFORM INFLOW GENERATED BY A 3-CYCLE WAKE SCREEN (J=0.83)

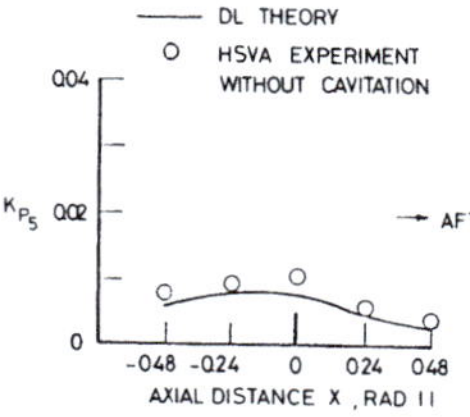

FIG. 9 TOTAL B-F PRESSURE AMPLITUDES AT POINTS ON A FLAT PLATE AT r=1.6 RADIUS AND ϕ=0, DUE TO LOADING AND THICKNESS OF 5-BLADED HSVA PROPELLER OPERATING IN A SHIPWAKE (J=0.80)

propeller to simulate the longitudinal wake of a specific hull. The effects of a strong wake "spike" are embodied in the DTNSRDC test results and DL theory compared in Figure 9. We can only wonder where the data would fall at the center where the theory predicts the maximum b.f. pressure amplitude. (A simple shift (relative to the propeller) of the plate (in which the gages were mounted) by half a gage spacing would have produced twice the number of measurements and one at the maximum--surely a good example of the need for cooperative effort between theorist and experimenter!)

4.3 Hull Pressures and Forces

The only sets of model hull pressures for non-cavitating propeller operation in the literature are those obtained by Huse [21] and by Mercier et al. [22]. Huse compared his measurements of pressure amplitudes at 35 points on a 6.1 m tanker model with those he calculated by a simplified quasi-steady lifting-line, thin-section theory, in order to secure a measure of the reflection or reaction effect of the hull at various locations. Ratios of the measured to the computed pressure amplitudes are summarized in Figure 10 where we observe wide variations

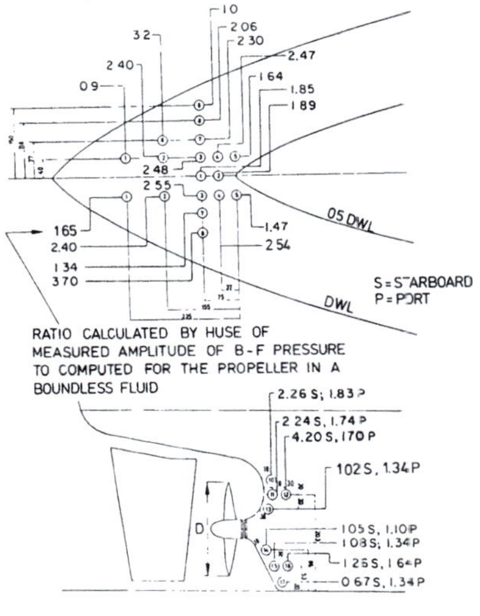

FIG. 10 ADAPTED FROM FIG. 36 OF REFERENCE [21] TO SHOW RATIOS OF MEASURED TO COMPUTED B-F PRESSURE AMPLITUDES

from the flat surface ratio of 2.0 used by most naval architects and researchers. The theory has been multiplied by 2.0 in the comparison shown on Figure 11 and good agreement is found because the hull is flat.

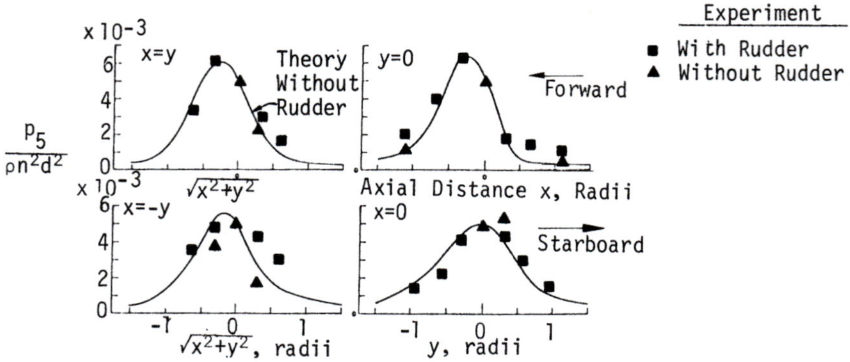

FIG. 11 BLADE-FREQUENCY PRESSURE AMPLITUDES INDUCED BY A 5-BLADE PROPELLER AT POINTS ON THE FLAT STERN OF A DESTROYER ESCORT MODEL FROM EXPERIMENT AND THEORY (J=0.82)

Hull force measurements carried out during the 1960's by Prof. F. M. Lewis at MIT were made with 3 m models built around a very stiff skeleton as depicted in Figure 12. To determine vertical "forces" he calibrated the model by exciting the model with a sinusoidally activated electro magnet at one point and measured the velocity response using a seismic pickup at another point. Effective vertical

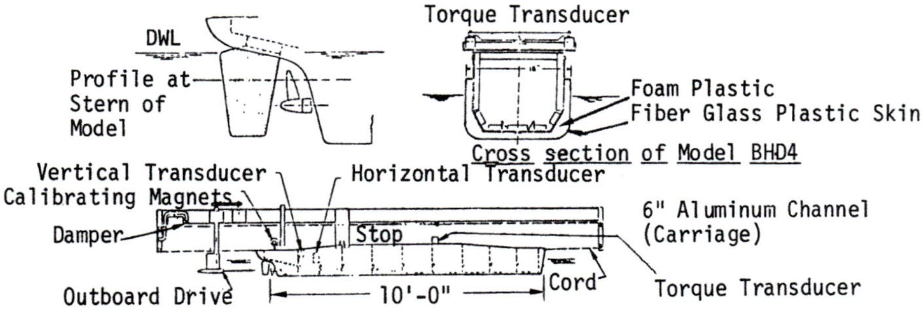

FIG. 12 PROPELLER VIBRATION FORCES, MODEL ARRANGEMENT AT M.I.T.

and horizontal forces were obtained with combinations of 4, 5 and 6-bladed propellers on a parent Series 60 hull and a variant. Bearing forces were obtained by subtracting measurements made with an outboard-driven propeller from those made with the propeller inboard-driven. He also instrumented rudders and determined the variation in horizontal forces as a function of rudder-propeller-hull clearance. In general, he found that the surface forces are larger than the total and the bearing forces and that the bearing forces generally are in opposition to the surface force.

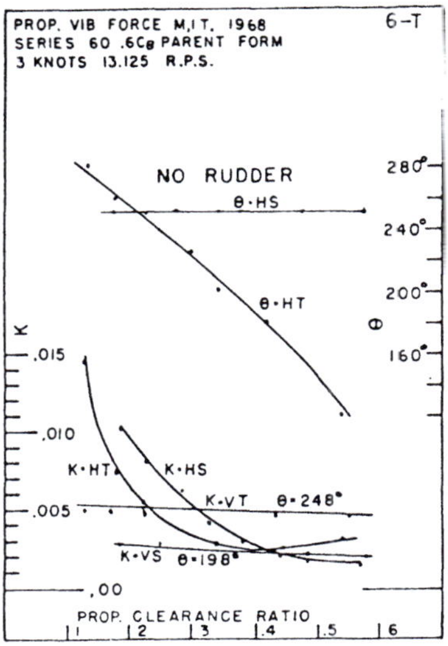

Propeller 6T: force, no rudder

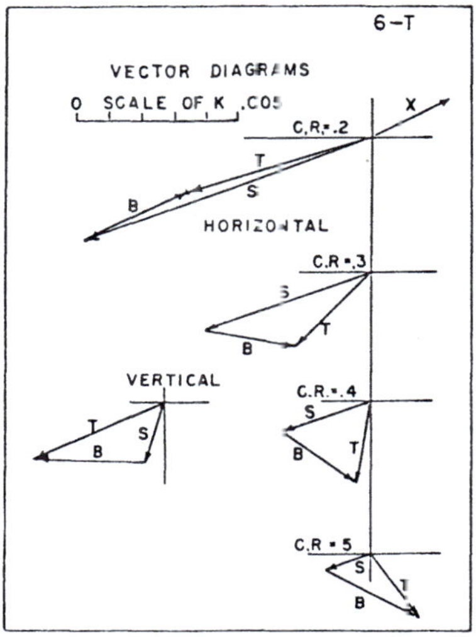

Vector diagram for Fig. 20

FIG. 13 VIBRATORY FORCES AND PHASES DETERMINED BY PROFESSOR LEWIS USING A 3mm MODEL FITTED WITH A 6-BLADED PROPELLER

A sample of his data is shown in Figure 13 (taken from Figures 20 and 21 of Reference 5) for a six-bladed propeller on the Series 60 parent form. Here we can observe the very rapid decay of the total horizontal force coefficient (K·HT) with increase in axial propeller clearance and the not-surprising virtual constance of the total vertical force coefficient (K·VT) with axial clearance. A replotting of his results for force variations with rudder location in Figure 14 (taken from Figures 19 and 20 of Reference 23) shows the strong dependence on rudder-propeller clearance and, moreover, that this exciting force can be reduced by more than 100 percent by proper selection of the clearance and *can be raised by increasing the clearance.* This harmonic behavior with propeller-rudder spacing is evident in the theory [24] and is due to the spacing of the helices shed by the propeller blades. In all, Lewis' experiments provided a means for comparative evaluation of the influence of several parameters and directed attention to the importance of surface forces even in single screw ships.

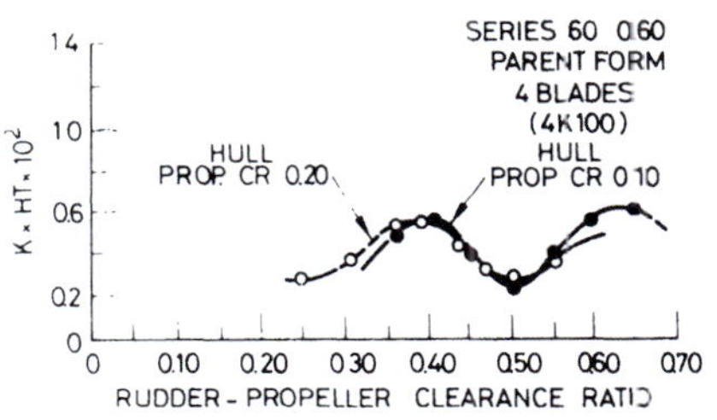

Transverse total force coefficient as function of rudder-propeller clearance for two propeller-hull clearances

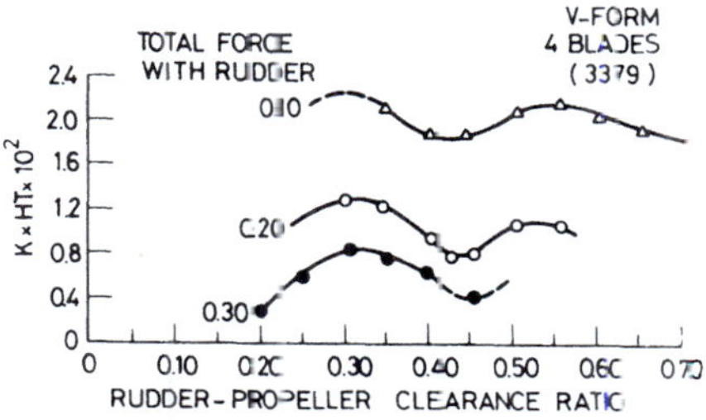

Transverse total force coefficient as a function of rudder-propeller clearance for various propeller-hull clearances

FIG. 14 EFFECT OF RUDDER-PROPELLER CLEARANCE ON TRANSVERSE BLADE FREQUENCY FORCES

A theory and a computer program for calculation of propeller vibratory forces and moments generated on the surface of hulls of *arbitrary* shapes has been developed at Davidson Laboratory (DL) and by Vorus at Michigan as noted previously. At DL, the hull surface is subdivided into quadrilateral elements over which are smeared sources of constant, but unknown, densities. The integral equation generated by requiring these sources, in aggregate, to annul the propeller-induced onset velocities normal to each and all elements is solved numerically by use of a generalized form of the Hess-Smith hull-surface flow program. A schematic of the source panel layout for a Series 60 hull is shown in Figure 15.

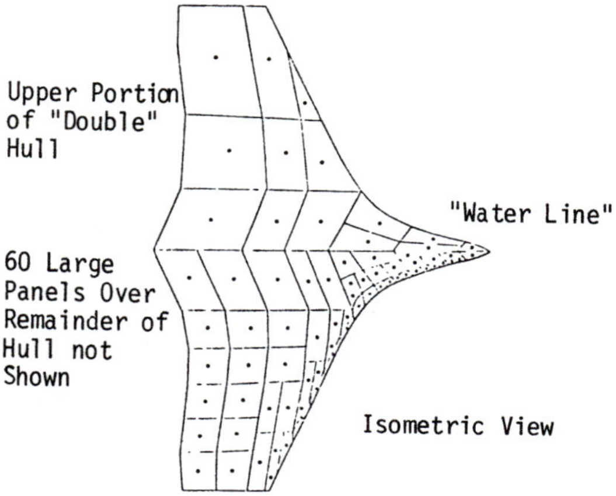

FIG. 15 SOURCE PANEL DISTRIBUTION ON "DOUBLE" SERIES 60 $C_B = 0.60$ HULL (V-SECTIONS AFT). ENTIRE HULL PLUS "REFLECTION IN AN INFINITE FLUID"

Application has been made to a spheroidal head form on which forces were measured at DTNSRDC for a configuration shown in Figure 16. Comparisons with DRNSRDC measurements, as shown in Figure 17, reveal close agreement. Vorus' method, which is quite different, also provides excellent agreement at design J.

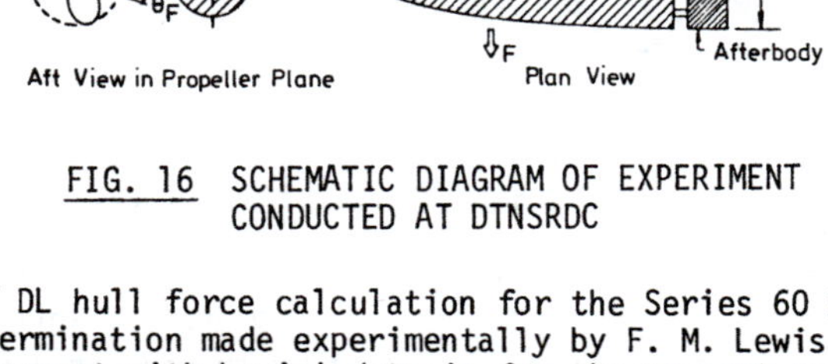

FIG. 16 SCHEMATIC DIAGRAM OF EXPERIMENT CONDUCTED AT DTNSRDC

Comparison of DL hull force calculation for the Series 60 hull with the "effective" force determination made experimentally by F. M. Lewis [5] is displayed in Figure 18. Agreement with Lewis' data is for the present not necessarily meaningful as he did not achieve a separation of the effects of force and moment. Further work is required to determine the relative responses of Lewis' model to force and moment.

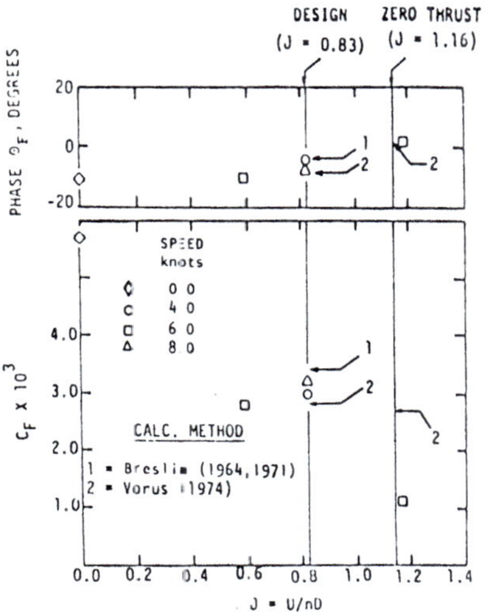

FIG. 17 CALCULATED AND MEASURED BLADE FREQUENCY FORCE FOR PROPELLER LOCATED AT $\ell = 16.0$ IN. WITH TIP CLEARANCE C = 3.0 IN.

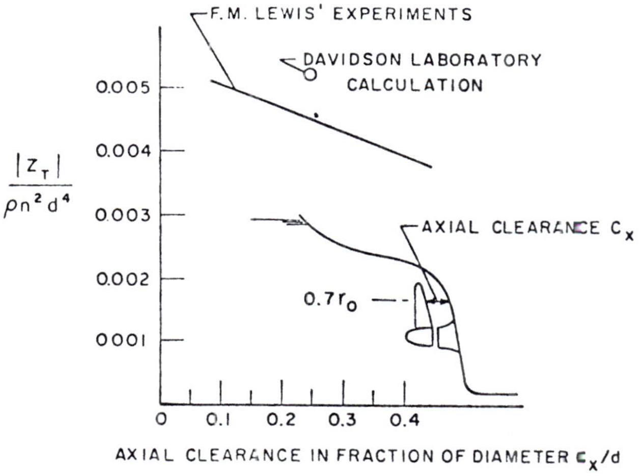

FIG. 18 COMPARISON OF TOTAL BLADE-FREQUENCY-VERTICAL-FORCE AMPLITUDE WITH PROFESSOR F. M. LEWIS' MODEL TEST DATA (SERIES 60 ($C_B = 0.6$) V-FORM, DTNSRDC 4-BLADED PROPELLER MODEL 3779)

5. FUTURE NEEDS

In spite of the quite successful comparisons exhibited, there still remains a paucity of b.f. excitation data for comparison with calculational methods. We plan to compare the DL propeller-hull program output with Huse's 36-point pressure measurements and, hopefully, upon the completion of a projected integration of the

MIT intermittently cavitating propeller-program with the DL propeller-hull program, one may embark on correlations with model and particularly full-scale pressures when cavitation is present. Det Norske Veritas has collected hull pressures on about 72 ships so there are full-scale data which might possibly be made available.

An outstanding problem is posed by the fact that calculation of propeller and hull vibratory forces are highly dependent on an accurate knowledge of the hull wake harmonics. These are currently deduced from 6-meter model wakes measured in the absence of the propeller, the so-called nominal wake. It is known, however, that the action of the propeller alters the hull flow ahead and the actual or effective wake harmonics seen by the propeller are therefore different from the nominal wake, particularly for classes of full-form ships where boundary layer separation is prevalent. In addition, the viscous flow is not scaled, the model Reynolds numbers being some 3 orders of magnitude smaller than those of the prototype. These wake questions are being studied, but it may be some time before they are resolved. Meanwhile, these coupled programs, together with finite element elastic hull programs could be used, hopefully early enough in the design stage, to give valid comparative evaluations of vibratory characteristics of alternate hull-propeller structural configurations.

It is to be appreciated that this summary is necessarily elliptic. For those interested in this topic, an up-to-date extensive bibliography can be secured from the Proceedings of the 15th International Towing Tank Conference (September 1978 at The Hague), published by the last host, The Netherlands Ship Model Basin, Wageningen, The Netherlands.

6. REFERENCES

1. Todd, F. H., "Ship Hull Vibration", Ed. Arnold Ltd., London, (1961).
2. Lewis, F. M., "Propeller Vibration", Trans. SNAME, Vol. 43, 252-285, (1935).
3. Lewis, F. M., "Propeller Vibration", Trans. SNAME, Vol. 44, 501-519, (1936).
4. Lewis, F. M. and Tachmindji, A. J., "Propeller Forces Exciting Hull Vibration", Trans. SNAME, Vol. 62, (1954).
5. Lewis, F. M., "Propeller Vibration Forces in Single Screw Ships", Trans. SNAME, Vol. 77, (1969).
6. Tachmindji, A. J. and Dickerson, M. C., "The Measurement of Oscillating Pressures in the Vicinity of Propellers", DTNSRDC Report 1130, (1957).
7. Denny, S. B., "Comparisons of Experimentally Determined and Theoretically Predicted Pressures in the Vicinity of a Marine Propeller", Naval Ship Research and Development Center, Report 2349, (May 1967).
8. Ramsay, J. W., "Aspects of Ship Vibration Induced by Twin Propellers", Trans. R.I.N.A., (1956).
9. Pohl, K. H., "Das Instationare Druckfeld in der Ungebung eine Schiffspropellers und die von ihm auf benachbarten Platten erzentzen periodschen Krafte", Schiffstechnik, 32, 107-116, (June 1959).
10. Breslin, J. P. and Tsakonas, S., "Marine Propeller Pressure Field Due to Loading and Thickness Effects", Trans. SNAME, Vol. 67, (1959).

11. Tsakonas, S., Jacobs, W. R., Ali, M. R., "Documentation of a Computer Program for the Pressure Distribution, Forces and Moments on Ship Propellers in Hull Wakes", Report SIT-DL-76-1863, Stevens Institute of Technology, (January 1976).

12. Van Gent, W., "Unsteady Lifting-Surface Theory for Ship Screws: Derivatives and Numerical Treatment of Integral Equations", Journal of Ship Research, Vol. 19, No. 4, 243-253, (Dec. 1975).

13. Hanaoka, T., "Hydrodynamics of an Oscillating Screw Propeller", Fourth Symposium on Naval Hydrodynamics, Ship Propulsion and Hydroelasticity, Washington, (August 27-31, 1962).

14. Kerwin, J., "Prediction of Steady and Unsteady Marine Propeller Performance by Numerical Lifting-Surface Theory", Presented by SNAME Annual Meeting, New York, N.Y., (November 1978).

15. Johnson, C. A., Søntvedt, T., "Propeller Excitation and Response of 230,000 TDW Tankers", Ninth Symposium on Naval Hydrodynamics, R. Brard and A. Castera, Editors, Vol. 1, Office of Naval Research, ACR-203, 581-669, (1972).

16. Noordzij, J., "Pressure Field Induced by a Cavitating Propeller", International Shipbuilding Progress, Vol. 23, No. 260, 93-105, (1976).

17. Vorus, W. S., "A Method for Analyzing the Propeller-Induced Vibratory Forces Acting on the Surface of a Ship Stern", Trans. SNAME, Vol. 82, 186, (1974).

18. Vorus, W. S., Breslin, J. P. and Tien, Y. S., "Calculation and Comparison of Propeller Unsteady Pressure Forces on Ships", Presented at Ship Vibration Symposium, Arlington, Va., (October 1978).

19. Breslin, J. P., Tsakonas, S. and Valentine, D., "A Method for Predicting Effects of Propeller-Hull Configurations on Vibratory Excitation of Ships", IAHR/IUTAM Symposium on Practical Experiences with Flow-Induced Vibrations Preprints, Karlsruhe, Germany, (September 1979).

20. Nelka, J. J., "Experimental Evaluation of a Series of Skewed Propellers with Forward Rake: Open-Water Performance, Cavitation Performance, Field-Point Pressures, and Unsteady Propeller Loading", NSRDC Report 4113, (July 1974).

21. Huse, E., "The Magnitude and Distribution of Propeller-Induced Surface Forces on a Single-Screw Ship Model", Norwegian Ship Model Experiment Tank Publication No. 100, The Technical University of Norway, (December 1968).

22. Jacobs, W. R., Mercier, J. and Tsakonas, S., "Theory and Measurements of the Propeller-Induced Vibratory Pressure Field", Report SIT-DL-70-1485, Stevens Institute of Technology, (December 1970); J. Ship Research, Vol. 16, No. 2, (June 1972).

23. Breslin, J. P., "Theoretical and Experimental Techniques for Practical Estimation of Propeller-Induced Vibratory Forces", Trans. SNAME, Vol. 78, 23-40, (1970).

24. Tsakonas, S., Jacobs, and Ali, M. R., "Propeller-Rudder Interaction Due to Loading and Thickness Effects", Journal of Ship Research, Vol. 19, No. 2, 99-117, (June 1975).

D2 CALCULATIONS, MEASUREMENTS AND RESULTING REMEDIES FOR VIBRATION PROBLEMS ON RO/RO SHIPS

Christian Nath
Research Engineer

Hans G. Payer
Head Structural Research Group

Eckhart A. Pleß
Head Shipbuilding Division

Germanischer Lloyd Hamburg, West Germany

SUMMARY

A class of large modern Ro/Ro (Roll on/Roll off) ships was investigated during design, construction and under service conditions. Natural frequencies were calculated for different cases, to theoretically analyse design alternatives. Moreover vibratory response was measured during tests with a mechanical exciter and for excitation in service as well. During these tests the influence of trim and ullage of double bottom tanks on the vibratory behavior was investigated.

Results of model tests and measurements on board the ship before and after application of flow accelerating fins above the propeller are described and comparisons are made. Vibrations were reduced even though the propeller tip clearance was decreased and the pressure amplitude there increased.

1. INTRODUCTION

Ro/Ro ships are designed to transport rolling cargo (e.g., trucks or trailers) using stern ramps for loading and unloading. Therefore, wide decks without intermediate support and large openings in the aft body are their common characteristics of these ships. One of the key problems in their design is to obtain a reasonably stiff aft body structure. This is of particular importance for larger ships of this type, since the unsupported span of the deck structure, which supports the deckhouse, may lead to fairly low natural freqency vibrations. In connection with increasing vibratory forces as a result of high power output, and possibly due to cavitation at the propeller, the vibration problems may be quite severe.

Knowing these problems, the ship dealt with in this paper was investigated very carefully, not only during the design phase, but also during construction, trial, and service.

2. CALCULATIONS AND MEASUREMENTS

During the design phase, finite element calculations were performed to determine the natural frequencies of the global aft body structure, deckhouse, and local systems. Some of the calculated eigenvalues were found to be in the range of the excitation frequencies (7.5 Hz first order of engine, 8.5 Hz propeller blade frequency).

To avoid the danger of resonance with the propeller, the substructure of the rudder and some structural components were locally stiffened. Newly calculated resonance frequencies for the deckhouse were checked with calculated excitation forces and gave no concern for excessive vibrations. The excitation forces were evaluated from a wake analysis assuming non-cavitating propeller operation.

After launching of the vessel, vibration tests were carried out with a mechanical exciter. Measured amplitudes for frequencies of 1.0 to 19.6 Hz were normalized with respect to the excitation forces of the exciter (up to 196 kN at 19.6 cps) to obtain the dynamic flexibility of the structure. This, together with the estimated excitation forces, showed acceptable levels of vibration (following ISO recommendations) to be expected for the ship.

Strong, global deckhouse vibrations nevertheless occurred during sea trials with the ship going at full power. Measured frequencies and mode shapes showed a good agreement with theoretical results; the magnitude of the amplitudes, however, indicated large excitation forces.

A sudden increase of vibration amplitude when exceeding a certain speed was an indication of the occurrence of cavitation. Two means of reducing the vibration level were investigated:

1. Detuning of the aft body structure by means of additional stiffening.
2. Decreasing of the propeller-induced excitation forces by tunnel or fins above the propeller, or by changing the number and/or the geometry of the propeller blades.

3. CURES

The first alternative was studied theoretically and put into practice. The effect of changes in the structural design was analysed using differing finite element models. Although no pronounced change in the frequency was found, the most promissing structural modification was employed when the foundation of the deckhouse was stiffened. The remedies were not sufficient to solve the problem since a direct support of the deckhouse was not possible without serious obstruction of the cargo passageway. A possible support in the centerline plane was investigated, and found to be inefficient due to the large deckheight below the deckhouse.

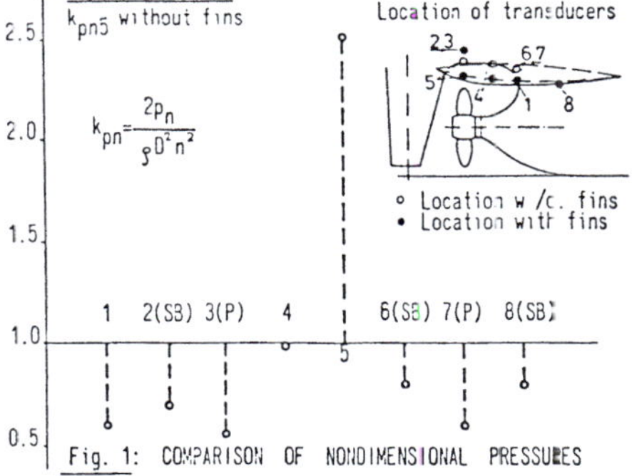

Fig. 1: COMPARISON OF NONDIMENSIONAL PRESSURES

Regarding a decrease of the excitation forces, flow accelerating fins above the propeller were tested in a cavitation tunnel with the model originally used for propulsion tests, sclale 1 : 28. Eight pressure transducers were installed on the hull of the model to measure the excitation forces. Tests were carried out with and without fins (Fig. 1).

A comparison showed cavitation to be circumferentially more uniform with the radial extent reduced and the peripheral extent being somewhat larger for the model with fins (Fig. 2).

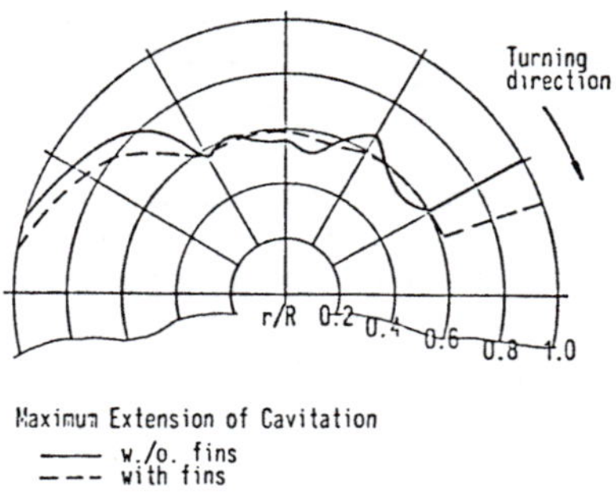

Fig. 2: EXTENSION OF CAVITATION OCCURRENCE

All pressure transducers, except the one directly above the propeller, showed a decrease in pressure fluctuation amplitude. The overall reduction of vibration amplitudes of the ship, as a result of fitting fins, was estimated to be 25 %. On the basis of these results it was decided to fit fins to one vessel of this class of Ro/Ro ships.

In addition, full scale measurements of the effect of fins were made. To achieve comparable boundary conditions, the ship was investigated for the same draughts shortly before and after application of the fins. As the measurement of pressures on the full scale hull is more difficult in practice, displacement spectra were recorded.

For trim and draught conditions where the fins were completely submerged (d_f = 4.2 m, d_a = 7.6 m) significant improvements in the vibratory behavior were found. The amplitudes of the panels in the deckhouse decreased by 40 to 80 %, whereas the combined amplitudes due to engine an propeller excitation (7.5 Hz and 8.5 Hz respectively) of the global deckhouse vibration decreased by about 20 %.

Pick-up No.	Amplitudes w./o. fins	Amplitudes with fins
1	37.1	36.4
2	44.6	33.4
3	58.0	50.5
4	57.0	50.4
5	51.2	48.2
6	36.7	28.5
7	31.2	24.3
8	28.8	22.8
9	31.2	25.9
10	47.9	37.3
11	52.2	47.6
15	101.1	73.7
16	110.1	86.0

Location of pick-ups

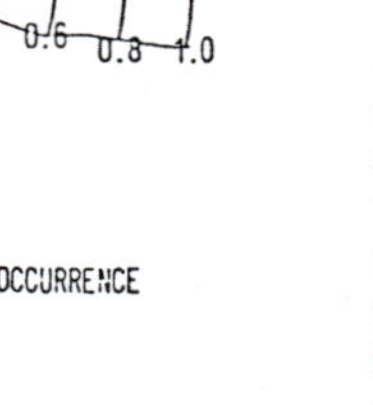

Fig. 3: MEASURED DISPLACEMENT AMPLITUDES

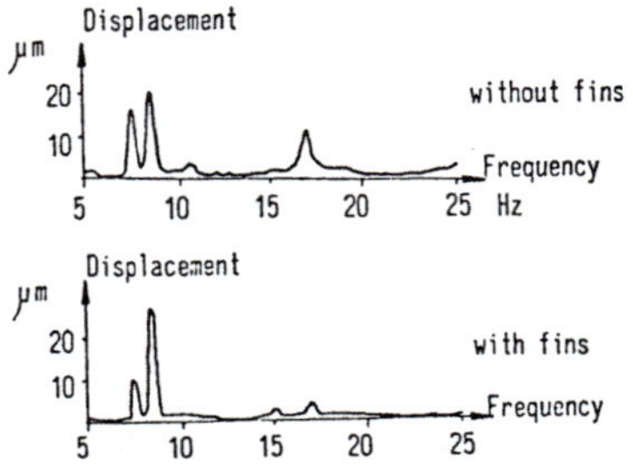

Fig. 4: MEASURED DISPLACEMENT SPECTRA OF LOCATION 1

In Fig. 3, selected results are shown for two comparable conditions, together with a layout of displacement pick-up points. Results for pick-up number 1, directly above the propeller, are presented in Fig. 4. As expected from the model tests (see Fig. 1), an increase at the propeller blade frequency (8.5 Hz) is evident. This increase is due to the increased pressure amplitudes, which are, in turn, associated with the decrease in propeller tip clearance. This increase of pressure, however, is locally confined. Remarkable is the amplitude reduction at the higher harmonics of the propeller blade frequency, a further indication of the improvement of the cavitation behavior due to the flow accelerating fins. Regarding the response at 7.5 Hz, a decrease of the displacement amplitude can be seen. This probably is due to a local detuning effect the stern floor structure caused by the fitting of the fins.

4. CONCLUSIONS

In conclusion, the effect of flow accelerating fins above the propeller resulted in pronounced improvement of the vibrational behavior of Ro/Ro ships. On the other hand, the results of model tests as well as speed and power measurements during trials, indicate a reduction of the propulsion efficiency of about 5 to 10 %. The analysis of the data is not sufficiently complete to allow a final evaluation of the results. The reduced efficiency may, however, be a price which must be paid to achieve improved comfort on this class of ships.

5. REFERENCES

Stiansen, S.G., "Propeller and Wave-Induced Hull Structure Vibrations", Ship Vibr. Symposium, Arlington, Va., (Oct. 1978).

Lindgren, H., Johnsson, C.A., "On the Influence of Cavitation on Propeller Excited Vibratory Forces and some Means of Reducing its Effect", International Symposium on Practical Design in Shipbuilding, Tokyo, (Oct. 1978).

Hylarides, S., "Some Hydrodynamic Considerations of Propeller Induced Ship Vibrations", Ship Vibration Symposium, Arlington, Va., (Oct. 1978)

Westram, A., "Experimentelle Schwingungsuntersuchung" - BMFT Vorhaben MTK 0058, Bericht STB-Nr. 700-1979, Germanischer Lloyd, Hamburg, Germany

Skaar, K.T., Carlsen, C.A., "Modelling Aspects for Finite Element Analysis of Ship Vibration", Conf. Eng. Appl. of the F.E.M., (May 1979), Oslo

Allanström, B., "Ro/Ro Ship-Model Tests with and without Fins in Cavitation Tunnel No. 2", Report No. 2374-1, Swedish State Shipb. Exp. Tank (SSPA) Göteborg, (1978).

6. ACKNOWLEDGEMENTS

The authors wish to acknowledge the good cooperation with the ship's owner Deutsche Dampfschiffahrts-Gesellschaft "Hansa", particularly Mr. Nehus, and the ship yard Howaldtswerke-Deutsche Werft AG, Kiel, who proposed and built the fins, and made the results of the model tests available.

RO/RO SHIPS

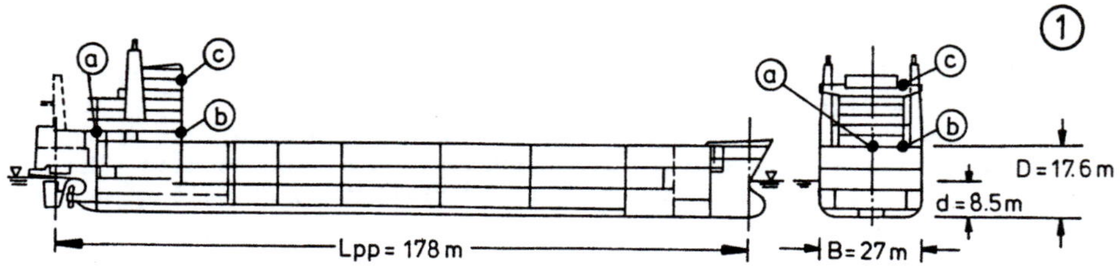

Problem: Deckhouse vibrations partly beyond acceptable limits (ISO-rules) occur at full speed due to propeller cavitation.

Geometry: See sketch 1. One controllable pitch propeller, four blades.

Flow conditions: Ship service speed v = 10 m/s. Model test in cavitation tunnel v = 5.5 m/s, σ_{cav} = 4.5, with turbulence stimulator scale 1 : 28.

Cavitation was observed for r/R≥0.6 between 10 and 2 o'clock position.

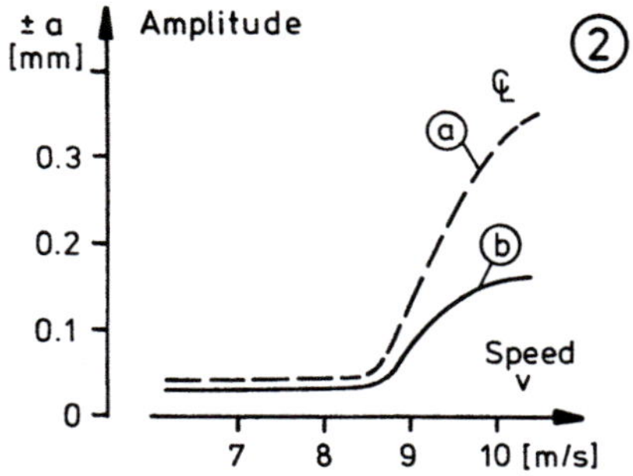

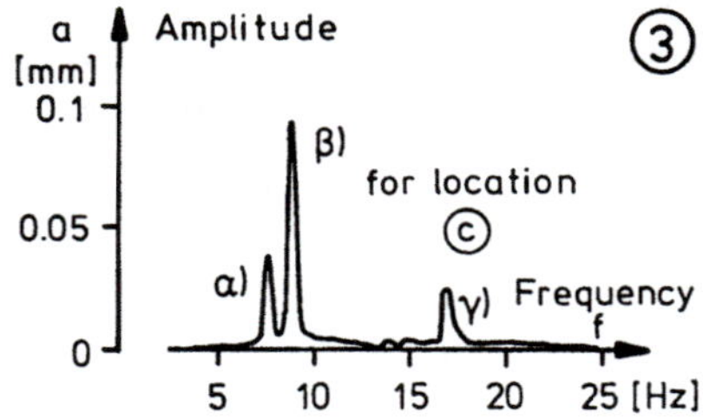

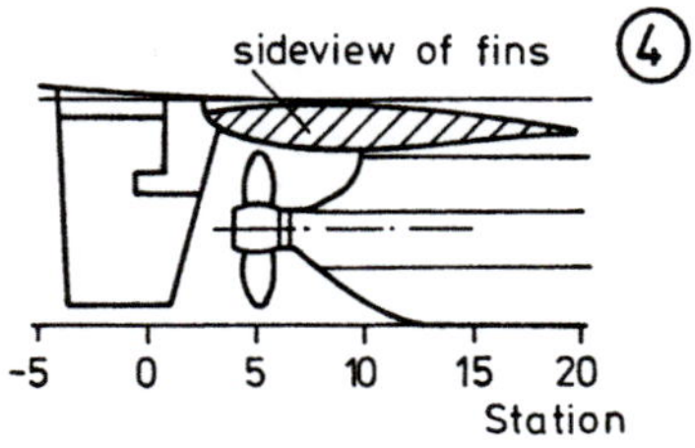

Operating conditions:
Vibratory response increased beyond 9 m/s ship speed (see sketch 2). Measured pressure amplitudes above propeller: ship, p_o = 25... 75 kPa; model, p_o = 27 kPa.

Structural dynamics:
Eigenmodes for several load cases were calculated. Higher torsional and bending mode in resonance with propeller blade frequency.

Vibrations:
Measured frequencies (see, e.g., (c) in sketch 1 and sketch 3) were: α) 7.5 Hz (first order of engine 7.5 rps), β) 8.5 Hz (propeller blade frequency), γ) 17 Hz (second order propeller blade frequency). Largest amplitudes were registered at location (c) in longitudinal direction.

Cures:
Structural stiffening was not sufficient as there was no possibility of direct support for the deckhouse. Application of fins (see sketch 4). Vibrations were reduced by 40 to 60 percent for load cases where fins were fully submerged, even though tip clearance was decreased and pressure amplitude there had doubled. The peripheral wake distribution and the cavitation boundary were somewhat steadier.

References: Hylarides, S., Some Hydrodynamic Considerations of Propeller Induced Ship Vibrations, Ship Vibr. Symp., Arlington, Va, Oct. 1978

D3

PROPELLER SINGING

F. Everett Reed
President and Technical Director

Littleton Research and Engineering Corp.
Littleton, Massachusetts
U.S.A.

SUMMARY

This paper discusses some practical experiences with the phenomenon of propeller singing, which consists of constant frequency noise generated by interaction between trailing vortices and complex high order modes of vibration. Cases of singing propellers on two different ships are discussed.

1. INTRODUCTION

Singing of propellers is characterized by a constant frequency tone generated by a propeller while operating over a range of speeds. Under normal circumstances, the range of speeds at which a tone is generated is small and the noise is not too objectionable. However, in war time, propeller singing is undesirable because it calls attention to a ship and precisely defines its location. Occasionally the noise becomes strong enough to be objectionable, and it is necessary to take steps to eliminate it.

The singing is generated by the coupling of von Karman vortices streaming off the trailing edge of a propeller with one of the higher natural modes of vibration of the propeller blade [1]. The frequencies of the blade are influenced by the virtual mass of the water vibrating with it, and so it is not possible to determine the natural frequencies by experiments in air. However, the general pattern of natural frequencies, which are reduced by the water mass, has been investigated by experiments in air and found to be similar to, but higher than, those observed in water.

The range of frequencies at which excitations are generated is related to the Strouhal number, $S = f_s\, b/U$, where f_s is the frequency of shedding, b is a somewhat indefinite characteristic length involving trailing edge thickness and boundary layer characteristics, and U is the flow velocity. Since the flow velocity, boundary layer thickness, and to some degree, the trailing edge radius vary over the length of the blade, a wealth of possible frequencies can be generated. If they excite a sympathetic mode of vibration, the frequency is sustained. There is also a wealth of possible vibration modes having closely spaced frequencies [2]. The coincidence of vibration modes with excitation frequencies appears to be greatest when the section thickness ratio is small; that is, when the blade is wide, or the maximum blade section thickness unusually small.

During the World War II years, the standard procedure for eliminating singing of propellers was to sharpen the tip and trailing edge to a sharp radius. By thus reducing the characteristic width b, the frequency of shedding can be raised out of coincidence with structural resonances having large amplitude at the trailing edge. This sharpening of the blade edge had its serious disadvantages because the edge was subject to damage from hitting floating objects when backing down, and a damaged edge is frequently found to provide the excitation required to initiate even stronger singing than the initial design.

Until recently, the general state of the art of avoiding propeller blade singing seemed to be that it was difficult to predict, but would be more apt to occur in propellers having wide regions of thin sections. Generally, it was not desirable to sharpen all propellers to a sharp V edge as a universal deterrent against singing because of the increased sensitivity of the edge to damage. If, during the preliminary trials of the ship, it was found that there was objectionable singing, the trailing edge of the propeller at the outer radii and over the tip were sharpened to sharp V edges.

Within recent years there has been some additional light thrown on the problem indicated by references [3,4,5,6,7]. An indication of these studies is that it is well to design the trailing edge so that the Von Karman vortex trails do not couple with the vibration of the propeller blade. This paper gives: an example where the traditional method of avoiding singing was unsuccessful and the more recent solution to the singing problem was successful; and an example of propeller singing that is believed to be unusual for the number of frequencies and their intensity. By modifying the trailing edge as indicated in references [3] to [7] all of the singing on this propeller was eliminated.

2. OBSERVATION OF VIBRATION

The first case, illustrated in a chart, is the twin propeller pilot boat NEW YORK, designed by M. Rosenblatt and Son, New York City, and built by Marionette Marine Corporation in Marionette, Wisconsin. The boat was admirably adapted to her service, providing a base for the New York-New Jersey Sandy Hook Pilots boarding and disembarking from ships being piloted into New York Harbor.

The pilot boat had one glaring fault for which nobody could be blamed. At a propeller speed extending from the lowest operational to about 155 RPM, a strong tone of about 120 Hz was generated by the boat. This tone was particularly loud in the region of the ship located over the propeller. The design of the ship placed the pilots' sleeping quarters on the lowest deck of the ship in the stern - part of them directly over the propellers and struts and all of them in the vicinity of the stern.

A noise attributed to singing propellers had been observed on the acceptance trials of the ship. Because of this noise, the trailing edge and the tips of the propellers had been tapered to a sharp trailing edge with distinct angles between the trailing wedge and the blade surface as shown in Modification A (see Chart). Since the tone was of constant frequency over its range, it was clearly self-excited. In fact, the same frequency tone could be observed when the boat was allowed to slow down from full speed, with the propellers kept from turning by shaft brakes.

Because sharpening the propeller edges was supposed to stop propeller singing, considerable effort was expended to no avail in attempting to find other sources for the noise: singing of the shaft struts; singing in the rudder; and friction in the shaft bearings. Finally, as a last resort, it was decided to modify the trailing edge as shown in Modification C (see chart). This modification was successful and the pilots are now well rested and alert when they bring ships into New York Harbor.

The second case, also shown on a Chart, involves a twin propeller tugboat whose propellers were designed by R. A. Stern and which was built at Peterson Builders, Inc., both in Sturgeon Bay, Wisconsin. This power unit, providing 3500 horsepower at 107 RPM on each of two shafts, is used to power a large barge to which it is rigidly connected at the stern. The propeller noise was so strong and extended over such a frequency range that it was clearly necessary to take some steps to eliminate the noise. Noise measurements were made in a ballast tank directly over the port propeller. As shown in the graph at the bottom of the second chart, the noise extended over the whole operating range of the ship. There are eight singing frequencies identified in these measurements. Several frequencies can be excited simultaneously. Some tones are excited at some shaft speeds, are lost as the speed is increased, and then reappear at higher shaft speeds.

To eliminate the noise, the trailing edge of the propeller was cropped to give the double sharp trailing edge shown in the following sketch.

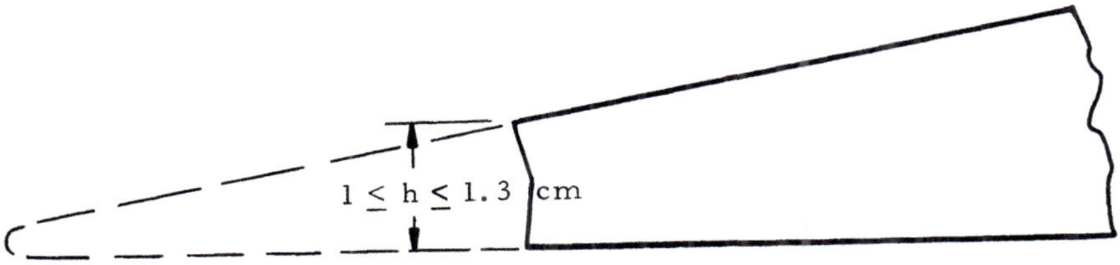

Because the blade was unusually thin, the value of h was specified to be not less than 0.4 inch (1 cm) and preferably 0.5 inch (1.3 cm) or more. Since the tugboat was too deep to be drydocked in the Great Lakes, the recommended propeller modification was not made until several months later in a shipyard in New Orleans. It is reported that propeller singing has ceased.

In closing, it should be mentioned that references [3] through [7] explain the importance of trailing edge shape on the self-excitation of high frequency vibrations of struts, runners and propellers.

3. REFERENCES

1. Cumming, R. A., "A Preliminary Study of Vortex-Induced Propeller-Blade Vibration and Singing", David Taylor Model Basin Report 1838, (Sept. 1965).

2. Slutsky, S. and Tamagno, J., "Propeller Blade Vibration Experiments", General Applied Science Laboratories, Inc. Tech. Report 213, AD254684, (June 1961).

3. Donaldson, R. M., "Hydraulic-Turbine Runner Vibration", Trans. ASME, Vol. 78, 1141-1147, (1956).

4. Heskestad, G. and Olberts, D. R., "Influence of Trailing Edge Geometry on Hydraulic-Turbine Blade Vibration", Trans. American Society of Mechanical Engineers, Series A, Journal of Engineering for Power, Vol. 82, 103-110, (Apr. 1960).

5. Toebes, G. H. and Eagleson, P. S., "Hydroelastic Vibrations of Flat Plates Related to Trailing Edge Geometry", ASME, Journal of Basic Engineering, Vol. 83, Series D, No. 4, (Dec. 1961).

6. Goldwag, E. and Berry, D. G., "Von Karman Hydraulic Vortexes Cause Stay Vane Cracking on Propeller Turbines at the Little Long Generating Station of Ontario Hydro", ASME, Journal of Engineering for Power, Vol. 90, Series A, No. 3, (July 1968).

7. Archibald, F. S., "Self-Excitation of an Acoustic Resonance by Vortex Shedding", Journal of Sound and Vibration, 38 (1), 81-103, (1975).

PROPELLER SINGING

Pilot Boat NEW YORK

The boat on which the pilots stay between taking ships in and out of New York Harbor.

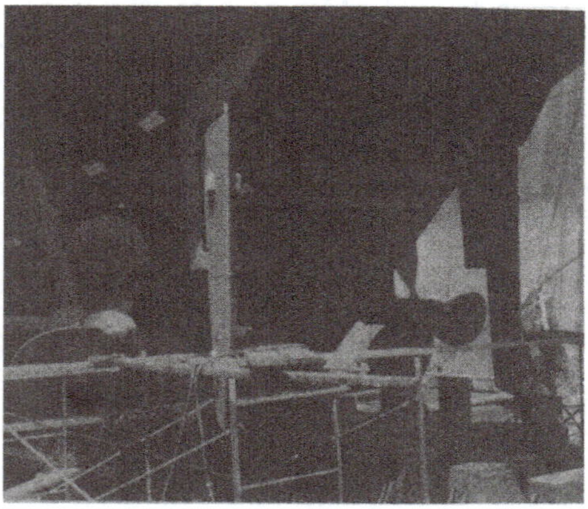

Problem: Propeller singing is constant frequency sound generated at particular rotating speeds of ship propellers. This is generally due to von Karman vortices off the trailing edge of the propeller, interacting with one or more of the complex higher natural modes of vibration of the propeller blade.

Measurements: Using a sound and vibration analyzer located inside the ship where the noise was strong, it was found that a strong tone of 121 Hz frequency was generated when two propellers, each having three blades, were operating with engine speeds from idling at 120 to 155 RPM and when only one propeller was operating over the engine RPM range 120 to 145 RPM. It was also noticed that a constant frequency sound at 275 Hz was generated when two engines were driving the ship at 190 - 200 RPM. The 121 Hz tone was observed when the ship was coasting down from full speed after the shafts were stopped by brakes.

Cure: Because singing (at defined frequencies) had been observed on earlier trials, the traditional cure for singing, sharpening the trailing edge of the blade, Sketch A, had been applied.

After investigating other possible sources of self-excited noise, the trailing edge was shaped as in Sketch B. The constant frequency noises are no longer noticeable.

Expanded Cylindrical Section

At 0.7R = 860 mm, Chord = 735 mm, Maximum thickness = 36 mm.

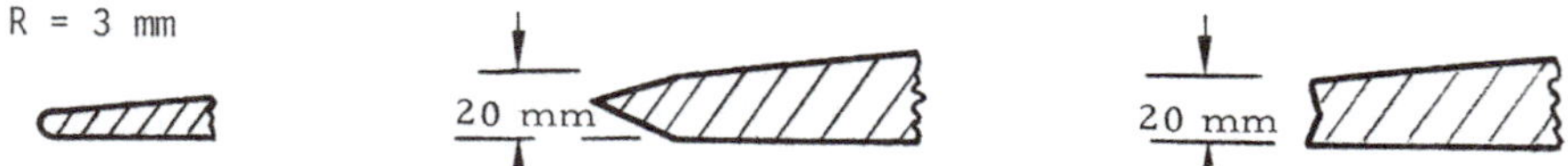

Initial Trailing Edge Modification A Modification B

PROPELLER SINGING
2000 HP TWIN-SCREW OCEAN TUG INTEGRATED WITH BARGE

The JAMIE E. BAXTER is a 7000 Horsepower Twin-Screw Ocean Tug which is used to propel a large barge to which it is rigidly attached. The tests were made without the barge.

Problem: Constant frequency noise from trailing vortices coupling with blade vibration.

Measurements: Measurements were made at various shaft speeds with a General Radio Sound and Vibration Analyzer, for which the microphone was located in an empty ballast tank directly above the port propeller. The propeller "sang" at a variety of frequencies over the whole operating range. The noise patterns at different propeller speeds are shown in the graphs at the bottom. It is believed that the large number of self-excited frequenices is due to the unusually thin propeller blade.

Cure: The trailing edge of the propeller was sharpened to the double "V"; tug is now quiet at all speeds.

At 0.7R = 1829 mm
Chord, C = 1433 mm
Max. thickness = 70 mm
Trailing edge:

Original, 3mm R
Modified, 10-12 mm

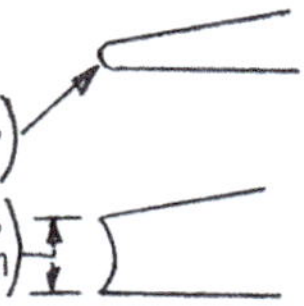

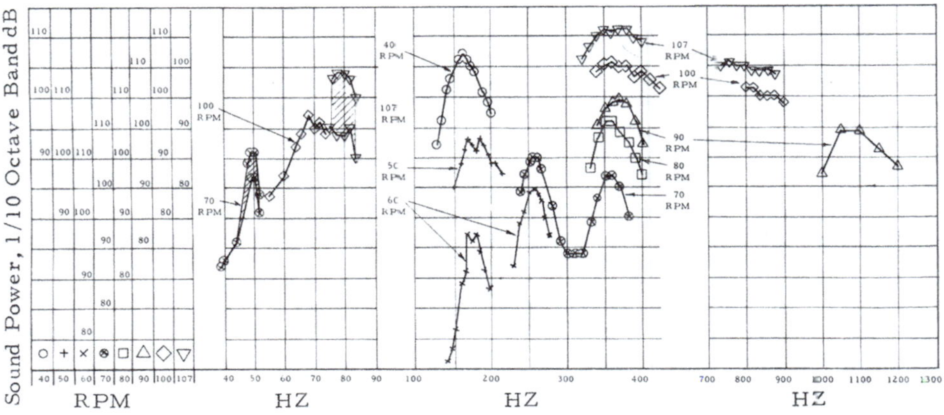

D4 PROPELLER EXCITING FORCES ESTIMATED FROM MEASURED ACCELERATION OF VERTICAL HULL VIBRATION OF SHIPS

Professor Toyoji Kumai
Technical Adviser

Mitsui Shipbuilding and Engineering Co. Ltd.
Tsukiji 5-6-4, Tokyo, Japan

SUMMARY

Propeller exciting forces are estimated using response factors, and compared with measured acceleration of vertical vibration at the stern of the upper deck during trial trips. Consideration is given to the difference between propeller exciting forces in four- and five-bladers.

1. INTRODUCTION

So far as 20,000 - 40,000 D.W.T. bulk carriers with five bladed propellers are concerned, there have frequently arisen severe fore and aft vibrations at the flying bridge. It seems that one of the causes of such vibration is resonance of the vertical main hull at the propeller blade frequency. On the other hand, the same or similar bulk carriers of four bladers are unlikely to exhibit such hull vibration and other local vibrations.

Many calculations of the propeller exciting forces of ships have been carried out and mainly confirmed by model tests. However, the propeller exciting forces can be approximately estimated from the measured vertical acceleration at the end of the upper deck on board ships during their trial trips, provided that the blade frequency response factors of the similar ships are obtained from the results of exciter tests.

In the present paper, the vertical exciting forces at the end of the hull in several four- and five-bladers are estimated by the above method using response factors. It is of interest that the vertical exciting forces in five-bladers are greater than that in four-bladers. One of the reasons for the difference in vertical exciting forces of four- and five-bladers is assumed to be due to the vertical wake bearing moment of the propeller in even and odd number bladers with conventional stern.

2. EXCITER TESTS, RESPONSE FACTORS AND FORCE ESTIMATES

In the full scale exciter tests for obtaining the response factors of the vertical vibration of ship hulls, the exciter was located near the after end of the upper deck. Acceleration pickups were provided for detailed measurements of the hull vibration at the higher frequencies. The results have been normalized in the form of the response factor, a/F, where a is the acceleration in cm/sec^2 measured at the position of the exciter, and F is the exciting force in tons. The response factors are recorded at the service blade frequency for each ship.

From the data of response factors mentioned above, the factors near the blade frequency of each ship are plotted for corresponding displacement as seen in Fig. 1. Since the response factor of the hull vibration at a given point is approximately inversely proportional to the displacement of the ship, an empirical formula will be given for obtaining the response factor:

$$\frac{a}{F} = \frac{20 \pm 6}{\Delta \times 10^{-4}} \quad , \qquad (1)$$

where, Δ = displacement in ton

The upper and lower limits of the above response factor give a meaning to the hump-hollow characteristic of the response factors of ship hull vibration (Ref.[1]).

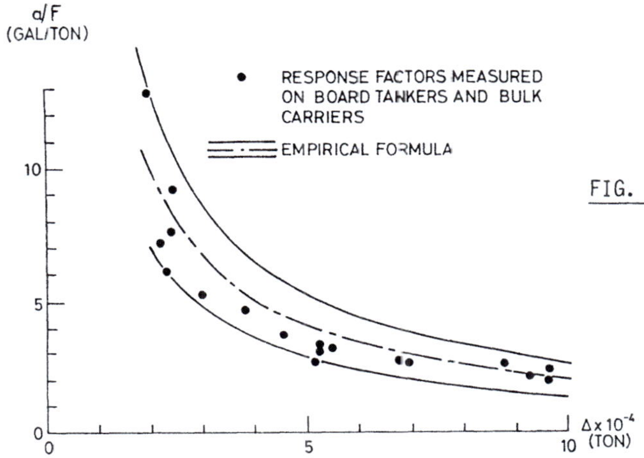

FIG. 1 RESPONSE FACTORS AT THE EXTREME STERN IN THE NODAL VIBRATION NEAREST TO THE BLADE FREQUENCY MEASURED FROM EXCITER TESTS ON BOARD SHIPS AND EMPIRICAL EXPRESSION (Ref.[1])

To obtain the exciting force F from the response factor, it is necessary to measure the acceleration a* at the stern of the hull during a trial trip of the ship at the same blade frequency as that of the exciter tests. The exciting force F is then obtained by dividing a* by a/F.

The mean value of acceleration at the after end of the deck measured on trial trips of sixteen bulk carriers with a four-bladed propeller is $a^* = 11$ cm/sec^2 at the blade frequency, whereas the average of measured acceleration in eight tankers and five bulk carriers with a five-bladed propeller is $a^* = 41$ cm/sec^2. Such accelerations measured on trial trips as mentioned above are approximately replaced by the exciting force F at the end of the hull using the formula (1). The force F will be considered to be the vibratory propeller exciting force composed of the wake bearing force and moment and the surface force at the blade frequency.

Fig. 2 shows propeller exciting forces per HP x 10^{-4} versus displacements of ships obtained from the above method. As can be seen in Fig. 2 it is of interest the mean value of the propeller exciting force in five-bladers is about five times greater than that in four-bladers.

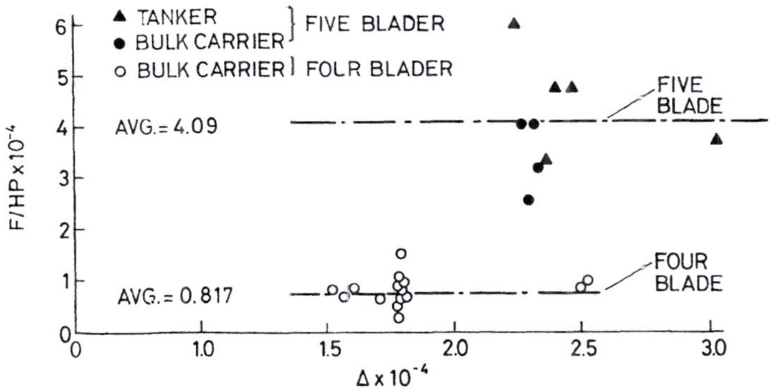

FIG. 2 PROPELLER EXCITING FORCES OBTAINED FROM MEASURED ACCELERATION ON TRIAL TRIPS

3. VERTICAL COMPONENTS OF PROPELLER EXCITING FORCES

The vibratory propeller forces mainly consist of the wake bearing force and the moment and surface forces. The bearing force and moment are induced by the ship wake distribution at the propeller disc. The vertical components of the moment and force are represented by Fourier components of the wake force distributions [2]:

$$M_v = M_o \frac{1}{2}(a_{z-1}+a_{z+1}) \cos z\omega t, \qquad (2)$$
$$P_v = P_o \frac{1}{2}(b_{z-1}-b_{z+1}) \sin z\omega t,$$

where,

M_o = mean thrust moment or $T_o r$

P_o = mean torque force or Q_o/r

r = 0.7R, R = radius of propeller

z = number of propeller blades

$a_{z\pm1}$ = $(z\pm1)$-th order Fourier components of thrust distribution curve at 0.7R

$b_{z\pm1}$ = $(z\pm1)$-th order Fourier components of torque distribution curve at 0.7R

Table I shows, as an example, Fourier components of thrust and torque curves of the propeller disc at 0.7R in the wake of the T-2 Tanker, based on the quasi-steady method.

At the end of the stern tube, the vibratory reaction due to a vertical moment applied at the center of the propeller disc is considered to be M_v/c by static beam theory, where c is the length of the stern tube. Then the total vibratory vertical force applied at the end of the stern tube is:

$$P_{vT}+P_v = \frac{T_o r}{2c}\{(a_{z-1}+a_{z+1})^2 + \beta^2(b_{z-1}-b_{z+1})^2\}^{1/2}\cos(z\omega t-\phi), \qquad (3)$$

where,

$P_{vT}=\frac{M_v}{c}$=reaction at the end of stern tube due to vertical thrust moment

P_v= vertical force component of torque force

$z\omega$= blade frequency

$$\beta = \frac{P_o c}{M_o} = \frac{Q_o c}{T_o r^2}, \quad \tan\phi = \frac{\beta(b_{z-1}-b_{z+1})}{(a_{z-1}+a_{z+1})}$$

TABLE I FOURIER COMPONENTS OF THRUST AND TORQUE CURVES AT 0.7R IN T-2 TANKER(Ref.[3])

z	a_z	a_z
0	0.9501	0.9744
1	0.2235	0.1703
2	0.2378	0.2028
3	0.0028	-0.0132
4	0.1152	0.0926
5	-0.0258	-0.0286
6	0.0720	0.0552
7	-0.0276	-0.0281
8	0.0521	0.0379
9	-0.0271	-0.0221
10	0.0373	0.0293
11	-0.0262	-0.0182
12	0.0288	0.0230
13	-0.0229	-0.0153

Since β and ϕ in the above expression are small compared with unity, the vertical vibratory force is approximately represented by

$$P_{vT}+P_v = P_{vT} = \frac{T_o r}{2c}(a_{z-1}+a_{z+1}) \cos z\omega t, \qquad (4)$$

The vertical force and the moment in a four-blader is excited by the 3rd and 5th orders of the wake harmonic variations, whereas a five-blader is excited by the 4th and 6th. The wake of a single screw ship has strong even harmonics, and weak odd orders, because of the strong top-to-bottom symmetry of the flow in the centre-plane aft of skeg region as shown in Table I. Thus, as the 4th order wake is larger than the 3rd order, the vertical moment of a five-bladed propeller is expected to be greater.

Numerical results obtained by the author [2] from the evaluation of a quasi-steady theory for the vertical moment on a propeller, in the wake of a T-2 Tanker, have been confirmed by model measurements reported by Van Manen et al [3]. For ships with conventional sterns, one can expect that the ratio of blade frequency vertical moment for a four blader to that for a five blader will be on the order of 1:8.

With regard to the hull surface force induced by the propeller, the vertical hull vibration is excited by the tip clearance force. Lewis' [4] results, obtained by measurements for the surface forces for a four-bladed propeller are 2.5 times larger than those for a five-blader on a fine-lined ship. The effect of cavitation on the surface force is, however, not considered in the present paper.

4. DISTRIBUTIONS OF BEARING FORCE AND SURFACE FORCE IN FOUR AND FIVE BLADERS

Since the ratios of the total propeller exciting forces in four- and five-bladers have been obtained, and the ratio of bearing moment and that of the surface force for both four-blader to five-blader have been assumed, the distributions of the wake bearing force induced by the moment and the surface force in the four-blader and five-blader can be estimated as follows.
If we put F_4, F_5 as total exciting forces in four and five bladers respectively, B_4, B_5 are vertical bearing forces induced by bearing moments in four- and five-bladers, respectively, and S_4, S_5 the surface forces in four- and five-bladers respectively, then we have the relation

$$\frac{F_4}{F_5} = \frac{B_4+S_4}{B_5+S_5}, \qquad (5)$$

From the above relation, the ratios S_z to B_z are obtained as

$$\frac{S_4}{B_4} = \frac{B_5/B_4-F_5/F_4}{F_5/F_4-S_5/S_4}, \quad (6) \qquad \frac{S_5}{B_5} = \frac{F_4/F_5-B_4/B_5}{S_4/S_5-F_4/F_5}, \quad (7)$$

The ratio of the propeller-induced forces in the four-blader to five-blader, for the case of 20,000 - 40,000 D.W.T. tankers or bulk carriers is F_4: F_5 = 1:5. This is the average value from the observations of vibratory accelerations at the stern of ships, shown in Fig. 2. The ratio B_4: B_5 = 1:8.14 is obtained by means of a quasi-steady method using Table I and eqn. (4), and S_4: S_5 is assumed 2.5:1 from Lewis' experiments. Then the distributions of B_z and S_z are obtained approximately by using eqns. (6) and (7), as given in the following table.

TABLE II DISTRIBUTIONS OF BEARING FORCE AND SURFACE FORCE INDUCED BY FOUR AND FIVE BLADED PROPELLERS

z	F_z	B_z	S_z
4	1	0.60	0.40
5	5	4.84	0.16

It is to be noted that large magnitude of the bearing force dominates the contribution to the vertical force for the five-blader with conventional stern, whereas the bearing force is a little larger than the surface force in the four blader.

5. CONCLUSIONS

The propeller-exciting forces have been estimated from measured acceleration at the stern of ships during trial trips. The exciting forces of ships with four-bladed propellers are smaller than those with five-bladed ones. Since one of the causes of such differences in four- and five-bladers is considered to be that of the vertical bearing moment of exciting forces in even-number bladed propellers and odd-number bladed ones, the results of observations were confirmed by using the results of the quasi-steady calculation of the wake bearing moment. As a result, it is to be noted that selection of the number of propeller blades at an early stage in design is one of the important problems for preventing severe hull vibration of ships.

6. ACKNOWLEDGEMENTS

The author expresses his thanks to Dr. J.P. Breslin, Director Session V, IAHR-IUTAM Symposium, for his advice and his help with preparation of this paper. The author is indebted to Professor D. Rockwell for preparing his manuscript. Thanks are due to Mr. Y. Sugimura, Chief Hull Design Section, Mitsui Engineering and Shipbuilding Co. Ltd., for supplying the measured data and permission to mention the results for the present paper.

7. REFERENCES

1. Kumai, T., "Virtual Inertia Coefficients and Response Factors in the Higher Modes of Main Hull Vibration." RINA, Suppl. Paper, 120 (1978).

2. Kumai, T., "Some Aspects of Propeller Bearing Forces Exciting Hull Vibration of Single Screw Ship" Schiffstechnik, (1962).

3. Van Manen, J.D. and Wereldsma, R., "Propeller Excited Vibratory Forces in the Shaft of Single Screw Tanker." I.S.P. 7, 11, (1960).

4. Lewis, F.M., "Propeller Vibration Forces in Single-Screw Ships" Trans. SNAME., (1969).

E1 DYNAMICS OF OFFSHORE STRUCTURES: PROBLEMS, SOLUTIONS, AND BENEFITS

Dr. L.R. Wootton

Atkins Research and Development

Epsom, Surrey, England

1. INTRODUCTION TO THE PROBLEMS

It is very difficult, and certainly uneconomic, to design most forms of offshore structures for deep water sufficiently stiff or massive to be completely insensitive to dynamic loading. Indeed, for many of the most advanced structures in the North Sea, the dynamics dominate the main design procedures, and by careful and diligent means, the adverse effects are mitigated. In fact, there are relatively few unacceptable occurrences of dynamic response of offshore structures though the potential problem always exists.

At the beginning of the development of North Sea Oil Platforms, dynamics was considered an aside to the main design process. The sea state was represented by equivalent static loading and the dynamic characteristics of the jacket only considered with respect to a (small) magnification of these loads. This was perfectly adequate as a design procedure of the BP Forties Field steel platforms in about 160 m water depth and for the Brent Concrete platforms installed in about the same water depth. Later, structures were designed for deeper water and it became necessary to consider dynamic properties and, particularly, their effect on fatigue life at the outset of the design process. The Conoco Murcheson steel jacket in 156 m of water represented an important step forward in this respect. At the same time, a number of compliant structures, designed to respond significantly to the waves, have been built and operate successfully. The major design problems associated with dynamics, therefore, that face an offshore structural designer are:

- The influence of the dynamic characteristics of fixed steel jackets on the fatigue life and the ability to withstand large wave loads.
- The dynamic response of floating moored rigs, buoyant towers, and other compliant structures to random sea states.
- The dynamic response of tethered buoyant structures in waves, with particular reference to the design of the moorings.
- The response of risers, moorings, and other tubular elements to vortex-induced forces in currents and waves.

In brief, the problems associated with dynamics are primarily those associated with analysis and the associated restrictions placed on design.

2. METHODS OF ANALYSIS

One of the earliest dynamic analyses carried out on a North Sea structure was that for the BP Forties Field and reported in Ref. 1 in 1975, in which the structure was represented by a single line of lumped masses and springs. This represented sway modes in the direction of the principle axes only; the torsional modes were not considered. Since that analysis was carried out, there have been enormous developments in techniques both in terms of modelling of the structure in three dimensions and in the representation of the wind and wave loads. There are now basically three techniques:

- Small dynamic effects: The natural frequencies and their associated modes of structure are calculated and found to be far removed from the frequencies of the main sources of energy input.

- Time domain solutions, in which representative time histories of the forces are fed into a simulation of the structure.
- Frequency domain solutions, in which the analysis is solved in the frequency domain.

The frequency-domain and time-domain methods have recently been critically examined by Vugts and Hayes. Summarising, the features of the two approaches are as follows:

a) Frequency-domain method. Constant coefficients such as mass, and frequency-dependent coefficients such as wave forces can be easily handled; non-linearities in the system such as soil stiffness or damping have to be represented by linear relationships; only the steady-state solution is obtained and this is derived directly the response of the structure and of its components to any particular form of input spectral force can be obtained easily; transient effects must be analysed by use of response-spectra techniques in which the envelopes of many time histories have previously been studied; it is not possible to derive phase relations between the responses in various modes; in general, frequency-domain methods are economic in terms of computing effort; water/structure interaction effects cannot be represented.

b) Time-domain methods. Constant coefficients can be treated simply using differential equations; frequency-dependent coefficients may also be accomodated; non-linearities can be represented realistically; local effects on the structure can be studied; transients, such as earthquakes and wave grouping can be studied by use of specific examples but it must be recognised that many examples may have to be taken to ensure that the envelope of response is obtained; steady-state solutions are obtained by running the time history for long enough: this may involve a large number of cycles; the accuracy of the result depends on the time interval used in the integration and great accuracy is likely to be very time consuming in computing effort; calculating the response of the structure to various forms of random excitation is likely to be a very expensive exercise in computing effort, - water/structure interaction effects can be represented.

It is worth noting that great analyses using the numerical techniques may not be justified because of the nature of some of the input properties. In the case of danamic response, the dominant factors are soil properties, mass, and damping. Often the soil properties are only known to a cursory extent at the time when the structure is being designed and analysed. When the installation has been started and piles are being driven or the gravity base being monitored, more reliable data is available and the structure can be reanalysed. The lack of precision in the soil properties has an overwhelming effect on the dynamic analysis of both steel piles and gravity base structures. Regarding the mass, there are two main components of mass that are difficult to predict, the added hydrodynamic mass (mainly due to the unknown influence of marine encrustation) and the deck load which inevitably increases throughout the design and installation phases and which cannot easily be measured. Concerning the damping, the combined and complex effects of soil, structure, and hydrodynamic damping are very difficult to estimate and resort must be made to global measurements to not discern the form of the individual components nor do they generally relate to large motions.

Despite these problems, large-scale dynamic analyses are carried out as an integral part of the design procedure (as described in Ref. 4, for example) and great reliance is placed on the results in determining the size of the members and the strength required of node assemblies. In the case of compliant structures, such as a Tethered Buoyant Platform, the dynamic characteristics dominate the overall design layout. The possible forms of motion include those where there is substantial water/structure interaction and which cannot be represented by linear simulations or in physical model tests and there must remain an element of doubt as to the validity of some of the forms of dynamic response suggested theoretically when no full-scale confirmation is available. This is the case for subharmonic oscillations of floating structures (i.e., natural-period motions in heave, pitch, and roll occurring in waves of quite different frequencies).

Perhaps one of the most important problems that cannot at present be analysed is that of buoys with a moored tanker. This problem is not particularly amenable to physical model testing because of scaling and the difficulty in reproducing a simulation of a random-sea state over a large area of a test tank. A time-history mathematical model simulation of the mooring buoy and its interaction with the tanker is technically feasible but practically impossible because of its demands in computing. As a result of this deficiency, the operational limits of tankers at particular moorings have to be set by experience during service. This may be simply based on visual observations or on measurements of the loading in the cables and the relative response of tanker and buoy (Ref. 6).

The tanker/buoy interaction problem also includes the influence of the relative motions of the structure in the water. In general this effect is not considered directly but, rather, is included in the assessment of the overall damping. Even for fixed structures this is inaccurate at very great water depths. The main problem is that if the relative-motion effects are significant then it is not at all obvious which sea states will cause the greatest stresses and loads. There is much activity at present in developing economic frequency-domain methods that will search for the worst sea states (Ref. 7).

Regarding the practical side, there has been a great deal of work on the use of measurements of vibration characteristics to detect failure of steel jackets (Ref. 9). Such schemes depend on a change in the natural frequencies and modes of the structure when an individual member looses stiffnes, due to a crack, or is completely severed. Although schematically possible, no such scheme has yet reached a level of reliability that justifies its use as a replacement for underwater inspection. The potential economic benefits of vibration monitoring techniques are very great but the technology has yet to be achieved. The real benefit that accrues from dynamics lies in proper integration into the design process. Often, the dynamic characteristics are analysed but the basic structural form is optimised in terms of static design loads. Only recently has it become possible to consider dynamics as part of the total design process.

3. REFERENCES

1. Gray, R.M., Berge, B. and Koehler, A.M., "Dynamic Analysis of the North Sea Forties Platforms", Offshore Technology Conference paper 2250, Houston, Texas, 1975.
2. Vugts, J.H. and Hayes, D.J., "Dynamic Analysis of Fixed Offshore Structures: A Review of Some Basic Aspects of the Problem", Engineering Structures, April 1979.
3. Ruhl, J.A., "Forced Vibration Tests on a Deepwater Platform", Offshore Technology Conference paper 3514, Houston, Texas, 1979.
4. Hallam, M.G., Heaf, N.J., Boudreaux, R.H. and Millman, D.N., "Dynamic and Fatigue Analysis of Mucheson Tower Structure", Offshore Technology Conference paper 3163, Houston, Texas, 1979.
5. Rainey, R.C.T., "The Dynamics of Tethered Platforms", Royal Institute of Naval Architects, Vol. 120, 1978.
6. Stewart, W.P., "Single Point Mooring: An Evolving Science", Dock and Harbour Authority, Vol. LVIII, no. 678/9, May and June, 1979.
7. Rainey, R.C.T. and Fish, P.R., "The Importance of Structural Motion in the Calculating of Motion of Work Loads on an Offshore Structure", to be published in Engineering Structures Journal.
8. Brown, D.R. and Huckvale, S.A., "Monitoring and Structural Response of Steel Offshore Structures", Europec Conference, London, October 1978.

E2 WAVE-INDUCED VIBRATION OF OFFSHORE STRUCTURES

T.L. Shaw

formerly of

University of Bristol, U.K.

SUMMARY

There exist few well documented records of the mechanics of wave-induced loading on structures in the sea. However, this subject has enormous economic consequences for many maritime countries, and is of world importance in the hydrocarbons market.

In order to consolidate and advance the state of knowledge of the fundamentals of wave action on the slender cylinders that make up many critical components of offshore structures, an international exchange of views formed from field, laboratory, and analytical experiences was timely in 1978. This Paper reviews those findings reported to that meeting that refer to practical experiences of some of the more important mechanisms involved.

Particular emphasis is now required for the function of vortex shedding from slender cylinders in waves and currents because, as in uni-directional flows, this is probably the principal mechanism that permits waves to induce these cylinders to vibrate. However, the coherence of periodic wakes in wave motion is basically much weaker, though the relative strength of currents and the orientation of cylinder axis to the vertical are important factors.

Information on these and related topics of importance is assembled in this Paper. There now exists the possibility of prescribing when wave-induced vibrations are possible, which should guide the many detailed studies remaining to be carried out before an adequately comprehensive understanding of this aspect of fluid loading is achieved.

1. INTRODUCTION

The study of the mechanics of wave-loading on offshore structures is relatively young compared with wind loading on land-based structures. Also, data recording is more difficult and expensive, and is in many cases more commercially sensitive as a consequence. Overall, therefore, the breadth of "practical experiences" is heavily biassed towards laboratory studies, though some comprehensive data of scientific value is now becoming available from a few well chosen installations.

It must be stressed immediately that offshore structures are seldom most seriously threatened by the action of waves alone, but rather by the combined effect of waves and currents (and perhaps also storm winds acting directly on superstructures).

Thus whereas even wave action will usually create hydrodynamically more complex pressure and force distributions over the slender cylindrical members used extensively for many offshore structures than are associated with wind loading on equivalent land-based structures, the addition of currents and winds, together with (in all cases) the occurrence of structural vibrations as the common mechanism for ultimate failure, establishes such complex motions as to make the comprehensive understanding of the important physical details still well beyond our grasp.

Nevertheless, the subject of wave loading obviously bears many similarities with that due to uni-directional flows, and it also has important economic overtones because of the dependence of the international market for hydrocarbons upon it. This has given the momentum to resolve problems more quickly than ever occurred for any type of land-based structure. (Even the principal phases of research activity that took place on suspension bridges, guyed towers and transmission cables were modest by comparison with the effort that has emerged in the last 10 years or so for offshore structures).

For these many reasons, a Symposium was held in Bristol in September 1978 under the title "Mechanics of Wave-Induced Forces on Cylinders". Although it was intended that the Symposium should concentrate on the physical processes of wave loading, it was recognised as valuable to include a number of presentations related directly to the practical experiences of appropriately detailed studies in both the laboratory and the field, though because of the physical complexity of the offshore loading processes referred to earlier, the emphasis of these experiences has been much more, though happily not totally inclined towards the small and sometimes misleading scale of the laboratory. These particular presentations are reviewed in this Paper. They are reported in full in the 768 pp Proceedings of the Symposium, published in 1979 by Pitman Publishing Ltd (1).

2. PRACTICAL EXPERIENCES OF WAVE LOADING FROM IAHR SYMPOSIUM, 1978

The first Paper to be considered is the lead Review Paper to the Symposium by Pearcey entitled "Some Observations on Fundamental Features of Wave-Induced Viscous Flows past Cylinders" (pp 1-54 of Ref.). In the case of wave loading on cylinders there immediately arises a second time factor in the list of relevant parameters, that is in addition to the period of vortex shedding there must be added the wave period. The similarity between these two periods may readily be seen to be relevant to whether a classic quasi-unidirectional flow wake will develop ($T_{wave} >> T_{vortex}$) or whether the wake will be of a form peculiar to wave-related flows. A range of intermediate wake forms is, of course, also possible.

Pearcey explains in detail "the shedding, convection and interaction of discrete vortices on all elements and components of discrete vortices on all elements and components of the forces. Many of the differences that occur for differently orientated cylinders seem to be associated with differences in the vortex effects". From Pearcey's explanation it is clear that even the wake structure of simple cylinders in regular waves holds as yet unresolved mysteries, which reflects on the following statements in the Preface to the Proceedings:

"Though much of the physics of wave and current loading on offshore structures bears closely on that for uni-directional (e.g., wind) flows, their many important differences make casual comparisons dangerous. The principal distinction is that because the fluid motion in waves has a circulatory form, the wake created during one half of a wave period becomes part of the incident flow during the next, and so on."

"The subject is further complicated by the fact that natural sea waves include a range of frequencies and heights, that they are seldom symmetric and may be breaking, that component waves are usually not directionally similar, and that their kinematics are nonlinearly affected by the presence of a current. The additional fact that wave-induced velocities decay roughly exponentially with depth (rather than with height from the ground as for uni-directional flows), suggests that the important load-creating wake forms for sea structures will be more three-dimensional, hence more difficult to predict". A similar statement of the now generally recognised complexity of this process is made in Ref. 2, and it is the methodical disassembly of the topic that forms the theme of Pearcey's Paper.

Much of the uncertainty about wave action on structures relates to the nature of the wake of slender rather than massive cylinders, for which vortex shedding may be expected to be related to that occurring in uni-directional flows. It remains to calculate for such flows the steady and fluctuating drag and lift coefficients as functions of both Reynolds number (based, say, on the maximum orbital wave velocity normal to the cylinder's axis), and of a parameter relating the periods of wave motion and vortex shedding (generally taken to be the 'Keulegan-Carpenter' number, K_c). Pearcey's approach to achieving this is based firstly on the "significant overlap between the set of viscous flow phenomena involved in periodic oscillatory flow and that for the steady, uni-directional flow". The limitations of attempting to demonstrate the wake structure of cylinders in orbital wave flows by experiments in planar oscillatory flows normal to the cylinder are now better understood, but the technique has value as Pearcey shows.

However, one of its limitations is that, in addition to the generally incorrect flow direction, it often gives data for motions that are unusually well correlated along the axis of the cylinder. Although this need not imply that the data so deduced are totally misleading, their application to the calculation of forces in waves must recognise that there is then much reduced correlation, i.e., only a short element of the cylinder is actually being represented. In terms of total forces, therefore, and more particularly the distribution of the component parts of these along the cylinder in space and time, the argument in favour of determining forces from an array of pressure sensors is particularly strong. (The same argument is, of course, also often advanced as justified for many studies in uni-directional flows, but because the general nature of these is both less complex and better understood than in the case of wave flows, the reasons for 'total force' measurements are much stronger).

Pearcey demonstrates from both laboratory and field measurements of pressures over cylinders that "All the evidence indicates a broad similarity between.......the two cases" (orbital and planar oscillatory flows), but that "There are important qualifications, notably that measurements of forces in the drag regime in waves are characterised by relatively large degrees of scatter and that there are still relatively few definite data for post-critical conditions. This sparsity is largely due to the difficulty of achieving post-critical conditions in laboratory tests combined with the fact that the scatter referred to above is especially evident in measurements made in real seas", (another well-known difficulty for the experimenter).

Pearcey therefore concludes that "the most satisfactory way of formulating design data.......will turn out to be in terms of statistical averages combined with defined variabilities or distributions, with average values and distributions depending on sea state as well as on other parameters such as K_c and Re.

A process more common to cylinders in the sea is that their surface roughness may change in time, often quite quickly. Pearcey notes that these typical changes can influence forces to a large degree. It is, of course, well known that a change in roughness will alter the range of Re values over which the transition from sub- to supercritical wake flow conditions occurs. However, predicting this range in waves is already difficult because of the inevitably three-dimensional and spatially varying nature of wake flows : the additional uncertainty due to changing roughness only magnifies this problem.

Pearcey goes on to explore in detail the components of the flow patterns in wave motion about horizontal, vertical and inclined cylinders. Experimental evidence is called upon where this is available to justify the detailed and informative conclusions about the velocity components hence the overall form of the wakes that may be expected in each case. This leads the author to state that "the shedding, convection and interaction of discrete vortices in turn dominates the viscous effects and differs substantially between one case and another of cylinder orientation". In consequence, Pearcey proposes that the orthogonal components of the resultant force on the cylinder may, in each case, be derived by linearly summing, for each element of the cylinder, the hydrodynamic force and its variations that would exist in potential flow, together with the effects of both viscosity and the influences associated with vortex shedding. This unusual approach adds to the clarity of the author's presentation of the hydrodynamic fundamentals and the associated force structure of wave action on cylinders. It should guide the thinking still needed before all the necessary studies of wave loading have been carried out. Even so, the author concludes by stating that the observations that he has been able to deduce from available evidence apply more certainly to sub-critical than to post-critical conditions; to sub-critical and post-critical plateau values and not to the intermediate transitional ones; to high Keulegan-Carpenter numbers (i.e., the drag regime) but not to the regime where the quadrature (inertia force) terms assume a major significance alongside the in-phase (drag) ones; to waves or currents separately, but not necessarily together and not necessarily to other modes of relative motion; to smooth cylinders but not to rough; to regular long-crested waves but not necessarily to irregular short-crested ones; to vertical and horizontal cylinders but not necessarily to those with other inclinations; to rigid, stationary but not necessarily to flexible, compliant cylinders; to isolated cylinders but not necessarily to those in groups or frames.

3. FIELD EXPERIENCES

Clearly, therefore, although the real problems remain to be resolved, the techniques whereby this can be achieved may now be available. Methodical investigations of specific topics are now needed, such as are given in many of the Papers to the Sympsoium. The report by Bishop (pp. 334-345) of experiments in the sea using equipment designed specially and exclusively for this purpose demonstrates how rapid progress may be achieved.

The equipment used by Bishop includes two vertical cylinders, of different diameters, which are instrumented to record local pressures, elemental forces and flow velocities along their lengths. Bishop argues that it is necessary to record the fundamental hydrodynamic processes in detail because although "statistical methods.......afford a good general assessment of the magnitude of the forces and their relations to the wave climate......., they do not relate individual force to the wave which causes it and so they may not be so well suited to predicting the maximum force for the design wave. Individual wave analyses are theoretically more suited to this and they are more appropriate as attempts are made to understand the hydrodynamics of the loading process. They do, however, suffer from problems with regard to sensitivity to errors in the experimental and analysis procedures. For instance, phase errors due to spatial separation of the force and wave measuring sensors can lead to errors in the derived force coefficients. Also, the method depends on accurate measurement of the detailed shape of the particle motion and force time histories. These problems are more acute in real sea conditions than they are in laboratory tests".

The analysis technique therefore developed by Bishop has been applied to the field data recorded during two test runs, and the results are presented in the Paper. It is worth stressing that each test run lasted for 20 minutes and although they were only 3 hours apart, the "current magnitude and direction had changed substantially" between them. (In fact, velocity measurements elsewhere show that major variations can occur on the timescale of minutes). Each set of data is analysed to give averaged drag (C_D) and inertia (C_M) force coefficients over different periods of time. Figure 1 (of the present Paper) shows how the choice of time-averaging period Δt_i influences the value of these coefficients for wave action on the more slender of the two cylinders, and how each value fluctuates as the scanning period moves forward in time. Compared with the equivalent results from the larger cylinders there is more variation in the values of the coefficients with Δt. Bishop explains how this variation may in part be attributed to the prevailing environmental conditions, and also to the problems of analysing a process that is sensitive to small data errors. The simultaneous action of currents is an additional influence on the pressure and force records that is difficult to allow for precisely.

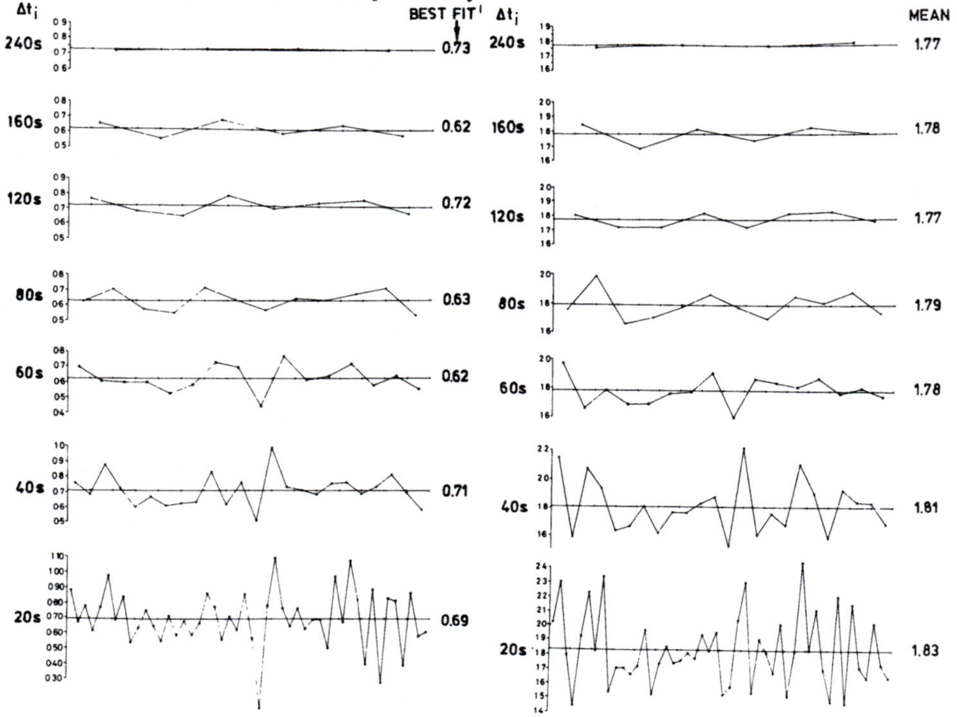

FIG. 1 EFFECT OF CHOICE OF Δt_i ON C_D AND C_M

Bishop therefore observes that the next step forward must be "to quantify the errors due to the experimental and analysis techniques at low values of Δt_i so that the remaining variation of the coefficients can be more confidently attributed to the hydrodynamic process".

A second Paper on field experience on wave loading on offshore structures reported results recorded off the coast of Australia. The Paper by Haritos, Sharp and Stevens ("A Study of a Simple Pile Structure in an Ocean Environment", pp. 346-358 of Ref. 1) deals with a cylinder of 0.41m diameter in approximately 13m depth of water. The instrumentation includes six wave height recorders, velocity probes and accelerometers on the structures.

Figure 2 "presents a sample set of traces for the resolved velocity components alongwave u, acrosswave v, and vertical w, obtained from the first set of drag wires (DW) 1m below MWL (mean water level). Included is the trace for the corresponding wave height from the capacitance probe and traces for velocity components associated with DW3 and DW4. DW3 was at the 1m level and DW4 at the 1.75m level below MWL, both being aligned in the same direction.

"It is observed from the traces that the u and w components are out of phase with respect to each other while u and waveheight appear to be in phase. Component v is observed to be of smaller relative magnitude compared with u or with w which appear to be of comparable size. In addition the form of the trace for the v component does not show the same characteristic cyclic variations as are observed in the u and w traces".

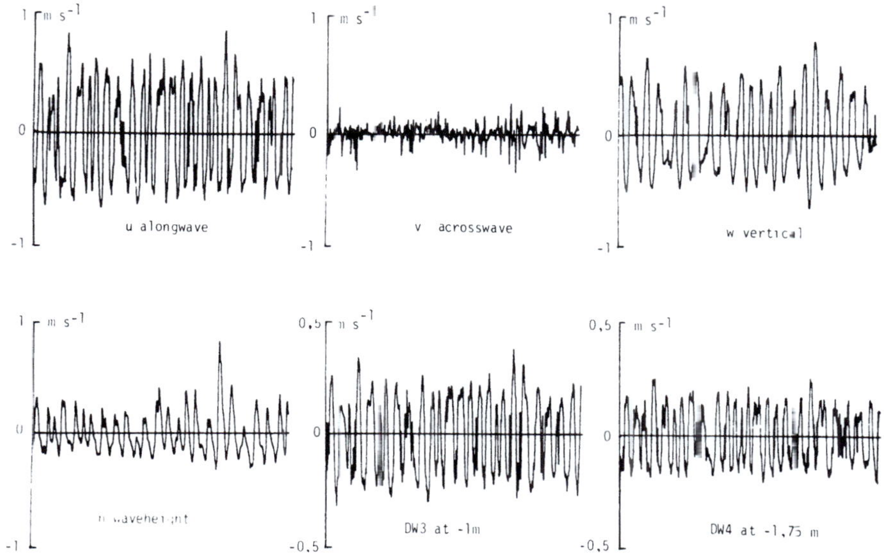

FIG. 2 SAMPLE SET OF VELOCITY COMPONENT MEASUREMENTS

The authors then compare their observed data for in-line transverse vibrations with the predictions of a mathematical representation. They conclude that in their particular case the damping ratio for transverse response is only 1% compared with nearly 6% for the in-line response.

4. LABORATORY STUDIES

From the many Papers reporting experimental data from laboratory studies, the paper by Matten, Hogben and Ashley entitled "A Circular Cylinder Oscillating in Still Water, in Waves and in Currents" (pp. 475-489) refers to particularly large-scale investigations. A cylinder of diameter 0.276m and length 1.6m could be towed along a wave tank and vibrated transversely at the same time. The pressures and forces on the cylinder were both recorded. Despite the title of this Paper, the content deals specifically with cylinder oscillations in currents. The authors are particularly concerned with the important problem of the effects on the imposed forces of strong axial correlation of wake motions during lateral vibrations. The tests reported include those made with and without end plates, and the results obtained are given in Table 1 (for Re = 0.27×10^6, K_c = 18). The authors conclude that "axial flows result in significant changes in the flow field around the cylinder and have a substantial effect on the resultant forces". However it is not immediately clear whether or not transitional Re effects may be significant here.

A related investigation was reported by Guilloud and Vignat, entitled "Motion and Deformation of a Pipe Submerged in Waves" (pp. 503-509). In this case the study relates to horizontal rather than to vertical pipes, and applies to the merits of the towing method of pipelaying as an alternative to welding sections at sea. A cylinder of diameter 0.762m was used in these tests, from which it is concluded (Table 2) that "large diameter pipes have been shown to stand up well to wave action".

The study by Yamamoto and Ciani ("Wave Forces on Armour Units for Submarine Cables", pp. 739-752) is of interest here because, like the Paper by Matten et al, it demonstrates the importance of axial correlation of fluid motions about a cylinder on the forces imposed on it. In this case, however, the cylinder is not prismatic; its outer profile includes frequent changes in section due to the flanges needed for the overall assembly of the armour units. For the whole structure, which in practice is close to the floor of the sea, it is shown that "The horizontal force increases 3 to 6 times as the flange angle increases from 0° to 90° (relative to the approaching waves). The vertical force increases up to 10 times as the flange angle varies from 0° to 45°". These results are, of course, related to the geometry of pipe and flanges, which is given in the Paper.

TABLE 1

PARAMETER	WITH END PLATES		WITHOUT END PLATES	
	Range	Mean	Range	Mean
Drag Coefficient	0.43-0.56	0.49	0.31-0.39	0.36
Added-Mass Coefficient	0.81-0.92	0.88	0.93-0.98	0.95
Overall Force Coefficient	1.01-1.07	1.05	1.08-1.11	1.09
Phase Angle between Max. Fluid Accn. and Max. Force	54° - 66°	60°	31° - 47°	41°
Lift Coefficient	0.73-1.50	1.30	0.47-0.71	0.59

TABLE 2

Tube (m)	0.254	1.016	0.508 B	0.762	0.762 B	1.016 B
Critical period of the pipe (sec)	8-9	10-12	8-10	12-14	10-12	12-14
Height at pipe failure (m)	10.5	12	8	16	9.5	13
Safe-load, limiting amplitude (m)	7.5	9	6	12	7	9.5
Relevant towing time (days)	6	7	6	6	7	8

A number of similarly specific studies are also reported in the Proceedings. Each includes both experimental and analytical work from which force and motion predictions are made for the various offshore structures considered in each case. These range from semi-submersible vessels to a cylindrical wave energy device. Of more direct relevance in the present context, however, are the results of several small-scale laboratory studies of wave action on simple cylinders, for example the presentation of White and Carr ("On the estimation of Morison coefficients in irregular waves", pp. 314-323). The so-called 'Morison' equation was proposed (in 1950) to give an expression for the instantaneous total force on an element of a structure according to the velocity and acceleration of the wave flow that occur at the cylinder at the time instant in question. Although criticised on many counts, this equation remains in wide use due to its relative simplicity and general verification in many contexts. The study by White and Carr concentrates on the value of the equation in the irregular wave climate of normal seas. Figure 3 shows their experimental results for inertia coefficients (C_M in irregular waves, C_M' in regular waves) when the imposed loading is dominated by (a) inertia, and (b) drag - i.e., at low and high K_c respectively. Despite the discrepancies apparent in the latter case, the authors conclude that it may well be possible to use "the wealth of information presently available from tests in irregular waves" for studies of wave loading on complex structures.

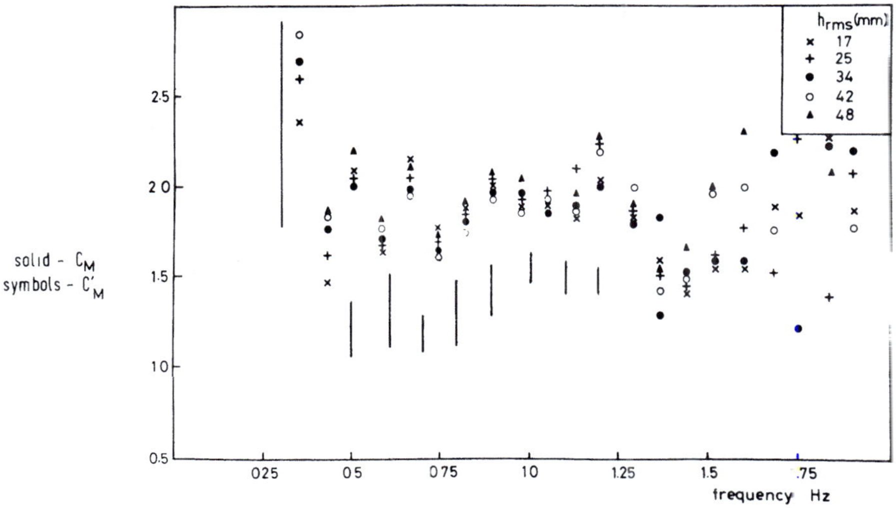

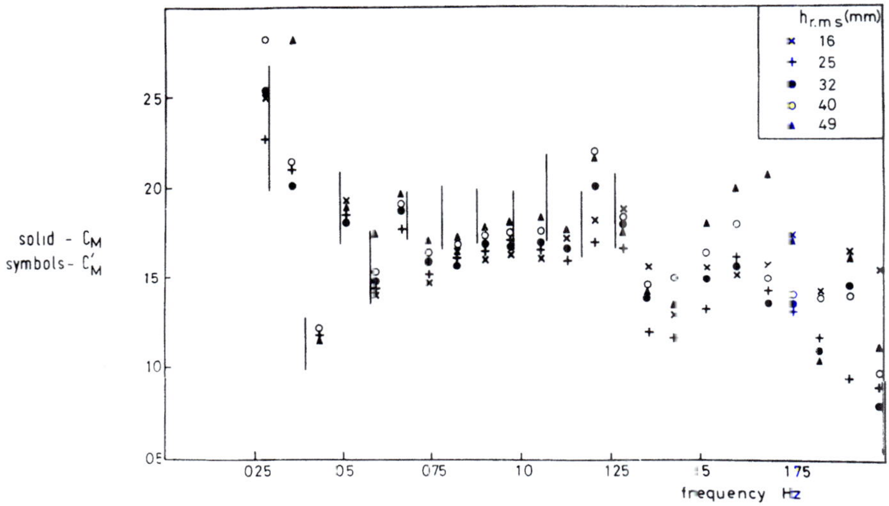

FIG. 3 INERTIA COEFFICIENTS IN IRREGULAR AND REGULAR WAVES WHEN LOADING DOMINATED BY INERTIA (TOP) AND DRAG (BOTTOM)

A similar conclusion is reached by Barnouin, Mattout and Sagner in their Paper "Experimental Study of the Validity Domain of Some Formulae for Hydrodynamic Forces for Regular and Irregular Flows", (pp. 393-405). This adds support for the results of tests in regular waves, for example that of Hutchinson (pp. 302-313) on instantaneous pressure distributions at different levels (z/d) around vertical cylinders (Figures 4a, b), where z/d is the proportion of the water depth downwards from the surface at which the pressure profiles are recorded.

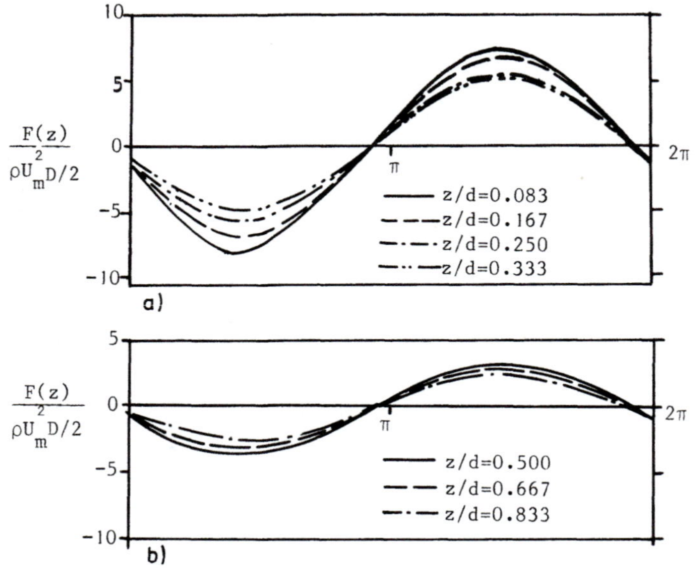

FIG. 4 PRESSURE VARIATIONS ALONG A VERTICAL CYLINDER IN A WAVE CYCLE

The complex situation of forces on partially submerged cylinders has applications to the members of structures lying in the wave zone, and to some types of wave energy device. The results of a theory developed by Dixon, Durrani and Greated (pp. 379-392) to describe this situation are compared with experimental data taken using irregular wave spectra. Although reasonable agreement is achieved, the authors state that "Work is continuing to improve the equations so that they are not so sensitive to the input parameters". This is, indeed, one of the most complex of all wave loading problems to understand, and unfortunately it often corresponds to the occurrence of maximum forces (from which the lesson is to minimise the number of structural components located in this zone).

5. SIMULATION OF COMPONENT HYDRODYNAMIC PROCESSES

It will be clear that there is a dichotomy between attempting to carry out fully representative studies of wave loading phenomena, and obtaining more readily, information from much simplified tests. The argument that simple studies can reveal the nature of important processes and so help with the understanding of comprehensive investigations lies behind the informative Papers by Grass and Kemp ("Flow Visualisation Studies of Oscillatory Flow past Smooth and Rough Circular Cylinders", pp. 406-420), and Bearman, Graham and Singh ("Forces on Cylinders in Harmonically Oscillating Flow", pp. 437-449). In both cases the orbital particle motion beneath waves is represented by a planar oscillatory flow, removing the input transverse hydrodynamic component. As Bearman et al show (Figure 5), the trend towards higher K_c gives C_D values that closely resemble those in uni-directional flow, i.e., the wake structures must then be closely similar. On this basis, for example, it is argued that the results obtained at lower K_c in planar oscillatory flows must be at least fairly representative of wave-induced motions, i.e., while the answers may not fully represent the axially varying flow induced by waves, they are not misleading and are more easily obtained.

The similarity at high K_c coincides with the zone in which vortex shedding may be expected to be most fully developed. Grass and Kemp show, from photographic studies in planar oscillatory flow, how this type of wake develops as the relative motion reverses and advances to and past the point of maximum velocity differential. From their photographs the authors have been able to

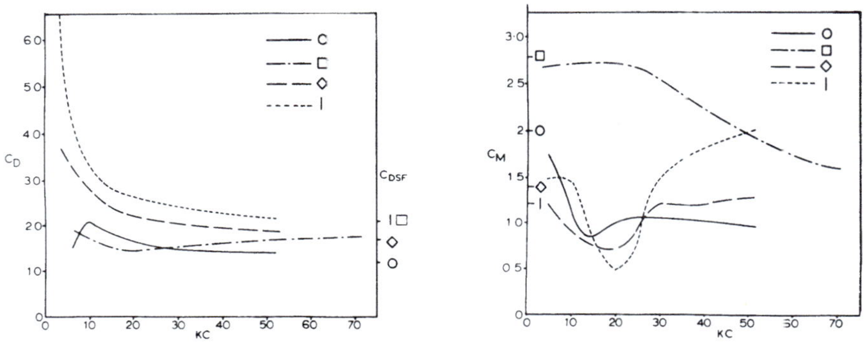

FIG. 5 VARIATIONS OF C_D AND C_M WITH K_c FOR DIFFERENT CYLINDERS

interpret how the angle of flow separation varies throughout the cyclic process (Figure 6) as the wave phase (t/T) advances from zero when the motion reverses. The authors conclude that "The present observations.......must somehow be incorporated into future numerical models if, for example, realistic drag force predictions are required". The implications of this statement are certainly endorsed by the numerical computing study of Stansby, reported in his Paper "Mathematical Modelling of Vortex Shedding from Circular Cylinders in Planar Oscillatory Flows, including Effects of Harmonics" (pp. 450-460).

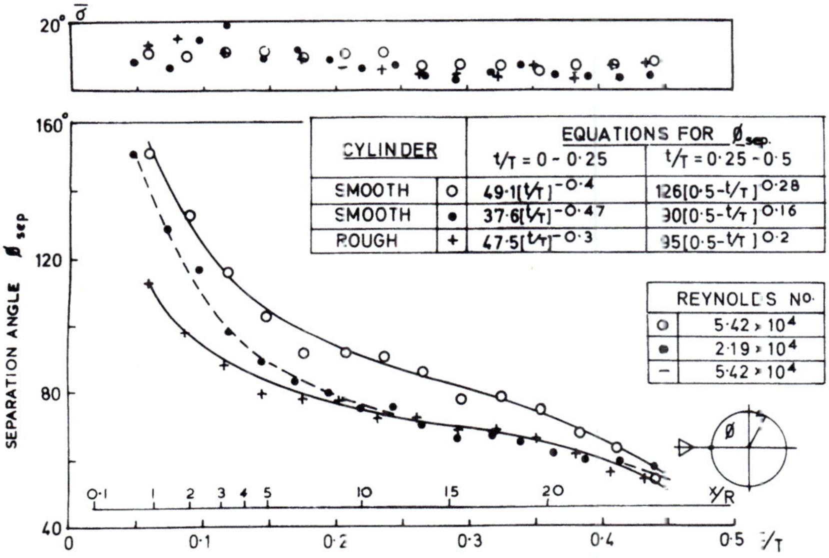

FIG. 6 VARIATIONS OF C_D AND C_M WITH K_c FOR DIFFERENT CYLINDERS

Further insight into the importance of the vortex shedding process on wave-induced forces at all values of K_c for which a significant drag component may be expected is provided by the analytical study of Graham ("Forces on Cylindrical Bodies in Oscillatory Flow at Low Keulegan-Carpenter Numbers", pp. 461-474). Graham acknowledges that the angle of flow separation relative to the shape of the cylinder in that region is fundamental in determining both the drag and inertia forces imposed on the cylinder, as far as vortex effects are concerned. The theoretical curves so deduced and shown in Figure 7 are compared with the planar oscillatory data given by Bearman et al in Figure 5. The close comparison again suggests that the sense and significance of vortex shedding at low K_c has been well represented.

6. PHYSICALLY COMPLEX SITUATIONS

Finally, the Paper by Sarpkaya ("Wave Loading in the Drag/Inertia Regime with Particular Reference to Groups of Cylinders", pp. 532-543) is an apt reminder that however complicated flows about a single cylinder may be, it is unusual for this to take place totally in isolation of influences from the flow about

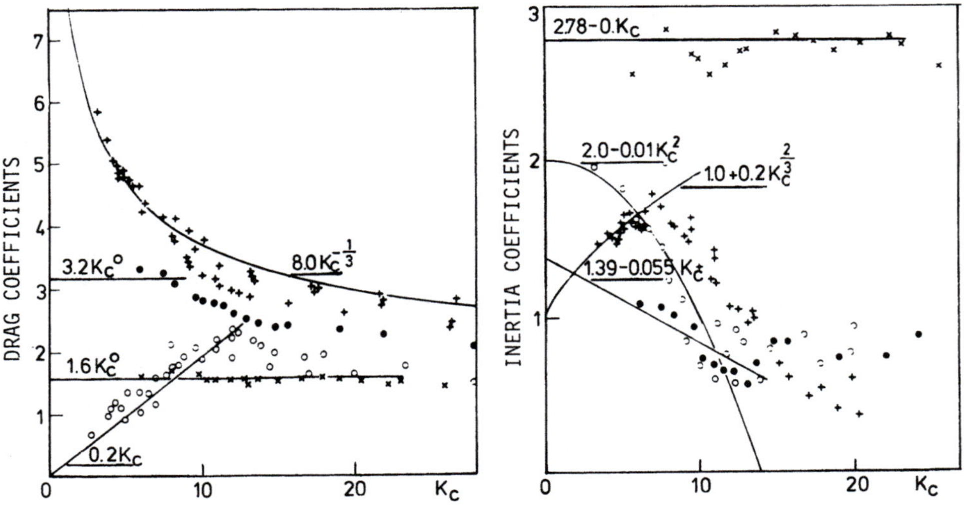

FIG. 7 DRAG AND INERTIA COEFFICIENTS FOR DIFFERENT CYLINDERS

neighbouring cylinders. Studying the configurations in Figure 8, Sarpkaya deduces the results shown. By comparison, potential flow theory suggests an inertia coefficient of 2.17 for both arrays. Sarpkaya concludes that "some fluid mass is entrapped within the array and that neither the potential theory nor the diffraction theory can adequately describe the behaviour of the complex separated flow". As a consequence, "It appears that experiments will continue to play a major rôle in the determination of the appropriate force-transfer coefficients for the type of tube bundles discussed".

7. CONCLUSIONS

In conclusion, therefore, it must be re-emphasised that the forces imposed on cylindrical structures by waves and currents are made up of particular complex distributions of pressures, that are continuously changing in time and space as the dynamics of the boundary layer, separation and wake motion adjust to basically oscillatory but often erratic motion of real sea waves. Figure 9 (Cokelet, pp. 287-301) vividly demonstrates the transient and spatially varying kinetic structure of water particles within the steep wave typical of design conditions, though in this case still limited to two-dimensions.

It need not be stressed that there exist few field data with sufficient detail to demonstrate even the rudiments of the interaction between typical waves and structures, though since 1975 there have been several moves to establish instructive offshore testing programmes. The first results of these are now becoming available. However, it will be many years before this source of information substantially displaces the need for good laboratory data. The compromise reported in the Proceedings and reviewed in this Paper nevertheless gives a sound assessment of the present state of understanding of the practical mechanics of wave-loading on cylinders.

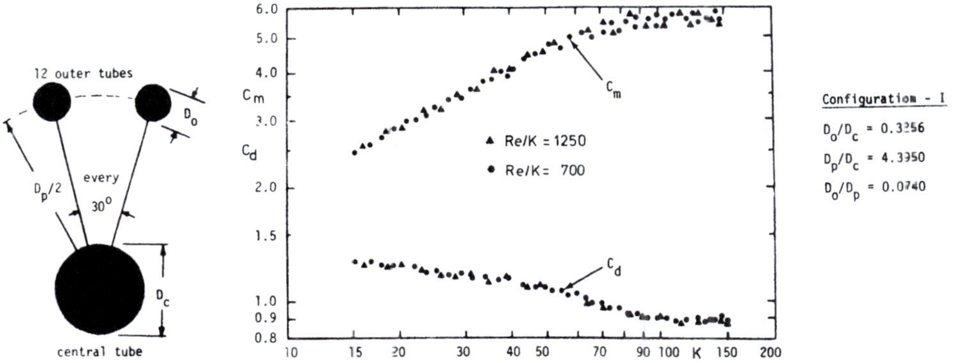

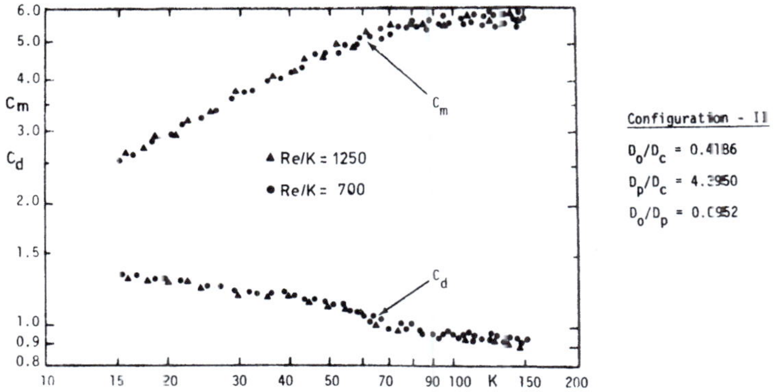

FIG. 8

DRAG AND INERTIA COEFFICIENTS FOR TWO RISER CONFIGURATIONS

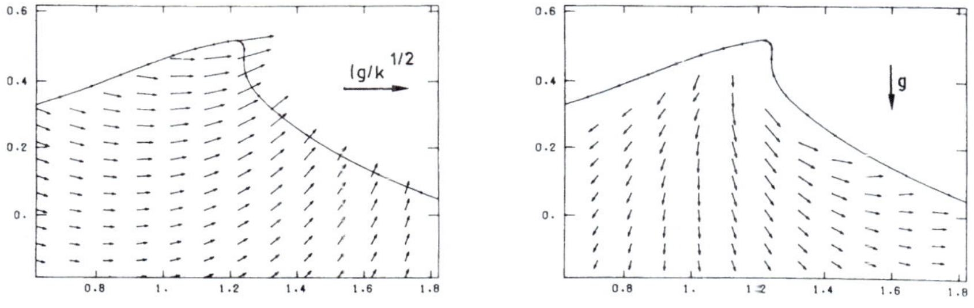

FIG. 9 THE VELOCITY (LEFT) AND ACCELERATION (RIGHT) FIELDS NEAR A WAVE CREST WHEN THE FACE OF A BREAKING WAVE BECOMES VERTICAL. SCALES ARE VALUES OF $2\pi x/L$ (HORIZONTAL) AND $2\pi y/L$ (VERTICAL, ABOVE STILL WATER LEVEL), WHERE L = WAVELENGTH

8. REFERENCES

1. Shaw, T.L.(Ed.), "Mechanics of Wave-Induced Forces on Cylinders", Pitman Publishing Ltd., 768 pp(1979).

2. Susbielles, G., "Wave Actions of Small Sized Obstacles", J. Soc. Underwater Technology, pp. 21-22, (March 1979).

E3 CURRENT-INDUCED OSCILLATIONS OF COGNAC PILES DURING INSTALLATION - PREDICTION AND MEASUREMENT

F. Joseph Fischer

Warren T. Jones

Shell Development Company, Houston, Texas, U.S.A.

Roger King

BHRA Fluid Engineering, Bedford, England

SUMMARY

Two phases of the underwater installation of foundation piles for Shell's Cognac platform in the Gulf of Mexico created situations where ocean currents could cause excessive oscillations of the piles due to vortex shedding. These phases were the platform pile-sleeve "stabbing" operation, when the cable-supported pile was a double pendulum, and the pile-driving operation, when the pile was a vertical cantilever supporting a massive hammer. Vortex-induced motions of the pendulum pile could delay or prevent stabbing or result in structural damage should pile-platform impact occur. Oscillations of the pile during driving could contribute to pile buckling or cause fatigue failure of the end-loaded cantilever pile. This paper describes the potential for such pile motions as assessed by scaled model tests and illustrates a means of attenuating these motions through the use of flexible splitter plates, or "flags". Pile and platform design modifications stemming from predicted pile motions are also discussed together with actual pile motions observed during installation operations.

1. POTENTIAL OPERATIONAL PROBLEMS

The underwater installation of the 24, 2.1-m diameter, 187-m long, foundation piles for the Cognac platform is shown schematically in Figure 1. The first phase of this operation would be the lowering of a vertical pile through the use of a four-part wire-rope system supported by a moored derrick barge. The critical aspect of this phase would be the insertion, or "stabbing", of the pile tip into the appropriate pile sleeve of the platform base section. The location of the pile tip relative to the top of the platform pile sleeve could be determined with an acoustic telemetry system and underwater television cameras. Quasi-static horizontal tip-sleeve offsets could be removed through barge repositioning.

Pile-stabbing operations would be complicated by dynamic tip motions. On the basis of in-house and published literature on the subject of flow-induced oscillations, e.g., Griffin 1 and Iwan 2, troublesome motions were judged to be possible. For a computed first-mode (rigid-body) frequency of 0.0154 Hz for the pendulum pile in still water, maximum pile-tip single amplitudes of 3.2-3.8 m for currents as low as 0.16 m/sec were predicted. These oscillations would be normal to the current flow and associated with nearly periodic shedding of vortices. Such motions could delay or prevent pile stabbing and greatly increase the risk of damaging critical structural members on the platform base section due to pile-platform impact.

After pile stabbing, the pile would be lowered through the pile sleeve and into the seafloor. Under its own weight the pile was predicted to penetrate the seafloor around 40 m. It must then be driven to a total penetration of 140 m with an underwater hammer. The pile in this cantilevered position was also judged to be susceptible to flow-induced oscillations. The natural frequency (in still water) of the 120-m long cantilevered pile segment with the hammer at its free end is around 0.0229 Hz and the pile could experience resonance for currents as low as 0.24 m/sec. For an unloaded pile the natural frequency increases to 0.0486 Hz and therefore a current speed of 0.52 m/sec might be required for large-scale transverse oscillations. These transverse oscillations could result in fatigue failure of the pile or could contribute to pile buckling. Associated with pile oscillations normal to current flow would be an increase in the effective drag coefficient for the pile in the direction of the flow; this would further aggravate the pile loading problem.

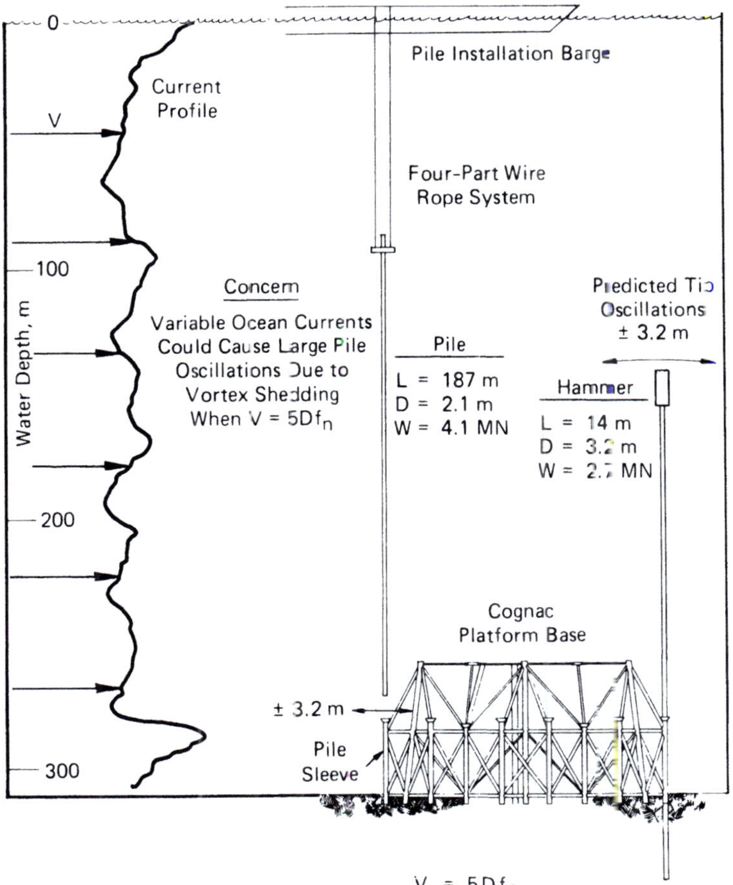

Configuration	f_n, Hz	$V = 5Df_n$, m/sec	Problem
Pile Being Stabbed, 90 m of Wire Rope	0.0154	0.16	Cannot Stab Pile. Pile Hits Platform
Pile Being Driven, 120 m Unsupported Above Pile Sleeve, Hammer Present.	0.0229	0.24	Pile Buckling.
Pile Being Driven, 120 m Unsupported Above Pile Sleeve, Hammer Absent.	0.0486	0.52	Pile Fatigue Failure.

FIGURE 1. COGNAC PILE INSTALLATION CONFIGURATION

Currents of magnitude sufficient to excite the pendulum pile during stabbing and the cantilevered pile being driven had been measured at the installation site. Storm conditions sufficient to drive the derrick barge to shore and thus lead to abandonment of a partially driven pile could cause currents of magnitude sufficient to excite a bare cantilevered pile in a resonant manner.

Effects of variability in current magnitude and direction over the length of the pile were not known. Large-scale oscillations reported by Wooton 3 and Albers 4 of relatively large cylindrical structures in natural environments have occurred, however, so further study in the form of model tests was proposed.

2. ASSESSMENT OF MOTIONS WHILE STABBING

For vortex-induced oscillations of flexible cylinders or flexibly supported rigid cylinders the dimensionless amplitude-to-diameter ratio, A/D, has been shown by Griffin 1 and King 5 to depend primarily upon the reduced velocity, V/Df_n, where f_n is the natural frequency of the cylinder measured in still water, the cylinder specific gravity with respect to water and the structural damping of the system. The dependence upon Reynolds number, if any, is usually slight and the major role of the elastic stiffness of a flexible system is in the determination of the natural frequency, f_n, of the system. The above observations stem from experimental and theoretical investigations and suggest the reasonableness of performing scaled model tests in an attempt to quantify prototype behavior. A comparison of some 1:30 and 1:27 scaled model tests with full-scale results has been given by King 6.

Scaled model tests pertinent to the Cognac pile installation were conducted in an attempt to refine motion predictions based on existing literature. In particular, models of the pile at the time of stabbing (double pendulum) were tested. These tests were conducted at Shell Development's Westhollow Research Center (WRC), at the Fluid Engineering Laboratory of the British Hydromechanics Research Association (BHRA) and at Offshore Technology Corporation (OTC).

The pendulum piles were also tested with "flags" attached. These flags were introduced during the tests in an attempt to reduce pile motions; they proved to be very effective.

Shell and BHRA tests were performed in current tanks wherein water flowed past a supported model and the OTC tests were conducted by towing a model pile through otherwise still water. Because of the nature of these facilities, all simulated current flow was unidirectional although not always uniform in magnitude, i.e., sheared flows were produced.

The physical model employed in the WRC tests is shown schematically and in photograph in Sketch 1 of Chart A. The model was a 1:168 scaled model and thus the 2.1-m diameter of the prototype Cognac pile was represented by a 1.27-cm diameter tube. Other dimensions were similarly scaled. The pile specific gravity, defined as

$$SG = \frac{\text{wt (pile + enclosed water)}}{\text{wt (displaced water)}}$$

was approximately 1.5 for the Cognac pile and this was preserved in the model.

Due to temporary flow-velocity limitations of the WRC current tank, the desired range for the reduced velocity, V/Df_n, was achieved by varying the pendulum-pile frequency, f_n. This was accomplished by varying a concentrated vertical force that was applied at the pile center of gravity. In essence, the gravitational restoring force was being modified. For ease of changing this "gravitational force" the actual model was inverted in the current tank.

The dimensionless amplitude ratio, A/D, and the dimensionless response frequency, f/f_n, for the pile with no flag are shown in Sketch 2 of Chart A as functions of the reduced velocity. The "jumps" in these quantities near the maximum response noted at $V/Df_n = 6$ suggest at least two possible stable oscillatory configurations. This complex phenomenon is known to be nonlinear in nature and hence multiple solutions are not unexpected.

The motions observed for the "bare" pile were greatly reduced with the introduction of a flexible splitter plate, or flag. The model flag was simply a thin sheet of plastic having a thickness of 0.00254 or 0.010 cm. Dimensionless response amplitudes and frequencies for pendulum piles with flags having different dimensions and positions relative to the pile tip are shown in Sketch 3 of Chart A. Flag dimensions and distances from the pile tip are given in terms of the pile diameter, D. The flags were effective in reducing the pile response to around 15 percent of the previously observed maxima at $V/Df_n = 5.5\text{-}5.7$. Apparently, the flags provided sufficient streamlining of the pile to eliminate the shedding of vortices in the vicinity of the pile over the height of the flag. This pile interval was not subjected to an oscillating force and thus provided substantial drag resistance to transverse pile oscillation.

Prototype-pile behavior can be estimated from model-test results using the prototype-pile diameter and natural frequency. In particular, the test result $A/D = 1.5$ at $V/Df_n = 6$ reported in Sketch 2 of Chart A, applied to the Cognac pile, becomes

$$A \text{ (prototype)} = 1.5D = (1.5)\,(2.1) = 3.2 \text{ m}$$

$$V \text{ (prototype)} = 6Df_n = (6)\,(2.1)\,(0.0154) = 0.19 \text{ m/sec}$$

A similar series of tests for pendulum piles was conducted by BHRA at the request of Shell. The models were real pendula with only gravitational restoring force but did not retain complete geometrical scaling. These tests, showing similar results to those quoted above, have been reported by King [7].

Both the WRC and BHRA model tests demonstrated the beneficial effects of a flag attached to a portion of the pile in reducing vortex-induced oscillations. However, these small-scale (1:168) tests did not focus on certain operational aspects that would arise if a flag were employed in prototype operations. In particular, the flags in these tests although "very flexible" were stiff enough to stand erect in air and hence in still water. This situation was realized to be unlikely for a flag of prototype dimensions, 21-m high and 11-m long. The droop of the flag could affect unfurling, effectiveness and total drag force. In order to assess these and other pile-flag characteristics, a series of flag tests were conducted at OTC in Escondido, California.

Currents flowing by a pendulum pile were simulated by towing a model pile through still water in OTC's basin of length 90 m. The 4.6-m depth of the basin dictated the pile length and hence maximum flag height to be used. The 4.3-m long, 28-cm OD "pile" shown in Figure 2 was used for these tests. As in other tests the pile specific gravity was the same as for the actual Cognac piles. The flag height was 3.6 m and its trailing length was 2.7 m or less. The flag was thus a 1:6 scaled model of the proposed prototype flag. Two flag materials were employed; one was neutrally buoyant, the other lost 82 percent of its weight upon submergence. The flags were

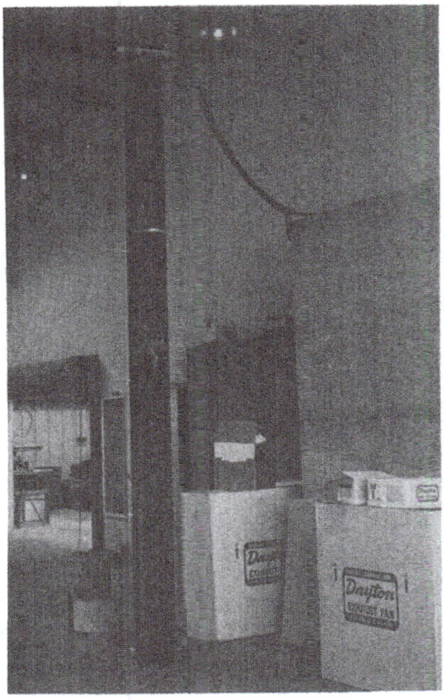

FIGURE 2. PIVOTED PENDULUM PILE USED FOR OTC FLAG UNFURLING TESTS
PILE: L = 4.3 m, D = 28 cm
FLAG: 4 m (HEIGHT) x 3 m

attached to the pendulum pile either at the front or back of the pile in a furled or unfurled state.

It should be mentioned that the model-pile characteristics were such that the model and prototype velocities were nearly the same at the same reduced velocity. This was fortunate since it was desirable (for similarity) to maintain the drag force-to-weight ratio which required equivalence of model and prototype velocities.

Some important observations of these tests are:

A. The apparent drag coefficient for the bare pile oscillating at resonance was 2.12 as compared with 0.93 for the same pile restrained from oscillation.

B. The flag retaining some of its weight upon submergence was superior to the neutrally buoyant flag with regard to ease of unfurling in a flow stream. Furthermore, its sag did not impair its performance in the suppression of pile oscillations.

C. Pleat folding of the flag (like an accordion) was superior to rolling the flag for the purpose of expediting unfurling.

D. Both flags were effective in reducing the oscillations to five percent of that for the bare pile when the flag was attached to the back (trailing edge) of the pile. A continuous time trace of pile response before, during and after flag unfurling is shown in Figure 3.

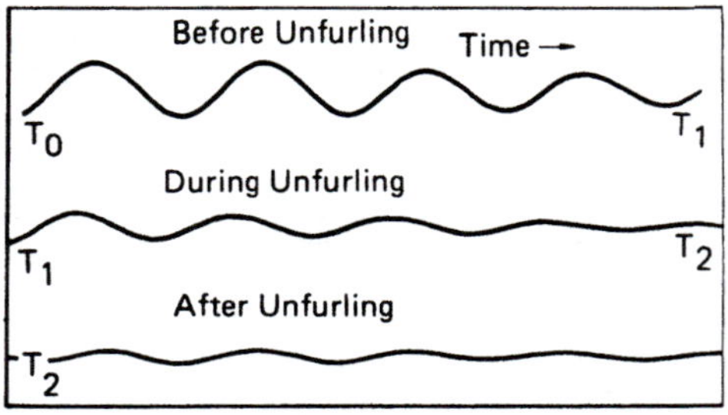

FIGURE 3. TRANSVERSE OSCILLATIONS OF OTC MODEL PILE DURING FLAG UNFURLING

E. Furled flags themselves were effective in reducing oscillation amplitudes to 50 percent of those of the bare pile.

F. Flags draping asymmetrically around one side of the pile, due to their attachment at the front (leading edge) of the pile, were not as effective as symmetric, trailing-edge flags. For flow velocities of bare-pile resonance, pile amplitudes were 30-50 percent of the bare-pile maximum. Maximum amplitude response was shifted to nearly twice the velocity yielding bare-pile resonance and was around 70 percent of the bare-pile maximum.

The Shell MARISE computer program for marine-riser dynamics was modified to accommodate the vortex fluid-force model due to Iwan and Blevins 8. This modified computer program was then used to predict the oscillatory behavior of the Cognac pendulum piles in various current profiles. A comparison of MARISE predictions with uniform-flow model-test results obtained by WRC, BHRA and OTC is given in Figure 4. Also shown is the theoretical maximum for a uniform pivoted rod due to Iwan 2.

Other important observations from the various test programs were:

A. Maximum oscillatory amplitudes of piles in limited shear-flow tests were not significantly reduced below those for uniform flows.

B. Impact-contact between the tip of the pile and a fixed object revealed almost instantaneous suppression of pile-tip motion.

3. ASSESSMENT OF MOTIONS WHILE DRIVING

Model tests pertinent to Cognac piles during driving were conducted by BHRA in order to refine response predictions based on literature data, e.g.,

Griffin 1 and Iwan 2, and to investigate the use of flags in attenuating such response. Models tested were vertical cantilever beams of circular cross section. Model lengths corresponded to the length of unsupported pile above the platform pile sleeve subsequent to stabbing the pile and lowering it into the seafloor under its own weight.

The cantilevers were either free-ended (no "hammer") or possessed an end mass (hammer). These two configurations were examined both with and without vortex-spoiling flags in uniform and sheared flows.

The model-test configuration is shown schematically in Sketch 1 of Chart B. Also shown in this sketch is the velocity profile for the sheared flow. As for the WRC pendulum tests, the model scale was 1:168 and the model specific gravity was 1.5.

The dimensionless pile-tip amplitude, A/D, has been plotted versus the reduced flow velocity, V/Df_n, in Sketch 2 of Chart B. For the sheared flow, V was chosen to be that near the midpoint of the model pile. The maximum responses for the two flow profiles are the same but more scatter in pile response occurs for the sheared flow.

The introduction of flags at the free end of the cantilever pile significantly reduced the amplitude of oscillation as indicated in Sketch 3 of Chart B. In no case did tip amplitudes exceed 0.2 pile diameters, as compared with 1.5 pile diameters without a flag.

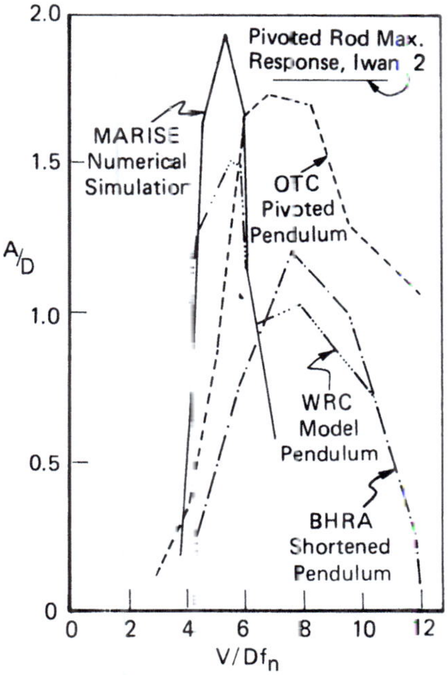

FIGURE 4. COMPARISON OF MODEL PENDULUM PILE RESPONSES

Amplitudes of transverse oscillation for a model cantilever pile fitted with a cylindrical end mass (hammer) are shown in Figure 5 versus reduced velocity. Results for several pile specific gravities are given. For the Cognac pile specific gravity of 1.5, results are also given for a "rough" hammer. This roughness was achieved by wrapping some wire mesh around the "smooth" hammer. This was done to simulate miscellaneous wires and hoses which surround the actual underwater hammer. The hammer roughness is seen to delay excitation and to reduce the response amplitudes. In all cases, the model-hammer specific gravity was equal to that of the prototype hammer, namely 2.5. Flags were also effective in reducing the amplitude of oscillation of the hammer-ended pile.

Measurements of the "steady" in-line deflection of the free-ended pile were made to assess any increase of the effective drag coefficient accompanying transverse oscillations. As before, an increase was noted by comparison of measured deflections with calculated ones based on a C_D of 1.2. This comparison is shown in Figure 6. A two-fold increase can be inferred in some instances.

Because of concern over the influence of model-pile structural damping on response results, BHRA were requested to conduct limited tests for a stainless steel cantilever pile; a PVC model had been used for the bulk of the tests. Results of these tests indicated a maximum amplitude of two pile diameters as compared with 1.5 pile diameters measured earlier with the PVC cantilever model. The logarithmic decrement (due to structural damping) of the PVC pile in air was 0.063 whereas that for the steel pile was around one-fourth of this value.

4. DESIGN/OPERATIONS CONSEQUENCES

The original plan for stabbing the Cognac piles called for the use of pile-sleeve cones (opening upward) having a large-end diameter of 5.5 m. WRC model pendulum tests suggested that peak-to-peak pile excursions of up to 6.4 m were possible for specific current conditions not unlike those measured at the Cognac location. When the pile diameter (2.1 m) is added to this centerline excursion, a "clear opening" for vertical stabbing of 8.5 m appeared to be needed.

Suffice it to say that lateral motions of the tip of the pile were anticipated to the extent that they could delay or prevent pile stabbing (vertical) into the 5.5-m diameter cones or possibly result in pile impact with the platform base section. Mr. B. E. Cox, Shell Cognac Installation Staff, thus proposed a side-entry, energy-absorbing pile-guide frame to replace the pile-sleeve cones. This guide frame with its 7.6-m "capture" width was intended to allow lateral control of the pile prior to lowering the pile into the sleeve. The 24 guides were also connected to one another by nylon rope thus forming a protective "fence" around vulnerable knee braces on the platform base section. These guide frames constituted the principal means of contending with potential pile-tip motions during stabbing operations. Although expensive, they were attractive in that they offered a passive solution to the problem.

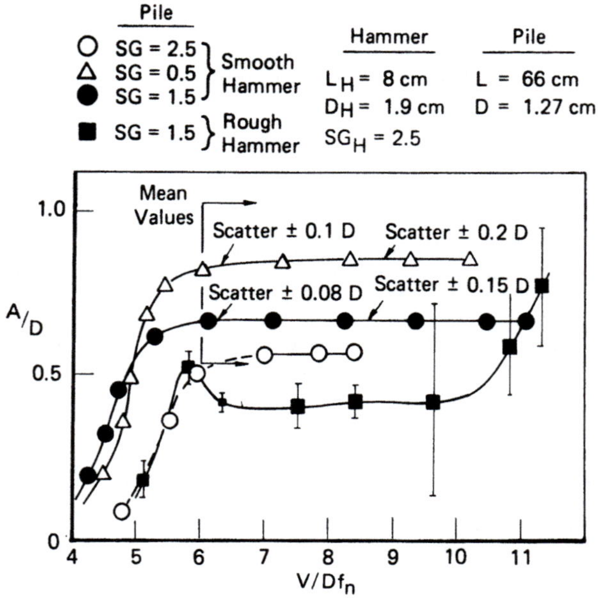

FIGURE 5. TRANSVERSE OSCILLATIONS OF BHRA MODEL CANTILEVER PILE WITH "HAMMER" END MASS

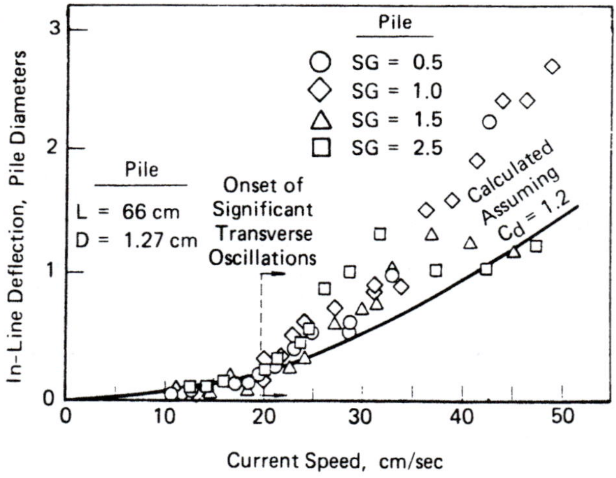

FIGURE 6. STEADY IN-LINE TIP DEFLECTION OF BHRA MODEL CANTILEVER PILE SHOWING INCREASE OF EFFECTIVE DRAG COEFFICIENT

A study of static and dynamic stress levels within Cognac piles in the driving mode was conducted by Dr. J. A. Ruhl, Shell Oil Company. For this study, in-line oscillations, transverse oscillations and drag-force amplifications due to transverse oscillations were considered in addition to more obvious sources of pile loading. His general conclusion was that "transverse oscillations triple calculated maximum stresses." This analysis provided the major impetus for using 414 MPa steel for much of the pile, i.e., that portion of the pile initially exposed above the pile sleeve subsequent to stabbing and initial lowering.

A study pertinent to possible fatigue failure of a cantilevered Cognac pile, without hammer, was also conducted by Ruhl. Using model test data he predicted failure of a pile in four days when exposed to a current of magnitude 0.46 m/sec. The exposed length of this pile was 130 m (likely maximum) and the wall thickness was 3.8 cm. Other assumptions included a stress concentration factor of unity and the AWS-X fatigue curve. On the basis of these and similar results for other current velocities and pile lengths, Ruhl recommended that piles be stabbed and driven one at a time to "keep exposure time to a minimum".

It was also recommended that a number of flags be readied for possible attachment to Cognac piles in the event of extreme pile oscillations. In addition to possibly expediting the stabbing of piles, use of a flag would allow continued pile driving in the event of large pile-hammer oscillations. The alternative might be the removal of the hammer and thus suspension of progress. The beneficial effect of a flag in suppressing flow-induced vibrations had been repeatedly demonstrated by model tests. The primary disadvantage to flag usage would be the requirement that divers would have to install and remove them.

5. FIELD EXPERIENCE

Flow-induced oscillations of Cognac piles during their installation did not cause any operational problems. In all but a few instances they appeared to be nonexistent. Significant pile motions did occur, however, while stabbing Pile No. 9.

The pile-installation configuration and the current profile at the time of stabbing Pile No. 9 are shown in Sketch 1 of Chart C. Although the currents were not uniform in magnitude (with significantly more shearing than examined in model tests) they were nearly unidirectional. With the current heading slightly west of the Cognac Y-axis, transverse pile oscillations should be predominantly along the Cognac X-axis. Such was the case by reference to the pile-tip amplitude time traces shown in Sketch 2 of Chart C. Typical peak-to-peak pile centerline excursions were around 3 m with several extreme excursions of nearly 6 m within a two minute interval. Typical periods of oscillation were around 30-40 seconds which were well below the periods of barge-drift oscillations, e.g., 200 seconds. Furthermore, barge-motion amplitudes were only around 0.3 m.

The motions observed for this pile were easily accommodated by the pile-sleeve side-entry guide frame shown in Sketch 3 of Chart C. Furthermore, these motions are in general agreement with those predicted on the basis of the WRC model tests. The cessation of pile motion upon contact with the guide, observed using underwater television cameras, was also consistent with model-test observations.

The positive-response data quoted above for Pile No. 9 are the most striking extracted from recorded data. By contrast, extremely small pile motions were usually observed during stabbing and driving operations. At this time, the lack of observed motion in particular current profiles is being attributed to a lack of unidirectionality of the flow or to a probable lack of vortex-shedding synchronization along the very long piles. The effect of variable current heading on flow-induced vibrations could not be quantified prior to field activities. After having been in the field, the effect is still not understood, but we now have some data/observations with which to improve our understanding.

6. CONCLUSIONS

Flow-induced oscillations of cylindrical systems such as piles, conductors, marine risers or cables can occur. Uniformity of current magnitude over the length of the cylinder does not appear to be necessary for oscillations to occur. This has been observed both in the model basin and in the field. Lack of unidirectionality of the flow over the cylinder length appears to greatly reduce the amplitudes of oscillations expected on the basis of current magnitude alone. Such reductions have been observed in the field during the installation of the Cognac piles.

7. REFERENCES

1. Griffin, O.M., Skop, R. A. and Ramberg, S. E., "The Resonant, Vortex-Excited Vibrations of Structures and Cable Systems", OTC Paper No. 2319, (1975).

2. Iwan, W. D., "The Vortex Induced Oscillation of Elastic Structural Elements", ASME Paper No. 75-DET-28, (1975).

3. Wooton, L. R., Warner, M. H. and Cooper, D. H, "Some Aspects of the Oscillations of Full-Scale Piles", Symposium on Flow Induced Structural Vibrations, Karlsruhe, Germany, (1972).

4. Albers, J. and DaSilva, M. L., "Innovative Risers Speed Drilling in the High Current Areas Off Brazil", Oil & Gas Journal, April 18, 74-81, (1977).

5. King, R., Prosser, M. J. and Johns, D. J., "On Vortex Excitation of Model Piles in Water", J. Sound and Vibration, 29 (2), 169-188, (1973).

6. King, R., "Vortex Excited Structural Oscillations of a Circular Cylinder in Steady Currents", OTC Paper No. 1948, (1974).

7. King, R., "Model Tests of Vortex Induced Motion of Cable Suspended and Cantilevered Piles for the Cognac Platform", Report RR1453, BHRA, Fluid Engineering, Bedford, (1978).

8. Iwan, W. D. and Blevins, R. D., "A Model for Vortex Induced Oscillation of Structures", J. Applied Mech., Sept., 581-586, (1974).

CHART A: WRC MODEL PENDULUM PILE

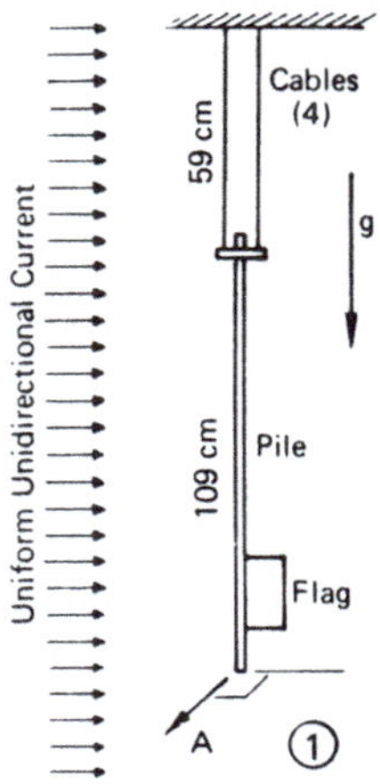

WRC MODEL PILE

SCALE = 1:168
D = 1.27 cm
SG = 1.5

Problem: Current-induced transverse oscillations of prototype underwater pendulum pile potentially delaying/preventing pile installation.

Geometry: See Sketch 1. Tip-oscillation amplitude, A, is perpendicular to current flow.

Boundary Conditions: See Sketch 1. Pile suspended vertically from fixed support by four parallel cables.

Flow Conditions: Unidirectional horizontal flow.

Operating Conditions: Model flow velocity around 0.26 m/sec. Reduced velocity, V/Df_n varied from 4 to 11.

Structural Dynamics: Mode of principal excitation is rigid-body pendulum mode wherein pile and cables are nearly collinear.

Vibration: For pile with no flag, see Sketch 2. Ratios of pile single amplitude over pile diameter, A/D, and pile response frequency over pile natural frequency plotted versus reduced velocity, V/Df_n. Strouhal shedding frequency denoted by f_s. Oscillation due to alternate shedding of vortices.

Cures: See Sketches 1 and 3. Flags of various dimensions attached near pile tip. Dimensionless amplitude and frequency responses given for reduced velocities. Flags act to streamline the pile, thus eliminating vortex shedding over the flag length along the pile.

References: Albers, J. and DaSilva, M. L. "Innovative Risers Speed Drilling in High Currents Off Brazil", Oil & Gas Journal, April 18, 74-81, (1977).

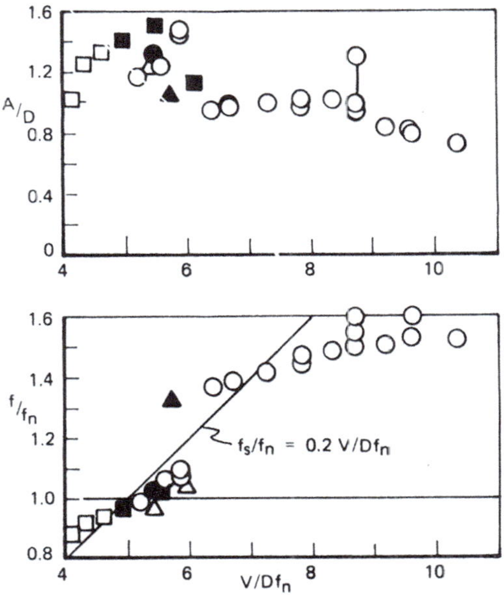

② TRANSVERSE OSCILLATIONS OF BARE MODEL PILE SUSPENDED BY 4 CABLES

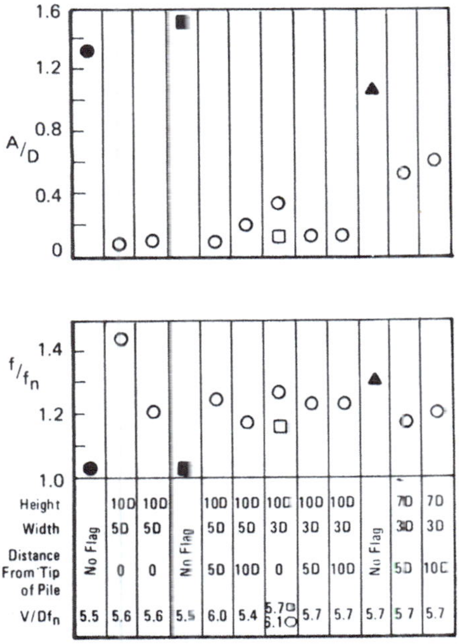

③ EFFECT OF FLAG ON TRANSVERSE OSCILLATIONS

CHART B: BHRA MODEL CANTILEVER PILE

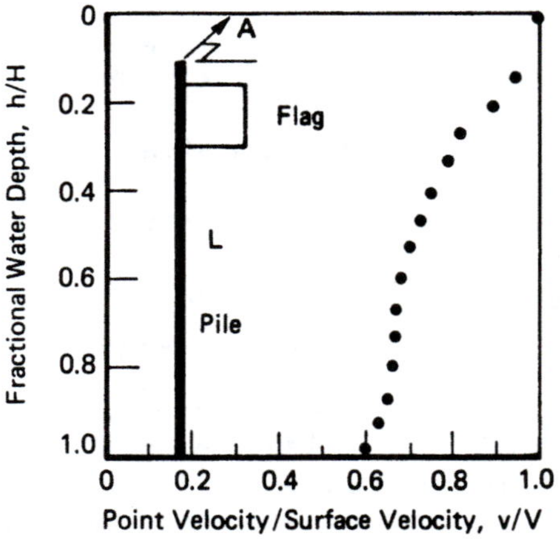

① BHRA Model Pile

Scale = 1:168
L = 66 cm
D = 1.27 cm
SG = 1.5

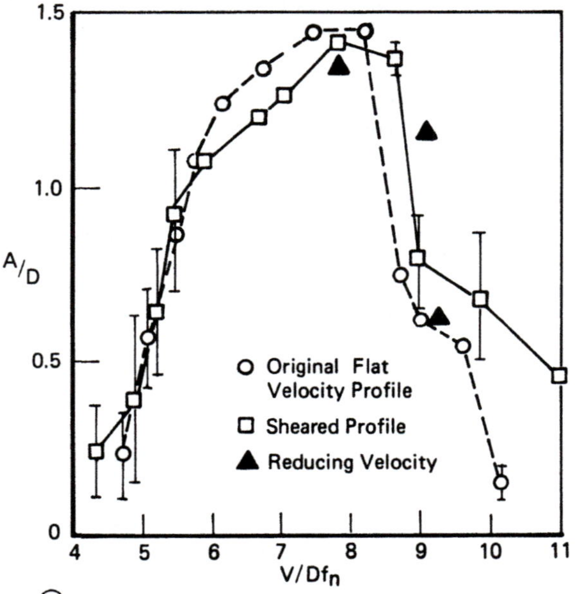

② TRANSVERSE OSCILLATIONS OF BARE PILE

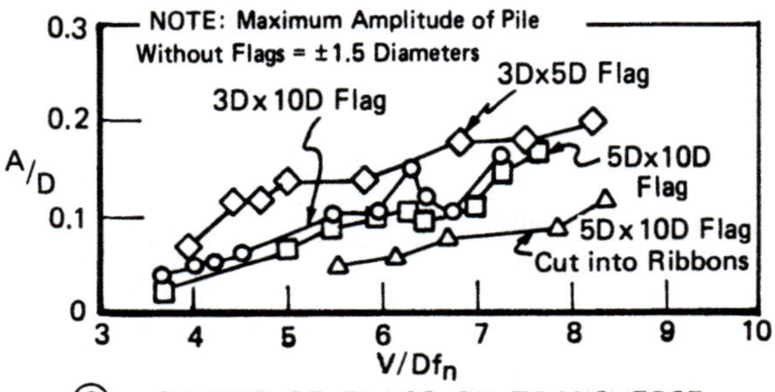

③ EFFECT OF FLAGS ON TRANSVERSE OSCILLATIONS

Problem: Current-induced transverse oscillations of prototype underwater cantilever pile potentially resulting in pile buckling or pile fatigue failure.

Geometry: See Sketch 1. Tip oscillation amplitude, A, is perpendicular to current flow.

Boundary Conditions: See Sketch 1. Vertical pile is cantilevered at base.

Flow Conditions: Uniform or sheared, unidirectional horizontal flow.

Operating Conditions: Reduced velocity, V/Df_n, for model varied from 4 to 11.

Structural Dynamics: PVC model bending stiffness, EI, was 2.83 $N\text{-}m^2$, resulting in a natural frequency in water of 3.85 Hz. Logarithmic decrement for structural damping was 0.063.

Vibration: For pile with no flag, see Sketch 2. Dimensionless single-amplitude ratio, A/D, is plotted versus reduced velocity, V/Df_n. Oscillations due to alternate shedding of vortices.

Cures: See Sketches 1 & 3. Flags of various dimensions attached near pile tip. Dimensionless amplitude ratio plotted versus reduced velocity.

References: Albers, J. and DaSilva, M. L., "Innovative Risers Speed Drilling in High Currents Off Brazil", Oil & Gas Journal, April 18, 74-81 (1977).

CHART C: COGNAC PILE IN OCEAN CURRENTS

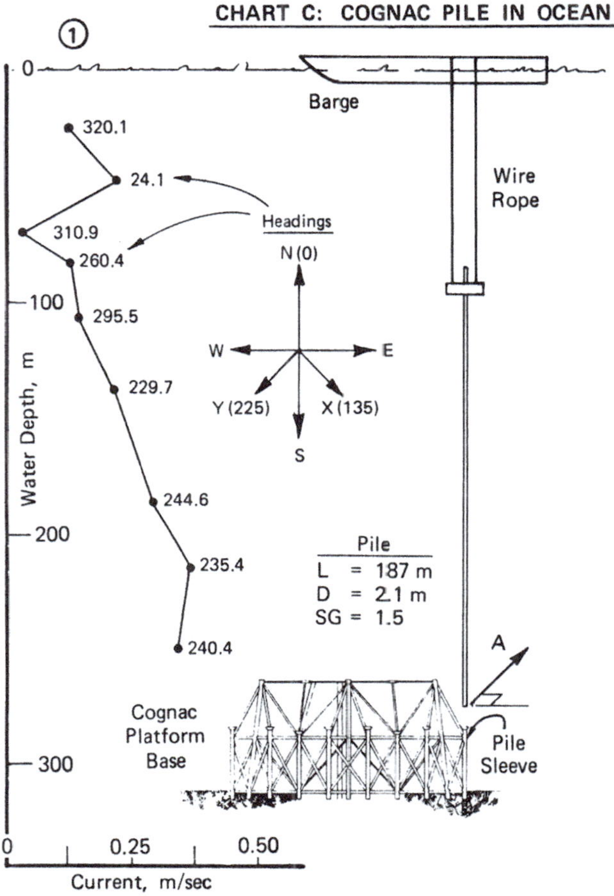

Problem: Current-induced transverse oscillations of underwater pendulum pile potentially delaying/ preventing pile installation.

Geometry: See Sketch 1. X and Y are specific Cognac horizontal coordinate axes. A is tip amplitude

Boundary Conditions: Pile suspended vertically from moored derrick barge by four parallel cables.

Flow Conditions: See Sketch 1. Current magnitude plotted with headings given by number.

Operating Conditions: Pile support point nominally fixed. Steady currents.

Structural Dynamics: First-mode rigid-body frequency for pendulum pile is 0.0154 Hz.

Vibration: See Sketch 2. Typical peak-to-peak excursion is 3 m and oscillation period is around 30–40 seconds. Believed due to alternate shedding of vortices.

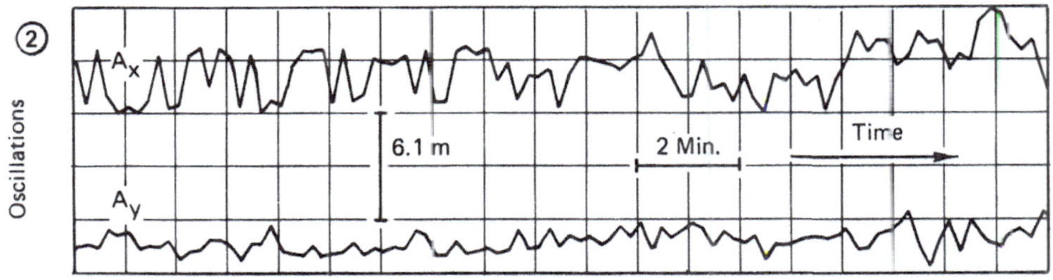

Cures: See Sketch 3. Side-entry pile-guide frames provided passive solution to pile-stabbing problem. Pile secured laterally before lowering vertically into pile sleeve.

E4 TRANSVERSE MOTION OF A BUOY BY SURFACE WAVE

Kunihiro Ogihara
Professor of Engineering

Toyo University
Department of Civil Engineering
Kawagoe, Saitama, Japan

SUMMARY

In this report, the transverse motion of a buoy is investigated by field and laboratory experiments as well as theoretical analysis. The results of field tests of a buoy constructed in Tokyo Bay by the Public Works Research Institute, Ministry of Construction, do not show clear transverse motion; the buoy moves in the direction of wave propagation and transverse to it. In laboratory model tests of the full-scale buoy transverse motion was clear, in addition to motion in the direction of wave propagation. In this paper, a clarification is attempted of the relations between the results in the field and model tests and the theoretical analysis.

1. FIELD TESTS

The field buoy shown in Fig. 1 has been constructed by the coastal engineering division of the Public Works Research Institute. This buoy has four instruments for measuring the wave parameters such as a pressure gauge for the wave height, an inclination meter to measure the wave direction, an accelerometer to measure the motion of buoy, and a gyrocompass. The data were recorded by tape recorder in the buoy and removed once a month for analysis.

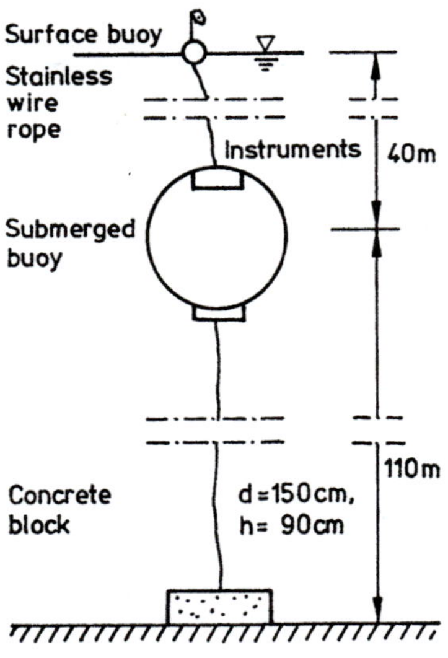

FIG.1 FIELD BUOY

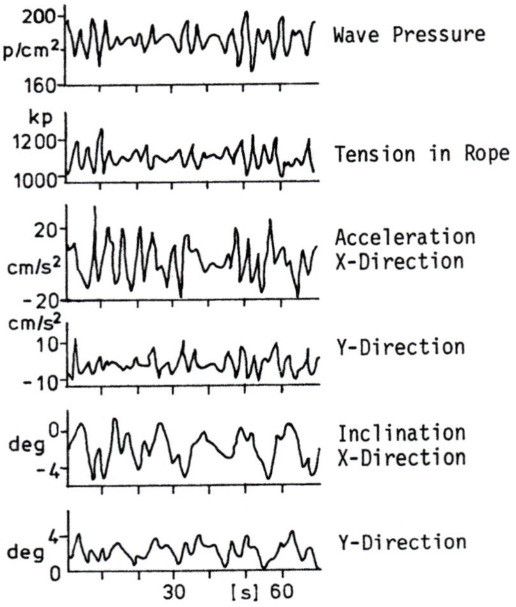

FIG.2 RECORDS

Fig. 2 shows typical records which were observed on the 12th November 1977. Typical evaluations are shown in Figs. 3 and 4. These points were calculated using the transverse-motion data from the inclination-meter; they show the peak vector displacement of the center of buoy in one cycle of motion.

The example in Fig. 4 shows the concentration of points in the direction of 300 degree (N.W. direction). For lower wave heights, in contrast, a stronger lateral motion was found (Fig. 3). It can thus be concluded that when the wave height is greater, the direction of motion is concentrated to one direction. It is not clear whether the motion shown in Fig. 3 arises from a directional spread of waves or from the phenomenon of transverse motion discussed below.

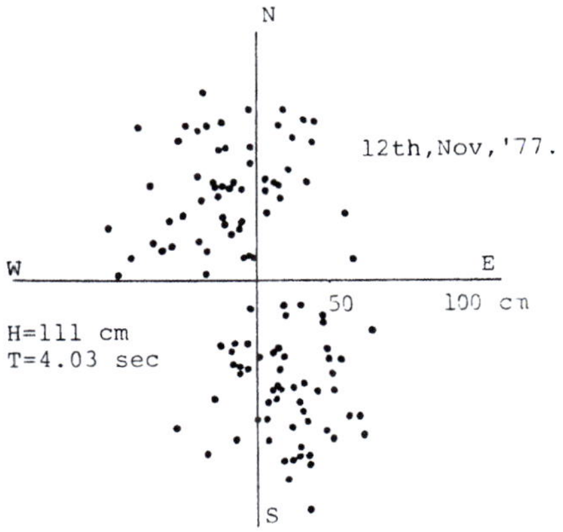

FIG.3 MOTION OF BUOY

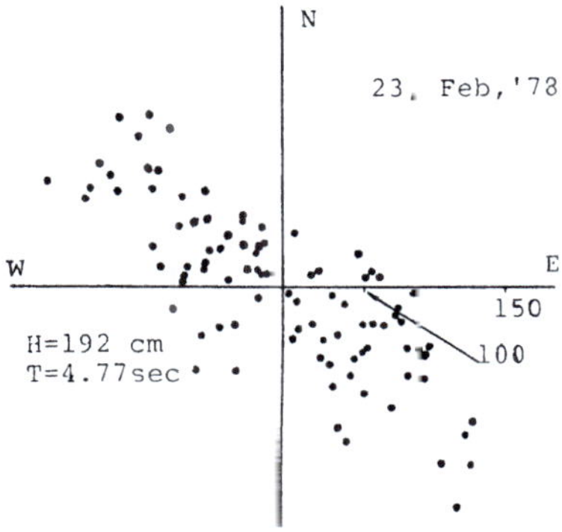

FIG.4 MOTION OF BUOY

2. MODEL TESTS

Before the field buoy was installed in Tokyo Bay, the author conducted model tests in which the motion of the buoy transverse to the wave direction arose under special conditions. In these tests two models were used; one is a ball of 13 cm diameter and 44.6 g mass, and the other is a cylinder of 6.5 cm diameter, 16.8 cm height, and 31.1 g mass (see Fig. 5).

Typical examples of the buoy motion are shown in Figs. 6 and 7. These two cases demonstrate that transverse motion arises at fairly large amplitude, as was reported by Dr. T. Sawaraghi [2] for the motion of cylindrical piers in model tests.

It is interesting and important to analyse how these motions arise under the wave action and to distinguish whether this oscillation occurs in the field or not.

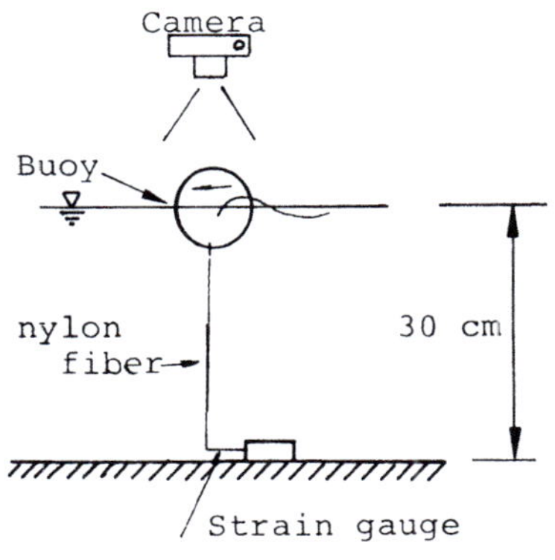

FIG.5 MODEL SCHEME

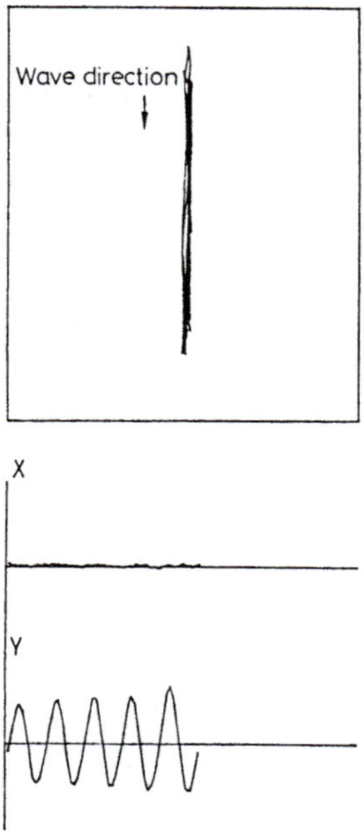

FIG.6 HORIZONTAL MOTION

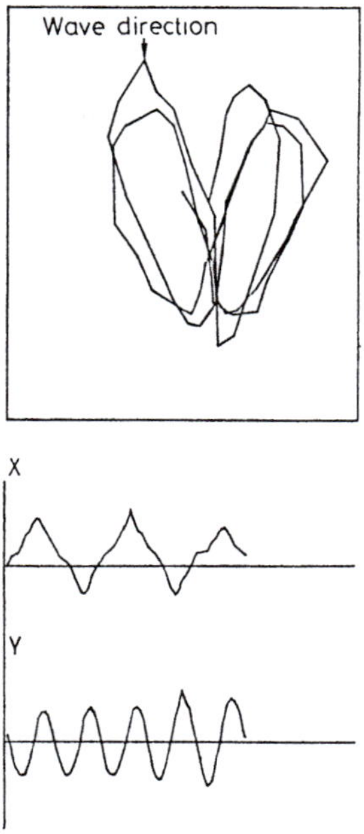

FIG.7 TRANSVERSE MOTION

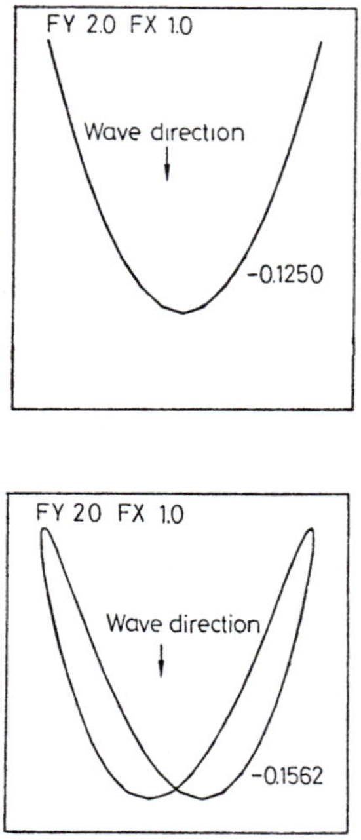

FIG.8 CALCULATED MOTION

3. THEORETICAL ANALYSIS

The following is a brief summary of the theoretical analysis described in more detail elsewhere. Consider a case such that the buoy is anchored with tensile stress in the anchor rope. The equations of motion of the buoy are in this case:

$$ml^2\theta + R\theta + k_\theta\theta = 0 \qquad m(l.\sin\theta)^2\phi + R\phi + k.\phi = 0 \qquad (1)$$

Here m is the virtual mass, k is a spring constant of deformation which is considered to be smaller than k_θ. The coordinate systems and notations are indicated in Fig. 9.

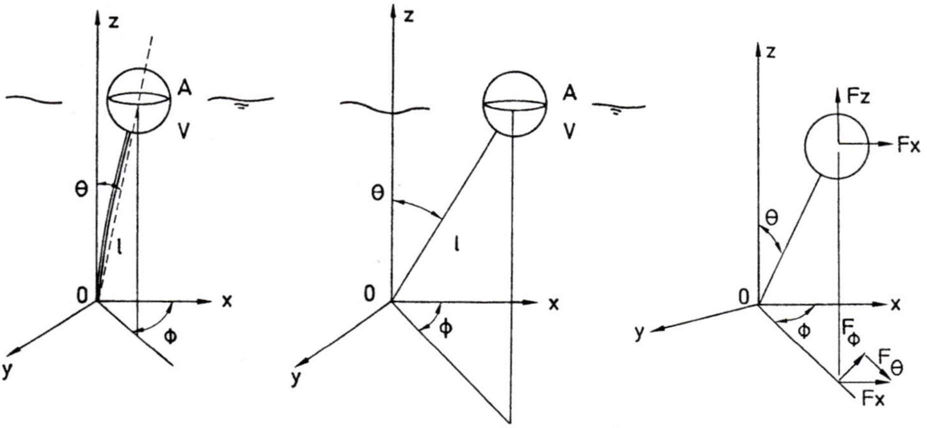

FIG.9 COORDINATE SYSTEMS AND NOTATIONS

As far as the wave forces on the buoy are concerned, Airy's equation is used for the wave hydrodynamics and Morison's equation for the wave force.

$$Fx = Cm\ .\ \rho V\ \frac{\partial}{\partial t}(u-x') + Cd.\rho A\ \frac{(u-x')}{2}\ |u-x'|$$
$$Fy = Cm\ .\ \rho V\ \frac{\partial}{\partial t}(v-z') + Cd.\rho A\ \frac{(v-z')}{2}\ |v-z'| \qquad (2)$$

The terms x' and z' are smaller than the terms u and v, and the second terms in equation (2) are neglected. For small ϕ and θ, moments about 0 are given by

$$M_\phi = -Fx.\phi.1\sin\theta_0 \qquad M_\theta = Fx.1\cos\theta - Fz.1\sin\theta_0 \qquad (3)$$

Neglecting terms of second order and using water particle velocities at the mean water level: $Fx = Ax(1).\cos\sigma t + Bx(1).\sin\sigma t$

$$Fz = Az(1).\cos\sigma t + Bz(1).\sin\sigma t \qquad (4)$$

where

$$Ax(1) = \frac{H}{2}\sigma^2\coth k'h.Cm.\rho V \qquad Bz(1) = -\frac{H}{2}\sigma^2 Cm.\rho V + wA\frac{H}{2}$$
$$Bx(1) = \left(\frac{\pi H}{T}\right)^2(\coth k'h)^2.Cd.\rho A\frac{4}{3\pi} \qquad Az(1) = \left(\frac{\pi H}{t}\right)^2 Cd.\rho A\frac{4}{3\pi}$$

The equations of motion are

$$ml^2\theta + R\theta + k_\theta\theta = 1[A_\theta(1).\cos\sigma t + B_\theta(1).\sin\sigma t]$$
$$m(1.\sin\theta_0)^2\phi + R\phi + k_\phi\phi = (-1\phi).[A_\phi(1).\cos\sigma t + B_\phi(1).\sin\sigma t] \qquad (5)$$

where

$$A_\theta(1) = Ax(1).\cos\theta_0 - Az(1).\sin\theta_0$$
$$A_\phi(1) = Ax(1).\sin\theta_0\ ,\ B_\theta(1) = Bx(1)\cos\theta_0 - Bz(1).\sin\theta_0 \qquad (6)$$
$$B_\phi(1) = Bx(1).\cos\theta_0 - Bz(1).\sin\theta_0$$

Defining

$$\frac{R}{ml^2} = 2\gamma,\ \frac{\theta}{ml^2} = \omega_0^2,\ h = \frac{\gamma}{\omega_0},\ u = \frac{\sigma}{\omega_0},\ F(u) = \frac{u^2}{\sqrt{(u^2-1)^2+4h^2u^2}},\ \tan^{-1}(\sigma\tau) = \frac{2\gamma\sigma}{(\omega_0^2-\sigma^2)} \qquad (7)$$

The solution in the θ direction is that of the forced oscillation of a linear system:

$$\theta = \frac{1}{ml}[A_\theta(1).\cos\sigma(t-\tau) + B_\theta(1).\sin\sigma(t-\tau)].F(u) \qquad (8)$$

The equation of motion for the ϕ direction is a standard-type Mathieu equation with damping, derived as follows:

$$m(1.\sin\theta_0)^2\ddot{\phi} + R\dot{\phi} + [k_\phi + 1.[A_\phi(1).\cos\sigma t + B_\phi(1).\sin\sigma t)]\phi = 0 \qquad (9)$$

This, by appropriate substitution (see Ref. 5) can be reduced to:

$$\ddot{\Phi} + (\delta + \varepsilon\cos\tilde{t})\Phi = 0 \qquad (10)$$

The solution of this equation can be seen in many books on vibrations, and it results in unstable conditions as shown in Fig. 10.

Analysis of model test. Typical traces of the horizontal motions of the buoy, observed by cine-camera are shown in Figs. 6 and 7. The first one is the case in which the transverse motion does not occur and the second one is the case that the transverse motion occurs to the same degree as the motion in the incident wave direction. The frequency of oscillation in the x and y directions in the second case are different, the former being half of the latter. This means, the parameter δ is 1/4 and this oscillation corresponds to the first unstable zone of Mathieu diagram in Fig. 10.

The traces derived by calculation, $A \sin 2\pi t/T$ for the x-direction and $A \sin(4\pi t/T+$ for the y-direction, are shown in Fig. 8 . The values -0.1562 and -0.125 are phase angles ψ by radian, these curves being similar to the second trace of Fig. 7. These phenomena about the circular pole structure are also observed by Dr. T. Sawaraghi of Osaka University and his data shown in Table 1 [3] .

Model	f_n	f_w	δ
8	1.075	0.913	1.386
		0.940	1.307
6	1.351	1.112	1.476
		0.885	2.33
		0.666	4.11
		0.436	9.60
3	1.914	0.625	9.37
		0.620	9.53

TABLE 1

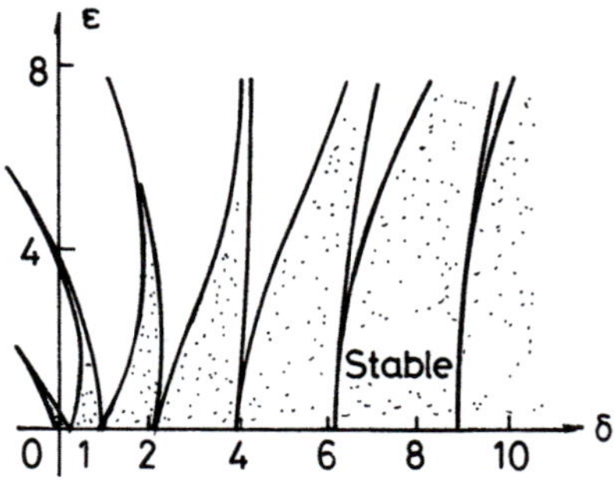

FIG.10 MATHIEU'S DIAGRAM

The value delta in this case is distributed 1.0, 2.0, 4.0, 9.0 - again corresponding to the unstable zone in Mathieu's diagram.

4. CONCLUSION

The transverse buoy motion in surface waves can be explained in terms of Mathieu's instability problem. There is still some doubt whether this phenomenon occurs in model tests only. Possible deviations between field and laboratory experiments may be due to the greater complexity of the field structure as compared to the model which consists, e.g., of a single cylinder or a single sphere. Moreover, the ratio delta between the natural frequency and the wave frequency may be much different from Mathieu's diagram. Therefore, the author intends to collect further field data on this phenomenon in the future.

REFERENCES

1. Tuboi, T.; "Oscillation", Gendaikogakusha.
2. Sawaraghi, G.; "Nonlinear Oscillation", Kyoritu Shutpan.
3. Sawaraghi, T., Nakamura, T., and Hideki Miki; Oscillation of circular cylinder under the wave", 23rd Congress Japanese Coastal Engineering, 1976 (in Japanese).
4. Hom-ma, M., Ogihara, K. and Emori, H.; "The motion of spherical buoy under the surface wave", Report of annual meeting of JSCE. No. 28, 1973.
5. Ogihara, K.; "Theoretical analysis of the transverse motion of a buoy under the surface wave", 24th Congress of Japanese Coastal Engineering, 1977 (in Japanese).
6. Hashimot, H. and Yamaguchi, O.; "Observation of surface wave by a submerged buoy in Tokyo Bay", 25th Congress of Japanese Coastal Engineering, 1978.

E5 SUPPRESSION OF FLOW-INDUCED MOTIONS OF A SUBMERGED MOORED CYLINDER

Dirse W. Sallet
Professor

The University of Maryland
Department of Mechanical Engineering
College Park, Md. 20742
U. S. A.

SUMMARY

A simply-moored, completely submerged cylindrical current meter will swing like a pendulum, cross-wise to the current at certain current velocities. These motions will falsify the velocity reading and should therefore be suppressed. This article deals with the effectiveness of different surface modifications (spoilers) and with the effectiveness of splitter plates to suppress these flow-induced motions. The article centers around the practical experiences gained from full-scale towing tests. The cylinder tested was 0.5 m in diameter, had an ℓ/d ratio of 5, and was exposed to flow in a Reynolds number range between 0.4×10^5 and 6.7×10^5. The cable length was variable up to 15 m, the positive buoyancy was 1419 N and the metacentric height was 0.30 m. It was found that splitter plates as well as certain spoilers will greatly reduce the flow-induced cross-wise motions.

1. INTRODUCTION

Flow-induced transverse vibrations will occur when an elastically supported cylindrical structure is subjected to cross flow. The structure in this study is a circular cylinder of 0.53 m diameter and 2.74 m length. The cylinder is a totally submerged buoy with positive buoyancy and moored by a single cable to the ocean floor. In a sense the physical system resembles an upside-down pendulum. In the presence of ocean currents the buoy will not only lean back due to the drag acting on the buoy but will also undergo transverse vibrations. These motions are undesirable as they falsify current velocity measurements taken by instruments within and attached to the buoy and also make sound monitoring much more difficult. The purpose of the tests described in this paper was to find methods which prevent or at least decrease the oscillating transverse motion of the buoy when it is exposed to cross flow due to ocean currents.

The cause of the flow-induced motions of the buoy is the alternating shedding of vortices from the surface of the buoy. The frequency of the vortex shedding is given by the Strouhal frequency, namely $f_S = SU/D$, where S is the Strouhal number, U is the velocity of the fluid relative to the buoy and D is the diameter of the buoy. The Strouhal number S is generally a function of the particular cylindrical shape of the elastically held body and the Reynolds number Re, ($Re = UD/\nu$, where ν is the kinematic viscosity). Flow-induced motions will occur when the Strouhal frequency is at or near the natural frequency of the elastic system, i.e., at or near the natural frequency of the pendulum like mooring cable - buoy system.

Three different types of solution methods in eliminating the detrimental buoy motions can be followed; one, holding the buoy in place by means of multipoint moors; two, detuning the system, i.e. insuring that none of the natural (resonant) frequencies of the vibrational system is at or near the Strouhal frequency; three, preventing the uniform, alternating vortex shedding on the buoy. The first method was not pursued here, because working with several mooring cables for the same buoy becomes difficult and is unpractical, particularly for long cables. The second solution method proved unfeasible because the detuning of the buoy-cable system had to be accomplished for the complete range of current velocities from 0 to 1.5 m/s. In other words, the physical parameters of the vibrational system such as cable length and metacentric height could not be adjusted within the given constraints that the lowest critical ocean current velocity is above 1.5 m/s, where the critical velocity is that velocity at which the vortex-shedding frequency equals one of the natural frequencies of the system. The third method of reducing the flow-induced buoy motions aims at preventing the uniform vortex shedding from the cylindrical buoy. This method was explored by attaching different types of spoilers and splitter plates to the buoy as is described in this article.

2. DESCRIPTION OF THE VIBRATIONAL SYSTEM

The vibrational system is similar to that of a compound pendulum as shown in Figure 1. The buoy is a circular cylinder 0.53 m in diameter and 2.74 m long. The length of the cable is variable between 3 and 15 m. The positive buoyancy ($W = B - w$) is 1419 N. The distance between the center of buoyancy and the center of mass of the buoy ($s = \ell_B - \ell_W$) is also variable between 0 and 0.30 m.

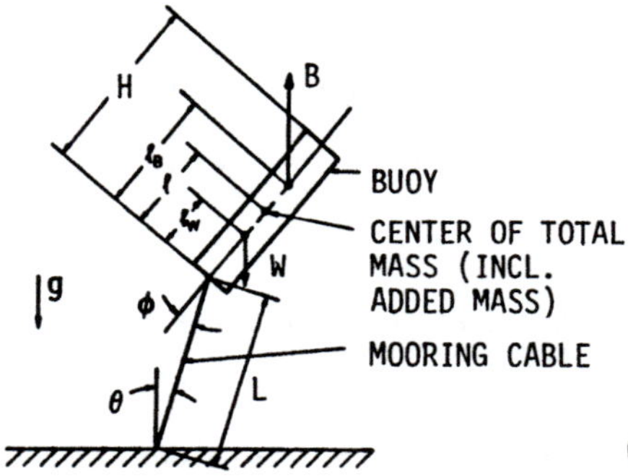

FIG. 1 END-PLATE MOORING OF BUOY (Flow Direction is Perpendicular to the Plane Shown).

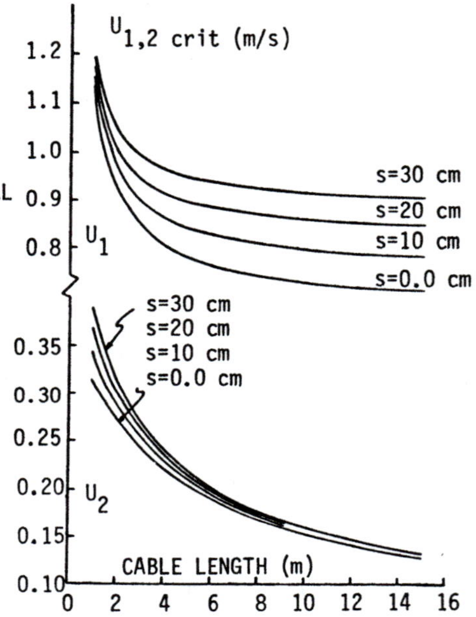

FIG. 2 LOWER AND HIGHER CRITICAL VELOCITIES AS FUNCTION OF CABLE LENGTH AND METACENTRIC HEIGHT ($s = \ell_B - \ell_W$) FOR SYSTEM OF FIGURE 1.

The undamped natural circular frequencies of the system were shown by Sallet and Berezow (Ref. 1) to be

$$\omega^2_{1,2} = \frac{\left\{ \frac{\ell_B B - \ell_W W}{m\ell^2 + I} + \frac{W}{mL} \right\} \pm \sqrt{R}}{2\left(1 - \frac{m\ell^2}{m\ell^2 + I}\right)} \quad (1)$$

where $\left(\frac{\ell_B B - \ell_W W}{m\ell^2 + I} + \frac{W}{mL}\right) - 4\left(1 - \frac{m\ell^2}{m\ell^2 + I}\right)\left(\frac{W(\ell_B B - \ell_W w)}{mL(m\ell^2 + I)}\right)$

The critical velocities near or at which flow-induced transverse vibrations for this system will start are:

$$U_{1,2}(\text{crit}) = \frac{(\omega_{1,2})D}{2\pi S} \quad (2)$$

The mass m and the moment of inertia include the added mass. Theoretical hydrodynamics shows that the added mass of a very long circular cylinder is equal to the mass of the displaced fluid, if the cylinder is in translation perpendicular to its center axis. This is the case at the start of the first mode of vibration, particularly when the mooring cable is long. The second mode of vibration, however, is essentially a rotational motion of the buoy around its center of total mass with the axis of rotation parallel to the flow direction. Compound pendulum tests with circular cylinders having the same length to diameter ratio as the here discussed buoy - cable system showed that equation (1) predicted the natural frequencies better when only 8/10 of the theoretical added mass was taken.

Figure 2 shows the lower and higher critical velocities of the buoy-mooring cable system. Experiments with the unmodified buoy which will be discussed in the next section are in good agreement with the estimated critical (or resonance) velocities. It is seen that it is not possible to shift the lowest critical velocity past 1.5 m/s by adjusting L or s or both within the given design criteria. For a study of the effect of the positive buoyancy and of the moment of inertia on $U_{1,2(crit)}$ the reader is referred to Reference 1.

3. SUPPRESSION OF FLOW-INDUCED BUOY MOTIONS USING SPOILERS

The rational for using spoilers is to attempt to break up or otherwise significantly influence the uniform vortex shedding from the surface of the buoy. Intuitively it seems plausible that generating many small vortices at right angles to the surface of the cylinder would prevent or at least to some degree inhibit the strong and uniform large vortex formation (vortex axis is parallel to buoy axis) which causes the flow-induced buoy motions.

The spoilers chosen were bonded to 1.59 mm thick neoprene sheets which were then taped to the surface of the buoy. Some of the spoilers tested are: [1] Lexan rods, 6.35 mm in diameter, 12.7 mm long, spaced equally 152.4 mm apart in columns which run parallel to the cylinder axis, successive columns are spaced 152.4 mm apart, successive columns are staggered by one half the column spacing (see Figure 3). [2] Same as 1. except the length of the lexan rods is 50.8 mm. [3] Same as 1. except the spacing within the column is 76.2 mm and the spacing between columns is 76.2 mm. [4] Rubber rods, 25.4 mm in diameter, 152.4 mm long, spaced equally 152.4 mm apart in columns which run parallel to the cylinder axis, successive columns are spaced 139.7 mm apart and are in line with the neighboring columns. [5] No modification. [6] Same as 1. except the lexan rods are 12.7 mm in diameter. [7] Same as 1. except the lexan rods are 12.7 mm in diameter and 50.8 mm long. [8] Same as 1. except the rods are 12.7 mm in diameter and the spacing is that of 3. [9] Same as 4., except that the spacing within the columns is 76.2 mm. [10] Nylon rope of 15.88 mm diameter arranged in a herring bone pattern; the rope is inclined 45°, the buoy is covered with 3 sections of parallel rope sections, the rope is separated 152.4 mm on centers (see Figure 4). [11] Same as 10., except that the rope is inclined 60° to the cylinder axis. [12] Same as 10., except that the rope is inclined 30° to the cylinder axis. [13] Same as 10. except that rope is inclined 30° to the cylinder axis and the rope separation is 76.2 mm on centers. [14] Inverted "T" extruded rubber spiral protruding 15.88 mm from the surface, six turns around the buoy. [15] Secondary vortex generators consisting of 25.4 mm by 25.4 mm by 3.18 mm rubber plates inclined 80° to the cylinder axis and irregularly spaced, 60 to 100 mm apart. [16] Rubber pieces, 50.8 mm long, with right triangular cross section, spaced 76.2 mm in a row and 127 mm in a column, flat sides facing each other in pairs along each column. [17] Aluminum angles, 19mm by 19 mm by 0.8 mm thick and 76.2 mm long, the angles are mounted length-wise parallel to the cylinder axis and are separated 76.2 mm from row to row and 101.6 mm within a column (parallel to the cylinder axis); see Figure 5.

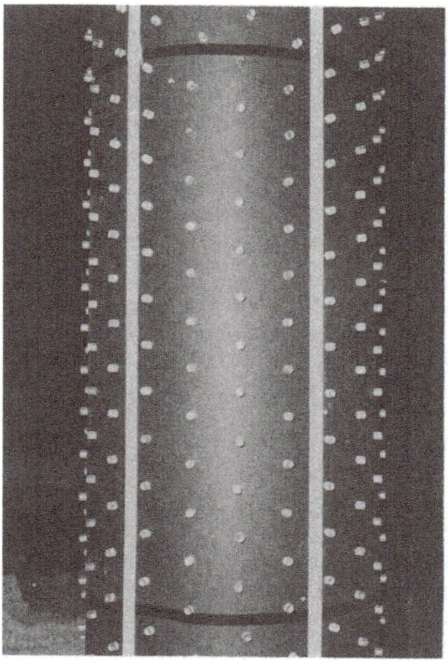

FIG. 3 SPOILER: ARRAY OF LEXAN STUDS

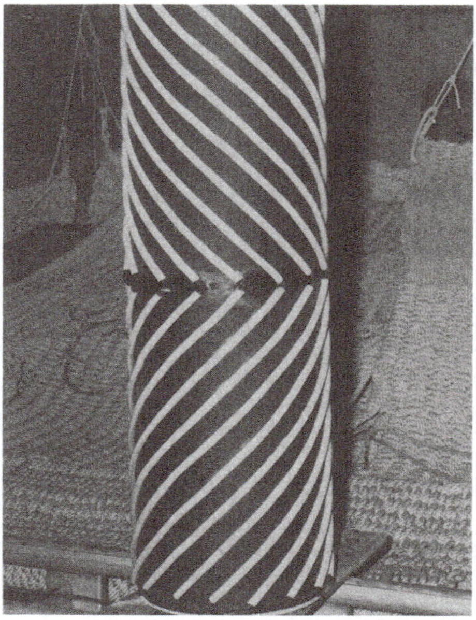

FIG. 4 SPOILER: NYLON ROPE, HERRING BONE PATTERN

FIG. 5 SPOILER: ALUMINUM ANGLES

The buoy was fitted with the described surface modifications and towed through the body of still water at different velocities. This was accomplished by attaching the mooring cable to a sled which moved on rails which in turn were mounted to the floor of a long towing tank. The motion of the buoy was measured photographically and is shown in Figure 6. The shown amplitude is the maximum horizontal displacement of the center of the buoy from its position at rest. Since the water was allowed to settle after each test run, the flow relative to the buoy is turbulence-free immediately ahead of the buoy.

4. SUPPRESION OF FLOW-INDUCED BUOY MOTIONS USING A SPLITTER PLATE

A more effective method of reducing the flow-induced motions of the buoy was found to be stabilization of the near wake by means of a splitter plate. The reason why splitter plates are effective in stabilizing the buoy and the necessary length of the plate was discussed by Sallet (Ref. 2). The vibrational system for the tests with splitter plates differed from that used for

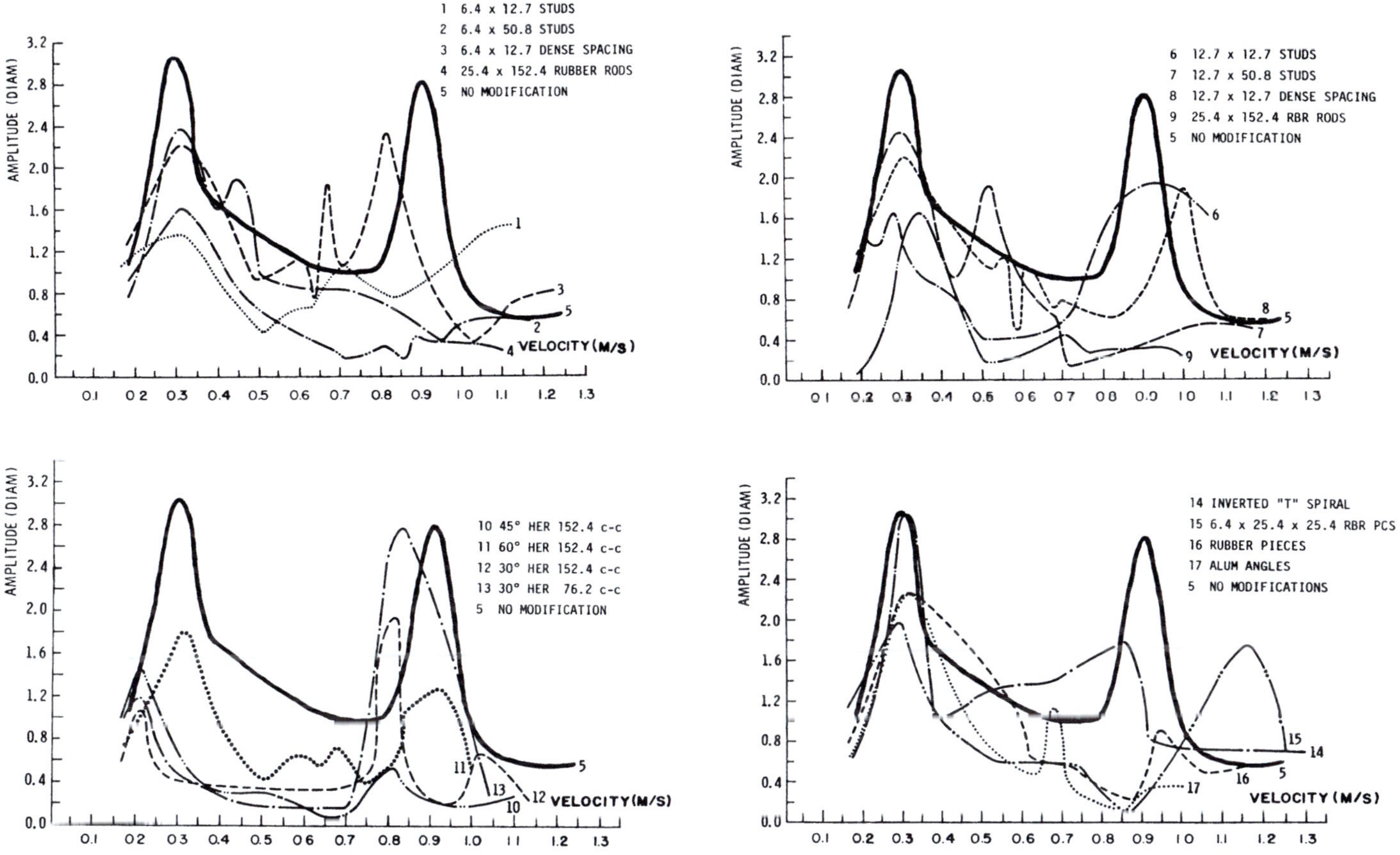

FIG. 6 EFFECT OF SURFACE MODIFICATION ON VIBRATION AMPLITUDE

spoilers. The mooring cable was attached to the buoy by means of a bridal through the center of drag (see Figure 7). This change eliminated the displacement of the buoy out of the initially vertical position due to steady drag alone.

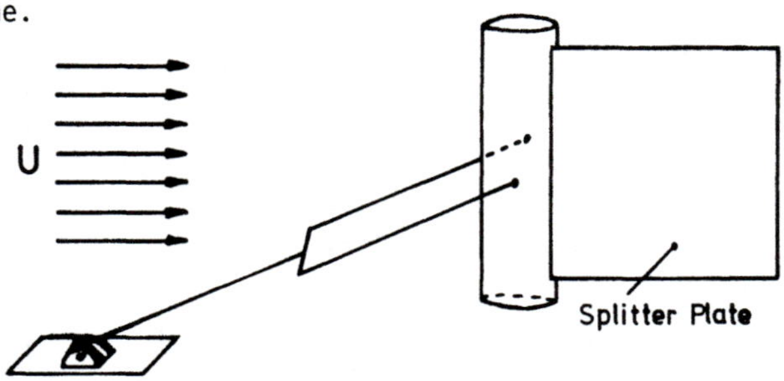

FIG. 7 STEADY CYLINDER WITH STABILIZING SPLITTER PLATE

The buoy was towed submerged through a tank of still water as was previously described. The recorded amplitude of the flow-induced pendulum-like oscillations is shown in Figure 8. Test series marked with A, B and C refer to

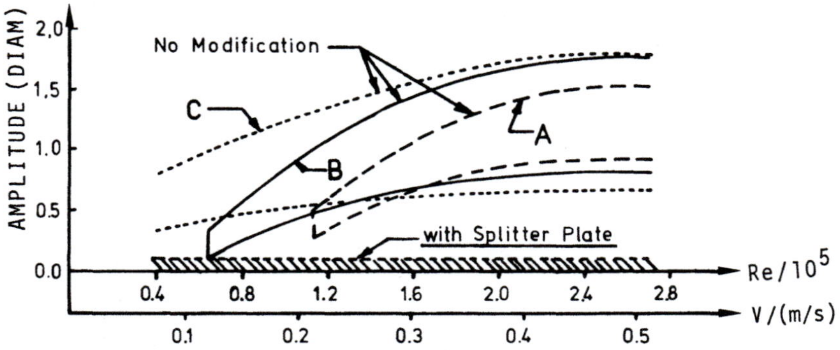

FIG. 8 AMPLITUDE VERSUS REYNOLDS NUMBER

tests of the unmodified buoy with mooring cable lengths of 4.57 m, 7.62 m and 11.23 m, respectively. For each test series an upper and lower value is given which brackets the observed values. The shown amplitudes for the buoy with a splitter plate include tests with all three cable lengths. The splitter plate was 3 buoy diameters long, extended 9/10 the length of the buoy and was recessed equally at both ends of the buoy. The material for the splitter plates was either 1.59 mm thick aluminum sheet or 3.18 mm thick plywood made neutrally buoyant with polystyrene and aluminum angles, respectively.

5. DISCUSSION AND CONCLUSION

The buoy was bottom moored with a single cable in two different ways; first, the cable was attached to the flat end of the buoy (see Figure 1) and second, the effective point of attachment was at the center of drag of the buoy by means of a bridal. The flow-induced transverse vibrations of the end-plate moored buoy resembled the oscillations of a compound pendulum. The two peaks in the amplitude-flow velocity plots (see solid line in Figure 6) are the excited natural (resonance) frequencies of the vibrational system. The velocities at or near which the peak amplitudes will occur can be estimated by equating the Strouhal frequency to the natural frequency of the vibrational system as was shown. This procedure yields an estimate only because of the following uncertainties and complications which were not taken into account:

1. The Strouhal number is Reynolds-number dependent. This dependency is particularly severe in that varying values for the Strouhal number have been observed in the range $2 \cdot 10^5 \leq Re < 3.5 \cdot 10^6$ which is encountered before the sec-

ond amplitude peak. A summary of applicable Strouhal numbers was given by Sallet (Ref. 3). The value taken for the calculations shown in Figure 2 was constant and equal to S = 0.2. The influence on lift and drag resulting from oscillatory motion was discussed by Sallet (Ref. 4).

2. The mass and the moment of inertia of the vibrational system have to include the added mass. The added mass depends not only upon the shape of the body undergoing motions in the fluid but also on the type of motion. The added mass of the here discussed finite circular cylinder was determined experimentally and found to be approximately 0.8 of the theoretical value for cylinders of infinite length in translation perpendicular to the cylinder axis.

3. The velocity for which the Strouhal numbers was determined as function of the Reynolds number is the velocity relative to the vortex shedding body. The velocities shown in Figure 2, 6 and 8 are the undisturbed free-stream velocities and are not adjusted for the fact that the buoy will be exposed to additional relative fluid velocity once the flow-induced motions start. Also, as a consequence, the excitation forces do not remain in a plane perpendicular to the free-stream velocity.

4. The end-plate moored buoy leaned back due to the drag thereby causing the flow not to be perpendicular to the cylinder axis. The velocity component perpendicular to the axis should be taken for the determination of the Strouhal number. It is seen that the errors introduced by not adjusting the critical velocity predictions for the last two effects partially offset one another.

The flow-induced motions of the bare cylinder when the center of drag mooring system was used did not exhibit the two distinct amplitude peaks of the double pendulum type mooring system. The critical velocities of this mooring configuration can be predicted using a simple pendulum as the vibrational model. Figure 8 shows the range of observed amplitudes over the tested velocities.

The tow tests showed that the splitter plate is the most effective means of preventing flow-induced vibrations. The splitter plate should be at least 3 cylinder diameters long and extend along the length of the cylinder; in addition the plate should be stiff enough so that it does not bend or have a flag-type flutter.

Surface attachments to the buoy (so-called spoilers) also reduce the flow-induced transverse motions. Figure 6 shows the effect of 12 different surface modifications. It is seen, that generally the most effective surface modification for damping the amplitudes at the lower critical flow velocities is not the most effective one for damping the amplitudes at the higher flow velocities. A herring-bone type spoiler such as modification 10, 11, 12 or 13 proved to be the best to prevent excessive flow-induced motions at the lower critical velocities while spoilers consisting of long rods protruding perpendicular from the surface of the buoy (modifications 7 and 9) were generally most effective at higher flow velocities. What speaks against application of spoilers, finally, is the possibility of mussle growth and the attachments of barnacles if the buoy is left in the sea for a longer period of time.

REFERENCES

1. Sallet, D.W. and Berezow, J., "Suppression of Flow-Induced Vibrations by Means of Body Surface Modifications", The Shock and Vibration Bulletin, Vol. 42, No. 4, 1972, pp. 215-228.
2. Sallet, D.W., "A Method of Stabilizing Cylinders in Fluid Flow", J. of Hydronautics, Vol. 4, No. 1, 1970, pp. 40-45.
3. Sallet, D.W., "On the Self Excited Vibrations of a Circular Cylinder in Uniform Flow", The Shock and Vibration Bulletin, Vol. 40, No. 3, 1969, pp. 303-309.
4. Sallet, D.W., "The Drag and Oscillating Transverse Force on Vibrating Cylinders Due to Steady Fluid Flow", Ingenieur-Archiv, Vol. 44, 1975, pp. 113-122.

SUBMERGED MOORED CYLINDRICAL BUOY

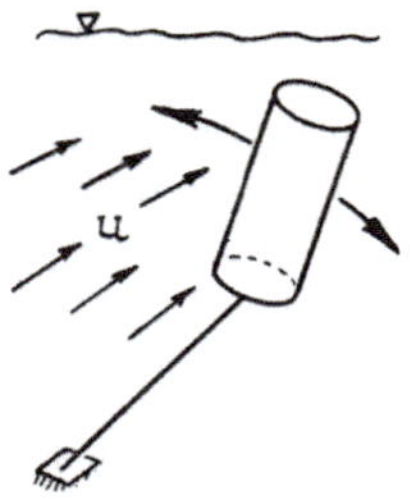

SKETCH 1

BUOY WITH SPOILERS

Problem: Flow-induced cross-wise oscillations of the buoy occur at certain ocean current velocities. These motions falsify current-velocity measurements and cause problems with sound monitoring devices.

Geometry: See sketch 1. The buoy is 0.53 m in diameter and 2.74 m long. The mooring cable is variable in length, up to 15 m.

Boundary conditions: Buoy is completely submerged; the mooring depth is such that surface effects do not occur.

Flow conditions: The current velocity can vary between 0 and 1.5 m/s. The approach flow is virtually turbulence free. These ocean conditions are simulated in full-scale tests in which the buoy is towed through a tank of still water.

Structural dynamics: The moment of inertia including the added mass is 598 mNs^2, the positive buoyancy is 1419 N and the metacentric height is 0.3 m.

Vibration: Peak amplitudes of approximately 3 cylinder diameters occur at absolute free-stream velocities of 0.3 m/s and 0.9 m/s for a system having a cable length of 6.1 m; these observed values are in line with predicted critical current-velocity estimates.

Cure: Attachment of spoilers to the surface of the buoy. Many different spoilers were investigated. One of the most effective spoilers results from bonding pieces of rope to the cylinder surface in a herring bone pattern (see photograph). The rope is 15.88 mm in diameter, inclined 45° and separated 152.4 mm on centers. The reduction in buoy oscillation is shown in the given amplitude versus free-stream velocity graph. An alternative cure consists in the attachment of a splitter plate. The splitter plate must be at least 3 diameters long and extend along the length of the cylinder.

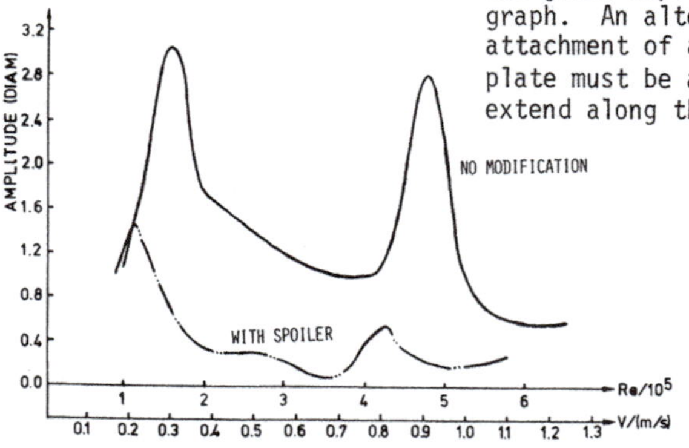

AMPLITUDE OF VIBRATION VERSUS CURRENT VELOCITY

F1 ON THE STATE OF STABILITY CONSIDERATIONS FOR SUSPENDED-SPAN BRIDGES UNDER WIND

Robert H. Scanlan

Department of Civil Engineering
Princeton University
Princeton, New Jersey, USA

SUMMARY

The paper briefly comments on the present state of the three outstanding problems of suspended-span bridges as they are affected by wind. These are vortex shedding, flutter, and buffeting. Each contains elements of self-excitation. In the first, the "lock-on" phenomenon, or the strong influence of structural motion upon concerted vortex shedding, is emphasized. The role of this phenomenon in the famous Tacoma Narrows incident is reviewed and is then disassociated with the more complex aeroelastic mechanism that instituted the final collapse of that bridge. This latter mechanism is better characterized as torsional, separated-flow flutter. This type of instability, associated with a distinctive and irreversible evolution of aerodynamic damping toward negative values as wind velocity increases, can be found to lie at the basis of the flutter instability of many bluff-section suspended-span bridges. However, this phenomenon, in turn, is to be sharply distinguished from "classical" or airfoil-type flutter, which may affect only very streamlined bridge types and usually is also associated with high flutter speeds. The total inappropriateness, however, of airfoil flutter theory to all cases of bridges is underscored. The necessary theory and experimental data for predicting bridge response to buffeting are discussed. The roles of sectional and three-dimensional wind tunnel models of bridges are assessed. The relatively modest state of full-scale testing is briefly commented upon. The paper acknowledges the work of many investigators over a four-decade period.

1. INTRODUCTION

For well over 100 years prior to the famous Tacoma Narrows incident in 1940 the failures of bridges under wind loading were scattered through the historical records. For many of those years the central question at issue was the assessment of the steady forces of the wind. This had been largely resolved, however, by 1940; and the remaining issue became the unsteady forces at play. Fig. 1 [4, Part I], [64] depicts an incident of 1836 that involved such forces. These have continued to be the focus of attention to the present time and in fact have formed an important segment of the larger study of bluff-body aerodynamics.

The rising tide of aerodynamic theory and experience had become strong enough by 1940 to permit some excellent insights into the aerodynamic mechanisms operating at Tacoma Narrows and converted that episode into a pivotal one of discovery. Yet today, some 4 decades later, further insights are still being sought, and newer, more analytical approaches to the problems of suspended-span stability are being developed, many through analogy with aeronautical results. Throughout, however, it has been necessary to emphasize the intrinsic differences that distinguish the pertinent bluff-body aerodynamics from the streamlined-body aerodynamics so typical of aeronautical applications.

FIG.1 BRIGHTON CHAIN PIER NOVEMBER 1836

Within the problems of unsteady, bluff-body aerodynamics there have been two important types -- those dealing with unsteadiness intrinsic to the flow, and those treating the consequencies of structural motion. In the first category lie the problems of vortex- and turbulence-initiation by the structure as well as buffeting of the latter by incident turbulence induced elsewhere. In the second category lie the full range of aeroelastic problems associated with fluid-structure interaction. After 1940 the preoccupation with bluff-body aeroelasticity grew as interest arose in new bridge designs.

Civil engineering structures, designed primarily with other major preoccupations than aerodynamics, have drawn necessary attention to the effects of wind upon bluff structures. Long-span bridges, bringing the additional factor of flexibility to this picture, have required the further consideration of the aeroelastic problems of such structures. For example, among the central concerns of a bridge project with a 2-mile center span for crossing the straits of Messina (Fig. 2 [1]) are its wind-driven dynamics. Under today's standards each new suspended-span bridge design has come to require proper analyses of its stability tendencies under wind. Some of the elements that appear in such analyses will be reviewed here.

FIG. 2 A STRAITS OF MESSINA PROJECT

2. VORTEX SHEDDING

Fixed, bluff objects in a two-dimensional flow shed vortices in a periodic fashion according to the well-known Strouhal relation. This is a phenomenon intrinsic to the flow itself, and it appears, according to recent observations (Fig. 3 [2]), to take place in organized fashion even up to Reynolds numbers exceeding 10^9, contrary to previously-held belief.

Mathematical characterizations of the phenomenon have been complex and, to date, few have been made that are completely adequate to the particular form of the problem that concerns bridges. However, bridges certainly shed vortices, particularly those with decks that have closed rectangular sections or other bluff forms such as the flattened H-section of the original Tacoma Narrows bridge. Over an extended, line-like structure in three dimensions, such as a long bridge of bluff cross-section, vortex shedding does not tend to develop completely coherently along the span. This is true even in very smooth flow, but the presence of turbulence in the natural wind breaks up vortex-shedding coherence even more.

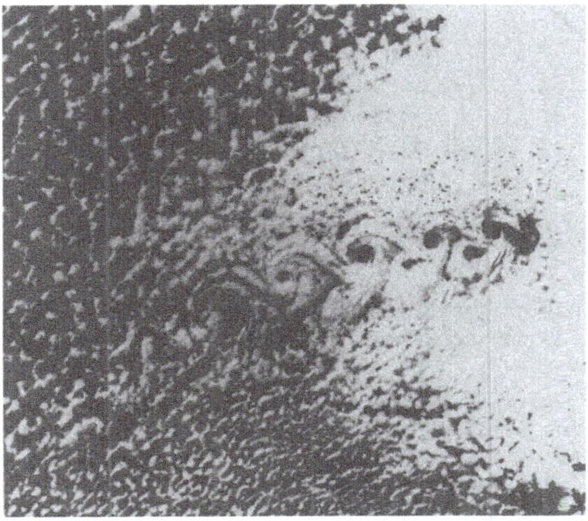

FIG. 3 CLOUD VORTEX PATTERN OVER JAN MAYEN ISLAND

However, quite coherent vortex discharge can still occur, given an important source of rhythm control, namely structural motion. When such motion occurs, most commonly at a natural frequency of the structure, the vortices shed by the wind may "lock on" to the rhythmic shift in local boundary position if the two natural rhythms -- fluid and structure -- are quite close together. In this circumstance the rhythm of vortex discharge ceases to conform exactly to the Strouhal rule and becomes instead the structural rhythm. The flow details further become considerably more complex than they are intrinsically when the flow alone dictates the frequency over the fixed structure. The phenomenon becomes aeroelastic.

Accompanying lock-on, it should be noted, is usually a narrow range of across-flow velocity, above and below which loss of the structural rhythmic control of vortex shedding occurs. Fig. 4 [2] depicts the typical effect of structural motion upon vortex rhythm. With bluff bodies like bridge decks in wind,

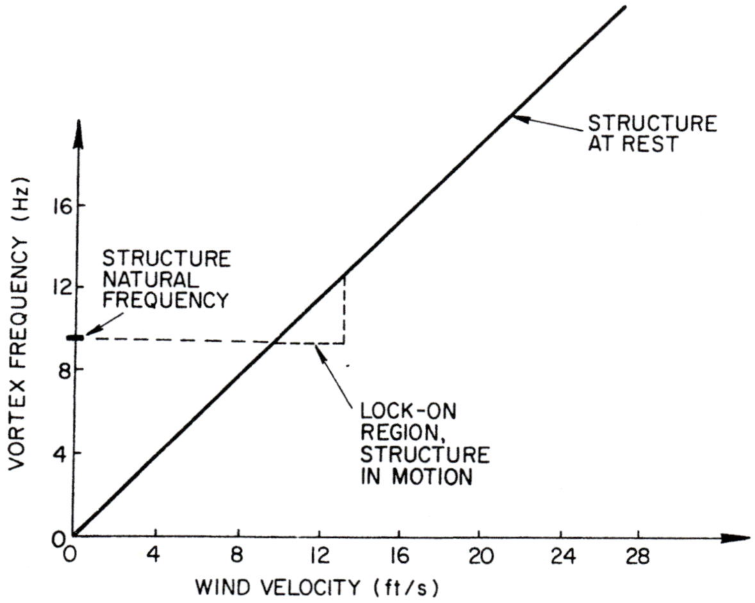

FIG. 4 VORTEX FREQUENCY CHARACTERISTICS WITH AND WITHOUT BODY MOTION

lock-on and such loss of synchronization occur mainly in connection with vertical structural motion, though occasionally also with very low-amplitude torsional motion.

A structure fixed in a steady flow develops alternating net forces and moments according to the Strouhal relation for natural vortex shedding. If the structure is also displaced it sheds additional wake vorticity and develops consequent surface pressures initiated by such movement. If this motion is periodic, the net resulting forces and moments will tend to contain components at two important frequencies: the Strouhal frequency and the frequency of mechanical motion. Which of these components predominates depends upon the degree to which the mechanical motion influences the local flow effects about the body. The relative importance of one or the other component may also simply depend upon their frequency separation and the frequency band upon which attention is focused for one reason or another. Fig. 5 [60] illustrates the relative amplitudes of transverse force components related to the Strouhal frequency f_s and the transverse motional frequency f_m for a square prism in two-dimensional flow at five different across-flow velocities.

For example, for large-amplitude, low-frequency vertical motion (as in "galloping") the relatively high frequency effects of vortex shedding can, in detail, be ignored, and only their average effects retained to explain the main aerodynamic forces influencing the phenomenon involved.

In the case of small-amplitude across-flow motion of the structure--motion of amplitude comparable to the thickness of the local boundary layer--the wake vorticity due to the motion can become confused with, and dominate, the naturally shed (Strouhal) vorticity, resulting in a single, structure-controlled, vortex-shedding rhythm; but this can only occur over a small range of near-coincidence of the two frequencies. Thereafter, should, for example, the Strouhal rhythm increase, the motional wake vorticity cannot so simply coalesce with the natural vorticity, and the lock-on pattern of the latter eventually ceases.

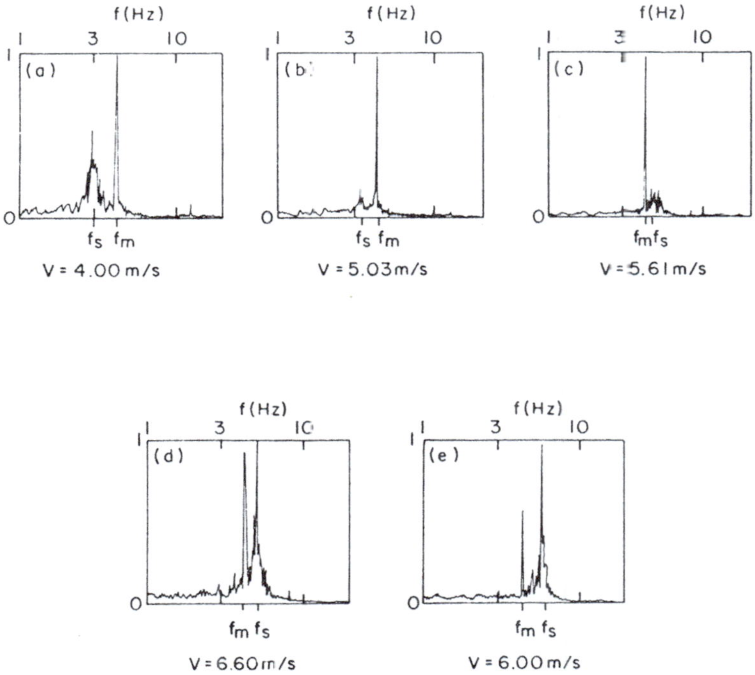

FIG. 5 FORCE SPECTRA ASSOCIATED WITH STROUHAL FREQUENCY AND BODY MOTION FREQUENCY

Wardlaw [3] has described a clear case of vortex-excited bridge deck vertical motion, the Long's Creek bridge, which was alleviated by judicious streamlining of the deck as an outcome of wind tunnel model studies. Fig. 6 depicts this modification.

FIG. 6 LONG'S CREEK BRIDGE AS MODIFIED

At Tacoma Narrows, classic vortex shedding played a prominent (though not the final catastrophic) role in the action of that bridge. What is strongly emphasized in the studies by Farquharson ([4], Part III, Ch. I-IV) and by Karman and Dunn ([4], Part III, Ch. VII), is that vortex xhedding became clear and well defined from that bridge only when deck motion occurred. The demise of that bridge has long been, according to conventional wisdom, laid simply to "vortex shedding", probably due in the main to the insistent use of that phrase by Karman and Dunn. However, a study of Ref. [4] reveals that an effect quite different from the simple lock-on of naturally shed vortices was associated with the final catastrophic motion at Tacoma Narrows.

All of the motion observed there, up to within about an hour of the collapse, was vertical deck motion only, and quite certainly vortex-induced. Low amplitude across-flow motion, however, does not have the strongly disruptive effect upon the general pattern of the local flow that torsional motion does. The collapse mechanism at Tacoma Narrows in a torsional mode has become the classic example of what is now called separated-flow torsional flutter. Fig. 7 [4, Part I] illustrates this mode just prior to the collapse.

FIG. 7 ORIGINAL TACOMA NARROWS BRIDGE IN ITS FAILURE MODE

3. SEPARATED-FLOW TORSIONAL FLUTTER

In the case of torsional motion within a flow of a bluff structure (such as an elongated rectangular prism, a flat H-section, or the original Tacoma Narrows deck section) the local flow becomes strongly modified. In particular, rotation of the leading surface upward is accompanied by an increased vortex discharge just above that surface; reversal of the motion creates a similar counterpart below the leading surface. These discharged vortices, close to the structure and not yet in its wake, pass along the surface giving rise to locally destabilizing pressures, i.e. pressure tending to aid the motion that initiated them. Fig. 8 (after [62], [7], [20]) illustrates progressively displacing vortices over a flat H-section that starts from rest and rotates to a position of maximum displacement.

Under the conditions described, the structural motion dominates the production of vorticity into the wake, and effects, if any, at the Strouhal frequency

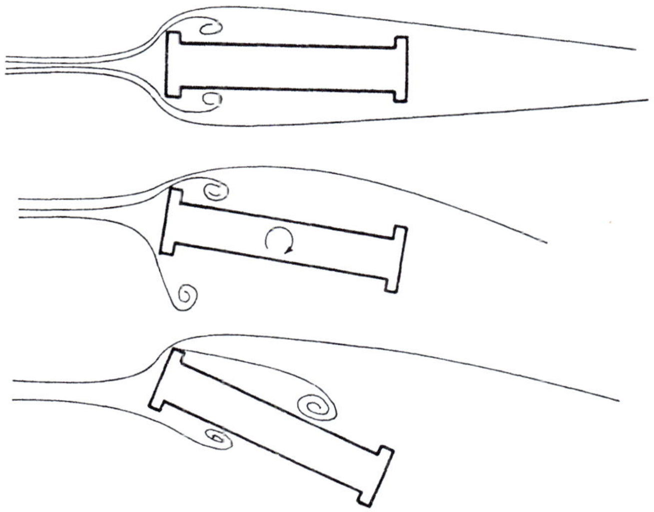

FIG. 8 VORTEX PATTERNS ACCOMPANYING TORSIONAL MOTION

become completely submerged in stronger effects. The mechanism described can, beyond some value of the reduced velocity (based on structural frequency), cause a torsional structural mode to auto-excite more and more violently with increasing wind velocity. Since natural Strouhal effects are now submerged, no eventual reversion to such effects (loss of synchronization) can occur. What has been described here no longer falls within the category usually understood as vortex shedding effects, but apperas to ceserve the new description of torsional flutter.

This is, in fact, the central phenomenon in most practical cases of bridge oscillatory instability under wind. The phenomenon has been characterized [5], [6], [7], as to overall effect, by a gradual diminution and eventual reversal in the sign of aerodynamic damping in torsion as wind velocity increases. Fig. 9 [5] depicts the evolution of a typical dimensionaless coefficient A_2^*, which is proportional to aerodynamic damping in torsion as the reduced velocity U/NB increases. (U = across flow velocity, N = oscillation frequency, B = deck width.) Fig. 10 [7] is a version of the same effect noted as early as 1941 during the Tacoma Narrows studies. (The ordinate δ/μ is the ratio of damping logarithmic decrement to a dimensionaless parameter comparing air density to structural density.) As has been suggested above, the flow mechanisms affecting the trends of these curves have their origins in body motion and thus have little or no relation to the natural shedding of wake vortices from a fixed body in the usual Strouhal-Karman-Roshko sense [8], [9].

The significant departures from ordinary wake vortex lock-on are the motion-related vortex discharge over the leading edge of the bluff body and the complete absence of evidence as to any loss-of-synchrony mechanism, permitting the phenomenon to proceed to catastrophic proportions with increasing wind velocity. During the occurrence the wake is still, of course, an array of alternating vortices, but one whose longitudinal spacing is not fixed naturally by the fluid. Rather, it is dictated by the reduced structural frequency only. In short, it is a flutter wake, not a Karman vortex street.

Most bridge deck sections that are oscillation-prone at elevated wind velocities can be shown by model studies to have separated-flow torsional flutter tendencies, even though in many cases some additional tendency toward torsion-bending aerodynamic coupling may be present. This has been commented on by Nakamura [10] and Scanlan [11]. In short, a strong intrinsic tendency toward single-degree

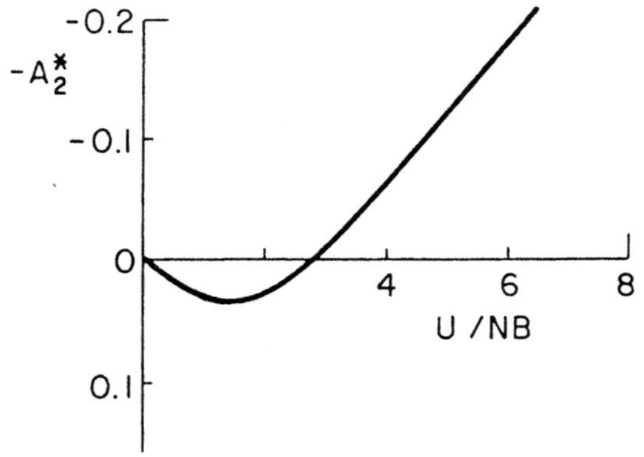

FIG. 9 TYPICAL TREND OF TORSIONAL AERODYNAMIC DAMPING COEFFICIENT FOR AN UNSTABLE BRIDGE

torsional flutter may "drive" the participation of vertical modes if some incidental aerodynamic coupling exists, as it often does [5].

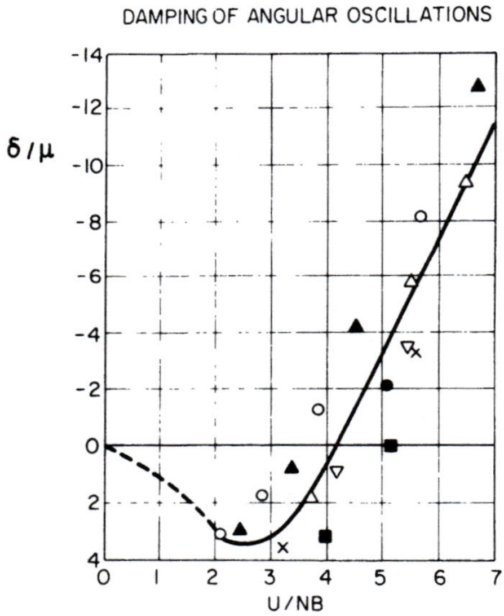

FIG. 10 KARMAN AND DUNN'S RESULT FOR TORSIONAL DAMPING, ORIGINAL TACOMA NARROWS MODEL

4. CLASSICAL FLUTTER

The usual flutter phenomenon of thin airfoils involving laminar, unseparated flows may be called "classical"; in most such cases both the elastic axis and the center of gravity of the airfoil section lie somewhere within its chord. In the case of bridges, where the elastic axis and c.g. are usually centered on the deck width, classical flutter with non-separated flow cannot occur without the coupling of bending and torsion degrees of freedom. When coupled flutter occurs the two degrees of freedom oscillate at the same frequency but with an important difference of phase between them, allowing an action of the system that extracts energy from the wind stream.

If a bridge deck is highly streamlined, such flutter may be possible at relatively high wind velocity, and occurring in a manner more strongly influenced by the stiffness terms of aerodynamic and structural origin in the equations of motion than by the damping terms. In fact, for both airfoils and well-streamlired bridge deck sections, individual vertical and torsional damping terms remain entirely on the side of stability for all values of the reduced velocity. For usual structural parameters, the phenomenon of classical flutter is possible only for a "high" velocity, so that streamlined bridge decks (Severn [12], Lillebaelt [13], Burrard Inlet [14], Humber [15], Bosporus [16], etc.) tend to exhibit high flutter speeds, often well beyond practically expected wind velocities at the bridge site. Fig. 11 presents three streamlined bridge sectional contours: they represent the Severn (L), Lillebaelt (L) and Burrard Inlet (B) project forms.

The theoretical framework of classical flutter has been a considerable guide, as to broad background questions of methodology, to researchers into questions of bridge flutter [5], [10], [21], [27], [28], [58].

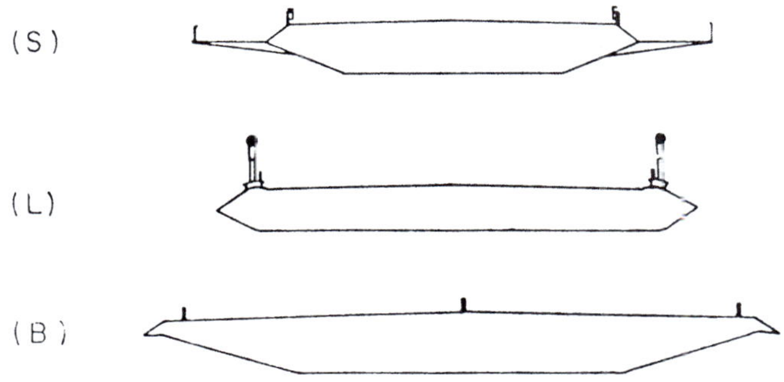

FIG. 11 THREE STREAMLINED BRIDGE DECK FORMS

5. THE COMPARISON OF SEPARATED-FLOW AND CLASSICAL FLUTTER

As has been pointed out, separated-flow, purely torsional flutter instability is a phenomenon, central to the bridge deck stability question, that is not to be confused either with natural vortex shedding, on the one hand, nor classical flutter on the other. In the course of a number of past bridge flutter investigations, such confusion has repeatedly occurred.

In the case of the linearized theory of classical flutter, the unsteady-motion lift and moment expressions are characterized by aerodynamic derivatives that are functions of reduced velocity, i.e., the dimensionless ratio defined as the velocity expressed in chord widths per second divided by the frequency of oscillation. Such terms also have alternative and equivalent formulations [17] as integrals over the entire motion that contain a kernel that is well known as the Wagner [18] or indicial-lift function. This is a smooth, monotonically increasing function that begins at half the eventual full lift value and proceeds, with time, asymptotically to unity, or the full-lift value (see Fig. 12).

On the other hand, in the case of separated-flow flutter, the Wagner function has been shown conclusively [17], [19], to be inapplicable, requiring

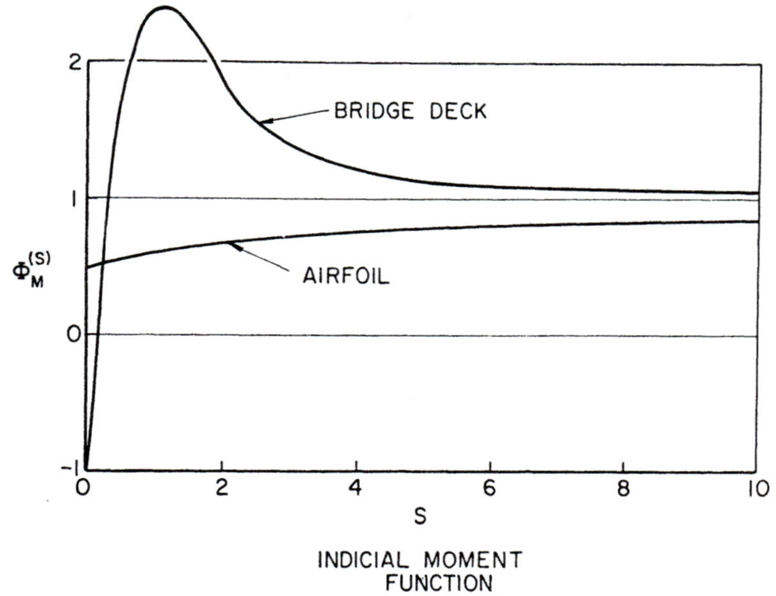

FIG. 12 COMPARISON OF INDICIAL MOMENT FUNCTIONS, BRIDGE AND AIRFOIL

replacement by a quite different indicial function characterized by an initial high "overshoot" followed by an asymptotic decay toward the steady unit value from above (rather than from below as in the Wagner case). Fig. 12 [17] presents a comparison between a typical bridge-deck indicial function and the Wagner function.

Thus, the two characterizations, mathematically equivalent respectively in frequency and time domains: the sign-reversing torsional aerodynamic derivative on the one hand, and the highly peaked indicial lift function on the other, have shown beyond any doubt that the majority of bridge instabilities are not cases of classical flutter.

A recent paper by Yoshimura and Nakamura [20] has further emphasized this point through a direct experimental investigation of the forms of the indicial functions for elongated rectangular bluff bodies. Selberg and Hjorth-Hansen have commented [21] on the limitations of flat-plate or classical flutter theory in the realm of bridge aerodynamics.

6. LINEARITY OF STABILITY THEORIES

There can be little doubt that the full phenomenon of any bridge deck instability is nonlinear, at least insofar as the details of the fluid flow are concerned. Yet, it is reasonable, and indeed experimentally supportable, to employ linear theories. This view is defended by the observations that, for design informational purposes: a) only small structural deflections are considered, b) only incipient dynamic phenomena are depicted, c) actual oscillations as viewed are often close to sinusoidal or to consist of a simple superposition of sinusoids, so that linear *post facto* theories can reasonably match observed dynamic activity.

Nakamura [22] has commented on the quite linear character of the aerodynamic responses of a bluff rectangular section for small motions.

On the other hand, certain workers, notably Miyata, Kubo, and Ito [23] have investigated some nonlinear effects. These may become pronounced for large amplitude excursions as in certain vortex-induced phenomena, high amplitude buffeting, or advanced stages of catastrophic flutter.

7. THE RESPONSE TO BUFFETING

The natural wind, stirred by surface obstacles, tends to be turbulent. Considerable meteorological effort has been devoted to ascertaining suitable representations of the spectral forms [2] that this turbulence takes. Fig. 13 presents typical nondimensional wind spectra in the form

$\frac{n\,S(n)}{u_\star^2}$ versus $f = \frac{nZ}{U}$ where n is frequency (Hz), Z is altitude, U is mean wind velocity, and $u_\star$ is a friction velocity equal approximately to $1/\sqrt{6}$ of the rms of fluctuating velocity. For applications to bridge engineering, information on longitudinal and vertical spectra, lateral cross-spectra, and scales are minimally necessary.

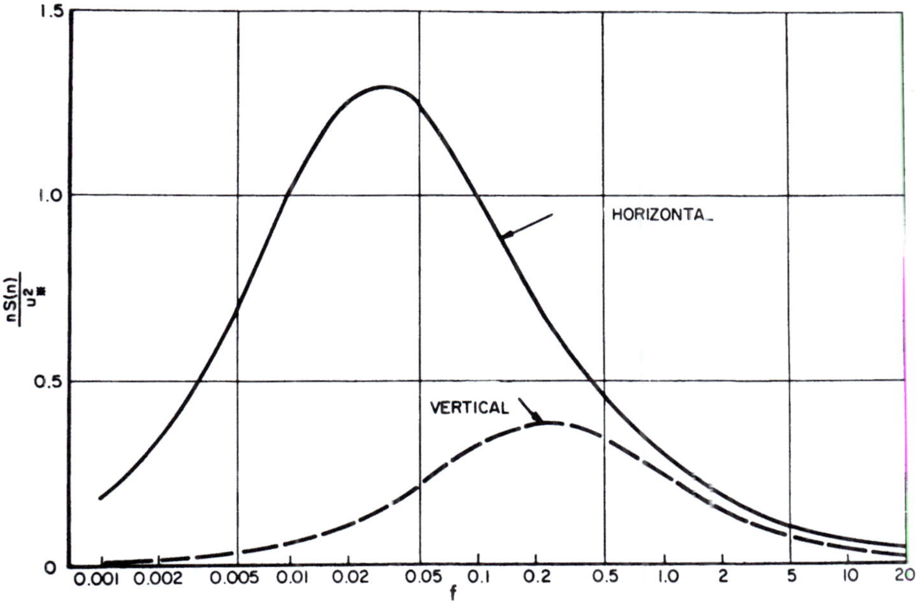

FIG. 13 SPECTRA OF LONGITUDINAL AND VERTICAL WIND COMPONENTS

It is important to recall that turbulence in the wind may both diminish certain undesired disturbances (such as vortex excitation and the onset of flutter) and increase others such as random motions. In the latter action it can be argued [24], [25] that the longitudinal pulsations of the wind are probably less disturbing than the vertical components. It is also important to comment that, in bridge design, it may on occasion be imprudent to count fully on the presence of natural turbulence to suppress an undesired structural response. Some observations [26] of natural wind at fair heights have found it unexpectedly laminar under certain meteorological conditions.

In the analytical formulation of the buffeting problem, a number of difficulties arise and the problem has therefore been a challenging one. The problem in principle requires a formulation that includes all aerodynamic excitation phenomena--from vortex excitation at "low" wind speeds to stability considerations (flutter) and external random-force excitations at higher wind speeds. It is usual to set aside the vortex-shedding question when using analytical approaches to relatively high-speed buffeting response. This may be completely justified if the bridge has either a very streamlined or an open-truss deck, and therefore sheds no strong vortices.

The stability problem alone received early attention by Bleich [27] and later Rocard [28] with what turned out to be essentially incorrect aerodynamics (use of airfoil theory out of context). Later work, of which [5], [29] indicate trends, based on schools of thought that required the basic aerodynamic data (flutter derivatives) to be experimentally acquired, was more successful, especially in documenting the aerodynamic ingredients of single-degree torsional flutter. Meanwhile, Davenport [30], [31] made notable first contributions to the bridge buffeting problem per se, especially to the input forces of the wind and to posing the problem as one in random vibrations. This work did not focus particular attention however upon those self-excitation forces that are ascribable to bridge motion. Most particularly, no provision was made for the outstanding proclivity of torsional damping to reverse sign to the side of instability in certain designs as reduced velocity increases. Recently, Irwin [32] for example, employing Davenport's theory, found it necessary to adjust torsional damping arbitrarily to get good agreement with wind tunnel buffeting results of a full model of Lion's Gate bridge. Other aspects of that study were then quite successful, as it served after modification to predict stresses in the bridge deck.

Work in Japan by a number of authors ([23], [33], [34], [35]) on the buffeting problem has produced useful methods of buffeting response prediction based on measured flutter derivatives and steady-state sectional force and moment coefficients.

Holmes [36], [37], has computed buffeting responses of the West Gate bridge by both statistical and Monte Carlo methods. As in Irwin's case [32], some adjustment for aerodynamic damping, based on experimental model evidence, was needed to bring calculated results into line with experiment. The net results exhibit excellent agreement with the wind tunnel results of Melbourne [51].

Béliveau, Vaicaitis, and Shinozuka [38] have performed exploratory buffeting calculations based on the Scanlan-Tomko [5] experimental flutter coefficients as well as the Scanlan-Budlong-Béliveau [17], [19], indical response functions.

Scanlan and Gade [25] have described a single-mode buffeting response estimate based on Scanlan-Tomko [5] type flutter derivative data and recently documented wind spectra [2]. The method ignores aerodynamic coupling between vertical and torsional motions. Scanlan [11] has recently extended this type of analysis in a theoretical study of multi-mode coupling, employing the full set of 6 self-excitation coefficients. Lin [39] has pointed out that the buffeting problem may properly require description as having time-varying aerodynamic coefficients and that this fact may yet be demonstrated to have an important effect upon calculated stability results, necessitating the introduction of stochastic notions of stability.

A few comments on some of the outstanding issues regarding the theory of bridge buffeting may be in order.

It has been usual to assume that the wind forces are of two separate kinds only: motion-dependent (self-excited) and time-dependent (proportional to fluctuating velocity components). The two are undoubtedly interconnected. The motion-dependent forces have usually been taken as functions of the mean reduced velocity. To what extent they are affected by the instantaneous velocity remains an open question. Another aspect of this is: How does turbulence affect the aerodynamic stability derivatives of a given deck section? Only tentative insights [40], [41], have to date been gained on this question.

Another unsettled question relative to buffeting by turbulence is the role of, and proper values for, the frequency-dependent aerodynamic admittance functions [42], [43], [44], which "correct" the analytic expressions for buffeting forces based on steady-state theory to agree with their statistically obtained values under turbulence. In fact, there are two kinds of frequency dependence involved in the parameters of the buffeting problem: the one concerned with structural frequencies and the other with the frequencies of the turbulence. Both are intimately related and no bridge response theory to date has completely clarified all the issues implicit in these relationships.

An interesting question remains unanswered relative to the exact effects of turbulence upon the onset of catastrophic flutter. Experimentally turbulence is observed to delay flutter onset, as this is usually observed under laminar flow conditions. Fig. 14 suggests qualitatively the striking difference in response characteristics of a full bridge when alternately subjected to laminar and turbulent cross flows. What elements in the flow are responsible for the retardation of flutter are not yet specifically understood. One theory holds that turbulence intrinsically alters the sectional flutter derivatives; another suggests that turbulence elicits, in the full bridge, many modes of response which, with their various aerodynamic and structural dampings, disperse the concerted energy of single-mode flutter, the various modes acting somewhat like vibration absorbers relative to one another.

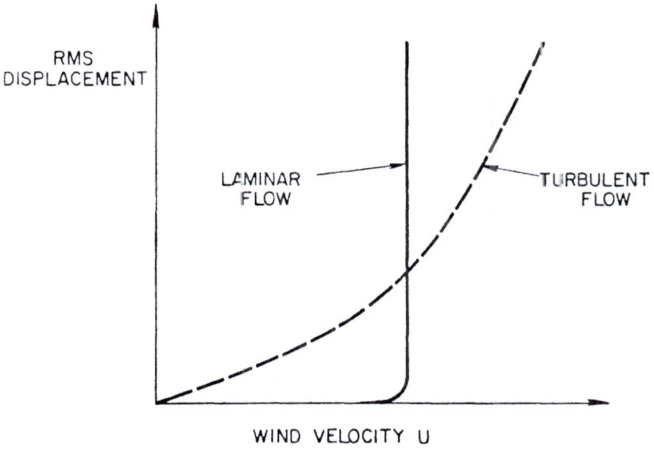

FIG. 14 TYPICAL DISPLACEMENT RESPONSE CHARACTERISTICS OF UNSTABLE BRIDGE: LAMINAR AND TURBULENT FLOW

It is further conceivable that should it become possible for the lowest single-mode flutter to be set up, by some circumstantial mechanism, even under turbulence, then that mode would in fact control the aerodynamic situation, notwithstanding the turbulent conditions, and proceed on to catastrophic flutter. This happens, in fact, to be a close description of what actually occurred in November 1940 at Tacoma Narrows. The final catastrophic flutter mode had in fact been observed before that day ([4], Part II, Ch. III). This possible circumstance may be a very serious argument in favor of laminar-flow flutter testing as a conservative design measure. Another such argument lies in the fact that, rather inexplicably, winds across high bridges may on certain occasions exhibit unusually laminar characteristics [65]. Thus one cannot definitively count on turbulence to relieve a laminar flow flutter tendency.

Work by Lin [39], as noted earlier, suggests that turbulence may also imply the need to consider stability stochastically.

8. STATE-OF-THE-ART IN EXPERIMENTAL BRIDGE AERODYNAMICS

The use of section models has, by now, accumulated 4 decades of experience. The dynamic section model remains the test device of first choice. Its primary service is to aid the indispensable determination of proper geometric form for the bridge deck section. Only after a reasonable geometry has been attained do other analytical or experimental studies become worthwhile. The determination of appropriate deck form, however, is part of the fine art of wind tunnel modelling, involving many details of experimental finesse that require the closest attention and skill of the aerodynamicist. Figs. 15 and 16 depict typical section models in position for wind tunnel testing. Fig. 15 is a model of the Humber bridge tested at the National Maritime Institute in England and Fig. 16 is a model of an open-truss bridge tested at the U.S. Federal Highway Administration Laboratory, near Washington, D.C.

Greatly oversimplifying, one may state that the general trends in results emerging from tunnel testing are the streamlining, or otherwise limiting of flow

FIG. 15 WIND STABILITY MODEL OF HUMBER BRIDGE

FIG. 16 BRIDGE DECK STABILITY MODEL IN FEDERAL HIGHWAY ADMINISTRATION TUNNEL, WASHINGTON, D.C.

separation effects, relative to single-roadway box girder decks; and the through-roadway venting, and occasional streamlining, of truss girder decks. Often the details of edge form, parapets, fences, roadway side or centerline barriers, etc., have surprisingly strong aerodynamic influence and must be modified. Of very characteristic importance are the "leading edge" geometric shapes first encountered by a side wind; for stability these apparently must tend to insure adherence of the flow, or at least its rapid reattachment, to the roadway surface, especially when a slight torsional deflection of the bridge is permitted. Fig. 17 [61] depicts typical geometric trends emerging from wind tunnel treatments of bridges.

Torsional motion lies at the heart of instability: in bluff bodies it can permit self-excited flutter; in streamlined ones, allowance of torsion can permit classical (coupled) flutter to occur. Hence, the concomitant structural byword is clear: the bridge must be as stiff as possible against torsion. Even if it is poor aerodynamically but stiff torsionally (as with a number of doubly-arrayed cable-stayed decks), a bridge may still be quite safe from catastrophic flutter. A case in point is the Sitka Harbor bridge [45]. The archetypical bad example of course remains the original Tacoma Narrows bridge: both aerodynamically very unstable and torsionally weak. The prototype of a fortunate "repair" is the Golden Gate bridge: aerodynamically mediocre but greatly improved torsionally by the addition of the bottom lateral truss system in 1954.

How should the section model be viewed? Essentially as an analog source of experimental aerodynamic information not obtainable by other means. Although conceived originally as a scaled dynamic model of a section of the prototype (and it still offers limited use to this end), it is more properly conceived of today as an analog computer of motional aerodynamic force and moment effects. To this end, the theories and experimental approaches for determining its flutter derivatives have been developed in Japan [29], the United States [5], [24], France [46] and elsewhere. The zero-frequency asmptotes of the flutter

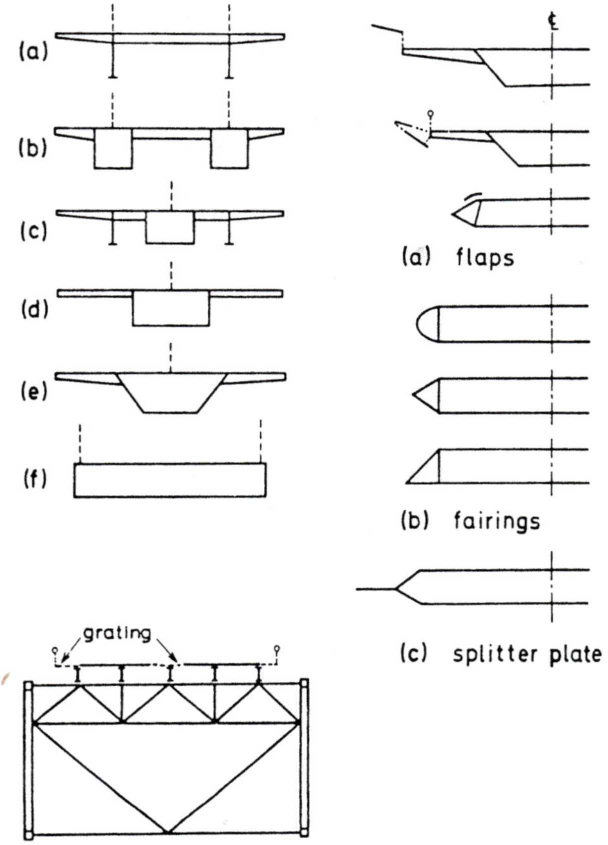

FIG. 17 TYPICAL CONTOUR ADJUSTMENTS AGAINST BRIDGE DECK INSTABILITY

derivatives constitute the steady-state force derivatives. With these data the analyst is armed to perform a variety of studies: vortex response, flutter, buffeting, etc., within which he can vary at will his hypotheses as to incident wind conditions and other parameters.

A final item of interest relative to the section model concerns the possibility of testing it under artificially induced "two-dimensional" turbulence, through the use of properly-driven mechanical flaps, creating turbulence exhibiting spectra and scales appropriate to those required by the model. Figs. 18a and 18b [63] depict an experimental form of such a device. In section model testing, geometric scale usually ranges from 1/100 to 1/25, whereas in boundary layer simulation wind tunnels it may be 1/500 to 1/100. A joint project of Colorado State University and Princeton University for the U.S. Department of Transportation is currently investigating the possibilities of actively-induced turbulence [47].

It is probable that bridges rendered aerodynamically very safe from catastrophic instability are well insured against bad performance under buffeting. Those that are only marginally safe, however, through having a flutter critical speed just beyond normal high winds for the site may be found to be distressed by buffeting in such winds because of progressive loss of torsional damping.

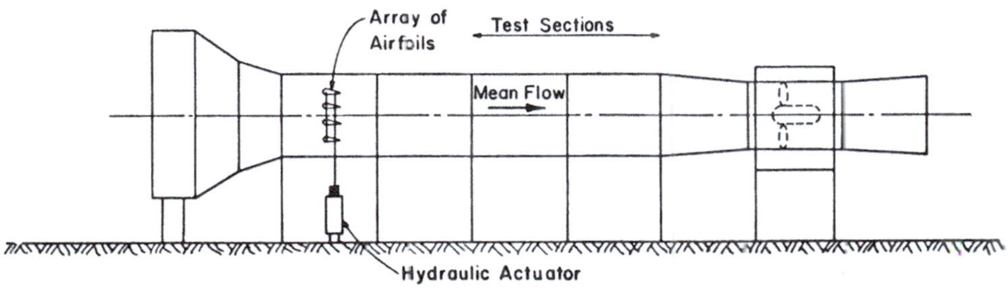

FIG. 18a LARGE-SCALE TURBULENCE GENERATOR

FIG. 18b AIRFOIL LATTICE

The experimental choices beyond the section model are three-dimensional: the partial span and full-span models. Davenport et al [48] have developed a partial-span reduced-scale ("taut-strip") model which approximately realizes two or more important modes and is relatively simple to build and test under three-dimensional scaled turbulence. Fig. 19 [41] depicts a taut-strip model under test at the University of Western Ontario.

Full models, the most expensive both in model and wind tunnel resources, have been realized on relatively rare occasions ever since the pioneering work of Farquharson [4] on the Tacoma Narrows and Scruton [49] on versions of the Severn and other bridges.

Full models, requiring an extra measure of modeler's skill, offer the possibility of direct modeling of all desired features of anticipated bridge conditions under wind. They become, in effect, a direct analog computer of the entire

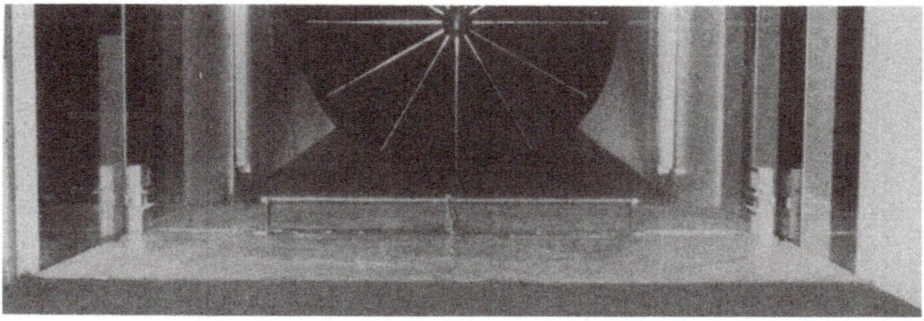

FIG. 19 TAUT STRIP MODEL IN UNIVERSITY OF WESTERN ONTARIO WIND TUNNEL

bridge dynamic response situation, not just of its aerodynamic forces. Recently, full models of the Heer-Agimont bridge in Belgium (Olivari and Thiry, Von Karman Institute) [50] (Fig. 20), the West Gate Bridge (Melbourne, Monash University, Australia [51], and the Lion's Gate Bridge (Irwin, National Research Council, Canada) [32] (Fig. 21) have been realized. Wardlaw [52] has recently given a cogent discussion of the various choices available in modeling bridge aerodynamics. Walshe [59] has presented rather complete descriptions of wind tunnel studies of a number of specific bridges.

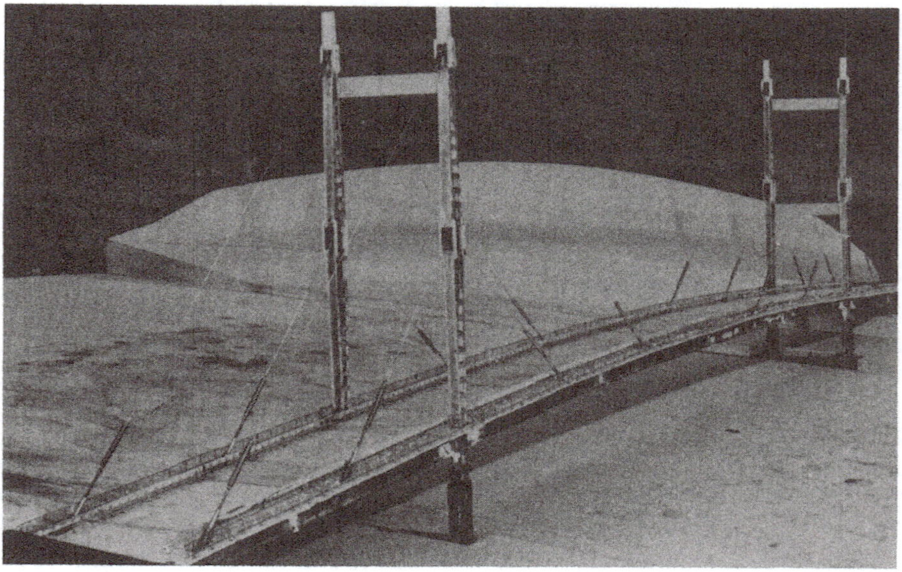

FIG. 20 FULL WIND TUNNEL MODEL OF HEER-AGIMONT BRIDGE

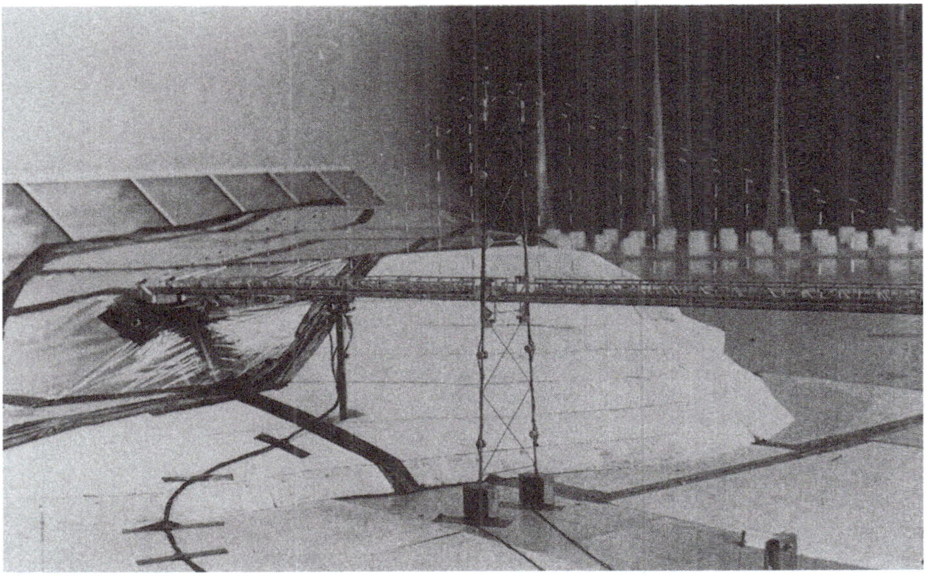

FIG. 21 FULL WIND TUNNEL OF LIONS' GATE BRIDGE

9. TESTS ON FULL-SCALE STRUCTURES

Tests on the aerodynamic response of prototype bridges have unfortunately been rare. On those occasions where some data have been gathered they have often been incomplete. Particularly lacking are bridge response data coupling simultaneous recording of both pertinent information on fluctuating wind and bridge motional response.

The well-known documentation by Farquharson [4] on the original Tacoma Narrows Bridge and by Vincent [53] on the Golden Gate were landmarks of this type even though incomplete.

In Canada, the National Research Council is presently sponsoring field measurements of wind and bridge response on the Lion's Gate Bridge, Vancouver. The new Pasco-Kennewick bridge in the state of Washington, U.S.A. is to be similarly studied for the U.S. Department of Transportation, as is the Luling bridge near New Orleans, Louisiana, U.S.A. Studies of full bridge response have been made in Japan by Narita and others [54]. Loiseau and Széchényi [55] have made some response and wind studies at the St. Nazaire bridge in France (see Fig. 22 [55]), Jensen and Hjorth-Hansen [56] have studied wind spectra and the dynamic action of the Sotra Bridge in Norway. Melbourne [57] has performed field response measurements on the West Gate Bridge in Australia.

10. ACKNOWLEDGEMENTS

The author is particularly grateful for personal communications received on

FIG. 22 SMOKE TESTS OF UNGUIDED AND DEFLECTOR-GUIDED WIND FLOW, ST. NAZAIRE BRIDGE

points touched on in this lecture, from Manabu Ito, University of Tokyo, Japan; Edmond Széchényi, ONERA, Châtillon, France; Erik Hjorth-Hansen, Norwegian Institute of Technology, Trondheim, Norway; R. L. Wardlaw, and H. P. A. H. Irwin, National Research Council, Ottawa, Canada; John D. Holmes, James Cook University, North Queensland, Australia; William H. Melbourne, Monash University, Melbourne, Australia; Alan G. Davenport, University of Western Ontario, London, Canada; Y. K. Lin, The University of Illinois, Urbana, U.S.A.; D. Olivari, Von Karman Institute, Rhode-Saint Génèse, Belgium; Leo Finzi, Technical University of Milan, Italy; Narahito Shiraishi, Kyoto University, Japan; Yuji Nakamura, Kyushu University, Japan; N. Narita, Public Works Research Institute, Japan; Frank H. Durgin, Massachusetts Institute of Technology, Cambridge, U.S.A.; Harold Bosch, Federal Highway Administration, Washington, D.C.; M. Shinozuka, Columbia University, New York City; F. J. Maher, Virginia Polytechnic Institute and State University, Balcksburg; Bogusz Bienkiewicz, Colorado State University, Ft. Collins; D. E. Walshe, National Maritime Institute, Teddington, England; and H. Teunissen, Atmospheric Engineering Service, Downsview, Ontario.

11. REFERENCES

1. Finzi, Leo, "Structural Problems for the Messina Narrows Bridge", Transportation Research Record 665, Vol. 2 TRB, National Academy of Sciences, Washington, D.C., 131-139, (1978).

2. Simiu, E. and Scanlan, R. H., Wind Effects on Structures, John Wiley and Sons, New York, (1978).

3. Wardlaw, R. L. and Goettler, L. L., "A Wind Tunnel Study of Modifications to Improve the Aerodynamic Stability of the Long's Creek Bridge", Report No. LTR-LA-8, NAE, National Research Council, Ottawa, Canada, (1968).

4. Farquharson, F. B., ed., Aerodynamic Stability of Suspension Bridges, University of Washington Engineering Experiment Station Bulletin No. 116, Parts I-V, (June 1949 - June 1954).

5. Scanlan, R. H. and Tomko, J. J., "Airfoil and Bridge Deck Flutter Derivatives", J. Eng. Mech. Div., ASCE, Vol. 97, No. EM6, 1717-1737, (Dec. 1971).

6. Nakamura, Y. and Mizota, T., "Torsional Flutter of Rectangular Prisms", J. Eng. Mech. Div. ASCE Vol. 101, No. EM2, 125-142, (1975).

7. Ammann, O. H., von Kármán, Th., and Woodruff, G. B., "The Failure of the Tacoma Narrows Bridge", Report to the Federal Works Agency, (March 28, 1941).

8. von Kármán, Th. and Rubach, H., "Ueber den Mechanismus des Fluessigkeits und Luftwiderstandes", Phys. Zeits. Vol. 13, (Jan. 15, 1912).

9. Roshko, A., "On the Wake and Drag of Bluff Bodies", J. Aeron. Sci., Vol. 22, No. 2, 124-132, (1955).

10. Nakamura, Y., "An Analysis of Binary Flutter of Suspension Bridge Deck Sections", J. Sound and Vibration, Vol. 57, 471-482, (1978).

11. Scanlan, R. H., "The Action of Flexible Bridges Under Wind, I: Flutter Theory, II: Buffeting Theory", J. Sound and Vibration, Vol. 60, No. 2, 187-199 and 201-211, (1978).

12. Roberts, G., "The Severn Bridge: A New Principle of Design," Proc. Symp. on Suspension Bridges, Laboratorio Nacional de Engenharia Civil, Lisbon, 629-639, (1966).

13. Ostenfeld, C., Hass, G. and Frandsen, A. G., "Motorway Bridge Across Lillebaelt: Model Tests for the Superstructure of the Suspension Bridge", Proc. Symp. on Suspension Bridges, Laboratorio Nacional de Engenharia Civil, Lisbon, 587-608, (1966).

14. Wardlaw, R. L., "Static Force Measurements of Six Deck Sections for the Proposed New Burrard Inlet Crossing", Report No. LTR-LA-53, NAE, National Research Council, Ottawa, Canada, (1970).

15. Walshe, D. E., "The Aerodynamic Investigation for a Box-Type Roadway Deck for a Suspension Bridge Proposed for the Humber Crossing", Special Report No. 012, National Physical Laboratory, Teddington, U.K., (1968).

16. Walshe, D. E., "The Aerodynamic Investigation for the Suspended Structure of the Proposed Bosporus Bridge", Special Report No. 020, National Physical Laboratory, Teddington, U.K., (1969).

17. Scanlan, R. H. and Budlong, K. S., "Flutter and Aerodynamic Response Considerations for Bluff Objects in a Smooth Flow", Proc. IUTAM-IAHR Symp. on Flow-Induced Vibrations, Karlsruhe, W. Germany, 1972, E. Naudascher (Ed.), Springer-Verlag, Berlin, 339-354, (1974).

18. Wagner, H., "Ueber die Entstehung des dynamischen Auftriebes von Tragfluegeln", Zeits. angew. Math. u. Mech., Vol. 5, 17-35, (1925).

19. Scanlan, R. H., Béliveau, J.-G., and Budlong, K. S., "Indicial Aerodynamic Functions for Bridge Decks", J. Eng. Mech. Div., ASCE, Vol. 100, No. EM4, 657-672, (Aug. 1974).

20. Yoshimura, T. and Nakamura, Y., "On the Indicial Aerodynamic Moment Responses of Bridge Deck Sections", Proceedings, 5th Int'l Conf. on Wind Engineering, Fort Collins, Colorado, (July 1979).

21. Selberg, A. and Hjorth-Hansen, E., "The Fate of Flat Plate Aerodynamics in the World of Bridge Decks", Proc. Theodorsen Colloq., Det Kongelige Norske Videnskabers Selskab, Trondheim, 101-113, (1976).

22. Nakamura, Y., "On the Aerodynamic Mechanism of Torsional Flutter of Bluff Structures", (to be published). Private communication.

23. Miyata, T., Kubo, Y., and Ito, M., "Analysis of Aeroelastic Oscillations of Long-Span Structures by Nonlinear Multidimensional Procedures", Proc. 4th Int'l. Conf. on Wind Effects on Buildings and Structures, London, 1975, Cambridge Univ. Press, U.K., 215-225, (1976).

24. Scanlan, R. H., "Recent Methods in the Application of Test Results to the Wind Design of Long, Suspended-Span Bridges", Report No. FHWA-RD-75-115, Federal Highway Administration, Office of Research and Development, Washington, D.C., (1975).

25. Scanlan, R. H. and Gade, R. H., "Motion of Suspended Bridge Spans under Gusty Wind", J. Struct. Div. ASCE Vol. 103, No. ST9, 1867-1883, (1977).

26. Durgin, F. H., Private communication.

27. Bleich, F., "Dynamic Instability of Truss-Stiffened Suspension Bridges under Wind Action", Proc. ASCE, Vol. 74, No. 8, 1269-1314, (1948); Vol. 75, No. 3, 413-416, (March 1949); Vol. 75, No. 6, 855-865, (June 1949).

28. Rocard, Y., L'instabilité en méchanique, Masson & Cie, Paris, Ch. VI, (1954).

29. Ukeguchi, N., Sakata, H., and Nishitani, H., "An Investigation of Aeroelastic Instability of Suspension Bridges", Proc. Int'l Symp. on Suspension Bridges, Laboratorio Nacional de Engenharia Civil, Lisbon, 273-284, (1966).

30. Davenport, A. G., "The Action of Wind on Suspension Bridges", Proc. Int'l. Symp. on Suspension Bridges, Laboratorio Nacional de Engenharia Civil, Lisbon, 79-100,(1966).

31. Davenport, A. G., "Buffeting of a Suspension Bridge by Storm Winds", J. Structure Div. ASCE Vol. 88, No. ST3, 233-268, (June 1962).

32. Irwin, H. P. A. H., "Wind Tunnel and Analytical Investigations of the Response of Lions' Gate Bridge to a Turbulent Wind", NAE, National Research Council, Report No. LTR-LA-210, Ottawa, Canada, (June 1977).

33. Konishi, I., Shiraishi, N., and Matsumoto, M., "Aerodynamic Response Characteristics of Bridge Structures", Proc. 4th Int'l Conf. on Wind Effects on Buildings and Structures, Heathrow, U.K. 1975, Cambridge University Press, 199-208, (1977).

34. Konishi, I., Shiraishi, N., and Matsumoto, M., "A Reliability Approach on Aeroelastic Responses of Long-Spanned Bridges", Reliability Approach in Structural Engineering, Maruzen, Tokyo, 329-343, (1975).

35. Okubo, T., Okauchi, I., and Murakami, E., "Aerodynamic Response of a Large-Scale Bridge Model Against Natural Wind", Reliability Approach in Structural Engineering, Maruzen, Tokyo, 315-328, (1975).

36. Holmes, J. D., "Prediction of the Response of a Cable-Stayed Bridge to Turbulence", Proc. 4th Int'l Conf. on Wind Effects on Buildings and Structures, Heathrow, U.K., September 1975, Cambridge University Press, 187-197, (1977).

37. Holmes, J. D., "Monte Carlo Simulation of the Wind-Induced Response of a Cable-Stayed Bridge", Wind Engng. Report 2/78, Dept. of Civil and Systems Engng., James Cook University, No. Queensland, Australia.

38. Beliveau, J-G., Vaicaitis, R. and Shinozuka, M., "Motion of a Suspension Bridge Subject to Wind Loads", J. Struct. Div., ASCE, Vol. 103, No. ST6, 1189-1205, (1977).

39. Lin, Y. K., "Stochastic Analysis of Bridge Motion in Large-Scale Turbulent Winds", Proc. 5th Int'l Conf. on Wind Engng., Colorado State University, Fort Collins, U.S.A., Paper 094, (July 1979).

40. Scanlan, R. H. and Lin, W.-H., "Effects of Turbulence on Bridge Flutter Derivatives", J. Eng. Mech. Div. ASCE, Vol. 104, No. EM4, 719-733, (1978).

41. Davenport, A. G., private communication.

42. Liepmann, H. W., "On Application of Statistical Concepts to the Buffeting Problem", J. Aeron. Sci., Vol. 19, No. 12, 793-800, 822, (Dec. 1952).

43. Davenport, A. G., "The Application of Statistical Concepts to the Wind Loading of Structures", Proc. Instn. of Civil Engrs., Vol. 19, 449-472, (Aug. 1961).

44. Shiraishi, N. and Matsumoto, M., "Aerodynamic Responses of Bridge Structures Subject to Strong Winds", Proc. Symp. of Engng. for Natural Hazards, Manila, (Sept. 1977).

45. Scanlan, R. H. and Gade, R. H., "Experimental Measurement and Interpretation of Aerodynamic Stability Coefficients for the Decks of Two Cable-Stayed Bridges", Dynamic Responses of Structures, Proc. of ASCE-EMD Spec. Conf., UCLA, Los Angeles, 478-486, (March 1976).

46. Loiseau, H. and Széchényi, E., "Etude du Comportement Aéroélastique du Tablier d'un Pont à Haubans", C.R. des journées AFPC, ONERA, Chatillon, France, (April 1974).

47. Cermak, J. E., Bienkiewicz, B., Peterka, J. and Scanlan, R. H., "Active Turbulence Generator for Study of Bridge Aerodynamics", Engineering Mechanics Division Specialty Conference, ASCE, Austin, Texas, (Sept. 17-19, 1979).

48. Davenport, A. G. and Tanaka, H., "Aerodynamic Study of Taut Strip Models in Turbulent Boundary Layer", Boundary Layer Tunnel Report BLWT-1-74, Univ. of Western Ontario, (Jan. 1974).

49. Scruton, C., "Experimental Investigation of Aerodynamic Stability of Suspension Bridges with Special Reference to Proposed Severn Bridge", Proc. in 5th Civil Engrs., London, Vol. 1, Part 1, No. 2, 189-222, (March 1952).

50. Olivari, D. and Thiry, F., "Wind Tunnel Tests of the Aeroelastic Stability of the Heer-Agimont Bridge", Technical Note No. 113, von Kármán Inst. for Fluid Dynamics, Rhode-Saint-Génèse, Belgium, (1975).

51. Melbourne, W. H., "West Gate Bridge Wind Tunnel Tests", Internal Report, Dept. Mechanical Engrg., Monash University, Clayton, Victoria, Australia.

52. Wardlaw, R. L., "Sectional versus Full Model Wind Tunnel Testing of Bridge Road Decks", DME/NAE Quarterly Bulletin No. 1978 (4), Ottawa, Canada, 25-47, (Jan. 1979).

53. Vincent, G. S., "Golden Gate Bridge Vibration Studies", J. Struct. Div., ASCE, Vol. 84, No. ST6, 1817-1 to 1817-40, (Oct. 1958).

54. Narita, N., Yokoyama, K. and Sato, H., "Field Observation of Aerodynamic Behavior of Long-Span Bridges, Paper 047, Proc. 5th Int'l Symposium on Wind Engng., Colorado State University, Fort Collins, U.S.A., (July 1979).

55. Széchényi, E. and Loiseau, H., "Mesures aérodynamiques et dynamiques sur le pont de Saint-Nazaire", T.P. No. 1978-26, ONERA, Chatillon, France, (June 1978).

56. Jensen, N. O. and Hjorth-Hansen, E., "Turbulence and Response Measurements at Sotra Bridge", Report STF 41, A 78003, SINTEF, Trondheim, Norway, (Feb. 1978).

57. Melbourne, W. H., "Model and Full Scale Response to Wind Action of the Cable-Stayed Box Girder West Gate Bridge", Proc. Symposium on Practical Experiences with Flow-Induced Vibrations, Karlsruhe, W. Germany, (Sept. 1979).

58. Kloeppel, K. and Weber, G., "Teilmodellversuche zur Beurteilung des aerodynamischen Verhaltens von Bruecken", Der Stahlbau, Vol. 4, 113-121, (1963).

59. Walshe, D. E. J., Wind-Excited Oscillations of Structures, National Physical Laboratory, H. M. Stationary Office, London, Ch. 7, (1972).

60. Otsuki, Y., Washizu, K., Tomizawa, H., and Ohya, A., "A Note on the Aeroelastic Instability of a Prismatic Bar with Square Section", J. Sound and Vibration, Vol. 34 (2), 233-245, (1974).

61. Ito, M. and Shiraishi, N., "Wind-Resistant Design Practices of Cable-Suspended Bridges in Japan", Proc. U.S. - S.E. Asia Symposium on Engineering for Natural Hazards Protection, Manila, 287-294, (1977).

62. Beeuwkes, R., Flow pictures taken at Case School of Applied Science, Cleveland (apparently private communication to von Karman), (1941).

63. Bienkiewicz, B., Private communication.

64. Reid, R. E. (presentation by Russell, J. S.), "On the Vibration of Suspension Bridges and the Means of Preventing Injury from this Cause", Trans. Royal Scottish Society of Arts, Vol. 1, (1841).

65. Teunissen, H., Private communication.

F2 VORTEX SHEDDING OSCILLATIONS OF BRIDGE DECK SECTIONS

Ichiro Konishi
Professor Emeritus, Dr. Eng.

Naruhito Shiraishi
Professor, Dr. Eng.

Masaru Matsumoto
Associate Professor, Dr. Eng.

Department of Civil Engineering
Kyoto University
Kyoto, Japan

SUMMARY

This paper describes the aerodynamic response characteristics in connection with vortex-induced oscillation of the longest span cable-stayed bridge in Japan. A series of wind-tunnel tests have been conducted in order to understand and improve the aerodynamic characteristics of a proposed deck section. Furthermore, the influence of unsteady angle of attack on the amplitude response are examined in two-dimensional wind-tunnel tests and with the use of an analytical method based on data obtained in steady state. Finally, the results are applied to evaluate the aerodynamic behavior of the prototype bridge.

1. INTRODUCTION

It has been pointed out that closed-form cross-sections of cable-stayed bridges provide advantageous mechanical resistance, which may, however, provoke vortex-induced oscillations at comparatively low wind velocity. In connection with construction plans of the Yamato River Bridge in Osaka, Japan, an investigation was carried out in order to clarify the following aerodynamic problems associated with wind design considerations: improvement of the proposed original bridge deck section to be aerodynamically stable so as to reduce response amplitude due to vortex-induced oscillation; consideration of influences of unsteady angle of attack on response amplitude; and evaluation of vortex-induced response of the prototype bridge from information obtained in wind tunnel tests.

2. THE YAMATO RIVER BRIDGE AND CHARACTERISTICS OF NATURAL WINDS

The Yamato River Bridge is to be constructed by the Hanshin Expressway Authority, Japan, as an urban express highway bridge obliquely crossing the Yamato River. Its configuration is illustrated in Sketch 1, i.e., 653 m in total length, 355 m in main span length, making it the longest cable-stayed bridge in Japan. In addition, it has a 30 m deck width and approximately 20 m clearance between deck and water level. The bridge deck section proposed by the Yamato River Bridge Technical Committee for the Hanshin Expressway Authority is a flat trapezoidal form, because of its mechanical and aerodynamic advantages as shown in Sketch 1. The deflectional and torsional natural frequencies of the lowest mode, as obtained analytically, are 0.319 Hz and 0.727 Hz, respectively.

The observation of natural winds at the site has been carried out in order to survey the wind characteristics during the last three years by means of a three component Gill Anemometer. The observed results of mean wind velocity, intensity of turbulence, and mean angle of attack, vary from 7 m/s to 20 m/s, 5% to 20% and -1° to 5°, respectively.

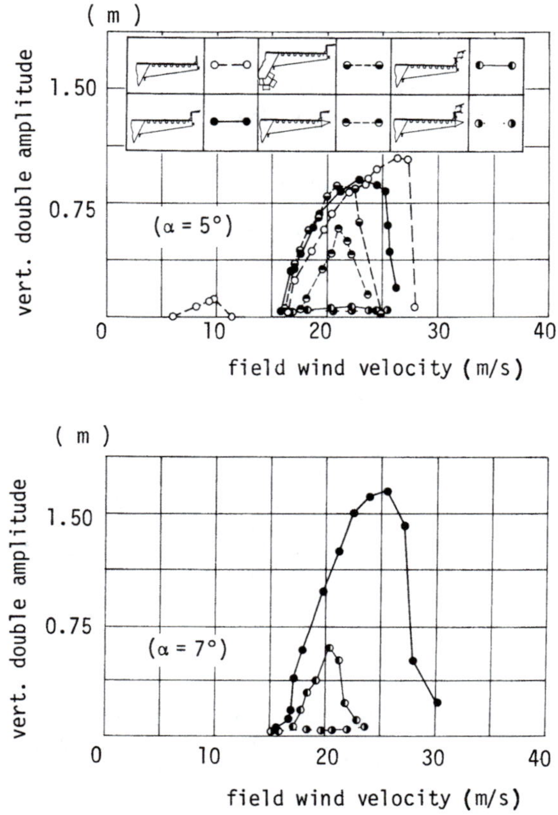

FIG. 1 AERODYNAMIC IMPROVEMENT BY MEANS OF SEVERAL ATTACHMENTS (HEIGHT OF CURB h = 3.8 mm IN MODEL SCALE)

3. AERODYNAMIC CHARACTERISTICS AND ITS IMPROVEMENT OF PROPOSED BRIDGE DECK SECTION

Some wind-tunnel tests have been conducted on a 1:75 scale sectional model in smooth flow in order to clarify its aerodynamic characteristics, with primary attention given to vortex-induced vertical and torsional oscillations. The model has a length of 0.93 m, a deck width of 0.40 m, a deck height of 0.048 m, and the working section of the wind-tunnel is 1.5 m x 2.5 m. Experimental conditions are carefully controlled and aerodynamic and vibrational similarity requirements are well satisfied. Vertical and torsional oscillations at $\alpha = 5°$ and 7° (α = angle of attack) of the proposed original bridge deck section without any curb have been observed. The presence of a curb makes the bridge deck section more unstable aerodynamically because of the generation of separated vortices from the leading edge. The configurations of several attachments to the proposed bridge deck section with a curb, having a height of 3.8 mm in the scale model (which is too high in practice), and their effectiveness with respect to vortex-induced oscillations at $\alpha = 5°$ and 7°, are illustrated in Sketch 2 and Fig. 1 respectively. As shown, it is possible to improve the rather unstable bridge deck section. The combination of bifacial flap-plates and fairings, or wind noses, is the most effective among the several attachments tried here. However, there may be mechanical disadvantages, due to high induced stress at the bracket of the bridge section. On the other hand, the amplitudes of vortex-induced vertical oscillations at $\alpha = 5°$ and 7° increase with the height of curb attached to the bridge deck having a single flap-plate, as shown in Fig. 2. Thus the height of the curb should be determined with care, so as not to exceed acceptable amplitudes of oscillation.

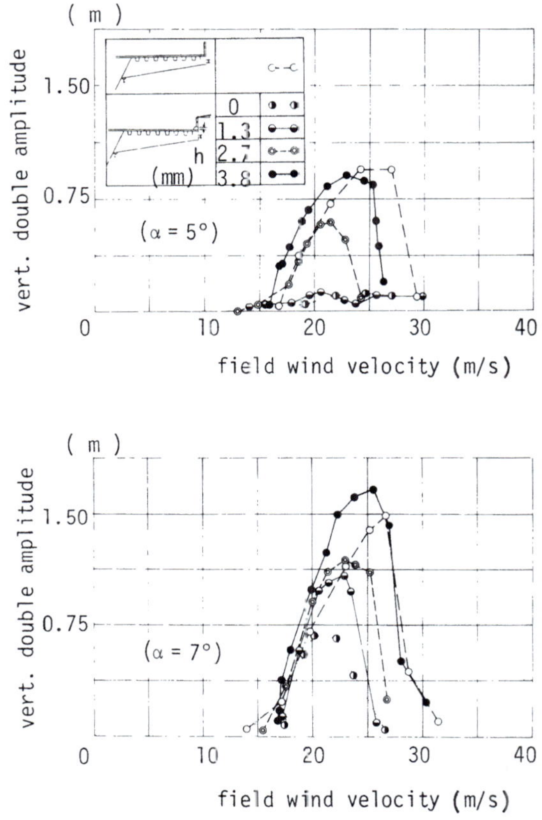

FIG. 2 INFLUENCES OF HEIGHT OF CURB (0 mm to 3.8 mm) IN SCALE MODEL, ON AERODYNAMIC VERTICAL RESPONSE CHARACTERISTICS

4. INFLUENCES OF UNSTEADY ANGLE OF ATTACK

Depending on the degree of sensitivity of vortex-induced oscillations to the steady angle of attack, an unsteady angle of attack of comparatively slow vertical wind velocity fluctuations could influence response by the introduction of a buffeting action. Wind tunnel tests in two-dimensional, slowly fluctuating flow and corresponding numerical analyses have been carried out. The calculation involves simulating the vibrations cycle by cycle using the concept of an aerodynamic derivative H_1^* defined by Scanlan [1], which is expressed as a function of vertical amplitude $(\bar{\eta})$, angle of attack $(\bar{\alpha})$, and reduced mean wind velocity $(\bar{V})$ in the following form:

$$H_1^* \, (\bar{\eta}, \bar{\alpha}, \bar{V}) = \sum_p \sum_q \sum_r \; C_{pqr} (2\,\bar{\eta})^p (\bar{\alpha})^q (\bar{V})^r \tag{1}$$

A numerical example and measured vertical response from wind-tunnel tests are illustrated in Fig. 3. The varying characteristics of measured and calculated amplitudes agree fairly well with each other; nevertheless, the calculated is slightly larger than the measured because of low intensity of turbulence, i.e., 3% of the air stream.

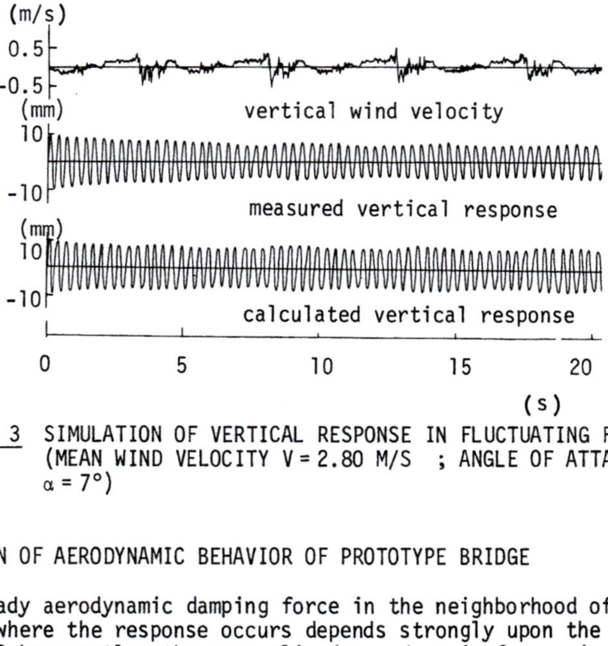

FIG. 3 SIMULATION OF VERTICAL RESPONSE IN FLUCTUATING FLOW (MEAN WIND VELOCITY V = 2.80 M/S ; ANGLE OF ATTACK α = 7°)

5. EVALUATION OF AERODYNAMIC BEHAVIOR OF PROTOTYPE BRIDGE

The unsteady aerodynamic damping force in the neighborhood of the wind velocity region where the response occurs depends strongly upon the amplitude of oscillation. Subsequently, the generalized aerodynamic forces in the j-th mode Lagrangian equation of the prototype bridge should be modified by accounting for the j-th modal shape. This procedure has been proposed by Vincent [2]. In other words, the aerodynamic damping forces may be linearized by means of the vibrating modal shape. Consequently, the response due to vortex-induced oscillation of the prototype bridge is evaluated to be larger than that obtained in the sectional model test if rather smooth and spanwise uniform wind acts on the bridge deck.

In practice the spanwise variation of mean wind velocity, the spanwise phase differences of vortex shedding, or rather strong turbulence, may diminish aerodynamic response. Calculated response of the first symmetrical deflectional (vertical) mode is compared with the measured response of the sectional model in Fig. 4.

6. CONCLUDING REMARKS

Vortex-induced oscillations at $\alpha = 5°$ and $7°$ were observed for the proposed original bridge deck section, but suitable attachments can reduce their response amplitudes. Furthermore, the height of curb strongly influences the magnitude of response. Concerning unsteady angle of attack, the calculated vertical response agrees fairly well with that measured in low turbulent slowly fluctuating flow. A calculated example of the first symmetrical deflectional (vertical) modal response of the prototype bridge is slightly larger than that measured on the sectional model.

The authors wish to express their thanks to members of the Yamato River Bridge Technical Committee and the staffs of the Hanshin Expressway Authority for their valuable advice and information, and also to Messrs. T. Takeuchi, H. Saeki and T. Shiotani for their assistance in wind tunnel tests.

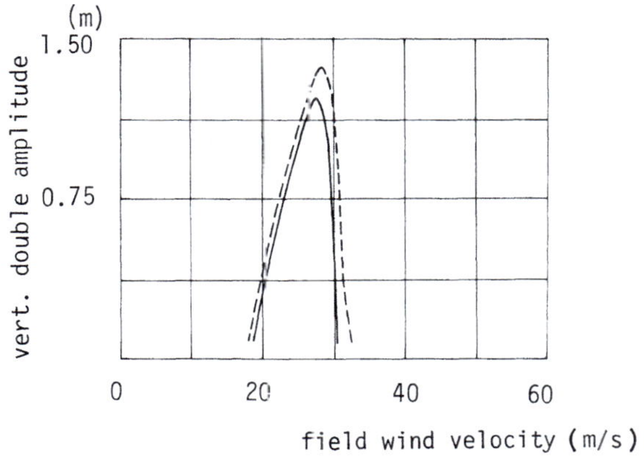

FIG. 4 COMPARISON OF MODIFIED VERTICAL AMPLITUDE OF VIBRATING MODE WITH DIRECTLY MEASURED AMPLITUDE IN MODEL TEST (IN PROTOTYPE SCALE: ANGLE OF ATTACK $\alpha = 7°$, HEIGHT OF CURB $h = 1.3$ mm)

7. REFERENCES

1. Scanlan, R. H. and Sabzebari, Ali, "Suspension Bridge Flutter Revisited", ASCE, Str. Eng. Conf., Seattle, Washington, (May 1967).

2. Farquharson, F. B., "Aerodynamic Stability of Suspension Bridges with Special Reference to The Tacoma Narrows Bridge", Part V, University of Washington, (1950).

BRIDGE DECK SECTIONS

Problem: Evaluation of the aerodynamic behaviour of the longest span cable-stayed bridge in connection with vortex-induced oscillation and improvement of aerodynamic instability of proposed original bridge deck section. From incident flow characteristics, including mean wind velocity and steady or unsteady angle of attack, and from deck sectional form, height of curb and suitable attachments, find conditions for improvement of aerodynamic instability.

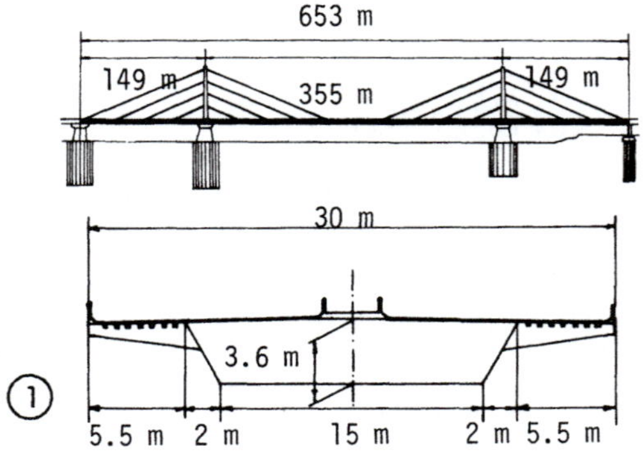

Geometry: Configurations of proposed bridge and its trapezoidal deck section are shown in Sketch 1.

Boundary Conditions: The bridge will be constructed at the mouth of the Yamato River and the clearance between deck and water level is around 20 m.

Flow Conditions: Simulated flow in wind tunnel is smooth or two-dimensional slowly fluctuating flow with low intensity of turbulence, and the range of angle of attack is -7° to 7°.

Structural Dynamics: Mass and mass inertia per unit length are $m = 1.705 \times 10^5$ kg/m and $I = 1.2072$ kg·m respectively, and the lowest natural frequencies are, respectively, 0.319 Hz for deflectional mode and 0.727 Hz for torsional mode.

Vibrations: Vertical oscillations with amplitudes of about 1.5 m to 1.0 m may occur over a limited range of wind velocity, i.e., 18 m/s to 30 m/s at a steady angle of attach of 7°.

Cures: Some suitable attachments shown in Sketch 2 make the proposed bridge deck section aerodynamically stable; also, decreasing height of curb promotes stability (numbers: dimensions in mm, model scale).

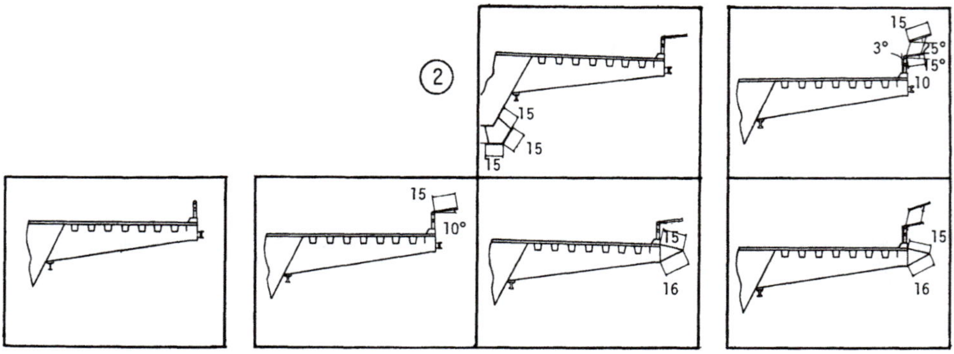

(a) Original design (b) Little effect (c) Successful

F3 MODEL AND FULL SCALE RESPONSE TO WIND ACTION OF THE CABLE STAYED BOX GIRDER WEST GATE BRIDGE

W.H. Melbourne
Professor of Fluid Mechanics
Department of Mechanical Engineering
Monash University, Melbourne, Australia

SUMMARY

In 1973, the Engineering Directorate for the redesign of the 848 m cable stayed box girder West Gate Bridge commissioned a programme of wind tunnel tests. The main part of this programme was concerned with the measurement of the deck response using a 1/150 scale, aeroelastic model of the full bridge, tested in two scaled models of the natural wind with turbulence intensities of 14% and 20%. The full model tests were further supported by sectional model tests in turbulent and smooth flow and some subsequent theoretical analyses, and five full model erection configurations were also tested. The predominant displacement of the deck of the service configuration was found to be in the first vertical bending mode with the peak vertical displacement at centre span, $\hat{z}$, having the following functional dependence,

$$\hat{z} = f\left[\left(\frac{\bar{u}}{n_o c}\right)^{2.5} \left(\frac{\sigma_u}{\bar{u}}\right)^{2.2} \left(\frac{1}{\zeta}\right)^{0.2} \cos\beta\right]$$

where $\bar{u}$ is the hourly mean longitudinal wind speed, n_0 is the fundamental frequency in the vertical deck bending mode, c is the deck chord length, $\sigma_u/\bar{u}$ is the turbulence intensity of the longitudinal velocity component, ζ is the critical damping ratio, and β is wind direction relative to the deck normal.

During 1978, the bridge was instrumented with wind recording equipment and two accelerometers at the end of one span during the final stages of construction and then at centre span since closure. Structural damping measurements were made at several construction stages, which showed the critical damping ratio to be about $\zeta = 0{\cdot}01$; unfortunately, these damping measurements could only be made at quite small amplitudes, decaying from ±2 mm at centre span.

To date, full scale measurements of deck response have been analysed for about ten wind storms, lasting more than three hours in given wind direction sectors. The full scale measurements show scattered, but generally good, agreement with the model measurements. In particular, the full scale response of the bridge to wind action shows a similar dependence on the turbulence intensity of the incident wind flow, providing confirmation that the cross-wind forcing mechanism for a slender box section is primarily due to resonant response to incident turbulent flow energy, via the section lift characteristics.

1. DESCRIPTION OF WEST GATE BRIDGE AND THE SITE

The West Gate Bridge connects the east and west banks of the lower Yarra River near the City of Melbourne. Being on a curve in the river, there is not a long fetch of open water normal to the bridge deck, hence the wind flow over the bridge site usually has turbulence characteristics of wind flow over suburban and industrial buildings. The site and approaches are topographically flat.

The steel spans of cable stayed plated box girder construction have a total length of 848 m with a centre span of 336 m and an average height above ground (water) level of 55 m. The full scale bridge is shown in Sketch 1, and a dimensional outline with some structural properties are given in Sketch 2.

2. THE MECHANISMS OF DECK RESPONSE TO WIND ACTION

The various mechanisms responsible for the cross-wind excitation of structures have been discussed in detail by Melbourne (1975). For bridge decks, the excitation mechanisms, which work primarily through the resonant response process at the lowest structural frequency, can be summarised as being due to

(i) incident turbulence inducing fluctuating forces via the sectional aerodynamic characteristics,
(ii) wake excitation, and
(iii) displacement dependent flutter mechanisms.

The box girder section used for West Gate Bridge is relatively streamlined having a depth (4·1 m) to chord (37·4 m) ratio of just over 10%, which implies the following aerodynamic characteristics with respect to bridge sections:

> low drag, high positive lift curve slope, negative lift at zero incidence (i.e. downwards), high positive pitching moment curve slope and low wake energy at zero incidence.

These characteristics plus relatively high torsional stiffness imply the following response to wind action.

(i) The low drag means that horizontal motions in the direction of the wind will be relatively small.
(ii) The low wake energy implies a reduced likelihood of significant wake induced critical velocity instabilities.
(iii) The high positive lift curve slope provides good aerodynamic damping for the vertical deck motions but increases the response of the deck to vertical components of the freestream turbulence.
(iv) The negative lift at zero incidence causes a mean downwards displacement.
(v) High torsional stiffness coupled with low pitch inertia removes onset of divergence, stall and classical flutter to higher wind speeds.

3. METEOROLOGICAL CONDITIONS

An estimate of the meteorological conditions at the site have been derived from long term wind records collected by the Bureau of Meteorology at Essendon Airport, about ten kilometres north of the site, and wind structure measurements made by Brook (1974) on a tower about a kilometre northeast of the site.

As the major response to wind action was expected to come from the effects of incident turbulence on the high lift curve slope characteristics of the deck section, particular attention was paid to obtaining full scale wind structure data for the site. In terms of a power law representation of mean wind speed profile, Brook found variations in the exponent from 0·15 to 0·20, and at a height of 55 m, turbulence intensity variations from 15% to 20%. It was decided that, for flows across the suburban roughness, the turbulence intensity could be as high as 20% and for flows down the river, about 14%.

4 THE MODEL TESTS

Only tests on the full service condition configuration of the bridge will be presented and discussed here. Other aspects covered by the model tests included profile and spectral details of the model wind flows, effects of turbulence on sectional models, five full model erection configuration tests, effects of changing bridge mass and stiffness, effects of change of edge conditions (a wind break configuration was tested), and traffic configurations, measurement of aerodynamic damping from the autocorrelation of response, wake characteristics, test to limit state wind speeds, and finally an integration of predicted response with the wind speed probability distribution.

4.1 Modelling Details

A photograph of the aeroelastic model of the full bridge service configuration is shown in Figure 1. The model construction for the deck, piers, cable towers and cables all consisted of a central structural spine which modelled the stiffness properties and about which were added segments to build up the correct aerodynamic shape and mass distribution.

The model to full scale scaling parameters were based on keeping constant the Froude number and the ratio of inertia to elastic forces; the latter ratio was kept constant for bending, torsion and cable elongation modes as relevant. The main scaling ratios in terms of model over full scale ratios were as follows:

Length,	$L_r = 1/50$	Mass moments of inertia	$I_{M_r} = L_r^5$
Velocity,	$V_r = \sqrt{L_r} = 1/12{\cdot}25$	Flexural and torsional stiffness	$(EI)_r = (GJ)_r = L_r^5$
Density,	$\rho_r = 1$		
Pressure,	$p_r = \rho_r V_r^2$	Axial stiffness	$(EA)_r = L_r^3$

Two turbulent shear flow natural wind models were developed in the wind tunnel to a length scale of 1/150, based on longitudinal turbulence scales, and with turbulence intensities, at scaled deck height, of 14% and 20% respectively.

FIG. 1 1/150 SCALE AEROELASTIC MODEL OF THE CABLE STAYED BOX GIRDER WEST GATE BRIDGE IN THE 4 m x 3 m WIND TUNNEL WORKING SECTION

4.2 Model Measurements

The first tests were concerned with an investigation of the modes excited under wind action and the probability distribution of peak displacements.

It was evident from the start that the main mode of oscillation was the 1st vertical bending mode. Spectral measurements of displacements at various positions were used to detect other bending and torsion modes, all of which were at least an order of magnitude less than the first vertical bending mode.

As part of this phase of the investigation, measurement of the spectral density of the longitudinal velocity component was made in the wake. This showed that energy distribution was still dominated by the freestream turbulence. However, it was evident that there were no marked peak effects which would have been necessary tp produce "critical velocity" effects. At no stage during the tests were any "critical velocity" effects detected.

From the decay of the autocorrelation the total damping in terms of percentage of critical was found to be

$\zeta = 3{\cdot}3\%$ and as $\zeta_{structural} = 1{\cdot}3\%$, the $\zeta_{aerodynamic} = 2{\cdot}0\%$, at $\bar{u}_{55} = 25$ m/s.

The aerodynamic damping, caused by the high sectional lift characteristics, is seen to be very significant.

As the problem of deck response under wind action became confined to the first vertical bending mode, the problem of obtaining a peak measurement, from which structural design parameters could be derived, became very much simplified. From measurements of upcrossing probability distributions, it was determined that the average maximum deck displacement was 4·0 (±0·02) standard deviations (σ_z) above the mean ($\overline{z}$). It was decided to use a peak factor of 4·0 to obtain the average maximum displacement per hour, z, in this study,

$$\text{i.e.}\quad \hat{z} = \overline{z} + 4{\cdot}0\,\sigma_z \qquad (1)$$

One must not lose sight of the fact that there is always a probability of getting higher peak values, for example there is an average probability per hour of about 10% of exceeding a value of Mean plus 4·8 σ.

Although measurements of deck response were made for a number of variations of the service configuration, only two sets of data relevant to the final full scale configuration will be presented in this report. In Figure 2 the vertical deck displacement at centre span is given as a function of wind direction (β) for two reduced velocities and one value of turbulence intensity. It can be seen that the peak deck response is dependent both on mean velocity and turbulence intensity to the power two or a little greater, and that for a given mean wind speed the respone varies approximately as the cosine of the angle between the wind direction and the deck normal, except at the larger angles. The structural damping of the model was varied using an oil dashpot damper and peak deck response was found to vary with critical damping ratio to the power -0·2. The strong effect of the aerodynamic damping is evident here as for purely mechanical assumptions the response would vary as damping to the power -0·5.

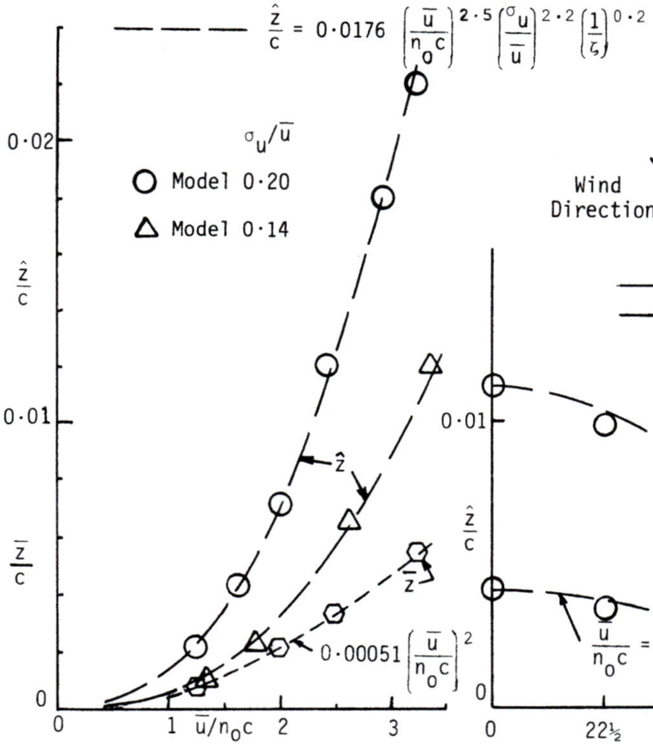

FIG. 2 MEAN AND AVERAGE HOURLY MAXIMUM CENTRE DECK VERTICAL DISPLACEMENT FROM THE MEAN AS A FUNCTION OF REDUCED VELOCITY, FROM MODEL MEASUREMENTS
$n_o = 0{\cdot}33$ Hz, $\zeta = 0{\cdot}013$, $\beta = 0°$

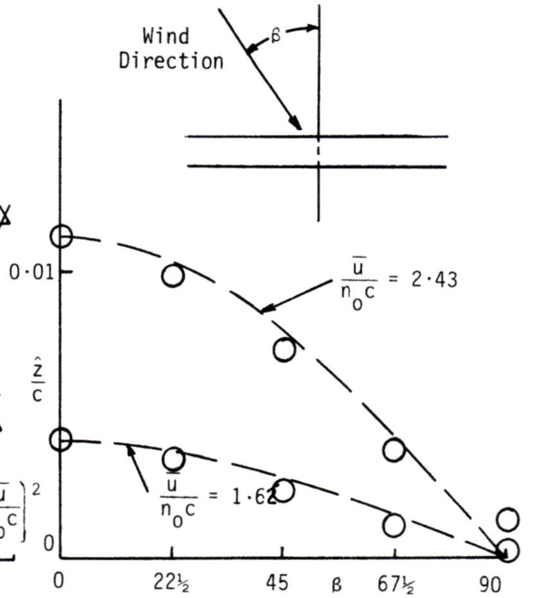

FIG. 3 AVERAGE HOURLY MAXIMUM CENTRE DECK VERTICAL DISPLACEMENT FROM THE MEAN AS A FUNCTION OF WIND DIRECTION FROM MODEL MEASUREMENTS
$n_o = 0{\cdot}33$ Hz, $\zeta = 0{\cdot}013$, $\sigma_u/\overline{u} = 0{\cdot}20$

To enable a comparison of model data with full scale data a function relationship for the peak vertical displacement at centre span was fitted to the model data as follows:

$$\frac{\hat{z}}{c} = 0{\cdot}0176 \left(\frac{\bar{u}}{n_o c}\right)^{2.5} \left(\frac{\sigma_u}{\bar{u}}\right)^{2.2} \left(\frac{1}{\zeta}\right)^{0.2} \cos\beta \qquad (2)$$

An evaluation of this function is shown for comparison with the model data in Figures 2 and 3.

5. FULL SCALE MEASUREMENTS

From mid 1978, just prior to bridge deck closure, an instrument package consisting of a cup anemometer and wind direction vane and two accelerometers was mounted at the 3rd box from the centre on the west side and subsequently all were moved to the centre span flange after deck closure. Nominally, continuous records from these instruments have been obtained since installation.

The bridge deck has oscillated primarily in the 1st vertical bending mode (symmetrical) in all wind storms to date. In fact, it has been difficult to detect other modes. One of the accelerometers was placed at the outer chord for a time and whilst on occasions slightly higher vertical accelerations were recorded compared with the accelerometer at mid-chord, the difference is small, and enough to conclude at this stage that the torsional response is very small. Similarly, one of the accelerometers was placed to measure horizontal acceleration for a time, and a small horizontal response was measured. As yet no measurements have been made with an accelerometer away from centre span to see if there is any significant response in the 2nd vertical vending mode (antimmetrical).

5.1 Frequency and Damping Measurements

Frequency and damping measurements have been made for several modes from a record of the decay in amplitude of deck oscillation following excitation initiated by driving a heavy mobile crane off the edge of a raised plank. These results are given in Table 1 along with model frequency values. Unfortunately, it has only been possible to measure the structural damping for very small amplitudes of oscillation, order ±2 mm, compared with peak oscillations due to wind action of ±50 mm which have been measured to date. The increased damping detected for the erection configuration is complicated by the fact that the outer temporary erection cables were still taking a significant load and the permanent outer cables were in various stages of tensioning.

5.2 Response at Centre Span

A continuous record of acceleration at centre span along with wind speed and wind direction was kept on a paper chart, such that maximum deck acceleration, mean and maximum gust wind speed and mean wind direction for each hour could be determined. From this information the peak displacement in each hour was calculated, assuming simple harmonic motion, $(\hat{z} = \hat{\ddot{z}}(2\pi n_o)^2$. The turbulence intensity was calculated by averaging the value of $\hat{u}/\bar{u}$ over three hours or more and using the relationship

$$\sigma_u/\bar{u} = 1/3{\cdot}5\ (\hat{u}/\bar{u} - 1) \qquad (3)$$

Evaluation of this full scale data for the three or more hours during eight storms is given in Table 2. It was thought originally that an average turbulence intensity could be determined for each wind direction. However, the variations of turbulence intensity for storms from the same direction were so sensitive to atmospheric stability conditions, and had such an influence on deck response, that the turbulence intensity was evaluated and averaged for each storm. A comparison is made with the measured response and that predicted by the model measurements, per Equation 2, using average values for each storm of reduced velocity, turbulence intensity and wind direction for a value of critical damping of $\zeta = 0{\cdot}01$. A further comparison between model and full scale response measurements is shown in

Sketch 3, where in particular the effect of turbulence intensity has been highlighted by adjusting the full scale values to a reduced velocity of 1·0 (using the functional relationship in Equation 2).

6 CONCLUSIONS

Both model and full scale measurements show that response to wind action is primarily in the first vertical bending mode, and that this response is dependent on the turbulence intensity squared (approximately). Such a dominant dependence on turbulence intensity confirms that the response mechanism is likely to be primarily due to resonant response to turbulent flow energy via the section lift characteristics.

There is generally good agreement between the model and full scale response measurements with the major sources of error being attributed to the determination in full scale of the freestream turbulence intensity and the structural damping at the larger amplitudes of oscillation.

There were no indications of any "critical velocity" type excitations of the deck either in the model or full scale measurements.

7 ACKNOWLEDGEMENTS

The author wishes to acknowledge the cooperation and support of the West Gate Bridge Authority to facilitate the full scale measurements and permission to publish the full and model scale data. The personal contributions by Mr. H.G. Wolfram, Dr. A.R. Toakley, Dr. J.D. Holmes, Professor B.J. Vickery and Messrs. M.G. Straker and D.B. Sharp and Mrs. I.H. Cody, are also gratefully acknowledged.

8 REFERENCES

1. Brook, R.R., "A Study of Wind Structure", Dept. of Sci., Bur. of Met. Paper No.27, (1974).
2. Davenport, A.G., "The Action of Wind on Suspension Bridges", Proc. Int. Symp. on Suspension Bridges, Laboratorio Nacional de Engenharia Civil, Lisbon, (1966).
3. Melbourne, W.H., "West Gate Bridge Wind Tunnel Tests", Consulting Report, (1973).
4. Melbourne, W.H., "Cross-Wind Response of Structures to Wind Action", 4th Int. Conf. on Wind Effects, pp 343-358, Cambridge University Press, (1975).
5. Vickery, B.J. & Watkins, R.D., University of Sydney, Civil Eng., Reports Nos S134 and S135, (1973).

Table 1 NATURAL FREQUENCY AND DAMPING MEASUREMENTS

Mode	Frequency, n Hz		Full Scale Critical Damping Ratio, ζ
	Model	Full Scale	
FULL BRIDGE			
1st Vertical Bending (Symmetrical)	0·33 (LYCA analytical estimate 0·34)	0·35	0·009 (±1·8 mm to ±0·7 mm)
2nd Vertical Bending (Antisymmetrical)	0·53 (LYCA analytical estimate 0·55)	Not detected as yet	
3rd Vertical Bending (Symmetrical)	0·97	1·0	
1st Torsion (Symmetrical)	0·80	0·87	0·003 (±0·006° to ±0·003°)
ERECTION CONFIGURATION JUST PRIOR TO CLOSURE			
1st Vertical Bending	0·30	0·38	0·012 (±3 mm to ±1·25 mm) 0·008 (±1·25 mm to ±0·75 mm)

Table 2 FULL SCALE CENTRE SPAN DISPLACEMENT MEASUREMENTS FOR EACH HOUR OF RECORD WITH RELATED WIND DATA AND MODEL COMPARISON

Wind Direction	Angle to deck normal β	Mean wind speed at deck $\bar{u}$ m/s	Maximum wind speed at deck u m/s	Turbulence Intensity $\sigma_u/\bar{u}$	Maximum vertical displacement $\hat{z}$ mm	$\hat{z}/c$	Reduced velocity $\bar{u}/n_o c$ (averaged)	Model $\hat{z}/c$ from Equation 2 for ζ = 0·01
NNE	15°	11·4 11·0 10·9	16·0 15·7 17·0	0·13	13·2 14·0 11·4	0·00035 0·00037 0·00030	0·90	0·00037
NNE	15°	14·4 15·3 15·2 14·9	20·7 22·3 23·5 23·0	0·14	45·7 42·8 37·2 44·7	0·00122 0·00114 0·00099 0·00120	1·22	0·00094
NNE	15°	12·9 13·2 13·4 14·8	19·1 18·1 18·7 19·9	0·11	15·2 20·9 16·0 20·9	0·00041 0·00056 0·00043 0·00056	1·09	0·00041
NNE	15°	13·1 15·5 14·5 12·0	19·1 21·4 20·7 17·4	0·12	19·3 22·1 21·3 20·9	0·00052 0·00059 0·00057 0·00056	1·12	0·00054
N	37°	15·4 13·2 12·8 12·8 13·9	23·6 20·1 19·9 18·9 21·2	0·15	25·0 21·3 19·3 15·2 29·4	0·00067 0·00057 0·00052 0·00041 0·00079	1·10	0·00069
NNW	50°	14·1 15·1 14·7 13·1	20·5 20·9 21·7 20·2	0·13	16·8 21·9 26·3 24·0	0·00045 0·00059 0·00070 0·00064	1·15	0·00045
W	53°	12·6 11·5 11·0 11·1 10·1	18·6 17·0 16·8 16·3 19·2	0·19	26·5 15·2 14·2 13·8 20·3	0·00071 0·00041 0·00038 0·00037 0·00054	0·91	0·00055
SSW	15°	14·3 14·2 13·2 12·1	19·7 17·8 18·1 17·1	0·11	23·3 29·4 19·3 15·0	0·00062 0·00079 0·00052 0·00040	1·09	0·00042
					Maximum Horizontal Displacement $\hat{x}$	$\hat{x}/c$		From model measurements $\hat{x}/c$
NNE	15°	11·4 11·0 10·9	16·0 15·7 17·0	0·13	2·0 3·0 2·0	0·00005 0·00008 0·00005	0·90	approx. 0·0001
NNE	15°	14·4 15·3 15·2 14·9	20·7 22·3 23·5 23·0	0·14	8·9 11·4 9·7 7·9	0·00024 0·00030 0·00026 0·00021	1·22	approx. 0·0002

THE CABLE STAYED BOX GIRDER WEST GATE BRIDGE

① FULL SCALE WEST GATE BRIDGE

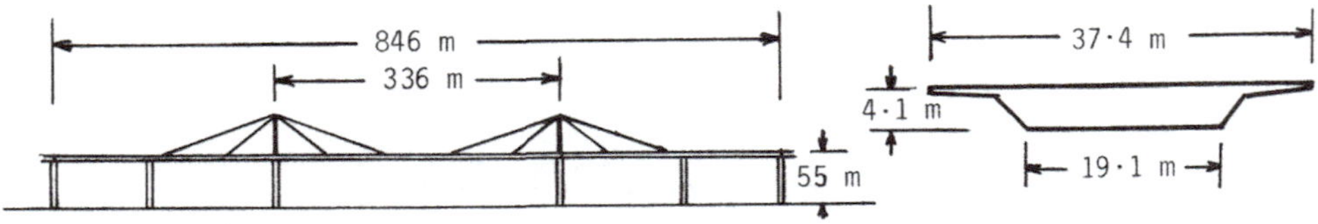

② AERODYNAMIC OUTLINE DIMENSIONS OF WEST GATE BRIDGE

Problem: Turbulent wind flow excites deck oscillations.
Geometry: Details of the bridge and deck are given in Sketch 2.
Flow condition: The incident flow is the natural wind approaching over terrain roughness varying from open sea to suburban/factory conditions. The turbulence intensity of the longitudinal component at deck height (55 m) varies from 10% to 20%.
Structural Dynamics: The average box mass is 2×10^4 kg/m. The average mass moment of inertia about the central axis of the box is $1{\cdot}8 \times 10^6$ kg m^2/s. The typical box section 2nd moments of area in bending are 100 and 3 m^4 respectively and in torsion 6 m^4. The mass of the outer and inner cables is 790 and 430 kg/m respectively. The fundamental natural frequency of the deck vertical bending mode is 0·35 Hz.
Vibration: The response of the deck to wind excitation has been measured and compared with measurements made on an aeroelastic wind tunnel model. A sample of these measurements is given in Sketch 3.

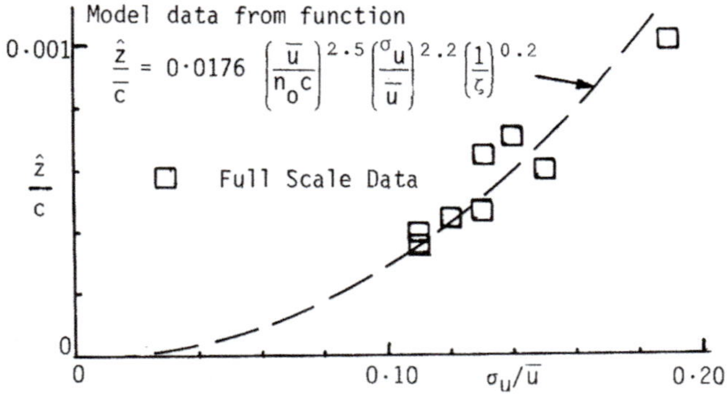

③ FULL SCALE AVERAGE HOURLY MAXIMUM CENTRE DECK VERTICAL DISPLACEMENT AS A FUNCTION OF THE TURBULENCE INTENSITY OF THE INCIDENT WIND FLOW, AT $\bar{u}/n_o c = 0{\cdot}1$, $\zeta = 0{\cdot}01$, $\beta = 0°$

F4 ON SUBSPAN OSCILLATION OF TWO COUPLED CONDUCTORS WITH ONE IN THE WAKE OF ANOTHER

Y. T. Tsui
Researcher

Institut de Recherche de l'Hydro-Québec
Varennes, Québec, Canada

SUMMARY

Skh.* 1 depicts a typical lift and drag distribution acting on the leeward conductor in the wake created by the windward conductor. For two-dimensional studies, this system can be simulated by a pair of rigid cylinders, mounted on a spring, as shown in Skh. 2.

As is well known, wake-induced flutter is referred to as subspan oscillation in power line parlance. For some years, two outstanding problems in wake-induced flutter, namely, flutter asymmetry and frequency coalescence mystified both researchers and practicing engineers. These problems were clarified by using the experimental set-up shown in Skh. 3 to study flutter of the leeward conductor in the wake of the fixed windward conductor [4]. These findings agree with recent theoretical results [2,3] and disagree with previous studies [5,10].

This paper is a generalization of the previous ones [2,3] in two aspects. First, the windward conductor, which is coupled to the leeward conductor through a spacer, is allowed to move. Second, damping is introduced. It is found that, as in the case of a fixed windward conductor, the wake flutter is symmetrical with respect to the horizontal axis of the wake if there is a change in sign of the $\varepsilon = k_{xy}/k_{xx}$. It is also found that, within our range of interest, the system is always unstable when the spacer spring constant, $K > 1.2\, k_{xx}$, where k_{xx} is the horizontal spring constant of the conductor suspension system. It appears that the spacer coupling ratio, K/k_{xx}, is the most important factor in determining stability. For practical cases, the system is almost always stable for a value of spacer coupling ratio in the neighborhood of 0.64, because the frequency ratio, $\kappa = \omega_y/\omega_x$, usually varies less than ten percent from unity for a typical transmission line. On the other hand, within our range of interest, damping has little or no effect on the stability of the coupled conductors. Nevertheless, it does have an influence on the stability of the leeward conductor when the windward conductor is fixed.

Although no experimental data for the general case of two-dimensional studies are available now, we are rather fortunate that Hydro-Quebec Research Institute does have a full-scale test line at the Magdalen Islands [11]. It is gratifying to report that our findings on the effect of damping, as well as of the spacer coupling ratio, on the stability of coupled conductors do agree qualitatively with the full-scale test line data [12].

1. INTRODUCTION

By and large, there are three kinds of flow-induced transmission line vibrations. One is galloping, which features very large amplitude in comparison with the conductor diameter, and occurs at its own natural frequency, usually below 1 Hz. The other is vortex-induced vibration, ascribable to vortex shedding. For transmission lines, it is known as aeolian vibration, which features low amplitude

*Skh. refers to sketch on Chart at end of text.

in comparison with conductor diameter and high frequencies of 10 to 50 Hz. To reduc corona loss, multiline conductors are commonly employed for modern high voltage transmission lines. As shown in Skh. 1, when the wind flows uniformly from the left the wake generated behind the windward conductor consists of very complex flow patterns. When such disturbed flow strikes the leeward conductor in the wake of the windward conductor, a third type of flow-induced oscillation can occur. This is known as wake-induced flutter or, in power line parlance, subspan or subconductor oscillation, which features a nearly horizontally oriented ellipse with frequency around 1 to 4 Hz.

2. STATEMENT OF THE PROBLEM

Skh. 1 depicts the wake behind a windward conductor in which the lift coefficient C_L and the drag coefficient C_D on the leeward conductor are referenced to the free-stream velocity. For two-dimensional analysis, this system can be simulated by a pair of rigid cylinders, mounted on spring suspension, with a single spacer in the form of another spring as shown in Skh. 2.

In reality, when subspan oscillations occur, both the windward and the leeward conductors will oscillate in a nearly horizontally elliptical orbit with opposite phase [13]. This paper is a generalization of [2,3] in two aspects. First, the windward conductor, which is coupled to the leeward conductor through a spacer, is allowed to move. Second, damping is introduced. The dynamic equations for both conductors, in the x and y directions, which have been idealized as two mass points are:

$$M\ddot{\bar{x}}_1 + d_{x_1}\dot{\bar{x}}_1 + k_{1111}\bar{x}_1 + k_{1121}\bar{y}_1 + k_{1112}\bar{x}_2 + k_{1122}\bar{y}_2 = F_{x_1} \tag{1a)'}$$

$$M\ddot{\bar{y}}_1 + d_{y_1}\dot{\bar{y}}_1 + K_{2111}\bar{x}_1 + k_{2121}\bar{y}_1 + k_{2112}\bar{x}_2 + k_{2112}\bar{y}_2 = F_{y_1} \tag{1b)'}$$

$$M\ddot{\bar{x}}_2 + d_{x_2}\dot{\bar{x}}_2 + k_{1211}\bar{x}_1 + k_{1221}\bar{y}_1 + k_{1212}\bar{x}_2 + k_{1222}\bar{y}_2 = F_{x_2} \tag{1c)'}$$

$$M\ddot{\bar{y}}_2 + d_{y_2}\dot{\bar{y}}_2 + k_{2211}\bar{x}_1 + k_{2221}\bar{y}_1 + k_{2212}\bar{x}_2 + k_{2222}\bar{y}_2 = F_{y_2} \tag{1d)'}$$

where the right hand side stands for the areodynamic forces, and the left hand side stands for the inertial forces (1st term), the mechanical damping forces (2nd term) and the elastic forces (3rd term through 6th term). The stiffness matrix is symmetric with ten elastic constants to be dtermined.

Since both the windward and the leeward conductors are identical, in so far as mass distribution and suspension system are concerned, it follows:

$$d_{x_1} = d_{x_2} = d_x \qquad k_{1111} = k_{1212} = k_{xx}$$

$$d_{y_1} = d_{y_2} = d_y \qquad k_{2121} = k_{2222} = k_{yy}$$

$$k_{1121} = k_{2111} = k_{1222} = k_{2212} = k_{xy}$$

The dynamical equations (1a)' through (1d)' can be simplified as follows:

$$M\ddot{\bar{x}}_1 + d_x\dot{\bar{x}}_1 + k_{xx}\bar{x}_1 + k_{xy}\bar{y}_1 - k_{1112}\bar{x}_2 + k_{1122}\bar{y}_2 = F_{x_1} \quad (1a)$$

$$M\ddot{\bar{y}}_1 + d_y\dot{\bar{y}}_1 + k_{xy}\bar{x}_1 + k_{yy}\bar{y}_1 - k_{2112}\bar{x}_2 + k_{2122}\bar{y}_2 = F_{y_1} \quad (1b)$$

$$M\ddot{\bar{x}}_2 + d_x\dot{\bar{x}}_2 + k_{1211}\bar{x}_1 + k_{1221}\bar{y}_1 + k_{xx}\bar{x}_2 + k_{xy}\bar{y}_2 = F_{x_2} \quad (1c)$$

$$M\ddot{\bar{y}}_2 + d_y\dot{\bar{y}}_2 + k_{2211}\bar{x}_1 + k_{2221}\bar{y}_1 + k_{xy}\bar{x}_2 + k_{yy}\bar{y}_2 = F_{y_2} \quad (1d)$$

Equations (1a) through (1d) form the basic equations in dealing with the aeroelastic stability of two coupled conductors, where one is in the wake of the other. In order to meet the spirit of this symposium on practical experiences, further theoretical treatment will be delegated elsewhere [1], in which the aerodynamic forces are linearized and the Routh-Hurwitz stability criteria are employed. Furthermore, in dealing with aerodynamic forces on coupled conductors in motion, the simple model is used because more sophisticated models, such as the Rawlins model and the Simpson-Flower model, are of academic interest at the present time due to paucity of required aerodynamic data.

3. RESULTS

In this paper, we shall present numerical results and elaborate on practical experiences. To be physically realistic, we choose the same parameters for numerical computations as in [2]:

Frequency ratio κ: 0.8, 0.91, 0.94, 0.97, 1, 1.03, 1.06, 1.09, 1.2 (9 cases)
Coupling coefficient ε: 0, $\pm$0.02, $\pm$0.04, $\pm$0.06, $\pm$0.08, $\pm$0.1 (11 cases)
Spacing to diameter ratio d/c: 10, 13, 16 (3 cases)
Reynolds number N_{Re}: 2.82 x 10^4, 5.64 x 10^4 (2 cases)
Incidence angle θ: 1° to 20° and -1° to -20° by 1° increment (40 cases)
Subspan length ℓ: 45.72m
Wt. per unit length M/ℓ: 1.596 kg/m
Diameter c: 3.048 x 10^{-2}m

In addition, we choose two more parameters: frequency ratio of spacer coupling ($\Omega/\omega_x = \sqrt{K/k_{xx}}$ where K/k_{xx} is spacer coupling ratio

$\frac{\Omega}{\omega_x}$ = 0, 0.8, 0.9, 1, 1.1, 1.2 (6 cases)

and critical damping ratio $\delta = (d_x/2M)\sqrt{k_{xx}/M} = (d_y/2M)\sqrt{k_{yy}/M}$
δ = 0, .005, .01, .02 (4 cases)

All together, we have 9 x 11 x 3 x 2 x 6 x 4 = 14,526 cases. For each case, we have 40 calculations for different incidence angles ranging from -20° to +200°. Since, in practice, the maximum value of critical damping of the conductor suspension system would hardly exceed one percent, we take the liberty of not tabulating the case of $\delta = 0.02$, even though the numerical results are available.

Based on our numerical results, it is found that the wake flutter symmetry is preserved, even for the spacer-coupled cases, if the sign of the static coupling coefficient, ε, is changed as shown in [2,3]. We shall take advantage of this symmetry property by tabulating only the negative coupling coefficients, ε, thus reducing the tabulation by one half. Also based on our numerical results, it is found that, for $\Omega = 1.1\ \omega_x$ and $\Omega = 1.2\ \omega_x$, the system is always unstable within our range of interest. We shall not tabulate these unstable cases.

Finally, all numerical results are condensed into 12 tables as given in [1] which will not be repeated here due to lack of space. Instead, we shall present some highlights.

Study of these tables reveals some important facts: it appears that the spacer coupling ratio, $K/k_{xx} = (\Omega/\omega_x)^2$, is the most dominant factor in determining stability. As this ratio increases from 0.64, the system becomes more and more unstable. When this ratio reaches and exceeds 1.2, the system is always unstable. For practical purposes, the system is always stable for $\Omega/\omega_x = 0.8$, since the frequency ratio, κ, hardly deviates 10 percent from unity. On the other hand, damping has little effect on the stability when $\Omega = 0.8\ \omega_x$ and $0.9\ \omega_x$; there is practically no effect when $\Omega = \omega_x$; and there is no effect whatsoever when $\Omega = 1.1\ \omega_x$ and $\Omega = 1.2\ \omega_x$. Only when $\Omega = 0$ does damping have some effect on the stability with more influence at $N_{Re} = 2.82 \times 10^4$ than at $N_{Re} = 5.64 \times 10^4$.

For the special case of no damping with a fixed windward conductor, the theoretical results are fully supported by the experimental findings [4]. We also resolved two outstanding problems of wake-induced flutter, namely, flutter symmetry and frequency coalescence [2]. Fortunately, the principle of flutter symmetry can be extended to the general cases of damping with a moving windward conductor provided that the sign of the static coupling coefficient of the suspension for the leeward conductor is changed. In view of imperfections of the simple model, one might raise the pertinent question: "Do we have any experimental results for the general case of damping with a moving windward conductor?" Our succint answer is no, simply because we have not performed such experiments yet, nor do we know of any. However, we are rather fortunate that Hydro-Quebec Research Institute does have a full scale test line facility at the Magdalen Islands [11,12].

4. PRACTICAL EXPERIENCES

All full scale tests were carried out at the Magdalen Islands test line facility which has been described in detail in [11]. Briefly, it consists of four portal type towers, 30 m. high by 15 m. wide, and two ground level anchoring bases making up three suspension spans of 274 m., 366 m. and 457 m., respectively, and two dead-end spans of 244 m., as shown in Skh. 4.

For these tests, two bundles of four Carillon conductors at 45.7 cm. spacing were strung in square configuration. Characteristics of the Carillon conductor are: stranding, 7 strands of 2.2 mm. diameter steel and 42 strands of 4 mm. diameter aluminum; overall diameter, 3.05 cm.; weight, 1.65 Kg/m; ultimate tensile strength, 117,565 N. The mechanical tension applied to the conductors are controlled to 20 percent U.T.S. within 1 percent.

Before proceeding further, it should be pointed out that both physical and geometrical properties of the test line, such as conductor weight and diameter, subspan length, etc., might vary somewhat from that used in two-dimensional simulation. Nevertheless, in both cases, typical values are employed. It is well known [14] that, in conventional bundles, each leeward conductor is influenced aerodynamically by only one windward conductor. Therefore, the same aerodynamic condition prevails for two conductors in simulation as well as for four conductors in a full scale test line.

Three basic types of articulated spacers have been tested. First is the commercial spacer-damper (herein called CSD) as shown in Skh. 5. Second is the articulated non-damping spacer (herein called ANDS) as shown in Skh. 6, which is the same as the first with their elastomeric elements removed. Third is the experimental spacer damper (herein called ESD) as shown in Skh. 7. As the name implies, ESDs are specially designed and made by Hydro-Quebec and consist of four subtypes with various physical and geometrical properties, as demonstrated in Skh. 8 and Table 1.

TABLE 1 SOME IMPORTANT PHYSICAL CHARACTERISTICS OF SIX SPACERS AND THEIR RELATIVE POINTS OF DEMERIT AFTER HARDY AND BOURDON [12].

SPACERS	CSD	ANDS	ESD1	ESD2	ESD3	ESD4
M (Kg) Frame mass	3.58	4.48	5.68	5.68	5.68	8.41
k_a (N/cm) Axial stiffness	1324	-	17206	10376	10376	10376
k_t (N/cm) Tangential stiffness	65.3	52.9	116.5	74.1	47.6	74.1
η Damping factor	.18	$\sim$0	.28	.28	.22	.28
Relative points of demerit obtained from test series		0.46 #1	2.30 #2	1.07 #3	0.46 #4	0.64 #5

Five series of tests were carried out. Referring to Skh. 4, there are two identical bundles, called #1 and #2.

In the first series, we employed CSD on bundle #1 and ANDS on bundle #2. The objective is to compare the performance of the commercial spacer-damper (CSD) and the articulated non-damping spacer (ANDS) for controlling subspan oscillations.

In the other four series, we employed CSD on bundle #1 and ESD1, ESD2, ESD3 and ESD4, respectively, on bundle #2. The objective is to establish the influence of both stiffness and mass of the spacer on the bundled conductor response to subspan oscillations.

As is well known, there is yet no standard method of measuring subspan oscillations. Hardy and Bourdon [12] devised a scheme called points of demerit (P.D.). In essence, points of demerit are proportional to the amplitude and frequency of subspan oscillations with weighting factor equal to 1 for sustained, and equal to 1/3 for unsustained oscillations. In five series of tests, bundle #1 always carries CSD and bundle #2 carries ANDS, EDS1, ESD2, EDS3 and EDS4. Therefore, in each series we divide the points of demerit of bundle #2 by the points of demerit of bundle #1, thus obtaining their relative points of demerit and listing them in the last row of Table 1.

Study of Table 1 reveals the following important findings: Comparison of columns 1 and 2 clearly shows that removing damping does attenuate subspan oscillations, while comparison of columns 1 and 3 clearly shows that increase of stiffness does aggravate subspan oscillations. The same trend prevails between column 1 and 4 and column 1 and 5 so far as tangential stiffness is concerned. Since spacer mass is not included in the two-dimensional simulation, we shall refrain from comparison of column 1 and 6. To sum up, there are very good qualitative agreement between full scale test line data and two-dimensional results.

5. REFERENCES

1. Tsui, Y. T. and Tsui, C. C., "On Wake Induced Flutter of a Circular Conductor in the Wake of Another", to appear at Flow-Induced Vibration Symposium at the 3rd National Congress on Pressure Vessel and Piping Technology, San Francisco, California, (June 25-29, 1979).

2. Tsui, Y. T., "On Wake-Induced Flutter of a Circular Cylinder in the Wake of Another", Journal of Applied Mechanics, pp. 194-200, (June 1977).

3. Tsui, Y. T., "Two Dimensional Stability Analyses of a Circular Conductor in the Wake of Another", Electric Power Systems Research, 1, pp. 87-95, (1977-78).

4. Watts, J. A. and Ko, R. G., "A Wind Tunnel Investigation Into the Aerodynamic Stability of Bundled Power Line Conductors for Hydro-Quebec, Part II: Dynamical Sectional Model Tests Using a Form Spring Orthogonal Suspension", LTR-LA-103, NAE, NRC, Ottawa, (Jan. 1973).

5. Simpson, A., "Stability of Subconductors of Smooth Circular Cross Section", Proceedings IEE, pp. 741-750, (April 1970).

6. Simpson, A., "On the Flutter of a Smooth Circular Cylinder in a Wake", Aeronautical Quarterly, pp. 25-41, (Feb. 1971).

7. Simpson, A., "Wake-Induced Flutter of Circular Cylinders: Mechanical Aspects", Aeronautical Quarterly, pp. 101-118, (May 1971).

8. Cooper, K. R. and Wardlaw, R. L., "Aeroelastic Instabilities in Wakes", Proceedings, Third International Conference on Wind Effects on Buildings and Structures, Tokyo, Japan, (Sept. 1971).

9. Scanlan, R. H. and Wardlaw, R. L., "Reduction of Flow-Induced Structural Vibration", Isolation of Mechanical Vibration, Impact and Noise, eds., Snowdon, J. C., and Ungar, E. E., ASME, N.Y., pp. 35-63, (1973).

10. Cooper, K. R., "Wake Galloping and Aeroelastic Instability", IUTAM-IAHR Symposium on Flow-Induced Vibrations, ed., Naudascher, E., Karlsruhe, Germany, (1972).

11. Cloutier, L. J., Hardy, C., and Brunelle, J., "The Magdalen Islands Test Line: A Research Tool for Studies of Bundle Conductor Dynamics", IEEE-PAS Conference Paper C 74 370-3, (1974).

12. Hardy, C. and Bourdon, P., "The Influence of Spacer Dynamic Properties in the Control of Bundle Conductor Motion", IEEE-PAS Conference paper to be presented at Vancouver, B.C., Canada, (July 15-20, 1979).

13. Sarkinen, C. F. and Wiitala, J. T., "Subspan Galloping Eased on EHV Line", Electrical World, pp. 34-37, (Dec. 1969).

14. Rawlins, C. B., "Fundamental Concepts in the Analysis of Wake-Induced Oscillation of Bundle Conductors", IEEE Transactions on PAS, Vol. 95, No. 4, pp. 1377-1393, (1976).

ON SUBSPAN OSCILLATIONS OF TWO COUPLED CONDUCTORS WITH ONE IN THE WAKE OF ANOTHER

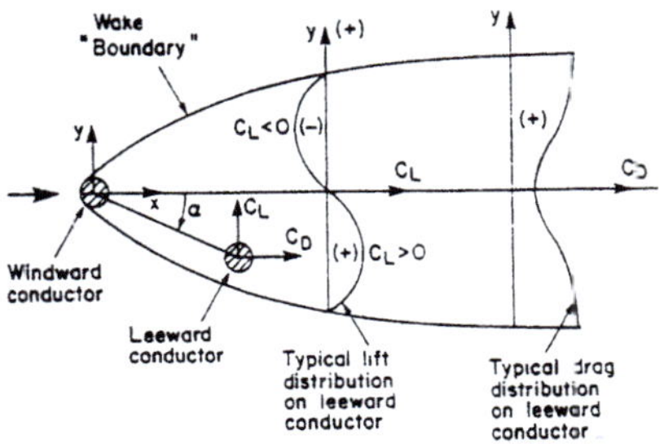

① Conductor diameter: 305mm. spacing to diameter ratio:10 to 16

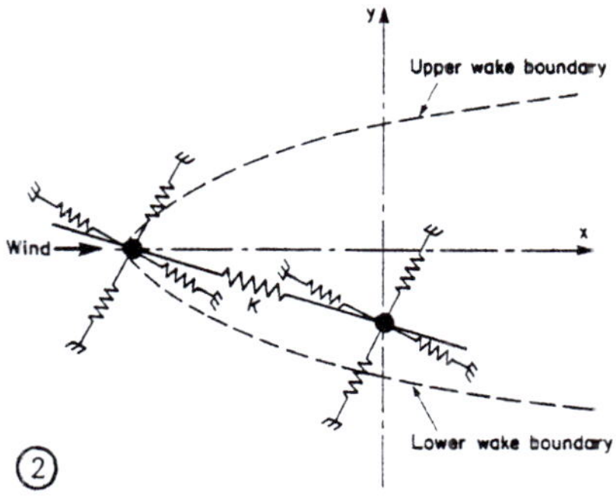

②

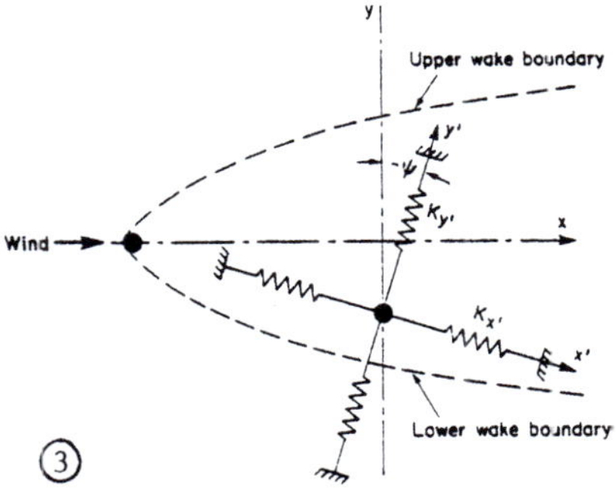

③

Problem: Subspan oscillations are wake-induced. Both conductors oscillate in horizontally oriented elliptical orbits with opposite phase at frequencies from 1 to 4 Hz. Damage to both conductors and spacers.

Geometry: See sketch 1. Subspan length around 46 meters.

Flow Conditions: See sketches 1 and 2. Uniform wind from left at Reynolds number 2.82×10^4 and 5.64×10^4.

Two Dimensional Simulation: See sketch 2. Two conductors, mounted on spring suspension, are connected by spring K, which simulates spacer.

Important Findings: (1) Subspan oscillation is symmetric with respect to centre-line of wake provided sign of static coupling is changed. (2) Vertical and horizontal frequency coalescence does not necessarily imply lack of oscillation. (3) Spacer coupling ratio, K/k_{xx}, is the most important factor in determining stability, having optimum value of 0.64. (4) Damping has little or no effect on stability of coupled conductors.

Items (1) and (2) have been verified experimentally for a special case of fixed windward conductor with K = 0, as shown in sketch 3. However, experimental results of a general case are not available. Look for full scale test data for qualitative verification.

Practical Experiences, Full Scale Test: Sketch 4 depicts side view of Magdalen Islands test line. It consists of four portal type towers and two ground level anchoring bases making up three subspans of 274 m, 366 m, and 457 m, respectively. Two bundles (#1 and #2) of four Carillon conductors at 45.7 cm. spacing were strung in square configuration. Main characteristics of Carillon conductor: diameter, 305 cm; weight 1.65 Kg/m/ ultimate tensile strength 117,656 N. The conductors are strung to 20 percent of U.T.S.

Three basic types of articulated spacer tested. Sketch 5: Commercial spacer damper (CSD). Sketch 6: Articulated non-damping spacer (ANDS), that is, CSD with their damping materials removed. Sketch 7: Experimental spacer damper (ESD) specially designed and made by Hydro-Quebec, which has four subtypes (ESD1, ESD2, ESD3, ESD4) with different stiffness and inertia properties as shown in sketch 8 and Table 1.

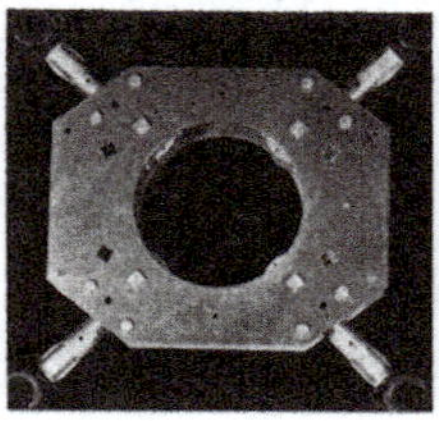

(5) Commercial spacer damper (CSD)

(6) Articulated non-damping spacer (ANDS)

(7) Experimenta[l] spacer samp[le] (ESD)

(4)

Table 1 Some important physical characteristics of six spacers and relative points of demerit after Hardy and Bourdon [2].

SPACERS	CSD	ANDS	ESD1	ESD2	ESD3	ESD4
M (Kg) Frame Mass	3.58	4.48	5.68	5.68	5.68	8.4
k_a (N/cm) Axial stiffness	1324		17206	10376	10376	1037
k_t (N/cm) Tangential stiffness	65.3	52.9	116.5	74.1	47.6	74.
η Damping factor	.18	~0	.28	.28	.22	.2
Relative points of demerits obtained from test series		0.46 #1	2.30 #2	1.07 #3	0.46 #4	0.6 #5

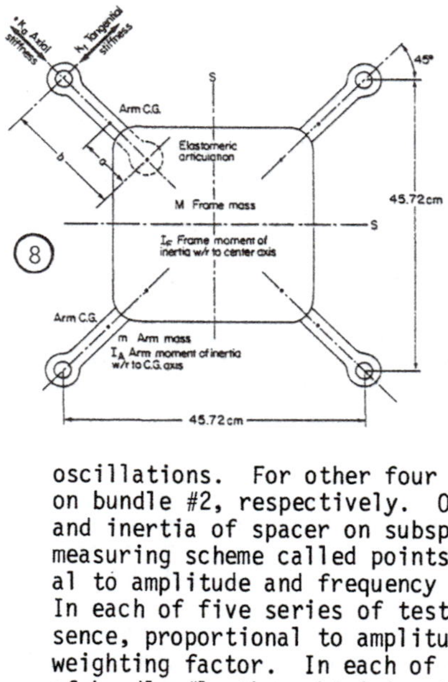

(8)

Five series of tests. For first series, CSD on bundle #1 and ANDS on bundle #2. Objective: to compa[re] performance of CSD and ANDS for controlling subspan oscillations. For other four series, CSD on bundle #1 and ESD1, ESD2, ESD3, ESD4 on bundle #2, respectively. Objective: to establish influence of both stiffness and inertia of spacer on subspan oscillations. Hardy and Bourdon devised a special measuring scheme called points of demerit (P.D.), which are, in essence, proportional to amplitude and frequency of subspan oscillations with certain weighting factor In each of five series of test, divide P.D. of bundle #2 by P.D., which are, in essence, proportional to amplitude and frequency of subspan oscillations with certain weighting factor. In each of five series of tests, divide P.D. of bundle #2 by P.D. of bundle #1, thus obtaining their relative points of demerit. It is shown that low stiffness is desirable in spacer-damper, but that damping is not essential for attenuation of subspan oscillation. This conclusion agrees qualitatively with findings from two dimensional simulation.

CURES: Spacers should have low stiffness, with optimium value of spacer coupling ratio of 0.64, and no damping, for attentuation of subspan oscillations.

REFERENCES:

1. Tsui, Y.T. and Tsui, C.C., "On Wake-Induced Flutter of a Circular Conductor in the Wake of Another". to appear at Flow-Induced Vibration Symposium at 3rd Natic al Congress on Pressure Vessel and Piping Technology, June 1979, San Francisco,
2. Hardy, C., and Bourdon, P., "The Influence of Spacer Dynamic Properties in the Control of Bundle Conductor Motion", IEEE-PAS, Summer Conference paper to be pre sented at Vancouver, B.C., Canada, July 1979.

Discussion: ON SUBSPAN OSCILLATIONS OF TWO COUPLED CONDUCTORS WITH ONE IN THE WAKE OF ANOTHER (Y.T. Tsui)

R.L. Wardlaw, National Research Council of Canada, Ottawa

In his Summary Dr. Tsui states, "For some years, two outstanding problems in wake-induced flutter, namely, flutter asymmetry and frequency coalescence, mystified both researchers and practicing engineers." He goes on to explain that these problems were clarified by the experimental work of Watts and Ko (Ref. [4] in the paper) and states that their results agree with his theoretical predictions but disagree with other published studies (Ref. [5-10] in the paper).

Unfortunately, while casting doubt on these latter references, Dr. Tsui does not make clear in the text of his paper either the essence of the mystery or the nature of the disagreement with other workers. In some early studies of the phenomenon the importance of the ratio, $\varkappa$, between the natural frequencies in the two degrees of freedom of the suspended cables, and of the angle of rotation, ψ, of the suspension system had not been examined. This may be what Dr. Tsui is referring to but it is unreasonable to suggest that this constitutes a disagreement. Subsequent work by Simpson (Ref. [5-7] in the paper) and by Watts and Ko (Ref. [4] in the paper) has clarified the nature of the coupling effects between the degrees of freedom and hence the behavioural sensitivity to $\varkappa$ and ψ. It would be useful if Dr. Tsui would clarify what the mystery and the disagreement are and eliminate any misunderstandings concerning References [5] to [10].

If there is some disagreement with other researchers, it would only be fair to them to reference their most recent and most appropriate publications. The references for Simpson's work (Tsui's Reference [5-7]) were published in 1970 and 1971. While it is appropriate to cite these, they do not give an up-to-date picture of that particular researcher's contributions to the understanding of the phenomenon and it would be desirable to include References [1] and [2] listed below. The work of Simpson and Flower (Ref. [2] below) was referred to in Dr. Tsui's text but a reference was not given. Tsui's References [8] to [10] are papers relating to work done at the National Research Council of Canada. These papers which were published in 1971, 1972 and 1973 represent an early stage in the NRCC investigations of aeroelastic instabilities in wakes and furthermore are general papers wherein the bundled conductor problem was only briefly discussed as one example of the phenomenon and therefore are quite unsuitable as references in the context of Tsui's paper. His Reference [4] published in 1974 represents a later stage in the NRCC work where more advanced experimental arrangements had been devised to permit careful examination of the effects of frequency ratio, suspension rotation angle and other parameters. The NRCC experimental and analytical program of study has been well documented in laboratory reports, while Reference [3] of this discussion is a comprehensive review and summary of the work that provides a suitable reference for Tsui's paper. Similarly, a more recent publication of Rawlins (Reference [4] below) should be cited. Also, I would like to draw attention to the work of Diana and his colleagues in Italy (References [5,6 and 7] below) who in parallel to other researchers have pursued successful wind tunnel and full scale investigations and analytical studies into the aeroelastic behaviour of bundles.

Finally, I would like to say that a full three-dimensional analysis of the aeroelastic behaviour of a multi-subspan bundle has been done by Ko (Reference [8] below) using finite element methods. His analysis includes multiple degrees of freedom of the spacer and the results demonstrate that careful choice of spacer arm stiffness may be a practical method of controlling motion. The aeroelastic behaviour appears to be sensitive to both the stiffness in the spacer plane and the out-of-place stiffness.

References

1. Simpson, A., Price, S.J., "On the Use of 'Damped' and 'Undamped' Quasi-Static Aerodynamic Models in the Study of Wake-Induced Flutter", IEEE Conference Paper C74-378-6, 1974.
2. Simpson, A., Flower, J.W., "An Improved Mathematical Model for the Aerodynamic Forces on Tandem Cylinders in Motion with Aeroelastic Applications", Journal of Sound and Vibration, Vol. 51 (2), pp. 183-217, 1977.
3. Wardlaw, R.L., Cooper, K.R., Ko, R.G. and Watts, J.S., "Wind Tunnel and Analytical Investigation Into the Aeroelastic Behaviour of Bundled Conductors", IEEE Transactions on Power Apparatus and Systems, Vol. PAS-94, No. 2, pp. 642-654, March/April 1975.
4. Rawlins, C.B., "Extended Analysis of Wake-Induced Oscillation of Bundled Conductors", IEEE Paper No. F 77 217-3, 1977.
5. Diana, G., Gasparetto, M., DiGiacomo, G., Nicolini, P., "Analytical Method for Computing Subspan Oscillation, Analytical and Experimental Results", IEEE Conference Paper C74-493 3, 1974.
6. Curami, A., Diana, G., Riva, R., DiGiacomi, G., Nicolini., P., "Wake-Induced Oscillations in Bundle Systems, Part I - Finite Element Method, Analytical Experimental Results", IEEE Paper A77 218-1, 1977.
7. Curami, A., Riva, R., DiGiacomi, G., Nicolini, P., "Wake-Induced Oscillations in Bundle Systems, Part II. Comparative Analysis of Various Bundle Configureations", IEEE Paper A77 219-9, 1977.
8. Ko, R., Wardlaw, R.L., "Three-Dimensional Analysis on the Wind Induced Subspan Oscillations of Bundled Conductors", IEEE Conference Paper C74 060-0, 1974.

Reply to Discussion:

I would like to thank Mr. Wardlaw for his interest and comments on my paper. By and large, the discussion offers two criticisms.

First, he states that I cast doubts on References [5 to 10] in the paper. This is certainly a misunderstanding. What I really mean, is to settle two outstanding problems in wake-induced flutter, namely flutter symmetry and frequency coalescence, once and for all. These two problems are solved simply by examining the behavior and relationship of the coefficients of the pertaining differential equations and the Routh-Hurwitz inequalities prior to any stability computation, and are valid for any kind of suspension system. This is the crux of difference between my work and Simpson's (and others), which has been spelled out in detail in References [2, 3] in the paper. In particular, the last section of Reference [3] dwells in length on previous findings, which, in my view, are unsatisfactory, misleading, or erroneous.

The second criticism is related to the most recent publications. References [1, 2] in the discussion do not deal with the above-mentioned two problems, at least explicitly. Nor are References [4 to 7] in the discussion. As noted by the discusser, Reference [3] in the discussion is a comprehensive review paper, the pertinent part of which is contained in Reference [4] in the paper. Besides, I have used up all available space for my Karlsruhe paper. As to chronological order, I obtained my theoretical results in early 1974, which were well documented in IREQ Report 1061. Several years delay in publishing my results is due to my inability to get through Cancam in 1974, IEEE-PAS in 1975, and Fluids Engineering in 1976.

Finally, I might add that a comprehensive two-dimensional study on this subject will appear in a forthcoming (tentatively March 8, 1980) issue of the Journal of Sound and Vibration and that three-dimensional analyses are now in progress at this Institute.

F5 WIND-INDUCED OSCILLATIONS OF SOME STEEL STRUCTURES

O. Mahrenholtz and H. Bardowicks
Institute of Mechanics, University of Hannover
Federal Republic of Germany

SUMMARY

The aeroelastic behavior of three steel structures - a floodlight mast, a pylon of a bridge and a steel chimney with five radial fins - is discussed. This behavior has been reproduced and explained by wind tunnel tests of sectional models of the structures. Lift and drag forces were measured. Further, the aerodynamic stability was studied using a forced oscillation method. Using this method, either vortex excitation or galloping could be observed.

1. INTRODUCTION

It is well known that steel structures, such as slender chimneys and masts, can be damaged by wind-induced oscillations caused by vortex excitation or galloping. This contribution reports on these types of aeroelastic vibrations for the cases of a floodlight mast, of a bridge pylon and of a chimney with radial fins. All three structures studied had similar properties. They were light constructions of welded steel with low internal damping, and had either stiff foundations or supports leading to low external damping. The aerodynamic behavior of these structures was tested using sectional models in a wind tunnel [1,2,3].

2. FLOODLIGHT MAST

In 1965, four floodlight masts were built in the Niedersachsenstadion in Hannover (Sketch 1 in Chart). The spectator area around the Niedersachsenstadion was a mound of rubble from bombed Hannover deposited there after 1945. Three long piles for each of the masts were driven into the loose ground. This very stiff foundation led to low external damping of aeroelastic vibrations of the masts. During their erection, the masts exhibited wind-induced oscillations. Hence, each of the masts were stayed by four cables as an effective measure against aerodynamic excitation.

A full model and some sectional models were built and studied in the wind-tunnel of the Institute of Mechanics. Figure 1 shows a partial model of the section mounted between end disks. The suspension allows vibrations of the model perpendicular to the smooth flow of the wind-tunnel having an open test section. Parallel to the springs and dampers, an electrodynamic shaker allows controlled oscillations of the model. The model technique used is known as the forced oscillation method or as the phase resonance method. In this case, the amplitude of vibration and the phase angle between the motion and the force of the shaker (viz. F_s), are regulated by control circuits [1].

For the model of unit length (see Fig. 1) the equation of motion can be written as:

$$m\ddot{y} + d\dot{y} + cy = F_a + F_s \quad . \tag{1}$$

For the aerodynamic force (viz. F_a) the approach given by Scruton [4] will be used:

$$F_a = \rho b^2 f\,(h_a\, f\, y + k_a\, \dot{y}) \quad , \tag{2}$$

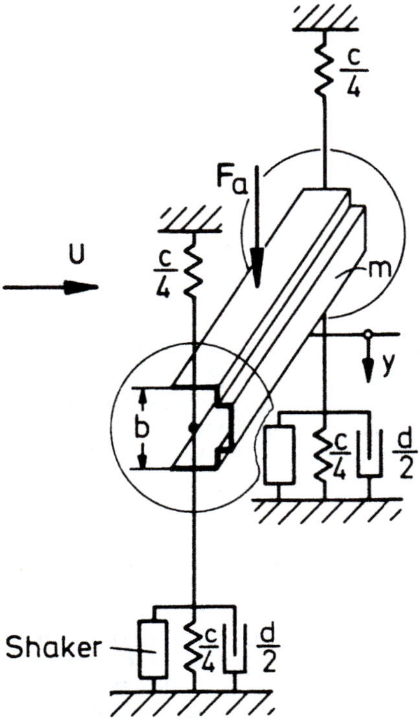

FIG. 1 SKETCH OF THE SECTIONAL MODEL

where ρ is the density, b is a characteristic length of the cross-section and is the frequency. The aerodynamic coefficients h_a and k_a depend on the wind speed U, the wind direction (i.e. angle of attack α), the reduced amplitude $\hat{\eta} = \hat{y}/b$, the Reynolds number $Re = Ub/\nu$ and the turbulence characteristics.

The force coefficient k_a describes the reduced input power to the model by the aerodynamic force:

$$k_a = m_r \frac{\int F_a \, dy}{2\,T} , \tag{3}$$

where

$$m_r = 2\,m / (\rho\, b^2) \tag{4}$$

is the mass ratio. T is the magnitude of the kinetic energy of the oscillation.

In the same way, the combined damping and mass parameter is defined as

$$k_s = m_r \frac{\int d\, \dot{y}\, dy}{2\,T} \tag{5}$$

An aeroelastic vibration becomes unstable if $k_s < k_a$.

The input power from eq. (3) can be determined by measuring the force of the electrodynamic shaker.

Lift and drag of the models were measured under the same boundary conditions as before. Using the quasi-steady approach given by den Hartog [5] and Parkinson and Smith [6], it can be expected from the lift and drag (see Sketch 3 of Chart with $C_L = 2L/(\rho b\, U^2)$ and $C_D = 2D/(\rho b\, U^2)$) that galloping occurs at $\alpha = 75°$, $\alpha = 98°$ and $\alpha = 162°$. This type of self-excited aeroelastic vibration was measured at $\alpha = 75°$ (see Sketch 4 of Chart).

The same plot shows vortex excitation at $\alpha = 85°$ with resonance conditions a

$$V_{r_{res}} = \frac{U}{f_v\, b} = \frac{1}{S} = 10 , \tag{6}$$

where f_v is the frequency of vortex shedding and S is the Strouhal number. Such vortex excitation was observed at several angles of attack with a similar magnitude of excitation of about $k_a = 60$ (reduced amplitude: $\hat{\eta} = \hat{y}/b = 0.09$).

The stiffening effect of the four cables is not high. As a remedy against aeroelastic vibrations, the cables working as external dampers should be effective.

3. LODEMANN BRIDGE PYLON

The Lodemann Bridge (see Sketch 1 of Chart) was built in 1965 near the Niedersachsenstadion in Hannover. During the heavy storm of November 13, 1972, with wind speeds about 40 m s^{-1}, the 30 m long pylon of the bridge collapsed [3]. Unfortunately no one witnessed this event. The crack appeared at a butt joint. The bending moment at this point, resulting from steady wind-loads, seems to be only a partial reason for the damage. In all likelihood, the failure of the pylon was caused by galloping. This could be reproduced by wind-tunnel tests of a sectional model, tested in the same way and under the same conditions as described in section 2. The prismatic model had a cross-section with a shape representative of the original pylon, which had a height of 17.5 m.

The lift and drag behavior is shown in Sketch 2. For $\alpha = 14°$, one obtains the slope of the normal force coefficient as

$$\frac{dC_{F_y}}{d\alpha} = - \left(\frac{dC_L}{d\alpha} + C_D \right) \approx 3 \quad . \tag{7}$$

Therefore violent galloping can be expected at this angle of attack. This quasi-steady approach is in good agreement with the measured behavior shown in Sketch 3 (dashed line). With the lowest eigenfrequency (in y-direction), $f_{o_1} \approx 1.1$ Hz, the highest reduced velocity

$$V_r = \frac{U}{f_{o_1} b} \approx 54 \quad , \tag{8}$$

was reached during a storm in November 1972. From Sketch 3, it can be seen that the self-excitation at $\alpha = 14°$ is greater than the vortex excitation, for instance, at $\alpha = 0°$ (dashed line). Vortex-induced vibrations should have appeared at wind speeds of about 6 m s^{-1}, but oscillations have not been witnessed at these low speeds. To this date, the pylon has not been re-erected.

4. STEEL-CHIMNEY WITH FIVE RADIAL FINS

The damaged steel-chimney which is shown in Sketch 1 of the Chart without its upper part, was fitted with five radial fins and a ladder extending from 4 m above the ground to the top of the chimney at a height of 28 m. The large, mat-like foundation was cast with direct coupling between the older foundations of the building and the adjacent 16 m high masonry chimney. This means that the external damping effect from the foundation is very small.

The five fins had been fitted with the hope that they would be the most effective "preventive measure" against vortex excitation. Under the best circumstances, DIN 4133 [7] allows the force coefficient (i.e. C_S) with respect to vortex excitation to be reduced by a factor of ten. However, for this chimney, the efficiency of the fins was overestimated.

Shortly after erection, oscillations of the chimney were observed at very low wind speeds. A short film of this event allowed determination of the reduced amplitude at the top of the chimney as

$$\hat{\eta} = \frac{\hat{y}}{b} = \frac{155 \text{ mm}}{612 \text{ mm}} = 0.25 \quad . \tag{9}$$

After a short time, several cracks were discovered at the lower part of the chimney and the upper part had to be removed as shown.

A sectional model of the chimney with 5 fins (relative height of the fins: $s/b \approx 0.1$), but without the ladder, was tested in a closed test section of a windtunnel [2]. The drag and lift coefficients are plotted in Sketch 2, and they show a magnitude of the drag coefficient $C_{D_{max}} = 2.7$ at $\alpha = 0°$ and $Re \approx 10^5$.

With this high drag, galloping cannot occur (cf. eq. (7)). This is verified by the results shown in Sketch 3. For reduced velocities greater than 12, oscillations of the model were damped by aerodynamic forces. Strong vortex excitation was measured at the same profile at $V_{r_{res}} = 7.9$ (i.e. $S = 0.13$, cf. eq. (6)). The Strouhal number is smaller than that of a plain circular cylinder.

The first eigenfrequency of the chimney was calculated as $f_0 = 0.93\ s^{-1}$. With these data, the critical wind speed of vortex shedding can be determined as

$$U_{crit} = V_{r_{res}} f_0 b \approx 4.5\ m\ s^{-1} \quad . \tag{10}$$

This is a low wind speed, but it is nevertheless dangerous for such a structure.

Five radial fins, in fact even 6 or 7 fins, when applied to the chimney, proved to be ineffective against vortex-induced vibrations.

5. REFERENCES

1. Bardowicks, H., Untersuchung der Einflüsse von Querschnittsform und Schwingweite auf aeroelastische Schwingungen scharfkantiger Prismatischer Körper. Dissertation, Univ. of Hannover, 1976.

2. Windrich, H., Maßnahmen zur Minderung wirbelerregter Schwingungen von Stahlschornsteinen - Windkanaluntersuchungen, Students work at the Institute of Mechanics of the Univ. of Hannover, 1978.

3. Mahrenholtz, O. and H. Bardowicks, "Aeroelastic Problems at Masts and Chimneys", presented at the 3 rd Colloquium on Industrial Aerodynamics, Aachen, 1978.

4. Scruton, C., "On the Wind-Excited Oscillations of Stacks, Towers and Masts", Proceedings of the 1st Conference on Wind Effects on Buildings and Structures, Teddington, 1963.

5. Den Hartog, J. P., "Tansmission Line Vibration Due to Sleet", Trans. AIEE 49, No. 444 (1930).

6. Parkinson, G. V. and J. D. Smith, "The Square Prism as an Aeroelastic Nonlinear Oscillator", Q. J. Mech. Appl. Math., XVII (2) (1964) 225-239.

7. DIN 4133, Schornsteine aus Stahl, 1973.

FLOODLIGHT MAST

<u>Problem:</u> During the erection of the floodlight masts at the Niedersachsenstadion in Hannover (Germany) in 1965 the masts showed wind-induced vibrations in forward and transverse directions.

<u>Geometry:</u> See Sketch 1 and Sketch 2 (cross-section of the lamp box). Height: 60 m, Length of the lamp box: 18 m.

<u>Boundary conditions:</u> 3 long piles as foundation.

<u>Flow conditions:</u> The vibrations occurred at wind speeds of about 19 m s^{-1}.

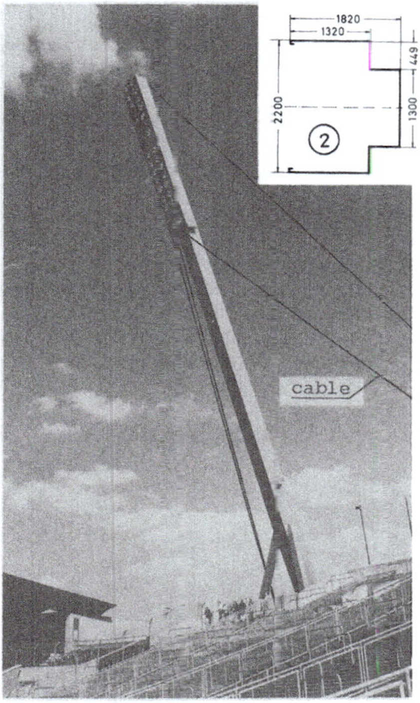

(1) FLOODLIGHT MAST

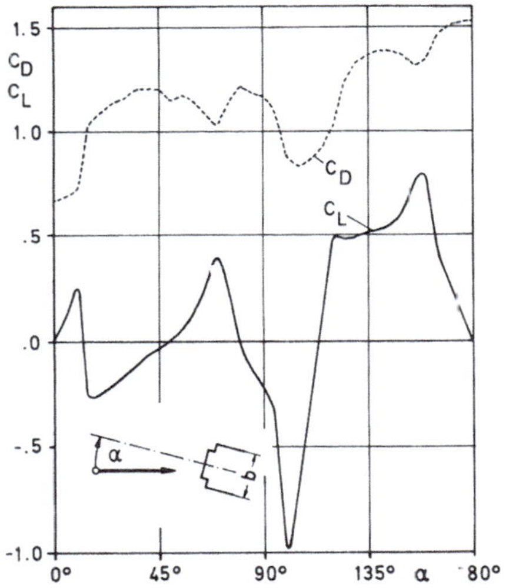

(3) LIFT AND DRAG OF THE LAMP BOX

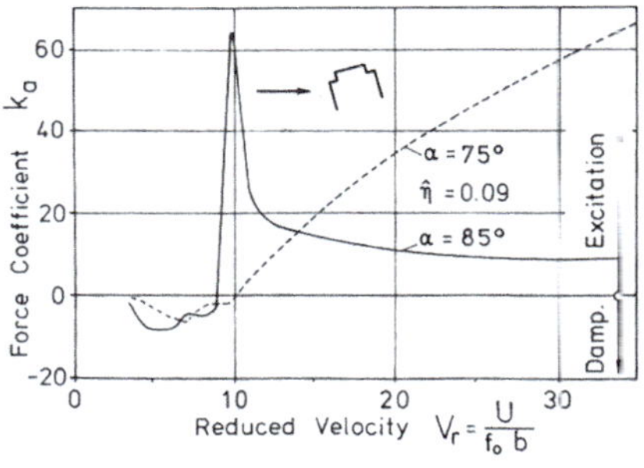

(4) STABILITY DIAGRAM

<u>Structural dynamics:</u> Eigenfrequencies and logarithmic decrements:
forward: f_n = 0.69 s^{-1}, Λ_n = 0.02,
transverse: f_t = 0.56 s^{-1}, Λ_t = 0.03.

<u>Vibrations:</u> Witnessed amplitudes at the top: forward: 0.5 m, transverse: 1.8 m.

<u>Wind-tunnel tests:</u> Full model and sectional model in smooth flow. Sketch 3 and Sketch 4 show results of a sectional model (cross-section: see Sketch 2). k_a (V_r, α) shows vortex excitation at α = 85° at S = $V_{r_{res}}^{-1}$ = 0.1 and galloping at α = 75°.

<u>Cures:</u> 4 cables at each of the masts (see Sketch 1).

<u>References:</u> Pestel, E. and O. Mahrenholtz Report of the Institute of Mechanics, Univ. of Hannover, 1965.
Bardowicks, H. Dissertation at the Univ. of Hannover, 1976.

PYLON OF THE LODEMANN BRIDGE

Problem: After the severe storm of Nov. 13, 1972 the pylon of the Lodemann Bridge fell down. Galloping seems to be the reason for this failure.

Boundary conditions: The crack appeared at a joint.

Flow conditions: Wind speeds of about 40 m s^{-1}.

Structural dynamics: Eigenfrequencies (calculated): $f_{oy} \approx 1.1\ s^{-1}$, $f_{oz} \approx 1.9\ s^{-1}$.

Wind-tunnel tests at a sectional model (cross-section: see Sketch 1). Examples of the results: Vortex excitation at $\alpha = 0^{o}$, galloping at $\alpha = 14^{o}$, aerodynamic damping at $\alpha = 25^{o}$.

References: Mahrenholtz, O. and H. Bardowicks "Aeroelastic Problems at Masts and Chimneys." Presentation at the 3rd Colloquium on Industrial Aerodynamics, Aachen, 1978.
Bardowicks, H. Dissertation at the Univ. of Hannover 1976.

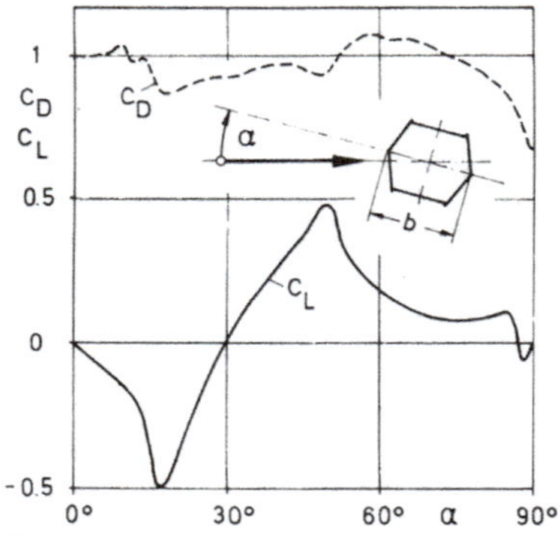

(2) LIFT AND DRAG

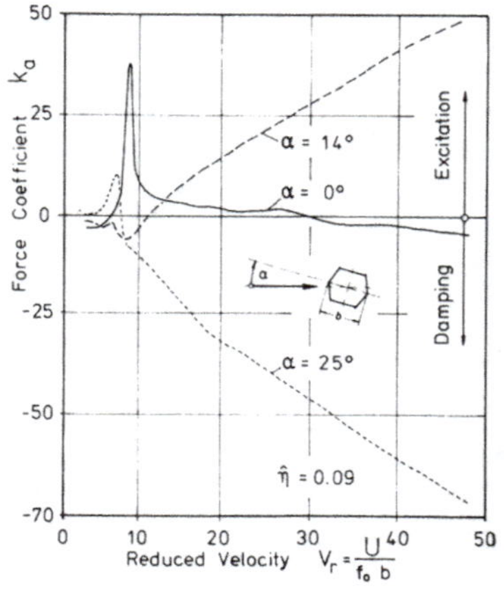

(3) STABILITY DIAGRAM

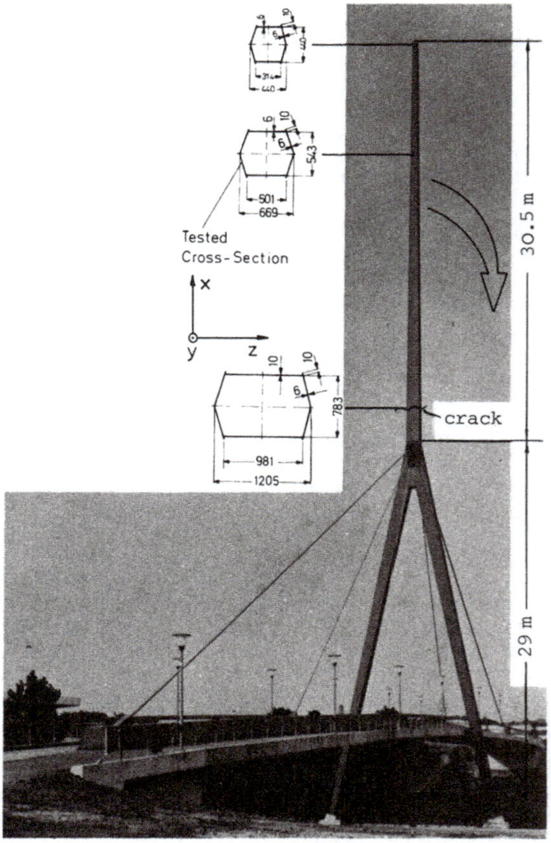

(1) LODEMANN BRIDGE

CHIMNEY WITH FIVE RADIAL FINS

Problem: Vortex excited aeroelastic vibrations leading to a crack at the lower part of the chimney shortly after its erection.

Boundary conditions: The mat-like foundation was casted between other foundations (high stiffnes, low damping effect).

Flow conditions: Low (calculated) wind speed at resonance conditions about 4.5 m s^{-1}.

Structural dynamics: Eigenfrequency (calculated): 0.93 s^{-1}.

Vibrations: Amplitude at the top of the chimney (taken from a film): 155 mm.

Wind-tunnel tests at a sectional model (cross-section of the chimney without the ladder) in a smooth flow. Example (see Sketch 3): Vortex excitation at $\alpha = 0^{\circ}$, Strouhal number: $S = V_{r\,res}^{-1} = 0.13$.

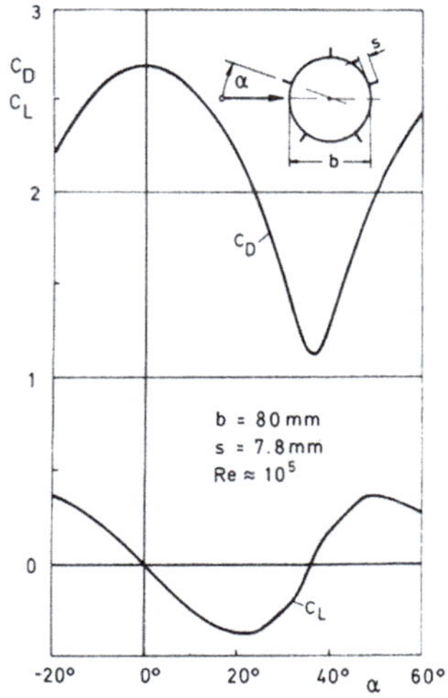

(2) LIFT AND DRAG

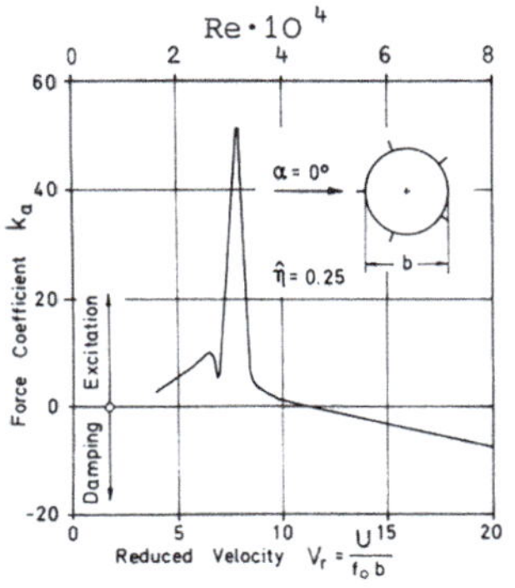

(3) STABILITY DIAGRAM

References: Mahrenholtz, O. and H. Bardowicks Aeroelastic Problems at Masts and Chimneys. Presentation at the 3rd Colloquium on Industrial Aerodynamics, Aachen, 1978.

Windrich, H. Students work at the Inst. of Mechanics, Univ. of Hannover, 1978.

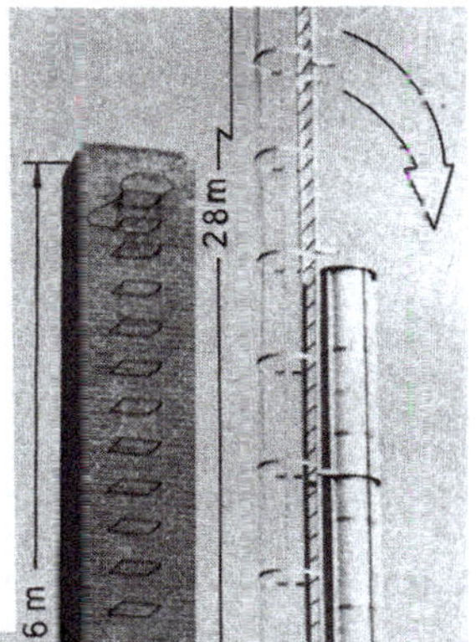

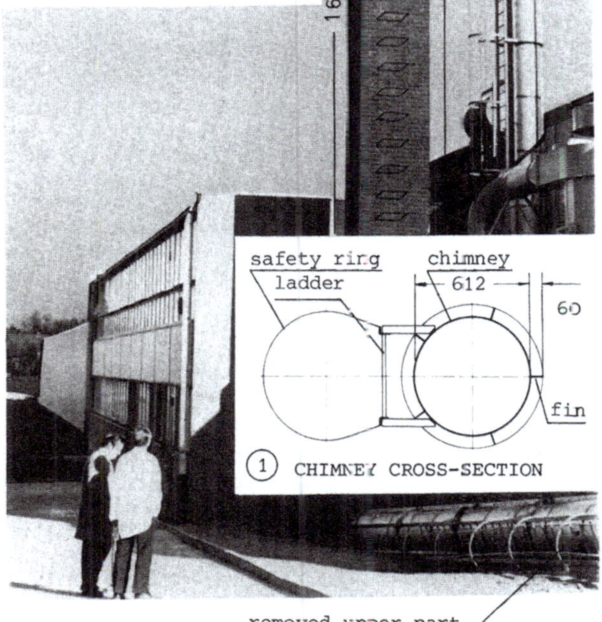

(1) CHIMNEY CROSS-SECTION

removed upper part

F6 APPROACHES TO THE SUPPRESSION OF WIND-INDUCED VIBRATIONS OF STRUCTURES

R.L. Wardlaw
Deputy Head
Low Speed Aerodynamics Laboratory

National Research Council of Canada
Ottawa, Canada

SUMMARY

This paper includes several Charts describing examples of wind-induced vibrations of slender structures that are caused by a variety of aerodynamic phenomena - vortex shedding, flutter, wake instability and galloping. The suppression or elimination of the vibrations has been achieved by changing the geometric shape, augmenting the system damping, or modifying the inherent elastic properties of the structure. The examples used are drawn from experiences in the wind tunnel investigation of long span bridges, towers and power cables.

1. INTRODUCTION

This presentation will deal with a variety of structures that are susceptible to vibration as a result of interaction with the earth's surface wind. These structures are typical of those encountered in the field of 'Wind Engineering' and are all characterized by being aerodynamically bluff in cross-section, as opposed to streamlined as characteristically met with in aircraft aerodynamics. The flow over these bluff shapes is, by contrast, much more complex and, consequently, much less amenable to analytical procedures. With the slender engineering proportions of long span bridges, power transmission cables, towers, poles and chimneys, the prospect of aerodynamic instability is always present. As advances in technology allow lighter and lighter structures to be built - by a factor of about three in recent decades - with lower damping and increased slenderness, they become more and more susceptible to the dangers of wind-induced vibrations.

Through past experiences, and from wind tunnel model investigations at reduced scale, the engineer is now able to recognize a variety of circumstances where vibrations are apt to arise and where steps should be taken to suppress the motion or to eliminate the cause. In some cases, such as the vortex shedding induced cross-stream vibration of circular structures such as chimneys, empirical data are available that allow for the prediction of critical wind speeds and acceptable estimates of motion amplitudes. However, with other structures, such as bridge road decks, the availability of empirical data is limited, and although analytical procedures are being developed, the stage has not been reached, particularly with regard to the effects of turbulence, where either vortex shedding response, onset velocities for flutter or buffeting response can be predicted with satisfactory precision. In these cases the engineer must resort to ad hoc wind tunnel model investigations in the design of the structure.

Unfortunately vibration problems sometimes are not anticipated in the design stage and only surface during erection or upon completion of the structure. More often the problem is identified earlier, but at a stage where the design concept is frozen and major design changes cannot be made. In these cases, "fixes" such as helical strakes for poles, edge fairings for bridge road decks or increased damping for towers or bridge road deck hangers must be incorporated. Most frequently, however, the prevention or avoidance of wind-induced vibration is considered at the preliminary design stage; and an appropriate stable aerodynamic shape is selected, adequate damping is designed-in, or by increasing structural stiffness, critical wind speeds are raised beyond site design wind speeds.

2. AUGMENTATION OF DAMPING

Vortex shedding-excited vibrations of structures is particularly sensitive to damping as demonstrated by the results of Scruton and Walshe (Ref. [1]) in Figure 1 for a circular cylinder. In this figure, the critical velocity in dimensionless form, V/fB, and amplitude, x/B, are plotted against the mass damping parameter $m\zeta/\rho B^2$ where

V = wind velocity

f = natural frequency

B = cylinder diameter

x = amplitude

m = mass/unit length

ζ = critical damping ratio

ρ = air density

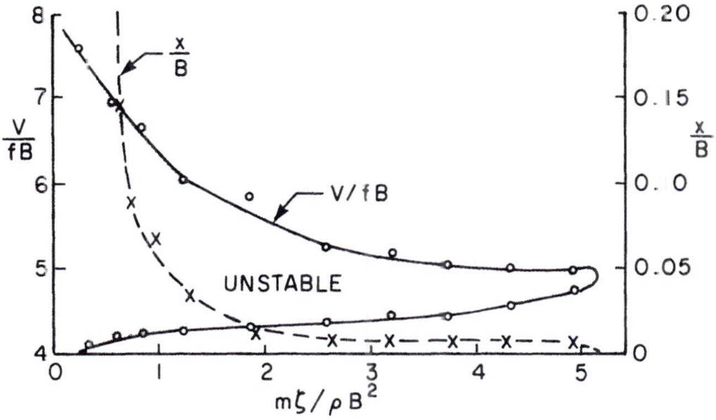

FIG. 1 RESPONSE OF A CIRCULAR CYLINDER TO VORTEX SHEDDING

It is seen that as the damping parameter is increased, the amplitudes drop sharply and that there is a maximum value of the parameter beyond which motion is eliminated. It has been generally found, for the vibration of structures, that only small increases in damping are required to suppress the motion.

The use of dynamic vibration absorbers, simply constructed with a viscoelastic rubber stem supporting a mass, for the purpose of suppressing motion is illustrated in the charts on the Antenna Towers, the Bras d'Or Bridge road deck hangers and the Commodore Barry Bridge I-beam truss members and the elevator shaft of the La Prade Heavy Water Plant. The use of vibration absorbers has been discussed in greater length in Reference [2]. Typically, it has been found in the laboratory that an absorber having a mass of about one percent of that of the structure is sufficient to control the motion, although in practice some of the installed dampers have been as much as 2 percent of the mass.

The damping sensitivity of the flutter onset velocity for bridge road decks depends on the cross-section geometry. For the proposed St. John's River cable-stayed crossing described in the accompanying chart, a 20 percent increase in the speed required to produce dangerous flutter amplitudes was achieved using a vibration absorber. At the same time, there was a 15-fold reduction of vortex shedding amplitude for torsional motion. A geometric fix developed for this road deck was the final solution. Inherently stable deck sections (Fig. 2) for a proposed, but not constructed, crossing of the Burrard Inlet at Vancouver on Canada's Pacific coast have been developed (Ref. [3]). In parallel investigations of other unstable sections studied, it was found that the use of vibration absorbers increased effective damping levels ten-fold for both torsion and bending while vortex excited motion was reduced 50-fold in torsion and 11-fold in bending.

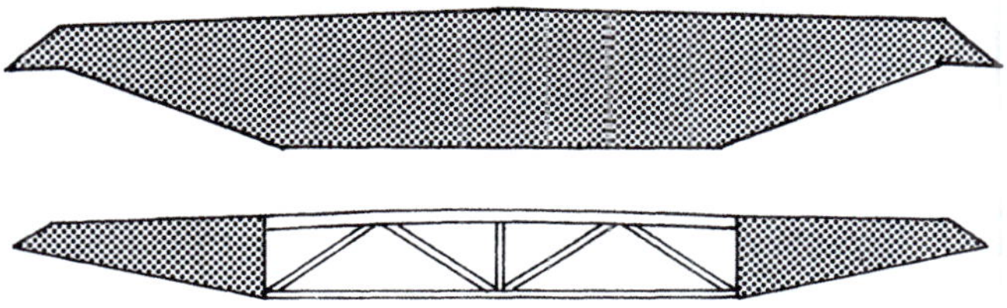

FIG. 2 BURRARD INLET PROPOSED DECK SECTIONS

FIG. 3 PASCO-KENNEWICK BRIDGE DECK SECTION

The augmentation of damping in tall buildings may be an effective and cost saving technique for the alleviation of buffeting in high winds. A computer-aided feasibility study (Ref. [4]) has shown that the probability of exceeding prescribed peak values was dramatically reduced by the use of vibration absorbers. For example, using a rigid mode model, it was found that the probability of exceeding an amplitude of 6×10^{-4} radians was reduced by a factor of 7 with a 0.1 percent damper ($\mu = 0.001$) and by a factor of 48 for a 1.0 percent mass ratio absorber ($\mu = 0.01$). These results are shown graphically in Figure 4.

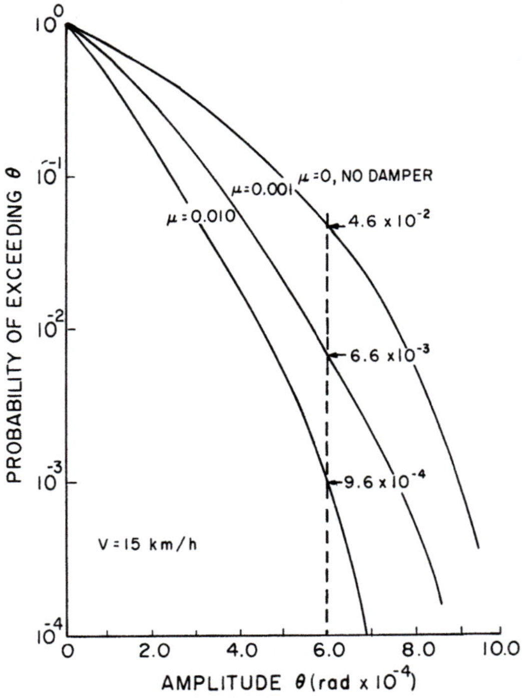

FIG. 4 BUILDING DAMPER STUDY

3. GEOMETRIC APPROACHES

The addition of triangular edge fairings to suppress vortex shedding excitation has been demonstrated both in the laboratory and at full scale for the Longs Creek Bridge as discussed in one of the accompanying charts. The idea of triangular edge fairings has also been used to develop aerodynamically stable shapes for the proposed new Burrard Inlet crossing already discussed and shown in Figure 2. A similar successful configuration using triangular box section edge beams and an open soffit was investigated and adopted for the now completed Pasco-Kennewick Bridge (Fig. 3) over the Columbia River (Ref. [5]). It was found for the Lion's Gate Bridge, as discussed in one of the charts, that a triangular edge fairing covering the central one-eighth of the full span, stabilized a torsional instability that occurred with a full model of the bridge in smooth flow. Subsequently, when tested in a simulated turbulent wind layer, it was found that the turbulence suppressed the instability, obviating the need for the fairing. Novel changes sometimes can be found for stabilizing road decks. For example it was found that large, powerful vortices were separating periodically from the cavity under the St. John's River Bridge. As pointed out earlier the resulting vibration was suppressed by the addition of solid baffle plates along the bridge at quarter-chord points that stabilized the vortex formations.

Galloping and vortex shedding excitation can be controlled by avoiding solid structures such as plate girders. The chart for the Similkameen River Ore Conveyor Bridge shows that openings in the conveyor cover have reduced vortex excited motion to acceptable levels. However, even with perforated webs the I-beam hangers on the Burton Bridge were not immune from severe motion during the erection phase. Stiffening struts were installed to increase frequencies and critical speeds. Later work on the Bras d'Or Bridge I-beam hangers showed that perforations would eliminate vortex shedding excitations but that galloping speeds were reduced to low values for wind blowing normal to the beam flanges.

The use of helical strakes has been demonstrated for steel grounding poles and the antenna towers. In the case of the antenna towers it was concluded that the vibration absorber provided the more attractive solution. A perforated shroud was found to reduce motion of the La Prade truss-supported elevator shaft, but again it was found that damping augmentation using vibration absorbers was the preferred solution.

The vibration behaviour of the bundled conductors described in one of the charts can be favourably altered by geometric factors. For example, increased separation of the conductors improves stability, and with a two-conductor bundle the instability can be controlled by tilting the bundle so that one conductor is not in the wake of another. This approach becomes less effective with bundles containing larger numbers of conductors.

4. ELASTIC PROPERTIES

Increasing modal frequencies will increase critical speeds and is often a practical solution as demonstrated with the hangers on the Burton Bridge. The use of stiffening ties on in-line towers such as are used at heavy water plants is more usbtle. The ties restrain the towers to move in unison and this appears to suppress tendencies toward instabilities of the wake-induced type and to reduce vortex shedding excitation.

Threshold speeds for subspan oscillations of bundled conductors can be increased by shortening the distance between spacers and thus increasing the subspan frequencies. Some degree of motion control is also achieved by influencing the elastic coupling between conductors by suitable design of the spacer elastic and damping properties.

5. REFERENCES

1. Scruton, C., Walshe, D.E.S., A Means for Avoiding Wind-Excited Oscillations of Structures with Circular or Nearly Circular Cross-Section, NPL Aero 335, (1957).

2. Wardlaw, R.L., Cooper, K.R., "Dynamic Vibration Absorbers for Suppressing Wind-Induced Motion of Structures", Proc. 3rd Colloquium on Industrial Aerodynamics, Aachen, p. 205 Part 2 (June 1978).

3. Wardlaw, R.L., Further Wind Tunnel Studies of the Aerodynamic Stability of Bridge Sections for the Proposed New Burrard Inlet Crossing, National Research Council of Canada, Report LTR-LA-54 (June, 1970).

4. Amyot, J.R., Wardlaw, R.L., Cooper, K.R. Van Blokland, G.P., Computer Studies of a Vibration Damper for Wind-Induced Motion of a Tall Building, National Research Council of Canada, Report LTR-AN-8, 28 (June, 1977).

5. Wardlaw, R.L., A Wind Tunnel Study of the Aerodynamic Stability of the Proposed Pasco-Kennewick Intercity Bridge, National Research Council of Canada, Report LTR-LA-163 (July, 1974).

6. ACKNOWLEDGEMENT

The author's colleagues, K.R. Cooper, H.P.A.H. Irwin and C.D. Williams have participated in most of the investigations described and have assisted in the preparation of the charts.

CHART A: LONGS CREEK BRIDGE ROAD DECK

Problem: Vortex shedding caused oscillation of road deck in first mode vertical bending.

Geometry: The bridge is cable-staged with an orthotropic steel deck and a plate girder beam at each edge. See Sketch 1.

Boundary conditions: Motion was observed when the bridge was 30m above the water. Subsequently the valley was flooded to 5m below the bottom of the road deck.

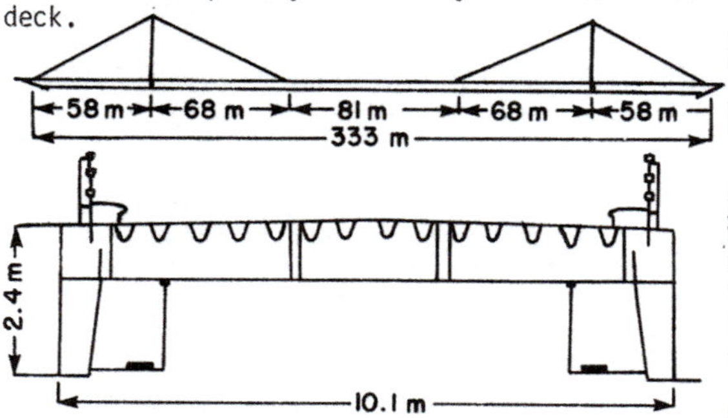

①

②

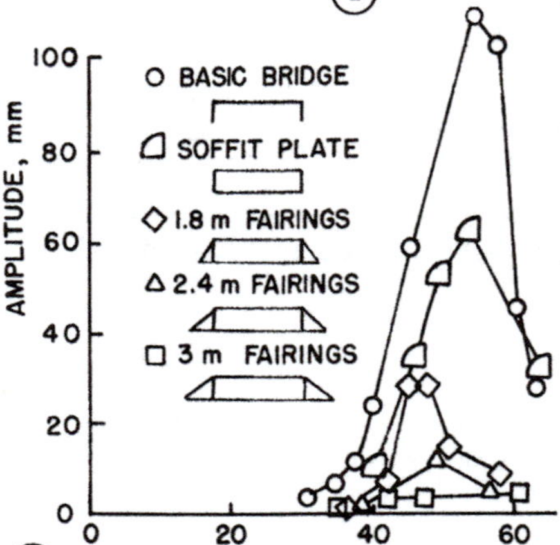

③

Flow conditions: The bridge is located in rolling terrain, partly wooded and partly farmland.

Operating conditions: Motion was observed over a range of dimensionless mean wind speed at the bridge height from $V/nB = 6$ to 9, $V = 9$ to 13m/s (B = girder depth, n = model frequency).

Structural dynamics: Frequency, $n = 0.59$ Hz; mass/unit length 4760 kg/m: moment of inertia/unit length = 4270 kg-m: critical damping ratio, $\zeta = 0.008$.

Vibrations: Motion was regularly observed. The largest amplitudes were about 0.3m.

Cures: The motion was eliminated by the temporary expedient of adding damping by suspending tote boxes in the water by cables from the road deck. The long term solution was the addition of soffit plating and 3m wide triangular edge fairings (Sketch 2) which were developed using 1:30 scale sectional wind tunnel models in smooth flow. Model results (Sketch 3) agreed with prototype observations. Optimum fairing design was sensitive to bridge height above water.

References: Wardlaw, R.L., Goettler, L.L., A Wind Tunnel Study of Modifications to Improve the Aerodynamic Stability of the Longs Creek Bridge, National Research Council, NAE LTR-LA-8, August 1968.

Wardlaw, R.L., Some Approaches for Improving the Aerodynamic Stability of Bridge Road Decks, Third International Conference on Wind Effects on Buildings and Structures, Tokyo, 1971.

CHART B: BRIDGE FLUTTER

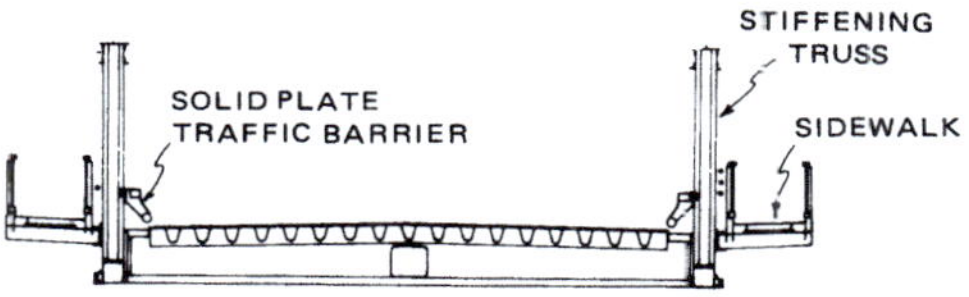

(1) CROSS-SECTION, DISTANCE BETWEEN TRUSSES = 12 m.

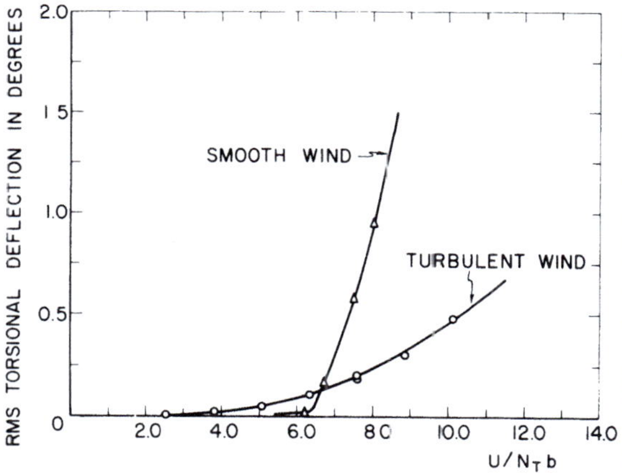

(2) TORSIONAL MOTION AT 1/10th SPAN

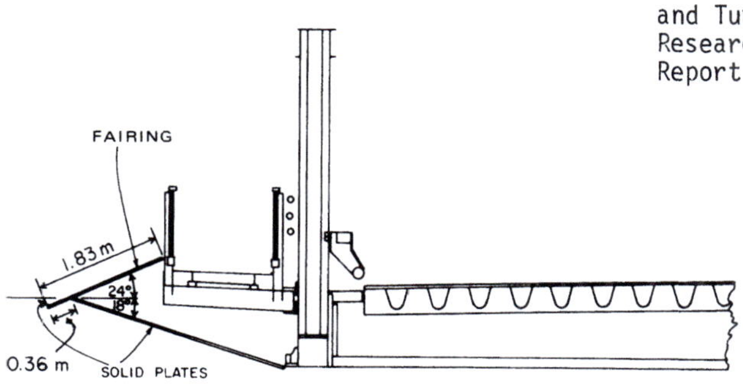

(3) CROSS-SECTION WITH FAIRING ADDED

Problem: An existing suspension bridge, with a main span of 472 m was to be modified by moving the sidewalk to the outside of the trusses (Sketch 1). Sectional model tests of the modified section indicated the flutter velocity was too low for the site. Because design changes were limited by having to use an existing structure the possibility of alleviation of flutter by wind turbulence was examined on a 1:110 scale full aeroelastic model in a 9 m x 9 m wind tunnel. A 2.1 m thick boundary layer was generated using the spire technique.

Cures: a) In 11.5% turbulence, with correctly scaled turbulence integral length scale, the flutter instability disappeared, Sketch 2. Lower turbulence levels were not tested. b) The instability could also be postponed to a high enough velocity even in smooth flow by adding the fairings shown in Sketch 3 to the central 1/8th of the full aeroelastic model length.

References: Irwin, H.P.A.H., and Schuyler, G.D., "Wind Effects on a Full Aeroelastic Bridge Model" ASCE Spring Convention, Pittsburgh April, 1978, Paper 3268.

Irwin, H.P.A.H. and Schuyler, G.D., "Experiments on a Full Aeroelastic Model of Lions Gate Bridge in Smooth and Turbulent Flow". National Research Council of Canada, N.A.E. Report LTR-LA-206, 1977.

CHART C: BRIDGE DECK - THE ST. JOHN'S RIVER BRIDGE

Problem: Vortex shedding excitation of vertical and torsional motion was detected during the wind tunnel phase of the design of a steel cable stayed bridge. Torsional motion exceeded 5.5 degrees amplitude at zero incidence on the 1:60 scale section model.

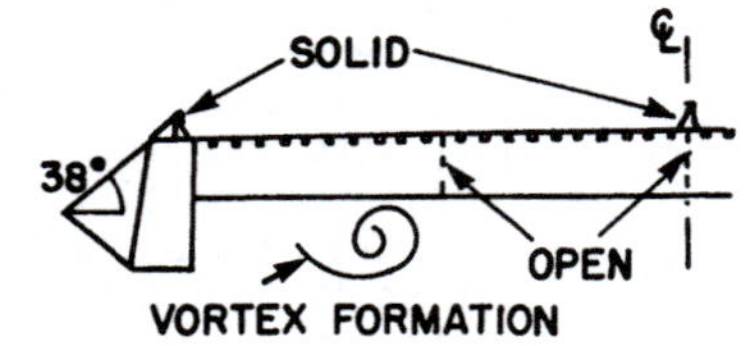

① INITIAL CONFIGURATION

Flow Mechanism: The predominant mechanism was vortex shedding. Strong vortices were observed, using smoke visualization, to form on the underside of the deck (Sketch 1).

Conditions: $\frac{M\zeta}{\rho b^2} = 0.05$, $\frac{I\zeta}{\rho b^4} = 0.004$ where M = mass per unit span, ζ = damping ratio, b = deck width (38m), ρ = air density and I = moment of inertia per unit span. Critical $\frac{V}{Nd} = 5.1$ for bending and 7.5 for torsion, where d = maximum depth of structure.

Cure: a) Installation of vertical baffle plates, to disrupt the underside vortex formation, at the quarter chord positions reduced the torsional amplitude to 1.8 degrees at zero incidence. b) A further reduction to 1.3 degrees resulted when the solid crash barriers were replaced by open railings. c) All motion was eliminated when, in addition, the triangular edge fairing angle was reduced to 26 degrees (Sketch 2).

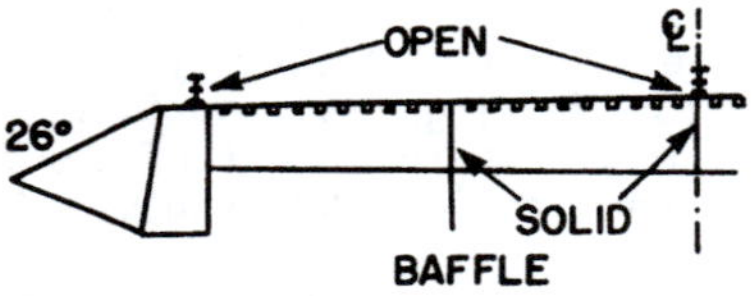

② FINAL CONFIGURATION

Reference: Irwin, H.P.A.H., Savage, M.G. and Wardlaw, R.L. "A Wind Tunnel Investigation of a Steel Design for the St. John's River Bridge, Jacksonville, Florida", National Research Council of Canada, N.A.E. Report LTR-LA-220, Feb., 1978.

CHART D: SUSPENDED ORE CONVEYOR OVER THE SIMILKAMEEN RIVER

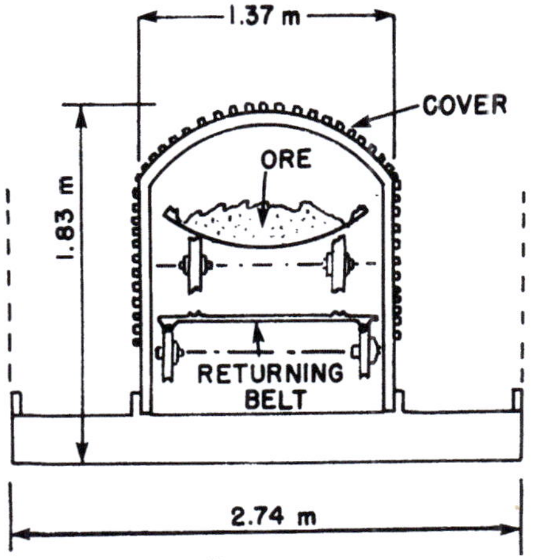

(1) CROSS-SECTION

Problem: During a wind tunnel investigation of a 1:12 scale sectional model of a proposed 396m span suspended ore conveyor, Sketch 1, vortex-induced vertical motion was observed. The peak to peak amplitude corresponded to 0.8m at full scale at a reduced velocity, V/Nd, of 5.6 where V = wind velocity, N = vertical frequency and d = vertical depth of structure = 1.83m. Mass = 600kg/m, damping ratio = 0.2%.

Cure: The problem was overcome by cutting slots in the cover over the conveyor. Sketch 2 shows the effect of the slots on peak amplitude and Sketch 3 shows a side view of the structure and the slot positions. Wind speed and pressure measurements around the conveyor showed the slots did not prevent the cover from performing its role of preventing the conveyor belt being blown off the line stands in high winds.

Reference: Irwin, H.P.A.H., and Savage, M.G., "A Wind Tunnel Investigation of a Suspended Ore Conveyor over the Similkameen River, British Columbia." National Research Council of Canada, N.A.E. LTR-LA-231, 1979.

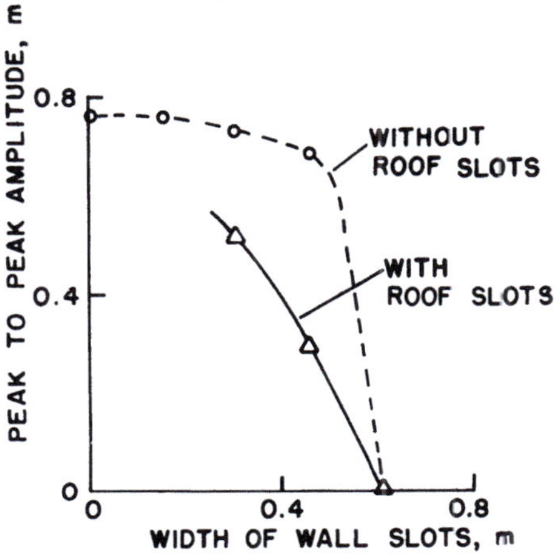

(2) EFFECT OF SLOT WIDTH

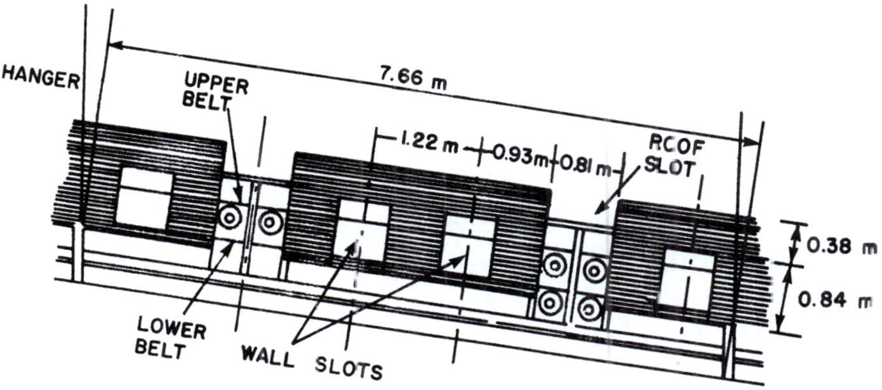

(3) SIDE VIEW SHOWING SLOTS

CHART E: ARCH HANGERS ON THE BURTON BRIDGE

Problem: To determine an effective method for controlling wind-induced oscillations of perforated H-section arch hangers on the Burton Bridge, using wind tunnel tests and full scale measurements.

Geometry: H-section dimensions 221 by 455 mm; hexagonal web perforating 200 mm across flats on 375 mm centers (20% open area ratio); hanger lengths from 10 to 22 m.

End conditions: HS bolts to gusset at upper end, welded to gusset at lower end, typical tensile load 450,000 N.

Flow conditions: Normal atmospheric turbulence over open water in wide river channel with low-lying land on either side; 100 year return period windspeed 82km/h .

①

Operating conditions: Hanger web perforations designed to reduce weak axis vortex-induced motions, however oscillations were observed in longer hangers during erection phase with tensile load 1/7th of design load; amplitudes of 20 mm observed at windspeeds greater than 32 km/h . Hairline cracks were found at the lower ends of the hangers. No oscillations observed at low windspeeds, once structure was completed.

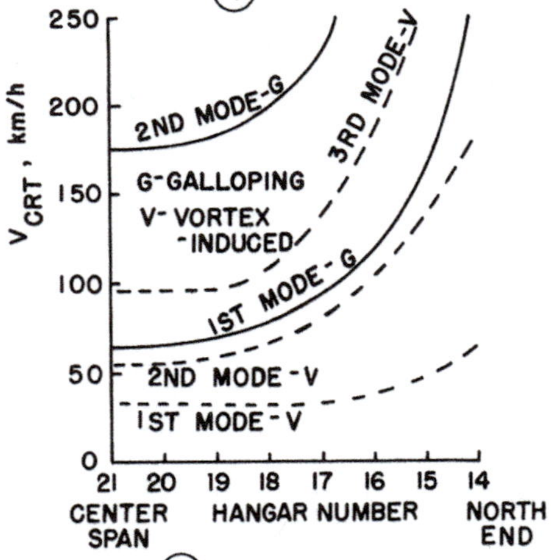

② CALCULATED CRITICAL SPEEDS FROM MEASURED FREQUENCIES

Measured structural dynamics: a) longest hanger: weak axis motion f_1 = 3.3 Hz, ζ_1 = 0.037, strong axis motion f_1 = 5.8 Hz, ζ_1 = 0.025, b) shortest hanger: weak axis motion f_1 = 6.7 Hz, ζ_1 = 0.065, strong axis motion f_1 = 25.7 Hz, ζ_1 = 0.005, c) road deck: vertical motion f_1 = 2.2 Hz, torsional motion f = 11 Hz, ζ = 0.004.

Vibrations: Wind tunnel tests on the longest hanger indicated (a) strong axis galloping began at 64 km/h , with an amplitude of 50 mm at 96 km/h . (b) weak axis vortex-induced motions occurred at 33 km/h with a maximum amplitude of 24 mm.

Cures: Horizontal channel section struts were installed at 5.9 m height, plus parallel 19 mm horizontal tensioned cables at 6.7 and 18.6 m and diagonal 19 mm horizontal tensioned cables at 13.4 m. Results (a) no galloping oscillations for windspeeds below 650 km/h . (b) vortex-induced oscillation amplitude reduced to 0.3 mm at 33 km/h in longest hanger.

③

CHART F: I-BEAM ROAD DECK HANGERS - BRAS D'OR BRIDGE

Problem: Fatigue damage due to crosswind weak axis bending vibration.

Geometry: 0.25mm × 0.36mm I-beam, 16mm flanges, 9.5mm web, up to 26m long.

Flow conditions: Wind exposure primarily of intermediate roughness, wooded hilly terrain.

(1) DAMPER INSTALLATION

Structural dynamics: I-beam mass = 91 kg/m, natural frequency, 5 Hz plus, depending on hanger length and mode.

Vibration: Amplitudes up to 127mm with motion beginning at 32 km/h, which is about the predicted critical speed for vortex-induced motion. Sectional wind tunnel experiments at 1:4 scale confirmed the full scale observations and identified galloping mode in strong axis bending at 153 km/h for critical damping ratio about 0.0015.

Cure: Horizontal tubular struts installed between hangers failed in fatigue. These have been removed and replaced by dynamic vibration absorbers consisting of a mass on a rubber stem mounted between beam flanges. Mass = ½% of beam mass was found effective in model. 1 and 2% mass dampers were installed on bridge. Dampers were found to retain damping capability for temperatures down to -23°C. The effectiveness of web perforations of 15, 25 and 35% of the web area was also investigated. This approach would eliminate vortex shedding excitation but would lower galloping speeds for weak axis bending.

References: Wardlaw, R.L., Cooper, K.R., Dynamic Vibration Absorbers for Suppressing Wind-Induced Motion of Structures, Proc. 3rd Colloquium on Industrial Aerodynamics, Part 2, Aachen, June 1978.

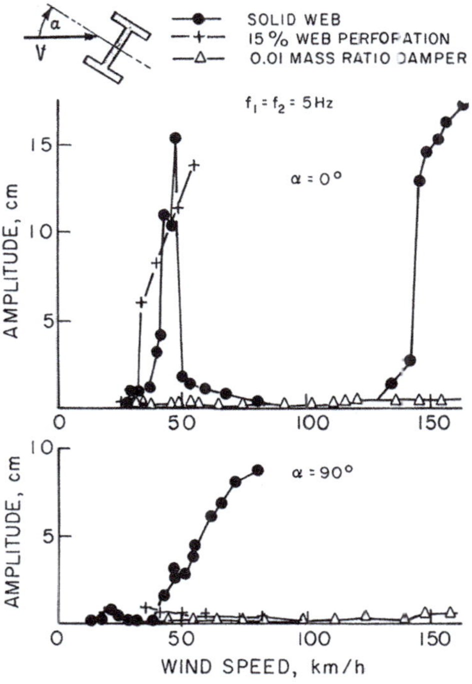

(2) SECTIONAL MODEL WIND TUNNEL RESULTS

CHART G: I-BEAM BRIDGE TRUSS MEMBERS

(1) Commodore Barry Bridge (500 m between piers)

(2) I-Beam Truss Members Showing Vibration Absorber

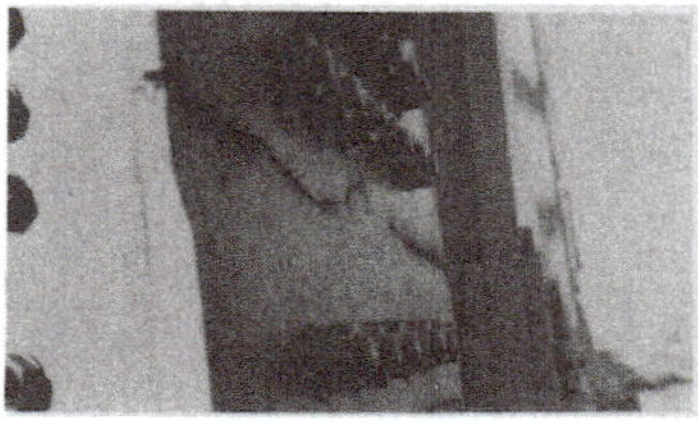

(3) Cracked Outer Flange

(4) 1:7.5 Scale Model in 4.5 m Wind Tunnel

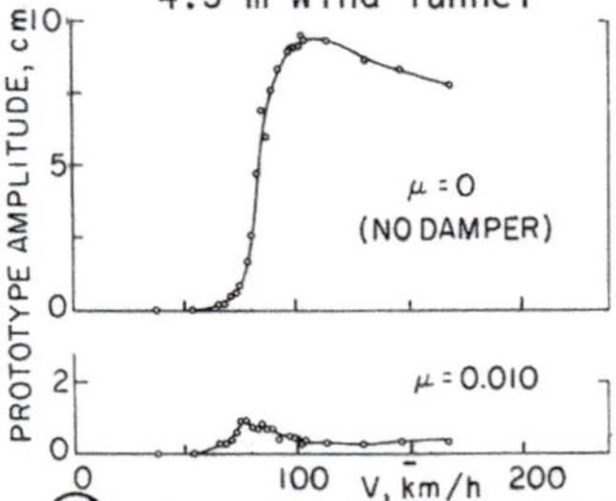

(5) Dynamic Response Without and With Absorber

Problem: Vortex shedding excitation led to the formation of serious cracks in I-section beams supporting the road-deck of a large truss bridge (Sketch 1) during a two day period of moderately high winds.

Geometry: Vertical and inclined I-beams with 81 cm flanges 16 mm thick, 71 cm webs 13 cm thick up to 34 m in length formed part of a Warren Truss (Sketch 2).

Boundary conditions: High strength steel beams were bolted to connections at each end to provide approximately fixed-fixed end conditions. The flanges were parallel to the bridge length.

Flow conditions: The beams were exposed to an approaching wind flow over water. The wind tunnel simulation wind used smooth flow.

Operating conditions: During a two day period winds of between 50 km/h and 65 km/h occurred at bridge road-deck height for many hours. Serious cracks were found in 15 members (Sketch 3). During erection, before the lower ends of the beam were attached to the deck torsional vibrations of some beams had been observed.

Structural dynamics: Beam geometry as outlined above. Beam weight was 274 kg/m. Depending on beam length and mode of vibration (weak axis bending, strong axis bending, torsion) the fundamental modes had frequencies of between 3 Hz and 6 Hz. $m\zeta/\rho b^2 = 1.54$.

Vibration: Wind tunnel investigation of the problem was undertaken using a 1:3 scale sectional model and a 1:7.5 scale elastic model (Sketch 4) of the longest, 34 m, beam. Motion was observed in weak-axis bending, strong-axis bending and torsion. The test results suggested that weak axis bending was most likely to have caused the damage observed.

Cures: Three cures were considered - stiffening the beams using intermediate cable ties, stiffening using structural steel ties, and the use of dynamic vibration absorbers. Based on economic, aesthetic and performance considerations the vibration absorbers were chosen. Typical results are shown in Sketch 5. A typical installation is indicated in Sketch 1.

References: Applications of Vibration Absorbers to Control Wind Induced Vibration of I-Beam Truss Members on the Commodore Barry Bridge. H.P.A.H. Irwin, K.R. Cooper, R.L. Wardlaw, National Research Council of Canada, LTR-LA-194. January 1976.

CHART H: OCTAGONAL STEEL GROUNDING POLES

Problem: Fatigue failure at base due to wind-induced galloping.

Geometry: Height 29 m, width between 0.41 m at base to 0.19 m at top, wall thickness 6.4 mm to 4.6 mm. Corner radii, 20 mm.

① WIND TUNNEL MODEL

Flow conditions: Wind exposure - open flat terrain.

Vibration: In first mode at 0.59 Hz, in wind of 40 km/h gusting to 56 km/h, elliptical motion of pole top had amplitudes of 0.76 m acrosswind and 0.46 m alongwind. Lower amplitudes were observed for higher modes. High value of dimensionless wind speed (about 60 to 70) and large amplitudes (about 2 × width) establishes phenomenon as galloping. 1:2 scale wind tunnel sectional model showed similar motion for wind angles between 8½ and 11½ degrees (0 degrees is normal to flat). Critical damping ratio for experiments, 0.0016. Den Hartog galloping can be predicted from force measurements. Motion has not occurred with luminaires on poles.

Cures: Doubling corner radius to 40 mm did not have a significant effect on amplitudes or critical speeds. A one-start tubular helical strake (height/width = 4.4%, pitch = 5 and 10 × width) eliminated galloping. A three-start flat plate helical strake (height/width = 17%, pitch = 6 × width) was also effective. Initial full scale installation of 3-start helix, with 5 × width pitch and height = 12%, over 25 percent of length was not satisfactory. Coverage of 67 percent of length is final solution.

References: Wardlaw, R.L., A Wind Tunnel Investigation of the Galloping Response of An Octagonal Cylinder, National Research Council of Canada, Report LTR-LA-90, December, 1971.

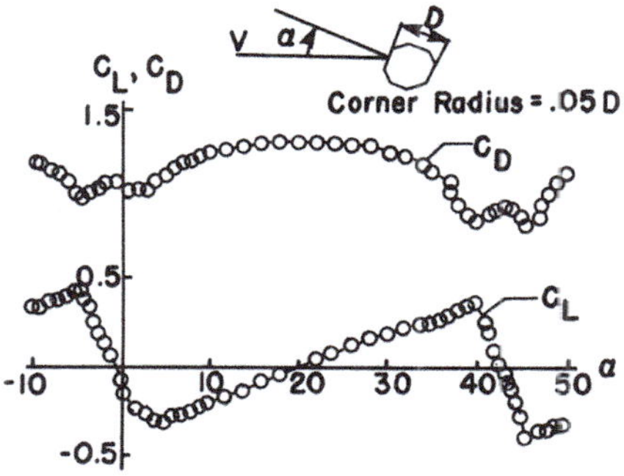

② AERODYNAMIC FORCE COEFFICIENTS

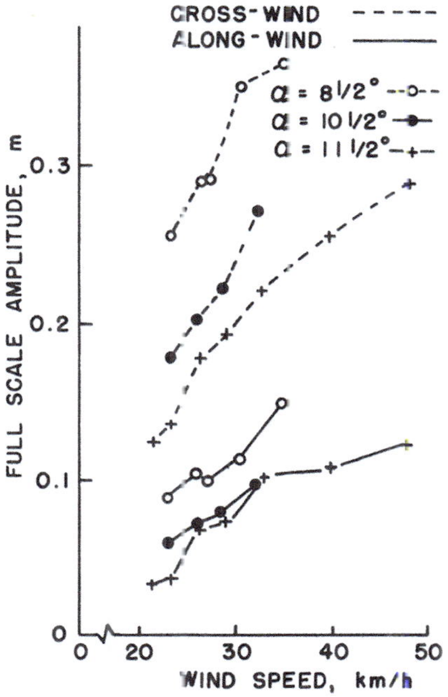

③ GALLOPING RESPONSE

CHART I: ANTENNA ARRAY

①

Problem: Fatigue damage of fibreglass insulators occurred as a result of vortex shedding.

Geometry: 120 aluminum towers, 7.4m high are uniformly spaced on the circumference of a 244m circle. The lower 2.6m is 406mm in diameter and the upper section is 150mm. Wall thickness is 6.3mm for both parts. They are connected structurally by two fibreglass plastic insulators. 1:9.6 scale wind tunnel model towers are shown in Sketch 1.

Flow conditions: Towers were located in intermediate terrain roughness.

Operating conditions: Motion was observed over a range of dimensionless mean wind speed V/nB = 3.0 to 6.6, V = 2.2 to 4.9m/s (B = diameter of upper cylinder, n = model frequency).

Structural dynamics: frequency, n = 5.0 Hz: mass/unit length of upper cylinder = 7.9kg/m: critical damping ratio ζ = 0.0046.

Vibrations: Vortex shedding is the cause of the vibrations. Prototype amplitudes up to 50mm have been reported for upstream towers. No motion has been reported for downstream towers. Wind tunnel results in smooth flow agree with prototype observations for upstream towers but higher amplitudes (120mm) were observed for towers on edge of circle.

Cures: Helical strakes do not provide adequate protection for towers on edge of ring where wake effects are strong but tip-mounted viscoelastic (natural rubber) dynamic vibration absorbers (Sketch 2) that were about 1.5 percent of the structure weight were effective. The damper critical damping ratio at room temperature was ζ = 0.016. The damping and stiffness characteristics of the absorber remained within suitable limits from -40°C to +20°.

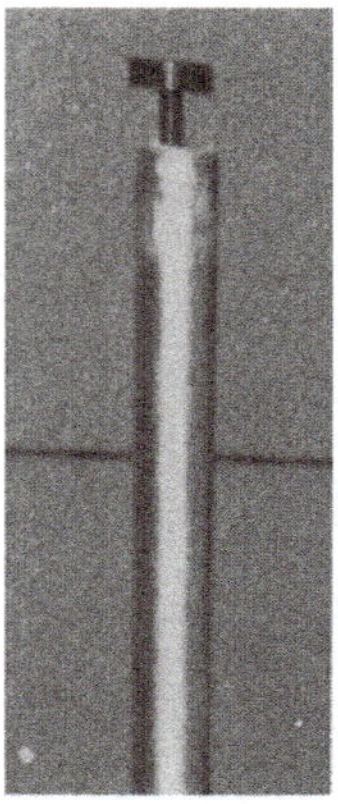

②

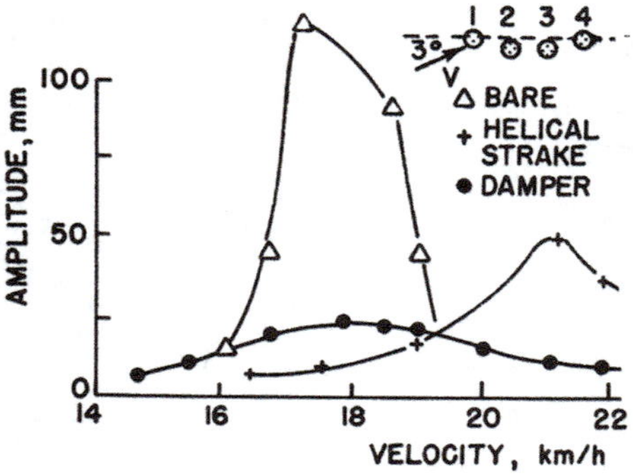

③ TOWER 4

References: Wardlaw, R.L., A Wind Tunnel Investigation of Devices to Suppress the Aerodynamic Instability of the Towers of a Wullenweber High Band Antenna Installation, National Research Council, NAE LTR-LA-309, May 1969.

CHART J: SLENDER TOWERS AT THE PORT HAWKESBURY HEAVY WATER PLANT

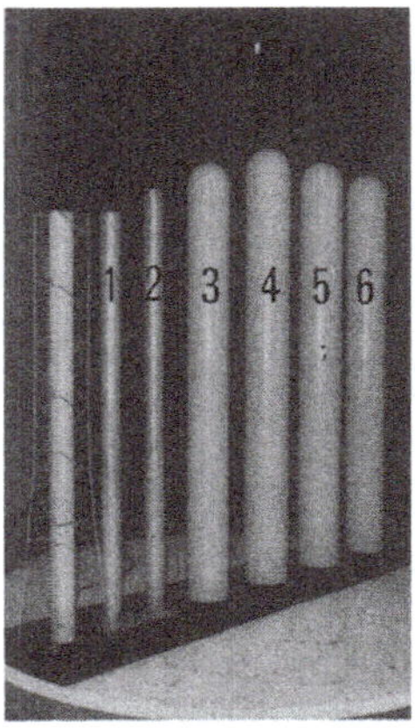

(1) HEAVY WATER PLANT

Problem: To investigate serious vibrations in two slender towers with an aeroelastic wind tunnel model. Damage to prototype foundation and tie-down bolts occurred in erection phase.

Geometry: Six towers and elevator shaft in-line: diameter - slender towers 3m, large towers 8.5m: height - slender towers 70m and 75m, large towers 81m and 85m: centre-line spacing - slender towers 9m, large towers 17m.

Boundary and flow conditions: Open-country-type mean velocity profile with 10% atmospheric turbulence; 7 towers in single concrete foundation.

Operating conditions: 100 year return period mean hourly windspeed 100km/h at 10m height. Vibrations observed of amplitude 0.3 - 0.6m, for gusts between 40 and 90km/h, for a wind blowing near to the line of the towers.

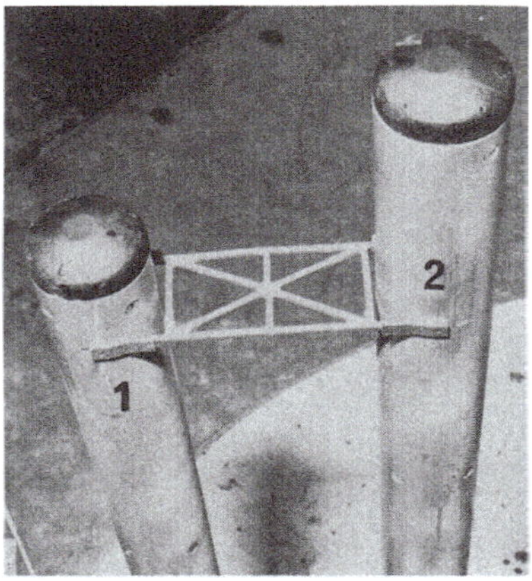

(2) RESTRAINING TIES

Structural dynamics: a) erection - $f_1 = 0.38$ Hz, $\zeta \simeq 0.005$, $m = 3400$ kg/m, $m\zeta/\rho D^2 = 1.5$ b) operating phase - $f_1 = 0.31$Hz, $\zeta \simeq 0.005$, $m = 5200$kg/m, $m\zeta/\rho D^2 = 2.4$.

Vibrations: Wind tunnel tests identified three causes of motion for the uncoupled towers as erected: vortex-induced, wake buffeting, wake galloping.

Cure: The two slender towers were coupled together with horizontal restraining ties.

References: K.R. Cooper, R.L. Wardlaw - A Wind Tunnel Investigation of Large Amplitude Vibrations of Slender Towers at the Port Hawkesbury Heavy Water Plant NRC-LTR-LA-35, July 1969.

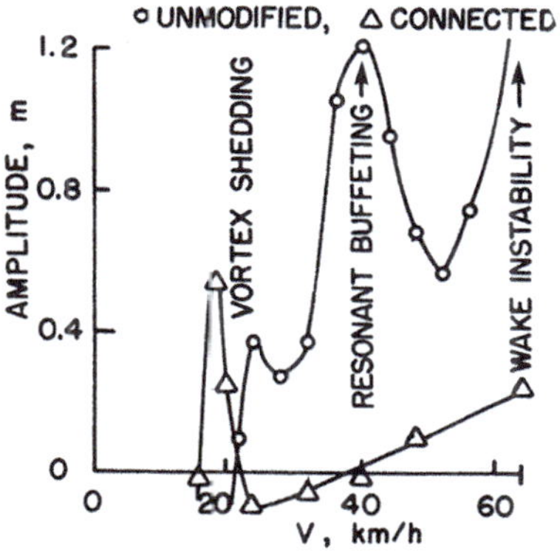

(3) RESPONSE OF TOWER NO. 2
$\alpha = 20°$ FROM LINE OF TOWERS,
TOWER NO. 1 UPSTREAM

CHART K: SLENDER TOWERS - LA PRADE HEAVY WATER PLANT

Problem: To investigate, with an aeroelastic wind tunnel model whether wind-induced oscillations of the 5 towers will occur during the erection or operating phases.

Geometry: Small towers - 75m height, 5m diameter, 10.7m spacing; large towers - 95m height, 8.8m diameter, 20.8m spacing, model scale 1:100.

Boundary and flow conditions: Open-country-type mean velocity profile with 15% atmospheric turbulence; 5 towers in a single concrete foundation.

Erection and operating conditions: 100 year return period mean hourly wind-speed at 10m is 82km/h.

Structural dynamics: a) small towers - erection phase f_1 = 0.81Hz, m = 5.3*10^3 kg/m; operating phase f_1 = 0.47Hz, m = 12*10^3kg/m. (b) large towers - erection phase f_1 = 0.90Hz, m = 16*10^3kg/m; operating phase f_1 = 0.55Hz, m = 35*10^3 kg/m. (c) critical damping ratio 0.002.

Vibrations: Wind tunnel measurements of wind-induced oscillations gave base bending moment and tip amplitude values at 80km/h of a) small towers - 437*10^6N-m, 1670mm, (b) large towers - 618 *10^6N-m, 140mm.

Cure: The towers were coupled together with horizontal restraining ties; the motion amplitudes at 80km/h were reduced to: a) small towers - 24*10^6N-m, 90mm, (b) large towers - 152*10^6N-m, 100mm.

References: H.P.A.H. Irwin, K.R. Cooper, R.L. Wardlaw - A Wind Tunnel Investigation of the Aeroelastic Behaviour of the Laprade Heavy Water Plant, NRC-LTR-LA-177, August 1975.

(1) HEAVY WATER PLANT

(2) RESTRAINING TIES

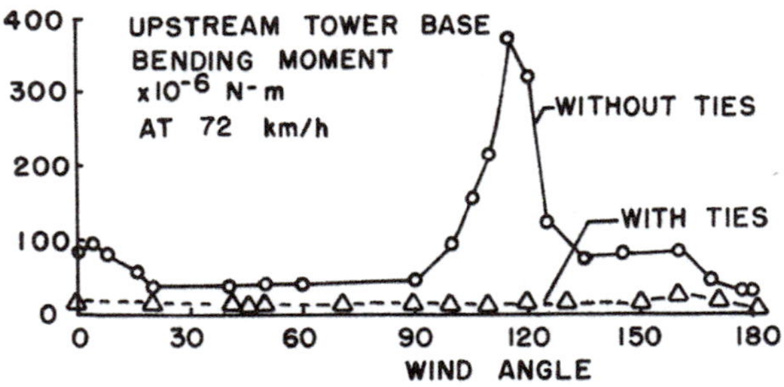

(3) BASE BENDING MOMENT FOR UPSTREAM SLENDER TOWER

CHART L: RECTANGULAR, TRUSS-SUPPORTED TOWER

(1) MODEL WITH ELEVATOR TOWER AT FAR END (UPSTREAM FOR ψ = 180°)

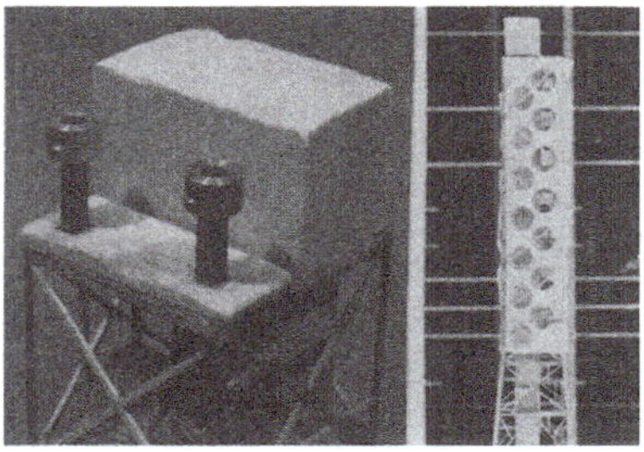

(2) MODEL DAMPERS (3) SHROUD

Problem: Unusual vortex-excited motion of wind tunnel model of a 100m tall, truss-supported, rectangular elevator shaft. Structural failure due to excessive dynamic loads probable.

Geometry: Open truss supporting solid 2.4m × 3.0m rectangular core. Located at end of row of 5 cylindrical towers. Truss base dimension 12.2m × 12.2m. Upper truss 6.1m × 6.1m. See Sketch 1.

Boundary conditions: Cantilevered truss tower mounted adjacent to larger cylindrical towers in open terrain near a river.

Flow conditions: Turbulent boundary layer simulation with gradient exponent of 0.16.

Structural dynamics: First lateral mode of vibration simulated using linear-mode-shape, rigid model mounted on pivot at ground level. Full scale structural frequency of 0.8 Hz and critical damping ratio of 0.01. $m\zeta/\rho b^2$ = 1.14.

Vibration: Largest motion occurred for winds blowing along line of towers with elevator tower upstream. Two response peaks in first mode observed. One due to shedding from central core, one due to shedding from core plus truss. Maximum base bending moments beyond design limit.

Cures: Aerodynamic cure tried first using 35% perforated shroud on upper parallel sided section of truss (Sketch 3). Insufficient amplitude reduction produced (Sketch 4). Dynamic tuned mass damper provided final solution. Two dampers each of 0.5% of the total tower mass were used (Sketch 2); although one alone will raise the damping ratio to 0.06 and suppress the motion.

References: A Wind Tunnel Study of the Aeroelastic Response of a Truss-Supported Elevator Shaft. K.R. Cooper, National Research Council of Canada, LTR-LA-190, Dec. 1975.

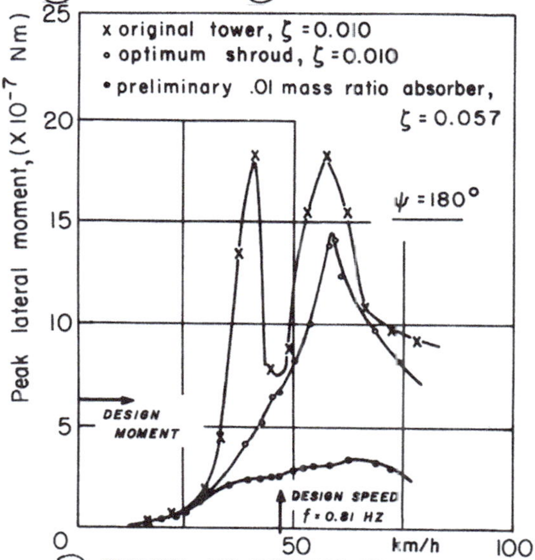

(4) ORIGINAL AND MODIFIED TOWER RESPONSE

CHART M: TRUSS-SUPPORTED CIRCULAR STACK

Problem: A wind tunnel test showed the possibility of large amplitude vortex-induced vibrations of a tall slender truss-supported tower.

Geometry: 151m high, 1.2m diameter welded steel stack supported by a truss to a height of 139m above ground. Truss tapers from 22m wide at ground to 6m wide at top. Upper 12m free of truss (Sketch 1).

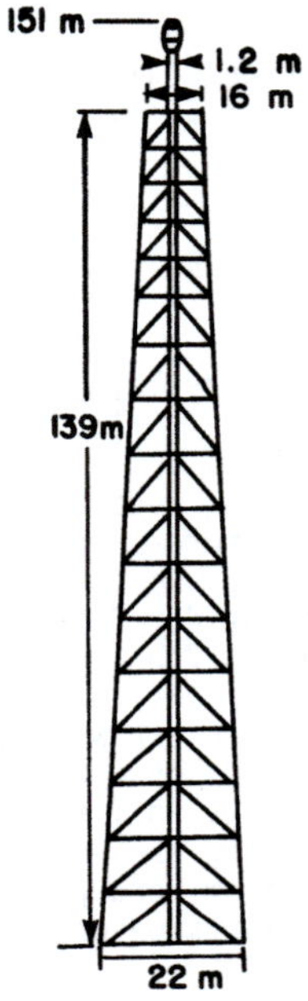

(1) SCHEMATIC VIEW OF TOWER

Boundary conditions: Free-standing tower supported by truss of low porosity.

Flow conditions: Surrounding terrain at site open country or water. Wind tunnel simulation employed smooth flow.

Structural dynamics: First three lateral modes having frequencies of 0.6 Hz, 1.8 Hz and 3.0 Hz, respectively. $m\zeta/\rho b^2 = 1.24$.

Vibration: Sectional wind tunnel model used (Sketch 2) to determine dynamic lift coefficients which were then used to compute tower response. Large amplitudes were found for the first mode. A typical test result is shown in Sketch 3.

Cures: Strakes were fitted to the model and completely eliminated any motion with the truss present and reduced amplitudes and rates of motion build-up for the tower alone (top portion of stack). Analysis of the test results indicated that a three-start helix, with a pitch length of 5 diameters and a strake height of 0.09 the tower diameter over the upper 30m of the stack would reduce tip amplitudes from 0.36m to 0.06m.

References: Wind Tunnel Tests on the La Prade Heavy Water Plant Flare Tower. K.R. Cooper, National Research Council of Canada. LTR-LA-225. July 1978.

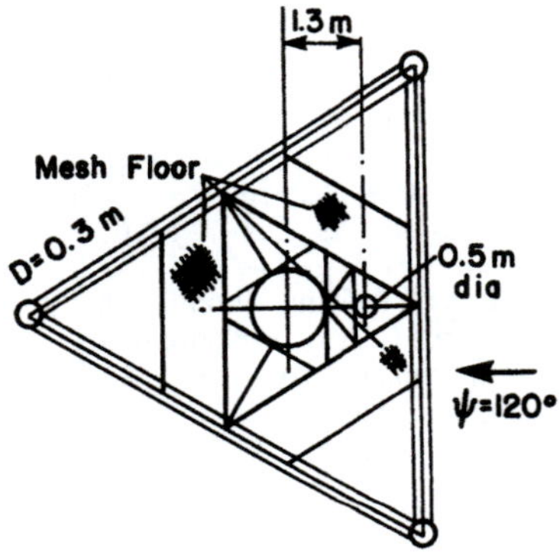

(2) TYPICAL CROSS-SECTION OF TOWER

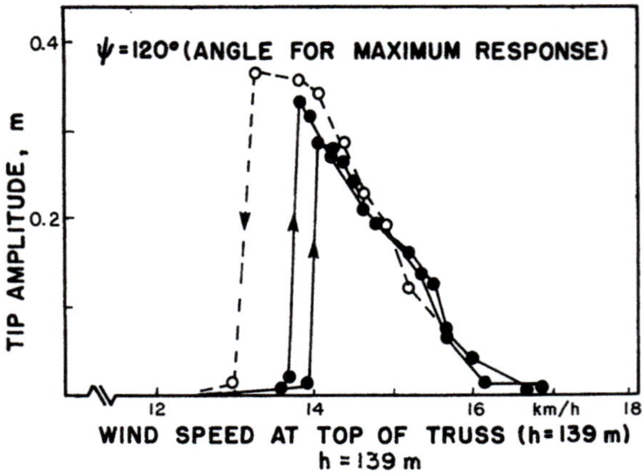

(3) CHARACTERISTIC RESPONSE

CHART N: ALUMINA DIGESTER TOWERS

Problem: To determine via a wind tunnel investigation whether wind-induced oscillations will occur in an in-line array of 3 welded steel towers, during erection and operating phases.

Geometry: Height 37 m, diameter 4.5m, spacing 12 m. Aeroelastic model scale 1:39.

Boundary and flow conditions: Open-country-type mean velocity profile with 15% turbulence intensity; 3 towers in single concrete foundation on bedrock.

Erection and operating conditions: 100 year return period mean hourly wind speed 137 km/h at 10 m height.

Structural dynamics:
Erection phase: $f_1 = 2.0$ Hz, $m = 11{,}000$ kg/m, $\zeta = 0.002$, $m\zeta/\rho D^2 = 0.9$
Operating phase: $f_1 = 1.13$ Hz, $m = 28{,}000$ kg/m, $\zeta = 0.002$, $m\zeta/\rho D^2 = 2.4$

Vibrations: Wind tunnel tests identified vortex-induced oscillations at low wind speeds (97 km/h), and motions due to turbulent buffeting at high windspeeds (194 km/h); maximum measured peak responses at 194 and 137 km/h (erection phase) - deflection 63 and 37 mm, base bending moment 36×10^6 and 21×10^6 N-m.

Cure: Restraining ties installed: maximum measured peak base bending moment (erection phase) reduced to 18×10^6 and 7×10^6 N-m at 194 and 137 km/h.

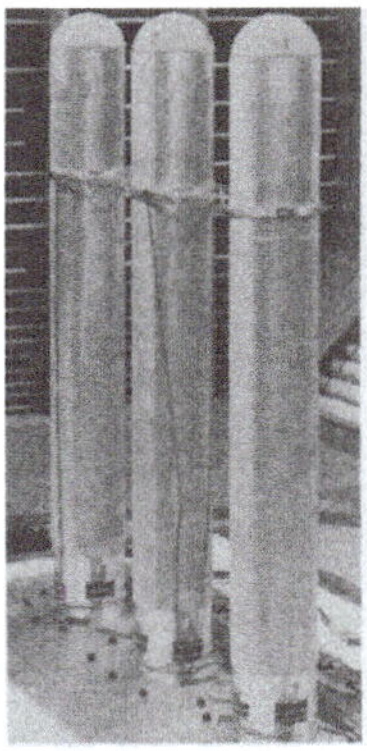

References: C.D. Williams: A Wind Tunnel Investigation of the Aeroelastic Behaviour of the Alcan Alumina Plant Digester Towers, San Ciprian, Spain, NRC-LTR-LA-217, November 1978.

C.D. Williams: A Wind Tunnel Investigation of the Aeroelastic Behaviour of the Alcan Alumina Plant Digester Towers, Shannon, Eire, NRC-LTR-LA-218, July 1978.

(1)

(2)

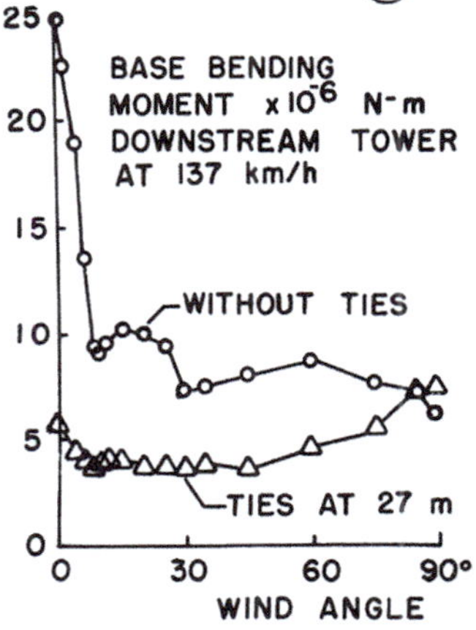

(3)

CHART O: BUNDLED OVERHEAD ELECTRIC POWER TRANSMISSION CABLES

Problem: Wake effects cause instability of downstream conductors. Power outages, conductor and line hardware damage occur.

(1) WIND TUNNEL FORCE MEASUREMENTS

Geometry: The cables are suspended in "bundles" of parallel conductors spaced 10 to 20 diameters apart, the bundle geometry between towers being maintained by spacers that divide each span into "subspans" of about 45m to 60m in length. At contemporary high voltages, 2, 3 or 4 conductors are used in a bundle. Conductor diameters are 25 to 50mm. The bundle geometry is chosen to optimize the effectiveness of the array in suppressing Corona discharge and by other practical electrical and mechanical considerations.

Flow conditions: Occurrences are restricted to smooth open terrain with low turbulence levels. This has been confirmed by sectional dynamic wind tunnel tests. Motion amplitude decreases for turbulence levels above 5 percent. Also, forces on downstream conductors are turbulence sensitive.

Operating conditions: The system is unstable above certain critical speeds that depend on bundle geometry, line dynamics, subspan length and bundle orientation to the horizontal. Instability occurs when downstream conduction is in lower half of wake of upstream conductor.

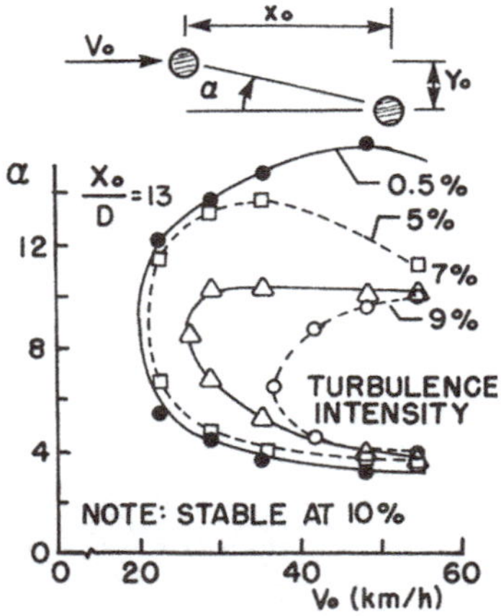

(2) CRITICAL STARTING SPEEDS

Structural dynamics: Mass/unit length of conductors is in range 1.5 to 6.0 kg/m. Dynamic coupling through spacers results in motion of upstream conductors as well as downstream conductors. Frequency = 1 to 4 Hz.

Cures: Motion is reduced by flexible spacers that incorporate damping elements. Other approaches are short subspans, large conductor separation and staggered subspan lengths. Novel approaches such as bundle twist could be pursued.

References: Wardlaw, R.L. et al, Wind Tunnel and Analytical Investigations into the Aeroelastic Behaviour of Bundled Conductors, IEEE Trans. PAS-94, No. 2, pp. 642-54, March/April 1975.

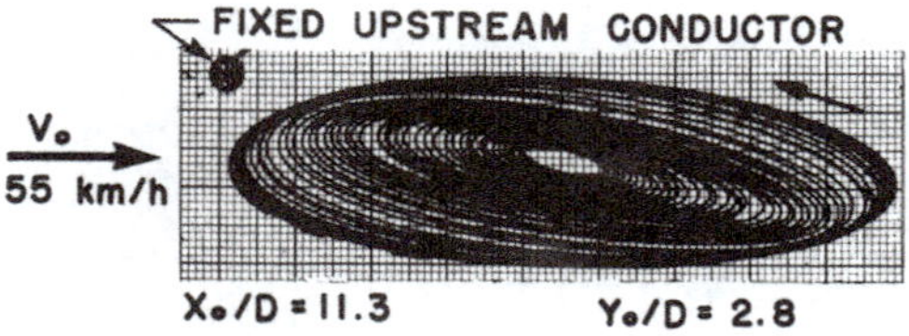

(3) MOTION BUILDUP

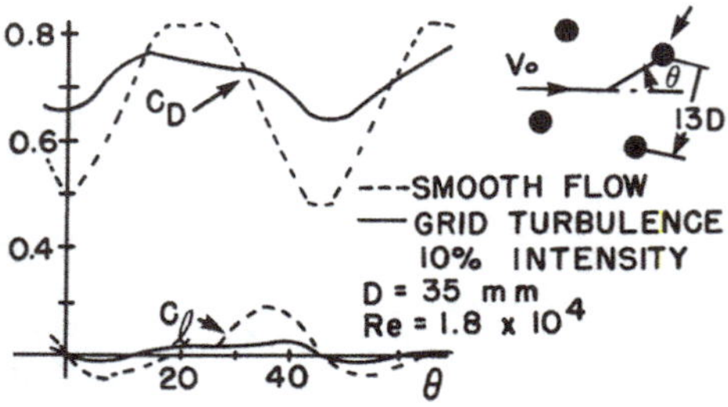

(4) FORCES ON DOWNSTREAM CONDUCTOR

CHART P: TETHERED BALLOON AND TOWED TARGET WITH SPHERICAL AFTERBODY

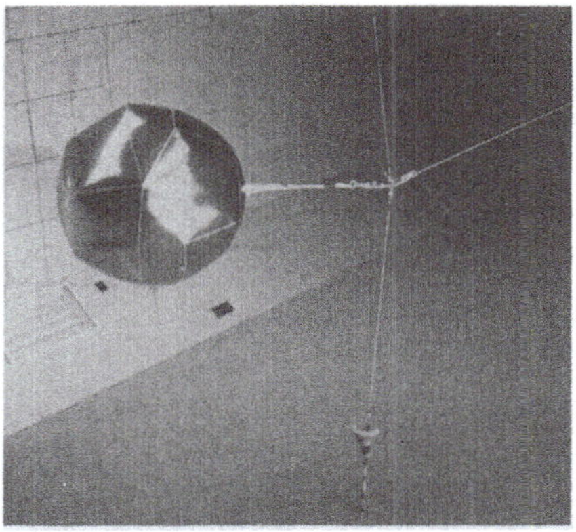

(1) BALLOON MARKER IN 9M x 9M WIND TUNNEL

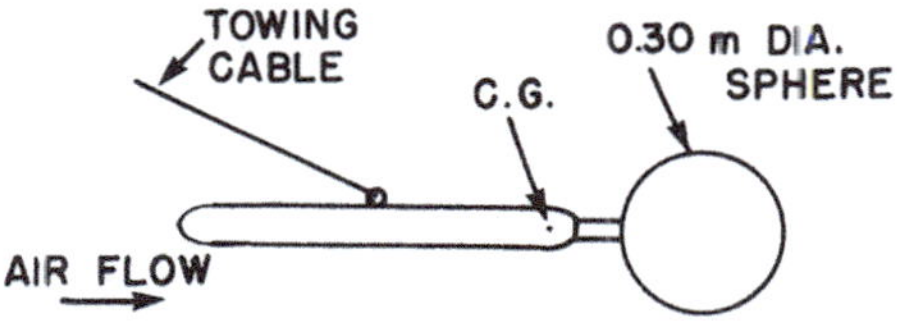

(2) TOWED TARGET AND MISS INDICATOR

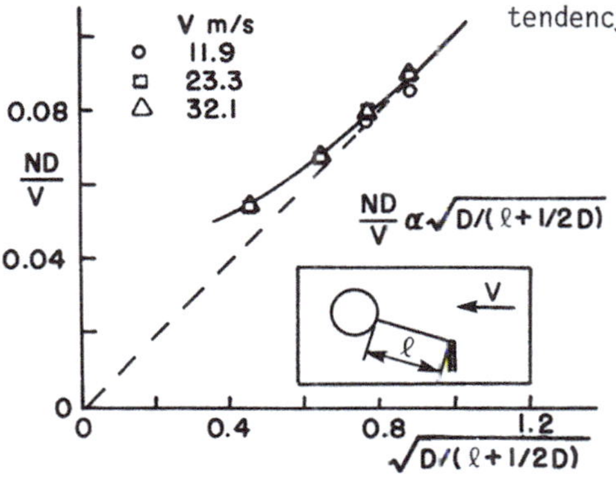

(3) ORBITAL FREQUENCY OF TETHERED BALLOON

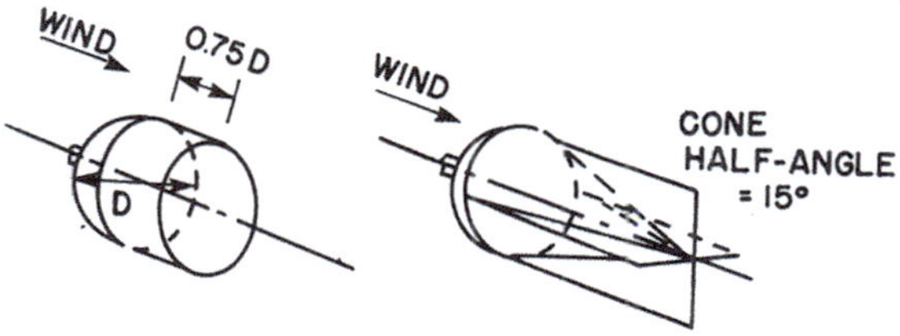

(4) AERODYNAMIC SOLUTIONS FOR TOWED TARGET

Problems: An air-filled tethered balloon marker, of 1.3 m diameter, 0.7 kg deflated mass, covered in reflecting material (Sketch 1). was wind tunnel tested up to 45 m/s in smooth uniform wind. At all speeds above 5 m/s it was found to orbit in a plane normal to the wind, the orbital diameter being approximately 2.3 balloon diameters, independent of tether length. The tests were in a 9 m x 9 m working section. An airborne towed target and miss indicator (Sketch 2) with a spherical radar reflector afterbody was found in flight trials to orbit violently on the end of the towing cable, the orbit being in a plane normal to the direction of travel.

Balloon tether force: For a tether length, ℓ, of 0.9m. Force in tether $/(\frac{1}{2}\rho V^2 A) = 0.8$ where ρ = air density, V = wind speed and A = frontal area of the balloon.

Orbital frequency: For the balloon the orbital frequency, N, was measured for various lengths, ℓ, of the tether. Sketch 3 shows the results plotted in the form ND/V versus $D/(\ell + D/R)$. For short tethers ND/V becomes proportional to D/R but for very long tethers there is a tendency for the orbital frequency to become insensitive to the tether length.

Cures: No cure was necessary for the marker, since the motion enhanced its visibility, and the oscillatory nature of the across-wind component of tether force on the tether post would be taken into account when designing the post. The marker was used during an extensive Arctic experiment in the summer of 1979.

Cures for the towed target are shown in Sketch 4. These were developed on a 1:8 scale model and were subsequently verified in flight trials.

CHART Q: AERODYNAMICALLY-INDUCED INSTABILITY OF A HIGH-SPEED MOTORCYCLE

Problem: Coupled oscillation in roll, yaw and side slip makes the machine unrideable at high speeds.

Geometry: Motorcycle of conventional lay-out with partial fairing typical of road racing machines (Sketch 1). Additional tail cone for operation in fully streamlined category (Sketch 2). Machine designed to set world speed records in the 125 cc partially streamlined and fully-streamlined classes at Bonneville Salt Flats, Utah, USA.

Boundary conditions: Motorcycle running in a straight line on a salt surface in still air conditions at an elevation of 1311m (Sketch 1).

Operating conditions: Smooth approach flow generated by motion over ground at speeds approaching 240 km/h. No damage as rider slowed machine before control lost.

Structural dynamics: Machine consists of two rigid bodies articulated through a steering axis inclined to the vertical (Sketch 3). Chassis elasticity effects are present, but of second order. Total motorcycle-rider weight of 160 kg.

Vibration: Coupled oscillation at about 2 Hz, beginning at 120 km/h and becoming uncontrollable at about 160 km/h. Instability only occurred when 1 kg, 1 m long tail cone in place (Sketch 2). Standard machine handling was excellent and speed of 222 km/h attained to set new record. Later tests demonstrated that effect of tail cone mass negligible. Instability attributed to unsteady aerodynamic forces.

Cures: Theoretical analysis of vehicle stability showed two options. The first was to shift the centre of gravity forward 130 mm. The second, which was used successfully, was to move the rear wheel back 150 mm.

References: The Effect of Aerodynamics on the Stability of High Speed Motorcycles. K.R. Cooper. Proc. Second AIAA Automotive Symposium. Los Angeles, Ca., May 1974.

(1) PARTIALLY-STREAMLINED CATEGORY

(2) FULLY-STREAMLINED CATEGORY

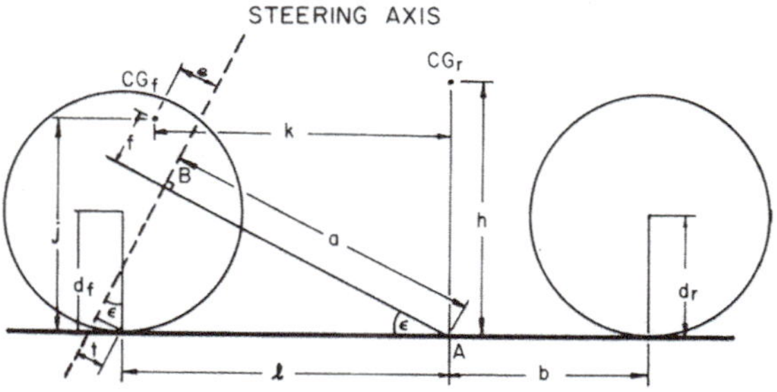

(3)

Discussion: APPROACHES TO THE SUPPRESSION OF WIND-INDUCED VIBRATIONS OF STRUCTURES (R. L. Wardlaw)

R. Blevins
General Atomic Co., San Diego, CA, U.S.A.

Thank you for your excellent paper. It is well known that you have perhaps more practical experience in resolving wind-induced vibration problems of bridges than anyone today. In his general lecture, Professor Scanlan aptly presented the flutter mechanism as produced by negative aerodynamic damping in torsion. In the three bridges discussed in this Session, one vibrated as a result of vortex shedding (N. Shiraishi et al.), one experienced only small buffeting oscillations (W. H. Melbourne), and one showed a potential negative torsion damping. In your experience, could you tell me, of the full-scale bridges which have come to your attention:

(1) approximately what percentage of the vibration problems were due to flutter, as described by Professor Scanlan, and what percentage were the problems due to either vortex shedding or turbulent buffeting;

and

(2) in those cases where there was flutter, did it appear that the vibrations were in one-degree-of-freedom (i.e., one principal mode) or two-degree-of freedom (i.e., two principal modes).

Reply to Discussion of: APPROACHES TO THE SUPPRESSION OF WIND-INDUCED VIBRATIONS OF STRUCTURES

R. L. Wardlaw

From my own experience and from the literature (see Ref. [1]), I am aware of eleven bridges where the road deck has been either seriously damaged or destroyed by wind and seven cases where adverse oscillations have been observed. With the exception of the original Tacoma Narrows Bridge, all those in the first category were constructed prior to 1900. The bridges in the second category are of relatively recent construction, the oldest of these being built in the 1930's. For most of the older bridges, detailed accounts of the events are unavailable, although outstanding observational reports have been made in the case of the Menai Straits Bridge in 1826, the Brighton Chain Pier in 1836 and the Wheeling Bridge in 1848 (see Ref. [1]).

These earlier experiences were not followed by the careful and exhaustive examination and research that is possible today that is exemplified by the investigations that arose following the collapse of the Tacoma Narrows Bridge in 1940 (Ref. [2]). This, of course, could not have happened in the 1800's because the development of practical aerodynamic testing facilities, in particular wind tunnels, has been a product of this century.

In many cases it is not possible to determine the causative phenomenon for the bridge motion. In one of the reported cases, the Longs Creek Bridge, it is quite clear that vortex shedding was the only cause of the oscillation and in another, the Tacoma Narrows Bridge, that it was one cause of motion. Evidence of flutter appears in six of the cases and it is tempting to conclude, for a variety of circumstantial reasons, that flutter was the culprit in some of the others. The influence of flutter cannot be established from the reports, although recently full model wind tunnel studies have been done in simulated turbulent wind and in the case of the Golden Gate Bridge the results have been compared to the behavior of the prototype (Ref. [3]).

In the context of this discussion the term flutter is being used loosely to include both one-degree-of-freedom and two-degree-of-freedom vibrations and in most cases it is impossible to make the distinction from available information. In some instances both torsional and vertical bending have been observed simultaneously and at a frequency between those of the two independent degrees-of-freedom. However the phase shift essential to classical flutter has not been reported. A coupled form of motion occurs in laboratory experiments that is essentially a rotational motion about a point forward to the bridge center-line so that the windward edge has a greater vertical displacement than the leeward edge. With respect to the section center of the principal elastic axes this would be an in-phase two-degree-of-freedom motion. Motion of this type is suggested by some descriptions of full-scale behavior.

Although simultaneous torsion and bending occur, it does not follow that the coupling is essential to the instability. This can be demonstrated in the laboratory by restraining the model to move in only one-degree-of-freedom, in which case motion will typically still occur. At the 1963 Symposium on Wind Effects on Buildings and Structures (Ref. [4]) Scruton reported this to be the case for the Tocoma Narrows Bridge and also that for all the bridges they had tested at the National Physical Laboratory there had been more where coupling between the degrees-of-freedom had been essential to instability.

REFERENCES

1. Podolny, W., Construction and Design of Cable-Stayed Bridges, John Wiley and Sons (1976).

2. Farquharson, F. B., Smith, F. C., Vincent, G. S., Aerodynamic Stability of Suspension Bridges, University of Washington Engineering Experiment Station Bulletin No. 116, Parts I to V (1941).

3. Davenport, A. G., Isyumov, N., Miyata, T., The experimental Determination of the Response of Suspension Bridges to Turbulent Wind, Proceedings, Wind Effects on Buildings and Structures, Tokyo (1971).

4. Scruton, C., Discussion on paper, D.E.J. Walshe, the Use of Models to Predict the Oscillatory Behavior of Suspension Bridges in Wind, Symposium No. 16, Wind Effects on Buildings an d Structures, Vol. II, National Physical Laboratory, Teddington (1963).

F7 TORSIONAL FLUTTER OF LARGE-SCALE MODEL OF SUSPENSION BRIDGE TRUSS DECK IN NATURAL TURBULENT WIND

Toshio Miyata
Associate Professor
Department of Civil Engg.
Faculty of Engg.
University of Tokyo
Tokyo 113, Japan

Isao Okauchi
Professor
Department of Civil Engg.
Faculty of Engg. Science
Chuo University
Tokyo 112, Japan

SUMMARY

Occurrence of torsional flutter vibrations is fatal to long-span suspension bridges. A large-scale model of a closed truss deck showed severe torsional flutter in natural turbulent wind. Evaluating the results of measurement, it was deduced that the occurrence of flutter was due to the negative damping effect of unsteady forces, and the torsional amplitude significantly responded to long-period wind fluctuations. Some open gratings installed in the highway deck eliminated the flutter completely.

1. INTRODUCTION

In connection with the wind-resistant design of long-span Honshu-Shikoku bridges proposed in Japan, a large-scale dynamic sectional model bridge was tested in the natural wind, aiming at a confirmation of the reliability of wind tunnel tests. The large-scale model tests consisted of response measurements associated with mean forces, transfer admittance functions between turbulent wind and fluctuating forces, buffeting vibrations and flutter vibrations. It was expected that such a model could be used to investigate the effects of turbulence, and these effects could be related closely to the response of prototype bridges.

In this contribution, the onset and persistence of torsional flutter observed on the model will be described, followed by a description of a means of attenuating the vibration by modifying the flow-structure interaction.

2. DESCRIPTION OF STRUCTURE AND FLOW CONDITIONS

The cross section of the large-scale model is shown in Sketch 1 in the Chart. The prototype is a proposed suspension bridge with a 1500 m center span length, which has a truss deck consisting of a six-lane highway in the upper deck and a double track railway in the lower deck. Severe flutter occurred in the case of the closed highway deck without any open gratings. This case is indicated as model A in Sketch 1 of the Chart. The case with open gratings is a modified deck for cures, named model B, where no flutter occurred. The structures are about 1/10 scaled rigid truss decks with 8 m length, which were elastically mounted as shown in Fig. 1 by using coil springs so as to satisfy dynamic similarity of flutter; that is, to allow oscillations in vertical translation and pitching (torsional) motions. The structural characteristics of the prototype and the model are given in Table 1.

The model rig was mounted on a turn-table which was adjusted to set the model axis at almost right angles with respect to the mean wind flow. The model was set on the beach where relatively strong seasonal winds were expected (See Sketch 2 in Chart). Approaching wind properties measured at a model height of about 6 m above ground were the mean wind velocity $\overline{U}$ = 5 to 20 m/s, the turbulence intensity I_u = 9 to 13%, the turbulence scale L_x = 25 to 100 m and the decay factor of lateral correlation with the order of k_y = 15 to 20 (in exp $(-k_y Ny/\overline{U})$), where N is the frequency and y the lateral distance. These were measured by sonic anemometers. The ratio of the turbulence scale to the streamwise deck width was assumed comparable to that of the scale of wind over open sea to the width of the prototype. The wind velocity for the prototype is calculated by $\overline{U}_p = \overline{U} \times 10.52\ (N_p/No) = 5.4 \times \overline{U}$ in the case of torsional vibration, where N_p and N_o are the natural frequencies of the prototype and the model respectively.

TABLE 1

	prototype	model
mass	45×10^3 kg/m	3.3×10^3 kg
polar moment of inertia	12.2×10^6 kgm^2/m	7.69×10^3 kgm^2
natural frequency vertical torsional	 0.11Hz 0.22Hz	 0.23Hz 0.43Hz
structural log. damping vertical torsional	 ——— ———	 $\sim$ 0.07 $\overline{-}$ 0.02

geometrical scale: 1/10.52

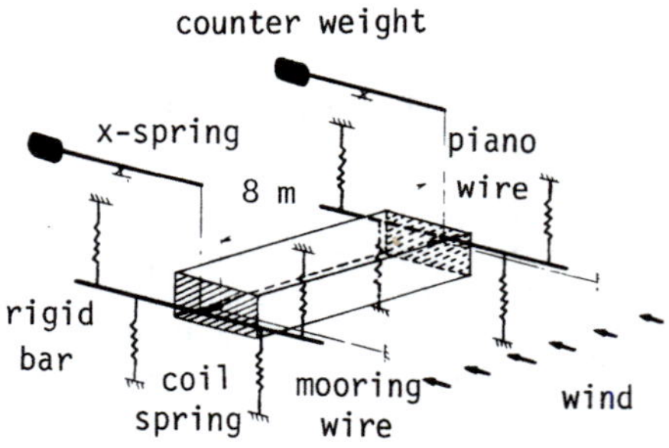

FIG. 1 MODEL MOUNTING SYSTEM

3. FLUTTER VIBRATIONS

Torsional flutter was clearly found to occur in the case of the closed highway deck without any open gratings (model A). Pitching (torsional) motion was predominant, accompanied by vertical translation. From records measured by potentio-displacement-meters, the frequencies of the flutter were nearly constant and equal to the natural frequencies. However, the torsional amplitude of flutter was influenced by the wind velocity fluctuations, particularly by comparatively long period fluctuations of the mean flow. Fig. 3 shows a time trace record near the onset of flutter. It can be seen that the envelope of torsional motion fluctuates extremely slowly, compared with the natural period. To describe the onset and persistence of torsional flutter, the root mean square (RMS) amplitude was calculated versus mean wind velocity, as shown in Sketch 3 in the Chart. The abrupt development of torsional response characteristically occurred for only a slight increase of mean wind velocity.

Such torsional flutter observed on the bridge deck is one of unstable self-excited vibrations which are induced by the aeroelastic, that is, airflow-structure interaction. It is said that the onset and persistence of flutter is due to a negative damping effect, caused by the interaction of the separated airflow from a leading edge of the deck structure with the torsional motion. In order to describe the flutter characteristics, it is useful to examine the aerodynamic damping due to the motion in relation to the wind velocity. Sketch 4 in the Chart shows the result of logarithmic aerodynamic damping at a specified amplitude, say 2°, versus mean wind velocity, in which the damping includes the structural value. The plotted data shows considerable scatter, although it can be seen that the onset of flutter was obviously due to the negative damping effect. Some damping values in Sketch 4, indicated by AR, were calculated by a parameter estimation method using an auto-regression model fitted to a fluctuating time trace of torsional motion. In this calculation, the equation of motion to describe the torsional flutter was assumed to be of the form

$$\theta(\ddot{\phi} + 2h_o\omega_o\dot{\phi} + \omega_o^2\phi) = M_R\phi + M_I\dot{\phi} + M_u u + M_w w \qquad (1)$$

where ϕ, $\dot{\phi}$ and $\ddot{\phi}$ are the torsional vibration, velocity and acceleration, h_o the structural damping constant, ω_o the natural circular frequency = $2\pi N_o$, θ the polar moment of inertia and the M's are coefficients of the pitching moment forces acting on the deck. $M_R\phi$ and $M_I\dot{\phi}$ are the unsteady self-excited force components proportional, respectively, to the torsional displacement and velocity, and $M_u u$ and $M_w w$ are the forced exciting forces due to the longitudinal and vertical components of fluctuating wind velocity.

Fig. 2 shows a comparison of aerodynamic damping data with that of wind tunnel tests for an about 1/10 scaled sectional model of the large-scale model. The large-scale model data are the same as in Sketch 4 in Chart. The wind tunnel

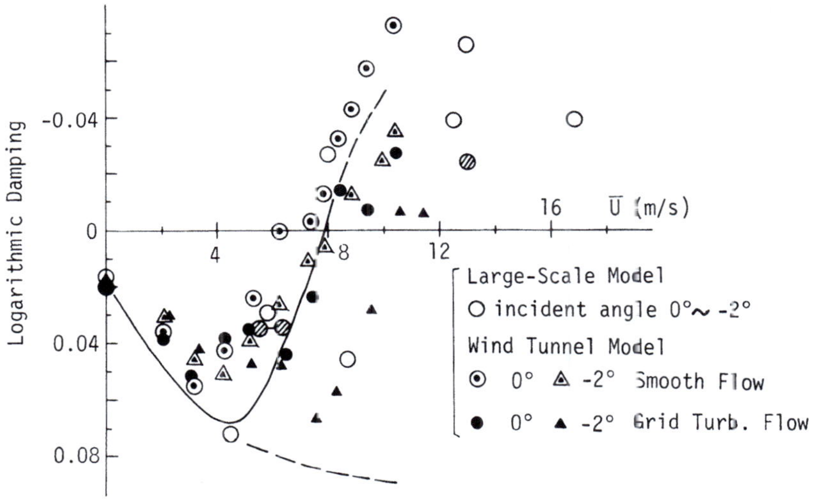

FIG. 2 COMPARISON OF AERODYNAMIC DAMPING OF LARGE-SCALE MODEL WITH THAT OF WIND TUNNEL TEST MODEL

tests were carried out in smooth flow as well as in a grid-generated turbulent flow, featuring an appropriate intensity, but a rather smaller scale than the model width. It can be seen that the wind tunnel data have a relatively smooth transition, but cannot well explain the scatter of the large-scale model data.

The scattered data in Sketches 3 and 4 may be attributed to wind velocity fluctuations in various forms. One of the effects of wind fluctuation is revealed in the envelope fluctuations of torsional flutter amplitude, which appear to be influenced by a comparatively long period fluctuation of longitudinal wind velocity. Fig. 3 is a typical time trace record that shows how the torsional flutter responds to wind velocity fluctuations, together with the accompanying random vibration in vertical translation. Considering three examples (1), (2) and (3) of wind fluctuations of relatively long period and the amplitude transition of pitching (torsional) motion indicated in Fig. 3:

(1) [0 to 100 sec, approx.] With decrease of wind velocity, the torsional amplitude, which has been gradually developing, begins to decrease after experiencing a peak.

(2) [300 to 400 sec. approx.] With increase of wind velocity, the amplitude starts to develop again and exhibits a flutter.

(3) [400 to 600 sec. approx.] Again wind velocity decreases, and the developing flutter tends to die away.

A common feature of these examples is a considerable time lag of the torsional flutter response and damping to long period fluctuations of wind velocity.

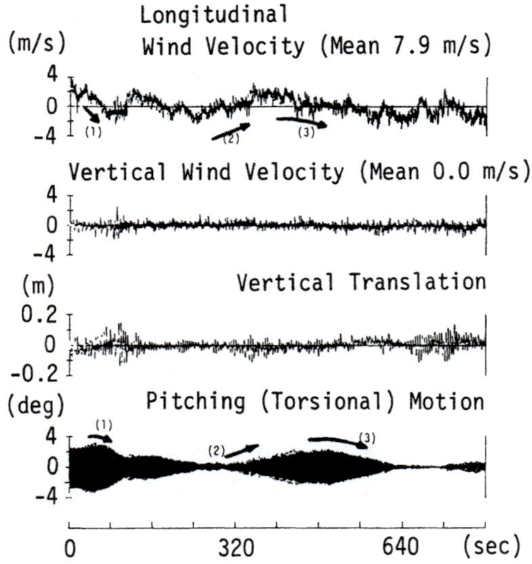

FIG. 3 RECORD OF TORSIONAL FLUTTER, ACCOMPANIED BY VERTICAL TRANSLATION, WITH LONGITUDINAL AND VERTICAL WIND VELOCITIES

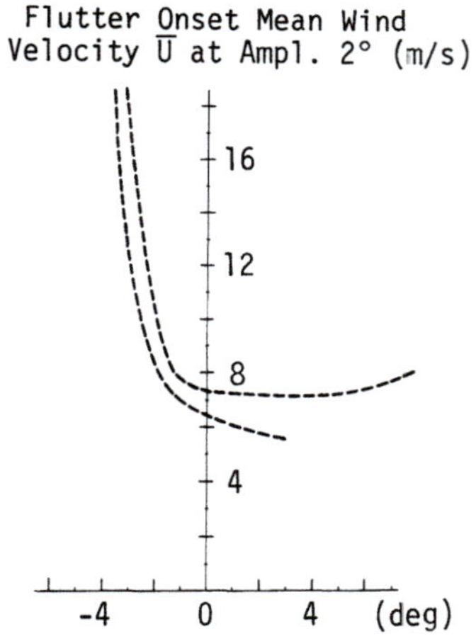

FIG. 4 FLUTTER ONSET MEAN WIND VELOCITY VS. EFFECTIVE INCIDENT ANGLE OF ATTACK OF WIND AT SPECIFIED AMPLITUDE

Torsional flutter is an unstable vibration induced by the negative damping effect of unsteady, self-excited forces. According to wind tunnel experiments in smooth flow, the flutter shows nonlinear characteristics with amplitude. A stationary vibration is usually maintained after its onset. To describe the flutter vibration, the formulation of eq. (1) is useful, in which the unsteady force coefficients M_R and M_I are assumed to be functions of nondimensional wind velocity U/NB and torsional amplitude ϕ_o, where U is wind velocity, N frequency and B width of the deck. That is,

$$M_R = M_R\ (U/NB,\ \phi_o) \quad , \quad M_I = M_I\ (U/NB,\ \phi_o) \qquad (2)$$

The torsional flutter observed in a turbulent wind is, as mentioned above, characteristic of response to long period wind fluctuation. Arguing this phenomenon in the form of eqs. (1) and (2), a value of mean wind velocity $\bar{U}$ must be chosen over such a time interval that behaviour (1), (2) and (3) in Fig. 3 is well described.

Fluctuations in the growth and decay of response amplitudes should also be expected from the time-varying angle of attack of the incident wind. These fluctuations arise from the dependence of the flutter velocity on wind inclination. Fig. 4 shows mean wind velocity at flutter onset versus the effective angle of attack of mean flow on the deck, where the two dotted lines indicate mean scatter of the data.

4. CURES OF FLUTTER

The introduction of gratings on model B produced positive aerodynamic damping (Sketch 4 in Chart). Thus, flutter was prevented and only small amplitude of buffeting vibrations were recorded (Sketch 3 in Chart).

5. ACKNOWLEDGEMENT

The large-scale bridge model plan was conducted by the Honshu-Shikoku Bridge Authority under a guidance of the technical committee on wind resistant design of Honshu-Shikoku bridges, Japan Society of Civil Engineers. This contribution is a presentation of the results of the plan.

6. REFERENCES

1. Technical Committee on Wind Resistant Design of Honshu-Shikoku Bridges, JSCE, Research Report on Large Scale Bridge Model Plan, No. 1, (1977), No. 2, (1978), (in Japanese).

2. Okauchi, I., Tajima, J. and Akiyama, H., "Response of the Large Scale Bridge Model to Natural Wind", Proc. 5th Int. Conf. on Wind Engineering, Colorado, USA, (1979).

SUSPENSION BRIDGE TRUSS DECK

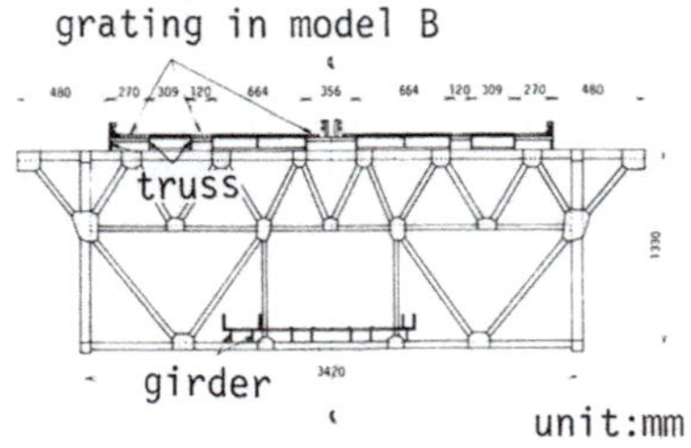

① MODEL A WITHOUT GRATINGS

②

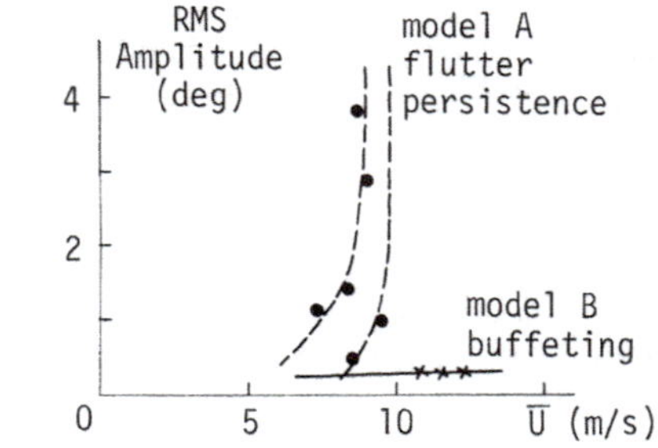

③ TORSIONAL RESPONSE

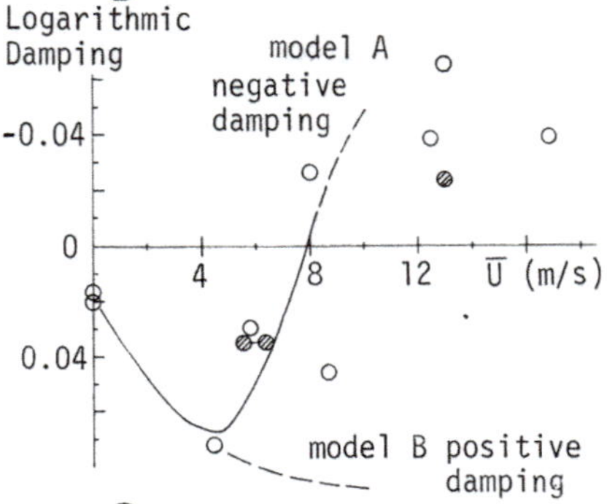

④ AERODYNAMIC DAMPING

Problem: Occurrence of torsional flutter oscillations is fatal to long-span suspension bridges. Large-scale model A of closed truss deck without gratings showed severe flutter in natural turbulent wind.

Geometry: See Sketch 1. 1/10 scaled rigid model (length 8 m) for prototype with 6-lane highway and double track railway.

Boundary conditions: Deck is mounted to allow to oscillate in vertical translation and pitching (torsional) motions by using coil springs.

Flow conditions: See Sketch 2. Approach flow of natural turbulent wind on beach with intensity Iu = 9∿13%, scale Lx = 25∿100 m and lateral correlation factor k_y = 15∿20 at model height 6 m.

Operating conditions: Exposed to mean wind velocity of $\bar{U}$ = 5∿20 m/s ($\bar{U}_p = 5.4 \times \bar{U}$ for prototype), acting at almost right angles to model axis.

Structural dynamics: Mass 3.3×10^3 kg. Polar moment of inertia 7.69×10^3 kgm^2. Natural frequency 0.43 Hz in pitching and 0.23 Hz in vertical translation. Log. damping 0.02 in pitching.

Vibration: Flutter was pitching (torsional) vibration, accompanied by vertical translation. Amplitude was influenced by wind velocity fluctuation. RMS amplitude versus mean wind velocity in Sketch 3. Frequency nearly equal to natural frequency. Log. aerodynamic damping versus mean wind velocity at specified amplitude 2° in Sketch 4 means flutter occurrence due to negative damping. (Data scattered by turbulence effect.)

Cures: Open gratings of model B in Sketch 1 eliminated flutter, as shown in Sketch 3, through turning to positive damping as shown in Sketch 4.

References: Okauchi, I. et. al., Response of the Large Scale Bridge Model to Natural Wind, Proc. 5th Int. Conf. Wind Engg., July 1979, Colorado, USA.

F8 TRAFFIC-INDUCED VIBRATIONS OF A MOTORWAY SIGN GANTRY

J S HAY

Transport and Road Research Laboratory
Crowthorne, Berkshire, England

SUMMARY

To assist in the writing of design rules on motorway sign gantries, wind tunnel experiments were carried out to determine the drag coefficient in winds from different directions. Since the flow was smooth rather than turbulent as in the atmosphere, a full scale sign gantry near TRRL was instrumented so as to obtain measurements to check the wind tunnel tests.

When the first recordings were made, it was immediately evident from signals of strain that the structure was subject to frequent bursts of oscillation at a frequency around 2.5 Hz. Observations of traffic soon established that the structure was being excited by an impulsive force whenever a high-sided vehicle passed underneath. Although it also responded to wind action, wind-induced vibrations never occurred at such a high frequency. As the intensity of vehicular buffeting is probably dependent on the relative speeds of wind and vehicle, it is unlikely to contribute much to the total loading in strong wind conditions. However, in view of its persistence, it merits consideration as regards fatigue, especially with lightly damped highway structures (although calculations suggest that fatigue is not a problem with the sign gantry used in the tests).

Since the Information Chart was prepared, the logarithmic decrement of the sign gantry has been derived from the shape of the spectrum near the natural frequency. The values obtained were 0.063 and 0.102.

TRAFFIC-INDUCED VIBRATIONS OF A MOTORWAY SIGN GANTRY

Problem: Transverse vibrations are induced by windage from high-sided vehicles. During periods of dense traffic, these vibrations may be nearly continuous.

Flow conditions: Structure on exposed motorwax site (see photo). Turbulence intensity 0.22 (at 7 m), probably including contribution from traffic. Lateral scale of turbulence estimated as 30 m.

Operating conditions: Vibrations initiated whenever highsided vehicle passes underneath.

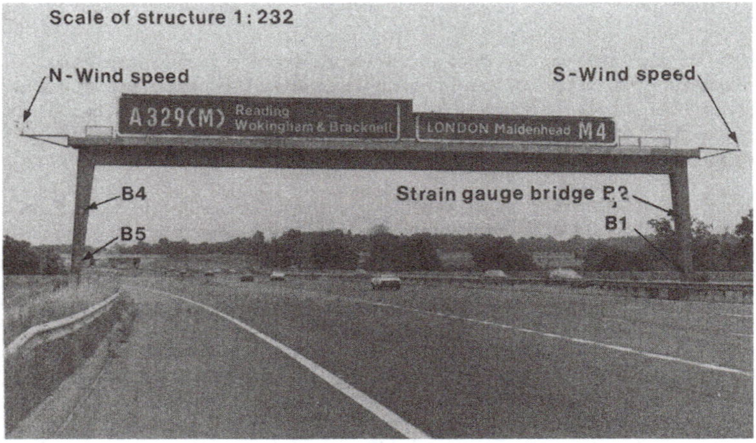

Structural dynamics: Values of damping (logarithmic decrement) derived from records ranged from 0.04 to 0.01 with a mean of 0.07. Second moment of area about neutral axis in each leg of gantry is 1.52×10^{-3} m^4 at 1 m and 1.37×10^{-3} m^4 at 3.5 m.

Vibrations: Frequency 2.49 Hz, indicated by spectra of strain measured in legs of structure. Examples of spectra shown in the sketch are of strain at 1 m (B1 and B5) and 3.5 m (B2 and B4) and of wind speed near the top of the legs (north and south ends of structure).

Cures: None attempted as cyclic stresses are low and assessment of fatigue indicates that problems are unlikely.

Reference: J A Hay. Forces on a sign gantry in normally directed winds. TRRL Report LR 820, Crowthorne, 1978 (Transport and Road Research Laboratory).

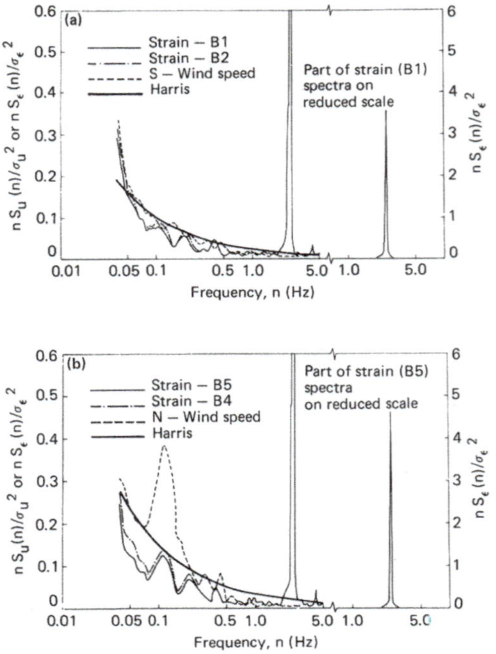

G1 THE INFLUENCE OF TURBULENCE ON THE AEROELASTIC RESPONSES OF TALL STRUCTURES TO WIND

A. G. Davenport

The University of Western Ontario
London, Ontario, Canada

SUMMARY

The paper considers the composition of the dynamic forces acting on a tall structure situated in a turbulent boundary layer flow. These can be broken down into turbulence induced and wake induced forces, and aeroelastic forces due to the structure's motion. The influence of turbulence on wake induced forces is discussed, as well as the key influence of aerodynamic damping terms. Examples of the latter for a variety of shapes are compared. An example showing the response of guyed masts is given.

1. RESPONSE OF TALL BUILDINGS TO TURBULENT WIND

1.1 Experiments on a Tall Building in Turbulent Flow

When the twin towers of The World Center in New York were built, over a decade ago, an unusual degree of attention was given to their behaviour in wind. At the time, wind tunnel studies were rarely carried out for buildings; if they were, they involved static pressure models tested in the normal uniform steady flow of aeronautical wind tunnels. Several aspects of the wind studies for The World Trade Center represented radical departures from previous practice; one was the use of aeroelastic models (hitherto used only for highly flexible structures such as bridges and chimneys) another was the modelling of the turbulent boundary layer flow characteristics of the natural wind, achieved partly through modelling the building topography of Manhattan.

The experimental study for the project, which the writer directed for the consulting engineers, employed three wind tunnels; the new micrometeorological wind tunnel at Colorado State University, designed by Cermak and Plate; the National Physical Laboratory by C. Scruton and the Boundary Layer Wind Tunnel Laboratory, The University of Western Ontario. The results affirmed the important role of dynamics in tall building response to wind and the important role of turbulence.

Experiments [1,2] carried out on a single square tower of similar proportions to the World Trade Center, the results of which are shown in Fig. 1, illustrate this. In these experiments the ratio of the height to width was 6.5, the mean density ratio of the building model to the air was approximately 182, the moderately rough boundary was representative of open country and the rough boundary represented a built-up urban area. The mean velocity and turbulence profiles are shown in Fig. 2. The responses are given in terms of the ratio of the rms and mean responses for the X and Y direction (σ_x, σ_y, X, Y respectively) to the building height H. The mean wind speed, U_o, at the top of the building is expressed through the velocity profile in Fig. 2 and the reduced velocity, U_g/f_oD, where f_o and D are the natural frequency and diameter of the model. Reduced velocities encountered in practice would seldom exceed 10 and are generally much below. When the experimental results are examined in detail the following points are noteworthy.

In the drag direction the slope of the mean response indicates the usual proportionality to the velocity squared. However, the dynamic rms drag response increases with a power slightly in excess of a square law; increases systematically

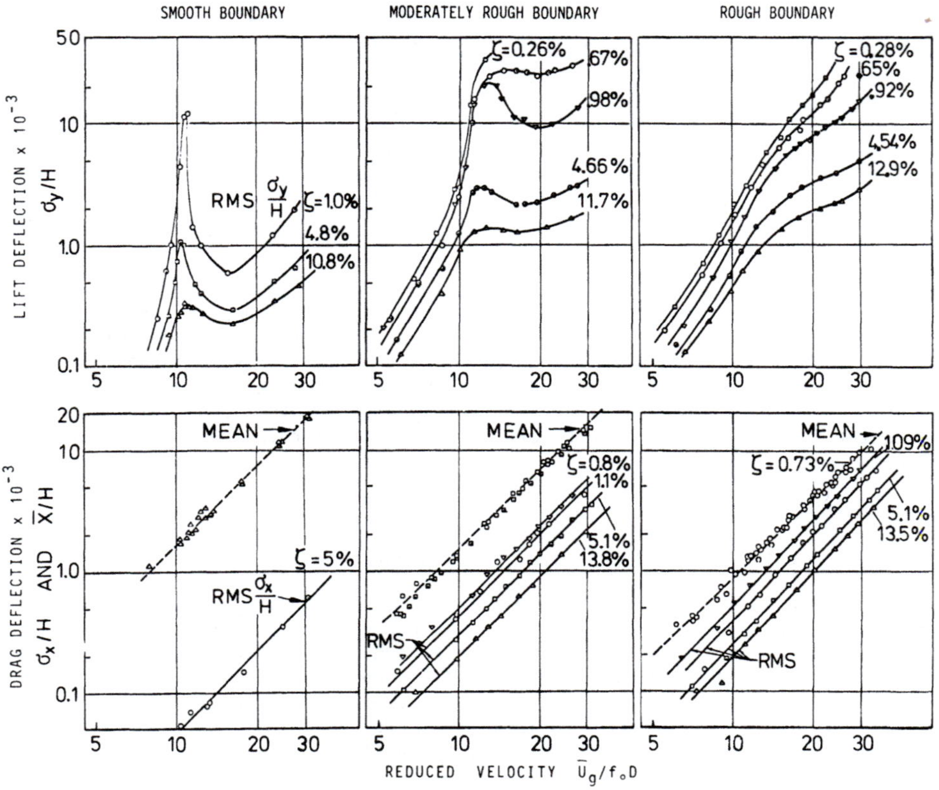

FIG. 1 DYNAMIC RESPONSE OF SQUARE TOWER (H=6.6D) TO UNIFORM SMOOTH FLOW AND TO TURBULENT BOUNDARY LAYER FLOW FOR VARIOUS CRITICAL DAMPING RATIOS

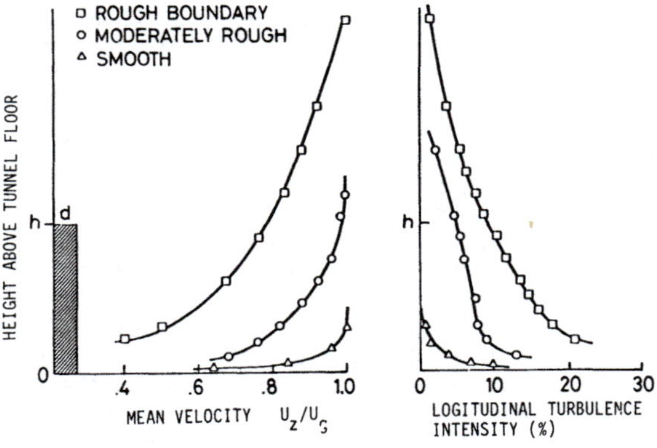

FIG. 2 MEAN VELOCITY AND TURBULENCE INTENSITY CHARACTERISTICS IN BOUNDARY LAYER FLOWS

with the roughness of the boundary; and gradually decreases with damping.

In the lift direction the dynamic response is considerably more complex. There is of course no mean response due to symmetry. With the smooth boundary (negligible turbulence) a marked peak occurs at a reduced wind speed of approximately 11, the amplitude of which collapses rapidly with increased damping. In the turbulent boundary layer flows the peak becomes blurred; the dynamic response of the off-peak wind speed increases, and the damping reduces the responses more gradually.

It is obvious from these results that the peak dynamic response, which may reach three times σ_x or σ_y, are highly significant compared to the mean and the influence of the turbulent boundary layer is dramatic. These facts have changed the direction of the wind tunnel testing and it would now be quite inappropriate to consider wind tunnel testing of buildings without modelling the boundary layer - at least approximately.

1.2 Theoretical Formulations

In recent years some progress has been made in characterizing these complicated aerodynamic processes. Some tentative models and explanations are now summarized.

A building or structure, such as The World Trade Center, experiences aerodynamic forces generated both by the upstream flow impinging on the structure, as well as by the wake behind the structure [3]. In addition there may be aeroelastic forces generated by motion of the body as it moves through the fluid in response to the external forces. Let the forces due to the oncoming flow, the wake and the aeroelastic effects be denoted respectively $F_T(t)$, $F_W(t)$ and $A(X, \dot{X}, \ddot{X}\,..)$, where $X, \dot{X}, \ddot{X}$ etc. are the vectors of displacement, velocity, acceleration etc.

Assuming a linear elastic system (mass M, damping C, stiffness K) the equation of motion is

$$M\ddot{X} + C\dot{X} + KX = F_T(t) + F_W(t) + A(X, \dot{X}, \ddot{X}\,..) \qquad (1)$$

This equation is quite general and may, if necessary, represent more than one degree of freedom (as in suspension bridge behaviour for example). Here we will concern ourselves primarily with uncoupled single degree of freedom behaviour, in which the response in each mode can be considered separately.

If amplitudes of motion are small the aeroelastic forces $A(X, \dot{X}, \ldots)$ can be expressed as a linear function of $X, \dot{X}$, etc. In the frequency domain this equation may be written

$$[-M\omega^2 + C\omega + K]\, X(\omega) = F_T(\omega) + [A(\omega) + i A^*(\omega)]\, X(\omega) \qquad (2)$$

where $X(\omega)$ is the component of the response at frequency ω; $F_T(\omega)$ and $F_W(\omega)$ are the frequency components of the turbulent and wake induced aerodynamic forces and $A(\omega)$ and $A^*(\omega)$ define the in-phase and out-of-phase components of the motion dependent forces.

Dividing by the stiffness, $K = M\omega_o^2$, we can rearrange equation (2) in the form

$$[(1 - \omega^2/\omega_o^2 - a^2) + 2i(\omega/\omega_o)(\zeta_s + \zeta_a)]\, X(\omega) = [F_T(\omega) + F_W(\omega)]/K \qquad (3)$$

where

$$a^2 = A(\omega)/K \qquad \zeta_a = -A^*(\omega)/(2M\omega_o) \qquad \zeta_s = C/(2M\omega_o)$$

The first of these terms, a^2, has the character of an aerodynamic stiffness ratio; it is, practically speaking, small compared to unity and usually can be neglected. By contrast, the term ζ_a, termed the aerodynamic damping, may be highly significant compared to the structural damping ζ_s (as discussed below).

1.3 Response to Turbulence

In turbulent flow the aerodynamic forces in equation (3) are random. The mean square generalized force, $\sigma^2_{GF_T}$, on the structure for the mode shape $\mu(z)$ is

$$\sigma^2_{GF_T} = \iint_A \sigma_p^2 (z, z') \; \mu(z) \; \mu(z') \; dz \, dz' \tag{4}$$

where $\sigma_p^2(z,z')$ is the covariance of the pressure at positions z and z' on the structural surface A. The corresponding spectrum of the generalized force is

$$S_{GF_T}(f) = \iint_A S_p(z, z'; f) \; \mu(z) \; \mu(z') \; dz \, dz' \tag{5}$$

The response to turbulence generally consists of background excitation at frequencies below resonance and a resonant peak. An approximation to the mean square fluctuating modal response, due to turbulence which is frequently suitable, is

$$\overline{X}^2 = [\sigma^2_{GF_T} + \frac{\pi}{4} \frac{f_o \, S_{GF_T}(f_o)}{(\zeta_s + \zeta_a)}] / K^2 \tag{6}$$

The terms in the [ ] represent the so called "background" and "resonant" excitation respectively.

For the type of problem at hand involving bluff bodies in boundary layer flow, the description of the covariance and cospectrum of pressure (σ_p and S_p) is difficult. Recent work by Hunt [4,5] and coworkers, using rapid distortion theory for determining the pressure fluctuations in the region upstram of the separation point has considerably enhanced understanding. It has so far not eliminated the need for approximate solutions in the practical instances. In instances where turbulence scales are larger than the structure, simple aerodynamic models involving direct projection of the flow fluctuations; correlations etc. onto the structure have worked surprisingly well.

For the prismatic structure considered in the experiments in Fig. 1, with a straight line mode shape, one such approach leads to the following approximations for variance and spectrum of the generalized force

$$\sigma^2_{GF_T} = (\rho \, \overline{U}_o^2 \, D \, H \, C_D)^2 I^2 \; \frac{3}{[(2+a)^2 + \frac{1}{2}(2a+3) \, H/L]} \; \frac{1}{[1 + \frac{1}{2} D/L]} \tag{7}$$

$$f_o \, S_{GF_T}(f_o) = (\rho \, \overline{U}_o^2 \, D \, H \, C_D)^2 I^2 \; \frac{84}{c^2} \left\{ \frac{\overline{U}_o}{f_o H} \right\}^{8/3} \frac{H}{D} \; \frac{1}{(7 + 8a)} \tag{8}$$

These approximations rely on a quasi-steady drag approach, a cross correlation of velocity of the form $\exp[-|z - z'|/L]$ where L is a turbulence scale, the assumption that the resonant frequency lies within the Kolmogorov inertial subrange, and a power law approximation to the mean velocity profile $(\overline{U} \propto z^a)$, c is the decay coefficient for narrow band correlation and has a value ~ 8.

Measurements of aerodynamic damping of prismatic bodies in the drag direction are not plentiful. However the values of the aerodynamic drag damping coefficient $C_D^*(\overline{U}/f D)$ for the flat plate shown in Fig. 3 may be indicative [6]. These coefficients are all positive and the value of the aerodynamic damping may be estimated from the relation

$$\zeta_a = \frac{1}{2\pi} \; \frac{\rho D^2}{M} \; \frac{\overline{U}}{f_o D} \; C_D^* \tag{9}$$

In the present case, with $M/\rho D^2 \approx 182$, $\overline{U}/f_o D \approx 10$, $C_D^* \sim 2.0$ we find $\zeta_a \approx .0175$. This value is clearly significant.

The expressions given above lend themselves to simple design formulations and are in fact employed in codes. The formulae lead to close agreement with the results shown in Fig. 1 as well as full scale experiments.

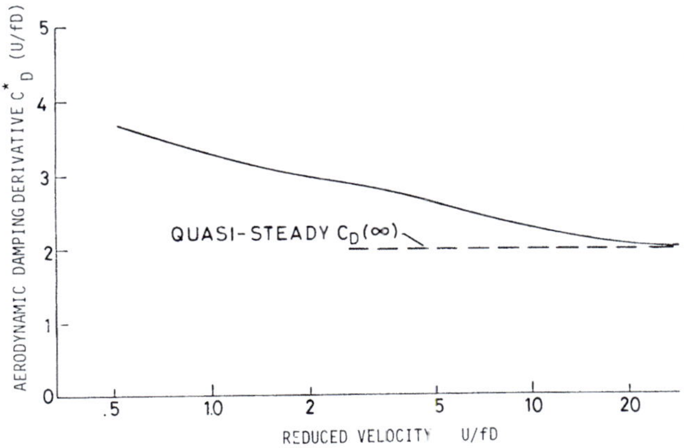

FIG. 3 AERODYNAMIC DERIVATIVE FOR DRAG DAMPING FOR FLAT PLATE

For the transverse lift direction a similar series of expressions can be derived for the excitation by turbulence in the oncoming flow. In this case the excitation is caused primarily by changes in the azimuth of the wind. Expressions for these transverse forces are similar to those in equations (7) and (8) but multiplied by the factor

$$[\partial C_{F_y} / \partial \alpha / 2 C_D]$$

and adjusted by the ratio or the turbulence intensities.

1.4 Response Due to Wake Forces

We now examine the transverse or lift components of the wake forces, $F_W(\omega)$. In uniform flow, the turbulent fluctuations are negligible, and the forces due to the wake will have the form

$$F_W(t) = \tfrac{1}{2} \rho \bar{U}^2 \sqrt{2}\, \tilde{C}_L D \epsilon^{i 2\pi f_e t} \tag{10}$$

where

$f_e = S\,\bar{U} / D$, the eddy frequency;
$\tilde{C}_L$ = r.m.s. lift coefficient;
S = the Strouhal number;
D = the frontal diameter.

Let $\bar{U}_c = f_o D/S$ define the critical windspeed for which the eddy frequency coincides with the natural frequency. The rms response of the system at frequency f, written $\tilde{X}(f)$, is found from equations (3) and (10) to be

$$\tilde{X}(f)/D = [(1 - \bar{U}/\bar{U}_c)^2 + 4(\zeta_s + \zeta_a)\,\bar{U}/\bar{U}_c]^{1/2} (\rho_a/\rho_s)(\bar{U}/\bar{U}_c)^2 \tilde{C}_L/(8\pi^2 S^2) \tag{11}$$

ρ_a/ρ_s is the ratio of air density to average structural density $(\frac{M}{D^2})$. The maximum value is when $\bar{U} = \bar{U}_c$ and

$$\frac{\tilde{X}(f)}{D}_{max} = \frac{(\rho_a/\rho_s)\,\tilde{C}_L}{16\pi^2(\zeta_s + \zeta_a)\,S^2} \tag{12}$$

If the flow is turbulent the fluctuations cause the eddy shedding frequency to vary. The energy instead, of being concentrated at one frequency is distributed over a spectrum. If the longitudinal velocity is $U(t) = \bar{U} + u(t)$ the lift force can be written

$$L(t) = \tfrac{1}{2} \rho \bar{U}^2 D \tilde{C}_L \exp[i(2\pi S/D)(\bar{U} + u)t]$$

The spectrum of the lift, or coefficient of lift can then be shown to correspond to the probability density of the velocity. If the latter is Gaussian the spectrum can be written

$$\frac{f S_{C_L}(f)}{C_L} = \frac{1}{\sqrt{2\pi}} \frac{f}{f_e} \exp[-\tfrac{1}{2}(f/f_e - 1)^2 / I_u^2] \tag{13}$$

where I_u is the intensity of turbulence. This form for the lift spectrum, suggested by Vickery [7] and observed at full scale by Ruscheweyh [8], is illustrated in Fig. 4.

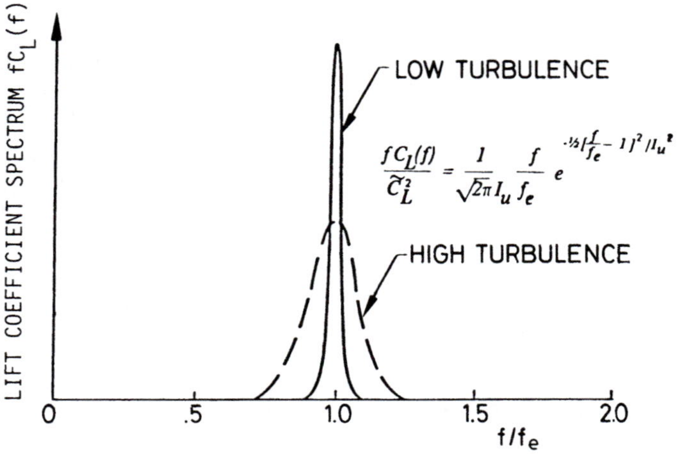

FIG. 4 EFFECT OF TURBULENCE ON LIFT COEFFICIENT SPECTRUM

Turbulence also affects the spanwise correlation. Vickery observed that even in smooth flow the spanwise correlation on a square has a spanwise scale of the order of 2-diameters; in turbulent flow this length scale is decreased. To account for this the co-spectrum of the force can be multiplied by $e^{-(\Delta z/\lambda D)}$ where Δz is the difference in the spanwise coordinates and λD is the transverse scale [11] ($\lambda \sim 1\text{-}2$). Similar results hold for the circular cylinder as has been found by Toebes [9] and Novak [10] for example. In most practical applications, it appears that the wake forces are locally correlated and the spectrum of mean square force per unit length may be reduced by the factor $2\lambda D/H$ to account for correlation.

In dealing with a tall structure, allowance must be made for variations of mean velocity and cross section with height and for the mode shape. Assume that these variables are $\bar{U}(z)$, $D(z)$ and $\mu(z)$ respectively and that reference values of the velocity and diameter are $\bar{U}_o$ and D_o (taken for example at the top of the structure). The spectrum of the generalized wake force is then

$$\frac{f\, S_{GF_W}(f)}{[\frac{1}{2}\rho \bar{U}_o^2 D_o H \tilde{C}_L]^2} = \frac{2\lambda D_o}{H^2} \int_0^H \left[\frac{\bar{U}(z)}{U_o}\right]^4 \left[\frac{D(z)}{D_o}\right]^3 \mu^2(z) \frac{1}{\sqrt{2\pi}\, I(z)} \frac{f}{f_e(z)} \exp\left[-\tfrac{1}{2}(f/f_e(z)-1)^2/I^2(z)\right] dz \qquad (14)$$

in which

$$f_e(z) = S\,\bar{U}(z)/D(z)$$

The form of this spectrum still retains the peaked shape but is significantly broadened by the variation with height of mean wind speed and diameter. For a highly damped structure the significant response will be concentrated at the resonant frequency f_o. From random vibration theory the mean square modal response $\bar{X}^2$ is

$$\frac{\bar{X}^2}{D_o^2} = \frac{\pi}{4} \frac{1}{(\zeta_s + \zeta_a)} \frac{f_o\, S_{GF_W}(f_o)}{(4\pi^2 f_o^2 D_o M)^2} \qquad (15)$$

where the generalized mass M is

$$M = \int_o^H m(z)\, \mu^2(z)\, dz$$

Substituting from equation (14) the rms response is

$$\sqrt{\overline{X^2}}/D_o = [\sqrt{\pi/16\,\pi^2}]\,[\zeta_s + \zeta_c]^{1/2}\,[\rho D_o^2\, H/M]\,[\tilde{C}_L/S^2]\,[2\lambda D_o/H]\, R(U^*) \qquad (16)$$

where

$$R^2(U^*) = [1/(\sqrt{2\pi}\,\bar{I}_o)]\, U^{*3} \int_o^H \theta_U^4(z)\, \theta_D^3(z)\, \mu^2(z)\, \exp[-\tfrac{1}{2}[\theta_U(z) - U^*]^2 [\theta_U(z)/U^* I_o]^2]\, dz \qquad (17)$$

and $$U^* = \bar{U}_o/\bar{U}_c = S\bar{U}_o/f_o D_o;\ \theta_D(z) = D(z)/D_o;\ \theta_U(z) = \bar{U}(z)/\bar{U}_o;\ I_o = I(z)/\theta_U(z) \qquad (18)$$

Equation (16) is the key equation for the modal response. U^* is directly proportional to the reference mean wind speed, $\bar{U}_o$ and the dependence of the response on windspeed is defined by $R(U^*)$ given in equation (17). This relationship is evaluated for the tall tower and shown in Fig. 5. It indicates a peaked form for the spectrum with maxima occurring in some cases at values of U^* significantly in excess of unity. Increases in turbulence intensity I_o, and variations in mean wind speed and diameter all have the effect of reducing the intensity of excitation but increasing the range of wind speeds over which excitation will take place.

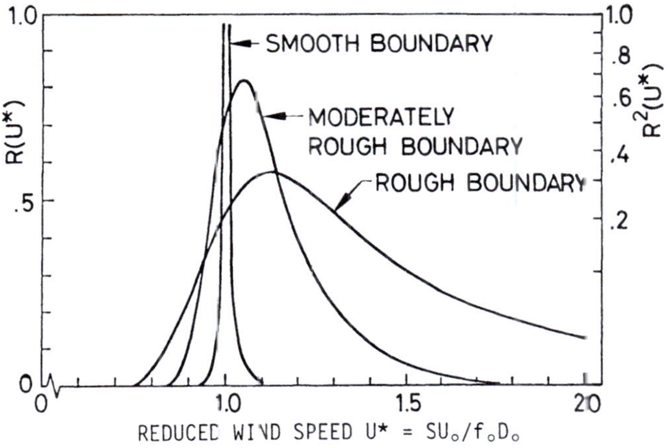

FIG. 5 VORTEX SHEDDING EXCITATION FUNCTION FOR SQUARE TOWER IN BOUNDARY LAYER FLOW

1.5 Aerodynamic Damping

The only component in equation (16) which has so far not been discussed is the aerodynamic damping. For large reduced velocities $(\bar{U}_o/f_o D_o)$ the value of aerodynamic damping can reasonably be expected to approach the quasi-steady value.

$$\zeta_a = \frac{\int_o^H \rho\, \bar{U}(z)\, D(z)\, \mu^2(z)\, (\partial C_{F_y}/\partial a)\, dz}{4\pi f_o M} \qquad (19)$$

$$= [1/4\pi]\,[\rho D_o^2\, H/M]\,[\bar{U}_o/f_o D_o] \int_o^H \theta_U(z)\, \theta_D(z)\, \mu^2(z)\, (\partial C_{F_y}/\partial a)\, dz/H$$

Here $\partial C_{F_y} / \partial a$ is the slope of the lift force at the origin.

It is clear that although the quasi-steady value gives good indications at high reduced velocities, at intermediate values the quasi-steady values cease to hold. This is particularly true close to the critical velocity for eddy shedding. This is illustrated in Fig. 6 which shows results by Washizu et al [12, 13] for rectangular prisms. (Note in these results the notation is such that the critical damping ratio is

$$\zeta_a = -[1/4\pi][\rho D_o^2 H/M][U_o/f_o D_o]^2 [C_L^* / X/D]$$

For the square cylinder these results show the dramatic transition in the damping at the reduced velocity of 8 - the critical value for eddy shedding. The results also show the quasi-steady value is negative as has been noticed often before.

It is apparent that the strongest negative damping lies slightly above the critical velocity - the region where generally the excitation is largest. This is clearly a dangerous combination.

These results refer to smooth flow and the question arises what happens in turbulent flow. Vickery [14] has suggested that for large scale turbulent flow the damping is 'blurred' by the constantly changing velocity. The effective damping is approximated by

$$\zeta_a(\bar{U}/fD)_{eff} = \int_o^\infty \zeta_a[(\bar{U}+u)/fD]\, p_u(u)\, du$$

where $p_u(u)$ is the probablity density function for u . This step clearly reduces the strongly negative peaks in the damping as illustrated in Fig. 7.

Although it is not possible to account for all of the quantitative aspects of the response, the above is nevertheless adequate to explain the trends in the response in Fig. 1. In the lift direction, the gradual flattening of the peak excitation is explained by the similar trend in Fig. 3 and the possible reduction in negative damping in Fig. 7. The increase in off-peak excitation is due to the response to turbulence.

Just as the application of quasi-steady aerodynamics to the galloping problem [23,24] has shown that nonlinearities are important so, it appears, these nonlinearities also exist at lower reduced velocities. The results shown in Fig. 6 for example only apply to small amplitudes. Even so this may be sufficient for many applications.

2. AERODYNAMIC DAMPING COEFFICIENTS

Although aerodynamic damping measurements (and other derivatives) have been measured for bridge decks, there is still no clear indications of the characteristics we should expect for bluff structures in general. It is instructive, therefore, to consider a number of illustrations of the damping forces on bluff bodies subject to transverse vibration.

The results in Fig. 6 also indicate the aerodynamic damping for two other rectangular shapes besides the square from experiments by Washizu et al [12,13]. The same pattern of damping in the square is repeated in the 1 x 4 rectangle. Note

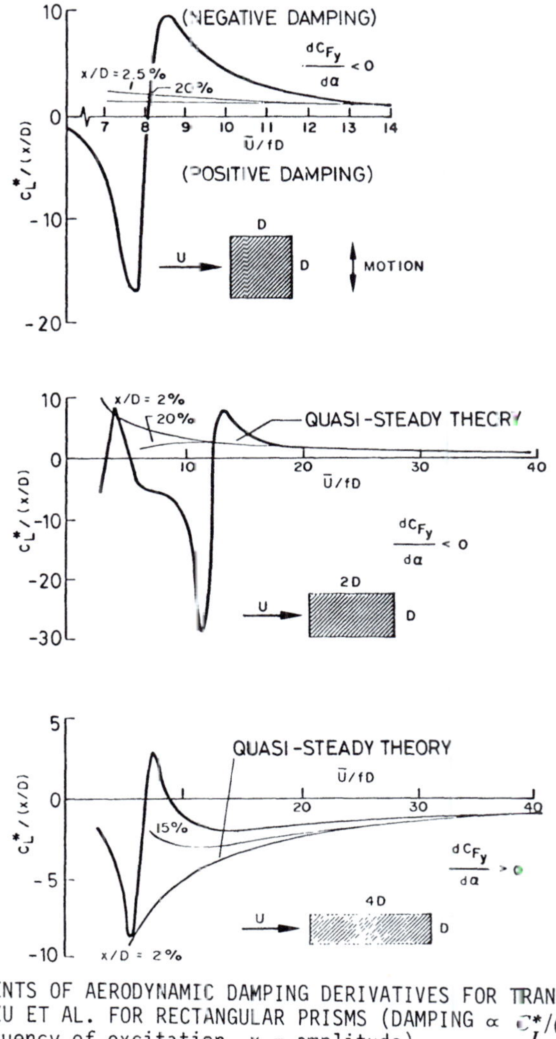

FIG. 6 MEASUREMENTS OF AERODYNAMIC DAMPING DERIVATIVES FOR TRANSVERSE MOTION BY WASHIZU ET AL. FOR RECTANGULAR PRISMS (DAMPING $\propto C_L^*/(x/D) \times U/f_o D$ f_o = frequency of excitation, x = amplitude)

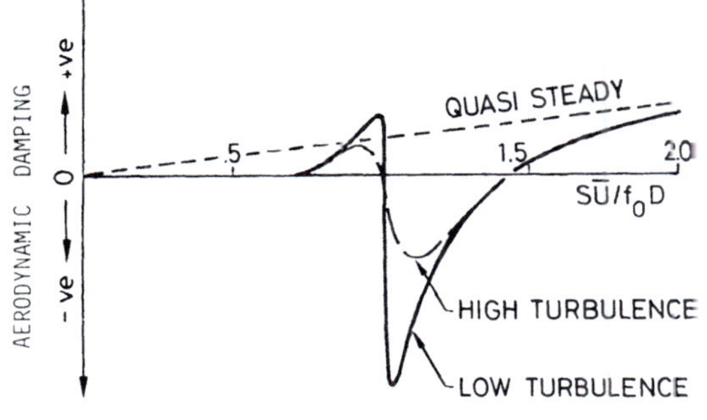

FIG. 7 EFFECT OF TURBULENCE ON AERODYNAMIC DAMPING

again that the quasi-steady value of damping is approached at reduced velocities approximately twice the critical. In this case the lift slopes and the damping values have the opposite sign. The 1 x 2 rectangle exhibits a double reversal, and not inconsistently, it also exhibits two typical vortex shedding frequencies - associated perhaps with reattachment.

Fig. 8 shows the aerodynamic damping for an equilateral triangle and, in Fig. 9, that for an H section bridge deck. The former were derived directly from pressure measurements by Twigge-Molecey [15] and Baines on an oscillating triangle. The latter selected from a series of bridge decks tested by Scanlan [16]. This had the bluffest cross-section, probably without reattachment taking place at the second side girder.

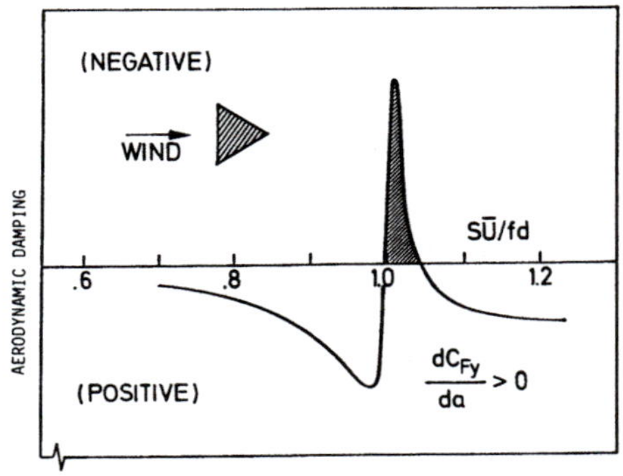

FIG. 8 AERODYNAMIC DAMPING COEFFICIENT FOR EQUILATERAL TRIANGLE IN HEAVING MODE (Pressure measurements - Twigge-Molecey and Baines)

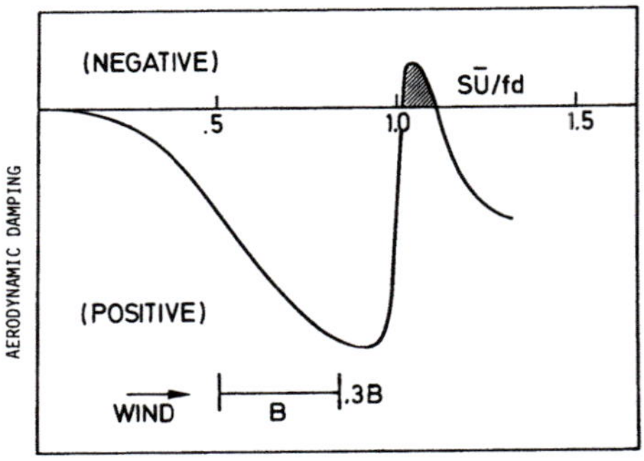

FIG. 9 AERODYNAMIC DAMPING COEFFICIENT FOR H-SECTION IN HEAVING MODE (Decay measurements-Scanlan)

Fig. 10 shows an example of the measurement of aerodynamic damping on circular cylinder by Scruton [17]. Also shown is a cylindrical tower with antennae on the outside. Exactly the same patterns are indicated. Direct measurement of the damping of circular cylinders has been carried out by Cincotta and Jones [18], Loiseau and Szecheny [19], Yano and Takahara [20] and by Nakamura, Kaku and Mingota [21].

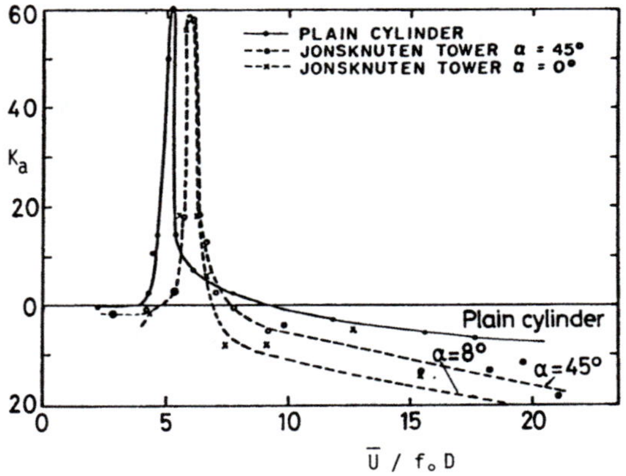

FIG. 10 AERODYNAMIC EXCITATION (DAMPING) OF CYLINDRICAL STRUCTURES (AFTER SCRUTON)

A summary of these results has been made by Vickery and is shown in Fig. 11. These have been normalized by the peak damping coefficient to avoid the problem of Reynolds number disparities.

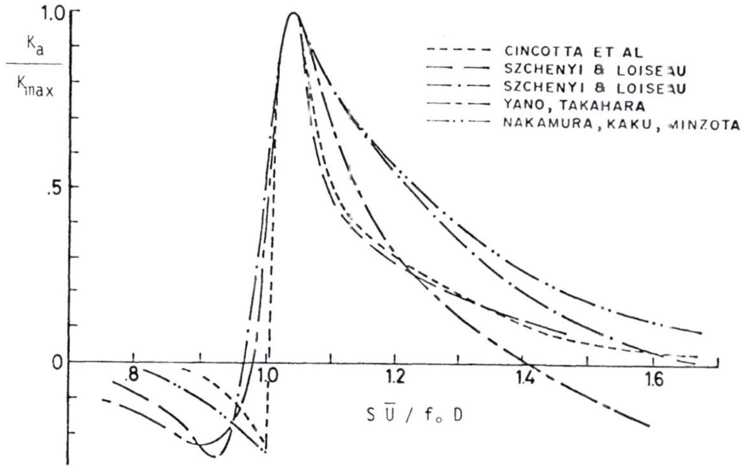

FIG. 11 SUMMARY OF EXPERIMENTAL ESTIMATES OF AERODYNAMIC DAMPING (AFTER VICKERY)

All of these dissimilar geometric shapes exhibit similar damping characteristics: namely a) positive damping for windspeeds below the critical; b) a rapid transition at the critical and c) a region of negative damping just above the critical velocity. Finally we note that the damping returns to the same sign as the quasi-steady damping implied by the lift slope $d\,C_{F_y}/da$. the reasons are not altogether clear. What is most significant from the practical viewpoint is the existence of the strong regions of negative aerodynamic damping which is a principal source of instability.

3. APPLICATION TO TALL GUYED MASTS

The discussion of circular cylinders brings us finally to consider tall cylindrical guyed masts in turbulent wind. These structures have in certain cases revealed a tendency to oscillate at low wind speed, in some instances with large amplitude. Figs. 12 and 13 show respectively a 203 metre mast and some measurements by van Koten [22] of comparatively low amplitude response. The general shape of the response curves in Fig. 13 is similar to the curves in Fig. 5, the extra breadth of the former being partly due to the varying diameter of the mast. The mode shapes are shown in Fig. 13 and the response patterns correspond to the excitation which occurs successively in the several modes.

A somewhat more surprising result rises from the comparison of responses of masts - A and B shown in Fig. 14. Both are just over 300 m in height, both are approximately the same mass and stiffness, and both are cylindrical or at least partially so. Where the masts differ is in the distribution and size of the cylindrical exteriors. Mast A is only partially cylindrical; Mast B is cylindrical throughout, but its diameter is slightly smaller. The responses, predicted using methods explained herein, are shown in Figs. 15 and 16. The magnitudes differ dramatically by a factor greater than 10. The principal explanation is in the coincidence of the large shroud on Mast A with the region of large amplitude deflection in Mode 3. From superficial appearances this large discrepancy in the movements would not have been anticipated.

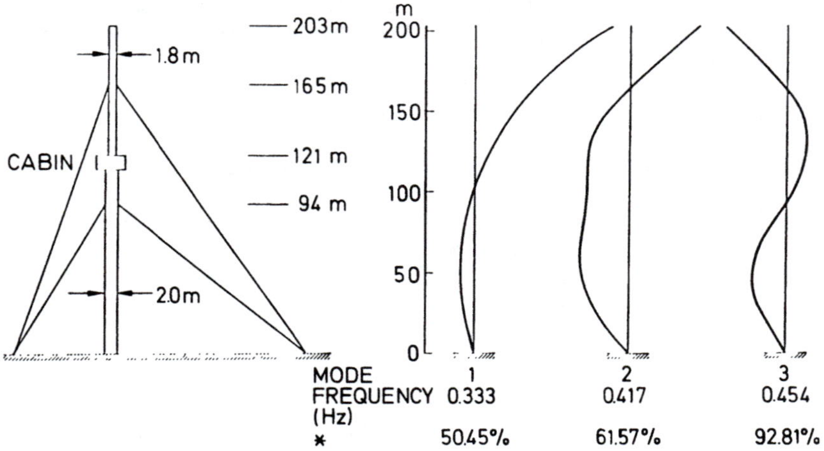

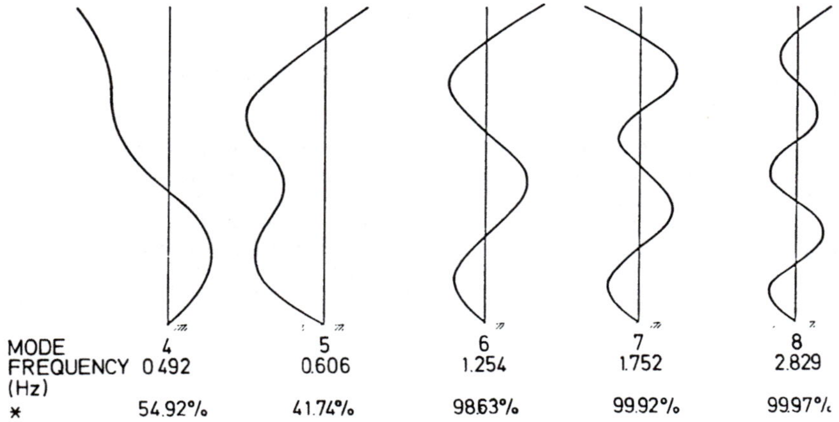

FIG. 12 CYLINDRICAL GUYED MAST (WIERINGERWERF) AND MODE SHAPES
(* CONTRIBUTION OF TOWER TO GENERALIZED MASS)

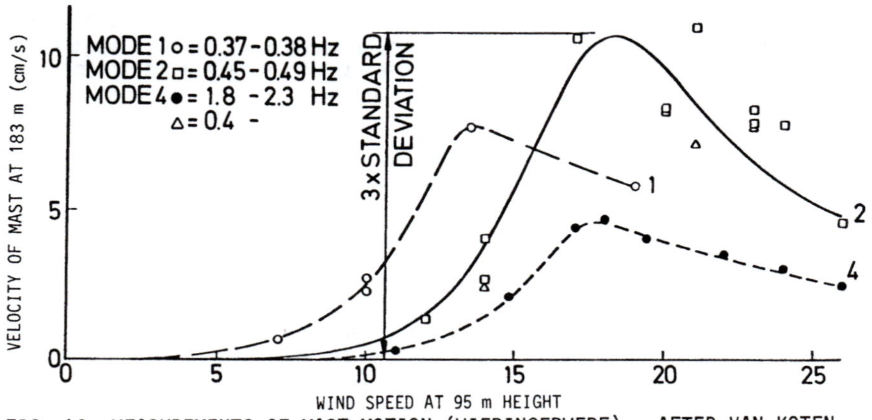

FIG. 13 MEASUREMENTS OF MAST MOTION (WIERINGERWERF) - AFTER VAN KOTEN

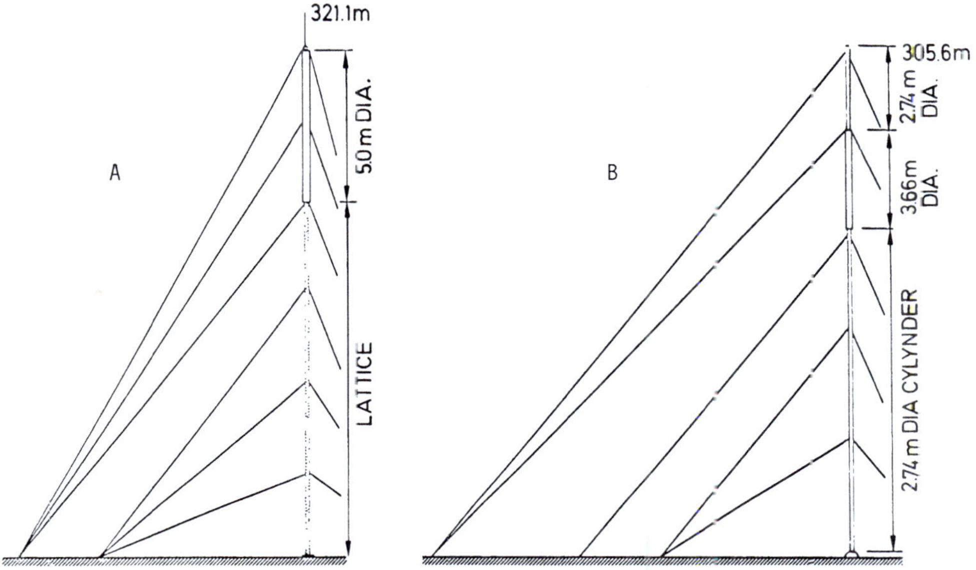

FIG. 14 LATTICE MAST WITH CYLINDRICAL SHROUD (A) AND CYLINDRICAL MAST (B)

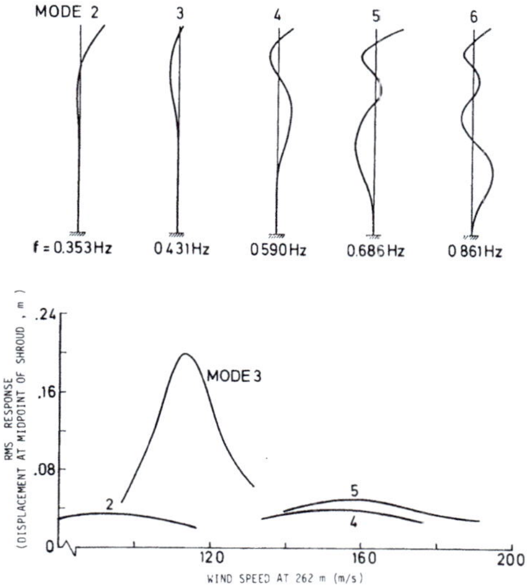

FIG. 15 RESPONSE OF LATTICE MAST WITH CYLINDRICAL SHROUD TO VORTEX SHEDDING (MAST A)

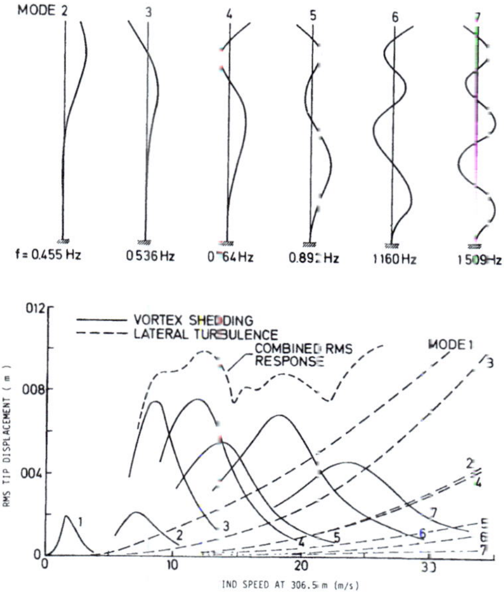

FIG. 16 RESPONSE OF CYLINDRICAL MAST TO VORTEX SHEDDING AND LATERAL TURBULENCE (MAST B)

4. CONCLUSIONS

It is apparent that trends to lightness, lower damping and reduced stiffness in buildings and structures can often lead to problems of wind excitation. Wind tunnel modelling of turbulent flow is now routine and should be considered as an essential component of building studies. Progress has been made in the understanding of turbulence effects on dynamic response. Renewed attempts to model the transverse excitation in turbulence are promising better predictions. Examples using tall guyed masts contrast the differing levels of response attainable with ostensibly similar structures.

REFERENCES

1. Davenport, A.G., "The Treatment of Wind Loading on Tall Buildings", Sym. Tall Build. Univ. of Southampton, (April 13-15, 1966), p. 441-482.

2. Rosati, P.A., "Wind Action on Finite Rectangular Cylinders in Turbulent Boundary Layer Flow", BLWT Res. Report, 1970-8 (M.E.Sc. Thesis supervised by A.G. Davenport).

3. Davenport, A.G., "The Prediction of Response of Gusty Wind", Int. Res. Seminar Safety of Structures under Dynamic Loading, June 1977, Trondheim, Norway (published by Tapir 1978).

4. Durbin, P.A., Hunt, J.C.R., "Fluctuating Surface Pressures on Bluff Structures in Turbulent Winds: Further Theory and Comparison with Experiment", Fifth Int. Conference on Wind. Eng., July 1979, Colorado State University, Fort Collins, Colorado.

5. Hunt, J.C.R., "A Theory for Fluctuating Pressures on Bluff Bodies in Turbulent Flows", IUTAM-IAHR, Karlsruhe, 1972 Symposium on Flow-Induced Vibrations (E. Naudascher), Springer-Verlag 1974.

6. Davenport, A.G., "The Application of Statistical Concepts to the Wind Loading of Structures", Inst. Civil. Engrs. Proc. V. 19 (Aug. 1961), p. 449-472.

7. Vickery, B.J. and Clark, A.W., "Lift or Across Wind Response of Tapered Stacks ASCE Jnl. Str. Div. Jan. 1972.

8. Ruscheweyh, H. and Hirsch, G., "Full Scale Measurements on Tower Shaped Structures", Proc. of Fourth Int. Conf. on Wind Effects on Buildings and Structures Heathrow, 1975.

9. Toebes, G.H., "Fluidelastic Features of Flow Around Cylinders", Intern. Res. Seminar Wind Effects on Build. & Struct., Ottawa.

10. Novak, M., Tanaka, H., "Pressure Correlations on Vibrating Cylinders", Proc. 4th Int. Conference on Wind Effects on Buildings and Structures", (Heathrow, U.K.), Ed. K.J. Eaton, Cambridge Univ. Press 1975.

11. Vickery, B.J., "Fluctuating Lift and Drag on a Long Cylinder of Square Cross-Section in a Smooth and in a Turbulent Stream", J. Fluid Mec. V. 25 Part 3 (1966), p. 481-494.

12. Washizu, K., Ohya, A., Otsuki, Y., Fujii, K., "Aeroelastic Instability of Rectangular Cylinders in a Heaving Mode", Jnl. Sound and Vibration (1978) vol. 59 (2), p. 195-210.

13. Otsuki, Y., Washizu, K., Tomizawa, H., Ohya, A., "A Note on the Aeroelastic Instability of a Prismatic Bar with Square Section", Jnl. of Sound and Vibration (1978), Vol. 34, p. 233-248.

14. Vickery, B.J., "Wind Loads on Chimneys", The University of Western Ontario, Research Report, BLWT-2-1979.

15. Twigge-Molecey, C.F.M., Baines, W.D., "Unsteady Pressure Distributions Due to Vortex Induced Vibrations of a Triangular Cylinder", IUTAM-IAHR Symposium Karlsruhe, 1972, Springer Verlag.

16. Scanlan, R.H., "An Examination of Aerodynamic Response Theories and Model Testing Relative to Suspension Bridges", Proc. Third Int. Conf. on Wind Effects on Buildings and Structures, Tokyo, 1971.

17. Scruton, C.S., "On the Wind Excited Oscillation of Stacks, Towers, and Masts", Proc. 1st Int. Conf. on Wind Effects on Buildings and Structures, Nat. Phys. Lab. Teddington, 1963.

18. Cincotta, J.J., "Experimental Investigation of Wind Induced Oscillation Effects on Cylinders in Two-Dimensional Flow at High Reynolds Numbers", NASA Langley Research Center.

19. Szechenyi, E., Loiseau, H., 'Portances sur Cylindre Vibrant Perpendiculaire ä Son Axe ä des Nombres de Reynolds Eleves: Onera (1975).

20. Yano, T. and Takahara, S., "Study on Unsteady Aerodynamic Forces Acting on an Oscillating Cylinder", Proc. Third Int. Conf. on Wind Effects on Buildings and Structures, Tokyo, 1971.

21. Nakamura, Y., Kaku, S. and Mizota, T., "Effect of Mass Ratio on the Vortex Excitation of Circular Cylinder", Proc. Third Int. Conf. on Wind Effects on Buildings and Structures, Tokyo, 1971.

22. Van Koten, H., "Meting van Bewegingsnelhelden aan de T.V. Mast te Wieringer Werf", T.N.O. Rapport N. B68-294/5/24.

23. Parkinson, G.V., "Mathematical Models for Flow Induced Vibrations of Bluff Bodies", IUTAM-IAHR Karlsruhe 1972, Symposium on Flow Induced Vibrations (E. Naudascher), Springer-Verlag 1974.

24. Novak, M.N. and Davenport, A.G., "Aeroelastic Instability of Prisms in Turbulent Flow", ASCE Eng. Mech. Div., Vol. 96, No. EM1, p. 17, Feb. 1970.

25. Surry, D., "Some Effects on Intense Turbulence on the Aerodynamics of a Circular Cylinder at Subcritical Reynolds Number, Jnl. F. Mech. 1972, Pt. 3, p. 543-563.

G2 ASSESSMENT OF WIND LOADS FOR GLAZING DESIGN

W.A. Dalgliesh
Division of Building Research
National Research Council of Canada
Ottawa, Canada K1A 0R6

SUMMARY

This paper develops a procedure for transforming detailed statistical descriptions of randomly fluctuating wind pressure into a single factor giving the equivalent loading effect on annealed glass panels in terms of whatever loading was used to generate the glass strength data. Such a transformation is considered indispensable for the proper application of data now available to architects and engineers through advanced wind tunnel tests.

1. INTRODUCTION

Wind loads for cladding design differ from those for the structure; this is now widely accepted, and is reflected in most, if not all, recent codes and standards. Detailed analysis of the interaction between the wind and the structure of very tall, slender buildings, usually with the aid of wind tunnel tests, has become commonplace and some codes even provide procedures requiring the input of such structural characteristics as size, fundamental frequencies, and damping. As size, frequency and damping of cladding elements are clearly different from those of the structure, it is evident that their response to wind action will also differ.

What has not been so readily appreciated, however, is the role played by the strength properties of the cladding material in a rational assessment of wind loads for cladding design. We are now better able to describe the statistical properties of the fluctuating wind pressures acting over panel-size elements of building surfaces and to predict the magnitude of extreme pressures and suctions, but little attention seems to have been given to their durations.

Fig. 1 GLASS CLADDING IN OTTAWA

Fig. 2 57-STOREY TOWER IN TORONTO

The widespread and growing use of glass as a cladding material testifies to its popularity from an architectural and functional point of view (Figs 1,2) and so it is appropriate to inquire whether the special nature of glass demands a more complete description of design wind pressures than just the extreme magnitude. A brief review will be provided of recent research shedding light on this question. The main objective of this paper is to demonstrate a method of describing and analyzing fluctuating pressures that takes into account not only storm duration but also the durations of extreme pressure "spikes" in order to supply an equivalent one-minute constant load as required for selecting glass thickness from strength data of North American glass manufacturers [1-3].

2. REVIEW OF PAPERS ON GLAZING RESPONSE TO WIND LOADING

Ishizaki et al [4] were among the first to report on a study to correlate the variations of glass strength with load duration and the variation of gust pressures with averaging time. Float glass plates with areas 1.5 m^2 and 2.5 m^2 were broken under constant pressures of different durations to determine an inverse relation of breaking pressure to the logarithm of load duration. The dependence of strength on load duration was then used in a computer simulation of wind loading on glass, using a scaled-up version of a wind pressure record measured on a high-rise building during a typhoon. The object of the computations was to determine how much the fluctuating wind pressure would weaken the glass before the failure situation is reached. The criterion for failure was based on the stress corrosion theory (to be described later). For the case studied, the authors concluded that the glass must be about 10 per cent stronger to withstand the cumulative effects of earlier gusts, than would be required just for the final gust pressure causing failure. They suggested an averaging time of 0.5 to 1 second for wind pressures for glass design. Ishizaki et al reported that there has been no trouble with wind-induced mechanical vibrations of window panes in Japan, where the natural frequencies range from about 10 to 50 Hz.

Mayne and Walker [5] obtained wind pressure measurements on two specially instrumented glass panels 1.6 m x 1.3 m x 4 mm mounted near the top of a 33 m high office building in Birmingham. They concluded that panels of this size responded to extreme gusts, with effective averaging times of 0.25 to 0.5 s, as if the pressure were uniformly distributed over the whole area. The authors found a more significant component of load from internal pressure than implied by the U.K. wind loading code, but concluded that the pressure coefficients in the U.K. glazing code give reasonable effective peak pressures when used with peak 3 second gust velocities. Fatigue effects, again investigated using a version of the stress corrosion theory, were considered adequately covered by the method of determining glass strength by loading to failure with steadily increasing pressure over a period of about one minute. Wind-induced vibrations, it was concluded, would not be a problem unless the natural frequency of the glass were less than 6 Hz.

Minor and Beason began research in 1970 into window glass failures caused by wind-borne missiles, to which they attribute a significant proportion of wind-induced glass breakage [6], then turned to more general studies of the response of window glass to wind effects. In a recent paper [7] they suggest that a constant pressure applied for the duration of the fastest-mile gust may not account for the sensitivity of glass strength to load duration. They have expressed reservations about the adequacy of tests forming the basis for design charts in wide use in the U.S.A., and advocate additional testing, analytical and experimental stress-field determinations in glass plates under load, and new analytical treatments of wind loads.

3. GLASS STRENGTH AND THE STRESS CORROSION THEORY

Considering the large amounts of glass used in buildings, surprisingly little information is available in the open literature on the strength of window-size glass panels to resist lateral loads [8-10]. Small glass rods and microscope slides, however, have been tested exhaustively to investigate the effects of duration and time variation of tensile stresses in bending on the breaking strength of the sample, and various theories have been advanced to explain the phenomena of static and dynamic fatigue [11-13].

Static fatigue refers to the progressive reduction in glass strength as the duration of a constant load producing tensile stresses is increased. In contrast to the high-cycle fatigue observed in some ductile materials such as steel, for glass the fatigue seems related only to the total duration of the load at the stress level in question, and not to the number of stress cycles. Dynamic fatigue seems to be the same phenomenon as static fatigue except that the loading is varied continuously, usually linearly increasing until failure occurs.

The stress corrosion theory states that the observed strength of glass is governed by tiny flaws on the surface in tension, which act as stress-raisers that grow progressively more severe during any period when the stresses are above some threshold level. The stress concentration factor at a flaw, of order 100, varies as the ratio of flaw depth to square root of flaw tip radius. When the flaw is in a region of high stress, flaw depth increases at a rate proportional to some large power of stress. Since the flaw tip radius is thought to remain essentially constant during this period of rapid flaw-depth growth, the stress concentration factor increases until the effective stress at the flaw is sufficient to cause sudden brittle failure.

More than ten years ago, W.G. Brown [14,15] undertook the task of reconciling apparent differences in the results of the few engineering strength tests then available of large glass plates [8-10]. To take care of the time variation of loading, he adopted the following simplified form of the stress-corrosion theory, as a cumulative damage criterion by which a time-varying loading of any sort could be equated to a simpler one, such as a constant load for a given period of time:

$$K = \int_0^T [p(t)]^s \, dt \qquad (1)$$

K = the damage caused by an applied load for a given glass panel surface condition (and edge condition if the edges are under stress in the test), size, aspect ratio, boundary stiffness and restraint and glass composition.

T = duration of load, s.

p(t) = load, varying as a function of time.

s = an exponent, assumed to range from 10 to 20 depending on many of the same factors affecting K.

In the case of heat-strengthened or fully-tempered glass, the surface compressive stress induced by the tempering process ranks with surface condition as the other major determinant of strength.

4. PEAK WIND PRESSURE COEFFICIENTS AND GLASS DESIGN

Design wind loads for cladding are generally determined in two stages, in recognition of the existence of two distinctly different scales of variation. First, a basic design pressure averaged over a period of at least 10 minutes, but usually equal to or less than an hour, is obtained from climatological records for the site in question, such that the probability of exceedance per year is acceptably low. Next, the smaller scale variations, called turbulence, are accounted for by multiplying by a gust effect factor, or alternatively, this factor can be incorporated with the external pressure coefficient which relates surface pressure on the building to the basic wind pressure at a suitable reference position, such as the roof height of the building. The peak pressure coefficient, with the gust effect factor already included, is a convenient starting point for developing a special pressure coefficient adapted for representing the cumulative effect of turbulent wind on glass. The peak pressure coefficient, $C_{\hat{p}}$ will be defined in terms of the mean and standard deviation pressure coefficients, and a peak factor:

$$C_{\hat{p}} = C_m + gC_{p'} \qquad (2)$$

$$C_m = [\bar{p} - p_o]/[0.5\ \rho v^2] \qquad (3)$$

$$C_{p'} = p'/[0.5\ \rho v^2] \qquad (4)$$

$$\bar{p} = \sum_{r=1}^{n} p_r/n \qquad (5)$$

where: p_o = internal "static" reference pressure, Pa
ρ = mass density of outside air, kg/m^3
v = mean reference speed (e.g., at roof height) m/s

$$p' = \left(\sum_{r=1}^{n} [p_r - \bar{p}]^2/[n[n-1]]\right)^{0.5} \qquad (6)$$

where: p_r = r^{th} reading of a digitally sampled pressure record, Pa
g = peak factor

5. PEAK FACTOR DERIVED FROM PRESSURE SPIKES

The most difficult parameter to specify is the peak factor, g, because it must account for the randomness inherent in the pressure fluctuations. The task is eased if the probability distribution of sequentially sampled pressures is Gaussian, as found on the windward side of a well-exposed building [16]. Unfortunately, on walls in the wake, or on walls nearly parallel to the wind direction, strong "spikes" of (usually negative) pressure occur at random intervals and tend to distort the tail of the probability distribution [17]. To cope with this problem, a study was made of pressure spikes, defined as excursions that are at least two standard deviations away from the mean and separated from adjacent spikes by returns toward the mean of at least two standard deviations.

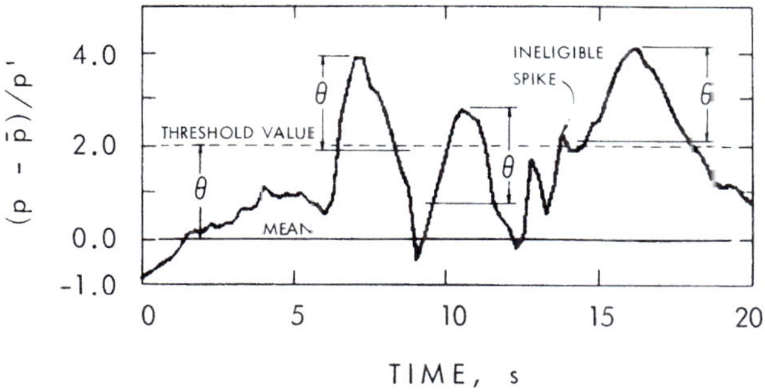

Fig. 3 TYPICAL TIME HISTORY WITH THREE SPIKES

Figure 3 shows a 20 s segment of a pressure record containing three spikes, from a tall office building in Toronto (Figure 2). The readings have been normalized by subtracting the mean value and then dividing by the standard deviation. The exponential distribution was found to describe pressure spikes reasonably well for a wide range of pressure tap locations and conditions, both for full- scale and for wind tunnel measurements [18] as shown in Figure 4. One useful result was that the most probable value for the peak factor was given by the following expression:

$$\tilde{g} = 2 + [\bar{g} - 2]\log_e n \qquad (7)$$

$\tilde{g}$ = the mode of the (non-dimensional) magnitude of the largest spike for a specified storm duration.

$\bar{g}$ = the mean magnitude of all pressure spikes during the storm period, assumed to apply to all such storm periods.

n = the average number of pressure spikes occurring in the specified duration.

Typical values for $\bar{g}$ were found to range from 2.3 to 3.3; for 5-minute periods, n ranged from 2 to 10. It is expected that n will increase with wind speed; the values stated here are for wind speeds of 20 to 25 ms^{-1}. The ratio of spike value in standard deviations to duration in seconds averaged for 80 of the largest positive-going spikes was 0.8 s^{-1}.

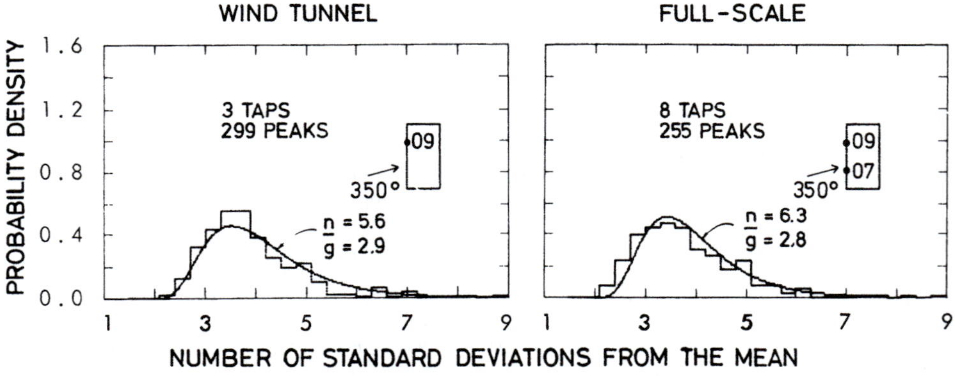

Fig. 4 PEAK PRESSURE HISTOGRAMS AND EXPONENTIAL CURVES

6.0 PEAK PRESSURE COEFFICIENT MODIFIED FOR GLASS DESIGN

The discovery that the exponential distribution gives a reasonable description of a population of pressure spikes, the largest of which is the peak for any given record, leads to a simple method for estimating their contributions to cumulative "weakening" of glass panels. Estimates of the most probable magnitudes of all n spikes are given by the following expression, the derivation of which is given in Appendix A:

$$\tilde{g}_r = 2 + [\bar{g} - 2]\log_e(n/[n-r+1]) \qquad (8)$$

where: $\tilde{g}_r$ = mode of rth smallest spike.

The average number of spikes per interval, n, depends on the length of the interval, the threshold value, θ, and the character of the pressure fluctuations at a particular spot on the building. For sampling intervals of five minutes, a threshold value of two standard deviations (the number 2 in Eqs 7 and 8) was found to produce at least one, but no more than about 10, spikes in most cases. Although somewhat arbitrary, this choice served to restrict attention to a conveniently small sub-set, including any excursions of a "non-Gaussian" nature.

The contribution of n trapezoidal loading pulses, K_e, each pulse consisting of the mean value plus a right triangular peak, can be summed as follows:

$$K_e = (q_{ref}C_m)^s(\sum_{r=1}^{n}[1 + g_r I_p]^{s+1} - n)/(dI_p[s + 1]) \qquad (9)$$

q_{ref} = basic reference pressure = $0.5 \cdot \rho \cdot v^2$, Pa

$I_p = C_p'/C_m$

d = spike value in standard deviations / duration s^{-1}

All other terms are as defined earlier. Equation (9) is derived in detail in Appendix B.

If the pressure record happens to be truly Gaussian, the cumulative damage function, K_g, can be computed as a function of q_{ref}, s, C_m and C_p' as follows:

$$K_g = (q_{ref}C_m)^s(1 + \sum_{r=1}^{s/2} {}_sC_{2r}[I_p^2/2]^r[2r]!/r!)T \qquad (10)$$

T = storm duration, s

${}_sC_{2r}$= the binomial coefficient

The derivation of Equation (10) is given in Appendix C, where certain simplifications are justified, such as taking s to be an even integer and neglecting a correction term.

Both K_e and K_g can be expected to underestimate the actual damage function, but for different reasons. K_e estimates the contributions of the n largest spikes but ignores the rest of the record, whereas K_g accounts for the whole record but may underestimate the largest spikes. The sum of the two taken to the $1/s^{th}$ power will be used to define a glass design coefficient, C_k, which will be slightly too large (no more than 5.9 per cent for s = 12).

A glass design coefficient, C_k, will therefore be defined to provide the magnitude of that constant, one-minute load that produces approximately the same value of the damage function (Eq. 1) as the entire storm characterized by C_m, C_p', g, n, T and d. Where charts or tables call for a constant one-minute load for glass design, the design wind load, P_d, is:

$$P_d = C_k q_{ref} = [(K_e + K_g)/60]^{1/s} \qquad (11)$$

The coefficient C_k matches the predicted load history during a wind storm with function of Equation (1) is in a sense a "characteristic time-to-failure" in seconds for a particular type of loading, and in Equation (11), the characteristic time for the actual storm, $(K_e + K_g)$ is related to the time for the glass strength referred to in a design chart (60 seconds).

7. C_k BY DIRECT INTEGRATION OF PRESSURE RECORDS

The modified pressure coefficient, C_K, is intended to serve as an estimate of the loading that best represents the effect of a particular period of fluctuating wind pressure on a glass panel. Depending as it does on five statistical parameters of the record and its length, C_k is in itself a statistical parameter that can be evaluated by direct integration of several similar pressure records using Equation (1) to define the necessary weighting function.

Pressure records from four sensors mounted on the 57-storey office tower shown in Figure 2 were integrated in this way as a check on the applicability of the assumptions made to define C_k. The records, referred to as Item "a" in Ref. 18, consisted of 27 blocks of 6000 samples, taken at a rate of 20 per second.

The integrations were performed 6 times for each sensor using two different values for s, (12 and 16) and three values for the mean pressure coefficient, C_m. The standard deviation pressure coefficient, $C_{p'}$, was set at 0.5. All four sensors were located on the windward side of the building, aligned vertically and on levels ranging from the 25th to the 50th floor. As their characteristics were reasonably similar, the records from the four sensors could have been combined but they have been treated separately to demonstrate the sensitivity of C_k to small variations in g (the average spike magnitude) and n_{300} (the number of spikes in the 300 s interval). The wind speed, at a reference height 46 m above the roof, was 23 ms^{-1}.

The results for the four sensors are given in Table I along with the values of g and n_{300} that were computed for Ref. 18. The observed C_k in columns 6 and 9 are averages for 27 record segments; coefficients of variation range from 11 to 22 per cent for C_m = 0.5, from 7 to 17 per cent for C_m = 1.0, and from 2 to 9 per cent for C_m = 2.5. Variance increases with s and g.

Table I
COMPARISON OF OBSERVED AND CALCULATED GLASS DESIGN COEFFICIENT, C_k

$\bar{g}$	n_{300}	$\tilde{\hat{g}}$	$\bar{\hat{g}}_{obs}$	C_m	s = 12 C_k obsv	 C_k calc	 calc/obsv	s = 16 C_k obsv	 C_k calc	 calc/obsv
2.58	5	2.93	3.20	0.5	1.47	1.56	1.06	1.57	1.68	1.07
				1.0	1.89	1.98	1.05	1.99	2.08	1.05
				2.5	3.32	3.42	1.03	3.34	3.44	1.03
2.65	8	3.35	3.57	0.5	1.58	1.60	1.01	1.68	1.71	1.02
				1.0	2.00	2.03	1.02	2.10	2.12	1.01
				2.5	3.40	3.47	1.02	3.45	3.49	1.01
2.92	6	3.65	4.01	0.5	1.73	1.65	0.95	1.85	1.76	0.95
				1.0	2.14	2.07	0.97	2.26	2.17	0.96
				2.5	3.50	3.48	0.99	3.58	3.52	0.98
2.95	8	3.98	4.22	0.5	1.78	1.73	0.97	1.91	1.85	0.97
				1.0	2.18	2.15	0.99	2.32	2.26	0.97
				2.5	3.52	3.54	1.01	3.62	3.59	0.99

Note: $C_{p'}$ = 0.5; d = 0.8 s^{-1}; T = 300 s

8. COMPARISON OF C_k AND C_p

Table II contains the results of a parameter study to determine the effect of using C_k instead of C_p for glass design for a one-hour storm. For the first three cases, for which the peak factor, g, = 3.0, C_k is approximately equal to C_p; even if the spike contributions were entirely omitted, C_k would stay within 1 per cent of the value given by Equation 11. For the last three cases (with peak factor 7.0) the situation is reversed; if the Gaussian contribution is omitted, C_k based on spike contributions alone is within 1 per cent of the defined value, and C_k is 20 to 25 per cent lower than the peak factor, C_p.

The first three cases correspond to situations in which the assumption of a Gaussian probability distribution for the pressure record is justified, and the last three are cases in which there is a strong tendency for non-Gaussian "spikes" to occur. The former show little difference between the peak factor and the C_k specially derived for glass design, but the latter suggest that when the peaks are extremely high, a substantial reduction may be in order.

Table II

GLASS DESIGN COEFFICIENT, C_k, COMPARED TO PEAK PRESSURE COEFFIENT, C_p'

					s = 12				s = 16			
$\bar{g}$	n_{300}	$\tilde{\bar{g}}$	C_m	$C_{\hat{p}}$	C_k	$C_k/C_{\hat{p}}$	% C_k with only K_e	% C_k with only K_g	C_k	$C_k/C_{\hat{p}}$	% C_k with only K_e	% C_k with only K_g
2.32	2	3.0	0.5	2.0	1.90	0.94	77	100	1.95	0.97	78	100
			1.0	2.5	2.41	0.96	79	100	2.41	0.96	81	100
			2.5	4.0	4.16	1.04	79	99	3.98	0.99	84	100
2.52	4	4.0	0.5	2.5	1.96	0.78	92	96	2.00	0.80	94	97
			1.0	3.0	2.47	0.82	91	97	2.47	0.82	94	97
			2.5	4.5	4.22	0.94	87	98	4.04	0.90	92	98
2.70	6	5.0	0.5	3.0	2.18	0.73	98	86	2.27	0.76	99	86
			1.0	3.5	2.66	0.76	97	90	2.72	0.78	99	89
			2.5	5.0	4.33	0.87	93	96	4.20	0.84	97	94
2.88	8	6.0	0.5	3.5	2.54	0.72	100	74	2.66	0.76	100	73
			1.0	4.0	2.98	0.74	99	80	3.09	0.77	100	78
			2.5	5.5	4.54	0.82	97	91	4.49	0.81	99	88
3.04	10	7.0	0.5	4.0	2.90	0.73	100	65	3.05	0.76	100	64
			1.0	4.5	3.34	0.74	100	72	3.47	0.77	100	69
			2.5	6.0	4.81	0.80	99	86	4.83	0.81	100	82

Note: $C_m = 0.5$; $d = 0.8\ s^{-1}$; $T = 3600$ s

9. CODES AND STANDARDS

If provision were to be made in codes and standards to take acount of the dependence of glass strength on loading rate, it would first be necessary to determine what wind storm characteristics are implied by the existing requirements. The National Building Code of Canada [19,20] requires a gust effect factor of 2.5 to be used for cladding; for a 50 m building, the combined gust factor and peak pressure coefficient for the worst case loading (assuming broken windows on the windward wall) could be as large as 5.25 (negative).

Referring to Table II for $C_m = 2$, $C_{p'} = 0.5$, such a value for C_p would imply a peak factor between 6 and 7. C_k used instead of C_p for glass design would mean a reduction of 20 per cent, to 4.2. This one example is not intended to be an argument for reducing wind loads in this particular code, but it does suggest that a closer study is required of those wind-storm characteristics of special importance in assessing the strength of glazing. More attention should be given to determining "effective" durations of peak pressures, and the average duration of wind storms. Storm duration emerges as an important parameter, as can be seen by comparing values of C_k from Table I (for 5-minute periods) to those from Table II (for 60-minute periods).

10. CONCLUDING REMARKS

Accounting for static fatigue through the cumulative damage criterion must be approached with caution, and it seems advisable to perform corroboratory tests on window-size panels. The stress corrosion theory suggests that some recovery of strength occurs during sufficiently long intervals between applications of high stress levels, a so-called "healing" phenomenon. It does not seem reasonable, therefore, to sum the damage over more than one storm. Research into the rate at which flaws in window-size panels are healed by ordinary corrosion (not under high stress) would be useful. The strength of glass in service as it ages has not yet been satisfactorily documented in the open literature.

When incidents of glass breakage occur in which wind is at least partly responsible, it is often difficult to rule out other factors, such as prior damage due to thermal stresses, or prestressing due to various combinations of thermal effects and improper glazing procedures. An accurate assessment of wind effects is required to facilitate investigation of the other factors, as well as to provide a sound basis for the specification of wind loads for glazing design.

11. ACKNOWLEDGEMENTS

The author wishes to thank his colleagues, particularly William R. Schriever and Joseph T. Templin, for their comments. This paper is a contribution of the Division of Building Research of the National Research Council of Canada and is published with the approval of the Director of the Division.

12. REFERENCES

1. Glass Product Recommendations: Wind Load Performance. Glass Division, PPG Technical Services, Report No. 101A, PPG Industries Inc., One Gateway Center, Pittsburgh, PA 15222. 1975

2. Glass for Construction, Libbey-Owens-Ford Company, 811 Madison Avenue, Toledo, Ohio 43675. 1976

3. PPG Glass Thickness Recommendations to Meet Architects' Specified One-Minute Wind Load. Technical Services/Flat Glass Division, PPG Industries, Inc., One Gateway Center, Pittsburgh, PA 15222. 1979

4. Ishizaki, H., Miyoshi, S., and Miura, T. On the Design of Glass Panes against Wind Loading. Proc., 4th International Conference, Wind Effects on Buildings and Structures, p. 655-61, Cambridge University Press. 1975

5. Mayne, J.R. and Walker, G.R. The Response of Glazing to Wind Pressure. Building Research Establishment Current Paper 44/76 Garston Watford WD27JR

6. Minor, J.E. and Beason, W.L. Window Glass Failures in Windstorms, Jnl. Struct. Div. ASCE Vol. 102 No. ST1 Paper 11834, p. 147-60. 1976

7. Beason, W.L. and Minor, J.E. Response of Window Glass to Wind Loads. Third U.S. Nat. Conference, Wind Engineering Research, Paper III-9, 1. 1-4 University of Florida, Gainesville. 1978

8. Orr, L. Engineering Properties of Glass. Conference on Windows and Glass in the Exterior of Buildings, Building Res. Inst. Pub. 478, Wash. D.C. 1957

9. Bowles, R. and Sugarman, B. The Strength and Deflection Characteristics of Large Rectangular Glass Panels Under Uniform Pressure. Glass Technology, Vol. 3 No. 5 p. 156-70. 1962

10. Hershey, R.L. and Higgins, T.H. Statistical Prediction Model for Glass Breakage from Nominal Sonic Boom Loads. Booz-Allen Applied Research Inc. NTIS-AD-763-594, Bethesda, MD. 1973

11. Hillig, W.B. Sources of Weakness and the Ultimate Strength of Brittle Amorphous Solids. Modern Aspects of the Vitreous State, Vol. 2, p. 152-94, edited by J.D. MacKenzie, Butterworths, London. 1962

12. Charles, R.J. Static Fatigue of Glass II. Journal of Applied Physics, Vol. 29 No. 11 p. 1554-60. 1958

13. Wiederhorn, S.M. Influence of Water Vapor on Crack Propagation in Soda-Lime Glass. Jnl. Amer. Cer. Soc. Vol. 50 No. 8 p. 407-14. 1967

14. Brown, W.G. A Practicable Formulation for the Strength of Glass and its Special Application to Large Plates, National Research Council of Canada, Division of Building Research, NRCC 14372, Ottawa, Canada. 1974

15. Brown, W.G. A Load Duration Theory for Glass Design, National Research Council of Canada, Division of Building Res., NRCC 12354, Ottawa. 1972

16. Dalgliesh, W.A. Statistical Treatment of Peak Gusts on Cladding, Jnl. Struct. Div. ASCE Vol. 97 No. ST9 p. 2173-87. 1971

17. Peterka, J.A. and Cermak, J.E. Wind Pressures on Buildings - Probability Densities. Jnl. Struct. Div. ASCE Vol. 101 No. ST6 p. 1255-67. 1975

18. Dalgliesh, W.A., Templin, J.T., Cooper, K.R. Comparison of Wind Tunnel and Full-Scale Building Surface Pressures with Emphasis on Peaks, Fifth Int. Conference on Wind Engineering, Colorado State University. 1979

19. National Research Council of Canada, Associate Committee on the National Building Code. The National Building Code of Canada 1977, NRCC 15555.

20. Ibid, Commentaries on Part 4 of the National Building Code of Canada 1977 Supplement No. 4 to the National Building Code, NRCC 15558, Ottawa.

APPENDIX A: MODE OF rth SMALLEST VALUE: EXPONENTIAL DISTRIBUTION

The probability density function* of the rth smallest value, x_r', among n independent random variables each having a standard exponential distribution is:

$$Px_r'(x) = \frac{n!}{(r-1)!(n-r)!} \exp[-(n-r+1)x] \cdot \left(1 - e^{-x}\right)^{r-1} \qquad \text{(A1)}$$

(where x is positive)

Let $$f(x) = \exp[-(n-r+1)x] \cdot \left(1 - e^{-x}\right)^{r-1} \qquad \text{(A2)}$$

Then the value of x for which Px_r' reaches a maximum is given by the solution to:

$$\frac{df(x)}{dx} = 0 \qquad \text{(A3)}$$

Thus the mode $\tilde{x}_r$ is given by:

$$\tilde{x}_r = \log_e \left| n/(n-r+1) \right| \qquad \text{(A4)}$$

The distribution of non-dimensionalized spike magnitude is put into standard exponential form as follows:

$$x_r' = (g_r - 2)/(\bar{g} - 2) \qquad \text{(A5)}$$

The mode, or most probable value for the rth smallest spike magnitude, is therefore obtained from Eqs. (A4) and (A5) as follows:

$$\tilde{g}_r = 2 + (\bar{g} - 2) \log_e \left| n/(n-r+1) \right| \qquad \text{(A6)}$$

which is Eq. (8). As a special case, the largest spike, or the peak, is given by Eq. (A6) for r = n:

$$\tilde{\bar{g}} = 2 + (\bar{g} - 2) \log_e n \qquad \text{(A7)}$$

which is Eq. (7).

---

* Johnson, N.I. and Kotz, S. Continuous Univariate Distributions - 1 Houghton Mifflin Co., Boston, 1970. p. 214.

APPENDIX B: CUMULATIVE DAMAGE CRITERION FOR n TRAPEZOIDAL SPIKES

Each spike is approximated as a rectangle of height $q_{ref} \cdot C_m$ (pascals) and base T_r (seconds) surmounted by a right triangle, itself of height $q_{ref} \cdot C_p' \cdot \tilde{g}_r$ (pascals). The resulting trapezoids represent n loading functions of time that can be inserted into Eq. (1) as follows:

$$K_{e_r} = \int_0^{T_r} [q_{ref}(C_m + C_p' \tilde{g}_r t/T_r)]^s \, dt \qquad \text{(B1)}$$

K_{e_r} = damage contribution of rth spike

Making the assumption that the base time, T_r, is related to spike height by having a common slope d, for all spikes, and making the substitution $I_p = C_p'/C_m$, Eq. (B1) becomes:

$$K_{e_r} = (q_{ref}\, C_m)^s \int_0^{\frac{\tilde{g}_r}{d}} [1 + dI_p\, t]^s \, dt \qquad \text{(B2)}$$

Evaluating the integral,

$$K_{e_r} = \frac{(q_{ref}\, C_m)^s}{s+1} \left([1 + \tilde{g}_r I_p]^{s+1} - 1 \right) \Big/ (dI_p) \qquad \text{(B3)}$$

The damage contribution summed for all n spikes is:

$$K_e = \frac{(q_{ref}\, C_m)^s}{s+1} \left[\sum_{r=1}^{n} (1 + \tilde{g}_r I_p)^{s+1} - n \right] \Big/ (dI_p) \qquad \text{(B4)}$$

which is Eq. (9).

APPENDIX C: CUMULATIVE DAMAGE CONTRIBUTION FOR A FLUCTUATING PRESSURE LOADING DESCRIBED BY A GAUSSIAN PROBABILITY DISTRIBUTION

The time series of varying pressures can be represented by a collection of sample values, each representing the magnitude, or height, of a rectangle of load of duration dt. If all rectangles of a given height are collected together, the sub-totals of times spent at each height form a histogram which should follow the Gaussian probability density function. The damage summation according to Eq. (1) can then be approximated as the rth moment of a linear function of normalized pressure:

$$K_g \approx \left[q_{ref}\ C_m\right]^s \left(\frac{1}{\sqrt{2\pi}} \int_{-1/I_p}^{\infty} (1 + I_p\ x)^s\ e^{-x^2/2}\ dx \right) T \qquad \text{(C1)}$$

$$I_p = C'_p / C_m \qquad\qquad x = (p_r - \bar{p})/p'$$

The lower limit of integration of the probability density function is necessary to avoid counting loadings acting in the opposite direction, i.e., when the load changes sign.

Because the dependence on the exact value of s is not very great, it can be rounded to the nearest even integer. The integral can then be approximated by applying the following relation for the $2r^{th}$ moment of a Gaussian distribution*:

$$\mu_{2r} = \frac{1}{\sqrt{2\pi}} \int_{-\infty}^{\infty} x^{2r}\ e^{-x^2/2}\ dx = \frac{(2r)!}{2^r r\ !} \qquad \text{(C2)}$$

Substituting Eq. (C2) into (C1) and simplifying,

$$K_g \approx \left[q_{ref}\ C_m\right]^s \left[1 + \sum_{r=1}^{s/2} {}_sC_{2r} \left(\frac{I_p^2}{2} \right)^r \frac{(2r)!}{r!} - R \right] T_s \qquad \text{(C3)}$$

$${}_sC_{2r} = \frac{s!}{(s-2r)!\ (2r)!} \qquad \text{(the binomial coefficient)}$$

$$R = \frac{1}{\sqrt{2\pi}} \int_{I/I_p}^{\infty} (1_p\ x-1)^s\ e^{-x^2/2}\ dx$$

The correction factor R turns out to be negligible for $I_p < 10$ and so has been dropped from Eq. (10).

---

* Johnson, N.I. and Kotz, S. Continuous Univariate Distributions -1. Houghton Mifflin Co., Boston, 1970. p. 45.

G3 WIND-INDUCED VIBRATIONS OF TOWERS AND STACKS

Hans Ruscheweyh

Institut für Leichtbau, Aachen Technical University,
Federal Republic of Germany

SUMMARY

The exciting forces due to vortex shedding have been investigated by full-scale measurements up to Reynolds numbers 2×10^7, and it was found that the regular vortex shedding stops after 2 to 4 cycles and restarts again after some seconds. - The correlation length of the exciting forces depends on the nondimensional vibration amplitude and the aspect ratio and the influence of this is estimated for circular cylindrical chimneys. The effective force coefficient is much lower than the local one. - The vibration of two in-line chimneys was investigated in a wind-tunnel test. Vortex excitation occurs at higher reduced velocities and with higher force coefficients. At small relative spacings interference galloping leads to very intense vibration. - A special cross section (cloverleaf) of a steel chimney tested in a wind-tunnel showed galloping-like instabilities which could be decreased by small variation of the geometry of the cross section and by adding strake strips. - Finally, damping measurements on a full-scale steel chimney, a steel mast, and a concrete natural cooling tower shell by means of different measurement techniques are being described.

1. INTRODUCTION

Although much research work has been done in the past to investigate the phenomenon of wind-induced vibrations, the problems are far from being solved. Special group arrangements of slender structures and their modern light-weight design may aggravate such problems substantially. A number of parameters of the flow, the structure, and the surroundings influence the mechanism of the wind-excited force. It is not possible to take all these parameters into account for a safe and rationalized design because of the lack of results of systematic investigations. Consequently, standard-code data for vibration problems are mostly too conservative and lead to uneconomic designs, or the data are not known at all such that certain vibration problems are not mentioned in the standard-code. In a lot of cases, therefore, consulting or ad-hoc investigations are necessary. One of the major problems is that the mechanism of the exciting forces is not well known. It would lead to effective progress in this field, if scientists and engineers would devote their research efforts to the clarification of the exciting flow field.

2. EXCITING WIND FORCES ON CYLINDRICAL STRUCTURES

Cylindrical or slender-cone structures such as stacks and television towers normally operate in the range of large (i.e., transcritical) Reynolds numbers. It is well known that in this range vortex separation becomes more stable and that vortex-excited oscillations will occur. Full scale measurements [1] on a concrete shaft of 10 m diameter (Reynolds number up to 2×10^7) give an impression of the dynamic behavior of the pressure distribution modified by the wind turbulence and the vortex separation (Sketch 1a). Sketch b shows the momentary pressure distribution and its modification by periodic vortex separation. The asymmetric distribution changes periodically from one side to the opposite, producing an alternating lift coefficient $\tilde{C}_{Lo}$ with an amplitude of $\tilde{C}_{Lo} = 0.36$ at a frequency given by the Strouhal number $S = 0.23$ for the case illustrated in Fig 1. This periodic change occurs when the separation point is fixed at φ^* which is an ultimate condition for stable vortex separation. We were able to show from the full-scale measurements that if this condition was not satisfied, that is if the separation point was not at a fixed position, the periodic change of pressure distribution did not occur any longer. This fact is shown in Sketch 1c: From t_1 to t_2 vortex shedding does occur, at t_2 the separation point moves from φ_{1*} to φ_{2*} and the vortex shedding stops. In the pressure distribution as shown in the two records of Sketch 1d, the regularly fluctuating part then vanishes. Some seconds later the vortex shedding starts again. In a continuous record of about 12 hours, many periods of vortex shedding each with two to four cycles were observed. No vortex period was found with more than four cycles! The full-scale measurements described above were taken at a very small relative oscillation amplitude y of the tower (i.e., $y/D < 10^{-3}$). At larger amplitude the locking-in effect tends to increase the stability of the vortex shedding, the number of cycles of regular vortices will increase, and the correlation length along the tower axis will increase also. As shown by Novak and Tanaka [2], the correlation length increases with the increase of the non-dimensional amplitude (see Sketch 2b). These relationships will be used in formulating the dynamic bending moment of a cylindrical stack as follows: Taking the correlation length L' (see Sketch 1a) over which the full aerodynamic forces are acting as a multiple of the diameter D, namely $L' = L \cdot D$ where $L = L'/D$ is the non-dimensional correlation length and further, approximating the tip effect which reduces the vortex shedding at the top of the stack [3] by the length of 2D, the effective moment arm to the base of the stack is $h = H - 2D - 2L'$ and when introducing the aspect ratio $\lambda = H/D$, this becomes $h = D \cdot \lambda (1 - 2(1+L)/\lambda)$.

Assuming that the local lift coefficient $\tilde{C}_{Lo}$ changes much less with the motion than the correlation length and can therefore be assumed constant, the dynamic base bending moment of the stack is

$$\tilde{M}_B = \tilde{C}_{Lo} \frac{\rho}{2} U^2_{crit} \, 2L \, D^3 \, \lambda \left(1 - \frac{2(1+L)}{\lambda}\right) V_{dyn} \tag{1}$$

where V_{dyn} = dynamic response factor and U_{crit} = critical wind speed.

In most of the standard codes, the dynamic-lift coefficient due to vortex shedding is given for the whole length of the stack, according to the formula (see sketch 2c).

$$\tilde{M}_B = \tilde{C}_L \frac{\rho}{2} U^2_{crit} \, D \frac{H^2}{2} V_{dyn} \tag{2}$$

Comparing eqns. (1) and (2), one obtains

$$\tilde{C}_L/\tilde{C}_{Lo} = (4L/\lambda)\,(1 - 2\,(1 + L)/\lambda) \tag{3}$$

In the equation (3) the dependency of the effective dynamic lift coefficient C_L from the aspect ratio $\lambda = H/D$ and the nondimensional correlation length $L = L'/D$ is expressed. The eqn. (3) of course is an approximation of a complicated problem ($\rightarrow$ generalized force), in particular the boundary of the integral for the correlation length is infinite while the stack has a finite length. Assuming that the correlation length will increase only little with spacing ratio $m > 2L$ the validity of eqn. (3) should be limited to $\lambda = 4L+2$. Sketch 2d shows the ratio C_L/C_{Lo} versus λ and one can see that the effective dynamic lift coefficient is only 15% to 45% of the local lift coefficient and that it decreases with increasing λ .

The practical consequence of eqn. (3) is as follows: The problem of calculating vortex-excited oscillation of a cantilever stack may be reduced to determining the local dynamic lift coefficient C_{Lo} at zero amplitude depending on Reynolds number, roughness, turbulence, etc., and to finding the non-dimensional correlation length L.

3. VIBRATION OF TWO IN-LINE CHIMNEYS

Stacks standing in a row have two different kinds of aerodynamic exciting mechanisms: Firstly, there is vortex excitation and, secondly, there is a dynamic instability similar to galloping. Whether the one or the other excitation will occur depends on the relative spacing of the two stacks and the wind direction (see Sketch 3a-e). For vortex excitation three points are of main importance.

i) The largest exciting forces act on the downstream chimney
ii) The aerodynamic coefficient $\tilde{C}_L$ of the latter is nearly twice as hight as the coefficient of the isolated stack.
iii) The reduced velocity U_r corresponding to the maximum coefficient $\tilde{C}_L$ is much larger than for the isolated stack.

The last two points entail an effective increase of vortex excitation by factors of 6.5 and of 3 at relative spacings a/D = 1.5 and 8 respectively (see sketch 3e). The wind direction β for maximum excitation varies from 40° at a/D = 1.5 to 16° at a/D = 8.

At a spacing a/D ≤ 2.5, a special instability occurs produced by two different flow conditions [4,5]. At a wind direction $\beta \simeq 0°$, the downstream stack is completely in the wake of the upstream one (flow condition „1"). At wind angle β = 8° to 10°, the flow suddenly begins to stream through the gap with high velocity (flow condition „2") (see sketch 3b). The pressure distributions of both stacks change dramatically. While the lift disappears at β = 0°, at $\beta \geq 10°$ a very high lift excits on the downstream stack. As an example the lift coefficient $\tilde{C}_L$ of the downstream cylinder is plotted in sketch 3c versus wind direction β for a/D = 1.5. One can see from the diagram that at β = 10° (critical β) $\tilde{C}_L$ jumps from zero to a maximum negative value and then increases continiously with increasing β to the well-known positive value at β = 90°. Turning the model back from the side position (β = 90°) to the length position (β = 0°), a hysteresis occurs between β = 82° and 94° as well as between 9° and 25°. This hysteresis is effected and enhanced by the vibration of the two stacks. If the two stacks vibrate out of phase with respect to each other near the critical angle β the flow changes from condition "1" to "2" and back. This change occurs with a phase shift with respect to the vibration which leads to an energy input to the system (see Fig. 1).

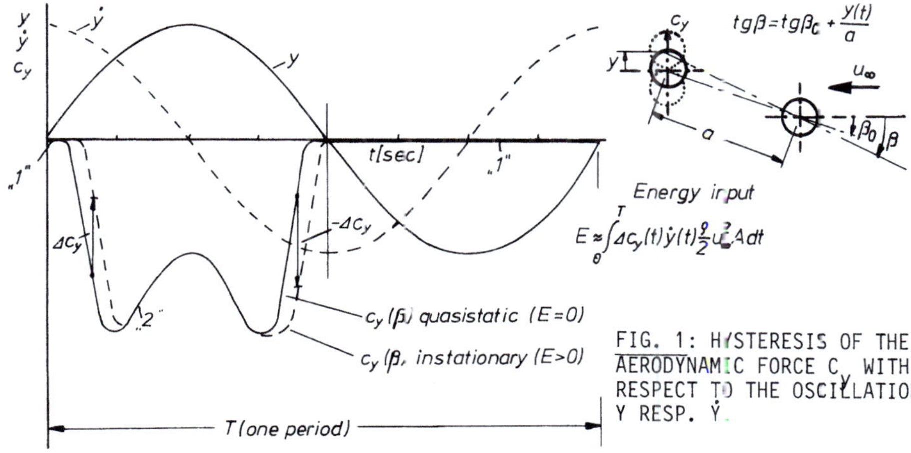

FIG. 1: HYSTERESIS OF THE AERODYNAMIC FORCE C_Y WITH RESPECT TO THE OSCILLATION Y RESP. $\dot{Y}$.

The stack motion increases to large amplitudes, and it increases further with increasing wind speed. Three different kinds of motions have been observed:

i) The upstream stack has relatively small amplitude oscillations, the downstream stack oscillates to very large amplitudes across the wind direction.
ii) The upstream stack vibrates across the wind direction, the downstream stack oscillates in a circle of an ellipse.
iii) Both stacks oscillate across the wind direction.

These vibration modes change with small variation of β in the range of critical angle β (see sketch 3d). The oscillations start at an onset wind speed U_0 and increase with increasing wind speed to extremly large amplitudes. The oscillations continue as the wind speed is reduced and they remain at speeds even lower than U_0. The characteristics near this speed are similar to the galloping phenomenon: There is an onset velocity, the amplitudes increase continuously with wind speed, and a hysteresis can be observed. All these effects are induced by the interference of the two cylinders. One can speak therefore of "interference galloping".

It is important to mention that this interference galloping is no classical "den Hartog"-galloping, because the aerodynamic exciting force is not primarily velocity-dependent but position-dependent. Therefore it is not possible to predict this instability by taking the slope of the static c_y-curve only into account. Dynamic investigations are necessary in this field.

Tests with helical strakes on both stacks at critical positions show the same effects with the only exception of higher onset velocities caused by the aerodynamic damping of the strakes. A successful device to avoid these dangerous instabilities is to interconnect the two stacks at the top or near the top. The relative motion of the two stacks which induces the periodic change from one flow condition to the other will then be prevented.

4. INSTABILITY OF A CLOVER-LEAF CROSS SECTION

A modern development in the steel-stack design is to connect two, three or four smoke flues to the outside of a supporting column. This measure leads to unconventional flow cross-sections.

Sketch 4a shows an example of such a design. It was tested in a wind tunnel by the author. A similar study was mentioned by Scruton [6]. Both vortex excitation and galloping-like instabilities occur as seen from sketch 4b, instabilities being indicated by dotted lines. In the model test, the wind speed had to be limited because of the extremely large amplitudes which developed near these instability points. Therefore, the characteristic of the instability could not be completely clarified. Sketch 4c shows the effective dynamic wind load on the full-scale prototype for two instability points. The logarithmic decrement of damping is assumed to $\delta = 0.07$. (Later on the assumed damping was confirmed by full-scale measurements on the prototype stack, obtaining δ=0.067).

While the vortex excitation is low and can be neglected, the galloping-like instability leads to extremly high dynamic loads. By increasing the combined mass damping parameter or/and the natural frequency of the system, the onset velocity U_{ro} may be increased to values higher than the probable wind speed. In other cases, aerodynamic or other damping devices are recommended.

5. DAMPING MEASUREMENTS ON FULL-SCALE STRUCTURES

Vortex-excited oscillations as well as the galloping depend on the mass-damping-parameter $2M\delta/(\rho D^2)$ in which M is the equivalent mass and δ the logarithmic decrement of damping. For the prediction of the vibration amplitude, a realistic estimation of damping is necessary. The damping depends on material hysteresis, stress level, structural design, connecting links, supporting elements, foundation, etc. and their influences are so complicated that a useful theoretical solution is not possible. Therefore, damping measurements on full-scale structures are necessary in order to collect empirical damping values of different structures.

Three measurement techniques may be distinguished:

i) Man-induced vibration; damping from decay curve of oscillation.

ii) System initially stressed by special devices (cables) and released; damping from decay curve of oscillation.

iii) Use of a random input force ($\rightarrow$ white noise) in the frequency range of natural frequency of the system; damping derived from the autocorrelation function.

Most of the structures exhibit increased damping with increasing vibration amplitude [7]. Therefore it is obvious that the damping-test amplitudes must be as large as possible. Method (i) permits only large amplitudes at very low-damped systems and is successfully applied to steel stacks without lining or steel antenna masts, etc. [8,9]. Sketches 5a,b show two examples: One is a four-flue steel stack with inner heat isolation. The damping of this type of stack increases rapidly with the amplitude. The second example is a welded conical steel antenna mast without any additional elements which could produce damping. Only the steel-material and the aerodynamic characteristics specify the damping in this case, and one can see in sketch 5c that the logarithmic decrement ist nearly constant up to nondimensional top amplitudes of η = 0.5 for the fundamental mode as well as for the first mode.

As an example for the second way of exciting a structure, a full-scale test on a natural-draught cooling tower is presented here. The concrete shell, shown in sketch 6a, can vibrate in different modes with different combinations of circumferential and meridian wave numbers [10]. A theoretical calculation of the natural frequencies is given in Sketch 6c. Because of the proximity of the frequencies for different modes it is difficult to assign the measured frequency to the correct vibration mode. Therefore, the measurements must include the measurement of the vibration modes. Steel bracing wires were used across the cooling-tower mouth (diameter 74 m, height 150 m) to produce three different modes: Diagonal to prestrain the mode n = 2, in an equilateral triangle to prestrain the mode n = 3, and in a square to prestrain the mode n = 4 (see sketch 6b). The wire tensions were 20 - 50 KN when the breaking shaft failed. Then the cooling-tower shell vibration comes into a transient state with the dominant frequency of the prescribed mode. Seven modes could be analysed (see sketch 6c), but because of disturbing influences of the initial shock of the breaking shaft, the vibration energy changed from one mode to another. Hence, the damping could not be analysed this way.

In a second test the correlation method could be applied successfully to measure the damping of the lowest frequency (mode n = 3, m = 1) of f_e = 1.27 Hz. The decay ratio of the autocorrelation function is shown in sketch 5d and the logarithmic decrement of damping is $\delta_{3,1}$ = 0.052 for the very small amplitude at the tower top of 0.1 mm.

6. REFERENCES

1. Ruscheweyh, H., "Statische und dynamische Windkräfte an kreiszylindrischen Bauwerken", Forschungsbericht NRW 2685, 1977, Westdeutscher Verlag.
2. Novak, M. - Tanaka, H., "Pressure Correlation on a Vibrating Cylinder" Proc. Wind Effects on Buildings and Structures, London 1975.
3. Ruscheweyh, H., "Tip Effect on Vortex-Excited Oscillation of a Model Stack with and without Efflux Stream. Flow-Induced Structural Vibrations", Springer-Verlag 1973.
4. Cooper, K.R. - Wardlaw, R.L., "Aeroelastic Instabilities in Wakes", Proc. Wind Effects on Buildings and Structures, Tokyo 1971.
5. Ruscheweyh, H., "Winderregte Schwingungen zweier engstehender Kamine", 3. Coll. Fachhochschule Aachen 1978.
6. Scruton, C., "Aerodynamics of Structures", Proc. Wind Effects on Buildings and Structures, Ottawa 1967.
7. Wootton, L.R., "The Estimation of the Structural Damping of Civil Engineering Structures", Atkin Research and Development 1971.
8. Hirsch, G., "Dämpfungsmaßnahmen zur Herabminderung winderregter Schwingungen schlanker Bauwerke", Haus der Technik-Veröffentlichungen Nr. 347, Essen, Vulkan-Verlag 1975.
9. Ruscheweyh, H., "Dynamische Torsionsbelastung von Hochhäusern unter Interferenzeinfluß und Dämpfungsmessungen an Originalbauwerken", DFG-Kolloquium Karlsruhe 1978, Institut für Beton und Stahlbetonbau, Universität Karlsruhe
10. Müller, F., "Zur Berechnung des Schwingungsverhaltens windbeanspruchter Kühltürme" DFG-Kolloquium, Mitt. d. Instituts f. Massivbau, Nr. 26, TH-Darmstadt 1977.

CHART 1: TELEVISION TOWER

(Full-scale measurements)

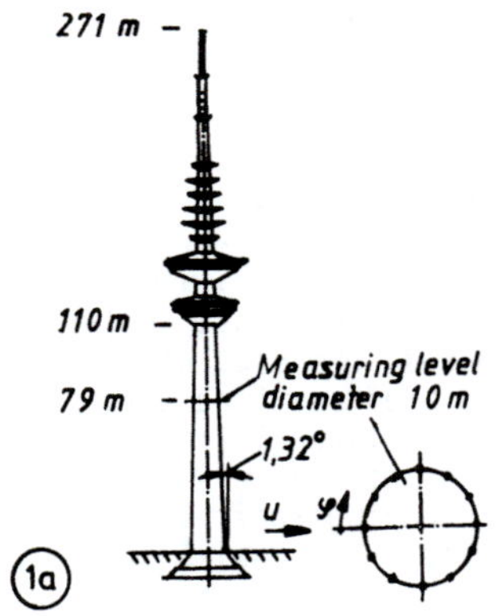

Problem: Regular exciting forces due to vortex shedding at transcritical Reynolds numbers and under turbulent wind condition. Stability of these regular forces.

Geometry: See sketch (1a). Concrete shaft of a Television Tower with small cone angle. Relative roughness k/D = 4 x 10^{-4}.

Boundary condition: City of Hamburg with surrounding buildings of a height of 20 m.

Flow condition: Natural wind over a city. Wind is measured by analysing the pressure distribution of the stagnation-point region of the tower shaft itself (⟶ cylindrical probe). The turbulence intensity is 12 % to 17 %.

Stability of vortex shedding: Vortex shedding occurs and produces asymmetric pressure distribution with high momentary lift forces. Stability of vortex shedding depends on constant separation point φ^* (see sketch 1b). Strouhal number S ≈ 0,23.
Vortex shedding stops if separation point moves from φ_1^* to φ_2^* (see sketch 1c).

Restart of regular vortex shedding after some seconds (see sketch 1d). No vortex period with more than 4 cycles!

Consequences: Effective vortex-shedding exciting forces are smaller than the maximum momentary local values.

References: [1] H. Ruscheweyh, Statische und dynamische Windkräfte an kreiszylindrischen Bauwerken. Forschungsbericht NRW 2685, 1977, Westdeutscher Verlag.

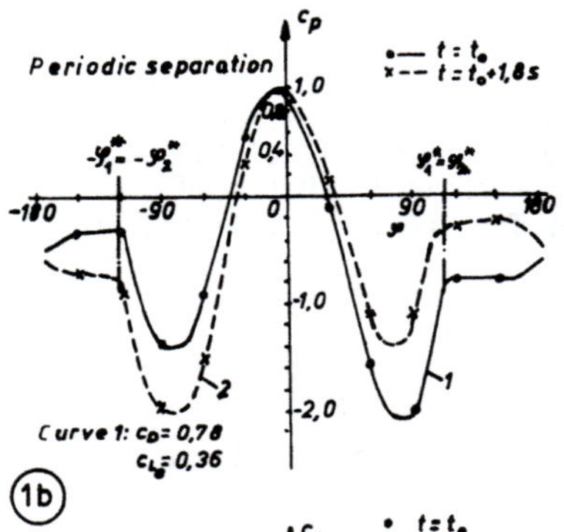

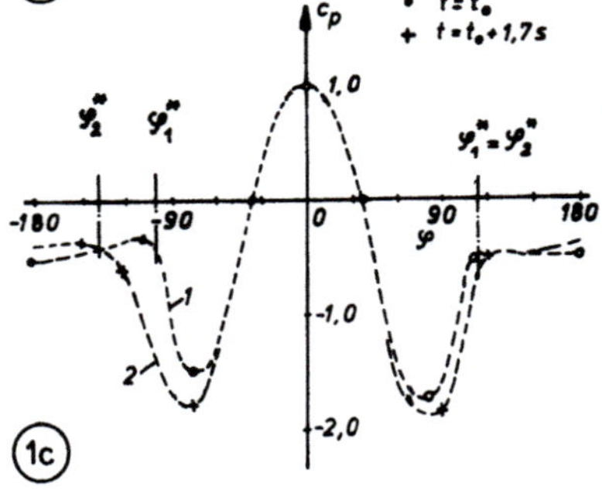

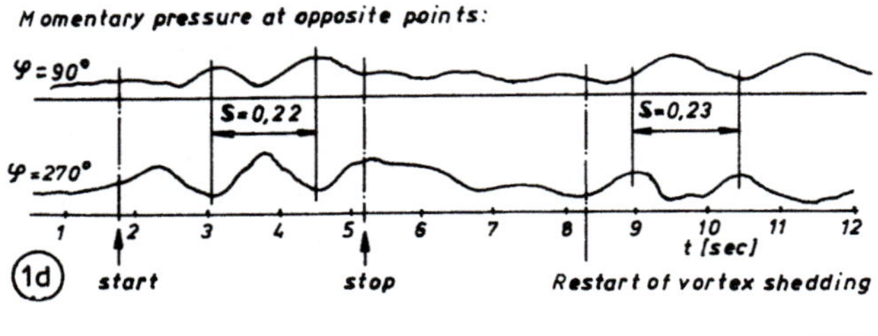

CHART 2: STACKS AND MASTS

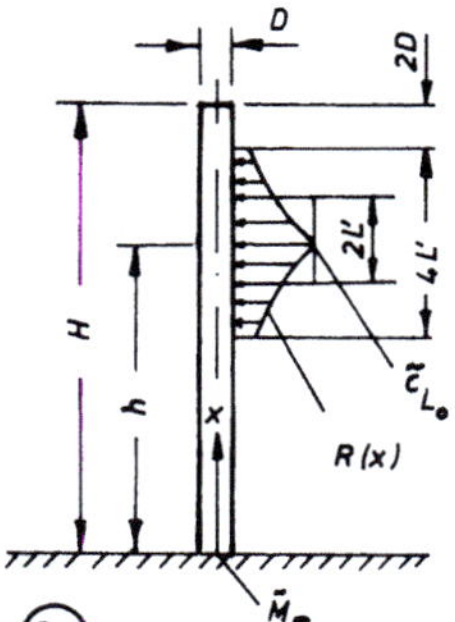

(2a)

Stack:
Aspect ratio $\lambda = \frac{H}{D}$

Correlation length of exciting force:

$$L' = \int_0^\infty R(x)\,dx = L \cdot D$$

Moment arm for base moment M:

$$h = D\,\lambda\left(1 - \frac{2(1+L)}{\lambda}\right)$$

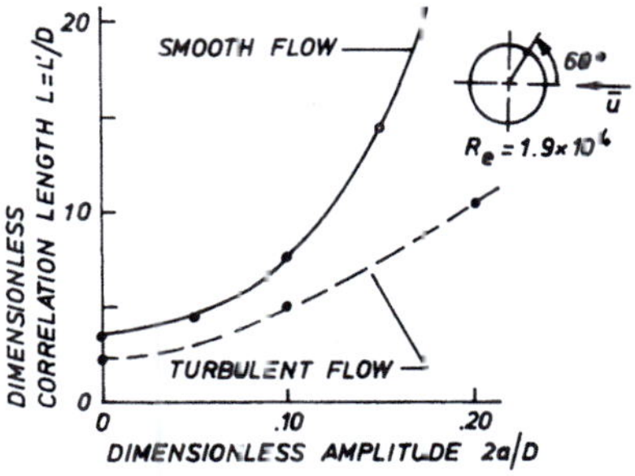

(2b)

Dynamic base moment $\tilde{M}_B = \tilde{C}_{L_0} \frac{\rho}{2} U^2_{crit} D \cdot 2 \cdot L' \cdot h \cdot V_{dyn}$ (1)

Problem: Vortex-excited oscillations of stacks. Prediction of resonance amplitude. Exciting force is influenced by the aspect ratio λ and the correlation length L: Estimation of the integral value $\tilde{C}_L$ for standard-code formulas.

Geometry: Circular cylindrical cantilever stacks or masts with aspect ratios $\lambda \geq 10$.

Flow conditions: All Reynolds number ranges, turbulent and smooth flow. Regular vortex shedding depressed at tip because of the tip effect.

Vibration: Vibration in fundamental mode at critical wind speed. Resonance factor $V_{dyn} = \pi / \delta$

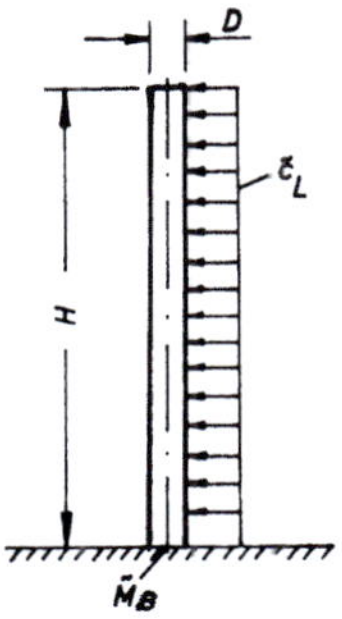

(2c)

Application to STANDARD CODES (Sketch 2c)

In standard codes:

$$\tilde{M}_B = \tilde{C}_L \frac{\rho}{2} U^2_{crit} D \frac{H^2}{2} V_{dyn} \quad (2)$$

Compare eqn. (1) with (2)

$$\boxed{\frac{\tilde{C}_L}{\tilde{C}_{L_0}} = \frac{4 \cdot L}{\lambda}\left(1 - \frac{2(1+L)}{\lambda}\right)}$$

$$\lambda \geq 4L + 2$$

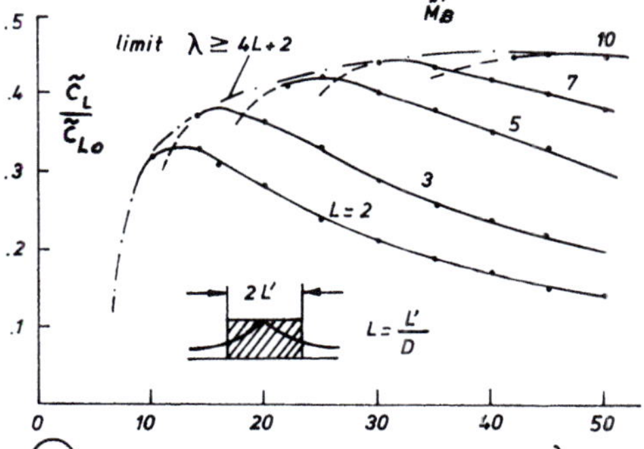

(2d)

References: [2] M.Novak, H. Tanaka, Pressure Correlation on a Vibrating Cylinder, Proc. Wind Effects on Build. and Struct., London 1975.

[3] H.Ruscheweyh, Tip Effect on Vortex-Excited Oscillation of a Modal Stack with and without Efflux Stream. Flow-Induced Structural Vibrations, 1973, Springer-Verlag.

CHART 3: TWO IN-LINE CHIMNEYS

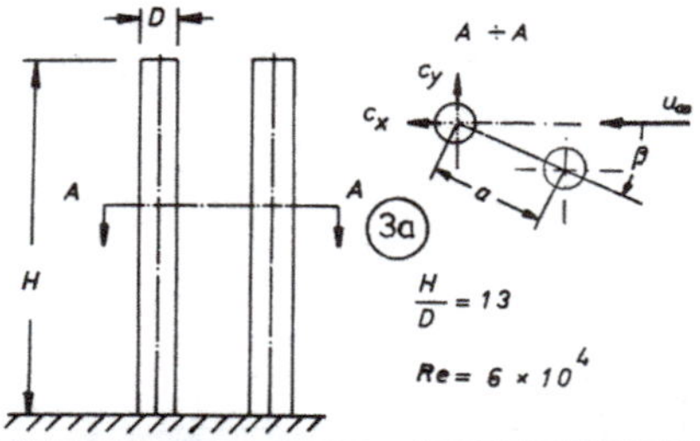

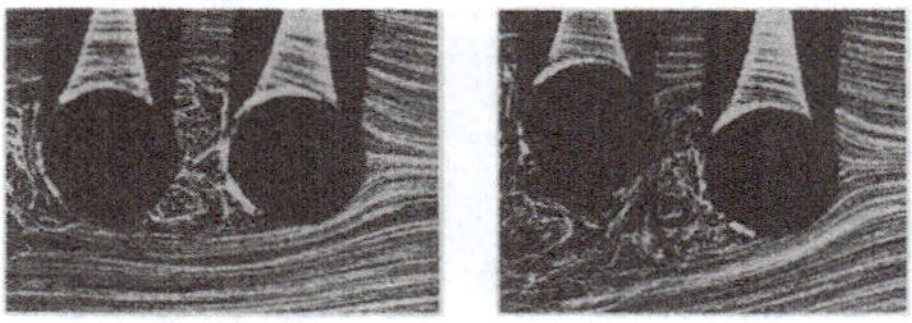

Flow condition „1" β = 0°
3b
Flow condition „2" β = 15°

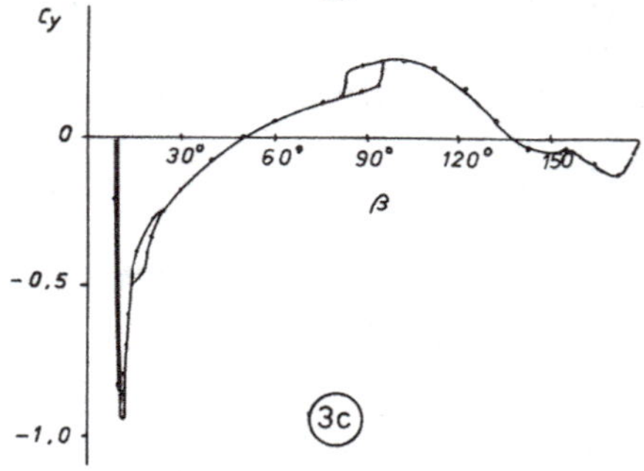

Problem: Aeroelastic instabilities have been observed to arise as a result of interference effects within arrays of several chimneys.

Geometry: See sketch 3a , two elastic models

$$1.5 \leq \frac{a}{D} \leq 8;\ 0 \leq \beta \leq 90°.$$

Flow condition: Smooth flow in a wind tunnel. At a slim angle of attack, the flow through the gap changes abruptly (see sketch 3b).

Side force on the downstream cylinder jumps from zero to a maximum negative value (see sketch 3c) with a Hysteresis.

Vibration: Vortex-excited vibration and interference galloping. Three types of motions. (See sketch 3d). Exciting-force coefficient varies with relative distance a/D and angle β . (See sketch 3e).

Cures: Connecting the chimney tops together reduces vortex-excited oscillation and avoids galloping.

References: [4] K.R. Cooper, R.L.Wardlaw, Aeroelastic Instabilities in Wakes. Proc. Wind Effects on Build. and Struct. 1971

[5] H. Ruscheweyh, Winderregte Schwingungen zweier engstehender Kamine, 3. Coll. Fachhochschule Aachen, 1978.

3d

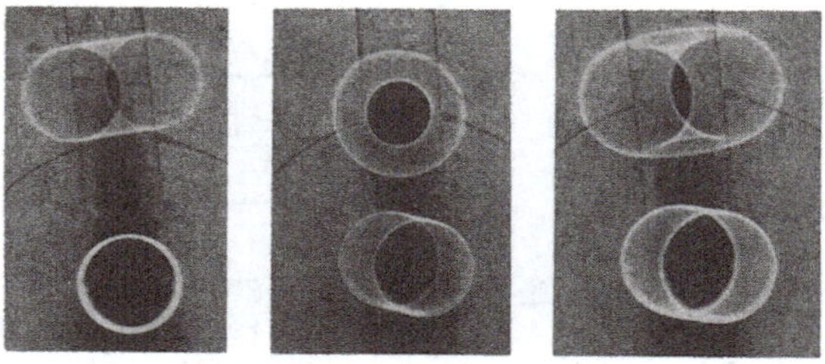

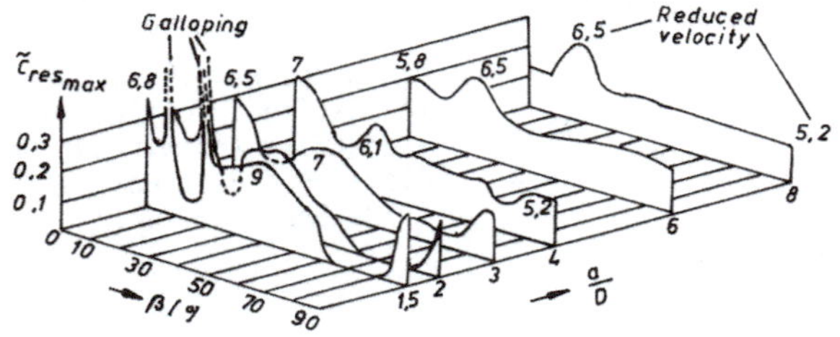

3e

CHART 4: STACK WITH CLOVERLEAF CROSS SECTION

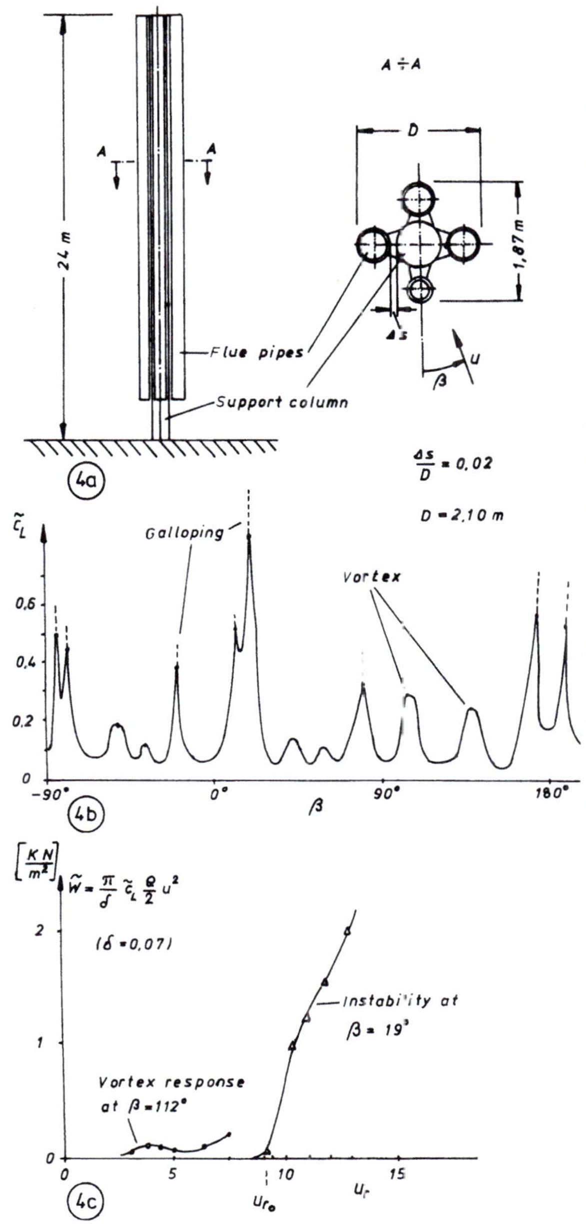

Problem: Steel stack with four smoke flues outside of the supporting column. Galloping-like instability occurs at different wind angles.

Geometry: See sketch 4a. Elastic model.

Flow condition: Smooth wind-tunnel flow.

Structural dynamics: Fundamental bending mode. Mass damping parameter

$$\frac{2\,M\,\delta}{\rho\,D^2} = 14 \text{ (model)}$$

and 8.8 (full-scale).

Vibration: Small vortex-excited vibration, high galloping-like oscillation at different wind direction β (see sketch 4b). Onset velocity at $\beta = 19°$:

$$U_{r_0} = \frac{U_0}{D \cdot f_e} = 9$$

Cures: Increasing the relativ gap $\Delta s/D$ to 0.05 reduces the number of instability points. Additional strake strips decrease the galloping intensity.

References: [6] C. Scruton, Aerodynamics of Structures. Proc. Wind Effects Buildings and Structures, Ottawa 1967.

CHART 5: DAMPING OF A STEEL STACK AND AN ANTENNA MAST

(Full-scale measurements)

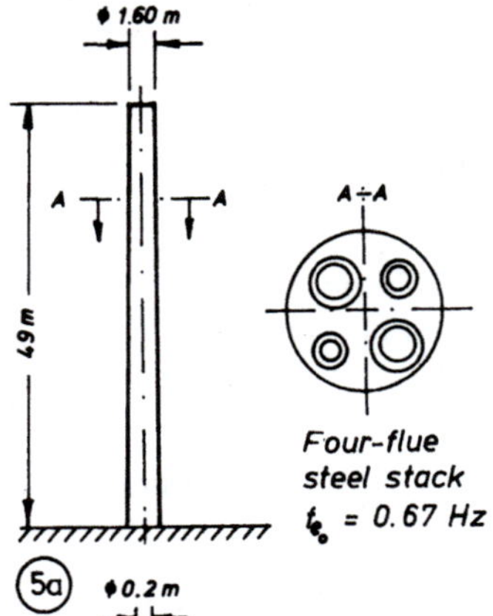

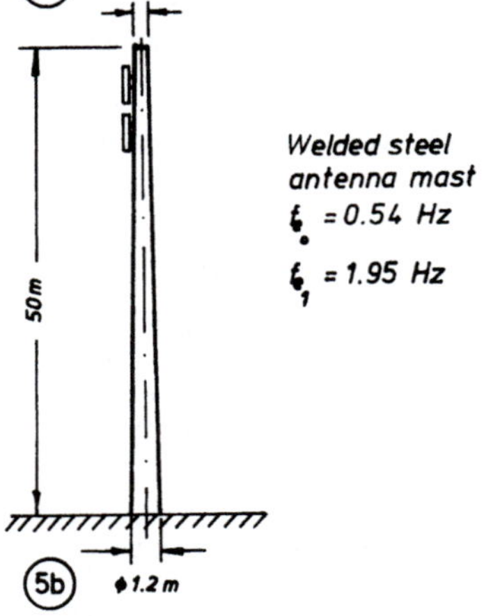

Problem: The damping of a structure cannot be predicted theoretically. It must be measured at the full-scale structures.

Geometry: See sketch 5a and 5b. Chimney with four inner flues.

Flow condition: Wind speed $U < 2$ m/s

Vibration: Man-induced vibration; logarithmic decrement of damping obtained from decay curve of oscillation.

Result: Damping of four-flue steel stack increases with increasing amplitude; damping of antenna mast is nearly constant up to top amplitudes of 10 cm (see sketch 5c) for the fundamental mode (f_{e_0}) as well as for the first mode (f_{e_1}).

References: [7] L.R. Wooton, The Estimation of the Structural Damping of Civil Engineering Structures, Atkin Research and Development, 1971

[8] G. Hirsch, Dämpfungsmaßnahmen zur Herabminderung winderregter Schwingungen schlanker Bauwerke, Haus der Technik-Veröffentlichungen, Nr. 347, Vulkan-Verlag Essen, 1975.

[9] H. Ruscheweyh, Dynamische Torsionsbelastung von Hochhäusern unter Interferenzeinfluß und Dämpfungsmessungen an Originalbauwerken, DFG-Kolloquium, Karlsruhe 1978, Institut für Beton und Stahlbetonbau, Universität Karlsruhe.

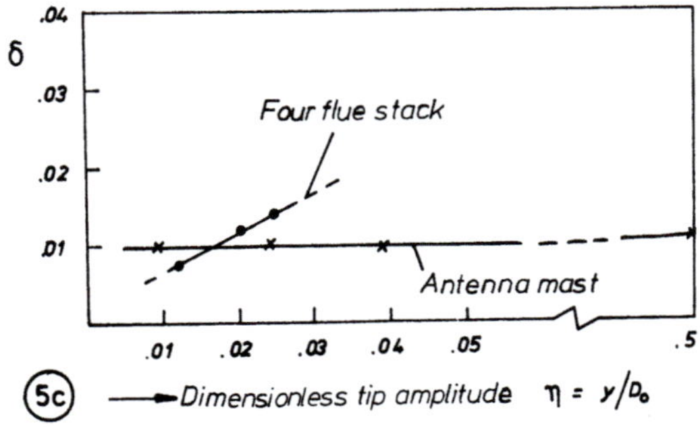

CHART 6: FREQUENCY AND DAMPING MEASUREMENT ON A NATURAL DRAUGHT COOLING TOWER

(Full-scale measurements)

Problem: Dynamic response estimation of a cooling-tower shell depends on the assumption of structural damping of the shell. Only few measured values are avaible.

Geometry: (see sketch 6a); concrete shell.

Flow condition: wind speed U < 2 m/s (first test) and U ≈ 6 m/s (correlation measurement).

Vibration: Man-induced vibration; top prestressed with cables in three different circumferential modes (see sketch 6b). Cable releases, shell vibrates dominantly in the prestressed mode.

Result: 7 frequences of the shell could be analysed. Comparison with theory see sketch 6c.
Damping for the lowest frequency (m = 1, n = 3, f_e = 1,27 Hz) from autocorrelation function R(K), see sketch 6d, (log. plot of autocorrelation envelope) is δ = 0,052 for a small amplitude.

- theory, SAP IV with conical shell element, shell without cooling installation

x m = 1, 0 m = 2, • m = 3: full-scale measurement, shell with cooling installations

---

References: [10] F.Müller, Zur Berechnung des Schwingungsverhaltens windbeanspruchter Kühltürme, DFG-Kolloquium, Mitteilung Institut für Massivbau TH Darmstadt, Nr. 26, 1977.

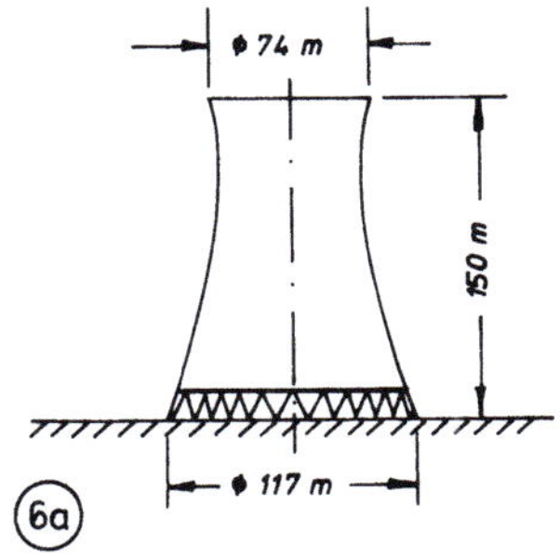

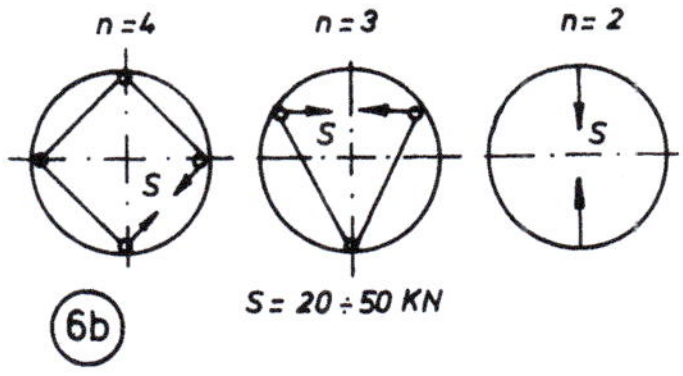

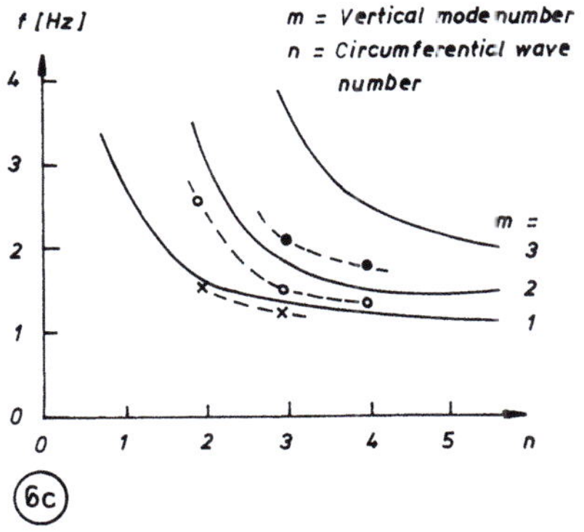

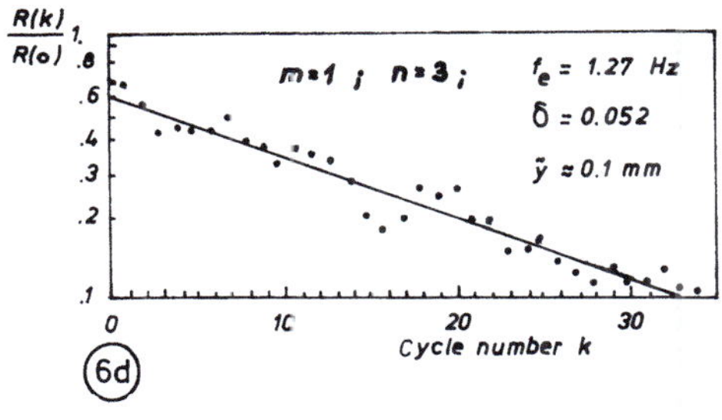

G4 EXPERIENCES WITH CROSS-WIND VIBRATIONS OF COLUMNS AND CHIMNEYS

H. van Koten
Institute TNO for
Building Materials and Building Structures
P.O. Box 49, 2600 AA Delft. Netherland

Three typical cases of cross-wind vibrations are mentioned on the charts added to this paper. Cross-wind vibrations are often seen at steel chimneys with a height above 20 m.
The first mentioned case was a lighting column with a height of 40 m. The calculations showed that cross-wind vibrations should not be important. Immediately after construction, the tower started to vibrate in the second mode. This case of vibration was not considered in the calculations. The accelerations in the column were so strong that attendance of persons inside the tower was not possible.
The vibrations were reduced by fitting helical strakes with a cross-section of 20 x 25 mm over the upper half of the tower. The cross-section of these strakes (2.5% D) are much smaller than indicated by Scruton (9% D), see Ref. given in Chart 1). This size of 2.5% D and the rectangular cross-section must be seen as a minimum. An earlier fitted strake with a round cross-section of 16 mm (1.6% D) was not effective at all. The difference between the results of Scruton and this full-scale experiment was probably caused by the Reynolds number. In the case of the tower, the Reynolds number was $Re = 7 \times 10^5$ (the critical wind speed was 10 m/s).

A second case of cross-wind vibrations was observed for a steel chimney with a height of 60 m and a cross-section of 1.60 m. The critical wind speed was 6 m/s. The first attempt to reduce the vibrations was a change in form of the chimney. The upper 23.5 m were replaced by a part with a diameter of 1.0 m. The critical speed for this part was 4 m/s. This reduction of critical speed and the two diameters were supposed to be sufficient to reduce the vibration amplitude. The effect was, however, negligable.
The amplitudes were reduced considerably after fitting helical strakes with a cross-section of 10 x 50 mm over the upper 23.80 m. The Reynolds number in this case was $Re = 2.8 \times 10^5$. The dimension of 50 mm (5% D) must again be seen as a minimum.

The last case of a vibrating chimney considered here, is a chimney with four different diameters and a height of 75 m. The dimensions are given in the Chart. The upper part of this chimney cracked 14 days after erection. Four helical strakes with a cross-section of 10 x 50 mm were fitted.

CONCLUSION

Different diameters over the height of chimneys are not effective for reduction of cross-wind vibrations. The minimum dimension in radial directing of helical strakes depends on the Reynolds number. Suggested is $d/D = (2 \times 10^4)/Re$. Four helical strakes with a length of 8 D and the above given dimensions are effective. No difference in behaviour was observed between a chimney with an octagonal or a round cross-section.

CHART 1: OCTAGONAL STEEL LIGHTING COLUMN

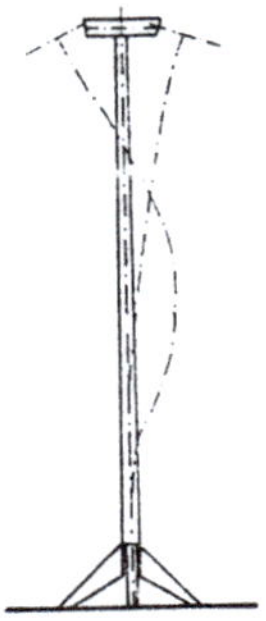

(1)

Problem: Large movements and stresses at a wind speed about 10 m/s; frequency of vibrations: 3.2 Hz (second-mode vibration of column).

Geometry: height : 40 m
diameter: 1 m

Situation: at an airport.

First cure without effect: (Sketch 2)
Four helical strakes with a round cross-section d = 16 mm = 1.6% of D.

Second effective cure: (Sketch 3)
Four helical strakes with a cross-section of 20 x 25 mm; d = 2.5% of D.

Vibrations: The movements before and after curing are given in Sketch 4.

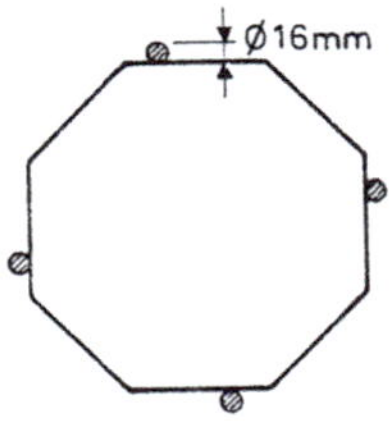

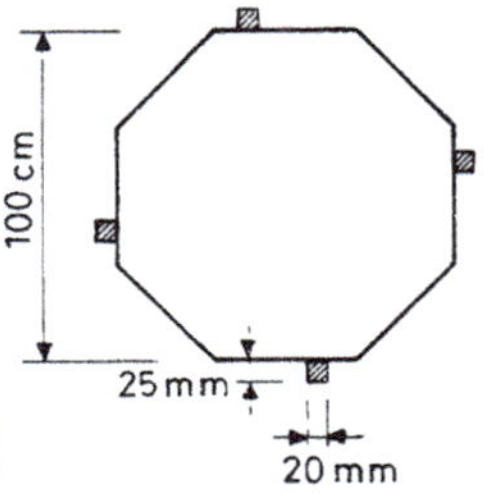

(3)

(2)

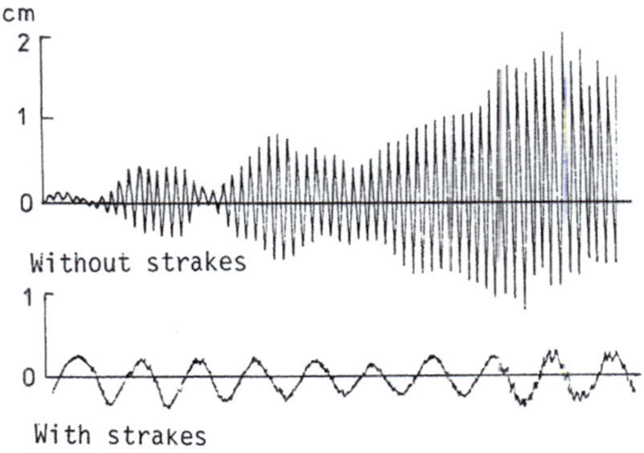

(4)

References: Koten, H. van,
An Application of Helical Strakes around a Tower to Reduce the Influence of Lateral Vibrations.
TNO-IBBC, Delft. Report no.: BI-74-15/08.1.130, 1974.

CHART 2: DAMAGE OF A STEEL CHIMNEY BY CROSS-WIND VIBRATION

① Chimney after erection

Problem: Cross-wind vibrations of a steel chimney of different diameters. The upper part was damaged two weeks after construction.

Geometry: height : 75 m
diameter: 14 m Ø 2.865 m
14 m Ø 2.220 m
10 m Ø 1.590 m
31 m Ø 9.555 m

Flow conditions: Turbulent wind conditions (build up area); V = 6 m/s.

Damage: The weld at a connection ring was cracked (Sketch 3). The upper part was taken down.

Cure: The connection ring was fitted again and welded on the inside of the chimney (see Sketch 2).

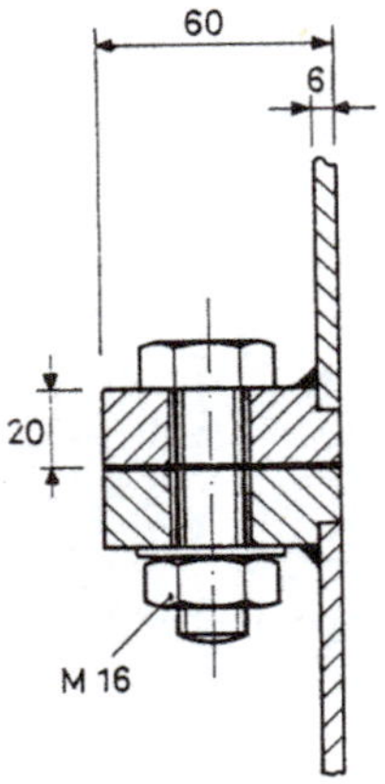

(2a) Original structure

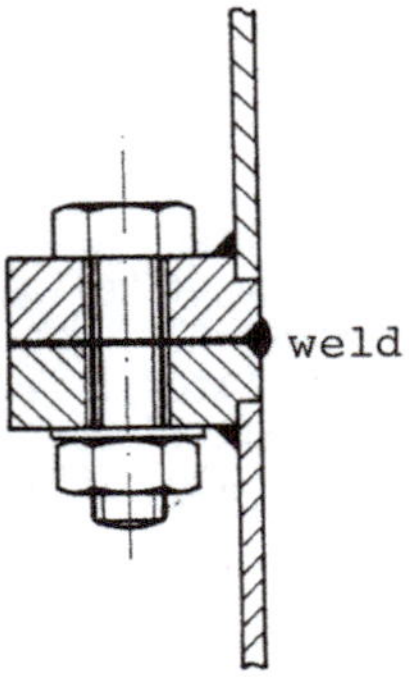

(2b) Cured structure

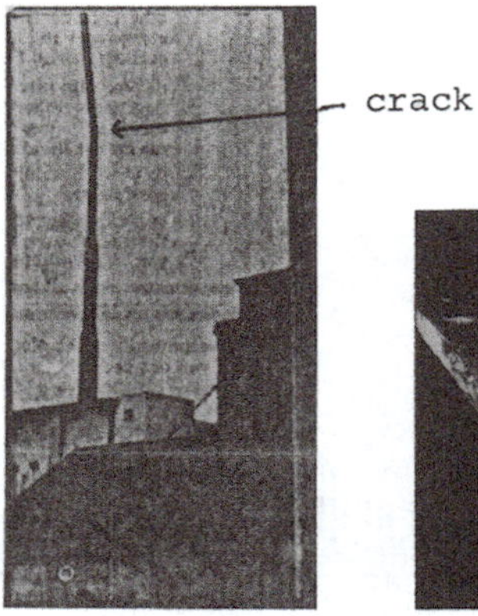

(3a) Chimney after cracking.

3b. Crack in weld.

CHART 3: CROSS-WIND VIBRATIONS OF A STEEL CHIMNEY

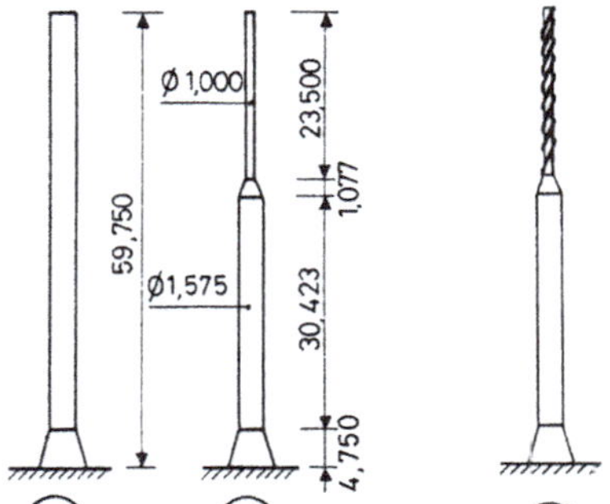

(1) (2) (3a)

(3b) Fitting of the strakes

Problem: Large cross-wind vibration occured at a cylindrical steel chimney.

Geometry: height : 60 m (Sketch 1)
diameter: 1.60 m

Flow conditions: Mean wind speed 4 m/s over open country. The statistical distribution of the wind speed at 30 m height during vibrations is given in Sketch 4. The statistical distribution of the top values of the stresses during vibration is given in Sketch 5.

First cure without effect: The upper part of the chimney was taken down and replaced by a part with a smaller diameter (Ø = 1.0 m) (Sketch 2). Expected was a considerable reduction of amplitudes because of the two critical speeds, but the top amplitude was still about 0.40 m.

Second effective cure: (Sketch 3)
Around the upper 23.50 m of the chimney, helical strakes with a cross-section of 10 x 50 mm (5% D) were fitted. The effect is seen in Sketch 6.

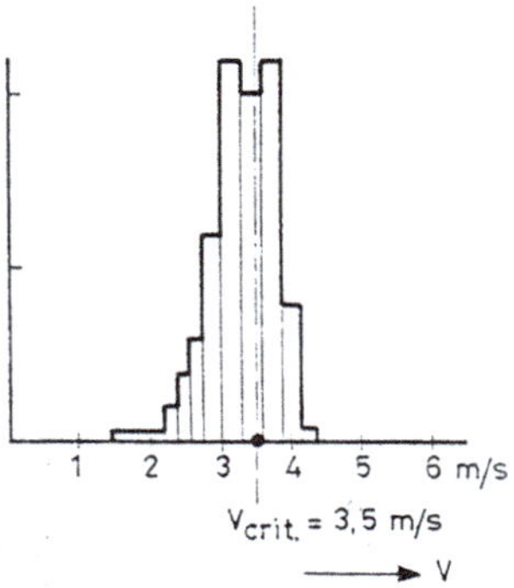

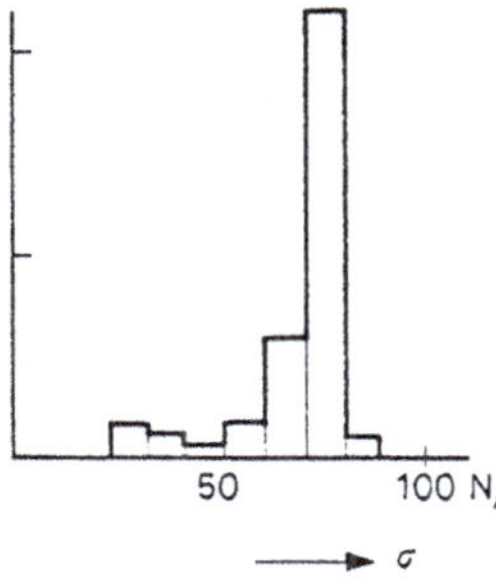

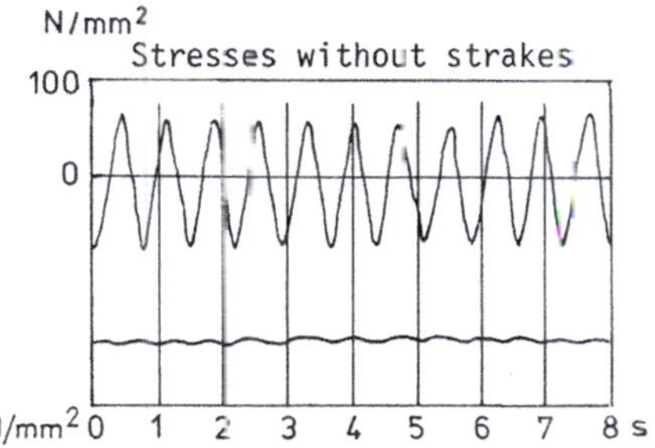

(4) Stat. distribution of the wind speed.

(5) Stat. distribution of the stresses.

(6) Stresses with strakes

G5 OSCILLATION DUE TO VORTEX SHEDDING OF REINFORCED CONCRETE CHIMNEYS OF RECTANGULAR CROSS-SECTION

T.A. Wyatt
Reader in Structural Design

Imperial College, London, U.K.

SUMMARY

Two reinforced-concrete chimneys 45 m high, of rectangular cross-section approximately 2200 x 2600 mm, have been observed to oscillate in wind due to vortex shedding. Measurement of the actual natural damping confirmed that the mass-damping parameter of these structures had the value anticipated by the designers, on which basis no significant oscillation was predicted. Site measurements and wind tunnel test results are presented and compared; the wind-tunnel tests include development of shroud-type aerodynamic spoilers.

1. OBSERVED BEHAVIOUR IN WIND

Two reinforced concrete chimneys of rectangular cross-section, 45 m high, were constructed at the North Armagh Hospital (near Portadown, Northern Ireland) during the summer of 1970. Scaffolding was removed in October of that year, and within a few weeks attention had been drawn to the occurrence of sustained oscillation in the fundamental mode during strong winds. During the ensuing period of a little over three months, significant oscillation was observed in five storms; 5 and 6 November 1970, 7 and 24 January and 12 February 1971.

On every occasion the mean wind direction was within $\pm 10^{\circ}$ of the direction 'A' shown on the key plan (Chart, Sketch 1), namely $231^{\circ} \pm 10^{\circ}$ true. The wind speeds were inferred from observations at the local weather station (Portadown) and checked against the observations at Aldergrove (Belfast Airport, about 30 km distant). The hourly mean speed corrected to a level towards the top of the chimney (40 m above ground, say) was about 17 m/s at the times when oscillation was observed.

Both chimneys oscillated, the predominant motions being perpendicular to direction 'A'. This corresponded to the less stiff principal direction for chimney 1, for which the frequency was 0.77 Hz, and to the more stiff direction for chimney 2, for which the frequency was 0.82 Hz. As closely as can be ascertained, these correspond to the still-air natural frequencies. It was not possible to measure the oscillations of the two chimneys simultaneously; in the storm of 24 January the two chimneys were observed in similar wind conditions and the amplitude was about 30% greater on the downwind chimney (No.2) than on the other.

The chimneys were of constant cross-section (2630 x 2170 mm chimney 1, 2720 x 2170 mm chimney 2) over the upper five-sixths of their height. The square root of the cross-section area will be used as the reference dimension, i.e. 2390 mm for chimney 1 and 2430 mm for chimney 2. The reduced velocities (V/nD) for maximum response were thus 9.2 and 8.5, respectively.

The oscillation of chimney 1 was studied in detail on 24 January by visual reading of deflection using a theodolite sighting on a target marked with quarter-inch (~6 mm) bands. The histogram of peaks derived from observations over 322 consecutive cycles is reproduced on the chart (Sketch 3). The Rayleigh distribution probability density curve based on the same r.m.s. value (namely 34 mm) is shown for comparison, making the appropriate conjunction of scales (the mark on the ordinate scale corresponds to a probability density of 1.0 per r.m.s. value, or 0.187 per $\frac{1}{4}$" band, and thus 60 observations in the histogram). It will be seen that the observations follow the Rayleigh distribution fairly closely, but with rather more large excursions.

The terrain on the approach to the site for wind direction A is open, such that the intensity of turbulence at level 40 m would be expected to be about 0.15. The buildings are simple in shape, the boiler house being only 8 m high; the fuel tanks are cylindrical, of similar height.

2. STRUCTURAL MEASUREMENTS AND CALCULATIONS

The natural frequencies and damping of chimney 1 in still air had been investigated by setting it in oscillation by means of a winch cable containing a 'fusible link'. The maximum force was 50 kN, which permitted satisfactory traces to be obtained for decay of amplitude in the first mode from 25 mm to 5 mm. The logarithmic decrement was found to be closely uniform over this range, with $\delta = 0.070$ for motion in the weak direction.

The static stiffness was also measured; the horizontal force at the top of the chimney to produce unit deflection in the weak direction was 1.65 kN/mm. The calculated mass of the structure was 7.0 t/m, effectively uniform over the region of interest. The calculated value of the natural frequency derived from these values agreed exactly (0.77 Hz) with the observations in still air and in wind. The actual value of the static stiffness is sensitive to foundation flexibility; the measured value assuming base fixity would imply an effective Young's modulus for the gross section of the reinforced concrete of 33 kN/mm^2. This is low for nominal 28 N/mm^2 concrete, which had an actual average 28-day cube strength of 43 N/mm^2, and suggests that foundation movement was significant.

The chimneys were lined with 4$\frac{1}{2}$" (~110 mm) refractory brickwork throughout their height; the cavity between lining and structure was filled with fibreglass mattress. The combination of this lining detail and the foundation compliance is presumed to account for the relatively high measured value of structural damping. The structure is believed not to have been subjected to sufficient load to cause tensile cracking by that time. The value of the combined mass and damping parameter was thus $m\delta/\rho D^2 = 70$.

Chimney 2 differed slightly by having large rectangular openings up the NW face. As a result the frequency in the direction of interest (the 'stiff' principal direction of this chimney) was only 6% higher than chimney 1.

3. DISCUSSION

Reinforced concrete chimneys in the 50 m height range are not normally regarded as seriously at risk from oscillations caused by vortex shedding. The high value of the combined mass and damping parameter referred to above is in accord with this expectation. Although it is well known that the square section is worse in this respect than the circular-section prism, the refined prediction methods available for circular sections make it interesting to estimate first the response of an isolated circular chimney with the same value of the mass-damping parameter.

The recent ESDU data item on this subject [2] has been applied to this purpose. With the given natural frequency, a circular chimney of the given dimensions proves to lie in the super-critical 'window' where resonant response to vortex-shedding is very unlikely to occur. Arbitrary adjustment of the effective Reynolds number as defined in the reference quoted to give developed subcritical or transcritical conditions, respectively, ($Ree = 3 \times 10^4$ or 3×10^6) would predict r.m.s. tip amplitudes of only 0.7 mm or 0.4 mm; these values are of course in the low-amplitude range where the correlation of vortex shedding is poor. The observed response was roughly 50 times larger.

Vortex shedding from an approximately square section has not been studied in such detail, especially in respect to behaviour in the natural wind. The leading reference [1] is based on a section model in smooth flow. It suggests that $m\delta/\rho D^2 = 44$ is sufficient to ensure stability. The term 'stability' here is perhaps somewhat ill-defined, but can be taken to mean that amplitudes are limited to small values, perhaps 0.01 D maximum, at which the correlation of shedding is poor even in the test conditions. The corresponding value quoted for a circular cylinder in the subcritical regime in the same reference is $m\delta/\rho D^2 = 30$. On this estimate the square section would be only modestly worse than the circular section. The suggested critical reduced velocity is $V/nD = 6.6$, which is some 25% less than the estimated worst condition at site.

The amplitude-damping curve in reference 1 suggests a rather more important difference at larger amplitudes, where the curve appears to approach the hyperbolic form corresponding to a constant value for the coefficient of fluctuating lift, $\hat{C}_L = 0.8$. Applying this value to $m\delta/\rho D^2 = 70$ for the real chimney, and including the correction for the mode shape in the conventional simple deterministic analysis, gives a maximum tip amplitude of 0.03 D, or 72 mm. This is approaching the 4σ value of 96 mm observed on chimney 1 (note that 34 mm was the observed r.m.s. peak value, so that the r.m.s. displacement value can be inferred to be $34/\sqrt{2} = 24$ mm). This would, however, be expected to be a very pessimistic approach, overlooking the fall-off from the hyperbolic relationship at the appropriate amplitudes even in the wind-tunnel test condition, as well as the expected effects of moving to the real wind condition. The distribution curve of the observations clearly does not support a deterministic analysis - but the magnitude of the observations appears much larger than would otherwise have been anticipated.

Proximity effects are well known to increase the risk of unsatisfactory behaviour. The chimney spacing was 5 D, known to be an unfavourable value, and the orthogonal orientation of the stiff principal directions of the two chimneys was introduced to minimise the risk of resonant interaction. It will be noted that the observations under discussion were made on the upwind chimney. Response of the downwind chimney is believed to have been up to 30% larger.

The rectangular section has also been extensively studied in relation to galloping, or negative lift slope instability [3]. It does not seem likely, however, that this significantly influenced the observed behaviour. The critical wind speed for this form of excitation predicted from reference [3] or reference [1] would be an order of magnitude higher than the observed critical value.

4. WIND-TUNNEL TESTS

Two series of wind-tunnel tests using an aeroelastic cantilever model of approximately 1/30 full size were conducted by the National Physical Laboratory [4]. A square section aluminium extrusion was used for convenience in modelling, as results were urgently required to aid decision on the need for (and possible nature of) remedial measures. The first series was carried out in smooth flow, the second series in small-scale turbulent flow. The turbulence intensity was 6.5% with longitudinal integral scale 2.2 D, produced by a grid.

The damping values of the two tests differed, being $m\delta/\rho D^2 = 16$ and 35, respectively. To facilitate comparison, the results have been reproduced on Figure 1 after scaling by factors 0.23 and 0.50, respectively; i.e. as a prediction for the full-size $m\delta/\rho D^2 = 70$, assuming inverse proportionality of amplitude to damping. It will be observed that small-scale turbulence enhances the second peak, at a velocity which is in better agreement with the estimated values of the speed at site than the section model tests discussed above. The predicted r.m.s. tip amplitude is now in good agreement with the observation, although it should be noted that the author would still have expected prediction on this basis to prove significantly pessimistic.

The tests included study of aerodynamic spoilers. It was found that the perforated shroud type which had been developed for circular chimneys was effective, but required to be fitted to a longer length than on circular chimneys. A shroud with rectangular perforations of pitch D/8 and 35% open area, mounted D/8 clear of the face of the chimney was recommended. If fitted to the upper 42% of the height, the vortex-shedding peak in response was effectively eliminated. Unlike spoilers on circular section chimneys, this caused negligible drag penalty.

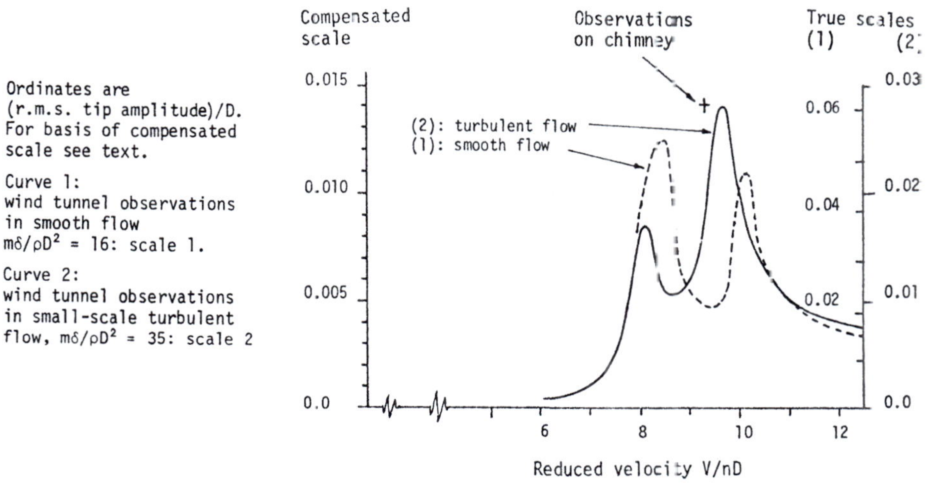

FIGURE 1: WIND TUNNEL TEST RESULTS

5. ACTION TAKEN

The wind-tunnel tests strongly reinforced the opinion that the observed movements fully represented the worst case that could be anticipated. The risk of structural damage in the long term, including fatigue effects, was considered with care, and it was decided that no short-term action was necessary. No spoilers were fitted. This confidence has been rewarded in that the chimneys have apparently settled to stability. No evidence can be presented for the reason for this change, although it may be guessed that cracking of the reinforced concrete section resulting from wind loading and/or thermal cycling has further augmented the damping.

6 ACKNOWLEDGEMENTS

The author was retained by the Consulting Engineers, Dr. I.G.Doran & Partners of Belfast, to make an independent assessment of the data that they had assembled. The observations on the behaviour of the chimneys in wind had been made by the Consulting Engineers' staff. The damping tests were carried out by Mr. L.R. Wootton and Mr. D.E.J. Walshe, then working at the National Physical Laboratory, and the wind tunnel tests were carried out by Mr. D.E.J. Walshe. Mr. C. Scruton had given advice and supervised the work by the NPL staff. The author has drawn heavily on these sources, and wishes in particular to thank Mr. J.A. Hill of I.G. Doran & Partners for making it possible to present this paper.

7 REFERENCES

(1) C. Scruton, "On the wind excited oscillations of stacks, towers and masts", Proc. 1st Int. Conf. Wind Effects on Buildings and Structures, London 1963 (HMSO London 1965).

(2) ESDU Data Item No. 78006 "Across-flow response due to vortex shedding", Engineering Sciences Data Unit, London, 1978.

(3) G.V. Parkinson, "Aeroelastic galloping in one degree of freedom", Proc. 1st Int. Conf. Wind Effects on Buildings and Structures, London 1963 (HMSO London 1965).

(4) D.E. Walshe, "The use of perforated shrouds for suppressing vortex excitation of cylinders of square section", NPL Mar. Sci. TECH. MEMO 2-71, 1971 (Oct.), and correspondence between Mr. Walshe and the Consulting Engineers, November 1971.

REINFORCED CONCRETE CHIMNEY OF RECTANGULAR CROSS-SECTION

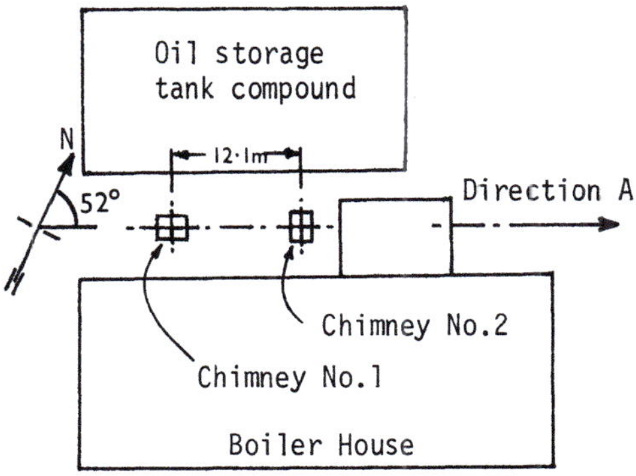

Sketch 1: Layout of site

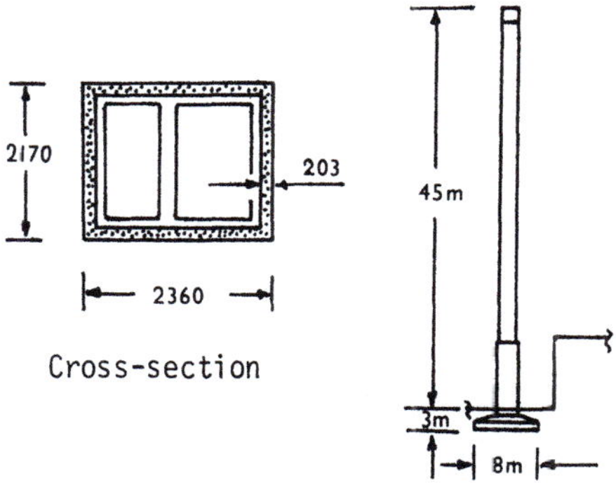

Cross-section

Sketch 2: Dimensions of Chimney No.1

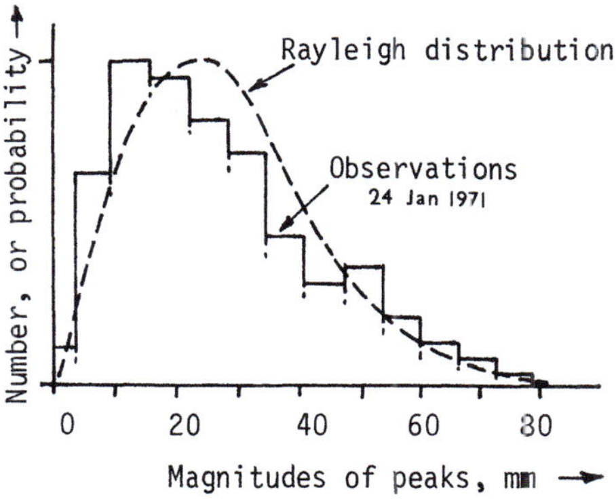

Sketch 3: Typical observations

Location: North Armagh Hospital, Northern Ireland.

Geometry: Two reinforced concrete chimneys 45 m high. Cross-section rectangular 2170 x 2630 mm; brick lining. Site plan as indicated by Skh.1. Note minor axis of No.1 is co-linear with the major axis of No.2.

Flow Conditions: Wind in direction A approached the site across open terrain with few obstacles and was not disturbed by local buildings.

Structural Dynamics: The lowest natural frequency of chimney No.1 in direction perpendicular to A ('weak' direction of cross-section) was 0.78 Hz. Damping was measured by decay of free oscillation giving logarithmic decrement 0.07 at amplitude 25 mm. The combined mass and damping parameter $m\delta/\rho BD = 70$.

Vibration: Significant oscillations of chimney No.1 at its lowest natural frequency in a direction perpendicular to A were observed during at least six wind storms in the first winter of the life of the structure. The wind was approximately in direction A, and maximum amplitudes were recorded when the hourly mean speed at height 40 m above ground was about 17 m/s. During the worst recorded period of oscillation, an analysis of the successive displacement peaks fitted closely a Rayleigh distribution with r.m.s. 34 mm (Skh.2).

Investigation: After wind-tunnel studies both in smooth flow and in small-scale turbulent flow, it was concluded that the observations represented the worst response conditions possible. This response was not damaging in the short term, and although designs for aerodynamic remedies were prepared, they were not found necessary. Monitoring was established on a long-term basis. After eight years there is no indication of continuing damaging vibration.

G6 FIELD TESTS ON A TALL TOWER - SUPPORTED STEEL STACK

Hatsuo Ishizaki
Professor, Kyoto University
Kyoto, Japan

Toshiaki Makihata
Researcher
Technical Institute

Masao Araki
Engineer
Off-shore Business Headquarters
Hitachi Shipbuilding & Engineering Co. Ltd.
Osaka, Japan

SUMMARY

Field experiments on a 200m high truss-stiffened stack were performed in order to ensure the structural safety of the stack against wind forces. The stack has three ducts supported together by a steel tower with a triangular cross-section. Wind speeds and directions, as well as lateral accelerations of the stack movements, were observed on two decks at the 180m and 90m levels, and the tilt angle of the stack was also measured on a deck at the 180m level. The vibrational amplitude of the stack caused by the wind was much smaller than the static or mean displacement. Large oscillations such as those frequently seen on self-supported steel stacks were not observed on this stack. It can thus be concluded that the dynamic effects were reduced by the supporting tower around the stacks.

1. INTRODUCTION

Over the last decade, a great number of tall stacks have been contructed in Japan. Some of these are made of concrete and others are steel stacks that are supported by surrounding truss towers. The self-supported steel stack is lighter in weight and smaller in energy absorption than corresponding concrete stacks. Accordingly, the vibration caused by wind frequently becomes a problem on the self-supported stack, particularly in the direction normal to the wind. On the other hand, the cross-wind vibration has not proved to be a serious problem with the tower-supported steel stack because the surrounding truss or frames will disturb the shedding of periodic vortices behind the stack. Thus, it seemed important to evaluate the fluctuating wind forces and the wind-induced vibrations of the tower-supported stack.

FIG.1 VIEW OF THE STACK

2. THE EXPERIMENTAL STACK AND THE INSTRUMENTATION

The stack is located at a power station near the sea coast. There is a building, 60m high, to the north of the stack and some storage tanks 30m high to the west. To the east and south, there are no obstacles which would disturb the wind flow from the sea.

A schematic diagram of the stack is shown in the Chart and Fig. 1 gives a photograph of the stack. The stack has three ducts, 200m high, supported together by a 180m-high steel tower with a triangular cross section. The ducts are connected at six levels with the tower so that the ducts are stiffened in the horizontal direction by the tower. The structural members of the stack were designed to resist the wind loads within the elastic range of the steel. The total weight of the structure is 3200 tons excluding the foundation.

Two propeller-type anemometers were installed on two decks at the 180m and 90m levels and two accelerometers on the same decks to measure the two orthogonal components of horizontal responses. The tilt angle of the stack was also measured by two clinometers at the 180m level. Fig. 2 shows the arrangement of the instruments on these two levels. The output from these instruments was stored on magnetic tape. The recorder was set in advance to start automatically when the wind speed became greater than a specific value.

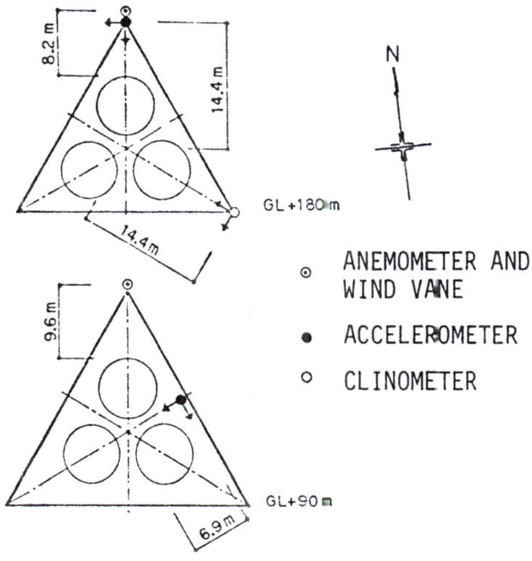

FIG. 2 ARRANGEMENT OF INSTRUMENTS

3. THE EXPERIMENTS

Small oscillations of the stack caused by impulsive ground motions were recorded under calm wind conditions in order to evaluate the natural frequencies and the structural damping of the stack. For this purpose, the structural characteristics of the stack were postulated as a discrete lumped-mass system. The calculated natural frequencies and the observed frequencies together with the logarithmic decrements of the damping are listed in Table 1.

The most detailed wind records were those of the seasonal winds during 1974 and 1975, and a record of Typhoon 7506 was also taken. The prevaileng directions of the seasonal winds were between north and west, while the wind direction of the typhoon moved from east to north and west.

The wind speed at the 180m level was slightly greater than the wind speed at the 90m level in some records; in other cases, the former wind speed was smaller than the latter. On the average, the wind speeds were almost

	THEORETICAL	EXPERIMENTAL	
MODE	F	F	δ
1st	0.660	0.612	.026
2nd	2.02	1.82	.028

F : FREQUENCY Hz

δ : LOGARITHMIC DECREMENT

TABLE 1 OBSERVED AND THEORETICAL DYNAMIC CHARACTERISTICS

equal. The largest value of the ten minutes wind speed at the 180m level in the records was 24.8m/s. From measurements at the 180m level it was found that, in the case of wind speeds less than 20m/s, the turbulence intensity is between 0.1 and 0.2, and for winds above 20m/s it is less than 0.1.

Sketch 2 in the Chart shows typical examples of movements of the stack which were obtained from the recorded accelerations at the 180m level. The figures display nearly elliptic orbits.

In order to determine the structural safety of the stack, the maximum displacement or amplitude caused by winds must be examined. The relation of the maximum amplitude in the cross-wind derection to the mean wind speed is shown by numbers in Fig. 3. When the wind blows from the direction approximately perpendicular to one side of the triangular section, as shown by numbers 4, 5 and 6 in the figure, the relation is comparatively distinct. Fig. 4 shows the comparison of two maximum amplitudes of the stack measured in two directions, parallel and normal to the wind direction. The amplitudes in

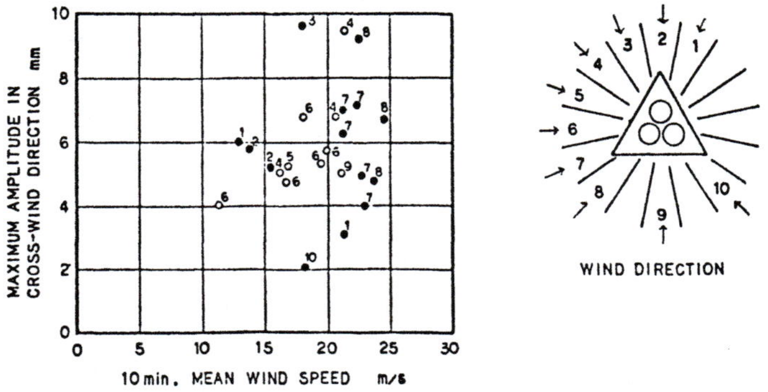

FIG. 3 MAXIMUM AMPLITUDES IN THE CROSS-WIND DIRECTION

the cross-wind direction are about 1.2 to 2.0 times the amplitudes in the wind direction.

To investigate the wind forces affecting the stack, the tilt angles of the stack at the 180m level together with the wind speeds and directions were measured for a short period every ten minutes. An example of the records obtained thereby is shown in Fig.5. It is evident from the figure that tilt angles are affected by sun radiation in the daytime. Fig. 6 shows the movements of the stack obtained from the observed tilt angles compared with the theoretical values shown by a dotted line. The theoretical values are calculated by postulating the temperature increase of the stack by sun radiation.

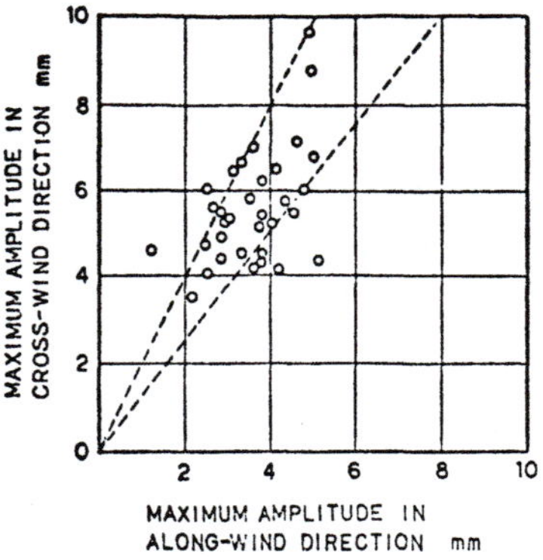

FIG. 4 COMPARISON OF MAXIMUM AMPLITUDES IN THE TWO DIRECTIONS, PARALLEL AND NORMAL TO THE WIND

To avoid these temperature effects and estimate the wind forces from the tilt angles, the records obtained in the night or on cloudy days were also considered. Fig. 7 is plotted from the records which were not greatly affected by temperature. The theoretical tilt angles caused by wind forces were also estimated and are shown in this figure. The theoretical drag coefficients for each stack and for the tower were postulated to be 0.6 and 2.0 respectively. As can be seen from this

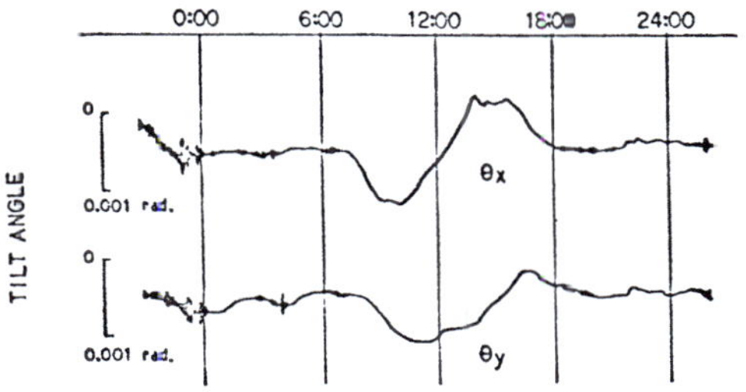

FIG. 5 RECORDS OF TILT ANGLE

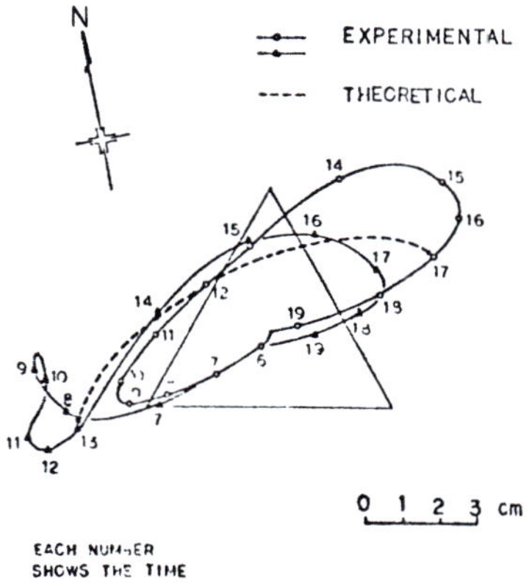

FIG. 6 MOVEMENTS OF THE STACK CAUSED BY SUN RADIATION

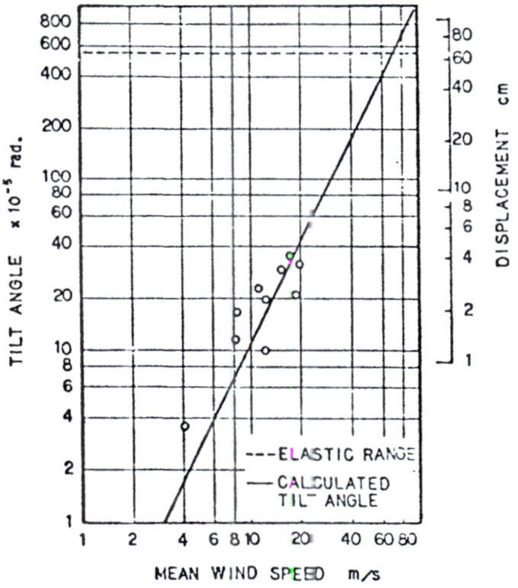

FIG. 7 TILT ANGLE OF THE STACK CAUSED BY WIND FORCES

figure, the observed angles coincide well with the theoretical values.

As shown in Fig. 7, the maximum tilt angle for the mean wind speed of 20 m/s is about 4×10^{-4} rad. or the displacements nearly 5cm. On the other hand, the maximum amplitude shown in Fig. 3 is smaller than 1cm in the cross-wind direction. The amplitude in the wind direction is about 1/2 to 1/1.2 of it. From these two amplitudes, the resultant amplitude of the vibration caused by winds is estimated to be less than 20% of the static or mean displacement. The value of 20% seems relatively small compared to values generally observed in wind-induced vibrations of self-supported stacks. Therefore, concerning the experimental stack, the dynamic effect caused by winds is concluded to be small for wind speeds less than 25m/s.

4. CONCLUSION

From the field experiments on a truss-stiffened stack, it is concluded that the vibrational amplitude of the stack induced by winds was much smaller than the static displacement. The corresponding reduction of the dynamic effects is caused by the increased rigidity due to the surrounding tower of the stack as well as the corresponding disturbance effects on the vortex shedding.

TALL STEEL STACK

Problem: Effect of truss-tower support on vibration of a steel stack 200m high.

Geometry: See Sketch 1.

Boundary conditions: The stack is located at a power station near the sea coast.

Structural dynamics: The period and damping of the stack were estimated from the records obtained by field measurements.
(i) Frequency (1st mode): 0.612 Hz
(ii) Logarithmic decrement of the damping: 0.026

Wind conditions: Records were obtained of the seasonal winds during the period from 1974 to 1975 and of Typhoon 7506. The prevailing directions of the seasonal winds were between north and west at velocities ranging from 10 to 25m/s.

Vibration: Typical examples of movements of the stack are shown in Sketch 2. The figure displays nearly elliptic orbits. The maximum amplitude of the stack at the 180m level was smaller than 1cm in the cross-wind direction. The amplitude in the wind direction is about 1/2 to 1/1.2 of it. The amplitude of the stack caused by the dynamic effects of winds was much smaller than the static or mean displacement.

Conclusion: Substantial reduction of dynamic effects was obtained as a result of the rigidity of the surrounding tower and the disturbance of the vortex shedding by the tower.

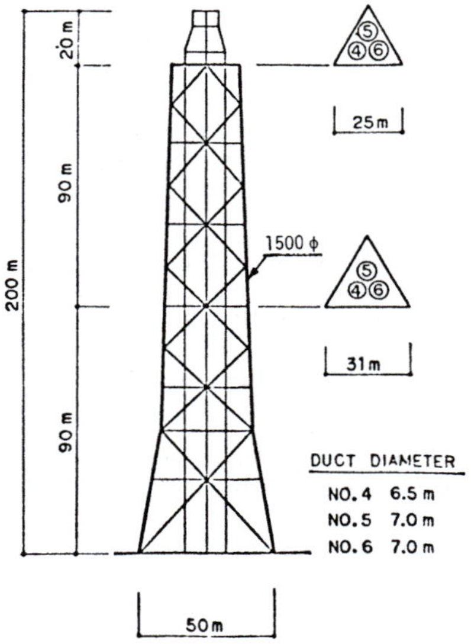

①

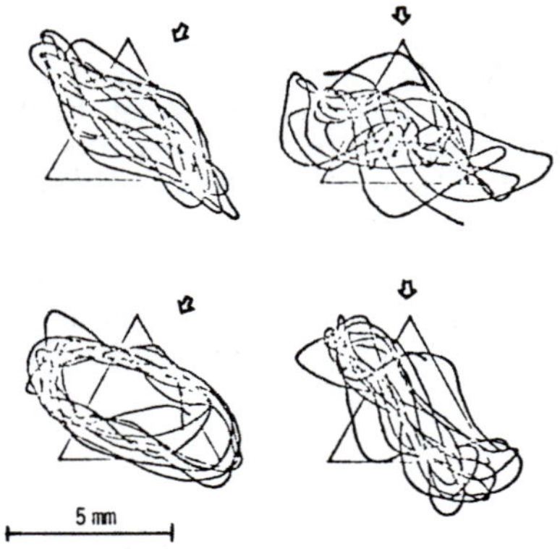

(1) (2)

②

G7

THE USE OF SLATS TO SUPPRESS VORTEX-INDUCED OSCILLATIONS OF SLENDER STRUCTURES

H. Y. Wong
Senior Lecturer

R. N. Cox
Senior Research Fellow

Department of Aeronautics and Fluid Mechanics
The University of Glasgow, Scotland.

SUMMARY

This paper describes an aerodynamic device for suppressing the flow-induced vortices which cause transverse oscillations on slender structures. By placing slats around the structure, the flow pattern is modified so as to reduce the amplitude of oscillations. A comparison is made with other aerodynamic damping devices, and practical experience with the slat system fitted to a chimney complex at the Edinburgh Royal Infirmary is described.

1. INTRODUCTION

Vortex-induced oscillations of slender structures having bluff cross-sections, such as chimney stacks or tall towers, can occur over a range of flow speed which varies with the structural damping and with the natural structural frequency. The transverse forces arise from the shedding of large vortices from alternate sides of the structure and if the vortex-shedding frequency is close to the natural frequency of the structure, large-amplitude oscillations can occur which may cause metal fatigue or may become divergent and lead to structural failure. The peak amplitude of oscillation can be reduced by increasing the structural damping. But if this is not practicable, aerodynamic damping systems can be employed. The present paper considers the use of longitudinal slats fitted round the structure (Sketch 1 of the Chart). The slats affect the vortex formation in such a way that the magnitude of the transverse exciting forces is considerably reduced and the amplitude of transverse oscillation is decreased.

2. DEVELOPMENT OF THE SLAT SYSTEM

The most efficient slat arrangement for a circularly cylindrical stack was determined in wind-tunnel studies reported by Wong (1977). Tests were carried out at a Reynolds number, $Re = VD/\nu$ (where V is the flow velocity, D the stack diameter and ν the kinematic viscosity) of the value up to 1.5×10^5 in a closed circuit low speed wind tunnel having a turbulent level of about 3%.

It was found that for all-round slats (which are necessary if the wind direction can vary, as is usually the case for an isolated stack) the maximum reduction in the amplitudes of transverse oscillation was obtained for the geometry shown in Sketch 1(b) with slats covering the upper half of the model stack, curve V of Sketch 2 shows the variation of the maximum peak-to-peak oscillation of the stack as the damping coefficient $k = 2M\delta_s/\rho D^2$ (where M is the mass per unit length, δ_s the logarithmic decrement of oscillation in still air, and ρ is the fluid density) is varied, and, for comparison, the curve for a circular cylinder with no damping device fitted is given as curve I. A considerable reduction of oscillation amplitude is obtained. If the slats are extended to the base of the stack, a further reduction in oscillation amplitude is obtained.

For unidirectional flow, such as might occur when the wind flow is channelled by other obstacles or by other parts of a structure (or in a tidal flow for marine structures), it was found that a further reduction in oscillation amplitude resulted from removing slats at the front and rear of the slat system to provide for larger openings there. The geometry involved is shown in Sketch 1(c) of the Chart, and the effect of varying the front and rear opening angles, θ, is given in Figure 1. The maximum effect on $\eta_{o_{max}}$ (the maximum non-dimensional amplitude of oscillation at the free end of the stack) is obtained for 2 slats (out of a total of 28 slats) removed at the front of the stack and between 2 and 4 at the rear. The value of $\eta_{o_{max}}$ obtained for a stack with slats over the upper 2/3 of the height, and with two slats removed at both front and rear of the model, is only 0.005, for a damping coefficient as low as k = 1.8! Further studies, which include the effect of varying the thickness of the slats were reported in Wong and Cox (1978).

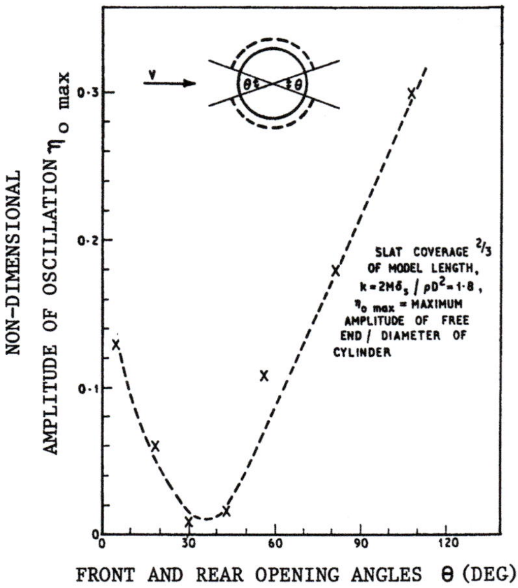

FIG. 1 MAXIMUM AMPLITUDE OF OSCILLATION OF CYLINDER FITTED WITH A SLAT SYSTEM AGAINST THE SIZE OF FRONT AND REAR OPENINGS

3. COMPARISON WITH OTHER AERODYNAMIC DEVICES AND FLOW MECHANISM

Two other aerodynamic damping devices for which test results are available are the helical strake, Sketch 1(d), developed by Scruton and Walshe (1957) and the perforated shroud, Sketch 1(e), used by Price (1956), and tested by Walshe and Cowdrey (1972). Two sets of results for stacks fitted with strakes are given in Sketch 2, curves II and IV. Not only are the strakes less effective in reducing flow-induced oscillation than the slats, but they also lead to an increased drag coefficient, and are also not adaptable to unidirectional flow situations.

The perforated shroud results given in curve III of Sketch 2 show that the optimised shroud system is about equivalent to the better of the two strake systems for which results are available, although their drag coefficient is lower. The slat system was also found to have a lower drag than the strakes, and further results comparing the three systems are given in Wong and Cox (1978). The effectiveness of the slat system in reducing vortex-induced oscillations depends on the modifications made by the slats to the flow field. Smoke studies have shown that high-velocity air entering between the slats facing the oncoming flow is channelled round the sides of the cylinder and is ejected through the slats over the rear portions of the cylinder. It is not possible for a vortex to grow between the cylinder and the slats, and any vortex growing on the outside of the slats is interfered with, and presumably weakened, by the flow through the rear slats. At the same time, the flow at the rear acts as a 'base-bleed' filling out the wake region, and leading to an increased base pressure. The flow mechanism is discussed in more detail in Wong and Cox (1978) and in Wong (1979).

Some recent studies have been made at Glasgow University of one cylinder in the wake of another when slats are fitted. With the slats appropriately arranged on the upstream cylinder, buffeting of the downstream cylinder can be effectively reduced for spacing up to 6 diameters of the cylinder, Wong (1980).

Tests have also been carried out on slatted cylinders in a towing tank at the University of Glasgow and Reynolds Numbers of up to 1.2×10^6 have been reached. The aim of the investigation was to find a cure for the riser vibration problem in tidal current. Even at such high values of Reynolds Number, the slats are still effective in suppressing the exciting forces caused by vortices.

4. EXPERIENCE WITH A CHIMNEY COMPLEX AT THE EDINBURGH ROYAL INFIRMARY

The main problem in the Infirmary Chimney Complex arose because of the geometrical arrangement of the five stacks as shown in Sketch 3. Flow incident normal to one side was liable to be diverted by the central stack so that it impinged at a high velocity on the two rear stacks. This is shown in the smoke photograph (see Chart) and tunnel tests confirmed that substantial oscillation did occur. By placing slats, as shown in Sketch 3, with one slat out of a total of nine (uniformly spaced) removed on the inside of each of the four corner stacks, it was considered that it would be possible to utilise the energy in the diverted flow for ejection to reduce the likelihood of periodic vortex excitation on the rear stacks, and thus the induced oscillation; this was confirmed by wind-tunnel tests to be the case. The arrangement of the slats was not based on the optimum configuration derived for an isolated cylinder because the chimney complex has a totally different cross-sectional geometry in which the channelling of flow through the stacks has a large part to play in vortex formation. To date, peak gust velocities of up to 30 m/s have been recorded. Without the slat device fitted, a maximum amplitude of oscillation of about 15 cm at the top of the chimney complex had been estimated for the reduced speed of about $V_r = V/ND' = 6.5$ (where V = wind speed, N = frequencies, and D' = distance between centres of two outer stacks) but the measurements showed that the maximum amplitude at the top of the chimney so far recorded was only 16 mm at a wind speed of about 15.5 m/s.

5. CONCLUSIONS

Model tests of a slat damping device on isolated stack models and test results from a chimney complex at Edinburgh have shown that the slat system is extremely effective in reducing the amplitude of vortex-induced oscillation on slender structures.

6. REFERENCES

1. Price, P., "Suppression of the Fluid Induced Vibration of Circular Cylinders", Journal of Engineering Mechanics Div., A.S.C.E., 82, Paper 1030, (1956).
2. Scruton, C. and Walshe, D.E.J., "A Means for Avoiding Wind-excited Oscillations of Structures with Circular or Nearly Circular Cross-section", National Physical Laboratory, Aero Report 335, (1957).
3. Wong, H.Y., "An Aerodynamic Means of Suppressing Vortex Induced Oscillation", Proc. Inst. Civil Engrs. Part 2, 63, 693-699, (1977).
4. Wong, H.Y., "The Suppression of the Effects of Vortices", Proc. 3rd U.S. Nat. Conf. on Wing Engng. Res., Florida, (1978).
5. Wong, H.Y. and Cox, R.N., "The Suppression of Vortex Induced Oscillation on Circular Cylinders by Aerodynamic Devices", Proc. 3rd Colloquium on Industrial Aerodynamics, Aachen, (1978).
6. Wong, H.Y., "A Means of Controlling Bluff Body Flow Separation", Journal of Industrial Aerodynamics, Vol. 4, No. 2, 183-201, (1979).
7. Wong, H.Y., "Vortex-Induced Wake Buffeting and Its Suppression", to be published in Journal Industrial Aerodynamics, (1980).

SLAT DAMPING DEVICE FOR STACKS

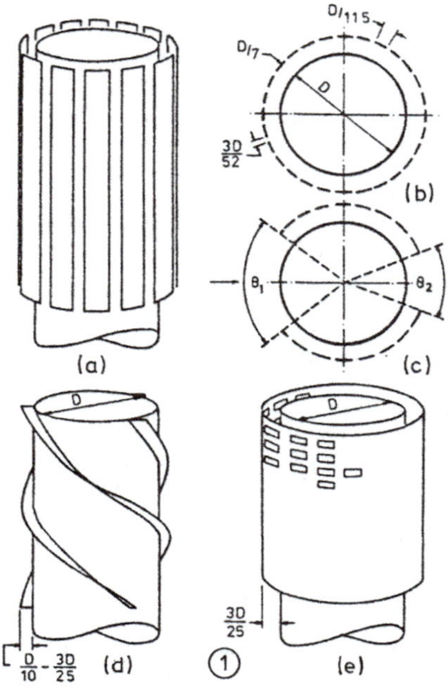

1

Problem: Vortex induced oscillation on slender structures.

Geometry: See Sketch 1(a), (b), (c) and 3.

Boundary Conditions: Flow multi-directional (1b -isolated stack) or uni-directional (1c)-wind channeled by neighbouring structures.

Flow Conditions: Natural atmospheric boundary layer.

Operating Conditions: Failure for isolated stacks following divergent oscillations can occur for V_r = 5 to 7 with k≃12 for typical structures. $V_r = V/ND$.

Cures: Use of slats over upper 1/3 to 1/2 of stack. Some slats can be removed in direction of most critical flow. Reduction of peak to peak from 14 cm (est.) to 1 cm for the Edinburgh Royal Infirmary chimney complex.

References: Wong H.Y., Proc. Inst. Civ. Eng. Pt. 2, 63, 693-699, 1977.
Wong H.Y., J. Industrial Aerodynamics Vol. 4, no. 2, 183-201, 1979.

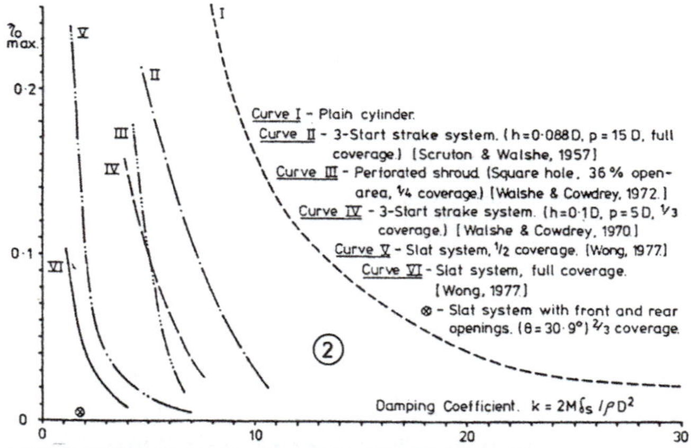

2

3

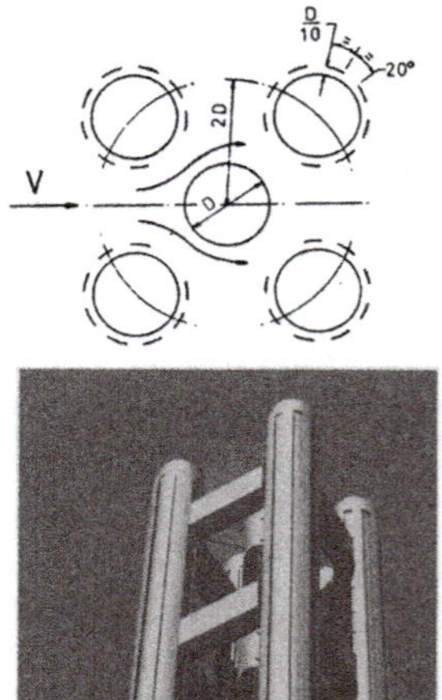

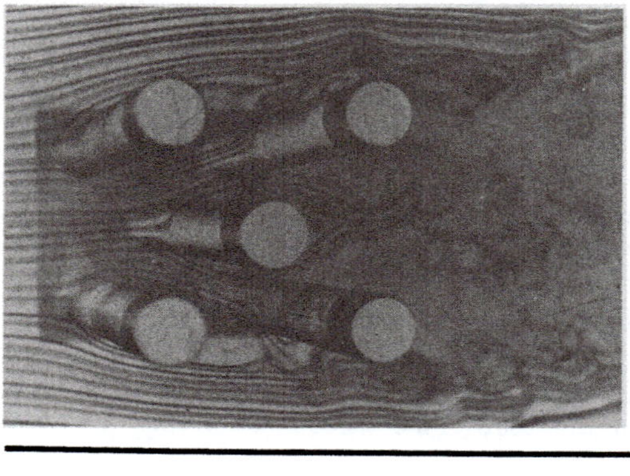

4

EDINBURGH ROYAL INFIRMARY CHIMNEY COMPLEX

G8 DAMPING OF STRUCTURES CONTAINING FLUIDS AND THEIR RELEVANCE TO WIND-INDUCED VIBRATIONS

David John Johns
Professor and Head

Department of Transport Technology
Loughborough University of Technology
Loughborough, England

SUMMARY

Results of experiments are reported on damping measurements of two classes of structure, oil jetty structural members and distillation columns. The tests were conducted with these structures empty and, also, partially or completely filled with a fluid. Significant increases in the logarithmic decrement were obtained by the addition of fluid.

Wind tunnel studies are also reported for models simulating oil jetty members and data are presented on the corresponding logarithmic decrements, vortex-excited dynamic responses and critical wind speeds. The data are compared with available theoretical data and encouraging agreement is obtained.

1. INTRODUCTION

The dynamic response of cylindrical structures to vortex excitation by a transverse external flow is dependent upon, inter alii, the value of the non-dimensional mass-damping parameter for the structure $k = 2m\,\delta/\varrho D^2$ (where D = diameter; m = mass per unit length; ϱ = air density).

The logarithmic decrement, δ, for typical structures is well known (Ref. [1]) and results are presented here for two further typical structural configurations, namely oil jetty structural members and distillation columns. Supplementary wind-tunnel studies of horizontal simply-supported tubes, with and without the addition of internal fluid, are also to be presented, and the data on vortex-excited vibration amplitudes, as a function of the parameter k, are to be compared with available predictions.

2. OIL JETTY STRUCTURAL MEMBERS

Large wind-excited oscillations were observed on the horizontal tie beams and some of the fuel pipes during the construction of a long oil jetty. The jetty structure comprised a series of portal frames linked by the tie beams, and Sketch 1 in Chart A shows an appropriate side-view of one tie beam and its pin-ended supports. The instrumentation used consisted of two accelerometers whose outputs were recorded directly and printed out on a Sanborn Recorder. Some of the tests made on the oil jetty are summarised in Table 1. Supplementary tests were performed on the shore and are summarised in Table 2. The details of the derivation of the logarithmic decrement, δ , from the test data are not discussed here as the techniques are well known, see Refs. [2] , [3] . Manual excitation was used on the jetty, and the tie beams set into resonance, by rhythmic movement of personnel on the member. f denotes natural frequency.

TABLE 1. JETTY INVESTIGATIONS

ELEMENT	LENGTH (m)	DIAMETER (mm)	FREQUENCY f (Hz) EXPT.	FREQUENCY f (Hz) THEORY S.S.*	δ
TIEBEAM 1	19.63	610	4	4.35	.022 (.015 to .026)
TIEBEAM 2	19.63	610	4.2	4.35	.029 (.014 to .036)

The values quoted for δ in Table 1 are, in general, the experimental averages, though in both cases the range of values is also shown. It was found that when one structural element was excited other adjacent elements could also be set into resonance. Thus, excitation of tie beam 1 caused out-of-phase vibration of the adjacent tie beam 2, with an amplitude 1/10th that of tie beam 1, and a corresponding excitation of the bowstring mode of the intermediate pile to which both beams were attached.

For the on-shore tests impulsive excitation was provided in which one end of each pipe or tie-beam section was lifted by crane and dropped onto a supporting wooden block.

TABLE 2. ON SHORE INVESTIGATIONS

ELEMENT	LENGTH (m)	DIAMETER (mm)	FREQUENCY f (Hz) EXPT.	FREQUENCY f (Hz) THEORY S.S.*	δ	COMMENTS
TIEBEAM 3 (a)	16.77	610	5.8	6.2	.063 (.048 to .074)	δ increases as vibration amplitude increases
(b)	16.77	610	5.8	6.2	.043 (.036 to .052)	
(c)	16.77	610	5.8	6.2	.035	
(d)	16.77	610	5.8	6.2	.022	
TIEBEAM 3 with 356 mm of water (e)	16.77	610	3.75	3.8	.25 (.23 to .29)	δ increases as vibration amplitude increases
(f)	16.77	610	3.75	3.8	.20 (.18 to .22)	

The values for δ quoted in Table 2 were less satisfactory in general than those in Table 1. A most significant result occurred when the horizontal tie beam 3 was just over half-full of water and a four-fold increase in δ was obtained. A corresponding reduction in frequency from 5.8 to 3.75 Hz was measured due to the addition of the water. It may therefore be concluded that

(a) values of δ for welded structures are dependent on vibration amplitude and can be as low as 0.014. However, a value of 0.03 is reasonably representative for design purposes [1] .

(b) the addition of water inside a welded tube can raise the value of δ fourfold. This, together with the consequential increase in m, the mass per unit length, significantly increases the value of the non-dimensional mass-damping parameter, k, though the natural frequency is lowered.

(c) for the particular empty tie-beam and using the measured values of natural frequency and logarithmic decrement it can be shown that vortex-excited oscillations could be anticipated and predicted. However, with a doubling in m and a fourfold increase in δ , i.e., with k eight times its original value, such oscillations would be unlikely.

*) S.S. = simply-supported ends.

3. WIND TUNNEL STUDY

The results obtained above prompted a detailed wind-tunnel study of the effects of an internal fluid on the wind-excited vibrations of a horizontal simply-supported tube representative of the oil jetty tie beams. The models were mounted (as in Chart A, Sketch 1) to be simply supported in the vertical plane only, and only vibrations in that plane were investigated with the airflow transverse to the model axis.

The wind tunnel to be used had an open, rectangular, working section of 1.06 m x 0.76 m and 1.60 m long. It is a closed-circuit wind tunnel with a useable speed range of 4 to 29 m/s. Tests were conducted for both uniform flow and artificially-induced turbulent flow conditions. Five separate mass and damping values were tested for each tube. Tables 3 and 4 describe the tubes and the results obtained. Strain gauges mounted at mid-length on each tube measured the lengthwise bending strain in the vertical plane and, by a suitable calibration a corresponding central, maximum transverse deflection was obtained from the strain data.

The results in Table 3 for δ showed, in general, a non-linear effect of vibration amplitude y on δ such that δ increased as y increased. This would be beneficial in practise, as it tends to be stabilising. The previously determined values of δ were used in conjunction with known values of m, the mass per unit length, to obtain the corresponding values of k. These results are superimposed on a plot of theoretical predictions in Chart A, Sketch 2. Clearly the addition of water into the horizontal tubes has a marked effect in increasing δ, m and, hence, k, as well as in reducing y/D. Reduction of y/D to about 1×10^{-2} requires a value of k of about 20.

TABLE 3. DATA ON CIRCULAR CYLINDRICAL TUBES

Model	Material	Overall Diameter (mm)	Wall Thickness (mm)	Fluid Content	Mass/ Length (Kg/m)	f (Hz)	δ	
							y/D= $.5\times10^{-2}$	y/D= $.7\times10^{-2}$
1	Steel	19.05	.25	Empty	.157	60.8	.017	.017
				1/4 full	.228	51.5	.051	.073
				1/2 full	.300	42.8	.100	.100
				3/4 full	.371	38.5	.085	.085
				Full	.442	35.1	.065	.065
2	Steel	25.40	.30	Empty	.245	73.3	.020	.020
				1/4 full	.371	56.5	.046	.055
				1/2 full	.496	45.8	.065	.074
				3/4 full	.625	42.4	.050	.056
				Full	.750	40.6	.047	.046
3	Aluminium	19.05	.315	Empty	.087	61.8	.019	.019
				1/4 full	.158	38.5	.044	.044
				1/2 full	.229	31.8	.083	.083
				3/4 full	.300	26.9	.063	.063
				Full	.371	24.1	.039	.039
4	Aluminium	25.40	.315	Empty	.115	84.6	.018	.018
				1/4 full	.241	48.5	.030	.030
				1/2 full	.368	38.1	.039	.040
				3/4 full	.495	31.8	.029	.029
				Full	.620	28.6	.021	.021

The experimental and theoretical values shown in Chart A, Skh. 2 are discussed in greater detail in Ref. [4] and apply specifically to the simply-supported tube configuration. The theoretical studies from Ref. [5] are based on a valuable compilation of such data by Blevins in Ref. [6]. Scruton's earlier experimental data (Ref. [7]) forms a resonable upper limit to the present experimental data and also agrees well with the predictions of the correlation model (Ref. [8]) which takes account of the fact that the vortex-induced oscillatory forces are (i) a non-linear function of vibration amplitude and (ii) not completely correlated along the length except, perhaps, at large vibration amplitudes y/D > 0.5.

TABLE 4. SUMMARY OF WIND-TUNNEL RESULTS

TUBE CONFIGURATION			UNIFORM FLOW				TURBULENT FLOW			
Material	Overall Diameter (mm)	Fluid Content	y/D $\times 10^2$	$\frac{fD}{U}$	k	U (m/s)	y/D $\times 10^2$	$\frac{fD}{U}$	k	U (m/s)
Steel	19.05	Empty	3.3	.22	14	5.2	1.9	.21	16	5.6
		1/4 full	2nd Mode only				.47	.21	51	4.8
		1/2 full					.21	.22	107	3.8
		3/4 full					.31	.22	112	3.4
		Full					.31	.22	96	3.0
Steel	25.40	Empty	1.40	.20	13	9.3	1.53	.21	14	9.1
		1/4 full	.55	.19	45	7.5	.49	.21	42	7.1
		1/2 full	.75	.18	95	6.6	.49	.18	82	6.4
		3/4 full	.50	.20	102	5.4	.35	.18	62	5.9
		Full	.41	.20	76	5.1	.31	.20	65	5.0
Aluminium	19.05	Empty	7.9	.17	8	6.4	6.3	.19	8	6.7
		1/4 full	Poor results				.84	.22	31	2.7
		1/2 full					1.05	.20	85	3.0
		3/4 full					Poor results			
		Full								
Aluminium	25.40	Empty	1.90	.2	6	10.4	.66	.21	5	10.3
		1/4 full	.98	.2	19	6.2	.43	.22	18	5.6
		1/2 full	.35	.2	36	4.8	.39	.22	36	4.4
		3/4 full	.19	.2	35	4.1	.09	.23	33	3.6
		Full	.24	.21	30	3.5	.19	.21	30	3.5

4. DISTILLATION COLUMNS

A potentially serious low-frequency sway motion caused by wind was observed on the tallest (61 m) of four in-line distillation columns, with a spacing of 6.1 m between the centres of each column. Supplementary wind-tunnel studies showed that this behaviour was most likely when the wind direction was at a shallow angle ($\simeq 15^\circ$) to the line of columns. The need was seen for investigations of the logarithmic decrement, δ , of the two tallest columns in both full and empty conditions being numbers 2 and 4 in line. (Chart B, Sketch 1).

All four columns consisted of welded steel tubes bolted together at intervals with flanges and having an insulation cladding. All four were fitted with helical strakes and their geometrical details are given in Chart B, Sketch 1. Two techniques were applied to excite the structures for the determination of the logarithmic decrement, δ , and natural frequency f:

(a) An impulsive loading followed by a decaying oscillation was achieved as in Ref. [2] . The initial input loading was applied by means of a hawser attached to the structure near its top with a mechanical fuse link (shear-pin assembly) in series. A condition was imposed that the maximum tip deflection should not exceed 150 mm. Sketch 1 in Chart B shows the loading directions in relation to the line of columns.

(b) Forced random excitation from the natural wind can also be analysed (Ref. [3]) assuming the response to be that of a linear system subjected to a white-noise input, to obtain the narrow band spectra, the autocorrelation function, and hence the logarithmic decrement, .

Measurement of wind speed and direction during the tests was accomplished by permanent on-site instrumentation at a height of 21.3 m. Two accelerometers were installed on each column near the top; one as close as possible to the line of the impulsive force and the other as close as possible to 90° to this line on the southern side. The four accelerometer signals, wind-speed and direction data, time, and a voice commentary were recorded on a tape recorder for later analysis, and the signals from the two accelerometers on the column under test were also recorded on a two-channel Sanborn chart recorder and oscilloscope for instant real-time display and preliminary analysis. The test results are summarised in Table 5.

The decaying oscillation traces obtained in all tests were not easy to analyse because of the interaction between the motions in two orthogonal directions as evidenced by the two accelerometer records. Ref. [2] discusses this difficulty, and the same analytical approach has been adopted here to obtain appropriate values for δ . Corresponding difficulties were experienced using the technique of Ref. [3] since the auto-correlation functions were not of the anticipated form (implying that the input was not that of a stationary random process). The two different analyses were conducted independently and by separate persons and it is therefore somewhat reassuring that the two sets of results are in such good agreement.

TABLE 5. DISTILLATION COLUMN TEST DATA

COLUMN	Wind Speed at 21.3m (mph)	Wind Direction	Wind Excited Tip Deflection (mm)	Test Pull Tip Deflection (mm)	Impulse δ	Wind Excited Damping δ	Estimated Range δ
2 Empty							
(i)	12-16	90-120°	0 - 30	135	.032-.047	-	.029-.047
(ii)	12-16	90-130°	0 - 30	125	.029-.041		
4 Empty							
(i)	11-13	80-110°	0 - 25	110	.017-.028	-	.017-.030
(ii)	11-16	80-110°	0 - 25	100	.109-.030		
2 Full							
(i)	14-16	230-250°	Small	150	.16 - .20	.12-.20	.10 - .20
(ii)	12-15	210-240°	Not Recorded	132	.10 - .18	.13-.18	
4 Full							
(i)	16-20	220-240°	Small	105	.19 - .36	.19-.33	.19 - .36
(ii)	14-18	220-240°	Not Recorded	120	-	.25-.36	

Sketch 2 in Chart B shows an interesting effect on Colum 4 (after an appropriate time delay) following an impulse on Column 2. This effect was not experienced with the same wind direction, etc., in the reverse situation, and could be aerodynamic in nature due to the energized wake shed from 2 - rather than ground borne. It may be concluded therefore, that

(a) the value of for the empty columns is
$0.029 \leqslant \delta \leqslant 0.047$ for Column 2 - Average 0.035
$0.017 \leqslant \delta \leqslant 0.030$ for Column 4 - Average 0.025

(b) the value of for the full column is
$0.10 \leqslant \delta \leqslant 0.20$ for Column 2 - Average 0.18
$0.19 \leqslant \delta \leqslant 0.36$ for Column 4 - Average 0.27

(c) there is aerodynamic coupling between vibrating columns which may cause significant responses in the downwind column. Thus, any wind-tunnel oscillation tests should be conducted with all columns being aeroelastically modelled.

5. CONCLUSIONS

New data is presented on the logarithmic decrements δ of various structural configurations showing also the effects of the inclusion of an internal fluid on δ. Corresponding wind-tunnel studies, showing the effects on δ and k and on the vortex-excited response of added internal fluid, are also believed to be novel. It may be concluded that added fluid can increase δ by a factor of between 4 and 10 and will then increase the mass-damping parameter k by a still greater factor. Whilst this may be a potent means of reducing large-amplitude vibrations in full-scale structures, it may also prove a simple effective means of varying k in model studies.

REFERENCES

1. Zorilla, E.P., "Predicting Aerodynamic Behaviour of Stacks", Oil and Gas Journal, pp. 69-76, November 1, 1971.
2. Johns, D.J. et al, "On Increasing the Structural Damping of a Steel Chimney", Earthquake Engineering and Structural Dynamics, Vol. 1, No. 1, pp. 93-100, 1972.
3. Jeary, A.P. and Winney, P.E., "Determination of Structural Damping of a Large Multi-Flue Chimney from the Response to Wind Excitation", Inst. of Civil Engineers, Tech. Note, 65, 1973.
4. Johns, D.J., and Skordilis, G., "Vortex Excitation by Transverse Flow of Tubes Containing Fluid", Paper submitted to the Third National Congress on Pressure Vessels and Piping Technology - Symposium on Flow-Induced Vibration, San Francisco, June 1979. (Also available as a Loughborough University of Technology Report, TT 78 R 05, November, 1978).
5. Milsted, M.G., and Johns, D.J., "On the Structural Damping of Steel Chimneys", Loughborough University of Technology Report TT 78 R 05, November 1978.
6. Blevins, R.D., "Vortex-Induced Vibration", Chap. of Flow-Induced Vibration, pp. 11-54. Van Nostrand, 1977.
7. Scruton, C., "On the Wind-Excited Oscillations of Stacks, Towers and Masts". Proceedings International Conference on Wind Effects on Buildings and Structures. N.P.L., Teddington, U.K., June 1963.
8. Blevins, R.D. and Burton, T.E., "Fluid Forces Induced by Vortex Shedding". Journal of Fluid Engineering, Vol. 95, pp. 19-24, 1976.

CHART A: OIL JETTY STRUCTURAL MEMBERS

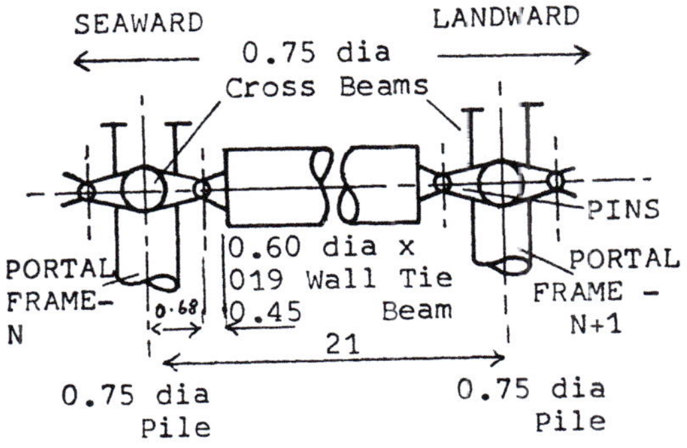

① Side view of horizontal tie beam linking adjacent portal frames (All dimensions in m)

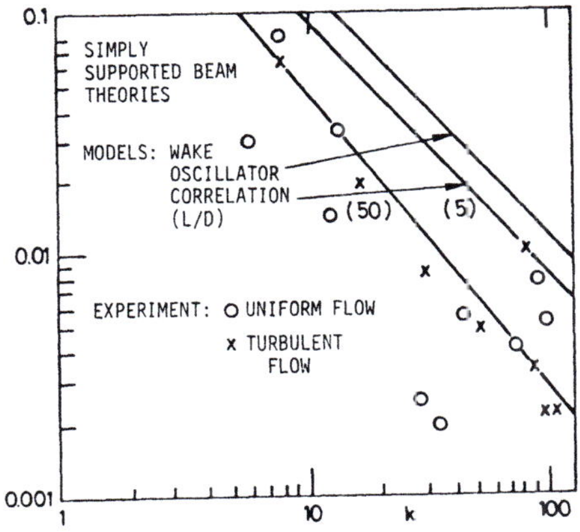

② Maximum resonant amplitude as a function of damping; theory and experiment

Problem: Horizontal tie beams and empty oil-bearing pipes excited by wind. Damage due to fatigue possible. Personnel on jetty react unfavourably. Phenomenon pronounced during construction phase.

Geometry: See Sketch 1

Boundary Conditions: Tie beams considered to have simply supported ends.

Vibration: Wind excitation occurred consistent with Strouhal number of 0.2.

Structural Dynamics:

Tie Beam: Length 19.63 m; diameter 0.61 m. Fundamental beam mode frequency 4 Hz (measured), 4.35 Hz (theory). Logarithmic decrement = 0.022.

Pipe A: Length 21.43 m; diameter 0.40 m. Fundamental beam mode frequeny 4 Hz (measured); 5.75 Hz (theory)clamped ends, 2.54 Hz (theory) simply supported ends. Logarithmic decrement = 0.04.

Cures:

1. Brace horizontal tie beams (structural members) to the adjacent oil bearing pipes (adopted) - effective increase of stiffness and added mass.

2. Add fluid to horizontal tubes to optimum amount to increase the mass damping parameter k = 2m D² (= air density). This was not proceeded with though explored experimentally subsequently. Addition of water half-full can raise fourfold and mass/unit length twofold. Hence 8 x increase in k is possible. Subsequent wind-tunnel studies confirmed fasibility (see Sketch 2).

Model	Material	Overall Diameter (mm)	Wall Thickness (mm)	Fluid Content	f (Hz)	δ
1	Steel	19.05	0.25	Empty	60.8	.017
				1/4 full	51.5	.051
				1/2 full	42.8	.100
				3/4 full	38.5	.085
				Full	35.1	.065

CHART B: FOUR IN-LINE DISTILLATION COLUMNS

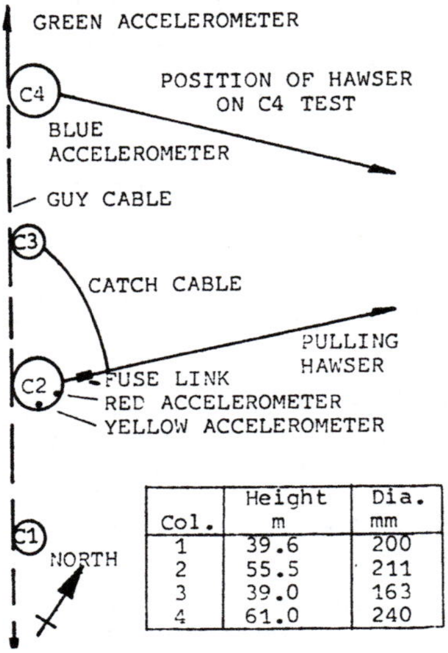

Col.	Height m	Dia. mm
1	39.6	200
2	55.5	211
3	39.0	163
4	61.0	240

(1)

Problem: Wind excitation observed on tallest (61 m) of four different inline distillation columns of basically circular section already fitted with helical strakes. Problem worst when columns are empty.

Geometry: See Sketch 1

Operating and Boundary Conditions: Maximum excitation known (from wind-tunnel studies) to occur on Column 2 and 4 with wind direction some 15° off-line of columns. All assumed to be free-standing vertical cantilevers which may be empty or have varying content of distillate fluid.

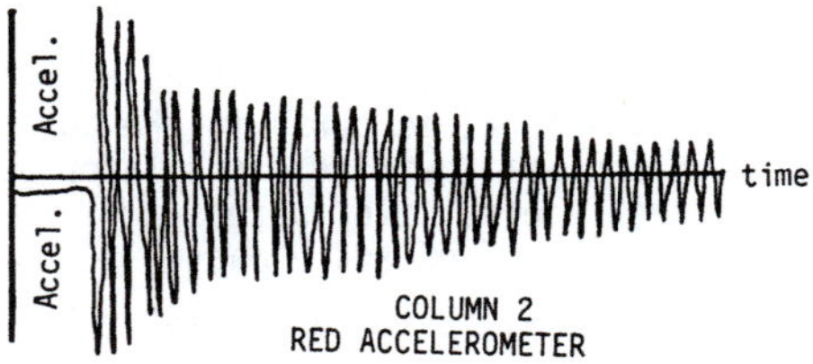

COLUMN 2
RED ACCELEROMETER

Structural Dynamics: Wind excitation as above treated as a white noise input and data analysed to derive logarithmic decrement δ . Impulse excitation (mechanical) also used to derive logarithmic decrement δ .

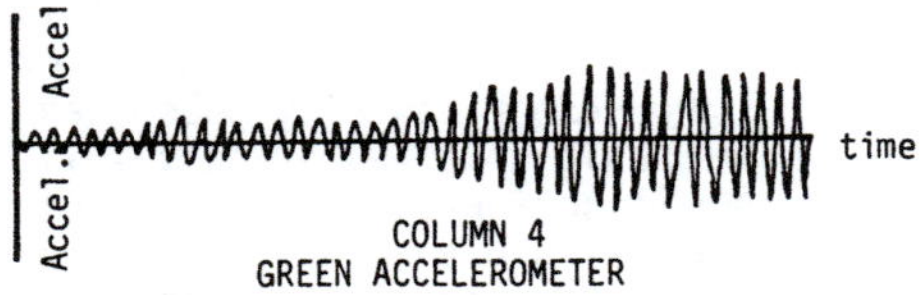

COLUMN 4
GREEN ACCELEROMETER

Vibrations: Sketch 2 shows interesting affect of aerodynamic coupling from Column 2 onto Column 4. With wind direction from 2 towards 4, an impulsive excitation of 2 develops (after appropriate time delay) an impulse on 4.

Conclusions:

(a) The value of δ for the empty column is
$0.029 \leq \delta \geq 0.047$ for Column 2 - Average 0.035
$0.017 \leq \delta \geq 0.030$ for Column 4 - Average 0.025

(b) The value of δ for the full column is
$0.10 \leq \delta \geq 0.20$ for Column 2 - Average 0.18
$0.19 \leq \delta \geq 0.36$ for Column 4 - Average 0.27

(c) There is aerodynamic coupling between vibrating columns which may cause significant responses in the downwind column. Thus any wind-tunnel oscillation tests should be conducted with all columns being aeroelastically modelled.

G9 THE WIND-INDUCED VIBRATION OF A TALL BUILDING

B. R. Ellis*, R. A. Evans**, A. P. Jeary* and B. E. Lee**

* Building Research Establishment, U.K.
** University of Sheffield, U.K.

SUMMARY

This paper describes how a series of tests, originally performed to measure the dynamic structural properties of a tall building, have been used both to determine the modal wind forces which act on the building and to suggest a cause for the failure of some of the building components. This failure involves large diagonal cracks which are present in the majority of the blockwork partition walls on the upper floors of the building, walls which in the terms of the design philosophy of the building would normally be considered to be 'non-structural'. Whilst a cure to the problem has been suggested to the building owners, the authors consider that problems of this type may continue to occur so long as our knowledge of the dynamic properties of buildings and of the generation of the dynamic wind forces produced by the flow-structure interaction remains at its present inadequate level.

1. INTRODUCTION

As part of an extensive research programme being carried out to determine the dynamic properties of tall buildings, a series of tests have been performed to determine the wind-induced modal forces acting on the Arts Tower at Sheffield University. This building is 80m high, 36m wide and 20m deep, and consists structurally of a central cast in-situ reinforced concrete core with deep floor slabs spanning between the core and external reinforced concrete columns. The building stands in a relatively exposed position, partially surrounded by low-rise structures.

Shortly after the Arts Tower was first occupied in 1966, a number of the occupants of the upper floors of the building claimed that they could clearly perceive movement of the building. Since then there have been few complaints until recently when, perhaps aware of the current test programme, one or two people have said that they thought that they had detected some movement. Apart from the perception of motion in the building the wind storms which have occurred over the last thirteen years have had other, more identifiable, effects. The major secondary effects concern the cracks which have appeared in the partition walls in the upper part of the building. These cracks are most noticeable on walls close to the periphery of the building and tend to run diagonally, descending to the periphery, across the walls for several metres. Since the cracks cause the surface plaster finish to fall out along their length, they represent a serious continuing maintenance problem.

2. FORCED-VIBRATION TESTS

To determine the dynamic properties of the building and to calibrate it in preparation for the wind response measurements, a series of forced-vibration tests were conducted in April 1976, using an eccentric mass rotary vibrator. The vibrator was mounted at a height of 70m and the response was monitored by fixed accelerometers at that level, and a further movable accelerometer which could be located anywhere in the building. Since known forces were applied at the modal frequencies and the subsequent response measured, it has been possible to calculate the values of modal mass, inertia and stiffness. The essential results of this series of tests are contained in Table 1. The elevational mode shapes of the fundamental modes were found to be acceptably approximated by straight lines.

A second series of forced-vibration tests were conducted in February 1979 using a set of four eccentric mass vibrators located at a height of 67m. These later tests, in which the input force levels were varied over a greater range than before, demonstrated that both the frequencies of the fundamental modes and their associated damping values vary with amplitude of motion, though these changes are not considered to be significant at the force levels associated with wind action on the building.

An inspection of the original design data for the building showed that the central core was intended to take all the wind load. Despite this, it contributes surprisingly little to the total stiffness of the building, 2 per cent in the north-south direction and 17 per cent in the east-west direction. Even when an allowance for the stiffness of the external columns is included, this only accounts for 25 per cent of the stiffness in the north-south direction and 60 per cent in the east-west direction. The additional stiffness is presumed to be provided by the internal partition walls and the external cladding panels.

TABLE 1 BUILDING PROPERTIES (measured April 1976)

MODE	N-S	E-W	θ
Fundamental natural frequency (Hz)	0.678	0.863	0.787
Modal mass (Kg)	6.35×10^6	5.84×10^6	-
Modal inertia (Kg/m^2)	-	-	8.14×10^8
Modal lateral stiffness (N/m)	1.15×10^8	1.70×10^8	-
Modal torsional stiffness (Nm/radian)	-	-	1.99×10^{10}
Damping (% critical)	0.86	0.86	0.90
Second mode natural frequency (Hz)	2.46	-	-

3. WIND RESPONSE MEASUREMENTS

Following the forced-vibration tests, the response of the building to the wind was measured using four fixed accelerometers at the 70m level. From the spectra of the accelerometer records the mean square accelerations in each mode were obtained for particular values of the incident wind speed and direction, using an ensemble averaging process. Since the values of modal mass, elevational mode shape, modal frequency and damping had already been determined from the forced-vibration tests, it was then possible to determine the mean square modal force at each of the fundamental frequencies. These results are shown in Table 2. The values of wind speed are equivalent hourly mean speeds at a height of 10m in open level country. Included in Table 2 are the corresponding values predicted by the method given in Ref. (1). This method predicts only the in-wind response and considers the worst case when the wind is blowing normal to a face. From the data in Table 2 the relationships between the wind speed and the r.m.s. modal forces were calculated. After averaging the relationships for all wind direction it was found that the $N\text{-}S_1$, the $E\text{-}W_1$ and the θ_1 modal forces are proportional to the velocity to the power 1.93, 2.08 and 2.30 respectively. The corresponding relationship given by (1) for all the modes, is velocity to the power 2.8. Using the calculated relationship for each wind direction the equivalent modal forces for a nominal wind speed of 10m/s were determined. Sketch 3 in the chart shows the variation of modal forces with wind direction for this wind speed.

TABLE 2 WIND CONDITIONS, FORCES AND BUILDING ACCELERATIONS, (1976-1977)

	Mean Wind Direction (degrees)	Mean Wind Speed (m/s)	$\bar{F}_r^2$		
			$N\text{-}S_1$ (KN^2/Hz)	$E\text{-}W_1$ (KN^2/Hz)	θ_1 (KN^2m^2/Hz)
	254	9.3	6.3	2.2	427
	254	11.1	8.5	1.7	1012
	275	8.5	3.4	1.8	324
	275	11.8	35.2	6.5	3749
	295	9.1	8.3	2.8	349
WIND EXCITATION RESULTS	295	11.8	24.1	6.5	1671
	305	7.0	3.4	1.5	174
	305	9.6	15.2	4.3	920
	305	11.4	23.4	6.1	1472
	305	13.3	27.0	6.7	1564
	338	10.0	3.4	0.3	133
	338	11.9	6.7	1.1	2.6
	338	13.3	8.4	1.7	400
	350	8.1	5.8	0.6	156
	270	10.0	--	1.2	--
	270	11.9	--	3.2	--
ESDU 76001 PREDICTIONS	270	13.3	--	5.9	--
	360	10.0	3.5	--	171
	360	11.9	9.3	--	452
	360	13.3	17.2	--	844

A number of conclusions are apparent from this sketch:

a) The variation of the $N\text{-}S_1$ and $E\text{-}W_1$ modal forces with wind direction are almost identical, where the magnitude of the $N\text{-}S_1$ modal forces is four times that of the $E\text{-}W_1$ modal forces. These translational modal forces exhibit a maximum when the wind direction is onto the corner of the building.

b) The variation of the torsional modal force shows two maxima, one when the wind is directed onto the narrow face of the building (270°) and one when the wind direction is onto the corner.

c) The modal forces are greater when the wind is normal to the narrow face than when it is normal to the wide face of the building.

4. THE FLOW-STRUCTURE INTERACTION

The modal forces, discussed in the previous section, can be considered to represent the fluctuating forces acting on the building, resolved into components and restricted to match the modal frequencies. In the terms of wind spectral content these modal frequencies may be considered to be approximately equal, and so the modal forces can be compared directly. However the variations of these modal forces cannot yield a complete picture of the flow-structure interaction since other important aerodynamic phenomena occurring at other frequencies have been discarded. Despite this restriction the conclusions of the previous section enable a tentative discussion to be held based on earlier experiments on fluctuating forces on bluff body wind tunnel models, see for example Ref. (2).

(i) Wind Direction 360°. Here the wind is normal to the wide face of the building and all the modal forces are small. For this direction the building presents a short afterbody arrangement to the flow and it is suggested that the separated flow from the front corners will not reattach onto the side walls. Hence it may be expected that only small fluctuating forces will act on the E and W side walls and the base, S wall, and that the major fluctuating forces will result from incident turbulence buffeting on the N face. This would then give rise to greater in-wind forces, N-S, than cross wind forces, E-W, and no mechanism would be available to promote torsional forces.

(ii) Wind Direction 270°. This wind direction now presents a long afterbody arrangement to the flow where it is likely that flow reattachment to the N and S side walls will occur intermittently over some proportion of the building height. This situation may then be expected to give rise to large cross wind forces and the incident buffeting effect would be small by comparison, a finding similar to that of Saunders and Melbourne, Ref. (3), who used wind tunnel models to investigate a similar flow-structure interaction. It is interesting to note that the value of the E-W modal force for a wind direction of 270° is approximately equal to that of the N-S modal force for a wind direction of 360°, when the relative face areas are taken into account, which suggests that the mechanism of incident flow buffeting remains relatively unchanged with flow configuration.

The large peak in the torsional modal force for the 270° wind direction may be explained by a poor correlation between the phase and position of the intermittent flow reattachment occurring on opposite side walls.

It is not considered profitable to extend this discussion on the aerodynamic aspects of the interaction since so little real evidence for the suggestions made here is available in the literature. There is clearly a need for more experimental data of the type given in Ref. (2) but for three dimensional bodies of various plan form shapes.

5. PARTITION WALL DAMAGE

There are three relevant points to be taken into account in considering the damage to the partition walls:

(a) The calculation of the actual stiffness of the building has suggested that the partition walls must be an integral part of the structure and thus stresses will be set up in them when external loads act on the building.

(b) Cracks are apparent in the majority of the partition walls in the upper part of the building. The surface finish of these walls is regularly re-decorated to cover the cracks, and yet they continue to appear. One particular instance of this was observed by one of the authors in January 1977 on a day during which a severe storm passed in the space of approximately 12 hours. A few days prior to this storm the room containing the recording apparatus had been re-decorated and the walls were covered in a fresh, uniform plaster surface, which had been painted. Within 6 hours of the start of the storm the diagonal cracks had re-appeared in the plaster, opening up to 3 mm wide at their centres.

(c) Large bending and torsional forces act on the building for wind directions in the range 270° to 300°. An analysis of the strong wind meteorological data for Sheffield has shown that a larger number of storms has occurred over this directional range since the completion of the building than from any other directional range, Ref.(4).

From a consideration of these points the hypothesis has emerged that the action of the wind on the building is responsible for the partition wall failure.

So far as a cure to the problem is concerned it has not been considered economically feasible to attempt to reduce the overall vibration level of the building. In order to attempt a local cure to individual walls it has been suggested to building owners that a small gap is made between the top of the wall and the underside of the floor slab to which it is, at present, connected. This may then help to reduce the build up of stresses due to the relative motion of adjacent floor slabs. However it should be pointed out that if this remedy were to be applied to all partition walls, the stiffness of the building may possibly reduce to a level at which the human perception of vibration during wind storms became a problem.

6. CONCLUSION

When the Arts Tower of the University of Sheffield was designed it easily satisfied the then current wind-loading code requirements. It gives the appearance of being a well proportioned building of solid construction. Yet were it not for the significant contribution of the so-called non-loadbearing elements of the building, the building would undoubtedly have been excessively flexible and would have had an unsatisfactory performance in the wind.

However the contribution of these 'non-structural elements to the building's structural properties has led to a large scale maintenance problem in dealing with the diagonal cracks in partition walls. Local cures for badly affected individual walls may be possible but a general solution to the problem is not yet in sight.

Acknowledgements

The authors wish to thank the University of Sheffield for permitting tests to be carried out on their building and also for the provision of services and laboratory accommodation. The authors are also grateful for the contributions to this work made by P.R.Sparks, Virginia State University, V.C.M. de Souza, supported by the National Research Council of Brazil, and by J. Plaquet of CEBTP, Paris.

7. REFERENCES

1. E.S.D.U., Response of Flexible Structures to Atmospheric Turbulence, Engineering Sciences Data Unit, Item 76001, London (1976).

2. Lee,B.E., The Effect of Turbulence on the Surface Pressure Field of a Square Prism, Jnl. Fl. Mech., Vol. 69, pp 263-282, (1975).

3. Saunders, J. and Melbourne, W. , Tall Rectangular Building Response to Cross-Wind Vibration, 4th Int. Conf. Wind Effects on Buildings and Structures, Heathrow, London (1975).

4. Lee, B.E., Extreme Wind Data and Building Design, Weather, Vol. 33, pp 416-423 (1978).

TALL BUILDING : FAILURE OF PARTITION WALLS

Problem: To determine whether wind action is responsible for diagonal cracking in non-structural blockwork partition walls of a high-rise(20-storey) office block. Main constructional elements are a central core and cantilevered floor slabs, with lightweight external cladding. Cracking is most serious on the upper six floors of the building on walls close to the periphery. Cracks in the surface plaster are causing a serious long-term maintenance problem.

Geometry: See Sketch 1. This is a typical floor plan. The building is 80m high.

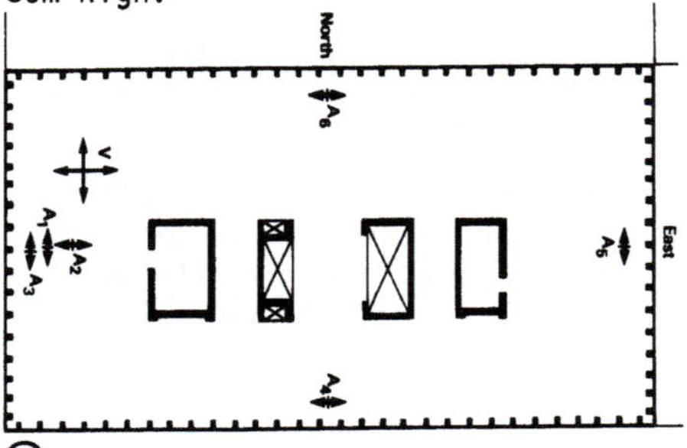

①

Boundary Conditions: The building stands in a predominantly suburban, low-rise terrain subject to strong winds from the west and north. Wind speeds are monitored at 6m above the roof, and in a nearby park at 10m above local roof height.

Flow Conditions: Cracks have been observed to form over the period of a single wind storm, average wind speed 15 m/s at 10m.

Structural Dynamics: Average frequencies of fundamental bending modes of vibration, N-S mode 0.678 Hz, E-W mode 0.863 Hz. Average frequency of fundamental torsional mode 0.787 Hz. From forced-vibration tests it has been found that both modal frequencies and associated values of damping vary with level of input force. See Sketch 2 for fundamental mode shapes.

Flow Measurements: The structural response of the building during high winds has been monitored. From these measurements the modal forces acting on the building have been calculated, for different wind directions. The variation of these forces with wind direction is shown in Sketch 3.

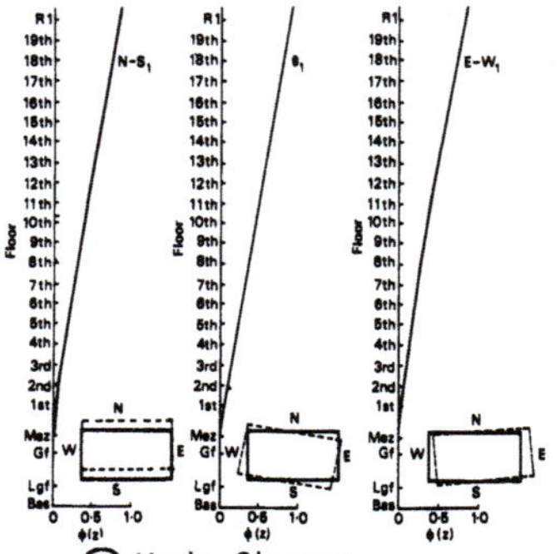

② Mode Shapes

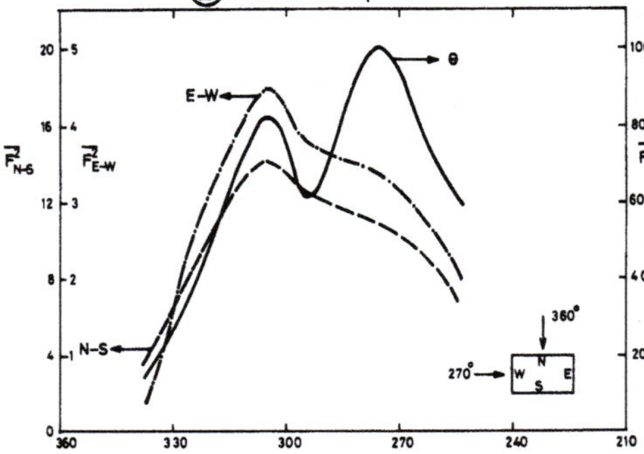

③ Modal Forces referenced to 10 m/s.

Findings: From forced-vibration tests it is apparent that the "non-structural" partition walls play a major role in providing the building's stiffness. The wind response measurements show large torsional forces acting on the building, particularly for oblique winds. This wind-induced predominantly torsion vibration is considered possibly to be responsible for the partition wall failure.

Cures: Attempts to reduce the overall vibration level of the building are not considered to be economically feasible. Cure to be attempted on some partition walls initially is to separate them at the top from the floor slab above by a small gap, thus permitting them to move with the lower slab.

Reference: Jeary, A.P., The Dynamic Behaviour of the Arts Tower, University of Sheffield. B.R.E. Curr. Pap. 48/78. (Editor). 1978.

G10

SEVERAL EXAMPLES OF WIND-INDUCED VIBRATIONS

Miroš Pirner
Associate Professor

Technical and Research Testing
Institute in Civil Engineering

15000 Prague 5
Czechoslovakia

SUMMARY

This work is concerned with wind-induced vibrations of Civil-Engineering structures. The first case deals with vortex-induced vibrations of road-bridge columns. As a cure, the columns were filled with granulated gravel, suppressing the amplitudes of induced vibrations to 1/20 of the original value. The second case is concerned with the collapse of a lighting tower on a sport stadium, where vibrations of all four lighting towers perpendicular to the wind direction were observed after their erection. At a wind speed of 30 m/s the vibrations started and bucklings of the thin wall of the shank eventually occurred. The cure consisted of using stiffening diaphragms. The third case is concerned with the damage of an aluminium cover on the flat roof of a swimming pool where vibration associated with suction behind the leading edge of the roof caused the damage of roof clamps, resulting in rolling-up of the roof. The fourth case is concerned with the elimination of wind-induced vibrations of TV towers in Czechoslovakia by using pendulum absorbers.

1. STEEL COLUMNS OF AN ARCH BRIDGE

Intensive vortex-excited vibrations in the lateral direction were observed in cylindrical columns of a road bridge across the Vltava [1] , as shown in the first Chart. The largest motions were observed in the longest columns (42 m long, o.d. of 1 m) whose first natural frequency was 1.6 Hz. The measured amplitudes at the midpoint of the columns were 11, 13, and 5.3 cm at wind velocities of 7, 8 and 11 m/s, respectively, [2] corresponding to Reynolds numbers in the supercritical range. The lateral vibrations at a mean wind velocity of 8 m/s caused the alternating stresses from nearly plus 5090 to minus 10570 N/cm². These self-induced vibrations of the columns did not couple with those of the roadway (fundamental frequency 3.4 Hz).

As a cure, it was decided to fill the columns with granulated gravel (6 to 15 mm) up to 3/4 of the column height. The intention of this measure was to increase the damping, decrease the first natural frequency and increase the mass. Other types of mechanical or aerodynamical dampers, as well as an increase of natural frequencies due to prestressing or bracing, were rejected. After filling, measurements of the wind-induced lateral vibrations gave the following results: decrease of the fundamental frequency to 0.7 Hz; no increase in damping (critical wind speed became 4 m/s); decrease in amplitudes (max. 6.5 mm at wind speed of 4 m/s).

2. LIGHTING TOWER

The second Chart shows one of four steel lighting towers around a sport stadium. At a wind speed of 30 m/s, the tower collapsed. There were no eye witnesses of the collapse, and therefore a detailed description of the vibration does not exist.The important aspect of the collapse was buckling of walls with large slenderness ratio: side walls (200); rear wall (150); and front wall (75). The buckling was combined with deformation of the shank cross-section; only cross-stiffening ribs (constructed from profile 80 x 80 mm) prevented deformation of the cross-section. The cross-stiffening ribs do not protect the wall which is under compression if, as in our case, the ratio a/b is larger than 0.7, where a, b are dimensions of the wall sides.

The cumulation of all effects (buckling, cross-section deformation, torsion, and bending of the shank) caused the generation of a plastic stability joint and the tower collapsed. All of these effects were studied on aeroelastic models in an aerodynamic tunnel with simulation of the atmospheric boundary layer.

The new lighting tower was stiffened so that the slenderness ratio did not exceed 50. The cross-stiffening ribs were replaced by diaphragms. The peripheral belt of the reflector was supplied with 14 circular openings, thereby increasing permeability. It was assumed that this increase would suppress vibrations in a plane perpendicular to the wind direction.

3. ROOF COVER OF SWIMMING POOL

As shown in the third Chart, the swimming-pool building in question is 20 m high (at the south edge) and rectangular in shape. The roof pitch is 5% from the south to the north. The cover (except rain channels, dilatations, and attics) is, from the view of the wind stream, quite smooth. The frame-plate of the roof cover has an overhang of nearly 2.5 m. From the view of aerodynamics, this overhang is undesirable. It allows the wind to penetrate the roof space from beneath, and controls the pressure distribution on the roof.

The surrounding buildings are generally about 10 to 15 m high with a relatively clear approach from the south and south west. The aluminium cover (0.8 mm thick) was first torn away in the zone behind the southern leading edge, at a wind speed of 23 m/s. After reconstruction, the second damage occurred at a wind speed of 21 m/s. The wind direction in both cases was the same.

For determining cladding loads due to the wind, pressure transducers were employed [3] . Based on these measurements, the pressure coefficients were found to be $C_p \simeq -2.0 \pm 0.25$ behind the leading edge , and the dominant frequency was 20 Hz at a wind speed of 15 m/s. The dynamic response of the aluminium cover was observed in this zone. The response caused the damage of cover clamps, followed by lifting of the cover.

The cure for the dynamic response was increasing the structural damping of the cover. Another solution (for instance the increasing of the cover weight) was rejected because of the small bearing-capacity reserve of the framework. The new aluminium cover (0.8 mm thick) was asphalted with a peripheral strip of 0.5 m width for increasing the structural damping.

4. TV TOWERS AND GUYED MASTS

In recent years, glass-reinforced plastic (GRP) cylinders have been used quite frequently as load-bearing structures of TV aerials transmitting in higher frequency bands on micro-waves. The circular cross-section of such structures (see fourth Chart) has the disadvantage that it is aerodynamically unstable, primarily in the plane perpendicular to that of the wind direction. For the elimination of these vibrations, various methods are recommended [3] .

As a rule, an increase of damping is relatively difficult to attain; however, in many cases it is possible to achieve significant damping by using dynamic vibration absorbers. In Czechoslovakia, such absorbers have been designed, mounted and tested on new masts, and a newly developed system of them is intended for installation on reconstructed masts. These absorbers are described in detail in [4] . If another mass is connected to a vibrating system of n degrees of freedom by means of an elastic element, a new system originates whose number of the degrees of freedom exceeds that of the former by one. A suitable selection of this additional mass and rigidity can result in improved behaviour of the new system.

Under aerodynamic excitation, various frequencies can arise in dependence on the wind velocity. If a mast or structure is to be safe under such circumstances, it is necessary to eliminate the resonance peaks in the frequency band of interest. This can be accomplished by means of a dynamic absorber only when - apart from an elastic element - a damping element is also inserted between the additional mass and the original structure. The theory of dynamic vibration absorbers is known from classical mechanics [3] .

The dynamic vibration absorbers for use on Czechoslovak masts and towers were developed specially for GRP cylinder having an inside diameter of 190 cm, closed with a steel lid. The absorber has the form of an annular frame supporting the necessary ballast and is suspended on 4 ties from the lid and provided with 4 hydraulic shock absorbers pushing against the lid. These absorbers are inclined at 30° with respect to the vertical. The tuning of the absorber can be changed by changing the length of the suspension ties, and the damping by the type of hydraulic shock absorbers.

The calculations and investigations of the structures carried out so far have shown that a pendulum vibration absorber can be applied to increase effective damping of high slender structures and eliminate undesirable vibrations.

REFERENCES

1. Pirner, M.: Limitation of self-induced vibrations. Proceedings of the Conference on Tower-Shaped Structures, IASS, The Hague, 1969.
2. Novák, M.: A statistical solution of the lateral vibrations of cylindrical structures in air flow. Acta Technica CSAV, No. 4, 1967.
3. Fischer, O. Koloušek, V.,Pirner, M.: Aeroelasticity in civil engineering (in Czech). Academia Prague, 1977.
4. Fischer, O: Some experience with the use of vibration absorbers on aerial masts. Acta Technica CSAV, No. 2, 1974.

CHART 1: STEEL COLUMNS OF AN ARCH BRIDGE

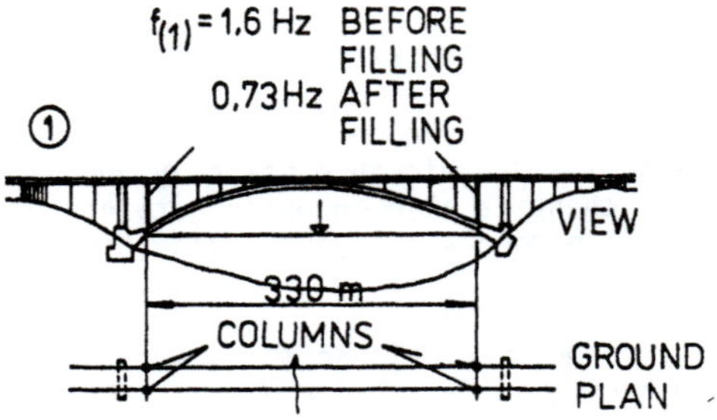

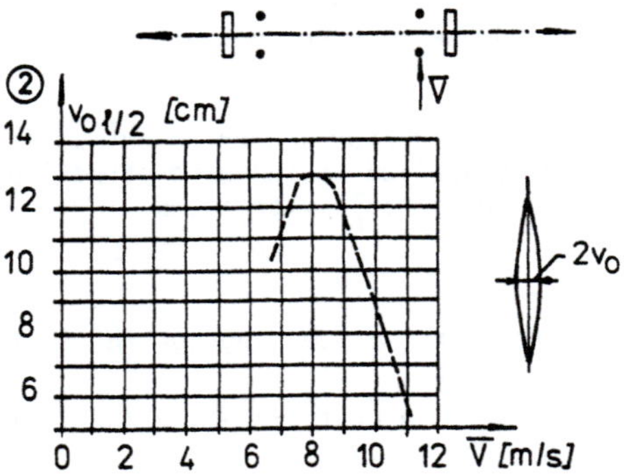

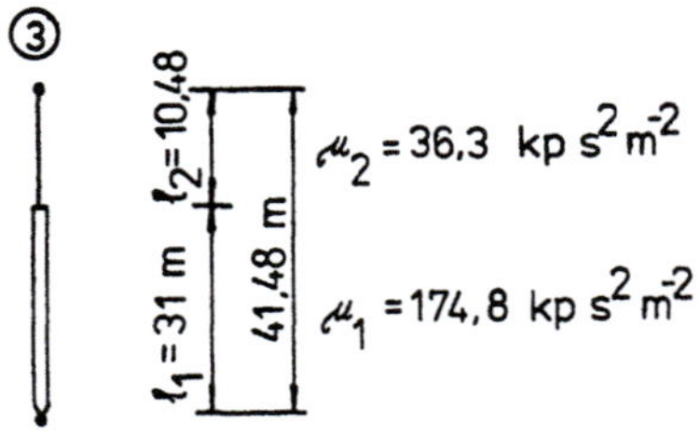

Problem: Column oscillation due to vortex shedding. Damage due to fatigue possible.

Geometry: See Sketch 1. Cross-section of columns is circular with diameter 1 m. Vertical distance between joints is 41.48 m.

Boundary conditions: Each end of column rigidly fixed.

Flow conditions:

$3.44 \times 10^5 < R_e < 8.25 \times 10^5$,

$5 < V < 12$ m/s, Strouhal number = 0.2 at V = 8 m/s.

Structural dynamics: Lateral vibration of columns having length of 41.48 m and shown in Sketch 3.

Vibration: Critical wind speed of 8 m/s. Dependence of amplitude on wind speed given in Sketch 2. Frequency = 1.6 Hz, max. amplitude = 130 mm. Oscillation of stress: -10570 N/cm² to +5090 N/cm² (due to dead load and wind).

Cures: All four columns filled with gravel (5-15 mm) to height of 31 m. First natural frequency lowered to 0.73 Hz and maximum amplitude reduced 20-fold. Critical wind speed of 5 m/s.

References:

Novák, M.: Über winderregte Querschwingungen der Ständer der Bogenbrücke über die Moldau, Der Stahlbau, 11, (1968).

CHART 2: ROOF COVER OF STADIUM

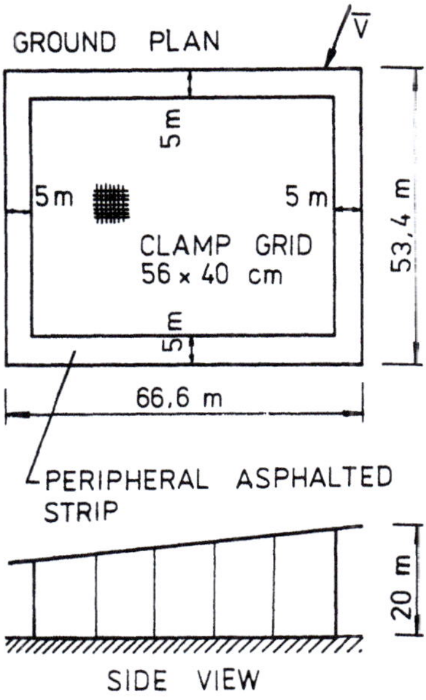

Problem: Damage of the aluminium cover due to wind oscillations.

Geometry: Flat roof, 66.6 x 53.4 m, inclined at a slope of 5%.

Boundary conditions: Aluminium cover, 0.8 mm thick, fastened by clamps with nails to cover boards. Clamp grid = 56 x 40 cm (see the ground plan). Clamps made from aluminium sheet 0.8 mm thick.

Flow conditions: Height of building nearly 20 m, neighborhood of hall is center of town. Wind speed at collapse = 23 m/s. Aerodynamic coefficient on roof measured in situ with pressure transducers.
$(C_p)_{max}$ = -2.0 ± 0.25 observed at leading edge. Fluctuation in C_p = ± 0.25. Dominant vortex frequency = 20 Hz at wind speed of 15 m/s.

Operating conditions: Dynamic wind effects (suction) at leading edge pulled nails out. Wind then "rolled" down aluminium cover.

Structural dynamics: First natural frequency of aluminium plate f_1 = 6.32 Hz; dimensions of plate given by the clamp grid.

Vibration: Amplitude of vibration ≈ 1.0 mm at wind speed of 12 m/s, observed at leading edge. Frequency of observed vibrations ≈ 6 Hz.

Cures: New aluminium cover asphalted with peripheral strip (see sketch) for increasing structural damping. Weight of aluminium cover, clamp grid, and clamps after reconstruction same as before.

References: Fischer, O., Koloušek, V., Pirner, M.: Wind effects on structures, Academia-Elsevier.

CHART 3: TV TOWERS AND GUYED MASTS

Problem: Vibrations due to wind loading. Damage due to fatigue possible. Disturbance of TV signal.

Geometry: General height of such structures: 100 to 300 m.

Boundary conditions: Top of structure generally furnished with cylindrical extension (nearly 20 m high) from glass-reinforced plastic, 1.9 m diameter.

Flow conditions: Combination of cross- and in-wind direction vibration caused by Karmān vortex street and incident turbulence.

Operating conditions: Stationary vibrations at low wind speeds (Strouhal number 0.2); random vibrations at higher wind speeds.

Structural dynamics: Systems with first natural frequency ≈ 0.2 Hz. Higher freq. up to 2 Hz. Structural damping = 0.02 (log. decrement).

Vibration: Motion in lower modes has elliptical trajectory. Motion of TV-tower top in horizontal plane at wind speed of 10 m/s shown in Sketch 1.

Cures: Pendulum vibration absorber (or absorbers) - see Sketch 2. Mass of each absorber is about 1/10 of generalised mass of construction for the corresponding natural mode. The mass has ring shape; hung on 4 supports. Necessary damping produced by 4 absorbers, adopted from car shock absorbers. Sketch 2 shows 2 vibration absorbers in two levels. Sketch 3 gives influence of absorber.

References: Fischer, O., Koloušek, V., Pirner, M.: Wind effects on structures, Academia-Elsevier

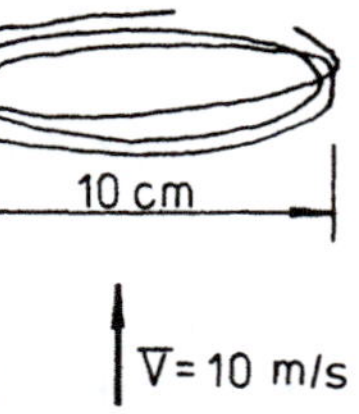

①

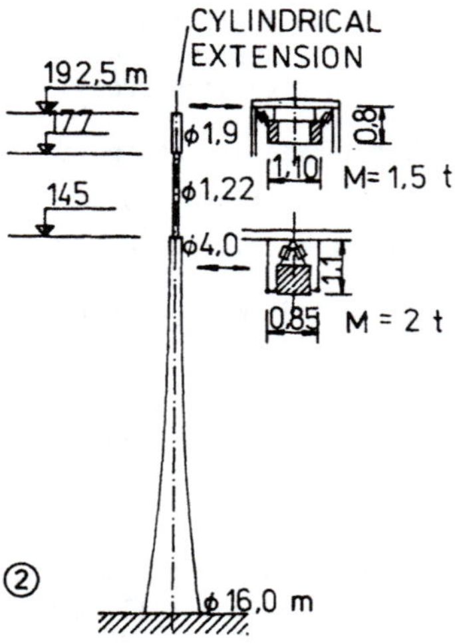

②

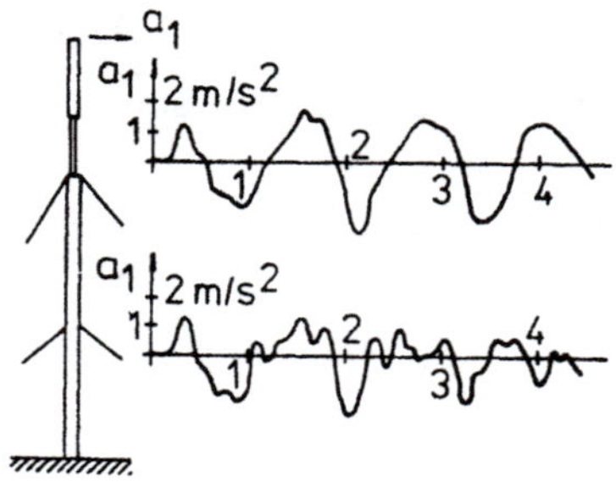

③

CHART 4: LIGHTING TOWER

①

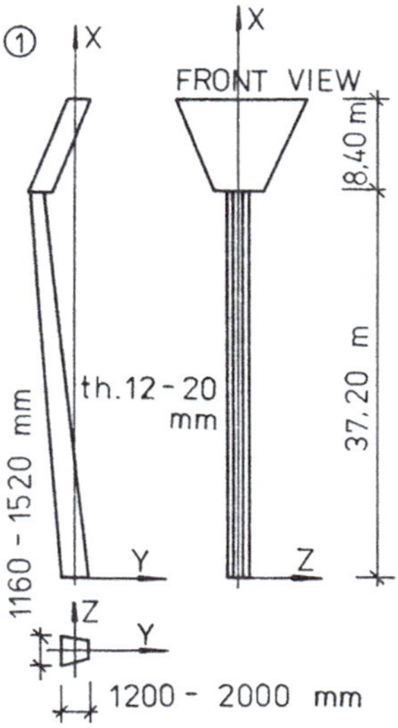

Problem: Vibrations in bending and torsional modes due to vortex shedding or galloping. Collapse of one of four towers (Sketch 3).

Geometry: see Sketch 1.

Boundary conditions: Thin wall steel profile (see Sketch 1). Area of reflector is permeable.

Flow conditions: In neighborhood of stadium is open country. Wind speed at time of collapse ≈ 30 m/s; wind direction at 45° to the axis Y (Sketch 1).

Structural dynamics: Moment of inertia at x = 4.5 m. J_y = 1384100 cm^4, J_z = 2756500 cm^4; thickness 10 mm. And at x = 18.6 m: J_y = 984800 cm^4, J_z = 1629500 cm^4, thickness 8 mm. Specific mass of the mast : μ = 420 kg/m^3.

Vibration: Natural theoretical freq. in direction Y: $f_{(1)}$ = 0.55 Hz; $f_{(2)}$ = 13.4 Hz. Direction Z: $f_{(1)}$ = 0.39 Hz; $f_{(2)}$ = 7.62 Hz. Torsional: f = 2.94 Hz. Motion before and at collapse not measured. Not yet determined whether vibration due primarily to flow-structure interaction of upper part of tower or support column.

Cures: Stiffening of mast. Peripheral rim of the reflector supplied with 14 circular openings (see Sketch 2).

②

③

G 11 TOWER CRANE IN TURBULENT WIND

Fétizon F, Jouannet J.C., Watremetz M.
Centre Technique des Industries Mécaniques (CETIM)
Senlis - France

SUMMARY

This paper describes an experimental study carried out on an aeroelastic model of a tower crane placed in a wind tunnel in which the boundary layer and the atmospheric wind spectrum are simulated. The mean and fluctuating components of the wind velocity and the appropriate forces and moments are determined. The fluid-structure interaction is shown for certain orientations of the crane in the wind : under its first modes (bending and torsion) the structure is sensitive to external forces due to spectral composition of the wind and, without attenuation and phase displacement, is generally sensitive to gustiness phenomena. A suitable increase in rigidity and an increased roughness, which enhances the aerodynamic damping phenomena, are cures for the vibration of the structure.

1. INTRODUCTION

In Europe, tower cranes are very widely used for the construction of buildings, silos, water towers, industrial and residential towers, etc. This type of structures is relatively flexible and lightly damped. Cranes can be installed on a variety of sites and the wind incidence is usually variable, making the determination of response characteristics more complicated. As far as the construction of cooling towers for nuclear stations is concerned, for example, tower cranes can be 100 m high and more. Considerations on the atmospheric boundary layer including wind spectra [1 to 3] are of importance, especially if one is interested in the dynamic aspect of the question. In the frequency band ranging from 0.02 to 10 Hz, the wind spectrum shows an exponentially decreasing trend [4]. The problem of the wind response [5] of tower cranes lies in the fact that the first modes of these structures are within this frequency band.

2. EXPERIMENTAL SET-UP

The experimental study was carried out in a model at a scale of 1/10. The actual dimensions of the structure are 31 m in height and 54 m for the jib and counterweight jib unit ; the total mass is equal to 67 t. Similarity was based on the Cauchy number $E/\rho V^2$. Based on the cross-dimension of the truss bars, the Reynolds number varies from 10^4 to 10^5, which corresponds to a clearly subcritical flow. The similarity concerning damping is less easily attained.

Tests have been carried out in the ETBS return-flow wind tunnel of 5.60 x 4.20 m^2 cross-section in Bourges. Two levels of turbulence were tested ; low-level turbulence (4 to 5 %) was that of the wind tunnel. The highly turbulent flow up to 20 % was generated by devices positioned upstream of the model : two beams of different dimensions producing fluctuations at the Strouhal frequencies, and a system comprising three bladed wheels which were rotated at variable speed filling the spectrum "gaps" in two ways : (a) by the blade passing frequencies and (b) by diffusion of turbulence energy around these spectrum lines.

In the highly turbulent flow, gusts of wind are simulated by the swivelling blades of the fan actuating the wind tunnel, which enables sudden accelerations and decelerations of the air stream ; thus, velocity variations ΔV of 4, 8 and

12 m/s corresponding to frequencies of 0,1, 0,04 and 0,02 Hz, respectively, have been realized at a maximum average velocity of 20 m/s.

The model was tested for 5 jib configurations corresponding to angles of incidence α of 0, 45, 90, 135 and 180° (sketch 1 in the chart), for each of three types of flow : (a) smooth flow ; (b) turbulent flow ; (c) turbulent flow with velocity variations ΔV. In each case the crane was subjected to various flow velocities.

The average flow velocities and fluctuations were measured by means of 5 Pitot tubes and 5 hot-wire probes positioned at various altitudes upstream of the model. The structural displacements were obtained by twice integrating the signals $\gamma_{x,y,z}$ delivered by the accelerometers positioned on the periphery of the structure and the forces were measured by an aerodynamic balance.

The structural modes and corresponding frequencies and damping coefficients of the model were determined by means of shocks exerted by a hammer fitted with a force pick-up.

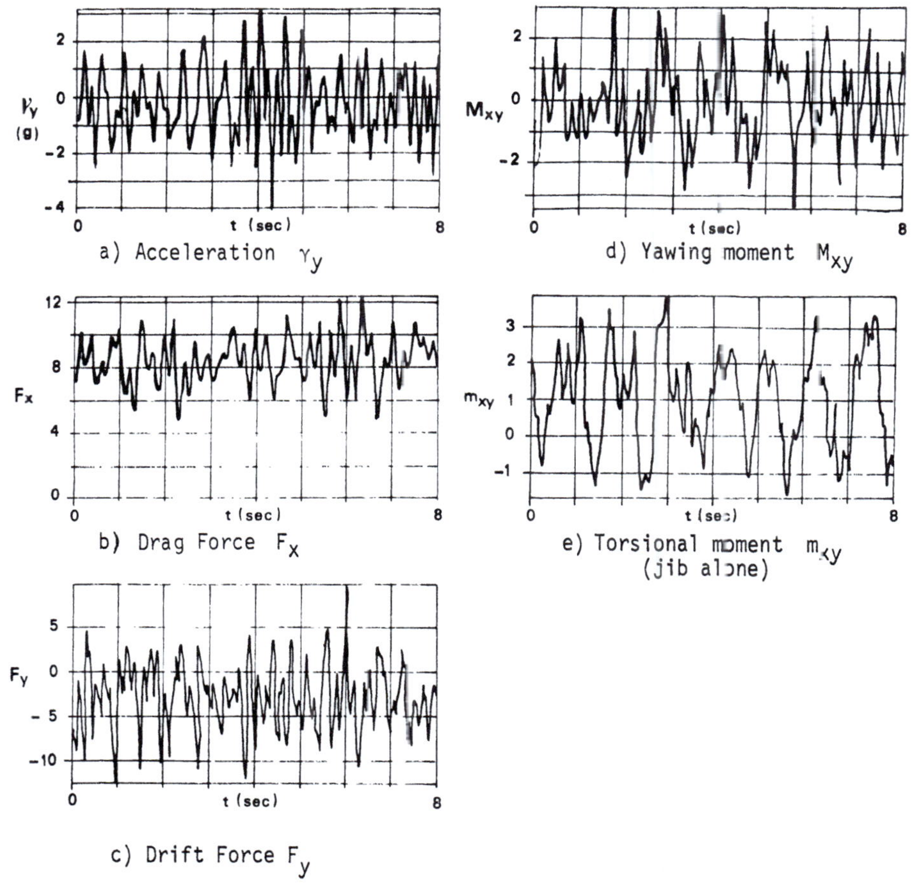

FIG 1 TIME REPRESENTATION FOR $V = 20$ m/s ; $\Delta V = 0$ m/s ; $\alpha = 180°$ (see Chart)

3. EXPERIMENTAL RESULTS

A typical example of the experimental results is shown in Fig.1 for the case of the jib in the wind direction. A marked periodicity was found for all signals, especially for the acceleration γ_y in the horizontal plane measured at the end of the jib (Fig 1a). The peak-to-peak displacements corresponding to the data shown in Fig.1 amounted to about 0.18 m. The force scales in that figure are arbitrary.

The spectral evaluations of the data of Fig.1 are represented in Fig.2. One will note good agreement between the simulated wind spectrum with those reported in the literature. Moreover, the evaluation shows that the structure in the vertical plane containing the jib is essentially excited in the first two modes of bending of the mast (Fig.2b). The level of energy exchange between wind and drag force is low (Fig.2c), the structure is sensitive to the "pink noise" of the wind excitation only in its first modes ; the energy extracted from the wind is therefore comparatively weak. This fact is confirmed by the analysis of the coherence function (Fig.2d). Except near 4.3 Hz where a bending resonance was observed.

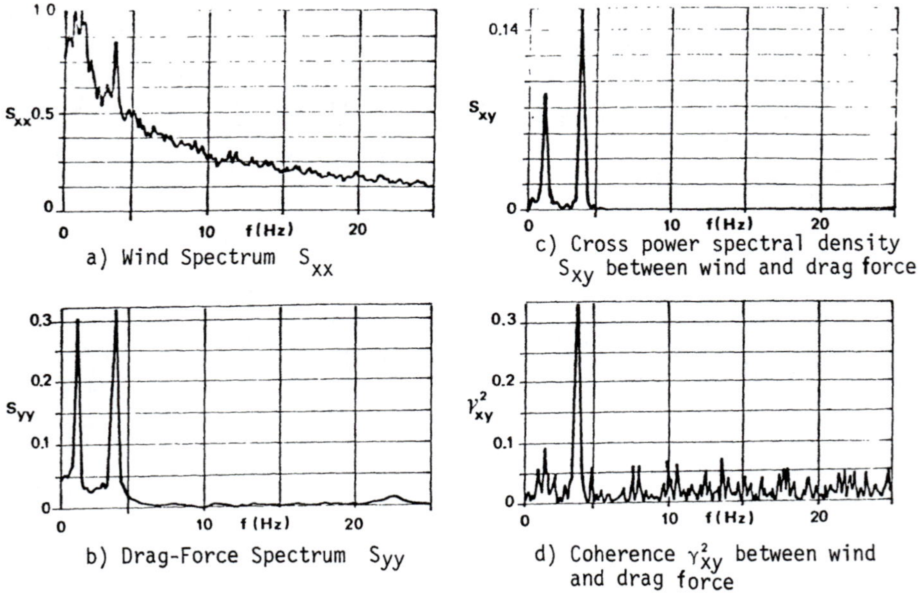

FIG 2 FREQUENCY REPRESENTATION FOR V = 20 m/s ; ΔV = 0 m, α = 0° (see Chart)

For the three types of flow studied, the increase in the RMS value of the drag varies parabolically with the increase in turbulence level. The forces pass through a maximum for a velocity fluctuation of ΔV = 4 m/s (Fig.3a). This can be interpreted as a filtered and distorded response of the structure near 4 m/s (0.10 Hz) fluctuation, whereas the 8 and 12 m/s fluctuations (0,04 and 0,02 Hz frequencies), brings about unfiltered responses of the structure.

The intensity variation of the force fluctuation f_x plotted against the intensity of turbulence shows a correlation of the linear type (fig 3b). A rough dimensional analysis of this result shows that, in order to obtain this type of correlation, the force should comprise a term of the form $F_x = k\, V\sqrt{\overline{v^2}}$. This result is confirmed in other respects as well : the average aerodynamic forces, especially the drag, increase with the intensity of turbulence for this type of structure and for the Reynolds range explored.

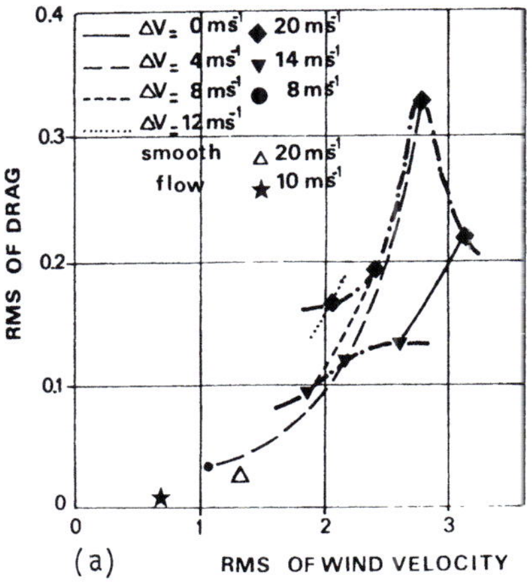

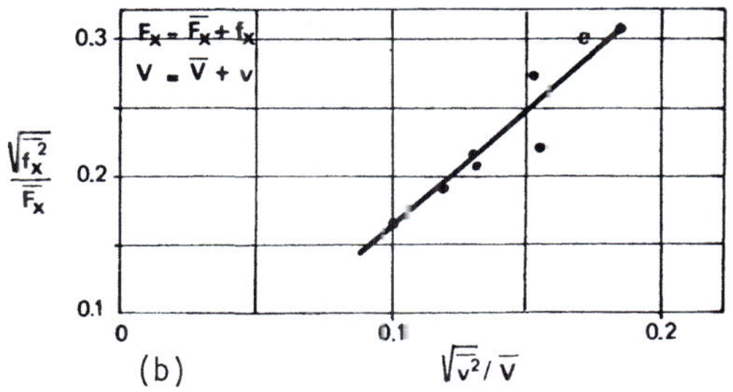

FIG. 3

(a) RMS VALUES OF THE FORCES PLOTTED AGAINST THE RMS WIND VELOCITY

(b) VALUES OF THE FORCE INTENSITY PLOTTED AGAINST TURBULENCE INTENSITY

The fluid structure interaction can be characterized in two ways : (a) To a wind spectrum such as that shown in Figure 2a, the structure responds in its first modes. This response is all the more important as the energy extracted from the wind is high ; it increases with the intensity of turbulence but decreases with frequency as the wind spectrum does. (b) To the gusts of wind the structure responds without filtration for frequencies less than 0.1 Hz ; it responds with filtration for higher frequencies, and this response is no longer linear if frequencies exceed 0.5 Hz. In smooth flow the crane responds in different modes, but the amplitudes remain limited. In turbulent flow with or without gust (ΔV), the interaction is largest when the jib is perpendicular to the wind. The structure is excited, through the jib, in its first torsional mode ; the amplitude increases, becomes very considerable and then decreases. This frequency modulation of about 0.2 Hz is characteristic of flutter phenomenon : the amplitude increases until the aerodynamic, and perhaps structural, damping becomes so considerable that it brings about a phase rotation and causes the vibration amplitude to decrease. In that case, the elastic limit is likely to be exceeded during these vibration of maximum amplitudes, or it may happen that the structure responds to a gust of wind which, by accident, agrees in phase with a mode of vibration.

The modeling of such a behaviour cannot be achieved with a linear model. An appropriate model would be that proposed by Wooton and Scruton [6] :

$$M \ddot{c} + (C_s + C_a) \dot{q} + K_s q = 0$$

where the coefficient of aerodynamic damping C_a is a function of the structural displacement, or the model proposed by Hartlen and Currie, [7] which comprises a relaxation term.

The measures against excessive vibrations for this type of structure are of two kinds : (a) The structure should be made more rigid, firstly, regarding the torsional mode of the mast to which the structure is by far most susceptible, and secondly, regarding mast bending. (b) the roughness of beams, plates, baffles, etc should be increased in order that the aerodynamic damping is raised; however, since an increase in roughness increases the mean forces which are detrimental to the static stability, this measure is to be executed with caution.

4. CONCLUSIONS

The study shows the considerable effect high-intensity turbulence has upon the aerodynamic behaviour of tower cranes :

(a) For this type of structure, turbulence increases the value of the mean wind loads.

(b) The structure responds essentially in different modes to turbulence and gusts of wind, except if the accelerations of the latter are small.

(c) The fluctuating parts of the forces represent a considerable amount of the mean forces and should therefore be taken into account in any calculation of possible fatigue failure or of the elastic limit being exceeded.

(d) The aerodynamic damping plays an important role in the attenuation of vibrations.

(e) Vortex excitation of the Strouhal type is possible only for each bar and not the truss in its entity (too small compactness). The result is an overall "white noise" excitation, which is not very dangerous to the structure.

(f) Higher bending and torsion rigidity of the mast as well as an increased roughness of the whole structure constitute satisfactory measures for an attenuation of these vibrations.

5. REFERENCES

1. DAVENPORT A.G. : The Spectrum of Horizontal Gustiness near the Ground in High Winds. Quarterly Journal of the Royal Meteorological Society, Vol 1, No 87, 1961

2. SHIOTANI, M. : Turbulence in the Lowest Layers of the Atmosphere. The Science Reports of the Thoroku University, Series 5, Vol.2, No 3, 1950.

3. HARRIS, R.I. : The Nature of the Wind. Proceedings of the Seminar : The Modern Design of Wind-Sensitive Structures - CIRIA, 18 June 1970.

4. BLEVINS, R.D. : Flow-Induced Vibration. Ed. Van Nostrand Reinhold - New-York- 1977.

5. UNGAR, E.E, MARK, W.D., BLISS, D.B. : Prediction of Wind - Responses of Tall Open Truss-Structures. Sound and Vibration, October 1974.

6. WOOTTON, LR. SCRUTON, C : Aerodynamic Stability. Proceedings of the Seminar : The Modern Design of Wind - Sensitive Structures - CIRIA - 18 June 1970

7. HARTLEN, CURRIE : Lift Oscillator of Vortex Induced Vibration. Journal of the Engineering Mechanics Division ASCE, October 1970.

TOWER CRANE IN TURBULENT WIND

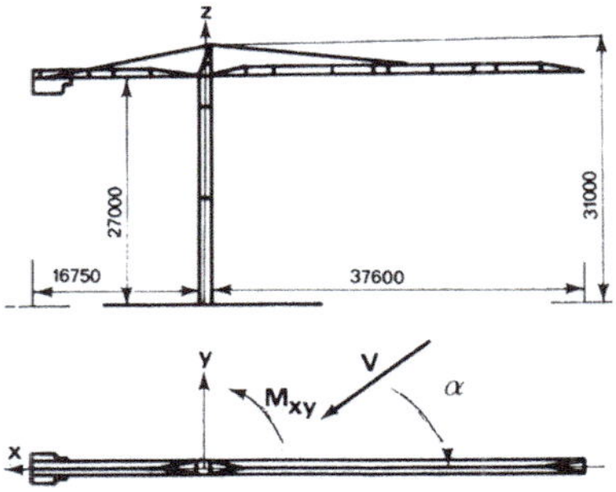

(1) All dimensions in mm

(2)

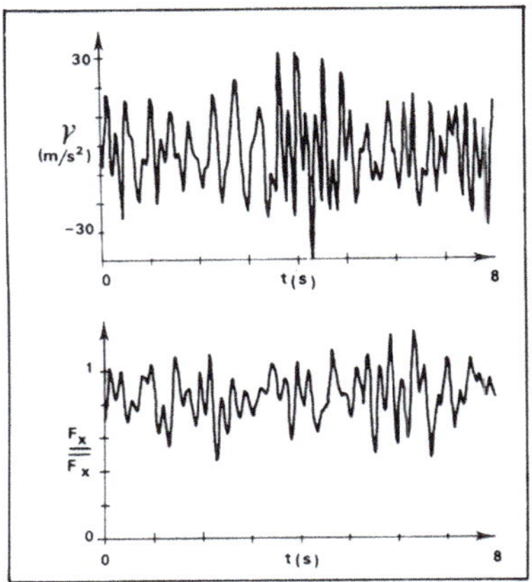

(3)

Problem: Effects of a turbulent wind on a crane. Possible resonant coupling obstacles of the same height.

Geometry: Sketch (1) shows full-size crane. Model shown in sketch (2) is to scale 1/10.

Boundary conditions: Flow in a 5.60x4.20 m wind tunnel with a flow disturbance system located 9 m upstream of the model. The thickness of the ground boundary layer which is 0.70 m is simulated by a floor.

Flow conditions: Approach flow has varying velocity fluctuations depending on variation of the pitch of the propeller in the disturbance system. Turbulence intensities of 20% attained.

Operating conditions: The incidence of the jib in the wind is variable. 5 configurations studied: α = 0, 45, 90, 135, 180°. Velocities of 0 to 20 m/s.

Structural dynamics: Lowest mode of structure is 0,5 Hz and highest analysed in range explored is 24 Hz. Damping coefficients 2% and 0,2% respectively.

Vibration: Typical traces of acceleration of tip of jib in y direction (γ_y) and drag force in x direction (F_x) for essentially smooth incidence flow, V^x = 20 m/s; incidence angle α = 180°. Maximum amplitude recorded at the end of the crane jib: 0,30 m on a torsional mode at 0,9 Hz and a bending mode at 1,6 and 4,3 Hz.

Cures: Increasing the lattice roughness, or securing a flat panel to the jib, makes it possible to increase the aerodynamic damping, but a compromise exists since the overall drag is increased.

References: Ungar, E.E., Mard, W.D., Bliss, D.B. Prediciton of wind-responses of tall open-truss structures. Sound and Vibration, Oct. 1974.

H1 INTRODUCTION TO THE WORKSHOP ON "PREDICTION AND MODELLING OF FLOW-INDUCED VIBRATIONS"

W.J. McCroskey

U.S. Army Aeromechanics Laboratory (AVRADCOM)
Ames Research Center, Moffett Field, California

Surveying the wide range of complicated flow-induced vibration problems that were described during the course of the symposium, one wonders whether there is much hope of calculating rigorously these demonic flow fields. Or, alternatively, are there approximate methods that can give insight and guidance in engineering design? The workshop sought to provide the participants with some assessments of the prediction and modelling techniques that currently exist. In particular, there were three main goals of the workshop session:

1. The identification of the common features of flow oscillations for the diversity of applications described in the symposium,
2. An assessment of the capabilities and limitations of various analysis and design methods, whether for prediction or for cures after problems arise,
3. Suggestions for future research topics that might lead to better, cheaper, more reliable, or otherwise improved design techniques.

The common denominator of many practical flow-induced vibrations is, of course, unsteady flow separation. The attendant fluctuating fluid dynamic forces can be almost completely stochastic, highly organized, or various combinations of random and periodic components. In studying such flows, the most popular model problem that has occupied researchers for over a hundred years is vortex shedding from a rigid circular cylinder. This seemingly simple flow problem remains unsolved today, and furthermore, by varying the Reynolds number, cross-sectional geometry, or elasticity in the lateral direction, all of the aforementioned types of fluctuating forces can be generated. Still another practical permutation on the classical cylinder problem is the addition of turbulent gusts or non-uniformities to the oncoming flow.

Figure 1, adapted from Ref. [1], shows the range of steady drag and fluctuating lift or side force that occurs on rigid, smooth circular cylinders. The shaded bands indicate the range of data that have been collected and published over the years. The drag data are generally better defined than the lift data, especially in the so-called subcritical Reynolds number regime. It is the fluctuating side forces that induce many of the vibration problems, and this quantity is a good barometer of the validity of prediction methods.

Attempts to estimate the fluid dynamic forces on bluff bodies fall into four categories: empirical correlations of wind tunnel measurements, Navier-Stokes calculations of the complete flow field, discrete vortex simulations of the periodic wake, and nonlinear oscillator models of the wake-body combinations. The experimental results were already described above, and the discrete symbols in Fig. 1 show a number of calculations, mostly Navier-Stokes. (The results of Chorin and of Sarpkaya and Shoaff are described and referenced in Dr. D.J. Maull's paper elsewhere in these proceeding; the other works are cited in Ref. [1]). Great strides are being made in computational fluid dynamics, and probably the overall impression created by the results shown in the figure is overly pessimistic with regard to present and future capabilities. However, the Navier-Stokes calculations are not likely to be attractive for routine engineering calculations for bluff bodies for some time, because of the large computer storage and CPU times required and because of the uncertainty in turbulence modelling. Therefore, this approach was not seriously considered in the workshop.

The discrete vortex approximation to the shed wake vorticity distribution is less rigorous than a complete flow field calculation, but it represents a promising compromise between rigor and practicality. As Dr. Maull stated in his lead-off lecture, this method, which is actively undergoing further development by several contemporary researchers, is reasonably successful in predicting the vortex shedding frequency and in determining whether oscillations are likely to occur. However, the magnitude and distribution of the fluctuating pressure field on a body are sensitive to certain details of the modelling, and consequently, quantitative force predictions are less certain at this time. It seems clear that further developments will have to include boundary layer calculations on the body and viscous diffusion in the vortex sheets.

A novel application of discrete vortex modelling of internal separated flows was presented by W.T. Ashurst and F. Durst, and their paper appears in these proceedings.

Professor G.V. Parkinson described the mathematical modelling of flow-induced vibrations of finite amplitude by means of a nonlinear oscillator, and Prof. E. Berger of the Technicsche Universtät Berlin made a short presentation on some significant refinements to the method. Also, Prof. E. Naudascher of the Universtät Karlsruhe showed that the approach is much more general than usually thought, being applicable in principle to any flow induced vibration problem. Several examples are described in his short note. Still other recent applications by Horváth [2] to turbomachinery and by St. Hilaire [3] to a musical reed may be mentioned in passing.

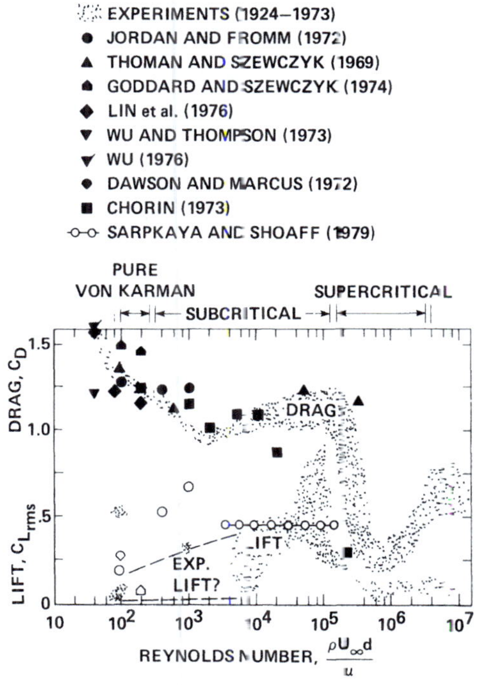

<u>FIG. 1</u> DRAG AND LIFT ON A RIGID CIRCULAR CYLINDER. SOLID SYMBOLS DENOTE CALCULATIONS OF STEADY DRAG, OPEN SYMBOLS FLUCTUATING LIFT

In the usual case of external flow over a bluff body, the wake-body combination is treated as a nonlinear system that is governed by an equation of the Rayleigh or van der Pol type, which admits resonance and limit-cycle behavior that resembles what has been observed. A considerate amount of ignorance regarding the basic flow behavior is glossed over and incorporated into empirical constants, and this helps to fuel the fires of controversy within the fluid dynamics community about the approriateness of the method [4]. Nevertheless, the nonlinear oscillator modelling offers a, albeit imperfect, tool for predicting the occurance of the so-called galloping and vortex-induced vibrations of bluff bodies and for estimating the maximum amplitudes as a function of system and flow parameters. Professor Parkinson's paper in these proceedings is, of course, mandatory reading on this subject.

A third major aspect of modern prediction methods for flow-induced vibrations was described by Prof. A.G. Davenport in a presentation entitled, "Response due to Incident Turbulence." Unfortunately, a complete paper does not appear in the proceedings, but the approach is described in References [5] and [6]. A very sketchy notion of the basic ideas is outlined below.

A flexible, spatially-extended structure is represented as a single degree of freedom spring-mass-damper system. The system is exposed to a wind that is comprised of randomly fluctuating velocities superimposed on a mean speed distribution in the atmospheric boundary layer. A quasi-steady drag coefficient, e.g. Fig. 1, is used to relate the instantaneous wind velocity to a steady-plus-fluctuating force that is applied to the system. Now since the fluctuating force is wind-velocity dependent and the mechanical damping of the system is structure-velocity dependent, the total system damping, and hence the dynamic response, depends upon both the mechanical properties of the structure and the spectrum of the atmospheric turbulence. Furthermore, the coupling between the elastic structures and the spatial structure of the atmospheric turbulence has to be considered. Despite the complexity of the problem, judicious combination of theory, experience, and empiricism has led to satisfactory predictions and cures for a number of practical applications. Unfortunately, the author is not qualified to make more definitive assessments of this approach.

In summary, considerable progress has been made during the past decade or so toward formulating approximate analyses that should provide much more insight and guidance to designers in the future. However, none of the techniques mentioned above are entirely satisfactory, nor are they really suitable yet for widespread application to most types of field problems. Instead, they should be viewed as valuable supplements to existing experimental methods and to practical experience.

Nevertheless, the discrete vortex and nonlinear oscillator models are advancing and maturing rapidly. Future refinements can be expected to resolve many, and perhaps most, of the remaining fluid dynamic limitations without an inordinate increase in complexity or computer requirements, thereby catching up with the structural dynamics aspects of many practical applications. Thus while the practical capability today leaves much to be desired, it seems evident that the forthcoming decade will see a considerable payoff of the research investment that has been made in the past decade.

REFERENCES

1. McCroskey, W.J.; "Some Current Research in Unsteady Fluid Dynamics - the 1976 Freeman Scholar Lecture," ASME Journal of Fluids Engineering, 99I, 8-39 (1977).

2. Horváth, A.J.T.; "Der Pumpvorgang von Verdichtern und Kreiselpumpen als nichtlineare Schwingung," Ph.D. Thesis, l'Ecole Polytechnique Fédéral de Lausanne, Comm. No. 3, Juris-Verlag, Zurich (1976).

3. St. Hilare, A.; "On Modeling Aerodynamically Induced Nonlinear Responses of Self-Excited Structures", Symposium on Unsteady Aerodynamics, Univ. Arizona, Tucson, 359-392 (1975).

4. Sarpkaya, T.; "Vortex-Induced Oscillations - A Selective Review," ASME Journal of Applied Mechanics, 46, 241 - 258 (1979).

5. Davenport, A.G.; "Wind Structure and Wind Climate" and "The Prediction of the Response of Structures to Gust Wind," Safety of Structures under Dynamic Loading, TAPIR Press, Norwegian Institute of Technology, Trondheim (1977).

6. Simiu, E., and Scanlan, R.H.; Wind Effects on Structures, Wiley Interscience, New York, Ch. 7 (1978).

H2

AN INTRODUCTION TO THE DISCRETE VORTEX METHOD

David J. Maull
Lecturer

Cambridge University Engineering Department,
Trumpington Street, Cambridge, England

SUMMARY

This paper presents an introduction to the discrete vortex method of calculating separated flows and jet flows. A discussion of the problems associated with using this method is given and some applications of the technique are described.

PREFACE

"....so that the true solution for an inviscous fluid presumably involves an infinite number of discontinuities, but otherwise does not differ essentially from that for a viscous case. Thus there is a second solution to every hydrodynamic problem, which does not imply steady motion and which entails an infinite number of discontinuous rotations. This is the solution we want for aerodynamics, and not that found by the ordinary mathematical method".

G.P. Thomson 1919.

"A mass of perfect fluid consisting of several distinct parts, each with its own velocity potential, and separated by thin sheets of transition (in the mathematical limit, vortex sheets) may be effectively redistributed..."

A.R. Low 1928.

1. INTRODUCTION

The idea of representing a vortex sheet by discrete vortices is not new, the first application having been given by Rosenhead[1] but the idea had been around for some years before. The advantage of this method over many others for solving fluid flow problems is that if the origin of the sheet in the flow is known then only the velocities on the sheet need to be calculated instead of, as in other methods, the entire flow field. This, however, can involve considerable computing effort and it is only relatively recently that large enough computers have become available to make the method of practical use. Who, nowadays, would contemplate doing Rosenhead's calculation, as he did, by hand?

2. BASIS OF THE METHOD

The flow round a body may be considered to be made up of three regions, the viscous, rotational boundary layer and wake and an inviscid outer region which is usually irrotational but could be rotational. In the method discussed here it is assumed that the flow is at a high Reynolds number and hence the diffusion of vorticity due to viscosity is small compared with the convection of the vorticity due to the flow. This does not mean, however, that viscosity is unimportant since it is due to viscosity that vorticity is generated on the body to be convected downstream in the wake.

The vorticity $\underline{\omega}$ is defined as

$\underline{\omega} = \nabla \times \underline{u}$, where $\underline{u}$ is the velocity and if we take the curl of the Navier-Stokes equation

$$\frac{\partial \underline{u}}{\partial t} + \underline{u}.\ \nabla \underline{u} = -\frac{1}{\rho}\nabla p + \nu\nabla^2 \underline{u}$$

we obtain the vorticity equation

$$\frac{\partial \underline{\omega}}{\partial t} + \underline{u}.\ \nabla \underline{\omega} = \frac{D\underline{\omega}}{Dt} = \nu\nabla^2 \underline{\omega} + \underline{\omega}.\nabla \underline{u} \qquad (1)$$

where $\frac{D\underline{\omega}}{Dt}$ is the usual substantive derivative. Equation (1) indicates that the rate of change of vorticity associated with a fluid element is made up of two terms, a diffusion of vorticity due to viscosity ($\nu\nabla^2\underline{\omega}$, analagous to the diffusion of heat) and a term ($\underline{\omega}.\nabla\underline{u}$) indicating an intensification of vorticity due to the stretching of vortex lines caused by the motion of the fluid.

For most of this discussion we will consider only two-dimensional motion, thus if the velocities are u and v in the x and y directions respectively, then $\underline{\omega}$ is a vector normal to the x,y plane with magnitude $\frac{\partial v}{\partial x} - \frac{\partial u}{\partial y}$. Therefore $\underline{\omega}.\nabla\underline{u}$ is zero and if we ignore the effects of viscous diffusion then equation (1) becomes simply $\frac{D\underline{\omega}}{Dt} = 0$, implying that the vorticity associated with a fluid element remains constant and moves with the fluid.

The basis of the method is then to follow the vorticity in the flow given that it is injected at a certain point (the separation point, in general there will be more than one of these) with a certain value at a particular time. The velocity of the fluid element is made up of two parts, firstly the velocity given by the basic irrotational flow around the body and secondly by the velocity induced by the vorticity in the flow including any images in the body necessary to satisfy the boundary conditions on the body. The basic irrotational flow round the body will be assumed to be known and the vorticity in the flow is assumed to lie along a sheet (s) of zero thickness with vortex strength $\gamma(s)$ per unit length of the sheet. The velocities at a point $z(= x + iy)$ due to this sheet are given by the Biot-Savart law

$$u - iv = \frac{i}{2\pi}\int_C \frac{\gamma(s)ds}{z - z(s)},$$ the integral being taken

the sheet contour c and $z(s)$ being the point on the sheet having vortex strength $\gamma(s)$. We will consider the velocity at a point on the sheet and this will be given by the irrotational flow field and the above integral if its Cauchy principal value is considered.

The problem is thus to calculate the evolution of the thin vortex sheets in the flow and other quantities of interest such as the forces on the body producing the sheets.

This introduction to the basic ideas has necessarily been brief and a more detailed treatment may be found in Clements [2] and Lin [3].

Before describing some of the uses that have been made of the method it is necessary to point out some of the problems that arise in applying the technique to practical situations.

3. DETAILS OF THE METHOD

The method may be described briefly as follows: the flow is started from rest and the irrotational flow round the body calculated which leads to vorticity being shed into the fluid from the separation points. This rate of production of circulation multiplied by the time step chosen for the calculation is represented in the flow by a vortex which convects with the fluid. After a further time step another vortex is generated and so on and eventually a distribution of vortices is formed which should represent a continuous vortex sheet.

There are, however, various problems and refinements that need to be considered before the method outlined above may be used successfully and these are dealt with below.

3.1 Discretisation of the sheet

Assume for the moment that the rate of transfer of circulation $(d\Gamma/dt)$ is given, thus in time Δt a vortex of strength $\Delta\Gamma = (d\Gamma/dt)\Delta t$ is formed which should represent the element of the continuous sheet formed during that time. This will be approximately the case if the time step Δt is small enough and $(d\Gamma/dt)$ is not varying greatly. The self-induced velocity of the vortex is zero but the short segment of the sheet that it represents has in general a non-zero self-induced velocity (i.e. as shown in figure 1). The short segment of the sheet has constant strength and thus the only point on the sheet with zero self-induced velocity is the mid-point and thus the single vortex representing this segment of the sheet should strictly be placed at the mid-point of the segment. As the flow evolves the sheet will stretch non-uniformly and thus the sheet should be "re-discretised" at each time step by replacing the current set of vortices with new ones equidistantly spaced and of a strength such that the total circulation is preserved. The mathematical argument leading to this re-discretisation is given by Fink & Soh [4]. This re-discretisation will obviously involve additional computing time but as shown by Fink & Soh [4] and Sarpkaya & Shoaff [5] it can lead to smooth rolling-up of the vortex sheet with none of the contorted vortex patterns evident in earlier calculations. However, Sarpkaya [6] states that in re-working his original calculation for the flow past an inclined flat plate, Sarpkaya [7], re-discretisation of the vortex sheets produced forces on the plate almost identical with those calculated without re-discretisation.

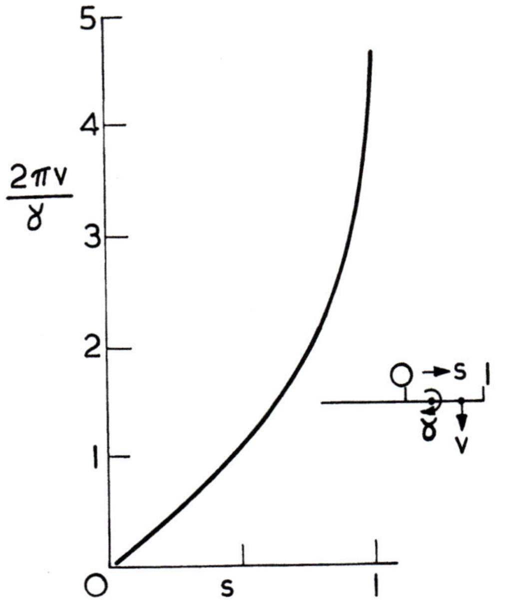

FIG. 1 SELF-INDUCED VELOCITY OF VORTEX SHEET.

Up until now most calculations have proceeded as outlined at the beginning of this section, that is without re-discretisation, the velocity induced at each vortex being calculated and the vortex moved to its new position at the beginning of each time step. One obvious difficulty is that if two vortices become too close then they will induce upon each other unrepresentatively high velocities. This can also occur if a vortex approaches a solid surface where it is now unduly influenced by its image. If re-distribution is not used other methods have been employed to stop these high velocities, for instance if the vortices approach too close to each other they are amalgamated (Clements & Maull[8]) or a vortex is given an inner core where the velocity varies like r and the usual potential velocity field varying like $1/r$ outside the core (Chorin & Bernard[9]). If a vortex approaches a surface then amalgamation effectively means that the vortex is removed from the flow (see for instance Clements & Maull[10]).

Two other methods have been proposed to compute the velocities when, for instance, the separation between a vortex sheet and a solid surface becomes small or when the sheet approaches another shear layer as can occur in the tightly rolled up portion of a wing trailing vortex. These are a 'sub-vortex' type of re-discretisation due to Maskew [11] and a cubic-spline interpolation procedure suggested by Clements [12] in an unpublished paper.

3.2 Initial placing of the vortices

There are three parameters involved in the placing of a vortex which represents the portion of the separated shear layer leaving the body in time Δt. They are the strength of the vortex $\Delta\Gamma$, its position and its convection velocity. In addition the Kutta-Joukowski condition should be satisfied although what this condition means and implies in detail is open to some discussion but it will at least involve the strength and position of the vortex. For a detailed discussion of the Kutta-Joukowski condition applied to the unsteady flow past aerofoils, see Basu & Hancock [13].

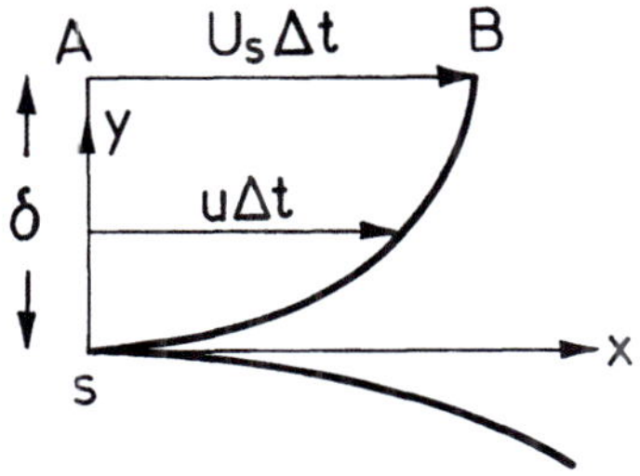

FIG. 2 SEPARATING BOUNDARY LAYER IN TIME Δt.

Consider the situation sketched in figure 2 which represents the boundary layer separating from a surface at a point s. The amount of fluid separating in time Δt is given by the area SAB, δ is the boundary layer thickness and the length AB is $U_s\Delta t$, U_s being the velocity at the edge of the boundary layer. Then the vortex replacing this portion of the shear layer must have a circulation equal to the circulation round the area SAB which equals the area times the vorticity, that is

$$\Delta\Gamma = \int_0^\delta u\Delta t\left(\frac{\partial u}{\partial y}\right)dy = \frac{1}{2}\,U_s^{\,2}\Delta t,$$

using the boundary layer approximations.

If a strip dy is considered, the circulation round it is $\frac{\partial u}{\partial y}\,dy\,u\,\Delta t$ convecting with a velocity u, so the total circulation times the convection velocity for the area SAB is

$\int_0^\delta u^2\Delta t\left(\frac{\partial u}{\partial y}\right)dy = \frac{U_s^{\,3}}{3}\,\Delta t$ and thus the mean convection speed of the replacement vortex should be $\frac{\Delta t\,U_s^{\,3}}{3}\Big/\frac{U_s^{\,2}\Delta t}{2} = \frac{2U_s}{3}$. As we shall see this convection speed, as far as one can tell, has not been used in any calculations.

As an example of the different methods used for placing the first vortex let us discuss one problem, that of the flow round sharp edges, for example flat plates at an incidence to a uniform stream.

An early application of the discrete vortex method was by Sachs, Lundberg & Hansen [14] who were interested in the flow past wing-body combinations, which involves, in the cross-flow plane, the flow sketched in figure 3(a). They placed in the plane of the wing a sheet of vorticity of strength γ per unit length with a length $\Delta s = v_i \Delta t$, v_i being the velocity induced at the tip by the vortices in the field. Satisfying the Kutta condition, that the flow leaves the plate smoothly, gave the value of the vorticity γ. The sheet was then replaced by a single vortex of strength $\gamma\Delta s$ at a point in the wing plane which again satisfied the Kutta condition and the vortex moved with the induced velocity at that point.

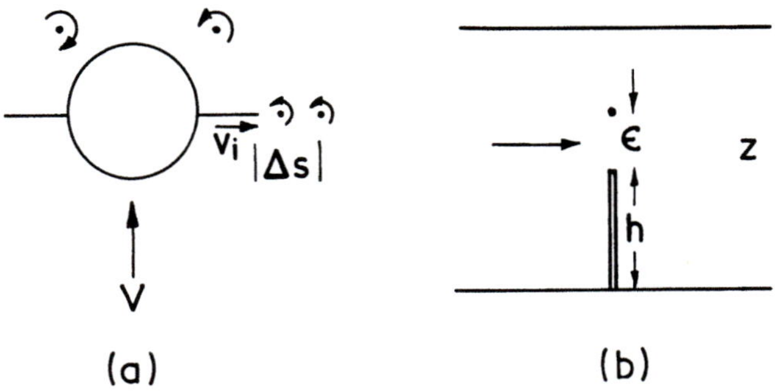

FIG. 3 APPLICATIONS OF THE METHOD.

Evans & Bloor [15] studied the flow past a flat plate in a channel as shown in figure 3(b). The vortex was placed at a point $z = i(h + \varepsilon/2)$ which is in the plane of the plate and the vortex strength was given by $\Delta\Gamma = \frac{1}{2}\left|\frac{dw}{dz}\right|^2 \Delta t$ where $\left|\frac{dw}{dz}\right|^2$ is the square of the velocity at the point $z = i(h + \varepsilon)$. This thus assumes that the vortex production is one half of the square of the velocity on the outer edge of the vortex sheet in contact with the main flow. Satisfying the Kutta condition enables the value of ε to be found and thus $\Delta\Gamma$. Again the vortex was moved with the induced velocity.

Kuwahara [16], Belotserkovski & Nisht [17], Sarpkaya [7] and Kiya & Arie [18] have all studied the problem of the flow about an inclined flat plate using a variety of schemes for placing a vortex at each step. Kuwahara [16] and Kiya & Arie [18] placed the vortex at a fixed point in the plane of the plate and determined the strength by satisfying the Kutta condition. Whereas Kuwahara placed the vortex at a point distance $0.0167b (\equiv a_s)$ away from the plate, whose half width was b, Kiya & Arie systematically investigated the effect of varying the ratio a_s/b and concluded that for the results to be representative a_s/b should lie in the range $0.005 < a_s/b < 0.0125$. Their results showed that the Strouhal number was not greatly changed by varying a_s/b, for instance a change of a_s/b from 0.009 to 0.05 produced a change in Strouhal number from 0.14 to 0.11 whereas the same change in a_s/b produced a change in drag coefficient from 2.38 to 3.72.

Sarpkaya [7], on the other hand, determined the strength of the new vortex by using the relation $\Delta\Gamma = \frac{1}{2} U_s^2 \Delta t$ and interpreted U_s as the mean of the velocities of the newest four vortices in the shear layer. The position of the vortex was then found by satisfying the Kutta condition. The value of U_s could have been found analytically at each time step and used instead of the mean velocity of the four vortices but the calculation is unstable when performed numerically.

The paper by Belotserkovski & Nisht [17] does not describe how the strength and position of the initial vortex at each time step was determined.

As pointed out by Kamemoto & Bearman [19] the 'best' position for introducing the new vortex will be a function of the time step Δt used in the calculation. They come to the conclusion that if the parameter $a_s/U_\infty \Delta t$ is too small, say less than 0.05, then the shear layer is not too well represented and find that the drag coefficient is very much a function of $a_s/U_\infty \Delta t$.

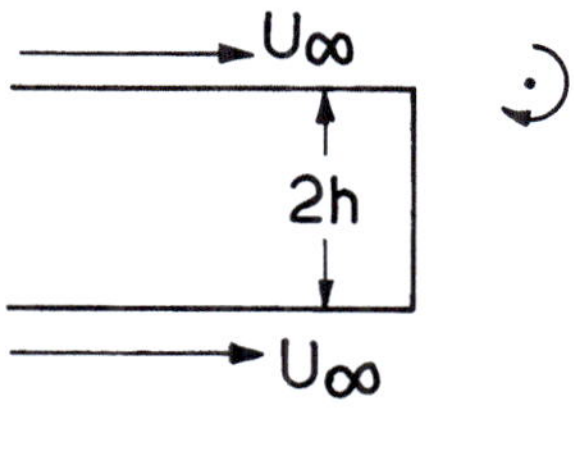

FIG. 4 APPLICATION TO BLUNT BASE WITH SHARP CORNERS.

Clements [20], see also Clements [21] and Clements & Maull [10], used two methods of introducing vortices into the flow around the body shown in figure 4. In the first case $d\Gamma/dt$ was put equal to $U_s^2/2$, U_s being the velocity at a point a small distance above the base and the vortices were introduced at the corners with a velocity equal to U_s. In the second case the Kutta condition was used (the corners being stagnation points in the transformed plane), the vortices being introduced at points a small distance above the corner with such a strength as to satisfy the Kutta condition. In both of these methods a single parameter, the distance from the corner, may be varied and investigation of this parameter showed that the Strouhal number was not greatly affected but that the base pressure was very sensitive to this parameter and to the time step used in the calculation.

In calculating the flow round a circular cylinder Stansby [22] used a modified form of the Kutta condition to determine the circulation round the new vortex. The separation point was assumed and a vortex placed on the normal from that point of sufficient strength to produce zero velocity at the separation point. The velocity (U_s) at the vortex was calculated and the strength put equal to $U_s{}^2\Delta t/2$ and a new position calculated to again give zero velocity at the separation point. A further iteration gave a new vortex velocity and strength. Chorin [23] also used this assumption of zero slip velocity at the separation point in calculating the flow past a circular cylinder. This no-slip condition at the separation point has also been used by Sarpkaya & Shoaff [5] for the flow around a circular cylinder together with the relationship that the vortex strength $\Delta\Gamma = U_s{}^2\ t/2$, thus enabling the position of the vortex to be calculated. In this case the vortex convects with the induced velocity at its point of introduction. A similar condition has also been used by Deffenbaugh & Marshall [24].

For the flow round a sphere, Leonard [25] uses a different condition to place the new vortex. He makes use of the result that the vortex strength is $U_s{}^2\Delta t/2$ and introduces it into the flow at a point such that its convection velocity is $U_s/2$.

Thus many different methods have been used to place a vortex in the flow at the start of a new time step in the calculation and some are better than others in modelling the physics of the flow. It does appear from the limited calculations done so far that the Strouhal number is relatively insensitive to the method by which the vortex is placed, but the force coefficients being directly proportional to the vortex strengths and their velocities (by Blasius' equation) are greatly influenced.

3.3 Use of transformations

The correct boundary condition on the surface of a circular cylinder in the presence of vortices is easily satisfied by using vortices at the inverse points in the cylinder. For other shaped bodies, however, a transformation may be required and it should be realised that the velocity of a vortex in one plane does not give the correct velocity when transformed to another plane. The correction required, first noted by Routh, is easily developed as follows.

Let two planes be related by the transformation $z = z(\zeta)$ with corresponding points z_i and ζ_i. The complex conjugate induced velocity of the vortex at

z_i is

$$q = \left[\frac{dF(z)}{dz} - \frac{i\Gamma}{2\pi(z - z_i)}\right]_{z=z_i}$$

where $F(z)$ is the complex potential in the z-plane including the vortex at z_i. The corresponding complex potential in the ζ-plane can be written as

$$F(\zeta) + \frac{i\Gamma}{2\pi}\ln(\zeta - \zeta_i)$$

where now $F(\zeta)$ is the complex potential without the vortex at ζ_i.

Thus $$\frac{dF(z)}{dz} = \frac{d}{d\zeta}\left\{F(\zeta) + \frac{i\Gamma}{2\pi}\ln(\zeta - \zeta_i)\right\}\frac{d\zeta}{dz} = \left\{\frac{dF(\zeta)}{d\zeta} + \frac{i\Gamma}{2\pi(\zeta - \zeta_i)}\right\}\frac{d\zeta}{dz}$$

so $$q = \left[\frac{dF(\zeta)}{d\zeta}\frac{d\zeta}{dz}\right]_{z_i,\zeta_i} + \frac{i\Gamma}{2\pi}\left[\frac{1}{\zeta - \zeta_i}\frac{d\zeta}{dz} - \frac{1}{z - z_i}\right]_{z_i\zeta_i}$$

where $\left[\frac{dF(\zeta)}{d\zeta}\right]_{\zeta_i}$ is the conjugate velocity of the vortex in the ζ-plane and the second term is the correction due to the transformation.

For example the transformation $z = \zeta - \frac{1}{\zeta}$ transforms a vertical plate in the z-plane to a circle in the ζ-plane. If the induced velocity of a vortex, strength Γ, at the point z_i is required then its complex conjugate velocity is

$$(u - iv) = (U - iV)\frac{\zeta_i}{1 + \zeta_i^2} + \frac{i\Gamma}{2\pi}\frac{\zeta_i}{(1 + \zeta_i^2)^2}$$

where $U - iV$ is the complex conjugate velocity of the vortex in the ζ-plane at the point ζ_i and the final term is Routh's correction.

3.4 Viscous effects

The calculation method, being a potential flow solution, is basically an infinite Reynolds number approximation but recently attempts have been made to incorporate into the model viscous effects to simulate finite Reynolds numbers. Viscous effects may be incorporated in two ways. Firstly by allowing the vortices to have a rotational core, the size of which is a function of the kinematic viscosity and the age of the vortex, that is an attempt is made to model viscous diffusion. This can be beneficial, as stated previously, since it allows vortices to approach each other without very large velocities being induced. Another method of modelling diffusion is by a random walk process as used by Chorin [23] applied to the flow round a circular cylinder for Reynolds numbers up to 10^5. The method has also been used by Ashurst [26].

The second incorporation of viscous effects is by performing at each time step of the main calculation a boundary layer calculation to determine the separation points and hence the strength and position of the next vortex to be introduced into the flow.

Deffenbaugh & Marshall [24] have calculated the flow round a circular cylinder using a boundary layer calculation to determine the separation points. The problem studied is the time development of the flow after an impulsive start with the Reynolds number in the high sub-critical regime ($10^4 < R < 10^5$). The predicted separation points after the flow has been established are too close to the front stagnation point resulting in unrepresentative Strouhal numbers and drag coefficients.

Sarpkaya & Shoaff [5], for the flow round a circular cylinder, have included viscous effects by making a boundary layer calculation at each step during the vortex calculation.

Chaplin [27] represented the boundary layer by vortices close to the body, their strength being calculated from laminar boundary layer theory.

It is known that the circulation found in vortex streets is less than that shed from the separation points. The dissipation of circulation (presumably due to three-dimensional effects in the experiments not modelled in the two-dimensional calculation method) has been taken into account by Sarpkaya & Shoaff [5], and by Nagano et al (unpublished) using an empirical dissipation function.

4. RESULTS

4.1 Flow round bodies

(a) Flat plates

Results of four calculations for the variation of the drag coefficient of a flat plate as a function of the angle of incidence are shown on figure 5. Those of Belotserkovski & Kuwahara are taken from published graphs and may not be very accurately plotted and the two values at 60° incidence due to Kiya & Arie are for different values of the parameter a_s/b that they investigated.

As can be seen all the calculated values over-estimate the drag coefficient indicating that the vortex strengths are also over-estimated. In Sarpkaya's [7] calculation up to 15% of the shed vorticity is lost by cancellation with vorticity of the opposite sign whereas it is generally accepted that up to 50% is lost in experimental investigations. The Strouhal number, based on the projected height of the plate, is almost independent of incidence and is found experimentally to be about 0.15. Sarpkaya [7] calculated a value of 0.154 and Kiya & Arie [18] a value of 0.14, so the Strouhal number is much better predicted than is the drag coefficient.

If the free stream is not uniform but sheared and hence rotational the discrete vortex method may still be used as shown by Kiya & Arie [28].

Author	R	C_d	C_d(expt)	S	S(expt)	Comments
Chorin [23]	2×10^2	1.18 - 1.30	1.30			
	10^3	1.15	1.0			
	2×10^3	1.04	0.96			
	10^4	1.09	1.12			
	2×10^4	0.87	1.15			
	2×10^5	0.29	0.35			
Deffenbaugh & Marshall [24]	$10^4 - 10^5$	0.9	1.15	0.16	0.18 - 0.19	Impulsive start, calculation only carried out until two vortices shed
Sarpkaya & Shoaff [5]	$10^4 - 10^5$	1.35 1.20 1.10	1.15	0.195 0.205 0.22	0.18 - 0.19	Function of dissipation rate
Stansby [22]		0.8	0.85	0.28	0.26	Separation fixed at 95°
Chaplin [27]		1.3	0.7 - 0.8	0.26	0.2	Separation fixed at 90°

TABLE 1. RESULTS FOR CIRCULAR CYLINDERS

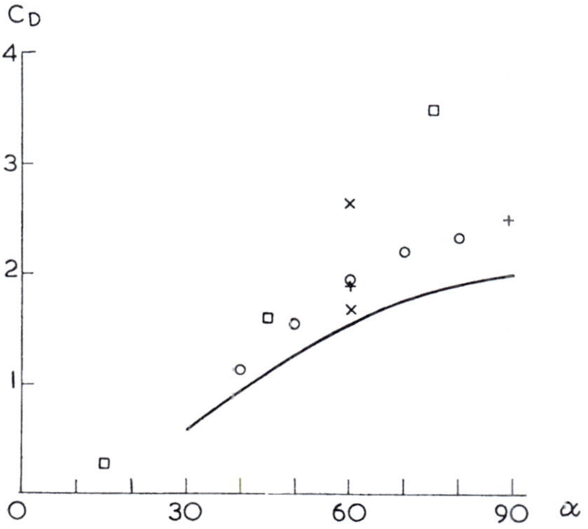

FIG. 5 DRAG OF FLAT PLATE VERSUS INCIDENCE.

o SARPKAYA [7], × KIYA & ARIE [18],
+ KUWAHARA [16], □ BELOTSERKOVSKI & NISHT [17],
— EXPERIMENT

(b) Circular cylinder

Table 1 shows results from calculations done by five workers for the flow past a circular cylinder. Again the Strouhal numbers are reasonably well predicted whereas large variations in the drag coefficients are found although at low Reynolds numbers Chorin's predictions are quite near to the experimental values.

(c) Other bodies

Sharp edged bluff bodies have been studied by Clements [20] and [21] and Clements & Maull [8], including the effects of base bleed, splitter plates and base cavities. For the basic body and for the body with base bleed the Strouhal numbers are reasonably well predicted but the base pressure is in general lower (higher drag) than that found experimentally. However, changes in base pressure with base bleed rate are relatively well predicted. This again points to the conclusion that the self-excitation of the shear layers caused by feedback to the separation points from the wake is reasonably independent of the strength of the shed vorticity but that the drag being very dependent upon the vortex strength is unduly influenced by the detailed modelling of the flow near the separation points and hence by the form of the Kutta condition used. The effect of a splitter plate and a base cavity were not well represented by the discrete vortex method. The splitter plate and the cavity walls act as vorticity sinks in these flows and for them to be adequately represented by this method some means must be found to take into account the sink effect.

The flow down a step is another case where a wall must act as a vorticity sink. Clements & Maull [8] found for this flow that to obtain results similar to those found experimentally up to 80% of the vortices put into the flow had to be removed when they approached the wall. It is difficult to see how this amount of vorticity, shed at the corner, is dissipated subsequently at the wall in practice and it is possible that most experiments on the flow down a step contain a three-dimensionality that removes vorticity. Vorticity is, of course, generated at the wall and this will usually be, for the flow down a step in the region of reattachment, of opposite sign to the vorticity shed at the separation point. Thus "dissipation" at the wall could also be considered as a partial cancellation of vorticity. The introduction of new vorticity at the wall to produce a no-slip wall condition has been used by Ashurst, [29].

The flow over a cavity has been modelled by Hardin & Mason [30] with the object of calculating the noise emission from the cavity. Results are given for the calculation but no comparison is given with experimental results.

If calculations involving three-dimensional vortex filaments are to be undertaken care must be exercised when taking into account the stretching of the filaments that can occur and this type of calculation has been reported by Leonard [31] where he investigated the interaction of aircraft trailing vortices and the self-induced perturbations of ring vortices. A natural extension of this work is to the flow round a sphere and this problem was tackled by Leonard [25] who showed that the method could qualitatively predict the asymmetric wake pattern found experimentally. Another interesting application of the method has been given by Leonard, [32], to the development of a turbulent spot in a laminar boundary layer where the boundary layer is modelled by vortex lines and the spot is the progressive time development of some of these lines.

4.2 Oscillating flow and response

An important application of the discrete vortex method would be to predict the hydroelastic response of a circular cylinder to vortex shedding. This was attempted by Ujihara [33] but the calculation was not taken far enough for any firm conclusions to be drawn. Sarpkaya, [34], has recently reported some calculations for an oscillating circular cylinder which qualitatively predict some of the experimentally observed features of this complex flow. It is worth pointing out here that the computer times required for this type of calculation are very long.

The more complicated problem of the response of a flexible cylinder in a stream which is also oscillating has been tackled by Stansby [35]. The results are in qualitative agreement with those found experimentally but until the hydroelastic response of a cylinder in a steady free stream can be adequately modelled it seems premature to tackle the more complicated problem.

The phenomenon of the 'locking-on' of the vortex shedding frequency of a bluff body to imposed transverse oscillations of the stream has been modelled by Clements & Maull [10] and the results are in qualitative agreement with experiment, showing a widening 'locking-on' range with increased amplitude of oscillation.

Longitudinal perturbations of the free stream at twice and four times the natural vortex shedding frequency can also cause 'locking-on' as shown by calculations by Clements [36]. Again the extent of the 'locking-on' is a function of the amplitude of the perturbation and in the 'locked-on' situation there is a significant fall in base pressure and hence a large rise in drag.

An oscillating free stream with a zero time mean occurs under waves and the forces on a body in this flow are functions of the Reynolds number and a parameter (the Keulegan-Carpenter number) which is proportional to the ratio of the amplitude of the fluid particle motion to the body dimension. This flow has been calculated by Stansby [22] and [35] and the results show the broad features found experimentally. We have seen that generally the vortex strengths predicted by the method tend to be high and in this application, where the vortices are swept back over the cylinder by the reversing free stream, it is to be expected that estimates of the forces are likely to be in error.

A free surface can be simulated, according to classical potential theory, by a distribution of circulation on the surface and this has been studied briefly by Zaroodny & Greenberg [37].

This does seem a useful application of the discrete vortex method and further work on this problem would be worthwhile. Zalosh [38] has extended the earlier work of Rosenhead [1] on the evolution of a vortex sheet to include effects of buoyancy and surface tension with an application to oil drop formation at an oil/water interface.

4.3 Jets and mixing layers

One important application of the method is to the calculation of the mixing between two streams, partly stimulated by the finding, in the early 1970's of large-scale eddy-like structures in these flows. This large scale structure and pairing of eddies has been demonstrated by Acton [39] in a preliminary paper and presented in more detail by her in a thesis, Acton [40]. The important point in this representation is that the shear layer, in order to produce pairing of the eddies, must have a finite thickness. It has been suggested that these large coherent structures play a significant role in noise production and Grabowski & Telste [41] used discrete vortex representation and acoustic theory to predict the noise from a shear layer. Whilst the vortex calculation showed large eddy formation similar to that found by Acton, the prediction of the noise was not good, mainly due to the influence, in the acoustic formulation, of the detailed motion of small-scale vortices.

The vortex growth in jets has been simply modelled by Beavers & Wilson [42] who studied the growth of a localised initial disturbance in a jet simulated by two vortex sheets. Significant features of the laboratory flows were reproduced in the calculations.

The stability of two vortex sheets was initially studied by Abernathy & Kronauer [43] and their work was extended by Boldman, Brinich & Goldstein [44] to the case where the velocities bounding the sheets were not necessarily equal. They found experimentally, and using the discrete vortex method, that as the velocity ratio was changed from unity vortex clouds became more diffuse.

Ashurst [26] used Chorin's random walk representation of viscosity and an ageing of the vortices in a study of a mixing layer. The interesting feature of these results is that the inclusion of the effect of ageing into the vortex velocity field improves the prediction of such quantities as the velocity fluctuations.

It has not been possible in this paper to review all the contributions to the discrete vortex method, further papers are discussed in Clements & Maull [10] and for a more general review of vortex dynamics the work of Saffman & Baker [45] is recommended.

5. CONCLUSIONS

The question to be answered is, 'What use is the method?'

There is little doubt that the discrete vortex method can be used to predict the gross features of some flows such as the vortex pairing in jets and the formation of a vortex street behind a bluff body. It thus helps to give an insight into the physics of quite complicated fluid flows. It is reasonably successful in predicting the vortex shedding frequency since this is dependent upon a feedback from the wake to the body and the strength of the feedback, as determined by the size of the vortices, does not seem to be important. However, the forces and pressures on a body are very dependent upon the strength of the vortices and they in turn are determined by what form of Kutta-Joukowski condition is used and what other parameters are used in the calculation. Since the forces are not well predicted by this method it is difficult to see how the hydroelastic oscillations of a body can be adequately modelled by this method as it stands at the moment. It should predict whether oscillations are likely to occur but not their magnitude.

Future developments to the method must take into account viscous effects by incorporating a boundary layer calculation as has been done by Sarpkaya & Shoaff [5] and viscous diffusion and ageing of the vortices as by Chorin [23] and Ashurst [26]. Flows which have regions where reattachment of the shear layer occurs are of practical importance, for instance in the flow down a step, and they do not seem to be well predicted unless gross assumptions are made about the destruction of vorticity and more work needs to be done on this aspect of the problem, although the key may be in introducing vorticity at the wall as has been done by Ashurst [26].

Free surface flows have not been studied in much detail and this seems an area where some application can be made of the method. Obviously one would eventually wish to calculate the flow around a three-dimensional body, possibly representing the body by vortex lattices but this will probably have to wait until further work is done on the two-dimensional problem to improve the prediction of the forces.

6. REFERENCES

1 Rosenhead, L. "Formation of vortices from a surface of discontinuity". Proc. Roy. Soc. London A, 134, 170-192 (1932)

2 Clements, R.R. "Flow representation, including separated regions, using discrete vortices". In Computational Fluid Dynamics AGARD Lecture Series No. 86, (1977)

3 Lin, C.C. "On the motion of vortices in two-dimensions". Univ. Toronto Studies, App. Math. Series No. 5, (1943)

4 Fink, P.T, & Soh, W.K. "A new approach to roll-up calculations of vortex sheets". Proc. Roy. Soc. London A 362, 195-209, (1978)

5 Sarpkaya, T. & Shoaff, R.L. "An inviscid model two-dimensional vortex shedding for transient and asymptotically steady separated flow over a cylinder". AIAA Journal, 17, 1193-1200, (1979)

6 Sarpkaya, T. "Comments on 'Theoretical study of lift-generated vortex wakes designed to avoid roll-up'". J. Am. Inst. Aero. Astro. 13, 1680-1681, (1975)

7 Sarpkaya, T. "An inviscid model of two-dimensional vortex shedding for transient and asymptotically steady separated flow over an inclined plate". J. Fluid Mech. 68, 109-128, (1975)

8 Clements, R.R. & Maull, D.J. "The rolling up of a trailing vortex sheet". Aero. J. Roy. Aero. Soc. 77, 46-51, (1973)

9 Chorin, A.J. & Bernard, P.S. "Discretisation of a vortex sheet with an example of roll-up". J. Comp. Phys. 13, 423-428, (1973)

10 Clements, R.R. & Maull, D.J. "The representation of sheets of vorticity by discrete vortices". Prog. Aerospace Sci. 16, 129-146, (1975)

11 Maskew, B. "Subvortex technique for the close approach to a discretized vortex sheet". J. Aircraft 14, 188-193, (1977)

12 Clements, R.R. "A refinement of the discrete vortex model". Univ. Bristol Rep. No. RRC3/77. Unpublished. (1977)

13 Basu, B.C. & Hancock, G.J. "The unsteady motion of a two-dimensional aerofoil in incompressible inviscid flow." J. Fluid Mech. 87, 159-178, (1978)

14 Sachs, A.H., Lundberg, R.E. & Hanson, C.W. "A theoretical investigation of the aerodynamics of slender wing-body combinations exhibiting leading-edge separation." NASA CR-719, (1967)

15 Evans, R.A. & Bloor, M.I.G. "The starting mechanism of wave-induced flow through a sharp-edged orifice". J. Fluid Mech. 82, 115-128, (1977)

16 Kuwahara, K. "Numerical study of flow past an inclined flat plate by an inviscid model". J. Phys. Soc. Japan 35, 1545-1551, (1973)

17 Belotserkovski, S.M. & Nisht, M.I. "Investigation of special features of flow over a flat plate at large angles of attack". Fluid Dynamics 5, 110-116, (1973)

18 Kiya, M. & Arie, M. "A contribution to an inviscid vortex-shedding model for an inclined flat plate in uniform flow". J. Fluid Mech. 82, 223-240, (1977)

19 Kamemoto, K. & Bearman, P.W. "The importance of time step size and initial vortex position in modelling flows with discrete vortices". Imp. Coll. Aero Note 78-108, (1978)

20 Clements, R.R. "An inviscid model of two-dimensional vortex shedding". J. Fluid Mech. 57, 321-336, (1973)

21 Clements, R.R. "Computer models of separated flows behind two-dimensional bluff bodies". Ph.D. Thesis, Cambridge University (1973)

22 Stansby, P.K. "An inviscid model of vortex shedding from a circular cylinder in steady and oscillatory far flows". Proc. Inst. Civ. Eng. 63, pt. 2, 865-880, (1977)

23 Chorin, A.J. "Numerical study of slightly viscous flow". J. Fluid Mech. 57, 785-796, (1973)

24 Deffenbaugh, F.D. & Marshall, F.J. "Time development of the flow about an impulsively started cylinder". J. Am. Inst. Aero Astro. 14, 908-913, (1976)

25 Leonard, A. "Simulation of three-dimensional separated flows with vortex filaments". Proc. 5th Int. Conf. Numerical Meth. Fluid Dyn. Springer-Verlag (1976)

26 Ashurst, W.T. "Numerical simulation of turbulent mixing layers via vortex dynamics". SAND 77-8613, Sandia Labs., Livermore, USA, also Proc. Conf. Turb. Shear Flows I, (1977)

27 Chaplin, J.R. "Computer model of vortex shedding from a cylinder". Proc. Am. Soc. Civil Eng. J. Hyd. Div. HY1, 155-165, (1973)

28 Kiya, M. & Arie, M. "An inviscid numerical simulation of vortex shedding from an inclined flat plate in shear flow". J. Fluid Mech. 82, 241-253, (1977)

29 Ashurst, W.T. "Calculation of plane sudden expansion flow via vortex dynamics". SAND 79-8679, also Proc. Conf. Turb. Shear Flows II, (1979)

30 Hardin, J.C. & Mason, J.P. "A vortex model of cavity flow". Am. Inst. Aero. Astro. Paper No. 76-524, (1976)

31 Leonard, A. "Numerical simulation of interacting, three-dimensional vortex filaments". Proc. 4th Int. Conf. Numerical Meth. Fluid Dyn. Springer-Verlag (1975)

32 Leonard, A. "Vortex simulation of 3-dimensional, spotlike disturbances in a laminar boundary layer". Proc. Conf. Turb. Shear Flows II, (1979)

33 Ujihara, B.H. "Hydroelastic analysis of a circular cylinder". North Am. Rockwell Corp. SD 68-996, (1968)

34 Sarpkaya, T. "Vortex-induced oscillations". J. Appl. Mech. 46, 241-258, (1979)

35 Stansby, P.K. "Mathematical modelling of vortex shedding from circular cylinders in planar oscillatory flows, including effects of harmonics and response". Proc. IAHR Synposium Mech. Wave Ind. Forces Cylinders. Bristol University, (1978)

36 Clements, R.R. "On locked vortex shedding in a flow with perturbations parallel to the main stream". J. Sound Vibr. 40, 563-564, (1975)

37 Zaroodny, S.J. & Greenberg, M.D. "On a vortex sheet approach to the numerical calculation of water waves'. J. Comp. Phys. 11, 440-446, (1973)

38 Zalosh, R.G. "Discretized simulation of vortex sheet evolution with buoyancy and surface tension effects". J. Am. Inst. Aero Astro. 14, 1517-1523, (1976)

39 Acton, E. "The modelling of large eddies in a two-dimensional shear layer". J. Fluid Mech. 76, 561-592, (1976)

40 Acton, E. "A modelling of large jet eddies". Ph.D. Thesis, Cambridge University (1977)

41 Grabowski, W.J. & Telste, J.G. "A discrete vortex simulation of a two-dimensional shear layer with prediction of the hydrodynamic noise". David Taylor Naval Ship Res. Dev. Center DTNSRDC-78/050, (1978)

42 Beavers, G.S. & Wilson, T.A. "Vortex growth in jets". J. Fluid Mech. 44, 97-112, (1970)

43 Abernathy, F.H. & Kronauer, R.E. "The formation of vortex streets". J. Fluid Mech. 13, 1-20, (1962)

44 Boldman, D.R., Brinich, P.F. & Goldstein, M.E. "Vortex shedding from a blunt trailing edge with equal and unequal external mean velocities". J. Fluid Mech. 75, 721-735, (1976)

45 Saffman, P.G. & Baker, G.R. "Vortex interactions". Ann. Rev. Fluid Mech. 11, 95-122, (1979)

H3 NONLINEAR OSCILLATOR MODELLING OF FLOW-INDUCED VIBRATIONS

G. V. Parkinson

Professor, Dept. of Mechanical Engineering
The University of British Columbia

SUMMARY

Two of the organized forms of flow-induced vibration of bluff prisms, galloping and vortex-induced, have been modelled mathematically as nonlinear oscillators. For galloping the quasi-steady theory is fairly complete, and can be used in design for the prediction of galloping behaviour based on the measurement of static fluid-dynamic force variation with incidence for the prism in the same kind of flow for which the galloping tendencies are being analysed. For vortex-induced vibration, the theory is still relatively primitive, and although mathematical models have been devised which reproduce most of the observed characteristics, the models require several empirical constants, and are of limited use for design prediction. However, their continued study in conjunction with flow-field models such as the discrete vortex models should lead to better understanding of the complex relationships between a vibrating prism and its vortex wake. In some important practical cases, critical wind speeds for galloping and vortex resonance lie close together, and while no complete theory is available for these complex situations, experimental evidence indicates that the galloping theory can provide boundaries.

1. INTRODUCTION

Since this lead-off paper is on the same topic as my general lecture for the 1972 Symposium [1], it is inevitable that some of the material will seem familiar, at least to those who heard or read the lecture. However, I will try to repeat myself only to the extent it is essential to the discussion of the themes for this Symposium. This will be the easier because the orientation of this Symposium is to the user -- the practising engineer or architect who has to worry about the possibilities of flow-induced oscillation of a structure.

First, the restrictions on the scope of the paper. As in Ref. 1, I will consider only transverse (plunging) oscillations of relatively long bodies of bluff cross-section, caused by galloping instability, vortex resonance, or a combination of them. This eliminates from consideration, in order to keep the discussion simpler and better focused, some important forms of flow-induced vibration, such as twisting motions, or combined plunging and twisting motions. Also, I will concentrate more on the physical than the mathematical aspects of the modelling, the latter having been treated in more detail in Ref. 1. Unlike Ref. 1, this will not attempt to be a survey of the current field, but merely a discussion of some important features.

Now, what can nonlinear oscillator modelling offer the practising engineer? Certainly, the interests of the researchers who have used such methods are often quite different with respect to flow-induced vibrations from those of the designer, who simply wants to know how he can avoid unacceptable vibrations of his structures. The researchers are more likely to be motivated by scientific curiosity about the vibration phenomena, and since the phenomena exhibit nonlinear characteristics, it becomes natural to use nonlinear oscillation theory in studying them. Well, the short answer to the above question is that nonlinear oscillator modelling offers the designer a hope of predicting, for galloping or vortex-induced vibrations, maximum amplitudes as a function of system and flow parameters, so that he learns the worst that can happen.

For the designer to make proper use of any nonlinear oscillator modelling technique, it is of course necessary for him to have some insight into the phenomena that are modelled, and the ultimate key to this insight is the detailed nature of the flow over and downstream of the body. Unfortunately for the analyst, the flow is unsteady, separated, and usually turbulent. Progress is being made in developing analytical methods of predicting fluctuating surface pressures for the unseparated parts of turbulent bluff-body flows [2], and potential flow methods can often predict effectively the mean pressures in such regions [3, 4]. However, for a long body of bluff cross-section experiencing galloping or vortex-induced oscillation, it is the part of the body cross-section exposed to the wake, hereafter called the section afterbody, that receives the dominant excitation through pressure loading from the flow field, and this excitation cannot be predicted at present by any exact flow-field modelling, although the discrete vortex methods described by Maull, in another lead-off paper of this Workshop, and in an earlier paper [5], hold promise. Fig. 1(a) shows typical pressure loading on the afterbody of a square section during galloping, and Fig. 1(b) shows typical fluctuating pressure loading on a circular section during vortex-induced oscillation, where it can be seen that the afterbody loading dominates (the mean separation line is at about θ = 80°).

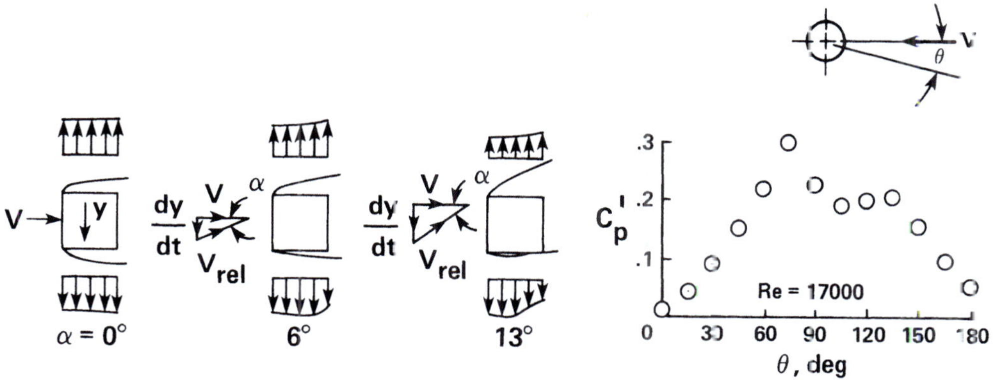

(A) GALLOPING SQUARE-SECTION PRISM

(B) CIRCULAR CYLINDER IN VORTEX-INDUCED OSCILLATIONS

FIG. 1 UNSTEADY PRESSURE DISTRIBUTIONS ON BLUFF BODIES.

The third complicating feature of the flow fields in galloping and vortex-induced vibrations, that they are unsteady, does not always present a difficulty. In the similar problem of vibrating airfoils, where the flow remains unseparated, an extensive unsteady flow theory exists, and it shows that if the vibration frequency, nondimensionalized by the airfoil chord and free-stream velocity, is considerably less than unity, the unsteady flow can be accurately represented by a quasi-steady model, in which the flow can be considered to pass through a sequence of steady states in the oscillation cycle, each one appropriate to the instantaneous apparent angle of attack of the airfoil to the approach flow, caused by the vector sum of the free-stream and vibration velocities. This situation is typical of galloping vibrations of bluff sections, in that the galloping frequency, always very close to the undamped natural circular frequency ω_n, when nondimensionalized by the transverse section dimension h and the free-stream velocity V, is usually considerably less than unity, so that a quasi-steady model is appropriate for pure galloping, and only the appropriate steady flows need be considered, as in Fig 1(a). Here we will use the reciprocal of reduced frequency, reduced velocity U, defined by

$$U = \frac{V}{\omega_n h}$$

so that for the quasi-steady model of galloping, we require $U \gg 1$. This has a physical significance for bluff-body flows not found in the airfoil flows, because of the Kármán vortex streets formed behind bluff sections, each with a characteristic Strouhal number S, defined by

$$S = \frac{f_v h}{V}$$

where f_v is the vortex formation cycle frequency in one shear layer. Thus, when

$$U = U_r = \frac{1}{2\pi S}$$

the wake vortex system and the body elastic system are in resonance, and since $0.1 < S < 0.2$ for sectional shapes of engineering interest then $0.8 < U_r < 1.6$, or $U_r = O(1)$, so that an equivalent requirement for quasi-steady modelling of galloping is $U \gg U_r$. The resulting modelling is considered in the next section. When U is close to U_r, vortex-induced vibrations of the body occur, and their nonlinear oscillator modelling is considered in the subsequent section. Finally, for a body capable of both galloping and vortex-induced vibrations, the more complex possibilities when the critical galloping flow velocity U_o, defined later, is near U_r, are considered in the last section.

2. GALLOPING

In the idealized form of galloping considered in Ref. 1 the long bodies are of constant section, the flow about them is two-dimensional, and the approach flow is uniform, although possibly homogeneously turbulent, and normal to the body span. The galloping is as a rigid body, in the one transverse degree of freedom. These, of course, do not represent field conditions for galloping; although they may be fairly closely met for horizontal structural members, the fact that such members vibrate as elastic beams requiring relative minor changes in the analysis. Nevertheless, although the idealized conditions are generally unrepresentative of field conditions, they provide a useful starting point, owing to their relative simplicity. Later we shall see how to adapt them to the three-dimensional world of wind engineering.

Galloping results from an instability to transverse disturbances. Thus, when the square section of Fig. 1(a) is given a small downward velocity $\mathring{y}$, the angle of attack α of the relative approach velocity causes the lower separating shear layer to lie closer to the underside of the section, increasing the suction there and creating a transverse force F_y in the direction of motion. The differential equation of the motion, in dimensionless form is

$$\ddot{Y} + 2\beta\dot{Y} + Y = C_y n U^2 \quad (1)$$

where

$$Y = y/h \ , \ n = \frac{\rho h^2}{2m} \ , \ \tau = \omega_n t$$

and (°) now means differentiation with respect to τ. The left side of Eq. (1) represents the assumed linear elastic system, of mass m per unit span, and with fraction β of critical viscous damping, and the right side represents F_y, expressed in terms of coefficient C_y for fluid of density ρ.

Eq. (1) becomes a nonlinear oscillator model when the quasi-steady assumption is invoked to represent the steady-state variation of C_y with α, obtained by wind tunnel measurement, as a polynomial in

$$\tan\alpha = \frac{\dot{y}}{V} = \frac{\dot{Y}}{U}$$

using Fig. 1(a). For the square section, an odd 7th degree polynomial is suitable:

$$C_y = A_1\left(\frac{\dot{Y}}{U}\right) - A_3\left(\frac{\dot{Y}}{U}\right)^3 + A_5\left(\frac{\dot{Y}}{U}\right)^5 - A_7\left(\frac{\dot{Y}}{U}\right)^7 \quad (2)$$

where A_1, A_3, A_5, A_7 are positive constants, the positive value of A_1 indicating the galloping instability. When Eq. (2) is substituted for C_y in Eq. (1), it becomes a nonlinear ordinary differential equation for Y, weakly nonlinear in the case of air flow, because of the small value of n. Then it can be solved by standard methods, described in Ref. 1, where the solutions are shown to give accurate predictions of the galloping amplitudes, as a function of wind speed and system parameters, both for the steady-state galloping and for the transient buildup. Thus, from static wind tunnel measurements of the C_y - α curve for a bluff section, one can predict the galloping behaviour of the section, particularly the maximum amplitudes to be expected. Also of interest to the designer is the critical wind speed for the initiation of galloping,

$$U_o = \frac{2\beta}{nA_1}$$

obtained from the linear terms in $\dot{Y}$ in Eq. (2). Again, this analysis is for two-dimensional conditions, although, as shown in Ref. 1, it continues to give good predictions when homogeneous turbulence is superimposed in the uniform approach flow, provided the C_y - α measurements are made in the same flow.

In most real situations for which the possibility of galloping is being considered, not only is the approach flow turbulent, but both flow and body must be considered three-dimensional. Nevertheless, if U>>Ur, the quasi-steady model of galloping can still be used to predict maximum amplitudes. As an example, we consider a tower of square section in the natural wind, a problem investigated recently in the author's laboratory [6, 7]. Novak [8] had previously considered such problems, and his formulation of the quasi-steady theory allowed for the lateral deflection mode of the tower, and for the variation with height of the wind speed. However, he used a lateral force coefficient averaged over the height of the tower. The analysis of Ref. 7 improved on Novak's method by including the measured variation with height of the local lateral force coefficient $C_y(Z)$.

That this variation is important can be seen from Fig. 2, showing $C_y(Z)$ vs $\tan\alpha$ for 5 different fractional values Z/ℓ of the tower height ℓ. The data are from tests in a large boundary-layer wind tunnel on a 1:550 scale model of a 400 m. tower of 28.5 m. square section in a typical 'suburban' wind exposure. Over the bottom third of the tower, both C_y and its initial slope are negative, (corresponding to A_1 negative in Eq.(2)), indicating stability against galloping. Over the upper two-thirds, the curves have the positive initial slopes indicating galloping instability, increasing to a maximum at about $Z/\ell = 0.7$, and then falling off as the top of the tower is approached, due to a tip flow effect.

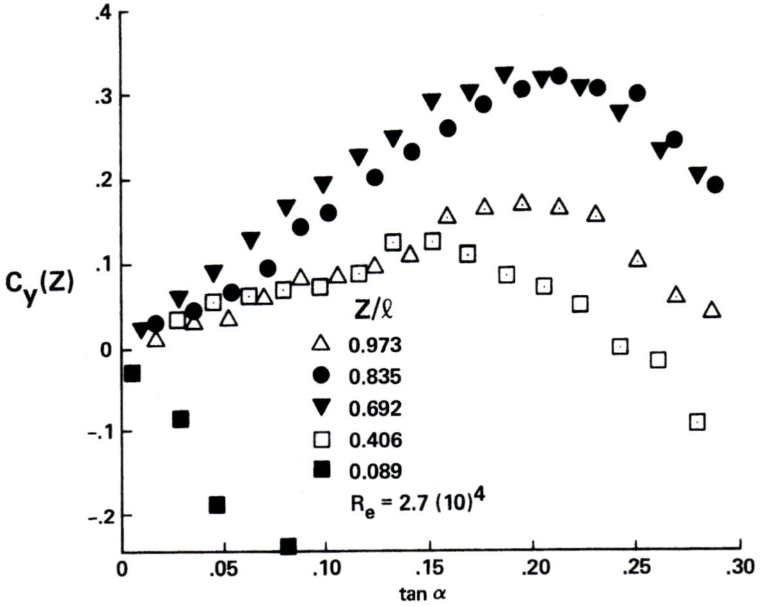

FIG. 2 LOCAL LATERAL FORCE COEFFICIENTS FOR TOWER OF SQUARE SECTION.

To incorporate this somewhat more complex behaviour of $C_y(Z, \alpha)$ into the quasi-steady galloping theory, a simplified procedure can be adopted, particularly if only the steady-state amplitudes are sought. Instead of a further appeal to the theory of weakly nonlinear differential equations, the requirement for an energy balance at steady-state galloping amplitudes between input from aerodynamic forces and dissipation by system damping is used. In this approach the difference between the local sectional equivalents of the terms $C_y nU^2$ and $2\beta\mathring{Y}$ from Eq.(1) is doubly integrated over the oscillation cycle period and over the height of the tower, and set equal to zero, with $Cy(Z, \alpha)$ represented as a double polynomial in $\mathring{Y}$, through the quasi-steady assumption

$$\tan\alpha = \frac{\mathring{Y}(Z)}{U(Z)}$$

as in Eq.(2), and in Z, from the data of Fig. 2. A suitable mode shape is assumed for Y(Z), and the incident wind velocity profile gives U(Z). As usual, for a weakly nonlinear system, the oscillation is assumed harmonic.

The integration yields an algebraic expression for the tip amplitudes $\bar{Y}_\ell$ of the tower as a function of tip wind speed U_ℓ and the system parameters, which can be plotted as a universal curve for a given tower shape and deflection mode shape exposed to a given wind profile. Fig. 3 gives a comparison of this predicted curve with experimental galloping data for several combinations of mass and damping parameters for the tower of Fig. 2, described earlier, with a linear deflection mode. The solid portion of the curve represents the predicted stable tip amplitudes, and the dotted portion the predicted unstable tip amplitudes, or the required initial tip amplitudes for oscillations to grow over that wind speed range (the stability is determined by methods similar to those described in Ref. 1). It can be seen that the theory gives good predictions of the observed amplitudes at the higher wind speeds, including a correct prediction of a small range of oscillation hysteresis for the highest damping level tested. The scatter in the data is largely due to uncertainty in the experimental values of system damping constant β.

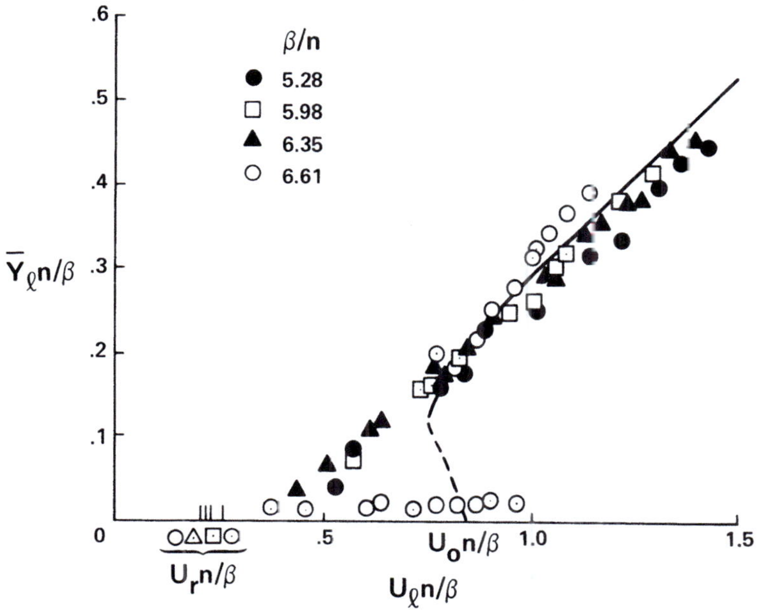

FIG. 3 REDUCED GALLOPING AMPLITUDES OF TOWER OF SQUARE SECTION.

It is no coincidence that the data for the highest damping level produces the best agreement with the theory. The reason is that U_o is then farthest removed from U_r, so the quasi-steady assumptions are more accurate. The facts that at lower damping levels U_r is closer to U_o, and that the towers vibrated with considerable amplitudes at wind speeds below the predicted range are related, and will be considered further in the last section of the paper. Thus, if the quasi-steady assumptions are reasonably well met, the galloping theory produces accurate predictions even in the real, turbulent, three-dimensional world, and is accordingly directly useful to the designer. We next turn our attention to the modelling of vortex-induced vibrations.

the oscillation hysteresis (two different possible variations of stable amplitude with wind speed) observed in some wind tunnel experiments at very low damping did not appear to be reproduced.

Now, the nonlinear term in the Hartlen-Currie model was chosen as the simplest form to produce the necessary self-excited oscillation of the lift coefficient for the stationary-cylinder case, corresponding to the normal formation of a Kármán vortex street. Although it is not commented on in Ref. 9, the model generates 3 real solutions for ω and ϕ, and, for certain choices of the system constants, for $\overline{Y}$ and $\overline{C}_y$ also, so that it is capable of displaying oscillation hysteresis. However, the multiple solutions for $\overline{Y}$ and $\overline{C}_y$ are unlike those observed. Perhaps a higher-order nonlinearity in the damping term of Eq.(3) would lead to a better reproduction of this oscillation hysteresis in $\overline{Y}$ and $\overline{C}_y$. (It is easily shown that such a change in Eq.(3) would not affect the variation of ω and ϕ, already fairly satisfactory in the Hartlen-Currie model). Berger [10] has used the greater freedom of such a higher-order nonlinearity to force agreement at more points with experimental data, and has produced quite close agreement with circular-cylinder data by Feng, (given in Ref. 1), on the details of the oscillation hysteresis in the $\overline{Y}$ - U variation during lock-in at low system damping, as shown in Fig. 4(a). On the other hand, at higher system damping, his model correctly predicts a $\overline{Y}$ - U variation without oscillation hysteresis, shown in Fig. 4(b).

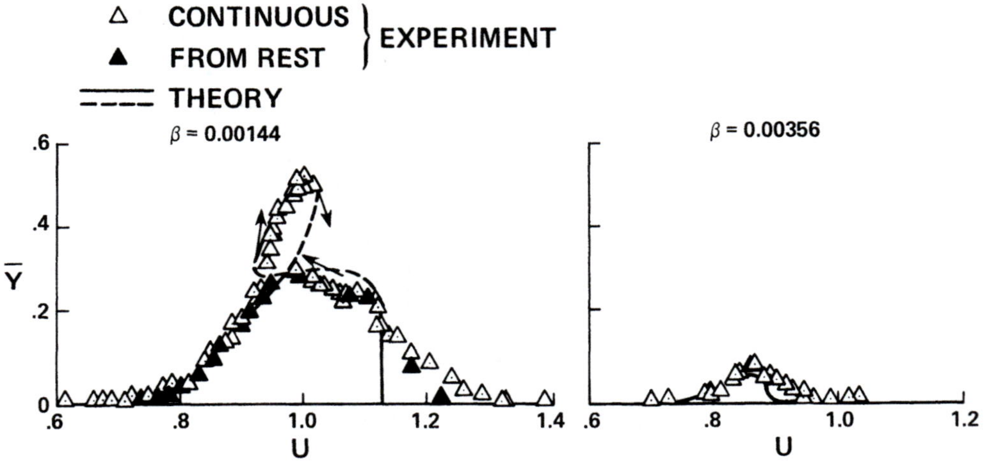

(A) LIGHTLY DAMPED (B) MORE HEAVILY DAMPED

FIG. 4 COMPARISONS OF MEASURED AMPLITUDE VS WIND SPEED WITH THEORY FOR VORTEX-INDUCED OSCILLATIONS OF A CIRCULAR CYLINDER [10].

The Berger and Hartlen-Currie models do not employ any flow-field theory in their derivation. Some investigators have attempted to put the nonlinear oscillator modelling on a more physical basis by deriving the equations using semi-empirical flow field theory. One model of this type, by Blevins and Iwan, is described by Blevins in his book on flow-induced vibration [11], and another, similar model was presented recently by Tamura and Matsui [12]. In the Blevins-Iwan model, the equivalent of Eq.(3) is a differential equation, again of van der

3. VORTEX-INDUCED VIBRATIONS

Nonlinear oscillator modelling of vortex-induced vibrations of long elastic bodies of bluff section is not as useful to the designer as is the corresponding modelling of galloping, for without any modelling he can assume that any bluff section with an appreciable afterbody is subject to the vibrations, that the range of wind speeds over which they will occur is a fairly small one containing U_r, and that although decreased system damping causes increased cylinder amplitudes, still the maximum amplitude $\bar{Y}_{max}$ will be at most of O(1). (The amplitude is observed to be self-limiting because of the changes in the wake vortex system with amplitude.) The chief benefit therefore of this form of nonlinear oscillator modelling is probably its contributions to satisfying scientific curiosity about aspects of the complex interactions between a vibrating cylinder and its vortex wake.

Nevertheless, the more successful models of this type do offer quantitative estimates of the variation of $\bar{Y}$ and other physical features of the vibration with U as a function of β, n, and other system parameters, almost exclusively for circular cylinders. The restriction is necessary, because in these models the numerical coefficients of key terms of the nonlinear equations are determined empirically by matching the models to experimental data for vibrating cylinders. In other words, the vortex-induced oscillator models really attempt to duplicate, rather than predict, behaviour. Since there is an abundance of data on vibrating circular cylinders, the theoretical models can be thoroughly tested, but again are less needed by the designer, who can turn to the original data. Another practical reason for the concentration on circular cylinders for vortex-induced vibration is that for non-circular sections the experimental separation of the effects from those of galloping may be difficult, since high-amplitude vortex-induced vibrations require low damping, and this causes a low value of U_o, which will then be close to U_r. Such combined effects are considered in the next section of the paper.

The most successful class of nonlinear oscillator models of vortex-induced vibration is that using two coupled ordinary differential equations for the displacement Y (Eq.(1), or a minor modification of it), and a flow variable related to the excitation of the cylinder. Only this second equation is nonlinear, and there only in the damping (first time derivative) term in the flow variable. The first of these models, which spawned the subsequent work, was devised for the circular cylinder by Hartlen and Currie [9]. In it the flow variable is the cylinder lift coefficient C_y, and the two equations are Eq.(1) and the equation for C_y, in which the nonlinearity is of the van der Pol type, selected because of its simplicity:

$$\ddot{C}_y - A\Omega_o\dot{C}_y + \frac{B}{\Omega_o}\dot{C}_y^{\,3} + \Omega_o^2 C_y = D\dot{Y} \qquad (3)$$

In Eq.(3), $\Omega_o = U/Ur$, and A, B, D are constants, selected to match the model to some of the data from wind tunnel experiments. Harmonic oscillation of Y and C_y are assumed at the same frequency ω, and Eqs.(1) and (3) are solved by harmonic balance. The principal characteristics that a good nonlinear oscillator model should reproduce are the synchronization, or lock-in, of the vortex formation frequency (i.e., the lift frequency) with the cylinder frequency, at the common value ω, which is always close to ω_n, over a range of U which depends on β/n, the variation of the phase angle φ between lift and cylinder displacement during lock-in, and the dependence of the amplitudes of the cylinder displacement $\bar{Y}$ and the lift coefficient $\bar{C}_y$ on U and β/n. The Hartlen-Currie model gave encouragingly good qualitative reproductions of the above characteristics, and the quantitative agreement was respectable, but there was clearly room for improvement. For example,

Pol type, in a 'hidden' flow variable defined so that its time derivative represents a weighted average of the transverse velocity within a two-dimensional control volume for which a momentum balance is sought. Parameters of the model again are fixed by matching data from circular-cylinder experiments. In the Tamura-Matsui model, a different version of a concept used by Nakamura, and described in Ref. 1, is employed. The concept is that of representing the effects of the vortex wake of a circular cylinder by a finite cavity attached to the cylinder and in torsional oscillation. In the equivalent of Eq.(3) for this model, the variable is angular position α, and the nonlinear term comes from an assumed oscillatory length of the cavity, which creates a nonlinear spring effect in the estimated aerodynamic restoring torque of the flow about the cylinder-cavity combination. Since the nonlinear term is linear in $\mathring{\alpha}$ and quadratic in α, the α-equation is again of the general van der Pol type.

The two models are quite similar mathematically, and although neither gives realistic estimates of oscillation hysteresis in $\bar{Y}$ and $\bar{C}_y$, both give quite good estimates of $\bar{Y}_{max}$ as a function of β/n, as can be seen from Fig. 5, in which calculated curves from both models are compared with a wide assortment of experimental data. The theoretical curves are nearly coincident below $\bar{Y}_{max} = 0.1$, but the Iwan model gives better agreement with higher-amplitude data.

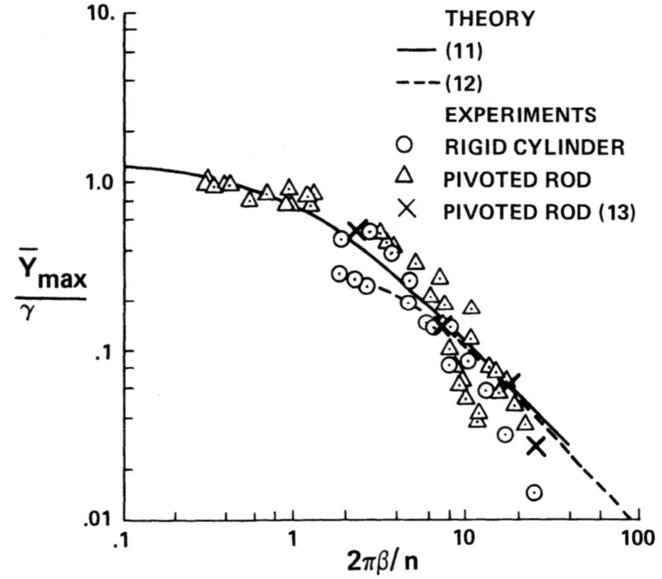

FIG. 5 MAXIMUM AMPLITUDE OF VORTEX-INDUCED OSCILLATION OF CIRCULAR CYLINDERS AS A FUNCTION OF MASS AND DAMPING PARAMETERS.

Finally, then, how can the designer benefit from the research on nonlinear oscillator modelling of vortex-induced vibrations? The use of circular sections for stacks, towers, and structural members is widespread, but the flows they experience are turbulent, and usually spatially nonuniform. The physical key in the flow to the strength of the cylinder vibrations is most probably the spanwise correlation of the vortex formation process, so that one might expect weaker vibrations in the field than in the laboratory, since the spanwise correlation should be worse. Also, since the oscillation hysteresis in $\bar{Y}$ and $\bar{C}_y$ seems likely to be caused physically by jumps in this spanwise correlation, again these phenomena seem more likely to occur in the laboratory than in the field. On the other hand, there is some evidence from experiments by Howell and Novak [13] that the higher amplitude vortex-induced vibrations of circular cylinders occurring with low system damping are not affected by turbulence in the incident flow, as shown by the inclusion of some of their data in Fig. 5. Therefore, the theoretical results of Fig. 5 may be directly helpful to the designer in indicating maximum amplitudes to be expected as a function of damping and mass parameters, even though the data on which the curves are based was obtained in smooth flow. In this context ordinates of the curve from the Blevins-Iwan model can be multiplied by mode shape factors γ to give maximum amplitudes for various elastic modes of vibration, in addition to the basic rigid body displacement ($\gamma = 1$).

4. COMBINED EFFECTS OF GALLOPING AND VORTEX RESONANCE

As a starting point for considering combined effects on flow-induced vibration of the galloping and vortex mechanisms, we examine the tower results of Fig. 3. Although the galloping theory predicts no galloping from rest below $U_{\ell o}n/\beta \cong 0.8$ this was found to be true experimentally only for the highest damping level. At the other three damping levels, oscillation started from rest at $U_\ell n/\beta \cong 0.4$ and the amplitude $\bar{Y}_\ell$ increased nearly linearly with U_ℓ, apparently as an extension of the linear upper portion of the galloping curve, for which the quasi-steady theory accurately predicted the amplitudes. Now, it can be seen in Fig. 3 that for the three lower damping levels, the wind speed range for tower vibration in each case begins slightly above U_{ℓ_r}, so that presumably the wake vortex excitation of the tower is playing a role. In these three examples, the ratio U_{ℓ_o}/U_{ℓ_r} ranges from 3.4 to 4.0. In a full-scale tower like the one modelled in the experiments of Fig. 3, using mass, elastic, and damping characteristics typical of modern construction, the ratio $U_{\ell o}/U_{\ell_r}$ might be even lower, perhaps from 1.5 to 3.0, so that interaction between galloping and vortex resonance effects might be expected to be even stronger. For the full-scale 400m tower, the actual value of V_{ℓ_r} would be about 35 m/s, corresponding to 13 m/s at the usual 10m reference height, and well within the design range of wind speeds for such a structure.

Thus, winds somewhat higher than V_{ℓ_r}, but still within the design range would from Fig. 3 be expected to cause much larger amplitudes of vibration of the tower than would the vortex excitation alone, even though the windspeeds were appreciably below the theoretical critical galloping value V_{ℓ_o}. It would clearly be desirable to know more about these interactions, and a recent paper [14] describes an experimental investigation of them in our laboratory. To reduce the number of influences to be interpreted, the experiments were carried out in smooth flow, under closely two-dimensional conditions. The cylinders, of square section, could have β and n changed so that Uc/Ur ranged from 0.15 to 7.56. Figure 6 shows some results of this investigation, measured variations of $\bar{Y}$ with U for two values of Uo/Ur. In Fig. 6(a), with Uo/Ur = 0.15, no vibration occurs until U = Ur, and then $\bar{Y}$ increases with U in an almost linear gradual approach to the theoretical galloping curve. In Fig. 6(b), with Uo/Ur = 1.49, there is again a sequence of stable amplitudes $\bar{Y}$ increasing nearly linearly with U, and approaching the upper branch of the theoretical galloping curve, in fact appearing as a continuation of it, as noted in Fig. 3. However, between U_r and U_o there are also two other sequences of $\bar{Y}$, one low sequence of stable amplitudes, and an intermediate sequence of unstable amplitudes, the two apparently forming a closed loop -- clearly a worthwhile object for nonlinear oscillator modelling.

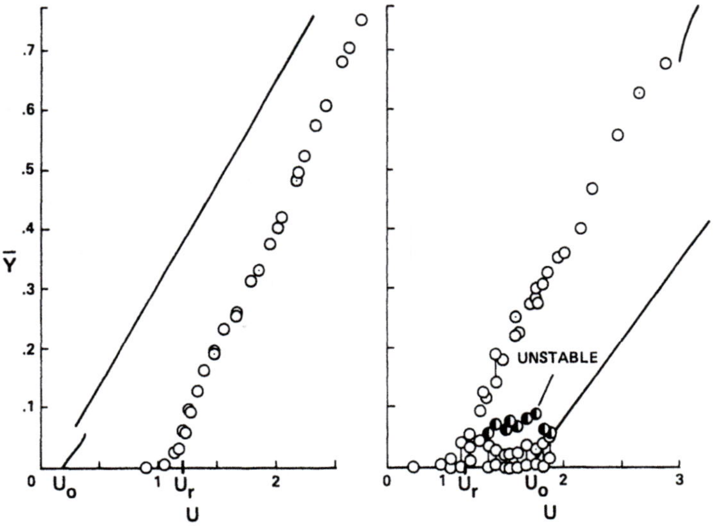

(A) $U_0/U_r = 0.15$ (B) $U_0/U_r = 1.49$

FIG. 6 AMPLITUDE-WIND SPEED VARIATION FOR SQUARE-SECTION CYLINDERS.

At appreciably higher ratios Uo/Ur, the vortex resonance and galloping effects became separated and appeared independent. The investigation, while suggesting some further explorations of the possibilities of nonlinear oscillator modelling, adds further support to the idea that long bodies of square section are subject to large-amplitude wind-induced vibrations at all wind speeds above U_r, and that the variation of $\bar{Y}$ with U can be given approximate limits by specifying a linear variation from zero at U_r to join the theoretical galloping curve, known to be accurate at high values of U.

5. REFERENCES

1. Parkinson, G.V.: "Mathematical Models of Flow-Induced Vibrations of Bluff Bodies", General Lecture, IUTAM-IAHR Symposium Flow-Induced Structural Vibrations, Karlsruhe, 1972, published Springer-Verlag, 81-127, (1974).

2. Durbin, P.A. and Hunt, J.C.R.: "Fluctuating Surface Pressures on Bluff Structures in Turbulent Winds: Further Theory and Comparison with Experiment", Proc. Fifth Int. Conf. Wind Engineering, Fort Collins, V-1-1 to V-1-17, (1979).

3. Roshko, A.: "A New Hodograph for Free Streamline Theory", NACA TN 3168, (1954).

4. Parkinson, G.V. and Jandali, T.: "A Wake Source Model for Bluff Body Potential Flow, Jour. Fluid Mech., 40, 3, 577-594, (1970).

5. Clements, R.R. and Maull, D.J.: "The Representation of Sheets of Vorticity by Discrete Vortices", Prog. Aerospace Sci., Pergamon, 16, 2, 129-146, (1975).

6. Parkinson, G.V. and Sullivan, P.P.: "Galloping Response of Towers", Proc. Third Colloq. Ind. Aero., Aachen, 2, 21-34, (1978). (To be published in Jour. Ind. Aero.)

7. Sullivan, P.P.: "Aeroelastic Galloping of Tall Structures in Simulated Winds", M.A.Sc. Thesis, U.B.C., (1977).

8. Novak, M.: "Galloping Oscillations of Prismatic Structures", Jour, Eng. Mech. Div., Proc. A.S.C.E., EM1, 27-46, (1972).

9. Hartlen, R.T. and Currie, I.G.: "Lift-Oscillator Model of Vortex-Induced Vibration", Proc. A.S.C.E., E.M.5, 577-591, (1970).

10. Berger, E.: "Some New Aspects in Fluid Oscillator Model Theory", Proc. Third Colloq, Ind. Aero., Aachen, 2, 19, (1978).

11. Blevins, R.D.: "Flow-Induced Vibration", van Nostrand, (1977).

12. Tamura, Y. and Matsui, G.: "Wake-Oscillator Model of Vortex-Induced Oscillation of Circular Cylinder', Proc. Fifth Int. Conf. Wind Engineering, Fort Collins, VIII-10-1 to VIII-10-10 (1979).

13. Howell, J.F. and Novak, M.: "Vortex Shedding from Circular Cylinders in Turbulent Flow", Proc. Fifth Int. Conf. Wind Engineering, Fort Collins, V-11-1 to V-11-11 (1979).

14. Wawzonek, M.A. and Parkinson, G.V.: "Combined Effects of Galloping Instability and Vortex Resonance", Proc. Fifth Int. Conf. Wind Engineering, Fort Collins, VI-2-1 to VI-2-12 (1979).

H4 USE OF OSCILLATOR MODELS FOR IDENTIFYING THE DANGER FOR FLOW-INDUCED VIBRATIONS IN A COMPLEX SYSTEM

E. Naudascher

Director, Institute of Hydromechanics
University of Karlsruhe

As limited as the oscillator-modelling method may be due to the difficulties of formulating the damping terms and to tracing them to the individual elements of the flow dynamics, it can be of great help in identifying and assessing the danger for excitation of flow-induced vibrations in a complex system. For illustration that this method may be far more universally applicable than we realize, consider the sluice gate shown in Fig. 1b, and the striking similarities in response of this gate with the response of a cylinder in crossflow (Fig. 1a), even as far as the lock-in behavior, including hysteretic jumps, or the dependence of maximum amplitude, on the system damping is concerned. In Fig. 1c, moreover, one may compare the same well-known cylinder response, for which oscillator models have been developed with great success, with the response of the standing-wave formation due to open-channel flow past a row of cylindrical piers. Again, a similar lock-in behavior can be discerned, despite the fact that it is an oscillating wave or water mass rather than an oscillating body which is being described in this diagram. It thus appears reasonable to seek differential equations for the analysis of the gate and wave responses which are similar to those formulated for the vortex-excited cylinder.

As a matter of fact, one can show that any flow-induced vibration can be viewed as the result of different actions and interactions of three types of oscillation in the system: (1) the "body" oscillator, represented by the elastically supported cylinder or gate in Fig. 1a,b; (2) the "fluid" oscillator, e.g.

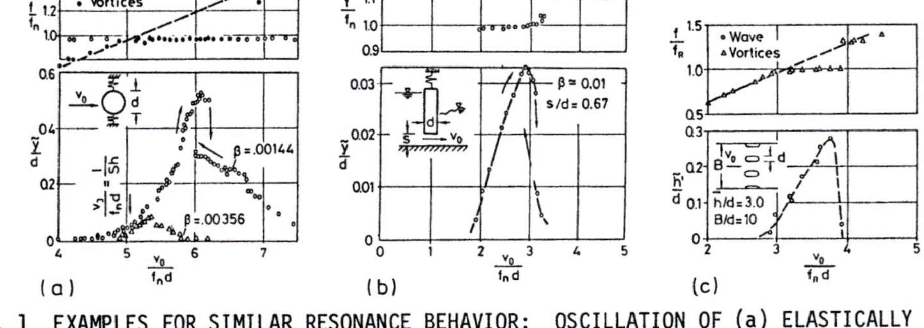

FIG. 1 EXAMPLES FOR SIMILAR RESONANCE BEHAVIOR: OSCILLATION OF (a) ELASTICALLY SUPPORTED CIRCULAR CYLINDER: (b) ELASTICALLY SUPPORTED SLUICE GATE: (c) STANDING WAVES BETWEEN PIERS IN AN OPEN CHANNEL

the oscillating water mass or wave in Fig. 1c; and (3) the "flow" oscillator, i.e. the alternating vortex shedding in all the examples depicted in Fig. 1. These three types of oscillators must not necessarily be present simultaneously for a given flow-induced vibration, nor must the number of oscillators be confined to one of a type.

To show what is gained by accepting this viewpoint, consider the example of Fig. 2. In this system, there is a body oscillator represented by the plug valve, a fluid oscillator represented by the surging water mass in the downstream tank; and a flow oscillator in the case that flow instability develops in the gap between valve and conduit as shown in Detail D1. Each of these oscillators, it turns out, can be described by a homogeneous differential equation when they are isolated, and the various types of flow-induced excitation which may develop in the overall system can be described by the same equations when the proper forcing or coupling terms are introduced (Fig. 3).

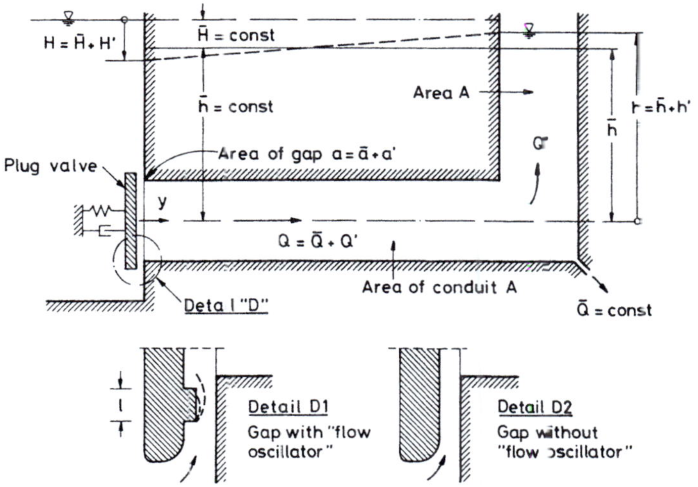

FIG. 2 PLUG VALVE WITH CONDUIT AND SURGE TANK - EXAMPLE FOR THE OSCILLATOR-MODEL APPROACH

For the particular system chosen here, these equations can be set up in a relatively simple fashion, and it is striking how much insight they provide for the variety of possible flow-induced vibrations. With only the body oscillation present, there can be self-excited vibration of the valve; with the flow oscillator active as well, the valve may also be susceptible to instability-induced vibration; and with the added effect of a fluid oscillator (i.e., the resonating surge in the example chosen), both these vibrations may be so substantially altered that neglect of this effect would lead to serious misjudgement of the system safety. (For more detail, see Naudascher, E., and Rockwell, D., "Oscillator-Model Approach to the Identification and Assessment of Flow-Induced Vibration in a System", to be published in the Journal of Hydraulic Research, January 1980).

The oscillator-model approach can thus be used as a tool for a lumped-parameter system analysis. What is important to note with regard to the practice-oriented theme of this Symposium is the fact that is a diagnosis of global assessment of dangerous spots and operating conditions of a system, along with recom-

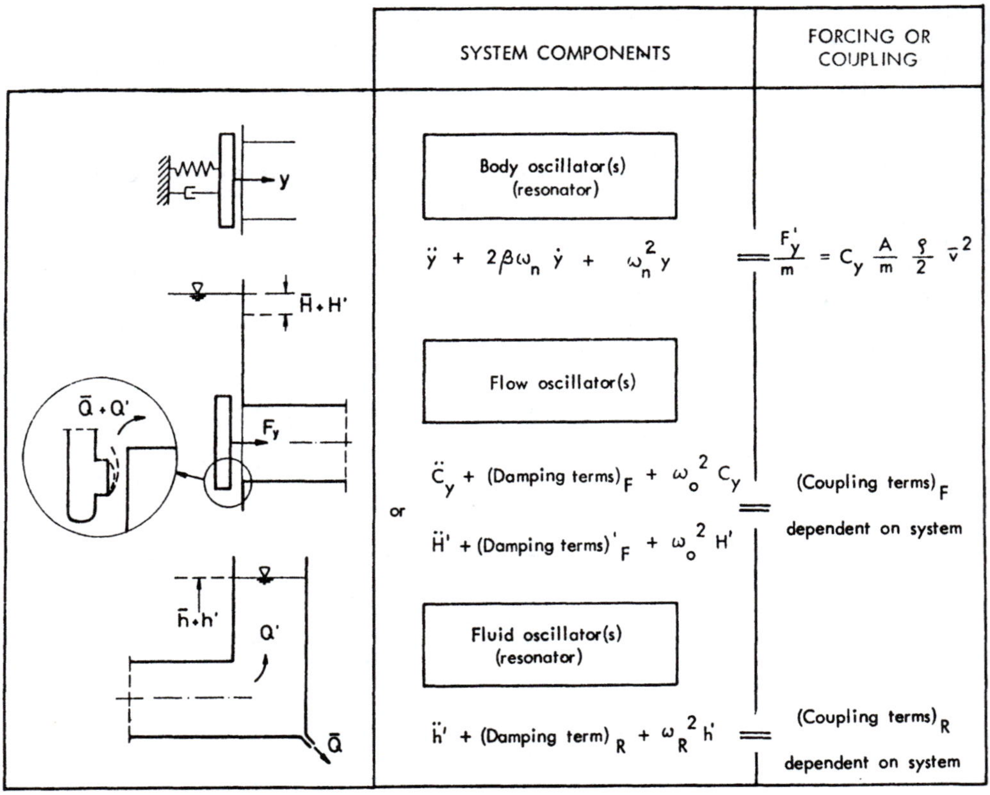

FIG. 3 SCHEME OF "BODY", "FLOW", AND "FLUID" OSCILLATORS, AND THE CORRESPONDING DIFFERENTIAL EQUATIONS, FOR THE OSCILLATOR-MODEL TREATMENT OF THE VALVE-CONDUIT-SURGE-TANK SYSTEM SHOWN IN FIG. 2.

mendations for cures, is all one needs, this oscillator-model system analysis will help greatly even if carried out in an approximate or conceptual fashion. It will help in safeguarding that no essential element of any of the great number of possible types of excitation, and none of their dangerous combinations, is overlooked. When investigating a problem on which information is scarce, moreover, this approach offers a basis for adapting, or at least benefiting from, methods and solutions developed in other fields. In certain cases, it may even yield quantitative prediction of the first-order system behavior.

H5 STUDIES OF FLOW INSTABILITIES IN TWO-DIMENSIONAL TEST SECTIONS WITH SUDDEN EXPANSIONS

W. T. Ashurst+ and F. Durst
Institut fur Hydromechanik,
University of Karlsruhe, W. Germany
+Sandia Laboratories, Livermore, California, USA

SUMMARY

Basic studies of flow instabilities are essential to advance the understanding of flow induced vibrations. Solutions to practical problems not only require field studies but also laboratory investigations and theoretical studies of the basic mechanisms involved in the cause of flow oscillations. The present paper describes experimental and theoretical studies of shear layer induced flow oscillations in a symmetric, two-dimensional, plane test section with sudden expansion geometry. The flow has been experimentally investigated with a laser-Doppler anemometer and numerically simulated with the discrete vortex dynamic method. The experimental and numerical results both show that the interaction of the two shear layers can produce asymmetric flow within a symmetric geometrical configuration with symmetric inlet and outlet velocity profiles.

1. INTRODUCTION

Flows in symmetric, two-dimensional plane test sections that have sudden expansion geometries exhibit symmetric flow properties at low Reynolds numbers. With increasing Reynolds number, asymmetric flows are observed in spite of the symmetric inlet and outlet velocity profiles and the symmetric boundary conditions. Both the symmetric and asymmetric flow fields have been experimentally studied by Durst et al [1] and Cherdron et al [2]. The same asymmetric flow field also exists in turbulent flow and has been investigated by Abbott and Kline [3] and more recently by Smyth [4]. Sparrow has carried out investigations of separated flows with mass transfer and demonstrated that the asymmetry of the flow is also reflected by wall transfer data [5].

Although investigations of flow with shear layer interactions are available in the literaure, detailed measurements in truly two-dimensional flows are still missing. Furthermore, combined experimental and theoretical studies have not been carried out to the extent required to deepen the physical understanding of the observed flow asymmetry within symmetric geometries. The present paper tries to remedy this situation by providing detailed velocity measurements carried out with a laser-Doppler anemometer system. Systems of this kind permit undisturbed, local velocity measurements in flow of transparent fluids. This paper also presents a comparison of experimental results with calculations using the discrete vortex dynamic method. In addition to a computer generated movie of the shear layer motion, the calculated time-averaged flow velocities and turbulent fluctuations are compared with experimental results. The agreement is found to be satisfactory. These results reveal that the unsteady vortex structures, which are amplified in the free shear layers, provide the mechanism for asymmetric flow within a symmetric geometry. The experimental results will be described followed by a description of the discrete vortex dynamic method.

2. EXPERIMENTAL INVESTIGATIONS

The present section summarizes the experimental investigations that have been carried out at the Institute of Hydromechanics of the University of Karlsruhe in order to obtain detailed measurements of the flow field in symmetric, two-dimensional, plane test sections with sudden-expansion geometries for flows at various Reynolds numbers. Detailed documentation of this work is presented in References 1, 2 and 6.

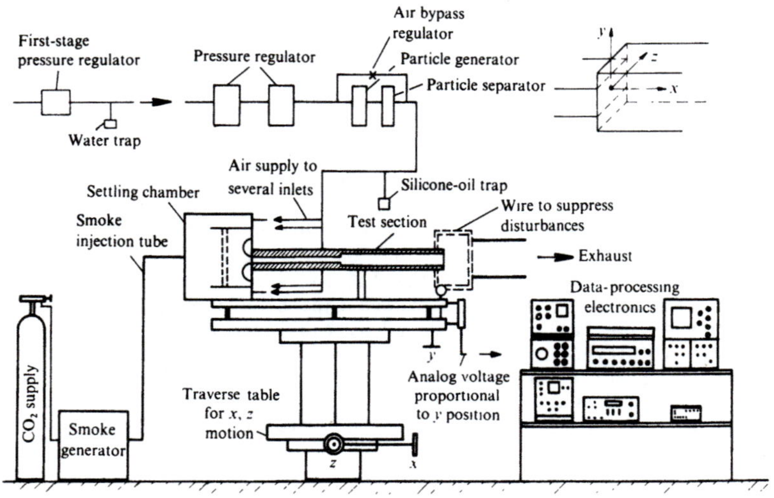

FIG. 1 SCHEMATIC REPRESENTATION OF EXPERIMENTAL TEST RIG.

Figure 1 indicates schematically the test section and experimental equipment that was employed to study internal gas flows of moderate Reynolds number in the vicinity of a sudden expansion. The symmetric sudden expansion is accomplished by a step change in the channel width.

Flow visualization has been used to provide understanding of the overall flow field structure. Smoke is induced in the upstream flow and long-time exposure photographs reveal the time-average recirculation regions while short-time exposures reveal the unsteady nature of the flow. In Figure 2 short-time exposures are given of a smoke streak (top) and of the instantaneous recirclation regions (bottom). From these pictures it is clear that the upstream wall boundary layers separate at the sudden expansion and become free shear layers. Disturbances in these layers are amplified and result in alternating vortex structures. It appears that the interaction of the vortex structures causes the asymmetric flow field.

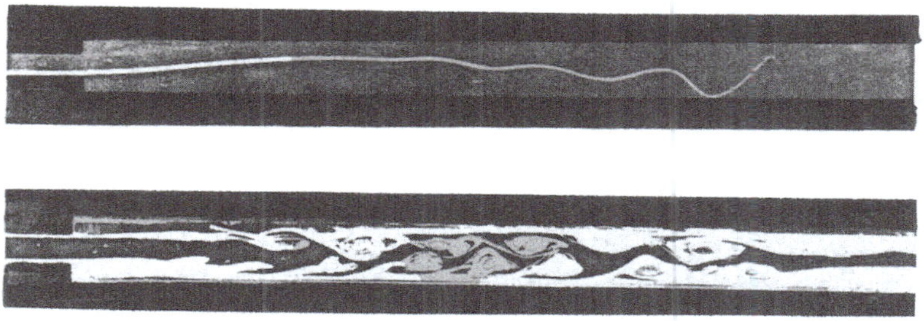

FIG. 2 SHORT TIME EXPOSURE OF INDUCED SMOKE IN SUDDEN EXPANSION (ON LEFT) FLOW.

For a more quantitative determination of the flow field a laser-Doppler anemometer (LDA) has been used. The LDA technique is well suited to flows with recirculation since the LDA system permits non-intrusive measurement of the velocity without the ambiguity of flow direction as would occur with the hot-wire method. A complete map of the time-average velocity profiles can be easily made. Figure 3 is an isometric view of the asymmetric, "almost" two-dimensional, flow field at two different Reynolds numbers.

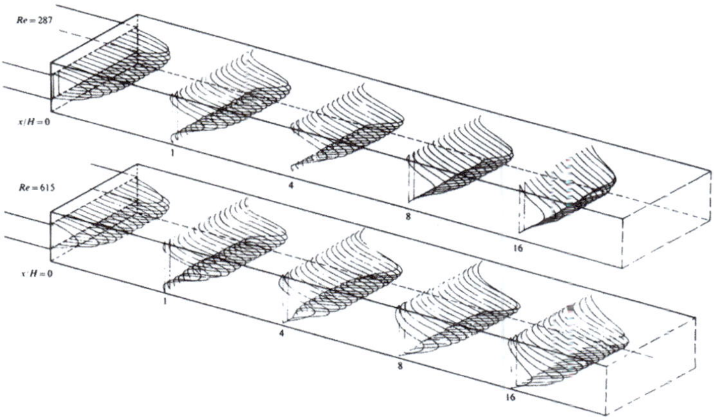

FIG. 3 SUDDEN EXPANSION MEAN VELOCITY PROFILES OBTAINED WITH A LASER-DOPPLER ANEMOMETER. NOTE THE SMALL SPANWISE EFFECT IN BOTH FLOWS AND THE LONGER RECIRCULATION REGION IN THE HIGHER REYNOLDS NUMBER FLOW.

The time resolution of the velocity record depends upon the number of particles that scatter the laser beam. It is possible to add particles to air flows and most water supplies already have plently of particles. With a finely resolved time-history velocity measurement it is possible to determine the frequency spectra by putting the LDA output signal into a fast Fourier analyser. The frequency of vortex shedding has been determined and is given in Reference [2].

3. THEORETICAL INVESTIGATIONS

In order to study the cause of the asymmetric flows that have been experimentally observed, theoretical studies were done using the discrete vortex dynamic method. This method has previously been used to numerically simulate a free shear or mixing layer. In that work [7] it was shown that the experimentally observed vortex structure in the mixing layer was also produced in the numerical results. Since the application of the discrete vortex dynamic method is somewhat new a brief description of its computer implementation will be given. The reader should also consult the review by Dr. D. Maull [8] which is concerned with exterior flow problems.

The discrete vortex dynamic method is a Lagrangian description of fluid flow. Vorticity has a kinematic definition as the vector curl of velocity $\underline{\omega} = \nabla \times \underline{u}$. In the vortex dynamic method the distribution of vorticity is discretized and the trajectory of each discrete vortex is calculated. The governing equation of motion is obtained by taking the vector curl of the Navier-Stokes equation, yielding:

$$\frac{D}{Dt}\left(\frac{\underline{\omega}}{\rho}\right) = \frac{\underline{\omega}}{\rho} \cdot \nabla \underline{u} + \nu \nabla^2 \underline{\omega}$$

when the density ρ is a function of pressure only. The symbol D/Dt is the material or Lagrangian derivative (time rate of change while moving with the fluid mass). The change in vorticity (over density) is given by two terms: a) vortex stretching and rotation $\underline{\omega}/\rho \cdot \nabla \underline{u}$ and b) vorticity diffusion by molecular viscosity $\nu \nabla^2 \underline{\omega}$. For two-dimensional flow the vorticity vector is normal to the velocity vector and thus $\underline{\omega} \cdot \nabla \underline{u}$ is always zero. For large Reynolds number flows (small ν) Chorin [9] has suggested that the diffusion effect be approximated with a random walk. Each discrete vortex is given a random component of motion (at each time step) which is taken from a Gaussian distribution of random numbers. The standard deviation of the random numbers is specified by the numerical time step Δt and the molecular viscosity ν (or Reynolds number) and is

$$\sigma = \sqrt{2\ \Delta t\ \nu} = L\ \sqrt{2\left(\frac{U\Delta t}{L}\right)\left(\frac{\nu}{UL}\right)}$$

To advance the vorticity to the next time step the velocity of each discrete vortex is required. There are two ways to determine the vortex velocity: direct vortex interaction or solution of the Poisson equation for the streamfunction. The direct interaction method is given by the Biot-Savart law which determines the velocity at a point from a volume integral over all regions which have non-zero vorticity. Only the discrete vortices have non-zero vorticity and this means each vortex velocity is given by direct interaction of all the other vortices. The calculational cost will be proportional to the number of vortex interactions which is (N - 1) times N/2 or $\alpha\ N^2$, for N discrete vortices. Note that only the vortex locations are needed to determine the velocity field. Thus spatial resolution of the flow field is automatically treated. There is no loss of resolution due to finite mesh spacing. Of course, the size of the discrete vortex provides a lower limit of spatial resolution and the finite time step provides an upper limit to resolvable frequencies.

The second way to calculate the velocity field requires a solution of a Poisson equation. Since any vector field may be described in terms of a scalar potential $\emptyset$ and a vector potential $\underline{A}$, the velocity field is written as $\underline{u} = \nabla \emptyset + \nabla \times \underline{A}$. From the kinematic relation of vorticity and velocity ($\underline{\omega} = \nabla \times \underline{u}$) the vector potential Poisson equation is obtained

$$\nabla^2 \underline{A} = - \underline{\omega} \text{ and in 2 dimensions } \nabla^2 A_z = - \omega_z$$

Thus the velocity is given as $\underline{u} = \nabla \times A_z$ where A_z is know as the stream function. Calculational cost of the Poisson solution is proportional to M Log M, where M is the number of mesh points. The cost of putting the discrete vorticity onto the mesh and finding the velocity from the stream-function will be proportional to N, the number of vortices. The cost of this method is M Log M + N, and for large N, this method will be faster than the direct method which cost N^2.

However the discrete numerical approximation of ∇^2 does not produce the point vortex velocity field. Instead of an infinite velocity at the vortex location with a r^{-1} decay (where r is the distance from the vortex) the velocity is zero at the vortex location and increases with r until a maximum is reached and then decays. The r^{-1} behavior is only achieved when r is large compared to the mesh spacing. The short range velocity is also non-isotropic. To remove these non-physical ,numerical effects the vorticity is artificially expanded to cover many mesh cells when the vorticity distribution is determined for the Poisson equation. This spreading of vorticity results in an isotropic velocity field which approaches the r^{-1} behavior within a few mesh cells (three cells in this work). The incorrect velocity behavior can be replaced with the true values for all interactions that occur within the short range region. During the creation of the vorticity distribution a linked list is made so that only those vortices that are near each other must be examined for the velocity correction. If all the pairs were examined the cost would become the same as the direct method, N^2.

The choice of which method to use for the velocity determination will also depend upon the flow problem being considered. For exterior flows (e.g., a bluff body shedding vorticity into a wake) the direct method allows easy treatment of the outer boundary conditions: nothing needs to be done. For interior flows (e.g., channel with flow separation) the outer boundary conditions can be included in the Poisson solution. In addition the solid wall boundaries act as sources and sinks of vorticity when the zero relative velocity condition is enforced. Thus for some exterior flows the direct method might be selected where as for interior flows the Poisson method will be faster.

The vortex dynamic method produces the velocity (and pressure from another Poisson equation) as a function of time at any desired spatial location. Thus the mean and fluctuation velocities (and pressures) can be directly calculated in terms of the flow geometry. As an example, the unsteady flow in a channel with a sudden expansion has been calculated and some of the results will be given here (further details are given in [10]).

The wall boundary layers that are upstream of the expansion location are used as inlet conditions. The continuous vorticity in these boundary layers is converted into several discrete vortices at each time step. New vorticity is also added along the channel side walls in order to maintain a no-slip velocity boundary condition. Selected frames from the computer generated movie of vortex motion are shown in Figure 4 where the line length corresponds to the vortex displacement for that time step. Initially the two shear layers have similar behavior and resemble the free mixing

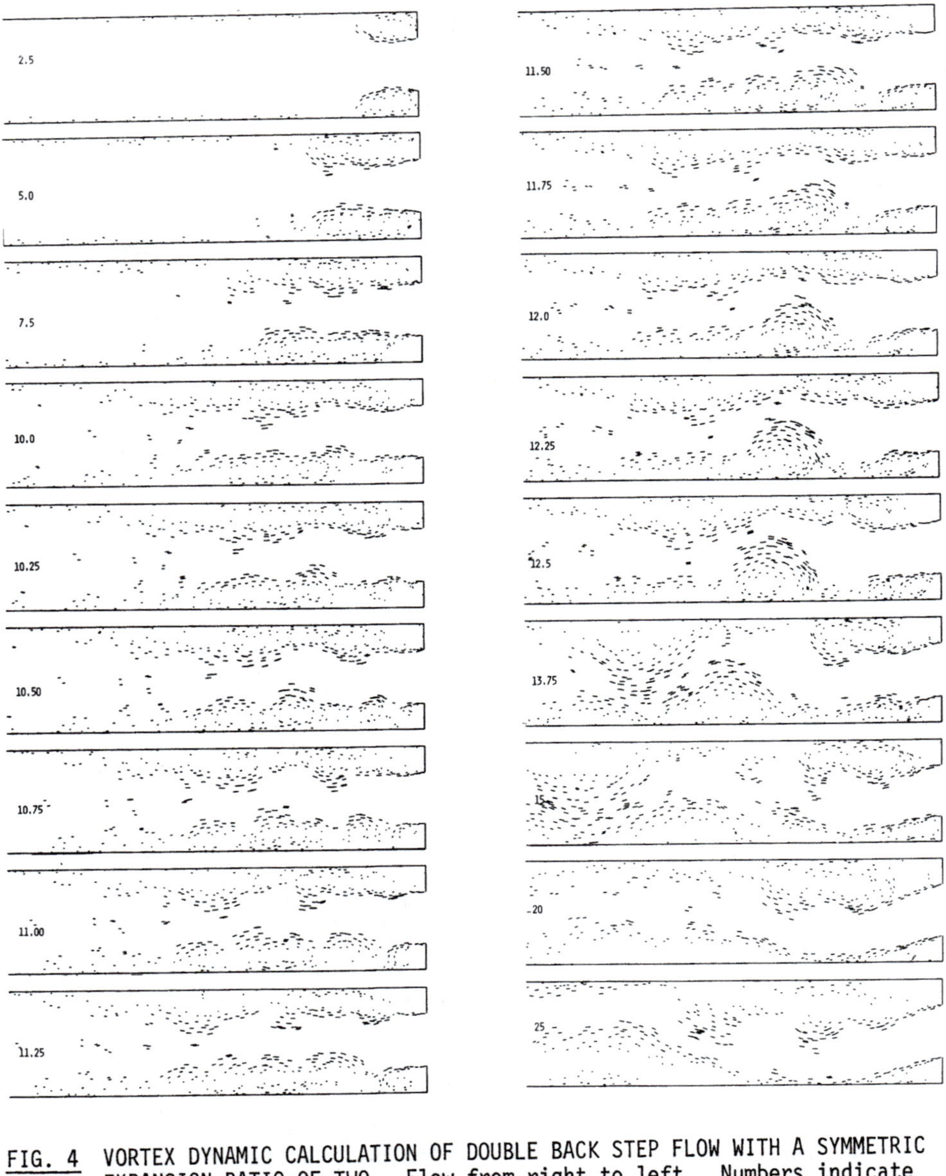

FIG. 4 VORTEX DYNAMIC CALCULATION OF DOUBLE BACK STEP FLOW WITH A SYMMETRIC EXPANSION RATIO OF TWO. Flow from right to left. Numbers indicate nondimensional time from impulsive start, L/Uin, where L = maximum channel width, Uin = entrance velocity. Dashed lines denote the discrete vortex displacements for the next time step, Δt = 0.05 L/Uin.

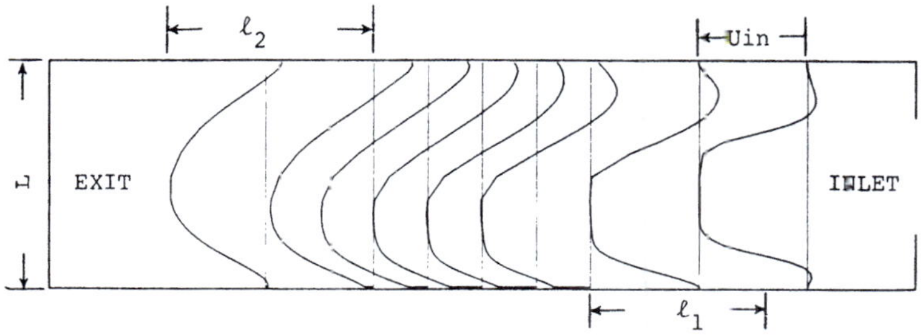

(A) CALCULATED TIME-AVERAGE STREAMWISE VELOCITY, U/Uin. Time period of 80 L/Uin; ℓ_1 and ℓ_2 denote reattachment regions from experiments [3].

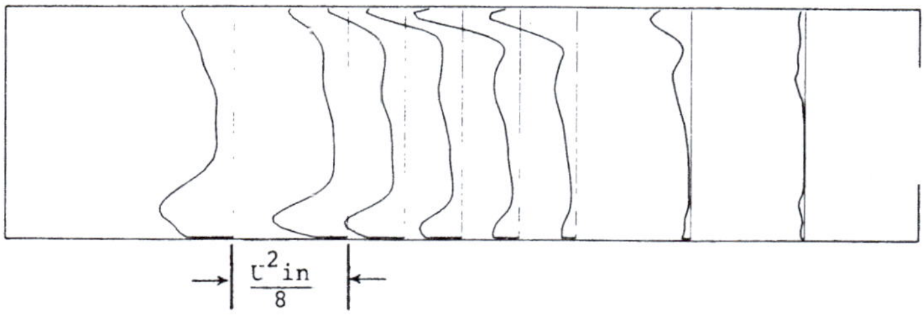

(B) CALCULATED TURBULENT NORMAL STRESS $(u/Uin)^2$.

FIG. 5 THE CALCULATED STREAMWISE VELOCITY AND TURBULENT NORMAL STRESS PROFILES AT EIGHT DIFFERENT STATIONS.

layer with clustering of vorticity. However, the pairing process produces larger and larger vorticity regions which begin to respond to the finite channel width. The wall boundary condition of zero normal velocity deforms the larger regions. A large cluster in one layer also affects the clustering sequence in the other layer. Over a period of time the lower layer is continually forced towards its wall and hence its vorticity is destroyed in a very short streamwise distance. Longer runs have not revealed any switching of the short recirculation region to the other wall. However, in a calculation using different initial conditions, the short recirculation region has appeared on the upper wall.

After the asymmetric flow field is established, the velocity at selected spatial locations is calculated each time step in order to determine the mean and fluctuation velocity values. Figure 5 presents the calculated velocity field using 248 locations (8 stations in the flow direction with 31 locations across the channel width). The asymmetric recirculation regions have good agreement with the experimental data of Abbott and Kline [3].

4. CONCLUSIONS

In hydraulic systems flows with separated regions often occur and exhibit a strong dependence upon the complete flow geometry and flow disturbances. Many of these system have produced large flow oscillations with great sensitivity to feedback effects. The flow in a channel with a sudden expansion is the example that has been experimentally and theoretically studied in this paper.

The theoretical method must be able to represent the unsteady flow effects that have been experimentally observed. A new result is the ability of the time-dependent vortex dynamic method to predict the asymmetric flow pattern that occurs in symmetric channels with sudden expansion. While the vortex dynamic method is expensive (problems require hours of computer time) it does provide unsteady flow simulation which can not be calculated by other available techniques.

The non-intrusive nature of the laser-Doppler anemometer (LDA) makes this technique well suited for flows with oscillations and flow reversal. The LDA technique has been applied to many flow geometries using a variety of fluids. The ability to measure frequency spectra of the flow field has also been demonstrated. These new experimental and numerical techniques will have many applications in the field of hydraulics.

5. ACKNOWLEDGMENT

W. T. Ashurst would like to thank Professors Naudascher and Durst for providing a summer visit at the Institute for Hydromechanics, University of Karlsruhe.

6. REFERENCES

1. Durst, F., Melling, A., and Whitelaw, J. H.; "Low Reynolds number flow over a plane symmetrical sudden expansion", J. Fluid Mech. 64, 111, (1974).

2. Cherdron, W., Durst, F., and Whitelaw, J. H., "Asymmetric flows and instabilities in symmetric ducts with sudden expansion", J. Fluid Mech., 84, 1, pp. 13-31, (1978).

3. Abbott, D. E., and Kline, S. J., "Experimental investigations of subsonic turbulent flow over single and double backward facing steps", J. of Basic Engineering, Trans. ASME D84, 317, (1962).

4. Smyth, R., "Experimental studies in turbulent, plane separated flows", Proc. ISL-AGARD Workshop on Laser-Anemometry, 233, (1976).

5. Private communication with F. Durst, 1979.

6. Durst, F., and Schonung, B., "Symmetric and asymmetric flows in two-dimensional test section with sudden expansions", SFB 80 report in preparation, University of Karlsruhe.

7. Ashurst, W. T., "Numerical simulation of turbulent mixing layers via vortex dynamics", Turbulent Shear Flow 1, Springer Verlag Heidelberg, 1979.

8. Maull, D. J., "An Introduction to the Discrete Vortex Methods", presented at this meeting.

9. Chorin, A. J., "Numerical study of slightly viscous flow", J. Fluid Mech. 57, 785-796, (1973).

10. Ashurst, W. T., "Calculation of plane sudden expansion flow via vortex dynamics", Proceeding of 2nd Symposium on Turbulent Shear Flows, Imperial College, London, July 1979.

J1 SUMMARY: PANEL DISCUSSION ON CRITICAL UNRESOLVED PROBLEMS

Donald Rockwell
Professor of Mechanical Engineering and Mechanics
Lehigh University
Bethlehem, PA, USA

1. INTRODUCTION

To set the stage for an open discussion between Panel members and the participants, six lead-off papers covering five basic disciplines were presented at the beginning of the Panel Discussion. These papers [1 to 7], which follow this summary, provide synopses of some critical unresolved problems in each of the respective disciplines. Taken together with the corresponding General Lectures and Invited Papers, as well as suggestions and ideas arising from open discussion with the participants, they bring forth an abundance of unresolved problems.

Whether a certain class of problems has remained "unresolved" obviously depends on the perspective of the engineer or scientist. The practicing engineer needs all essential information at his disposal in order to identify and prevent disastrous scenarios. In this setting, the issue of whether a problem remains unresolved is not open for contention: on-the-spot prevention or cure is the single measure of success. For the long run, however, we should strive for attaining prevention/cure at minimum expense, financial or otherwise. Moreover, occurrence of flow-induced vibrations is often intimately linked to noise generation, excessive pressure drop and losses within the system, alteration of heat transfer characteristics, etc., making an in-depth understanding of the basic mechanisms even more essential.

From the standpoint of one who has an experimental modelling facility at hand, overall simulation of the full-scale structural response may be regarded as effectively resolving the problem. Indeed, the flow and structural complexities of many situations make careful simulation mandatory. On the other hand, if one theoretically or experimentally achieves knowledge of the basic features of a given flow-structure interaction, thereby allowing extension to other problems of a similar sort, there would appear to be resolution of the problem in a more universal sense. The need for deeper insight into these basic features was underscored at the outset of the Panel Discussion, when experts characterized the uncertainties in engineering prediction of flow-induced vibrations. For example, in the area of tall buildings and structures (in wind), use of an appropriate wind tunnel is likely to allow prediction within 30%, though for vortex-shedding excitation, within 100% is considered to be an accomplishment. For the area of heat exchanger and reactor components, estimates were: internal axial flow configurations within 50%, cross flow within 100% and external axial flow within an order of magnitude; it was noted that special testing could significantly improve these estimates. Although in some cases significant uncertainty is associated with the structural aspects, such as damping, most participants would agree that, in general, the intimidating complexity of the fluid mechanics is the major obstacle.

The fact that the major share of flow excitation mechanisms inherently involve unstable, transitional, or turbulent flows greatly pleases researchers, insuring challenges for decades to come. Up until about ten years ago, most workers habitually characterized turbulence on a purely time-averaged statistical basis; since then, several breakthroughs have strongly suggested an underlying organization, or coherence, of turbulence (e.g., Roshko [8]). For workers in the area of flow-induced vibrations, this trend is particularly significant, since the coherence of the flow is likely to be most obvious in the near wake region just downstream of separation. In this context, the long recognized reappearance of organized vortex shedding at a high "turbulent" Reynolds number ($Re_D \approx 3.5 \times 10^6$, e.g., see Morkovin [9]) does indeed invite all of us to take a fresh look at separated flows, using recently developed techniques that allow examination of discrete flow events (Kovasznay [10]), compared to more conventional (and less fascinating) time averaging.

By no means is it within the scope of this brief review to survey all unresolved problems brought forth during the Panel Discussion and in the contributions to the Symposium. The reader is encouraged to study the overviews [1 to 7] that follow this summary, as well as the range of contributions in these Proceedings. With the intention of complementing points made therein, and emphasizing common features of the underlying fluid mechanics, selected types of excitation mechanisms are discussed below.

2. VORTEX SHEDDING

Although classical vortex shedding from bluff bodies immersed in essentially turbulence-free flow might be considered as well understood, there are apparently a number of open questions. The influence of end effects seems to be often overlooked, though investigation of any other feature of vortex shedding should require consideration of this aspect, involving characterization of the degree of spanwise coherence. Even the apparently simple issue of end plates on bluff bodies is a deceptive one; their role in promoting spanwise coherence is not obvious (Graham [11]). Consequently, further consideration of three-dimensionalities at the bases and tips of tall structures having various aspect ratios, as well as for tubes in confined systems, and their influence on the character of the flow dynamics along the entire structure, is worthy of further study.

Workers in the field of ocean structures, where there can be alternating flow past cylinders in the presence of little or no mean flow component, called for a more detailed understanding of vortex shedding mechanisms. Concerning vortex shedding in the presence of a mean-flow component without oscillations of the free stream, an interesting type of "in-line shedding" has been observed in water flows (Wooton et al. [13]), though its mechanism is not well understood. It was felt that this type of shedding is more likely to occur in water than air, if one considers that the response at resonance is proportional to the mass of fluid displaced divided by the structural mass.

In the area of blading and stay vanes in hydraulic machinery and equipment, where vortex shedding is, in some cases, felt to occur downstream of the trailing edge, it is not clear how the characteristic length (or more appropriately width) should be formulated for the dimensionless Strouhal number; use of the plate thickness,together with the displacement thicknesses of the separating boundary layers,does not seem to provide an adequate relation (Chen [12]). Although it seems to be well accepted that the trailing edge shape can strongly influence the shedding process, a detailed physical understanding is lacking. Moreover, the occurrence of transitory or fully developed stall on the surface of a vane or blade may exert an overriding influence on the very existence of shedding; if it does persist, the frequency and amplitude are likely to be strongly altered.

Of course, fluid-elastic effects and free-stream unsteadiness provide an additional dimension of complexity for all the above-mentioned aspects. With regard to the aforementioned stay vanes, guide vanes, and runners in hydraulic machinery, it would seem important to characterize the organized nature of the downstream wake, as well as separated flow zones occurring near the surface at high incidence. If fluid-elastic effects are present, or if there is unsteadiness in the incident velocity field, vortex shedding involving generation of major vortices from the leading edge of the body, as well as vortices in the separated shear layer above the body, may occur, as shown in Figure 1 (McAlister and Carr [14]). Here, the major vortex shed from the leading edge of the airfoil has a larger scale, and consequently lower frequency than the smaller vortices due to hydrodynamic instability of the shear layer. In assessing the dynamic loading, one must carefully distinguish between these various sources of vortex shedding.

The types of vortex shedding associated with hydrodynamic instability of a shear layer,such as that occurring in jets and mixing layers, may well be the essential feature of organized oscillations in certain valves, valve headers, elbows and other geometries discussed at the Symposium. A common aspect of nearly all oscillations reported was the strong resonant coupling between a turbulent separated flow region (i.e. valve, etc.) and the acoustic modes of a long steam, water, or oil pipeline. In recent years, there has been considerable interest in vortex development in turbulent jets excited by acoustic signals (Crighton [15]), and the self-excited oscillations of turbulent jets passing through cavities (Schachenmann and Rockwell [16]) and sudden expansions (Hill and Greene [17]). The type of resonant coupling occurring in these latter two cases is analogous to that studied in detail for coupling between turbulent vortex shedding from the trailing edge of a flat plate(s) and acoustic modes of the test section (Parker [18]). Moreover, the possible similarities between this resonant coupling mechanism and that occurring in tube banks, where an underlying flow instability is enhanced by coupling with an acoustic mode of the tube bank, deserves investigation.

3. GALLOPING

It is well known that galloping-type oscillations are a strong function of the structural cross-section, reflected in variations of the lift coefficient of the structure with incidence angle of the oncoming wind. Consequently, it is not surprising that the formation of ice on one side

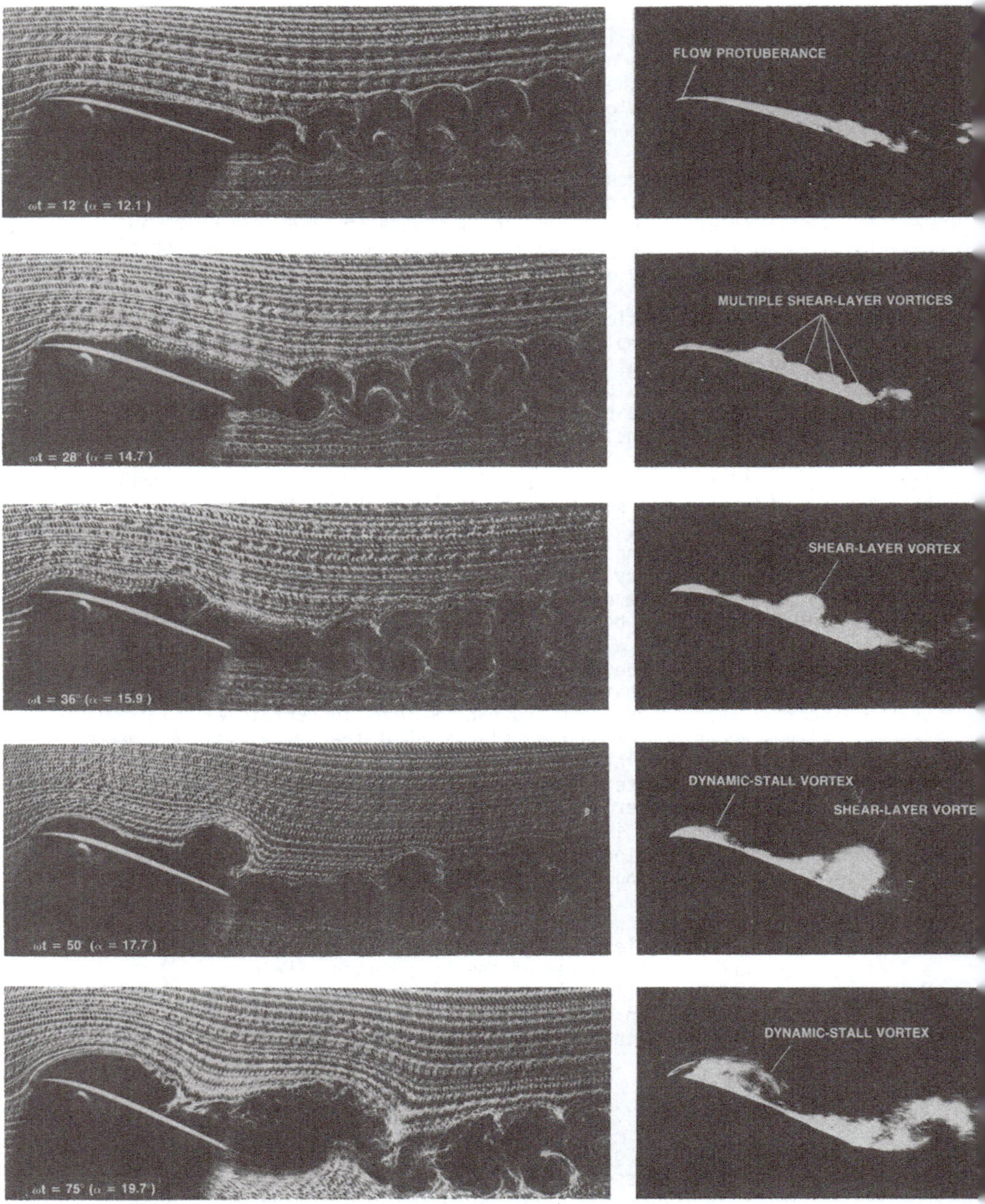

(a) (b)

<u>FIG. 1</u> CRITICAL STAGES OF DYNAMIC STALL (a) Free stream hydrogen bubble electrode only (b) Model electrodes only (Courtesy of K.W. McAlister and L.W. Carr [14] and American Society of Mechanical Engineers)

of a power conductor makes it highly susceptible to galloping - what could be more challenging than a "weather dependent" cross section of a structure! It has been pointed out that a simple one-dimensional model is inappropriate, requiring consideration of the torsional mode of oscillation (Wardlaw [1]). In the area of bridge decks, the torsional degree of freedom is the essential feature of self-excited vibration of most bluff cross-sections, leading to separated-flow torsional flutter, sometimes called torsional stall flutter. Scanlan [19] points out that this mechanism is sometimes confused with vortex-shedding excitation, both producing wakes with coherent vortices. It would seem that experimental methods used in studying dynamic stall of, for example, helicopter rotor blades undergoing large pitch variations (McAlister and Carr [14]), could lead to an improved understanding of the fluid mechanical aspects of torsional stall flutter. Such information would be particularly enlightening in the case of more streamlined cross-sections (e.g., certain bridge decks), where the extent of separation, or stall, on the surface can be expected to vary during the oscillation cycle.

4. EFFECTS OF INCIDENT TURBULENCE

A difficulty in characterizing the effects of incident turbulence is, in fact, adequately defining the nature of the turbulence itself. Inadequate meteorological information was put forth as a major unresolved problem. Tornadic winds, thunderstorms, and flow past open-water fetches and other types of terrain all produce peculiarities that call for more detailed characterization of the lower atmospheric wind (Wardlaw [1]). In an analogous sense, the nature of turbulence impinging on the upstream parts of various heat-exchanger and reactor components, and stay vanes and blading in turbomachines, is strongly influenced by the type of approach flow configuration, requiring re-definition for each change in configuration. Even in seemingly well-controlled wind-tunnel experiments, there is question, in some cases, as to whether turbulence generators have produced turbulence of proper spectral intensity, scale, and degree of isotropy. In addition, attempts to effectively simulate the turbulence scale can lead to other complications, such as scaled Reynolds numbers of the order 10^3 to 10^5, at which there is large sensitivity of the lift coefficient of a cylinder to not only small changes in Reynolds number but also to small changes in turbulence intensity (Vickery [2]).

A part of the Discussion addressed the most important scale of the incident turbulence, relative to a characteristic length of the structure. Obviously, in the case of pure buffeting of the upstream side of the structure, for example, larger scales produce large values of spatial pressure correlation, resulting in substantial loading of a structural component or the structure itself. In the case of vortex shedding, if the incident wavelength is of the order of the shedding wavelength, the shedding process itself may be altered, in turn yielding changes in magnitude of the fluctuating lift forces. Clearly, in this case, turbulent buffeting and vortex shedding cannot be viewed as separate and uncoupled phenomena. Detailed surface pressure measurements, in conjunction with unsteady velocity measurements of the approach and wake flow are required. Important first steps in this direction have already been taken. As pointed out by Brunn and Davies [20], it should be possible, to first order, to experimentally sort out the effects of turbulence and vortex shedding via narrow band spatial correlations of surface pressure, provided the longitudinal length scale of the incident turbulence is larger than the cylinder diameter. Surry [21] has found that vortex shedding from a cylinder is "not seriously disrupted" by intense turbulence intensities (>10%) and longitudinal length scales ranging from 0.36 to 4.4 diameters, at $Re_D = 4x10^4$. For very long incident wavelengths, corresponding to relatively large-scale turbulence, it has been suggested (Vickery [2]) that forces can be effectively simulated by considering the turbulence to modulate the smooth flow forces. On the other hand, it was noted that incident small-scale turbulence can alter the entrainment rate of separated shear layers, thereby altering reattachment lengths; as a consequence, the mean lift coefficient of certain structures might be changed, producing different galloping characteristics.

Distortion of the turbulence approaching a body is an important consideration if, for example, one wishes to deduce the correlation pattern of unsteady pressure on the body surface from corresponding upstream estimates or measurements of the incident turbulence. In addition, possible changes in turbulence intensity as the body is approached must be accounted for. Bearman [22] has addressed certain features of distortion, and some of his findings have been theoretically described by Hunt [23]. In essence, Bearman's experiments show that, if L_x is the longitudinal scale of the incident turbulence, and D is the characteristic body width: for $L_x/D >> 1$, a quasi-steady approach is appropriate and the streamwise intensity attenuates as the mean flow approaching the body, whereas for $L_x/D << 1$ it is actually amplified because of vortex stretching. For $L_x/D \sim 1$, there is a combination of these effects.

5. PROXIMITY EFFECTS

Even at separation distances of fifty diameters, adjacent antennae have experienced enhancement of vortex shedding, relative to a single, isolated antenna (Wardlaw [1]). This observation strongly suggests the possible dynamic interaction between structures and not only their own wakes but also adjacent wakes. In most cases, the mechanisms of interaction do not seem to be well understood. For the situation of structures located relatively far apart, where classical hydrodynamic/aerodynamic coupling effects would appear to be negligible, one must still contend with the unsteady pressure field generated by movement of an adjacent structure, its wake, or even the wake of the structure under consideration. It is well known that even small amplitude perturbations of appropriate frequency can influence the near field of separating shear layers, through interaction of pressure perturbations with the extremely sensitive region near separation. If the ratio of the characteristic distance to the acoustic wavelength is small, locations away from concentrations of vorticity instantaneously experience changes in velocity through the Biot-Savart law; for an essentially incompressible field, this also means an instantaneous change in pressure (Ffowcs Williams [24]). The practical consequence is that, aside from possible perturbations arising from adjacent structural motion or unsteady impingement on structures (Lighthill [29]), changes in vorticity concentration resulting from vortex distortion/interaction in neighboring wakes or shear layers may exert significant influence - even, of course, in the upstream direction. In fact, for the case where a structure is located downstream of another, favorable phasing of unsteady events between the two structures can lead to strongly coherent oscillations of the shear layer between them (Rockwell and Naudascher [25]).

The challenges posed by flow through various kinds of tube banks and other groups of structures, involving coupling effects not adequately accounted for by added mass, and the complexities imposed by fluid-elastic effects, are discussed by Paidoussis [3], Blevins [4,7], and Wardlaw [1]. It would be generally agreed that there is need for fundamental experimental investigations to better define the basic mechanisms; instrumentation and experimental techniques have reached a sophistication that allows correlations of velocity, pressure, and displacement within groups of structures, providing a direct assessment of present models and formulation of new ones.

Relative to the difficulties of examining this class of problems, workers in the field of turbomachinery would seem to have been granted a license of luxury in studying cascades of blades. However, they are confronted with mistuning of blades (Whitehead [26]) whereby groups of blades around the cascade tend to vibrate together, meaning that the amplitude and phase of each blade relative to the others can be drastically different than expected on the basis of first considerations. In addition, subsonic stall flutter, involving separation of flow along a portion of, and downstream of, a blade can occur (Yashina and Tanaka [27]). For this situation, there may be some parallels with, for example, coupled vibration of a single row of cylinders spaced across the flow. Investigators would be well-advised to follow corresponding experimental and theoretical developments.

6. CONFINED SWIRLING FLOWS

Important for a wide variety of internal-flow systems is the possible occurrence of unstable oscillations associated with swirling, or vortical, flows where the primary vector of the swirl (vorticity vector) tends to be oriented in the direction of, or opposite to that of, the mean flow. Flow in draft tubes of hydraulic machines, turbomachine exit chambers, cyclone separators, corners of hydraulic structures, and pump intake arrangements can exhibit severe, and coherent, unsteadiness. It appears that investigations of the type recently carried out by Escudier and Merkli [28] should be pursued. Insightful measurement and visualization of these three-dimensional and unsteady vortices, often involving backflow through the central region of the vortex, is essential. In doing so, the possible occurrence of features such as bistable behavior of the fluctuating pressure field and simultaneous existence of several oscillation modes [28] should be characterized.

7. CLOSING COMMENTS

Among the classes of problems not discussed herein, but of critical significance, are: excitation mechanisms involving two phase (liquid-vapor) flows, where the vapor phase is formed as a result of cavitation, flashing, or boiling processes in internal-flow systems; high speed internal- or external-flow systems where compressibility effects can play a key, if not overriding role in the excitation mechanism; a variety of problems associated with slender-body aerodynamics and blading in turbines, fans, and compressors such as self-excited classical flutter, transonic, and supersonic flutter; and forced excitation due to inlet distortion, and other related problems.

In addition, from the mechanical or structural perspective, unresolved aspects of characterizing the inherent structural damping, as well as design of suitable damping systems for various sorts of vibrating structures, have, due to the limited scope of the Panel Discussions,not been assessed.

In summary, it is clear that, insofar as possible, we should all strive for a better physical understanding of the excitation mechanisms leading to vibrations. To this end, practicing engineers and research scientists have a great deal to gain from interacting with one another.

8. REFERENCES

1. Wardlaw, R.L. "Unresolved Problems: Bridge Decks, Beams, and Cables", Proceedings of 1979 IAHR/IUTAM Karlsruhe Symposium on Practical Experiences with Flow-Induced Vibrations, Naudascher, E. and Rockwell, D., eds., Springler-Verlag (1980).

2. Vickery, B.J., "Wind Effects on Buildings and Structures: Critical Unresolved Problems" ibid.

3. Paidoussis, M.P., "Flow-Induced Vibrations of Heat Exchanger and Reactor Components: Critical Unresolved Problems", ibid.

4. Blevins, R.D., "Critical Unresolved Problems in Flow-Induced Vibrations of Heat Exchangers", ibid.

5. Douma, J.H., "Hydraulic Structures: Critical Unresolved Problems", ibid.

6. Chaix, B., "Critical Unresolved Problems of Flow-Induced Vibrations in Hydraulic Machinery and Equipment", ibid.

7. Blevins, R.D., "Critical Unresolved Problems in Flow-Induced Vibration of Ocean Structures", ibid.

8. Roshko, A., "Structure of Turbulent Shear Flows: A New Look", AIAA J. 14, 1349-1357 (1976).

9. Morkovin, M.V., "Flow Around a Circular Cylinder-A Kaleidoscope of Challenging Phenomena", Symposium on Fully Separated Flows, Hansen, A.G., ed., ASME, New York, 102-118 (1964).

10. Kovasznay, L.S.G., "Measurement in Intermittent and Periodic Flow", Proceedings of the Dynamic Flow Conference 1978, P.O. Box 121, DK-2740 Skovlunde, Denmark (1978).

11. Graham, J.M.R., "The Effect of End Plates on the Two-Dimensionality of a Vortex Wake", The Aeronautical Quarterly, XX, 237-247, (August, 1969).

12. Chen, Y.N., "Flow-Induced Vibrations of Plates, Single Cylinders, and Tube Bundles", ibid., reference [1].

13. Wooton, L.R., Warner, M.H. and Cooper, D.A., "Some Aspects of the Oscillations of Full-Scale Piles", Proceedings 1972 IUTAM/IAHR Symposium, Karlsruhe, Naudascher, E., ed. Springer-Verlag, 587-601 (1974).

14. McAlister, K.W. and Carr, L.W., "Water Tunnel Visualization of Dynamic Stall", ASME Publication, Symposium on Nonsteady Fluid Dynamics, Winter Meeting, Crow, D.E. and Miller, J.A., eds., 103-111 (1978).

15. Crighton, D. "Basic Principles of Aerodynamic Noise Generation", Progress in Aerospace Sciences 16, 31-96 (1975).

16. Schachenmann, A. and Rockwell, D., "Self-Sustained Oscillations of Turbulent Pipe Flow Terminated by an Axisymmetric Cavity", to be published in J. Sound and Vibrations (1980).

17. Hill, W.G. and Greene, P.R., "Increased Turbulent Jet Mixing Rates Obtained by Self-Excited Acoustic Oscillations", J. Fluids Engineering, 99, 520-525 (1979).

18. Parker, R. "Resonance Effects in Wake Shedding from Parallel Plates; Some Experimental Observations", J. Sound and Vibration 4, 62-72 (1966).

19. Scanlan, R., "On the State of Stability Considerations for Suspended Span Bridges Under Wind", ibid., reference [1].

20. Bruun, H.H. and Davies, P.O.A.L., "An Experimental Investigation of the Unsteady Pressure Forces on a Circular Cylinder in a Turbulent Cross Flow", J. Sound and Vibration 40(4), 535-559 (1975).

21. Surry, D. "Some Effects of Intense Turbulence on the Aerodynamics of a Circular Cylinder at Subcritical Reynolds Number", J. Fluid Mechanics 52(3), 543-563 (1972).

22. Bearman, P., "Some Measurements of the Distortion of Turbulence Approaching a Two-Dimensional Bluff Body", J. Fluid Mechanics 53(3), 451-467 (1972).

23. Hunt, J.C.R., "A theory of Turbulent Flow Round Two-Dimensional Bluff Bodies", J. Fluid Mechanics 61(4), 625-706 (1973).

24. Ffowcs Williams, J.E. "Hydrodynamic Noise", Annual Review of Fluid Mechanics 1, 197-222 (1969).

25. Rockwell, D. and Naudascher, E. "Self-Sustained Oscillations of Impinging Free Shear Layers", Annual Review of Fluid Mechanics 11, 67-94 (1979).

26. Whitehead, D.S., "Effect of Mistuning on the Vibration of Turbomachine Blades Induced by Wakes", J. Mechanical Engineering Science 8, 15-21 (1966).

27. Yashima, S. and Tanaka, H. "Torsional Flutter in Stalled Cascade", ASME Paper 77-GT-72 (1977).

28. Escudier, M. and Merkli, P. "Observations of the Oscillatory Behavior of a Confined Ring Vortex", AIAA Journal 17(3), 253-260 (1979).

29. Lighthill, J., Waves in Fluids, Cambridge University Press (1979).

J2

UNRESOLVED PROBLEMS

BRIDGE DECKS, BEAMS AND CABLES

R.L. Wardlaw
Deputy Head
Low Speed Aerodynamics Laboratory

National Research Council of Canada
Ottawa, Canada

SUMMARY

An attempt is made to identify some of the problems where more research into the development of experimental and analytical techniques would be beneficial to the design of structures against wind action. Included for discussion are proximity effects, the galloping of iced power conductors, the effects of turbulence on the aeroelastic behaviour of slender structures, and the development of methods of predicting behaviour in turbulent wind from sectional model measurements in smooth flow. Emphasis is placed on the importance of developing practical methods of augmenting system damping and the concept of added damping being incorporated into the structure at the design stage rather than as a 'fix' after the structure is completed.

1. INTRODUCTION

One is hard-pressed to think of wind-induced vibration problems that are unresolved. In our practical experience, problems that have arisen have been resolved or partially resolved. What must be discussed, then, is the need for finding the most practical and cost effective solutions, the need to learn more about some of the aerodynamic phenomena and the need to hone our analytical and experimental skills for predicting behaviour and stress levels.

On the other hand, it would be wrong to sit back and think that there are no problems ahead, waiting to catch us by surprise. It is frequently the case that problems occur, not as a consequence of the engineer overlooking some known wind-prone situation, but as a result of technological advances that lead into new and unknown aerodynamic circumstances where the vibration would neither have been anticipated or predictable in terms of contemporary knowledge. This was the case with the original Tacoma Narrows Bridge that collapsed in 1940 (Ref. [1]). Innovative advances incorporated into the design of the bridge resulted in a road deck structure of unusually light weight, low torsional stiffness and an aerodynamically bluff cross-section, all of which enhanced its susceptibility to oscillation.

Another example is the bundled conductor (Ref. [2] Chart 0 and Ref. [3]), conceived to suppress the Corona discharge, a consequence of high transmission line voltages that are now approaching megavolt levels. At the time of their introduction two decades ago, it is doubtful if the aerodynamicist, even had he been asked, would have been apprehensive about their behaviour, and, in retrospect, it is clear that he would not have had the knowledge required to predict the severe and damaging aerodynamic instabilities that have been encountered.

2. EFFECTS OF TURBULENCE

The earth's surface wind is a turbulent boundary layer that varies from 300 m to 600 m in depth depending on the roughness of the local terrain. It is well-known that the wind turbulence may have a profound effect on the aerodynamic stability of slender structures. For example, it has been demonstrated in the laboratory, with full models of the Halifax Harbour Bridge (Ref. [4]) and the

Lions' Gate Bridge (Ref. [2], Chart B), that aero-elastically unstable behaviour observed in smooth flow was no longer present in a simulated turbulent wind. It has also been observed in wind tunnel experiments that turbulence suppresses the subspan oscillation of bundled conductors. This appears to be confirmed at full scale by the observation that transmission lines exposed to wind blowing over smooth, open terrain will oscillate but those exposed to wind blowing over rough terrain will not.

Unfortunately, turbulence does not provide universal relief from wind-induced vibrations and many examples could be given such as the galloping of iced electric power transmission lines, the failure of the original Tacoma Narrows Bridge and the vortex-shedding excited motion of the Longs Creek Bridge (Ref. [2], Chart A).

Why does the turbulence suppress the vibrations in some cases and not in others? To what extent does the behaviour depend on turbulence scale, turbulence intensity, the size of the structure etc.? The answers to these questions are not always clear, and a better understanding of how the turbulence affects the response would be beneficial for several reasons.

Testing in simulated turbulent wind is not always practical because of constraints prescribed by the size of the wind tunnel and the size of the model. For example, full spans of power transmission lines could not be satisfactorily modelled at the small scales at which we can model the full boundary layer in conventionally sized wind tunnels. However, if we had a better understanding of the influence of turbulence it might be possible to predict, by analytical means, their behaviour in turbulent flow using measurements made on sectional models in smooth flow. Alternatively it may be that satisfactory results could be obtained by testing sectional models in a wind that modelled only the small scale end of the turbulence spectrum.

Similar arguments can be made for slender towers, where the design and construction of small scale models with accurate elastic and mass similarity becomes difficult. In this case larger scale full models could be constructed for testing in either smooth flow or a partial boundary layer simulation, if the full scale behaviour could be satisfactorily inferred from the measurements.

For bridge road decks, the sectional model has the advantage over full models of low cost, large scale, short lead time for model construction and ease of incorporating experimental modifications of section geometry. It would be useful to be able to predict the aerodynamic stability and buffeting response of the full bridge from sectional model tests done in either smooth flow or partial boundary layer simulation. With further development, the analytical concepts discussed by Scanlan (Ref. [5]) may soon be successfully implemented for this purpose. Even when such tools become available, the importance and value of the full model study done in a complete wind simulation should not be downplayed, and investigations of a large bridge could include both sectional and full model testing.

Finally, more attention must be given to the measurement of the statistical properties of the turbulence in the lower part of the atmospheric wind. This is particularly important for open water fetches where unexpectedly low turbulence intensities have recently been measured. This is important for long span bridges where the danger exists that assessments of the aerodynamic stability of bridge road decks at higher turbulence intensities may be misleading.

3. PROXIMITY EFFECTS

In recent years there have been many examples documented in the literature of severe vibration problems arising where parallel slender structures are in close proximity. In some instances the proximity effect is such that the level of vibration of the single structure is worsened by the presence of the additional parallel structure. In others, aeroelastic instabilities exist entirely as a consequence of their being multiple parallel structures.

An example of the former is the response of the slender antennae described in Chart I of Reference [2] where vortex shedding excitation is magnified even at separations of fifty diameters. Examples of the latter type are the bundled conductors already discussed and the heavy water plant towers described in Chart J of Reference [2] and the towers described by Ruscheweyh in Reference [6].

In general the proximity effects that have been documented are, at the moment unpredictable in terms of critical wind speeds or motion amplitude. An exception is the bundled conductor behaviour first analyzed by Simpson (Ref. [7]). There is a need, therefore, to systematically research these phenomena and to develop practical solutions such as stiffening ties, spacers and damping augmentation; and to codify the results so that they are available in a useful form to the practicing engineer. As well it is important to further the development of analytical procedures, as has been done for bundled conductors.

4. GALLOPING OF ICED CONDUCTORS

This is a very destructive instability caused by a thin crescent of ice deposited on one side of a power conductor during freezing rain storms. The problem has been around for a long time and serious and extensive research on the subject has been underway for more than forty years. This may be the exception to my opening remark that one is hard-pressed to think of wind-induced vibration problems that are unresolved. Although line damage and blackouts still plague utility companies in ice-prone regions, there are, in fact, some devices available that offer limited relief. However, the utilities are still striving for a more definitive understanding of the problem and the development of better means of controlling motion.

It now appears that the phenomenon is not adequately described by the classical den Hartog single degree-of-freedom model of galloping but that the torsional degree-of-freedom is essential to the observed behaviour. It is also becoming clear that susceptibility to galloping is sensitive to turbulence intensity and terrain roughness. Much productive research has been done at the Ontario Hydro (Ref. [8]) which has led to promising, but unproven, new concepts for damping devices, one of which is described in Reference [9].

5. PREDICTION OF RESPONSE

Much progress has been made in recent years relating to the prediction of buffeting response and aerodynamic stability by both experimental and analytical means. New techniques such as the taut strip model, developed by Davenport (Ref. [10]), for studying road decks in turbulent wind, fill a gap on the experimental side while analytical procedures for buffeting calculations continually improve as demonstrated by the work of Irwin on the Lions' Gate Bridge (Ref. [11]).

However novel approaches should be sought out and developed and other methods, now in the research phase, should be pursued, for example, the techniques proposed by Scanlan for using measurements made with sectional models to predict full scale behaviour. Research underway at the Colorado State University, aimed at the development of large scale longitudinal two-dimensional "turbulence" should lead to a useful research and development tool for sectional models that will enhance our understanding of turbulence effects and improve our ability to predict response.

It is important that all analytical and experimental procedures be validated by comparisons of results to full scale measurements. Emphasis must, therefore, be placed on continued programs of comparative full scale measurements such as those done by Dalgleish on Commerce Court, Toronto (Ref. [12]) and other buildings.

6. DAMPING AUGMENTATION

There are many instances on record where damping has been added after the completion of a structure in order to control unexpected vibration problems.

Several of these have been described in reference [2] (see Charts A, F, G, I). More recently the use of dampers as designed-in features has become a reality and an example is the large scale tuned mass damper installed in the Citicorp Center, Manhattan (Ref. [13]).

The availability of effective damping systems makes possible large reductions in the probability of large amplitude motions of buildings with the potential of important savings in structural material. Although similar schemes have not yet been applied to bridge road decks, it is possible that their use would lead to more efficient structural design, wherein the section geometry would not have to be compromised for reasons of aerodynamic stability or vortex shedding excitation.

Much research and development effort has been directed towards the use of dynamic vibration absorbers, and their practicality has now been amply demonstrated. However it is possible that other more effective and economical approaches could be developed and research to this end is warranted. One possibility that has already been successfully applied to the isolation of railway vibrations from buildings (Ref. [14]) is the use of viscoelastic layers applied to structural members. This approach eliminates the need for tuning and would be less sensitive than absorbers to the extreme temperature ranges encountered with outdoor applications. The application of this technique to buildings and perhaps to a greater degree to bridge road decks appears to warrant more attention.

7. REFERENCES

1. Farquharson, F.B., Smith, F.C., Vincent, G.S.,"Aerodynamic Stability of Suspension Bridges with Special Reference to the Tacoma Narrows Bridge", University of Washington, Engineering Experiment Station Bulletin No. 116, Parts I to V (1950-54).

2. Wardlaw, R.L.,"Approaches to the Suppression of the Wind-Induced Vibrations of Structures, Proc. Symposium on Practical Experiences with Flow-Induced Vibrations",IAHR/IUTAM, Karlsruhe, Germany, Sept. 3-6, 1979.

3. Tsui, Y.T.,"On Subspan Oscillation of Two Coupled Conductors with One in the Wake of Another, Proc. Symposium on Practical Experiences with Flow-Induced Vibrations",IAHR/IUTAM, Karlsruhe, Germany, Sept. 3-6, 1979.

4. Davenport, A.G., Isyumov, N., Miyata, T.,"The Experimental Determination of the Response of Suspension Bridges to Turbulent Wind",Proc. Third International Conference on Wind Effects on Buildings and Structures, Tokyo, pp. 1207-1219 (Sept. 1971).

5. Scanlan, R.H.,"The Action of Flexible Bridges Under Wind, I: Flutter Theory, II: Buffeting Theory, Journal of Sound and Vibration",Vol. 60, No. 2, pp. 187-199 and 201-211 (1978).

6. Ruscheweyh, H.,"Wind-Induced Vibrations of Towers and Stacks, Proc. Symposium on Practical Experiences with Flow-Induced Vibrations',IAHR/IUTAM, Karlsruhe, Germany, Sept. 3-6, 1979.

7. Simpson, A., Flower, J.W.,"An Improved Mathematical Model for the Aerodynamic Forces on Tandem Cylinders with Aeroelastic Applications",Journal of Sound and Vibration, Vol. 51, No. 2, p. 183-217 (1977).

8. Chadha, J.,"A Dynamic Model Investigation of Conductor Galloping",Ontario Hydro Research Division, Report No. 73-399-K.

9. Nigol, O., Clarke, G.J.,"Conductor Galloping and Control Based on Torsional Mechanism",IEEE Paper C74-016-2 (1974).

10. Davenport, A.G.,"The Use of Taut-Strip Models in the Prediction of the Response of Long Span Bridges to Turbulent Wind",Proc. Symposium on Flow-Induced Structural Vibrations, IUTAM/IAHR, Karlsruhe, Germany, Aug. 14-16, 1972.

11. Irwin, H.P.A.H.,"Wind Tunnel and Analytical Investigations of the Response of Lions' Gate Bridge to a Turbulent Wind", National Research Council of Canada, NAE Report LTR-LA-210 (June, 1977).

12. Dalgleish, W.A., Templin, J.T., Cooper, K.R.,"Comparisons of Wind Tunnel and Full-Scale Building Surface Pressures with Emphasis on Peaks", Proc. Fifth International Conference on Wind Engineering, Colorado State University, Fort Collins, Colorado, July 8-14, 1979.

13. Petersen, N.R.,"Design of Large Scale Tuned Mass Dampers", ASCE Preprint 3578, Boston, April 2-6, 1979.

14. Grootenhuis, P.,"Sandwich Damping Treatment Applied to Concrete Structures", Royal Society, Phil. Trans., Series A 263, pp. 455-59 (1968).

J3 WIND EFFECTS ON BUILDINGS AND STRUCTURES - CRITICAL UNRESOLVED PROBLEMS

B. J. Vickery

University of Western Ontario
London, Canada

1. INTRODUCTION

In discussing unresolved problems it is useful to draw a distinction between problems that can be identified and quantified by theoretical means without recourse to testing, and those which require the construction and testing of a model. While the last decade or so has seen a considerable advance in both theoretical and experimental approaches to the identification of wind-induced vibrations, there is no doubt that it is in the theoretical area that most problems remain.

Even in areas where existing theoretical approaches yield reasonable engineering estimates of the structural response, our understanding of the fluid mechanics and our ability to provide correct solutions is at times weak. A notable example is the application of the "gust factor" approach to the prediction of the along-wind response of buildings: the methods proposed, for example, by Davenport [1], Vickery [2], and Simiu [3], give reasonable estimates, but the theoretical model on which all are founded is an extreme simplification of the actual fluid flow. The work of Hunt [4] over the past years has pointed the way to a better understanding of the flow field, but predictions of behavior based on this work remain distant.

In the area of across-wind motions, progress in theoretical methods of prediction has been slower, although the work of Novak [5], for example, has shown that some instabilities at high values of reduced velocity can be well predicted both qualitatively and quantitatively. The theoretical prediction of motions due to interference effects in turbulent shear flow, perhaps the most significant influence for very tall buildings in a city environment, is not conceivable at the present time.

While theoretical models present many unresolved problems, the methods that are available and the data that have been accumulated do enable the identification of most situations in which wind-induced vibrations may, or will, be a major design consideration. In such cases, recourse to wind tunnel testing will commonly provide the necessary data, although it would be overly optimistic to consider the wind tunnel as an infallible predictor of structural response. Although full scale data gathered over the past decade has shown that the wind tunnel can provide good engineering estimates of full scale behavior (i.e. Davenport [6], Dalgleish [7,8], Melbourne [9] and others [10]), problems remain.

Among these is the difficulty of dealing with large discrepancies in Reynolds number for structures with curved surfaces; the adequacy of modelling atmospheric flows, i.e., directional changes with height; thunderstorms and tornadic winds; experimental limitations, i.e., response time of pressure tubing; and the difficulties of dealing with future changes in the environment of buildings. Apart from these problems, which are associated with the actual test program, there are problems in the estimation of the structural parameters, most notably damping, and problems in the adequate definition of wind climate. Davenport [11] has demonstrated the importance of the directional distribution of wind speed but in many situations, the necessary meteorological data is not available or, if available, is in a form which requires considerable processing before it can be incorporated into the design process.

Many of the problems alluded to in this introduction are exemplified in the papers presented at this meeting and some of those raised, particularly those in Session E, will be discussed in further detail.

2. UNRESOLVED PROBLEMS: RELATION TO SYMPOSIUM CONTRIBUTIONS

In the general lecture, Dr. Davenport drew attention to a number of areas in which there is a clear need for further work. One such area is the torsional motion of buildings, a subject which was also treated by Ellis et al. in a later paper of that Session. Data for the prediction of torsional motions is extremely limited and, furthermore, it has not been uncommon to omit the torsional mode from the more simple aero-elastic models. It is perhaps significant that two major buildings which have suffered permanent, structural damage as the result of wind action (Mayer-Kaiser, Miami; Great Plains Life, Lubbock, Texas) exhibited marked permanent deformations in torsion.

While not dealing specifically with vibration problems, the paper by Dalgleish raises some questions in regard to techniques for determining the response of small elements such as windows. Reference is made to the work of Mayne and Walker and the conclusions that vibrations may be a problem at frequencies below 6 Hz and that window panels respond to extreme gusts with effective averaging times of 1/4 to 1/2 second. These conclusions imply a need to examine pressure fluctuations with averaging times of the order of 0.2 seconds (full scale) or about 2 milliseconds at typical model scales. If such a need is in fact real, then a severe problem in experimental method exists, since the conventional approach using a single transducer with a scanning pressure switch and tubing yields a response time of the order of 10 milliseconds.

While flush diaphragm or closely coupled transducers could be employed, such an approach would be prohibitively uneconomical in practice. The work of Dalgleish suggests that the response time need not be so rapid, but the full scale study by Eaton and Mayne [12] did show remarkable changes in peak pressure as the averaging time was reduced to 30 milliseconds (full scale). It would be a significant help to the experimental fraternity if the need for a rapid responding pressure measuring system could be clearly defined. It is the view of the writer that the common practice of accepting a response time of about 10 ms at model scale will introduce errors, but that these are unlikely to be significant in relation to other sources of error, particularly errors associated with the definition of wind climate and those introduced due to inadequacies in modelling atmospheric flows.

In the invited paper on the wind-induced vibration of stacks and towers, Ruscheweyh states "although many research works have been done in the past the .... the problem is as up to date as before". The writer tends to agree with this appraisal of the situation. The work by Iwan and Blevins [14] and Hartlen and Currie [13] and others in devising models for shedding in smooth flow has, unfortunately, done little to aid in the reliable prediction of full scale vortex induced motions in turbulent shear flow.

The prediction of vortex-induced motions of circular structures was also discussed by van Koten, Ishizaki, both of whom presented valuable full scale data, and by Pirner. Other contributions by Wong and by Johns, give a total of six out of ten papers that deal significantly with vortex-induced motions of circular structures. There is little doubt that it is a problem which continues to attract attention, and hopefully the comment of Ruscheweyh will be invalid in the near future.

The major problems in predicting tower motions are, on the structural side, the difficulty of accurately defining the damping and, on the experimental side, the inability to reproduce the Reynolds number. While there has been considerable work done at high values of Re, it has, of necessity, been conducted in essentially smooth flow. If the turbulence scale is to be reproduced in existing boundary layer tunnels, the linear scale will be of the order of 1:300 and the Reynolds number commonly in the range 10^3 to 10^5. Unfortunately, it is in this range that Gerrard [15] and Vickery [16] have shown that there exists a remarkable sensitivity, not only to changes in Re, but to minor changes in turbulence level (Fig. 1). It is therefore very difficult, if not impossible, to assess the true influence of turbulence on the full scale structure.

The problem is particularly significant when interference effects are being examined. In the case of buffetting by an upstream structure, it is not at all clear as to whether the greatly augmented response is a true reflection of the buffetting action or is due to an increased lift coefficient associated with the aerodynamic changes induced by turbulence in the flow, and peculiar to that range of Reynolds number.

In an attempt to deal with the problem of response to shedding in large scale turbulent flow, the writer has been examining the use of a numerical simulation method. In this model, the forces in smooth flow are represented by a narrow band random force and a non-linear aerodynamic damping term which is a function of wind speed. The forces are simulated by regarding the turbulence as a modulator of the smooth flow forces. Some simulated response records are shown in Fig. 2, for damping ratios of 2%, 0.5% and 0.2% of critical. In each case, the mass parameter $(m/\rho d^2)$ is 100.

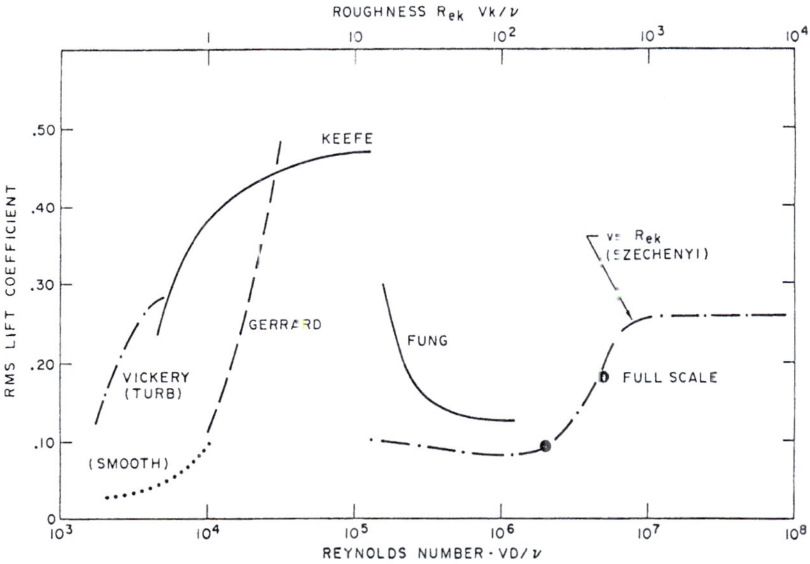

FIG. 1 SENSITIVITY OF RMS LIFT COEFFICIENT TO Re AND TURBULENCE

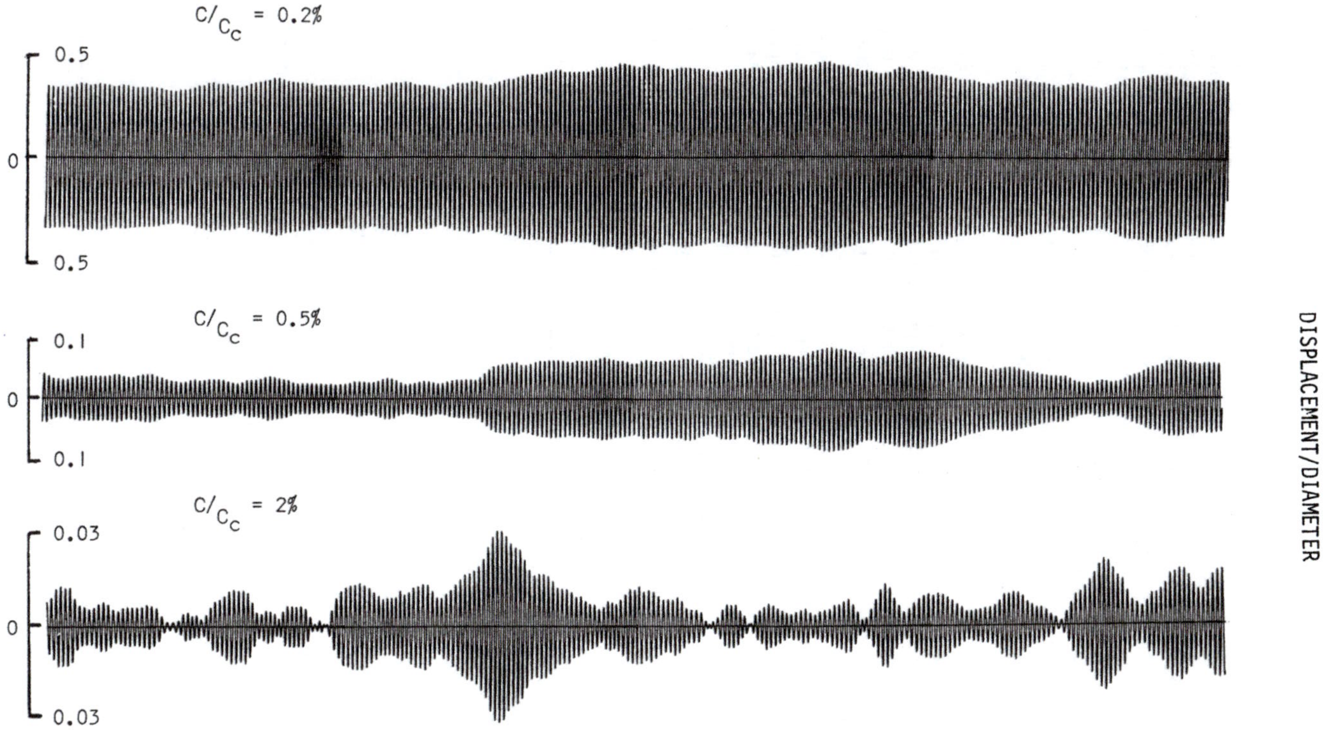

FIG. 2 NUMERICALLY SIMULATED RESPONSE TO SHEDDING IN LARGE SCALE (L/D > 100) TURBULENCE

$C_L = 0.20$, $i = 0.10$, $M/\rho d^2 = 100$

The encouraging feature of these results is the very close qualitative correspondence with experimental data obtained, for example, by Kwok [17]. The transition from a narrow-band Gaussian response to an almost sinusoidal response is clearly evident. The comparison with Wooton's [18] observations (Fig. 3) shows again the remarkable similarity between the simulated and experimental response traces.

In concluding this discussion on unresolved problems it is worthwhile to note that the proof of theoretical and experimental methods of prediction lies in the comparison with observed full scale behavior. In this regard the paper by Melbourne in Session D warrants some comment. The agreement observed here is indeed encouraging even though individual observations show discrepancies of 50 percent

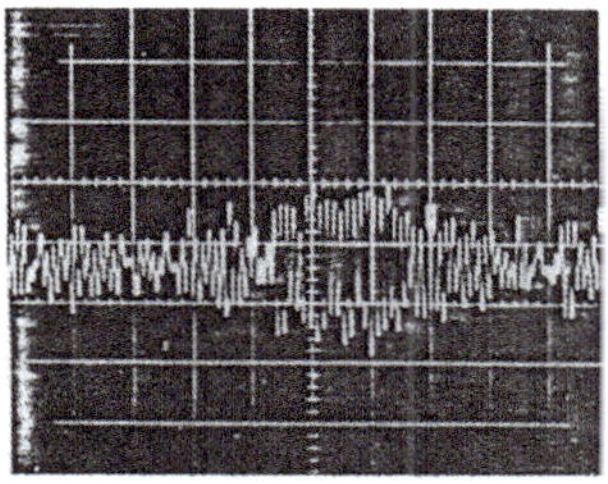

(a) High Damping

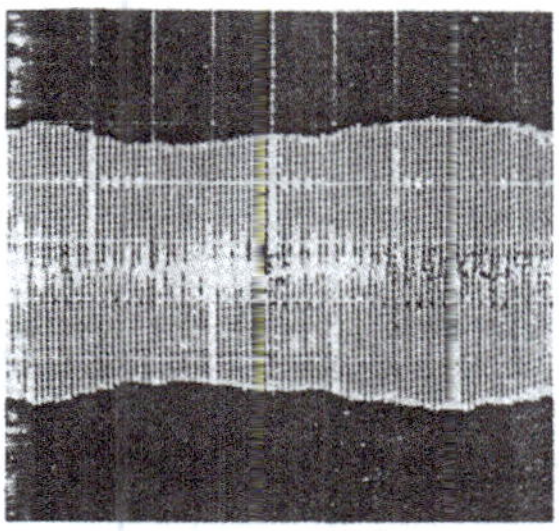

(b) Low Damping

FIG. 3 TYPICAL OBSERVED RESPONSE TRACES AT HIGH AND LOW VALUES OF DAMPING

and more. For the most part, these discrepancies result from the stochastic nature of the measured response and the difficulty of obtaining adequate sample lengths in full scale experiments. The variability in the measured response corresponds to a coefficient of variation of 10 percent and when this is taken into account, the root-mean-square discrepancy between model and prototype of 19 percent demonstrates that the wind tunnel was, in this instance, a most useful predictor.

3. REFERENCES

1. Davenport, A. G., "Gust Loading Factors", Journal Struct. ASCE 93, No. St3, 11-34, (1967).

2. Vickery, B. J., "On the Reliability of Gust Loading Factors", Proc. Techn. Meeting on Wind Loads on Building Structures, Bldg. Sc. Series 30, N.B.S., Washington, 93-106, (1970).

3. Simiu, E., "Gust Factors and Along-wind Pressure Correlations," Journal Struct. Div. ASCE 99, St4, 773-783, (1973).

4. Durbin, P. A. and Hunt, J. C. R., "Fluctuating Surface Pressures on Bluff Structures in Turbulent Winds: Further Theory and Comparison with Experiment", Proc. 5th Int. Conf. on Wind Engng., V-1-1 - 17, Colorado (1979).

5. Novak, M., "Galloping Oscillations of Prismatic Structures", J. Eng. Mech. Div. ASCE 98, No EM1, 27-46, (1972).

6. Davenport, A. G., et al., "Wind Induced Response of Suspension Bridges - Wind Tunnel Model and Full Scale Observations", Proc. 5th Int. Conference on Wind Engineering, VII 1-1 - 18, Colorado, (1979).

7. Dalgleish, W. A., Templin, J. T. and Cooper, K. R., "Comparisons of Wind Tunnel and Full Scale Building Surface Pressures with Emphasis on Peaks", Proc. 5th Int. Conference on Wind Engineering, V. 6 1-13, Colorado, (1979).

8. Dalgleish, W. A., "Comparison of Model and Full Scale Wind Pressures on a High-rise Building", J. Ind. Aero., Vol. 1, 55, (1975).

9. Melbourne, W. H., "Model and Full Scale Response to Wind Action of the Cable Stayed Box Girder West Gate Bridge", Proc. Symp. on Practical Experiences with Flow Induced Vibrations, Session D, 187-194, Karlsruhe, (1979).

10. "Symposium on Full Scale Measurements of Wind Effects on Tall Buildings and Other Structures", University of Western Ontario, July, (1974).

11. Davenport, A. G., "The Dependence of Wind Action on Wind Direction", Proc. ASCE-EMD Conference, University of Waterloo, Ontario, May, (1976).

12. Eaton, K. J. and Mayne, J. R., "The Measurement of Wind Pressures on Two Storey Houses at Aylesbury", Bldg. Res. Est., Current Paper CP 10/14, U.K., (1974).

13. Hartlen, R. T. and Currie, I. G., "Lift-Oscillator Model of Vortex Induced Vibration", J. Eng. Mech. Div. ASCE 96, 577-591, (1970).

14. Iwan, W. D. and Blevins, R. D., "A Model for the Vortex Induced Oscillation of Structures", J. Appl. Mech., ASME 41, 581-585, (1974).

15. Gerrard, J. H., "A Disturbance-Sensitive Reynolds Number Range of the Flow Past a Circular Cylinder", J. Fluid Mech., Vol. 22, Part 1, 187-196, (1965).

16. Vickery, B. J., "A Model for the Prediction of the Response of Chimneys to Vortex Shedding", Proc. 3rd Int. Symp. for Design of Industrial Chimneys, 157-162, Munich, (1978).

17. Kwok, K., "Cross-Wind Response of Structures due to Displacement Dependent Excitations", Ph.D. Thesis, Monash University, Melbourne, (1977).

18. Wooton, L. R., "The Oscillation of Large Circular Stacks in Wind", Proc. Inst. Civ. Eng., Vol. 43, 573-598, (1969).

J4 FLOW-INDUCED VIBRATIONS OF HEAT-EXCHANGER AND REACTOR COMPONENTS - CRITICAL UNRESOLVED PROBLEMS

M. P. Païdoussis
Professor and Chairman

Department of Mechanical Engineering
McGill University
Montreal, Québec, Canada

1. INTRODUCTORY COMMENTS

The observations to be made here will, in many ways, be complementary to the author's General Lecture on this topic, and will, therefore, be very brief.

In the General Lecture, the practical cases of flow-induced vibration encountered in heat exchangers and nuclear reactors were classified according to whether they have been caused by cross, axial, or annular flow, for cases where the flow was external to the vibrating component, or by internal flow. This simply reflects the fact that most research work in this area may be classified into these four broad categories; in no way should it be thought to suggest that flow in heat exchangers and reactors, or in parts thereof, is purely axial, or transverse, let us say. Indeed, although such idealizations make good sense for

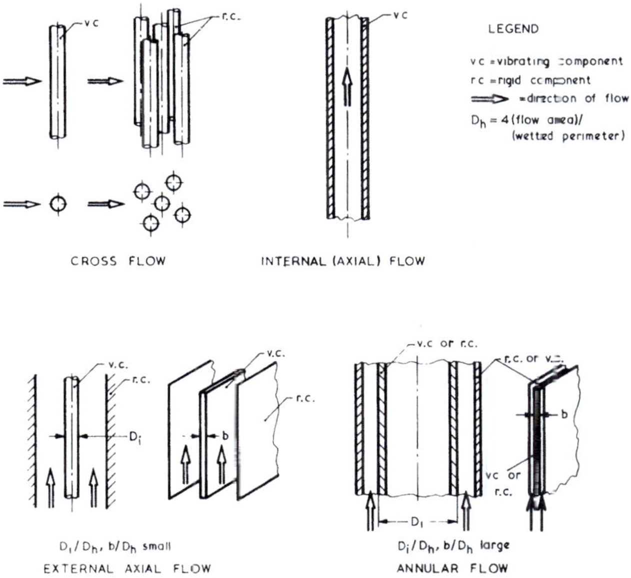

DEFINITION OF IDEALIZED CLASSES OF FLOW

constructing tractable mathematical or physical models, in real situations, the flow is invariably more complex. A good example is the flow in a typical Pressurized Water Reactor. The coolant enters the reactor from several ports, mixes and simultaneously flows down two parallel coaxial annular passages, formed by the pressure vessel, the thermal shield and the core barrel; it then flows into the lower part of the reactor, turns around 180°, and flows up into the reactor core. Flow visualization studies show that large vortices and flow instabilities are possible in the annular flow — hardly an idealized, purely axial, annular flow.

The danger is that occasionally, by presuming the problem to be caused by one or the other of the aforementioned idealized classes of flow-induced vibration, one may overlook the very detail which is actually responsible for the problem — throwing away the baby with the bathwater, so to speak (*vide*, e.g., Chart XLI of the General Lecture). However, fortunately, the nature of flow-induced vibration problems is not usually so perverse, and one may, at the very least, utilize this classification to discuss the state of the art in this area.

Another observation that may usefully be made is that, whenever fluid flow is in contact with some structural component, there is bound to be *some* vibration. Vibration is not bad *per se;* in fact, it may be quite innocuous; in each case the *effect* of the vibration should be assessed. For example, in cases of axial-flow-induced vibration, the vibration amplitudes are often very small, and it is only because of wear and fretting with closely adjacent structures that they become worrisome. Hence, designers are really interested in cumulative wear over the life-span of the component — just as they are sometimes interested in fatigue life — rather than the details of vibratory behaviour. Therefore, the link between impact level and fretting wear for different materials under different conditions — or between vibration and fatigue — is an additional and essential block of knowledge, necessary for preventing failures due to flow-induced vibration.

2. PROBLEMS INVOLVING INTERNAL FLOWS AND EXTERNAL, AXIAL OR ANNULAR, FLOWS

Leaving annular-flow-induced vibration aside for the moment, it is true to say that most vibration problems of components subjected to internal flow or to external axial flow are caused by near- or far-field noise, the effect of which is magnified either by structural resonance or through other amplification mechanisms — e.g. acoustical resonance. I.e., the problems are caused by perturbations to the mean flow, rather than the mean flow itself. The current widespread practice in analyzing vibration is to ignore the mean flow almost entirely, taking but two of its effects into account: (i) the reduction of *in vacuo* structural frequencies by the hydrodynamic ("added") mass effect, and (ii) the generation and transmission of velocity and pressure perturbations by the mean flow, which constitute the excitation field. Thus, in this scheme of things, the problem is transformed to one of structural mechanics, involving a structure "softened" by the hydrodynamic mass effect and submitted to a prescribed stochastic excitation field.

In this analytical framework, the principal difficulty has been the designer's inability to realistically predict the excitation field for a given physical system, without the aid of *in situ* measurements. Recently, however, this has become possible, at least for some simple cases. E.g., for flow in piping, it is now possible to predict the noise characteristics of an elbow or a valve at a certain distance from the component of interest. However, further work is necessary in this direction, eventually leading to a codified data-bank of excitation characteristics of all commonly used piping and other components.

The approach described above, in cases where the characteristics of the excitation field could either be predicted well or measured, has been spectacularly successful. However, in general, this is not possible; e.g., except under special circumstances, one's ability to predict amplitudes of axial-flow-induced vibration is typically within one order of magnitude.

Several weaknesses in this approach have become evident. Thus, in internal and annular flow problems it is not always possible to treat the excitation field as completely unaffected by vibratory movement. Also, in axial and annular flows, there often exists important hydrodynamic coupling of the various structural elements, not wholly taken into account by the hydrodynamic added mass of the still fluid. Finally, in annular flows, flow-induced vibration problems may but very rarely be treated successfully by this approach: the mean flow is usually important, as is the leakage mechanism. Clearly, further work is needed, to define the practical limits of applicability of existing methods and to develop more refined approaches, when such are necessary.

3. PROBLEMS INVOLVING CROSS FLOW

Cross flow usually implies the existence of separated flow regions, rendering analysis — and understanding of what goes on — much more difficult.

For relatively simple geometries, e.g. a cylinder in cross flow, the state of the art is quite good, as demonstrated by the relative rarity of serious failures in recent times, indicating that the response of the structure may be predicted with reasonable confidence. Recent work has concentrated in filling the gaps in existing knowledge. Sometimes, this work, initially perceived as only providing such auxiliary information, gives new insight into the overall problem. This is the case with the study of the excitation and vibration of a cylinder in the supercritical and transcritical Reynolds number régimes at high turbulence intensities; it showed that, at least at high turbulence levels, it is wrong to consider turbulent buffeting and vortex shedding as independent of each other and linearly superposable phenomena.

When it comes to banks of cylinders, which are of great practical interest for heat exchangers of all types, the situation is much more confused. The distinction between vortex shedding excitation and buffeting, referred to above, is even more unclear in this case. Moreover, the very important mechanism of fluidelastic instability, which has decimated heat exchangers in the past fifteen years (cumulative damage of the order of $\$10^8$) is insufficiently well understood. The weaknesses here have been discussed, at some length, in the General Lecture. The main problem is that insufficient *fundamental* experimental work has been undertaken to date; most of the experiments done aimed at obtaining information of immediate usefulness in preventing the problem in specific designs, which, moreover, necessitated that the parameters used be in very narrow bands, close to particular types of designs. Despite this, design guides do exist, which may be used with some confidence. Insufficient understanding, nevertheless, may lead to over-conservative design, with the usual financial and efficiency penalties.

A great deal of fundamental work remains to be done for tube-bank vibration, especially experimental work, which should provide at once (i) the means of assessing existing semi-analytical models and (ii) the building blocks for improving them or devising new ones.

4. CONCLUSION

In general, the greatest difficulties in this area arise from insufficient physical understanding of the problems, which translates to insufficient understanding of the fluid mechanics of the problems. This becomes painfully clear by the cures applied to the many reported cases of flow-induced vibration, which, although they cannot be castigated as being of the "brute force and ignorance" type — indeed they are good engineering solutions, yet do not display the finesse often so evident in cures in the aircraft industry, for instance.

Laconically, one may say that (i) the state of the art for internal-flow-induced vibration is good, but not all of it is available in the open literature; (ii) for external axial-flow-induced vibration, there exist good analytical tools, and good potential for improvement, to bring prediction capabilities to the level of internal flows; (iii) cross-flow-induced vibration is plagued by inadequate physical understanding which is essential for improving available prediction capability; and (iv) for annular-flow-induced vibration the state of the art is insufficiently advanced, and a great deal remains to be done.

Finally, it is worth stressing that most of the above apply to single phase flows. Understanding fluid-solid interactions in two-phase flows, despite their importance and near ubiquity, is truly in its infancy.

J5

CRITICAL UNRESOLVED PROBLEMS IN FLOW-INDUCED VIBRATION OF HEAT EXCHANGERS

R. D. Blevins
Staff Engineer
General Atomic Co.
San Diego, California, U.S.A.

1. INTRODUCTION

Since the demonstration of an interactive mechanism for fluid-elastic whirling of a tube array in cross flow in 1971, there has been considerable investigation of this phenomenon. In essence, the phenomenon results in large amplitude vibration 'whirling' of closely spaced tubes exposed to mean flow perpendicular to the tube axis, once a critical flow velocity is attained. This critical velocity is specified by an instability constant K. The critical velocity has been correlated very successfully with industrial heat exchanger tube failures. A large number of references on this subject are cited in [1,2,3].

2. UNRESOLVED PROBLEMS

The present theoretical models of the whirling instability only provide a framework for the analysis, as they apply only to a single tube row. The details of the interactions between multiple tubes cannot be accurately predicted at present. A number of questions remain unresolved and are fruitful areas for research.

a.) Typical values of the instability constant are K = 2.5 or K = 3 for closely spaced tube arrays. Yet there must be some dependence on tube array spacing since K must approach infinity as the tube array spacing becomes large. At present, the influence of K on the array spacing is obscured by variations in the definition of onset of instability. Some authors have used the point at which organized motion is detectable as the onset of instability, while the more common definition is the point at which the tube amplitude reaches 0.2 mm. There is a basic need for consistent data to predict the instability constant as a function tube array geometry.

b.) It is a theoretical result that the instability should sharply increase with frequency differences between tubes because tubes can exchange energy efficiently only if their bandwidths overlap. Yet it has been demonstrated experimentally that detuning of tubes does not generally produce the predicted increases in the velocity for onset of instability. Clearly, the tubes are not interacting in the manner modeled by the theory due to: i) omitted terms in the equation, or ii) the presence of additional fluid phenomena.

c.) Experimental data indicate that the theoretically predicted dependence of critical velocity on tube damping is not always obtained. Structural damping appears to influence the instability to a smaller degree than is predicted by the theoretical model [4]. What fluid mechanism is responsible for this and how does one interpret these effects theoretically?

3. REFERENCES

1. Blevins, R.D., Flow-Induced Vibration, Van Nostrand Reinhold, NY., London, 1977.

2. Paidoussis, M.P., "Flow-Induced Vibrations in Nuclear Reactors and Heat Exchangers," Symposium on Practical Experiences with Flow-Induced Vibrations, Karlsruhe, Germany, 1979.

3. Blevins, R.D., "Flow-Induced Vibration in Nuclear Reactors: A Review," Progress in Nuclear Energy, to be published Fall 1979 or Spring 1980.

4. Weaver, D.S., and L.K. Grover, "Cross-Flow-Induced Vibration in a Tube Bank," J. Sound Vib. 59, 277-294 (1978).

J6

HYDRAULIC STRUCTURES - CRITICAL UNRESOLVED PROBLEMS

Jacob H. Douma
Consulting Hydraulic Engineer

SUMMARY

The purpose of this lead-off paper is to present a brief review of hydraulic structure vibration problems which have been satisfactorily solved and set the stage for discussion of some unresolved problems.

1. INTRODUCTION

The Symposium papers under Session F, Hydraulic Structures, present much valuable information on many important flow-induced vibration problems experienced in the field of hydraulic structures.

The General Lecture by Mr. Kolkman contains outstanding discussions of vibration sources, applicable to structural mechanics vibration theories, and design criteria for remedying vibration problems of most types of gates and valves in general use today. The invited paper by Mr. Pickett is devoted primarily to experience of the United States Corps of Engineers with flow-induced vibration of gates and valves. In this invited paper, Mr. Petrikat largely discusses flow-induced gate vibration caused by improperly designed seals. He presents modifications in seal and gate design, based on both model tests and field experience, to eliminate or minimize gate vibration. The invited paper by Mr. Falvey covers, in addition to gate seals and vibration of fixed-cone valves, several other kinds of structural vibration experienced by the United States Bureau of Reclamation. Spillway nappe vibration and stilling basin wall vibration are of particular interest, and they will be discussed in more detail later in this lead-off paper.

Many of the additional Symposium papers deal with structural vibration of various kinds of gates. They provide valuable contributions to the general knowledge of solutions for gate vibration problems.

2. RESOLVED VIBRATION PROBLEMS

It is not surprising that so much has been presented on flow-induced vibration of gates and valves. For many years, this has been the most important vibration problem pertaining to hydraulic structures. Many design and research engineers have devoted much effort during the past 40 years to developing gates, valves and seals that are free of excessive vibration.

The Symposium papers indicate that most vibration problems pertaining to gates, valves and seals have been solved. This is verified by favorable field experience in most instances. Other flow-induced structural vibration problems, cited in the papers, have generally been cured by change in design, providing greater structural rigidity, of by scme other means of eliminating the vibration problems pertaining to hydraulic structures have been solved satisfactorily.

3. UNRESOLVED VIBRATION PROBLEMS

In spite of the fact that there is an abundance of information available on resolved flow-induced structural vibration problems, there are some important vibration problems that have not been studied sufficiently to provide design criteria insuring protection against excessive structural vibration. Several of these problems will be described as a basis for further discussion.

3.1 Vortex Shedding at Gates

The problem of local vortex shedding at the ends of gates is not well understood. These vortices are developed in the corners formed by approach conduit walls and the gate face. For high-head outlet works, vortex velocities are sufficiently high to produce cavitation pressures within the vortex. After passing underneath part-open gates, cavitation zones within vortices collapse on downstream conduit or chute inverts. Pulsating pressures in collapse areas can cause excessive vibration of steel liners and cavitation erosion of concrete floors.

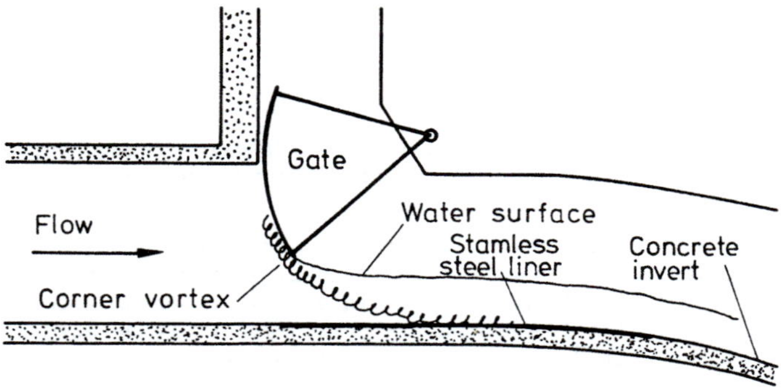

ELEVATION - OUTLET STRUCTURE

FIG. 1 VORTEX SHEDDING IN CORNER OF PART-OPEN GATE

At a large dam, initial operation of a 16-foot wide by 24-foot high tainter gate under heads up to about 300 feet, indicated that vortex shedding at the gate ends did not produce excessive gate vibration, but the downstream stainless-steel liner was severely damaged and excessive cavitation erosion occurred in the concrete chute invert downstream of the liner. Part of the liner was torn out and the remainder was loosened.

It appeared that pulsating pressures within high-velocity flow underneath the gate, particularly in the side vortices, caused the steel liner to vibrate sufficiently to crack welds near the upstream end of the liner. Cracked welds permitted pulsating pressures to occur underneath the liner, causing excessive vibration of liner plates which resulted in fatigue failure of anchors.

The steel liner was repaired by replacing the missing parts, providing much greater anchorage, and pressure grouting underneath the liner to eliminate all voids between steel and concrete. Liner rigidity was thereby significantly improved and the gate has operated successfully for several years under heads up to about 400 feet.

3.2 Stilling Basin Floor and Walls

Another flow-induced structural vibration problem for which adequate design criteria has not been developed stems from hydraulic jump action in large stilling basins. When flows with velocities of 100 feet per second, or more, enter a stilling basin to form a hydraulic jump, large pressure surges occur in the upstream part of the basin. These surges are produced in the process of energy dissipation through hydraulic jump action. Associated therewith are large pressure fluctuations which produce rapidly varying hydrodynamic loads on the stilling basin floor and walls. If the floor and walls are not sufficiently rigid, these pulsating loads will cause them to vibrate and eventually fail in fatigue.

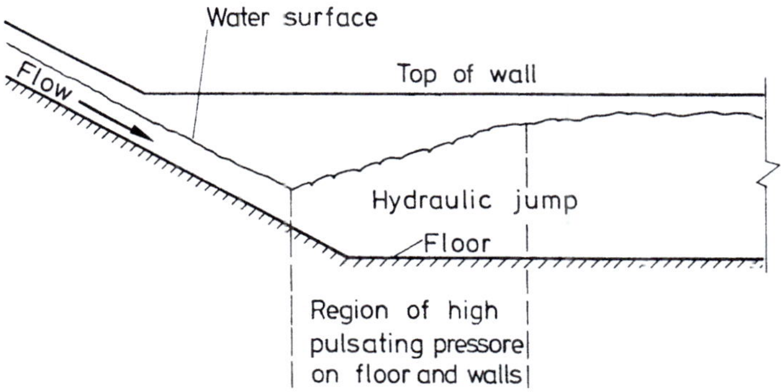

ELEVATION - STILLING BASIN

FIG. 2 VIBRATION OF STILLING BASIN FLOOR AND WALLS

The usual practice is to design stilling basin floors and walls with a high safety factor against vibration failure. Nevertheless, there have been failures which likely would not have occurred if adequate design criteria were available.

Mr. Falvey cites the case of a center wall failure for a hollow-jet valve stilling basin. He states that the solution to the vibration problem was to eliminate the center wall. This may have been possible for that particular project, but at some projects a similar center wall is essential for successful stilling basin operation. A full-length center wall may be required when only one gate or valve is in operation to assure good hydraulic jump performance, prevent stones from being transported into the basin by return eddies, and permit inspection and maintenance repairs in the inoperative half of the basin.

3.3 Plunge Pools

Pulsating pressures in large plunge pools produce flow-induced vibration problems somewhat similar to those encountered in stilling basins. A high-velocity jet deflected by a flip bucket into a plunge pool creates unknown large pressure fluctuations at the bottom and sides of the plunge pool. When the plunge pool is pre-excavated and completely concrete lined, the lining must be designed sufficiently rigid to prevent flow-induced pressure pulsations from producing excessive vibration.

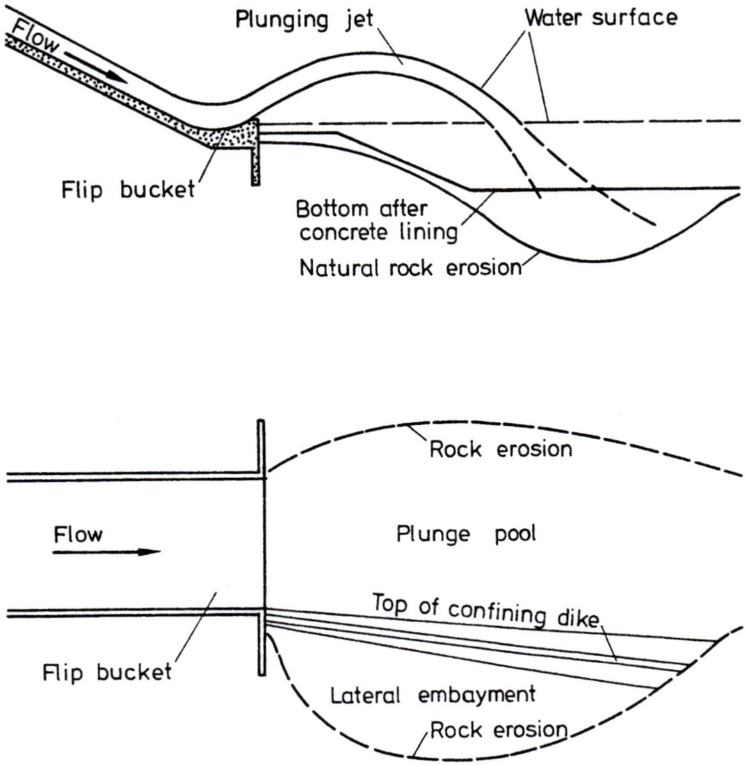

FIG. 3 PLUNGE POOLS

In natural plunge pools, the depth and extent of lateral rock erosion depends on the magnitude of pressure pulsations, as well as the erodibility characteristics of the rock. Sufficient information is not available to correlate plunge pool pressure pulsations and rock erodibility characteristics for predicting ultimate plunge pool sizes.

At the previously referred to large dam, extensive lateral erosion occurred on the right side of the natural plunge pool for the service spillway which threatened to undermine the flip bucket. The spillway has operated several months each summer during the past five years. Flows up to about 320,000 cubic feet per second have discharged into the plunge pool with jet velocities up to about 150 feet per second.

A confining concrete dike is being constructed on the right side of the plunge pool to separate the lateral embayment from the main plunge pool, thereby preventing additional lateral erosion and improving plunge pool action. The confining dike will be constructed of a gravity rollcrete section faced with 10-foot thick structural concrete slabs. The face slabs are being rigidly anchored to insure that plunge pool pulsating pressures will not cause excessive vibration and possible failure of the slabs. Pressure flucturations at the face slabs were measured in a model since generalized criteria for determining fluctuating pressures at a fixed boundary adjacent to a plunge pool were not available.

3.4 Low-Ogee Crest

Several papers have discussed vibration caused by flow over flap gates and overflow weirs. Flap-gate vibration is easily remedied by using spoilers to supply sufficient air underneath the overflow jet, thereby eliminating jet oscillation and structural vibration. However, more study needs to be given to the problem of vibration of low-ogee crest structures. These structures are often used for navigation and small conservation dams and for high-level chute spillways of large dams.

When placed on poor rock or alluvial soils, the rigid crest structure is founded on a relatively elastic foundation with a comparatively low elastic modulus. The mass of water over the crest at maximum head may be several times larger than the mass of the structure itself. Consequently, a significant part of the system comprised of the water flowing over the crest, crest structure, and foundation, is in motion. Should some flow instability or hydroelastic control be present, structural vibration of the crest becomes a distinct possibility as the head on the crest becomes progressively greater.

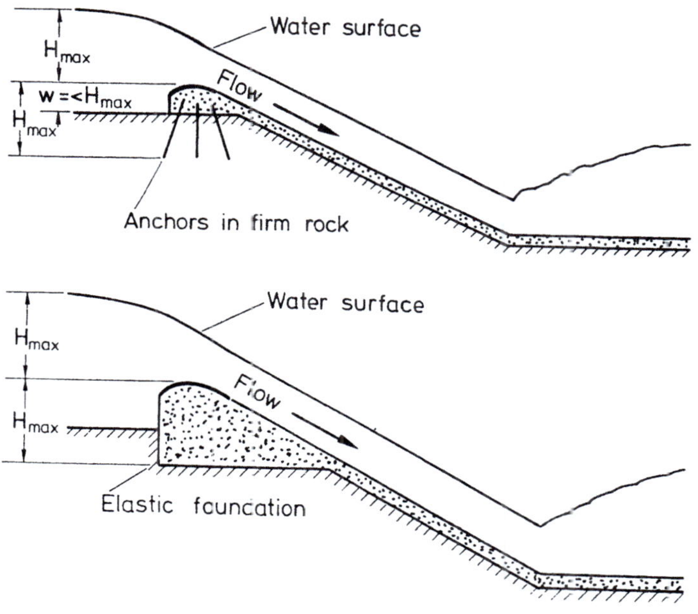

FIG. 4 LOW-OGEE SPILLWAY CREST

Prototype experience has indicated generally that excessive structural vibration does not occur when the crest structure mass is equal to or larger than the mass of water over the crest. In practice, when a low crest is founded on good rock anchors are provided so that the combined mass of concrete and rock engaged by the anchors exceeds the mass of overlfow water at maximum head. When a low crest is founded on poor rock or soils, it is not possible to engage sufficient foundation by anchors. Therefore, generalized design criteria are needed on the relationship between height of low crest on an elastic foundation and limiting overflow head which will prevent excessive flow-induced structural vibration.

4. REQUIRED RESEARCH

Each of the aforementioned unresolved flow-induced structural vibration problems requires the completion of a research program to establish reliable design criteria which will solve the problem.

Large-scale laboratory tests should be made to study the mechanics of vortex shedding at gates. Pressure pulsations are needed in vortex collapse areas for reservoir heads of 150 to 450 feet to establish requirements for steel liner rigidity and length to insure against excessive flow-induced vibration of steel liners and cavitation erosion of adjacent downstream concrete inverts.

Although laboratory measurements of fluctuating pressures have been made on stilling basin floors and walls for some specific projects, generalized design criteria are not available to insure that the floor and walls of large stilling basins will not be subjected to excessive structural vibration. A comprehensive series of large-scale model tests should be made to determine the magnitude of fluctuating pressures over stilling basin floor and wall areas for an appropriate range of hydraulic jump conditions. These pressures should be translated into equivalent hydrodynamic forces on increment areas for design of floors and walls with sufficient rigidity to withstand structural vibration.

The unresolved problem of structural vibration of plunge pool elements requires large-scale model tests to develop design criteria similar to that described for stilling basins. In this case, fluctuating pressures are required at boundaries which are located near the plunging jet. Tests should be made with various depths and widths of plunge pool lining and alignment of confining dike.

Research is required for the unresolved problem of vibration of a low-ogee crest on an elastic soils foundation when subjected to high-head discharge. While small-scale model tests have been made with the model crest placed on a rigid foundation, large-scale laboratory tests are needed with the crest placed on appropriate springs to simulate an elastic soils foundation. The tests should be conducted for several heights of low-ogee crest to determine for each at which head flow begins to induce structural vibration. The results of these tests will provide design criteria for the structural height of a low-ogee crest for any specific design discharge head over the crest.

5. CLOSING REMARKS

In closing, it should be mentioned that a classification of vibration problems according to common excitation mechanisms and basic flow categories, as has been proposed by Professor Naudascher, together with an index of pertinent references, would be of great value to engineers who design hydraulic structures which may be subject to flow-induced structural vibration.

J7 CRITICAL UNRESOLVED PROBLEMS OF FLOW-INDUCED VIBRATIONS IN HYDRAULIC MACHINERY AND EQUIPMENT

Bernard Chaix
Professor

Institut fur Flussigkeitstechnik
(Hydraulic Machinery and Equipment)
ETH - Zurich

SUMMARY

This contribution to the Panel Discussion first recalls fundamental features of a hydro-electric plant as an introduction for specialists from other fields. Most vibration phenomena reviewed are not limited to water turbines and pumps, calling for interdisciplinary work. Future investigations are suggested, all of them aiming at more precise data from tests and theoretical models.

1. INTRODUCTION

The papers presented in Session C show that almost all of those flow-induced vibration problems have been effectively attenuated in practice. Critical is the fact that most of these problems were tackled after putting the plant into operation. What about vibration prevention at the design stage?

Actually, from the very beginning, the manufacturer is designing and computing in terms of transients and vibrations; tests on models are increasingly analyzed taking these aspects into account. Although flow patterns and vibration modes of mechanical parts are often known, the prediction of all flow-induced vibrations is difficult due to the complexity of interactions; hence the need for engineers trained to collaborate with specialists in mechanical vibrations, fluid mechanics and system theory.

The designer not only needs appropriate methods, but also reliable numbers to work with. The more precise the prediction, the more competitive the design. The latter must meet increasingly severe safety requirements against ever growing risks. This problem is well known in chemical and nuclear plants; the trend for more stringent requirements in the field of hydroelectric machinery (see Fig. 1) stresses the importance of future quantitative investigations.

2. EXAMPLE OF A HYDROELECTRIC SCHEME

In Figure 2, elements representing some classical categories of flow-induced vibrations or transients are depicted. The names refer to contributions to Session C; those in brackets deal with the same type of problem in other machines or elements.

Concerning methods for solving our problems, two recommendations emerge from consideration of Figure 1:
1) Take into account non-linear relationships. As an example, the above mentioned throttle acts in proportion to the square of the flow rate—and the coefficient may vary in the ratio 1:20 depending on the flow direction.

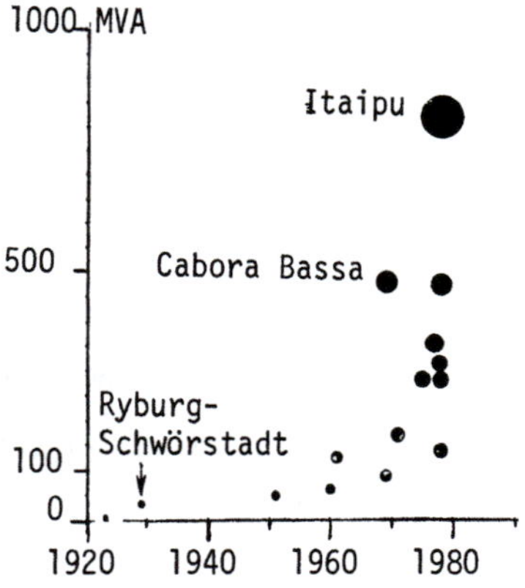

FIG. 1 UNIT POWER OF GENERATOR IN HYDRO-ELECTRIC PLANTS AS A FUNCTION OF DATE OF ORDER (FROM REF. 1)

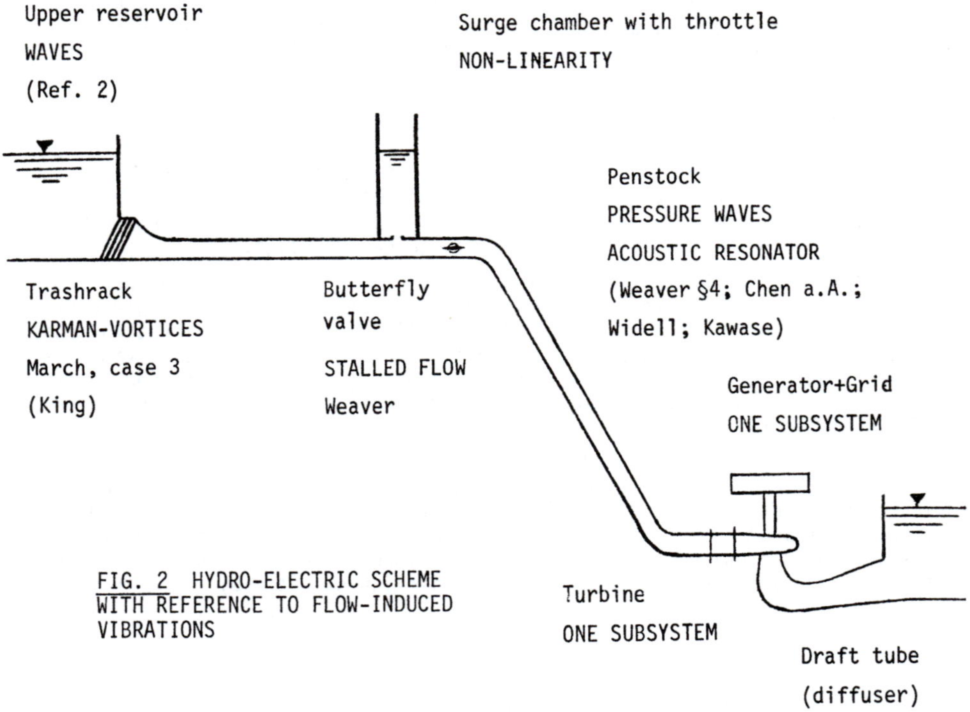

FIG. 2 HYDRO-ELECTRIC SCHEME WITH REFERENCE TO FLOW-INDUCED VIBRATIONS

2) Take advantage of the methods of system theory, as used in automatic control. Although a block diagram does not bring new information, it helps overlooking complex mechanisms (Widell*). Fig. 1 may be considered as a system having at least 4 subsystems, from which the penstock is rarely omitted. In such a system, a stability analysis using the describing function has the further advantage of handling non-linear systems (Glattfelder, Grein & Huser).

With regard to the category of hydraulic excitation, we note examples of separated flow: Karman vortices build up in the detached shear layer; throttles induce losses via a separated jet, as do valves while closing. Future investigations should provide the designer with methodical spectral analysis of pressure fluctuations in such flows, including sharp bends and diffusers.

3. TURBOMACHINERY COMPONENTS

Unlike vanes, and certain reactor components, turbines and pumps are designed for minimum losses. The risk of vibrations is accordingly lower under optimum efficiency conditions: the flow does not break away from the streamlined surfaces until near the trailing edge of the blades. Depending upon the shape of the blade and of the boundary layer, an organized Karman street or a turbulent wake is produced, as shown in Fig. 3. An unresolved problem arises from the specific geometry of the blading. For isolated profiles, the amplitude of the fluctuating forces, linked with Karman vortices, are known (Chen). Moreover, we are still missing precise information for radial cascades, in accelerating or decelerating flow.

Another problem is the interference between blade rows, whether the wakes are organized or not. Most important with respect to vibrations is the relative motion between guide vanes and runner blades, depicted in Fig. 4. Future research on such interferences should build on work done for thermal turbomachines (See VDI-Symposium Düsseldorf 1980). Under design conditions, water leaves the runner of a turbine approximately in the axial direction. At part load (or overload) a turbine like the one in Fig. 4 has an outflow with strong peripheral velocity. The eccentricity of the resulting swirl flow may induce vibrations (Kubota). That is why future investigations should go further into the three-dimensional computation of eccentric swirl flow. A broad definition of this category of problems would include: the flow upstream of axial-flow runners (Eichler); tip vortices of ship propellers; the vortex in the inlet of a pump (Wang, Elder); and the confined vortex at the exit of some compressors (Escudier).

4. ADDED MASS AND DAMPING

Just as in a mechanical oscillator, flow may act well as a spring, a damper, or a mass, in addition to its role as an excitation mechanism ("negative damping," according to Ref. 3). The following table, similar to Kolkman's, gives examples of hydraulic forces in phase with the displacement, Y, or its derivatives. Future research topics are given in capital letters. The added mass is a practical concept, which works astonishingly well, although the computation assumes potential flow (P. Kolkman, Session F). This kind of approach for the fluctuating flow still holds when a potential flow perpendicular to the oscillations is superimposed; it is not necessarily valid when viscous effects produce separated shear layers. An oscillating cylinder shedding Kármán vortices experiences hydraulic forces which

*Names refer to corresponding papers in the Symposium Proceedings.

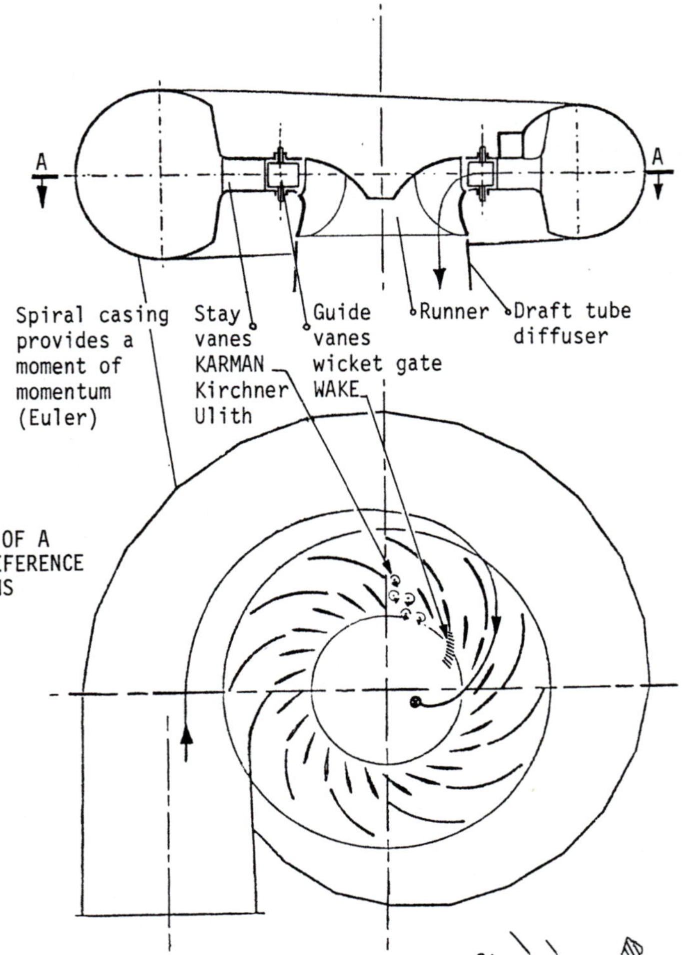

FIG. 3 TYPICAL ELEMENTS OF A HYDRAULIC TURBINE WITH REFERENCE TO FLOW-INDUCED VIBRATIONS

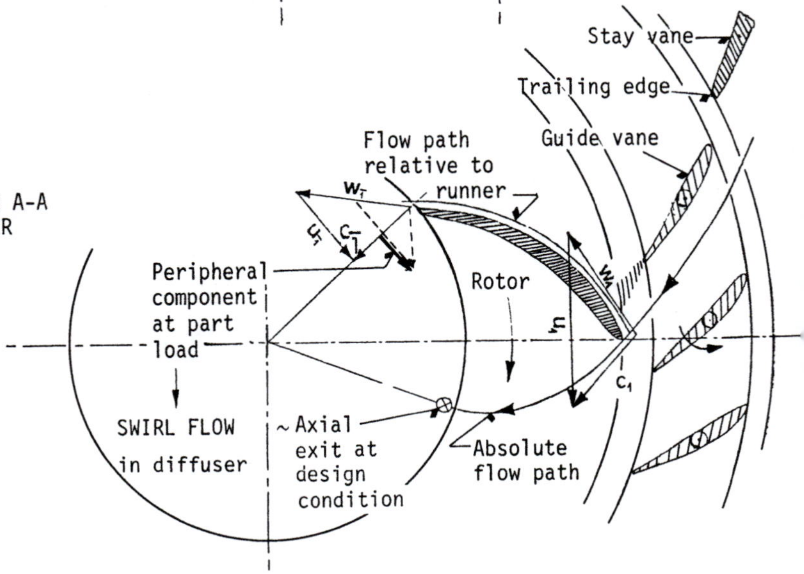

FIG. 4 SECTION A-A WITH WAKE/RUNNER INTERFERENCE AND PART LOAD SWIRL

Y	$\dot{Y}$	$\ddot{Y}$
m	m/s	m/s^2
oil film in bearing		"added mass"
THREE-DIMENSIONAL TEMPERATURE FIELD (Ref. [5])		WITH DETACHED FLOW (Ref. [4])
	leakage flow through labyrinth seal	
	LOMAKIN - EFFECT (Ref. [6])	

correspond to the so-called added mass. In this example, between Strouhal numbers 0.1 to 0.3, the force in phase with the acceleration makes a swing typical for resonance (resonance with respect to the frequency of Kármán vortices).

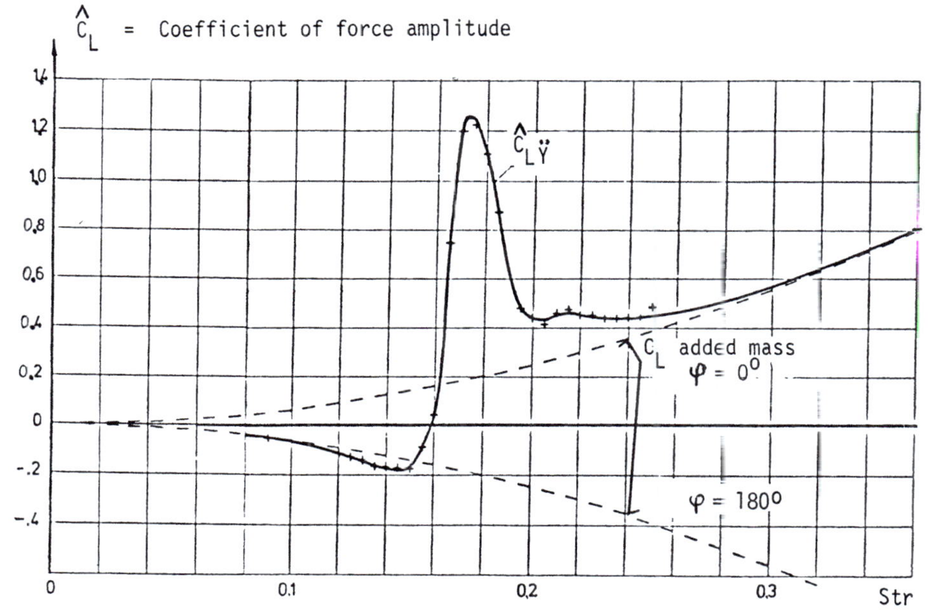

FIG. 6 FORCE IN PHASE WITH ACCELERATION OF OSCILLATING CYLINDER

Abscissa: $\text{Str} = \text{Strouhal number} = \dfrac{\text{Frequency of oscillation x Diameter}}{\text{Velocity of incident flow}}$

Ordinate: $\text{Force coefficient} = \dfrac{\text{Hydraulic force per unit length}}{\text{Dynamic pressure of incident flow x Diameter}}$

Solid curve = Measured force in phase with acceleration $\ddot{Y}$

Dotted line = Force computed as ρ_{Fluid} x Volume of cylinder x $\ddot{Y}$

Further Notations: ϕ = Phase angle

$c_{L\ddot{Y}} = c_L(\phi = 0°)$

$c_L(0°) \equiv -c_L(180°)$

$Re = 6 \times 10^4$

Amplitude of oscillation = $\pm$ 0.1 x Diameter

5. CONCLUSIONS

In the future, vibration problems should be approached using teamwork, i.e. interdisciplinary collaboration of mathematicians, specialists in mechanics and fluid-mechanics, and occasionally with designers of thermal turbomachines or ship propellers. The tools should include the methods of control theory. Non-linearity should be taken into account.

The subjects of future research, which are discussed in this paper, are not new. However, demands for accuracy are new, and will continue to grow.

6. REFERENCES

1. Krick, N., "Der Export von Wasserkraft-Generatoren und Gesamtanlagen", Wasser, Energie, Luft, 70, 11/12, 309-314, (Nov./Dec. 1970).

2. Plichon, J. N. and Lescure, A., "Instabilités hydrauliques dans les centrales", S.H.F.-Sessions des 15 et 16 novembre 1978, (Société Hydrotechnique de France, 119 rue de Grenoble, F-75 007 Paris).

3. Naudascher, E., "Flow-Induced Vibrations - A Unified Approach", IAHR Invited Paper, IAHR Congress, Baden-Baden, 1977.

4. Staubli, Th., "Hydrodynamic Forces on an Oscillating Prifile in a Uniform Flow", work in progress at the Institute für Flüssigkeitstechnik, ETH Zurich.

5. Fust, A. and Odermatt, R., "Three-Dimensional Temperature Distribution in Oil Film in Axial and Radial Bearings", work in progress at the Institute für Grundlagen der Maschinenkonstruktion, ETH Zurich (previous work: Hunter, 1960).

6. Priesnitz, Chr., "Der Einfluss von Ein- und Dreispaltdichtungen auf die Lage der kritischen Drehzahl und die Laufruhe von Pumpenwellen", VEB WTZ, Pumpen und Verdichter, Halle (DDR) (1967).

J8

CRITICAL UNRESOLVED PROBLEMS IN FLOW-INDUCED VIBRATION OF OCEAN STRUCTURES

R. D. Blevins
Staff Engineer
General Atomic Co.
San Diego, California, U.S.A.

1. INTRODUCTION

The vast majority of analyses of wave loads on ocean structures has been based on a static structure whose deformation is negligible (see, for example, [1,2,3]). The maximum wave force is applied to the structure and the resultant stresses are computed from static principles. This static model is appropriate as long as the fundamental natural frequency of the structure is well above the wave frequency so that the dynamic amplification is very small. In practical terms, the highest frequency waves with significant energy content have periods of about 5 seconds, corresponding to a frequency of 0.2 Hz. Thus the fundamental natural frequency of the structure would have to be at least three times this value, about 0.6 Hz, in order for the static response of the structure to be an accurate estimation of the deformation produced by ocean wave loading.

Many large ocean structures, or partially constructed ocean structures such as unsupported pilings, have natural frequencies below 0.6 Hz and the static approximation would not appear appropriate for these structures [3]. These low frequency structures can resonate with either the fundamental wave frequency or with a higher harmonic of the wave frequency, and significant dynamic amplification can be produced. Moreover the structure may resonate with vortex shedding either due to wave action or due to a tide [4]. Thus, the problem of the dynamic response of ocean structures to wave loading is of considerable practical importance.

2. UNRESOLVED PROBLEMS

Some fundamental unresolved problems in the dynamic analysis of ocean structures are as follows:

a.) The wave loads on cylindrical structures are the sum of an inertial and a drag component, which are in turn characterized by an inertia and a drag coefficient. Experimental data are becoming available for these coefficients for single cylinders in a sinsuscidal flow (Ref. [5]). What are these coefficients for a complex sea state and for a cylinder in the presence of other cylinders as in a piling array?

b.) While a theoretical analysis is available for the dynamic in-line response of cylinders in ocean waves, there are essentially no data to validate the analysis [6]. Moreover, there are very little data available to predict the vortex-induced response perpendicular to the flow. Methods for predicting the dynamic response of structures to ocean waves need to be qualified by comparison with experiments on elastic structures exposed to ocean waves.

3. REFERENCES

1. Foster, E.T., "Model for Nonlinear Dynamics of Off Shore Towers," ASCE J. Mechanics Div. EM1, 41-66 (1970).

2. Wirsching, P.H. and Prasthofer, P.H., "Preliminary Dynamic Assessment of Deepwater Platforms," ASCE J. Structural Div., St 7, 1447-1462 (1976).

3. Zienkiewicz, O.C. (editor); Numerical Methods in Offshore Engineering, John Wiley, NY., 1978.

4. Blevins, R.D., Formulas For Natural Frequency and Mode Shape, Van Nostrand Reinhold, NY., 1979.

5. Sarpkaya, T., "In-line and transverse Forces on Cylinders," J. Ship Research 2 (4), 200-216 (1977).

6. Blevins, R.D., Flow-Induced Vibration, Van Nostrand Reinhold, N.Y., London, 1977.

FILM SESSION

P. Ulith — "Flow Around Stay Vanes in a Spiral Case of a Model Turbine". - The flow in front of and around the stay vanes in a spiral casing of a medium specific speed model turbine is visualized.

T. Kubota — "Swirling Flow with Vortex Breakdown in the Draft Tube of a Francis Turbine".- The behavior and breakdown of cavitating vortex cores generated in the center of upper region of the draft tube of a model Francin turbine with a runner speed of 1,000 rpm is illustrated.

U. Schumann, W. Olbrich — "Core Barrel Oscillations During Blowdown of a Pressurized Water Reaktor". - A computer generated movie shows the pressure field and the core barrel oscillations of a typical PWR, both with and without accounting for fluid-structure interactions.

M.M.Zdravkovich — "Flow-Induced Vibration of an Aero Elastic Tube Bank". - The vibrational response of tube tips of a 10-row deep bank with 19 tubes per row is depicted. In the first film, the tubes are flexibly mounted cantilevers arranged in a triangular pattern at a pitch of 1.375 diameter. In the second film, different vibration of a 3-row deep in-line tube bank is shown.

M. Ishii — "Model Tests on the Dynamic Instability of Tainter Gates". - Six model tests on the dynamic instability of Tainter-gates are shown: free oscillation; intense self-excited oscillation; periodic oscillation; damping oscillation; intense self-excited oscillation.

M. Gerodetti — "Gate Vibration at the Weir of Schinznach, Switzerland". - The vibration of gates caused by underflow with an insignificant gate opening of only a few centimeters is shown.

F. Büsching	"Model Tests for Stoplogs Subjected Simultaneously to Overflow and Underflow". - Some tests with stoplog models of different dimensions and positions are shown; vibration forces are addressed.
R.L. Wardlaw, H.P.A.H. Irwin	"Wind-Tunnel Investigation of a 1:110 Scale Model of the Lion's Gate Bridge in Smooth and Turbulent Flow". - In smooth, uniform flow the bridge develops a catastrophic torsional oscillation. This motion does not occur when the bridge is in a simulated atmospheric wind layer; however large amplitude random motion is observed that is caused by turbulent buffeting.
A.G. Davenport	"Guyed Mast Vibration". - A variety of oscillations of tall slender guyed masts, including vortex shedding and guy cables is shown.
E. Berger, T. Kobayashi	"Oscillator-Model Theory". - The film (16 mm, 2 min.) shows 3 sequences of the near wake visualized by smoke with a phase controlled stroboscope: cylinder at rest; cylinder oscillating with small amplitude in "hard" excitation; and cylinder oscillating with large amplitude at the Strouhal resonance.
W.T. Ashurst	"Prediction of Unsteady Flow by the Discrete-Vortex Technique". - The unsteady flow behind a symmetric sudden expansion in a channel has also been calculated. This computer movie shows the time-development of asymmetric flow that is established within the symmetric channel.
D. Rockwell, C. Knisely	"Vortex Impingement on Corner of Cavity". - Separated flow past a cavity gives rise to coherent oscillations, characterized by impingement of vortices on the downstream corner of the cavity. The hydrogen bubble technique is used to visualize the flow dynamics.

IUTAM Symposia

International Union of Theoretical and Applied Mechanics

Verformung und Fließen des Festkörpers. Deformation and Flow of Solids. 1955.

Grenzschichtforschung. Boundary Layer Research. 1957.

Creep in Structures. 1960.

Dynamics of Satellites / Dynamique des Satellites. 1962.

Kreiselprobleme. Gyrodynamics. 1962.

Symposium Transsonicum. 1962.

Stress Waves in Anelastic Solids. 1963.

Applied Mechanics. 1964.

Rheology and Soil Mechanics. Rhéologie et Mécanique des sols. 1964.

Trajectories of Artificial Celestial Bodies as Determined from Observations. Trajectoires des corps célestes artificiels déterminées d'aprés les observations. 1965.

Irreversible Aspects of Continuum Mechanics. – Transfer of Physical Characteristics in Moving Fluids. 1966.

Theory of Thin Shells. 1967.

Mechanics of Generalized Continua. 1967.

Thermoinelasticity. 1968.

Dynamics of Satellites. 1969.

Instability of Continuous Systems. 1969.

Creep in Structures. 1970.

Flow-Induced Structural Vibrations. 1972.

Optimization in Structural Design. 1973.

Photoelastic Effect and Its Applications. 1973.

Buckling of Structures. 1974.

Satellite Dynamics. 1974.

Dynamics of Rotors. 1974.

Mechanics of Visco-Elastic Media and Bodies. 1974.

Symposium Transsonicum II. 1975.

Dynamics of Multibody Systems. 1977.

High Velocity Deformation of Solids. 1977.

Physics and Mechanics of Ice. 1979.

Mechanics of Sound Generation in Flows. 1979.

Practical Experiences with Flow-Induced Vibrations. 1979.

Laminar-Turbulent Transition. 1979.

Springer-Verlag
Berlin
Heidelberg
New York

GPSR Compliance

The European Union's (EU) General Product Safety Regulation (GPSR) is a set of rules that requires consumer products to be safe and our obligations to ensure this.

If you have any concerns about our products, you can contact us on ProductSafety@springernature.com

In case Publisher is established outside the EU, the EU authorized representative is:

Springer Nature Customer Service Center GmbH
Europaplatz 3
69115 Heidelberg, Germany

Batch number: 10620157

Printed by Printforce, the Netherlands